AF741149

Frontiers in Guided Wave Optics and Optoelectronics

Edited by

Bishnu Pal

Frontiers in Guided Wave Optics and Optoelectronics

Edited by Bishnu Pal

Republished by InTech

Janeza Trdine 9, 51000 Rijeka, Croatia

Copyright © 2010 InTech

Abstracting and non-profit use of the material is permitted with credit to the source. Statements and opinions expressed in the chapters are these of the individual contributors and not necessarily those of the editors or publisher. No responsibility is accepted for the accuracy of information contained in the published articles. Publisher assumes no responsibility liability for any damage or injury to persons or property arising out of the use of any materials, instructions, methods or ideas contained inside. After this work has been published by the Intech, authors have the right to republish it, in whole or part, in any publication of which they are an author or editor, and the make other personal use of the work.

Technical Editor Teodora Smiljanic

Cover Designer Dino Smrekar

Additional hard copies can be obtained from orders@intechopen.com

Frontiers in Guided Wave Optics and Optoelectronics, Edited by Bishnu Pal
p. cm.
ISBN 978-953-7619-82-4

We are IntechOpen, the world's largest scientific publisher of Open Access books.

3,250+
Open access books available

106,000+
International authors and editors

112M+
Downloads

151
Countries delivered to

Our authors are among the
Top 1%
most cited scientists

12.2%
Contributors from top 500 universities

CLARIVATE ANALYTICS BOOK CITATION INDEX INDEXED

WEB OF SCIENCE™

Selection of our books indexed in the Book Citation Index
in Web of Science™ Core Collection (BKCI)

Interested in publishing with us?
Contact book.department@intechopen.com

Numbers displayed above are based on latest data collected.
For more information visit www.intechopen.com

For EU product safety concerns:
IN TECH d.o.o., Prolaz Marije Krucifikse Kozulić 3, 51000 Rijeka, Croatia,
info@intechopen.com or visit our website at intechopen.com.

Meet the editor

Bishnu P. Pal is Professor of Physics at Mahindra École Centrale Hyderabad India since July 1st 2014 after retirement as Professor of Physics from IIT Delhi; Ph.D.'1975 from IIT Delhi; Fellow of OSA and SPIE; Senior Member IEEE; Honorary Foreign Member Royal Norwegian Society for Science and Arts; Member OSA Board of Directors (2009- 11); Distinguished Lecturer IEEE Photonics Society (2005- 07).

Preface

I have great pleasure in introducing this e-book "Frontiers in Guided Wave Optics and Optoelectronics" under the new series Advances in Lasers and Electro Optics being published by IN-TECH Publishers. Guided wave optics and optoelectronics are at the heart of optical communications, optical signal processing, miniaturization of optical components, biomedical optics, defense applications, and so on. The most recent recognition of the importance of this subject has been acknowledged through the conferral of the Noble Prize in Physics for 2009 to Dr. Charles Kao "for groundbreaking achievements concerning the transmission of light in fibers for optical communication". The charter of the Noble Prize states that it is given to " .... who shall have conferred greatest benefit on mankind". Optical communication in the last two decades has revolutionized the way information is transferred in terms of instant transmission as well as access especially the Internet. Optical fiber networks are now taken for granted regardless of the scale of access, be it inter-city, inter-continental, metro or local. Transmission loss in silica fibers has been reduced to nearly the lowest possible limit, dispersion of signals in a telecom grade fiber can be highly controlled to a level where signals are transmitted over long distances without any significant impairment, and nonlinearity-induced distortions can be reduced drastically through appropriate fiber design. With all these remarkable achievements, it appeared for a while in the mid-1980s that there would be no further scope for research in optical fibers. However it turned out that new demands arose for specialty fibers, in which dispersion and nonlinearity could be tailored to achieve transmission properties, which are otherwise impossible in conventional fibers. Around this time, the concept of photonic crystals (PhC) was put forward, which eventually led to the development of microstructured optical fibers (MOF), which are also referred by some as photonic crystal fibers (PCF).

Chapter 1 forms the subject matter of specialty optical fibers such as large negative dispersion coefficient fibers, which are known to serve as dispersion compensating fibers. Inherently gain flattened erbium doped fibers, which do not require any gain flattening filters as a cost effective solution for the metro optical networks are also described in this chapter. The so-called photonic bandgap guided fibers, which fall within the class of microstructured fibers, constitute a completely new type of fiber waveguides, whose functional principle for light guidance and confinement is totally different from that of conventional fibers. In certain forms, these fibers are capable of guiding light in an air core thereby enabling the propagation of light with ultra low loss. As a corollary, these fibers can also serve as a conduit for high-energy pulses without causing any material damage within the fiber or impairing the signal due to nonlinearities induced at high powers. Microstructured fibers also form a versatile platform for generating supercontinuum (SC) light, which is a highly spatially coherent, intense broadband light with imminent applications that range from biomedical imaging and engineering to laser development and the generation of stable frequency combs.

Chalcogenide glass-fibers, which form another example of application-specific specialty fibers, are described in chapter 2. Fibers drawn from As-S-Se-like glass systems exhibit large nonlinearities. The authors, who are from a leading group in the US on this topic, have highlighted the utility of such fibers for building up compact Raman as well as Brillouin amplifiers, SC light generators in the near and mid-infrared wavelength regions, fast optical switches, optical regenerators for high speed telecommunication systems, efficient slow light realization and several other issues.

Measurements on nuclear radiation response of a variety of commercially available radiation hardened fibers are described in chapter 3. In nuclear environments, optical fibers are often employed for data collection as hybrid sensors and transmission of the same to a remote date processing center, as well as light-guides for control and diagnostics. Through detailed studies on the gamma, beta and neutron irradiation effects on these fibers, the authors have described the role and mechanism behind the formation of different types of colour centers as a result of the irradiation. These colour center formations are attributed to molecular bonding between different basic constituents at the atomic level. Such studies are important from the point of view of choosing material systems in the fabrication of radiation-hardened fibers.

Chapter 4 (which is a collaborative effort by authors from Russia and Scotland) describes programmable all-fiber techniques (e.g. linearly chirped fiber Bragg grating) for the synthesis and control of fast (pico to sub-pico second) temporal optical pulses for applications in nonlinear optical switching and wavelength conversion devices. The control of short, parabolic flat-top optical pulses is also important for self-phase modulation-induced supercontinuum generation experiments. Such optical signal processing techniques are important in e.g. ultrahigh-bit-rate optical communications, coherent control of atomic and molecular processes, and generation of ultra-wideband RF signals.

In chapter 5, the authors deal with the topical phenomenon of *slow light* (SL), which has garnered enormous attention in recent years. SL essentially refers to the significant slowing down of the group velocity of light in a medium near its resonance i.e. at frequencies where propagating light resonantly interacts, e.g. for gain or absorption, with atoms or molecules. Slow light has several potential applications as delay lines, optical buffers, equalizers etc, especially from the optical communication perspective. Unfortunately the topic essentially remained an academic curiosity for a long time because at a resonance the gain or absorption of light in the media is far too strong to be fruitfully fully exploited for device realization. In late 1990s it was shown that through *nonlinear* resonance interactions, such as in *stimulated Brillouin scattering (SBS)*, group velocities could be brought down to few tens of meters per second because $\mathrm{d}n/\mathrm{d}\omega$ could be large and yield sharp resonances. The authors argue that realizing this effect in an all-fiber form, e.g. as a delay line, is all the more attractive because as a device, this could be seamlessly integrated into an optical fiber network through techniques such as splicing. The authors make a detailed study of SBS and the physical mechanism by which a Stokes pulse can be slowed in an optical fiber.

In chapter 6, optical amplification characteristics of Bismuth doped glass and fibers in the O-band spanning from 1260 nm to ~ 1360 nm are described. Optical amplification through erbium doped fibers in the wavelength band spanning the C- and L-band across 1530 ~ 1610 nm has matured as a present technology. However, since the demand for bandwidth is ever-increasing, utilization of additional potential amplifier bands of high-silica fibers for wavelength division multiplexed transmission is always attractive. Though Fiber Raman amplifiers and Praseodymium-doped glass fiber amplifiers could work at the O-band, their gain bandwidth is rather limited. Bismuth doped fiber amplifiers can potentially offer a much wider gain bandwidth, which is the topic of study of this chapter. Since zero dispersion wavelengths of conventional fibers like ITU specified G-652 fibers (example, SMF-28 of Corning Inc.) fall within the O-band, development of wide band amplifiers (as well as lasers) for this band is very attractive because it would amount to adding more transmission bandwidth and could complement fiber amplifiers already studied for the C-, L-, and S- bands. One important finding of their studies is the necessity of co-doping with Aluminium for achieving broadband luminescence at the infrared wavelengths when pumped with commercially available 810 nm high power laser diode.

New demands for broadband wireless in access networks, radar related data processing, hybrid fiber radio (HFR), mm wave and THz generation systems, etc have given rise to a new application-specific research area of optical fibers known as Radio-over-Fibre (RoF), which forms the subject matter of chapter 7. This subject area is also often referred to as microwave photonics, in which a radio signal typically in the millimetre wave band is transmitted through optical fibers employing laser sources and electro-optical devices. HFR, in some sense, is similar to hybrid-fibre coaxial (HFC) network, in which a combination of digital and analog channels are distributed in the last mile through coaxial cables with the main difference that in HFR, the 'last-mile' distribution is done wirelessly. The authors state that in the RoF technology the required RF signal processing functions in one shared Central Office and then optical fibers are used to distribute the RF signals to the remote access units (RAU), which make this approach very cost effective. This enables the incorporation of advanced network features such as dynamic allocation of resources.

As mentioned in chapter 1, nonlinear aspects of microstructured specialty fibers (also referred to as photonic crystal fibers) is exploited to dramatic effect for broadband SC light generation. These fibers also offer engineerable dispersion profiles, which facilitate control over the nature of the generated SC light. SC generation has been known to result from soliton fission processes as well as Raman effects and some phase matching phenomena, which occur on these fibers under the right conditions. The dynamics and interactions between solitons as well as the phenomenon of dispersive wave generation are dealt with in chapter 8 and provide insight into how propagating femtosecond pulses generate new frequencies. This topic is also extensively investigated experimentally by deploying the so-called sum frequency generation (SFG) X-FROG (Cross correlation frequency resolved optical gating) technique, which is a *time-spectral* visualisation technique. Besides conventional MOFs, the authors have also investigated SC from soft glass like SF_6-based extruded MOFs, in which nonlinearity could be much stronger than high-silica glass and they could show quite remarkable agreement between theory and experiment.

In chapter 9, the authors describe three varieties of dispersion compensating devices, which play important role in the generation, amplification and propagation of dispersion-free femtosecond pulses. Various such devices include chirped mirrors, grating and prism pairs and are discussed in this chapter, out of which grating pair-based dispersion compensation approaches have been discussed in greater detail, which is of critical importance for the field of ultra-fast optics.

Dispersion tailored photonic crystal or microstructured fibers form the subject of study in chapter 10. Though there is some overlap of the discussions with those discussed in chapter 1, this chapter provides greater details on dispersion characteristics of MOFs. The introduction contains a concise review of the state-of-the-art of various fiber designs. The authors present their analysis of the dispersion profiles of photonic band gap (PBG) fibers including solid-core and hollow-core photonic crystal fibres, whose periodicity is modified by applying resonant Gires-Tournois interferometric (GTI) resonant layers around the core, which induces a frequency-dependent phase shift of the light. This concept is then utilized to design Bragg fibers with the layers immediately adjacent to the core having parameters that are distinct from the rest of the high and low index layers in the periodic cladding, which form the GT layers. The same concept is then extended to an all-glass PCF as well as a hollow-core Bragg fiber and a hollow-core PCF with honeycomb cladding for tailoring their dispersion properties.

Chapter 11 is concerned with a detailed study on the origin and influence of refractive index change (RIC) induced due to population inversion in resonant fiber structures such as Ytterbium (Yb)-doped fiber lasers, which operate in the 1 μm spectral region. The results presented in this chapter indicate that far-resonance electronic transitions in the UV rather than near-resonant IR transitions are responsible for the RIC. By extending their arguments, the authors propose exploiting this effect to develop a simple all-fiber solution for coherently beam combining rare-earth-doped fiber amplifiers through active phase control in an all-fiber spliced configuration.

In chapter 12, current research and applications of periodically poled lithium niobate (PPLN)-based optoelectronic devices, which include tunable wavelength filters, polarization controllers, electro-optical and various switches are reviewed. Since lithium niobate is used extensively for configuring integrated optical waveguides, this study, in principle, can be extended to waveguide geometries. Theoretical analyses and designs of integrated optical wavelength-selective switches (with potential operating speeds of *a few tens of picoseconds or faster)* in the form of a cascaded Mach-Zehnder interferometer (MZI) are discussed in the subsequent chapter (Chapter 13). Cascaded MZI's are an excellent candidate for configuring integrated optical wavelength interleavers due to their inherent strong wavelength selectivity. In chapter 13, the authors propose a combination of Raman waveguide amplifiers integrated in the arms of the first stage MZI and the integration of an attenuator in one of the arms of the second stage MZI as a means to achieve control of light amplitude through stimulated Raman scattering. The authors also describe an alternative architecture for such a switch in which three Raman amplifiers are placed in the lower arms of a three-stage cascaded MZI. The efficiency of its switching operation is verified through computer simulation by employing a finite difference beam-propagation-method.

Chapter 14 discusses nonlinear integrated optical device platforms based on high-index doped-silica materials. Often referred to as silicon photonics, this platform offers a compromise between the linear optical properties of single-mode fibers and those of semiconductors and other nonlinear glasses. In particular, measurement techniques to characterise the linear and third order nonlinearities, with specific applications to parametric four wave mixing (FWM) are described. Potential applications such as narrow line width, and/or multi-wavelength sources, on-chip generation of correlated photon pairs, as well as sources for ultra-low power optical hyper-parametric oscillators are also projected.

Advances in inscription of waveguides and micro/nano-photonic devices through the use of high-power, focused femtosecond laser pulses, generally referred to as femtosecond micromachining, are reviewed in chapter 15. In this chapter, the authors highlight the interactions between a suitable material (typically glass) and short pulses that result in permanent changes in the physical, chemical and optical properties of the material on a sub-micron scale. State-of-the-art femtosecond laser sources, machinable materials, and some of the current applications for this type of technology are also reviewed. Interestingly this technique has been used to produce phase masks, which are essential for fabricating wavelength selective fiber devices such as in-fiber Bragg gratings. This technology can be used to realize high aspect ratio, micron-scale channels as microfluidic lab-on-chip devices such as for measuring a specific particle to particulate sorting and counting. Due to the high point density that can be achieved through the spatial confinement of the femtosecond pulse-material interaction, this technology has been also used in high density data storage and retrieval applications, for creating sub-micron features in polymers through polymerization, producing photonic crystal structures and even for fabricating medical stents. The authors state that the industrialisation of micromachining processes will be of great significance in the future of solar cell and flexible organic light emitting diodes (OLEDs) or manufacturing techniques which require highly localised and fast creation of complex, machinable patterns.

Magneto-optical materials for integrated optical waveguides, which form the topic of chapter 16, are attractive for optoelectronics because of their unique characteristics like non-reciprocity and retention of memory. The most common example of one such component, which is required in an optical communication network, is an optical isolator, which is invariably used as an integral component in a fiber amplifier to prevent it from lasing. The authors of this chapter report on the growth of (Cd,Mn)Te waveguides on GaAs substrate for realizing magnetooptical integrated optical isolator with a high isolation ratio of 27 dB, a low optical loss of 0.5 dB/cm, and a high magneto-optical figure-of-merit of 2000 deg/dB/kG over a 25-nm wavelength range. They have also utilized the magnetization reversal of nano-magnets through spin-polarized photo-excited electrons for realizing non-volatile, high-speed optical memory. Metal clad magnetooptical waveguides have also been described in this chapter.

The subject of the next chapter 17 is based on hollow optical waveguides, known as Bragg reflectors, for integrated optics. These waveguides confine light by Bragg reflectors oriented transversely to the direction of propagation. A widely tunable Bragg reflector is introduced which demonstrates on-chip polarization control for adjustable compensation of polarization mode dispersion (PMD). Owing to a weak dependence of air on temperature, the phase delay suffered by the light confined in hollow core waveguides (which have been introduced in Chapter 1), is nearly temperature-insensitive, which is of significant advantage in waveguide-based sensors. By incorporating MEMS-based actuators on either of the multilayer Bragg mirrors, the air gap between the mirrors can be tuned to achieve tunable propagation characteristics of such waveguides. Chapter 18 is also devoted to Bragg gratings in which the grating is located exclusively along the longitudinal direction of a fiber's core – the so-called Fiber Bragg Grating (FBG) - with a focus on increasing the operating temperature ranges of these FBGs. The authors refer to these devices as regenerated fiber gratings, which can withstand temperatures in excess of 1200°C. These gratings have a number of potential applications, which include their role in monitoring furnace temperatures in various situations, and their utility as a component in high peak power fibre lasers. In the following chapter 19, optical deposition of carbon nano tubes (CNTs) onto optical fibers to realize CNT-based fiber devices is described including fundamental properties of CNTs, their fabrication, and CNT-based optical devices. Utility of CNT as a passive mode-locker, or as a saturable absorber for ultrashort pulse generation is well known for sometime now. Authors developed a method, which enables area-selective deposition of CNTs only onto the core region of an optical fiber end. Evanescent coupling between CNTs and propagation mode of a microfiber is one way to realize a polarization insensitive CNT device. The authors demonstrate a passively mode-locked fiber laser having optically deposited CNTs (to serve as saturable absorber) circumscribing a microfiber, which's tapered (realized through heat and stretch method) waist was ~ 6 μm.

Thulium (Tm^{3+}) doped fibers initially attracted attention from the point of view of their use as fiber amplifiers at the S-band. Chapter 20 describes high power Tm^{3+} fiber lasers and their utility as pump for chromium doped ZnSe (Cr^{2+}:ZnSe) lasers. In Tm^{3+}-doped fiber lasers, output at the mid-infrared wavelengths ~ 2 μm is realizable, which is extremely important from the point of view of laser microsurgery due to high absorption by water in this spectral region. Thus these lasers could provide high-quality laser tissue cutting and welding in biomedical optics besides potential applications in environment monitoring, LIDAR, optical-parametric-oscillation (OPO) OPO-based pump sources, and so on. Authors discussed a variety of double clad fiber structures to configure Tm^{3+} doped fiber lasers. This longest chapter spread over 70 pages in the book contains details of several issues e.g. spectroscopy, fabrication, scalability, nonlinear optical effects, wavelength tunability, self-pulsing, Q-switched operation, etc. of Tm^{3+} fiber lasers.

In chapter 21, development of ~ 2 μm wavelength emitting lasers in the form of crystal lasers, fibre lasers and semiconductor lasers are discussed. For the crystal and fiber lasers, authors focus on thulium and holmium doped systems, in which output powers close to 1 kW and slope efficiencies of up to 68 % have been reported. The chapter also describes latest improvements of GaSb-based laser diodes and ends by indicating their potential applications in spectroscopy, sensing, surgery, and material processing.

Chapter 22 focuses on the design and realisation of photonic crystal (PhC)-based micro resonators for lasers. In view of the versatility afforded by PhCs, optical properties of such resonator can be manipulated without almost any restrictions. Two different schemes for designing PhC laser resonators were discussed by the authors. The first one uses a bulk active region, which is surrounded by a PhC-mirror and the second type uses the PhC directly as the gain medium. Incorporation of PhCs allows for a full control of the dispersion relation of a resonator, and hence enables newer designs of resonators. Surface plasmonic waveguides, in which the active semiconductor region is sandwiched between two metallic layers of gold, were also used to realize THz quantum cascade lasers.

Ceramic lasers form the subject matter of chapters 23 and 24. Due to their short fabrication period and being mass-producible, the cost of ceramic laser materials could be much lower than that of single crystals. Furthermore, no complex facilities and critical techniques are required for the growth of large sized ceramics. At a low doping concentration, efficiency of a diode-end-pumped Nd:YAG ceramic laser was found to be higher than that of a Nd:YAG single crystal laser. A high-power (6.8 W) tunable Yb:YAG ceramic laser with a slope efficiency as high as 72% has been demonstrated at room temperature. A diode pumped passively mode-locked Yb:YAG ceramic laser was also demonstrated for the first time by the authors in chapter 23. Broad tunability in the spectral range from 994.35 to 1098.87 nm and from 992.52 to 1110.83 nm in two different cases at room temperature were also obtained in such Yb:YAG ceramic lasers. In chapter 24, polarization properties of laser diode pumped micro-chip Nd:YAG ceramic lasers is presented. The author of this chapter points out that ceramic lasers, which consist of randomly distributed single crystals surrounded by grain boundaries, are interesting for studying lasing properties in random media. Observed segregations into multiple local-modes due to field interference effect among the local-modes and the associated variety of dynamic instabilities that occur in laser diode-pumped Nd:YAG ceramic samples were discussed.

Chapter 25 is concerned with surface-emitting circular Bragg lasers. Surface-emitting lasers have attracted attention because of their salient features such as low-threshold current, single-mode operation, and wafer-scale integration. Their low-divergence surface-normal emission also facilitates output coupling and packaging. Vertical Cavity Surface Emitting Lasers (VCSELs) though available commercially, their single-modedness and good emission patterns are guaranteed only over a very small mode area (diameter of ~ μm). Attempts to further increase the emission aperture have failed mostly because of the contradictory requirements of large-area emitting aperture and single modedness. From this perspective, circular Bragg lasers could circumvent this problem. This chapter contains a comprehensive and systematic theoretical study on the surface-emitting Hankel-phased circular Bragg lasers in various geometries. According to the authors, these lasers should find applications in ultra-sensitive biochemical sensing, all-optical gyroscopes, and coherent beam combining, and as high-power, high-radiance sources in communications and display technology.

The integration of optical and wireless mode of communication in access networks is the subject matter of chapter 26. It essentially implied the convergence of two conventional technologies - radio frequency (RF) for wireless access in the last tens of meters and optical fiber for wired transmission across long-range links. RoF techniques have been discussed in an earlier chapter. Authors here argue that service providers for next-generation access networks are expected to offer end users greater choice, convenience and variety in an efficient way, which would require delivery of voice, data and video services with mobility feature to serve the fixed and mobile users in a unified networking platform. Ultra uncompressed HD Video with (UHDV) 7680 x 4320 pixels (33 Mega pixels) plus 22.2 sounds (24 channels in three layers) would require 40 GB/s or higher data speed. The chapter contains details of a variety of enabling technologies to accomplish such convergence including results from a test bed.

Photonic crystal (PhC)-based optical multiplexers (MUX)/demultiplexers (DMUX) are discussed in chapter 27, in which the authors initially dwell on definition and functional principle of PhCs including role of defects in periodic structures of this kind for realizing optical components and devices. Two different geometries of PhC were studied in detail as MUX/DMUX of 1310/1550 nm wavelengths. According to them integration of PhC with planar lightwave circuits would play an important role in integrating photonic components and devices to an optical network.

Maximum transmission distance in a fiber optical community antenna TV (CATV) system is still limited by RF parameters. A number of techniques like split-band schemes, light-injection and dispersion compensation have been applied to extend this bottleneck. Authors of chapter 28 discuss these techniques with the target to make a cost effective fiber optic CATV system, for which they argued that it is important to combine fiber optical CATV systems with other applications, e.g. Internet access and WiMAX services.

Active optical alignment of optoelectronic components in any optical communication system is an important issue especially in high-speed optical communication systems such as 10 Gb/s or higher. In most situations, passive techniques are employed. Chapter 29 describes a bi-directional MEMS-based optical beam steerer as an active solution to minimize such technological issues. This was achieved through use of a silicon optical bench containing a set of micro-machined silicon components comprising of the substrate, RF feed-through, hermetic sealing and optical alignment functions.

As the editor, I feel extremely happy to present to the readers such a rich collection of chapters authored/co-authored by a large number of experts for this book covering the broad field of guided wave optics and optoelectronics from around the world. It was indeed a monumental task to edit such a large volume. Most of the chapters are state-of-the-art on respective topics or areas that are emerging. Several authors narrated technological challenges in a lucid manner, which was possible because of individual expertise of the authors on their own subject specialties. I have no doubt that the book would be useful to graduate students, teachers, researchers, and practicing engineers and technologists and they would love to have it on their book shelves for ready reference at any time.

I thank Professor Vedran Kordic for inviting me to edit this book, which was a delightful experience for me in view of the ocean of knowledge and information that is contained in each chapter. In fact Professor Kordic contacted individual chapter lead authors. On behalf of Prof. Kordic and the publisher of this e-book, I would thank all the contributors for their scholarly contributions. Finally I would like to thank my daughter Parama Pal, presently with the Wellman Laboratory of Photomedicine at the Harvard University Medical School in Boston, USA and also my wife Subrata for helping me with the composition of the preface and editing of the book.

January 27, 2010

Editor

Bishnu Pal
Physics Department,
Indian Institute of Technology Delhi
New Delhi, India

Contents

1

Application Specific Optical Fibers

Bishnu P. Pal
Indian Institute of Technology Delhi
Physics Department
New Delhi: 110016
India

1. Introduction

Optical fiber technology was considered to be a major driver behind the information technology revolution and the huge progress on global telecommunications that has been witnessed in recent years. Fiber optic telecommunication is now taken for granted in view of its wide-ranging application as the most suitable singular transmission medium for voice, video, and data signals. Indeed, optical fibers have now penetrated virtually all segments of telecommunication networks - be it trans-oceanic, transcontinental, inter-city, metro, access, campus, or on-premise [Pal, 2006]. Initial R&D revolution in this field had centered on achieving optical *transparency* in terms of exploitation of the *low-loss* and *low-dispersion* transmission wavelength windows of high-silica optical fibers. The earliest optical fiber communication systems exploited the first low loss wavelength window centered on 820 nm of silica with graded index multimode fibers as the transmission media. However, primarily due to unpredictable nature of the bandwidth of jointed multimode fiber links, since early-1980s the system focus had shifted to single-mode fibers. This was accentuated by the discovery of the zero material dispersion characteristic of silica fibers, which occurs at a wavelength of 1280 nm [Payne & Gambling, 1975] in close proximity to its second low loss wavelength window centered at 1310 nm. The next revolution in lightwave communication took place when *broadband* optical fiber *amplifiers* in the form of erbium doped fiber amplifiers (EDFA) were developed in 1987 [Mears et al, 1987], whose operating wavelengths fortuitously coincided with the lowest loss transmission window of silica fibers centered at 1550 nm [Miye et al, 1979] and heralded the emergence of the era of *dense wavelength division multiplexing* (DWDM) technology in the mid-1990s [Kartapoulos, 2000]. Recent development of the so-called low water peak fibers like AllWave™ and SMF-28e™ fibers, which are devoid of the characteristic OH^- loss peak (centered at 1380 nm) extended the low loss wavelength window in high-silica fibers from 1280 nm (235 THz) to 1650 nm (182 THz) thereby offering, in principle, an enormously broad 53 THz of optical transmission bandwidth to be potentially tapped through DWDM technique! State-of-the-art technology has already demonstrated exploitation and utilization of 25 ~ 30 THz optical bandwidth. The emergence of DWDM technology has also driven the development of various specialty fibers and all-fiber components for seamless growth of the lightwave communication technology. These application-specific fibers were required to address new issues/features such as broadband dispersion compensation, realization of specialized components such as fiber couplers for multiplexing pump and signal wavelengths required in configuring fiber amplifiers, erbium doped fibers for realizing EDFAs, realization of wavelength sensitive in-fiber grating-based components, low sensitivity to nonlinear impairments with propagation, and so on. There has also been a resurgence of interest amongst researchers to design and fabricate an exotic class of special fibers - fibers in which transmission losses of the material would not be a limiting factor while nonlinearity and dispersion characteristics could be conveniently tailored to achieve certain application-specific fibers, not necessarily for telecommunication applications only. Research targeted towards such fiber designs led to the emergence of a new class of fibers, broadly referred to as *microstructured optical fibers* (MOF), which are characterized by wavelength-scale periodic refractive index features across its physical cross-section resulting in *photonic bandgaps* when appropriately designed. These features could be periodically located air holes/low refractive index regions in the cladding region, which surround the central core region, which could be of higher or lower refractive index than the average refractive index of the cladding region. Due to the large degree of design freedom and flexibility and also strong dependence on wavelength of the mode effective index, microstructured fibers have opened up a variety of new applications such as spectral broadening of a short pulse due to extreme nonlinear effects after propagating through a MOF resulting in generation of supercontinuum light, wide band transmission, high power delivery, endlessly single mode, very large or very small mode effective area, low-loss guidance of light in an air core and so on. In this chapter we would attempt to describe several application-specific specialty fibers and include some results of our own work in this direction.

2. Optical transparency

2.1 Loss spectrum

Loss spectrum, dispersion and *nonlinear propagation effects* are the three most important propagation characteristics of any signal transmitting single-mode optical fiber in the context of modern optical telecommunication. An illustrative example of the loss spectrum of a state-of-the-art commercially available conventional ITU[1] recommended standard G.652 type of single-mode fiber is shown in Fig. 1 [Pal, 2006]. Except for portion of the loss spectrum around 1380 nm at which a peak appears due to absorption by minute traces of OH^- present in the fiber, the rest of the spectrum in a G.652 fiber could be well described through a wavelength dependence as $A\lambda^{-4}$ meaning thereby that signal loss in such fibers is essentially dominated by Rayleigh scattering; A is the Rayleigh scattering coefficient. With GeO_2 as the dopant and relative core-cladding index difference $\Delta \sim 0.37\%$, where $\Delta \sim [(n_c - n_{cl})/n_{c,cl}]$ estimated Rayleigh scattering loss in a high-silica fiber is about 0.18 - 0.2 dB/km at 1550 nm; $n_{c,cl}$ correspond to core and cladding refractive indices, respectively. Superimposed on this curve over the wavelength range 1360 ~ 1460 nm (often referred to as the E-band) is a dotted curve without the peak but overlapping otherwise with rest of the loss spectrum; this modified spectrum represents a sample corresponding to a low water peak fiber like AllWave™ or SMF-28e™ fiber. In real-world systems, however, there are other sources of loss, which are required to be accounted for, on case-to-case basis,

[1]ITU stands for International Telecommunication Union

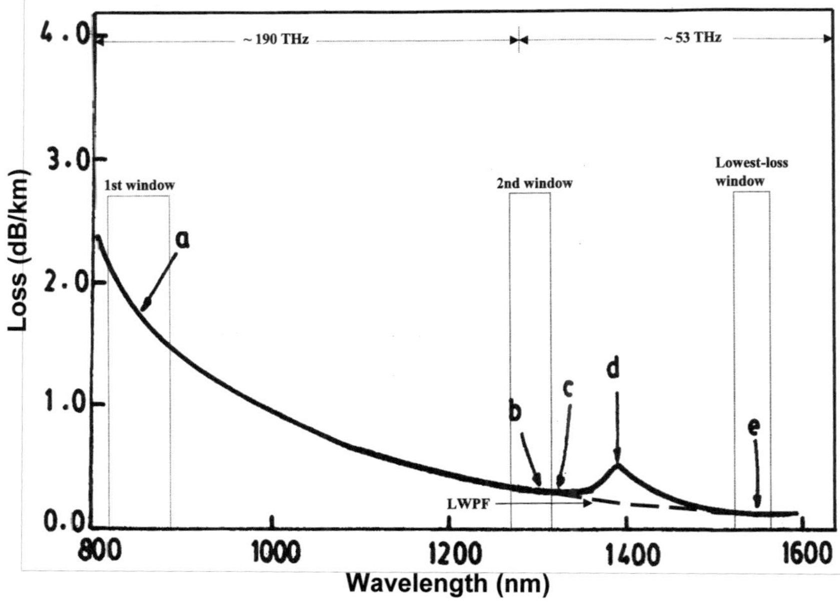

Fig. 1. Loss spectrum (full curve) of a state-of-the-art G.652 type single-mode fiber, e.g. SMF-28 (adapted from Corning product catalogue©Corning Inc.): (a) 1.81 dB/km at 850 nm, (b) 0.35 dB/km at 1300 nm, (c) 0.34 dB/km at 1310 nm, (d) 0.55 dB/km at 1380 nm, (e) 0.19 dB/km at 1550 nm. The dashed portion of the curve corresponds to that of a low water peak fiber due to reduction of the OH^- peak in enhanced SMF; theoretical transmission bandwidths at different low loss spectral windows are also shown [After Pal, 2006; ©2006 Elsevier Press].

and which could add up to more than the inherent loss of the fiber. Examples of these potential sources are cabling-induced losses due to microbending, bend-induced losses along a fiber route due to wrong installation procedures, losses due to splices and connectors including those involving insertion of various components in a fiber link. Several of these also depend to a certain extent on the refractive index profile of the fiber in question. Detailed studies have indicated that these extraneous sources of loss could be addressed by optimizing mode field radius (W_P, known as Petermann spot size/mode field radius) and effective cut-off wavelength (λ_{ce}) [Pal, 1995]. The parameter W_P effectively determines transverse offset-induced loss at a fiber splice as well as loss-sensitivity to microbends and λ_{ce} essentially determines sensitivity to bend-induced loss in transmission. For operating at 1310 nm, optimum values of these parameters were found to range between 4.5 and 5.5 μm and between 1100 and 1280 nm, respectively. An indirect way to test that λ_{ce} indeed falls within this range is ensured if the measured excess loss of 100 turns of the fiber loosely wound around a cylindrical mandrel of diameter 7.5 cm falls below 0.1 dB at 1310 nm and below 1.0 dB at 1550 nm [Pal, 1995].

2.2 Dispersion spectrum

Chromatic dispersion, whose very name implies that it is dependent on wavelength and whose magnitude is a measure of the information transmission capacity of a single-mode fiber, is another important transmission characteristic (along with loss) and it arises because of the dispersive nature of an optical fiber due to which the group velocity of a propagating signal pulse becomes a function of frequency (usually referred to as *group velocity dispersion* [GVD] in the literature), which limits the number of pulses that can be sent through the fiber per unit time. This phenomenon of GVD induces frequency chirp to a propagating pulse, meaning thereby that *leading edge* of the propagating pulse differs in frequency from the *trailing edge* of the pulse. The resultant frequency chirp [i.e. $\omega(t)$] leads to inter-symbol interference (ISI), in the presence of which the receiver fails to resolve the digital signals as individual pulses when the pulses are transmitted too close to each other [Ghatak & Thyagarajan, 1998]. Thus the signal pulses though started, as individually distinguishable pulses at the transmitter, may become indistinguishable at the receiver depending on the amount of chromatic dispersion-induced broadening introduced by the fiber after propagating through a given length of the fiber (see Fig. 2). For quantitative purposes, dispersion is expressed through dispersion coefficient (D) defined as [Thyagarajan & Pal, 2007]

$$D = \frac{1}{L}\frac{d\tau}{d\lambda_0} = -\frac{\lambda_0}{c}\frac{d^2 n_{eff}}{d\lambda_0^2} \tag{1}$$

where τ is the group delay and n_{eff} is the mode effective index of the fundamental mode in a single-mode fiber. It can be shown that the total dispersion coefficient (D_T) is given to a very good accuracy by the algebraic sum of two components $D_T \cong D_M + D_{WG}$ where D_M and D_{WG} correspond to material and waveguide components of D, respectively. A plot of typical dispersion spectrum is shown in Fig. 3. It can be seen from it that D_{WG} is all along negative and while sign of D_M changes from negative to positive (going through zero at a wavelength of ~ 1280 nm [Payne & Gambling, 1975]) with increase in wavelength. The two components of D cancel each other at a wavelength near about 1300 nm, which is referred to as the *zero dispersion wavelength* (λ_{ZD}) and it is a very important design parameter of single- mode fibers. Realization of this fact led system operators to choose the operating wavelength

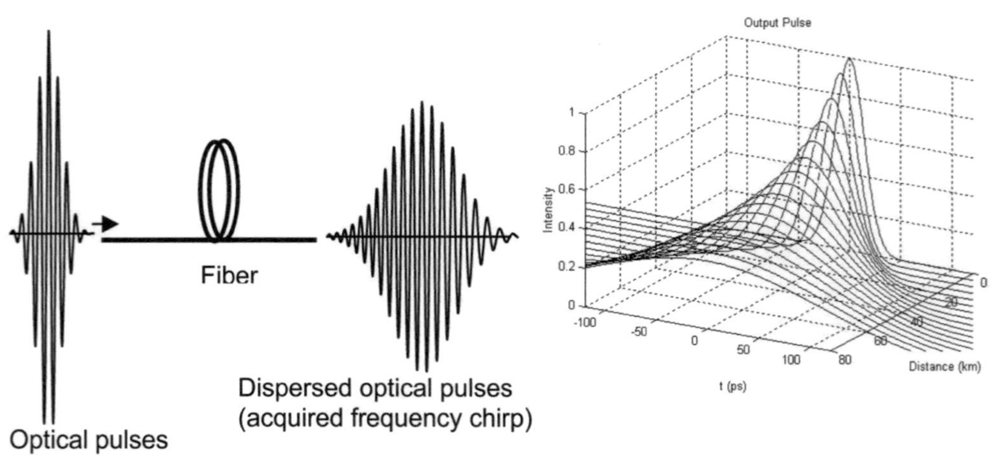

Fig. 2. a) Schematic showing dispersion of a Gaussian input pulse with propagation through a dispersive medium e.g. an optical fiber; frequency chirp is apparent in the dispersed pulse; b) Calculated dispersion induced broadening of a Gaussian pulse [Results courtesy Sonali Dasgupta]

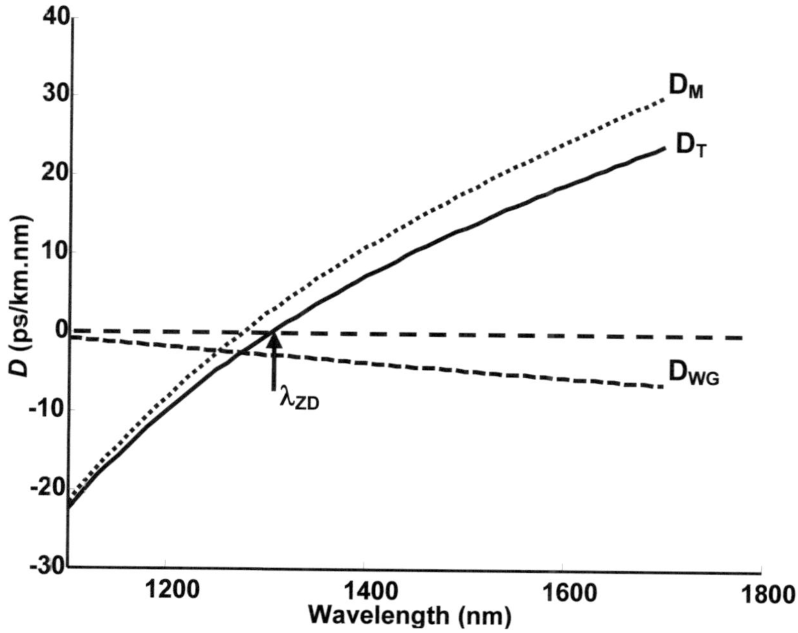

Fig. 3. Dispersion coefficient, D as a function of wavelength of a typical G.652 type fiber; D_M and D_{WG} correspond to material and waveguide components, respectively of total D while λ_{ZD} refers to wavelength for zero total dispersion [After Thyagarajan and Pal, 2006; ©2006 Springer Verlag].

of first generation single-mode fibers as 1310 nm. These fibers optimized for transmission at 1310 nm are now referred to as G.652 fibers as per ITU standards; millions of kilometers of these fibers are laid underground all over the world. Though it appears that if operated at λ_{ZD} one might get infinite transmission bandwidth, in reality zero dispersion is only an approximation (albeit a very good approximation) because it simply signifies that only the second order dispersive effects would be absent. In fact as per ITU recommendations, SMF-28 type of standard G.652 fibers qualify for deployment as telecommunication media provided at the 1310 nm wavelength, its D_T is < 3.5 ps/(nm.km). At a wavelength around λ_{ZD} third order dispersion would determine the net dispersion of a pulse. In the absence of second order dispersion, pulse dispersion is quantitatively determined by the dispersion slope S_0 at $\lambda = \lambda_{ZD}$. A knowledge of D and S_0 enables determination of dispersion (D) at any arbitrary wavelength within a transmission window e.g. EDFA band in which D in G.652 fibers varies approximately linearly with the operating wavelength λ_0 [Pal, 2006]. Besides pulse broadening, since the energy in the pulse gets reduced within its time slot, the corresponding signal to noise ratio (SNR) will decrease, which could be compensated by increasing the power in the input pulses. This additional power requirement is termed as *dispersion power penalty* [Thyagarajan & Pal, 2007]. For 1 dB dispersion power penalty at the wavelength of 1550 nm, we can write the following inequality as a design relation for estimating maximum allowed dispersion

$$B^2.D.L \leq 10^5 \text{ Gb}^2\text{.ps/nm} \quad (2)$$

where B is measured in Gbits, D in ps/(nm.km) and L in km [Pal, 2006; Thyagarajan & Pal, 2007]. Based on Eq. (2), Table I lists the maximum allowed dispersion for different standard bit rates for tolerating a dispersion power penalty of 1 dB.

Data rate (B)	Maximum allowed dispersion ($D.L$)
2.5 Gb/s (OC-48)*	~ 16000 ps/nm
10 Gb/s (OC-192)	~ 1000 ps/nm
40 Gb/s (OC-768)	~ 60 ps/nm

*OC stands for Optical Channels, which are standards for DWDM systems

Table I. Maximum allowed dispersion at different standard signal transmission rates

2.3 Nonlinear propagation effects

With the emergence of DWDM transmission, a transmission fiber is typically required to handle large power density and long interaction lengths due to transmission of multiple channels over long fiber lengths. In view of this, nonlinear propagation effects have become yet another important transmission characteristic of an optical fiber. It is well known that under the influence of intense electromagnetic fields (comparable to the inter-atomic fields that exist in the propagating medium) a fiber may exhibit certain nonlinear effects [Agrawal, 2007]. This nonlinear response of the fiber could be mathematically described through the following nonlinear relation between **P** and **E**, where **P** represents polarization induced by the electric field **E** of the propagating em field:

$$P = \varepsilon_0 \left(\chi^{(1)} E + \chi^{(2)} EE + \chi^{(3)} EEE + ...\right) \qquad (3)$$

where ε_0 is vacuum permittivity and $\chi^{(n)}$ is the nonlinear susceptibility of the medium. In silica, $\chi^{(2)}$ is absent while $\chi^{(3)}$ is finite, which is the major contributor to nonlinear effects in silica-based optical fibers. A consequence of this finite $\chi^{(3)}$ is that it leads to an intensity dependent refractive index of silica as

$$n^{NL} = n_1 + n_2 I \qquad (4)$$

where n_1 is the linear part of the fiber core refractive index and n_2 (~ 3.2 x 10^{-20} m^2/W in silica as measured at 515 nm through self-phase modulation (SPM) technique; at the 1550 nm wavelength region it is usually less by ~10% [Agrawal, 2007])) is the nonlinear component of the nonlinear refractive index, related to $\chi^{(3)}$ through [Ghatak & Thyagarajan, 1998]

$$n_2 = \frac{3}{4} \frac{\chi^{(3)}}{c \varepsilon_0 n_1^2} \qquad (5)$$

and I is the intensity of the propagating em wave. For describing nonlinear effects in a fiber, often a parameter known as γ, which is defined through

$$\gamma = \frac{2\pi}{\lambda} \frac{n_2}{A_{eff}} \qquad (6)$$

is used, where [Agrawal, 2007]

$$A_{eff} = \frac{\left[\int_0^{\infty}\int_0^{2\pi} E^2(r) r dr d\varphi\right]^2}{\int_0^{\infty}\int_0^{2\pi} E^4(r) r dr d\varphi} \tag{7}$$

is known as the mode effective area of a fiber. The parameter γ ~ 1/(W.km) in a standard fiber like SMF-28 (of Corning Inc.). Its value could be enhanced either through smaller A_{eff} (obtainable through appropriate fiber designs) or choice of the glass materials or both. Values as large as 20 times that of SMF-28 has been reported in certain fibers through these routes [Okuno et al, 1999; Monro, 2006]. Often the fundamental LP_{01} mode of a single-mode fiber is approximately describable by a Gaussian distribution with a spot size (or mode field radius) of ω_0; in that case $\pi\omega_0^2$ would yield A_{eff} [Ghatak & Thyagarajan, 1998; Agrawal, 2007]. For simple estimations, we may consider ω_0 to be ~ 5 μm for a typical single-mode fiber at the operating wavelength of 1550 nm and hence A_{eff} ~ 75 - 80 μm^2 ⇒ for a propagating power of ~ 100 μW across this cross-section of the fiber, net intensity would be ~ 106 W/m^2, resulting in a net change (= n_2I) in refractive index of the fiber ~ 3.2 x 10^{-14}, which though small in magnitude could lead to substantial nonlinear effect in a fiber. This is because the typical fiber lengths encountered by a propagating optical signal could be few tens to hundreds of kilometers. Nonlinear effects in a fiber could be broadly classified into two domains depending on the physical mechanism involved: refractive index related effects like self-phase modulation (SPM), cross-phase modulation (XPM), and four wave mixing (FWM) or stimulated scattering effects like stimulated Brillouin scattering (SBS), stimulated Raman scattering (SRS), and soliton self frequency shift [Agrawal, 2007]. For example, FWM could lead to generation of new waves of different wavelengths in a DWDM link. If there are N propagating signal channels, number of new waves generated [Li, 1995] due to FWM would be ~ $N^2(N - 1)/2$. In view of this, cross coupling of power between these waves/side-bands could lead to depletion of power from the signal channels and hence could result in lowering of signal to noise ratio in a particular signal channel. Detailed analyses show that FWM effect could be substantially reduced if the fiber is so designed that signals experience a finite dispersion within the fiber and/ or channel spacing is large in a DWDM stream. As an example, for a single-mode fiber having average loss of 0.25 dB/km at 1.55 μm, FWM induced power penalty in terms of the ratio of generated power [at the new frequency ω_1 (= $\omega_3 + \omega_4 - \omega_2$)] to the output power after a length of 100 km of the fiber for a channel spacing of 1 nm is ~ - 25 dB for D = 0 ps/(nm.km), ~ - 47 dB for D = 1 ps/(nm.km), and ~ - 70 dB for D = 17 ps/(nm.km); if the channel spacing is reduced to 0.5 nm, the corresponding power penalties are ~ - 25 dB, ~ - 35 dB, and ~ - 58 dB, respectively [Li, 1995]. Thus non-linear optical effects are important issues required to be accounted for in a DWDM communication systems of today. As per ITU standards, in a DWDM stream, signal channels are required to be equally spaced. One of the important parameters often considered in designing DWDM systems is known as spectral efficiency (*SE*), which is defined as the ratio of bit rate to channel spacing. Since, bit rate cannot be increased arbitrarily due to constraints imposed by electron mobility, *SE* could be enhanced through smaller and smaller channel spacing and hence in that context nonlinear effects like FWM becomes an important issue. For DWDM applications, fiber designers came up with new designs for the signal fiber for low-loss and dedicated DWDM signal transmission at the 1550 nm band, which were generically named as non-zero dispersion shifted fibers (NZ-

DSF). For a while, in between there was a move to deploy single-mode fibers with $D \sim 0$ at the 1550 nm band to take simultaneous advantage of lowest transmission loss in silica fibers in that band. Such fibers were known as dispersion shifted fibers (DSF). However, they were found to be of no use for DWDM transmission because nonlinear effects would be significantly high due to D being 0 in these fibers! These fibers were designed to substantially suppress nonlinear effects like FWM by allowing each of the DWDM signals to experience a finite amount of dispersion during propagation [Pal, 2006]. ITU has christened such fibers as G.655 fibers, which should exhibit dispersion $2 \leq D$ (ps/(nm.km)) ≤ 6 in the 1550 nm band in order to detune the phase matching condition required for detrimental nonlinear propagation effects like four-wave mixing (FWM) and cross-phase modulation (XPM) to take place during multi-channel transmission of DWDM signals.

3. Emergence of amplifiers

3.1 Fiber amplifiers

In the late 1980s typical state-of-the-art repeater-less transmission distances were about 40-50 kms @ 560 Mb/s transmission rate. Maximum launched optical power into a fiber was below 100 μW, it was difficult to improve system lengths any further and use of electronic repeaters became inevitable. At a repeater, the so-called 3**R**-regeneration functions (*re-amplification, retiming, and reshaping*) are performed in the electric domain on the incoming attenuated as well as distorted (due to dispersion) signals after detection by a photo-detector and before the revamped signals are fed to a laser diode drive circuits, wherefrom these cleaned optical pulses are re-injected in to next section of the fiber link. However, these complex functions required unit replacement in case of network capacity upgrades because electronic components are bit rate sensitive. By mid-1980s, it was felt that *an optical amplifier is needed to bypass this electronic bottleneck.* Fortuitously in 1986, the Southampton University research group in England reported success in incorporating rare earth trivalent erbium ions into host silica glass during fiber fabrication [Mears et al, 1986]; Erbium is known to have strong fluorescence at 1550 nm. Subsequently, the same group demonstrated that excellent noise and gain performance is feasible in a large part of the 1550 nm window with erbium-doped silica fibers [Mears et al, 1987], which ushered in the era of erbium doped fiber amplifiers (EDFA) and DWDM networks; EDFAs went commercial by 1996. Absorption bands most suitable as pump for obtaining amplification of signals at the 1550 nm wavelength region are 980 nm and 1480 nm wavelengths although 980 is a more popular choice; in Fig. 4 a schematic of an EDFA is shown. When pumped at either of these wavelengths, an erbium doped fiber was found to amplify signals over a band of almost 30 ~ 35 nm extending from 1530 ~ 1565 nm, which is known as the C-band fiber amplifier.

Thus, a single EDFA can be used to amplify several channels simultaneously required for dense wavelength division multiplexing (DWDM). EDFAs are new tools that system planners now almost routinely use while designing high capacity optical networks. Practical EDFAs with output power of around 100 mW (20 dBm), 30 dB small signal gain, and a noise figure of < 5 dB are available commercially. We may note that core diameter of an EDF is typically smaller (almost half) of the standard single-mode fibers like SMF-28 and require special program(s) in a fiber splice machine to achieve low-loss splice of these special fibers with signal carrying standard fibers like SMF-28. Further, it could be seen from Fig. 4 that an integral component of an EDFA is a fused fiber coupler, which multiplexes 980 nm pump and 1550 nm band of signal wavelengths at the input of the EDF. In order for the fiber

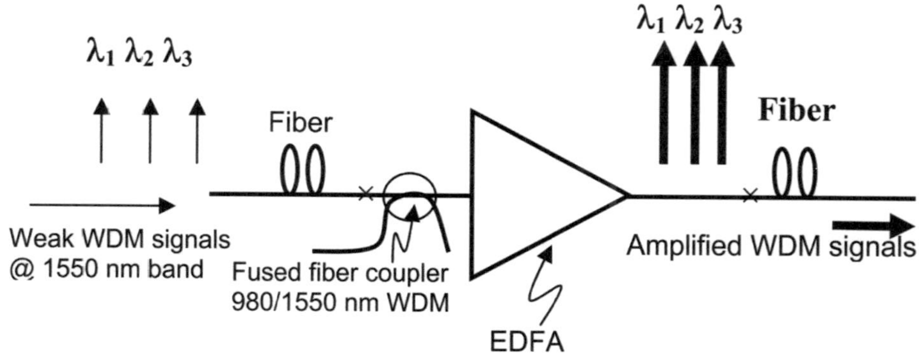

Fig. 4. Schematic of an EDFA in which 980 (or 1480) nm wavelength is used as a pump, which creates population inversion in Er^{+3} ions in the EDF and the weak WDM signals at the 1550 nm band get amplified as it propagates through the population inverted EDF

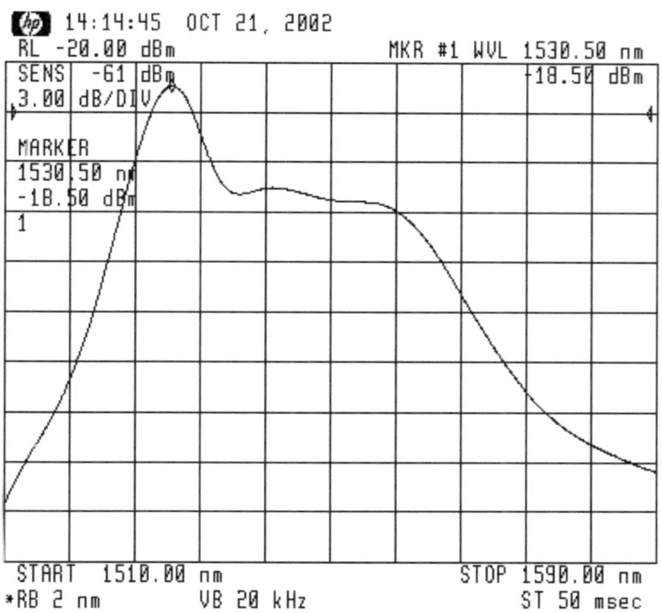

Fig. 5. Measured ASE spectrum (power vs wavelength) of an EDF as an example

used to configure such a multiplexer to be single moded at both the pump and signal wavelengths, λ_{ce} of the fiber should be < 980 nm. This led to development of special single-mode fibers known as SMF 980™ and Flexcor™ fibers, both being registered trademark of two different companies. Figure 5 shows typical amplified spontaneous emission (ASE) spectrum of an EDF when pumped with a diode laser at 980 nm. It could be seen from the figure that the spectrum is non-uniform, this caharacteristic in conjunction with the saturation effects of EDFAs cause increase in signal power levels and decrease in the optical signal-to-noise ratio (OSNR) to unacceptable values in systems consisting of cascaded chains of EDFAs [Srivastava & Sun, 2006]. These features could limit the usable bandwidth of EDFAs and hence the amount of data transmission by the system. Accordingly various schemes of gain equalizing filters (GEF) such as Mach-Zehnder filter [Pan et al, 1995], acousto-optic filter [Kim et al, 1998], long-period fiber-grating [Vengsarkar et al, 1996], fiber-loop mirror [Li et al, 2001; Kumar et al, 2005], side-polished fiber based filter [Varshney et al, 2007] and so on have evolved in the literature. However, in the design of certain special networks like a metro network, one of the major drivers is low installation cost in addition to achieving low maintenance/repair costs. Naturally one of the routes to achieve these objectives would be to use fewer components in the network. Use of an intrinsically gain flattened EDFA would cut down the cost on the GEF head. This motivated some investigators [e.g., Nagaraju et al, 2009] to investigate design of an inherently gain flattened EDFA by exploiting a wavelength filtering mechanism inherent in a co-axial dual-core fiber design scheme. An example of the design of such a gain-flattened EDFA is shown in Fig. 6, which was based on a highly asymmetric dual-core concentric fiber design (see Fig. 6), whose inner core was only partially doped with erbium [Nagaraju et al, 2009]. Refractive index profile (RIP) of the designed inherently gain flattened EDFA and the RIP of the corresponding fabricated preform are shown in Fig. 6(a) and Fig. 6(b), respectively; r_d refers to ER^{+3} doping radius. The RIP was measured using a fiber analyzer. The so realized RIP was close to the designed one except for small profile perturbations typical in fibers fabricated by the MCVD process. Figure 7 shows the measured gain and noise figure as a

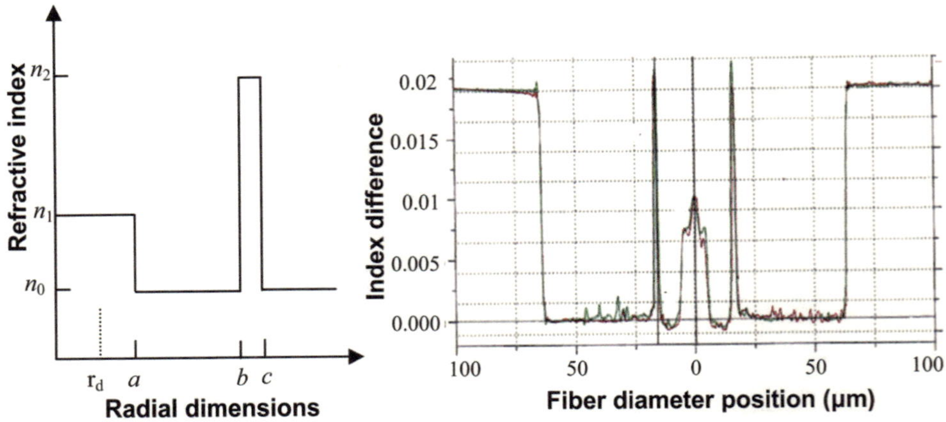

Fig. 6. a) Schematic of the theoretically designed refractive index profile of an inherently gain-flattened EDFA; b) Corresponding refractive index profile of the fabricated fiber preform (After Nagaraju et al, 2009; ©2009 Elsevier Press)

function of wavelength for the fabricated coaxial fiber; some improvements in the noise figure at longer wavelengths could be seen due to the increased overlap between the pump and the signal modes at longer wavelengths. Gain variation across the C-band was found to be more than the designed one, which is attributable to small variations in the fabricated fiber RIP parameters from the one that was designed. A very precise comparison is, in any case, difficult due to lack of sufficient precision inherent in measurement instruments for estimating various parameters of the fiber RIP and the dopant level. Phase-resonant optical coupling between the inner low index contrast core and the outer high index contrast narrow ring that surronds the inner core was so tailored through optimization of its refractive index profile parameters that the longer wavelengths within the C-band experience relatively higher amplification compared to the shorter wavelengths

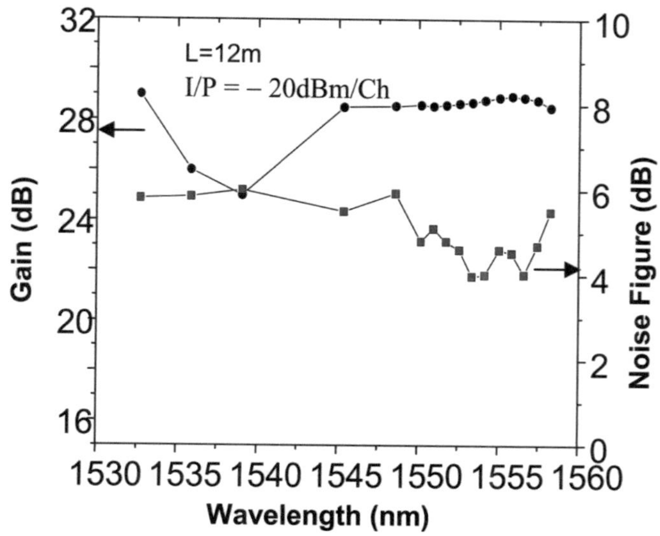

Fig. 7. Experimental results for measured gain (•) and noise figure (■) with wavelength on a 12 m long fabricated EDF (After Nagaraju et al, 2009; ©2009 Elsevier Press)

thereby reducing the difference in the well-known tilt in the gains between the shorter and longer wavelength regions. The fabricated EDFA exhibited a median gain ≥ 28 dB (gain excursion below ± 2.2 dB within the C-band) when 16 simultaneous standard signal channels were launched by keeping the I/P level for each at -20 dBm/channel. We may mention that another variety of EDFA is known as L-band EDFA, L standing for long wavelength band (1570 ~ 1610 nm) [Sun et al, 1997]. With suitable optimization of EDFAs, C-band and L-band amplifiers could be used as two discrete amplifiers to simultaneously amplify 160 signal wavelength channels by separating the transmission of two bands of weak signals for amplification by these amplifiers.

In addition to EDFA, another kind of fiber amplifiers that is commonly used is known as Raman fiber amplifiers (RFA). One of the nonlinear scattering processes namely, stimulated Raman scattering (SRS) is responsible for Raman amplification, which is quite broadband (up to 40 THz) with a broad peak appearing near 13.2 THz in bulk silica [Agrawal, 2007]. Classically this scattering process is described as an inelastic scattering process in which frequency of a propagating pump light beam of energy $\hbar\omega_P$ in a molecular medium shifts to a lower frequency i.e. suffers red shift to generate lower energy photon of energy $\hbar\omega_S$, which is determined by the molecular vibrational levels of the medium, in this case the doped silica fiber, according to the well known fundamental process known as the Raman effect. Without going in to the details, quantum mechanically it can be shown that the process leads to an amplification of a co-propagating signal as long as the frequency difference $\omega_P - \omega_S$ lies within the bandwidth of the Raman gain spectrum [Agrawal, 2007]. The most important parameter that characterizes the amplification process in a RFA is known as the Raman gain efficiency (C_R), defined as the ratio of coefficient (γ_R) to A_{eff} of a fiber. Figure 8 depicts a plot of C_R as a function of the difference $\omega_P - \omega_S$ in THz for three different variety of specialty fibers: NZDSF (A_{eff} ~ 55 μm^2), super large area special fiber (SLA of A_{eff} ~ 105 μm^2), and a dispersion compensating fiber (DCF of A_{eff} ~ 15 μm^2); pump wavelength was 1.45 μm and signal wavelength was ~ 1.55 μm [Bromage, 2004]. It can be seen that C_R is largest in the case of DCF, which could be attributed to its A_{eff} that was the least and also it had a high Germania content compared to other two fiber varieties. Difference in the peak value of the Raman gain curve could be attributed to difference in A_{eff} and degree of overlap between the pump and the signal transverse modes [Urquhart & Laybourn, 1985; Bromage, 2004]. It is known that GeO_2 molecules exhibit larger peak gain (than silica) near 13.1 THz [15]. Dispersion compensating fibers, which represent one class of specialty fibers, are described in the next section; these are typically characterized with large Δ (typically ~ 2%) and small A_{eff}. In view of the large Raman gain exhibited by a typical DCF, these are often used as RFA. In certain DWDM systems, EDFA as well as RFA are used as a hybrid broadband amplifier for DWDM signals.

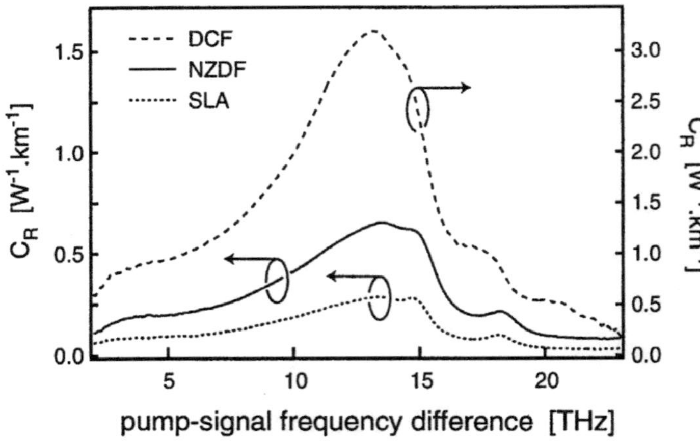

Fig. 8. Mesured C_R for three different variety of germanosilicate fibers having different 1550 nm A_{eff} and each pumped with 1450 nm diode laser (After Bromage, 2004; ©2004 IEEE).

3.2 Dispersion compensating fibers

To counter potentially detrimental nonlinear propagation effects in a DWDM link since a finite (albeit low) D is deliberately kept in NZDSF fibers, signals would accumulate dispersion between EDFA sites! Assuming a D of 2 ps/(nm.km), though a fiber length of about 500 km could be acceptable @ 10 Gbit/s before requiring correction for dispersion, @ 40 Gbit/s corresponding un-repeatered span would hardly extend to 50 km [see Eq. (2)]. The problem is more severe in G.652 fibers for which @ 2.5 Gbit/s though a link length of about 1000 km would be feasible at the 1550 nm window, if the bit rate is increased to 10 Gbit/s, tolerable D in this case over 1000 km would be hardly ~ 1 ps/(nm.km)! Realization of this fact triggered development of some dispersion compensating schemes in mid-1990s, which could be integrated to a single-mode fiber optic link so that net dispersion of the link could be maintained/managed within desirable limits. Three major state-of-the-art fiber-based optical technologies available as options for dispersion management are: *dispersion compensating fibers* (DCF) [Ramachandran, 2007], *chirped fiber Bragg gratings* (CFBG) [Kashyap, 1999], and *high-order-mode* (HOM) *fiber* [Ramachandran, 2006].

In order to understand the logic behind dispersion compensation techniques, we consider the instantaneous frequency of the output pulse from a single-mode fiber, which is given by [Thyagarajan & Pal, 2007]

$$\omega(t) = \omega_c + 2\kappa\left(t - \frac{L}{v_g}\right) \tag{8}$$

where ω_c represents carrier frequency, the center of the pulse corresponds to $t = L/v_g$, where v_g is the group velocity, and κ is a parameter, which depends on $d^2\beta/d\lambda_0^2$. Accordingly the leading and trailing edges of the pulse correspond to $t < L/v_g$ and $t > L/v_g$, respectively. In the normal dispersion regime (where operating wavelength < λ_{ZD}) κ is positive, thus the leading edge of the pulse will be down-shifted i.e *red-shifted* in frequency while the trailing edge will be up-shifted i.e. *blue-shifted* in frequency with respect to the center frequency ω_c. The converse would be true if the signal pulse wavelength corresponds to the anomalous dispersion region (operating $\lambda > \lambda_{ZD}$) where κ is negative. Hence as the pulse broadens with propagation due to this variation in its group velocity with wavelength in a dispersive medium like single-mode fiber it also gets chirped. If we consider propagation of signal pulses through a G.652 fiber at the 1550 nm wavelength band at which its D is positive, it would exhibit *anomalous* dispersion. If this broadened temporal pulse were transmitted through a DCF, which exhibits *normal* dispersion (i.e. its dispersion coefficient D is negative) at this wavelength band, then the broadened pulse would get compressed with propagation through the DCF. Thus if the following condition is satisfied:

$$D_T L_T + D_c L_c = 0 \tag{9}$$

the dispersed pulse would recover back its original shape; subscripts T and c stand for the transmission fiber and the DCF. Consequently if a G.652 fiber as the transmission fiber is operated at the EDFA band, corresponding DCF must exhibit negative dispersion at this wavelength band ⇒ its D_{WG} (negative in sign) must be large enough to be more than D_M in magnitude. Large negative D_{WG} is achievable through appropriate design tailoring of the refractive index profile of the fiber so that at the wavelengths of interest a large fraction of its modal power rapidly spreads into the cladding region for a small change in the propagating wavelength. The *first generation* DCFs relied on narrow core and large Δ (typically ≥ 2 %) fibers to achieve this task and hence necessarily involved relatively large insertion loss. Accordingly, a parameter named as figure of merit (FOM) is usually ascribed to any DCF defined through

$$\text{FOM} = \frac{-D_c}{\alpha_c} \tag{10}$$

These DCFs were targeted to compensate dispersion in G.652 fibers at a single wavelength and were characterized with a $D \sim -50$ to -100 ps/(nm.km) and a positive dispersion slope. Since DWDM links involve multi-channel transmission, it is imperative that ideally one would require a broadband DCF so that dispersion could be compensated for all the wavelength channels simultaneously. The key to realize a broadband DCF lies in designing a DCF, in which not only that D versus λ is negative at all those wavelengths in a DWDM stream, but also its dispersion slope is negative. Broadband dispersion compensation ability of a DCF is quantifiable through a parameter known as relative dispersion slope (*RDS*), which is defined through [Pal, 2006; Thyagarajan & Pal, 2007]

$$RDS = \frac{S_c}{D_c} \tag{11}$$

Values of *RDS* (nm^{-1}) for well-known NZ-DSF's like LEAF™, TrueWave-RS™, and Teralight™ are 0.0202, 0.01, and 0.0073, respectively. Thus if a DCF is so designed that its *RDS* matches that of the transmission fiber then that DCF would ensure perfect compensation for all the signal wavelengths. Such DCFs are known as dispersion slope compensating fibers (DSCF). *RDS* for G.652 fibers at 1550 nm is about 0.00335 nm^{-1}. One recent dual-core DCF design, whose refractive index profile was similar to that shown in Fig. 6 with the difference that it had a high refractive contrast for the central core and lower refractive contrast for the outer wider ring core, had yielded the record for largest negative D (– 1800 ps/(nm.km) at 1558 nm) in a DCF. The two cores essentially function like a directional coupler. Since these two concentric fibers are significantly non-identical, through adjustments of index profile parameters their mode effective indices could be made equal at some desired wavelength called phase matching wavelength (λp), in which case the effective indices as well as modal field distributions of the normal modes of this dual core fiber exhibit rapid variations with λ around λp [Thyagarajan & Pal, 2007]. Further research in this direction has led to designs of dual-core DSCFs for broadband dispersion compensation in G.652 as well as G.655 fibers within various amplifier bands like S-, C- and L- bands [Pal & Pande, 2002; Pande & Pal, 2003]. Mode effective areas of dual core DSCFs could be designed to attain A_{eff}, which are comparable to that of the G.652 fiber (≈ 70 - 80 μm^2). The net residual dispersion spectra of a 100 km long G.652 fiber link along with so designed DSCFs (approximately in ratio of 10:1) at each of the amplifier band are shown in Fig. 9. It could be seen that residual average dispersion is well within ± 1 ps/(nm.km) within all the three amplifier bands.

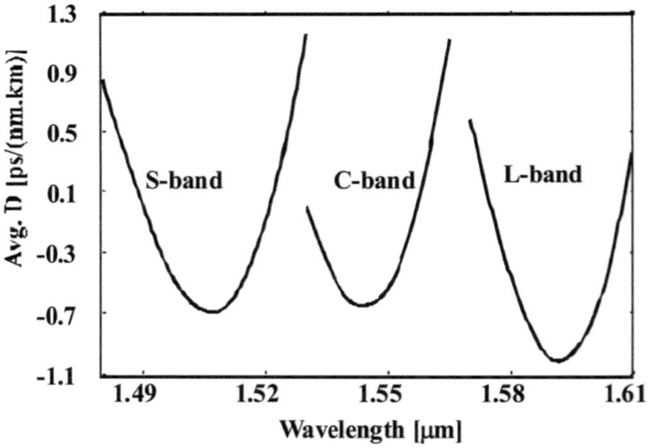

Fig. 9. Net residual dispersion at different amplifier bands of a dispersion compensated link design consisting of 100: 10 (in kms) of G.652 fiber and 10 km of a dual-core DSCFs (After Pal & Pande, 2004; ©2004 Elsevier Press).

3.3 Fibers for metro networks [Pal, 2006]

During the IT bubble burst, there has been a slowing down of business in optical communication due to the so-called huge fiber glut in the long haul networks (typically trans-oceanic). However, the gap between the demand and supply of bandwidth has been much less in the metro sector and in recent years metro optical networks have attracted a great deal of attention due to potentials for high growth. A metro network provides generalized telecommunication services transporting any kind of signal from one point to another in a metro, usually running a couple of hundred kilometers in length. In transport, DWDM is the key enabling technology to expand the capacity of existing and new fiber cables without optical-to-electrical-to-optical conversions. Accordingly the network trend in the metro sector has been to move towards transparent rings, in which wavelength channels are routed past or dropped off at the nodes [Ryan, 2003]. Gigabit Ethernet is fast evolving as a universal protocol for optical packet switching. Thus, in addition to voice, video, and data a metro network should be able to support various protocols like Ethernet, fast Ethernet, and 10 Gbit/s Ethernet. DWDM in a metro environment is attractive in this regard for improved speed in provisioning due to possibility of allocating dynamic bandwidth on demand to customers and for better-cost efficiency in futuristic transparent networks running up to 200 km or more. Legacy metro networks relied heavily on directly modulated low-cost FP-lasers in contrast to the more expensive externally modulated DFB lasers. However directly modulated lasers are usually accompanied with a positive chirp, which could introduce severe pulse dispersion in the EDFA band if the transmission fiber is characterized with a positive D. The positive chirp-induced pulse broadening can be countered with a transmission fiber if it is characterized with normal dispersion (i.e. negative D) at the EDFA band. This is precisely the design philosophy followed by certain fiber manufacturers for deployment in a metro network e.g. MetroCor™ fiber of Corning Inc. For typical transmission distances encountered in a metro network, a DCF in the form of a standard SMF with positive D is used to compensate for dispersion in a MetroCor™ kind of fiber. However, due to a relatively low magnitude of D in a standard SMF, long lengths of it are required, which increase the overall loss budget and system cost. An alternative type of metro fiber has also been proposed and realized, which exhibits positive $D \sim 8$ ps/(nm.km) at 1550 nm [Ryan, 2002]. The argument in favor of such positive dispersion metro-fibers is that dispersion compensation could be achieved with readily available conventional DCFs of much shorter length(s) as compared to standard SMFs that would be necessary to compensate for the negative dispersion accumulated by the metro fibers [Ryan, 2002].

4. Microstructured Optical Fibers (MOF)

4.1 Discovery of the concept of photonic crystals

The state-of-the art in silica-based optical fiber technology could be described as

- Loss close to theoretical limit (0.14 dB/km)
- Dispersion could be tailored close to zero anywhere at a wavelength ≥ 1310 nm but not below 1200 nm unless fiber core is significantly reduced through tapering, for example [Birks et al, 2000]
- Minimum nonlinear impairments over distances ≥ 100 km
- High quality fiber amplifiers with low noise to compensate for whatever be the transmission loss; noise figure could be close to theoretical minimum of 3 dB
- Demonstration of hero experiments at transmission rates > terabit/s over a single fiber through modulation techniques like CSRZ-DQPSK and polarization mode division multiplexing.

With so much of development it appeared for a while that there was no further research scope for development of newer fibers. However it became increasingly evident in the early 1990s that there is a need to develop specialty fibers in which material loss is not a limiting factor and fibers in which nonlinearity and/ or dispersion could be tailored to achieve propagation characteristics, which are otherwise impossible to achieve in conventional fibers. Two research groups - Optoelectronic Research Center in University of Southampton in UK (and soon after University of Bath) and MIT in USA exploited the concept of Photonic Crystals (PhC), which were proposed for the first time independently in two papers, that appeared simultaneously in the same issue of Physical Review Letters in 1987 [Yablonovich, 1987; John, 1987], to develop a completely new variety of specialty fibers broadly known as microstructured optical fibers. Yavlonovitch and John in their 1987 papers (coincidentally EDFA was also discovered in the same year!) showed independently that a lattice of dielectrics with right spacing and different optical properties could generate a photonic bandgap (analogous to electronic band gap in a semiconductor). A square lattice of periodic air holes in a higher index dielectric like silica shown in Fig. 10 could be cited as an example. They showed that wavelength scale structuring of such lattices in terms of refractive index features could be exploited as a powerful tool to modify their optical properties. Light wavelengths falling within the structure's characteristic band gap will not propagate i.e. would not be supported in that medium. If, however, a defect is created in the lattice (e.g. through removal of one hole at the center so that the region becomes a solid of same material as the rest of the medium) to disrupt its periodicity (akin to change in semiconductor properties by dopants), the same forbidden band of wavelengths could be supported and localized in that medium within the defect region. Thus if one were to extend this concept to a cylindrical geometry like a fiber, the defect region would mimic the core of an optical fiber with the surrounding 2D-periodic arrays of air holes in silica like PhC as forming a cladding of a lower average refractive index. This indeed formed the functional principle of "holey" type of MOFs. Due to strong interaction of the propagating light in the bandgap with the lattice, n_{eff} becomes a strong function of the propagating wavelength, which yields an additional functionality, which could be exploited to form a fiber that could function as an endlessly single-mode fiber over a huge bandwidth [Birks et al, 1997]. In view of this large bandwidth, holey fibers are usually not referred to as a photonic band gap structure while the other variety of microstructured fibers are often referred to as photonic bandgap guided fibers as described in the next sub-section.

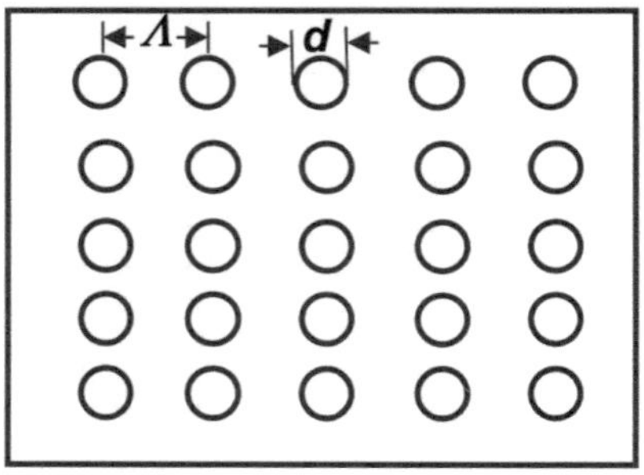

Fig. 10. A square lattice of periodic air holes in a higher index dielectric; Λ is the pitch and d is the diameter of the holes

4.2 Holey and photonic bandgap MOFs

Wavelength scale periodic refractive index features across its cross-section, which run throughout the length of the fiber, characterize MOFs [Monro, 2006]. These special fibers opened up a lot of application potentials not necessarily for telecom alone. In contrast to the fundamental principle of waveguidance through total internal reflection in a conventional fiber, waveguidance in a MOF is decided by two different physical principles - *index* guided and *photonic bandgap* guided (PBG). In index guided MOFs like holey fibers (see Fig. 11), in which the central defect region formed by the absence of a hole yielding a material of refractive index same as the surrounding solid, light guidance in these MOFs could be explained by a *modified total internal reflection* due to the average refractive index created by the presence of PhC cladding consisting of 2D-periodic arrays of air holes in the silica matrix, which is lower than the central defect region. This average index of the cladding depends on the relative distribution of the modal power supported by the silica and air hole lattice, which vary with wavelength. As the wavelength decreases, more and more power gets concentrated within the high index region, the cladding index increases and effective relative refractive index difference between the core and the cladding decreases. As a result the normalized frequency remains relatively insensitive to wavelength. Accordingly, over a broad range of wavelengths the fiber functions as a single-mode fiber. In fact, if the ratio of air hole diameter to the pitch of crystal is kept below 0.45, the fiber remains endlessly single moded [Knight et al, 1996; Birks et al, 1997]. For larger d/Λ, it supports higher order modes as long as λ/Λ is < a critical value.

a) b)

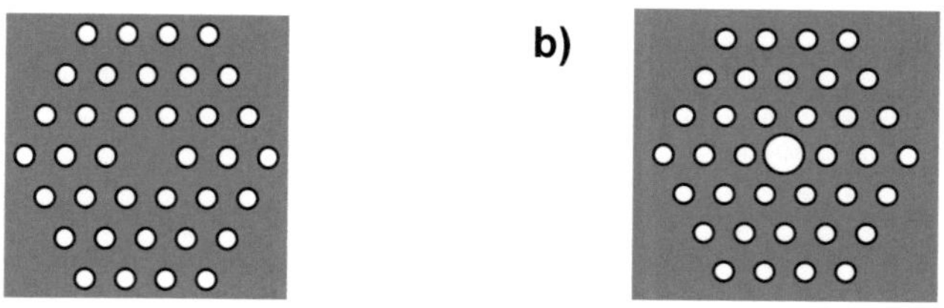

Fig. 11. Schematics of MOFs with white regions representing air or low index medium a) index guided holey type; b) Photonic bandgap guided fiber.

In contrast, in a photonic band gap guided MOF, the central defect region is of a lower refractive index (usually air), which forms the core; typically it is larger in diameter than the low index regions of the PhC cladding. The central core region could have a refractive index same as that of the low index region of the periodic cladding. Functionally light within the photonic bandgap is confined in the central lower index region due to multiple Bragg reflections from the air-silica dielectric interfaces, which add up in phase. Though periodicity is not essential in case of index-guided structures, periodicity is essential in case of PBG guided structures.

4.3 Dispersion tailored holey fibers

Availability of MOFs offered a huge design freedom to fiber designers because a designer could manipulate propagation effects in such a fiber through either or all of the parameters such as lattice pitch (Λ), air hole size (d) and shape, refractive index of the glass, pattern of the lattice, core size and refractive index. We depict some sample results based on use of the software CUDOS* (of University of Sydney, Australia). It is evident from Fig. 12 that D

---

* CUDOS is license free software for simulating index-guided MOFs that is available from University of Sydney, Australia.

becomes a stronger function of wavelength as pitch decreases for a fixed size of the hole and also for larger holes when pitch is kept fixed. Figure 13 depicts D versus wavelength for an index guided DCF, in which the air hole size of one of the outer rings has been reduced, thereby effectively functioning as an outer core of smaller index contrast. Its effective refractive index profile is analogus to that of a dual concentric core fiber discussed earlier in the section on DCF. Figure 13 representing its dispersion spectrum shows broadband DCF nature of such a structure when appropriately optimised. Our deeper studies on such a broadband DCF have revealed that D vs λ curve becomes flatter and confinement loss becomes smaller when the outer core i.e. the ring with smaller sized holes is nearer to the core. Dispersion slope of the designed DCF is ~ – 3.7 ps/km.nm^2, and accordingly its RDS is 0.00357 nm^{-1}, which matches well with the RDS value for standard G.652 type of single-mode fibers (0.0036 nm^{-1}). Figure 14 depicts effect of varying d_2 on D vs λ curve for the dual core holey fiber of the kind shown on Fig. 13. The confinement loss was found to reduce from ~ 10^3 to ~10^{-1} dB/m as the number of cladding rings increases from 5 to 8.

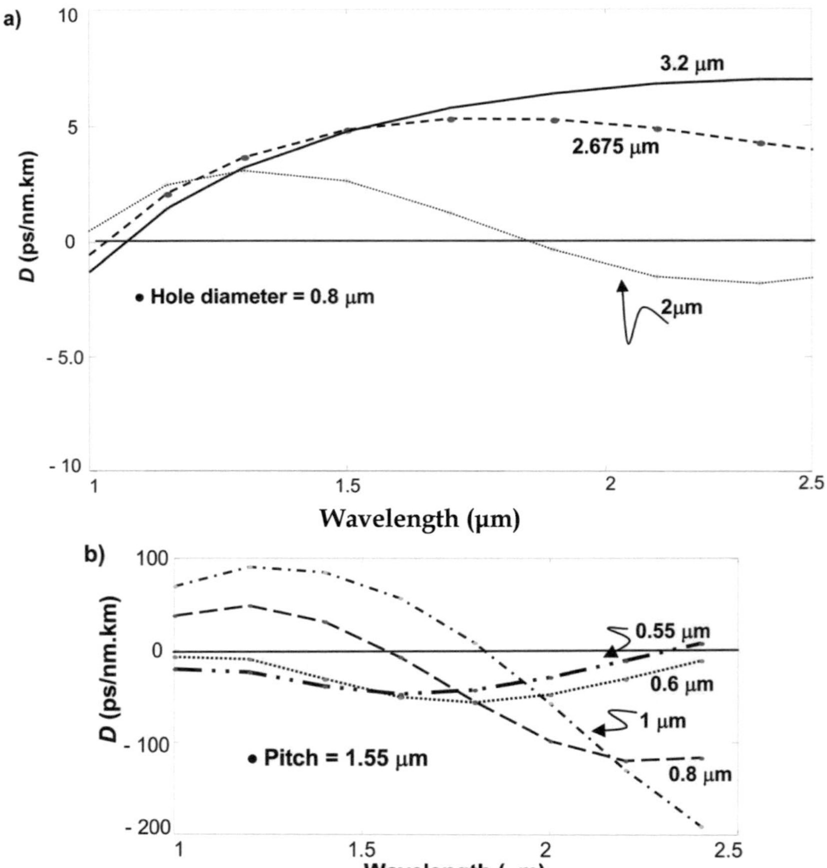

Fig. 12. Dispersion spectra in a holey fier having three cladding rings; a) for different pitch and constant hole diameter; b) foe different hole diameter and constant pitch [After Mehta, 2009]

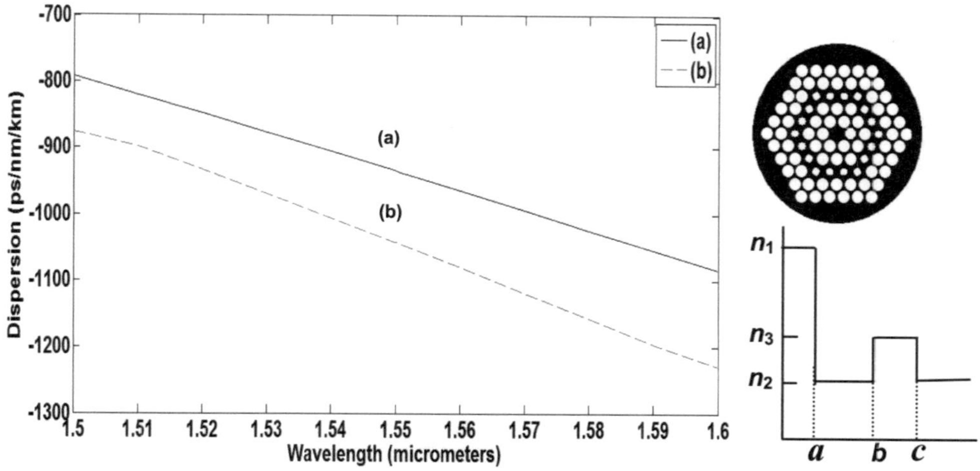

Fig. 13. Dispersion spectra of index guided DCF (shown on the side), whose pitch (Λ) is 0.8 μm and $d_1/\Lambda = 0.85$; a) for d_2= 0.58 μm, b) for d_2= 0.56 μm; d_1 corresponds to diameter for the bigger air holes in the cladding rings while d_2 ($<d_1$) correspond to diameter of the air holes in the third outer ring. Equivalent effective refractive index profile is also shown in the figure [After Mehta, 2009].

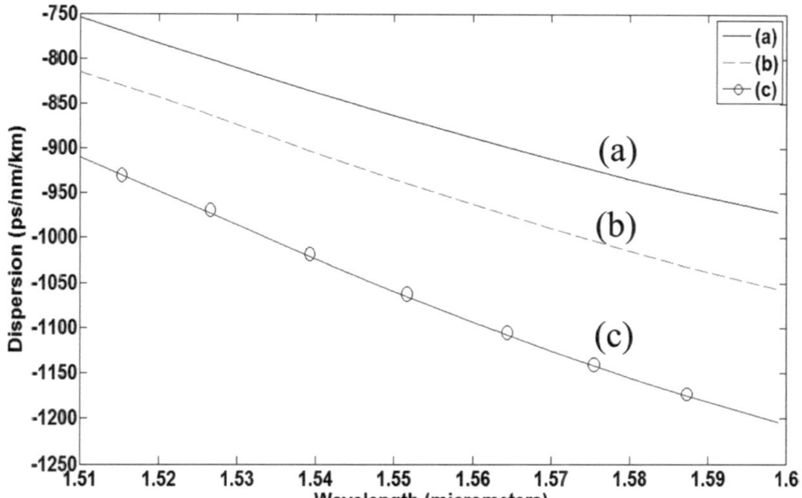

Fig. 14. Effect on dispersion spectrum for variation in d_1 while d_2 and pitch were kept fixed at 0.3 μm and 1 μm, respectively; (a) d_1 = 0.85 μm, (b) d_1 = 0.87 μm, (c) d_1 = 0.9 μm. [After Mehta, 2009]

Confinement loss is an important characteristic parameter of MOFs. It is defined as (Kuhlmey et al, 2002)

$$\text{Confinement loss (dB/m)} = \frac{20\times10^6}{\ln 10}\frac{2\pi}{\lambda}\mathrm{Im}\left(n_{eff}\right) \tag{12}$$

where n_{eff} is effective index of the fiber mode.

4.4 Birefingent microstructured optical fiber

Birefringence could be easily generated in a MOF by increasing diameter of the holes adjacent to the core along one direction as shown in Fig. 15. Birefringence (B) is defined through Eq. (12). In Fig. 16, we depict sample of a typical variation of B with the ring

$$B = n_x - n_y = \frac{\lambda(\beta_x - \beta_y)}{2\pi} \tag{12}$$

location in which the diametrically opposite larger sized pair of holes on a specific ring in the cladding. It is apparent that birefringence reduces as the pair of diametrically opposite air holes is located farther away from the core of the holey fiber.

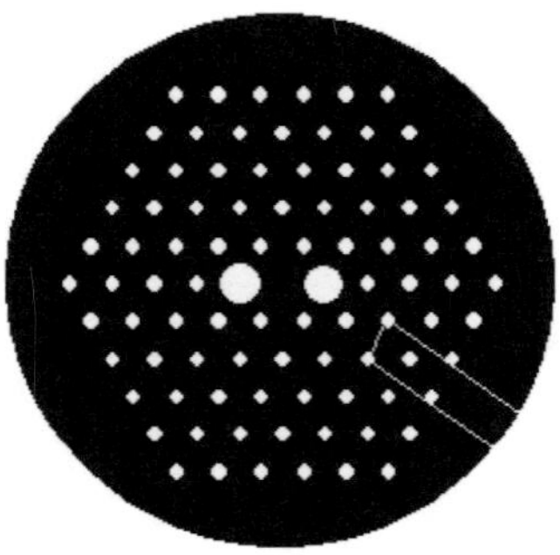

Fig. 15. Cross section of a sample birefringent index guided holey MOF

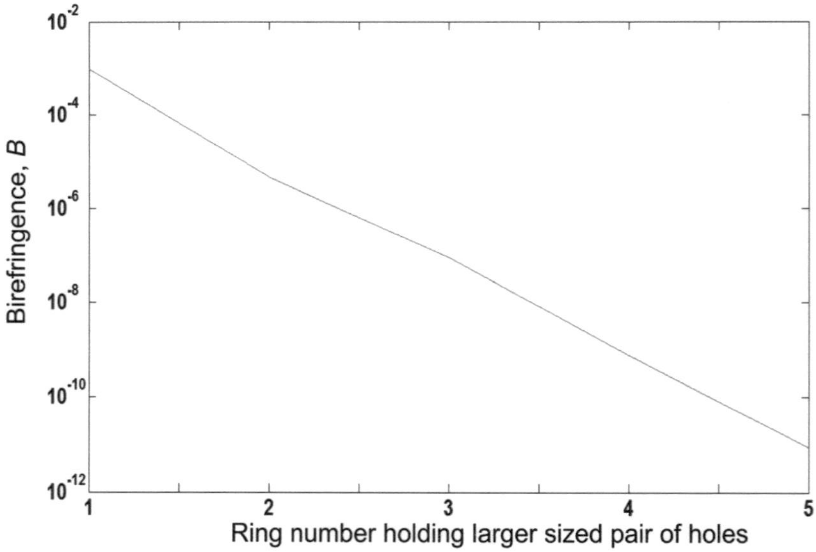

Fig. 16. Birefringence as a function of location of the larger sized pair of holes with respect to the core of the holey fiber [After Mehta, 2009]

5. One dimensional photonic bandgap fibers: Bragg fiber

5.1 Dispersion compensating Bragg fibers

Bragg fibers represent an 1D-PBG MOF, which was first proposed in 1978 [Yeh et al, 1978], almost 10 years before the concept of PhC emerged; although the authors missed out in identifying the inherent photonic band gap characteristic feature of these fibers. These fibers consist of a low-index central region (serving as the core) that is surrounded by concentric layers of alternate high and low refractive index materials. Although in the first proposal the refractive indexes of the cladding bilayers were assumed to be higher than that of the core (see Fig. 17), Bragg fibers could also be designed such that only one of the cladding bilayers has a refractive index higher than that of the core [Katagiri et al, 2004].

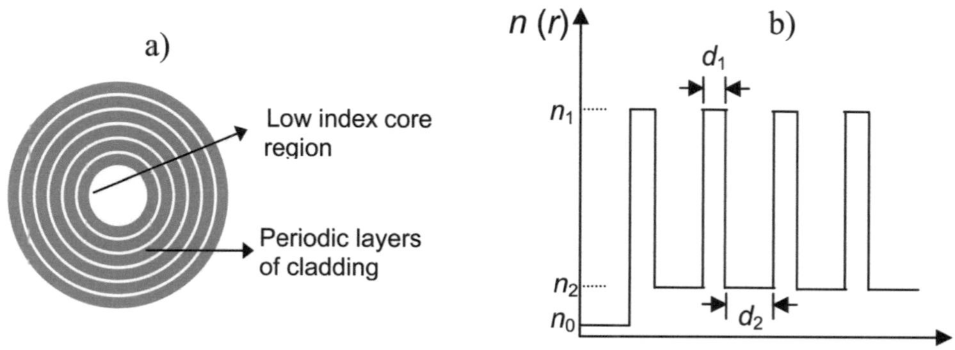

Fig. 17. a) Cross sectional view of a Bragg fiber; b) Corresponding refractive index profile

In either case, the refractive index periodicity in the cladding spawns a PBG. In mid-1980s, air-core Bragg fibers attracted some attention due to a possibility of guiding light in an air core, thereby offering the possibility of attaining a very low transmission loss [Doran & Blow, 1983]. With the emergence of the concept of PBGs in the context of photonic crystals there has been a resurgence of interest in air-core Bragg fibers as an alternative PBG fiber platform for a variety of applications. The first report on a solid silica-core Bragg fiber appeared in 2000 [Brechet et al, 2000], which aimed to achieve zero GVD at wavelengths shorter than conventional telecommunication wavelength windows; in addition, liquid-filled polymer-based Bragg fibers [Cox et al, 2006; Pone et al, 2006], and hollow/low-index core Bragg fibers [Argyros et al, 2006; Skorobogatiy, 2005] also attracted attention because they exhibit diverse propagation characteristics. Bragg fibers with relatively large refractive index difference between its core material and cladding layer materials could essentially be modeled like planar stacks of periodic thin films of alternating materials, similar to an interference filter [Dasgupta et al, 2006]. Inherently the Bragg fibers are leaky in nature due to its refractive index profile, however the loss can be minimized by appropriately choosing the cladding layer parameters. For a given number of cladding layers, the following *quarter-wave stack condition* minimizes the radiation loss of the TE modes supported by a Bragg fiber

$$k_0 n_1 d_1 = k_0 n_2 d_2 = m\pi/2 \tag{19}$$

where n_1, n_2 and d_1, d_2 are the refractive indices and thickness of odd and even layers, respectively, k_0 is free space wave number at the operating wavelength, which is the central wavelength of the band gap that is characteristic of the Bragg fiber and the integer m (≥ 1)

represents order of the quarter wave stack condition. Satisfying this condition implies that round-trip optical phase accumulated from traversal through a pair of claddings is 2π. By choosing $m = 3$, design of a dispersion compensating Bragg fiber (DCBF) of extremely large figure of merit has been reported [Dasgupta et al, 2005]. Figure 18 depicts radiation loss and dispersion spectra of the designed air-core high index contrast Bragg fiber based on a polystyrene-tellurium (PS-Te) material system. It could be seen that this DCBF exhibits dispersion and radiation losses of −1245 ps/(nm.km) and 0.006 dB/km (with 20 cladding bi-layers), respectively, at 1550 nm. These values amount to an effective FOM of 200,000 ps/(nm.dB), which are 2 orders of magnitude larger than that of conventional DCFs. The average dispersion of the DCBF is −1800 ps/(nm.km) across the C-band of an EDFA, with an estimated average radiation loss of 0.1 dB/km.

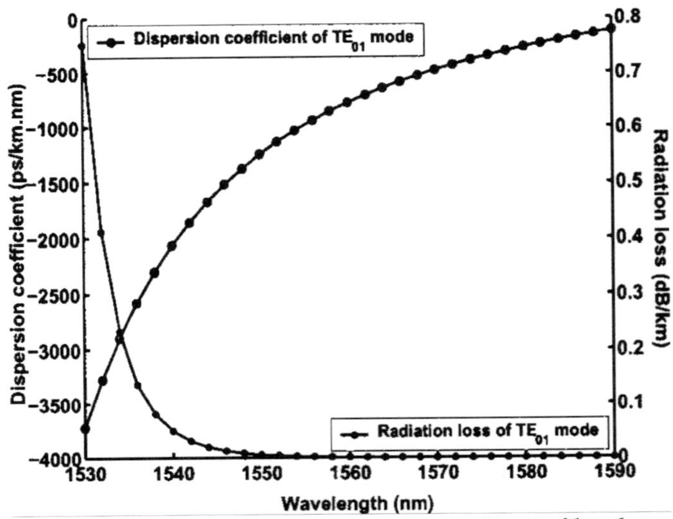

Fig. 18. Dispersion and radiation loss spectra of an air-core Bragg fiber-based DCF [After Dasgupta et al, 2005; ©2005 OSA]

5.2 Bragg fiber for metro networks

Exact modeling of Bragg fibers as reported in [Yeh et al, 1978] was extremely computation intensive because of the large number of material interfaces involved. However, a semi-asymptotic matrix approach involving much less complexity and computational time has been reported in [Xu et al, 2003], according to which the field in the first few of the inner cladding layers are assumed to be exact Bessel solutions while the field in the rest of the outer layers is assumed to be describable by the asymptotic form of Bessel functions. The Bloch theorem is then applied to set up the eigen value equation, solutions of which give the propagation constants of the guided modes. This formalism has been applied to design an air core metro centric Bragg fiber [Pal et al, 2005]. Figure 19 shows the dispersion spectrum of the lowest order TE mode of such a metro-centric Bragg fiber. Its dispersion characteristics are very close to that of the metro fiber of AlcaLuc, average dispersion of the fiber across the C band being 10.4 ps/(nm.km) with a dispersion slope of 0.17 ps/(nm^2.km) at 1550 nm. This should enable a dispersion-limited fiber reach before dispersion compensation is required of ~ 96 km in the C band at 10 Gbits/s, assuming dispersion power penalty of 1 dB. Its average dispersion across the L-band was 15.9 ps/(nm.km) with a dispersion slope of 0.09 ps/(nm^2.km) at 1590 nm. Inserting a conventional dispersion compensating fiber or another Bragg fiber-based dispersion compensator discussed in the previous section can easily compensate the accumulated positive dispersion of this NZDSF in a metro-like network system. The strong confinement of the TE_{01} mode within the air-core allows the material-related losses in the cladding to be ignored at the design stage as compared to the radiation loss. The average radiation loss of the TE_{01} mode across the C band was estimated to be only 0.03 dB/km while it was nearly an order of magnitude lower across the L band. To achieve efficient optical coupling, this Bragg fiber, which had a core diameter of ~ 10 µm, can be tapered to attain significant modal overlap between its TE_{01} mode and the LP_{01} mode of the SMF-28 single-mode fiber pigtail of a laser diode. Additionally, the air-core of the Bragg fiber enhances the threshold for optical nonlinearity while simultaneously allowing a large mode effective area as compared to conventional fibers - both these features are attractive for large power throughput with low sensitivity to detrimental nonlinear effects. Large power throughput serves to offset the signal distribution losses at the nodes of a typical metro and/ regional network rings, which is advantageous from the point of view of power budgeting options. Such a metro fiber, if realized should allow a fiber span length of ~ 100 km without requiring any dispersion compensating device and amplifier. These features, to a large extent, satisfy the required features like reduced installation and operational costs, and complexity of a metropolitan network.

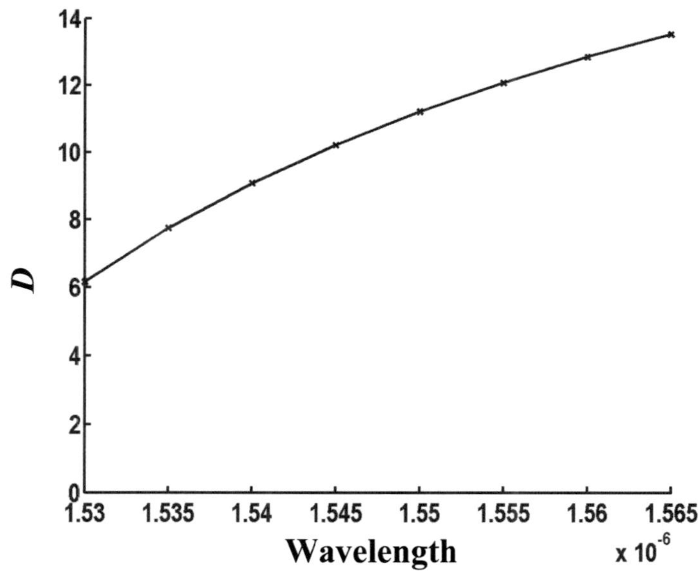

Fig. 19. Dispersion spectrum for the TE_{01} mode of a positive dispersion flattened air core Metro centric Bragg fiber [After Pal et al, 2005 ; ©2005 OSA]

5.3 Nonlinear spectral broadening in solid core Bragg fibers

Solid core Bragg fibers could be designed to have very small (~ 3 µm^2) or very large mode area (~ 500 µm^2). The first report on realization of a solid core Bragg fiber was reported in [Brechet et al, 2000]. Solid core Bragg fibers are attractive from fabrication point of view and also for investigating nonlinear effects in such fibers. As in index guided and photonic bandgap guided MOFs fibers, zero dispersion in a Bragg fiber could be tailored to wavelengths below 1200 nm. An important additional advantage offered by a solid core Bragg fibers is that it could be fabricated by the widely well-known and mature process of modified chemical vapor deposition (MCVD) technology unlike the complex process involved in fabricating other varieties of MOFs. We have studied the phenomenon of supercontinuum generation centered at 1.05 μm in a solid core dispersion decreasing Bragg fiber (DDBF) [Pal et al, 2006; Dasgupta et al, 2007]. Since a silica-based solid core fiber amenable to fabrication by the MCVD technology would be a low index contrast fiber, evolution of a short high power pulse in a silica core Bragg fiber having 10 bi-layers of cladding was designed by modeling it through a matrix method applicable to LP modes of an optical fiber [Pal et al, 2006]. By simulating the propagation of different order soliton pulses, we observed that the shortest distance at which such a pulse is maximally compressed is ~ $z_0/(N-1)$, where z_0 is the soliton period, and N is the whole integral value for $N \geq 2$. Beyond this propagation distance, the pulse begins to break up, and the corresponding frequency spectrum does not exhibit any further broadening. Pulse evolution with propagation in the frequency domain as modeled through solution of the nonlinear Schroedinger equation [Agrawal, 2007] and taking in to consideration higher order group velocity dispersion effects. Maximum broadening of the pulse in the frequency domain was found to occur in a dispersion decreasing fiber (through up-taper along its length) of length ~ 70 cm. The nonlinearity induced broadening of the pulse as a supercontinuum pulse at the output of the dispersion decreasing Bragg fiber is shown in Fig. 20; it could be seen that the 25 BW of the output pulse is ~ 152 nm. It could be broadened further but at the expense of worsening of the flatness of the spectrum. Still broader spectrum would be possible if a lower dispersion slope Bragg fiber could be designed. More recently experimental demonstration of super continuum light from a solid core Bragg fiber fabricated through the MCVD method has been reported [Bookey et al, 2008] (as shown in Fig. 21).

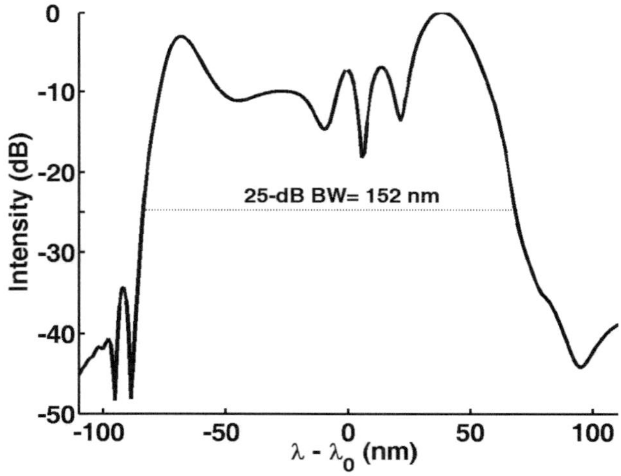

Fig. 20. Simulations demonstrating broadened frequency spectrum achieved through a dispersion decreasing solid core Bragg fiber for a 100 fs input pulse of peak power 4 kW [Calculations courtesy Dasgupta, 2006]

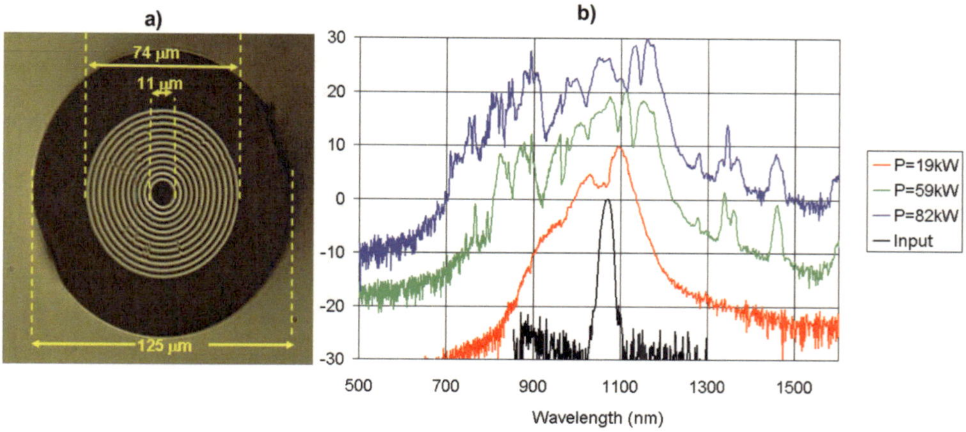

Fig. 21. a) Optical micrograph of the cross section of a solid core Bragg fiber fabricated through MCVD technology; b) Nonlinear spectral broadening in 3 cm of this Bragg fiber showing the input spectrum and for 19 kW, 59 kW, and 82 kW of launched peak powers from an optical parametric amplifier (OPA). OPA was tuned to 1067 nm and FWHM of the launched pulse was 120 fs (After Bookey et al, 2009; ©2009 OSA).

6. Conclusion

In this chapter we have attempted to provide a unified summary description of the most important propagation characteristics of an optical fiber followed by discussion on several variety of special fibers for realizing fiber amplifiers, dispersion compensating fibers, microstructured optical fibers, and so on. Even though huge progress has been made on development of optical fibers for telecom application, a need for developing special fibers, not necessarily for telecom alone, has arisen. This chapter was an effort to describe some of these special fibers. Detailed discussions are given on our own work related to inherently gain-flattened EDFA, DCFs of large mode effective area, index-guided MOF and Bragg fibers for realizing dispersion compensation, for metro network centric applications, and for generating super continuum light.

7. Acknowledgement

The author acknowledges many interesting discussions and exchange of ideas in the course of gathering cumulative knowledge in this field with his colleagues Ajoy Ghatak, M. R. Shenoy, K. Thyagarajan, and Ravi Varshney. He is also grateful to his graduate students namely, Sonali Dasgupta, B. Nagaraju, and Kamna Pande, for many fruitful discussions during their thesis work, which led to several publications with them on specialty fibers, which are referred to in this chapter. Manu Mehta carried out and executed many of the design calculations as part of her M.Tech. Dissertation at our Institute on application specific index guided holey fiber structures, which were based on use of the CUDOS software, made available to us by B. Eggleton and Boris Kuhlmey from University of Sydney. This work was partially supported by our ongoing Indo-UK collaboration project on Application Specific Microstructured Optical Fibers under the UKIERI scheme sponsored by the UK Government and the Indo-French Network collaboration project on Specialty Optical Fibers and Amplifiers sponsored by DST (Govt. of India) and French Ministry of Research.

8. References

Agrawal, G. P. (2007), *Nonlinear Fiber Optics*, Fourth edition, Academic Press, San Diego.

Agrawal, G. P. (2006a), Fiber optic Raman amplifiers in *Guided Wave Optical Components and Devices: basics, Technology, and Applications*, B. P. Pal (Ed.), pp. 1-25, Elsevier Academic Press, Burlington & San Diego.

Argyros, A., Eijkelenborg, M. V., Large, M. and Basset, I. (2006), Hollow core microstructure polymer optical fiber, *Opt. Lett.*, Vol. 31, pp. 172-174.

Auguste, J. L., Jindal, R., Blondy, J. M., Clapeau, M., Marcou, J., Dussardier, B., Monnom, G., Ostrowsky, D. B., Pal, B. P. and Thyagarajan K. (2000), -1800 ps/(nm.km) chromatic dispersion at 1.55 μm in a dual-core fiber, *Electron. Lett.* Vol. 36, pp. 1689-1691.

Birks, T. A., Knight, J. C., and Russel, P. St. J. (1997), Endlessly single-mode photonic crystal fiber, *Opt. Lett.*, Vol. 22, pp. 961-963.

Birks, T. A., Wadsworth W. A., and Russel, P. St. J. (2000) Supercontinuum generation in tapered fibers, *Opt. Lett.* Vol. 25, pp. 1415-1417.

Bookey H. T., Dasgupta, S., Bezawada, N., Pal, B. P., Sysoliatin, A., McCarthy, J. E., Salganskii, M., Khopin, V., and Kar, A. K. (2009), Experimental demonstration of spectral broadening in an all-silica Bragg fiber, *Opt. Exp.* Vol. 17, pp. 17130-17135.

Brechet, F., Roy, P., Marcou, J. and Pagnoux, D. (2000), Single-mode propagation in to depressed-core-index photonic bandgap fiber designed for zero-dispersion propagation at short wavelength, *Electron. Lett.*, Vol. 36, pp. 514-515.

Bromage, J. (2004), Raman amplification for fiber communication systems, *IEEE J. lightwave Tech.*, Vol. 22, pp. 79-93.

Cox, F.M., Argyros, A. and Large, M .C. J. (2006), Liquid-filled hollow core microstructured polymer optical fiber, *Opt. Exp.*, Vol. 14, pp. 4135-4140.

Dasgupta, S., Pal, B. P. and Shenoy, M. R. (2005), Design of dispersion compensating Bragg fiber with ultrahigh figure of merit, *Opt. Lett.*, Vol. 30, pp. 1917-1919.

Dasgupta, S. (2006), Personal Communication.

Dasgupta, S., Pal, B. P. and Shenoy, M. R. (2006), Photonic bandgap guided Bragg fibers in *Guided Wave Optical Components and Devices: basics, Technology, and Applications*, B. P. Pal (Ed.), pp. 1-25, Elsevier Academic Press, Burlington & San Diego.

Dasgupta, S., Pal, B. P. and Shenoy, M. R. (2007), Nonlinear spectral broadening in solid core Bragg fibers, *IEEE J. Lightwave Tech.*, Vol. 25, pp. 2475-2481.

Doran, N. J. and Blow, K. J. (1983), Cylindrical Bragg fibers: a design and feasibility study for optical communications, *IEEE J. Lightwave Tech.*, Vol. LT-1, pp. 588-590.

Ebendorff-Heidepriem, H., Petropoulos, P., Asimakis, S., Finazzi, V., Moore, R. C., Frampton, K., Koizumi, F., Richardson, D. J. and Monro, T. M. (2004), Bismuth glass holey fibers with high nonlinearity, *Opt. Exp.*, Vol. 12, pp. 5082-5087.

Ghatak, A. and Thyagarajan, K. (1998), *Introduction to Fiber Optics*, Cambridge University Press, Cambridge.

John, S. (1987), Strong localization of photons in certain disordered dielectric superlattices, *Phys. Rev. Letts.*, Vol. 58, pp. 2486-2489.

Johnson, S. G., Ibanescu, M., Skorobogatiy, Weisberg, O., Engeness, T. D., Solgacic, M., Jacobs, S. A. and Joannopoulos, J. D. (2001), Low loss asymptotically single-mode propagation in large core omniguide fibers, *Opt. Exp.*, Vol. 9, pp. 748-779.

Kartapoulos, S. K. (2000), *Introduction to DWDM Technologies*, SPIE Press, Bellingham, Washington & IEEE Press, Piscataway, NJ.

Kashyap, R. (1999), *Fiber Bragg Gratings*, Academic Press, San Diego.

Katagiri, T., Matsuura, Y. and Miyagi, M. (2004), Photonic bandgap fiber with a silica core and multiplayer dielectric cladding, *Opt. Lett.*, Vol. 29, pp. 557-559.

Kim, H. S., Yun, S. H., Kim, H. K., Park, N., and Kim, B. Y. (1998), Actively gain-flattened erbium-doped fiber amplifier over 35 nm by using all-fiber acousto-optic tunable filters, *IEEE Photon. Tech. Lett.*, Vol. 10, pp. 790-792..

Knight, J. C., Birks, T. A., Russel, P. St. J., and Atkin, D. M. (1996), All-silica single-mode optical fiber with photonic crystal cladding, *Opt. Lett.*, Vol. 21, pp. 1547- 1549.

Kuhlmey, B. T., White, T. P., Renversez, G., Maystre, D., Botten, L. C., de Sterke, C. M., and McPhedran, R. C., 2002, Multipole method for microstructured optical fibers,.II. Formulation, *J. Opt. Soc. Am. B*, Vol.19, pp. 2331-2340.

Kumar, N., Shenoy, M.R., Pal, B.P. (2005), A standard fiber-based loop mirror as a gain-flattening filter for erbium-doped fiber amplifiers, *IEEE Photon. Tech. Lett.*, Vol. 17, pp. 2056-2058.

Li, S., Chiang, K. S., Gambling, W. A. (2001), Gain flattening of an erbium-dped fiber amplifier using a high-birefringence loop mirror, *IEEE Photon. Technol. Lett.*, Vol. 13, pp. 942-944.

Li, T. (1995), The impact of optical amplifiers on long-distance lightwave communications, *Proc. IEEE*, Vol. 81, pp. 1568-1579.

Mehta, M. (2009), M.Tech. (Optoelectronics and Optical Communication) dissertation *Studies on Microstructured Optical Fibers*, IIT Delhi.

Mears, R.J., Reekie, L., Poole, S. B. and Payne, D. N. (1986), Low-threshold tunable cw and Q-switched fiber laser operating at 1.55 μm, *Electron. Lett.*, Vol. 22, pp. 159-160.

Mears, R. J., Reekie, L., Jauncy, I. M. and Payne, D. N., (1987), Low-noise fiber amplifier operating at 1.54 μm, *Electron. Lett.*, Vol. 23, pp. 1026-1027.

Miya, T., Terunume, Y., Hosaka, T. and Miyashita, T. (1979), An ultimate low-loss single-mode fiber at 1.55 μm, *Electron. Lett.*, Vol. 5, pp. 106-108.

Monro, T. M. (2006), Microstructured Optical fibers in *Guided Wave Optical Components and Devices: basics, Technology, and Applications*], B. P. Pal (Ed.), pp. 1-25 Elsevier Academic Press, Burlington & San Diego.

Nagaraju, B., Paul, M. C., Pal, M., Pal, A., Varsheny, R. K., Pal, B. P., Bhadra, S. K., Monnom, G. and Bernard, D. (2009), Design and fabrication of an intrinsically gain flattened Erbium doped fiber amplifier, *Opt. Comm.*, Vol. 282, pp. 2335-2338.

Okuno, T., Onishi, M., Kashiwada, T., Ishikawa, S. and Nishimura, M. (1999), Silica-based functional fibers with enhanced nonlinearity and confinement loss trade-offs, *IEEE J. Select. Top. Quantum Electron.*, Vol. 5, pp. 1385-1391.

Pal, B. P. (1995), Optical transmission in *Perspective in Optoelectronics*, S. S. Jha (Ed.), pp. 195-297, World Scientific, Singapore.

Pal, B. P. (2006), Optical fibers for broadband lightwave communication: evolutionary trends in designs in *Guided Wave Optical Components and Devices: basics, Technology, and Applications*, B. P. Pal (Ed.), pp. 1-25, Elsevier Academic Press, Burlington & San Diego.

Pal, B. P. and Pande, K. (2002), Optimization of a dual-core dispersion slope compensating fiber for DWDM transmission in the 1480-1610 nm band through G.652 single-mode fibers, *Opt. Comm.*, Vol. 201, pp. 335-344.

Pal, B. P., Dasgupta, S. and Shenoy, M. R. (2005), Bragg fiber designs for transparent metro networks, *Opt. Exp.*, Vol. 13, pp. 621-626.

Pal, B.P., Dasgupta, S. and Shenoy, M. R. (2006), Supercontinuum generation in a Bragg fiber: a novel proposal, *Optoelectron. Letts.*, Vol. 5, pp. 342-344.

Pan, J.Y., Ali, M. A., Elrefaie, A. F., and Wagner, R. E. (1995), Multi-wavelength fiber amplifier cascades with equalization employing Mach-Zehnder optical filter, *IEEE Photon. Technol. Lett.*, Vol. 7, pp. 1501-1503.

Pande, K. and Pal, B. P. (2003), Design optimization of a dual-core dispersion compensating fiber with high figure of merit and a large mode effective area for dense wavelength division multiplexed transmission through standard G.655 fibers, *App. Opt.*, Vol. 42, pp. 3785-3791.

Payne, D. N. and Gambling, W. A. (1975), Zero material dispersion in optical fibers, *Electron. Lett.*, Vol. 11, pp. 176-178.

Pone, P., Dubois, C., Guo, N., Gao, Y., Dupuis, A., Boismenu, F., Lacroix, S. and Skorobogatiy, M. (2006), Drawing of the hollow all-polymer Bragg fibers, *Opt. Exp.*, Vol. 14, pp. 5838-5852.

Ramachandran, S. (Ed.) (2007), *Fiber-based Dispersion Compensation*, Springer-Verlag, Berlin.

Ramachandran, S. (2006), Dispersion-tailored higher order mode fibers for in-fiber photonic devices in *Guided Wave Optical Components and Devices: Basics, Technology, and Applications*], B. P. Pal (Ed.), pp. 291-310, Elsevier Academic Press, Burlington & San Diego.

Ryan, J. (2002), Fiber considerations for metropolitan networks, *Alcatel Telecom. Rev.*, Vol. 1, pp. 52-56.

Skorobogatiy, M. (2005), Efficient ant-guiding of TE and TM polarizations in low index core waveguides without the need of omnidirectional refelector, *Opt. Lett.*, Vol. 30, pp. 2991-2993.

Sun, Y, Sulhoff, J. W., Srivastava, A., Zysking, J. L., Srasser, T. A., Pedrazzani, J. R., Wolf, C., Zhou, J., Judkins, J. B., Espindola, R. P., and Vengsarkar, A. M. (1997), 80 nm ultra wideband erbium doped silica fiber amplifier, *Electron. Lett.*, Vol. 33, pp. 1965-1967.

Srivastava, A. and Sun, Y. (2006), Erbium-doped fiber amplifiers for dynamic optical networks in *Guided Wave Optical Components and Devices: Basics, Technology, and Applications*], B. P. Pal (Ed.), pp. 181-204, Elsevier Academic Press, Burlington & San Diego.

Thyagarajan, K., Diggavi, S., Taneja, A. and Ghatak, A. K. (1991), A simple numerical technique for the analysis of cylindrically symmetric refractive index profile optical fiber, *Appl. Opt.*, Vol. 30, pp. 3877-3879.

Thyagarajan, K. and Pal, B. P. (2007), Modeling dispersion in optical fibers: applications to dispersion tailoring and dispersion compensation in *Optical Fiber Communication Reports*, Ramachandran, S. (Ed.), Vol. 4, pp. 173–213, Springer-Verlag, Berlin.

Urquhart, W. P. and Laybourn, P. J. (1985), Effective core area for stimulated Raman scattering in single-mode optical fibers, *Proc. Inst. Elect. Eng.*, Vol. 132, pp. 201–204.

Xu, Y., Yariv, A., Fleming, J. G. and Lin S. Y. (2003), Asymptotic analysis of silicon based Bragg fibers, *Opt. Exp.*, Vol. 11, pp. 1039-1049.

Varshney, R. K., Nagaraju, B., Singh, A., Pal, B. P., and Kar, A. K. (2007), Design and Realization of an All-Fiber Broadband Tunable Gain Equalization Filter for DWDM Signals, *Opt. Exp.*, Vol. 15, pp. 13519-13530.

Vengsarkar, A. M., Lemaire, P.J., Judkins, J. B., Bhatia, V., Erdogan, T., Sipe, J. E. (1996), Long period fiber gratings as band rejection filters, *IEEE J. Lightwave Tech.*, Vol. 14, pp. 58-65.

Yablonovitch, E. (1987), Inhibited spontaneous emission in solid-state physics and electronics, *Phys. Rev. Letts.*, Vol. 58, pp. 2059-2062.

Yeh, P., Yariv, A. and Marom, E. (1978), Theory of Bragg fiber, *J. Opt. Soc. Am.*, Vol. 68, pp. 1196-1201.

2

Nonlinear Properties of Chalcogenide Glass Fibers

Jas S. Sanghera, L. Brandon Shaw,
C. M. Florea, P. Pureza, V. Q. Nguyen,
F. Kung, Dan Gibson and I. D. Aggarwal
Naval Research Laboratory
USA

1. Introduction

Chalcogenide glasses are based on the chalcogen elements S, Se and Te with the addition of other elements such as Ge, As and Sb to form of stable glasses (Borisova, 1981). Due to their large IR transparency, fibers fabricated from these glasses are ideal for transmission of high power IR light. Several applications of chalcogenide fibers for IR transmission have been documented (Sanghera et al., 2005a). Also of interest is the high nonlinearity of these glass compositions. The high $\chi^{(3)}$ nonlinearities of chalcogenide glasses make them excellent candidates for applications such as all optical processing, Raman amplification, parametric amplifiers and supercontinuum generation.

2. Glass preparation

Chalcogenide glasses are melted directly in quartz ampoules using chemicals purified via distillation/sublimation (Sanghera et al., 1994a). Typical melt temperatures range from 600°C to 900°C, depending upon composition. The liquids are quenched and the glass rods annealed at temperatures around the appropriate softening temperatures. The optical fibers are obtained by the double crucible (DC) process (Sanghera et al., 1995). The DC process enables adjustments to be made in the core/clad diameter ratio during fiber drawing by independent pressure control above each melt. Therefore both multimode and single mode fibers can be drawn with relatively few processing steps.

3. Fiber properties

Figure 1 compares the losses routinely obtained for a couple of chalcogenide glasses along with the lowest ("champion") losses reported in the literature (Sanghera et al., 1994b; Churbanov, 1992). Depending upon composition, the sulfide, selenide and telluride based fibers transmit between about 0.8-7 μm, 1-10 μm, and 2-12 μm, respectively. Therefore, the practical applications dictate the type of fiber to be used. As-S fibers loss routinely achieved is about 0.1-0.2 dB/m in fiber lengths of about 500 meters. Losses for As-Se fibers typically range from 0.5 to 1 dB/m in the near IR around 1.5 μm.

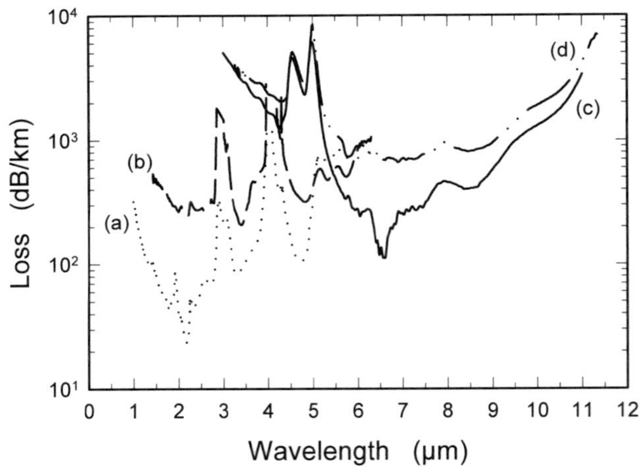

Fig. 1. Transmission loss spectra of (a) lowest loss sulfide fiber, (b) typical sulfide fiber, (c) lowest loss telluride fiber, and (d) typical telluride fiber.

4. Nonlinear properties

It is well established that the values of $\chi^{(3)}$ for chalcogenide glasses are about two orders of magnitude larger than silica (Nasu et al, 1989; Richardson et al, 1998). More recently, glasses have been reported with non-linearities approaching 1000 times silica (Lenz et al., 2000; Harbold et al., 2002). These large nonlinearities would allow small compact low power devices for telecommunications. The subpicosecond response of these nonlinearities is ideal for high data rate telecommunication devices.

For efficient nonlinear devices utilizing the optical Kerr effect, the nonlinearity must be high and the nonlinear absorption must be low. A figure of merit FOM = $n_2/(\beta\lambda)$ can be defined as a useful metric to determine optimum compositions, where n_2 is the nonlinear index and β is the nonlinear absorption. For isotropic medium, one and two photon resonant processes dominate the third-order susceptibility. For frequencies approximately half of the material resonance, two photon processes resonantly enhance the nonlinear index n_2. Normally, however, the two photon resonance enhancement is accompanied by two photon absorption which competes with the nonlinear index n_2. In the case of amorphous materials such as chalcogenide glass, an exponential Urbach tail exists and its absorption edge extends below the half gap. This edge leads to two photon absorption (TPA) below the half gap and thus n_2 may increase faster than TPA absorption in this region. Consequently, the best performance in terms of nonlinear index strength vs. TPA (FOM) will occur just below the gap. Figure 2 shows the bandgap of the As-S-Se system vs. Se concentration.

Here, the bandgap is defined at the point of 10^3 cm^{-1} absorption. In the graph, Se content of 0 at. % corresponds to pure $As_{40}S_{60}$ while Se content of 60 at. % corresponds to pure $As_{40}Se_{60}$. The bandgap of the glass system decreases with Se content. For operation at 1.55 µm (0.8 eV), we would expect an optimum composition of $As_{40}Se_{60}$ where $E_g/h\nu \sim 0.45$. This is borne out by experimental data.

Spectrally resolved two beam coupling measurements of As-S-Se system have been performed to determine the magnitude of the nonlinear index n_2 and the two photon

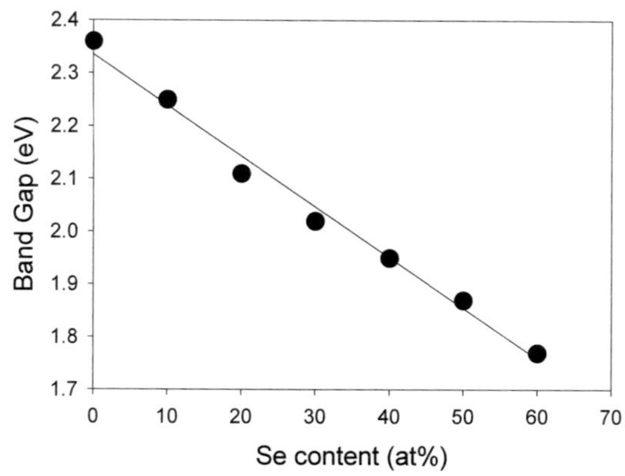

Fig. 2. Bandgap of As-S-Se glass system (defined at the point of 10^3 cm^{-1} absorption).

absorption coefficient. Details of these measurements can be found in (Harbold et al., 2002). Figure 3 shows the results of these measurements. Values for As-S were found to be ~220 times higher than for silica at 1.55 μm and increased with Se substitution of S to a value of ~930 times higher than silica for As-Se. Likewise, two photon absorption also increases with increasing Se content. This data can be used to calculate the FOM for the As-Se system (Figure 4). As expected, the glasses with the largest FOM for operation at 1550 nm occurs for $E_g/h\nu$ at ~0.45 which is the As-Se composition (Slusher et al., 2004).

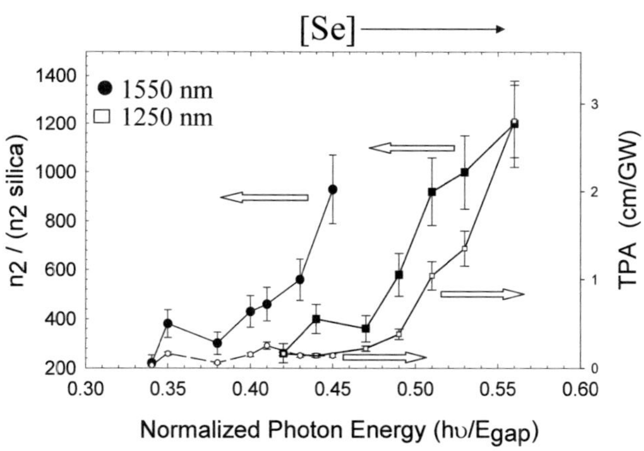

Fig. 3. n_2 and TPA absorption of As-S-Se glass system.

High speed optical processing has been demonstrated by exploiting these high nonlinearities in chalcogenide glass fiber and waveguides. Earlier work on all optical switching in chalcogenide fiber was performed by Asboe (Asobe et. al. 1993) who demonstrated switching of an 80-GHz pulse train in a 2 meter length of As_2S_3 based fiber using an optical kerr shutter configuration. More recently, 640 Gb/s demultiplexing has been demonstrated in a 5 cm long chalcogenide rib waveguide on silicon by utilizing FWM (Galili et. al. 2009). 40 Gb/s all optical wavelength conversion has also been demonstrated in chalcogenide tapered fibers (Pelusi, et. al. 2008). Here, a CW laser at the conversion wavelength was modulated by XPM with the co-propagating 40 Gb/s signal.

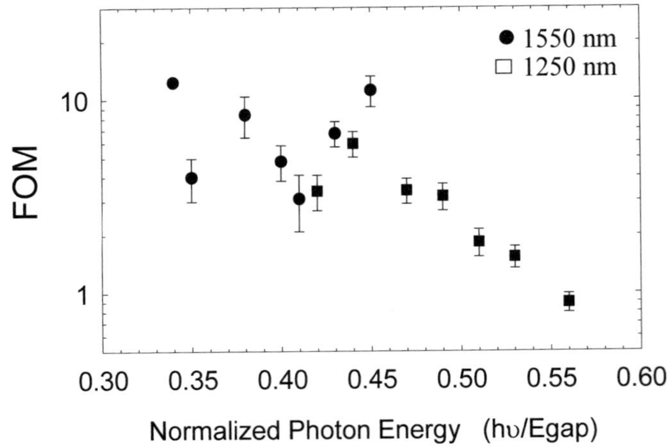

Fig. 4. FOM for As-S-Se glass system.

5. Raman amplification

Figure 5 shows the normalized Raman spectra of $As_{40}S_{60}$, $As_{40}Se_{60}$, and silica. $As_{40}Se_{60}$ glass has a much narrower Raman line (~60 cm^{-1}) than silica glass (~250 cm^{-1}). In addition, the Raman shift for $As_{40}Se_{60}$ glass is much smaller (~240 cm^{-1}) than the Raman shift of silica glass (~440 cm^{-1}) due to the heavier atoms present in the chalcogenide glass. Previous studies have looked at stimulated Raman scattering in $As_{40}S_{60}$ glass, a very similar glass system to $As_{40}Se_{60}$ (Asobe et al., 1995). These studies found the Raman gain coefficient of $As_{40}S_{60}$ to be almost two orders of magnitude higher than that of silica. It was also found that this enhancement in the Raman gain roughly corresponded to the enhancement in the nonlinear index, n_2. Consequently, one might expect to see an even larger Raman gain coefficient in $As_{40}Se_{60}$ since the selenide glass has shown an even larger nonlinearity and also a narrower Raman spectrum.

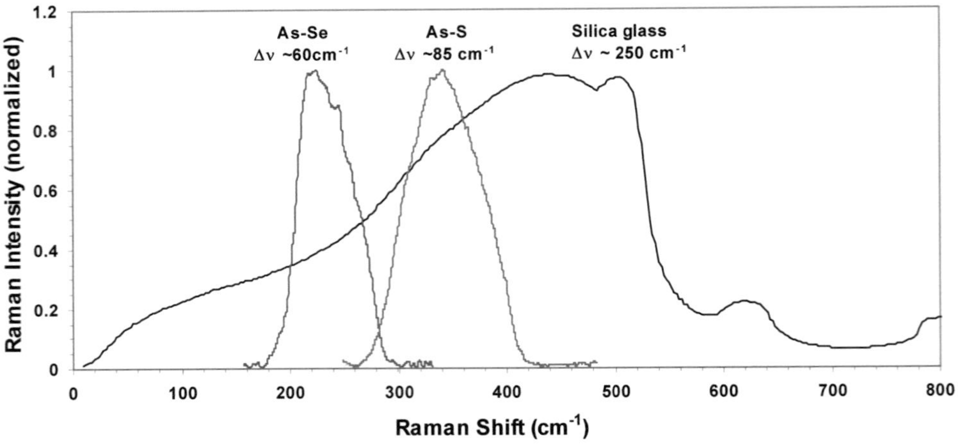

Fig. 5. Raman spectra of As_2S_3 and As_2Se_3 glass. Silica glass is shown for reference.

Raman amplification at 1.55 μm has been demonstrated in small core As-Se fiber (Thielen et al., 2003a). The results of the Raman amplification experiment are shown in shown in Figure 6. Over ~23 dB of gain was achieved in a 1.1-meter length of fiber pumped by a nanosecond pulse of ~10.8 W peak power at 1.50 μm. The peak of the Raman gain was shifted by ~230 cm^{-1} to 1.56 μm. The Raman gain coefficient was estimated to be ~300 times silica in this experiment. More recent measurements of the Raman gain coefficient show a value of about 780x greater than that of silica (Slusher et al. 2004).

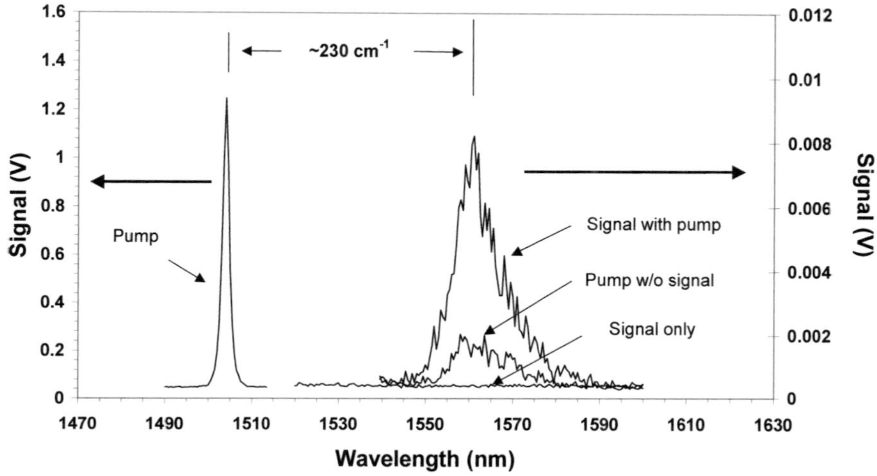

Fig. 6. Raman amplification in As-Se fiber. Shown is amplifier output with signal and no pump, pump and no signal (showing background stimulated Raman scattering (SRS) resulting from pump), and amplified signal with pump.

The large Raman gain coefficient of chalcogenide glass coupled with its large IR transparency show promise for lasers and amplifiers in the near and mid-IR. The potential for Raman lasers and amplifiers can be assessed by defining a figure of merit (FOM). The expression for single pass gain, G_A, in a Raman fiber laser is given by [1]:

$$G_A = \exp\left(\frac{g_R P_0 L_{eff}}{A_{eff}}\right) \quad (1)$$

Where g_R is the Raman gain coefficient, P_0 is the pump power, A_{eff} is the fiber effective area and L_{eff} is the fiber effective length. The fiber effective length is given by

$$L_{eff} = \frac{1}{\alpha}\left(1 - e^{-\alpha \cdot L}\right) \approx \frac{1}{\alpha} \quad (2)$$

Where α is the fiber loss. For long lengths, L_{eff} is approx $1/\alpha$. From these equations, the gain is proportional to *exp* $(-g_R/\alpha)$ for long fiber lengths. Thus, the value g_R/α can be used as a rough FOM for Raman amplification. Table 1 compares the performance of an As-Se Raman fiber laser or amplifier operating at 4 μm to a silica Raman fiber laser or amplifier operating in the telecommunications band at 1.5 μm. Here, the Raman gain coefficient of As-Se, g_R, which is measured to be 780x silica at 1.5 μm is extrapolated to it value in the mid-IR since the Raman gain coefficient scales inversely with wavelength. α is the fiber loss. For silica, a loss of 0.2 to 0.3 dB/km is typical of telecommunication grade fiber. For As-Se, two losses are given. The loss of 200 dB/km is typical of "champion losses" achieved at NRL for As-Se fiber while the loss of 3 dB/km is theoretical loss for As-Se fiber (Devyatykh et al., 1992). For the loss of 200 dB/km, g_R/α for an As-Se fiber Raman amplifier operating at 4 μm is about 0.38 compared to 1.1 for a silica fiber Raman amplifier. For the theoretical loss of 3 dB/km, g_R/α for As-Se fiber operating at 4 μm is 23 times that of silica fiber operating at 1.5-μm.

	λ (μm)	g_R (cm/W)	Loss (dB/km)	α (cm⁻¹)	FOM ($10^{-6}W^{-1}$)
Silica Fiber	1.5	0.65×10^{-12}	0.2-0.3	$\sim 6 \times 10^{-7}$	1.1
As-Se Fiber	4	1.7×10^{-10}	200	5×10^{-4}	0.34
			3	7.5×10^{-6}	23

Table 1. Figure of merit for Raman amplification in As-Se fiber at 4-μm compared Raman amplification in silica fiber at 1.5-μm. The loss value of 200 dB/km (a) for As-Se is typical of a "champion" loss value. The loss value of 3 dB/km (b) is theoretical loss.

A Raman laser has been demonstrated in As-Se fiber by Jackson (Jackson et. al. 2000). They generated 0.64 W of first Stokes at 2062 nm with a slope efficiency of 66% under 2051 nm pumping in a 1 meter length 6 μm core, 0.19 NA fiber. Reflection off the endface of the fiber (~22% at normal incidence) was used for feedback at the output end of the fiber while a broadband Au-coated mirror was used as a back reflector. Note that the braodband nature of the cavity reflectors allowed the Raman laser to oscillate on a number of vibrations. The line at 2062 nm was attributed to interlayer vibrations of As_2Se_3. Raman output at 2102 from bond bending vibrations and at 2166 nm for bond stretching vibrations were also observed. Stimulated Raman scattering (SRS) has been observed in the mid- IR. Figure 7 shows the SRS in a ~ 1m length of As-Se fiber under CW CO laser pumping at ~ 5.4 μm. The SRS is seen at ~ 6.1 μm. Raman laser operating in the wavelength range of from 6.1 to 6.4 μm would have applications in laser surgery. These wavelengths correspond to amide bands in

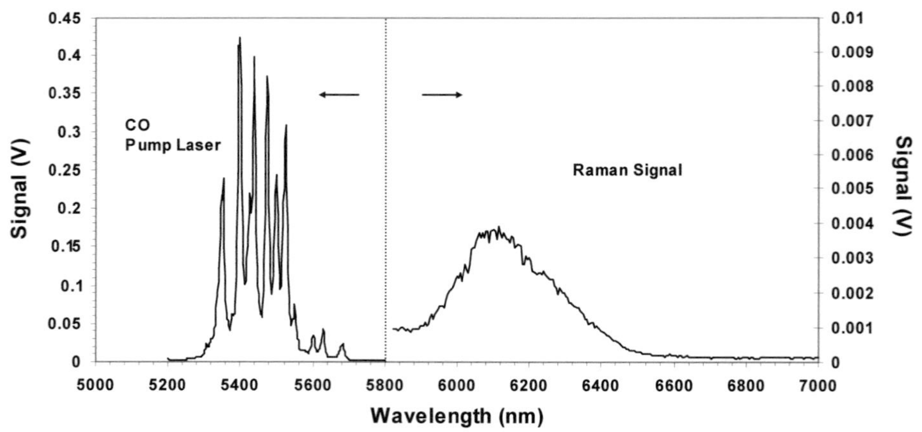

Fig. 7. SRS signal observed at 6.1 μm under ~5.4 μm CO laser pumping.

tissues and studies have shown that ablation of soft tissue is possible at these wavelengths with minimal collateral damage, thus accelerating healing (Edwards et al. 1994). Modeling of a Raman laser operating at 6.45 μm under CO laser pumping at 5.59 μm has shown high slope efficiencies and moderate threshold power operation is possible (Thielen et al. 2003b).

6. Supercontinuum generation

Supercontinuum generation has been demonstrated between 2 to 3 μm in small core sulfide and selenide fibers as well as photonic crystal selenide fibers (PCF) (Shaw et al., 2005). The 1 meter length of fibers were pumped with a Ti:sapphire pumped OPA laser at a wavelength of 2.5 μm using 100 fs pulses and 100 pJ/pulse. The outputs from the fibers are shown in figure 8. The sulfide and selenide fibers were 7 μm core diameter, while the PCF fiber had a 10 μm core diameter. In all cases, pumping was in the normal dispersion region of the fibers and much of the broadening can be attributed to self phase modulation (SPM) with some broadening to the red due to Raman (Hu et. al., 2008).

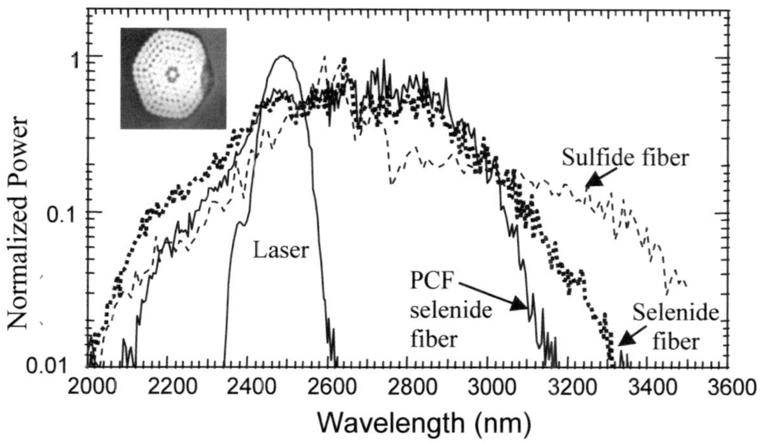

Fig. 8. Supercontinuum generation in small core chalcogenide fibers. The insert shows the cross-sectional view of the selenide PCF fiber.

By using chalcogenide glass PCF, the dispersion of the fiber by can controlled and the zero dispersion wavelength can be shifted to the near-IR making it feasible to pump in the anomalous dispersion region of the fiber with conventional near-IR fiber laser pumps. Modeling has shown that very broad supercontinuum bandwidths can be generated with properly designed chalcogenide PCF fiber and proper pump (Hu et. al. 2009)

7. Poling of chalcogenide glass

Isotropic materials such as glasses lack a center of inversion symmetry and thus have no second order nonlinear susceptibility ($\chi^{(2)}$) they should not exhibit second harmonic generation (SHG) (Dianov et al., 1989). However, undoped and Pr-doped GaLaS glasses have exhibited SHG (De Aruajo et al., 1996) through optical pumping. This SHG may be due to crystallization or the effect of frozen-in electric fields. The latter arises from the relationship $\chi^{(2)} = E_{dc}\chi^{(3)}$, where E_{dc} is the frozen-in electric field (Dianov et al., 1989). Electric

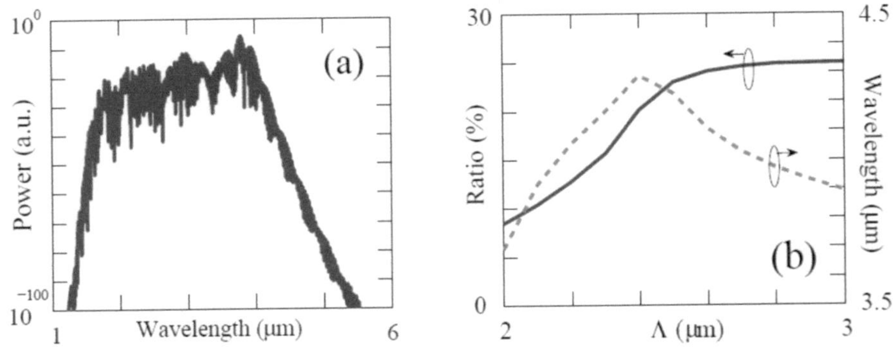

Fig. 9. (a) Modeled supercontinuum spectrum in As-S Photonic crystal fiber with $\Lambda = 3$ μm under 2 μm, 500 fs, 1 kW peak power pumping. (b) The central wavelength of the soliton with the largest power (dashed curve) and the ratio of the power generated between 3 μm and 5 μm to the total input power as a function of the pitch at the end of the tapered PCF (solid curve) (Hu, et. al. 2009)

poling has been successfully used to produce SHG in silica based fiber systems (Kazansky et al., 1997). It is not unreasonable to expect similar results in chalcogenide fibers.

Since $\chi^{(3)}$ is about 2 to 3 orders of magnitude larger in chalcogenides compared with silica, we expect larger SHG efficiencies in electrically poled chalcogenide glasses. However, the question arises as to whether the electric fields can be frozen-in for chalcogenide glasses. Second harmonic generation has been observed at 780 nm using electrically poled arsenic sulfide glass when pumping a 1 mm thick arsenic sulfide glass disk at 1560 nm as shown in Figure 10. The sample was electrically poled at 100°C for 5 hours under nitrogen gas atmosphere. At the present time the magnitude appears comparable to silica glass but the mechanism is unknown.

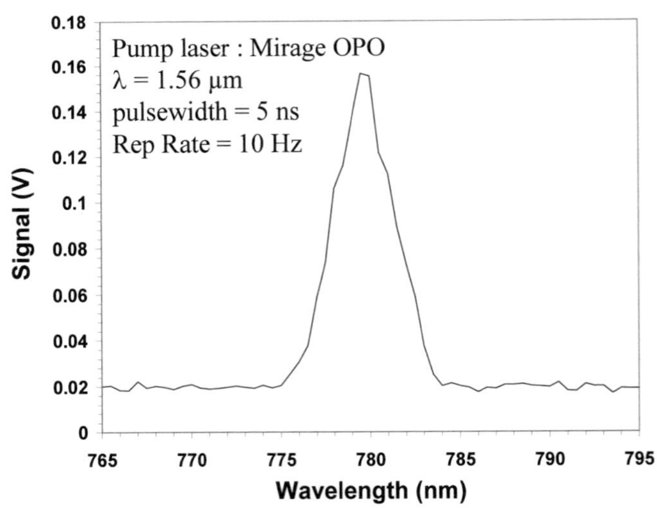

Fig. 10. Second harmonic generation in poled As-S glass. Glass was pumped at 1.56 μm. Shown is the 780 nm SHG signal.

8. Brillouin scattering

In order to estimate the Brillouin gain coefficient, the threshold power of the stimulated Brillouin scattering (SBS) process can be measured using the experimental setup detailed in Fig. 11. The threshold power is easily determined by measuring the amount or monitoring the spectrum of the reflected light using a high-resolution optical spectrum analyzer (OSA) as sampled by the circulator. The fibers can be coated with liquid gallium on 10-cm lengths on each end to eliminate the radiation leaking into the cladding. In the example provided, the fiber ends were not anti-reflection coated and hence cavity effects were significant due to the high refractive index of the fiber. The losses in the fiber and in the coupling optics are all taken into account when estimating the amount of pump launched into the core. A 45% coupling efficiency was estimated in the As_2S_3 case, and 37% in the As_2Se_3 case. These values can be optimized and hence the SBS threshold power can be reduced, which is desired trend from a system design perspective.

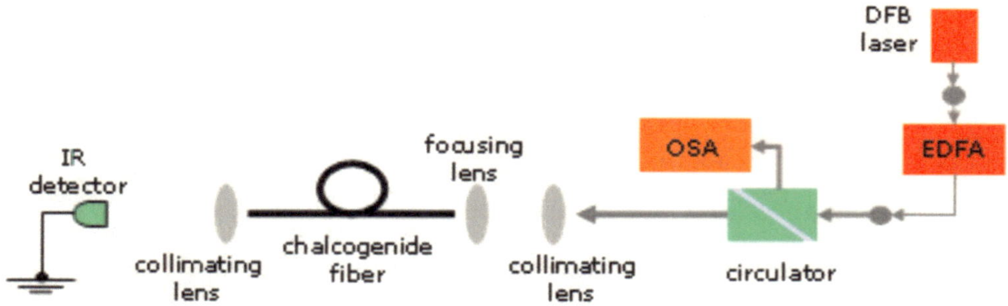

Fig. 11. Experimental setup used for SBS threshold measurements.

The spectral changes of the backward wave propagating through the chalcogenide fiber, as sampled by the circulator, are shown in Fig. 12 for the As_2S_3 fiber, and in Fig. 13 for the As_2Se_3 fiber, respectively. The cavity effects reduced the accuracy of the threshold measurement as indicated in the captions. Nevertheless, the threshold is easily identified by the significant jump in the peak of the Brillouin-shifted signal monitored on the OSA. Additionally, clamping of the pump output power was observed, once the threshold was reached, since most of the pump power wastransfered to the Stokes wave (Ruffin, 2004).
The numerical aperture (NA) of a fiber determines the mode-field diameter and hence the effective area of the fundamental mode, with direct implications on the threshold power estimation for the SBS process. It also determines the number of modes supported by the fiber at a given wavelength, λ. The V-number for a step-index fiber is a function of NA as given in Eq. 3, where d is the core diameter:

$$V = \frac{\pi d}{\lambda} NA \tag{3}$$

A value of V=2.405, or lower, indicates single mode behavior. The V-number for the As_2S_3 fiber used was ~2.8 During the experiments, the mode field pattern was monitored by imaging the output on a Vidicon camera to make sure only the fundamental mode was launched. Using the NA and V-number values, the Mode Field Diameter (MFD), d_{1/e^2}, for the fundamental mode will be given by Eq. 4 and is listed in Table 2:

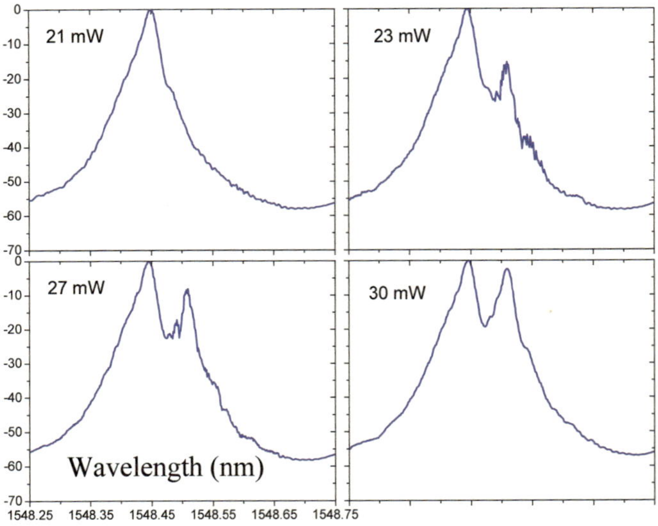

Fig. 12. Typical spectra of the reflected light sampled by the circulator for different launched pump powers into the As_2S_3 fiber core. Fiber length was 10.0 m. Estimated SBS threshold: (27 ± 3) mW. Tick labels shown only on one plot for clarity.

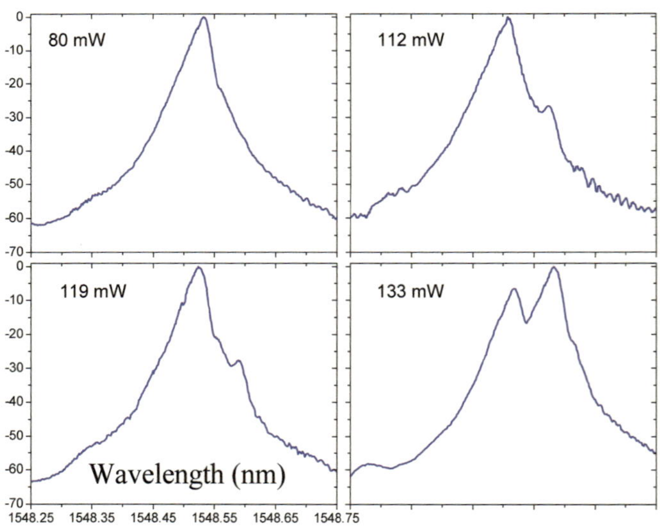

Fig. 13. Typical spectra of the reflected light sampled by the circulator for different launched pump powers into the As_2Se_3 fiber core. Fiber length was 5.0 m. Estimated SBS threshold: (127 ± 7) mW. Tick labels shown only on one plot for clarity.

$$d_{1/e^2} = d \times (0.65 + \frac{1.619}{V^{1.5}} + \frac{2.879}{V^6}) \quad (4)$$

The propagation loss is also an important parameter as it defines the effective interaction length for the Brillouin scattering process. The values reported in Table 2 represent relatively low losses for both singlemode fibers at 1.56 μm. However, it should be possible to lower the losses even further by improved fiber drawing and glass fabrication processes.

Fiber	Core dia. [μm]	Clad dia. [μm]	Core Refractive Index	NA	V-number	$d_{1/e2}$ MFD [μm] (calculated)	Loss [dB.m^{-1}]
As_2S_3	4.2	142.0	2.45	0.33	2.8	4.2	0.57
As_2Se_3	6.5	175.0	2.81	0.14	1.8	9.0	0.90

Table 2. Chalcogenide fiber parameters (at wavelength of 1.56 μm).

From the experimentally determined threshold power values (Pth) shown in figures 12 and 13, one can estimate the Brillouin gain coefficient (g_B) using Eq. 5 (Song et al, 2006; Ippen and Stolen, 1972):

$$P_{th} \cong 21 \frac{A_{eff}}{L_{eff} g_B k} \quad (5)$$

In the Eq. 5, k is a constant which reflects whether the polarization is maintained constant throughout the interaction (k = 1) or not (k = 0.5, our case). Also, the Aeff and Leff are the effective area of the fundamental mode, and the effective interaction length, respectively. These are given by Eq. 6 and Eq. 7, where L is the fiber length, α is the propagation loss, and the mode-field diameter is determined by Eq. 3 above.

$$A_{eff} = \frac{\pi d_{1/e^2}^2}{4} \quad (6)$$

$$L_{eff} = \frac{1}{\alpha}\left(1 - e^{-\alpha L}\right) \quad (7)$$

Using Eqs. 5-7, the parameters from Table 3, and the fiber lengths and pump threshold values indicated in Fig. 11 and Fig. 12, The Brillouin coefficient is estimated to be (3.9 ± 0.4) x 10^{-9} m.W^{-1} for the As_2S_3 and (6.75 ± 0.35) x 10^{-9} m.W^{-1} for As_2Se_3. The value for the As_2Se_3 is close to the only other previously published result for this composition (Song et al, 2006). The value for the As_2S_3 fiber, although lower than the one for As_2Se_3, is still two orders of magnitude higher than that for fused silica (~4.4 x 10^{-11} m.W^{-1}) (Song et al, 2006; Ogusu et al., 2004).

9. Slow light

The slow-light technique based on stimulated Brillouin scattering (SBS) in optical fibers has attracted interest as it allows a very simple and robust implementation of tunable optical pulse delays, using mostly standard telecom components. Especially important are non-silica-based fibers with higher nonlinearity since these require lower powers and shorter lengths for practical implementations.

To date, there have been reports of slow-light generation in Bi-oxide high-nonlinearity fiber (Jáuregui, C. et al., 2006), telluride fiber (Abedin, K., 2008) and of very efficient slow and fast light generation in As_2Se_3 chalcogenide fiber (Song, K. et al., 2006). Additionally, the SBS process has been studied in As_2S_3 glass fibers (Florea et al., 2006). The very large Brillouin gain coefficient presents the chalcogenide fibers as alternatives to silica fiber for slow-light applications. A figure of merit (FOM) has been proposed (Song et al., 2006) in order to quantify the usefulness of a given fiber for slow-light based applications. The Brillouin gain is considered a positive factor while the length, the refractive index, and the power are considered as negative factors impacting the response time and the onset of additional nonlinear effects in the system. The FOM (Song et al., 2006) requires knowledge of the actual Brillouin gain which has to be measured, and takes into account the effective length not the total length of fiber. One can re-write the FOM such as to reduce it to the primary quantities describing the fiber (effective area, length and propagation loss, refractive index, and Brillouin gain coefficient expressed in dB):

$$FOM \equiv \frac{Gain[dB]}{P_p nL} = \frac{10 \times \log(\exp(g_B k \frac{P_p}{A_{eff}} L_{eff}))}{P_p nL} \tag{8}$$

The FOM can be further reduced to:

$$FOM = 4.34 \frac{g_B k L_{eff}}{n A_{eff} L} \tag{9}$$

It is important to keep in mind that this FOM essentially determines what length and power are needed in a system to achieve a certain gain, and hence a certain time delay. The FOM as defined above in Eq. 9 tends to be a quantity which obscures the physical meaning contained in Eq. 8. Actually, the theoretical gain (G_{th}), expressed in dB, as given by Eq 10, could be used instead to compare different fibers, if one considers a standard fiber length of 1 m and a standard pump power of 1 mW. Then, the theoretical gain is given by Eq. 10:

$$G_{th}[dB] = 4.34 \frac{g_B k \times 1mW \times L_{eff}|_{L=1m}}{A_{eff}} \tag{10}$$

One can use this last, fairly simple expression to compare the most representative fibers considered so far: silica (Song et al., 2005; Ruffin et al., 2005), high-nonlinearity bismuth fiber (Jáuregui et al., 2006; Lee et al. 2005), As_2Se_3 fiber (Song et al., 2006), along with the results reported here. The comparison is provided in Table 3, with all the data reported for experiments without polarization control (k=0.5). Also included is the FOM as defined above for completion. One can easily notice the significant increase in the theoretical gain (or FOM) for the As_2S_3 fiber due to its smaller core, lower loss and slightly reduced refractive index.

A typical experimental setup for slow light demonstration using chalcogenide fiber is detailed in Figure 14. The components contained within the dashed contour lines were only employed for the delay measurements. The output of a DFB laser (at 1548 nm) was split in two components, one which will serve as a pump while the other will serve as a counter-propagating signal.

	Silica [a]	Bi-HNL [b]	As_2Se_3 [c]	As_2Se_3	As_2S_3
N	1.47	2.22	2.81	2.81	2.45
Aeff [m^2]	6.78x10^{-11}	0.3x10^{-11}	3.94x10^{-11}	6.31x10^{-11}	1.39x10^{-11}
loss [dB.m^{-1}]	0.001	0.91	0.84	0.90	0.57
L [m]	2.0	2.0	5.0	5.0	10.0
Leff [m]	2.0	1.63	3.23	3.1	5.6
g_B [m.W^{-1}]	4.40x10^{-11}	6.43 x10^{-11}	6.10x10^{-9}	6.75x10^{-9}	3.90x10^{-9}
G_{th} [dB]	**0.076**	**0.003**	**1.084**	**0.719**	**3.398**
FOM [dB.W^{-1}.m^{-1}]	**1**	**17**	**77**	**51**	**139**

[a] Song et al.,2005; Ruffin et al., 2005 ; [b] Jáuregui et al., 2006 ; [c] Song et al., 2006

Table 3. Comparison of figure of merit for slow-light based applications at 1.56 μm.

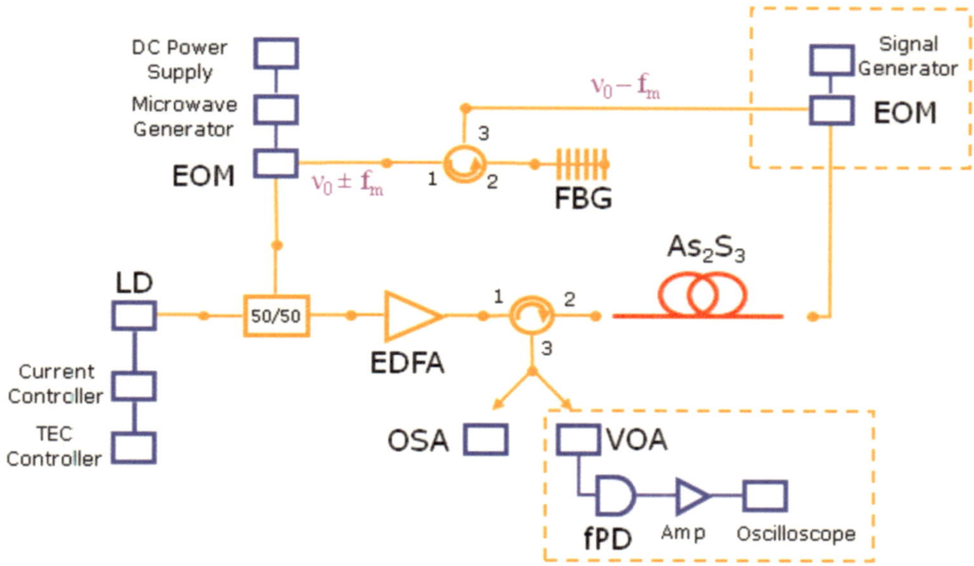

Fig. 14. Experimental setup used for gain and delay (dashed contour line) measurements. Abbreviations: LD - laser diode, EOM - electro-optical modulator, FBG - fiber Bragg grating filter, EDFA - Er-doped fiber amplifier, VOA - variable optical attenuator, fPD - fast photodiode, Amp - electrical amplifier.

The signal component is frequency shifted by a certain amount (f_m) such as to match the Brillouin shift. Using a $LiNbO_3$ modulator and a signal generator one can generate two side-bands which are then separated by using a fiber Bragg grating (FBG) filter. The center frequency is suppressed through DC biasing. For the gain measurements, the signal is coupled into the chalcogenide fiber and the output is monitored with an OSA. For the delay measurements, the signal, prior to being coupled into the fiber, is modulated (sine wave at 25 MHz) with a $LiNbO_3$ modulator and a DS345 signal generator. The output is then passed through a variable optical attenuator (VOA) and detected with a fast photodiode and an amplifier on an oscilloscope. The VOA allowed us to control the signal on the detector such

that we maintained the same signal (as low as possible) throughout the gain measurement to avoid any electronics-induced time response.
The pump is amplified with a standard EDFA and passed through a circulator before being coupled into the chalcogenide fiber, counter-propagating with the signal. The circulator allows us the signal to be picked off and sent to the detector.
The in-house drawn fiber used in this work was similar to the one used in previous work (Florea et al., 2006) but this time the fiber was cabled and both ends were antireflection coated. The fiber had a core of 5.2 μm diameter and a clad of 150 μm diameter, while the loss at 1550 nm was measured to be 0.138 m^{-1} (0.6 $dB.m^{-1}$). The effective area of the fundamental mode was measured and the critical power, P_{th}, for a 10-m length of fiber was determined, directly from the variation, with pump power, of the counter-propagating signal generated through Brillouin scattering. This was done in order to check the previous estimate of the g_B coefficient (Florea et al., 2006), which was obtained by rather qualitatively analyzing the spectral changes of signal. By using A_{eff} and P_{cr} to determine g_B as detailed below, this approach follows the method used in previous work (Song et al., 2006; Abedin, 2006) although a more exact analysis was proposed elsewhere (Ogusu, 2002).
The effective area (A_{eff}) was measured by imaging the fiber output on a vidicon camera using an appropriate microscope objective. A_{eff} was measured directly rather than use a theoretical estimate (Song et al., 2006) due to the fact that the fiber had a very high NA (greater than 0.30) making it possible for a second, higher order mode to contribute to the fundamental mode field. The measuring system was calibrated by also imaging a patch of SMF28 fiber with well known mode-field diameter (MFD) of 10.4 ± 0.8 μm at 1550 nm. The MFD for the chalcogenide fiber was thus determined to be 5.2 ± 0.4 μm.
The critical power was measured by monitoring the intensity of the Brillouin scattered signal versus the launched, counter-propagating pump power. A more precise analysis is usually performed in silica fibers (Ruffin et al., 2005). The coupling efficiency was estimated from fiber throughput measurements. The reflected signal was collected using a circulator, and the values of the Brillouin peak were read directly from the optical spectrum analyzer (OSA). Several measurements were made which yielded an average P_{th} of 29 ± 6 mW, which is close to the previously reported value (Florea et al., 2006) of 27 ± 3 mW. A typical data set is shown in Figure 15.

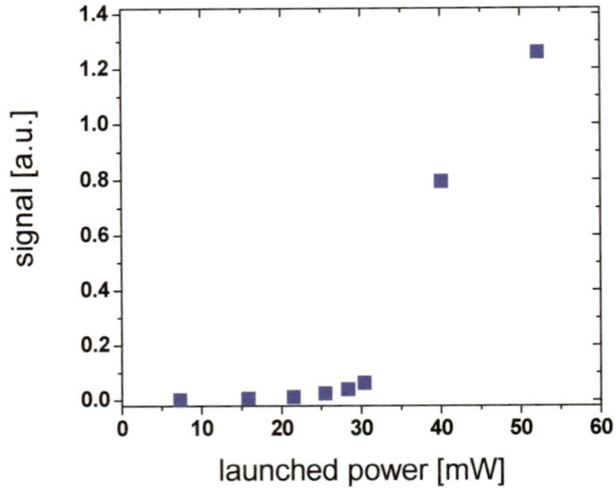

Fig. 15. Brillouin scattered signal in As_2S_3 fiber versus launched pump power.

Using Equation 7, in which α is the fiber loss and L is the fiber length, an estimate of the effective fiber length (L_{eff}) can be obtained, giving a value of 5.4 m. Finally, one can use these values for A_{eff}, P_{th}, and L_{eff}, to estimate the Brillouin scattering coefficient using Equation 5, where k = 0.5, in this case. Using proper error analysis, the Brillouin scattering coefficient was determined to be (5.7 ± 2.0) x 10^{-9} m.W^{-1} for the As_2S_3 fiber.

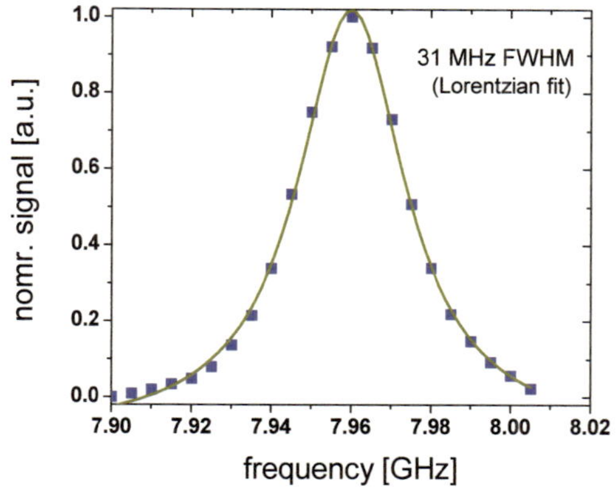

Fig. 16. Typical linewidth of the Brillouin signal at low pump power.

Additionally, the linewidth of the Brillouin signal was measured using a small probe (~ 8 μW) launched counter-propagating through the fiber. The Brillouin shift was identified to be 7.96 GHz while the linewidth of the Brillouin shift was found to be 31 MHz with typical data being represented in Figure 16. The linewidth was measured at low pump powers. Linewidth narrowing was observed for higher powers with linewidths as small as 19 MHz being recorded.

Gain and delay measurements using a small signal (~ 8 μW) have been performed in the chalcogenide fiber. For the gain measurement, the signal peak values were read from the OSA for different pump powers. For the delay measurement, the relative shift of the sine wave was read from the oscilloscope. Typical set of traces is shown in Figure 17.

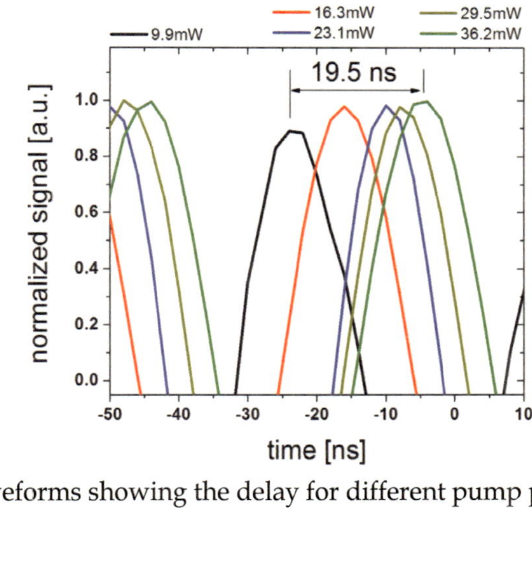

Fig. 17. Typical waveforms showing the delay for different pump powers.

The observable gain and delay were limited by the damage threshold of the AR coating, which unfortunately was lower than the threshold for the bare As_2S_3 glass. A slow variation of the amplified signal was observed which perhaps was due to the lack of polarization control in the setup. The overall results are represented in Figure 18.

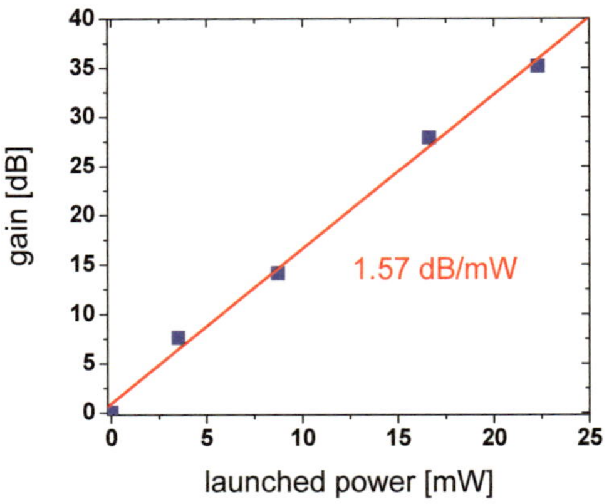

(a)

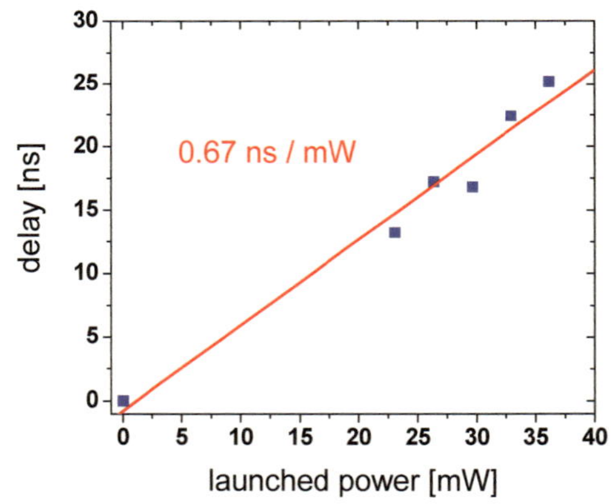

(b)

Fig. 18. (a) Gain and (b) pulse delay measurements in 10-m long As_2S_3 fiber at 1548 nm.

The slope of gain-versus-power is twice as large as the best previously reported result (Abedin, 2008). This was expected based on the analysis of the figure of merit (FOM) for the SBS process in these fibers (Song et al., 2006; Florea et al., 2006). However, the gain slope falls short of the theoretical estimate. Using the undepleted pump approximation, the small-signal gain is given theoretically by Equation 11:

$$G_{th}[dB] = 4.34\frac{g_B \times k \times L_{eff}}{A_{eff}} \times P \qquad (11)$$

Using the experimentally determined values for the involved parameters along with the associated uncertainties, Equation 11 gives us a slope in the range [1.8 ... 5.0] $dB.mW^{-1}$. Inhomogenities in the fiber core diameter which we have noticed, the potential presence of a second mode and the pump depletion approximation can be viewed as factors contributing to the discrepancy.
The same factors can also influence the delay data. Once again one can predict theoretically how much the peak of the signal pulse would be delayed (Δt) assuming an undepleted pump. The group velocity (given by $v_g = c/n_g$, c - speed of light, n_g total fiber group index) determines the time that a given pulse will take to travel the effective length of fiber. In the presence of the pump, the group velocity at the peak of the Brillouin gain will be modified according to Equation 12 (Okawachi et al., 2005), where $\Delta\upsilon$ is the linewidth (full-width half-maximum) of the Brillouin shift:

$$\frac{1}{v_g} = \frac{n_{fg}}{c} + \frac{G / L_{eff}}{2\pi \times \Delta\nu} \tag{12}$$

For a narrow linewidth pulse the delay, that is difference between the transit times required by the pulse with and without the pump, will then be given by Equation 5 (Okawachi et al., 2005):

$$\Delta t \approx \frac{G}{2\pi \times \Delta\nu} = \frac{g_B \times k \times L_{eff}}{A_{eff} \times 2\pi \times \Delta\nu} \times P \tag{13}$$

Using the experimentally determined values for the involved parameters along with the associated uncertainties, Equation 13 gives us a slope in the range [2.1 ... 5.9] $ns.mW^{-1}$. For this type of fiber, Equation 13 indicates theoretically that delays on the order of 100 ns or more can be obtained for reasonable powers.
While in practical terms a 19 ns delay was obtained for only 31 mW of pump power, which is marginally better than the result in the As_2Se_3 fiber (Abedin, 2008), these experimental values fall short of the theoretical expectations. The choice of the 25 MHz frequency for modulation of the signal was unfortunate since it turned out to be too close to the Brillouin linewidth, especially at low powers. Future work will try different modulation parameters and will also provide a study to gain an insight into the nature and origin of fiber imperfections and the role of polarization which can negatively influence the performance of this system. This understanding will pave the way forward for delays of the order of 20 ns with as little as 10 mW of launched power.

10. Conclusions

The large nonlinearities and fast response of the nonlinearity of the As-S-Se system make fibers drawn from these glasses well suited for optical switches, optical regenerators for high speed telecommunication systems. Use of these materials will allow compact devices cm's in length with optical powers <1W peak power (1 pJ in 1 ps pulses). The large Raman gain of the As-S-Se fibers coupled with the large IR transparency make these well suited for compact Raman amplifiers for telecommunications as well as fiber lasers and amplifiers in the mid-IR. These high nonlinearities also allow efficient supercontinuum generation which

is useful for broadband sources in the near and mid-IR. Finally these materials can be poled to induce an effective $\chi^{(2)}$, opening up the potential of waveguide parametric amplifiers.
The stimulated Brillouin scattering process was studied in As_2S_3 and As_2Se_3 single mode fibers. Values of the Brillouin gain coefficient were measured to be $(3.9 \pm 0.4) \times 10^{-9}$ m.W^{-1} and $(6.75 \pm 0.35) \times 10^{-9}$ m.W^{-1}, respectively. An analysis of the figure of merit for slow-light based applications indicates that the smaller core As_2S_3 fiber performs best due to the lower loss, reduced core size and slightly lower refractive index. The configuration using the small-core As_2S_3 fiber yields a figure of merit which is about 140 times larger, or a theoretical gain about 45 times larger, than the best silica-based configurations reported to date.
The continued improvement of chalcogenide materials will make such devices feasible in the near term.

11. References

Abedin, K. (2006). Observation of strong stimulated Brillouin scattering in single-mode As_2Se_3 chalcogenide fiber. *Opt. Express* 13, pp. 10266-10271.

Abedin, K.; Lu, G. and Miyazaki, T. (2008). *Electron. Lett* 44 , pp. 16.

Asobe, M.; Kanamori, T.; Naganuma, K.; Itoh, H. and Kaino, T. (1995). Third-order nonlinear spectroscopy in As_2S_3 chalcogenide glass fibers. *J. Appl. Phys.* 77, pp. 5518-5523.

Asobe, M; Kobayashi, H; Itoh, H; and Kanamoir, T; (1993) Laser-diode-driven ultrafast all-optical switching by using highly nonlinear chalcogenide glass fiber, *Opt Lett*, 18, pp. 1056-1058.

Borisova, Z. (1981). *Glassy Semicondutors*, Borisova, Z. (Ed.), Plennum Press, NY, USA.

Churbanov, M. (1992). *J. Non-Cryst. Solids*, Vol. 140, pp. 324.

De Aruajo, M.; Vermelho, M.; Gouveia-Net, A.; Sombra, A. and Medeiros Neto, J. (1996). *IEEE Photonics Technol. Letts.*, Vol. 8, pp. 821.

Devyatykh, G.; Churbanov, M.; Scripachev, I.; Dianov, E. and Plotnichenko, V. (1992). *International Journal of Optoelectronics*, Vol. 7, pp. 237-254.

Dianov, E.; Kazansky, P. and Stepanov, D. (1989). *Sov. J. Quant. Electron.*, Vol. 19, pp. 575.

Edwards, G.; Logan, R.; Copeland, M.; Reinisch, L.; Davidson, J.; Johnson, B.; Maciunas, R.; Mendenhall, M.; Ossoff, R.; Tribble, J.; Wekhaven, J. and O'Day, D. (1994). *Nature*, Vol. 371, pp. 416-418.

Florea, C.; Bashkansky, M.; Dutton, Z.; Sanghera, J. & Aggarwal, I. (2006) Stimulated Brillouin scattering in single-mode As_2S_3 and As_2Se_3 chalcogenide fibers. *Opt. Express*, 14, 12063-12070.

Galili, M; Xu, J; Mulvad, H; Oxenlowe, L;, Clausen, A; Jeppesen, P; Luther-Davies, B; Madden, S; Rode, A; Choi, D; Pelusi, M; Luan, F; and Eggleton, B (2009) Breakthrough switching speed with an all-optical chalcogenide glass chip: 640 Gbit/s demultiplexing, *Opt. Exp*, 17, pp. 2182-2187

Harbold, J.; Ilday, F.; Wise, F.; Sanghera, J.; Nguyen, V.; Shaw, L. and Aggarwal, I. (2002) *Optics Letters*, Vol. 27, pp. 119.

Hu, J.; Menyuk, C., Shaw, L., Sanghera, J., and Aggarwal, I., (2009) Generating mid-IR source using As_2S_3-based chalcogenide photonic crystal fibers, Proc. *Conference on Lasers and Electro-Optics (CLEO) 2009*, Optical Society of America

Hu, J.; Menyuk, C., Shaw, L., Sanghera, J., and Aggarwal, I., (2008) Raman response function and supercontinuum generation in chalcogenide fiber, *Proc. Conference on Lasers and Electro-Optics (CLEO) 2008*, Optical Society of America

Ippen, E. and Stolen, R. (1972). Stimulated Brillouin scattering in optical fibers. *Appl. Phys. Lett.*, 21, pp. 539-541.

Jackson, S; and Anzueto-Sanchez, G; (2006) Chalcogenide glass Raman fiber laser, *Appl. Phys. Lett.*, 88, p. 221106

Jáuregui, C.; Ono, H.; Petropoulos, P. and Richardson, D. (2006). Four-fold reduction in the speed of light at practical power levels using Brillouin scattering in a 2-m bismuth-oxide fiber. *OFC*, paper PDP2.

Kazansky, P.; Russell, P. and Takabe, H. (1997). *J. Lightwave Tech.*, Vol. 15, pp. 1484.

Lenz, G.; Zimmermann, J.; Katsufuji, T.; Lines, M.; Hwang, H.; Spalter, S.; Slusher, R.; Cheong, S.; Sanghera, J. and Aggarwal, I. (2000). *Optics Letts.*, Vol. 25, pp. 254.

Lee, J.; Tanemura, T.; Kikuchi, K.; Nagashima, T.; Hasegawa, T.; Ohara, S. and Sugimoto, N. (2005). Experimental comparison of a Kerr nonlinearity figure of merit including the stimulated Brillouin scattering threshold for state-of-the-art nonlinear optical fibers. *Opt. Lett.* 30, 1698-1700.

Nasu, H.; Ibara, Y. and Kubodera, K. (1989). *J. Non-Cryst. Solids*, Vol. 110, pp. 229.

Ogusu, K. (2002). Analysis of Steady-State Cascaded Stimulated Brillouin Scattering in a Fiber Fabry–Pérot Resonator. *IEEE Photon. Tech. Lett.* 14, pp. 947-949.

Ogusu, K.; Li, H. and Kitao, M. (2004). Brillouin-gain coefficients of chalcogenide glasses. *J. Opt. Soc. Am. B.*, 21, pp. 1302-1304.

Okawachi, Y.; Bigelow, M.; Sharping, J.; Zhu, Z.; Schweinsberg, A.; Gauthier, D.; Boyd, R. and Gaeta, A. (2005). *Phys. Rev. Lett.* 94, pp. 153902.

Pelusi, M; Ta'eed, V; Fu, L.; Magi, E.; Lamont, M; Madden, S; Choi, D; Bulla, D; Luther-Davies, B; Eggleton, B; (2008) Applications of highly-nonlinear chalcogenide glasss devices tailored for high-speed all-optical signal processing, *IEEE J. Sel. Top in Quant. Electron.*, 14, pp. 529-539.

Richardson, K.; McKinley, J.; Lawrence, B.; Joshi, S. and Villeneuve, A. (1998). *Opt. Mats.*, Vol. 10, pp. 155.

Ruffin, A. (2004). Stimulated Brillouin Scattering: An overview of measurements, system impairments, and applications. *NIST Symposium on Optical Fiber Measurements, Technical Digest*, pp. 23-28.

Ruffin, A.; Li, M.; Chen, X.; Kobyakov, A. and Annunziata, F. (2005). Brillouin gain analysis for fibers with different refractive indices. *Opt. Lett.* 30, pp. 3123-3125.

Sanghera, J.; Busse, L. and Aggarwal, I. (1994a). *J. Appl. Phys.*, Vol.75, pp. 4885 .

Sanghera, J.; Nguyen, V.; Pureza, P.; Kung, F.; Miklos, R. and Aggarwal, I. (1994b). *J. Lightwave Tech.*, Vol. 12, pp. 737.

Sanghera, J.; Aggarwal, I.; Busse, L.; Pureza, P.; Nguyen, V.; Miklos, R.; Kung, F. and Mossadegh, R. (1995). *SPIE*, Vol.2396, pp. 71.

Sanghera, J.; Aggarwal, I.; Busse, L.; Pureza, P.; Nguyen, V.; Kung, F.; Shaw L. and Chenard, F. (2005). Chalcogenide Optical Fibers Target Mid-IR Applications. *Laser Focus World*, April Vol. 41 No. 4, pp. 83-87.

Shaw, L.; Thielen, P.; Kung, F.; Nguyen, V.; Sanghera, J. and Aggarwal, I. (2005). IR supercontinuum generation in As-Se photonic crystal fiber. *Proceedings of Advanced Solid State Photonics*, TOPS vol. 98, Optical Society of America.

Slusher, R.; Hodelin, J.; Sanghera, J.; Shaw, L. and Aggarwal, I. (2004). *JOSA-B*, Vol. 21, pp. 1146.

Song, K.; Herráez, M. and Thévenaz, L. (2005). Observation of pulse delaying and advancement in optical fibers using stimulated Brillouin scattering. *Opt. Expr.* 13, pp. 82-88.

Song, K.; Abedin, K.; Hotate, K.; Herráez, M. and Thévenaz, L. (2006). Highly efficient Brillouin slow and fast light using As_2Se_3 chalcogenide fiber. *Opt. Expr.* 14, pp. 5860-5865.

Thielen, P.; Shaw, L.; Pureza, P.; Nguyen, V.; Sanghera, J. and Aggarwal, I. (2003a). Small-core As-Se fiber for Raman amplification. *Optics Letters*, 28, pp. 1406-1408.

Thielen, P.; Shaw, L.; Sanghera, J. and Aggarwal, I. (2003b). *Optics Express*, Vol. 11, pp. 3248-3253.

3

Irradiation Effects in Optical Fibers

Sporea Dan[1], Agnello Simonpietro[2] and Gelardi Franco Mario[2]
[1]National Institute for Lasers, Plasma and Radiation Physic, Laser Metrology Laboratory,
[2]University of Palermo, Department of Physical and Astronomical Sciences
[1]Romania,
[2]Italy

1. Introduction

Intrinsic and extrinsic optical fiber-based sensors are promising devices to be used in very different and complex environments, by their very nature: capabilities to work under electromagnetic fields; possibility to carry multiplexed signals (time, wavelength multiplexing); small size and low mass; ability to handle multi-parameter measurements in distributed configuration; possibility to monitor sites far away from the controller. In the case of the optical fibers, the possibility to be incorporated into various types of sensors and actuators, free of additional hazards (i.e. fire, explosion), made them promising candidates to operate in adverse conditions as those required by space applications and terrestrial nuclear facilities (Alam et. al.[a], 2006; Alam et al.[b], 2006; Berghmans et al., 2008; Ott, 2002). In nuclear environments optical fibers found an application niche in optical communication links, embedded into various all-fiber or hybrid sensors or as light-guides for control and diagnostics (Alfeeli et al., 2007; Ahrens et al., 2001; Fernando et al., 2005; Fielder et al., 2005; Florous et al., 2007; Gan et al. 2008; Henschel et al., 2001; Kimurai et al. 2002; O'Keeffe et al. 2008; Reichle et al., 2007; Troska et al., 2003). For applications related to fusion installations the requirements are quite demanding because of the exposure to (Campbell, 2005; Griscom, 1998; Hodgson, 2006; ITER Physics Expert Group on Diagnostics, 1999; Shikama, 2003; Zabezhailov, 2005): ionising radiation, high temperature, and high electromagnetic disturbances.
One of the major drawbacks for optical fibers use under ionizing radiation is related to the development of colour centres, which affect dramatically the optical transmission in UV-visible-NIR spectral ranges (Griscom, 1998; Karlitschek, 1995). For this reason, optical fibers are by more than 30 years in the focus of colour centres research (Friebele, 1976; Kaiser, 1974).
Research on radiation induced colour centres in pure and doped bulk silica materials has a long history of over 50 years (Weeks, 1956), but it is still actual (Radiation effects, 2007; Devine et al., 2000; Pacchioni et al., 2000), as new materials and devices (optical fibers, waveguides, multiplexers, or fiber lasers) are continuously devised and evaluated. Apart from the diversity of the investigated materials and devices new challenges are presented by the various irradiation conditions to which such materials and devices are subjected.
The complexity of the colour centres dynamics lead to the use of complementary methods to individuate these centres (electron paramagnetic resonance: EPR, luminescence) besides the

traditional optical absorption spectroscopy. New insights are also provided by in-situ experiments, when optical absorption changes are monitored during the irradiation, aiming to evaluate the irradiation effects disentangled from post irradiation relaxation and with the target to go deeper into the mechanisms of defects generation.
This chapter reports original results concerning the investigations carried out on optical fibers irradiated by gamma-ray, beta-ray, and neutron, using comparative measurements of optical absorption, EPR and luminescence. The optical measurements were performed mostly in the UV spectral range as the most effective colour centres are generated there. The EPR measurements were used to detect the presence of the intrinsic point defects E′ centers (Devine et al., 2000; Pacchioni et al., 2000), as well as H related paramagnetic point defects typically observed in H loaded fibers (Li et al, 1996; Karlitschek et al., 1998). Finally, the photoluminescence measurements were applied to reveal the presence of oxygen deficient defects (ODC(II)) induced by irradiation (Skuja, 1998; Skuja et al.; 2005).

2. The physics of colour centres

Up to now, most of the investigations on the use of optical fibers in radiation environments were done in relation to their use for sensing and data transfer. One approach of interest is related to the use of optical fibers for periscopes (light transmitting pipe-lights) for plasma diagnostics in fusion installations. Within such a task, the spectral behaviour of optical fibers in the UV-visible spectral range is of interest (Brichard, 2005).
Many years of study on bulk silica and on preforms or silica fibers have enabled to evidence optical absorption effects from IR up to vacuum-UV induced by the devices manufacturing or by their subsequent use during high intensity light propagation or in irradiation environments (Devine et al., 2000; Pacchioni et al., 2000; Skuja, 1998). The relevant role of the absorption in affecting the optical efficiency of the devices has lead to isolate the contributions of various absorption bands. Many efforts have been done also to assign these bands to given point defects and to determine specific procedures that could enable to prevent the same defect formation and/or could let their conversion into optically inactive species. A short review of the main intrinsic absorption bands occurring in the UV spectral range will be given in the following.
The ubiquitous and most easily induced absorption features in the UV range is the band at about 213 nm (5.8 eV), attributed to the E′ centers (Pacchioni et al., 2000; Skuja, 1998). This band is characterized by an oscillator strength, *f*, of about 0.2, and can be well described by a Gaussian profile with full width at half maximum (FWHM) of about (0.7±0.1) eV. The point defect responsible for this absorption is characterized by a microscopic structure consisting in an unpaired electron in an approximately sp^3 hybrid orbital of a dangling bond of a threefold coordinated silicon atom: O≡Si$^\bullet$, where the symbol ≡ represents single bonds to three oxygen atoms and the symbol • stays for the unpaired electron (Pacchioni et al., 2000).
The electronic excitation process associated to this absorption band has been strongly debated and, recently, support to its attribution to an electronic transition from the defect related bonding states to the unpaired electron level has been given (Agnello et al., 2008).
It is noteworthy to observe that this defect is also paramagnetic and is characterized by an EPR signal (Weeks 1956) well correlated to the optical absorption (Weeks et al., 1963) with principal g-values (Weil et al., 1994, Pacchioni et al., 2000): g_1=2.0003, g_2=2.0006, g_3=2.0018.

Another relevant absorption band, diffusely observed in the UV range in various forms of irradiated silica, is the one peaked at about 258 nm (4.8 eV) (Pacchioni et al., 2000; Skuja, 1998). This band can be described by a gaussian profile with FWHM of 1.05 eV and oscillator strength f=0.03 (Cannas et al., 2006). Excitation within this absorption originates a well known red luminescence band at about 650 nm (1.9 eV) that has an additional excitation channel at 620 nm (2.0 eV). This overall optical activity has been attributed to the Nonbridging Oxygen Hole Centers (NBOHC) (Pacchioni et al., 2000). The microscopic model associated to this defect consists in a non-bridging oxygen atom: $O{\equiv}Si\text{-}O^{\bullet}$, having an unpaired electron in a non-bonding oxygen orbital (Pacchioni et al., 2000). The electronic transitions responsible for the absorption and emission of this defect are still debated but, usually, they are related to electron promotion from bonding states of the $O{\equiv}Si\text{-}O^{\bullet}$ moiety to the non bonding states of the dangling O and, in particular, to that of the unpaired electron (Bakos et al., 2004). Due to the presence of an unpaired electron in the oxygen orbital, this defect is responsible also for an EPR signal (Pacchioni et al., 2000; Skuja, 1998). By studying the dependence of this signal as a function of temperature the following main g values have been determined: g_1=1.9999, g_2=2.0095, g_3=2.078 (Pacchioni et al., 2000). However, the identification of NBOHC by the EPR technique is complex due to the overlap with other signals and to the resonance line being spread over a wide magnetic field (about 80 G at 9.8 GHz microwave frequency, X-band) that prevents its detection at low defect concentrations (typically below 10^{16} centers/cm^3).
Other defects responsible for the absorption increase around 258 nm (4.8 eV) are interstitial ozone molecule (O_3) and the peroxy radical: $O{\equiv}Si\text{-}O\text{-}O^{\bullet}$ (Skuja et al.; 2005). Both these defects should be observed as product of irradiation in oxygen rich materials or as a consequence of oxygen displacements processes induced by irradiation. Their detection is however difficult due to the overlap with the NBOHC band and, in particular, the attribution of an absorption band to the peroxy radical in this spectral range is still a matter of debate (Skuja et al.; 2005).
A defect of particular relevance for the UV optical absorption in SiO_2 is also the oxygen deficient center: ODC(II) (Pacchioni et al., 2000; Skuja et al.; 2005; Skuja, 1998). This defect is usually associated to an absorption band peaked at 248 nm (5.0 eV) and gives rise to two emission bands at 460 nm (2.7 eV) and 280 nm (4.4 eV) (Skuja, 1998). The ODC(II) can be induced during fiber fabrication but it is also relevant in relation to the photosensitivity of silica and in connection to fiber photon writing processes (Pacchioni et al. 2000; Karlitschek et al., 1998). In this respect, the connection of the ODC(II) with another defect usually named H(I) center is of particular importance due to the use of hydrogen for radiation hardening of fibers. This latter defect is paramagnetic and is characterized by a doublet of EPR lines split by 74 G (Skuja, 1998). The structural model for the H(I), confirmed by many experimental investigations, consists in a silicon atom bonded with two oxygen and one hydrogen and having an unpaired electron on silicon: $O{=}Si^{\bullet}\text{-}H$ (Pacchioni et al., 2000; Skuja, 1998).
The existence of a conversion between ODC(II) and H(I) has been evidenced by studies on surface defects on silica and has been used to support the model of the ODC(II) as consisting in a two-fold coordinated silicon: $O{=}Si^{\bullet\bullet}$ (Radzig, 1998, Skuja, 1998).
Despite the almost clear identification of many absorption features in the UV spectral range, the processes of defect formation and defect stability are still matter of investigation, both experimentally and theoretically (Griscom, 2001; Mashkov et al. 1996; Kajihara et al., 2008). In particular, it has been shown that the defect formation processes feature a complex

dynamics related to the presence of precursor of the defects and also affected by the inhomogeneity intrinsic to the amorphous nature of the material. Furthermore, the stability of irradiation induced defects is affected by the presence of species present in the material and, in particular, it has been evidenced that H related species (H, H_2, H_2O) usually affects the formation and thermal removal of many defects as NBOHC, E′ and ODC(II) (Griscom et al., 1998; Devine et al., 2000; Pacchioni et al. 2000; Nuccio et al. 2009; Messina et al. 2009).

3. Experiment

The goals of the here reported investigations were: to further improve the optical absorption measurements in the UV-visible spectral range in fibers; to evaluate the possibility to use alternative methods for identifying colour centres in fibers (i.e. EPR, luminescence); to check through an inter-laboratory comparison the capabilities of two laboratories located in Italy and Romania in performing EPR tests on irradiated optical fibers.
The investigated optical fibers were commercially available products. The characteristics of some of the investigated optical fibers are summarised in Table I.

Characteristics / Optical fiber sample nickname	Core diameter (μm)	Cladding diameter (μm)	Maximum operating temperature (°C)	Jacket type
S42-1	1000	1035 (TECS)	-	Tefzel
S43-1	400	440 (doped silica)	150	Tefzel
S44-1	600	660 (doped silica)	300	Polyimide
S45-1	600	660 (doped silica)	300	Polyimide
S46-1	600	660 (doped silica)	300	Polyimide
S47-1	400	500	-	Tefzel

Table I. The irradiated optical fibers.

The optical fibers are either solarization resistant optical fibers, optical fibers with enhanced UV transmission or radiation hardened optical fibers developed for UV applications.
The length of the samples for the optical absorption measurements was 120 mm. For the case of EPR and luminescence tests five pieces from the same optical fibers samples, without jacket and having a length of 10 mm each, were measured in the same time.

3.1 Set-ups

The general set-up for the off-line optical absorbance measurements is similar to that previously used (Sporea, 2005; Sporea, 2007), but, for the purpose of this investigation, it has a better S/N ratio (1,000:1 full signal), 16 bits A/D conversion resolution, a dynamic range of 25,000:1, a greater quantum efficiency in the UV range (65 % at 250 nm), spectral resolution 1.2 nm, a sensitivity of 0.065 counts/e^-, and a minimum OD detection level of 0.4 for the optical set-up used (this value is determined by two factors: first, the core of the connecting optical fibers and the core of the samples are different, and second, the sample optical fibers have no fixed connectors, hence, a biasing level which limits the set-up lowest detectable OD). Such a detecting scheme makes possible a better tracking of the colour centre development in the UV spectral range and enables a higher range of absorption levels to be detected (O.D. of 4.4). For the reported optical absorption curves the signal was averaged over three detected acquisitions with a value of 2 for the box car parameter. Irradiation and off-line measurements were carried out at room temperature.

Depending on the case, various set-ups were used for on-line measurements. During the gamma, beta and neutron irradiations tests were carried out both at room temperature and with the optical fiber heating. For these purposes specially designed set-ups were built (Figure 1), to make possible the simultaneous irradiation at room temperature (upper plate) and with the heating of the optical fibers (lower plate). In the case of neutron and beta irradiation a small oven was constructed which made possible the irradiation of the optical fibers in a linear geometry. For on-line measurements during gamma irradiation a circular geometry was used, the optical fiber samples (items 4 and 7 in Fig. 1 b) were coiled with a diameter of about 130 mm, and the gamma source was placed at the coil centre (item 6 in Fig. 1 b).

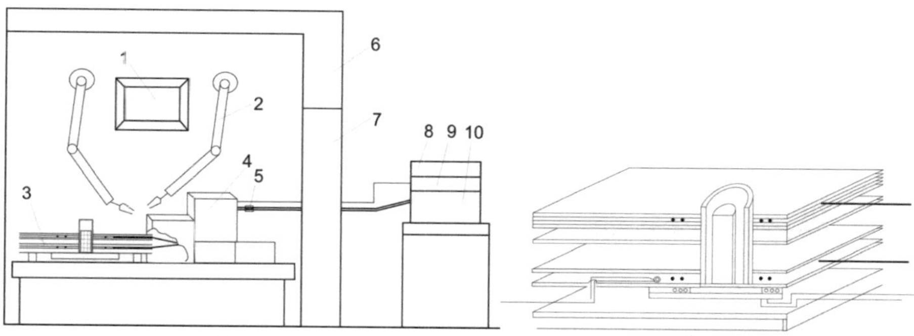

Fig. 1. The set-up for on-line measurements of gamma-irradiated optical fibers at room temperature and under heating conditions: a – the sketch of the irradiation hot-cell (1 – hot - cell window; 2 – remote manipulators; 3 – set-up for the optical fibers positioning; 4 – shielding; 5 – connecting optical fibers; 6 – hot-cell room; 7- hot cell access door; 8 – optical fiber mini spectrometer; 9 – broad band stabilized light source; 10 – laptop); b – detail of the set-up for the optical fiber positioning and heating.

The design of the special set-up made possible the irradiation at room temperature of one sample while a second sample was heated up to 240 °C. The possibility to have the sample heated during the irradiation enables the study of the colour centres dynamics under temperature stress.

3.2 Irradiations and measurements

The ^{60}Co gamma-ray off-line irradiations have been performed at room temperature by using the SVST-Co-60/B, tote-box type, multipurpose irradiation facility of the National Institute for Physics and Nuclear Engineering - Horia Hulubei (IFIN-HH). The SVST-Co-60/B irradiator has today an activity of 267000 Ci, and the irradiations were performed in the irradiation chamber, at a dose rate ~ 700 Gy/h.

For the absorbed dose measurements two dosimetric systems are used: ALANINE / EPR dosimetry system, calibrated for 20 Gy - 10 kGy range, and ECB dosimetric system, calibrated for 1 kGy - 100 kGy ranges. The measured absorbed doses for both the systems are traceable to RISO High Dose Reference Laboratory - Denmark, and their performances were tested by international dosimetric intercomparisons. The uncertainty of absorbed dose measurement is around 3%.

The on-line irradiations were performed in a hot-cell at the ^{60}Co gamma-ray irradiation facility of the "Horia Hulubei" National Institute of R&D for Physics and Nuclear Engineering-IFIN-HH. The dose rate can be modified by changing the distance of the optical fiber sample from the gamma source (i.e. a dose rate of 5.3 kGy/ h at 6 cm distance form the source).

The beta irradiation conditions were the following: the mean electron energy: 6 MeV, the electron beam current: 1 μA, the pulse repetition rate: 100 Hz, the pulse duration: 3.5 μs, the beam diameter: 10 cm, the spot uniformity: +/- 5 %. In order to have Bremsstrahlung, the electron beam collided on a tungsten target, and gamma-rays were generated. To separate the electron flux from the gamma-rays two Al foils, one having a thickness of 3 mm, and one of 4 mm were placed at the distances of 10 mm and respectively at 120 mm away from the target.

The neutron irradiation was performed at the Cyclotron accelerator facility of the National Institute of R&D for Physics and Nuclear Engineering "Horia Hulubei", Magurele, based on the reaction 9Be + d →n + X, using a deuteron beam (13 MeV) and a thick beryllium target of 165 $mg.cm^{-2}$. To obtain the desired neutrons fluencies the samples are located downstream at the distances from 10 to 40 cm from the Be target. The neutrons flux above 1 MeV is estimated with a relative error of about 20 %. The maximum neutron flux achievable in our setup, at a distance of 10 cm from the target, is 2.10^9 $n.cm^{-2}.s^{-1}.\mu A^{-1}$, corresponding to a deuteron beam intensity of 10 μA. In practice, a neutron fluence of 10^{13} $n.cm^{-2}$ can be obtained in about 1-6 days of irradiation, depending on the position of the samples in respect to the Be target. It is estimated that the neutron flux was uniform over 10 mm length of the optical fiber sample, with a spatial uniformity of 3%.

The EPR measurements were performed in Romania at the IRASM irradiation facility (National Institute for Physics and Nuclear Engineering - "Horia Hulubei") using an X band Magnettech MiniScope MS 200 spectrometer provided with a rectangular TE_{102} resonant cavity and 100 kHz modulation frequency.

Both non-irradiated and irradiated samples have been introduced into fused quartz tubes (1 cm sample length, 5 mm outer tube diameter) and fixed into the resonant cavity always in the same position. The mass of measured samples varied between 20 and 130 mg. The EPR spectra have been registered under the experimental conditions, detailed in the following figures, and with the magnetic field sweep rate of 18 G/min. The g-factors of the observed EPR signals have been determined by using a ZnS: Mn^{2+} standard. Similar measurements were done at the University of Palermo on a Bruker EMX-micro working in X-band (about 9.8 GHz) with a rectangular TE_{102} resonant cavity and at 100 kHz magnetic field modulation using an EPR tube, 10 mm outer diameter, for sample mount. For specific colour centres detection, a preliminary calibration of the two instruments (the Magnettech MiniScope MS 200 spectrometer in Romania and the Bruker EMX-micro in Italy) was run using a gamma ray irradiated sample of high purity silica to compare the sensitivity of the two instruments. Based on this calibration, additional parallel measurements were performed in the two laboratories and with the equipment of Palermo a concentration estimate has been carried out.

The luminescence effects were investigated at this stage only in Palermo with a Jasco FP6500 spectrofluorometer using a 150 W Xenon lamp. Excitation light impinged on the fiber samples perpendicularly to their axis using an artwork mount at 45° back-scattering configuration, and emission light was collected at 90° with respect to the excitation direction.

4. Results

Figures 2-4 illustrate the optical absorbance of the irradiated optical fibers as they were subjected to various total gamma irradiation doses. These spectra evidence that irradiations mainly induce the formation of bands below 300 nm that could be associated to E'-Si centres and Non-Bridging oxygen hole centres (Skuja, 1998), whose amplitude increase on increasing the irradiation dose. At higher wavelength negligible absorption is found. It can be observed that the fibers S44, S45 and S46 are more radiation resistant than the others.

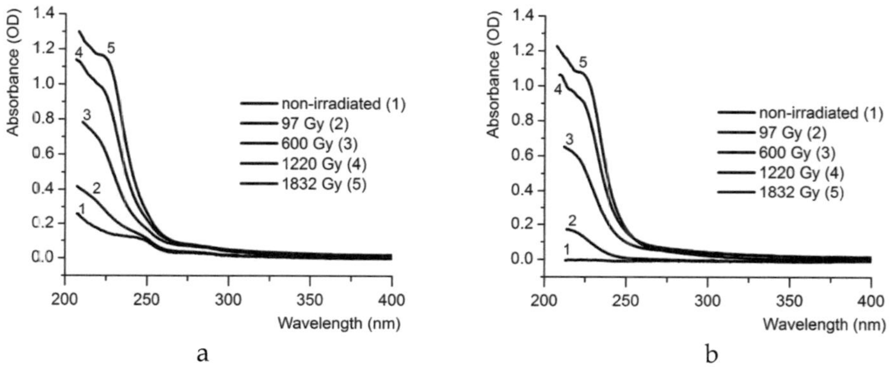

Fig. 2. Optical absorption spectra in optical density units (OD) for: a - sample S42-1; b - sample S43-1.

Optical fibers S44-1, S45-1 and S46-1 are hydrogen loaded type optical fibers, so they exhibit a higher radiation hardening than the other samples. Optical fibers S42-1 and S43-1 are optical fibers with an enhanced UV response and have a high OH content. The optical fiber S47-1 was designed for UV laser beam delivery applications with high fluorine content.

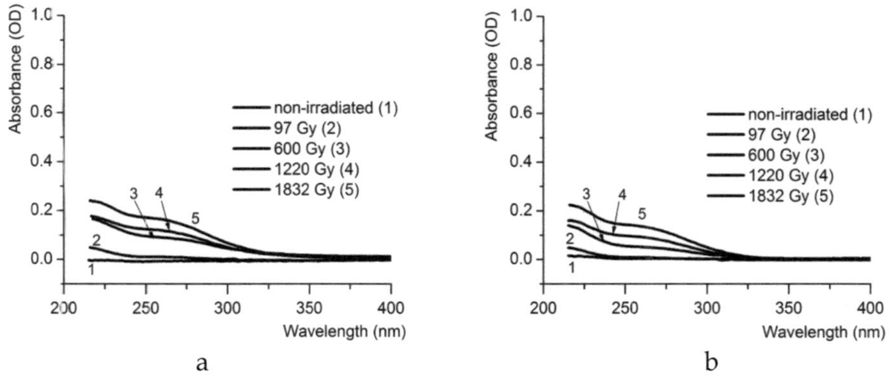

Fig. 3. Optical absorption spectra in optical density units (OD) for: a - sample S44-1; b - sample S45-1.

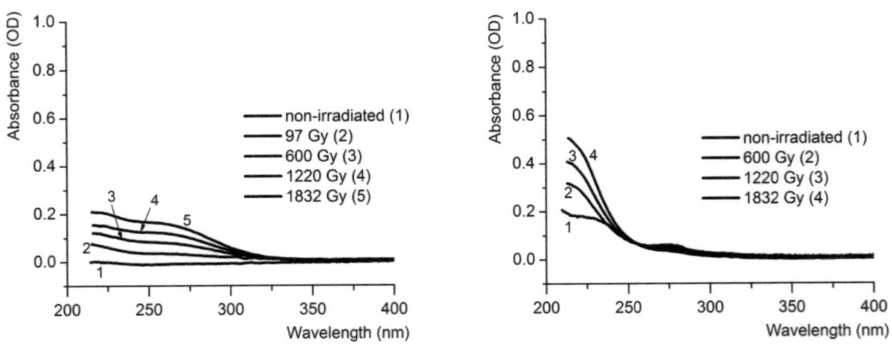

a b

Fig. 4. Optical absorption spectra in optical density units (OD) for: a -sample S46-1; b - sample S47-1.

Decomposition of the spectra as a function of energy obtained from the data of fibers S43, S44 and S45 (Figs. 2, 3) has been carried out using Gaussian bands. This decomposition enabled to further explore the presence of specific point defects in the irradiated fibers.

In Fig. 5 an example of the decomposition result is reported for the fiber S45 at the various irradiation doses. The found bands are characterized by peak position (4.8 ± 0.1) eV and FWHM (1.0 ± 0.1) eV, and by (5.7 ± 0.1) eV and FWHM (0.8 ± 0.1) eV, in reasonable agreement with the bands attributed to the NBOHC and to the E′ center (Pacchioni et al., 2000; Skuja, 1998), respectively.

It is found also that more than 70% of the absorption spectra is attributed to the absorption bands of the E′ center and of the NBOHC. A further band peaked at 198 nm (6.25 eV), whose origin is unknown, is necessary to describe the absorption below 210 nm (at energy larger than 5.9 eV). Usually this band contributes less than 25% to the overall decomposition. It was also found that these Gaussian amplitudes increase for all investigated dose evidencing that the induced absorption is related to the processes of these defects generation.

Similar results are found for the other investigated fibers. As can be observed the spectra of the S43 sample (Fig. 2) feature a low contribution at about 258 nm (4.8 eV) with respect to the component at 213 nm (5.8 eV), at variance to the samples S44, S45 and S46 (Fig. 3, 4). These aspects are confirmed by the spectra decomposition. Since the ratio between the composing bands, and in particular those related to the E′ center and to the NBOHC, is not constant and depends on the fiber, it can be guessed that the microscopic process of defects generation depends on the specific preparation method and could be affected by hydrogen loading (Devine et al., 2000, Karlitschek et al., 1998).

In Fig. 6 the change of the optical attenuation for an optical fiber with enhanced UV transmission, 400 μm core diameter, and Tefzel jacketing (sample similar to S43-1) under neutron irradiation at a total fluence of 1.18 x 10^{13} n/cm^2 is given as an example.

It is observed that also in this case the absorption increases non-linearly with the fluence for wavelengths below 400 nm. Due to the instrumental reduced sensitivity below 235 nm no specific feature can be associated to point defects in this case.

A solarization resistant 600 μm core diameter, Polyimide jacketed optical fiber (sample similar to S46-1) was subjected to beta irradiation and separately to Bremsstrahlung (Fig. 7a), at room temperature. During an additional experiment the same type of optical fiber

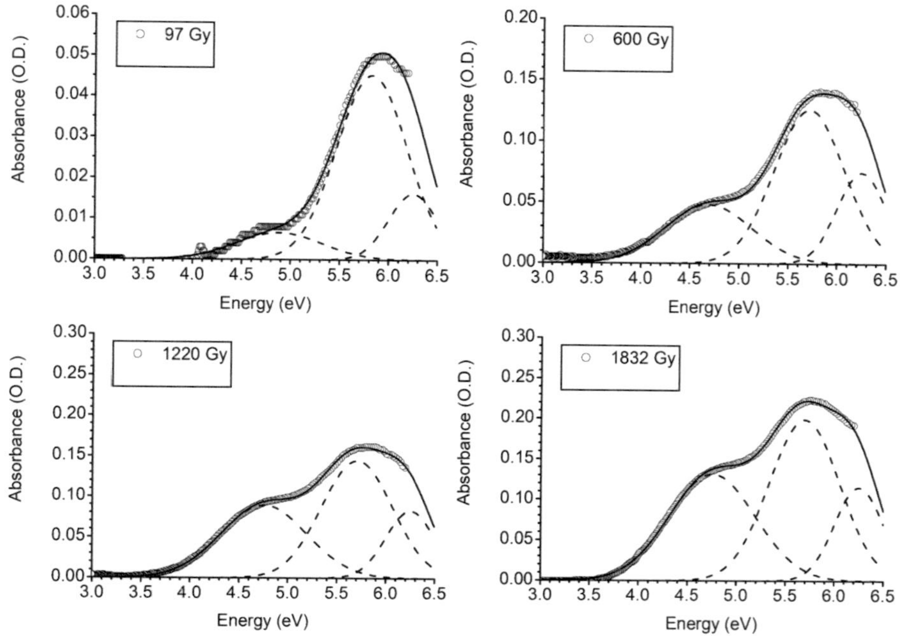

Fig. 5. Spectral decomposition using Gaussian bands of the spectra of sample S45-1 irradiated at various doses; the circles represent experimental points, the dashed lines the component Gaussian bands (peaked at about 4.8 eV, 5.8 eV and 6.25 eV) and the full line the sum of dashed lines.

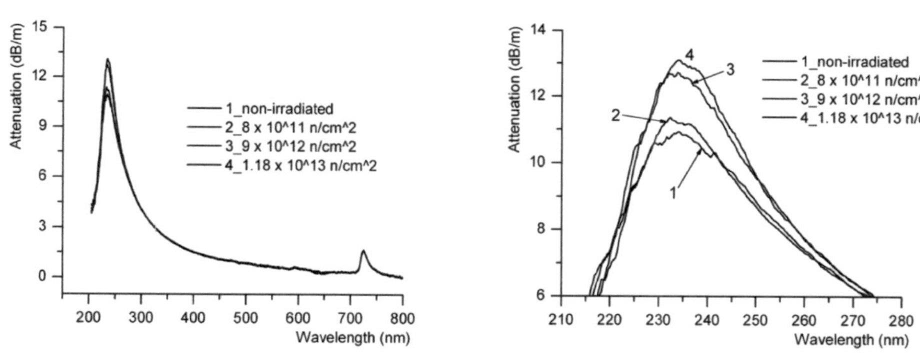

Fig. 6. The optical absorption spectra in the UV-visible range for an optical fiber with enhanced UV transmission, subjected to neutron irradiation: a - the UV-visible spectrum; b - detail of the attenuation spectrum in the 210 nm - 280 nm spectral range.

was also heated during the irradiation (Fig. 7b). The tested optical fibers were of hydrogen loaded type and exhibit a lower radiation hardening under heating conditions, as the temperature contributes to hydrogen diffusion.

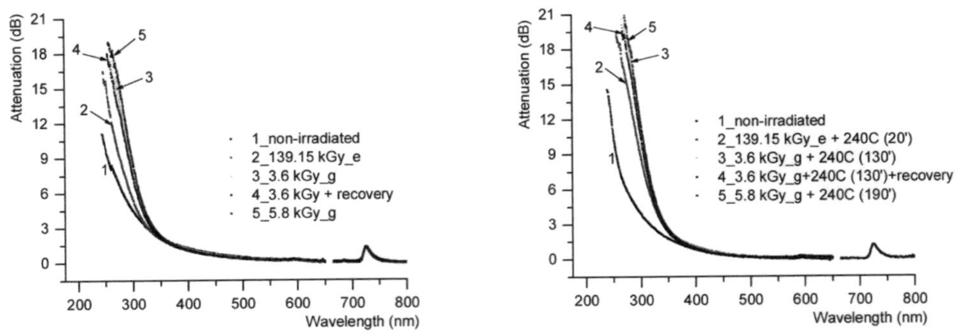

a b

Fig. 7. The change of the optical attenuation of an optical fiber subjected to beta irradiation (total dose of 139.15 kGy_e) and to Bremsstrahlung (total dose of 5.8 kGy_g): a - at room temperature; b – heated at 240 °C during the irradiation.

Investigation of the effects of irradiation on these fibers has been also carried out by EPR measurements. As shown in Fig. 8, these measurements evidenced the presence of resonance signals in some non-irradiated samples. A detailed investigation carried out in Palermo has shown that these signals do not arise from intrinsic defects of silica.

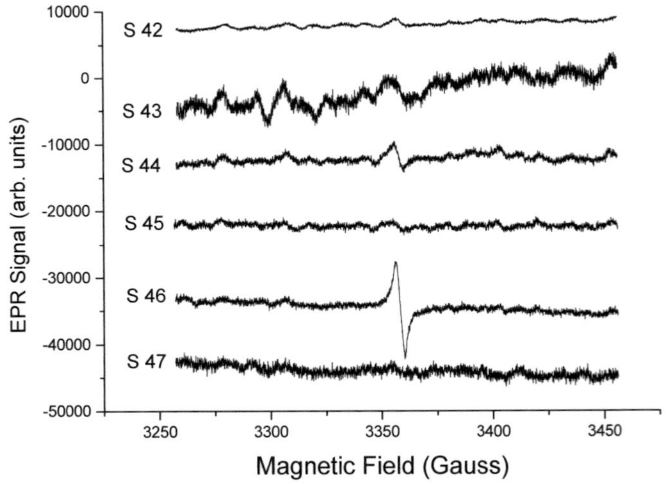

Fig. 8. The EPR signals for the non-irradiated samples corrected for sample mass (Power = 3.2 mW; modulation amplitude = 4 G; scanning time = 655 s; receiver gain = 9 x 10^2). Spectra are arbitrarily vertically shifted.

The EPR measurements carried out on the irradiated fibers are reported in Figs. 9 and 10. It is worth to note that in these figures the spectra recorded for the irradiated samples subtracted by spectra recorded for the respective pristine materials are reported, and show that the defects contents increase by irradiation. Furthermore, the higher radiation resistance of the samples S44, S45, S46 with respect to the other investigated fibers, already evidenced by optical measurements, is confirmed also by the EPR measurements.

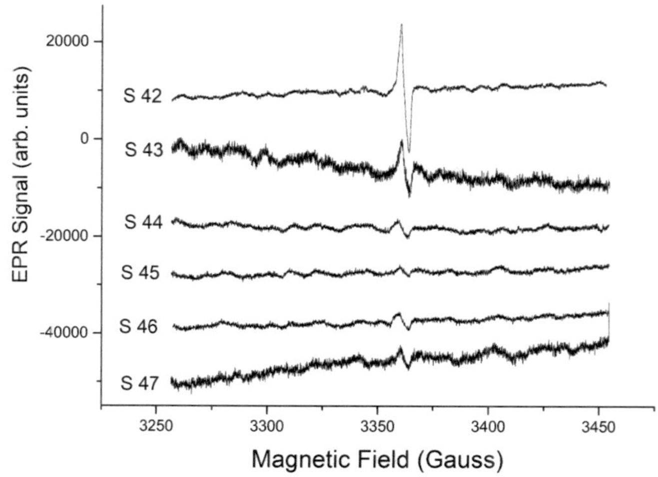

Fig. 9. The EPR spectra for 618 Gy irradiated samples corrected for sample mass, after non-irradiated signal subtraction (Power = 3.2 mW; modulation amplitude = 4 G; scanning time = 655 s; receiver gain: 1 x 10^2 for S 42 and 9 x 10^2 for the others). Spectra are arbitrarily vertically shifted.

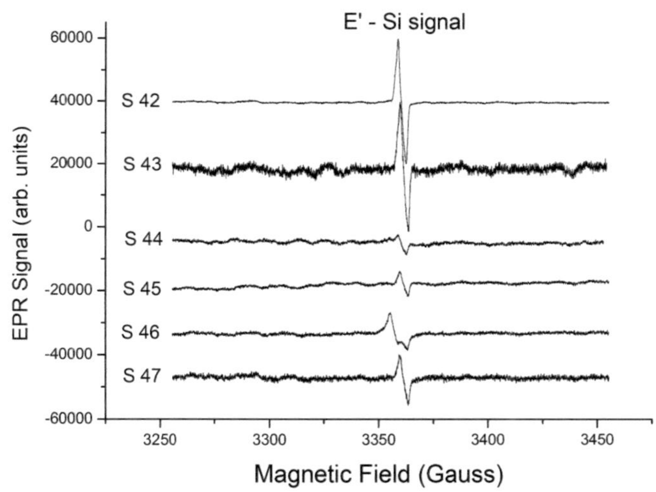

Fig. 10. The EPR spectra for 1800 Gy irradiated samples corrected for sample mass, after non-irradiated signal subtraction. (Power = 3.2 mW; modulation amplitude = 4 G; scanning time = 655 s; receiver gain: 1 x 10^2 for S 42 and 9 x 10^2 for the others). Spectra are arbitrarily vertically shifted.

In addition to the optical fiber specified in Table I, several irradiated samples (S7.7; S7.9; S1-440B and S3-440B) were investigated through EPR for comparative studies both in Bucharest and Palermo. The results of EPR measurements performed in Bucharest are given in Figs. 11-13 (spectra are arbitrarily vertically shifted and are not normalized for the samples weight).

These measurements evidence the presence of E'-Si centres (Pacchioni et al., 2000; Karlitschek, 1995) and of the H(I) defects in some of the irradiated samples (Skuja, 1998). The presence of H(I) defects, identified by the doublet of lines split by about 74G, is evidenced by the magnified spectrum.

The same measurements have been done in Palermo showing qualitatively the same results, and supporting the calibration procedure to individuate specific point defects induced by irradiation. Furthermore, the quantitative analysis carried out in Palermo, enabled to estimate the concentrations of E'-Si and H(I) point defects as reported in Table II.

Optical fiber sample nickname	Sample mass (g)	E'-Si defects (centers/cm³)	H(I) defects (centers/cm³)
S7.7	0.124	8.2×10^{16}	1.2×10^{15}
S7.9	0.132	5.7×10^{16}	1.0×10^{15}
S1_440B	0.023	3.3×10^{16}	Not detected
S3_440B	0.020	3.1×10^{16}	Not detected
SSU1.1	0.052	Not detected	Not detected

Table II. The concentration of E'-Si and H(I) point defects.

For some non-irradiated and irradiated samples luminescence measurements were performed (Fig. 14). These measurements have shown that in the non-irradiated samples no PL associated to intrinsic point defects of silica could be detected.

As shown in Fig. 14, in the irradiated fibers S7.9, S1_440B and S3_440B, the luminescence centred at about 4.4 eV excited at 5.0 eV associated to the oxygen deficient ODC(II) defects is detected (Skuja, 1998).

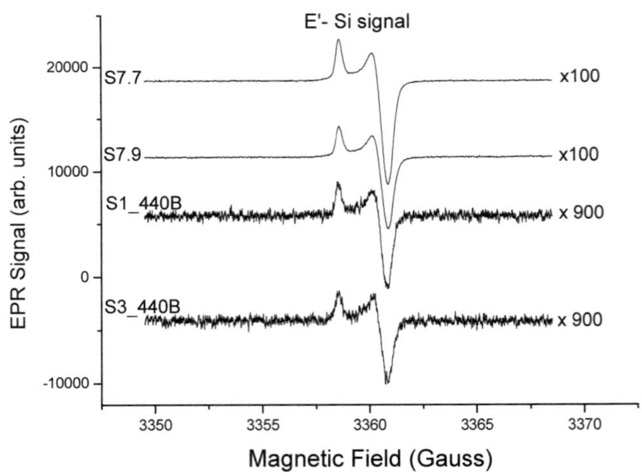

Fig. 11. The EPR signals for some irradiated samples at 113 kGy (samples S7.7 and S7.9) and at 40 kGy (samples S1-440B and S3-440B). Power = 0.1 mW; modulation amplitude = 0.1 G; scanning time = 655s; receiver gain = 1×10^2 for S7.7 and S7.9 and 9×10^2 for S1_440B and S3_440B.

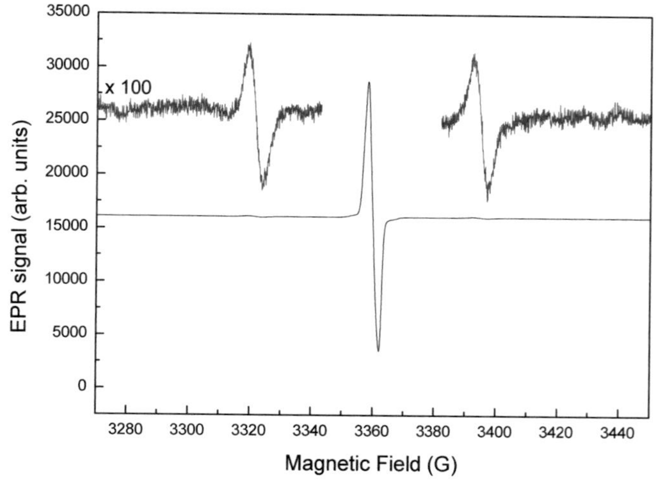

Fig. 12. The EPR signals recorded in Bucharest for the S7.7 sample irradiated at the dose of 113 kGy.

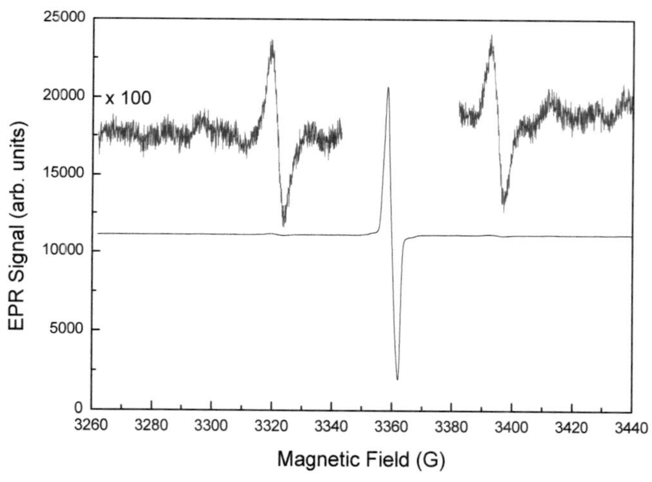

Fig. 13. The EPR signals recorded in Bucharest for the S7.9 sample irradiated at 113 kGy.

Sample S.7.7 was a common, 1000 µm core diameter optical fiber, gamma irradiated up to 113 kGy, in 4 steps, with sample heating at 100 ^{o}C for 4 h between each irradiation step. The optical fiber S.7.9 was a solarization resistant optical fiber of 1000 µm core diameter, gamma irradiated to 113 kGy. Samples S1-400B and S3-440B were of the same type (enhanced UV transmission), irradiated by gamma-rays up to a total dose of 40 kGy, the first one was also subjected to temperature stress (140 ^{o}C for 4 h, between the irradiation steps). In Fig. 14, the SSU1.1 fiber, not showing emission, is reported for comparison. Photoluminescence excitation spectra confirmed these attributions by evidencing an excitation band centred at 5.0 eV, as reported in literature for ODC(II) (Skuja, 1998). The presence in the sample S7.9 of the ODC(II) emission band agrees with the observation of the H(I) defects that, as reported in the paragraph 2, can be considered a product of the reaction between ODC(II) and H species dissolved in the sample (Skuja, 1998). As regards the samples S3_440B and S1_440B, the absence of H(I) centers signal is also compatible with the much lower amplitude of the emission band of ODC(II), proving a lower content of these latter defects.

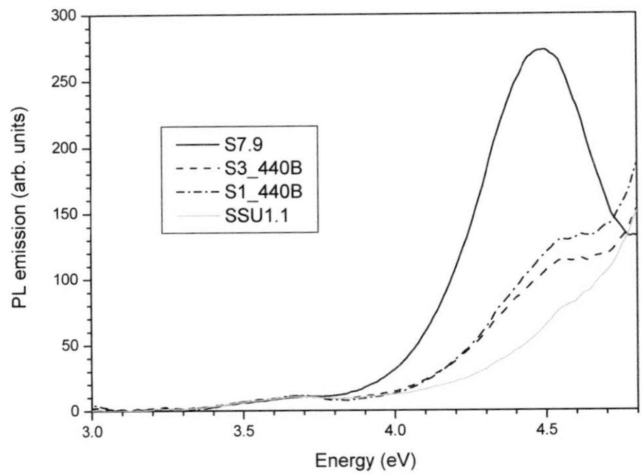

Fig. 14. Photoluminescence spectra excited at 5.0 eV of fibers irradiated at 113 kGy (sample S7.9), at 40 kGy (samples S1-440B and S3-440B) and at 500 kGy (sample SSU1.1).

5. Future work

Further investigations on the generation/recovery of colour centres in the UV-visible spectral range are of interest as new radiation hardened optical fibers are developed and more accurate measuring means are available. We shall focus in the near future on the evaluation of the irradiation effects on commercially optical fibers from different vendors in order to assess the role played in such cases by the dopants, core/cladding ratio, temperature stress. On-line measurements have to be extended to have a better understanding of the colour centre dynamics. Tests at higher irradiation dose and dose rates will be run for gamma and beta rays irradiation. To complement the inter-laboratory comparison on EPR measurements a supplementary inter-comparison dedicated to photoluminescence investigations will be organized between laboratories in Romanian and Italy.

6. Conclusions

Eleven types of commercially available optical fibers operating in the UV spectral range were investigated as they were subjected to gamma, beta, and neutron irradiation. The evaluation of the irradiation induced colour centres was done by off-line optical absorption measurements, EPR, and luminescence. E′-Si and H(I) defects were revealed by EPR, the

ODC(II) diamagnetic centres were identified by photoluminescence. From the optical absorption measurements it is observed that the induced absorption increases for all the investigated doses suggesting a growth of the concentration of induced colour centres, with a prominent role of the E′ centers and of the NBOHC. A dependence of irradiation effects on the materials is found.

7. Acknowledgements

The work was done in the frame of the bilateral collaboration existing between the National Institute for Laser, Plasma and Radiation Physics in Bucharest, and the Department of Physical and Astronomical Sciences of the University of Palermo. The Romanian authors acknowledge the financial support of the Romanian Ministry for Education, Research and Innovation in the frame of the research grant 12084/2008. The Italian authors acknowledge the financial support by the University of Palermo grant CoRI 2006. The authors thank to the following colleagues, for their competent technical assistance and advice: Dr. Adelina Sporea for taking care of the optical absorption measurements and samples preparations, Ms. Rodica Gerogescu for gamma irradiations and some of the EPR measurements, Mr. I. Vata for neutron irradiation, Dr. C. Oproiu for beta irradiations, Mr. A. Alessi and Dr. L. Nuccio for part of EPR and photoluminescence measurements.

8. References

Agnello, S.; Buscarino G.; Gelardi, F.M. & Boscaino, R. (2008). Optical absorption band at 5.8 eV associated with the E'_γ centers in amorphous silicon dioxide: Optical absorption and EPR measurements, *Physical Review B*, Vol. 77, No.19 (May 2008), (195206 1-7), ISSN 1098-0121

Ahrens, R. G.; Jaques, J.J.; Luvalle, M.J.; Digivanni, D. & Windeler, R.S. (2001). Radiation effects on optical fibers and amplifiers, *Proceedings of Testing, Reliability, and Applications of Optoelectronic Devices*, pp. 217-225, SPIE Vol. 4285, ISBN 9780819439635, May 2001, SPIE

Alam[a], M.; Abramczyk, J.; Manyam, U.; Farroni, J. & Guertin, D. (2006). Performance of optical fibers in space radiation environment, *Proceedings of ESA/CNES,ICSO, 6th Intl. Confernce on Space Optics*, ISBN 92-9092-932-4, Noordwijk, June 2006, ESA Publications Division, Noordwijk

Alam[b], M.; Abramczyk, J.; Farroni, J.; Manyam, U. & Guertin, D. (2006). Passive and active optical fibers for space and terrestrial applications, *Proceedings of Photonics for Space Environment XI*, pp. 1-14, SPIE Vol. 6308, ISBN 9780819463876, San Diego, August 2006

Alfeeli, B.; Pickrell, G.; Garland, M.A. & Wang, A. (2007). Behavior of random hole optical fibers under gamma ray irradiation and its potential use in radiation sensing applications", *Sensors*, Vol. 7, No. 7, (2007), (676-688) EISSN 1424-8220

Bakos, T.; Rashkeev, S. N. & Pantelides, S.T. (2004). Optically active defects in SiO_2: The nonbridging oxygen center and the interstitial OH molecule. *Physical Review B*, Vol. 70, No.7 (August 2004), (075203 1-9), ISSN 1098-0121

Berghmans, F.; Brichard, B.; Fernandez Fernandez, A.; Gusarov, A.; Van Uffelen, M. & Girard, S. (2008). An introduction to radiation effects on optical components and fiber optic sensors, In: *Optical Waveguide Sensing and Imaging,* NATO Science for Peace and Security Series B: Physics and Biophysics, Bock, W.J.; Gannot, I. & Tanev, S. (Eds.), (127-165), Springer Netherlands, ISBN 978-1-4020-6950

Brichard, B. & Fernandez Fernandez, A. (2005). Radiation effects in silica glass optical fibers, Short Course Notebook, *New Challenges for Radiation Tolerance Assessment, RADECS 2005*, pp. 95-138, Cap d'Agde, Septembert 2005, RADECS Association

Campbell, D.J.; Ingesson, L.C.; Cecconello, M. & Ciattaglia, E. (2005). ITER Diagnostics, *Presented at ITER - Opportunities for European industry Workshop,* Barcelona, December 2005

Cannas, M.; Vaccaro, L. & Boizot, B. (2006). Spectroscopic parameters related to non-bridging oxygen hole centers in amorphous-SiO_2, *Journal of Non-Crystalline Solids,* Vol. 352, (January 2006), (203-208), ISSN 0022-3093

Devine, R.A.B.; Duraud, J.P. & Dooryhée, E. (2000). *Structure and Imperfections in Amorphous and Crystalline Silicon Dioxide,* John Wiley & Sons, ISBN 978-0-471-97536-6, Chicester

Fernando, W.; Gan, K.K.; Jackson, P. D.; Johnson, M.; Kagan, H.; Kass, R.; Rahimi, A.; Rush, C.; Smith, S. & Ter-Antonian, R. (2005). Radiation-Hard Optical Link for the ATLAS Pixel Detector, Presented at *IEEE05,* Puerto Rico

Fielder, R.S.; Duncan, R. & Palmer, M. (2005). Recent advancements in harsh environment fiber optic sensors: An enabling technology for emerging nuclear power applications, Luna Innovations, Inc., 2005

Florous, N.J.; Saitoh, K.; Murao, T. & Koshiba, M. (2007). Radiation dose enhancement in photonic crystal fiber bragg gratings: towards photo-ionization monitoring of irradiation sources in harsh nuclear power reactors, *Proceedings of Conference on Lasers and Electro-Optics, CLEO 2007,* pp. 1 – 2, ISBN: 978-1-55752-834-6, Baltimore, May 2007, Institute of Electrical and Electronics Engineers, Inc.

Friebele, E.; Sigel, G. & Griscom, D. (1976). Drawing-induced defect centers in a fused silica core fiber, *Applied Physics Letters.* Vol. 28, No. 9, (1976), (516-518), ISSN DOI:10.1063/1.88839

Gan, K.K.; Abi, B.; Fernando, W.; Kagan, H.P.; Kass, R.D.; Law, A.; Lebbai, M.R.M.; Rau, A.; Rizardinova, F.; Skubic, P.L. & Smith, D.S. (2008). Radiation-hard optical link for SLHC, *J. of Physics: Conference Series,* Vol. 110, (2008), doi:10.1088/1742-6596/110/9/092011

[a]Griscom, D. L. (1998). Radiation hardened fiber optics for fusion reactor diagnostics, *Final Technical Report,* Naval Research Laboratory, Department of the Navy

[b]Griscom, D.L. & Mizuguchi, M. (1998). Determination of the visible range optical absorption spectrum of peroxy radicals in gamma-irradiated fused silica, *Journal of Non-Crystalline Solids,* Vol. 239, No. 1-3, (1998), (66-77), ISSN 0022-3093

Griscom, D.L. (2001). Fractal kinetics of radiation-induced point-defect formation and decay in amorphous insulators: Application to color centers in silica-based optical fibers *Physical Review B,* Vol. 64, No.17 (October 2001), (174201 1-14), ISSN 1098-0121

Henschel, H.; Körfer, M. & Wulf, F. (2001). Fibre optical radiation sensing system for TESLA, *Proceedings DIPAC 2001 – ESRF,* pp. 73-75, Grenoble, May 2001, http://epaper.kek.jp/d01/papers/CT04.pdf

Hodgson, E.R.; Decreton, M.; Cecconello, M.; Ingesson, C. & Campbell D.J. (2006). Radiation-hard ceramic materials for diagnostic and heating and current drive systems for ITER, *Proceedings of the 33rd EPS Conference on Plasma Phys.*, Rome, June 2006, http://epsppd.epfl.ch/Roma/pdf/P1_119.pdf

ITER Physics Expert Group on Diagnostics (1999). ITER Physics basis, *Nucl. Fusion*, Vol. 39, (1999), (2541-2575)

Kaiser, P. (1974). Drawing-induced coloration in vitreous silica fibers, *JOSA*, Vol. 64, No. 4, (1974), (475-481), ISSN doi:10.1364/JOSA.64.000475

Kajihara, K.; Hirano, M.; Skuja, L. & Hosono, H. (2008). Intrinsic defect formation in amorphous SiO_2 by electronic excitation: Bond dissociation versus Frenkel mechanisms, *Physical Review B*, Vol. 78, No.9 (September 2008), (094201 1-8), ISSN 1098-0121

Karlitschek, P.; Hillrichs, G. & Klein, K.-F. (1995). Photodegradation and nonlinear effects in optical fibers induced by pulsed uv-laser radiation, *Optics Communications*, Vol. 116, No. 1-3, (1995), (219-230), ISSN 0030-4018

Karlitschek, P.; Hillrichs, G. & Klein, K.-F. (1998). Influence of hydrogen on the colour center formation in optical fibers induced by pulsed UV-laser radiation: Part 2: All-silica fibers with low-OH undoped core, *Optics Communications*, Vol. 155, (October 1998) (386-397), ISSN 0030-4018

Kimurai, E. Takada, Y. Hosono, M. Nakazawa, Takahashi, H. & Hayami, H. (2002). New Techniques to Apply Optical Fiber Image Guide to Nuclear Facilities, *J. of Nuclear Sci. and Techn.*, Vol. 39, nº 6, (603-607), ISSN 0022-3131

Li, J.; Lehman, R. L. & Sigel, G. H. Jr (1996). Electron paramagnetic resonance characterization of nonuniform distribution of hydrogen in silica optical fibers, *Applied Physics Letters*, Vol. 69, No. 14, (September 1996), (2131-2133), ISSN 0003-6951

Mashkov, V. A.; Austin, Wm. R. ; Zhang, Lin & Leisure, R. G. (1996). Fundamental role of creation and activation in radiation-induced defect production in high-purity amorphous SiO_2, *Physical Review Letters*, Vol. 76, No.16 (October 1995), (2926-2929), ISSN 0031-9007

Messina, F.; Agnello S.; Cannas, M. & Parlato, A. (2009). Room temperature instability of E'_γ centers induced by γ irradiation in Amorphous SiO_2, *The Journal of Physical Chemistry A*, Vol. 113, No.6, (December 2008), (1026-1032), ISSN 1520-5215

Nuccio, L.; Agnello, S. & Boscaino, R. (2009). Role of H_2O in the thermal annealing of the E'_γ center in amorphous silicon dioxide, *Physical Review B*, Vol. 79, N. 12, (March 2009), (125205 1-8) ISSN 1098-0121

O'Keeffe, S.; Fitzpatrick, C.; Lewis, E. & Al-Shamma'a, A.I. (2008). A review of optical fibre radiation dosimeters, *Sensor Review*, Vol. 28, No. 2, (136 – 142), ISSN 0260-2288

Ott, M.N. (2002). Radiation effects data on commercially available optical fiber: database summary, *Proceedings of 2002 IEEE Radiation Effects Data Workshop*, pp. 24 – 31, ISBN 0-7803-7544-0, Phoenix, July 2002, Institute of Electrical and Electronics Engineers, Inc.

Pacchioni, G.; Skuja, L. & Griscom, D. L. (2000). *Defects in SiO_2 and related dielectrics: science and technology*, Kluver Academic Publishers, ISBN 0-7923-6686-7, Dordrecht

Radiation effects, The 10 th Europhysical Conference on Defects in Insulation Materials, *Phys. Status Solidi*, Vol. 4, No. 3 , (2007), (1060-1175)

Radzig, V. A. (1998). Point defects in disordered solids: Differences in structure and reactivity of the (O≡Si-O)$_2$Si: groups on silica surface, *Journal of Non-Crystalline Solids*, Vol. 239, (1998), (49-56), ISSN 0022-3093

Reichle, R.; Brichard, B.; Escourbiac, F.; Gardarein, J.L.; Hernandez, D.; Le Niliot, C.; Rigollet, F.; Serra, J.J.; Badie, J.M.; van Ierschot, S.; Jouve, M.; Martinez, S.; Ooms, H.; Pocheau, C.; Rauber, X.; Sans, J.L.; Scheer, E.; Berghmans, F. & Decréton, M. (2007). Experimental developments towards an ITER thermography diagnostic, *J. of Nuclear Materials*, Vol. 363-365, (June 2007), (1466-1471), doi:10.1016/j.jnucmat.2007.01.207

Shikama, T.; Nishitani, T.; Kakuta, T.; Yamamoto, S.; Kasai, S.; Narui, M.; Hodgson, E.; Reichle, R.; Brichard, B.; Krassilinikov, A.; Snider, R.; Vayakis, G. & Costley A. (2003). Irradiation test of diagnostic components for ITER application in a fission reactor, Japan materials testing reactor, *Nuclear fusion*, Vol. 43, No. 7, (2003), (517-521), ISSN 0029-5515

Skuja, L. (1998). Optically active oxygen-deficiency-related centers in amorphous silicon dioxide, *Journal of Non-Crystalline Solids*, Vol. 239, No. 1-3, (1998), (16-48), ISSN 0022-3093

Skuja, L.; Hirano, M.; Hosono, H. & Kajihara, K. (2005). Defects in oxide glasses, *Physica Status Solidi (c)*, Vol. 2, No.1, (2005), (15-24), DOI:10.1002/pssc.200460102

Sporea D. & Sporea, R. (2005). Setup for the in situ monitoring of the irradiation-induced effects in optical fibers in the ultraviolet-visible optical range, *Review of Scientific Instruments*, Vol. 76, No. 11, (2005), DOI:10.1063/1.2130932

Sporea, D.; Sporea, A.; Agnello, S.; Nuccio, L.; Gelardi, F.M. & Brichard, B. (2007). Evaluation of the UV optical transmission degradation of gamma-ray irradiated optical fibers, *CLEO/ The 7th Pacific Rim Conference on Lasers and Electro-Optics*, Seoul, August 2007

Troska, J.; Cervelli, G.; Faccio, F.; Gill, K.; Grabit, R.; Jareno, R.M.; Sandvik, A.M. & Vasey, F. (2003). Optical readout and control systems for the CMS tracker, *IEEE Trans. Nuclear Sci.*, Vol. 50, No. 4, Part 1, (2003) (1067-1072), ISSN 0018-9499

Weeks, R.A. (1956). Paramagnetic resonance of lattice defects in irradiated quartz, *J. Appl. Phys.*, Vol. 27, No. 11, (1956), (1376-1381), DOI:10.1063/1.1722267

Weeks, R. A. & Sonder, E. (1963). The relation between the magnetic susceptibility, electron spin resonance, and the optical absorption of the E_1'center in fused silica, In: *Paramagnetic Resonance II*, W. Low (Ed.), 869-879, Academic Press, LCCN 63-21409, New York

Weil, J. A.; Bolton, J. R. & Wertz, J. E. (1994) *Electron Paramagnetic Resonance*, John Wiley & Sons, 0-471-57234-9, New York

Zabezhailov, M.O.; Tomashuk, A.L.; Nikolin, I.V. & Plotnichenko, V.G. (2005). Radiation-induced absorption in high-purity silica fiber preforms, *Inorganic Materials*, Vol. 41, No. 3, (2005), (315–321), ISSN 0020-1685

4

Programmable All-Fiber Optical Pulse Shaping

Antonio Malacarne[1], Saju Thomas[2], Francesco Fresi[1,2], Luca Potì[3], Antonella Bogoni[3] and Josè Azaña[2]

[1]*Scuola Superiore Sant'Anna, Pisa,*
[2]*Institut National de la Recherche Scientifique (INRS), Montreal, QC,*
[3]*Consorzio Nazionale Interuniversitario per le Telecomunicazioni (CNIT), Pisa,*
[1,3]*Italy*
[2]*Canada*

1. Introduction

Techniques for the precise synthesis and control of the temporal shape of optical pulses with durations in the picosecond and sub-picosecond regimes have become increasingly important for a wide range of applications in such diverse fields as ultrahigh-bit-rate optical communications (Parmigiani et al., 2006; Petropoulos et al., 2001; Oxenlowe et al., 2007; Otani et al., 2000), nonlinear optics (Parmigiani et al., 2006 b), coherent control of atomic and molecular processes (Weiner, 1995) and generation of ultra-wideband RF signals (Lin & Weiner, 2007). To give a few examples, (sub-)picosecond flat-top optical pulses are highly desired for nonlinear optical switching (e.g. for improving the timing-jitter tolerance in ultrahigh-speed optical time domain de-multiplexing (Parmigiani et al., 2006; Petropoulos et al., 2001; Oxenlowe et al., 2007)) as well as for a range of wavelength conversion applications (Otani et al., 2000); high-quality picosecond parabolic pulse shapes are also of great interest, e.g. to achieve ultra-flat self-phase modulation (SPM)-induced spectral broadening in super-continuum generation experiments (Parmigiani et al., 2006 b). For all these applications, the shape of the synthesized pulse needs to be accurately controlled for achieving a minimum intensity error over the temporal region of interest. The most commonly used technique for arbitrary optical pulse shaping is based on spectral amplitude and/or phase linear filtering of the original pulse in the spatial domain; this technique is usually referred to as 'Fourier-domain pulse shaping' and has allowed the programmable synthesis of arbitrary waveforms with resolutions better than 100fs (Weiner, 1995). Though extremely powerful and flexible, the inherent experimental complexity of this implementation, which requires the use of very high-quality bulk-optics components (high-quality diffraction gratings, high-resolution spatial light modulators etc.), has motivated research on alternate, simpler solutions for optical pulse shaping. This includes the use of integrated arrayed waveguide gratings (AWGs) (Kurokawa et al., 1997), and fiber gratings (e.g. fiber Bragg gratings (Petropoulos et al., 2001), or long period fiber gratings (Park et al. 2006)). However, AWG-based pulse shapers (Kurokawa et al., 1997) are typically limited to time resolutions above 10ps. The main drawback of the fiber grating approach (Petropoulos et al., 2001; Park et al. 2006) is the lack of programmability: a grating device is designed to realize a single pulse shaping operation over a specific input pulse (of prescribed wavelength and bandwidth) and once the grating is fabricated, these specifications cannot be later modified. Recently, a simple and practical pulse shaping technique using cascaded two-arm interferometers has been reported (Park & Azaña, 2006). This technique can be implemented using widely accessible bulk-optics components and can be easily reconfigured to synthesize a variety of transform-limited temporal shapes of practical interest (e.g. flat-top and triangular pulses) as well as to operate over a wide range of input bandwidths (in the sub-picosecond and picosecond regimes) and center wavelengths. However, this solution presents all the drawbacks due to a free-space solution where it is needful to strictly set the relative time delay inside each interferometer in order to "program" different obtainable pulse shapes. Therefore the pursuit of an integrated (fiber) pulse shaping solution, including full compatibility with waveguide/fiber devices, which can be able to provide the additional functionality of electronic programmability, manifests to be useful for a lot of different application fields. For this reason a programmable fiber-based phase-only spectral filtering setup has been recently introduced (Azaña et al., 2005; Wang & Wada, 2007). In the next section the working principle of this spectral phase-only linear filtering approach is discussed and an improvement of the solution reported in (Azaña et al., 2005) is presented and widely investigated.

2. Programmable all-fiber optical pulse shaper

A pulse shaper can be easily described in the spectral domain as an amplitude and/or phase filter. Using linear system theory it is possible to consider an input signal $e_{in}(t)$ whose frequency spectrum is $E_{in}(\omega)$ as reported in Fig. 1, and the corresponding output spectrum $E_{out}(\omega)$. The pulse shaper is represented by a filter transfer function $H(\omega)$ so that:

$$E_{out}(\omega) = E_{in}(\omega) \cdot H(\omega) = \Im\{e_{out}(t)\} \tag{1}$$

where $H(\omega)$ is found out so that the output temporal shape $e_{out}(t) = u(t)$, with $u(t)$ the desired target intensity profile.

Previous solutions are based on amplitude-only filtering (Dai & Yao, 2008), amplitude and phase filtering (Petropoulos et al., 2001; Weiner, 1995; Park et al., 2006; Azaña et al., 2003), or phase-only filtering (Azaña et al., 2005; Wang & Wada, 2007; Weiner et al., 1993). In term of power efficiency phase filtering is preferred since the energy is totally preserved with respect to amplitude only or amplitude and phase filtering where some spectral components are attenuated or canceled. Avoiding any amplitude filtering, in principle we may achieve an energy lossless pulse shaping. Moreover, if only the output temporal intensity profile is targeted, keeping its temporal phase profile unrestricted, a phase-only filtering offers a higher design flexibility, even if obviously it rules out the possibility to obtain a Fourier transform-limited output signal or an output phase equal to the input one. Then, with phase-only filtering we are able to carry out an arbitrary temporal output phase but with a programmable *desired* temporal output intensity profile.

In this case the system is represented by a phase-only transfer function $M(\omega) = K e^{j\Phi(\omega)}$, where the design task is to look for $\Phi(\omega)$ such that:

$$\Im^{-1}\{M(\omega) \cdot E_{in}(\omega)\} = u(t) \tag{2}$$

The very interesting fiber-based solution for programmable pulse shaping proposed in (Azaña et al., 2005) and used in (Wang & Wada, 2007) is based on time-domain optical phase-only filtering. This method originates from the most famous technique for programmable optical pulse shaping, based on spatial-frequency mapping (Weiner et al., 1993).

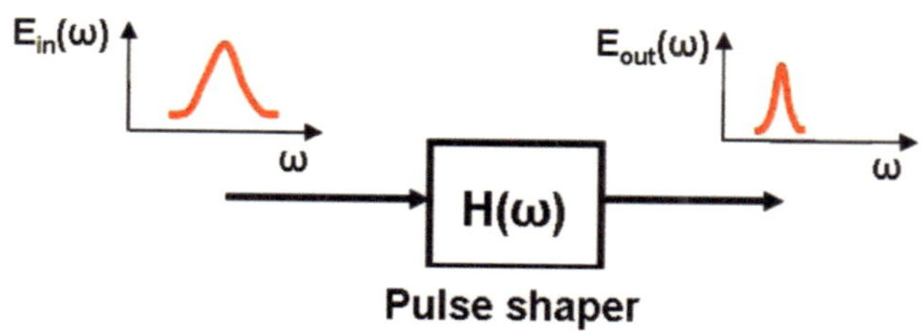

Fig. 1. Transfer function for a pulse shaper

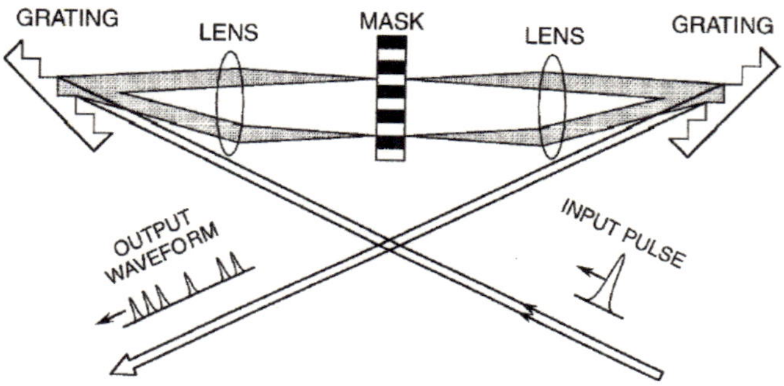

Fig. 2. Spatial-domain approach for shaping of optical pulses using a spatial phase-only mask

The scheme is shown in Fig. 2: a spatial dispersion is applied by a grating on the input optical pulse, then a phase mask provides a spatial phase modulation and finally a spatial dispersion compensation is given by another grating. Its main drawback consisted in being a free space solution with all the problems related to a needful strict alignment, including significant insertion losses and limited integration with fiber or waveguide optics systems. For these reasons we looked for an all-fiber solution that essentially is a time-domain equivalent (Fig. 3) of the classical spatial-domain pulse shaping technique (Weiner et al., 1993), in which all-fiber temporal dispersion is used instead of spatial dispersion.

To achieve this all-fiber approach we started from a different solution based on the concept concerning a time-frequency mapping using linear dispersive elements (Azaña et al., 2005). As shown in Fig. 3 (top), applying an optical pulse at the input of a first order dispersive medium, we obtain an output signal $e_{disp}(t)$ dispersed in time domain corresponding to the spectral domain of the input pulse. In this way, a temporal phase modulation $\varphi(t)$ applied to the dispersed signal coming out from the dispersive medium corresponds to a spectral phase modulation $\Phi(\omega)$ applied to the input spectrum (Fig. 3, bottom). For a given first order chromatic dispersion coefficient β_2, the correspondence between temporal and spectral phase modulations is:

$$\varphi(t) = \Phi(\omega = t/\beta_2) \tag{3}$$

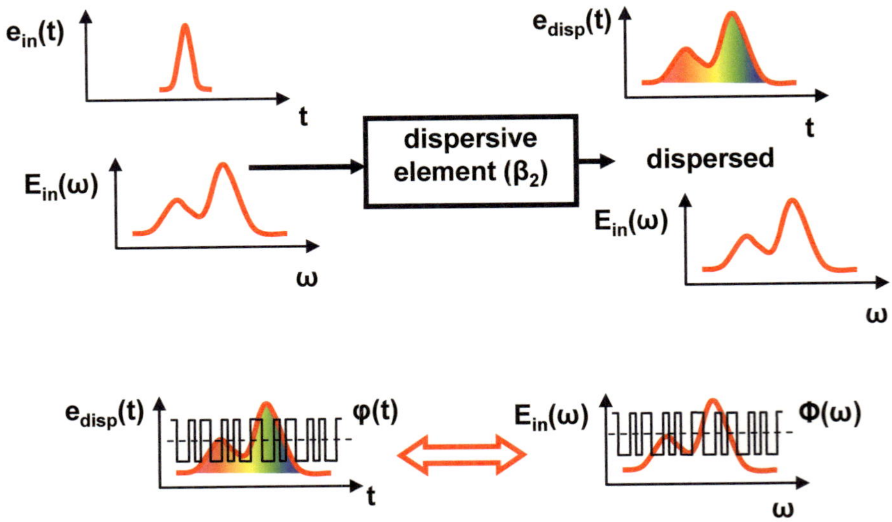

Fig. 3. Principle of time-frequency mapping for the time-domain pulse shaping approach. β_2: first order dispersion coefficient; $\varphi(t)$: temporal phase modulation applied to the dispersed signal; $\Phi(\omega)$: spectral phase modulation applied to the input spectrum, corresponding to $\varphi(t)$

To apply the mentioned phase modulation an electro-optic (EO) phase modulator will be used. As it will be more clear afterwards, *any* $\Phi(\omega)$ that satisfies Eq. 2 will not be practical in terms of design and implementation. Therefore we restrict $\Phi(\omega)$ to a binary function with levels $\pi/2$ and $-\pi/2$ and a frequency resolution determined by practical system specifications (input/output dispersion and EO modulation bandwidth). It is possible to demonstrate that with such a binary phase modulation with levels $\pi/2$ and $-\pi/2$, the re-shaped signal is symmetric in the time domain. The temporal resolution of the binary phase code, similarly to Eq. 2, is related to the corresponding spectral resolution this way:

$$\omega_{pix} = T_{pix} / \beta_2 \qquad (4)$$

Finally, to achieve the inverse Fourier-transform operation on the stretched, phase-modulated pulse, such a pulse is compressed back with a dispersion compensator providing the conjugated dispersion of the first dispersive element (Fig. 4).

As reported in Fig. 4, the binary phase modulation is provided to the EO-phase modulator by a bit pattern generator (BPG) with a maximum bit rate of 20 Gb/s.

Dispersion mismatch between the two dispersive conjugated elements has a negative effect on the performance of the system and for obtaining good quality pulse profiles it is critical to match these two dispersive elements very precisely. In our work, this was achieved by making use of the same linearly chirped fiber Bragg grating (LC-FBG) acting as pre- and post-dispersive element, operating from each of its two ends, respectively (Fig. 5); this simple strategy allowed us to compensate very precisely not only for the first-order dispersion introduced by the LC-FBG, but also for the present relatively small undesired higher-order dispersion terms.

As reported in Fig. 6, reflection of the LC-FBG acts as a band-pass filter applying at the same time a group delay (GD) versus wavelength that is linear on the reflected bandwidth. In particular the slope of the two graphs of Fig. 6 (left) represents the applied first-order dispersion coefficient, respectively +480 and -480 ps/nm for each of the two ends of the LC-FBG.

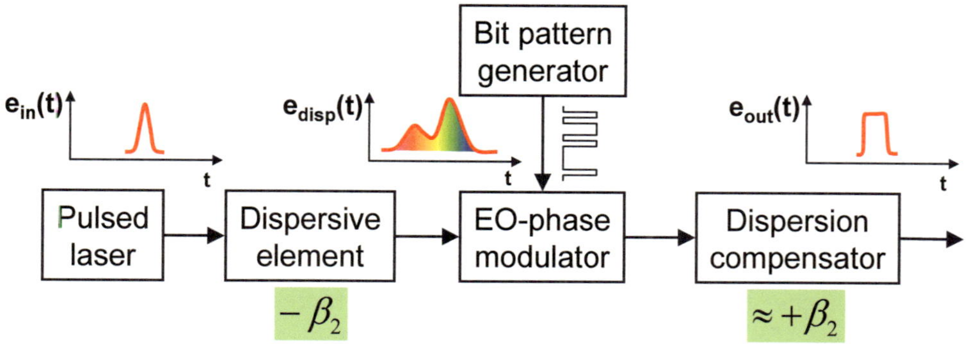

Fig. 4. Schematic of the pulse shaping concept based on time-frequency mapping and exploiting a binary phase-only filtering

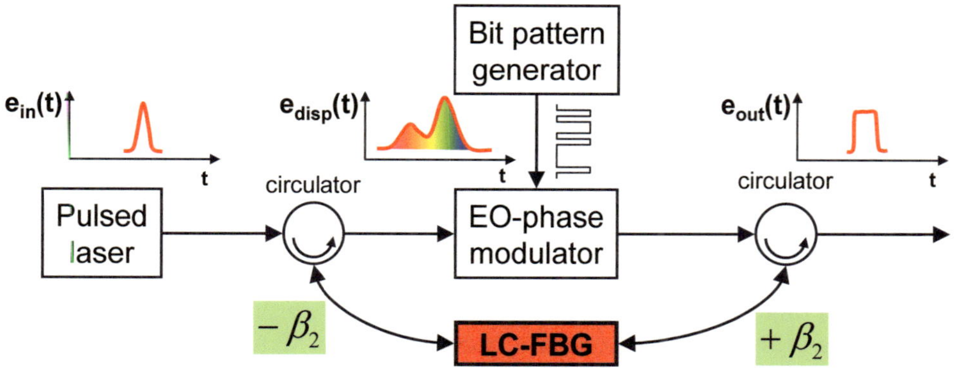

Fig. 5. Schematic of the pulse shaping concept based on time-frequency mapping exploiting a single LC-FBG as pre- and post-dispersive medium

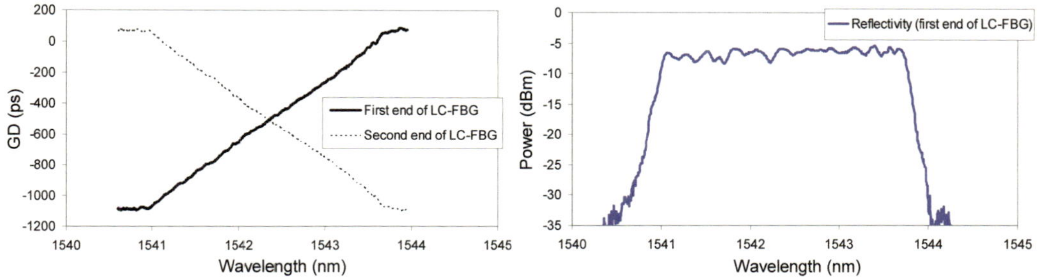

Fig. 6. Reflection behavior of the LC-FBG. (left) Group delay over the reflected bandwidth for both the ends; (right) reflected bandwidth of the first end

Similarly to any linear pulse shaping method, the shortest temporal feature that can be synthesized using this technique is essentially limited by the available input spectrum. On the other hand, the maximum temporal extent of the synthesized output profiles is inversely proportional to the achievable spectral resolution $\omega_{pix.}$

2.1 Genetic algorithm as search technique

To find the required binary phase modulation function we implemented a genetic algorithm (GA) (Zeidler et al., 2001). A GA is a search technique used in computing to find exact or approximate solutions to optimization and search problems. GAs are a particular class of evolutionary algorithms that use techniques inspired by evolutionary biology such as inheritance, mutation, selection, and crossover (also called recombination), and they've been already exploited for optical pulse shaping applications (Wu & Raymer, 2006). They are implemented as a computer simulation in which a population of abstract representations (called *chromosomes*) of candidate solutions (called individuals) to an optimization problem evolves toward better solutions. Traditionally, solutions are represented in binary as strings of logic "0"s and "1"s. The evolution usually starts from a population of randomly generated individuals and happens in generations. In each generation, the fitness of every individual in the population is evaluated, multiple individuals are stochastically selected from the current population (based on their fitness), and modified (recombined and possibly randomly mutated) to form a new population. The new population is then used in the next iteration of the algorithm. Commonly, the algorithm terminates when either a maximum number of generations has been produced, or a satisfactory fitness level has been reached for the population. If the algorithm has terminated due to a maximum number of generations, a satisfactory solution may or may not have been reached.

In our case we use GA to find a convergent solution for phase codes corresponding to desired output intensity profiles (targets), starting from an input spectrum nearly Fourier transform-limited. First we code each spectral pixel with '0' or '1' according to the phase value ($\pi/2$ or $-\pi/2$, respectively). Each bit pattern producing a phase code is a chromosome. We start with 48 random chromosomes. We select the best 8 chromosomes in terms of their fitness (in terms of *cost function*, explained later). We obtain 16 new chromosomes from 8 pairs of old chromosomes (all of them chosen within the best 8) by *crossover* (2 new chromosomes from each pair). Then we obtain 24 new chromosomes from 24 random old chromosomes (1 new chromosomes from each) by *mutation*. Then we have 48 chromosomes again ("the best 8" + "16 from crossover" + "24 from mutation"). This iteration can be repeated a certain number of times. For our simulations we've chosen 10÷30 iterations corresponding to elaboration times in the range of 5÷15 seconds (10 iterations for flat-top and triangular pulses generation, 20÷30 iterations for bursts generation).

The fitness of each chromosome is indicated by its corresponding *cost function*. Each *cost function* C_i generally represents the maximum deviation in intensities between the predicted output signal $e_{out}(t)$ and the target $u(t)$ in a time interval $[t_i,t_{i+1}]$:

$$C_i = \max\left\{\left|e_{out}(t) - u(t)\right|\right\}; \quad t \geq 0 \; and \; t \in [t_i, t_{i+1}] \tag{5}$$

while the *total cost function* C_{tot} is defined as sum of the partial cost functions C_i, each of them with a specific weight w_i :

$$C_{tot} = \sum_i C_i w_i \tag{6}$$

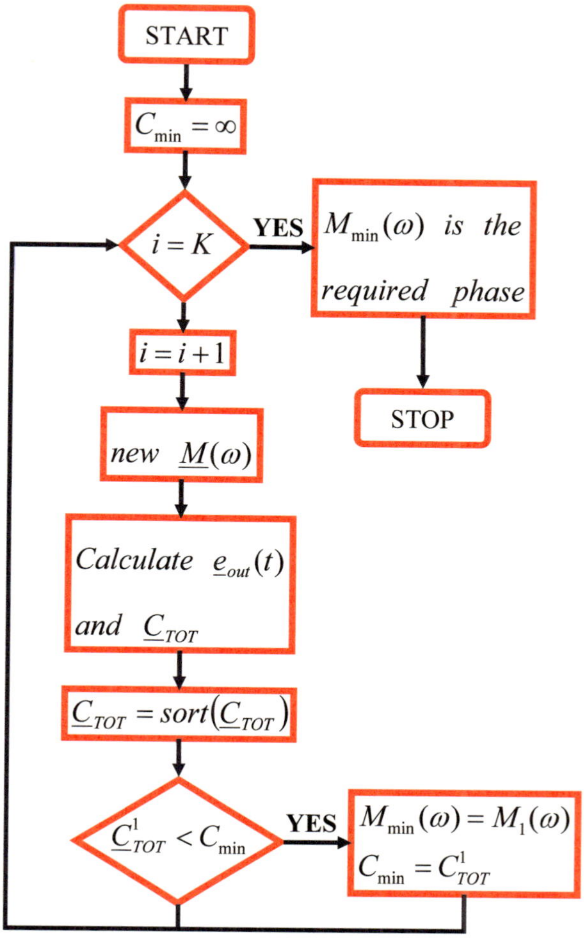

Fig. 7. Flow chart of the applied optimization technique

During each iteration, thanks to GA we move in a direction that reduces the total cost function. This way we derived the particular phase code so as to obtain the desired output temporal intensity profile, whose deviation from the target hopefully is within an acceptable limit. After a sufficient number of iterations, the obtained phase profile can be then transferred to the experiment. In Fig. 7 the flow chart for a general optimization technique is shown. In our case within the block where we calculate the new array of transfer functions $\underline{M}(\omega)$, we apply GA through crossover and mutation as explained above.

To better understand what a cost function is, we report here a couple of examples concerning the cost functions used for single flat-top pulse and pulsed-burst generations. In Fig. 8 (left) the features taken into account for a flat-top pulse generation are shown. Since the generated signal is symmetric in the time domain, we considered just the right half of the output profile.

Three time intervals correspond to three cost functions: the first one (C_1) is related to the flatness in the central part of the pulse, the second one (C_2) concerns the steepness of the falling edge, whereas the last one (C_4) is related to the pedestal amplitude. In particular, in Fig. 9(a) we report the comparison between the simulated temporal profile carried out through GA and its relative theoretical target for the case of a flat-top pulse. In this case, the defined total cost function was $C_{tot}=5C_1+C_2+C_4$.

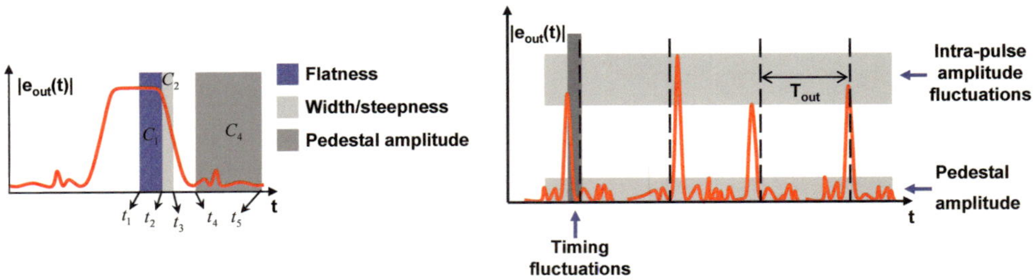

Fig. 8. (left) Cost functions for a single flat-top pulse generation. (right) Features taken into account with cost functions for a pulsed-burst generation

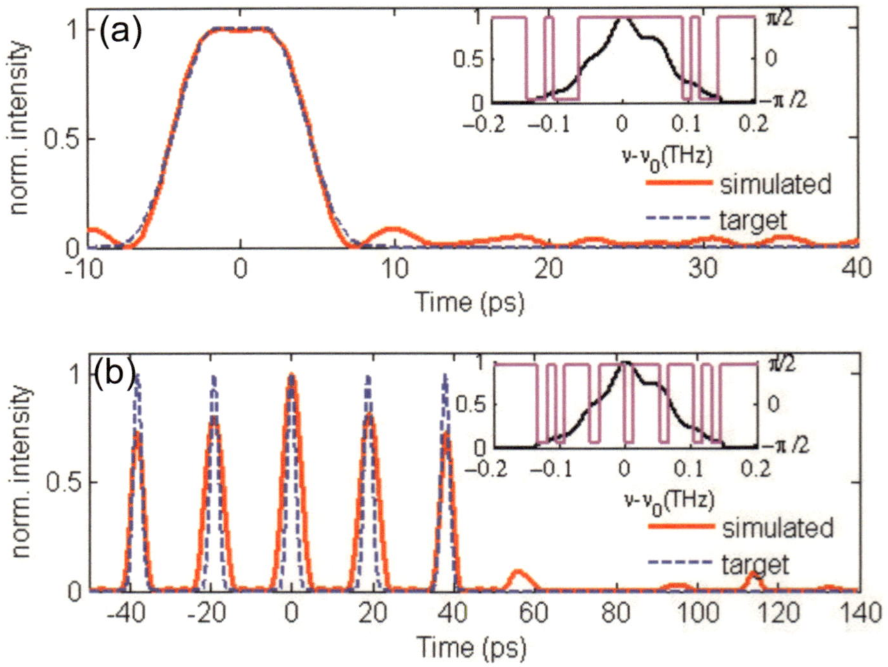

Fig. 9. Simulated and target profiles for a flat-top pulse (a) and a 5-pulses sequence (b). The used phase codes are shown in the insets (solid) together with the input pulse spectrum (dashed)

In Fig. 8 (right) another example considering a pulsed-burst as target shows the considered features: the intra-pulse amplitude fluctuations, the timing fluctuations and the pedestal amplitude again. In particular, Fig. 9(b) shows the comparison between the simulated temporal profile and its relative theoretical target for the case of a 5-pulses sequence. In this case, even though we weighted the partial cost functions in order to obtain a sequence with flat-top envelope, because of the limited spectral resolution, the simulated sequence is not so equalized (inter-pulse amplitude fluctuations ≈ 25%) as the theoretical target.

To demonstrate the programmability of the proposed scheme, we targeted shapes like flat-top, triangular and bursts of 2, 3, 4 and 5 pulses with nearly flat-top envelopes, defining a specific total cost function for each case.

2.2 Experimental setup

As shown in the experimental setup in Fig. 10, the exploited optical pulse source was an actively mode-locked fiber laser producing nearly transform-limited ~ 3.5 ps (FWHM) Gaussian-like pulses with a repetition rate of 10 GHz, spectrally centered at λ_0 = 1542.4 nm. The source repetition rate was decreased down to 625 MHz, corresponding to a period of 1.6 ns, using a Mach-Zehnder amplitude modulator (MZM) and a bit pattern generator (BPG 1) producing a binary string with a logic "1" followed by fifteen logic "0".

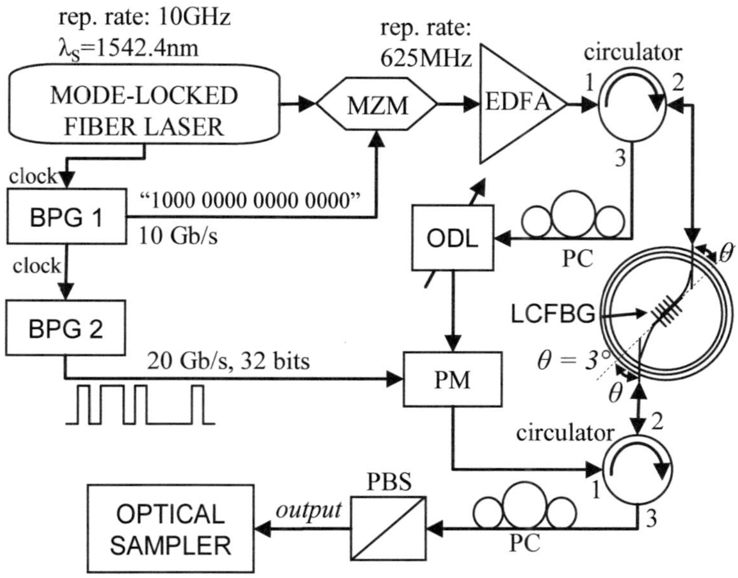

Fig. 10. Experimental setup of the programmable all-fiber pulse shaper

In order to temporally stretch the optical pulses, they were reflected in a LC-FBG, incorporated in a tunable mechanical rotator for fiber bending, which allowed us to tune the chromatic dispersion coefficient by changing the stretching angle θ (Kim et al. 2004). Such tunable dispersion compensator will be deepened and described in next Section (Section 2.2.1). Setting $\theta = 3°$, we obtained a first-order dispersion coefficient of 480 ps/nm ($\beta_2 \approx$ -606 ps^2/rad) over a 3dB reflection bandwidth of 2.3 nm (centered at the laser wavelength λ_0=1542.4nm). The dispersed pulses (port 3 of first circulator), each extending over a total duration of ~1.6 ns, were temporally modulated using an EO phase modulator (PM) driven by a second bit pattern generator (BPG 2), generating 32-bit codes, each with a bit rate of 20 Gb/s and a period of 1.6 ns, according to the designed codes obtained from the GA. To accurately synchronize the phase code and the stretched pulse we employed an optical delay line (ODL) together with shifting bit by bit the code generated from BPG 2. In order to precisely compensate for the previously applied chromatic dispersion value, we used the same LC-FBG operated in the opposite direction, thus introducing the exact opposite dispersion (-480 ps/nm). At port 3 of the second circulator we obtained the desired output pulse together with a small amount of the input pulse transmitted through the grating. The desired output was discriminated using a polarization controller (PC) and a polarization beam splitter (PBS). Finally, the output temporal waveform was monitored by a commercial autocorrelator first, and then acquired by a quasi asynchronous optical sampler prototype (Section 6 of Fresi's chapter) based on four wave mixing (FWM), with a temporal resolution of ~ 100 fs.

2.2.1 Tunable dispersion compensator based on a LC-FBG

Referring to (Kim et al. 2004), a method to achieve tunable chromatic dispersion compensation without a center wavelength shift is based on the systematic bending technique along a linearly chirped fiber Bragg grating (LC-FBG). The bending curvature along the LC-FBG corresponding to the rotation angle of a pivots system can effectively control the chromatic dispersion value of the LC-FBG within its bandwidth. The group delay can be linearly controlled by the induction of the linear strain gradient with the proposed method. Based on the proposed method, the chromatic dispersion could be controlled in a range typically from ~100 to more than 1300 ps/nm with a shift of the grating center wavelength less than 0.03 nm over the dispersion tuning range.

In our particular case, to "write" the LC-FBG prototype exploited in the experiment presented in Section 2.2, we used a setup where a UV laser with a wavelength of 244 nm was employed. Its light beam was deflected by a sequence of mirrors; the last mirror was fixed on a mechanical arm, whose position was automatically driven by a proper LabView software, so as to hit a phase mask. Such a mask divided the input beam in two coherent beams so as to create interference fringes through beating. Such fringes had the task to photo-expose the span of fiber in order to realize the LC-FBG. In this case the linear chirp (periodicity linearly increasing/decreasing along the fiber) was directly introduced by the phase mask.

In Fig. 11 the measured reflection spectrum and the group delay (GD) of a typical LC-FBG are reported, showing excellent results in terms of amplitude ripples (< 0.5dB) (Fig. 11(a)) and linear behavior of the GD versus wavelength (Fig. 11(b)). The main difference between the LC-FBG described in this section and the one employed in Section 2.2 is the central wavelength (1542.4 nm instead of 1550.4 nm).

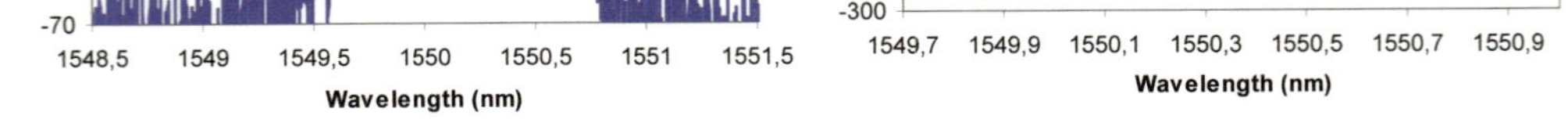

Fig. 11. (a) Reflection spectrum of a typical LC-FBG. (b) GD of the same LC-FBG

The LC-FBG was carefully attached to the cantilever beam fixed on the rotation stage in order to compose the dispersion-tuning device (Kim et al. 2004). Through the device a certain tunable bending angle is applied on the metal beam where the grating is attached.

Both the bandwidth and the chromatic dispersion value (the derivative of the graph in Fig. 11(b)) of the grating change with the bending angle applied to the grating. In particular, increasing the rotation angle it is possible to decrease the chromatic dispersion and to increase the reflection bandwidth.

In Fig. 12(a),(c) variation of reflection spectra with the rotation angle are shown, whereas in Fig. 12(b),(d) variation of GD with the rotation angle are reported. As shown in Fig. 12(a),(c), the central wavelength of the reflection bandwidth is fixed and equal to ~1550.4 nm. In Fig. 13 variation of the chromatic dispersion (left) and the -3dB-bandwidth (right) with the rotation angle are reported.

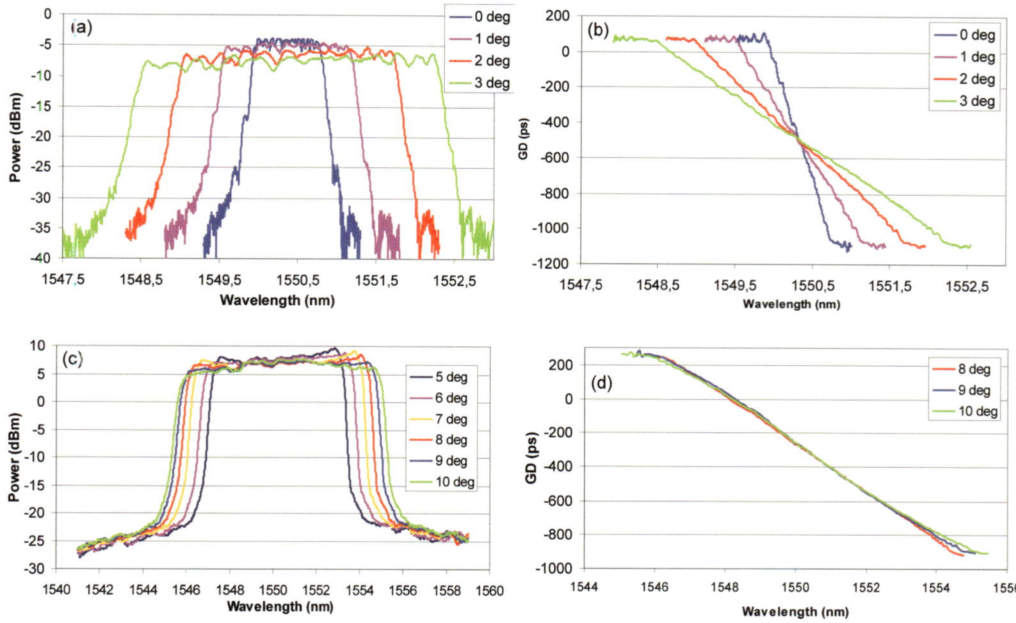

Fig. 12. Measured results of the variation of (a),(c) the reflection spectra and (b),(d) the group delay of the tunable dispersion compensator with the rotation angle

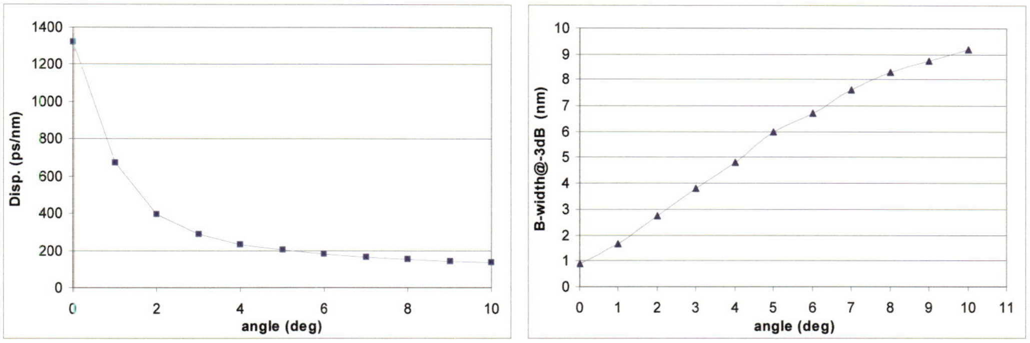

Fig. 13. Variation of chromatic dispersion (left) and -3dB-bandwidth (right) of the tunable dispersion compensator with the rotation angle

Concluding, in the example reported in this section a LC-FBG has been fabricated through a proper setup and it has been employed with a mechanical rotator in order to compose an all-fiber tunable chromatic dispersion compensator able to provide a chromatic dispersion in the range (±134.4;±1320.4) ps/nm. The sign of the applied chromatic dispersion depends on which end of the grating we employ. Furthermore, adding a circulator on an end of the LC-FBG, we obtain a system where port 1 and port 3 of the circulator represents the input and the output of the tunable dispersion compensator respectively (Fig. 14).

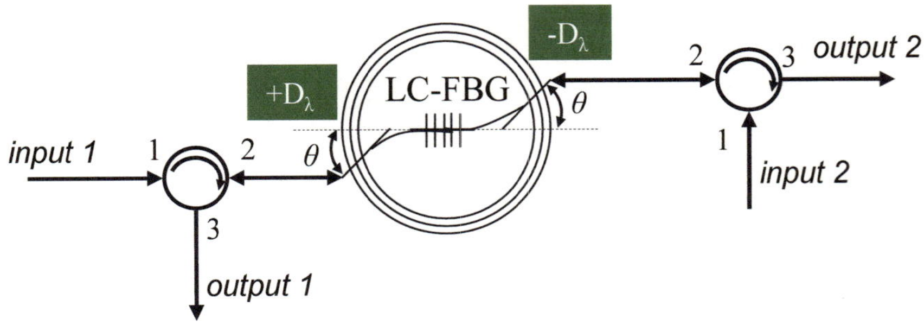

Fig. 14. Tunable chromatic dispersion compensator scheme. From input 1/output 1 a positive chromatic dispersion is provided whereas from input 2/output 2 a negative chromatic dispersion is provided

2.3 Experimental results

The capabilities of our programmable picosecond pulse re-shaping system were first demonstrated synthesizing the flat-top optical pulse related to Fig. 9(a) and the 5-pulses sequence related to Fig. 9(b), monitoring the temporal profile of the output signal through a commercial autocorrelator (Fig. 15), then the experiment has been repeated monitoring the output optical signal by an optical sampler. In particular we synthesized five different temporal waveforms of practical interest (Petropoulos et al., 2001; Park et al., 2006; Azaña et

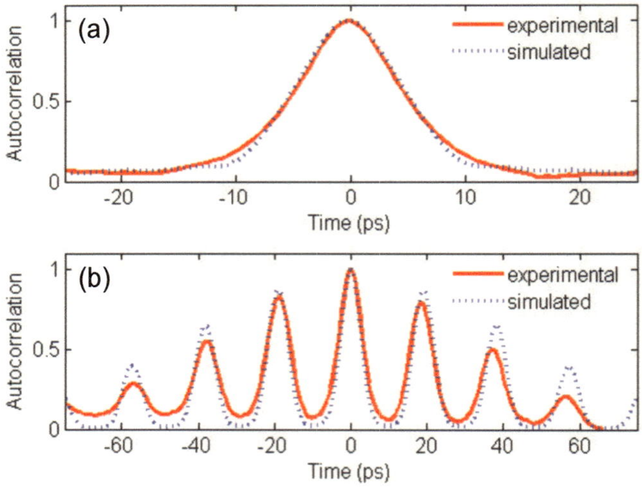

Fig. 15. Experimental and simulated autocorrelation curves for the flat-top pulse (a) and the 5-pulses sequence (b)

al., 2003) (see Fig. 16), namely a 9-ps (FWHM) flat-top optical pulse (Fig. 16(a)), a 8.5-ps (FWHM) triangular pulse (Fig. 16(b)), and three pulse sequences with flat-top envelopes, respectively a "11" (Fig. 16(c)), a "111" (Fig. 16(d)) and a "101" (Fig. 16(e)) sequence, with ~ 20-ps bit spacing.

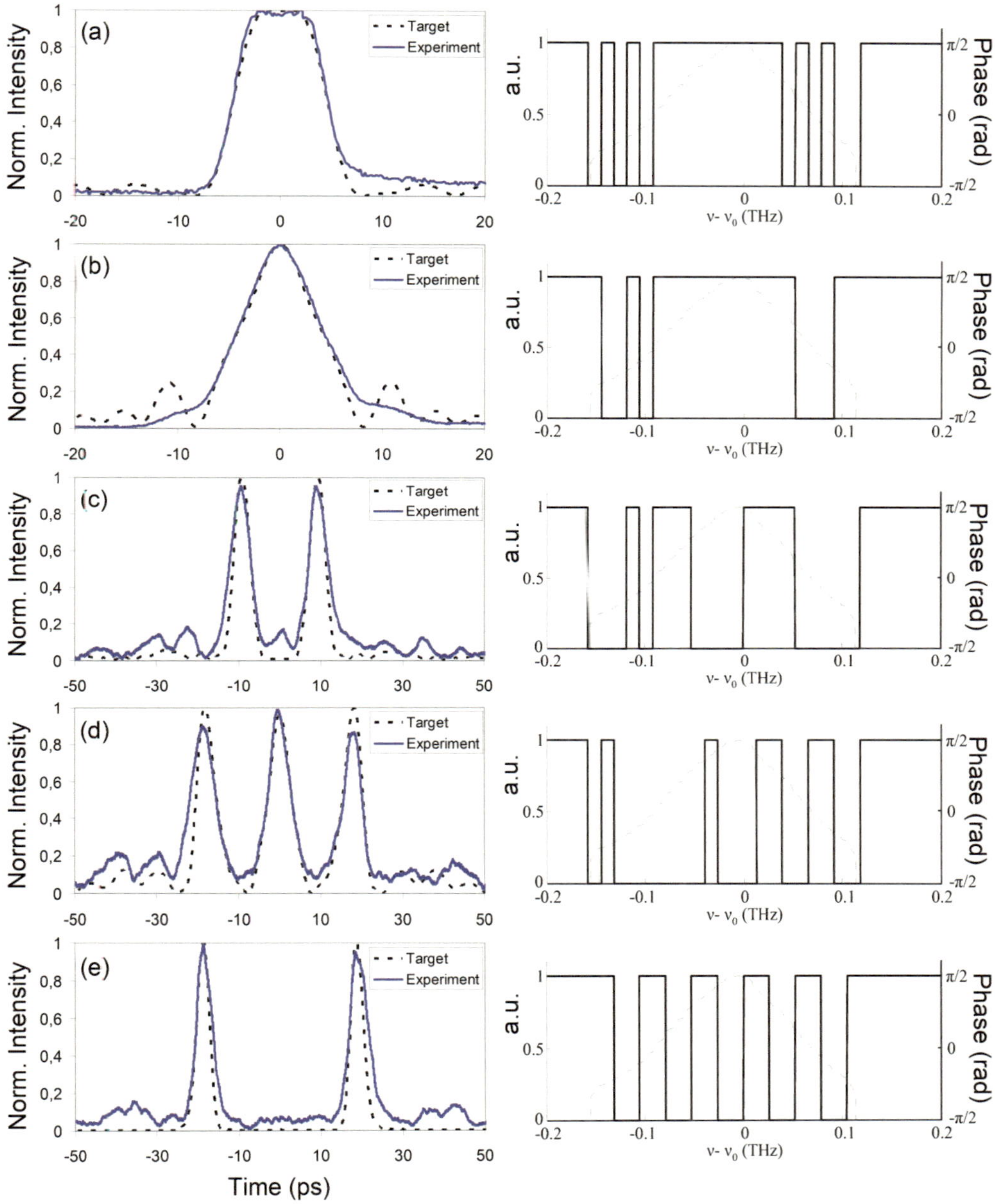

Fig. 16. Target and experimental profiles for a flat-top pulse (a), a triangular pulse (b), a "11" sequence (c), a "111" sequence (d) and a "101" sequence. The respective used phase codes are shown on the right (solid) together with the input pulse spectrum (dashed)

Fig. 16 shows the traces of the five synthesized pulse shapes experimentally acquired by the optical sampler in comparison with the simulated pulse shapes (the required binary codes to synthesize each of the target shapes are shown on the right of each graph), showing an excellent agreement between theory and experiments in all cases. Based on the values of the temporal pixel (the bit period of the BPG 2 was T_{pix} = 20 ps) and first-order dispersion used in our setup (D_λ = 480 ps/nm), we estimated a spectral resolution (Eq. 4) of ~ 13.1 GHz, which restricted the extension of the synthesized waveforms to ~ 76 ps, limiting the number of pulses per synthesized pulse burst, each with a repetition period of ~ 20 ps, to a maximum of three consecutive pulses.

To show the behavior of the system working on targets with a temporal extent larger than the above mentioned maximum, in Fig. 17 we report the comparison between simulated targets and experimental output temporal profiles acquired by the optical sampler, for cases with a temporal extent larger than 80 ps. In the first case (Fig. 17(a)) even though the agreement between simulation and experiment is quite good by the amplitude peaks of the target, the pulse shaper is not able to maintain the pedestal amplitude within an acceptable level, especially by the logic "0"s of the sequence. Moreover, in the target of the sequence "1001" two side residual peaks are already present due to a limited spectral resolution.

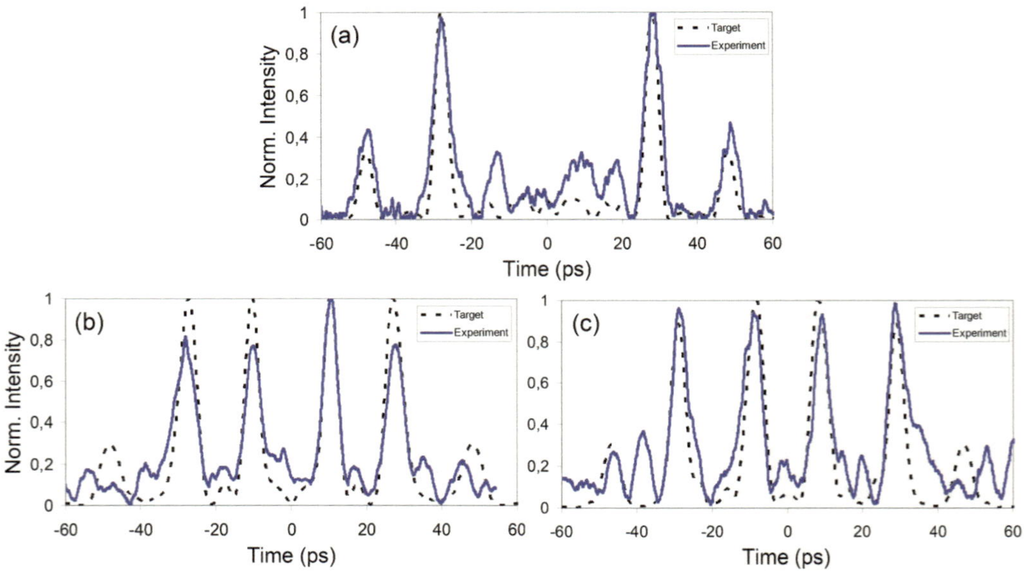

Fig. 17. Target and experimental profiles for a "1001" sequence (a), a "1111" sequence with an equalized target (b) and a "1111" sequence with a non-equalized target (c)

This limitation is due to the limited chromatic dispersion imposed by the LC-FBG (with a dispersion more than 480 ps/nm the reflection bandwidth would be narrower than the input signal bandwidth giving rise to unacceptable distortions on the output signal) and to the bit rate of the BPG 2 (20 Gb/s is the maximum value).

If we consider all the features mentioned in Section 2.1 about a pulsed-burst (acceptable pulses amplitude fluctuations, timing fluctuations, pedestal amplitude), having a look on Fig. 17(b)-(c) it is possible to notice bad performances in particular for the equalization and the pedestal level of the pulsed sequence. Moreover, the mismatch between simulated targets and experimental results increased if compared with all the cases shown in Fig. 16, confirming the non-correct working condition.

Considering the frequency bandwidth of the output pulses from the pulse shaper (FWHM ≈ 4.5 ps corresponding to a bandwidth ≈ 222 GHz), the reported setup provided a fairly high time-bandwidth product > 16.

As indicated by Eq. 4, a higher spectral resolution (i.e. longer temporal extension for the synthesized waveforms) can be achieved by increasing the bit rate of BPG 2 or by use of a higher dispersion. Using a higher dispersion would however require to decrease the repetition rate of the generated output pulses (assuming the same input pulse bandwidth). Other experimental non-idealities affecting the system performance include spectral fluctuations of the input spectrum, the non-perfect squared shape of the electric binary code produced by the BPG 2 and undesired higher order dispersion terms introduced by the LC-FBG.

3. Conclusion

In conclusion, we have demonstrated a fiber-based time-domain linear binary phase-only filtering system enabling arbitrary temporal re-shaping of picosecond optical pulses. Flat-top and triangular pulses together with two and three pulse-bursts have been synthesized from the same input pulse by properly programming the bit pattern code driving an EO phase modulator.

4. References

Azaña, J.; Slavik, R.; Kockaert, P.; Chen, L.R.; LaRochelle, S. (2003). Generation of customized ultrahigh repetition rate pulse sequences using superimposed fiber Bragg grating. *IEEE Journal of Lightwave Technology*, Vol. 21, No. 6, (June 2003) 1490-1498, 0733-8724

Azaña, J.; Berger, N. K.; Levit, B.; Fischer, B. (2005). Reconfigurable generation of high-repetition-rate optical pulse sequences based on time-domain phase-only filtering. *Optics Letters*, Vol. 30, No. 23, (December 2005) 3228-3230, 0146-9592

Dai, Y.; Yao, J. (2008). Arbitrary pulse shaping based on intensity-only modulation in the frequency domain. *Optics Letters*, Vol. 33, No. 4, (February 2008) 390-392, 0146-9592

Kim, J.; Bae, J. K.; Han, Y. G.; Kim, S. H.; Jeong, J. M.; Lee, S. B. (2004). Effectively tunable dispersion compensation based on chirped fiber Bragg gratings without central wavelength shift. *IEEE Photonics Technology Letters*, Vol. 16, No. 3, (March 2004) 849-851, 1041-1135

Kurokawa, T.; Tsuda, H.; Okamoto, K.; Naganuma, K.; Takenouchi, H.; Inoue, Y.; Ishii, M. (1997). Time-space-conversion optical signal processing using arrayed-waveguide grating. *Electronics Letters*, Vol. 33, No. 22, (October 1997) 1890-1891, 0013-5194

Lin, I.S.; Weiner, A.M. (2007). Hardware Correlation of Ultra-Wideband RF Signals Generated via Optical Pulse Shaping, *IEEE International Topical Meeting on Microwave Photonics, 2007*, pp. 149-152, 1-4244-1168-8, Victoria, BC, Canada, October 2007

Otani, T.; Miyazaki, T.; Yamamoto, S. (2000). Optical 3R regenerator using wavelength converters based on electroabsorption modulator for all-optical network applications. *IEEE Photonics Technology Letters*, Vol. 12, No. 4, (April 2000) 431-433, 1041-1135

Oxenlowe, L.K.; Slavik, R.; Galili, M.; Mulvad, H.C.H.; Park, Y.; Azana, J.; Jeppesen, P. (2007). Flat-top pulse enabling 640 Gb/s OTDM demultiplexing, *Proceedings of European Conference on Lasers and Electro-Optics, 2007 and the International Quantum Electronics Conference. CLEOE-IQEC 2007*, CI8-1, 978-1-4244-0931-0, Bourgogne, France, June 2007

Park, Y. and Azaña, J. (2006). Optical pulse shaping technique based on a simple interferometry setup, *Proceedings of 19th Annual Meeting of the IEEE Lasers & Electro-Optics Society, 2006*, pp. 274-275, 9780780395558, Montreal, QC, Canada, November 2006

Park, Y; Kulishov, M; Slavík, R; Azaña, J. (2006). Picosecond and sub-picosecond flat-top pulse generation using uniform long-period fiber gratings. *Optics Express*, Vol. 14, No. 26, (December 2006) 12670-12678, 1094-4087

Parmigiani, F.; Petropoulos, P.; Ibsen, M.; Richardson, D.J. (2006). All-optical pulse reshaping and retiming systems incorporating pulse shaping fiber Bragg grating. *IEEE Journal of Lightwave Technology*, Vol. 24, No. 1, (January 2006) 357-364, 0733-8724

Parmigiani, F.; Finot, C.; Mukasa, K.; Ibsen, M.; Roelens, M. A.; Petropoulos, P.; Richardson, D. J. (2006). Ultra-flat SPM-broadened spectra in a highly nonlinear fiber using parabolic pulses formed in a fiber Bragg grating. *Optics Express*, Vol. 14, No. 17, (August 2006) 7617-7622, 1094-4087

Petropoulos, P.; Ibsen, M.; Ellis, A.D.; Richardson, D.J. (2001). Rectangular pulse generation based on pulse reshaping using a superstructured fiber Bragg grating. *IEEE Journal of Lightwave Technology*, Vol. 19, No. 5, (May 2001) 746-752, 0733-8724

Wang, X. and Wada, N. (2007). Spectral phase encoding of ultra-short optical pulse in time domain for OCDMA application. *Optics Express*, Vol. 15, No. 12, (June 2007) 7319-7326, 1094-4087

Weiner, A. M.; Oudin, S.; Leaird, D. E.; and Reitze, D. H. (1993). Shaping of femtosecond pulses using phase-only filters designed by simulated annealing. *Journal of the Optical Society of America A*, Vol. 10, No. 5, (May 1993) 1112-1120, 0740-3232

Weiner, A. M. (1995). Femtosecond optical pulse shaping and processing. *Progress in Quantum Electronics*, Vol. 19, No. 3, (1995) 161-237, 0079-6727

Wu, C.; Raymer, M.G. (2006). Efficient picosecond pulse shaping by programmable Bragg gratings. *IEEE Journal of Quantum Electronics*, Vol. 42, No. 9, (September 2006) 873-884, 0018-9197

Zeidler, D.; Frey, S.; Kompa, K.-L.; and Motzkus, M. (2001). Evolutionary algorithms and their application to optimal control studies. *Physical Review A*, Vol. 64, No. 2, (August 2001), 1050-2947

5

Physical Nature of "Slow Light" in Stimulated Brillouin Scattering

Valeri I. Kovalev[2], Robert G. Harrison[1] and Nadezhda E. Kotova[2]
[1]*Department of Physics, Heriot-Watt University, Edinburgh,*
[2]*PN Lebedev Physical Institute of the Russian Academy of Sciences, Moscow,*
[1]*UK*
[2]*Russia*

1. Introduction

It is well known that the velocity of a light pulse in a medium, referred to as the group velocity, is smaller than the phase velocity of light, c/n, where c is the speed of light in vacuum, and n is the refractive index of the medium. The difference between phase and group velocity of light is a result of two circumstances: a pulse is generically composed of a range of frequencies, and the refractive index, n, of a material is not constant but depends on the frequency, ω, of the radiation, $n = n(\omega)$. A group index $n_g(\omega) = n + \omega(dn/d\omega)$ is used to quantify the delay (or advancement), Δt_g, of an optical pulse, $\Delta t_g = n_g L/c$, which propagates in a medium of length L, where c/n_g is called the group velocity (Brillouin, 1960).
For about a century studies of this phenomenon, now topically referred to as slow light (SL), were mostly of a scholastic nature. In general the effect is very small for propagation of light pulses through transparent media. However when the light resonantly interacts with transitions in atoms or molecules, as for gain and absorption, the effect is greatly enhanced. Fig. 1 shows the gain (inverted absorption) spectral profile around a resonance together with its refractive index dispersion profile, the gradient of which results in $n_g(\omega)$.

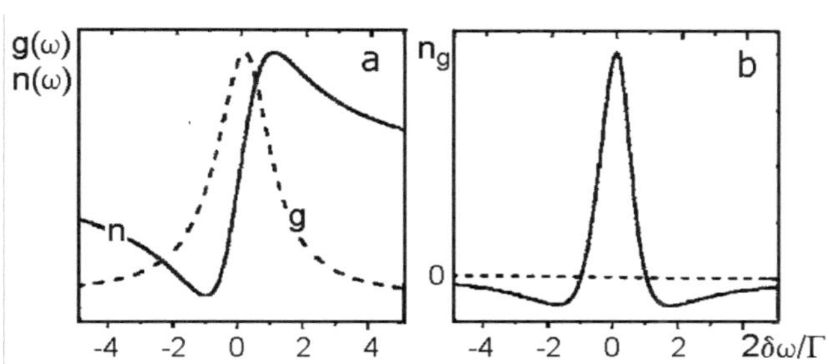

Fig. 1. a), Normalized dispersion of the gain coefficient, $g(\omega)$, (dashed line), the refractive index, $n(\omega)$, and b),the group index, $n_g(\omega)$, for a gain resonance.

As seen in the figure $n_g(\omega)$ peaks at line centre and it is here that the group delay is a maximum. However in reality for a meaningful delay the gain required must be high and this leads to competing nonlinear effects, which overshadow the slowing down (Basov et al., 1966). On the other hand in the vicinity of an absorbing resonance the corresponding absorption is much too high to render the group effect useful. An exciting breakthrough happened in the early nineties when it was shown that group velocities of few tens of meters per second were possible with *nonlinear* resonance interactions (Hau et al., 1999). Two important features of nonlinear resonances make this possible: substantially reduced absorption, or even amplification, of radiation at a resonance, and sharpness of such resonances; the sharper a resonance, the higher $dn/d\omega$ and so the stronger the enhancement of group index, and hence the greater the pulse is delayed.

Widely ranging applications for slow light have been proposed, of which those for telecommunication systems and devices (optical delay lines, optical buffers, optical equalizers and signal processors) are currently of most interest (Gauthier, 2005). The essential demand of such devices is compatibility with existing telecommunication systems, that is they must be of wide enough bandwidth (≥10 GHz) and able to be integrated seamlessly into such systems.

Of the various nonlinear resonance mechanisms and media, which allow sufficiently long induced delays, stimulated Brillouin and Raman scattering (SBS and SRS) in optical fiber are deemed to be among the best candidates. Currently SBS is the most actively investigated and many experimental and theoretical papers on pulse delaying via SBS in optical fiber have been published in the last few years, see the review paper (Thevenaz, 2008) and references therein. In this process the pulse to be delayed is a frequency down-shifted (Stokes) pulse. This is transmitted through an optical fiber through which continuous wave (CW) pump radiation is sent in the opposite direction to prime the delay process. It is supposed that the Stokes pulse is amplified by parametric coupling with the pump wave and a material (acoustic) wave in the medium (Kroll, 1965), and the amplification is characterised by a resonant-type gain profile. The dispersion of refractive index associated with this profile (which is similar to that in Fig.1) can then be used to increase the group index for optical pulses at the Stokes frequency (Zeldovich, 1972).

Along with obvious device compatibility, there are several other advantages of the SL via SBS approach for optical communications systems: slow-light resonance can be created at any wavelength by changing the pump wavelength; use of optical fibre allows for long interaction lengths and thus low powers for the pump radiation, the process runs at room temperature, it uses off the shelf telecom equipment, and SBS works in the entire transparency range of fibers and in all types of fiber. Currently a main obstacle to applications of this approach is the narrow SBS gain spectral bandwidth, (Thevenaz, 2008), which is typically ≈ 120-200 MHz in silica fiber in the spectral range of telecom optical radiation (~1.3-1.6 μm) (Agrawal, 2006).

This chapter reviews our ongoing work on the physical mechanisms that give rise to pulse delay in SBS. In section 2 the theoretical background of the SBS phenomenon is given and the main working equations describing this nonlinear interaction are presented. In section 3 ways by which the SBS spectral bandwidth may be increased are addressed. Waveguide induced spectral broadening of SBS in optical fibre is considered as a means of increasing the bandwidth to the multi-GHz range. An alternative way widely discussed in the literature, (Thevenaz, 2008), is based on spectral broadening of the pump radiation. However it is shown through analytic analysis of the SBS equations converted to the frequency domain that pump radiation broadening by any reasonable amount has only a negligible effect on increasing the SBS bandwidth. Importantly in this section we show that, irrespective of the nature of the broadening considered, the SBS gain bandwidth remains centred at the Brillion frequency which is far removed from the centre frequency of the Stokes pulse. Consequently the associated group index, which is enhanced at and around the SBS gain centre, cannot lead to group index induced delay of a Stokes pulse as claimed in the literature (Thevenaz, 2008). In section 4 the actual physical mechanisms by which a Stokes pulse is delayed through SBS are examined. Analytical analysis of the equations in the time domain shows that the SBS amplification process does not amplify an external the Stokes pulse and so again cannot induce group delay of this pulse. Rather the delay is shown to be predominantly a consequence of SBS gain build-up determined by inertia of the acoustic wave excitation. Finally in section 5 conclusions are drawn from this work in regard to current understanding of SL in SBS.

2. Theory of stimulated Brillioun scattering

In SBS, the resonance in a medium's response occurs at the Brillouin frequency, Ω_B, which is the central frequency of the variation of density in a medium, $\delta\rho(z,t) = 1/2\{\rho(z,t)\exp[-i(\Omega_B t+qz)] + c.c.\}$. This density variation is resonantly induced by an electrostrictive force resulting from interference of two plane counter-propagating waves, the forward-going (+z direction) Stokes and backward-going (-z direction) pump optical fields, $E_S(z,t) = 1/2\{E_S(z,t)\exp[-i(\omega_S t-k_S z)] + c.c.\}$ and $E_p(z,t) = 1/2\{E_p(z,t)\exp[-i(\omega_p t+k_p z)] + c.c.\}$, respectively, where $\rho(z,t)$, $E_S(z,t)$ and $E_p(z,t)$ are the amplitudes of the acoustic wave and of Stokes and pump fields, with $\Omega = \omega_p - \omega_S$ and $q = k_p + k_S$, ω_S and k_S, and ω_p and k_p being their radian frequencies and wavevectors and c.c. is the abbreviation for complex conjugate. In an isotropic medium $\delta\rho(z,t)$ is described by the equation, (Zeldovich et al., 1985),

$$\frac{\partial^2\delta\rho}{\partial t^2} - v_s^2\nabla^2\delta\rho - A\nabla^2\frac{\partial\delta\rho}{\partial t} = -\rho_0\frac{\partial\varepsilon}{\partial\rho}\frac{1}{16\pi}\nabla^2\left|E(z,t)\right|^2, \tag{1}$$

where v_s is the speed of a free acoustic wave, A is its damping parameter, $\nabla^2 \equiv \partial^2/\partial z^2$ in the chosen plane wave model, ε and ρ_0 are the dielectric function and equilibrium density of the medium, and $E(z,t) = E_p(z,t) + E_S(z,t)$. Since the amplitude $\rho(z,t)$ is supposed to be slowly varying in space, then $\nabla^2\delta\rho(z,t) \cong -q^2\delta\rho(z,t)$ and Eq.(1) is usually reduced to

$$\frac{\partial^2\delta\rho}{\partial t^2} + \Omega_B^2\delta\rho + \frac{\Gamma_B}{2}\frac{\partial\delta\rho}{\partial t} = -\rho_0\frac{\partial\varepsilon}{\partial\rho}\frac{1}{8\pi}\frac{\partial^2}{\partial z^2}\left[E_p(z,t)E_S(z,t)\right]. \tag{2}$$

This is then the equation for the induced acoustic wave. It is a typical equation for an externally driven damped resonant oscillator, in which the right-hand side is the driving force, $\Omega_B = qv_s = 2v_s\omega_p/(c/n+v_s) \cong 2nv_s\omega_p/c$ is the resonant frequency of the oscillator, known as the Brillouin frequency, Γ_B is the FWHM spectral width of the resonant profile with $2/\Gamma_B$ being the decay time of the acoustic wave.

The pump field reflected by the induced acoustic wave is a new Stokes field, which in turn interacts with the pump field to further electrostrictively enhance the acoustic wave and so the Stokes field and so forth. Increase of the Stokes field in SBS is therefore a direct consequence of increase of reflectivity of the acoustic wave for the pump field. As such, so called "SBS gain" characteristics are determined by the reflectivity, spectral characteristics and dynamics of the acoustic wave. In the approximation that the CW pump radiation is not depleted over the interaction length, L, the spatial/temporal evolution of the Stokes signal is described by the nonlinear wave equation,

$$\frac{\partial^2 E_S}{\partial z^2} - \frac{\varepsilon}{c^2}\frac{\partial^2 E_S}{\partial t^2} = \frac{1}{c^2}\frac{\partial \varepsilon}{\partial \rho}\frac{\partial^2}{\partial t^2}\left[\delta\rho(z,t)E_p(z,t)\right]. \tag{3}$$

Eqs (2) and (3) are the basic equations, which describes the SBS phenomenon in an optically lossless medium in the small signal plane wave approximation. Since the density and Stokes field amplitudes, $\rho(z,t)$ and $E_S(z,t)$, vary slowly in both space and time and the acoustic wave in SBS attenuates strongly, their evolution is usually reduced to two well known first order equations: from Eq.(2) the relaxation equation for $\rho(z,t)$,

$$\frac{\partial \rho}{\partial t} + \left(\frac{\Gamma_B}{2} + i\delta\Omega\right)\rho = -i\rho_0 \frac{\partial \varepsilon}{\partial \rho}\frac{\Omega_B}{8\pi v_s^2} E_p(t)E_S^*(z,t), \tag{4}$$

which describes the amplitude of the driven damped resonant oscillator, and from Eq.(3) the partial differential equation for $E_S(z,t)$,

$$\frac{\partial E_S}{\partial z} + \frac{n}{c}\frac{\partial E_S}{\partial t} = -i\frac{\omega_S}{2cn}\frac{\partial \varepsilon}{\partial \rho}\rho^*(z,t)E_p(t). \tag{5}$$

Here $\delta\Omega = \Omega - \Omega_B$ is the difference between the acoustic drive frequency, Ω, and the resonant Brillouin frequency and asterisk, *, marks complex conjugate. The right-hand side of Eq. (5) is a source of the Stokes emission.

3. Spectral broadening of SBS

In the literature on group index induced slow light it is argued that rate at which optical pulses may be delayed is ultimately determined by the spectral bandwidth of the resonance responsible for slow light generation in the material (Boyd & Gauthier, 2002). So, the narrower the bandwidth the larger is the delay. On the other hand, to minimize pulse distortion the bandwidth must exceed substantially that of the optical pulse to be delayed and consequently determines a lower limit for the duration of the optical pulse. This argument is correct for systems in which a resonance in the material is in resonance with the optical pulse to be delayed, such as those based on electromagnetically induced transparency and coherent population oscillation (Boyd & Gauthier, 2002). However as shown below this does not apply to SBS since the resonance occurs around the Brillouin frequency, Ω_B, which is far from the frequency of the Stokes pulse to be delayed. This point has been overlooked in the literature on SL via SBS and as a consequence has led to misinterpretation of experimental findings of Stokes pulse delay in SBS. This issue is considered in some detail in section 4 where it is shown the Stokes delay arises from the inertial build up time of SBS and not group index delay as has been claimed throughout the literature. Nevertheless it is still of academic interest to consider ways in which the spectral bandwidth of SBS may be increased and this is considered below.

The physical mechanism responsible for Γ_B is attenuation of the Brillioun acoustic wave, in liquids and solid optical media this is predominantly due to viscosity (Zeldovich et al., 1985). Such spectral broadening is homogeneous in nature. For bulk silica, Γ_B, scales with pump radiation wavelength, λ, as $\Gamma_B \cong 2\pi 40/\lambda^2$ MHz (Heiman et al. 1979), where λ is in μm. It is evident from this expression that the shorter the radiation wavelength the wider the spectrum, so for radiation in the short wavelength transmission window of silica, $\lambda \cong 0.2$ μm, Γ_B is expected to be ~2π GHz compared to ~ 20 MHz at telecom wavelengths, $\lambda \cong 1.3$-1.6 μm. The SBS gain bandwidth in fibers may also be broadened through varying fiber design, doping concentration, strain and/or temperature (Tkach et al., 1986; Shibata et al., 1987; Azuma et al., 1988; Shibata et al., 1989; Yoshizawa et al., 1991; Tsun et al., 1992; Yoshizawa & Imai, 1993; Shiraki et al., 1995; LeFloch & Cambon, 2003). However the highest achieved line-width enhancement factor, compared to Γ_B is ~5, (Yoshizawa et al., 1991). A potentially attractive solution to increasing Γ_B is by waveguide induced spectral broadening (Kovalev & Harrison, 2000), which is discussed in some detail below (Sect. 3.1). Spectral broadening of the pump radiation has also been proposed (Stenner et al., 2005, Herraez et al., 2006) as a means for broadening Γ_B and is currently a subject of considerable activity (Thevenaz, 2008). However, as shown below (see Sect. 3.2) the effect is in fact negligible.

3.1 Waveguide induced spectral broadening of SBS

Due to the waveguiding nature of beam propagation in optical fiber and its effect on the SBS interaction, such propagation has been shown to render the Stokes spectrum inhomogeneous (Kovalev & Harrison, 2000), the bandwidth of which is massive in fibers of high numerical aperture, NA (Kovalev & Harrison, 2002). The nature of the broadening arises from the ability of optical fiber to support a fan of beam directions within an angle $2\theta_c$ (Fig. 2), where θ_c is the acceptance angle of the fiber, defined as

$$\theta_c = \arcsin\left\{1 - \frac{n_{cl}^2}{n_{co}^2}\right\}^{1/2} = \arcsin\left[\frac{NA}{n_{co}}\right], \tag{6}$$

where $n_{cl,co}$ are the refractive indices of the fiber cladding and core, respectively.

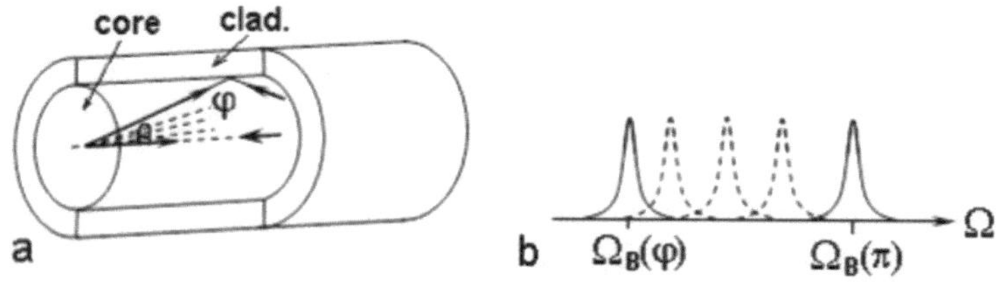

Fig. 2. Sketch showing the nature of waveguide induced broadening of SBS gain spectrum; a), schematic of fiber, and, b), homogeneously broadened spectral profiles for different angles of scattering, φ.

The frequency shift of the Stokes depends on the angle, φ, between the momentum vectors of the pump and scattered radiation through the relation $\Omega_B(\varphi)=4\pi n v_s \sin(\varphi/2)/\Lambda$. So the range of $\Omega_B(\varphi)$ in a fiber will be from $\Omega_B(\pi) = 4\pi n v_s/\Lambda$ to $\Omega_B(\pi-2\theta_c) = 4\pi n v_s \cos\theta_c/\Lambda$. For every $\Omega_B(\varphi)$ there corresponds a homogeneously broadened line of the form

$$\gamma_h(\Omega,\Omega_B) = \frac{\Gamma_B^2}{\left[4(\Omega_B - \Omega)^2 + \Gamma_B^2\right]}. \tag{7}$$

The Stokes spectrum, broadened by guiding, is then the convolution of frequency-shifted homogeneously broadened components, each generated from a different angular component of the pump and Stokes signal (such broadening is inhomogeneous by definition). The shape of the broadened Brillouin linewidth is described by the equation (Kovalev & Harrison, 2002),

$$\gamma_i(\Omega,\theta_c)=\frac{\Gamma_B}{2[\Omega_B(\pi)-\Omega_B(\pi-2\theta_c)]}\left[\tan^{-1}\left(2\frac{\Omega_B(\pi)-\Omega}{\Gamma_B}\right)-\tan^{-1}\left(2\frac{\Omega_B(\pi-2\theta_c)-\Omega}{\Gamma_B}\right)\right], \tag{8}$$

where θ_c is linked to NA through Eq.(6). Fig. 3 shows $\gamma_i(\Omega,\theta_c)$ for five values of NA.

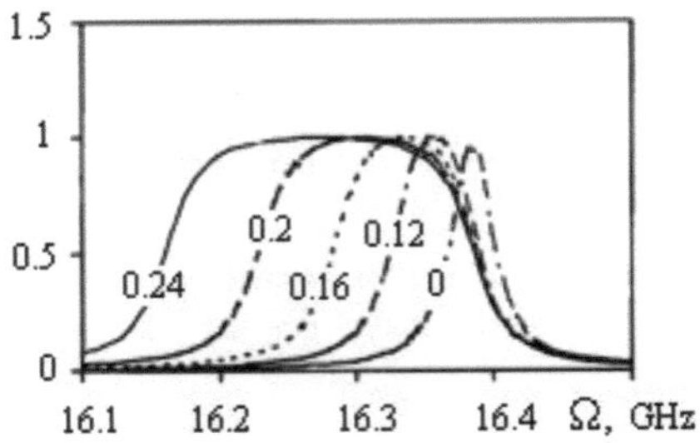

Fig. 3. Line shape of the Brillouin gain spectrum in optical fiber for several NA values (Kovalev & Harrison, 2002).

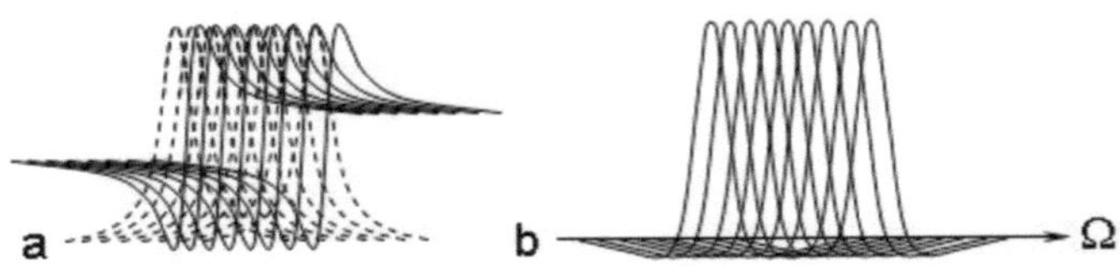

Fig. 4. Relation between, a), the gain profile (dashed) and the phase index (solid) and, b), group index, for SBS in optical fibers.

Intuitively, the dispersion and group index profiles, which are associated with the convolutionally broadened SBS gain spectrum (dashed lines in Fig.4a), are expected to also be convolutionally broadened (solid lines in Fig.4a and Fig.4b). As seen (Fig.4b) the maximum n_g is expected to be more or less constant and so it's value for each and all the homogeneous spectral components that contribute to the group index profile is the same.
The shape of the group index spectrum is determined more precisely by numerical simulation (Kovalev et al., 2008). The group index in the case of the SBS resonance in optical fiber can be expressed as (Okawachi, 2005; Kovalev & Harrison, 2005),

$$n_g(\Omega_B,\Omega)=n-\frac{g_0 I_p \lambda \Gamma_B}{2\pi}\left\{\frac{(\Omega_B-\Omega)}{4(\Omega_B-\Omega)^2+\Gamma_B^{\,2}}+\omega\frac{4(\Omega_B-\Omega)^2-\Gamma_B^{\,2}}{\left[4(\Omega_B-\Omega)^2+\Gamma_B^{\,2}\right]^2}\right\}, \tag{9}$$

where g_0 is the value of the SBS gain coefficient at the exact Brillouin resonance and I_p is the pump radiation intensity. When several resonant frequencies, Ω_B, exist in the medium, and the distribution of their relative amplitudes over the range from $\Omega_B(\pi)$ to $\Omega_B(\pi\text{-}2\theta_c)$ is some function $F(\Omega_B)$, the spectrum of the "broadened" group index, $n_{gb}(\Omega)$, is described by the convolution integral,

$$n_{gb}(\Omega) = \int_{\Omega_B(\pi-2\theta_c)}^{\Omega_B(\pi)} F(\Omega_B) n_g(\Omega_B, \Omega) d\Omega_B \,. \tag{10}$$

Results of calculations for the case when $F(\Omega_B) = 1$ in the range from $\Omega_B(\pi)$ to $\Omega_B(\pi-2\theta_c)$ are presented in Fig.5. For the sake of illustration the results are centred by shifting the limits of integration in such a way that the lower limit is $\omega_1 = \omega_0 - m\Gamma_B/2$ and $\omega_2 = \omega_0 + m\Gamma_B/2$, where m, which is called the rate of broadening, varies from 2 to 100. As seen the width of the profile increases continuously with increasing m. The original shape of the profile (individual components in Fig 4b) is retained for a broadening of $m \leq 2$. Beyond this, the profile becomes top-hat, at $m \cong 4$, and for $m > 4$ it exhibits a dip, the depth of which increases with increasing m. For $m > 20$ the dip tends to becomes flat-bottomed. It can therefore be seen that the value of group index can stay constant over a broad range of frequencies, especially when $m > 60$. However the price for this is a reduced magnitude of the group index. However, as seen in Eq.(9), this may be compensated for by increase of the pump intensity.

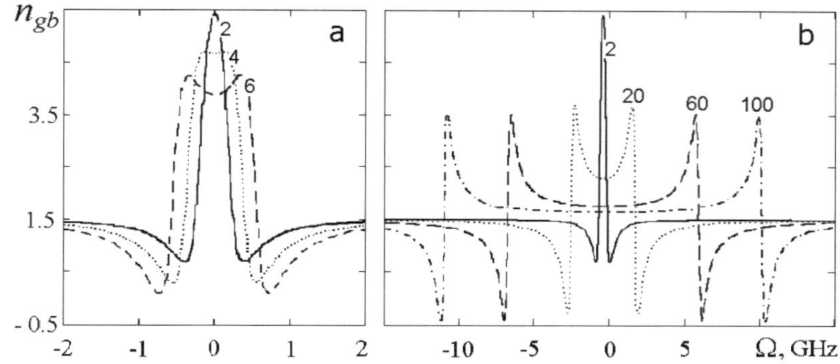

Fig. 5. Shape of group index, n_{gb}, spectrally broadened due to waveguiding nature of fiber, a) for m = 2-6, and b), for m = 2-100.

Earlier work has shown that waveguide induced broadening is dependant on the numerical aperture of fiber through the equation (Kovalev & Harrison, 2002),

$$\Gamma \cong \sqrt{\Gamma_B^2 + \Omega_B^2 \frac{(NA)^4}{4n_{co}^4}} \,. \tag{11}$$

It follows from Eq.(11) that in the calculations above, m = 2 corresponds to NA = 0.12, which is standard for single-mode telecom fiber. However, it is now readily possible to realise single-mode fiber with much higher NA, ~0.8 (Knight et al., 2000). For such fiber the broadening is ~15 GHz, which is comparable with the needs of telecom devises. As noted above this analysis assumes that the homogeneously broadened Brillouin gain contributions to the inhomogeneous profile are uniformly distributed, $F(\Omega_B) = 1$. It is relatively straight forward to account for alternative distributions by introducing their appropriate shape function $F(\Omega_B)$ into Eq.(10).

These considerations therefore show that waveguide induced spectral broadening of the gain bandwidth of SBS in optical fiber is potentially massive (> 10 Gb/s), and readily achievable.

3.2 On the effect of the pump spectral width on spectral broadening of SBS

It has been proposed in (Stenner et al., 2005) that spectral broadening of the pump radiation may lead to comparable spectral broadening of the Stokes pulse in SBS. This approach has since been the focus of many publications, (Minardo et al., 2006; Shumakher et al., 2006; Zhu & Gauthier, 2006; Zadok et al., 2006; Chin et al., 2006; Schneider et al., 2006; Kalosha et al., 2006; Zhu et al., 2007; Song & Hotate, 2007; Lu et al., 2007; Zhang et al., 2007-1 & -2; Yi et al., 2007; Shi et al., 2007; Pant et al., 2008; Ren & Tomita, 2008; Sakamoto et al., 2008; Wang et al., 2008; Schneider et al., 2008; Cheng et al., 2008), aimed at high data rate applications of SL. In this section the validity of this assertion is examined and it is shown that the effect is in fact negligible.

This may be seen from examining the spectral features of the medium's response and the Stokes emission through Fourier transformation of Eqs (4) and (5) using the following basic properties of Fourier transforms, $F(\omega) \equiv S[f(t)]$, (Korn & Korn, 1967),

$$S[f(t)] = \int_{-\infty}^{\infty} f(t)e^{-i\omega t}dt,\ S[\frac{d}{dt}f(t)] = i\omega S[f(t)],\ and\ S[f_1(t)f_2(t)] = \frac{1}{2\pi}\int_{-\infty}^{\infty} F_1(\nu)F_2(\omega-\nu)d\nu\,, \tag{12}$$

where $f(t)$ is a function of time. Here ω and ν are Fourier transform frequencies, which are the difference frequencies of the acoustic, Stokes and pump signals from their respective line centers. Eqs (4) and (5) then give

$$\left(\frac{\Gamma_B}{2} - i(\omega + \delta\Omega)\right)\tilde{\rho}^*(z,\omega) = i\rho_0 \frac{\partial\varepsilon}{\partial\rho}\frac{\Omega_B}{8\pi v_s^2}\frac{1}{2\pi}\int_{-\infty}^{\infty} \tilde{E}_p^{\prime *}(\nu)\tilde{E}_S^{\prime}(z,\omega-\nu)d\nu\,, \tag{13}$$

and

$$\frac{d\tilde{E}_S(z,\omega)}{dz} + i\omega\frac{n}{c}\tilde{E}_S(z,\omega) = -i\frac{\omega_S}{2cn}\frac{\partial\varepsilon}{\partial\rho}\frac{1}{2\pi}\int_{-\infty}^{\infty} \tilde{E}_p(\nu)\tilde{\rho}^*(z,\omega-\nu)d\nu\,. \tag{14}$$

Note that here we have primed the Stokes and pump fields within (13), which are responsible for inducing the acoustic wave, to distinguish them from the generated Stokes field, (LHS of (14)), and from the pump field, which generates the new Stokes field, (the field $\tilde{E}_p(\omega)$ under the integral in (14)). $\delta\Omega$ is a detuning parameter, which can influence only the strength of the medium's response to the drive force at frequency, Ω. (In the case of non-monochromatic pump and Stokes fields $\delta\Omega$ can be considered as the difference between the central frequencies of their bandwidths.)

Consider the case of a typical SBS slow light experiment in which the spectrum of the Stokes signal corresponds to that of a temporally smooth pulse and the spectrum of the pump radiation is the Fourier-transform of a continuous wave field the amplitude of which is randomly fluctuating in time. As seen, the right-hand sides of these equations are proportional to the convolution integrals of spectra $\tilde{E}_p^{\prime *}(\omega)$ and $\tilde{E}_S^{\prime}(z,\omega)$ in Eq.(13), and $\tilde{E}_p(\omega)$ and $\tilde{\rho}^*(z,\omega)$ in Eq.(14), respectively.

Eq. (13) is an algebraic equation, the solution of which gives the spectrum of the medium's response

$$\tilde{\rho}^*(z,\omega) = i\rho_0 \frac{\partial \varepsilon}{\partial \rho} \frac{\Omega_B}{8\pi v_s^2} \frac{F_\rho(z,\omega)}{\left(\Gamma_B/2 - i(\omega + \delta\Omega)\right)}, \tag{15}$$

where

$$F_\rho(z,\omega) = \frac{1}{2\pi} \int_{-\infty}^{\infty} \tilde{E}_p^{'*}(\nu) \tilde{E}_S^{'}(z,\omega-\nu) d\nu \tag{16}$$

is the function which determines the spectrum of the driving force for the medium's response. The spectrum of the medium's response is then given by the modulus of $\tilde{\rho}^*(z,\omega)$, $|\tilde{\rho}(z,\omega)|$.

The spectrum of the Stokes field is described by the first order differential equation, Eq.(14), the solution of which is

$$\left|\tilde{E}_S(z,\omega)\right| = \left| e^{-i\frac{\omega n}{c}z} \left[\tilde{E}_S(0,\omega) + i\frac{\omega_S}{2nc}\frac{\partial \varepsilon}{\partial \rho} \int_0^z F_E(x,\omega) e^{i\frac{\omega n}{c}x} dx \right] \right|, \tag{17}$$

where $\tilde{E}_S(0,\omega)$ is the spectrum of an input Stokes signal at z = 0 and

$$F_E(z,\omega) = \frac{1}{2\pi} \int_{-\infty}^{\infty} \tilde{E}_p(\nu) \tilde{\rho}^*(z,\omega-\nu) d\nu \tag{18}$$

is the function which determines the spectrum of the source of the generated Stokes field. Equations (15)-(18) describe the spectral features of the SBS-induced material response and Stokes field when both optical fields, pump and Stokes, are non-monochromatic. Note that solution Eq.(17) in the *spectral* domain is entirely consistent with the analytical solution of Eqs. (4) and (5), previously obtained in the *temporal* domain for stimulated scattering induced by non-monochromatic pump and monochromatic Stokes fields in (Kroll, 1965, Charman et al., 1970, Akhmanov et al., 1971, Akhmanov et al., 1988), and for monochromatic pump and non-monochromatic Stokes fields in (Kovalev et al., 2009).

It is easily seen that the solution (17) for the Stokes field differs substantially from that usually deduced in textbooks from (4) and (5) in the steady state approximation (that is when both pump and Stokes fields are considered *monochromatic*),

$$E_S(z) = E_S(0) e^{\frac{g_0 |E_p E_p^*| z}{2(1-i2\delta\Omega\Gamma_B^{-1})}}, \tag{19}$$

which results from the equation for the Stokes field of the form, (Zeldovich et al., 1985),

$$\frac{dE_S(z)}{dz} = \left[\frac{g_0 |E_p E_p^*|}{2(1-i2\delta\Omega\Gamma_B^{-1})} \right] E_S(z). \tag{20}$$

Here it is again important to remember that δΩ is a detuning parameter, the value of which is the difference between the frequencies of the monochromatic pump and Stokes fields as chosen, $\Omega = \omega_p - \omega_S$, and the resonant Brillouin frequency, Ω_B, $|E_pE_p^*|$ is the pump radiation intensity, and g_0 is the standard steady state SBS gain coefficient. Equation (20) means that the Stokes field is amplified in the medium with gain proportional to the pump radiation intensity. To appreciate the difference between this case and our case let us consider both pump and Stokes fields in Eqs.(13)-(18) to be *monochromatic*. The spectra of the amplitudes of driving forces and of the medium's response then reduce to δ-functions at $\omega = 0$ and these depend on z only. The equation for the Stokes field is then,

$$\frac{dE_S(z)}{dz} = \left[\frac{g_0(E_p^{'*}E_S^{'}(z))}{2(1 - i2\delta\Omega\Gamma_B^{-1})} \right] E_p, \tag{21}$$

The physical meaning of Eq.(21) is substantially different from that of Eq.(20), though their mathematical forms may look similar. Equation (21) describes the spatial evolution of the amplitude of the Stokes field, which results from reflection of the pump field by the induced acoustic wave. Since the Stokes field on the RHS of Eq.(21) is responsible for creating the acoustic wave, it is not the same as the reflected Stokes field on the LHS of this equation and therefore it is distinguished by its prime. Though for this monochromatic case the Stokes fields have the same frequency, their roles still remain physically distinct as in our general treatment above. Such distinction is not made in the text-book treatment that leads to Eq.(20) and to its familiar exponential solution, Eq.(19), which displays "gain" and a "modified propagation constant" for the Stokes field (see (Zhu et al., 2005)). Evidently the solution for Eq.(21) cannot be the same. As such, though the RHS of this equation has both real and imaginary parts (in the case of non-zero detuning, δΩ), this does not modify the propagation constant for the reflected Stokes field and therefore it can have no bearing on changing the refractive and group index for this field.

Returning now to the solutions (15)-(18) of Eqs. (13) and (14) for the general case in which either one of the fields or both have nonzero bandwidth, they display three important features of the SBS interaction: i) the external input Stokes signal, as seen in Eq.(17), propagates through a non-absorbing medium without gain or measurable loss (its energy loss for creating the acoustic wave is usually negligible), ii) the SBS-generated Stokes signal is a result of reflection of the pump radiation by the acoustic wave, which is created by the pump and the original Stokes fields (see Eqs. (4) and (5)), and iii) each spectral component of the generated Stokes signal arises from a range of spectral components of the non-monochromatic pump and Stokes fields, see Eqs. (17) and (18).

To see the consequences of this, consider the case when the growth of the Stokes field along z is small. As such, the z dependence of $\tilde{E}_S(z,\omega)$ and $\tilde{\rho}^*(z,\omega)$ is dropped. While this approximation does not account for gain narrowing of the SBS spectrum, typical for higher amplification, (Zeldovich et al., 1985), it still captures reasonably well the trends in the spectral features of the Stokes field, $\tilde{E}_S(z,\omega)$, the medium response, $|\tilde{\rho}(\omega)|$, and so the SBS gain and the modified refractive and group indices. It then follows from Eq.(17) that the output spectrum of the Stokes signal, $|\tilde{E}_S(\omega)|$, is the sum of the Fourier spectra of the input Stokes signal, $|\tilde{E}_S(0,\omega)|$, and the convoluted spectrum of the pump field and medium's excitation, characteristics which are described by the function $F_E(\omega)$ (see Eq. (18)). The latter is the SBS-induced contribution to the Stokes signal, and is determined by which of $\tilde{E}_p(\omega)$ or $\tilde{\rho}^*(\omega)$ is spectrally broadest. These spectral features of the output Stokes radiation are consistent with those predicted earlier in the low gain approximation using the *temporal* domain treatment of stimulated Raman scattering, (Akhmanov et al., 1971, Akhmanov et al., 1988), and SBS, (Zeldovich et al., 1985), with non-monochromatic pump fields.

In contrast to the spectral features of the Stokes emission, those of the medium's response, $|\tilde{\rho}(\omega)|$, have to date received little attention. According to Eq.(15) this case is not as straightforward as that of the Stokes spectrum since a Lorentzian shape multiplier of bandwidth Γ_B appears in addition to the convolution integral, $F_\rho(\omega)$. From Eq. (15), $|\tilde{\rho}(\omega)|$ is given as

$$|\tilde{\rho}(\omega)| = \rho_0 \frac{\partial \varepsilon}{\partial \rho} \frac{\Omega_B}{8\pi v_s^2} \frac{|F_\rho(\omega)|}{\sqrt{\Gamma_B^2/4 + (\omega + \delta\Omega)^2}} . \qquad (22)$$

Examples of the spectra, $|\tilde{\rho}(\omega)|$, are shown in Fig. 6 by solid curves for $\delta\Omega = 0$ (Fig. 6(a)) and $\delta\Omega = 0.1$ GHz (Fig. 6(b)).

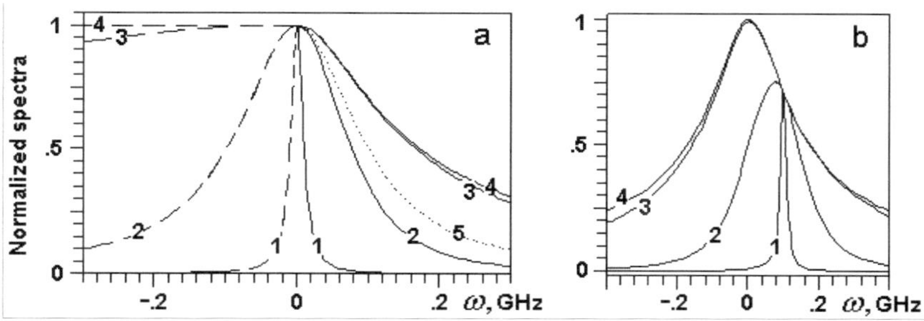

Fig. 6. a), LHS and RHS halves of the (symmetrical) spectra of $|F_\rho(\omega)|$ (dashed) and of $|\tilde{\rho}(\omega)|$ (solid lines) respectively for spectral widths of $F_\rho(\omega)$ 0.02(1), 0.2(2), 2(3) and 20(4) GHz, when $\Gamma_B = 0.2$ GHz and $\delta\Omega = 0$. The dotted line 5 is the Lorentzian-shaped spectrum with $\Gamma_B = 0.2$ GHz; b), spectra of $|\tilde{\rho}(\omega)|$ for the same spectral widths of $F_\rho(\omega)$ when $\delta\Omega = 0.1$ GHz.

It follows from Eq.(22) that when the spectral width, $\delta\omega_p$, of the driving force, $|F_\rho(\omega)|$ (shown by dashed curves in Fig. 6(a)), is narrower than Γ_B, the width, Γ, and centre frequency of $|\tilde{\rho}(\omega)|$ is determined by $|F_\rho(\omega)|$, (the convolution of $\tilde{E}_p^*(\omega)$ and $\tilde{E}_S(\omega)$ as given in Eq. (16)), the central frequency of which is detuned by $\delta\Omega$ as shown by curve 1 in Fig. 6(b). Nonzero detuning $\delta\Omega$ results also in decreased amplitude of the medium's response (compare solid curves 1 of Figs. 6(a) and 6(b)). As $\delta\omega_p$ increases, Γ grows and for $\delta\omega_p > \Gamma_B$ it saturates at $\sim 1.7\Gamma_B$ (solid curves 3 and 4 in Fig. 6(a)), and the effect of the detuning $\delta\Omega$ on the features of the medium's spectrum becomes negligible (compare solid curves 3 and 4 in Figs. 6(a) and 6(b)). In essence, this means that irrespective of how broad the bandwidth of the broadband pump and/or Stokes emission is/are, the bandwidth of the material response spectrum is predominantly determined by the features of the material and can never be much greater than that of the medium's resonant response ($\sim\Gamma_B$). This is exactly the features expected from an externally driven damped resonant oscillator (see Eq.(2)).

From Eq. (22) the SBS induced dispersion of refractive index, $\Delta n(\omega) \cong (\partial\varepsilon/\partial\rho)|\,\tilde{\rho}(\omega)\,|/2n_0$, and its corresponding induced group index at the frequency of the Stokes radiation, $n_g(\omega_S) = n_0(\omega_S) + \omega_S(d\Delta n/d\omega)|_{\omega_S}$, may be directly determined. To do so let us replace the relative frequency ω in the equations above with the absolute frequency, ω', and consider the spectrum of $|F_\rho(\omega')|$ to be Gaussian in shape centred at frequency Ω_B, $F_\rho(\omega') = F_{\rho a}\exp[-(\omega'-\Omega_B)^2/\delta\omega_\rho^2]$, where $F_{\rho a} = \sqrt{I_p I_S}$ is the spectral amplitude and I_S is the Stokes signal intensity. The induced group index at absolute frequency ω' is then

$$n_g(\omega') = -B\frac{\omega'(\omega'-\Omega_B)}{\delta\omega_\rho^2}\left[\frac{\Gamma_B^2}{2}+2(\omega'-\Omega_B)^2+\delta\omega_\rho^2\right]\left[\frac{\Gamma_B^2}{4}+(\omega'-\Omega_B)^2\right]^{-3/2} e^{-\frac{(\omega'-\Omega_B)^2}{\delta\omega_\rho^2}}, \qquad (23)$$

where $B = \rho_0\left(\frac{\partial\varepsilon}{\partial\rho}\right)^2\frac{\Omega_B\sqrt{I_p I_S}}{16\pi n_0 v_s^2}$.

The dispersion of the SBS-induced group index, which is described by Eq. (23), has a maximum at $\omega' = \Omega_B - \Delta$ and a minimum at $\omega' = \Omega_B + \Delta$, where Δ is functionally dependent on $\delta\omega_\rho$ and is $< \Gamma_B/2$. These maximum and minimum, in principle, may be of very high amplitude, however it is critically important to note that they are located at $\omega' \approx \Omega_B$. At the Stokes frequency, $\omega' = \omega_S$, since $\omega_S >> \Omega_B > \Gamma_B$ ($\omega_S \approx 1.7\cdot10^4\Omega_B$ in silica), Eq.(23) reduces to

$$n_g(\omega_S) = -\frac{2B\omega_S}{\delta\omega_\rho^2}e^{-\frac{\omega_S^2}{\delta\omega_\rho^2}}, \qquad (24)$$

which shows that the SBS-induced $n_g(\omega_S)$ is independent of both Γ_B and Ω_B, and is negligibly small for any reasonable bandwidth, $\delta\omega_\rho$, of $|\tilde{\rho}(\omega)|$. As such the acoustic resonance can have next to no effect in enhancing or modifying the natural group index in the medium for the Stokes signal. Consequently spectral broadening of pump radiation by any reasonable amount results in only minute increase of the spectral width of the material's resonant excitation in SBS, and furthermore because this resonance is far away from a Stokes optical signal frequency to be delayed the approach cannot be effective in modifying a natural group index of a medium.

This may also be seen using the following less rigorous simple argument. The resonant Brillouin frequency, Ω_B, that is the frequency of the acoustic wave, for Stokes radiation excited by a monochromatic pump radiation, is $\Omega_B = 2n v_s\omega_p/c$. Broadening of the pump radiation spectrum by $\Delta\omega_p$ then results in pump induced broadening of the acoustic wave spectrum, $\Delta\Omega_B = 2n v_s\Delta\omega_p/c$. It follows that for $\Delta\omega_p$ of ~$2\pi\cdot$(12-25) GHz (Zu et al., 2007; Song & Hotate, 2007), which is of order of Ω_B for SBS excited by 1.55 μm pump radiation in silica fiber ($n \cong 1.45$, $v_s \cong 6$ km/s), $\Delta\Omega_B \cong 2\pi\cdot$(0.6-1.5) MHz << $\Gamma_B \cong 2\pi\cdot16$ MHz. Clearly then pump induced broadening of the acoustic wave spectrum is negligible compared to its homogeneous spectral width, Γ_B, for any reasonable value of the pump spectrum width up to few tens of GHz.

4. Effect of acoustic wave inertia on Stokes pulse delay in SBS

In this section the underlying physical processes that give rise to Stokes pulse delay in SBS are addressed.

In typical SBS-based slow light experiments the CW pump power is kept below the value at which the SBS interaction experiences pump depletion. The pump power is therefore constant throughout the interaction length (in lossless media). This is also an underlying reason why the contribution of spontaneous scattering to the SBS interaction is considered sufficiently small to be ignored in theoretical treatments of this problem (Song et al., 2005; Okawachi et al., 2005; Zhu et al., 2005). Equations (4) and (5) with appropriate boundary conditions are therefore sufficient for describing the evolution of a Stokes pulse in a medium.

It is convenient to introduce the new temporal coordinate $t' = t - zn/c$ and suppose that the centre frequency of the Stokes pulse spectrum coincides with the resonant Brillouin Stokes frequency, that is $\delta\Omega = 0$. In terms of the new variables Eqs (4) and (5) can be rewritten as

$$-\frac{\partial E_S}{\partial z} = -i\frac{\omega_S}{2cn}\frac{\partial \varepsilon}{\partial \rho}(\rho(t'))^* E_p, \quad (25)$$

and

$$\frac{\partial \rho}{\partial t'} + \frac{1}{\tau}\rho = -i\rho_0 \frac{\partial \varepsilon}{\partial \rho}\frac{\Omega_B}{8\pi v_s^2} E_p E_S^*(t'). \quad (26)$$

This set of equations has an analytic solution, which can be obtained using Reimann's method (Bronshtein & Semendyaev, 1973). Consider the case addressed in typical SL experiments, in which the duration of the Stokes pulse is much less than its transit time in the medium and the pump is CW monochromatic radiation. Assuming that there are no acoustic waves in the medium before a Stokes pulse enters, and $E_S(t' \leq 0) = 0$, $E_S(z=0,t') = E_{S0}(t')$ and $E_p(z,t') \equiv E_p = \text{const}$, the solutions for the Stokes field and the density variation are then

$$E_S(z,t') = E_{S0}(t') + \frac{g\mathrm{I}_p z}{\tau}\int_0^{t'}\left(e^{-\frac{t'-\vartheta}{\tau}}\right) I_1\left(\sqrt{2g\mathrm{I}_p z\frac{t'-\vartheta}{\tau}}\right)\sqrt{2g\mathrm{I}_p z\frac{t'-\vartheta}{\tau}} E_{S0}(\vartheta)d\vartheta \quad (27)$$

and

$$\rho(t') = -i\rho_0 \frac{\partial \varepsilon}{\partial \rho}\frac{\Omega_B}{8\pi\tau v^2} E_p \int_0^{t'}\left(e^{-\frac{t'-\vartheta}{\tau}}\right) I_0\left(\sqrt{2g\mathrm{I}_p z\frac{t'-\vartheta}{\tau}}\right) E_{S0}^*(\vartheta)d\vartheta . \quad (28)$$

Here $\mathrm{I}_p = |E_p|^2$ is the pump radiation intensity in $[W/cm^2]$, $I_{0,1}(x)$ are the Bessel functions of imaginary argument x, and g is the SBS gain coefficient,

$$g = 10^7 \frac{\omega_S^2 \rho_0 \tau}{4nc^3 v_s}\left(\frac{\partial \varepsilon}{\partial \rho}\right)^2 [cm / W]. \quad (29)$$

Suppose that the input Stokes signal is an optical pulse, the time dependent intensity of which is given by

$$\mathrm{I}_S(z=0,t) = |E_S(z=0,t)|^2 = \mathrm{I}_{S0}(3.5t/t_p)^2 e^{-3.5t/t_p}, \quad (30)$$

where I_{S0} is the intensity at the peak of the pulse, t_p is the FWHM pulse duration. The shape of the pulse is shown in Figs. 7 and 8 (curves for $G = gI_pL = 0$) and it is a good approximation for pulses actually used in experiments (Pohl & Kaiser, 1970). Since the SBS exponential gain, G, in the fiber is supposed to be below the SBS threshold, this, according to (Kovalev & Harrison, 2007), limits G to ≤12 for standard silica fibers of > 0.1 km length.

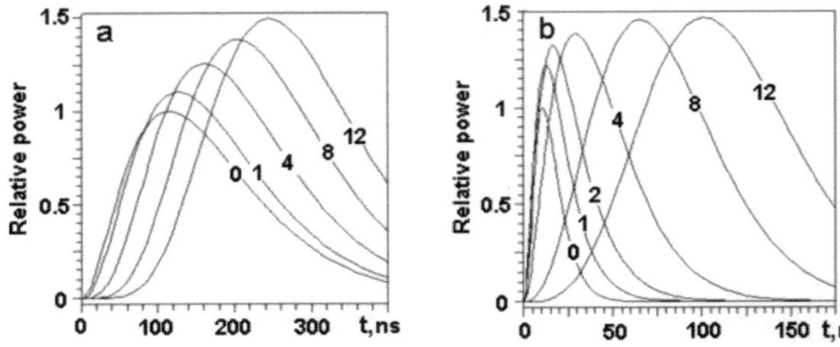

Fig. 7. Output Stokes pulse shapes for a), t_p = 200 ns (>>τ) and G = 0 (1), 1(0.42), 4(0.026), 8(0.0006), and 12(0.000012) and b), for t_p = 18 ns (=τ) and G = 0(1), 1(0.9), 2(0.6), 4(0.2), 8(0.09), and 12(0.00026), where numbers in brackets are the amplitude magnification factors. The numbers on curves are Gs.

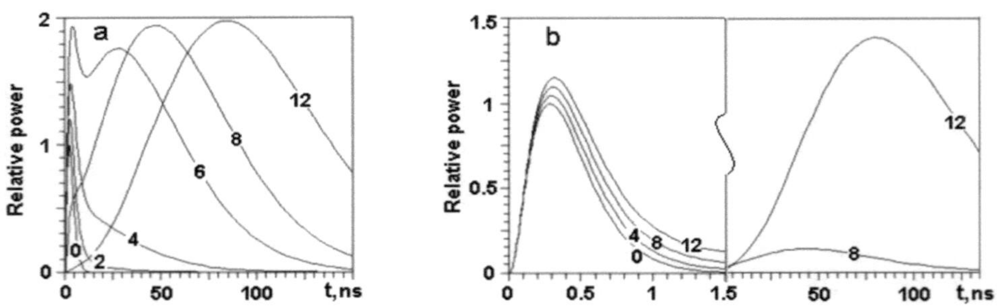

Fig. 8. Output Stokes pulse shapes for a), t_p = 4 ns (<τ) and G = 0(1), 2(1), 4(1), 6(1), 8(0.2), and 12(0.007), and b), for t_p = 0.5 ns (<<τ) and G = 0(1), 4(1), 8(1), and 12(1 at the first peak and 0.3 at the tail). LHS of b) is temporally stretched to show profiles.

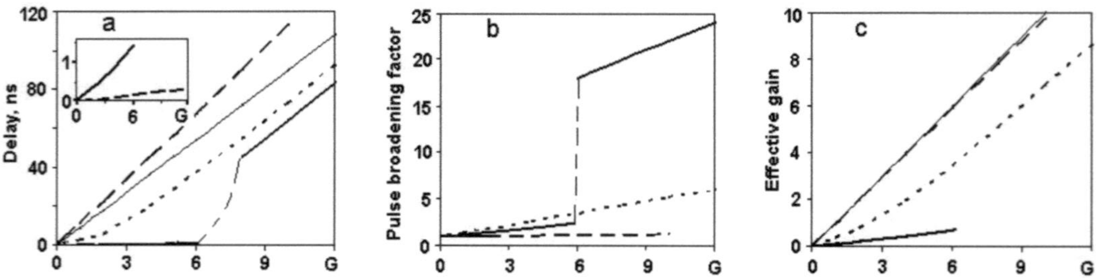

Fig. 9. Output Stokes pulse a), delay, b), broadening factor, and c), effective exponential gain at peak of output pulse vs gain G for pulse durations t_p = 200 (dashed), 18 (dotted), 4 (thick solid in the main graph and in the insert), and 0.5 ns (dashed in the inset). Thin solid lines in a) and c) are ΔT_d = 9G ns and G_{ef} = G dependencies respectively.

Figs 7 and 8 show the calculated relative output Stokes pulse powers, $P_S(t) = |E_S(t)|^2 S$ (S is the effective area of fiber-mode cross section), shapes, amplitudes and delays for four input pulse durations, t_p, and different G. Here the decay time, τ, of the hyper-sound wave in silica is taken to be $\tau = 18$ ns at a pump radiation wavelength ~1.55 μm. It should be emphasized that the decay time of the acoustic wave, which is determined by the viscosity of fused silica (Pinnow, 1970), is used in these calculations rather than the SBS spectral width. This is because the latter depends on intrinsic characteristics of the fiber (core/cladding design, doping type and concentration, numerical aperture, etc. and the ambient environment (mechanical, thermal and electromagnetic fields (see Sec. 3), all of which can result in substantial variation of the resonant Brillouin frequency, Ω_B. However they do not appreciably effect the viscosity of the medium.
It follows from Figs 7 and 8 that the induced delay of the output Stokes pulse, its duration and peak power, which is estimated to be $P_{S0}e^{G_{ef}}$, where G_{ef} is the effective SBS exponential gain, in all cases increase with increase of G. Rates of these growths depend substantially on the ratio of pulse duration to acoustic wave decay time, t_p/τ, as shown in Fig. 9.
For a long input Stokes pulse, $t_p = 200$ ns that is $t_p/\tau = 200/18 >> 1$, the output pulse, on increase of G, remains similar in form (Fig. 7(a)), and is increasingly delayed following $\Delta T_d/G \cong 11$ ns as shown by the thick dashed line in Fig. 9(a). Additional calculations have shown that it may coincide with $\Delta T_d \cong \tau G/2 \cong 9G$ ns (shown by the thin solid line in Fig. 9(a)), but this happens only when $t_p/\tau \cong 2.5\pm0.5$ and >100. For $t_p/\tau < 2$ the growth of ΔT_d falls below the value $\Delta T_d \cong 9G$ ns. The duration of the output pulse compared to that of the input is slightly broadened with increase of G (see dashed line in Fig. 9(b)), and G_{ef} decreases very slightly compared to G (dashed line in Fig. 9(c)).
For $t_p = 18$ ns (Fig. 7(b)), that is $t_p/\tau = 1$, ΔT_d for the output pulse again increases with G though not linearly; at lower G (0 to ~2) with slope $\Delta T_d/G \cong 3$ ns, and at higher G (> 2) with slope $\Delta T_d/G \cong 9$ ns (dotted line in Fig. 9(a)). The pulse broadening in this case increases substantially with G, by a factor ≥ 5 at G > 10 (dotted line in Fig. 9(b)) and G_{ef} decreases notably, by a factor of ~3 at lower G to ~1.5 at G ≥ 10 (dotted line in Fig. 9(c)).
For the short pulses, $t_p \leq 4$ ns (Fig. 8), that is for $t_p/\tau < 1$, significant new features appear. For G < 4 the output pulses approximately retain their shape with only a slight increase of ΔT_d with G ($\Delta T_d/G \cong 0.1$-0.15 ns for $t_p = 4$ ns and $\Delta T_d/G \cong 0.03$ ns for $t_p = 0.5$ ns, as is shown in inset of Fig. 9(a), solid and dashed lines respectively). In both cases G_{ef} for the leading peak decreases substantially, by a factor ~10. For G between 4 and 6 there is substantial growth of the power in the tail of the pulses and for G > 8 the maximum of the pulse shifts to the tail (see Figs. 8(a) and 8(b)). This is because of the long decay of the acoustic wave excited by the short Stokes pulse interacting with the CW pump. The dependence of ΔT_d on G for G > 8 then follows the linear relation $\Delta T_d \cong (9G - 25)$ ns (thick solid line in Fig. 9(a)). The pulse broadening factor is then ~20 for $t_p = 4$ ns (solid line in Fig. 9(b) at G > 6) and ~200 for t_p = 0.5 ns.
It is interesting to note that analytical results presented in Figs. 7 and 8, which are obtained in the small signal limit, give dependencies of pulse delay and broadening on G quite similar to these obtained numerically both for long pulses, $t_p/\tau \cong 15$ and 5 in (Zhu et al, 2005), and for short pulses, $0.1 < t_p/\tau < 2$ in (Kalosha et al., 2006). There are however some quantitative differences, which are important to highlight. The numerical modelling in (Zhu et al., 2005) gives $\Delta T_d = \tau G/2$ dependence for $t_p/\tau \cong 15$, while calculations here predict this

only for t_p/τ in the region ~2.5±0.5, while for $t_p/\tau > 5$ the slope $\Delta T_d/G$ is a factor 1.2-1.3 steeper. Also the shape of the ΔT_d upon G dependence for $t_p/\tau \cong 5$ in ((Zhu et al, 2005)) is similar to that in these calculations but for $t_p/\tau \cong 1$. As to the numerical results presented in (Kalosha et al., 2006), the difference with findings here is most probably because all their results were obtained for an input Stokes power well above that for onset of pump depletion (input Stokes power is of 1 mW compared with pump power of ≤ 20 mW).

To understand the underlying nature of the behaviour described above consider through Eq. (28) the temporal and spectral characteristics of the complex dielectric function variation in the medium, $\Delta\varepsilon = (\partial\varepsilon/\partial\rho)\rho = \Delta\varepsilon' + i\Delta\varepsilon''$, induced by the interaction of the pump and Stokes signals; $\Delta\varepsilon(\omega)$ results in the SBS gain and $\Delta\varepsilon'(\omega)$ is responsible for modification of the refractive index, $\Delta n(\omega) \cong \Delta\varepsilon'(\omega)/2n_0$, of the medium, where n_0 is the refractive index of a medium without SBS. In the limit of small gain, G < 1, the Bessel function $I_0(x)$ in Eq. (28) may be set to unity, and analysis is greatly simplified. While this approximation does not allow us to describe gain narrowing of the SBS spectrum typical for higher G, it still captures reasonably well the trends in the temporal and spectral features of ρ, which determine those of the SBS gain and modified refractive index.

When $I_0(x) = 1$, the integral in Eq. (28) can be taken for $E_{S0}(t)$ given by Eq. (30). It results in the following analytic expression for $\rho(t)$,

$$\rho(t) = A\left\{\frac{2}{(2b-\tau)^2}\left[2\tau b\left(e^{-\frac{t}{\tau}} - e^{-\frac{t}{2b}}\right) + (2b-\tau)te^{-\frac{t}{2b}}\right]\right\}, \tag{29}$$

where $b = t_p/3.5$, and $A = -i\rho_0 \frac{\partial\varepsilon}{\partial\rho}\frac{\Omega_B}{8\pi v^2}E_pE_S^*(0)$. The spectral characteristics of the acoustic wave amplitude follow from the Fourier transform of Eq.(31),

$$\rho(\omega) = A\sqrt{\frac{2}{\pi}}\frac{4\tau b}{(2b-\tau)^2}\left[\tau\frac{1+i\tau\omega}{1+(\tau\omega)^2} - 2b\frac{1+i2b\omega}{1+(2b\omega)^2}\right] + \sqrt{\frac{2}{\pi}}\frac{8b^2}{(2b-\tau)}\left[\frac{1+i4b\omega-(2b\omega)^2}{[1+(2b\omega)^2]^2}\right]. \tag{30}$$

The temporal dynamics of the induced acoustic wave amplitudes and their spectra for values of t_p/τ considered above are shown in Fig. 10 (the dynamics for t_p = 0.5 ns is not shown on the timescale of Fig. 10(a)). The curves in Fig. 10(a) represent different characteristic types of SBS interaction: curves 1 give an example of quasi-steady state interaction when $t_p/\tau >> 1$, curves 3 demonstrate transient type interaction when $t_p/\tau << 1$, and curves 2 are for an intermediate case when $t_p/\tau \cong 1$.

In the first case of a long Stokes pulse, that is $t_p/\tau >> 1$, the shape of the acoustic wave pulse shown in Fig. 10(a) as the solid curve 1 almost reproduces the shape of the input Stokes pulse (dotted curve 1). The spectrum of the excited acoustic wave in this case, Fig. 10(b), reproduces the spectrum of the input Stokes pulse, shown by the solid and dashed curves 1 respectively. These are both narrower than the Lorentzian-shaped spectrum corresponding to τ = 18 ns (dotted curve 5).

This is to be expected since Eq. (2) is in essence the equation for the amplitude of a driven damped oscillator; for such a system the spectrum of the induced oscillations is fully determined by the spectrum of the driving force when it is narrower than the reciprocal decay time of the oscillator.

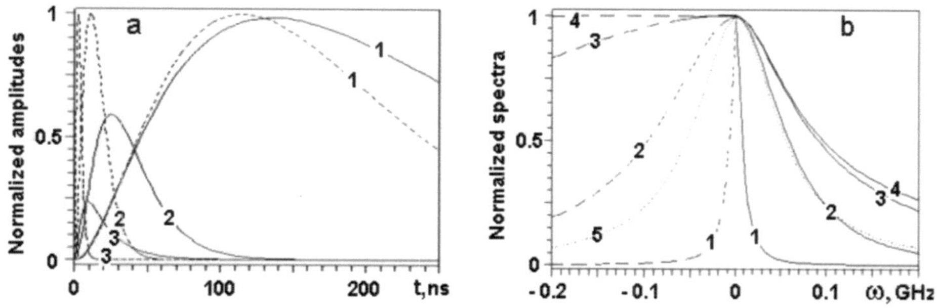

Fig. 10. a), Dynamics and b), spectra of the normalised acoustic wave amplitude (solid lines) for Stokes pulses of the shape given by Eq. (30) (dashed lines) with t_p = 200(1), 18(2), 4(3) and 0.5 ns(4), τ = 18 ns. The dotted line 5 in b), is the Lorentzian spectrum with τ = 18 ns. Curves in b) show half spectra of pulses at LHS and of corresponding medium's response at RHS.

In the case of shorter Stokes pulses, $t_p/\tau \leq 1$, (dashed curves 2 and 3 in Fig. 10(a)) the spectrum of the driving force, that is that of the input Stokes pulse, is broad band, (dashed curves 2 and 3 and also curve 4 for t_p = 0.5 ns pulse in Fig. 10(b)). As seen the dynamics of the medium's response (solid curves 2 and 3 in Fig. 10(a)) and its spectra (solid curves 2, 3 and 4 in Fig. 10(b)) differ substantially from those of the Stokes pulses and their spectra. In the temporal domain the maximum amplitude of the induced ρ(t) decreases with decrease of pulse duration and a long tail appears after a Stokes pulse, which decays exponentially with a decay time of τ. The spectra of ρ(t) in these cases are narrower, increasingly so for shorter pulses, than the spectra of the driving force (input Stokes pulse). Their width is then determined predominantly by the reciprocal decay time of the oscillator (dotted curve 5 in Fig. 10(b)) as is to be expected for a damped oscillator, which is driven by a broad-band force. Reduction of amplitude of the induced ρ(t) with decrease of pulse duration results in reduced amplitude of the output Stokes pulse and its effective G as shown in Fig. 9(c).

The imaginary part of $\rho(\omega)$ described by Eq. (13) gives the refractive index of the medium modified by the SBS interaction, $n(\omega) = n_0 + \Delta n(\omega) \cong n_0 + (\partial\varepsilon/\partial\rho)\rho(\omega)/2n_0$ and its corresponding group index, $n_g(\omega) = n_0 + \omega[d\Delta n(\omega)/d\omega]$ (Okawachi et al., 2005, Zhu et al., 2005). Spectra of the Stokes pulses and the group indices induced by these pulses are shown in Fig. 11 for the four different t_p/τ. It is important to remember (see section 3.2, Eqs (23) and (24)) that the group index profile is frequency shifted from the centre frequency of the Stokes pulse by ~$1.7\cdot10^4\Omega_B$ and therefore has negligible effect on the Stokes pulse delay. In the literature on SL via SBS, it is assumed that their centre frequencies coincide. However even if this were so it is clear from Fig. 11 that the spectral width of a Stokes pulse is bigger than that of the SBS induced group index regardless of t_p/τ. As such, in this hypothetical case, while some, central, part of the input spectrum may experience a group delay, other parts of the spectrum will experience group advancement or neither delay or advancement (see Figs. 11(c) and 11(d)).

It therefore follows that regardless of the pulse length of the input Stokes pulse the pulse delays associated with SBS amplification of a Stokes pulse, as described above, cannot be attributed to SBS induced group delay. They are predominantly a consequence of the phenomenon of SBS build-up.

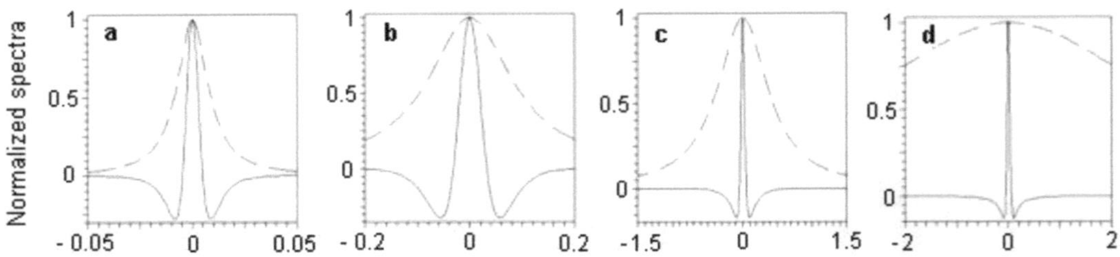

Fig. 11. Relation between the spectra of a Stokes pulse and of the group index induced by this pulse for t_p = 200 a), 18 b), 4 c), and 0.5 ns d). Horizontal scales are in GHz.

5. Conclusions

The results presented in this chapter raise the question of whether slow light, as first discussed in (Zeldovich, 1972), can be realised. To answer this recall the nature of the group delay effect. It is a linear phenomenon exhibited by a pulse propagating through a medium with normal dispersion of refractive index (Brillouin, 1960). The effect is greatly enhanced in the vicinity of a medium's resonance, which for a gain medium is normally dispersive. For SBS the maximum of this resonantly enhanced dispersion is centred around the Brillioun frequency, Ω_B, and it is all but negligible at the Stokes frequency. As such the acoustic resonance can have next to no effect in enhancing or modifying the natural group index in the medium for the Stokes signal. Consequently pulse delay associated with Stokes pulse induced SBS cannot be attributed to SBS induced group delay. It is predominantly a consequence of the phenomenon of SBS build-up, which arises from the inertia of the medium in responding to the optical fields. Also, spectral broadening of the pump radiation by any reasonable amount has next to no effect on the SBS spectral bandwidth of the excited acoustic wave in the medium, which is commonly believed to determine the SBS gain bandwidth (Thevenaz, 2008).

6. References

Agrawal, G. P. (2006). *Nonlinear Fiber Optics*, Academic Press, Boston.

Akhmanov, S. A., Drabovich, K. N., Sukhorukov, A. P. & Chirkin, A. S. (1971). Stimulated Raman scattering in a field of ultrashort light pulses, *Sov. Phys. JETP*, Vol. 32, 266-273.

Akhmanov, S. A., Dyakov, Yu. E. & Chirkin, A. S. (1988). *Introduction to Statistical Radiophysics and Optics*, Springer, Berlin.

Azuma, Y., Shibata, N., Horiguchi, T. & Tateda M. (1988). Wavelength dependence of Brillouin-gain spectra for single-mode fibres, *Electronics Letters*, Vol. 24, 250-252.

Basov, N.G., Ambartsumian, R.V., Zuev, V.S., Kryukov, P.G. & Letokhov, V.S. (1966). Nonlinear amplification of a light pulse. *Sov. Phys. JETP*, Vol. 23, 16-24.

Boyd, R. W. & Gauthier, D. J. (2002). "Slow" and "fast" light, *Progress in Optics*, Vol. 43, 497-530.

Brillouin, L. (1960). *Wave Propagation and Group Velocity*, Academic Press, New York.

Bronshtein, I. N. & Semendyayev, K. A. (1973). *A guide book to mathematics*. Verlag Harri Deutsch, Zurich.

Carman, R. L., Shimizu, F., Wang, C. S. & Bloembergen, N. (1970). Theory of Stokes pulse shapes in transient Raman scattering, *Phys. Rev. A*, Vol. 2, 60-72.

Cheng, A., Fok, M. P. & Shu, C. (2008). Wavelength-transparent, stimulated-Brillouin-scattering slow light using cross-gain-modulation-based wavelength converter and Brillouin fiber laser, *Optics Letters*, Vol. 33, 2596-2598.

Chin, S., Herraez, M. G. & Thevenaz, L. (2006). Zero-gain slow and fast light propagation in an optical fiber, *Optics Express*, Vol. 14, 10684-10692.

Gauthier, D. J. (2005). Slow light brings faster communications, *Physics World*, Vol. 18, 30-32.

Hau, L. V., Harris, S. E., Dutton, Z. & Behroozi, C. H. (1999). Light speed reduction to 17 meters per second in an ultracold atomic gas, *Nature*, Vol. 397, 594-596.

Heiman, D., Hamilton, D. S. & Hellwarth, R. W. (1979). Brillouin scattering measurements in optical glasses, *Phys. Rev.* Vol.19, 6583-6592.

Herraez, M. G., Song, K. Y. & Thevenaz, L. (2006). Arbitrary-bandwidth Brillouin slow light in optical fibers, *Opt. Express*, Vol. 14, 1395-1400.

Kalosha, V. P., Cheng, L. & Bao, X. (2006). Slow and fast light via SBS in optical fibers for short pulses and broadband pump, *Optics Express*, Vol. 14, 12693-12703.

Knight, J. C., Arriaga, J., Birks, T. A., Ortigosa-Blanch, A., Wadsworth, W. J. & Russel, P. St. (2000). Anomalous dispersion in photonic crystal fibre, *IEEE Photonics Technology Letters*, Vol. 12, 807-809.

Korn, G. A. & Korn, T. M. (1967). *Manual of mathematics*, McGraw-Hill, New York.

Kovalev, V. I. & Harrison, R. G. (2000). Observation of inhomogeneous spectral broadening of stimulated Brillouin scattering in an optical fiber, *Phys. Rev. Lett.* Vol. 85, 1879-1882.

Kovalev, V. I. & Harrison, R. G. (2002). Waveguide-induced inhomogeneous spectral broadening of stimulated Brillouin scattering in optical fiber, *Optics Letters*, Vol. 27, 2022-2024.

Kovalev, V. I. & Harrison, R. G. (2005). Temporally stable continuous-wave phase conjugation by stimulated Brillouin scattering in optical fiber with cavity feedback, *Optics Letters*, Vol. 30, 1375-1377.

Kovalev V. I. & Harrison, R. G. (2007). Threshold for stimulated Brillouin scattering in optical fiber, *Optics Express*, Vol. 15, 17625-17630.

Kovalev, V. I., Harrison, R. G., Knight, J. C. & Kotova, N. E. (2008). Waveguide induced spectral bandwidth enhancement of slow light group index caused by stimulated Brillouin scattering in optical fiber," *Laser and Particle Beams*, Vol. 26, 319-322.

Kovalev, V. I., Kotova, N. E. & Harrison, R. G. (2009). Effect of acoustic wave inertia and its implication to slow light via stimulated Brillouin scattering in an extended medium, *Optics Express*, Vol.17, 2826-2833.

Kroll, N. (1965). Excitation of hypersonic vibrations by means of photoelastic coupling of high-intensity light waves to elastic waves, *J. Appl. Phys.* Vol. 36, 34-43.

Le Floch, S. & Cambon, P. (2003). Study of Brillouin gain spectrum in standard single-mode fiber at low temperatures (1.4-370 K) and high hydrostatic pressures (1-250 bars), *Optics Communications*, Vol. 219, 395-410.

Lu, Z., Dong, Y. & Li, Q. (2007). Slow light in multi-line Brillouin gain spectrum, *Optics Express*, Vol. 15, 1871-1877.

Minardo, A., Bernini, R. & Zeni, L. (2006). Low distortion Brillouin slow light in optical fibers using AM modulation, *Optics Express*, Vol. 14, 5866-5876.

Okawachi, Y., Bigelow, M. S., Sharping, J. E., Zu, Z., Schweinsberg, A., Gauthier, D. J., Boyd, R. W. & Gaeta, A. L. (2005). Tunable all-optical delays via Brillouin slow light in an optical fiber, *Phys. Rev. Lett.* Vol. 94, 153902.

Pant, R., Stenner, M. D., Neifeld, M. A. & Gauthier, D. J. (2008). Optimal pump profile designs for broadband SBS-based slow light systems, *Optics Express*, Vol. 16, 2764-2777.

Pinnow, D. A. (1970). Guide lines for the selection of acoustooptic materials, *IEEE J. Quantum Electron.* Vol. QE-6, 223-238.

Pohl D. & Kaiser, W. (1970). Time-resolved investigations of stimulated Brillouin scattering in transparent and absorbing media: Determination of phonon lifetimes, *Phys. Rev. B*, Vol.1, 31-43.

Ren L. & Tomita, Y. (2008). Reducing group-velocity-dispersion-dependent broadening of stimulated Brillouin scattering slow light in an optical fiber by use of a single pump laser, *J. Opt. Soc. Am. B*, vol. 25, 741-746.

Sakamoto, T., Yamamoto, T., Shiraki, K. & Kurashima, T. (2008). Low distortion slow light in flat Brillouin gain spectrum by using optical frequency comb, *Optics Express*, Vol. 16, 8026-8032.

Schneider, T., Junker, M. & Lauterbach, K.-U. (2006). Potential ultra wide slow-light bandwidth enhancement, *Optics Express*, Vol. 14, 11082-11087.

Schneider, T., Henker, R., Lauterbach, K.-U. & Junker, M. (2008). Distortion reduction in Slow Light systems based on stimulated Brillouin scattering, *Optics Express*, Vol. 16, 8280-8285.

Shi, Z., Pant, R., Zhu, Z., Stenner, M. D., Neifeld, M. A., Gauthier, D. J. & Boyd, R. W. (2007). Design of a tunable time-delay element using multiple gain lines for increased fractional delay with high data fidelity, *Optics Letters*, Vol. 32, 1986-1988.

Shibata, N., Waarts, R. G. & Braun R. P. (1987). Brillouin gain spectra for single-mode fibers having pure-silica GeO_2-doped, and P_2O_5-doped cores, *Optics Letters*, Vol. 12, 269-271.

Shibata, N., Okamoto, K. & Azuma Y. (1989). Longitudinal acoustic modes and Brillouin-gain spectra for GeO_2-doped-core single-mode fibers, *JOSA B*, Vol. 6, 1167-1174.

Shiraki, K., Ohashi, M. & Tateda, M. (1995). Suppression of stimulated Brillouin scattering in a fiber by changing the core radius, *Electronics Letters*, Vol. 31, 668-669.

Shumakher, E., Orbach, N., Nevet, A., Dahan, D. & Eisenstein, G. (2006). On the balance between delay, bandwidth and signal distortion in slow light systems based on stimulated Brillouin scattering in optical fibers, *Optics Express*, Vol. 14, 5877-5884.

Song, K. Y., Herraez, M. G. & Thevenaz, L. (2005). Gain-assisted pulse advancement using single and double Brillouin gain peaks in optical fibers, *Optics Express*, Vol. 13, 9758-9765.

Song K. Y. & Hotate, K. (2007). 25 GHz bandwidth Brillouin slow light in optical fibers, Optics Letters, Vol. 32, 217-219.

Stenner, M. D., Neifeld, M. A., Zu, Z., Dawes, A. M. C. & Gauthier D. J. (2005). Distortion management in slow-light pulse delay, *Opt. Express*, Vol. 13, 9995-10002.

Thevenaz, L. (2008). Slow and fast light in optical fibres, *Nature Photonics*, Vol. 2, 474-481.

Tkach, R. W., Chraplyvy, A. R. & Derosier, R. M. (1986). Spontaneous Brillouin scattering for single-mode optical-fibre characterisation, *Electronics Letters*, Vol. 22, 1011-1013.

Tsun, T.-O., Wada, A., Sakai, T. & Yamauchi, R. (1992). Novel method using wight spectral probe signals to measure Brillouin gain spectra of pure silica core fibres, *Electronics Letters*, Vol. 28, 247-249.

Wang, S., Ren, L., Liu, Y. & Tomita, Y. (2008). Zero-broadening SBS slow light propagation in an optical fiber using two broadband pump beams, *Optics Express*, Vol. 16, 8067-8076.

Yi, L., Zhan, L., Hu, W. & Xia, Y. (2007). Delay of broadband signals using slow light in stimulated Brillouin scattering with phase-modulated pump, *IEEE Photon. Technol. Letters*, Vol. 19, 619-621.

Yoshizawa, N., Horiguchi, T. & Kurashima, T. (1991). Proposal for stimulated Brillouin scattering suppression by fibre cabling, *Electronics Letters*,Vol. 27, 1100-1101.

Yoshizawa, N. & Imai, T. (1993). Stimulated Brillouin scattering suppression by means of applying strain distribution to fiber with cabling, *Journal of Lightwave Technology*, Vol. 11, 1518-1522.

Zadok, A., Eyal, A. & Tur, M. (2006). Extended delay of broadband signals in stimulated Brillouin scattering slow light using synthesized pump chirp, *Optics Express*, Vol. 14, 8498-8505.

Zeldovich, B. Ya. (1972). Time of establishment of stationary regime of stimulated light scattering, *JETP Lett.* Vol. 15, 158-159.

Zeldovich, B. Ya., Pilipetskii N. F. & Shkunov V. V. (1985). *Principles of phase conjugation*, Springer Verlag, Berlin.

Zhang, B., Yan, L., Fazal, I., Zhang, L., Willner, A. E., Zhu, Z. & Gauthier, D. J. (2007)-1. Slow light on Gbit/s differential-shift-keying signals, *Optics Express*, Vol. 15, 1878-1883.

Zhang, B., Zhang, L., Yan, L.-S., Fazal, I., Yang, J.-Y. & Willner, A. E. (2007)-2. Continuously-tunable, bit-rate variable OTDM using broadband SBS slow-light delay line, *Optics Express*, Vol. 15, 8317-8322.

Zhu, Z., Gauthier, D. J., Okawachi, Y., Sharping, J. E., Gaeta, A. L., Boyd, R. W. & Willner, A. E. (2005). Numerical study of all-optical slow-light delays via stimulated Brillouin scattering in an optical fiber, *J. Opt. Soc. Am. B*, Vol, 22, 2378-2384.

Zhu Z. & Gauthier, D. J. (2006). Nearly transparent SBS slow light in an optical fiber, *Optics Express*, Vol. 14, 7238-7245.

Zhu, Z., Dawes, A. M. C., Gauthier, D. J., Zhang, L. & Willner, A. E. (2007). Broadband SBS slow light in an optical fiber, *J. Lightwave Technol*. Vol. 25, 201-206.

6

Bismuth-doped Silica Fiber Amplifier

Young-Seok Seo and Yasushi Fujimoto
Korea Atomic Energy Research Institute
Korea

1. Introduction

To accommodate rapidly increasing quantities of information and communication, optical fiber transmission technology with high-speed high-capacity is demanded. The development of an optical gain medium and fiber amplifiers to cover the 1250~1650 nm region, which is the entire optical telecommunication windows of silica fiber, becomes an important issue for ultra-wide broadband optical communication. Bismuth-doped glasses exhibit a broadband luminescence in the near infrared region. Thus, they are potential gain media for extending the spectral bandwidth of the current erbium-doped silica fiber amplifiers. There are several reports on an infrared luminescence from bismuth-doped glasses such as germanate, phosphate, borate et al. (Meng et al., 2005ab; Peng et al., 2005abc; Suzuki & Ohishi, 2006). According to their research, bismuth-doped glasses are therefore very promising for creating broadband amplifiers for fiber telecommunication lines and tunable or femto-second lasers.
There are two useful wavelengths for optical communication. One is the 1550 nm, erbium-doped fiber amplifiers (EDFAs) working wavelength, which has minimum losses. EDFA developments within the third telecommunication window have contributed to the rapid growth of wavelength division multiplexing (WDM) transmission systems. The L-band (1570~1605 nm) of EDFA can be used in WDM systems in conjunction with C-band (1530~1560 nm). However, efforts to use WDM techniques to exploit this capability have been hampered by nonlinear fiber effects such as four-wave mixing. In addition, the amplification bandwidth of silica-based EDFA is as small as ~70 nm (Yamada et al., 1998). A broadband amplifier with a gain bandwidth of more than 70 nm was reported by the integration of EDFA with thulium-doped fiber amplifiers or fiber Raman amplifiers (FRAs) (Yamada et al., 1998).
The other useful wavelength (1300 nm) for optical communication is O-band, between 1260 and 1360 nm, which is the natural zero-dispersion region of silica glass fiber where the temporal distortion of transferred optical pulses can be minimized. For example, praseodymium-doped fluoride fiber amplifiers (PDFFAs) and FRAs have been successfully used for 1300 nm amplification, but it also suffer from narrow bandwidth (~25 nm) in operating wavelength and low efficiency (Miyajima et al., 1991; Whitley, 1995). In addition, a PDFFA normally made from fluoride glass, which is very brittle and cannot be fusion spliced to the silica glass fiber. An alternative core fiber material and fiber amplifier for 1300 nm amplification is expected.

High-average-power lasers are widely used for material processing, thin film fabrication with ablation, generation of extreme ultraviolet light sources for lithography, and various scientific research purposes. As there are so many applications of high-power lasers, demand is growing rapidly for more powerful lasers; therefore, the requirements for thermally resistant laser media are becoming tougher. Some of the absorption energy created in a laser medium by excitation with flash lamps or laser diodes is extracted as an output of laser energy, but most of the absorption energy generates heat accumulation in the medium. This heat accumulation causes a thermal lensing effect and a thermal birefringence in the medium; of course, the excess heat accumulation leads to destruction of the medium. Therefore, laser media must be operated under strict conditions during excitation.

Silica glass is one of the most attractive materials for high-peak-power or high-average-power lasers because it has favourable thermal and mechanical toughness, high optical transmittance from the ultraviolet to the infrared regions, and a low nonlinear refractive index compared to the other commercial laser glasses. These properties are indispensable to the laser driver for inertial confinement fusion because phosphate laser glasses have such weak thermal shock toughness that they cannot work at a high repetition rate (Fujimoto et al., 1999).

Bismuth-doped silica glass (BiSG) is a new material that emits a broadband fluorescence peak at around 1250 nm with a bandwidth over 300 nm. We suggested BiSG which can complement the problems of the other optical amplifiers (Fujimoto & Nakatsuka, 2001; 2003). It is important to understand the optical properties of the BiSG for technological applications such as optical fiber amplifier and fiber laser. It is well known that bismuth ions contribute to increase the refractive index of the glass and give no specific absorption band in the visible and near infrared region. However, BiSG has three excitation bands at 500, 700 and 800 nm showing the luminescent bands at 750, 1120 and 1250 nm respectively. Its 800 nm absorption band makes this material have a potential to be pumped by commercialized powerful laser diodes. In addition, cw lasing has been obtained in the spectral region between 1150 and 1300 nm in a bismuth-doped aluminosilicate glass fiber (Dianov et al., 2005; Dvoyrin et al, 2006; 2007).

We previously reported a new infrared luminescent BiSG that had a possibility of being a high-power laser material and a possibility of being an optical fiber amplifier. Also, we achieved optical amplification in a BiSG and bismuth-doped silica fiber(BiDF) at 1310 nm with 810-nm excitation. BiSG has many attractive features, which make it suitable as a core fiber material of an optical fiber. The near-infrared spectral regions with a wide luminescence in the range from 1000 ~ 1600 nm and a long lifetime of about 100~600 μs of luminescence make such a fiber promising for the development of lasers and amplifiers.

In this chapter, we demonstrate an optical amplification at the 1260~1360 nm region (O-band) in various BiSG and fiber. The optical amplification was observed at five different wavelengths between 1260 and 1360 nm with 810 nm excitation. The optical gain profile is similar to the fluorescence spectrum and the amplification bandwidth is greater than 75 nm in the 1300 nm region. The laser diode pumped fiber amplifier at 1310 nm showed gain characteristics with a 5.0, 8.0-cm length BiDF and a wide-band tuned amplification through an over 100-nm bandwidth. We also report on a simultaneous amplification results for BiSG and fiber at two different wavelengths in the 1300 nm region. Simultaneous amplification of 25 nm bandwidth in four different wavelength regions was obtained. This technique can be useful for WDM optical amplifiers at the second telecommunications window.

2. Optical amplification in a BiSG

2.1 Sample preparation and experimental setup

The sample compositions (mol %) of BiSG are shown in Table 1. In the case of sample A, bismuth-oxide (Bi_2O_3; 99.9 %), aluminum-oxide (Al_2O_3; 99.99 %) and silica powder (SiO_2; 99.8 %) were mixed and kneaded in a mortar. The mixture compound of Bi_2O_3: 1.0 mol %, Al_2O_3: 7.0 mol %, and SiO_2: 92 mol % was put in an alumina crucible and then melted at 1750 °C for 50 hour in the air atmosphere. In the case of sample B, lithium oxide (Li_2O) was added and the mixture was put in an alumina crucible and then melted at 1700 °C for 30 hours. The composition of the sample B was Bi_2O_3 (1.0 mol %), Al_2O_3 (7.0 mol %), SiO_2 (92-x mol %) and Li_2O: (x = 1, 5, and 10 mol %). The samples were cut in the pieces (0.5 cm × 0.5 cm × 0.24 cm) and optically polished. The color of the prepared samples was reddish brown but transparent, with a transmittance of ~90 % at 1300 nm region.

Sample	Bi_2O_3 (mol %)	Al_2O_3 (mol %)	SiO_2 (mol %)	Li_2O (mol %)
A	1	7	92	
	1	7	91	1
B	1	7	87	5
	1	7	82	10

Table 1. Glass composition of bismuth-doped silica glass and added in lithium oxide, Li_2O.

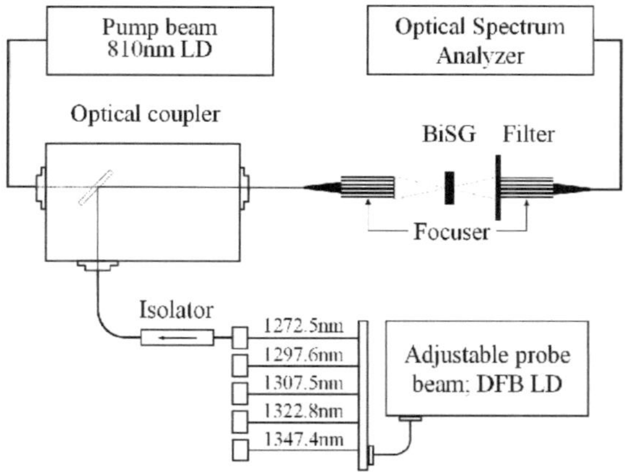

Fig. 1. The experimental setup for the optical amplification in a bismuth-doped silica glass.

The experimental setup for optical gain measurement is shown in Fig. 1. The 810 nm semiconductor laser (Unique Mode: UM2500-50-15) was used as an excitation source. Distributed feedback (DFB) semiconductor lasers (Mitsubishi Electric Corp.: FU-436SDF-EW41Mxx) with five different wavelengths were used as probe beams. The wavelengths of the five DFB semiconductor lasers were 1272.5, 1297.6, 1307.5, 1322.8 and 1347.4 nm, and these probe beams can be adjusted using a single laser driver. The probe laser beams were combined with the excitation beam using an optical coupler (OFR: 810/1310 nm), and then focused into the multi-mode fiber (MMF) with an output focusing lens (focuser). The

combined beam was focused into the BiSG sample surface by the focuser and the probe beam was detected by an optical spectrum analyzer (Ando: AQ6317B) using a pigtail focuser. The excitation beam is manually chopped to make the state of a probe beam without excitation I_0 and a probe beam with excitation I. The optical gain coefficient was defined as $g = (1/l)\ln(I/I_0)$, where l is the length of the BiSG sample.
The transmittance and the fluorescence spectra were measured using U-4100 spectrometer (Hitachi Ltd.) and SS-25 spectrometer (JASCO Corp.) with 150 W Xe lamp excitation. The transmittance of the new sample in the region of 1000 to 2500 nm was 90 % at maximum, which is higher than previous result (Fujimoto & Nakatsuka, 2001). The fluorescence spectra of BiSG samples with 800 nm excitation have a peak at ~1250 nm and the broad emission with a FWHM of ~300 nm. The strong absorption at 300 nm is considered to be the absorption edge of the Bi_2O_3-containing glass system (Sugimoto et al, 1996). However, we could not find any luminescence derived from previously reported Bi^{3+} luminescence and the infrared emission was hardly observed from the sample without aluminum (Parke & Webb, 1973; Weber & Monchamp, 1973).

2.2 Gain characteristics of the bulk-type BiSG

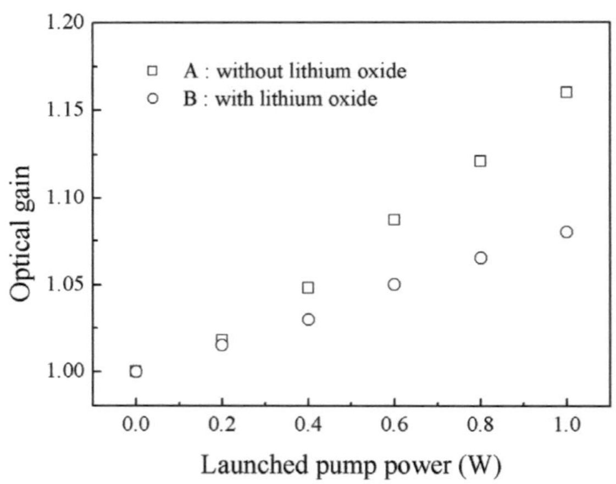

Fig. 2. Optical gain profiles of each sample at thickness 0.24cm.

We have measured the gain characteristics of the BiSG samples over a useful part of the 1300 nm optical communication window. The probe beam (1307.5 nm) and the excitation beam (810 nm) were focused onto the sample surface. Between the two MMF pigtail focuser, we measured amplified signals in BiSG samples with the maximum launched pump power (1.0 W) at 810 nm. Figure 2 shows the optical gain profile of two samples at 1307.5 nm. The optical gain increases linearly with excitation power up to 1.0 W. In the case of sample A, the maximum optical gain and the gain coefficient was 1.16 and 0.62 cm^{-1}, respectively. The absorbed energy was calculated as 42 % of excitation power, which is 0.42 W at 1.0 W excitation for a 0.24 cm thick sample. If the excitation length such as fiber core material becomes longer, the absorption will increase then the gain will be larger than that is demonstrated in this experiment. In the case of sample B, the maximum optical gain and the gain coefficient was 1.08 and 0.32 cm^{-1}, respectively.

The spectral dependence of the optical gain is important if optical fiber amplifier is to be incorporated in broadband WDM system. As shown in Fig. 3, the measured spectral dependence of the optical gain shows a close resemblance to the fluorescence spectrum, as shown by the dotted line. The optical gain was observed at five wavelengths, 1272.5, 1297.6, 1307.5, 1322.8 and 1347.4 nm. The optical gain spreads widely between 1260 and 1360 nm. A maximum optical gain and gain coefficient was 1.18 and 0.70 cm^{-1} respectively at the 1272.5 nm wavelength region, which is closed to the peak wavelength of the fluorescence spectrum. Moreover, the broad fluorescence spectrum suggests that optical gain in excess of 1.09 (0.35 cm^{-1}) should be available over a bandwidth as large as ~75nm (Seo et al., 2006a).

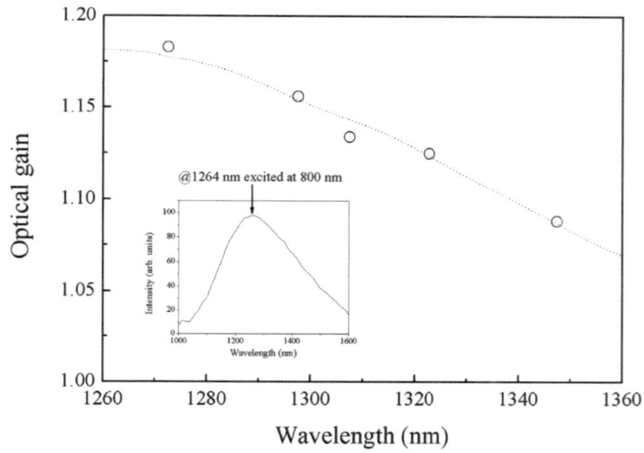

Fig. 3. Optical gain as a function of different signal wavelength 1272.5, 1297.6, 1307.5, 1322.8 and 1347.4 nm. Points and curve represents experimental measurements and fluorescence spectrum at 800 nm excitation, respectively.

3. Optical amplification in a 6.5-cm-long BiSG

3.1 Sample preparation and experimental setup

The composition of this sample was Bi_2O_3 (1.0 mol %), Al_2O_3 (7.0 mol %), and SiO_2 (92.0 mol %). The BiSG can be drawn by pulling the melted material upward using an alumina bar. The fabricated sample was cut in the same shape of 6.5-cm-long with an elliptical cross section (250 × 450 μm). The BiSG was optically polished at both ends to provide good optical transmission. The experimental setup for optical amplification in a BiSG is shown in Fig. 4. For the probe beam, a 1308 nm distributed feedback (DFB) semiconductor laser (Afonics: FXP0034, peak at 1308 nm) and four DFB semiconductor lasers (Mitsubishi Electric Corp.: FU-436SDF-EW41Mxx) with different wavelengths 1272.5, 1297.6, 1322.8, and 1347.4 nm respectively, were combined using a singlemode fiber (SMF) WDM coupler. And the probe beams can be adjusted using a single driver. The DFB semiconductor laser of 1308 nm was used as the anchor wavelength. The probe laser beam was combined with the excitation beam in the optical coupler (OFR: 810/1310 nm) and then focused into the MMF with an output focusing lens (focuser). The combined beam was focused into the BiSG sample by the focuser, and the gained probe beam was detected by an optical spectrum analyzer (Ando: AQ6317B) using another pigtail focuser.

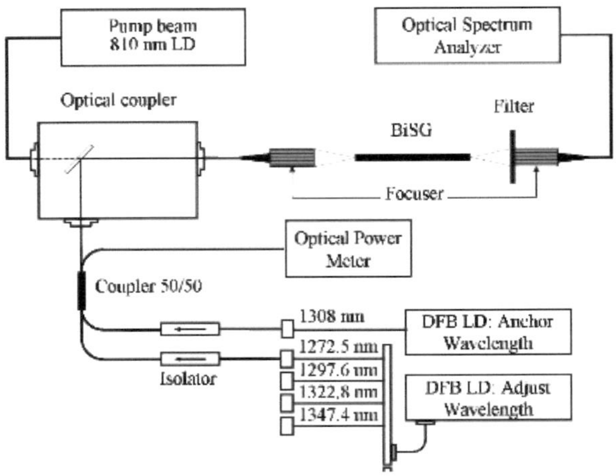

Fig. 4. Experimental setup for optical amplification in a 6.5-cm-long bismuth-doped silica glass

3.2 Gain characteristics of the rod-type BiSG

We measured the gain characteristics of the BiSG at 1300 nm optical communication window. By pigtail focuser, the probe beam (1308 nm) and excitation beam (810 nm) were focused onto the sample surface. The maximum optical gain was calculated to be 4.79 (6.80 dB) and therefore, gain coefficient g was 0.24 cm^{-1}. The optical gain increases linearly with excitation power up to 360 mW. The measured signal input/output power for BiSG was -8.3 dBm / -1.5 dBm (@ 1308 nm), exhibiting much lower power conversion efficiency. The optical gain in the previous result with a bulk-type BiSG was 1.19 with excitation power of 2.0 W, though the sample thickness was 0.26 cm (Fujimoto & Nakatsuka, 2003). Because the optical gain coefficient of bulk sample was 0.13 cm^{-1} at 360 mW excitation power, the fiber shape clearly affects the gain increment due to beam mode matching between the pump and the probe. If the excitation length is longer, absorption power will increase greatly, and then the gain will be larger than that demonstrated in this experiment.

Between the two MMF pigtail focusers, we measured optical amplification in the BiSG with a threshold of 360 mW launched pump power at 810 nm. Figure 5 shows simultaneous amplification at two wavelengths near 1300 nm second telecommunication window. The wavelength of the excitation beam was 810 nm, and the sample length was 6.5 cm. The probe wavelengths to measure the simultaneous amplification in the BiSG sample were adjustable wavelengths (1272.5 nm, 1297.7 nm, 1322.8 nm and 1347.4 nm) and anchor wavelength (1308 nm). For the four cases of amplification experiments, the maximum optical gain at these wavelengths were 3.30 (5.19 dB), 3.18 (5.02 dB), 3.05 (4.84 dB) and 2.87 (4.58 dB) respectively and 3.12 (4.94 dB) at the anchor wavelength was obtained. The optical gain shown in simultaneous measurements of two wavelengths suggests that it is possible to realize WDM optical fiber amplifiers in O-band (1260~1360 nm) by using this gain material.

The spectral dependence of optical gain is an important consideration if a fiber amplifier is to be incorporated in a broadband WDM system. As shown in Fig. 6, the measured spectral dependence of the optical gain coefficient appears to closely resemble the fluorescence

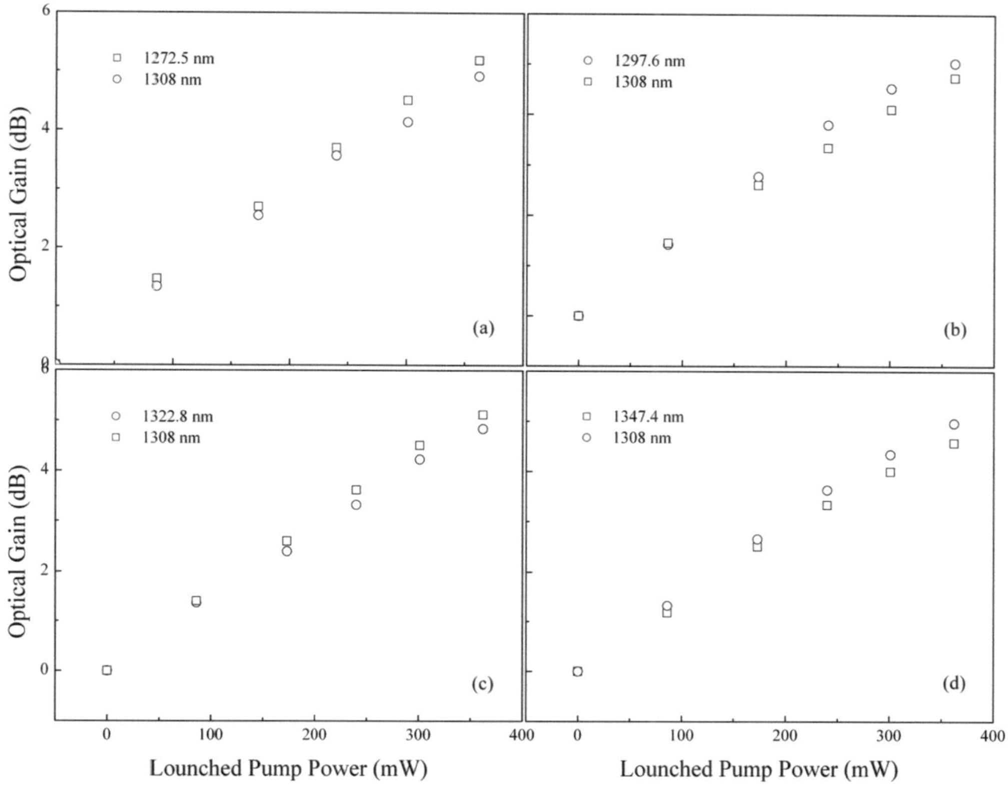

Fig. 5. Simultaneously optical amplification properties at two wavelengths of 1300 nm range: (a) 1308 nm and 1272.5 nm; (b) 1308 nm and 1297.6 nm; (c) 1308 nm and 1322.8 nm; and (d) 1308 nm and 1347.4 nm.

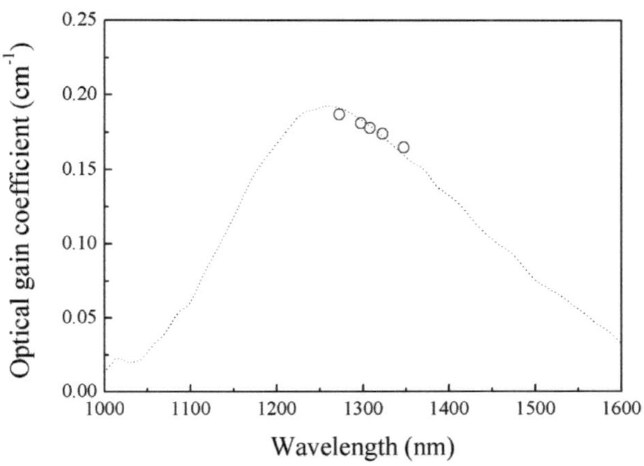

Fig. 6. Optical gain coefficient as a function of different signal wavelengths: 1272.5, 1297.6, 1308.0, 1322.8 and 1347.4 nm. Points and curve represent experimental measurements and fluorescence spectrum at 800 nm excitation, respectively.

spectrum, as represented by a dotted line. The optical gain spread widely from 1250 to 1360 nm. The maximum optical gain was obtained at 1272.5 nm. This wavelength is close to the peak wavelength of the fluorescence spectra. Moreover, broad fluorescence spectrum suggests that optical gain coefficient exceeding 0.165 cm^{-1} should be available over a bandwidth as large as 75 nm (Seo et al., 2006b).

4. Optical amplification in a multi-mode BiDF

4.1 Sample preparation and experimental setup

BiDF sample was fabricated by modified rod-in-tube method. The mixture compound of this sample was Bi_2O_3: 1.0 mol %, Al_2O_3: 7.0 mol %, and SiO_2: 92 mol %. Glass core was inserted into a tube of glass cladding to form a preform, which was drawn by heating in the drawing furnace. The fiber with refractive index difference, Δn~0.017, was drawn from the perform. The 8.0 cm long sample was optically polished at both ends to provide good optical transmission. Its core diameter and outer diameter was 13 μm and 230 μm, respectively.

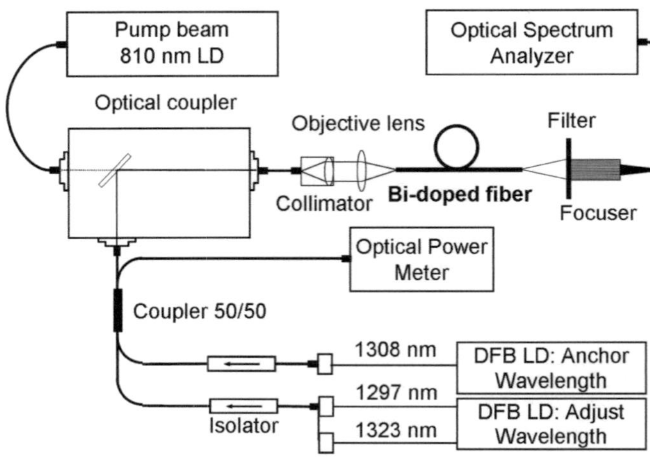

Fig. 7. Schematic diagram for optical gain measurement in bismuth-doped silica fiber.

The experimental setup for optical gain measurement in BiDF is shown in Fig. 7. An 810-nm CW laser diode was used as the pump source. A 1308 nm distributed feedback (DFB) laser diode and two laser diodes with different wavelengths were used as probe beam. The wavelengths of the two laser diodes were 1297 nm and 1323 nm, respectively, and these probe beams can be adjusted using a single driver. The probe laser beam was combined with the excitation beam in the optical coupler (OFR 810/1310 nm), and focused on an end of the fiber by an objective lens. The amplified probe signal was observed with an optical spectrum analyzer, and an optical low path filter cut out the unused pump light.

4.2 Gain characteristics of the multi-mode BiDF

Glass core was the optical amplification in an 8.0 cm long BiDF sample at single wavelength, 1308 nm, is shown in Fig. 8. The maximum optical gain was calculated to be 3.76 (5.8 dB) and therefore, gain coefficient was 0.038 dB/mW. The optical gain increased linearly with

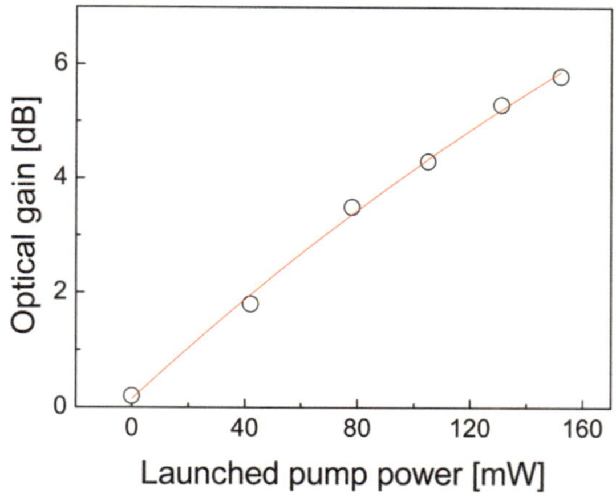

Fig. 8. Optical gain profile of an 8.0 cm long bismuth-doped silica fiber.

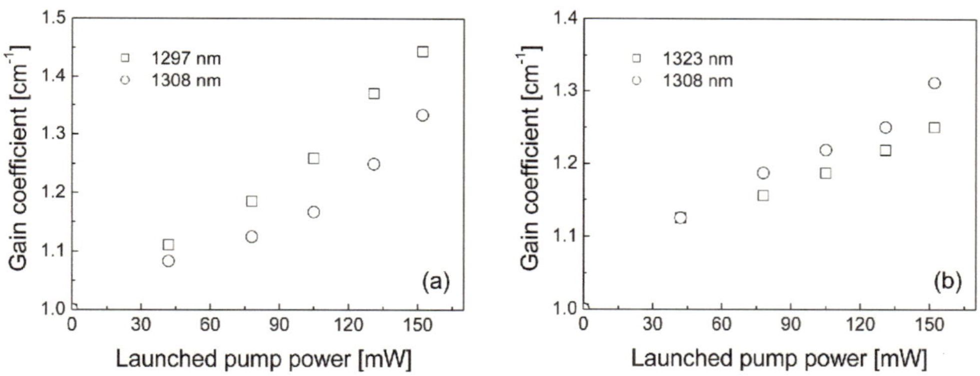

Fig. 9. Simultaneous optical amplification properties at two wavelengths of 1300-nm range: (a) 1308 nm and 1297 nm; (b) 1308 nm and 1323 nm.

excitation power up to 152 mW. The measured signal input/output power for BiDF was -29 dBm / -23 dBm (@ 1308 nm), exhibiting much lower power conversion efficiency. The optical gain in the previous result with a bulk-type BiSG was 1.16 with excitation power of 1.0 W, though the sample thickness was 0.24 cm. Because the specific gain coefficient of the bulk sample was 0.62 cm^{-1}/W, the fiber shape affects the gain increment due to beam mode matching between the pump and the probe.

Figure 9 shows simultaneous amplification at two wavelengths near the 1300 nm region. The signal wavelengths to measure the simultaneous amplification in the BiDF were adjustable wavelengths (1297 and 1323 nm) and anchor wavelength (1308 nm). For the two cases of amplification experiments, the maximum gain coefficients at the adjustable wavelengths (0.046 and 0.036 cm^{-1}) and the anchor wavelength (0.036 and 0.034 cm^{-1}) were obtained. The optical gain shown in simultaneous measurements of two wavelengths suggests that it is possible to realize WDM optical fiber amplifiers in O-band (1260~1360 nm). Performance of fiber amplifier largely depends on the fiber specification. Optical gain in this experiment was much smaller than that for EDF or PDFFA. The smaller core cross section gives promising potential to the practical gain performance. The gain characteristics will be further improved by optimizing the fiber structure, such as a partially doped core structure and deformed shape of first clad layer for efficient pumping (Seo et al., 2007a).

5. Optical amplification in a single-mode BiDF

5.1 Sample preparation and experimental setup

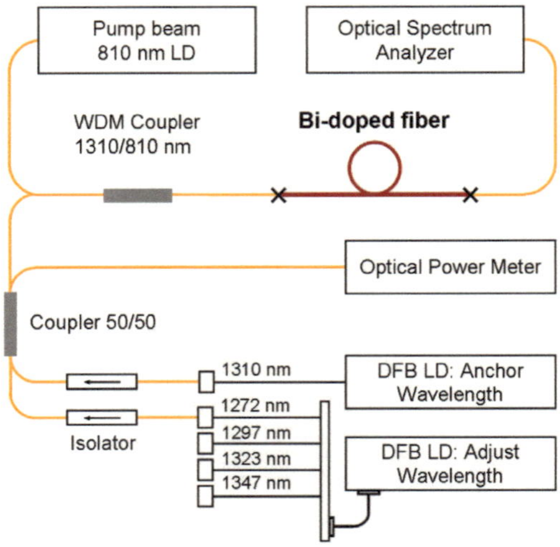

Fig. 10. Schematic diagram for the optical gain measurement of a bismuth-doped silica fiber.

BiDF with a refractive index difference, Δn~0.017, was drawn from the preform. Its core diameter and outer diameter was 1.54 µm and 125 µm, respectively. The experimental setup for the optical gain measurement in the BiDF is shown in Fig. 10. An 810-nm cw laser diode was used as a pump source. A 1310 nm distributed feedback (DFB) laser diode and four laser diodes with different wavelengths were used as a probe beam. The wavelengths of the four DFB laser diodes were 1272, 1297, 1323 and 1347 nm, and these probe beams can be adjusted by using a single laser driver. The probe laser beams were combined with an excitation beam using a WDM coupler (810/1310 nm). BiDF was fusion spliced (FSM-40PM, Fujikura) to output coupler with a SMF. The combined beam goes through into the BiDF, and the gained probe beam was detected by an optical spectrum analyzer (Ando: AQ6317B).

5.2 Gain characteristics of the single-mode BiDF

The optical amplification in a 5.0-cm long BiDF sample at a single wavelength, 1310 nm, is shown in Fig. 11. The maximum optical gain was calculated to be 9.6 dB and therefore, the gain coefficient was 0.442 cm^{-1}. The optical gain increased linearly with an excitation power up to 100 mW. The measured signal input/output power for the BiDF was -30.0 dBm / -20.4 dBm (@ 1310 nm), exhibiting a much lower power conversion efficiency. The optical gain in the previous result with a bulk-type BiSG was 1.16 with an excitation power of 1.0 W, though the sample thickness was 0.24 cm (Seo et al, 2005; 2006a). Because the specific gain coefficient of the bulk sample was 0.62 cm^{-1}/W, the fiber shape clearly affects the gain increment due to a beam mode matching between the pump and the probe. If the excitation length is longer, absorption power will increase greatly, and then the gain will be larger than that demonstrated in this experiment.

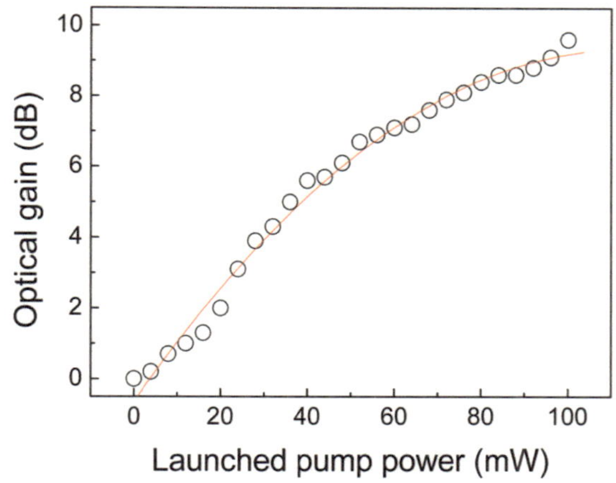

Fig. 11. Optical gain profile of the bismuth-doped silica fiber at a length of 5.0 cm.

Figure 12 shows a simultaneous amplification at two wavelengths near the 1300 nm region. The wavelength of the excitation beam was 810 nm, and the sample length was 5.0-cm. The signal wavelengths to measure the simultaneous amplification in the BiDF were adjustable wavelengths (1297 nm and 1323 nm) and an anchor wavelength (1310 nm). For the two cases of amplification experiments, the maximum optical gain at these wavelengths were 7.99 dB and 7.17 dB respectively and 8.69 dB at the anchor wavelength was obtained. The optical gain shown from the simultaneous measurements of two wavelengths suggests that it is possible to realize WDM optical fiber amplifiers in the O-band (1260~1360 nm). One of the most fundamental parameters, the pump efficiency is defined as the net gain per unit pump power. Net gain is obtained for over a 100-mW pump power. The pump efficiency is 0.095 dB/mW (Seo et al., 2007b).

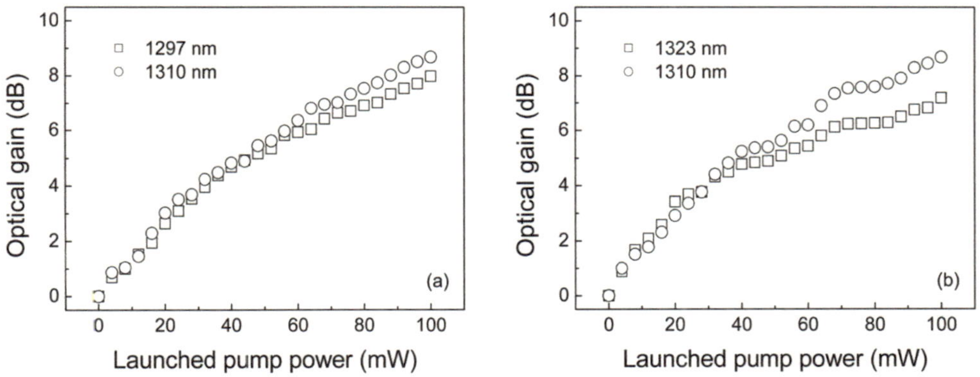

Fig. 12. Simultaneous optical amplification properties at two wavelengths of the 1300-nm range: (a) 1310 nm and 1297 nm; (b) 1310 nm and 1323 nm.

6. Conclusion

The BiSG can complement other optical fiber amplifiers by solving such problems as fusion splicing and wideband amplification. Understanding the optical properties of BiSG is important due to its potential technological applications in optical fiber amplifiers and fiber lasers. The spectroscopic properties of the investigated BiSG are different from those of the previously reported luminescence from Bi^{3+} ions ((Parke & Webb, 1973; Weber & Monchamp, 1973). The peaks of absorption and emission spectra of BiSG exist in the visible and infrared regions, respectively. The absorption cross section of the $^1S_0 \rightarrow {}^3P_1$ allowed transition for Bi^{3+} ions was estimated to be 2.4×10^{-17} cm^2; that for bismuth ions of BiSG is 1.2×10^{-20} cm^2 at best. The Stokes shift between absorption and emission in a BiSG sample is much larger than rare earth ions. Because of the large Stokes shift, population inversion can be realized, and optical amplification occurs via the stimulated emission process.

The origin of light emission in a BiSG, which is still unclear, is the valence electrons of the bismuth ions. We are considering more important reasons why aluminum ions are needed to generate BiSG luminescence. Aluminum ions are needed to generate BiSG luminescence. Aluminum is expected to have a special role in the formation of Bi luminescent center. Therefore, discovering the aluminum status in BiSG, especially the aluminum coordination state, will help us understand the unknown luminescent center. Aluminum coordination state can be investigated by using ^{27}Al-NMR and XAFS. Co-doping of Al and Bi is indispensable for the broadband infrared luminescence of BiSG. The aluminum ion has to roles in BiSG: assisting the configuration of the peculiar luminescent center of Bi ion with some coupling effect, and increasing compatibility with the silica network (Fujimoto & Nakatsuka, 2006; Ohkura et al., 2007).

There are many attractive characteristics of BiSG for a laser application, such as, a long emission lifetime (over 600 μs) and a high absorption cross-section which enable an effective pumping and a short pulse generation. It is important to measure the quantum yield, because the performance of a laser amplification and oscillation is very sensitive for the quantum yield. The quantum yield is defined as the rate of the emitted photon number to the absorbed photon number. The quantum efficiency of the BiSG was measured as 60-70 % (Fujimoto et al, 1999).

BiSG has many attractive features that make it suitable for use in an optical fiber amplifier. First, because BiSG is a silica-based material, it can be transformed to an optical fiber and easily fusion connected to the silica fibers of a network system. Second, the peak wavelength is at 1250 nm, and the bandwidth is 300 nm of FWHM, which is five or six times wider than erbium-doped silica fibers. Third, as the simultaneous amplifying signals near the 1300 nm region (~75 nm), BiSG fiber amplifier is considered a more effective solution for WDM broadband systems. Fourth, a common commercial semiconductor laser (~808 nm) can be used for excitation of this medium. As discussed above, BiSG is a promising candidate for the core fiber material of an optical amplifier at a natural zero-dispersion wavelength, 1300 nm, of silica glass fiber (Fujimoto & Nakatsuka, 2001; 2003).

In conclusion, we have demonstrated optical amplification in a bismuth-doped silica glass and BiDF at second telecommunication window. The amplification was obtained at five different wavelength and the amplification bandwidth is greater than 75 nm (1272.4 ~ 1347.4 nm) in a 0.24 cm BiSG. And we demonstrated simultaneous optical amplification at two wavelengths of the 1300 nm region in a BiSG. Simultaneous amplification was obtained at four different wavelengths, and the amplification bandwidth was greater than 75 nm. This technique can be useful for WDM optical amplifiers at 1300 nm second telecommunication windows. Therefore, this gain medium is expected to be useful for applications in ultrawide broadband optical communication.

Also we reported optical amplification in a BiDF. The optical gain obtained in an 8.0-cm-long multi-mode BiDF at 1308 nm with 810-nm excitation and discussed simultaneous amplification at two wavelengths of the 1300 nm region. And we have demonstrated on the optical amplification phenomenon in a 5.0-cm-long single-mode BiDF at 1310 nm and discussed a simultaneous amplification at two wavelengths of the 1300 nm region. These spectroscopic characteristics and the amplification observed at the 1300 nm range have shown that such fibers are good candidates for cw and pulsed fiber lasers and fiber amplifiers for a spectral range of 1100~1400 nm.

7. References

Dianov, E. M.; Dvoyrin, V. V.; Mashinsky, V. M.; Umnikov, A. A.; Yashkov, M. V. & Guryanov, A. N. (2005). CW bismuth fibre laser, *Quantum Electronics,* Vol. 35, 1083-1084, ISSN 1063-7818

Dvoyrin, V. V.; Mashinsky, V. M.; Bulatov, L. I.; Bufetov, I. A.; Shubin, A. V.; Melkumov, M. A.; Kustov, E. F.; Dianov, E. M.; Umnikov, A. A.; Khopin, V. F.; Yashkov, M. V. & Guryanov, A. N. (2006). Bismuth-doped-glass optical fibers—a new active medium for lasers and amplifiers, *Opt. Lett.,* Vol. 31, No. 20, 2966-2968. ISSN 0146-9592

Dvoyrin, V. V.; Mashinsky, V. M. & Dianov, E. M. (2007). Yb-Bi pulsed fiber lasers, *Opt. Lett.,* Vol. 32, No. 5, 451-453, ISSN 0146-9592

Fujimoto, Y.; Murata, K. & Nakatsuka M. (1999). New material for high average power laser based on silica glass, *Fusion Eng. Des.,* Vol. 44, 431-435, ISSN 0920-3796

Fujimoto, Y. & Nakatsuka, M. (2001). Infrared luminescence from bismuth-doped silica glass, *Jpn. J. Appl. Phys.,* Vol. 40, No. 3B, L279-L281, ISSN 0021-4922

Fujimoto, Y. & Nakatsuka, M. (2003). Optical amplification in bismuth-doped silica glass, *Appl. Phys. Lett.,* Vol. 82 No. 19, 3325-3326, ISSN 0003-6951

Fujimoto, Y. & Nakatsuka, M. (2006). ^{27}Al NMR structural study on aluminum coordination state in bismuth doped silica glass, *J. Non-Cryst. Solids,* Vol. 352, 2254-2258, ISSN 0022-3093

Meng, X.; Qui, J.; Peng, M.; Chen, D.; Zhao, Q.; Jiang, X. & Zhu, C. (2005). Infrared broadband emission of bismuth-doped barium-aluminum-borate glasses, *Opt. Express,* Vol. 13, No. 5, 1635-1642, ISSN 1094-4087

Meng, X.; Qui, J.; Peng, M.; Chen, D.; Zhao, Q.; Jiang, X. & Zhu, C. (2005). Near infrared broadband emission of bismuth-doped aluminophosphate glass, *Opt. Express,* Vol. 13, No. 5, 1628-1634, ISSN 1094-4087

Miyajima, Y.; Sugawa, T. & Fukasaku, Y. (1991). 38.2dB amplification at 1.31 μm and possibility of 0.98 μm pumping in Pr^{3+}-doped fluoride fibre, *Electron. Lett.,* Vol. 27, No. 19, 1706-1707, ISSN 0013-5194

Ohkura, T.; Fujimoto, Y.; Nakatsuka, M. & Seo, Y. (2007). Local structures of bismuth ion in bismuth-doped silica glasses analyzed using Bi LIII X-ray absorption fine structure, *J. Am. Ceram. Soc.,* Vol. 90, No. 11, 3596-3600, ISSN 0002-7820

Parke, S. & Webb, R. S. (1973). The optical properties of thallium, lead and bismuth in oxide glasses, *J. Phys. Chem. Solids,* Vol. 34, 85-95, ISSN 0022-3697

Peng, M.; Qiu, J.; Chen, D.; Meng, X.; Yang, I.; Jiang, X. & Zhu, C. (2004). Bismuth- and aluminum-codoped germanium oxide glasses for super-broadband optical amplification, *Opt. Lett.*, Vol. 29, No. 17, 1998-2000, ISSN 0146-9592

Peng, M.; Qiu, J.; Chen, D.; Meng, X. & Zhu, C. (2005). Broadband infrared luminescence from Li_2O-Al_2O_3-ZnO-SiO_2 glasses doped with Bi_2O_3, *Opt. Express*, Vo. 13, No. 18, 6892-6898, ISSN 1094-4087

Peng, M.; Qiu, J.; Chen, D.; Meng, X. & Zhu, C. (2005). Superbroadband 1310 nm emission from bismuth and tantalum codoped germanium oxide glasses, *Opt. Lett.*, Vol. 30, 2433-2435, ISSN 0146-9592

Peng, M.; Wang, C.; Chen, D.; Jiang, X. & Zhu, C. (2005). Investigations on bismuth and aluminum co-doped germanium oxide glasses for ultra-broadband optical amplification, *J. Non-Cryst. Solids*, Vol. 351, 2388-2393, ISSN 0022-3093

Seo, Y.; Fujimoto, Y. & Nakatsuka, M. (2005). Amplification in a bismuth-doped silica glass at second telecommunication windows, *Conference on Lasers and Electro-Optics*, CThR6, ISBN 1-55752-7709, 22-27 May 2005, Baltimore, Maryland, U.S.A., Optical Society of America

Seo, Y.; Fujimoto, Y. & Nakatsuka, M. (2006). Optical amplification in a bismuth-doped silica glass at 1300 nm telecommunications window, *Opt. Commun.*, Vol. 266, No. 1, 169-171, ISSN 0030-4018

Seo, Y.; Fujimoto, Y. & Nakatsuka, M. (2006). Simultaneous amplification at two wavelengths near 1300 nm in a 6.5-cm-long bismuth-doped silica glass, *IEEE Photon. Technol. Lett.*, Vol. 18, No. 18, 1901-1903, ISSN 1041-1135

Seo, Y.; Lim, C.; Fujimoto, Y. & Nakatsuka, M. (2007). 9.6 dB Gain at a 1310 nm Wavelength for a Bismuth-doped Fiber Amplifier, *Journal of the Optical Society of Korea*, Vol. 11, No. 2, 63-66, ISSN 1226-4776

Seo, Y.; Lim, C.; Fujimoto, Y. & Nakatsuka, M. (2007). Bismuth-doped silica glass as a new laser material, *J. Korean Phys. Soc.*, Vol. 51, No. 1, 364-367, ISSN 0374-4884

Sugimoto, N.; Kanbara, H.; Fujiwara, S.; Tanaka, K. & Hirao, K. (1996). Ultrafast response of third-order optical nonlinearity in glasses containing Bi_2O_3, *Opt. lett.*, Vol. 21, No. 20, 1637-1639, ISSN 0146-9592

Suzuki, T. & Ohishi, Y. (2006). Ultrabroadband near-infrared emission from Bi-doped Li_2O–Al_2O_3–SiO_2 glass, *Appl. Phys. Lett.*, Vol. 88, 191912-1-191912-3, ISSN 0003-6951

Weber, M. J. & Monchamp, R. R. (1973). Luminescence of $Bi_4Ge_3O_{12}$: Spectral and decay properties, *J. Appl. Phys.*, Vol. 44, No. 12, 5495-5499, ISSN 1066-7814

Whitley, T. J. (1995). A review of recent system demonstrations incorporating 1.3-μm praseodymium-doped fluoride fiber amplifiers, *J. Lightwave Technol.*, Vol. 13, No. 5, 744-760, ISSN 0733-8724

Yamada, M.; Mori, A.; Kobayashi, K.; Ono, H.; Kanamori, T.; Oikawa, K.; Nishida, Y. & Ohishi, Y. (1998). Gain-Flattened Tellurite-Based EDFA with a Flat Amplification Bandwidth of 76 nm, *IEEE photon. Technol. Lett.*, Vol. 10, No. 9, 1244-1246, ISSN 1041-1135

Yamada, M.; Ono, H. & Ohishi, H. (1998). Low-noise, broadband Er^{3+}-doped silica fibre amplifiers, *Electron. Lett.*, Vol. 34, No. 15, 1490-1491, ISSN 0013-5194

7

Radio-over-Fibre Techniques and Performance

Roberto Llorente and Marta Beltrán
Universidad Politécnica de Valencia
Spain

1. Introduction

Radio-over-fibre (RoF) techniques have been subject of research during the last decades and find application in optical signal processing (photonic analogue-to-digital converters, photonic-microwave filters, arbitrary waveform generation), antenna array beamforming, millimetre-wave and THz generation systems, or photonic up- and down-converting links for applications such as broadband wireless access networks, electronic warfare and RADAR processing, imaging and spectroscopy or radio-astronomy (Seeds & Williams, 2006; Capmany & Novak, 2007). In these applications a radio signal typically in the millimetre-wave band is transmitted through optical fibre employing laser sources and electro-optical devices.

The use of optical fibre links to distribute telecommunication standards is the more successful application of RoF technology, usually known as hybrid fibre-radio (HFR) networks (Jager & Stohr, 2001). HFR networks have been deployed in the last decade due to the increasing demand of high-bitrate communication services in today's access network. This demand is based on the steady market introduction of services requiring the transmission of massive data quantities, like high-definition movie distribution, on-line gaming and rich Internet experience by example (Merill Lynch, 2007).

The HFR concept applied to the enhancements of community antenna television (CATV) networks reflected in the so-called hybrid-fibre coax (HFC) network, in which a combination of digital and analogue channels is distributed from a central location to many users distributed geographically (Darcie & Bodeep, 1990; Wilson et al., 1995). In HFC networks the last mile connection is provided through coaxial cable whilst in HFR networks the last mile connection is always a wireless link. This is not a minor difference, as the wireless environment is much more hostile than cable imposing restrictive RoF link performance requirements in terms of linearity, noise and power handling capabilities, key parameters to guarantee a spurious free dynamic range (SFDR) for the whole link high enough to cope with geographical dispersion of users and complex modulation formats used by current wireless standards. A simplified schematic of a HFR network is shown in Fig. 1.

RoF technology allows centralising the required RF signal processing functions in one shared location (Central Office, CO) and then to use optical fibre to distribute the RF signals to the remote access units (RAU). This allows important cost savings as the RAUs can be simplified significantly, as they only need to perform optoelectronic conversion and filtering and amplification functions. It is possible to use wavelength multiplexing techniques (WDM) in order to increase capacity and to implement advanced network features such as dynamic allocation of resources. This centralised and simplified RAU scheme allows lower cost system operation and maintenance, which are reflected into major system OPEX savings, especially in broadband wireless communication systems where a high density of RAUs is necessary.

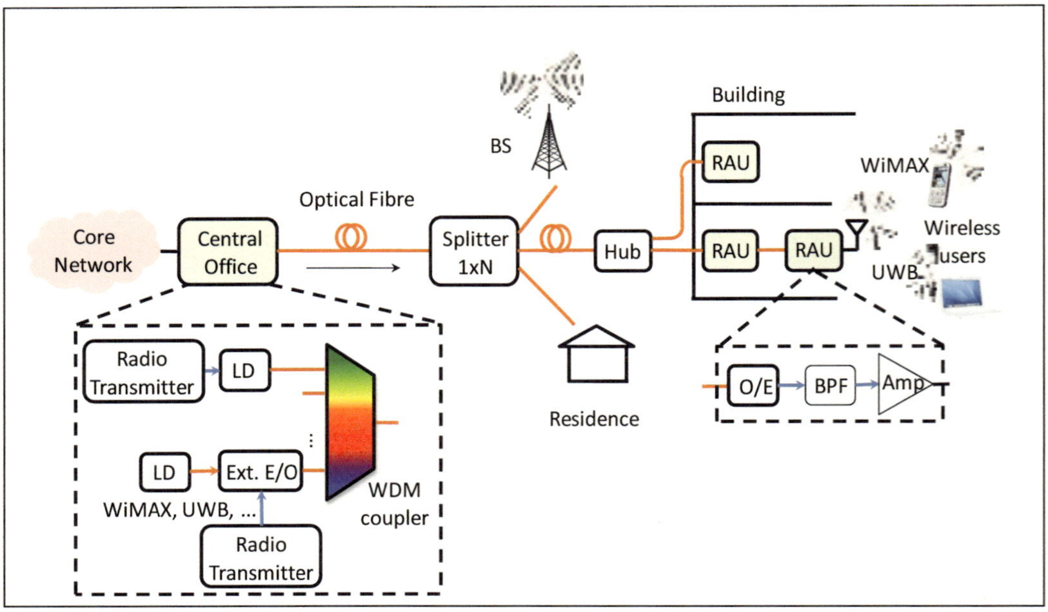

Fig. 1. Simplified schematic of a RoF system. LD: Laser diode. BS: Base Station. RAU: Remote Access Unit. BPF: Band-pass filter. Amp: Electrical amplifier.

The CO and the RAU perform electro-optical (E/O) and opto-electronic (O/E) conversion of wireless signals respectively. E/O conversion is achieved employing either directly modulated laser sources or external electro-optic modulators. O/E conversion is done employing photodetectors or photoreceivers (Seeds & Williams, 2006). Regarding the RF transport, when the signal is transported directly at the frequency of operation there are benefits regarding cost, complexity and upgradeability, as there is no need for complex RF signal processing at the RAU involving up/down conversion or base-band mux/demux (Capmany & Novak, 2007; Jager & Stohr, 2001).

RoF techniques and complete transmission systems have been demonstrated for frequencies up to 120 GHz (Hirata et al., 2003). As mentioned before, the most successful application of RoF technologies has been the transmission of wireless standards over optical fibre links in centralized architectures, also known as distributed antenna systems (DAS) for both indoor and outdoor applications. The broad bandwidth of the optical fibre facilitates standard-independent multiservice operation for cellular systems, such as GSM (Owaga et al., 1992), UMTS (Persson et al., 2006), wireless LAN (WiFi 802.11 a/b/g/n) (Chia et al., 2003; Niiho et al., 2004; Nkansah et al., 2006) and also for emerging technologies WiMAX (Pfrommer et al., 2006) and Ultra-wideband (UWB) (Llorente et al., 2007). Available commercial systems however are typically limited to frequency ranges between 800-2500 MHz. Demonstrations of such DAS systems include their deployment to provide uniform wireless coverage in important sportive events such as the 2000 Olympic games and 2006 world cup (Rivas & Lopes, 1998; Cassini & Faccin, 2003). For indoor applications where picocell configurations are envisaged, advanced multi-function devices such as waveguide electro-absorption modulator (Wake et al., 1997) or polarization independent asymmetric Fabry-Perot modulators (Liu et al., 2003; Liu et al., 2007) are used as detector/modulator.

Two key factors limiting the overall transmission performance in RoF systems are the optical source and the electro-optic modulation technique employed. Regarding the laser source, at frequencies used for major wireless standards (GSM, WiFi 802.11 a/b/g, UMTS) and also WiMAX up to 5-6 GHz, directly modulated semiconductor lasers are preferred due to lower cost (Qian et al., 2005). For higher frequencies, the required performances can be satisfied only by externally modulated transmitters. Devices with bandwidth handling capabilities in excess of these required by near-term WIMAX deployments, in particular distributed feedback (DFB) lasers offering the required bandwidth and performances, exist commercially, but normally at a high cost taking into account the number of devices required for typical applications. Recently, a lot of research efforts have been devoted to the development of low-cost/high-performance transmitters, for instance uncooled lasers (Ingham et al., 2003; Hartmann et al., 2003) or vertical-cavity surface-emitting lasers (VCSEL) (Persson et al., 2006; Chia et al., 2003). Probably, the most restrictive requirement for wireless services provision over RoF systems is the SFDR. Nowadays SFDRs in excess of 100 dB·$Hz^{2/3}$ have been demonstrated experimentally, providing enough dynamic range to be employed in real applications (Seeds & Williams, 2006).

2. Ultra-wideband radio-over-fibre

2.1 Optical generation

The basic elements of RoF systems are broadband laser sources either employing direct or external modulation, a suitable transmission media such as multi-mode fibre (MMF), single-mode fibre (SMF) or plastic optical fibre (POF), and broadband photodetectors or photoreceivers (Seeds & Williams, 2006; Capmany & Novak, 2007; Dagli, 1999). The laser source and modulation method is the key element in the performance of RoF applications.

The generation of the optical signal to be transmitted in the RoF system is of special difficulty in the case of UWB signals. UWB is a radio technology intended for cable replacement in home applications within a range of tens of meters (picocell range), with high-definition video and audio communications a potential application (Duan et al., 2006). UWB is also attractive in many other applications including medicine, sensor networks, etc. UWB radio offers: High data rate capability (>1 Gbit/s), low radiated power spectral density (PSD) minimising the interference, low-cost equipment commercially available. UWB is available in two main implementations: Multi-band orthogonal frequency-division multiplexing (MB-OFDM) and impulse radio. The ECMA standard (ECMA-368, 2007) uses MB-OFDM in 528 MHz individual sub-bands, whilst the impulse-radio implementation employs short pulses (in the range of hundreds of picoseconds) modulated in amplitude, time, polarity or shape to fill a desired bandwidth. MB-OFDM generally shows superior performance to the impulse-radio approach in terms of multi-path fading and intersymbol interference (ISI) tolerance, whilst impulse-radio is able to provide simultaneously communications, localization and ranging to a sub-centimetre resolution.

Currently, UWB uses the unlicensed band from 3.1 to 10.6 GHz mainly for indoor communications (FCC 04-285, 2004; ECMA-368, 2007) and the 24 GHz band for vehicular short-range radar applications (SARA Group, 2009), with a bandwidth larger than 20% of the centre frequency or a 10-dB bandwidth of at least 500 MHz as in FCC regulation (FCC

04-285, 2004) or at least 50 MHz as in ETSI regulations (ETSI, 2008) and a maximum radiated PSD of -41.3 dBm/MHz to guarantee spectral coexistence with other wireless narrowband services complementary in terms of range and bitrate such as WiMAX. Nevertheless, the whole UWB band 3.1-10.6 GHz is not available worldwide due to coexistence concerns (WiMedia, 2009). Outside the United States, available effective bandwidth is 1.5 GHz which only supports hundreds of Mbit/s. However, the unlicensed 60 GHz band enables UWB multi-Gbit/s wireless communications worldwide, as shown in Fig. 2, while challenges related to wireless channel and transceiver design have to be addressed (Daniels & Heath, 2007).

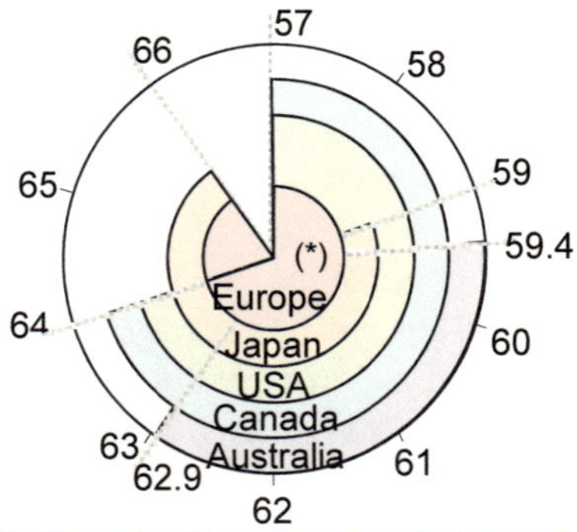

Fig. 2. International frequency allocations in the 60 GHz band (as of January 2009). (*) (ECC, 2009)

RoF distribution of UWB signals, termed UWB-over-fibre, has received great interest to extend the UWB range exploiting the advantages of the broad bandwidth, low loss, light weight, and immunity to electromagnetic interference offered by optical fibres.

In this section, different techniques for generating impulse-radio UWB signals in the optical domain is reported, featuring frequencies ranging from baseband up to millimetre-wave bands, including 24 GHz and 60 GHz. Some laser source characteristics are also discussed.

2.1.1 Impulse-radio ultra-wideband baseband

For UWB-over-fibre systems, it is desirable to generate UWB signals directly in the optical domain, avoiding the use of additional E/O and O/E conversions and exploiting the advantages provided by optics such as broadband processing, light weight, small size, and immunity to electromagnetic interference. Many techniques have been proposed to generate impulse-radio UWB signals in the 3.1-10.6 GHz band in the optical domain. These techniques have mainly focused on generating Gaussian monocycle and doublet pulse shapes, which have been demonstrated to provide better bit error rate (BER) and multipath performance among different pulse types (Chen & Kiaei, 2002).

RoF distribution of UWB signals in the band from 3.1 to 10.6 GHz for high-definition audio/video broadcasting in optical access networks, e.g. in fibre-to-the-home (FTTH) networks has been proposed (Llorente et al., 2008). The performance of both MB-OFDM and impulse-radio UWB implementations at 1.25 Gbit/s is experimentally analysed and compared for different SMF links, ranging from 25 km up to 60 km. Both UWB implementations exhibit error-free operation (BER$< 10^{-9}$) up to 50 km without dispersion compensation. The impulse radio technology exhibits degraded performance compared with OFDM although other optimized impulse-radio generation and detection schemes could lead to different results. OFDM-UWB degrades quickly with fibre length, due to the carrier suppression effect (Schmuck, 1995).

Several demonstrations on optical generation of impulse-radio UWB in the band from 3.1 to 10.6 GHz including fibre and/or wireless transmission have been reported achieving 65 cm wireless distance at 500 Mbps data employing on-off keying (OOK) amplitude modulation (Abtahi et al., 2008); 20 cm at 1.025 Gbit/s OOK-modulated data after up to 10 km of dispersion-compensated SMF (Hanawa et al., 2009); 5 cm at 1.6875 Gbit/s OOK-modulated data after 24 km of SMF (Pan & Yao, 2009a); at 1.625 Gbit/s data employing pulse-position modulation (PPM) after up to 200 m of SMF (Shams et al., 2009a), or 37 km of SMF with no wireless transmission (Shams et al., 2009b); and at 781.25 Mbit/s data employing binary phase-shift keying (BPSK) modulation after 30 km of SMF (Yu et al., 2009). In addition, techniques have been reported capable of pulse shape modulation (PSM) (Dong et al., 2009), or reconfigurable for multiple modulation formats (Pan & Yao, 2009b).

Photonic generation of Gaussian monocycle pulses based on balanced photodetection of data Gaussian pulses has been proposed (Hanawa et al., 2007; Beltrán et al., 2008). Data Gaussian pulses are first generated by intensity modulation of an electrical data sequence with optical Gaussian pulses from a pulsed laser. This technique is shown in Fig. 3. Optical

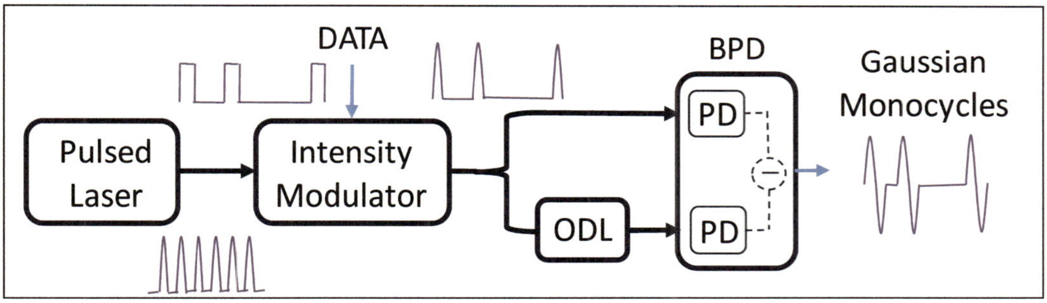

Fig. 3. Gaussian monocycle pulse generation based on balanced photodetection. ODL: Optical delay line. BPD: Balanced photodetector. PD: Photodetector.

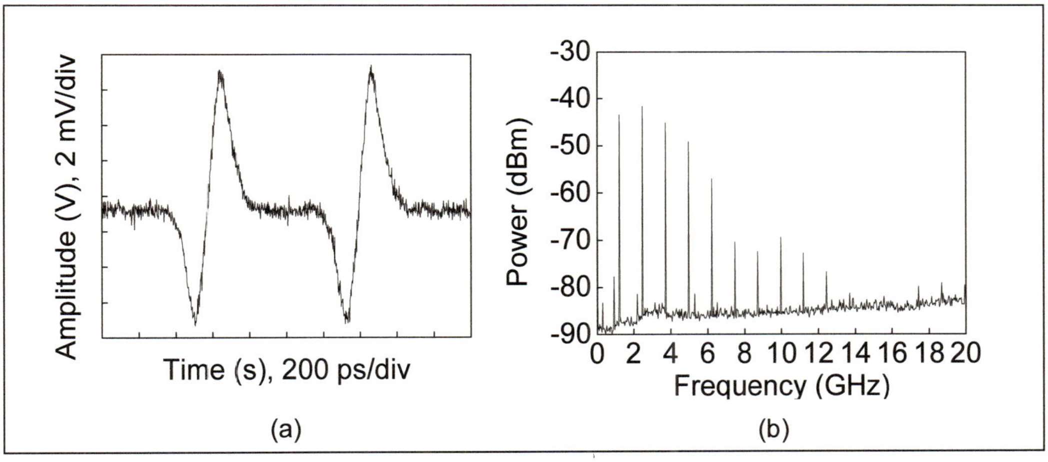

Fig. 4. Monocycle pulses generated employing the technique in Fig. 3; (a) the temporal waveform and (b) its spectrum (resolution bandwidth: 30 kHz).

data pulses are split into two equal parts to drive the two inputs of the balanced photodetector. Optical delay is employed to adjust the relative time delay between the two signals. The pulse width of Gaussian pulses and the time-delay difference are adjusted so as to generate the desired UWB bandwidth. This approach has been experimentally demonstrated employing an actively mode-locked fibre laser and a Mach-Zehnder modulator (MZM) (Beltrán et al., 2008). To control the pulse width, a spool of standard SMF is included after the MZM. Fig. 4 shows monocycles generated based on balanced photodetection exhibiting a UWB 10-dB bandwidth of 6 GHz at 1.25 Gbit/s.

Gaussian monocycle pulses can also be generated based on differential photoreception of data Gaussian pulses (Beltrán et al., 2009b) targeting to reduce cost and complexity. Fig. 5 shows this technique. Again, data Gaussian pulses are first generated by intensity modulation of an electrical data sequence with optical Gaussian pulses from a pulsed laser. Optical data pulses are photodetected and amplified by an electrical amplifier providing complementary outputs. The two outputs are combined after adjusting their relative time delay to generate monocycles. The pulse width of Gaussian pulses and the time-delay difference are adjusted so as to generate the desired UWB bandwidth. This approach has been experimentally demonstrated employing an actively mode-locked fibre laser and a

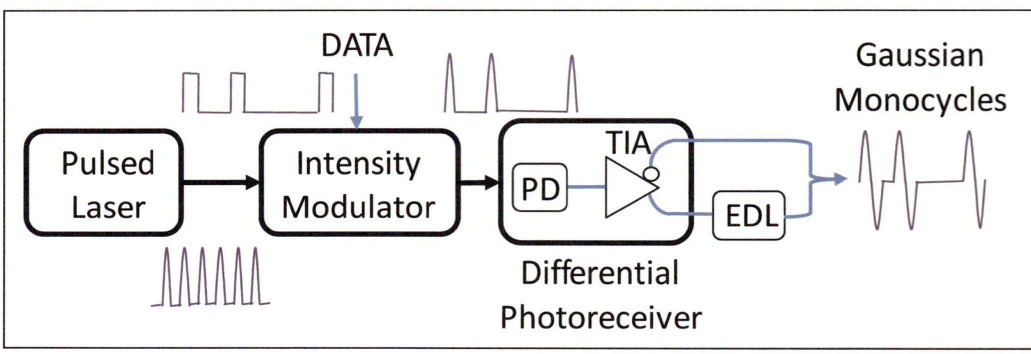

Fig. 5. Gaussian monocycle pulse generation based on differential photoreceiver. PD: Photodetector. TIA: Transimpedance amplifier. EDL: Electrical Delay Line.

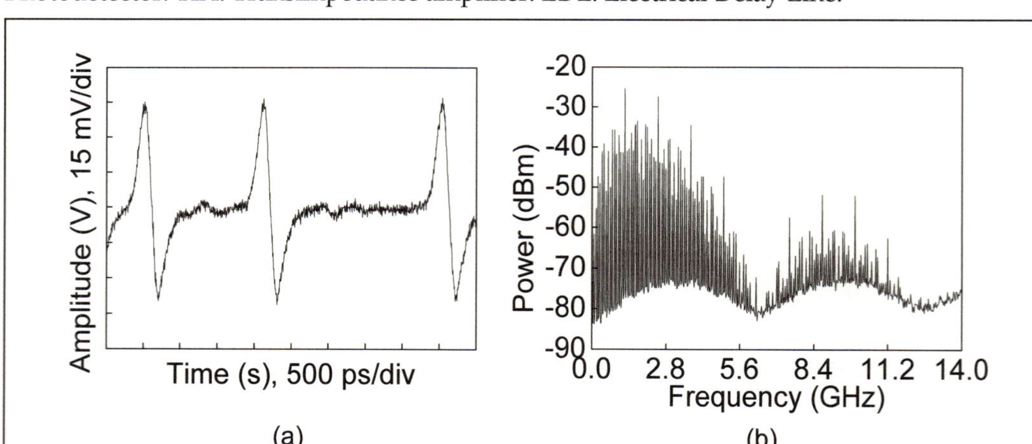

Fig. 6. Monocycle pulses generated employing the technique in Fig. 5; (a) the temporal waveform and (b) its spectrum (resolution bandwidth: 300 kHz).

MZM. To control the pulsewidth, a spool of standard SMF is included after the MZM. Fig. 6 shows monocycles generated based on differential photoreception exhibiting a UWB 10-dB bandwidth of 3.8 GHz at 1.244 Gbit/s.

The photonic techniques for Gaussian monocycle generation shown in Fig. 3 and Fig. 5 are capable of providing high-quality pulses covering the whole UWB band with simple and flexible configuration compared with other techniques employing custom fibre Bragg gratings, nonlinear optical processes or spectrum shaping components.

UWB generation requires a pulse width in the order of hundreds of picoseconds to generate a suitable UWB bandwidth and a multi-gigahertz pulse repetition rate equal to the target data rate of the system. Gain-switched laser diodes, as used in (Hanawa et al., 2007; Kaszubowska-Anandarajah et al., 2008), passive or active mode-locked fibre lasers, e.g. (PriTel, 2009), and emerging optically pumped passively mode-locked vertical-external-cavity surface-emitting lasers (VECSEL) can provide repetition rates suitable for UWB systems. In contrast to mode-locked lasers, gain switched lasers are simpler and more compact. However, timing jitter and also fluctuations of other pulse parameters are larger than for mode-locked lasers. Optically pumped VECSELs have the potential for very compact and cheap. In addition, suitable pulses for UWB applications could also be generated by modulating a continuous-wave light source, e.g. (Wu et al., 2007).

2.1.2 Impulse-radio ultra-wideband in the millimetre-wave band

For UWB optical transmission in RoF operating in the millimetre-wave bands, 24 GHz and 60 GHz, broadband optical frequency up-conversion centralized in the CO appears as a cost-effective solution instead of employing broadband electrical mixing at each RAU. A number of techniques have been reported for millimetre-wave impulse-radio UWB signal generation based on optical up-conversion. These techniques have been demonstrated in the 24 GHz band regulated for vehicular radar and also used in communications. One approach (Kuri et al., 2006) up-converts electrical rectangular pulses based on a complex self-heterodyne technique employing an arrayed waveguide grating (AWG) and a special MZM with high extinction ratio to suppress the residual RF carrier to meet the UWB emission mask. Another approach (Guennec & Gary, 2007) up-converts monocycle or doublet pulses generated by electrical transmitters commercially available by modulation in a MZM biased in nonlinear regime. This technique is simple, however it requires high-frequency electro-optic devices which make it difficult to be upgraded to higher frequency bands. The approaches demonstrated in (Fu et al., 2008; Li et al., 2009) serve as both frequency up-conversion and optical amplification and up-convert monocycle pulses generated employing electrical Gaussian pulses and a frequency discriminator. The former method is based on nonlinear polarization rotation in a semiconductor optical amplifier (SOA) exhibiting limited performance at high frequencies, whilst the latter method employs a more complex architecture based on a fibre optical parametric amplifier (OPA) but can be extended to higher frequency bands. An approach based on up-conversion of optical pulses in a MZM biased in non-linear regime has also been demonstrated (Chang et al., 2008). Optical monocycle and doublet pulses are generated by driving a dual-parallel MZM with electrical Gaussian pulses. In this demonstration, the performance of the millimetre-wave UWB signal after fibre transmission is analyzed showing the doublet pulse has better tolerance to fibre dispersion than the monocycle pulse.

Gaussian monocycle pulses do not meet the FCC spectrum mask in the 3.1-10.6 GHz band. However, these pulses have been demonstrated to be suitable for further frequency up-conversion (Guennec & Gary, 2007). Optical up-conversion to 20 GHz of the generated baseband monocycles shown in Fig. 4 based on the technique in (Guennec & Gary, 2007) was also demonstrated in (Beltrán et al., 2008). A simpler approach to generate UWB monocycles in the millimetre-wave band based on frequency up-conversion of data Gaussian optical pulses in a MZM at the CO and monocycle shaping at RAUs has been proposed (Beltrán et al., 2009a; Beltrán et al., 2009c). This technique is depicted in Fig. 7.

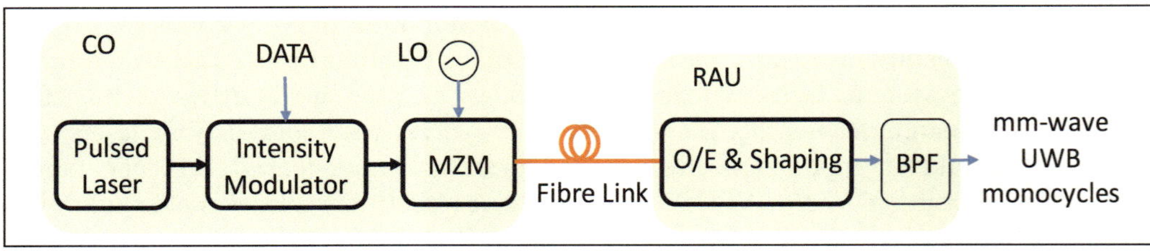

Fig. 7. Millimetre-wave impulse-radio UWB-over-fibre based on up-conversion in a Mach-Zehnder modulator at the CO and monocycle shaping at the RAU. LO: Local oscillator.

This method has been demonstrated in two proof-of-concept experiments by modulating Gaussian optical pulses from an actively mode-locked fibre laser with data in a MZM to generate data Gaussian pulses whose pulse width is subsequently controlled by standard SMF. A second MZM driven by a local oscillator (LO) signal and biased at quadrature point (linear regime) generating an optical double-sideband signal with carrier is employed for up-conversion. The millimetre-wave signal is so obtained after photodetection and filtering at the RAU. In order to verify appropriate operation, the millimetre-wave signal is down-converted with the same LO signal employed for up-conversion in an electrical mixer (conventional homodyne detection) and further low-pass filtered, with no fibre and no air transmission.

In the first experiment (Beltrán et al., 2009a), monocycle shaping is based on balanced photodetection as shown in Fig. 3. UWB monocycles are generated at 19 GHz, exhibiting a single-sideband 10-dB bandwidth of 2.5 GHz at 622 Mbit/s, as shown in Fig. 8. Also shown

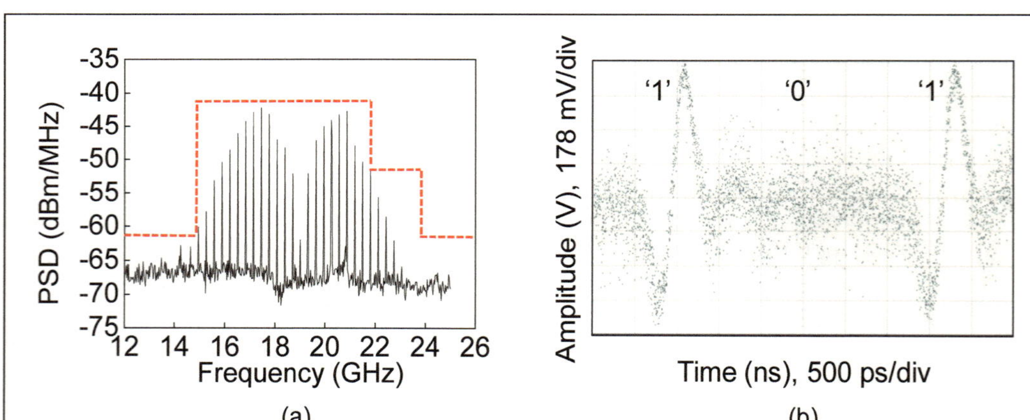

Fig. 8. (a) RF spectrum of UWB monocycles at 19 GHz generated in the setup in Fig. 1 with monocycle shaping as in Fig. 3. The FCC UWB mask is shown as a dashed line translated to 19 GHz; (b) down-converted data monocycles.

in Fig. 8 is the down-converted signal. The monocycles bear OOK-modulated data so that they are suitable for simultaneous vehicular radar and communications in the 24 GHz band to provide traffic safety applications.

Monocycle shaping based on differential photoreception as shown in Fig. 5 is performed in the second experiment (Beltrán et al., 2009c). This technique does not increase significantly the complexity of RAUs. UWB monocycles are generated at 16.85 GHz, exhibiting a single-sideband 10-dB bandwidth of 2.5 GHz and bearing OOK-modulated data at 1.244 Gbit/s, as shown in Fig. 9. Also shown in Fig. 9 is the down-converted signal.

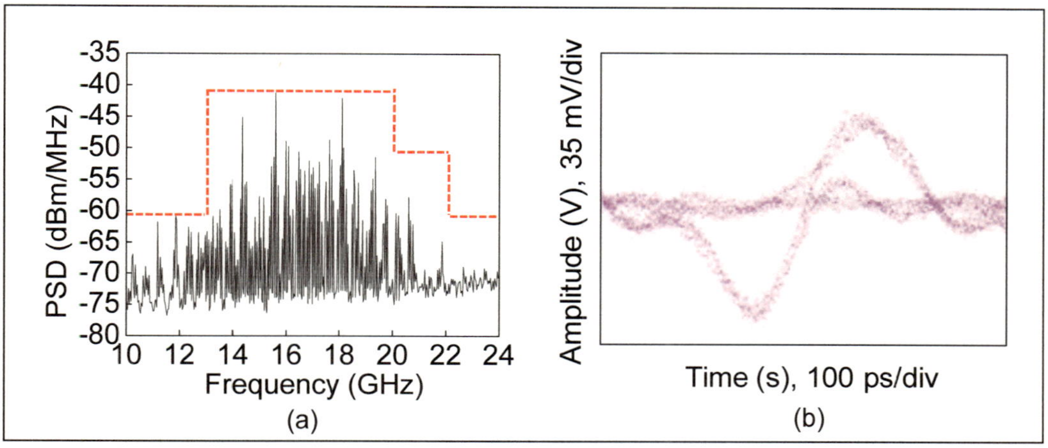

Fig. 9. (a) RF spectrum of UWB monocycles at 16.85 GHz generated in the setup in Fig. 7 with monocycle shaping as in Fig. 5. The FCC UWB mask is shown as a dashed line translated to 16.85 GHz; (b) eye diagram of down-converted data monocycles.

As can be observed in Fig. 8 and Fig. 9, the RF residual carrier does not limit the UWB emission mask. This enables the simultaneous wireless transmission of the two spectral sidebands improving receiver sensitivity at expense of reduced spectral efficiency. The two sidebands could be also filtered separately enabling a simultaneous dual-band generation.

Fig. 10 shows a UWB-over-fibre system where optical data monocycles are frequency up-converted in a MZM in nonlinear regime. This approach has been proposed and demonstrated in a proof-of-concept experiment for millimetre-wave UWB generation in the 60 GHz band (Beltrán et al., 2009b). Electrical Gaussian monocycles are converted to optical domain by external modulation in a MZM to generate optical data monocycles. In the experiment, electrical OOK-modulated monocycles are generated as shown in Fig. 5. Fig. 11 (a) shows the so-obtained optical data monocycles. A low-frequency LO of 14.25 GHz multiplied by 2 is applied to a second MZM biased at minimum transmission point to generate an optical double-sideband signal with a suppressed carrier (optical carrier suppression modulation), resulting in UWB monocycles at 57 GHz after photodetection and filtering at the RAU. The millimetre-wave signal is down-converted by electrical homodyne detection employing the LO signal multiplied by 4 and further low-pass filtered to verify appropriate operation, with no air transmission.

UWB monocycles are generated at 57 GHz, exhibiting a 10-dB bandwidth of 3.8 GHz at 1.244 Gbit/s, as shown in Fig. 11 (b). In this technique it is required to filter the residual RF carrier frequency for wireless transmission in practice. Further transmission over 100 m of standard SMF is demonstrated with no performance degradation. Fig. 11 (c) shows the demodulated signal after fibre transmission. This UWB RoF system has been proposed for multi-Gbit/s high-definition video/audio distribution within in-vehicle networks, e.g. in aircrafts, where also fibre interconnects RAUs along the vehicle. The impulse-radio UWB approach offers also ranging and localization functionalities of special interest for localization of users potentially interfering and for radio tagging and passenger identification applications.

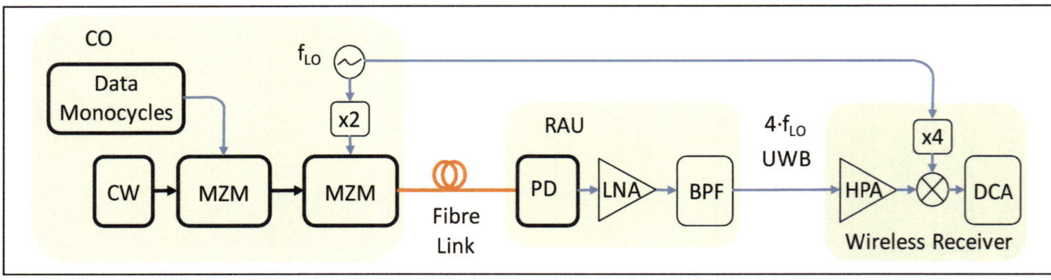

Fig. 10. Setup of the photonic 60 GHz impulse UWB-over-fibre based on up-conversion of data optical monocycles in a Mach-Zehnder modulator in nonlinear regime. CW: Continuous-wave laser. LO: Local oscillator. PD: Photodetector. LNA: Low-noise amplifier. BPF: Band-pass filter. HPA: High-power amplifier. DCA: Digital communications analyzer.

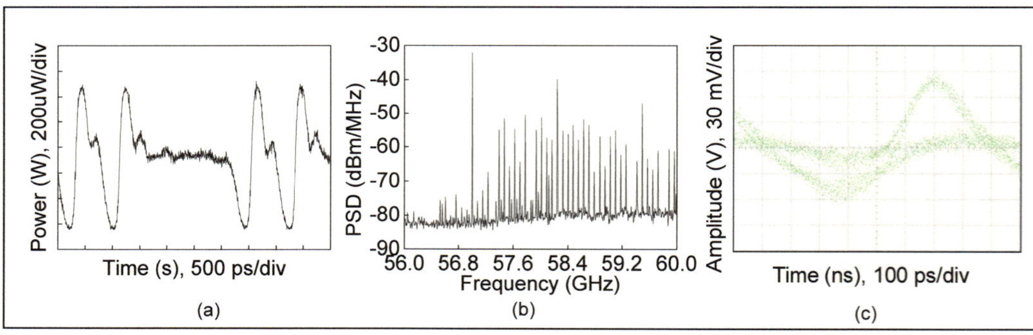

Fig. 11. Measurements in the setup in Fig. 10; (a) optical data monocycles; (b) RF spectrum of UWB monocycles at 57 GHz; (c) eye diagram of down-converted data monocycles.

In the techniques in Fig. 7 and Fig. 10, biasing the MZM employed for up-conversion at minimum transmission point requires half of the LO frequency and reduces the RF power fading effect due to fibre chromatic dispersion (Ma et al., 2007), however higher power is required in the system to not degrade performance with respect to bias at quadrature point. In addition, in both techniques baseband signal is also available after photodetection, which could be provided via a wired connection and a user could employ a simple, low-cost receiver to detect the signal by filtering out the millimetre-wave signal. Also, the baseband signal could meet the UWB mask in the 3.1-10.6 GHz band and be radiated employing an antenna with a suitable frequency response (Pan & Yao, 2009a).

2.1.3 Pulsed laser source characterization

As described in the previous section, optical generation of impulse-radio signals can be achieved employing pulsed laser sources. The overall RoF performance depends directly on the characteristics of the specific pulsed laser employed. In particular, polarization stability i.e. the variation of polarization over time is of special importance when external modulation is employed.

The polarization stability of a pulsed laser source can cause spectrum distortion. The experimental setup for the characterization of the polarization stability of a femtosecond pulsed laser is depicted in Fig. 12. A computer controls the process of capture and storage of data.

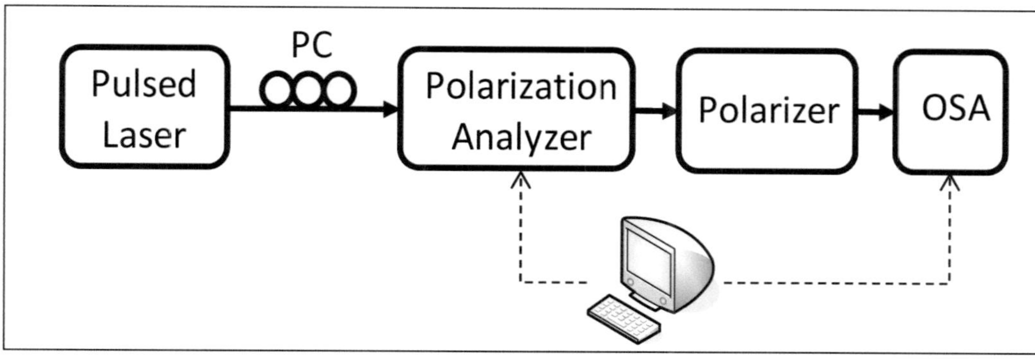

Fig. 12. Experimental setup to characterize the polarization stability of pulsed lasers. PC: Polarization controller. OSA: Optical spectrum analyzer.

The polarization stability and the distortion of the laser spectrum are evaluated for linear horizontal polarization (LH) with an orientation of 0° (launched polarization) adjusted by the polarization controller in Fig. 12. In practice, the orientation adjusted is ~3.6° (LH+3.6°). The optical spectrum analyzer captures the spectrum with 0.05 nm resolution bandwidth and -80 dBm sensitivity. The evaluation is performed at different wavelengths at which the spectrum gets distorted for different launched polarizations. The measurement time is 24 h. Fig. 13 (a) shows the orientation ψ (LH+ψ) calculated from the normalized Stokes vector (S1, S2, S3) given by the polarization analyzer as a function of time. Abrupt changes in the behaviour are due to abrupt temperature changes (disconnection/connection of conditioned air) in the laboratory measurement environment. Fig. 13 (b) is a plot of Poincare sphere showing the evolution over time of the normalized Stokes vector. From the Stokes vector other parameters characterizing polarization such as the degree of polarization (DOP),

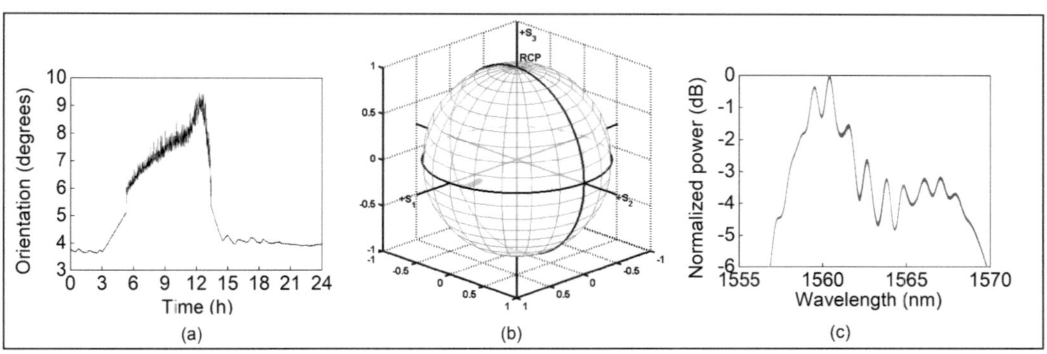

Fig. 13. Characterization of the polarization stability of a femtosecond pulsed laser; (a) Orientation; (b) Poincare sphere; (c) spectra over 1 h.

degree of linear polarization (DOLP), degree of circular polarization (DOCP) and ellipticity can also be calculated. Fig. 13 (c) shows the evolution of spectrum over time for a measurement time of 1 h over a zone in which the temperature is stable. The polarization stability, expressed as the standard deviation of the normalized Stokes parameters, is lower than 0.001 at 1 h independently on the wavelength at which is evaluated.

2.2 Transmission performance

In RoF systems, analog radio signals are modulated on the intensity of the optical signals (E/O conversion) to be transmitted over an optical fibre link employing either directly modulated lasers or external modulators, as shown in Fig. 1. Directly modulated semiconductor lasers such as DFB lasers and VCSELs are preferred due to lower cost whilst for high frequencies the required performance can be satisfied only by externally modulated transmitters.

It has been shown that a VCSEL has higher RF to optical power conversion efficiency than an external MZM and a DFB laser diode for the same output optical power (Gamage et al., 2008a; Gamage et al., 2008b). In addition, in case of the bandwidth is not a limiting factor direct modulation of a DFB laser leads to less distortion on UWB signals than external modulation with MZM because of its less nonlinearity, for the same output power (Jazayerifar et al., 2008). In addition, impulse-radio UWB signals are more sensitive to nonlinear distortion and less sensitive to noise than OFDM UWB signals for the same transmitted energy due to the higher peak-to-peak power. In practice, higher modulated power can be obtained with MZM but increasing the input optical power. The impact of laser chirp on UWB signals is almost negligible, but it affects the amount of dispersion when the UWB signal is transmitted over fibre.

Most of the UWB RoF systems have focused on SMF which is best suited for long-distance access applications. RoF in combination with MMF fibres can be deployed within homes and office buildings for baseband digital data transmission supporting 3.5 GHz wireless signals. The large core diameter of MMF fibres (typically 50 µm or 62.5 µm) offers easier installation and maintenance in within-building environments and reduced cost compared to SMF (Koonen & Garcia, 2008). Note that MMF is also widely used in within-building fibre installations for baseband data transmission systems at far more than 10 Gbit/s. Compared to silica MMF, graded-index plastic optical fibres (GI-POF) offer further advantages such as smaller bending radius, better tolerance to tensile load and stress, and simpler connectorization.

A RoF system employing VCSEL direct modulation of impulse-radio UWB signals in the 3.1-10.6 GHz band has been demonstrated over 100 m MMF (Jensen et al., 2009). Error-free operation employing FEC is achieved at a wireless distance of 8 m at 2.5 Gbit/s or 4 m at 4 Gbit/s. Impulse-radio UWB generation employing DFB direct modulation and transmission over 100 m GI-POF has also been recently demonstrated (Abraha et al., 2009).

In this section, an analysis of impulse-radio UWB propagation on standard SMF is presented. The analysis compares two modulation schemes: External modulation in a MZM at 1550 nm and direct modulation in a VCSEL at 1310 nm. Fig. 14 shows the two UWB radio-over-fibre implementations considered. The analysis targets to evaluate the impact of the modulation index on the reach and has been performed employing the commercial simulation tool VPITransmissionMaker™ (version 7.5).

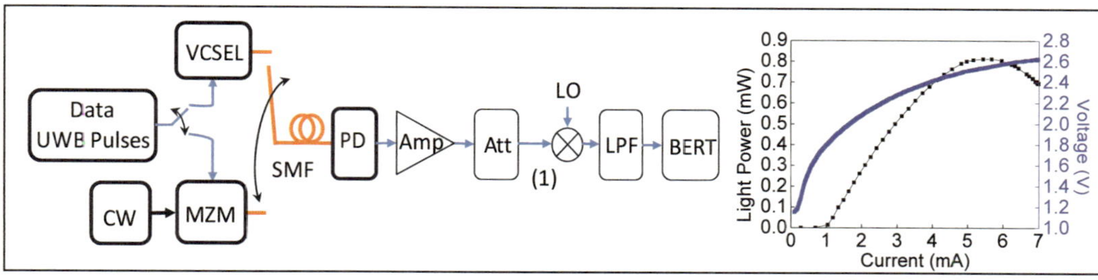

Fig. 14. Impulse-radio UWB RoF optical link configuration and VCSEL characteristics. CW: Continuous-wave laser. PD: Photodetector. Amp: Amplifier. Att: Variable attenuator. LPF: Low-pass filter. BERT: BER test.

Table 1 summarises the parameters of the components shown in Fig. 14 employed in the analysis. Typical parameters of commercially available components are considered.

CW param	**value**	**MZM param**	**value**	**Photodiode PIN param**	**value**
Laser power	10 dBm	Chirp	0	Responsivity	0.65 A/W
RIN	-130 dB/Hz	Vπ DC	5 V	Dark current	5 nA
Amplifier param	**value**	Vπ RF	5V	Thermal noise	10pA/√Hz
Gain	40 dB	Insertion losses	6 dB	**VCSEL param**	**value**
Noise factor	3.8 dB	Extinction ratio	35 dB	Core radius	$2 \cdot 10^{-6}$ m
LO param	**value**	**LPF param**	**value**	Active region Thickness	$0.3 \cdot 10^{-6}$ m
Power	7 dBm	LPF type	Bessel 4th order	Confinement factor	0.03
Frequency	3.75 GHz	Bandwidth	3 GHz		

Table 1. UWB radio-over-fibre parameters employed in the setup in Fig. 14.

The impulse-radio UWB signal consists of fifth-derivative Gaussian pulses at 5.4 GHz 10-dB bandwidth, which are compliant with current UWB regulation in the 3.1-10.6 GHz band. This signal is OOK-modulated with 2^{14}-1 pseudo-random bit sequence (PRBS) data at 1.25 Gbit/s resulting the data UWB pulses to be modulated. Both MZM and VCSEL are biased in linear regime, MZM at its quadrature bias point and VCSEL at the centre of the linear zone of its L-I curve, i.e. at $(I_{th} + I_{sat})/2$ where I_{th} is the threshold current of 1 mA and I_{sat} is the saturation current of 3.5 mA which corresponds to the 1-dB compression point. Fig. 14 shows the V-I and L-I curves of the commercial VCSEL at 27 °C in the analysis. At each modulation index, the attenuator sets a maximum PSD of -41.3 dBm/MHz at the radiation point (1) in Fig. 14. This power level would meet the UWB mask in current regulation employing a transmitter antenna with 0 dBi gain. Down-conversion to baseband is performed by electrical mixing (6.5 dB conversion losses) to obtain a suitable eye diagram for BER performance evaluation. Optical noise is predominant at receiver so that a *Chi2* method is employed for BER estimation. BER performance at different fibre lengths is compared with the back-to-back (B2B) configuration (with no fibre transmission) to evaluate the fibre degradation. Fig. 15 shows the BER results. External modulation gives a maximum reach of 50 km at error-free operation (BER< 10^{-9}) at 0.25 modulation index. Direct modulation gives a maximum reach of 25 km at 0.16 modulation index. Higher modulation indexes than these shown in Fig. 15 result in distortion which makes the UWB spectrum non-compliant with regulation. Longer reach could be achieved employing a pre-amplifier at receiver or forward error correction codes (FEC) (BER< $2.2 \cdot 10^{-3}$) at expense of increased complexity. Fig. 16 and Fig. 17 show examples of UWB signal and down-converted signal for external and direct modulation, respectively.

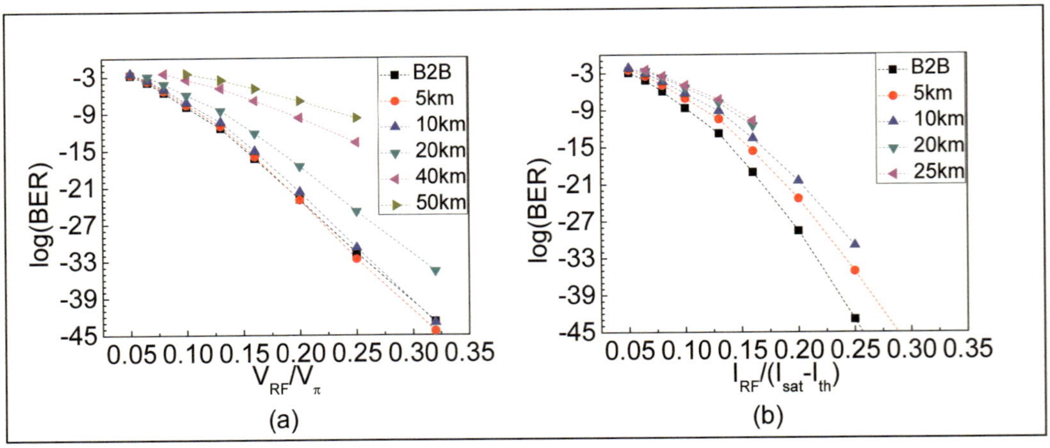

Fig. 15. Performance of impulse-radio UWB over SMF as a function of the modulation index; (a) External modulation in a MZM; (b) Direct modulation in a VCSEL.

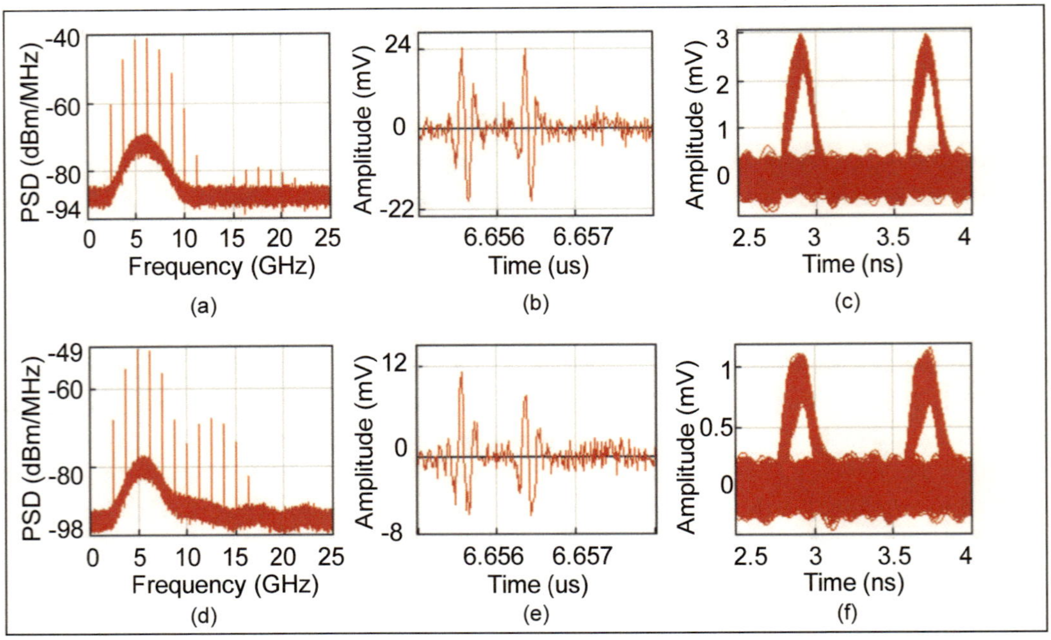

Fig. 16. Impulse-radio UWB signal at point (1) in Fig. 14 (a) RF spectrum and (b) time-domain, (c) down-converted eye diagram at 0.2 external modulation index for B2B configuration. (d-f) the same for 40 km SMF.

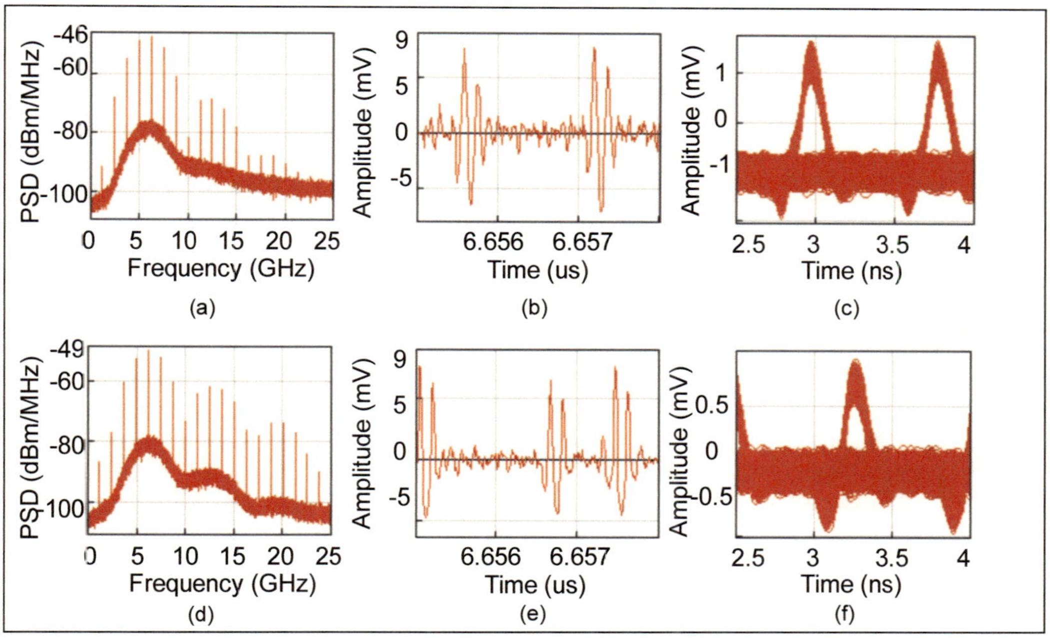

Fig. 17. Impulse-radio UWB signal at point (1) in Fig. 14 (a) RF spectrum and (b) time-domain, (c) down-converted eye diagram at 0.16 direct modulation index for B2B configuration. (d-f) the same for 20 km SMF.

3. Conclusion

In this chapter, the principles and state-of-the-art of RoF technology has been presented. The use of optical sources for the generation of impulse-radio UWB RoF, as one of the most challenging applications to date, has been described and the expected performance after optical transmission has been presented. A technique based on frequency up-conversion of optical UWB signals in a MZM in nonlinear regime has been presented. This technique permits the optical generation of UWB monocycles at 57 GHz bearing data at 1.244 Gbit/s. The generation and further transmission of impulse-radio UWB over 100 m of SMF has been demonstrated with good quality pulses. Another technique performing up-conversion in a MZM of optical pulses with subsequent electrical UWB shaping has been demonstrated at 16.85 GHz bearing data at 1.244 Gbit/s with good quality. The polarization stability of a pulsed laser has also been presented as key factor limiting system performance. Finally, the RoF distribution of impulse-radio UWB has been analyzed over long-distance SMF links suitable for access networks. The longest transmission reach is achieved employing external modulation in a MZM.

4. References

Abraha, S.T.; Yang, H.; Tangdiongga, E. & Koonen, A.M.J. (2009). Novel generation and transmission of FCC-compliant impulse radio ultra wideband signals over 100-m GI-POF. *Proceedings of International Topical Meeting on Microwave Photonics*, pp. 1-4, ISBN 978-84-8363-424-0, Valencia, Spain, October 2009

Abtahi, M.; Mirshafiei, M.; LaRochelle, S. & Rusch, L.A. (2008). All-optical 500-Mb/s UWB transceiver: an experimental demonstration. *IEEE Journal Lightwave Technology*, Vol. 26, No. 15, August 2008, pp. 2795–2802, ISSN 0733-8724

Beltrán, M.; Morant, M.; Pérez, J.; Llorente, R. & Martí, J. (2008). Photonic generation and frequency up-conversion of impulse-radio UWB signals, *Proceedings of IEEE Lasers and Electro-Optics Society*, pp. 498-499, ISBN 978-1-4244-1931-9, Newport Beach, California, United States, November 2008

Beltrán, M.; Llorente, R.; Sambaraju, R. & Martí, J. (2009a). 24 GHz UWB-over- fibre system for simultaneous vehicular radar and communications, *Proceedings of ICT-MobileSummit*, pp. 1-6, ISBN 978-1-905824-12-0, Santander, Spain, June 2009

Beltrán, M.; Llorente, R.; Sambaraju, R. & Martí, J. (2009b). 60 GHz UWB-over-fiber system for in-flight communications, *Proceedings of IEEE MTT-S International Microwave Symposium*, pp. 5-8, ISBN 978-1-4244-2803-8, Boston, United States, June 2009

Beltrán, M.; Sambaraju, R.; La Porta, A.; Llorente, R. & Perez, J. (2009c). Photonic generation and envelope detection of millimeter-wave ultra-wideband impulse-radio employing Mach-Zehnder modulators, *Proceedings of IEEE International Conference on Ultra-Wideband*, pp. 428-432, ISBN 978-1-4244-2931-8, Vancouver, Canada, September 2009

Capmany, J. & Novak, D. (2007). Microwave photonics combines two worlds. *Nature Photonics*, Vol. 1, June 2007, pp. 319-330

Cassini, A. & Faccin, P. (2003). *Proceedings of International Topical Meeting on Microwave Photonics*, pp. 123–128, ISBN 978-1-4244-2168-8, Budapest, Hungary, September 2003

Chang, Q.; Tian, Y.; Ye, T.; Gao, J. & Su, Y. (2008). A 24-GHz ultrawideband over fiber system using photonic generation and frequency up-conversion. *IEEE Photonics Technology Letters*, Vol. 20, No. 19, October 2008, pp. 1651-1653, ISSN 1041-1135

Chen, X & Kiaei, S. (2002). Monocycle shapes for ultra-wideband systems, *Proceedings of International Symposium on Circuits and Systems*, pp. 597-600, ISBN 0-7803-7448-7, Scottsdale, Arizona, United States, May 2002

Chia, M. Y. W.; Luo, B.; Yee, M. L. & Hao, E. J. Z. (2003). Radio over multimode fibre transmission for wireless LAN using VCSELs. *IEEE Electronics Letters*, Vol. 39, No. 15, July 2003, pp. 1143–1144, ISSN 0013-5194

Dagli, N. (1999). Wide-Bandwidth Lasers and Modulators for RF Photonics. *IEEE Transactions on Microwave Theory and Techniques*, Vol. 47, No. 7, July 1999, pp. 1151-1171, ISSN 0018-9480

Daniels, R. C. & Heath, R. W. (2007). 60 GHz wireless communications: emerging requirements and design recommendations. *IEEE Vehicular Technology Magazine*, Vol. 2, No. 3, September 2007, pp. 41-50

Darcie, T.E. & Bodeep, G.E. (1990). Lightwave subcarrier CATV transmission systems. *IEEE Transactions on Microwave Theory and Techniques*, Vol. 38, No. 5, May 1990 pp. 524-533, ISSN 0018-9480

Dong, J.; Zhang, X.; Huang, D. & Rosas-Fernández, J.B. (2009). All-optical ultra-wideband doublet generation and non-degraded transmission over optical fiber, *Proceedings of Optical fiber communication conference*, pp. 1-3, ISBN 978-1-4244-2606-5, San Diego, California, United States, March 2009

Duan, C.; Pekhteryev, G.; Fang, J.; Nakache, Y-P.; Zhang, J.; Tajima, K.; Nishioka, Y. & Hirai, H. (2006). Transmitting multiple HD video streams over UWB links, *Proceedings of Consumer Communications and Networking Conference*, pp. 691-695, ISBN 1-4244-0085-6, Las Vegas, United States, January 2006

ECC/REC/(09)01, Use of the 59 - 64 GHz frequency band for point-to-point fixed wireless systems, January 2009

ECMA-368 International Standard, High rate ultra wideband PHY and MAC Standard, 2007.

ETSI EN 302 065, Ultra WideBand (UWB) technologies for communication purposes; Harmonized EN covering the essential requirements of article 3.2 of the R&TTE Directive, February 2008

FCC 04-285, Revision of part 15 of the commission's rules regarding ultra-wideband transmission systems, December 2004

Fu, S.; Zhong, W.; Wen, Y.J. & Shum, P. (2008). Photonic monocycle pulse frequency up-conversion for ultrawideband-over-fiber applications. *IEEE Photonics Technology Letters*, Vol. 20, No. 12, June 2008, pp. 1006-1008, ISSN 1041-1135

Gamage, P.A.; Nirmalathas, A.; Lim, C.; Wong, E.; Novak, D. & Waterhouse, R. (2008a). Multi-services distribution using power-efficient low-cost VCSELs, *Proceedings of International Topical Meeting on Microwave Photonics*, pp. 169–172, ISBN 978-1-4244-2168-8, Gold Coast, Australia, October 2008

Gamage, P.A.; Nirmalathas, A.; Lim, C.; Wong, E.; Novak, D. & Waterhouse, R. (2008b). Performance comparison of directly modulated VCSEL and DFB lasers in wired-wireless networks. *IEEE Photonics Technology Letters*, Vol. 20, No. 24, December 2008, pp. 2102–2104, ISSN 1041-1135

Hanawa, M.; Nakamura, K. & Nonaka, K. (2007). High-quality electrical Gaussian-monocycle pulse generation by electrical-optical hybrid signal processing, *Proceedings of Optoelectronics and Communications Conference*, pp. 138-139, Kanagawa, Japan, July 2007

Hanawa, M.; Mori, K.; Nakamura, K.; Matsui, A.; Kanda, Y. & Nonaka, K. (2009). Dispersion tolerant UWB-IR-over-fiber transmission under FCC indoor spectrum mask, *Proceedings of Optical Fiber Communication conference*, pp. 1-3, San Diego, California, United States, March 2009

Hartmann, P.; Webster, M.; Wonfor, A.; Ingham, J.D.; Penty, R.V.; White, I.H.; Wake, D. & Seeds, A.J. (2003). Low-cost multimode fibre-based wireless LAN distribution system using uncooled directly modulated DFB laser diodes, *Proceedings of European Conference on Optical Communication*, pp. 804–805, Rimini, Italy, September 2003

Hirata, A.; Harada, M. & Nagatsuma, T. (2003). 120-GHz wireless link using photonic techniques for generation, modulation, and emission of millimeter-wave signals. *IEEE Journal of Lightwave Technology*, Vol. 21, No. 10, October 2003, pp. 2145-2153, ISSN 0733-8724

Ingham, J.; Webster, M.; Wonfor, A.; Penty, R.; White, I. & White, K. (2003). Wide-frequency-range operation of a high linearity uncooled DFB laser for next-generation radio-over-fibre, *Proceedings of Optical Fiber Communication conference*, pp. 754–756, ISBN 1-55752-746-6, Atlanta, Georgia, United States, March 2003

Jager, D. & Stohr, A. (2001). Microwave Photonics, *Proceedings of European Microwave conference*, pp. 1-4, London, United Kingdom, September 2001

Jazayerifar, M.; Cabon, B. & Salehi, J.A. (2008). Transmission of multi-band OFDM and impulse radio ultra-wideband signals over single mode fiber. *IEEE Journal Lightwave Technology*, Vol. 26, No. 15, August 2008, pp. 2594–2603, ISSN 0733-8724

Jensen, J.B.; Rodes, R.; Caballero, A.; Yu, X.; Gibbon, T.B. & Monroy, I.T. (2009). 4 Gbps impulse radio (IR) ultra-wideband (UWB) transmission over 100 meters multi mode fiber with 4 meters wireless transmission. *Optics Express*, Vol. 17, No. 19, September 2009, pp. 16898–16903, ISSN 113626

Kaszubowska-Anandarajah, A.; Perry, P.; Barry, L.P. & Shams, H. (2008). An IR-UWB photonic distribution system. *IEEE Photonics Technology Letters*, Vol. 20, No. 22, November 2008, pp. 1884–1886, ISSN 1041-1135

Koonen, A.M.J. & Garcia, L.M. (2008). Radio-over-MMF techniques Part II: Microwave to millimeter-wave systems. *IEEE Journal Lightwave Technology*, Vol. 26, No. 15, August 2008, pp. 2396–2408, ISSN 0733-8724

Kuri, T.; Omiya, Y.; Kawanishi, T.; Hara, S. & Kitayama, K. (2006). Optical transmitter and receiver of 24-GHz ultra-wideband signal by direct photonic conversion techniques, *Proceedings of International Topical Meeting on Microwave Photonics*, pp. 1-4, ISBN 1-4244-0204-2, Grenoble, France, October 2006

Le Guennec, Y. & Gary, R. (2007). Optical frequency conversion for millimeter-wave ultra wideband-over-fiber systems. *IEEE Photonics Technology Letters*, Vol. 19, No. 13, July 2007, pp. 996-998, ISSN 1041-1135

Li, J.; Liang, Y. & Kin-Yip Wong, K. (2009). Millimeter-wave UWB signal generation via frequency up-conversion using fiber optical parametric amplifier. *IEEE Photonics Technology Letters*, Vol. 21, No. 17, September 2009, pp. 1172-1174, ISSN 1041-1135

Liu, C.; Seeds, A.; Chadha, J.; Stavrinou, P.; Parry, G.; Whitehead, M.; Krysa, A. & Roberts, J. (2003). Bi-directional transmission of broadband 5.2 GHz wireless signals over fibre using a multiple-quantum-well asymmetric Fabry–Pérot modulator/photodetector, *Proceedings of Optical Fiber Communication conference*, pp. 738–740, ISBN 1-55752-746-6, Atlanta, Georgia, United States, March 2003

Liu, C.; Polo, V.; Van Dijk, F.; Pfrommer, H.; Piqueras, M.A.; Herrera, J.; Martinez, A.; Karlsson, S.; Kjebon, O.; Schatz, R.; Enard, A.; Yichuan Yu; Tsegaye, T.; Chin-Hsiu Chuang; Seeds, A.J. & Marti, J. (2007). Full-Duplex DOCSIS/WirelessDOCSIS Fiber–Radio Network Employing Packaged AFPMs as Optical/Electrical Transducers. *IEEE Journal Lightwave Technology*, Vol. 25, No. 3, March 2007, pp. 673–684, ISSN 0733-8724

Llorente, R.; Alves, T.; Morant, M.; Beltran, M.; Perez, J.; Cartaxo, A. & Marti, J. (2007). Optical distribution of OFDM and impulse-radio UWB in FTTH networks. *Proceedings of Optical Fiber Communication conference*, pp. 1–3, ISBN 978-1-55752-856-8, San Diego, California, United States, February 2008

Llorente, R.; Alves, T.; Morant, M.; Beltran, M.; Perez, J.; Cartaxo, A. & Marti, J. (2008). Ultra-wideband radio signals distribution in FTTH networks. *IEEE Photonics Technology Letters*, Vol. 20, No. 11, June 2008, pp. 945–947, ISSN 1041-1135

Ma, J.; Yu, J.; Yu, C.; Xin, X.; Zeng, J. & Chen, L. (2007). Fiber dispersion influence on transmission of the optical millimetre-waves generated using LN-MZM intensity modulation. *IEEE Journal of Lightwave Technology*, Vol. 25, No. 11, November 2007, pp. 3244-3256, ISSN 0733-8724

Meryll Lynch (2007). Telecom services-wireless/cellular. Available at: http://www.scribd.com/doc/7656725/ML-Data-Growth

Niiho, T.; Nakaso, M.; Masuda, K.; Sasai, H.; Utsumi, K. & Fuse, M. (2004). Multi-channel wireless LAN distributed antenna system based on radio-over-fibre techniques, *Proceedings of IEEE Lasers & Electro-Optics Society*, pp. 57-58, ISBN 0-7803-8557-8, Rio Grande, Puerto Rico, November 2004

Nkansah, A.; Das, A.; Lethien, C.; Vilcot, J.-P.; Gomes, N.J.; Garcia, I.J.; Batchelor, J.C. & Wake, D. (2006). Simultaneous dual band transmission over multimode fibre-fed indoor wireless network. *IEEE Microwave and Wireless Components Letters*, Vol. 16, No. 11, November 2006, pp. 627-629, ISSN 1531-1309

Ogawa, H.; Polifko, D. & Banba, S. (1992). Millimetre-wave fibre optic systems for personal radio communication. *IEEE Transactions on Microwave Theory and Techniques*, Vol. 40, No. 12, December 1992, pp. 2285–2293, ISSN 0018-9480

Pan, S. & Yao, J. (2009a). Photonic generation of chirp-free UWB signals for UWB over fiber applications, *Proceedings of International Topical Meeting on Microwave Photonics*, pp. 1-4, ISBN 978-84-8363-424-0, Valencia, Spain, October 2009

Pan, S. & Yao, J. (2009b). A photonic UWB generator reconfigurable for multiple modulation formats. *IEEE Photonics Technology Letters*, Vol. 21, No. 19, October 2009, pp.1381-1383, ISSN 1041-1135

Persson, K.-A.; Carlsson, C.; Alping, A.; Haglund, A.; Gustavsson, J.S.; Modh, P. & Larsson, A. (2006). WCDMA radio-over-fibre transmission experiment using singlemode VCSEL and multimode fibre. *IEEE Electronics Letters*, Vol. 42, No. 6, March 2006, pp. 372–374, ISSN 0013-5194

Pfrommer, H.; Piqueras, M. A.; Polo, V.; Herrera, J.; Martinez, A. & Marti, J. (2006). Radio-over-Fibre Architecture for Simultaneous Feeding of 5.5 and 41 GHz WiFi or 4G Access Networks, *Proceedings of IEEE MTT-S International Microwave Symposium*, pp. 301–303, Chicago, United States, June 2006

PriTel, Inc. 1550-nm Actively Mode-locked Fiber Lasers [Online]. Available: http://www.pritel.com

Qian, X.; Hartmann, P.; Ingham, J.D.; Penty, R.V. & White, I.H. (2005). Directly-modulated photonic devices for microwave applications, *Proceedings of IEEE MTT-S International Microwave Symposium*, pp. 4, ISBN 0-7803-8845-3, Long Beach, California, United States, June 2005

Rivas, I. & Lopes, L.B. (1998). A microcellular DCA scheme using variable channel exclusion zones, *Proceedings of IEEE Vehicular Technology Conference*, pp. 1395–1399, ISBN 0-7803-4320-4, Ottawa, Canada, May 1998

SARA Group, Strategic Automotive Radar frequency Allocation, 24 GHz UWB SRR frequency allocation [Online]. Available: http://www.sara-group.org/official_information/specific_decisions_for_different_countries/

Schmuck, H. (1995). Comparison of optically millimeter-wave system concepts with regard to chromatic dispersion. *IEEE Electronic Letters*, Vol. 31, No. 21, October 1995, pp. 1848–1849

Seeds, A.J. & Williams, K.J. (2006). Microwave photonics. *IEEE Journal of Lightwave Technology*, Vol. 24, No. 12, December 2006, ISSN 4628-4641

Shams, H.; Kaszubowska-Anandarajah, A.; Perry, P.; Barry, L.P. (2009a). Optical generation, fiber distribution and air transmission for ultra wide band over fiber system, *Proceedings of Optical fiber communication conference*, pp. 1-3, ISBN 978-1-4244-2606-5, San Diego, California, United States, March 2009

Shams, H.; Kaszubowska-Anandarajah, A.; Perry, P.; Barry, L.P. & Anandarajah, P. M. (2009b). Fiber distribution of IR-UWB signals based on an externally injected gain switched laser, *Proceedings of European Conference on Lasers and Electro-Optics 2009 and the European Quantum Electronics Conference*, pp. 1-1, ISBN 978-1-4244-4079-5, Munich, Germany, June 2009

Wake, D.; Johansson, D. & Moodie, D.G. (1997). Passive picocell—A new concept in wireless network infrastructure. *IEEE Electronics Letters*, Vol. 33, No. 5, February 1997, pp. 404–406, ISSN 0013-5194

Wilson, B.; Ghassemlooy, Z. & Darwazeh, I. (Eds.) (1995), *Analogue Optical Fibre Communications*, Institution of Engineering and Technology (15 Aug 1995), ISBN 978-0852968321, London, United Kingdom

WiMedia Alliance, Worldwide regulatory status [Online]. Available: http://www.wimedia.org

Wu, X.; Christen, L.; Yang, J.; Nuccio, S.R.; Willner, A. & Paraschis, L. (2007). 40-GHz CSRZ optical pulse generation using a 10-GHz Mach-Zehnder modulator and a 25-ps delay line interferometer, *Proceedings of IEEE Lasers and Electro-Optics Society*, pp. 882-883, ISBN 978-1-4244-0925-9, Florida, United States, October 2007

Yu, X.; Gibbon, T. B. & Monroy, I. T. (2009). Experimental demonstration of all-optical 781.25-Mb/s binary phase-coded UWB signal generation and transmission. *IEEE Photonics Technology Letters*, Vol. 21, No. 17, September 2009, pp. 1235-1237, ISSN 1041-1135

8

Time-Spectral Visualization of Fundamental Ultrafast Nonlinear-Optical Interactions in Photonic Fibers

Anatoly Efimov
Center for Integrated Nanotechnologies, Los Alamos National Laboratory
USA

1. Introduction

Recent renewed interest to nonlinear optical processes in waveguiding structures was sparked in large part by the original reports on broadband supercontinuum (SC) generation in photonic crystal fibers (PCF) and uniform fiber tapers (Ranka et al., 2000; Birks et al., 2000) introduced a few years earlier by the group in Bath (Knight et al., 1996; Knight, 2003). The significance of these first reports stemmed from the fact that it was sufficient to use femtosecond oscillators only (nanojoule pulse energies) to produce octave-spanning continua in the visible. The fact that the broadband spectra emitted from a PCF pumped with femtosecond pulses centered at 800 nm covered all of the visible range and thus were easily observed with the naked eye as a spectacular rainbow of colors must have had something to do with the subsequent popularity of this research area. This is because in the previous decade and even much earlier a number of works reported broadband continuum generation in the near-infrared (Lin et al., 1978; Morioka et al., 1994), which remained unnoticed by the wider optics community. Those older studies were usually performed with pump sources at 1550 nm and were targeted towards telecom applications at that time. Subsequent studies unveiled important differences in SC generation mechanisms and their properties depending on the fiber or PCF parameters and the duration of the pump pulse. One very important advantage of a SC generated with femtosecond pulses is its high degree of coherence over the whole or most of the generated spectrum (Dudley et al., 2006), making it useful for a number of advanced applications, such as absolute phase stabilization of ultrashort pulses (Jones et al., 2000) and frequency metrology (Udem et al., 2002).
Importantly, it was soon realized that PCFs and other engineered structures allow for much greater flexibility in dispersion design, in addition to high effective nonlinearities due to strong confinement of the transverse optical mode. In large part it was the particular shape of the dispersion profile of the PCF as a function of the wavelength, which fortuitously allowed for broadband SC generation in the early experiments. The importance of dispersion should not be underestimated as it controls various phasematching conditions for nonlinear optical processes, such as emission of Cherenkov continuum (responsible for the visual portion of the typical SC) by solitons (Wai et al., 1990; Karpman, 1993; Akhmediev & Karlsson, 1995). Solitons themselves play an important role in the process of SC generation and thus must be studied carefully. However, spectroscopic measurements are

not sufficient in many situations to uncover the fundamental physics involved. The same example of SC generation clearly demonstrates this: By looking at the spectrum recorded at the output of a long PCF it is virtually impossible to extract any useful information on the underlying dynamical mechanisms in play behind its generation, even though the excitation pulse had a well-defined temporal structure, usually a transform-limited pulse of a few tens of femtoseconds in duration.

Temporal resolution in the measurements, thus appears as important as spectral one and a number of time-resolved techniques exist, which are applicable to waveguiding geometries. Importantly, however, any linear interferometric method, such as spectral interferometry, can not be applied to the highly nonlinear propagation regimes of interest, because the output spectrum differs significantly from the input one. Thus, nonlinear methods (featuring ultrafast nonlinear gating) must be used. Two encompassing domains can be identified: auto-correlation and cross-correlation methods. The autocorrelation approaches are very well suited for the characterization of unknown pulses, e.g. the output pulse from a femtosecond laser (Trebino, 2000). Here, the unknown pulse gates itself, so to speak, to produce a nonlinear autocorrelation trace of a specific shape, depending on the particular nonlinear mechanism used, which carries information about the amplitude and the phase of the unknown pulse, which can be extracted. Often times some fundamental limitations apply, such as time-reversal symmetry for second harmonic generation (SHG) frequency-resolved optical gating (FROG). In addition, technical difficulties arise if complex or broadband pulses are being measured, such as phasematching bandwidth requirements for SHG FROG or spectral phase interferometry for direct electric field reconstruction (SPIDER), or sensitivity in polarization-gating FROG and others. This is why the very first time-resolved images of supercontinuum generated in silica PCFs were obtained using cross-correlation methods—the sum-frequency generation (SFG) X-FROG (Dudley et al., 2002) and cross-correlation sonogram measurements (Hori et al., 2004). In a typical cross-correlation approach the signal pulse is gated with a different reference pulse, usually a replica of the input pump pulse split off before the nonlinear medium under study and appropriately delayed. While sonogram measurements require both delay and spectral filter scanning to obtain the 2D time-frequency image of the pulse (Taira & Kikuchi, 2001), the SFG X-FROG approach requires only one delay scan with wavelength resolution achieved by spectrally resolving the upconverted sum-frequency signal, and thus is simpler to implement with off the shelf components.

For our intended work on visualizing weak and broadband signals the optimal measurement technique must be sensitive and broadband itself, as well as yield the most intuitive images of the output pulse possible. Arguably, the SFG X-FROG fits the bill very well. In this second-order nonlinear cross-correlation type measurement technique a portion of the pulse from the laser is used to gate the signal pulse. The total two-dimensional spectrogram is built by scanning the delay between the signal and reference pulses and acquiring the spectrum of the SFG signal (Linden et al., 1998):

$$I_{X-FROG}^{SFG}(\omega,\tau) = \left| \int_{-\infty}^{+\infty} E_{sig}(t)\, E_{ref}(t-\tau)\, e^{-i\omega t} dt \right|^2 , \tag{1}$$

where $I_{X-FROG}^{SFG}(\omega,\tau)$ is the intensity of the X-FROG signal as a function of optical frequency ω and delay τ produced by cross-correlating the signal $E_{sig}(t)$ with the delayed reference

$E_{ref}(t-\tau)$ electric field pulses. Expression (1) is written specifically for sum-frequency $\omega_{SF} = \omega_{sig} + \omega_{ref}$ generation X-FROG where the product of two fields is obtained through SFG in a thin crystal with instantaneous quadratic nonlinearity and the Fourier transform is obtained using a spectrometer. The focal plane detector of the spectrometer measures the spectral intensity, hence the spectrogram is proportional to the modulus squared of the spectrum of the sum-frequency generated signal. Special reconstruction algorithms can be used to recover the complete temporal behavior of the electric field, or equivalently, the amplitude and the phase of the optical pulse in either temporal or spectral domains. Most importantly, however, the information conveyed by the 2D X-FROG spectrogram itself allows one to grasp the whole picture of a complex ultrashort pulse visually. The fact that a spectrogram may be a much better representation of a complex pulse, than a pair of phase and amplitude curves in either time or spectral domains, was recognized by Treacy back in 1971 (Treacy, 1971). Moreover, highly intuitive nature of the (nearly) constrain-free X-FROG spectrograms is becoming increasingly attractive for theorists as well, who often find it to be the most suitable way to represent data from numerical simulations (Hilligsøe et al., 2004; Skryabin & Yulin, 2005). This fact facilitates direct visual comparison between theory and experiment, as is shown below.

X-FROG can be used to study a wide variety of propagation effects ranging from well-established processes to novel nonlinear regimes with ultrashort optical pulses, continuous waves (!) and their interactions. In what follows, our X-FROG setup developed for experimental work at telecom wavelength with arbitrary fibers is briefly described first (Efimov & Taylor, 2005a). Several illustrative examples of soliton formation in 100 meters of regular fiber and Raman soliton dynamics in standard cobweb PCFs (Efimov & Taylor, 2005b) are presented next, followed by more interesting cases of Cherenkov continuum generation by stabilized solitons in the vicinity of a second zero-dispersion (2ZD) point of a small-core PCF (Efimov et al., 2004). Resonant interactions of solitons and dispersive waves (Efimov et al., 2005c; Efimov et al., 2006) are illustrated next where excellent agreement with theoretical predictions was obtained. Further, experimental measurements and numerical simulations of SC in highly-nonlinear soft glass SF6 PCF are compared, where again we find excellent agreement, except for the solitonic part of the trace, where temporal jitter of solitons smears the experimental picture, revealing the effect of subtle instabilities present at the output of the pump laser used in the experiment (Efimov & Taylor, 2008). Finally, recent results on soliton dynamics and continuum generation across the bandgaps of an all-glass photonic bandgap fiber (PBF) are summarized before concluding this chapter.

2. X-FROG system and simple visualization examples

Performing cross-correlation type measurements requires matching optical propagation path lengths for both the signal and the reference arms. This is easily done with mechanical delay stages for relatively short samples, up to a meter or so long. Longer fiber samples will require more clever approaches. One interesting solution was offered in (Nishizawa & Goto, 2001) where the reference pulse is derived from the part of the generated signal as the red-most soliton, filtered out by a long-pass filter. An adjustable delay line was still needed after the fiber sample to provide delay scanning. Hori, et al. built a completely electronically controlled delay scanning using a separate fiber in the reference arm of their cross-correlation sonogram system, where an acousto-optic modulator was used to vary the input power launched in the reference fiber, thus adjusting the delay of the propagating Raman soliton (Hori et al., 2003). In this approach the delay variation is intimately connected with soliton's central frequency. Although certainly creative, these delay scanning tactics are somewhat complicated and not universally applicable to any experimental situation. Thus, in our system we used a simple free-air delay stage and relied on the fact that, aside from the carrier-envelope phase (which is not measured by X-FROG) all the pulses in a pulse train produced by the laser are identical and therefore different pulses can be used as signal and reference.

The X-FROG system designed for use with arbitrarily long fibers is shown in Fig. 1. An optical parametric oscillator (OPO) generates a train of 100 fs pulses tunable in the range 1400-1600 nm with repetition rate of 82 MHz and average power up to 350 mW. A polarizing beam splitter separates the signal and the reference pulses. The former is coupled into the fiber under study using an aspheric lens and the latter propagates in a free-space 4-meter delay stage, equipped with precision retroreflecting corner cubes. The delay line of the X-FROG system is designed to provide optical delays in excess of the time spacing between individual pulses in the pulse train, thus allowing cross-correlation of different pulses within the pulse train generated by the OPO. The jitter of pulse temporal positions within the pulse train does not influence the results even for 100 meter-long fibers. The reference and the signal pulses are subsequently mixed in a 200 μm thick BBO crystal to generate sum-frequency signal, which is spectrally resolved in a grating spectrometer. The phasematching bandwidth of the nonlinear crystal at this particular wavelength region is sufficient to accurately convert more than an octave of signal spectrum, due to favorable material dispersion in the crystal.

Phasematching bandwidth available for SFG is, in fact, one of the most critical parameters of the whole system. From the long history of autocorrelation measurements of ultrashort pulses it is known that sufficiently thin crystals are required since the phasematching bandwidth is inversely proportional to the length of the nonlinear medium (Weiner, 1983). Interestingly, chirped pulses may appear shorter if measured with thick nonlinear crystals. This is because the narrow phasematching bandwidth effectively cuts the wings of the

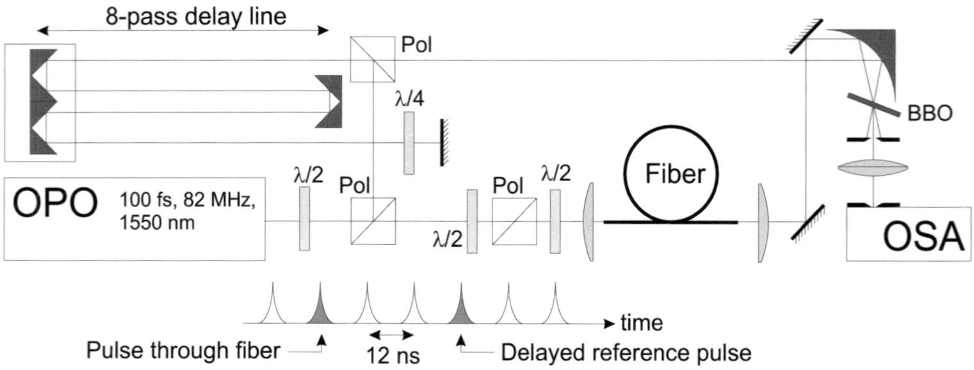

Fig. 1. SFG X-FROG system for studying pulse dynamics in fibers of arbitrary length. Schematically shown at the bottom is the pulse train generated by the laser system with separate reference and signal pulses shaded. Pol—polarizer, λ/2 and λ/4 are the waveplates.

pulse's spectrum, which for a chirped pulse, correspond to the leading and trailing edges of the pulse in time, making the autocorrelation trace appear shorter. For a uniaxial crystal, such as BBO, there are two phasematching possibilities for SHG: Type 1 $o+o \rightarrow e$, and Type 2 $o+e \rightarrow e$, where o and e denote "ordinary " and "extraordinary" rays respectively. For the SFG process, however, used in X-FROG there exist two distinct Type 2 processes: $o_{sig}+e_{ref} \rightarrow e_{SF}$ or $e_{sig}+o_{ref} \rightarrow e_{SF}$ because the signal and the reference are not identical. Using the "ordinary" and "extraordinary" refractive index data for the nonlinear crystal used in the system, all three cases may be examined for phasematching bandwidth, Fig. 2, and the most optimal configuration selected for the experiment. In Fig. 2 all three phasematching cases are represented for the reference wavelength of 1.55 μm and 200 μm thick BBO crystal. Clearly, more than an octave bandwidths are possible in all three cases. Most of the experiments described below were performed around the pump wavelength of 1.55 μm so the most optimal case for us is the Type 1 process, which also provides almost 2000 nanometers of bandwidth.

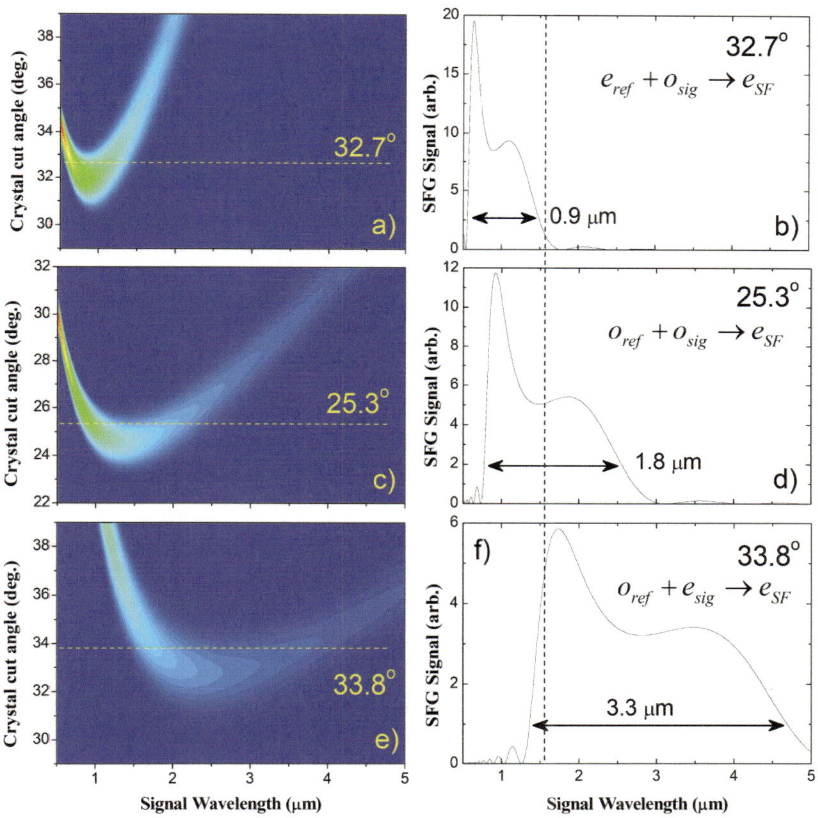

Fig. 2. Phase matching curves for non-collinear sum-frequency generation of a fixed wavelength reference pulse at 1.55 μm and variable wavelength signal pulse in a 200 μm BBO crystal. Left: density plots for crystal cut angle vs. signal wavelength. Dashed lines show most optimal cut angles for obtaining large phasematching bandwidths, and corresponding phasematching curves are plotted on the right for three possible arrangements. Vertical dashed line indicates λ = 1.55 μm central wavelength.

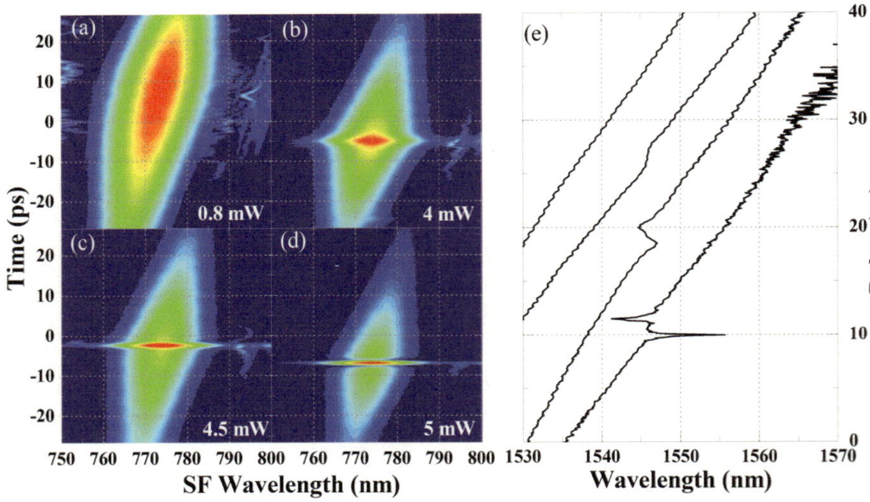

Fig. 3. X-FROG traces at the output of 100 meters of 3M FS-PM-7811 fiber. Average input power increases sequentially from (a) 0.8 mW, to (b) 4 mW, to (c) 4.5 mW to (d) 5 mW. Color scale mapping is logarithmic with the SFG signal intensity. (e) Wavelength-dependent group delay for several input powers obtained directly from respective X-FROG traces in (a-d). Curves are offset for clarity. Power increases from top to bottom.

As a simple demonstration of SFG X-FROG sensitivity and very intuitive traces produced, consider the problem of soliton formation in a regular single-mode fiber. Fig. 3 displays a series of traces taken at the output of a 100 meter-long piece of single-mode fiber (3M FS-PM-7811). Sensitivity of the system is sufficient to obtain more than 40 dB of signal-to-noise ratio even in the linear pulse propagation regime, Fig. 3a. As the input power is increased, soliton formation is observed with linear waves present on the wings, Fig. 3b-d. Naturally, blue (red) spectral components of the linear wave lead (trail) the soliton. As the input power is increased, a larger fraction of the input pulse energy contributes to the soliton.

It is interesting to observe the dynamics of the wavelength-dependent group delay as a function of the input power directly from the X-FROG trace, i.e. without applying reconstruction algorithms, as was suggested in (Treacy, 1971). The data shown in Fig. 3e is obtained by following the "center of mass" of each spectrum slice from the corresponding spectrogram in time. Each curve in the figure in sequence from top to bottom is obtained from the respective X-FROG trace from Fig. 3a-d. Formation of group delay features near the soliton position is clearly visible. Qualitatively, the behavior of the group delay versus wavelength is easy to understand when the soliton is almost formed: At the wings away from soliton the linear chirped wave gives rise to the wavelength-dependent delay due to dispersion in the fiber. The soliton, however, being a nonlinear wave maintains all its wavelength components at the same delay (no chirp) which corresponds to the near-horizontal section of the curve at the location of the soliton.

Soliton propagation at telecom wavelengths in a standard cobweb PCF presents another interesting example for visualization of the nonlinear dynamics. The fiber used for these experiments is a high-delta silica microstructured fiber with a 2.5 μm guiding core, 63 mm in length. Fig. 4 shows a series of X-FROG traces recorded at the output of the PCF for varying input powers of the pump pulse. The input wavelength is located well within the

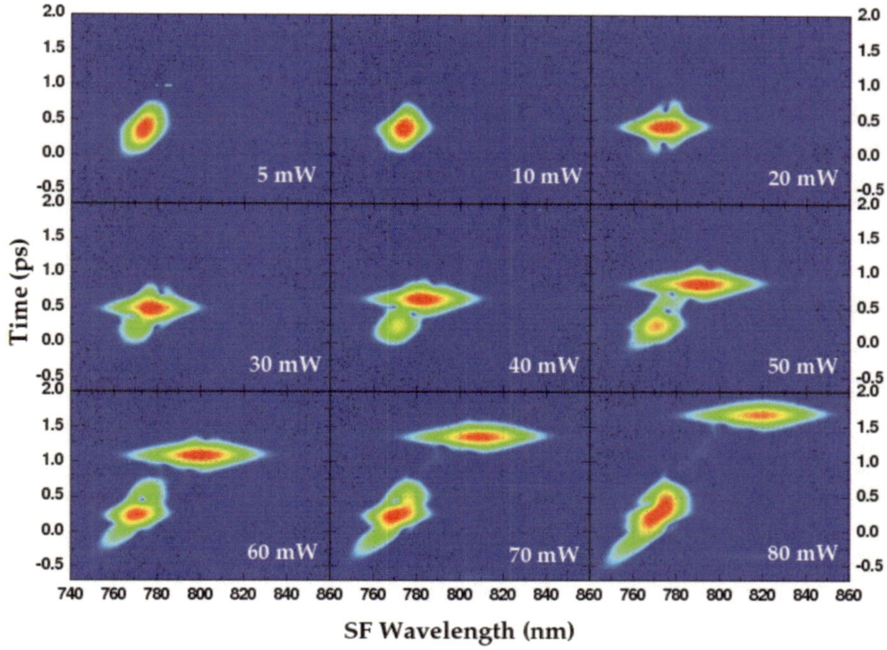

Fig. 4. Power-dependent X-FROG traces at the output of 2.5 μm-core cobweb PCF, 63 mm in length. Formation of a Raman-shifting soliton is clearly observed. Horizontal axis corresponds to sum-frequency signal wavelength. Color scale is logarithmic

anomalous dispersion region of this fiber so no broadband supercontinuum generation is observed. At high powers, however, intermodally phase-matched third harmonic generation occurs in which the Raman soliton dynamics plays crucial role (Efimov & Taylor, 2005b). At 5 mW the propagation of the fundamental pulse is nearly linear, which can be inferred from the elliptical shape of the trace as well as parabolic phase from X-FROG reconstruction, Fig. 5, in both temporal and spectral domains. Soliton formation sets in at about 10 mW average power, at which point the X-FROG trace assumes the required diamond-like shape, characteristic to the Sech(t) temporal amplitude profile. Spectral and temporal phase functions, Figs. 5a and 5b show substantial flattening, as expected, for 10 mW and 20 mW input powers. Using the parameters of this silica PCF (effective area $A_{eff} = 5\mu m^2$, nonlinear index $n_2 = 3 \cdot 10^{-16} cm^2 W^{-1}$, dispersion $D = 200 ps \cdot nm^{-1} km^{-1}$) and input pulse duration of 100 fs we can estimate the average power, corresponding to the fundamental soliton, to be about 17 mW, which agrees well with the experiment if the fiber coupling efficiency of ~60% is taken into account.

With further increase of the input power pulse splitting and Raman self-frequency shift of the main soliton can be clearly seen. The frequency shifting soliton maintains the diamond-like X-FROG shape which is a clear evidence for the solitonic Sech(t) amplitude profile. From the reconstructed spectral phase functions we see that the Raman soliton remains flat-phased (the tilt of the phase function corresponds to linear shift on the time axis), whereas the pulse remaining in the vicinity of the pump wavelength is, in fact, chirped, as evidenced by the curved phase in Fig. 5b. The wavelength shift of the Raman soliton is accompanied by the delay offset in accordance with the anomalous dispersion of the fiber. With increasing

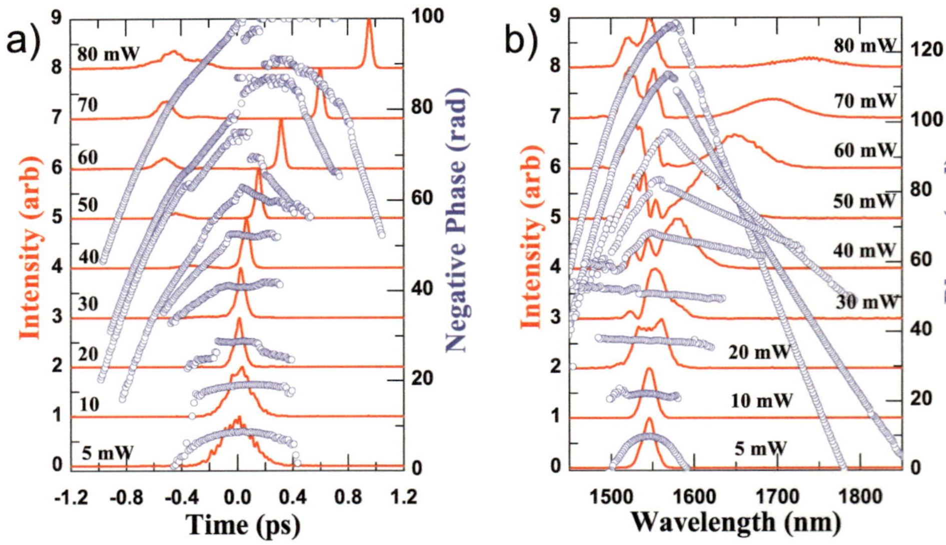

Fig. 5. Intensity (colid curves) and phase (circles) of the pulse in temporal, (a), and in spectral, (b), domains at the output of the PCF reconstructed from the experimental X-FROG spectrograms of Fig. 4. Reconstruction error does not exceed 0.002 on 256 × 256 grid for all power levels. Input power increases from bottom to top as indicated.

input power the spectral width of the Raman soliton increases and at the same time the temporal duration of the soliton reduces since the Raman soliton constitutes a transform-limited pulse. At the 80 mW input power the intensity full-width at half-maximum of the Raman soliton shrinks to 42 fs, which gives the soliton compression ratio of ~2.5 times.

3. Soliton dynamics near the second dispersion zero in a small-core PCF

Novel regimes of nonlinear soliton dynamics can be realized in strongly guiding waveguides. When the waveguide is designed for strong confinement of the mode, usually by using large index difference between the core and the cladding, the waveguide contribution to the total dispersion may become compatible and even exceed the material dispersion. For such waveguides with core diameters on the order of the wavelength a second zero dispersion (2ZD) point may exist at a longer wavelength. The dispersion slope $\frac{\partial D(\lambda)}{\partial \lambda}$ in the vicinity of the 2ZD point is negative, whereas it is positive near the first zero-dispersion point, typical for the regular fibers and previous generations of PCFs (Ranka et al., 2000; Dudley et al., 2002; Dudley et al., 2006). Theoretical predictions (Skryabin & Yulin, 2005; Biancalana et al., 2004) and previous experiments (Harbold et al., 2002; Skryabin et al., 2003; Genty et al., 2004) suggested unusual soliton dynamics in the vicinity of the 2ZD point. These effects may have far-reaching consequences since the wavelength of the 2ZD point can be tuned to the telecommunications region around 1550 nm through careful control of the fiber structure during manufacturing.

Figure 6 shows the dispersion and the optical loss for two of the fibers used in our experiments. The 2ZD points for two fibers shown are located at 1510 and 1560 nm and measured dispersion slopes are large and negative. The dispersion of a regular fiber is

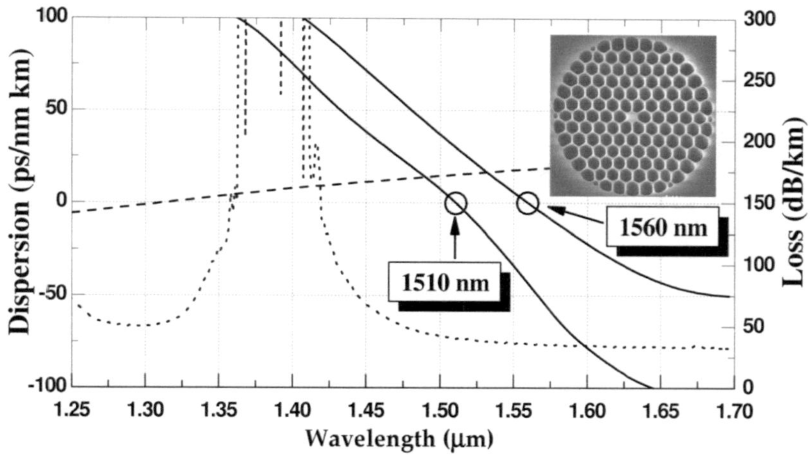

Fig. 6. Optical properties of small-core PCFs. Solid lines—dispersion curves for two distinct fibers; dotted curve—attenuation. For comparison the dispersion of a regular telecom fiber is shown with the dashed curve. Inset—scanning electron microscope image of one of the PCFs; the central guiding core diameter is about 1.2 μm.

shown for comparison in the figure with a dashed curve. The small core of the fiber also provides for high peak intensities for the propagating optical pulses. The nonlinear parameter for these PCFs is estimated to be $\gamma \sim 100 \cdot W^{-1}km^{-1}$, about two orders of magnitude larger than in standard telecom fibers. Note that in the vicinity of the 2ZD point, anomalous dispersion regime is realized at shorter wavelengths whereas normal dispersion lies at longer wavelengths. Thus, solitons launched on the anomalous dispersion side are pushed toward the 2ZD by the Raman effect, rather than away, as would be the case near the first ZD point. One interesting consequence of this fact is that the soliton gets stabilized near the 2ZD through spectral recoil while emitting the Cherenkov continuum across the 2ZD point at longer wavelengths (Skryabin et al., 2003) according to the phase-matching condition (Lu et al., 2005):

$$\Delta\beta \equiv \beta(\omega_s) - \beta(\omega_{cher}) + \frac{\omega_{cher} - \omega_s}{v_g} + \gamma P_s = 0 , \qquad (2)$$

where β is the propagation constant, ω_s and ω_{cher} are the soliton and the Cherenkov continuum frequencies, respectively; v_g is the soliton's group velocity; γ is the fiber's nonlinear coefficient, and P_s is the soliton peak power. Since the soliton never crosses the 2ZD point, it is clear, also, that the soliton can potentially transfer all of its energy to the Cherenkov continuum (Chen et al., 2009).

Detailed time-spectral dynamics of a soliton propagating in the vicinity of 2ZD point of a small-core PCF, observed with X-FROG is shown in Fig. 7. The PCF is pumped in the anomalous dispersion region at 1440 nm (λ_{SF} = 720 nm). Under these conditions most of the pump energy contribute to form a soliton, which subsequently self-frequency shifts towards the red and approaches the 2ZD point. When sufficient portion of the soliton's spectrum extends across the 2ZD point ($\lambda_{2ZD} = 1560$ nm , which corresponds to SF wavelength of 750 nm) into the normal dispersion region, efficient generation of Cherenkov continuum

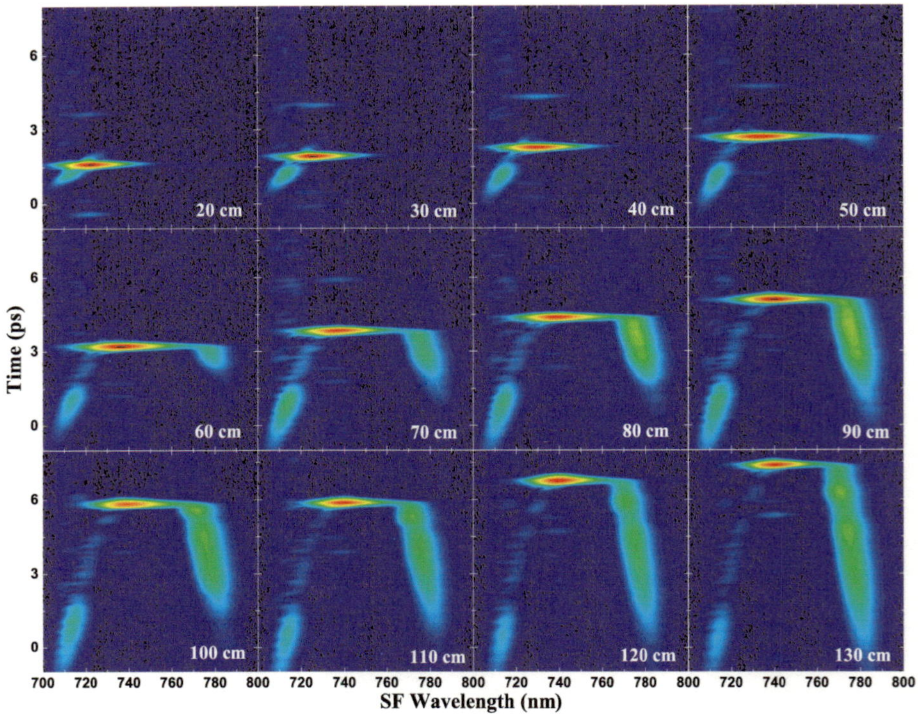

Fig. 7. Experimental X-FROG spectrograms detailing the dynamics of femtosecond soliton propagation in a small-core PCF obtained using a cut-back technique. Input power is held constant corresponding to the fundamental soliton. Raman self-frequency shift seizes as soon as appreciable amount of continuum is generated at about 70 cm. Soliton is slowly loosing energy to the Cherenkov continuum which causes it's spectrum to shrink in width. Color scale is logarithmic.

occurs. In the experimental X-FROG traces of Fig. 7 this continuum manifests itself through a long tail at SF wavelengths longer than 760 nm (1600 nm fundamental) in the normal dispersion region of the fiber. Spectral recoil from the continuum stabilizes the soliton at a wavelength near to the 2ZD point, (after 70 cm of propagation). With further propagation the central wavelength of the soliton remains virtually unchanged due to the appearance of a recoil "force" acting on the soliton's spectrum and balancing the Raman self-frequency shift. Note that there is a continuous energy flow from the soliton to the continuum through uninterrupted spectral overlap (Akhmediev & Karlsson, 1995). Because the soliton is loosing energy to Cherenkov continuum, it's duration must adiabatically increase in order to maintain $N = 1$ soliton number:

$$N^2 \sim \frac{\gamma \cdot E \cdot T_0}{|\beta_2|}, \tag{3}$$

where N is the soliton number, E and T_0 are soliton energy and duration, and β_2 is the dispersion parameter of the fiber. This increase in duration is manifested in spectral width reduction of the soliton in Fig. 7, even though the 100 fs long reference pulse used in these measurements limits the direct temporal resolution of the effect.

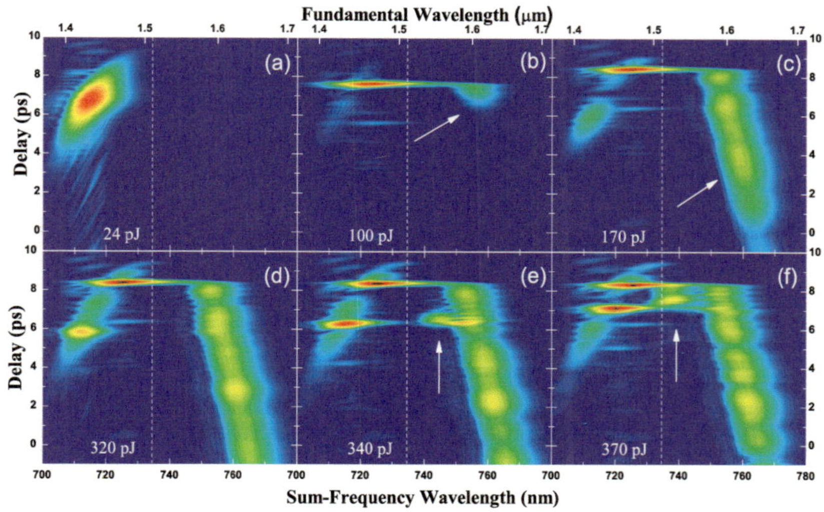

Fig. 8. Power-dependent X-FROG spectrograms measured at the output of a 130 cm-long PCF with $\lambda_{2ZD} = 1510nm$. The fiber is pumped at 1430 nm, i.e. in the anomalous dispersion region. At lower input powers formation of the primary strong soliton and its Cherenkov radiation are observed, arrows in (b) and (c). Subsequently, a second weaker soliton is formed which interacts with the continuum emitted by the first soliton. The scattering of the Cherenkov radiation by the second soliton leads to energy transfer into a spectral band located between the solitons and the Cherenkov radiation band, arrows in (e) and (f). Vertical dashed line marks 2ZD. Color scale is logarithmic.

More interesting things begin to happen if we increase the input power to the level when two solitons can form from the initial input pulse, Fig. 8. It appears that, contrary to the common belief, the solitons can strongly interact with continuous waves the result of which seems to be the generation of new signals at some non-degenerate frequencies. There is no big surprise here, however, because solitons do not inelastically interact with cw only in integrable systems, described by the standard nonlinear Schrodinger (NLS) equation. Large perturbations either due to dispersion or birefringence change the picture dramatically. As described below, nonintegrability leads to observable and quite spectacular scattering of continuous waves on solitons.

4. Scattering of continuous waves on solitons

Interaction of solitons and continuous waves has been extensively studied in the past. In the framework of standard NLS equation the soliton-cw mixing results in small changes in soliton's position, amplitude and phase. When the cubic dispersion term is included in the propagation equation, the soliton can emit Cherenkov continuum, as described previously. The most recent theoretical analysis (Skryabin & Yulin, 2005), however, predicts the existence of two more phasematched processes in which mixing of a soliton and a cw leads to generation of new spectral components. Physically, the four-wave mixing (FWM) between cws (linear waves) and solitons (nonlinear waves) is different from the textbook case of all-cw FWM (Agrawal, 2001) because the soliton exchanges momentum with other waves as a whole. Therefore, we can no longer associate a particular momentum with a certain frequency component of the soliton as is possible for a monochromatic cw. From a practical standpoint, the frequency components comprising the soliton's spectrum are dispersionless (see constant group delay within soliton's spectrum in Fig. 3e). Specifically, the wavenumber associated with a certain frequency component of a soliton is a linear function of frequency within the soliton's spectrum $k_{s/\omega}(\omega) = k(\omega_s) + \frac{\omega - \omega_s}{v_g(\omega_s)} + \gamma P_s$, where ω_s is the carrier frequency of the soliton. This same expression enters the phasematching condition for Cherenkov continuum emission, Eq. (2), which in turn is only one of three possible resonances in a more general set of equations for the frequencies of new signal waves generated as a result of soliton-cw interaction:

$$k_{s/sig}\left(\omega_{sig}\right) + J\left[\beta_{cw}\left(\omega_{cw}\right) - k_{s/cw}\left(\omega_{cw}\right)\right] = \beta_{sig}\left(\omega_{sig}\right) \tag{4}$$

where J can take on three values: 0, -1, and +1. In equation (4), derived using the perturbation theory applied to the generalized NLS equation of propagation, the J = 0 case describes the Cherenkov continuum emission and does not include any separate cw pump. The other two nonzero J cases describe the resonances involving a soliton *and* a separate cw pump. Solving these equations yields the frequencies ω_{sig} of newly generated signals. A number of examples were illustrated in the theoretical work (Skryabin & Yulin, 2005) and are not repeated here.

In addition to dispersion, inelastic scattering of a cw on a soliton can be induced by birefringence of the fiber, provided the soliton and the cw pump propagate in orthogonal eigenpolarization states. To account for birefringence, the $J \neq 0$ equations are easily generalized:

$$J = +1: \quad k_{s/sig}^{(x)} - k_{s/cw}^{(x)} = \beta_{sig}^{(y)} - \beta_{cw}^{y} \quad \text{phase-insensitive} \tag{5}$$

$$J = -1: \quad k_{s/sig}^{(x)} + k_{s/cw}^{(x)} = \beta_{sig}^{(y)} + \beta_{cw}^{y} \quad \text{phase-sensitive} \tag{6}$$

where superscripts (x) and (y) denote separate eigenaxes. Physically, the interaction between the orthogonally polarized fields is mitigated by cross-phase modulation terms in the pair of NLS equations written for each eigenpolarization. The phase-matching equations (5) and (6) above are also marked as phase-sensitive and phase-insensitive to highlight the important difference between these two resonances: For the phase-insensitive resonance, both phase and group velocity differences cancel out on the left- and right-hand side of Eq. (5). On the other hand, the phase-sensitive resonance (6) retains the dependence not only on the difference in group velocities, but also the difference in phase velocities between the two eigenaxes of the fiber.

In the experiment of Fig. 8 the signal marked with the arrow in Fig. 8e and f is exactly the result of the resonant interaction of the second soliton with the continuum emitted by the first soliton. Close inspection reveals that the J = +1 phase-insensitive resonance is responsible for the emission of the signal at 1.53 μm between the solitons and the Cherenkov tail. The results of numerical modeling for the conditions of the experiment (not shown) were found to match the experimental results nearly perfectly, which further validated the theory of resonant scattering of continuous waves on solitons outlined above.

It is interesting, however, to perform soliton-cw interaction experiments in the most pristine manner, with separate lasers as sources of solitons and continuous waves, thus directly testing the predictions of the theory. To do this, a small-core PCF with 2ZD point at 1510 nm was chosen for its high effective nonlinearity and large dispersion slope to render the system non-integrable due to high-order dispersion. The length of the fiber used was 90 cm. The OPO served as the source of ~100 fs solitons, while the cw field at 1546 nm was generated by a *separate* temperature-controlled laser diode followed by an erbium-doped fiber amplifier. The cw signal is thus located in the normal dispersion region, and the power coupled to the fiber is estimated at ~0.1 W.

The results of this experiment are shown in Fig. 9. The soliton peak power increases from 1.7 to 4.4 to 7 kW for the 1st, 2nd, and 3rd columns respectively. The bottom row in Fig. 9 shows experimental X-FROGs for the case when no cw pump was launched into the fiber. From the top two rows, one can see that as the soliton power increases its interaction with the cw radiation becomes more and more pronounced. There are two reasons for this: The first is that, for this particular phase-insensitive interaction the FWM term responsible for the interaction between the soliton and cw pump is proportional to the modulus squared of the soliton field and only to the first power of the (unconjugated) cw field (Efimov et al., 2005c). Second, and more important, is that, with the increase of the pump power, the soliton

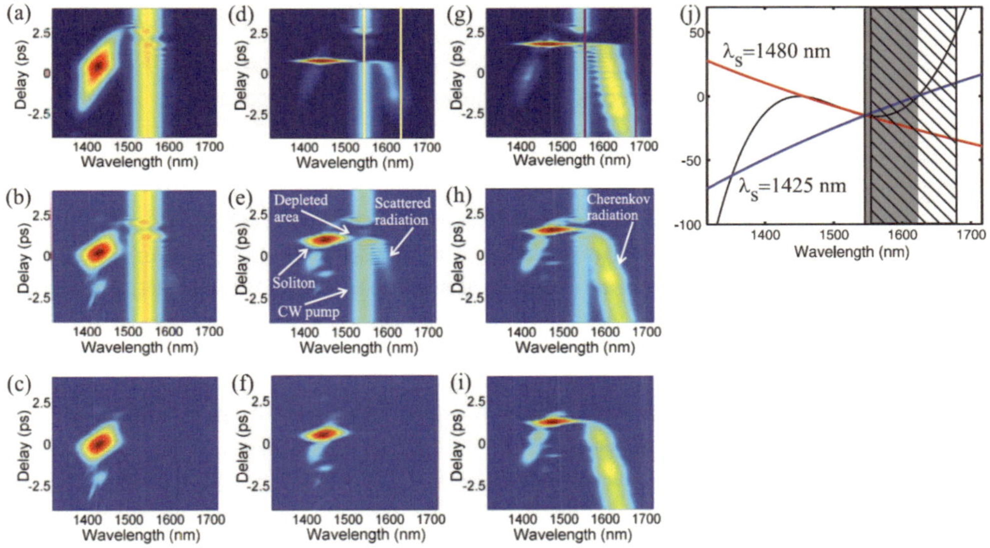

Fig. 9. Interaction between the soliton and cw. First row of spectrograms, (a, d, g)-numerical modeling, second row, (b, e, h)—experimental measurements, and. third row, (c, f, i) shows experimental measurements with the cw pump switched off. 1st, 2nd, and 3rd columns correspond to the soliton peak power 1.7, 4.4, and 7 kW, respectively. Propagation length is 90 cm and cw power is 0.1 W for all the panels. Panel (j) shows the graphical solution to the wave vector matching condition (4) for the two values of the soliton central wavelength. Grey area shows the FWM band, dashed area shows the Cherenkov band. Color scale used in spectrograms is logarithmic.

becomes temporally shorter, and, therefore, the Raman effect brings the soliton closer to the cw frequency, which strongly enhances the efficiency of the FWM process.

Comparison of theoretical and experimental results unambiguously indicates that the frequency component to the longer wavelength side of the cw line in Figs. 9d and e is generated as a result of the $J = +1$ phase-matched FWM between the soliton and the cw pump. When the cw pump is switched off, Fig. 9f, the FWM signal disappears. Another important observation which can be made here is the appearance of the hole in the cw pump; see Figs. 9e and h. This hole is formed because the energy taken from the cw pump is transferred (scattered) by the soliton to the new spectral component.

Conveniently normalized and rescaled geometrical representation of the wave vector matching (4) is shown in Fig. 9j, where the nearly straight lines represent the left-hand side of Eq. (4) for the two different values of the soliton wavelength, and the curved line is the right-hand side of Eq. (4). Since the soliton wavelength is being shifted by the Raman effect, the matching point of the FWM process also shifts with both propagation distance and pump power. The shaded region in Fig. 9j and the white lines in Fig. 9d show the boundaries of the FWM band. The excitation efficiency of the other FWM resonances located to the left of the shaded region is low, and they are not observed in the experiment.

If the input soliton power is made large (third column in Fig. 9) the efficient emission of the Cherenkov continuum by the soliton itself overwhelms the FWM signal, because for this particular PCF the Cherenkov and FWM resonances overlap in large part (see the overlap between the dashed and shaded regions in Fig. 9j). Comparison of Figs. 9g–i clearly shows that this strong radiation is indeed the Cherenkov one, because its emission occurs independently of the cw pump. Thus, in the case when the soliton and cw pump have the same polarizations, observation of the newly generated wave can be achieved only for a relatively small power range of the soliton pump. Other experiments performed using birefringent PCFs help to reveal more details about soliton-cw scattering as well as to observe the other, *phase-sensitive* resonance, described by equation (6).

If the soliton and cw pumps are orthogonally polarized, propagating in a birefringent fiber, then the difference between the phase-sensitive and phase-insensitive scattering should be particularly revealing. This is because the condition of the phase-sensitive resonance depends strongly on the phase velocity difference of the two polarizations, and therefore spectral location of this resonance can be expected to depend strongly on the choice of polarizations for the cw and the soliton pumps. Thus, in what follows we used a commercially available birefringent PCF with a 2.5 μm core. This fiber is anomalously dispersive in the wavelength range of interest around 1.55 μm with third- and higher-order dispersion terms being not important at all. Therefore, the non-integrability of the system, as the key requirement for existence of inelastic soliton-cw scattering, stems here from birefringence, rather than dispersion.

Again, we use separate lasers as sources of solitons and cw. Figure 10 shows the graphical representation of phase-matching conditions (5) and (6) along with experimental and numerical X-FROG spectrograms. Analytical theory predicts in Fig. 10a that the locations of phase-sensitive and phase-insensitive resonances differ significantly depending on whether cw propagates in the fast fiber axis and the soliton—in the slow axis, or vice versa. In the experiment the cw and femtosecond beams were orthogonally polarized and launched into a 24 cm long piece of PCF by use of a polarizing cube. At the output of the PCF the cw eigenaxis is selected with another polarizer, thus strongly, but not completely damping the cross-correlation signal from the soliton. Figure 10b shows the X-FROG trace for the case

when the soliton central wavelength is 1430 nm and the input power is adjusted such that the stimulated Raman scattering shifts the soliton spectrum by no more than 10 nm.
Let us assume that the soliton is x-polarized and the cw is y-polarized. The theory predicts that the waves generated as a result of the FWM interaction between the orthogonally polarized soliton and the cw retain the cw polarization. The source of the FWM signal for the phase-insensitive resonance (5) is the $\left|A_x^2\right|A_y$ term, whereas the phase-sensitive resonance (6) is due to the $A_x^2 A_y^*$ term in the generalized NLS equation:

$$\begin{aligned}&\left[i\partial_z + D_{x,y}(i\partial_t)\right]A_{x,y} + \gamma\left[\left|A_{x,y}\right|^2 + \frac{2}{3}\left|A_{y,x}\right|^2\right]A_{x,y} + \\ &+\frac{\gamma}{3}A_{y,x}^2 A_{x,y}^* + \gamma A_{x,y}\int dt' R(t') I(t-t',z) = 0,\end{aligned} \tag{7}$$

where $A_{x,y}$ are the field amplitudes in orthogonal polarizations, $I = \left|A_x\right|^2 + \left|A_y\right|^2$ is the intensity, γ is the nonlinear coefficient, $R(t)$ is the standard Raman response (Agrawal, 2001), and $D_{x,y}(i\partial_t)$ are the dispersion operators, specific for each eigenaxis where constant (phase velocity) and linear (group velocity) Taylor expansion terms are appropriately included.
The graphical solutions, Fig. 10a, to the wave-vector matching equations (5) and (6) indicate that for a soliton at 1430 nm central wavelength, both resonances are located near 1350 nm for one set of launching conditions. If, however, the eigenaxes are swapped, the phase-sensitive resonance disappears, while phase-insensitive resonance remains in the vicinity of the soliton. Experiments indeed reveal the presence of the signal at 1350 nm for one set of input launching conditions. If the eigenaxes are swapped, then the signal disappears from the picture, Fig. 10d. The numerically computed X-FROG trace showing the FWM signal, Fig. 10c matches nearly perfectly the experimental one, Fig 10b. These and other evidence (Efimov et al., 2006) verify the observation of phase-sensitive resonance in scattering of a continuous wave on a soliton in a typical birefringent PCF.
It is important to note that the observation of a subpicosecond structure in a truly cw wave, as the result of resonant interaction with a soliton, or otherwise, is highly nontrivial from the experimental standpoint and has not been demonstrated before. X-FROG appears to be a very potent technique for a wide spectrum of time-frequency-resolved measurements, including measuring the temporal structure of separate continuous wave beams. As described below, careful broadband X-FROG measurements can resolve previously obscure features in supercontinua, generated in PCFs. There, we will see the effects of inelastic soliton-cw interaction as an important element of supercontinuum dynamics during its formation and propagation in the fiber.

5. Supercontinuum generation in a highly nonlinear soft glass PCF

The success of silica photonics crystal fibers, which offered new degrees of freedom to manipulate dispersion and nonlinear properties, inspired the exploration of other glass materials as the host for these structures. Currently, such materials as polymers (van Eijkelenborg et al., 2001), highly-nonlinear glasses (Ravi Kanth Kumar et al., 2002) and even quantum dot-doped matrices (Yu et al., 2007) are being used to manufacture PCFs. The

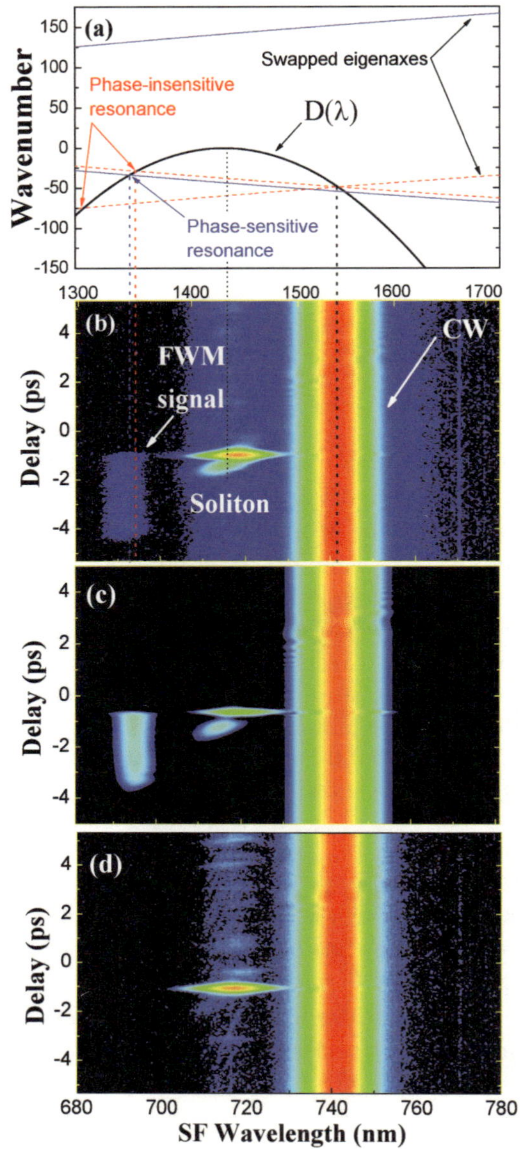

Fig. 10. (a) Graphical solutions to the wave-vector matching equations (5) and (6). Wavelengths of the expected FWM signals are given by the intersections of the quasi-parabolic function, describing the dispersion of linear waves, with the straight lines. The solid (dashed) curves correspond to the phase-sensitive (phase-insensitive) resonance; see the text for details. (b) Experimental X-FROG spectrogram showing generation of the new spectral component near 1350 nm. (c) Numerically generated X-FROG diagram corresponding to the experimental conditions of (b). (d) Experimental X-FROG diagram showing no new spectral components when polarizations of the soliton and cw are swapped. The blob between the signal and the cw in (b)-(d) is a small part of the soliton power coupled to the orthogonal polarization.

diversity of the modern PCF family is truly spectacular. The urgent need to study the dynamics of optical signals in these structures in time-frequency can be addressed with X-FROG.

Recently developed extrusion-based PCFs made from soft glasses, such as Schott SF6, can be about an order of magnitude more nonlinear as compared to their silica counterparts. Indeed, recent supercontinuum generation experiments in these SF6 PCFs indicate that a multioctave continuum can be generated efficiently in only millimeters of fiber using 100 fs pulses at telecom wavelengths and nanojoule energies (Omenetto et al., 2006). Previous measurements of supercontinua in time-frequency suffered from low signal-to-noise ratio and thus could not be directly compared to numerical results (Dudley et al., 2002). Nevertheless, low-noise broadband measurements with X-FROG are possible and may look almost indistinguishable from the numerically generated spectrograms.

Figure 11 shows theoretically and experimentally produced spectrograms of a supercontinuum pulse generated in a 11 cm long sample of extruded SF6 PCF using the X-FROG system described in Section 2. Input pump pulse is of typical 100 fs duration centered at 1550 nm. The fact that the broadband radiation in general lies on the parabola, which corresponds to the group delay as a function of wavelength, immediately tells us that all of the supercontinuum spectrum is generated at the very beginning of the fiber, supporting the previous findings (Omenetto et al., 2006). Subsequent propagation is, in fact, nearly purely dispersive. If the normal dispersion side of the experimental spectrogram (shorter wavelengths) matches the numerical one nearly perfectly, we see a clear discrepancy on the anomalous side of the picture, where solitons are formed. It appears that the experimental solitons are elongated in time. Notwithstanding the fact that the red-most numerical soliton is also stretched (due to steep confinement loss beyond 1700 nm), the explanation is, in fact, the temporal jitter associated with the presence of slight envelope modulations on the pump pulse (Efimov & Taylor, 2008).

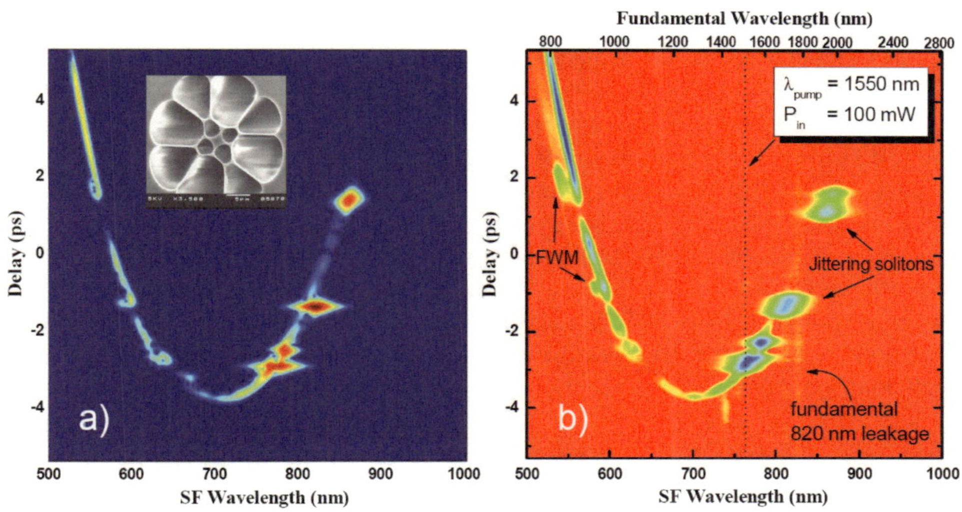

Fig. 11. Numerical (a) and experimental (b) X-FROG spectrograms of a supercontinuum pulse at the output of 11 cm long piece of SF6 PCF. Lower axis is sum-frequency and upper axis in (b) – fundamental wavelength of the signal. Color scale is logarithmic covering 30 dB of signal intensity. Inset shows the SEM image of the fiber tip

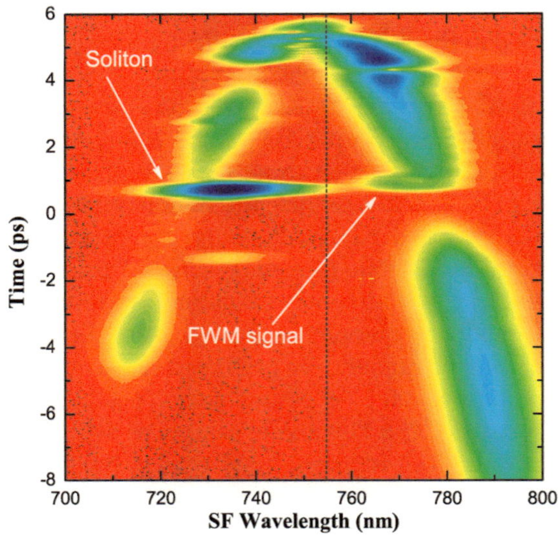

Fig. 12. Experimental X-FROG spectrogram showing the dynamics of the continuum in the vicinity of the 2ZD point of a small-core *silica* PCF from Fig. 6. The PCF is pumped exactly at 2ZD point (vertical dashed line). Dispersion is anomalous at shorter wavelengths and normal at longer wavelengths. A soliton forms on the anomalous side and interacts with the quasi-cw continuum on the normal dispersion side, which rresults in FWM signal generation (arrow). Color scale is logarithmic.

Nevertheless, the agreement between the theory and experiment is remarkable. On the normal dispersion side (shorter wavelengths) the quasi-continuum is pierced with holes directly opposite to solitons. The missing radiation is scattered in the FWM signals as the result of soliton-cw interaction, described in the previous section. As is mentioned in (Skryabin & Yulin, 2005) the group velocity-matched propagation of a soliton and a cw, located on both sides of ZD point, leads to the formation of a temporal hole in the cw, directly opposite to soliton. In the current experiment, however, the soliton undergoes slow Raman self-frequency shift during propagation so that it's carrier wavelength becomes progressively longer and thus group velocity—slower. Careful consideration of the phase-matching condition (4) shows that the FWM signal will appear at wavelengths shorter than cw, as is observed in both experimental and numerical spectrograms in Fig. 11.

Interestingly, if we were to study the continuum formation in the vicinity of the second dispersion zero (of a different PCF), a similar effect is observed, Fig. 12, except that in this case the soliton shifts towards inside the "parabola" and the FWM signal also appears inside, in total agreement with the analytical theory and numerical computations.

6. Solitons and continuum generation in all-glass bandgap PCF

Continuing technical advancement in the area of complex fiber design recently resulted in the development of a new class of all-glass PCFs (Feng et al., 2003; Luan et al., 2004). These fibers consist of an arrangement of glass rods embedded in a background matrix of a different glass and no air channels. Depending on the index difference between the rods and the background these fibers will display either index guiding or resonant bandgap guiding characteristics (Litchinitser et al., 2002). For the latter case the rods would have higher index (e.g. Ge-doped silica) than the background (e.g. pure silica). Such a design offers an alternative to liquid-filled PCFs (Larsen et al., 2003). The transmission spectra of these resonant-guiding PBFs consist of multiple low-loss bandgaps separated by high-attenuation regions. Importantly, in each bandgap the dispersion varies from normal to anomalous with increasing wavelength. Thus, each bandgap can support both solitons and dispersive waves (Fuerbach et al., 2005).

Moreover, intricate cross-bandgap phase-matching conditions can be realized, e.g. for the process of Four-Wave Mixing (FWM) as was recently predicted (Rasmussen et al., 2008). These intricate interactions can be revealed using time-spectral visualisation techniques as described previously. One complication arises, however: Typically, all-solid PBF designs are not well-suited for studies of nonlinear-optical interactions because of relatively large core sizes with effective mode areas on the order of tens of square micrometers. Thus, at best a single soliton formation can be observed with just oscillator pumping (Fuerbach et al., 2005). The observation of nonlinear cross-bandgap processes will thus require amplified pulses.

The fiber used in our experiments is similar to the one described in (Wang et al., 2006) and consists of a triangular array of Ge-doped glass rods embedded in a pure silica background (index difference $\Delta n = 0.033$). The cladding pitch and rod diameter are $\Lambda = 6.5\ \mu\text{m}$ and $d = 3.0\ \mu\text{m}$ respectively. The guiding region is defined by six rods surrounding a pure silica central region. The fiber structure in the vicinity of the core and its spectral transmission are shown in Fig. 13.

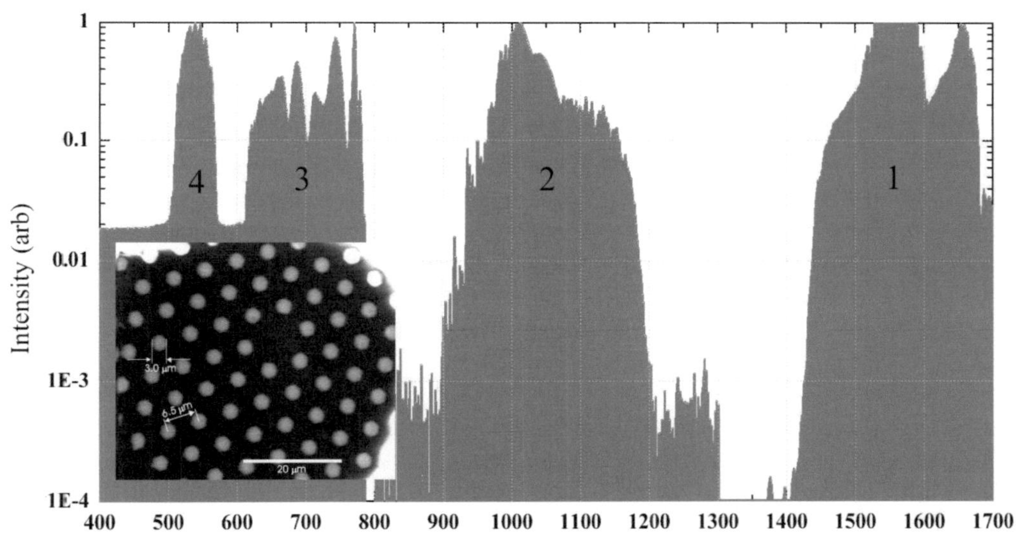

Fig. 13. Spectral transmission of the all-glass PBF. The structure of the transmission bands reflects the non-uniform spectral shape of the continuum used for their measurement. Bandgaps marked 1, 2, 3, and 4, and their edges can be clearly identified. Inset shows the SEM image of the central region fiber cross-section with rod diameter d and pitch Λ identified

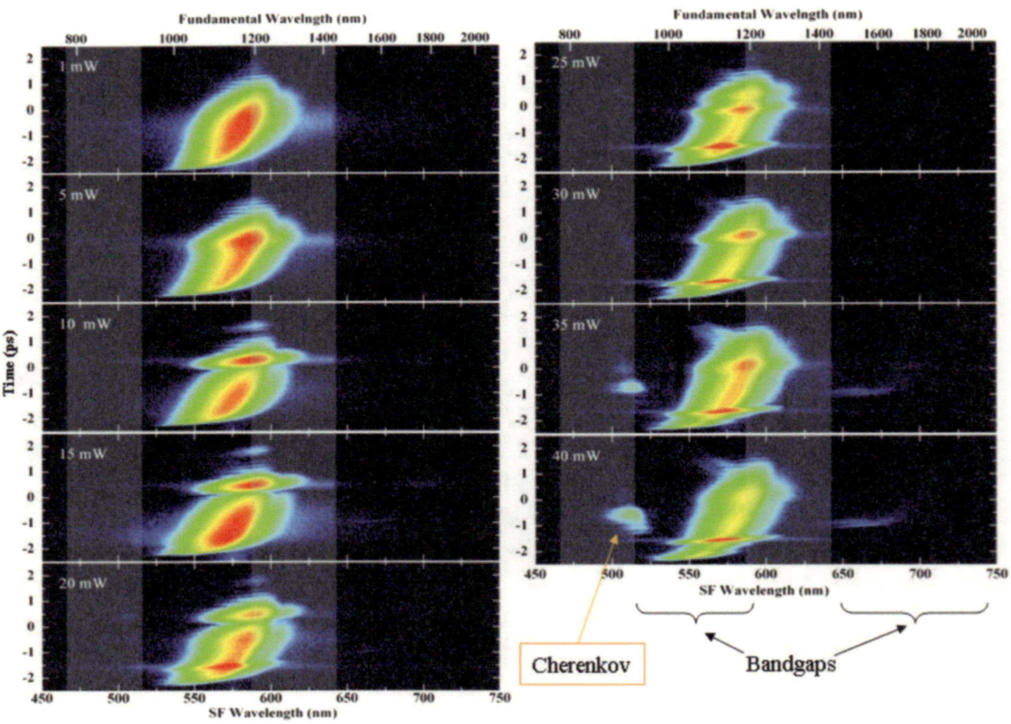

Fig. 14. Sequence of experimental X-FROG traces for varying input pump powers ranging from 4 nJ(1 mW input average power) to 160 nJ (40 mW) as indicated. Semi-transparent regions mark the high-loss interbandgap space. Color scale is logarithmic spanning 40 dB of sum-frequency signal.

To be able to access higher pump pulse intensities the fiber was pumped with 50 fs pulses at 1150 nm central wavelength from a laser, regenerative amplifier and optical parametric amplifier system (250 kHz Coherent Mira-RegA-OPA). At the output of 21 cm long fiber sample the optical pulses are analyzed with a SFG X-FROG system similar to the one described in Section 2, except that a KDP crystal was used instead of BBO in order to increase even further the phase-matching bandwidth to cover 500 to 2000 nm wavelength span.

Previous FROG measurements in a liquid-filled PCF revealed the formation of a soliton and generation of a phase-matched Cherenkov wave in the same bandgap (Fuerbach et al., 2005). Fig. 14 shows a sequence of X-FROG traces measured at the output of our all-glass PBF for different input pulse energies. Because the dispersion varies substantially over the bandwidth of the input pulse, only a fraction of the input spectrum on the red side contributes to the first soliton. This soliton forms very close to the edge of the bandgap and being pushed to longer wavelength by the stimulated Raman scattering, responsible for the soliton self-frequency shift, efficiently looses energy and disappears. The behavior of the second soliton is very similar: it forms at 25 mW input power, then elongates at 35 mW and finally disappears at 40 mW. The fact that these solitons do not extend their spectrum over to the normal dispersion region at wavelengths shorter than 1000 nm explains why the Cherenkov continuum appears only when the third soliton forms. This soliton propagates with its spectrum situated further away from the bandgap edge so that it is not affected by the loss as much. As a result, its duration is much shorter than that of the first two solitons and the blue edge of its spectrum extends over to the normal dispersion side making it possible to emit the Cherenkov continuum.

The result of this analysis shows that the presence of a substantial loss on one side of the soliton's spectrum, e.g. due to the bandgap edge, limits the extent of not only the affected side, but also the opposite side of the spectrum. This is expected because within a soliton, being a nonlinear wave, the spectral components are not independent, as is the case for a dispersive wave. The immediate consequence of this result is that the Cherenkov emission in a bandgap fiber is less efficient than that in more traditional waveguides. Only when pumped very close to the dispersion zero the radiation of the Cherenkov wave can be effective.

At higher pump powers one can also observe generation of visible radiation around 650 and 550 nm, which reside within the next two higher-order bandgaps. Fig. 15 presents the X-FROG spectrogram under these conditions. Here the pump wavelength is set to 1125 nm, i.e. closer to ZDW so the Cherenkov continuum is more prominent. More importantly, however, the time-spectral structure of the visible components in the two higher-order bandgaps is uncovered. The shape of the trace in the 550 nm bandgap, as well as the nearly-uniform illumination of the fiber suggests that the emission occurred at the very beginning of the fiber. Group delay differences between different bandgaps can be directly determined from the trace of Fig. 3 and show expected dependence on the bandgap number: Shorter

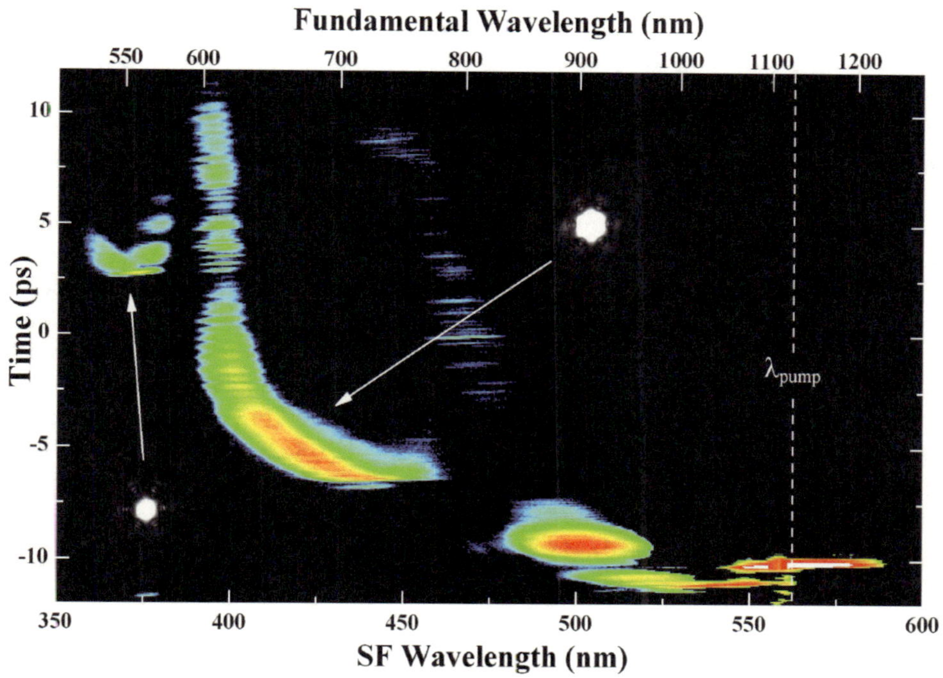

Fig. 15. Experimental X-FROG trace showing time-spectral properties of the signals in three bandgaps (number 2, 3, and 4 in Fig. 13) simultaneously. Insets show the near-field intensity distribution of the modes nonlinearly generated in bandgaps 3 and 4, as marked by arrows. Gray/color scale is logorithmic spanning 35 dB of sum-frequency signal.

wavelength bandgaps are delayed with respect to the longer ones (Rasmussen et al., 2008). Insets show the optical near-field images of the generated visible modes and indicate fundamental mode profiles at both wavelengths.

The observed interbandgap generation may involve two mechanisms, working separately or simultaneously. First, the input pulse may generate a broadband continuum due to the usual mechanisms (self-phase modulation, soliton compression and fission, dispersive wave generation) at the very beginning of the fiber and the resulting spectral components fill in the bandgaps, where they propagate dispersively with minimal loss. In addition, phase-matched inter-bandgap FWM generation may play a role in the observed behavior (Rasmussen et al., 2008). Previously, intermodally phase-matched FWM (Stolen & Leibolt, 1976; Akimov et al., 2003) as well as third harmonic generation (Efimov & Taylor., 2005b) were observed in both regular fibers and PCFs. Inter-bandgap phase-matched FWM is a very similar process and thus appears likely to contribute to the generation mechanism. Complete modeling of the PBF fiber structure as well as further experiments are needed to fully understand the observed phenomena.

7. Conclusion

Thorough understanding of fundamental nonlinear processes occurring as optical signals propagate along a waveguide require obtaining as much experimental information as possible. It was recognized a long time ago (Treacy, 1971) that spectrogram representation of optical pulses, that is 2D plots of signal amplitude in time and frequency showing relative temporal positions of different frequency components comprising the pulse, is, in fact, the most intuitive and rich way to describe the pulse. Simultaneous temporal and spectral resolution, limited only by the fundamental Fourier relation, gives unprecedented clarity of images, substantially simplifying the interpretation of the data and allowing visual comparison with theory and numerical modeling. An experimental tool such as X-FROG being slightly more involved in implementation than spectral or autocorrelation-type systems will reward one with the most dramatic images of nonlinear evolution of ultrashort pulses in optical fibers and waveguides. As often happens, however, when new advances in experimental techniques mature, they bring to light not only answers to long-standing questions, but create new puzzles to challenge both theorists and experimentalists alike.

In this work the emphasis was placed on visualizing and studying various aspects of soliton dynamics in regular and photonic crystal fibers and their interaction with dispersive waves. Recent analytical theory explains well all the resonant soliton-cw scattering processes observed in the experiments described in sections 3 and 4. This scattering of cw on solitons in non-integrable systems, either due to high-order dispersion or birefringence, leads to generation of new non-degenerate signals, which were observed and confirmed numerically. These processes are important, for example, in supercontinuum generation, where quasi-cw radiation propagating in the normal dispersion region interacts with Raman-shifting solitons formed on the anomalous side (section 5). Broadband and high dynamic range X-FROG measurements of supercontinua show excellent agreement with numerics, as well as directly show experimental artifacts, such as timing jitter of solitons.

We also obtained most clear images of well-known nonlinear processes such as soliton formation in long lengths of regular fiber and Raman solitons in silica cobweb photonic crystal fibers (section 2). Among other things, this brings about the educational value of X-FROG measurements and presentation of numerical results.

Finally, X-FROG measurements of multiple soliton formation and nonlinear cross-bandgap generation in all-glass PBFs challenge the current state of affairs in our understanding of very complex nonlinear-optical interactions occurring in waveguiding structures. Further discoveries can be expected by applying the X-FROG technique to the investigation of ultrafast nonlinear optical processes in other types of photonic crystal structures, complex waveguides and other nonlinear systems.

8. References

Agrawal, G. P. (2001). Nonlinear Fiber Optics. Academic Press, San Diego.

Akhmediev, N. & Karlsson, M. (1995). Cherenkov radiation emitted by solitons in optical fibers. *Physical Review A* Vol. 51, No. 3, pp. 2602-2607.

Akimov, D. A.; Serebryannikov, E. E.; Zheltikov, A. M.; Schmitt, M.; Maksimenka, R.; Kiefer, W.; Dukel'skii, K. V.; Shevandin, V. S. & Kondrat'ev, Yu. N. (2003). Efficient anti-Stokes generation through phase-matched four-wave mixing in higher-order modes of a microstructure fiber. *Optics Letters* Vol. 28, No. 20, pp. 1948-1950.

Biancalana, F.; Skryabin, D. V. & Yulin, A. V. (2004). Theory of the soliton self-frequency shift compensation by the resonant radiationin photonic crystal fibers. *Physical Review E* Vol. 70, No. 1 016615-1:9.

Birks, T. A.; Wadsworth, W. J. & Russell, P. St. J. (2000). Supercontinuum generation in tapered fibers. *Optics Letters* Vol. 25, No. 19, pp. 1415-1417.

Chen, Z.; Taylor, A. J. & Efimov, A. (2009). Coherent mid-infrared broadband continuum generation in non-uniform ZBLAN fiber taper. *Optics Express* Vol. 17, No. 7, pp. 5852-5860.

Dudley, J. M.; Gu, X.; Xu, L.; Kimmel, M.; Zeek, E.; O'Shea, P.; Trebino, R. Coen, S. & Windeler, R. S. (2002). Cross-correlation frequency resolved optical gating analysis of broadband continuum generation in photonic crystal fiber: simulations and experiments. *Optics Express* Vol. 10, No. 21, pp. 1215-1221.

Dudley, J. M.; Genty, G. & Coen, S. (2006). Supercontinuum generation in photonic crystal fiber. *Reviews of Modern Physics* Vol. 78, No. 4, pp. 1135-1185.

Efimov, A.; Taylor, A. J.; Omenetto, F. G.; Yulin, A. V.; Joly, N.; Biancalana, F.; Skryabin, D. V.; Knight, J. C. & Russell, P. St. J. (2004). Time-spectrally-resolved ultrafast nonlinear dynamics in small-core photonic crystal fibers: Experiment and modelling. *Optics Express* Vol. 12, No. 26, pp. 6498-6507.

Efimov, A. & Taylor, A. J. (2005a). Cross-correlation frequency-resolved optical gating for studying ultrashort-pulse nonlinear dynamics in arbitrary fibers. *Applied Optics* Vol. 44, No. 20, pp. 4408-4411.

Efimov, A. & Taylor, A. J. (2005b). Spectral-temporal dynamics of ultrashort Raman solitons and their role in third-harmonic generation in photonic crystal fibers. *Applied Physics B* Vol. 80, No. 6, pp. 721-725.

Efimov, A.; Yulin, A. V.; Skryabin, D. V.; Knight, J. C.; Joly, N.; Omenetto, F. G.; Taylor, A. J. & Russell, P. St. J. (2005c). Interaction of an Optical Soliton with a Dispersive Wave *Physical Review Letters* Vol. 95, No. 21, pp. 213902-1:4.

Efimov, A.; Taylor, A. J.; Yulin, A. V.; Skryabin, D. V. & Knight, J. C. (2006). Phase-sensitive scattering of a continuous wave on a soliton. *Optics Letters* Vol. 31, No. 11 1624-1626.

Efimov, A. & Taylor, A. J. (2008) Supercontinuum generation and soliton timing jitter in SF6 soft glass photonic crystal fibers. *Optics Express* Vol. 16, No. 8, pp. 5942-5953.

van Eijkelenborg, M. A.; Large, M. C. J.; Argyros, A.; Zagari, J.; Manos, S.; Issa, N. A.; Bassett, I.; Fleming, S.; McPhedran, R. C.; de Sterke, C. M. & Nicorovici1, N. A. P. (2001). Microstructured polymer optical fibre. *Optics Express* Vol. 9, No. 7, pp. 319-327.

Feng, X.; Monro, T. M.; Petropoulos, P.; Finazzi, V. & Hewak, D. (2003). Solid microstructured optical fiber. Optics Express Vol. 11, No. 18, pp. 2225-2230.

Fuerbach, A.; Steinvurzel, P.; Bolger, J. A. & Eggleton, B. J. (2005). Nonlinear pulse propagation at zero dispersion wavelength in anti-resonant photonic crystal fibers *Optics Express* Vol. 13, No. 8, pp. 2977-2987.

Genty, G.; Lehtonen, M.; Ludvigsen, H. & Kaivola, M. (2004). Enhanced bandwidth of supercontinuum generated in microstructured fibers. *Optics Express* Vol. 12, No. 15, pp. 3471-3480.

Harbold, J. M.; Ilday, F. O.; Wise, F.; Birks, T. A.; Wadsworth, W. J. & Chen, Z. (2002). Long-wavelength continuum generation about the second dispersion zero of a tapered fiber. *Optics Letters* Vol. 27, No. 17, pp. 1558-1560.

Hilligsøe, K. M.; Andersen, T. V.; Paulsen, H. N.; Nielsen, C. K.; Mølmer, K.; Keiding, S. R.; Kristiansen, R.; Hansen, K. P. & Larsen, J. J. (2004). Supercontinuum generation in a photonic crystal fiber with two zero dispersion wavelengths. *Optics Express* Vol. 12, No. 6, pp. 1045-1054.

Hori, T.; Nishizawa, N.; Goto, T.; & Yoshida, M. J. (2003). Wideband and nonmechanical sonogram measurement by use of an electronically controlled, wavelength-tunable, femtosecond soliton pulse. *Journal of the Optical Society of America B* Vol. 20, No. 11, pp. 2410-2417.

Hori, T.; Nishizawa, N. & Goto, T. (2004). Experimental and numerical analysis of widely broadened supercontinuum generation in highly nonlinear dispersion-shifted fiber with a femtosecond pulse. *Journal of the Optical Society of America B* Vol. 21, No. 11, pp. 1969-1980.

Jones, D. J.; Diddams, S. A.; Ranka, J. K.; Stentz, A.; Windeler, R. S.; Hall, J. L. & Cundiff, S. T. (2000). Carrier-Envelope Phase Control of Femtosecond Mode-Locked Lasers and Direct Optical Frequency Synthesis. *Science* Vol. 288, No. 5466, pp. 635-639.

Karpman, V. I. (1993). Radiation by solitons due to higher-order dispersion. *Physical Review E* Vol. 47, No. 3, pp. 2073-2082.

Knight, J. C.; Birks, T. A.; Russell, P. St. J. & Atkin, D. M. (1996). All-silica single-mode optical fiber with photonic crystal cladding. *Optics Letters* Vol. 21, No. 19, pp. 1547-1549.

Knight, J. C. (2003). Photonic crystal fibers. *Nature* Vol. 424, No. 6950, pp. 847-851.

Larsen, T. T.; Bjarklev, A.; Hermann, D. S. & Broeng, J. (2003) Optical devices based on liquid crystal photonic bandgap fibres. *Optics Express* Vol. 11, No. 20, pp. 2589-2596.

Lin, C.; Nguyen, V. T. & French, W. G. (1978). Wideband near-IR continuum (0.7-2.1 μm) generated in low-loss optical fibres. *Electronics Letters* Vol. 14, No. 25, pp. 822-823.

Linden, S.; Giessen, H. & Kuhl, J. (1998). XFROG - A New Method for Amplitude and Phase Characterization of Weak Ultrashort Pulses. *Physica Status Solidi (b)*, Vol. 206, No. 1, pp. 119-124.

Litchinitser, N. M.; Abeeluck, A. K.; Headley, C. & Eggleton, B. J. (2002) Antiresonant reflecting photonic crystal optical waveguides. *Optics Letters* Vol. 27, No. 18, pp. 1592-1594.

Lu, F.; Deng, Y. & Knox, W. H. (2005). Generation of broadband femtosecond visible pulses in dispersion-micromanaged holey fibers. *Optics Letters* Vol. 30, No. 12, pp. 1566-1568.

Luan, F. George, A. K.; Hedley, T. D.; Pearce, G. J.; Bird, D. M. Knight, J. C. & Russell, P. St. J. (2004). All-solid photonic bandgap fiber. *Optics Letters* Vol. 29, No. 20, pp. 2369-2371.

Morioka, K.; Mori, K.; Kawanishi, S. & Saruwatari, M. (1994). Pulse-width tunable, self-frequency conversion of short optical pulses over 200 nm based on supercontinuum generation. *Electronics Letters* Vol. 30, No. 23, pp. 1960-1962.

Nishizawa, N. & Goto, T. (2001). Experimental analysis of ultrashort pulse propagation in optical fibers around zero-dispersion region using cross-correlation frequency resolved optical gating. *Optics Express* Vol. 8, No. 6, pp. 328-334.

Omenetto, F. G.; Wolchover, N. A.; Wehner, M. R.; Ross, M.; Efimov, A.; Taylor, A. J.; Ravi Kanth Kumar, V. V.; George, A. K.; Knight, J. C.; Joly, N. Y. & Russell, P. St. J. (2006). Supercontinuum Generation in Sub-Centimeter Lengths of High-Nonlinearity Photonic Crystal Fibers. *Optics & Photonics News* Vol. 17, December 2006, p. 35.

Ranka, J. K.; Windeler, R. S. & Stentz, A. J. (2000). Visible continuum generation in air-silica microstructure optical fibers with anomalous dispersion at 800 nm. *Optics Letters* Vol. 25, No. 1, pp. 25-27.

Rasmussen, P. D.; Laegsgaard, J. & Bang, O. (2008). Degenerate four wave mixing in solid core photonic bandgap fibers. *Optics Express*, Vol. 16, No. 6, pp. 4059-4068.

Ravi Kanth Kumar, V. V.; George, A. K.; Reeves, W. H.; Knight, J. C.; Russell, P. St. J.; Omenetto, F. G. & Taylor, A. J. (2002). Extruded soft glass photonic crystal fiber for ultrabroad supercontinuum generation. *Optics Express* Vol. 10, No. 25, pp. 1520-1525.

Skryabin, D. V.; Luan, F.; Knight, J. C. & Russell, P. St. J. (2003). Soliton Self-Frequency Shift Cancellation in Photonic Crystal Fibers. *Science* Vol. 301, No. 5640, pp. 1705-1798.

Skryabin, D. V. & Yulin, A. V. (2005). Theory of generation of new frequencies by mixing of solitons and dispersive waves in optical fibers. *Physical Review E* Vol. 72, No. 1, pp. 016619-1:10.

Stolen, R. H. & Leibolt, W. N. (1976). Optical fiber modes using stimulated four photon mixing. *Applied Optics* Vol. 15, No. 1, pp. 239-243.

Taira, K. & Kikuchi, K. (2001). Optical sampling system at 1.55 μm for the measurement of pulse waveform and phase employing sonogram characterization. *IEEE Photonics Technology Letters* Vol. 13, No. 5, pp. 505-507.

Treacy, E. B. (1971) Measurement and Interpretation of Dynamic Spectrograms of Picosecond Light Pulses. *Journal of Applied Physics* Vol. 42, No. 10, pp. 3848-3858.

Trebino, R. (2000). Frequency-Resolved Optical Gating: The Measurement of Ultrashort Laser Pulses. Kluwer, Boston.

Udem, Th.; Holzwarth, R. & Hänsch, T. W. (2002). Optical frequency metrology. *Nature* Vol. 416, No. 6877, pp. 233-237.

Wai, P. A.; Chen, H. H. & Lee, Y. C. (1990). Radiations by ''solitons'' at the zero group-dispersion wavelength of single-mode optical fibers. *Physical Review A* Vol. 41, No. 1, pp. 426-439.

Wang, A.; George, A. K. & Knight, J. C. (2006). Three-level neodymium fiber laser incorporating photonic bandgap fiber. *Optics Letters* Vol. 31, No. 10, pp. 1388-1390.

Weiner, A. M. (1983). Effect of group velocity mismatch on the measurement of ultrashort optical pulses via second harmonic generation. *IEEE Journal of Quantum Electronics* Vol. 19, No. 8 pp. 1276-1283.

Yu, H. C. Y.; Argyros, A.; Barton, G.; van Eijkelenborg, M. A.; Barbe, C.; Finnie, K.; Kong, L.; Ladouceur, F. & McNiven, S. (2007). Quantum dot and silica nanoparticle doped polymer optical fibers. *Optics Express* Vol. 15, No. 16, pp. 9989-9994.

9

Dispersion Compensation Devices

Lingling Chen, Meng Zhang and Zhigang Zhang
Institute of Quantum Electronics, State Key Laboratory of Advanced Optical Communication Systems and Networks, School of Electronics Engineering and Computer Science, Peking University
P.R.China

1. Introduction

Dispersion compensation plays a key role in generation, amplification and propagation of femetosecond pulses. In the dispersive medium, the pulse can be broaden or compressed depending on the sign of chirp and dispersion. To obtain the ultrashort pulses, the pulse group delay (GD) should have about frequency independence after the dispersion compensation. Especially to compress a pulse to near the transform limit one should not only compensate the GD but also eliminate the high order dispersion term. In this chapter, we briefly review the calculation formulae and design rules of the various dispersive devices including grating and prism pairs and chirped mirrors. An integrative optimization of chirped mirror and its function in mode locked Ti:sapphire laser is described in details. Finally, a useful tool — dispersion map is introduced.

2. Grating pairs with negative and positive dispersion

A parallel grating pair as Treacy demonstrated (Fig. 1) can offer a negative group delay dispersion (GDD) (Treacy, 1969). The commonly used phase shift formula through the parallel grating pair is

$$\varphi = \frac{\omega}{c} b(1+\cos\theta) - \frac{2\pi G}{d}\tan(\gamma-\theta)$$

where b and G are the slant and the perpendicular distance path length between the two gratings respectively, d the grating constant, c the speed of light in vacuum, and γ and θ the incident and refraction angle respectively.

For a grating pair with the positive dispersion, Zhang et al (Zhang et al. 1997) proposed a grating- mirror model in contrast to the grating-lens model of Martinez (Martinez, 1987), to demonstrate that the phase shift of a grating-mirror system is a conjugation of that of parallel grating pair. In the model, a mirror with the radius of curvature of R is placed so that the incident point on the grating is right at the circular center of the mirror, as shown in Fig. 2. The simple ray-tracing would result a phase shift as

$$\varphi = \frac{\omega}{c}[4R - b(1+\cos\theta)] + \frac{2\pi G}{d}\tan(\gamma-\theta)$$

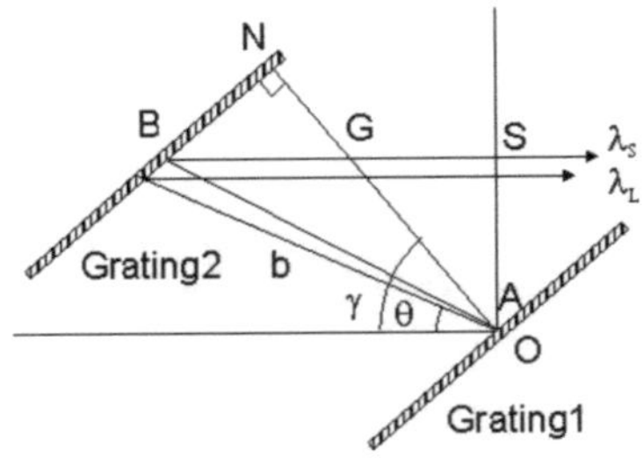

Fig. 1. Treacy grating pair with negative dispersion (adapted from Treacy, 1969)

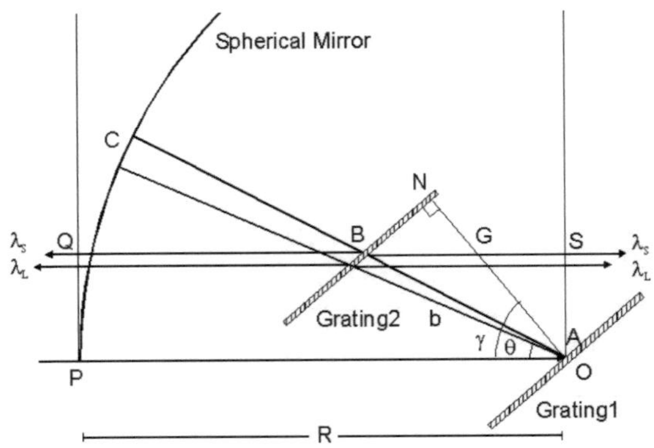

Fig. 2. Graing-mirror with positive dispersion

The subsequent differentiation with respect to frequency results in GD, GDD and the third order dispersion (TOD). Obviously, the only difference between the phase shifts of those two configurations is a constant (4R, depends on the starting point of calculation) and a opposite sign before the term of $b(1+\cos\theta)$. The differentiation would vanish the constant term and leaves an opposite sign. This proves that the grating-mirror configuration is a phase conjugation of the parallel grating pair and can be used as a pulse "stretcher".

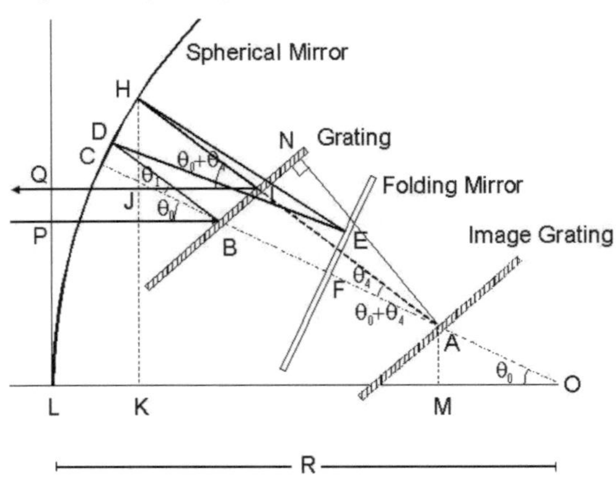

Fig. 3. Folded graing-telescope with positive dispersion

However, this grating-mirror stretcher cannot be realized because, in practice, the incident beam must have a finite size, while the single-mirror system gives no beam collimation. Therefore a telescope is required for the beam collimation.

To derive the similar formula to Equation for the telescope stretcher, we would prefer to start with the folded-grating-telescope stretcher and to adapt it in the coordinate shown in Fig.3. For explicitness, only the incident and the output rays are displayed, the folding mirror and some optical paths have been omitted.

$$\varphi(\omega)=\frac{\omega}{c}(C+A)-\frac{\omega}{c}b\times\left[1+\cos(\theta_0+\theta_4)\right]+\frac{2\pi G}{d}\tan(\gamma-\theta_0-\theta_4)+\frac{2\pi}{d}(G_0-G)\tan(\gamma-\theta_0),$$

where $C=2R-(R-s_1)\cos\theta_0$ is a constant, s_1 =BC, A is an angle dependent variable

$$A=R\left\{\sin(\theta_1-\phi_1)\left(\frac{1}{\sin\theta_1}+\frac{1}{\sin\theta_2}\right)+\sin(\theta_3-\phi_3)\left(\frac{1}{\sin\theta_3}+\frac{1}{\sin\theta_4}\right)\right\}-R\frac{\sin\phi_4}{\sin\theta_4}\cos\theta_0$$

All variables are defined in reference (Zhang et al. 1997).It is noticed that, although the formula is complicated, it takes the form of:

$$\frac{\omega}{c}b\times\left[1+\cos(\theta_0+\theta_4)\right],$$

indicating that it is indeed the phase conjugation of the pulse compressor, except for the mirror induced stigmatism. Similarly, we can have the formulism for Öffner type stretcher, which is (Jiang et al. 2002)

$$\varphi(\omega)=\frac{\omega}{c}(C+A)-\frac{\omega}{c}b\times\left[1+\cos(\theta_0+\theta_6)\right]+\frac{2\pi G}{d}\tan(\gamma-\theta_0-\theta_6)+\frac{2\pi}{d}(G_0-G)\tan(\gamma-\theta_0),$$

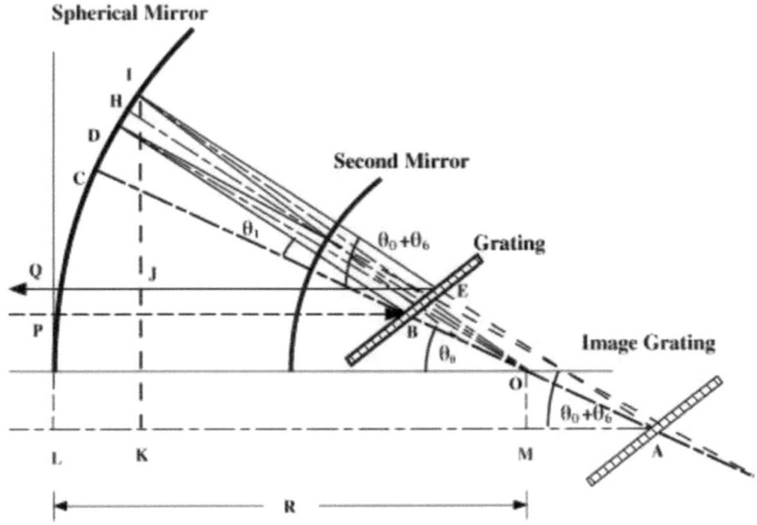

Fig. 4. Öffner type stretcher (adapted from Jiang et al. 2002)

The only difference with Martinez stretcher is the subscript of the angles. All variables are defined in reference (Jiang et al. 2002).

The above formulas, although they are two dimensional, are very useful in optimization of chirped pulse amplification systems.

3. Prism pairs

Prism pairs have also been widely used for dispersion control inside laser oscillators since they can be very low loss in contrast to grating pairs. In 1984 Fork et al. derived formulism of GDD for pair of Brewster-angled prisms (Fork et al. 1984). Nakanuma and Mori noticed that the commonly used formulas developed by Fork were not convenient to use in the case where the adjustable prism insertion is required. They proposed to separate the prism separation and prism insertion (Nakanuma & Mori, 1991) so that those parameters can be easily measured. The simplified formulism is found in reference (Zhang & Yagi, 1993). The GDD and TOD can be expressed to functions of prism separation and prism insertion:

$$\varphi''(\omega) = D_{2L}L + D_{2x}x\,,\quad \varphi'''(\omega) = D_{3L}L + D_{3x}x$$

where the coefficients are approximately

$$D_{2L} = -\frac{2\lambda}{\pi c^2}\lambda\left(\frac{dn}{d\lambda}\right)^2,$$

$$D_{2x} = \frac{2\lambda}{\pi c^2}\lambda^2\frac{d^2n}{d\lambda^2}\frac{1}{n},$$

$$D_{3L} = \frac{3\lambda}{\pi^2c^3}\left(\lambda\frac{dn}{d\lambda}\right)\left(\lambda\frac{dn}{d\lambda}+\lambda^2\frac{d^2n}{d\lambda^2}\right),$$

$$D_{3x} = -\frac{3\lambda^2}{2\pi^2c^3}\left(\lambda^2\frac{d^2n}{d\lambda^2}+\frac{\lambda^3}{3}\frac{d^3n}{d\lambda^3}\right)\frac{1}{n}.$$

where n is the refractive index of the prism material. An important advantage of using the above formulism is that they can be presented in a so called dispersion map, as described in section 5.

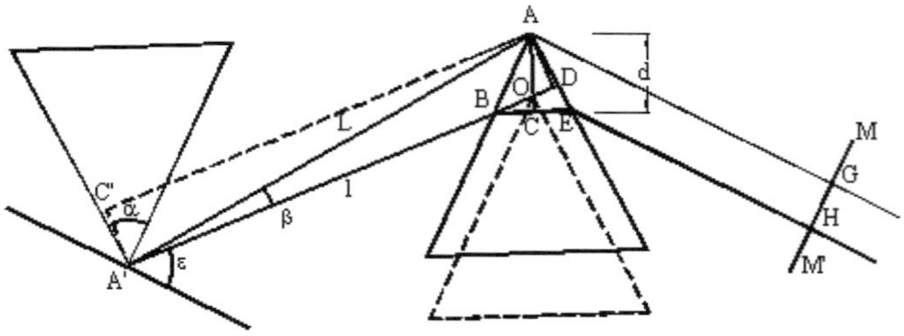

Fig. 5. A typical prism pair arrangement.

4. Chirped mirrors

Although grating and prism pairs are widely used as the compressor schemes, they suffer from high order dispersion. By contrast, chirped mirrors (CMs) which are dispersive multilayer structures designed by optimizing the initial design are usually used for pulses compression by providing dispersion without much material in the beam path. They can provide the precisely controlled GD, GDD, even higher order dispersion such as third order dispersion (TOD), as well as the high reflectivity in a broadband or ultra-broadband wavelength range. They are key components for extremely short pulse compression and especially have played a critical role in the development of octave-spanning lasers and the generation of mono-cycle pulses. In 2007, 1.5-octave chirped mirrors have been proposed and demonstrated, for extra-cavity compression to 3 fs pulses (Pervak et al. 2007). Furthermore, they are important for compression of attosecond soft-x-ray pulses (Wonisch et al. 2006).

There are always ripples in GDD curves of broadband chirped mirrors, due to the interference between the reflections from different layers of a multilayer structure. Although the ripples can be minimized by the optimization procedure such as with commercially available software, it is difficult to reduce the ripples to acceptable level when the design of a single chirped mirror has to operate over the wavelength ranges up to an octave spanning or more. Therefore, several other approaches to reduce GDD ripples have been proposed, such as double-chirped mirrors (Kärtner et al. 1997; Matuschek et al. 1998; Kärtner et al. 2001), complementary CM pairs (Pervak et al. 2005; Pervak et al. 2007). By pairing the CMs, the dispersion ripples of CMs are opposite to each other. Thus, the residual ripple of the CM pairs could be very small.

The design rule of CMs consists of two main issues. The first issue is the initial multilayer design. The design procedure of chirped mirrors is usually to start with a favorable initial structure, then to perform optimization. In this procedure, the initial design is very crucial. If the initial design is not close enough to the target, the optimization procedure is impossible to reach a satisfactory result. The commonly used initial designs are double chirp (DC) and modulated layer thickness (MLT) (Tempea et al. 1998).

The second issue is the optimization process including the choices of targets and optimization methods. Although a CM is characterized by a certain value of GDD that is the second derivative of the phase shift on reflection with respect to the angular frequency, GD was mostly chosen as the target, because the second derivative of the phase has two problems: the accuracy and the time consumption in the optimization. However, we found that just taking GD as the target cannot ensure a smooth GDD. Besides the targets setting, we should also consider the optimization methods. Although paring two CMs is a good idea to eliminate the ripples but in some cases such as generating octave-spanning in four-mirror Ti:Sapphire ring cavity it is not enough. Then we introduce the intracavity optimization.

4.1 Initial design of chirped mirrors

Here, we summarize two different designs as the initial layer structure for the optimization and compare the differences. First was the DC, and the second was the MLT.

The conception of DC mirrors has been proposed by Käertner (Kärtner et al. 1997; Matuschek et al. 1998) where in addition to the chirp of the Bragg wave number, the thickness of the high refractive index layers is also chirped.

Assuming the double chirped mirrors composed of 2N index steps with high refractive index medium such as TiO_2 and low refractive index medium such as SiO_2. The thicknesses of low index layer are linearly chirped while the thicknesses of high index layer are given by

$$d_{H,i}(x) = \frac{\lambda_{\max}}{4n_H}\left(\frac{N+1-i}{N}\right)^{\eta} \quad \text{where } \eta = 1, 2 \quad i=1,2,\ldots,N+1$$

Where η=1, 2 for linear and quadratic chirp. The clear reduction of oscillations in the GD by the double chirped technique is visible (Matuschek et al. 1998). Moreover, the index-matching can be extended by a properly high quality AR-coating.

MLT method (Tempea et al. 1998) is based on the recognition that in the two component structure, the effects of modulating the layer thickness is similar to that of a corresponding modulation of refractive index. Composing the mirror with narrow bandwidth filters with gradually shifted central wavelength may avoid some resonant interference. The general formula described the variation of the layer thickness is given by

$$t(x) = t_0(x) + A(x)\sin(2\pi\frac{x}{\Lambda(x)})$$

where x is the distance of the respective layer from the substrate and the number of layers within a modulation period Λ must be kept constant,

$$\Lambda(x) = \alpha t_0(x)$$

where the parameter α determines the number of layers within a modulation period and can be set as $\alpha = 5$ (i.e.), equal to its minimum permitted value.

For an early constant negative GDD in the high reflectance range we need

$$t_0(x) = \frac{1}{4}(\frac{\lambda_{\min} - \lambda_{\max}}{d}x + \lambda_{\max})$$

where d is the total optical thickness of the coating. One problem in CM design is to obtain negative GDD at the shortest wavelength of the high-reflectance range. This difficulty is related to that the shortest wavelength components have a limited penetration depth which tends to result in a constant GD. This problem can be alleviated by increasing the amplitude of the modulation depth by

$$A(x) = \frac{A_2 - A_1}{d}x + A_1$$

where A_1, A_2 (>A_1) are limits between which the amplitude is allowed to vary.

The above equations consist of a complete recipe of the design of chirped mirror by using MLT method.

We simulated two chirped mirror pairs to compare these two above initial designs—DC (pair 1) and MLT (pair 2) (Chen et al. 2007). The target was CM pairs providing negative GDD of around -50 fs^2 and TOD of around -35 fs^3 at 800 nm with the reasonable residual GDD ripples and high reflectivity in wavelength range of 600-1200 nm. In the DC model, we chose the first order. In the MLT model, the number of layer within a modulation period was 5, the minimum modulation amplitude was 12.5 nm and maximum modulation amplitude was 62.5 nm, same as in the reference (Tempea et al. 1998). The total layer number of each mirror in a pair for both models was 80.

After optimization, from Fig. 6 and 7, we can see that both of the CM pairs meet the target at 800 nm with the residual GDD ripples <15 fs^2 and the reasonable high reflectivity. Meanwhile, we find that the time consuming on optimization for pair 1 is little longer than pair 2, the GDD curve is closer to what we required and the GD is matched better in the spectral range. Particularly, the reflectivity is relatively higher.

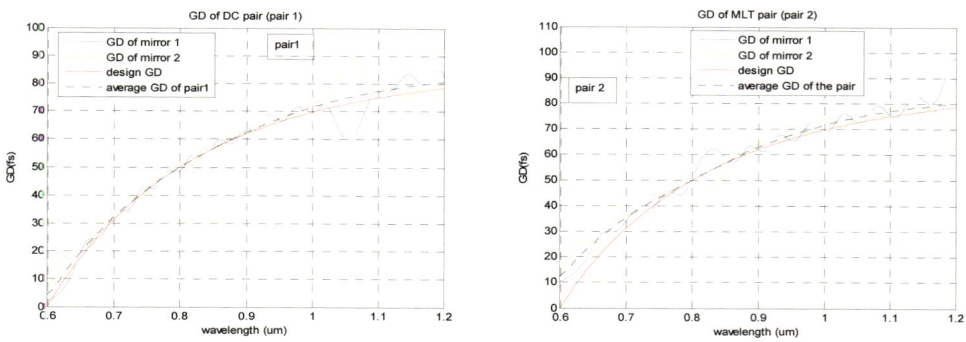

Fig. 6. GD for different design DC mirrors (pair1) and MLT CMs (pair2)

It is noticed that, from Fig. 8 we can see that the reflectivity of pair 1 is above 99.4% in most of spectral range except the 1170~1200 nm. However, reflectivity of pair 2 is unexpectedly low (< 99%) in the range of 810~910 nm. If these CM pairs are going to be used inside laser cavity, the loss of mirrors is one of the most important factors we should consider. In this regard, pair 1 is better than pair 2.

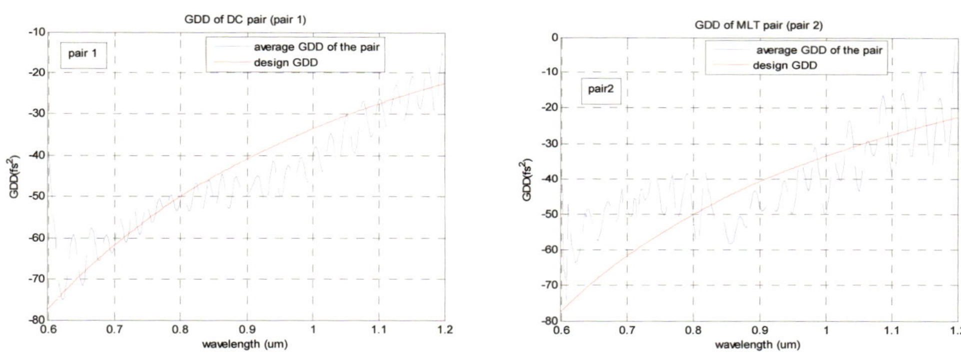

Fig. 7. GDD for different design DC mirrors (pair1) and MLT CMs (pair2)

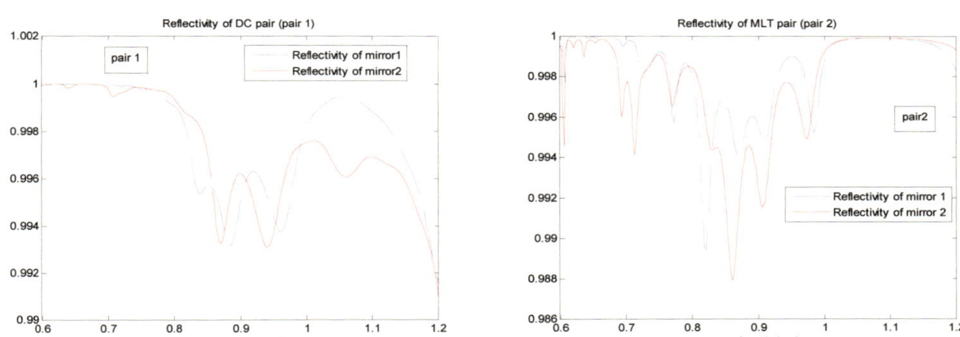

Fig. 8. Reflectivity for different design DC mirrors (pair1) and MLT CMs (pair2)

The above simulation is just one example to design the CM pairs. We can see each initial design has its own advantages and disadvantages; thus, the choice mainly depends on the specific necessities.

4.2 Optimization of chirped mirrors

4.2.1 Optimization targets

It is noticed that in some cases, the two CMs can be better paired with each other with GD as the target than with GDD as the targets (Pervak et al. 2007). For complementary pair mirrors, the merit function is modified as below when we take GD as the target:

$$F = \sum_{i=1}^{n} \left\{ \left[\frac{R_1(i) - R_t(i)}{\Delta R} \right]^{\alpha} + \left[\frac{R_2(i) - R_t(i)}{\Delta R} \right]^{\beta} + \left\{ \frac{[GD_1(i) + GD_2(i)]/2 - GD_t(i)}{\Delta GD} \right\}^{\chi} \right\} \quad (1)$$

where the subscripts j = 1, 2 represents each individual mirror. R_t (i) and GD_t(i) are the desired values at the wavelength λ_i. R_j(i) and GD_j(i) are the calculated instantaneous values. ΔR and ΔGD are the corresponding tolerances and n is the number of wavelength points. The weighting parameters α, β and χ, have to be adjusted according to the error in the process to handle the optimization.

We should also notice that CMs are finally characterized by the GDD. Therefore, in the final optimization step, GDD can be used to be the target. The merit function thus takes the following form in this case:

$$F = \sum_{i=1}^{n} \left\{ \left[\frac{R_1(i) - R_t(i)}{\Delta R} \right]^{\alpha} + \left[\frac{R_2(i) - R_t(i)}{\Delta R} \right]^{\beta} + \left\{ \frac{[GDD_1(i) + GDD_2(i)]/2 - GDD_t(i)}{\Delta GDD} \right\}^{\chi} \right\} \quad (2)$$

where the variables are the same as in Eq. (1), except GDD in the place of GD.

Considering these two aspects, we find that the optimization process can be divided into three steps. In the first and second optimization, the GD and GDD are set to be the optimization target respectively, as described in Eq. (1) and Eq. (2), where the derivative of the phase is calculated by the Birge and Kaertner method (Birge et al. 2006), since this method reduces the calculation time. During the third optimization, the exact differential process is applied for fine adjustment for the target of GDD.

4.2.2 Optimization methods

As we said before, to minimize the ripples in GDD curves of broadband chirped mirrors, especially ones operate over the wavelength ranges up to an octave spanning or more, people developed complementary CM pairs. However, in the octave-spanning lasers, particularly in the high repetition rate (~1 GHz) ring cavity lasers (Fortier et al. 2006; Nogueira et al. 2006), there are three CMs and one output coupler. Two of the CMs are usually paired, but the other one has no mirror to pair with. Therefore, the total dispersion is usually heavily oscillated, and the compensation is incomplete in such a laser (Fortier et al. 2006; Nogueira et al. 2006). The same difficulty also happens to the linear cavity, where the end mirrors offer one bounce reflection while other CMs experience twice which can be designed in pair. Silver mirror has been used as the end mirror to avoid this problem (Matos et al. 2004). However, the high reflection loss and the weak durability of the silver mirror make the laser inefficient and short-lived. Though octave spanning spectrum has been obtained by some groups with such a cavity, it requires quite a critical alignment. Therefore, more precise overall intracavity dispersion compensation scheme has to be developed.

In 2008, a new cavity for octave-spanning laser is proposed by the MIT group (Crespo et al. 2008), where two pairs of chirped mirrors (four chirped mirrors) were designed and combined to obtain quite a low ripple of the intracavity GDD. The output coupler, however, was performed by the coating on a fused-silica wedge. Sub-two-cycle pulses were obtained. In 2009, another method for precise intracavity dispersion compensation: the integrative optimization of all cavity mirrors, provided that the output coupler has a flat GDD (Chen et al. 2009) is proposed by us. Both GDD and TOD were considered. This technique allows minimizing the ripples of the whole GDD for ultra-broadband femtosecond lasers.

Take an example as in four-mirror Ti:Sapphire ring cavity for this intracavity optimization to obtain the octave spanning. The four-mirror ring cavity for optimization is shown in Fig. 9. In an octave-spanning Ti:Sapphire laser, one or two mirrors have to have a low reflectivity at 532 nm for the pumping beam to go through while to maintain a high reflectivity from 600 nm to 1200 nm. This constrict makes the oscillation of the GDD larger than the other mirrors. Therefore, we first designed this pump mirror, and then matched the dispersion of this mirror with the other two mirrors. After that, we optimized these three mirrors together with all requirements (reflection and GDD). In Fig. 9, M1 and M2 are concave mirrors. M1 is the pump mirror. M3 may be a convex mirror (Fortier et al. 2006; Nogueira et al. 2006). The optimization procedure for two mirrors was similar to that in reference (Crespo et al. 2008), but with more steps as described below.

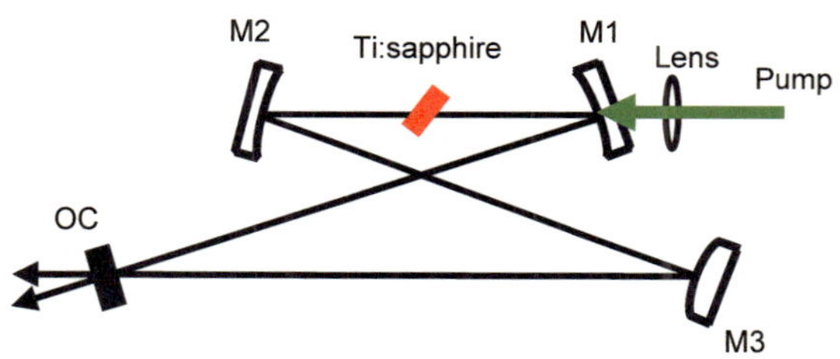

Fig. 9. Schematic of four-mirror ring cavity. M1, M2 and M3 are chirped mirrors for integrative dispersion optimization; OC is the output coupler.

The reflectivity of the pump mirror M1 was designed to be less than 20% in the wavelength range of 500-540 nm, particularly less than 5% at 532 nm. The reflectivity of the other two mirrors was set without the restriction on the transmission at 532 nm. The reflectivity of all the mirrors was maintained to be more than 99.8% in the desired spectral range of 600-1200 nm. The target GDD and TOD of each CM were set to be -50 fs^2 and -35 fs^3 at the central wavelength of 800 nm so as to compensate the dispersion employed by the 2.3 mm Ti:Sapphire crystal (The GDD employed by the output coupler is assumed to be zero). For all other wavelengths the GD and GDD were calculated respectively, as for the optimization target of 351 points in the wavelength range of 500-1200 nm. The initial design of all mirrors was adapted from the double chirped mirror. The optimization process was adapted as we said in 4.2.1, which was divided into 3 stages and merit functions were modified as below:

$$F=\sum_{i=1}^{n}\left\{\left[\frac{R_1(i)-R_t(i)}{\Delta R}\right]^{\alpha}+\left[\frac{R_2(i)-R_t(i)}{\Delta R}\right]^{\beta}+\left[\frac{R_3(i)-R_t(i)}{\Delta R}\right]^{\gamma}+\left\{\frac{[GD_1(i)+GD_2(i)+GD_3(i)]/3-GD_t(i)}{\Delta GD}\right\}^{\chi}\right\}$$

$$F=\sum_{i=1}^{n}\left\{\left[\frac{R_1(i)-R_t(i)}{\Delta R}\right]^{\alpha}+\left[\frac{R_2(i)-R_t(i)}{\Delta R}\right]^{\beta}+\left[\frac{R_3(i)-R_t(i)}{\Delta R}\right]^{\gamma}+\left\{\frac{[GDD_1(i)+GDD_2(i)+GDD_3(i)]/3-GDD_t(i)}{\Delta GDD}\right\}^{\chi}\right\}$$

The final optimization results for all CMs are shown in Fig. 10 and Fig. 11. Figure 10 shows that the reflectivity of M1 has a high transmission at 532 nm, and all three mirrors have the reflectivity more than 99% for the wavelength range of 600-1200 nm. It also shows that the combined GD of the three mirrors agrees well with the desired GD curve. This in turn results in a best fit with the GDD curve (Fig. 11). The GDD ripples of the combined three mirrors are less than 25 fs^2 for most wavelengths from 600 to 1200 nm. This is in comparison with the unbalanced mirror dispersion in reference (Fortier et al. 2006), where ripples of net GDD were as high as more than 100 fs^2.

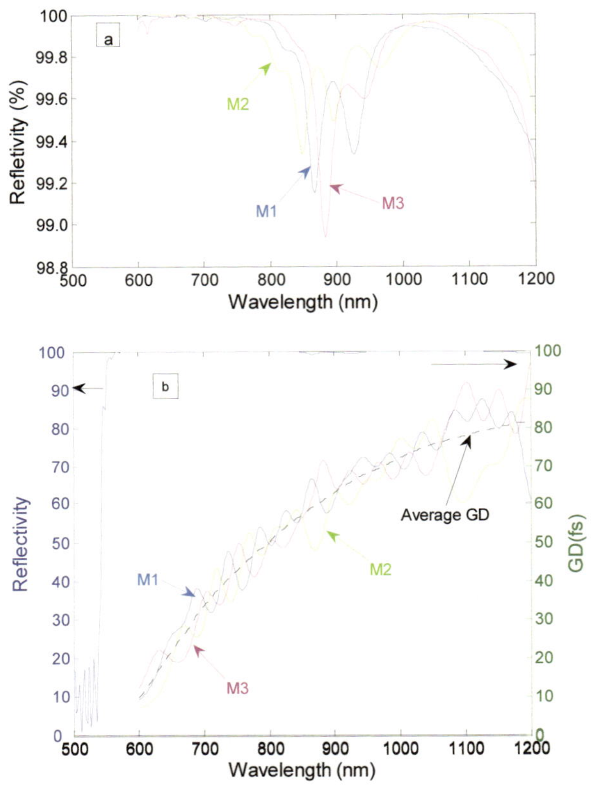

Fig. 10. (a) Reflectivity of each CM. (b) Individual and combined average GD (GD1+GD2+GD3)/3) of three CMs.

Although the combined GDD is smooth and flat, the large ripple of GDD in each individual mirror is still a concern that may destroy the pulse evolution. To testify the pulse profile changes on those mirrors, we introduce a pulse on the three mirrors in sequence and look at the pulse evolution. The pulse is assumed to be secant hyperbolic and chirp-free with spectrum of 696–904 nm, corresponding to a transform-limited pulse of 8 fs (FWHM). For each CM, a constant GDD is added so that the GDD at 800 nm is based to be 0 fs^2. In this way, the effect of the GDD ripples on the pulse could be clearly seen. When this pulse goes to those three mirrors in the order of M1, M2, and M3, the pulse duration becomes 8.21 fs, 8.80 fs, 9.31 fs and the corresponding energy in the main pulse is 97.14%, 92.89%, and 92.24% of the initial value respectively (Fig.12). When three mirrors are combined together, the

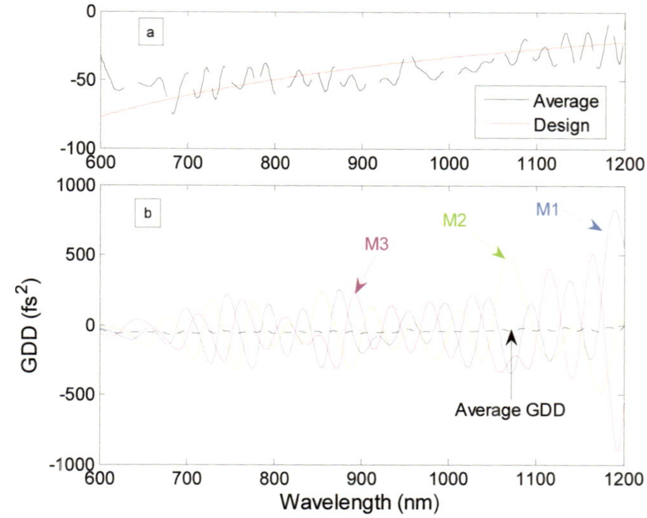

Fig. 11. (a) Combined and averaged GDD ((GDD1+GDD2+GDD3)/3) of three CMs. The solid curve is the designed GDD. (b) Individual and averaged GDD of three mirrors in a larger scale.

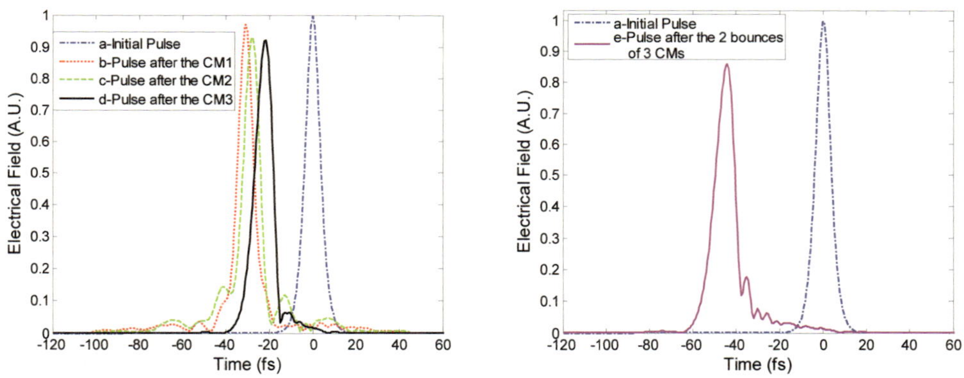

Fig. 12. Temporal analysis of the reflection of an incident secant hyperbolic-shaped free-chirp 8-fs pulse (a) from the CMs. The curve (b), (c), (d) correspond to the envelope of the pulse reflected from M1, M2, M3 orderly after one bounce, (e) reflected from the three CMs after two bounces. The temporal shift of the curves is artificial. Notice that a constant GDD is added to each CM to base the GDD at 800 nm to be 0 fs^2, so that the effect of the GDD ripples on the pulse could be clearly seen.

pulse width becomes 9.31 fs after one bounce and 10.53 fs after two bounces. The energy in the main pulse is 85.86% of the initial value after two bounces. It is noticed that slight satellite pulses appear after two bounces but the pulse still maintains its shape (Fig. 12), indicating that the residual GDD ripples of these three CMs do not destroy the pulse.

In a laser cavity, the pulse is circulating and the accumulated GDD ripples may destroy the pulse formation. However, the amplitude modulation caused by Kerr lens mode locking plays also an important role in pulse shaping. We simulated the pulse evolution in the Ti: Sapphire laser cavity with the master equation (Haus et al. 1992). The Kerr lens mode locking was modeled as a saturable absorber described by

$$q(a) = q_0 / (1 + |a|^2 / P_A) \text{ (Kärtner et al. 2005),}$$

where a is the electrical field, q_0 the modulation depth, and P_A the saturation fluence of the absorber. By solving the master equation, we obtained a stable solution for the pulse profile. The intracavity pulse spectrum expands from 600 nm to 1200 nm at about 20 dB, shown in Fig. 13. The simulation demonstrated that the ultrashort pulses in the Ti:Sapphire laser would ultimately build up in such a cavity without pulse splitting or collapse and the intracavity GDD would be compensated well with these three CMs.

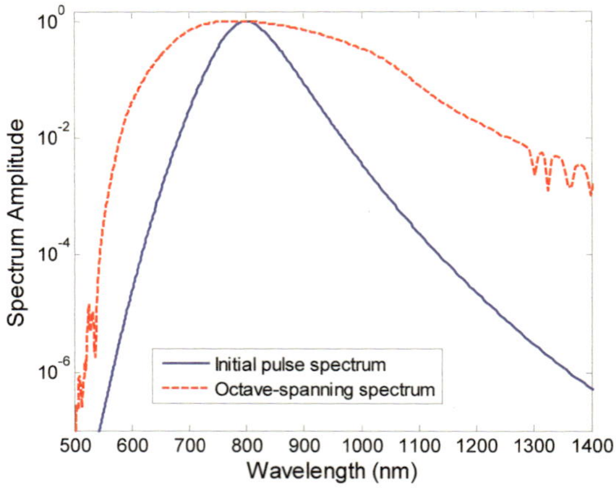

Fig. 13. The shaping of a pulse iteratively propagated in the ring cavity of Ti:Sapphire laser in frequency domain.

The inconvenience of this scheme in practice is that one additional mirror has to be coated separately, in addition to the chirped mirror pair. This will increase the manufacturing cost as well as the risk of phase errors resulted from imperfect coating. Nevertheless, the new scheme should be able to significantly reduce uncontrolled ripples and hopefully to achieve octave-spanning spectrum easier than using unbalanced mirrors.

In summary, we noticed from the simulation that this multi-mirror simultaneous optimization technique can reduce the total cavity dispersion ripples in a ring cavity. This technique allows minimizing the ripples of the whole GDD for ultra-broadband femtosecond lasers. This technique is able to easily extend to the design of chirped mirrors in linear cavities, where the number of bounces is difficult to match for paired mirrors. Another logical extension of this technique is the design of multi-chirped-mirror set for over one octave-spanning spectrum for extra-cavity dispersion compensation.

5. Mapping dispersion for compensation

In order to show how the dispersion is compensated in a laser system, or in any other extracavity dispersion compensation scheme, Nakanuma et al introduced a so-called dispersion "map" where the second order and the third order dispersion are expressed in a orthogonal coordinates (Nakanuma et al. 1991). The second order dispersion is the horizontal while the third order dispersion is in the vertical coordinate respectively. This map is very useful in selecting dispersion compensation schemes.

Let us take an example. In a Ti:Sapphire laser, if we only consider the dispersion introduced by the sapphire crystal, its second and third order dispersion are fs^2 and fs^3 respectively. In the map, the dispersion can be expressed in a vector with the horizontal and vertical coordinates of (Nakanuma et al. 1991). The purpose of using this map is to basically bring the vector back to zero, by introducing prism pairs. The prism pair dispersion depends on the prism separation and insertion, which are not independent. However, by the geometry introduced in section 3, for the crystal length of 20 mm, the dispersions could be

$$D_2 = -26.7l + 471.6d$$

$$D_3 = -63.2l + 320.9d$$

that is, in the map, the prism pair dispersion can be expressed by two vectors: one represent the prism separation dependent and the other is the prism insertion dependent. One can chose the prism separation and the prism insertion to make the total dispersion zero. As indicated in reference (Zhang et al. 1997), this can be made by solving equations

$$D_{2l}l + D_{2d}d + D_{2g} = 0$$

$$D_{3l}l + D_{3d}d + D_{3g} = 0$$

To obtain finite and positive solutions of l and d, D_{2t}, D_{3t}, D_{2d} and D_{3d} should satisfy either condition 1:

$$R_l > R_g \text{ and } R_d > R_l$$

or condition 2:

$$R_l < R_g \text{ and } R_d < R_l$$

where $R_l = D_{3l} / D_{2l}$ and $R_d = D_{3d} / D_{2d}$

In Ti:sapphire crystal, at the wavelength of 800 nm, the slope of the crystal is 0.7 fs; if we chose SF10 glass prism pair (A_1-B_1-C_1), the slope R_l and R_d are 1.71 fs and 0.69 fs and respectively. They cannot satisfy either condition, therefore the vector is open. One can choose prism material with R_l and R_d as small as possible, such as fused silica (A_1-B_2-C_2). However, because the unit dispersion is smaller too, the required prism separation would be enormously long that is not acceptable in a laser cavity. The only choice is to shorten the crystal length so that the residual third order dispersion is minimized (A_2-B_3-C_3) for the shortest pulse generation.

However, as the laser operation wavelength changes, the corresponding material dispersion changes, as well as the dispersion slope does. In this case, the map can help choosing the prism materials, if a prism pair is preferred. For example, a Cr:forsterite laser, which the operation wavelength is 1300 nm, the slope of the dispersion is 3.8 fs, much higher than at

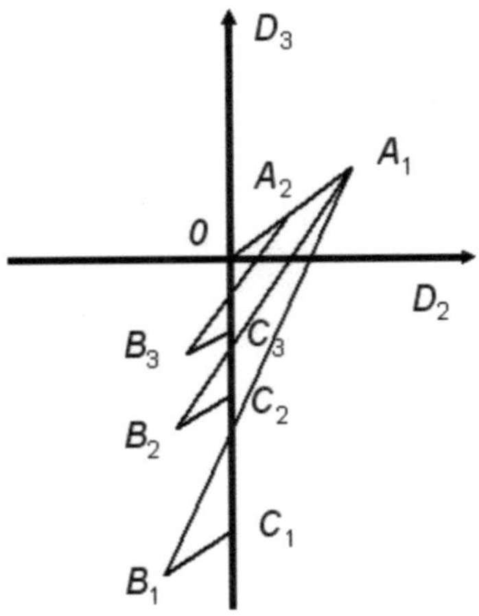

Fig. 14. Dispersion mapping for Ti:Sapphire crystal

800nm. Figure x shows a typical map for forsterite at 1290 nm. The gain material dispersion D_{2g} and D_{3g} are drawn as a vector *OA*, which shows a significantly steep slope in the map. By investigation, most of the prism materials available do not fulfill condition 1: $R_l > R_g$, but fit to $R_l < R_g$. Therefore, we have to search for the materials which satisfy the second part of condition 2: $R_d < R_l$. Some of the materials do satisfy condition 2 such as SFS01, SF58, and SF59. In the map, the dispersion of a SFS01 prism pair is drawn as another vector AB and BC which bring the gain dispersion vector back to origin so that the net D_2 and D_3 are simultaneously zero, to complete a triangle loop. The prism separation *l* and the insertion *d* can be derived from section 3. The dispersion vector of the SF14 prism pair is also plotted in dashed lines in Fig. 15 as a comparison. Apparently it is not possible to complete the triangle using the SF14 prism pair at this wavelength and a net positive TOD will remain in the cavity. The vectors for SF6 prisms (not plotted) will also end up with a net positive TOD because of $R_d \approx R_l$. Checking the wavelength dependence of the dispersion material, we found that SF14 at a wavelength shorter than 1230 nm satisfies the condition 2: $R_d < R_l$. Therefore, it is reasonable to choose SF14 prisms as the dispersion compensator for the laser operated at a shorter wavelength than 1230 nm. If either condition 1 or condition 2 can be satisfied, there should be no severe restriction on the gain crystal length. This is particularly good for Cr:forsterite because its low thermal conductivity does not allow to use a thin crystal with high doping density. It is reasonable to use a relatively long crystal to ease the thermal problem (Ivanov et al 1995). This is in contrast to the case of Ti:sapphire lasers where neither of the conditions can be fulfilled by available prism materials so that thin crystal is the only choice.

In the wavelength of 1550 nm, the dispersion slope of the Cr:YAG crystal increases to >11 fs. There are no more glass materials making the prism pair have such a high slope. One can draw the map as in Fig. 15 and will find that any glass material will leave a positive and high TOD. Rather, simply a piece of fused silica itself would do the compensation of second order only (Ishida & Naganuma, 1994), because its bulk material has a negative second order and positive third order, while the fused silica prism pair does not offer a negative third order by increasing the prism separation. Therefore, either prism pair or bulk material will ultimately make the third order positive.

It should be noted that, in an inhomogeneously broadened laser medium, such as Nd:glass, this map does not work well, because it requires much more dispersion to make it mode locked (Lu et al. 2001).

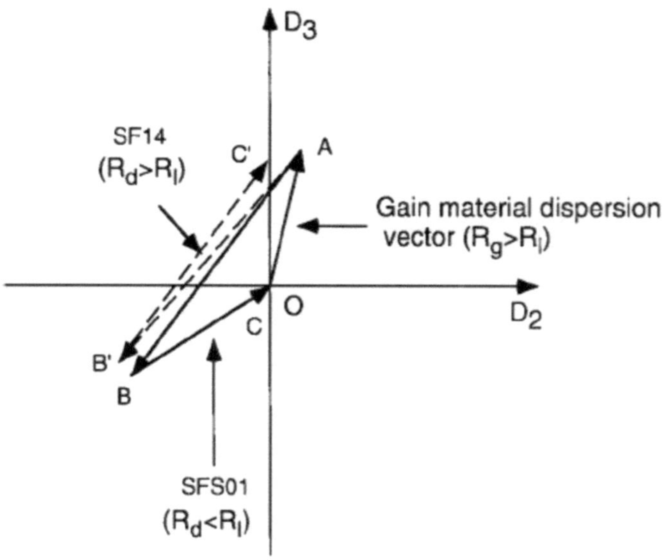

Fig. 15. Intracavity dispersion mapping for Cr:forsterite crystals

The similar method can be applied into the chirped pulse amplifier (CPA). In a grating stretcher (O-A_1) compressor (A_2-A_3) system with identical incident angle and slant distance,

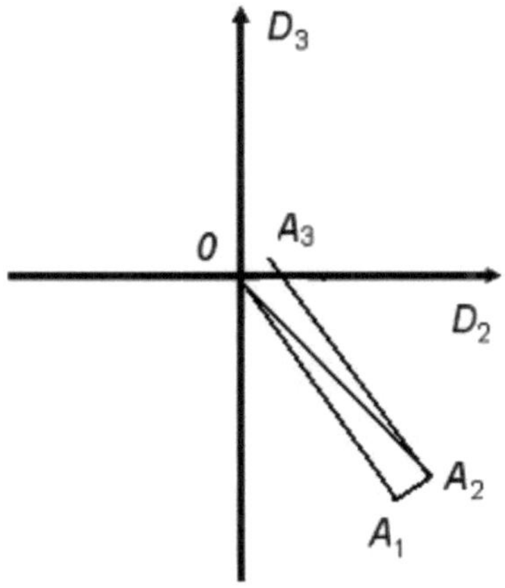

Fig. 16. Dispersion mapping for a CPA system

the material dispersion (A_1-A_2) plays a key role in breaking the balance (i.e., bring the dispersion to the origin) of dispersion introduced by grating pairs. The necessity of compensating the third order dispersion (A_2-O) can be very obvious in the map.

6. Conclusions

We have reviewed three important dispersion compensation devices including grating pair, prism pair and chirped mirrors. We also summarized the detailed calculation formulism for readers' convenience. The emphasis was given to the novel design of chirped mirrors. Finally, a dispersion map was introduced for conveniently mapping intracavity and extracavity dispersions of a femtosecond laser or an amplification system.

7. References

Birge, J. R.; & Kaertner, F. X. (2006). "Efficient analytic computation of dispersion from multilayer structures". *Appl. Opt.* 45(7): 1478~1483, ISSN: 0003-6935

Chen, L.; & Zhang, Z. "Design of broadband chirped mirrors with different initial designs". *Conference on Lasers and Electro-Optics/Pacific Rim 2007,* paper TuB1_5, ISBN: 978-1-4244-1173-3, Korea, Aug. 2007, Seoul.

Chen, L.; Yang, W.; Wang, X. & Zhang, Z. (2009). "Integrative optimization of chirped mirrors for intracavity dispersion compensation". *Opt. Commun.* 282(4): 617~620, ISSN: 0030-4018

Crespo, H. M.; Birge, J. R.; Falcão-Filho, E. L.; Sander, M. Y.; Benedick, A. & Kärtner, F. X. (2008). "Nonintrusive phase stabilization of sub-two-cycle pulses from a prismless octave-spanning Ti:sapphire laser". *Opt. Lett.* 33(8): 833~835, ISSN: 0146-9592

Fork, R. L.; Martinez, O. E. & Gordon, J. P. (1984). "Negative dispersion using pairs of prisms". *Opt. Lett.* 9(2):150~152, ISSN: 0146-9592

Fortier, T. M.; Bartels, A. & Diddams, S. A. (2006). "Octave-spanning Ti:sapphire laser with a repetition rate >1 GHz for optical frequency measurements and comparisons". *Opt. Lett.* 31(7): 1011~1013, ISSN: 0146-9592

Haus, H. A.; Fujimoto, J. G. & Ippen, E. P. (1992). "Analytic theory of additive pulse and Kerr lens mode locking ". *IEEE J. Quantum Electron.* 28(10): 2086~2096, ISSN: 0018-9197

Ishida, Y. & Naganuma, K. (1994). "Characteristics of femtosecond pulses near 1.5 μm in a self-mode-locked Cr4+:YAG laser". *Opt. Lett.* 19(23): 2003~2005, ISSN: 0146-9592

Ivanov, A. A.; Minkov, B. I.; Jonusauskas, G.; Oberle, J. & Rulliere, C. (1995). "Influence of Cr4+ in concentration on cw operation of forsterite laser and its relation to thermal problems". *Opt. Commun.* 116(1-3): 131~135, ISSN: 0030-4018

Jiang, J.; Zhang, Z. & Hasama, T. (2002). "Evaluation of chirped-pulse-amplification systems with Offner triplet telescope stretchers". *J. Opt. Soc. Am. B*, 19(4): 678~683, ISSN: 0740-3224

Kärtner, F. X.; Matuschek, N.; Schibli, T.; Keller, U.; Haus, H. A.; Heine, C.; Morf, R.; Scheuer, V.; Tilsch, M. & Tschudi, T. (1997). "Design and fabrication of double-chirped mirrors". *Opt. Lett.* 22(11): 831~833, ISSN: 0146-9592

Käertner, F. X.; Morgner, U.; Schibli, T. R.; Ippen, E. P.; Fujimoto, J. G.; Scheuer, V.; Angelow, G. & Tschudi, T. (2001). "Ultrabroadband double-chirped mirror pairs for generation of octave spectra". *J. Opt. Soc. Am. B* 18(6): 882~885, ISSN: 0740-3224

Kärtner, F. X.; Ippen E. P. & Cundiff, S. T. (2005). "Femtosecond Laser Development,", In: "*Femtosecond Optical Frequency Comb: Principle, Operation, and Applications,*" Ye, J. and Cundiff, S. T. (Ed.), 54~75, Springer, ISBN: 0-387-23790-9, Boston.

Lu, W.; Yan, L. & Menyuk, C. R. (2001). "Kerr lens mode locking of Nd:glass laser," *Opt. Commun.* 200(1-6): 159~163, ISSN: 0030-4018

Martinez, O. E. (1987). "3000 times grating compressor with positive group velocity dispersion: Application to fiber compensation in 1.3-1.6 μm region". *IEEE J. Quantum Electron.* QE-23(1): 59~64, ISSN: 0018-9197

Matos, L.; Kleppner, D.; Kuzucu, O.; Schibli, T. R.; Kim, J.; Ippen, E. P. & Kaertner, F. X. (2004). "Direct frequency comb generation from an octave-spanning, prismless Ti:sapphire laser". *Opt. Lett.* 29(14): 1683~1685, ISSN: 0146-9592

Matuschek, N.; Kärtner, F. X. & Keller, U. (1998). "Theory of double-chirped mirrors". *IEEE J. Select. Topics Quantum Electron.*4(2): 197~-208, ISSN: 1077-260X

Naganuma, K. & Mogi, K. (1991). "50-fs pulse generation directly from a colliding-pulse mode-locked Ti:sapphire laser using an antiresonant ring mirror". *Opt. Lett.* 16(10): 738~740, ISSN: 0146-9592

Nogueira, G. T. & Cruz, F. C. (2006). "Efficient 1 GHz Ti:sapphire laser with improved broadband continuum in the infrared". *Opt. Lett.* 31(13): 2069~2071, ISSN: 0146-9592

Pervak, V.; Naumov, S.; Tempea, G.; Yakovlev, V.; Krausz, F. & Apolonski, A. (2005). "Synthesis and manufacturing the mirrors for ultrafast optics". *Proc. SPIE* 5963 (59631P). ISSN: 0277-786X

Pervak, V.; Tikhonravov, V.; Trubetskov, M. K.; Naumov, S.; Krausz, F. & Apolonski, A. (2007). "1.5-octave chirped mirror for pulse compression down to sub-3 fs". *Appl. Phys. B* 87(1): 5 ~12, ISSN: 0946-2171

Tempea, G.; Krausz, F.; Spielmann, Ch. & Ferencz, K. (1998). "Dispersion control over 150 THz with chirped dielectric mirrors". *IEEE J. Selec. Topics in Quan. Electron.*, QE-4(2): 193~196, ISSN: 0018-9197

Treacy, E.B. (1969). "Optical Pulse Compression with Diffraction Gratings". *IEEE J. Quantum Electron.* QE-5(9), 454~458, ISSN: 0018-9197

Wonisch, A.; Neuhäusler, U.; Kabachnik, N. M.; Uphues, T.; Uiberacker, M.; Yakovlev, V.; Krausz, F.; Drescher, M.; Kleineberg, U. & Heinzmann, U. (2006). "Design, fabrication, and analysis of chirped multilayer mirrors for reflection of extreme-ultraviolet attosecond pulses". *Appl. Opt.* 45(17): 4147~4156, ISSN: 0003-6935

Zhang, Z.; Yagi, T. & Arisawa, T. (1997). "Ray-tracing model for stretcher dispersion calculation". *Appl. Opt.* 36(15): 3393~3399, ISSN: 0003-6935

Zhang, Z. & Yagi, T. (1993). "Observation of group delay dispersion as a function of the pulse width in a mode locked Ti:sapphire laser". *Appl. Phys. Lett.* 1993, 63(22): 2993~1995, ISSN: 0003-6951

Zhang, Z.; Torizuka, K.; Itatani, T.; Kobayashi, K.; Sugaya, T. & Nakagawa, T. (1997). "Femtosecond Cr:forsterite laser with mode locking initiated by aquantum-well saturable absorber". *IEEE J. Quantum Electron.* QE-33(11): 1975~1981, ISSN: 0018-9197.

10

Photonic Crystal Fibre for Dispersion Controll

Zoltán Várallyay[1] and Kunimasa Saitoh[2]
[1]*Furukawa Electric Institute of Technology Ltd.*
[2]*Hokkaido University*
[1]*Hungary*
[2]*Japan*

1. Introduction

The idea to prepare photonic crystal fibres (PCF) as their name show, goes back to the birth of photonic crystals. The ability to tailor structures on the micro and nano scale range, in the late 1980s, provided the opportunity to investigate the relation between the structure of matter and light. Three dimensional photonic crystal structures are periodic, dielectric structures in which light may behave similar way as electron waves in a crystal lattice: under suitable circumstances, the periodic potential presented by the crystal may open up a frequency band in which the propagation of electron waves with certain energies are prohibited. This frequency band in the case of photonic crystals is generally called photonic bandgap (PBG) as a nomenclature borrowed from solid state physics. Within this framework, the photonic crystal emerged and became an extensively studied scientific area since 1987 (Yablonovitch, 1987; John, 1987).

To introduce two dimensional photonic crystal waveguides into the practice which implies a waveguide being uniform along the propagation direction holding a two dimensional periodic photonic crystal lattice around the core as shown in Fig. 1, was justified by the novel optical properties of these waveguides compared to traditional optical fibres. One may distinguish two type of photonic crystal PCFs. One type is index guiding PCF (Knight et al., 1996) which means that the core refractive index is larger than the average index of the cladding therefore the propagation occurs due to the total internal reflection on the core-cladding interface and only the fibre structure recalls the structure of photonic crystals. The other type of PCFs is bandgap guiding fibres (Knight, Broeng, Birks & Russell, 1998; Cregan et al., 1999) where the fibre core has lower refractive index than the average index of the surrounding cladding region and the propagation occurs due to the bandgap guidance and not due to the total internal reflection which would not be possible in such a structural arrangement. Both type of PCFs hold very interesting new properties for manipulating and controlling the propagating light or light pulses.

Index guiding photonic crystal fibres with small-core sizes can have a large waveguide contribution to the material dispersion, therefore anomalous dispersion can be achieved well below the 1.27 μm region where material dispersion alone is negative (Mogilevtsev et al., 1998). This allows the dispersion management of the fibres even in the visible wavelength range with a suitable cladding design by changing the hole size and hole-to-

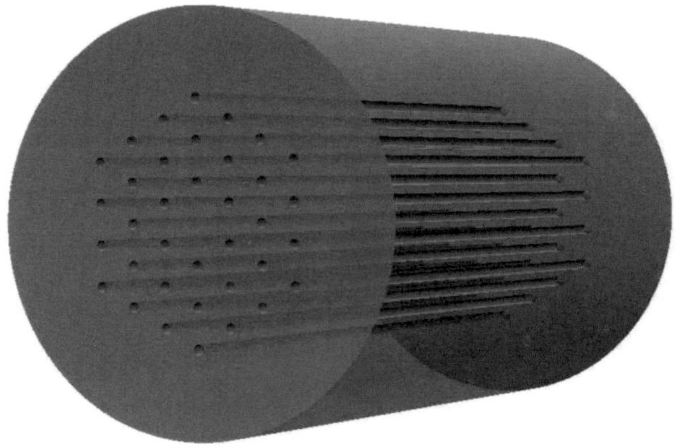

Fig. 1. General scheme of a piece of photonic crystal fibre. Crystal lattice with three periods is illustrated with the missing center most node forming the core. Crystal nodes may consist of holes or low index materials for index guiding fibres and high index glass or liquids for bandgap fibres.

hole spacing (Ferrando et al., 2000; Saitoh et al., 2003). These types of fibres opened up new areas in nonlinear optics such as ultra-broad supercontinuum generation in silica microstructured fibres even in the visible range (Ranka et al., 2000). The process of continuum and super-continuum generation had been known for many years but initial experiments used very high power lasers with large pulse energies (above micro-Joule) and ultra short pulses (< 1 ps), focused into glass, sapphire or even water. The breakthrough provided by photonic crystal fibre in 2000 was the design of the fibre that allowed the use of much lower pulse energies (~1-2 nJ) to produce the continuum effect and the zero dispersion wavelength of the fibre could be close to the pump wavelength of Ti:Sapphire. The continuum is also particularly broad, often spanning over two optical octaves. These novel light sources are cheap and effective sources for spectroscopy, frequency metrology and optical coherent tomography.

Dispersion tailored PCFs also made possible to use these fibres in Ytterbium fibre laser oscillators as a dispersion compensating element because other type of silica glass fibres have normal dispersion at around 1 micron. The obtain pulse widths was approximately 100 fs this way with pulse energy of 1 nJ in an all-fibre arrangement (Lim et al., 2002).

Ultrashort pulse generation via nonlinear compression is also reconsidered at low pulse energies (< 1 nJ) around 800 nm (Várallyay et al., 2007) where the 10 fs region in pulse duration was achieved experimentally by the utilization of dispersion tailored PCF having a zero dispersion wavelength at 860 nm and effective area of 5 μm^2. Theory showed that under optimal conditions the sub-6 fs region can be reached with the same pulse energies and fibre.

Index guiding photonic crystal fibres with sub-micron core diameter was also prepared. These waveguides are usually called nano-wires (Tong et al., 2003). These fibres can provide suitable conditions for ultra broadband soliton-effect compression of femtosecond pulses. In an experiment, 6.8 fs, few-cycle duration was demonstrated starting from 70 fs pulses around 800 nm (Foster et al., 2005) and simulations predicted that self-compression down to single-cycle duration is possible. This compression technique using photonic nanowires provides a simple method for the self-compression of sub-nJ pulses to few-cycle durations without any additional optical elements.

Further useful property of index guiding PCFs that the core size can be scaled well above the core size of single mode step-index fibres without loosing the single-mode operation. Large-mode area photonic crystal fibres yields the possibility of carrying only the fundamental mode in a large mode area decreasing this way the nonlinearity in these fibres for high energy pulse transmission (Knight et al., 1998; Birks et al., 1997; Mortensen et al., 2003). Using ytterbium doped, large mode area PCF higher than 1000 μm^2 effective core area was reported without losing the single mode operation in a fibre amplifier (Limpert et al., 2004). Exceeding the 2000 μm^2 effective core area in a chirped pulse amplifier system was also demonstrated still having the diffraction limited output (Limpert et al., 2006). Doped microstructured fibres are excellent candidates for solely fundamental mode amplification due to the escaping higher order modes between the holes running along the fibre cladding. Bending loss however can be significant and the fibre has to be kept straight. The dispersion of these fibres can not be tailored using structural modifications in the cladding since the waveguide contribution becomes negligible at large core sizes, therefore the dominant dispersion is material dispersion.

Bandgap guiding fibres with a hollow core was first demonstrated in 1998 (Knight et al., 1998). Photonic bandgap fibres enable light guidance in a low refractive index material such as air, vacuum or gas that way that the surrounding cladding is made of a higher index dielectric material such as silica glass. Significantly high percentage of the light energy is guided in the hollow core in the bandgap if surface modes or leaking modes do not affect the propagation. This may offer a significantly reduced nonlinearity with respect to silica core fibres (Lægsgaard et al., 2003; Ouzonov et al., 2003).

Lower loss than in conventional solid core fibres are theoreticaly also achievable (Roberts et al., 2005) as well as light transmission at wavelengths where the material absorption would otherwise be prohibitive (Stephens et al., 2004; Shephard et al., 2005). These properties are uniq features of these waveguides which would not be available without the existance of photonic bandgap effect.

The low nonlinearity and low loss is advantageous for high intensity pulse transmission which is an intended property in laser physics for delivering high power, focused laser beams through fibres. A suitable dispersion profile is however indispensable for ultrashort pulse applications where pulse broadening and pulse distortions may occur very quickly due to improper dispersion management. Photonic bandgap fibres have a typical third order function-like dispersion function (Müller et al., 2002; Jasapara et al., 2003) which increases continuously from the normal dispersion region to the anomalous one with an inflection point in the middle of the bandgap. In order to see this property of bandgap fibres we calculated the dispersion function in a few types of fibres plotted in Fig. 2. This behavior is independent of the constituting materials but it is a basic feature of the bandgap guidance (Bouwmans et al., 2003). The dispersion function of bandgap fibres usually show an anomalous dispersion around the center of the bandgap and it crosses zero close to the short wavelength edge of the bandgap. The monotonically increasing function from the short to long wavelengths shows a significant positive dispersion slope anywhere in the bandgap which limits the compression ratio (Ouzounov et al., 2005) or the quality of the compressible, ultra short pulse (Lægsgaard & Roberts, 2008).

The shape of the dispersion function of photonic bandgap fibres is however affected by the waveguide contribution at small core sizes and by resonances at even large core sizes (Jasapara, Bise, Her & Nicholson, 2003; Jasapara, Her, Bise, Windeler & DiGiovanni, 2003). Harmful effects such as mode anti-crossing (Saitoh et al., 2004; Humbert et al., 2004), surface and leaking modes (Várallyay et al., 2008) may also affect the dispersion function of PCFs.

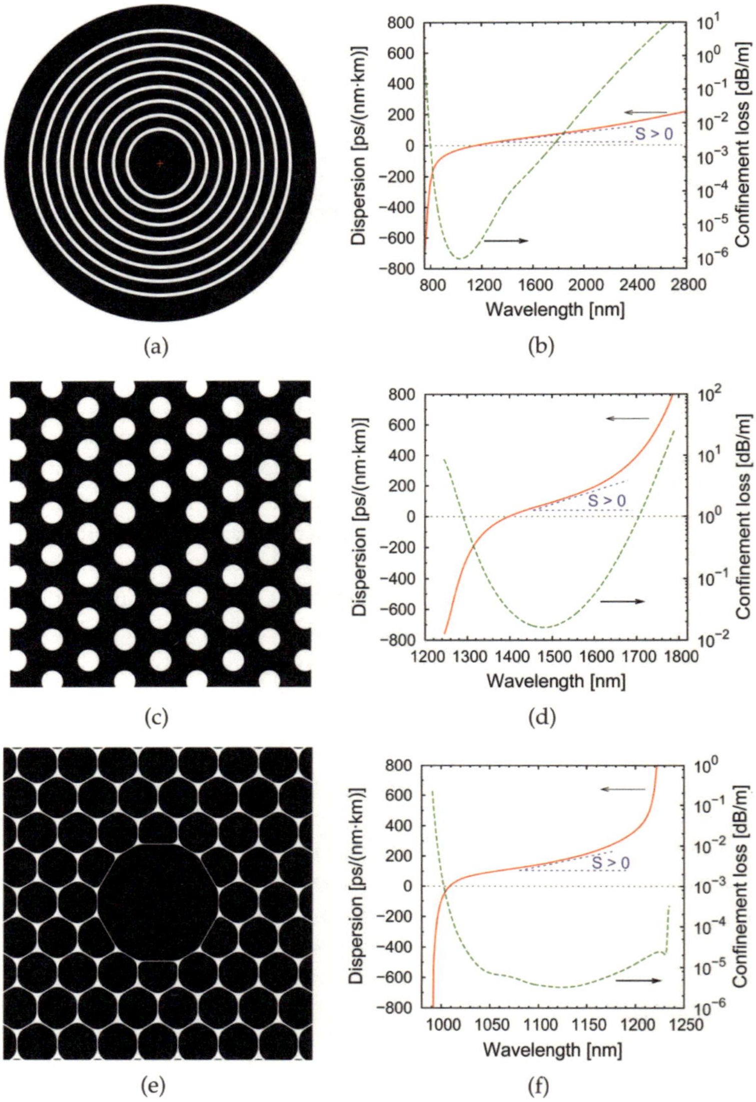

Fig. 2. Structure and dispersion function of different photonic bandgap fibres calculated with the finite element method. *S* refers to the dispersion slope which is clearly positive in all cases. (a) Solid core Bragg fibre with silica (low index regions) and doped silica (high index, white regions). The optimized structure is detailed in Sec. 3.1, (b) obtained dispersion and confinement loss curves with an extra broad bandgap of the annular structure having a full (C_∞) symmetry, (c) silica microstructured fibre where the holes are filled with a high-index liquid. The structure is same as measured in (Jasapara, Bise, Her & Nicholson, 2003): pitch Λ= 5.5 μm, hole diameter d = 2.2 μm and the refractive index of the liquid was n_H = 1.59, (d) obtained dispersion and loss functions for the high index liquid filled PC fibre with a close agreement to the measurements, (e) Structure of a hollow-core fibre with honey-comb cladding and high air-filling fraction. This fibre is used in later calculations and the structure is described in Sec. 4 and (f) the obtained dispersion and loss profiles of the hollow-core fibre.

Namely, an ideal fibre compressor or dispersion compensation unit should posses the following properties to fulfil the requirements in high power, ultra-short pulse compression which is applicable in fibre lasers (Nielsen et al., 2006; Ruehl et al., 2007) as well as in fibre amplifier technology (de Matos et al., 2003):

- low loss
- low nonlinearity (air-guidence or large core size)
- anomalous dispersion (where dispersion is normal otherwise)
- negative or flat dispersion slope in a broad wavelength range.

The last item is believed that it can be realized by structural modifications in photonic bandgap fibres such as introducing resonant layers (Várallyay et al., 2009) which resembles to that of Gires-Tournois interferometers (theory is explained in Sec. 2).

The modification of the dispersion function of bandgap guiding PCFs due to the above mentioned advantageous properties are in high interest. Introducing geometrical defects and resonances in the perfectly periodic structure of a PBG cladding results in a modified groupdelay dispersion and even negative dispersion slope was investigated first in hollow-core OmniGuide fibres (Engeness et al., 2003). Dispersion modification in all-glass PBG fibres was also initiated by the introduction of defect rods in the cladding region (Fang et al., 2007) which theoretically results in resonance behavior. This modification shifted and changed the magnitude of the dispersion function but it preserved the canonical form of the dispersion profile. It was also shown that dispersion modification did work for hollow-core PBG fibres by changing the first period of the fibre suppressing this way the anti-crossing events within the bandgap (Roberts, 2007). The resulted dispersion function is a nearly flat function over a wide wavelength range with very small positive dispersion slope. Negative dispersion slope for hollow-core fibres is also predicted by resonant layers changing the hole size and the shape of the holes in the first and second periods of a hollow-core fibre (Lægsgaard et al., 2007). We showed however that detune the parameters of the first period compared to other periodes of a hollow-core, all-silica Bragg fibre may result in a reversed dispersion slope ($S < 0$ instead of $S > 0$) which can be theoretically considered as an equivalent of a thin-film Gires-Tournois interferometer (Várallyay et al., 2008b; Várallyay et al., 2008). The same was shown for solid core PBG fibres where the freedom of introducing resonant layers is completed by the ability of changing the refractive index of a particular layer (Várallyay et al., 2008a).

In this chapter we present a complete analysis on the possibilities of modifying the dispersion of PBG fibres including solid-core and hollow-core photonic crystal fibres by applying two dimensional, resonant Gires-Tournois interferometers around the fiber core.

2. Theory

We aim to apply resonant structures in PBG fibre claddings resembling to that of Gires-Tournois interferometers (GTI) in solid-core and hollow-core fibres taking the analogy from one dimensional PBG structures. GTIs have an arrangement similar to Fabry-Pérot interferometers. Whereas Fabry-Pérot cavity is enclosed by two high-reflecting mirrors, GTI consists of both a high and a low reflector which results in a highly frequency dependent phase shift on the light field falling on the low-reflectivity surface (Kuhl & Heppner, 1986).

In order to provide a qualitative description of resonances we start with one dimensional photonic bandgap structures which can be associated by dielectric mirrors. The assembly of alternating high and low index thin films can be characterised by the characteristic matrix of the arrangment having N layers and neglecting the absorbance the following expression can be given (Macleod, 2001)

$$\begin{bmatrix} B \\ C \end{bmatrix} = \left\{ \prod_{q=1}^{N} \begin{bmatrix} \cos(\delta_q) & i\sin(\delta_q)/\eta_q \\ i\eta_q \sin(\delta_q) & \cos(\delta_q) \end{bmatrix} \right\} \begin{bmatrix} 1 \\ \eta_m \end{bmatrix} \tag{1}$$

where

$$\delta_q = kn_q d_q \cos(\theta_q) \tag{2}$$

with $k = 2\pi/\lambda$ the free-space wavenumber, n_q and d_q are the refractive index and physical thickness of the qth layer, respectively and η_q is the addmittance of the same layer at oblique incidence for the p and s-polarization components can be written as

$$\begin{array}{ll} \eta_q = n_q Y/\cos(\theta_q) & \text{for } p\text{-polarisation (TM)} \\ \eta_q = n_q Y \cos(\theta_q) & \text{for } s\text{-polarisation (TE)} \end{array} \tag{3}$$

where $Y = \sqrt{\varepsilon_0/\mu_0}$ is the admittance of the free space. In Eq. (1), η_m is the admittance of the substrate or emergent medium. The amplitude reflection coefficient can be calculated from the characteristic matrix as follows

$$\rho = \frac{\eta_0 - A}{\eta_0 + A} \tag{4}$$

where $A = C/B$ can be obtained from Eq. (1). The phase shift on the reflecting wave is now given as a quotient of the imaginary and real part of Eq. (4)

$$\tan(\varphi) = \frac{\Im\rho}{\Re\rho} = \frac{-2\eta_0 \Im A}{\eta_0^2 - (\Re A)^2 - (\Im A)^2} \tag{5}$$

where $\Im$ and $\Re$ stand for the imaginary and real parts, respectively.

Resonant layers can be introduced in the cladding of a particular bandgap fibre by the modification of one or necessarily more layers.

2.1 One resonant layer

Gires-Tournois (GT) interferometer can be considered as a thin film layer in a simple case with 100% reflection at one side where the high reflectance is ensured by the fibre cladding in a PCF and low reflection at the other side: core and thin-film interface as shown in Fig. 3. The amplitude reflection of this layer is given by Eq. (4) where η_0 is the admittance of the incident medium (core) and A is the admittance of the GT layer which can be obtained from the characteristic matrix (Eq. (1)) yielding

$$A = \frac{\eta_m \cos(\delta) + i\eta_1 \sin(\delta)}{\cos(\delta) + i(\eta_m/\eta_1)\sin(\delta)} \tag{6}$$

where η_m is the admittance of the subsequent layers (the cladding) and δ is the phase factor of the positive going wave and now it is given by $\delta = kn_1 d_{GT}\cos(\theta_1)$ using the notation of Fig. 3. Here, d_{GT} is the GT layer thickness, n_1 is the refractive index of the resonant layer and θ_1 is the angle of refraction.

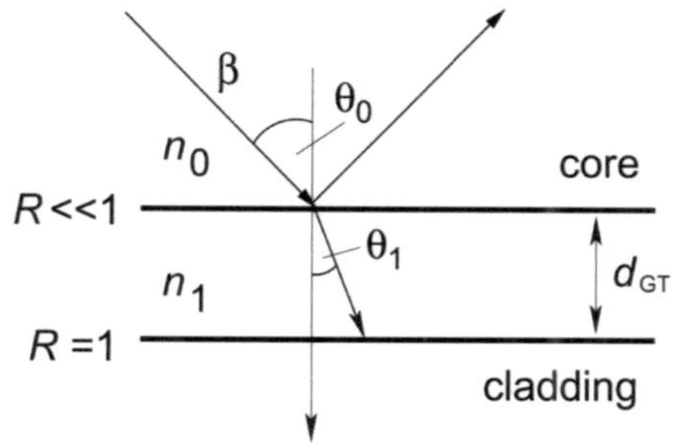

Fig. 3. Schematic illustration of a GT interferometer between core and cladding in a PBG fibre.

η_m can be assumed to be zero in Eq. (6) because the reflectance is 100% = $(\eta_1 - \eta_m)/(\eta_1 + \eta_m)$ at this side of the TF which is a good approximation at the close center of the bandgap.
Writing back Eq. (6) into Eq. (4) with the above mentioned simplification (η_m = 0) the amplitude refraction of a GT interferometer can be given by

$$\rho = \frac{\eta_0 - i\eta_1 \tan(\delta)}{\eta_0 + i\eta_1 \tan(\delta)} \tag{7}$$

The real and imaginary part of Eq. (7) is yielding the phase shift on the incident field as follows

$$\tan(\varphi) = \frac{-2\eta_0\eta_1 \tan(\delta)}{\eta_0^2 + \eta_1^2 \tan^2(\delta)} \tag{8}$$

The group-delay (GD) ($d\varphi/d\omega$), the group-delay dispersion (GDD) ($d^2\varphi/d\omega^2$) and higher order dispersions can be easily derived from Eq. (8). Here, we only give the calculated GD function derived using Eqs. (2) and (8)

$$\frac{d\varphi}{d\omega} = \frac{2\eta_0\eta_1(\eta_0^2 - \eta_1^2)\tan^2(\delta)}{\cos^2(\delta)\left[\eta_0^4 + 6\eta_0^2\eta_1^2\tan^2(\delta) + \eta_1^4\tan^4(\delta)\right]}\frac{d\delta}{d\omega} \tag{9}$$

where $d\delta/d\omega = n_1 d_{GT}\cos(\theta_1)/c$.
The calculation with oblique incidence is essential in the approximation of fiber behavior due to the high propagation angle of light in fibres which prerequisite can be obtained from the corresponding eigenvalue of the investigated mode:

$$\theta_0 = \arcsin\left(n_{\text{eff}}/n_0\right) \tag{10}$$

where n_{eff} is the effective refractive index and n_0 is associated with the core index.
For the Bragg fibre investigated in Sec. 3.1, the propagation angle obtained from the Helmholtz eigenvalue equation is approximately 87° for the fundamental mode. The

refractive index of GTI is chosen to be n_1 =1.46 which is 0.01 higher than that of the core index which is n_0 = 1.45 at around 1.03 μm. We plot the GD and GDD functions of the GTI based on Eq. (5) in Fig. 4 applying thicknesses that show the nonlinear phase shift and resonance behavior at around 1 μm. We found that the thickness of the GTI must be d_{GT} = 1.34 μm (zeroth-order resonance) or 4.02 μm (first-order resonance). The mediate thicknesses shift the position of the resonance peak as it is indicated by the black solid line in Fig. 4. The reversed dispersion region of the GDD function is broader for the lower-order resonance and sharper for the higher-order one.

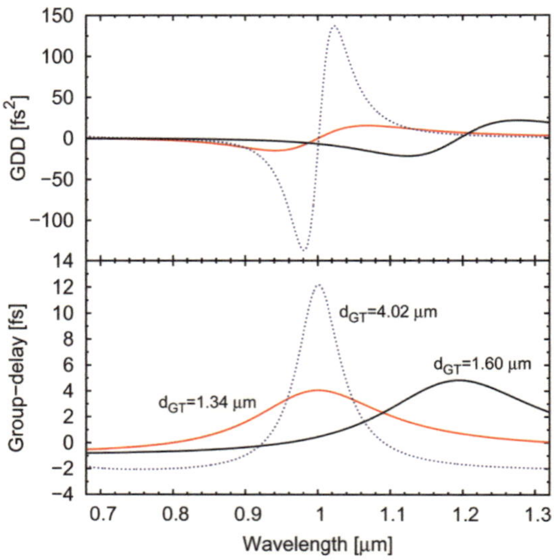

Fig. 4. Zeroth- and first-order resonances on the group-delay and group-delay dispersion curves of Gires-Tournois interferometer for oblique incidence with d_{GT}=4.02 μm and 1.34 μm thicknesses, n_0 = 1.45, n_1 = 1.46 and θ_0 = 87°. Calculated from the derived Eq. (5). Black solid line is a shifted resonance with d_{GT} = 1.6 μm.

2.2 Two resonant layers

There are fibre structures where the introduction of an annular layer around the core having a refractive index close to the core index is not accessible. All-silica hollow core PBG fibres are typical examples where slightly higher index dielectric material than the air is not available, therefore the resonant structure must be set up by changing the thickness of the high index glass and low index air-spacer layers in the same time. Namely the first periode around the fibre core has to be readjusted. This modification as it is shown below must result in the same physical effect as above. A schema of the thin-film model of this arrangement is shown in Fig. 5.

Using the notation of Fig. 5 the characteristic matrix can be written for the two layers in the following form

$$\begin{bmatrix} B \\ C \end{bmatrix} = \begin{bmatrix} \cos(\delta_1) & i\sin(\delta_1)/\eta_1 \\ i\eta_1 \sin(\delta_1) & \cos(\delta_1) \end{bmatrix} \begin{bmatrix} \cos(\delta_2) & i\sin(\delta_2)/\eta_2 \\ i\eta_2 \sin(\delta_2) & \cos(\delta_2) \end{bmatrix} \begin{bmatrix} 1 \\ \eta_m \end{bmatrix} \tag{11}$$

where $\delta_{1,2} = \omega / c n_{1,2} d_{1,2} \cos(\theta_{1,2})$ and $\eta_{1,2} = n_{1,2} Y / \cos(\theta_{1,2})$ for the p-polarization.

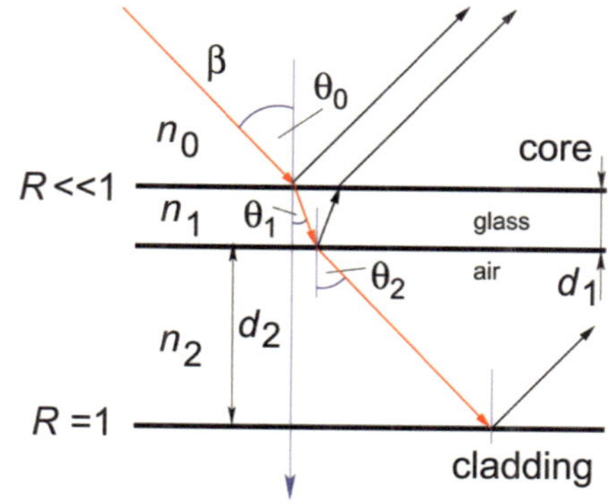

Fig. 5. Schematic of the high and low index layers (first periode) between core and further parts of the cladding having two adjustable thicknesses: d_1 is the physical thickness of the glass layer and d_2 is the thickness of the air-spacer layer.

Performing the operations in Eq. (11) and taking into consideration that $\eta_m = 0$ same as previously, $A = C/B$ is obtained as

$$A = i\frac{\eta_1 \sin(\delta_1)\cos(\delta_2) + \eta_2 \cos(\delta_1)\sin(\delta_2)}{cos(\delta_1)\cos(\delta_2) - \eta_2/\eta_1 \sin(\delta_1)\sin(\delta_2)} \tag{12}$$

Writing back Eq. (12) to Eq. (4) and deriving Eq. (5), one can obtain the phase shift of the incident field upon one reflection given by

$$\tan(\varphi) = \frac{-2\eta_0 \Im A}{\eta_0^2 + A^2} \tag{13}$$

where it was exploited that A has only imaginary part in this particular case.
Assuming a simplified air-silica Bragg structure with annular silica and air layers around the holow core which is analyzed in Sec. 4.1 where we found that the quater-wave stack condition around one micron is fulfiled if the high (silica) and low (air) index layers have the thicknesses of d_H =0.25 μm and d_L =3.9 μm, respectively. The corresponding angle of incidence obtained from Eq. (10) at around 1.05 μm is $\theta_0 = 86°$. We are looking for resonant peaks in the group-delay and reversed GDD function around the 1 μm wavelength region changing the thickness of the first high d_1 and first low d_2 index layers. We plotted three different resonance curves in Fig. 6(a) and 6(b).
Figure 6(a) shows broad resonances with a low index layer thickness $d_2 < d_L$ where d_L is the thickness of the original quater-wave stack for the air layers. Figure 6(b) shows much narrower resonance curves with much sharper peaks having low index thicknesses $d_2 > d_L$. The high index layer was chosen as 0.1, 0.2 and 0.3d_H. One can see that the thiner the glass layer the broader the resonance is. Changing the thickness of the silica layer changes the magnitude of the reflection on the first facet of this combined Gires-Tournois interferometer. Figure 6 is a good illustration for showing that narrower the resonance the steeper the corresponding dispersion curve is.
The phase properties of photonic bandgap fibres can be well approximated with one dimensional dielectric structures (Fekete et al., 2008) using the p-polarization component of the light at grazing incidence hence, in the followings, we demonstrate that the calculated GTI works well for solid core as well as for hollow core PBG fibres.

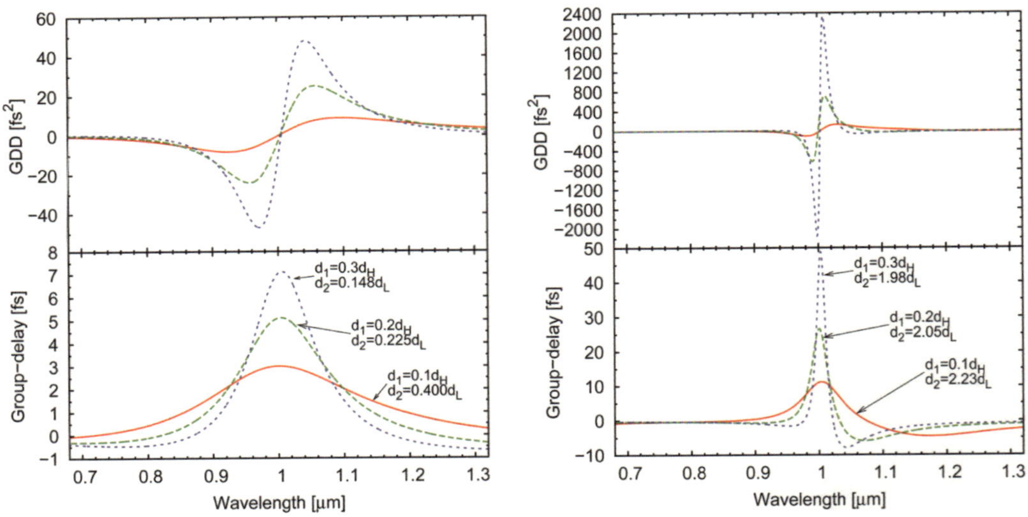

(a) (b)

Fig. 6. (a) zeroth order and (b) first order Gires-Tournois resonances achived by adjusting the first and second high and low index layers assuming a perfectly reflecting underlying cladding ($\eta_m = 0$). The applied thicknesses are expressed in terms of the original quater-wave stacks $d_H = 0.25\ \mu$m and $d_L = 3.9\ \mu$m.

3. Solid core fibres

Bragg fibres analyzed below are not photonic crystal fibres due to the annular arrangement of the alternating high and low index layers not forming a crystal lattice-like pattern . But light guidence is achieved by the bandgap effect similarly to those fibres where a microstructured arrangement ensures the bandgap guidence. Namely the average index of the cladding is higher than the core medium. It is also notable that the first proposition and theoretical analyses on these fibres (Yeh et al., 1978) were presented well before the appearence of photonic crystals and photonic crystal fibres. Bragg fibre can be imagened as a round bent dielectric mirror with properly adjusted high and low index layers for the large angles of incidence of light. In this case the core is an air core and the cladding is formed of the dielectric structure. It is also known however that the loss properties of Bragg fibres are the most advantageous if the core index is identical to the refractive index of the low index layer (Dianov et al., 2009). Therefore the analyses presented below are carried out for such Bragg fibres where the core is a solid, glass core having the same refractive index as the low index layer (same structure was presented in Fig. 2(a)). The introduced resonant layer appeares around the core followed by 8 period of alternating high and low index layers.
We also investigate the properties of solid core photonic crystal fibres with a similar arrangement has shown in Fig. 2(c).

3.1 Solid core Bragg fibre

We realize the same GTI structure in a solid-core Bragg fibre than it is obtained from the analytical calculations in Sec. 2.1. The used structure has shown in Fig. 2(a) and the modified one in Fig. 7. We found that if we use realizable high and low index materials for cladding

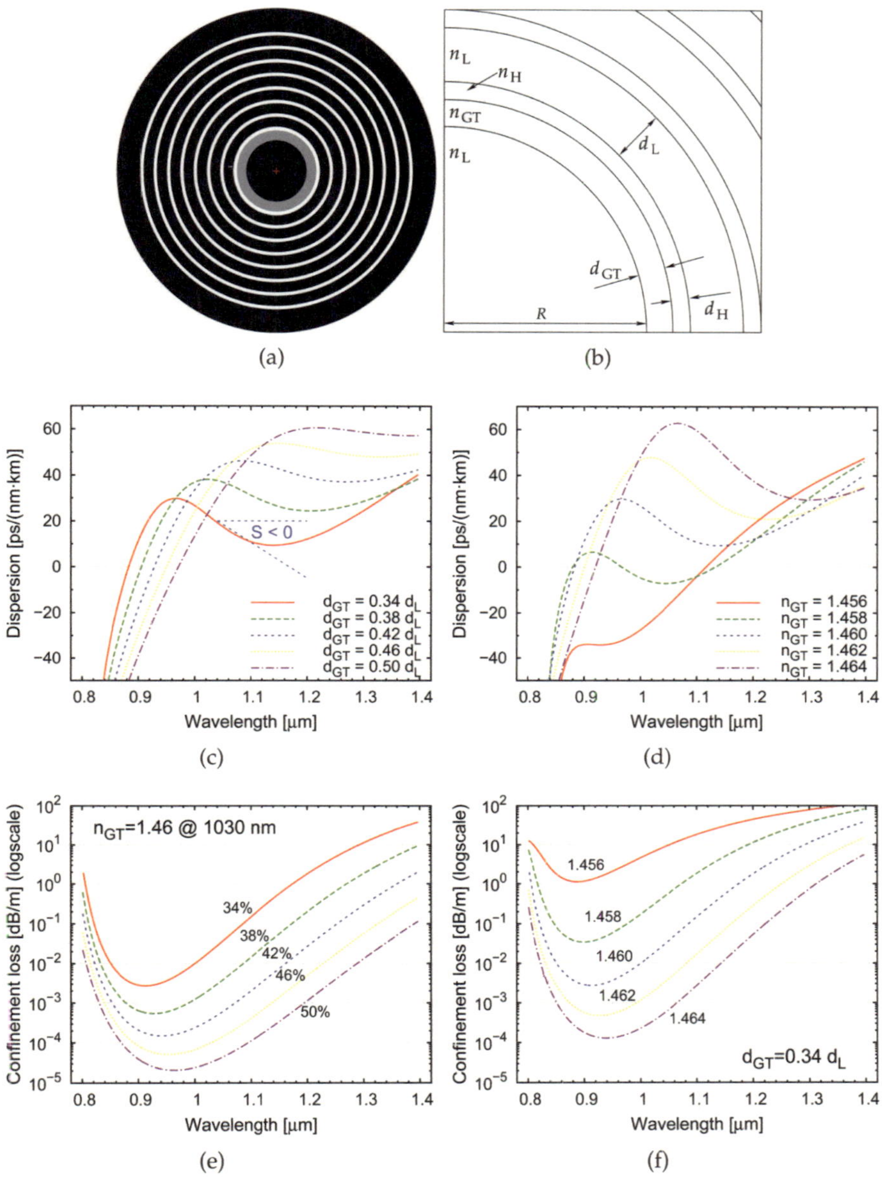

Fig. 7. (a) Solid core Bragg fibre with a resonant first layer just around the core having a refractive index of n_{GT} and thickness of d_{GT}. The original structure is shown in Fig. 2(a), (b) parameters of the fibre: R: core radius, subscripts L and H indicate the low and high index properties, respectively, while GT stands for the resonant GT layer. (c), (d), (e) and (f) shows the obtained dispersion and confinement loss profiles of different GTIs realized around the core in the above Bragg PBG structure. (c) Dispersion functions with different thicknesses of GTI, (d) dispersion functions with GTI having different refractive indices. Dispersion slope (S) is clearly smaller than zero in a broad wavelength range. (e) confinement loss belongs to the case changing GTI thicknesses and (f) confinement loss with different GTI refractive indices.

with n_L =1.45 and n_H =1.5, than the optimum structure which shows a broad and regular bandgap with 10^{-6} dB/m confinement loss must have low index and high index layer thicknesses of 3.8 μm and 0.95 μm, respectively.

Two parameters can be adjusted in connection with the reversed dispersion slope (RDS); the refractive index and the thickness of the GT layer. We set the refractive index of the GT layer to 1.46, and the thickness of it to $0.34d_L$ which results in a 180 nm RDS region from 960 nm to 1140 nm. Increasing the thickness of this resonant layer shifts the RDS region to longer wavelengths (Fig. 7(c)), while the increase in it's refractive index causes stronger resonances and steeper dispersion function due to the decreasing refraction on the first surface (core-GT interface). Figure 7(d) however, shows a high sensitivity on the changes of the GTI index. 0.1- 0.2% index difference may cause almost 100 nm shift and some changes in the steepness of the dispersion slope. This shows that small deviations from the ideal conditions at manufacturing process may result an essentially different dispersion profile.

Nearly zero dispersion profile is also demonstrated in Fig. 7(d) for n_{GT} = 1.458, where the dispersion fluctuation is 12 ps/(nm·km) over a 280 nm wavelength range. Similarly, for d_{GT} = $0.5d_L$ (Fig. 7(c)) the changes of the dispersion function is less than 4 ps/(nm · km) over an almost 200 nm wavelength range. The loss profiles show an acceptable loss (< 0.1 dB/m) for using these fibres in real applications. In Fig. 7(e) and Fig. 7(f), one can see that the wavelength range where dispersion tailoring is achieved falls on the increasing tail of the loss profile which indicates that the mode has continuously stronger resonant coupling with the GT layer, resulting in higher loss. In order to see the changes in the mode profile due to the resonance coupling we plotted the fundamental core mode at the short and long wavelength edge and in the middle of the RDS region in Fig. 8, corresponding to the GTI with d_{GT} = $0.34d_L$ and n_{GT} = 1.46.

The higher the light energy stored in the GTI the larger group-delay of the propagating pulse can be expected as it is already demonstrated at one-dimensional PBG structures (Szipőcs et al., 2000). This is the reason that the introduced resonant layer is able to reverse the dispersion function resulting in a wavelength dependent standing field in the GTI. The mode at the long wavelength edge of the reversed slope looks like more an LP_{02} than an LP_{01} mode due to the resonant coupling of the fundamental mode with the resonant layer (Fig. 8).

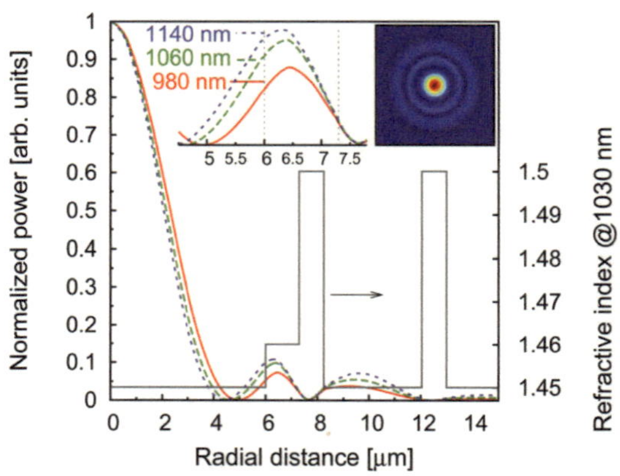

Fig. 8. Propagating fundamental mode profiles at 980 nm, 1060 nm and 1140 nm wavelengths. The embedded graph zooms to the close surrounding of GTI. Vertical black lines show the region where the refractive index n_{GTI} = 1.46. Inset shows the 2D distribution of the propagating fundamental mode at 1060 nm.

3.2 All-glass photonic crystal fibre

All-glass photonic bandgap fibres have pure silica glass core and the cladding consist of cylindrical high-index inclusions forming a crystal lattice. The high-index glass can be dopped silica glass applying germanium oxide. The significance of these fibres lies in the similar property to their index guiding analogons: single mode guidance of the light can be ensured in a broad wavelength range even if the core size is extremaly large (Egorova et al., 2008) and the loss of these fibres are tolarable small for laser applications.

A solid core PCF with photonic bandgap guidence can be an ideal choice to preserve the single modeness at large core sizes keeping low nonlinearity and for the possibility of introducing resonance layers to tailor the dispersion. These fibres are serving many opportunity to introduce the resonant layer in the cladding region, for example, changing the size of the high-index inclusions (d) changing the pitch (Λ), changing the refractive index of the high index glass (n_H) and all the combinations of these changes are available.

In our calculations, we consider an all-silica PCF with a triangular lattice having an index difference between silica and GeO_2 doped silica as large as Δn =0.015. The pitch is Λ=7.4 μm and the hole size is selected to be d =2.47 μm which result in a mode field diameter of ~12 μm around one micron. This structure provide a broad, fundamental bandgap around the 1 μm wavelength range.

The investigated resonant layer is basically that one relates to the theory of GTI with two layers (Sec. 2.2). Namely, the diameter of high index inlaying glass (d_1) and the distance of the first and second periode of the high index regions (Λ_2) are considered as variable parameters. Any other solution such as changing the refractive index of the high index circular inclusions along with varying the geometry is considered in later publications.

Figure 9 shows the dispersion profile and loss profile of the fibre choosing d_1 = 0.741 μm = 0.3d and Λ_2 = 3.11 μm. RDS is achieved between the 1390 nm and 1620 nm wavelengths with an exponentially increasing loss in this range. At the short wavelength edge of the RDS, the confinement loss is 6.8 $\cdot 10^{-3}$ dB/m, at 1550 nm 5 dB/m and at the long wavelength edge, it is almost 100 dB/m. This loss can be decreased by shifting the bandgap to longer

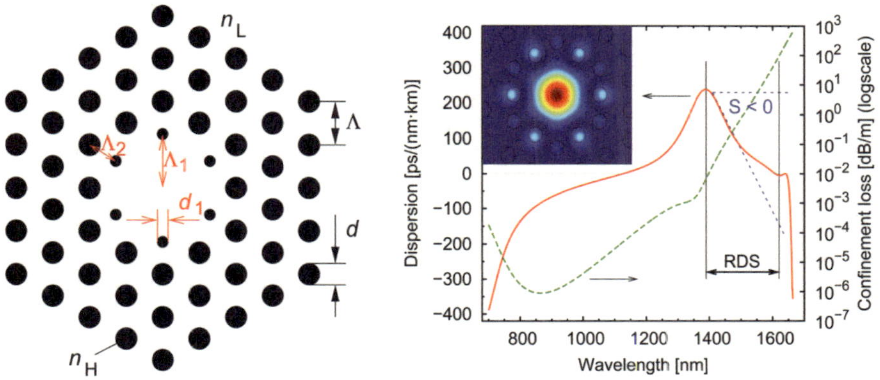

(a) (b)

Fig. 9. (a) Scheme of the structure having a modified first periode in the crystal lattice with smaller diameter of high index region $d_1 < d$ and modified pitch so that $\Lambda_1 > \Lambda_2$ but $\Lambda_1 + \Lambda_2 = 2\Lambda$ (b) Dispersion profile and the corresponding confinement loss curve. RDS is achieved around the telecommunication wavelengths. Inset shows the electric field norm at 1500 nm.

wavelengths by increasing the value of d and Λ and keeping the parameters of the resonant layer for the same value.

The topic to achieve reversed dispersion profile through resonant structures in all-glass photonic crystal fibres is a hot topic in our recent investigations and more detailed analysis on these structures will be presented in later publications.

4. Hollow core fibres

In the followings, we intend to introduce resonant layers in hollow-core, air-silica fibres for the analogy described in the theory in Sec. 2.2. By adjusting the thickness of the first high index layer (core wall) and the following low index layer properly the wavelength dependent nonlinear phase shift on the propagating light may cause a dispersion slope with opposite sign of the conventional bandgap guidence (See Fig. 2(b), 2(d) and 2(f)).

We investigate a simplified, hollow-core Bragg fibre by neglecting the silica struts between the glass annular layers in Sec. 4.1. This type of fibre is intended to combine the advantageous property of Bragg fibres with extra broad bandgaps and the light guidence in a low nonlinearity hollow core region (Vienne et al., 2004).

Due to the significance of small details in PBG structures which are crucial in the design of reliable PBG waveguides, a more accurate model is presented with a honey-comb cladding structure in Sec. 4.2. These type of fibres can be leaking-mode free in a relatively wide wavelength range (Saitoh et al., 2007) therefore the modified dispersion properties can be demonstrated freely. The leaking or surface mode free guidence of light in the case of hollow-core Bragg fibres with support bridges in the low index cladding regions (Foroni et al., 2007) has not been resolved yet. The simplified model however can demonstrate that the theoretical model was presented in Sec. 2.2 is valid for the two-dimensional calculations and explains the underlaying physics.

4.1 Simplified hollow-core Bragg fibre

We use three periods of high (glass) and low (air) index layers in our hollow-core Bragg fibre model (similar to Fig. 2(a) if the dark regions are considered to be the air and white regions the glass). We neglected the effect of silica struts between the glass layers in a first approximation, though these small structural parts modify the effective index of the low index layers by a few percent which requires the modification of the fibre cross-section parameters (Fekete et al., 2008). Silica struts also introduce some mode anti-crossing events partialy due to the lowering of symmetry from full cylindrical (C_∞).

After some optimization of the glass and air layer thicknesses at around one-micron (Várallyay et al., 2008) the critical angle of incidence, the thickness of the low and high index layers were found to be $\theta_0 = 86.16°$, $d_L = 3.92\ \mu$m and $d_H = 0.25\ \mu$m, respectively. The core radius is $R = 10\ \mu$m.

The GTI layer is formed by readjusting the first high and low index layers (first periode) around the core. The thin glass layer plays the role of a partial reflector, and behind the resized low index layer (cavity) cladding acts as a high reflectivity mirror of the two-dimensional Gires-Tournois interferometer. In our present simulation, we choose three different silica (high index) layer thicknesses for partial reflector layers around the core, which are 10, 20 and 30% of the thickness of the high index layers forming the quarter-wave stack Bragg mirror design. The obtained results of the computations in Fig. 10(a) clearly

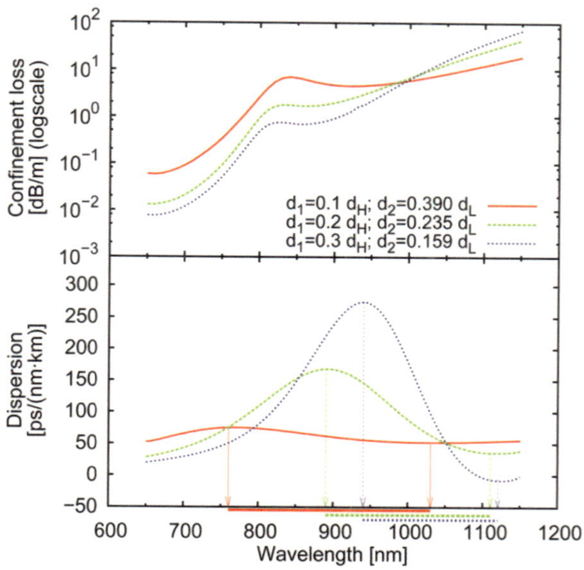

(a)

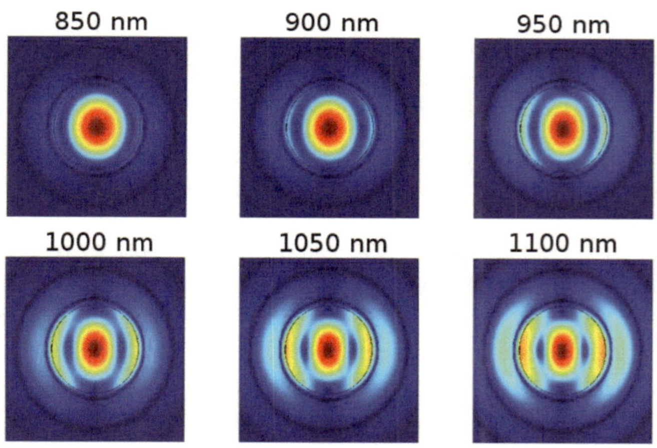

(b)

Fig. 10. (a) Dispersion and confinement loss of the fundamental core mode propagating in hollow-core Bragg fibre corresponding to three different GTI arrangement. Physical thicknesses of the high index partial reflector layers (d_1) and the low index air spacer layers (d_2) forming the resonant layer are expressed in terms of the high and low index quarter-wave stacks. (b) Fundamental mode field distribution with resonant coupling with the GTI in a hollow-core Bragg fibre having a readjusted first periode and design parameters $d_1 = 0.3d_H$ and $d_2 = 0.159d_L$ (blue curve in Fig. 10(a)).

show that RDS can be obtained in a relatively wide wavelength range. In our models, the d_2 spacing between the cladding and the thin partial reflector layer had to be set to 0.39, 0.235 and 0.159 times the spacing between the fused silica layers in the quarter-wave stack design (d_H). The RDS regims corresponding to the different partial reflector layers are 260, 220 and 180 nm, respectively.

The used parameters in the theory (Sec. 2.2) predicted the RDS around one micron accurately compared to the two-dimensional model presented here. Layer thicknesses for GTI are almost exactly the same values predicted analytically and obtained numerically.
In order to have an insight into the physical effect of the cylindrical GTI layer, what we can regard as a realigned first period we computed the mode field distributions at different wavelengths which results are shown in Fig. 10(b). The superimposed GTI layer contribute to the effective index of the LP_{01} mode in a resonant way through the frequency dependent mode field distribution. The ratio between the peak of the transversal mode distribution and side peaks vary around the resonance wavelength similar way was shown in Fig. 8, leads to the desired RDS.

4.2 Hollow core fibre with honey-comb structured cladding

In the case of all-silica hollow-core PBG fibres the resonant GT layer can be presented by the same way was discussed in the previous subsection changing the thickness of the first air layer whose reflectance can be further reduced by decreasing the thickness of the core wall (first high index layer). The high reflector is the properly designed microstructured cladding where the high and low index layers satisfy the quarter wavelength condition. The used model of hollow-core PBG fibre with a honey-comb structured cladding and the necessary geometrical changes are shown in Fig. 11(a) and the corresponding parameters are represented in Fig. 11(b). In a 7-unit-cell hollow-core-PBG fibre, core radius is determined by the hole-to-hole spacing Λ, the core wall thickness t and the introduced core expansion coefficient E

$$R_c = (E+1)(1.5\Lambda - t/2) \tag{14}$$

where the role of E is clearly to provide a larger core and a detuned first period with thin air-spacer layer (parameters are in Fig. 11(b)). The pitch was selected to $\Lambda = 2.85\ \mu$m to obtain a bandgap centered around 1100 nm wavelength range and the hole size and pitch ratio to $d/\Lambda = 0.98$ which indicates a very high air-filling fraction in order to obtain large PBG in the cladding and lower the leakage losses (Saitoh & Koshiba, 2003). The core wall thickness was $t = T(\Lambda - d)$ where T is a parameter usually smaller than unity. We can suppress the existence of surface modes which localize energy at the core wall by selecting $T < 0.5$ (Saitoh et al., 2007) at the RDS wavelengths.

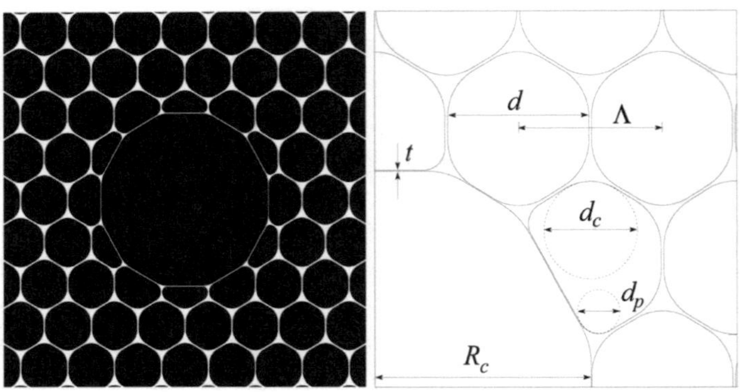

(a) (b)

Fig. 11. (a) the modified fibre structure with an applied core expansion based on the structure was shown in Fig. 2(c) and (b) the model parameters.

In Fig. 12(a), and 12(b) we show the calculated dispersion profiles for the LP_{01}-like fundamental air-core mode for changing the *T* and *E* parameters, respectively, in a 7-cell-core PBG fibre. If $T = 0.3$ and $E = 0.18$, the obtained dispersion function is the one having a RDS around 1050 nm and it ranges from 1012 nm to 1106 nm. This 94 nm wide reversed slope region presented for the LP_{01}-like fundamental mode can be raised to 120 nm by setting $T = 0.45$ (see Fig. 12(a)). This increase in the core wall thickness introduces stronger resonance in the phase-delay of the light due to the increasing reflectance of the first facet light reaches during the propagation and consequently results in a steeper dispersion slope than those belonging to smaller *T* values. Decreasing the magnitude of the expansion coefficient (Fig. 12(b)) will shift the reversed slope region to longer wavelengths without changing the magnitude of the negative slope range. Namely, varying the thickness of the first low index layer will determine the position of the resonance property.

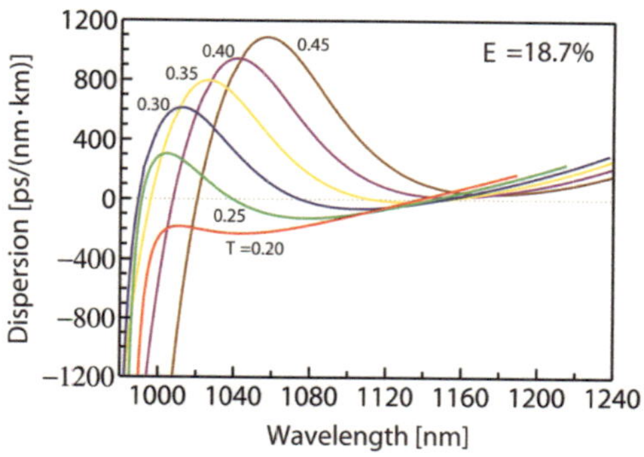

(a)

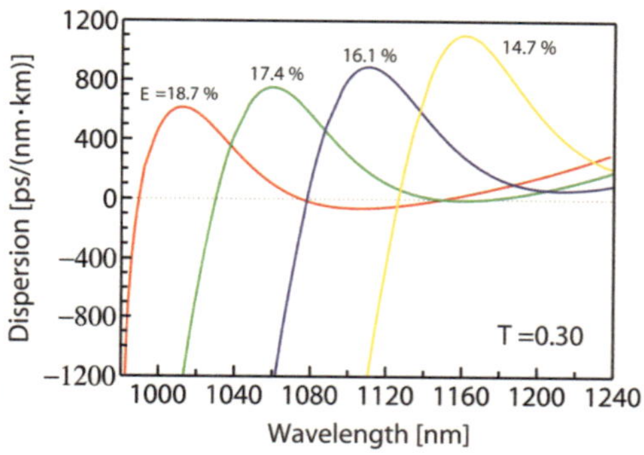

(b)

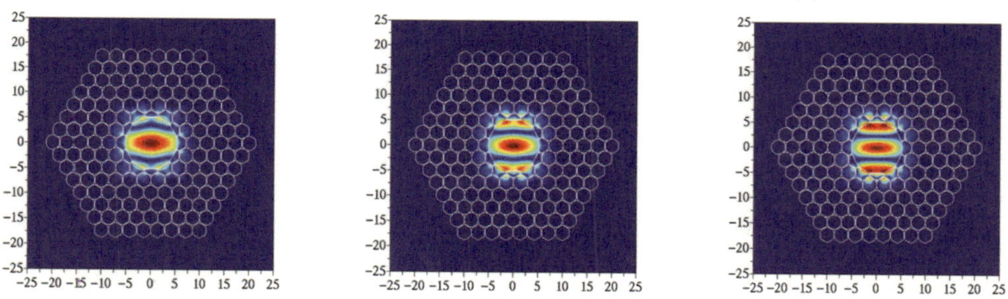

(c)

Fig. 12. Dispersion functions of the designed hollow-core PBG fibre with (a) different core wall thicknesses and (b) different core expansion coefficients. (c) Mode distribution for one particular polarization component of LP_{01} at 1000 nm, 1050 nm and 1100 nm.

Because of the large index difference between the low and high index materials constituting the fibre cladding the resonant coupling with the GTI is also larger than in the case of solidcore fibres with smaller index differences, resulting in a much more distorted mode profile. Figure 12(c) shows the mode field distribution at 1000, 1050 and 1100 nm.

Figure 13 is intended to show the wavelength-dependence of the confinement loss for the fundamental air-core mode with and without GTI in hollow-core PBG fibres using 6 periods of air-holes in the cladding region. *T* is kept at 0.3 in both cases without core expansion as well as with $E = 0.187$. Due to the presence of resonances for $E = 0.187$, what we obtained is an elevated loss profile by almost three magnitudes, but still resulting in less than 0.01 dB/m which is acceptable for the purpose of dispersion management.

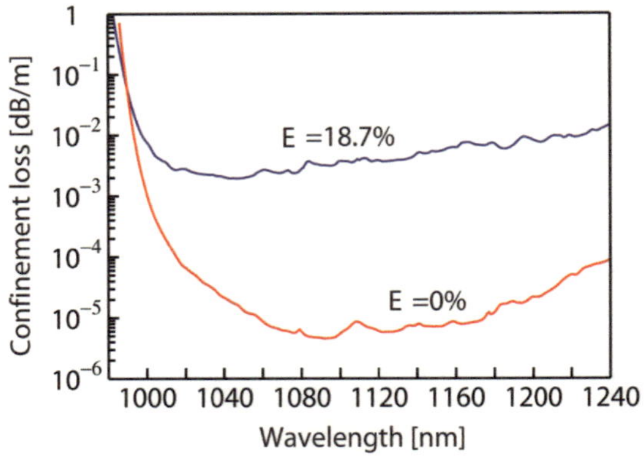

Fig. 13. Confinement losses of hollow-core PBG fibres with $E = 0$ and $E = 0.187$ core expansion factors.

5. Fibre properties

5.1 Coupling loss

Due to the mode distortions it is reasonable to investigate how large coupling loss can be expected when one splices the investigated fibre in Sec. 3.1 to an other Bragg or a step-index profiled fibre. We computed the overlap integral between the modes of the modified cladding structure and the modes of the fibre without GTI:

$$\mu(\lambda) = \int_0^{2\pi}\int_0^{\infty} \Psi^{(1)}(r,\varphi,\lambda)\Psi^{(2)^*}(r,\varphi,\lambda) r dr d\varphi \tag{15}$$

where $\Psi^{(k)}$, $k = 1,2$ is the normalized electric field distribution in the kth fibre, and the symbol asterisk stands for complex conjugate. r and φ are the variables of polar coordinates in Eq. (15). The results are summarized in Fig. 14(a) where we present the loss due to coupling the light from one fibre to another one between 900 and 1200 nm. First we tested the coupling between the unmodified Bragg fibre ($d_L = 3.8$ μm, $d_H = 0.95$ μm $n_L = 1.45$ and $n_H = 1.5$) and a step-index fibre having the same 6 μm core radius and an index difference between the core and cladding: $\Delta n = 0.0036$. The cladding refractive index follows the index of fused silica obtained from the Sellmeier equation at any wavelength. The overlap factor shows 81% coupling at 1040 nm which is equivalent with a 0.9dB loss (curve with full circles in Fig. 14(a)). By the introduction of the GT layer around the core in the Bragg fibre ($d_{GT} = 0.34d_L$, $n_{GT} = 1.46$ at 1030 nm) and investigate the coupling to a Bragg fibre with the same cladding structure except the GTI, the overlap factor yields a value of 30%which corresponds to 5.2dB loss (full squares in Fig. 14(a)). This large coupling loss becomes even larger if we couple the distorted mode field into the step-index fibre (open circles in Fig. 14(a)) where the loss is more than 7.2 dB at 1050 nm.

Fortunately, this large loss can be optimized. If we take a look at the effective indices of Bragg fibres with and without GTI we can see that the difference can be considered

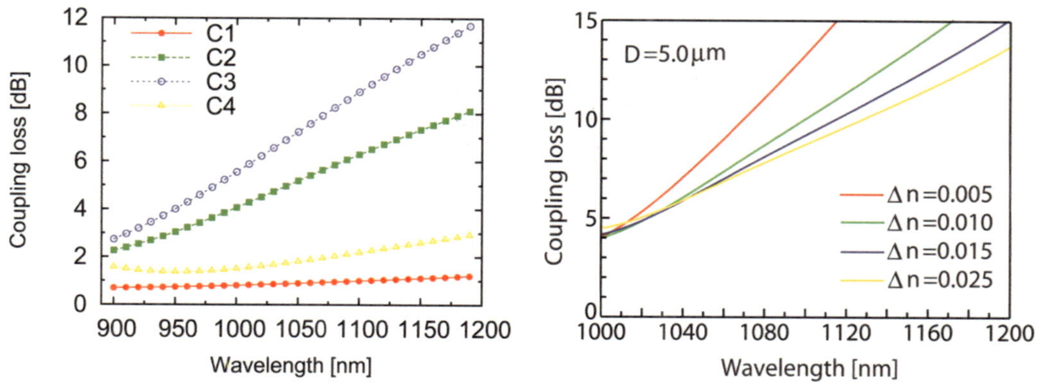

(a) (b)

Fig. 14. (a) Coupling loss calculated from the overlap integral between LP_{01} modes of different fibres. "C1" refers to the coupling between the Bragg fibre without GTI and a stepindex fibre; "C2" is the coupling between two Bragg fibres with and without GTI and both R_c is 6 μm; "C3" is the coupling between step-index and Bragg fibre with GTI and "C4" is the improved coupling between two Bragg fibres with and without GTI by changing the core size of the Bragg fibre without GTI. (b) Coupling loss as a function of wavelength between hollow-core RDS and step index fibres with different index profiles using a core diameter of 5 μm.

significant and should be moved closer in order to obtain a better coupling efficiency. The effective refractive indices are 1.447329 and 1.448562 at 1040 nm for a Bragg fibre with and without GTI, respectively. If we start to decrease the core size of the regular Bragg fibre without GTI, we can decrease the effective index as well, since the wave vector of the propagating mode will enclose a smaller incident angle (see Eq. (10)). For a 4.1 μm core radius of the normal Bragg fibre the effective index is 1.447351 for LP_{01} at 1040 nm and this way the coupling loss becomes less than 1.7dB with the distorted mode (open triangles in Fig. 14(a)).

In the case of hollow-core PBG fibre attaching simply the modified and unmodified structures will result in 4-7 dB loss along the RDS region. Using step-index fibres the loss between the hollow-core RDS and these fibres are even larger. Fig. 14(b) shows the coupling loss between the step-index fibre with different index profiles and the hollow-core RDS fibre where the expansion factor is set to E = 18.7%. The calculations presented in Fig. 14(b) was performed with a step-index fibre having a core size of 5 μm. The coupling loss is found to be between 4 and 9 dB over the RDS range.

We can see also in Fig. 14(b) that the coupling loss can be decreased by increasing the index difference of step-index fibre. Larger mode field diameter is presented for longer wavelengths in index guiding fibres but this larger mode field diameter can be decreased using a fibre with larger index difference. This larger Δn however gains the possibility of multimode operation. The calculated fibre structures are multi-moded which indicate that coupling may occur to higher order modes of the RDS fibre and may degrade the beam quality of the output if it is used for a pulse compressor. Analyses and solutions on this issue require further investigations of these fibres.

5.2 Propagation loss

The propagation loss in optical fibres can be evaluated as a sum of different type of losses such as confinement loss and scattering loss. In the case of using the fibre in a wavelength range where the bandgap is surface-mode and leaking-mode free the dominant loss will be the scattering loss (Roberts et al., 2005).

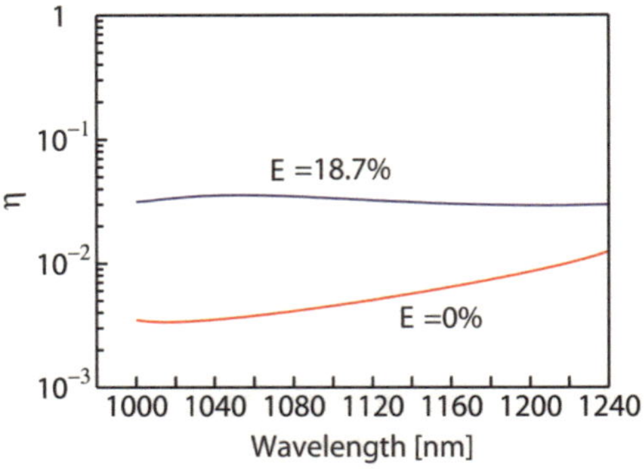

Fig. 15. The η-factor of the fundamental air-core mode as a function of wavelength with $d/\Lambda = 0.98$, $dc/\Lambda = 0.70$, $d_p/\Lambda = 0.30$, $\Lambda = 2.85\ \mu$m, $T = t/(\Lambda - d) = 0.30$, and the expansion coefficient is set to E=0% and 18.7%.

Figure 15 shows the η-factor (Murao et al., 2008) of the fundamental air-core mode in hollow-core fibre as a function of wavelength with $d/\Lambda = 0.98$, $d_c/\Lambda = 0.70$, $d_p/\Lambda = 0.30$, $\Lambda = 2.85\ \mu$m, $t = 0.3(\Lambda - d)$, and the expansion coefficient is set to E=0% and 18.7% (see parameters in Fig. 11(b)), where the η-factor is defined as the normalized overlap integration of the power in the silica-ring as follows:

$$\eta = \frac{\int_{\text{glass annulus}} (\mathbf{E} \times \mathbf{H}^*) \cdot \hat{\mathbf{z}} dA}{\int_{\text{cross-section}} (\mathbf{E} \times \mathbf{H}^*) \cdot \hat{\mathbf{z}} dA} \tag{16}$$

The η-factor, which is proportional to the scattering loss of hollow-core PBG fibres due to the surface roughness of the silica-ring around the air-core. In the hollow-core PBG fibre with GTI, it is about one order of magnitude larger than the scattering loss in the hollow-core PBG fibre without GTI.

5.3 Fibre nonlinearity

The fibre nonlinearity of hollow-core PBG fibres can be almost one thousand times smaller than that of single mode fibres ($\gamma \approx 10^{-6}$ $(\text{Wm})^{-1}$). This value was reported theoretically (Lægsgaard et al., 2003) as well as experimentally (Hensley et al., 2007). The utilization of resonant structures however may increase this value due to the increased guided mode overlap with silica.

In Fig. 5.3, we show the nonlinear coefficient γ as a function of wavelength which is a crucial quantity in high power light delivery. We use the following definition of the nonlinear coefficient (Vincetti et al., 2006)

$$\gamma = \gamma_{\text{air}} + \gamma_{\text{silica}} = \frac{2\pi n_{2_{\text{air}}}}{\lambda A_{\text{eff}_{\text{air}}}} + \frac{2\pi n_{2_{\text{silica}}}}{\lambda A_{\text{eff}_{\text{silica}}}} \tag{17}$$

where n_2 is the nonlinear refractive index coefficient and A_{eff} is the effective mode area calculated for air and silica, respectively, defined as (Lægsgaard et al., 2003)

$$A_{\text{eff}_i} = \frac{\left(\int_{\text{cross-section}} (\mathbf{E} \times \mathbf{H}^*) \cdot \hat{\mathbf{z}} dA\right)^2}{n_i^2 \varepsilon_0^2 c^2 \int_{A_i} |\mathbf{E}|^4 dA} \tag{18}$$

where i = air or silica, ε_0 is the permittivity and c is the speed of the light in vacuum. It can be seen clearly from the results in Fig. 16 that γ is increased for the PBG fibre with GTI with an order of magnitude. The increased nonlinearity may still ensure the validity of using these fibres for high power pulse compression but the available power level with fibres using this type of resonant structure is clearly lower than using low nonlinearity hollow-core fibres. The improvement of nonlinearity is essential in these fibres to introduce them in high intensity pulse compression though we present some solutions which still allow the generation of ultra-short and high energy pulses (Sec. 6).

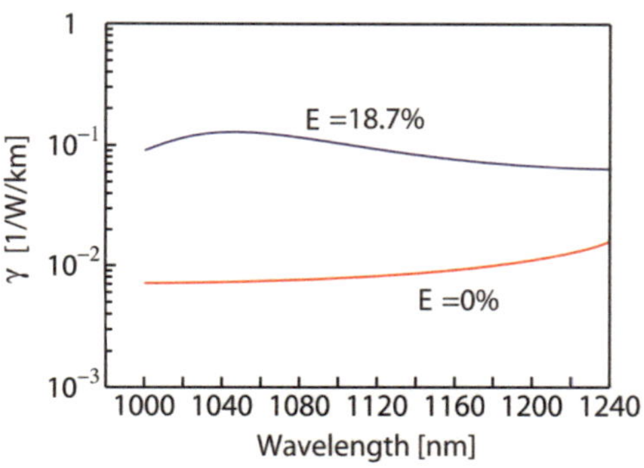

Fig. 16. The nonlinear coefficient γ as a function of wavelength in a 7-cell hollow-core PBG fibre with and without core expansion.

In the case of solid-core fibres we have to account for only the second term of Eq. (17) since we can assume that the nonlinear refractive index of the different glasses are nearly the same. From the calculated $A_{eff}(\lambda)$ which increases with the increase of the wavelength, solid-core fibre nonlinearity shows an opposite behavior than hollow-core fibre (Várallyay et al., 2009). This is due to the expansion and larger penetration of the mode into the cladding region as a function of wavelength. This phenomenon is the same in the two types of fibres but the larger fraction of pulse power in the cladding of hollow-core fibres results in the significantly larger silica contribution to γ increasing it's value.

We obtained that the nonlinear coefficient of hollow-core RDS fibre is still more than a magnitude smaller than in solid-core PBG fibres at around 1050 nm (Várallyay et al., 2009). The increase of core size may lower the nonlinearity in solid-core fibres but this changes limit the wavelength range of the resonance since the propagation angle of the mode gets closer to the right angle.

6. Propagation analysis

We investigate a pulse compressor consisting of two serially connected hollow-core fibres with and without RDS. In the case of linear propagation, the necessary fibre lengths for compensating the chirp on a broadband pulse up to the third order can be estimated from the following equation-system

$$\beta_2^{(1)} L^{(1)} + \beta_2^{(2)} L^{(2)} + GDD_0 = 0 \tag{19}$$

$$\beta_3^{(1)} L^{(1)} + \beta_3^{(2)} L^{(2)} + TOD_0 = 0 \tag{20}$$

where $\beta_i^{(j)}$ is the ith order term in the Taylor series of the dispersion of the jth fibre, $L^{(j)}$ is the length of the jth fibre and GDD_0 and TOD_0 are the pulse initial GDD and third-order dispersion (TOD), respectively. It can be seen from Eq. (19) and (20) if the input chirps of the pulses are zero the pulse shape will be preserved when the ratio of the third and second order dispersion terms are same for the two fibres.

Self-phase modulation adds some additional chirp on the pulse in the case of nonlinear propagation which depends on the actual pulse power. The propagation is treated by the generalized nonlinear Schrödinger equation (Agrawal, 2007)

$$\frac{\partial E(z,t)}{\partial z} + \left(\sum_{k=2}^{N} \beta_k \frac{i^{k-1}}{k!} \frac{\partial^k}{\partial t^k} \right) E + \frac{\alpha}{2} E = i\gamma \left(|E|^2 E + \frac{i}{\omega_0} \frac{\partial}{\partial t} \left(|E|^2 E \right) - T_R E \frac{\partial |E|^2}{\partial t} \right) \tag{21}$$

where E is the complex envelope function, z is the coordinate in the propagation direction, t is the retarded time, α is the attenuation coefficient, β_k is the kth order dispersion from the Taylor-series of the propagation constant, γ is the nonlinear coefficient and T_R is the first moment of the nonlinear Raman response. Eq. (21) is solved by the split-step Fourier method (Agrawal, 2007) such a way that the maximal phase change of the propagating pulse remained below 0.01 radian.

The GDD and 3rd order dispersion values of the regular and RDS fibres are −0.07748 ps^2/m, 0.00036 ps^3/m and −0.05847 ps^2/m, −0.00309 ps^3/m, respectively. The nonlinear coefficient (γ) of the two fibres are $7 \cdot 10^{-6}$ and 10^{-4} 1/(Wm) based on the calculations presented in Sec. 5.3. Detailed simulation parameters including higher order dispersion values, pulse and additional fiber parameters are same as used in (Várallyay et al., 2009).

The first guess for the lengths of the fibres is taken from Eq. (19) and (20) and a brute-force optimization is used to find the highest peak power around the initial guess which target yields the possible shortest pulses with the possible highest quality (Várallyay et al., 2007).

We calculate the propagation and compression of Gaussian pulses with $5 \cdot 10^5$ fs^2 linear input chirp and over 16 nm bandwidth. The resulted input pulse width is 11.1 ps. This magnitude of input chirp can be a realistic value for pulses coming out from fibre oscillators or amplifiers. The dispersion parameters up to the fifth order is taken into consideration for both type of hollow-core fibres (see (Várallyay et al., 2009)). Dispersion properties of hollow-core RDS is fitted to the dispersion function obtained from the calculations which belong to the parameters E = 0.187 and T = 0.3 (see Fig. 12). The dispersion function of the regular hollow-core PBG fibre is fitted to a HC-1060 type fibre (see, for instance, in Ref. (Saitoh et al., 2003)). The β_3 parameter has a different sign in the two fibres and an approximately ten times larger absolute value for hollow-core RDS fibre than HC-1060. The coupling loss between the two fibres is assumed to be 5 dB.

The calculations are shown in Fig. 17(a) and 17(b). In this case the combined fibre compressor yields 1.5 times shorter pulses with significantly better pulse quality (QF=90.6%)

compared to the performance of regular hollow-core fibres. The HC-1060 type fibre can compress the 100 nJ pulses with 74.3% QF after 6.3 m length of propagation.

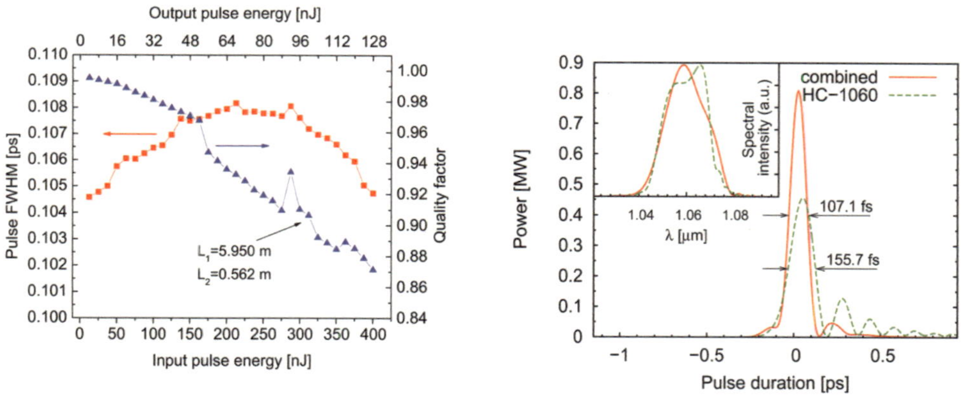

(a) (b)

Fig. 17. (a) Pulse full width at half maximum (FWHM) evaluation as a function of pulse energy with $GDD_0 = 5 \cdot 10^5$ fs^2 input chirp; black arrow shows the regular (L_1) and RDS fiber lengths (L_2) at 90% quality factor, (b) pulse shapes with combined and HC-1060 type fibre compressors.

7. Conclusion

We have given a qualitative explanation of adopting one-dimensional Gires-Tournois interferometers (GTI) in two-dimensional photonic bandgap (PBG) structures. We showed that by properly adjusting one or two resonant layers around the fibre core in solid- or hollow core bandgap fibers may result a tailored dispersion profile even reversed dispersion slope (RDS). Finite element analyses confirmed that the dispersion profile of PBG fibres can be tailored effectively by the application of resonant layers for the propagating fundamental mode. And this works for higher order modes as well. We calculated the dispersion properties of solidcore Bragg and all-silica photonic crystal fibers as well as hollow-core Bragg and photonic crystal fibres with honey-comb cladding structure. Negative dispersion slope compared to the canonical form of bandgap guidance as well as flat and nearly zero dispersion profiles are obtained with relatively low losses. Fiber properties were also analyzed. After structural optimization, low coupling losses between two fibres, with and without GTI, also demonstrated. The obtained dispersion profiles showing negative dispersion slope over 100 nm wavelength regions may yield new perspectives in dispersion management even if the increased nonlinearity and propagation loss of hollow-core RDS fibres somewhat decrease the transmittable intensities. Further optimization and investigation of the presented structures may improve their physical properties. The propagation and compression of pre-chirped, broadband Gaussian pulses confirmed the validity of further developments on RDS PBG fibres.

8. References

Agrawal, G. P. (2007). *Nonlinear Fiber Optics, fourth edition*, Academic, San Diego, CA.

Birks, T. A., Knight, J. C. & Russell, P. S. J. (1997). Endlessly single-mode photonic crystal fiber, *Opt. Lett.* 22: 961–963.

Bouwmans, G., Luan, F., Knight, J., Russell, P. S. J., Farr, L., Mangan, B. & Sabert, H. (2003). Properties of a hollow-core photonic bandgap fiber at 850 nm wavelength, *Opt. Express* 11: 1613–1620.

Cregan, R. F., Mangan, B. J., Knight, J. C., Birks, T. A., Russell, P. S. J., Roberts, P. J. & Allan, D. C. (1999). Single-mode photonic band gap guidance of light in air, *Science* 285: 1537-1539.

de Matos, C., Taylor, J., Hansen, T., Hansen, K. & Broeng, J. (2003). All-fiber chirped pulse amplification using highly-dispersive air-core photonic bandgap fiber, *Opt. Express* 11: 2832–2837.

Dianov, E. M., Likhachev, M. E. & Février, S. (2009). Solid-core photonic bandgap fibers for high-power fiber lasers, *IEEE J. Sel. Topics Quantum Electron.* 15: 20–29.

Egorova, O. N., Semjonov, S. L., Kosolapov, A. F., Denisov, A. N., Pryamikov, A. D., Gaponov, D. A., Biriukov, A. S., Dianov, E. M., Salganskii, M. Y., Khopin, V. F., Yashkov, M. V., Gurianov, A. N. & Kuksenkov, D. V. (2008). Single-mode all-silica photonic bandgap fiber with 20 μm mode-field diameter, *Opt. Express* 16: 11735–11740.

Engeness, T., Ibanescu, M., Johnson, S., Weisberg, O., Skorobogatiy, M., Jacobs, S. & Fink, Y. (2003). Dispersion tailoring and compensation by modal interactions in OmniGuide fibers, *Opt. Express* 11: 1175–1196.

Fang, Q., Wang, Z., Jin, L., Liu, J., Yue, Y., Liu, Y., Kai, G., Yuan, S. & Dong, X. (2007). Dispersion design of all-solid photonic bandgap fiber, *J. Opt. Soc. Am. B* 24: 2899–2905.

Fekete, J., Várallyay, Z. & Szipőcs, R. (2008). Design of high-bandwidth one- and two-dimensional photonic bandgap dielectric structures at grazing incidence of light, *Appl. Opt.* 47: 5330–5336.

Ferrando, A., Silvestre, E.,Miret, J. J. & Andrés, P. (2000). Nearly zero ultraflattened dispersion in photonic crystal fibers, *Opt. Lett.* 25: 790–792.

Foroni, M., Passaro, D., Poli, F., Cucinotta, A., Selleri, S., Lægsgaard, J. & Bjarklev, A. (2007). Confinement loss spectral behavior in hollow-core Bragg fibers, *Opt. Lett.* 32: 3164–3166.

Foster, M. A., Gaeta, A. L., Cao, Q. & Trebino, R. (2005). Soliton-effect compression of super-continuum to few cycle durations in photonic nanowires, *Opt. Express* 13: 6848–6855.

Hensley, C. J., Ouzounov, D. G., Gaeta, A. L., Venkataraman, N., Gallagher,M. T. & Koch, K.W. (2007). Silica-glass contribution to the effective nonlinearity of hollow-core photonic band-gap fibers, *Opt. Express* 15: 3507–3512.

Humbert, G., Knight, J. C., Bouwmans, G., Russell, P. S. J., Williams, D. P., Roberts, P. J. & Mangan, B. J. (2004). Hollow core photonic crystal fiber for beam delivery, *Opt. Express* 12: 1477–1484.

Jasapara, J., Bise, R., Her, T. & Nicholson, J. W. (2003). Effect of mode cut-off on dispersion in photonic bandgap fibers, *Optical Fiber Communication Conference, Technical Digest*, Optical Society of America, p. ThI3.

Jasapara, J., Her, T. H., Bise, R., Windeler, R. & DiGiovanni, D. J. (2003). Group-velocity dispersion measurements in a photonic bandgap fiber, *J. Opt. Soc. Am. B* 20: 1611–1615.

John, S. (1987). Strong localization of photons in certain disordered dielectric superlattices, *Phys. Rev. Lett.* 58: 2486–2489.

Knight, J. C., Birks, T. A., Cregan, R. F., Russell, P. S. J. & de Sandro, J. P. (1998). Large mode area photonic crystal fibre, *Electron. Lett.* 34: 1347–1348.

Knight, J. C., Birks, T. A., Russell, P. S. J. & Atkin, D. M. (1996). All-silica single-mode optical fiber with photonic crystal cladding, *Opt. Lett.* 21: 1547–1549.

Knight, J. C., Broeng, J., Birks, T. A. & Russell, P. S. J. (1998). Photonic band gap guidance in optical fibres, *Science* 282: 1476–1478.

Kuhl, J. & Heppner, J. (1986). Compression of femtosecond optical pulses with dielectric multilayer interferometers, *IEEE Trans. Quant. Electron.* QE-22: 182–185.

Lægsgaard, J., Mortensen, N. A., Riishede, J. & Bjarklev, A. (2003). Material effects in air-guiding photonic bandgap fibers, *J. Opt. Soc. Am. B* 20: 2046–2051.

Lægsgaard, J. & Roberts, P. J. (2008). Dispersive pulse compression in hollow-core photonic bandgap fibers, *Opt. Express* 16: 9628–9644.

Lægsgaard, J., Roberts, P. J. & Bache,M. (2007). Tailoring the dispersion properties of photonic crystal fibers, *Optical and Quantum Electronics* 39: 995–1008.

Lim, H., Ilday, F. & Wise, F. (2002). Femtosecond ytterbium fiber laser with photonic crystal fiber for dispersion control, *Opt. Express* 10: 1497–1502.

Limpert, J., Liem, A., Reich, M., Schreiber, T., Nolte, S., Zellmer, H., Tünnermann, A., Broeng, J., Petersson, A. & Jakobsen, C. (2004). Low-nonlinearity single-transverse-mode ytterbium-doped photonic crystal fiber amplifier, *Opt. Express* 12: 1313–1319.

Limpert, J., Schmidt, O., Rothhardt, J., Röser, F., Schreiber, T., Tünnermann, A., Ermeneux, S., Yvernault, P. & Salin, F. (2006). Extended single-mode photonic crystal fiber lasers, *Opt. Express* 14: 2715–2720.

Macleod, H. A. (2001). *Thin-film optical filters third edition*, J WArrowsmith Ltd., Bristol.

Mogilevtsev, D., Birks, T. A. & Russell, P. S. J. (1998). Group-velocity dispersion in photonic crystal fiber, *Opt. Lett.* 23: 1662–1664.

Mortensen, N. A., Nielsen, M. D., Folkenberg, J. R., Petersson, A. & Simonsen, H. R. (2003). Improved large-mode-area endlessly single-mode photonic crystal fibers, *Opt. Lett.* 28: 393–395.

Müller, D.,West, J. & Koch, K. (2002). Interferometric chromatic dispersion measurement of a photonic band-gap fiber, *in* A. K. Dutta, A. A. S. Awwal, N. K. Dutta & K. Okamoto (eds), *Active and Passive Components for WDM communications II*, Vol. 4870, SPIE, pp. 395–403.

Murao, T., Saitoh, K. & Koshiba, M. (2008). Structural optimization of air-guiding photonic bandgap fibers for realizing ultimate low loss waveguides, *J. Lightwave Technol.* 26: 1602–1612.

Nielsen, C. K., Jespersen, K. G. & Keiding, S. R. (2006). "a 158 fs 53 nJ fiber-laser system at 1μm using photonic bandgap fibers for dispersion control and pulse compression, *Opt. Express* 14: 6063–6068.

Ouzonov, D. G., Ahmad, F. R., Müller, D., Venkataraman, N., Gallagher, M. T., Thomas, M. G., Silcox, J., Koch, K. W. & Gaeta, A. L. (2003). Generation of megawatt optical solitons in hollow-core photonic band-gap fibers, *Science* 301: 1702–1704.

Ouzounov, D., Hensley, C., Gaeta, A., Venkateraman, N., Gallagher, M. & Koch, K. (2005). Soliton pulse compression in photonic band-gap fibers, *Opt. Express* 13: 6153–6159.

Ranka, J. K., Windeler, R. S. & Stentz, A. J. (2000). Visible continuum generation in air-silica microstructure optical fibers with anomalous dispersion at 800 nm, *Opt. Lett.* 25: 25- 27.

Roberts, P. J. (2007). Control of dispersion in hollow core photonic crystal fibers, *Conference on Lasers and Electro-Optics (CLEO 2007)*, CLEO, p. CWF2.

Roberts, P. J., Couny, F., Sabert, H., Mangan, B. J., Williams, D. P., Farr, L., Mason, M. W., Tomlinson, A., Birks, T. A., Knight, J. C. & Russell, P. S. J. (2005). Ultimate low loss of hollow-core photonic crystal fibres, *Opt. Express* 13: 236–244.

Roberts, P. J., Williams, D. P., Mangan, B. J., Sabert, H., Couny, F., Wadsworth, W. J., Birks, T. A., Knight, J. C. & Russell, P. S. J. (2005). Realizing low loss air core photonic crystal fibers by exploiting an antiresonant core surround, *Opt. Express* 13: 8277–8285.

Ruehl, A., Prochnow, O., Engelbrecht, M., Wandt, D. & Kracht, D. (2007). Similariton fiber laser with a hollow-core photonic bandgap fiber for dispersion control, *Opt. Lett.* 32: 1084–1086.

Saitoh, K., Florous, N. J., Murao, T. & Koshiba, M. (2007). Realistic design of large-hollow-core photonic band-gap fibers with suppressed higher order modes and surface modes, *J. Lightwave Technol.* 25: 2440–2447.

Saitoh, K. & Koshiba, M. (2003). Leakage loss and group velocity dispersion in air-core photonic bandgap fibers, *Opt. Express* 11: 3100–3109.

Saitoh, K., Koshiba, M., Hasegawa, T. & Sasaoka, E. (2003). Chromatic dispersion control in photonic crystal fibers: application to ultra-flattened dispersion, *Opt. Express* 11: 843- 852.

Saitoh, K., Mortensen, N. A. & Koshiba, M. (2004). Air-core photonic band-gap fibers: the impact of surface modes, *Opt. Express* 12: 394–400.

Shephard, J., MacPherson, W., Maier, R., Jones, J., Hand, D., Mohebbi, M., George, A., Roberts, P. & Knight, J. (2005). Single-mode mid-ir guidance in a hollow-core photonic crystal fiber, *Opt. Express* 13: 7139–7144.

Stephens, T. J., Maier, R. R., Barton, J. S. & Jones, J. D. C. (2004). Fused silica hollow-core photonic crystal fibre for mid-infrared transmission, *Conference on Lasers and Electro-Optics (CLEO 2004)*, CLEO, p. CPDD4.

Szipőcs, R., Kőházi Kis, A., Lakó, S., Apai, P., Kovács, A. P., DeBell, G., Mott, L., Louderback, A. W., Tikhonravov, A. V. & Trubetskov, M. K. (2000). Negative dispersion mirrors for dispersion control in femtosecond lasers: chirped dielectric mirrors and multi-cavity Gires-tournois interferometers, *Appl. Phys. B* 70: S51–S57.

Tong, L.M., Gattass, R. R., Ashcom, J. B., He, S. L., Lou, J. Y., Shen, M. Y., Maxwell, I. & Mazur, E. (2003). Subwavelength-diameter silica wires for low-loss optical wave guiding, *Nature* 426: 816–819.

Várallyay, Z., Fekete, J., Bányász, Á. & Szipőcs, R. (2007). Optimizing input and output chirps up to the third-order for sub-nanojoule, ultra-short pulse compression in small core area PCF, *Appl. Phys. B* 86(4): 567–572.

Várallyay, Z., Fekete, J. & Szipőcs, R. (2008a). Higher-order mode photonic bandgap fibers for dispersion control, *CLEO Conference, San Jose, CA*, OSA, p. JThA48.

Várallyay, Z., Fekete, J. & Szipőcs, R. (2008b). Higher-order mode photonic bandgap fibers with reversed dispersion slope, *OFC/NFOEC Conference 2008, San Diego, CA*, OSA, p. JWA8.

Várallyay, Z., Saitoh, K., Fekete, J., Kakihara, K., Koshiba, M. & Szipőcs, R. (2008). Reversed dispersion slope photonic bandgap fibers for broadband dispersion control in femtosecond fiber lasers, *Opt. Express* 16: 15603–15616.

Várallyay, Z., Saitoh, K., Szabó, Á. & Szipőcs, R. (2009). Photonic bandgap fiberswith resonant structures for tailoring the dispersion, *Opt. Express* 17: 11869–11883.

Vienne, G., Xu, Y., Jakobsen, C., Deyerl, H. J., Jensen, J., Sorensen, T., Hansen, T., Huang, Y., Terrel, M., Lee, R., Mortensen, N., Broeng, J., Simonsen, H., Bjarklev, A. & Yariv, A. (2004). Ultra-large bandwidth hollow-core guiding in all-silica Bragg fibers with nano-supports, *Opt. Express* 12: 3500–3508.

Vincetti, L., Maini, M., Poli, F., Cucinotta, A. & Selleri, S. (2006). Numerical analysis of hollow core photonic band gap fibers with modified honeycomb lattice, *Opt. and Quantum Electron.* 38: 903–912.

Yablonovitch, E. (1987). Inhibited spontaneous emission in solid-state physics and electronics, *Phys. Rev. Lett.* 58: 2059–2062.

Yeh, P., Yariv, A. & Marom, E. (1978). Theory of Bragg fiber, *J. Opt. Soc. Am.* 68: 1196–1201.

11

Resonantly Induced Refractive Index Changes in Yb-doped Fibers: the Origin, Properties and Application for all-fiber Coherent Beam Combining

Andrei A. Fotiadi[1,3], Oleg L. Antipov[2] and Patrice Mégret[1]
[1]*Université de Mons,*
[2]*Institute of Applied Physics RAS,*
[3]*Ioffe Physico-Technical Institute RAS,*
[1]*Belgium*
[2,3]*Russian Federation*

1. Introduction

For years, solid-state lasers have replaced traditional systems in numerous laser applications. Now, fiber lasers bring the ultimate in solid-state reliability and operating convenience to applications in materials processing, telecom, sensing, and medicine and so on. Various configurations of continuous wave (cw), mode-locked, Q-switched, or single-frequency lasers have been reported, offering a large variety in performance characteristics. Ytterbium-doped silica fiber lasers operating near 1 μm as well as erbium fibers at 1.5 μm have been among the most successful fiber lasers creating a great impact on many areas of technology. Current progresses in fiber lasers are associated with further scaling of the output power. In particular, such power scaling could be achieved through coherent combining of the beams delivered by a number of regular fiber sources. In this perspective, apart from the use of active fibers as a gain media, it has appeared that the unique and rather special properties of Yb-doped fibers discussed in this Chapter have still to be exploited in novel high-brightness all-fiber laser systems.

The Chapter is dedicated to the effect of refractive index changes taking place in standard commercial Yb-doped optical fibers under diode pumping, and also to applications of this effect for coherent combining of fiber lasers. Being associated with changes of population of different ion states, the effect of refractive index changes (RIC) is essentially a side effect of the population inversion mechanism that is responsible for light amplification in lasers. The electronic and thermal RIC mechanisms, which are important both for laser crystals and glasses, are intensively investigated for several years. Correct understanding of these phenomena is vital for Yb-doped fiber lasers and amplifiers, whose performance characteristics rely on physical processes in the active medium.

The Chapter includes remarkable amount of original experimental studies, theoretical modeling and demonstration of advanced fiber configurations. The structure of the Chapter is logical and easy to follow. Part 2 contains judicially selected background material to enable those who are not expert in the field to appreciate the subject of the research. We give a clear physical insight into the electronic and thermal RIC mechanisms, and present important estimations helping the reader to clearly understand the scientific context, motivation and aims of this work.

The main body of the work is centered around two experimental studies (Fotiadi et al., 2008c; Fotiadi et al., 2009c), reproduced in parts 3 and 4. Part 3 is dedicated to the fundamental properties of the Yb-doped fibers sensitive to the diode pumping at 980 nm. In this part we show our original RIC experiments performed with different samples of aluminum-silicate fibers and report on the dynamical characteristics of the RIC effect measured in the spectral range 1460 - 1620 nm. Detailed analysis of these experimental data has brought us to the conclusion that the process is predetermining by the electronic RIC mechanism that can be quantitatively described by a two-level population inversion model with the polarizability difference (PD) of the Yb-ion in the excited and ground states as the only material model parameter determined from the experimental data. The absolute PD value measured in the experiment is found to be independent on the testing wavelength and the fiber specification parameters, i.e. the cross-section profile, the size, the length and the Yb-ion concentration. Our results ensure a predominant contribution to the RIC effect from far-resonance UV electronic transitions rather than near-resonant IR transitions.

Part 4 of the Chapter focuses on potential applications of the RIC effect in advanced fiber laser systems implemented through our original concept of all-fiber coherent beam combining. The concept employs Yb-doped fibers as optically controllable phase shifters to be conjugated with Er-doped fiber amplifiers operating at ~1.55 μm. The coherent combining of 500-mW Er-doped amplifiers successfully demonstrated in our experiment confirms the validity of the concept. The performance characteristics of the multichannel fiber laser system associated with the proposed power scaling method are also under discussion.

2. RICs in Yb-doped optical fibers: the origins and fundamentals

In this part we summarize fundamentals of the electronic and thermal mechanisms of the refractive index changes (RIC) in single-mode ytterbium-doped optical fibers induced by optical pulses at 980 nm and we report the general equations describing RIC evolution. In particular, the electronic RIC dynamics is shown to follow the dynamics of the population of the excited/unexcited ion states with a factor proportional to their polarizability difference (PD). The thermal contribution to the RICs induced in standard Yb-doped fibers is shown to be negligible under condition of all-fiber low-power experiments discussed in the Chapter.

2.1 Electronic and thermal RIC mechanisms caused by nonlinear pump absorption

Ytterbium-doped silicate glass fibers are widely used for high-power, high-beam-quality laser sources due to their impressing gain efficiency, single transverse mode quality, relaxed thermal management problems, and overall robustness to environmental disturbances. Resulting from the confined mode structure and typically long propagation distances, nonlinear phenomena such as stimulated Brillouin scattering, stimulated Raman scattering, optical Kerr effect, inter-mode four-wave mixing, soliton and supercontinuum generation are observed in fiber lasers at relatively low power due to the high field intensity in the fiber core (Agrawal, 1989). In particular, the optical Kerr effect can cause refractive index changes (RIC) leading to self-focusing, self trapping, self phase modulation, and soliton formation in the fiber cavity. It has been shown, however, that power-dependent RICs other than the optical Kerr effect can take place in ytterbium-doped fibers (Arkwright et al., 1996). These specific phenomena are attributed to the population inversion mechanism that is also responsible for optically induced light amplification and saturating absorption in the active fiber media (Digonnet et al., 1997). The propagation of an optical beam through the absorbing (or amplifying) fiber media induces changes in the refractive index, which could cause self- and cross- phase modulation, light-induced gratings, spectral hole-burning effects, i.e. effects contributing dynamical behavior of the fiber lasers. In the particular case of Yb-doped optical fibers, one could account for thermal and non-thermal RIC effects driven by the population inversion mechanism in an indirect and direct manner, respectively. The thermal RIC effects are caused by pump thermalization of the absorbed energy into heat. The electronic RICs are explained by the polarizability difference (PD) between the excited and ground states. We have to distinguish the electronic RIC effect from the regular Kerr nonlinearity that does not relay on the population inversion. In many experiments, it is not straightforward to separate electronic and thermal contributions to refractive index change because they usually have similar response times. In the next sections of this part, we present an analytical treatment and then rough estimations in order to evaluate the RIC signal and determinate the fractions of the electronic and thermal contributions to this signal in the experiments discussed further in the Chapter.

2.2 Fundamentals of electronic RICs in Yb-doped fibers

Silica glass, the most common material for the production of fibers, is a good host for Yb-ions. The spectroscopy of the Yb-ion is simple compared to other rare-earth ions (figure 1) (Paschotta et al., 1997). For amplification in optical spectrum range, only two level manifolds are important: the ground-state manifold ($^2F_{7/2}$) and the excited-state manifold ($^2F_{5/2}$). They consist of four and three sublevels, respectively. The transitions between sublevels are smoothed by strong homogeneous and inhomogeneous broadening. As a result, ytterbium fibers are able to provide an optical gain over the very broad wavelength range from 975 to 1200 nm with a range of possible pump wavelengths from 860 nm to 1064 nm. In the spectrum range outside these resonances, ytterbium fibers are optically transparent and are the subject to pure RIC effect measurements.

The electronic-population RIC contribution is owned to the polarizability difference (PD) between the excited and ground states (Born & Wolf, 2003). The electronic polarizability is defined as the ratio of the dipole moment induced in an ion to the electric field that produces this dipole moment. The polarizability of the Yb-ion on some electron level q at the given testing optical frequency ν_T is determined by the probabilities of all possible transitions from this level to other levels $i \neq q$ and is expressed as the following:

$$p_q(\nu_T) = \frac{e^2}{4\pi^2 m} \sum_i \frac{f_{qi}(\nu_{qi}^2 - \nu_T^2)}{(\nu_{qi}^2 - \nu_T^2)^2 + (\nu_T \Delta\nu_{qi})^2} \tag{1}$$

where e and m are the electron charge and mass; f_{qi} is the oscillator force of the transition between levels q and i; ν_{qi} and $\Delta\nu_{qi}$ are the resonance frequency and the linewidth.

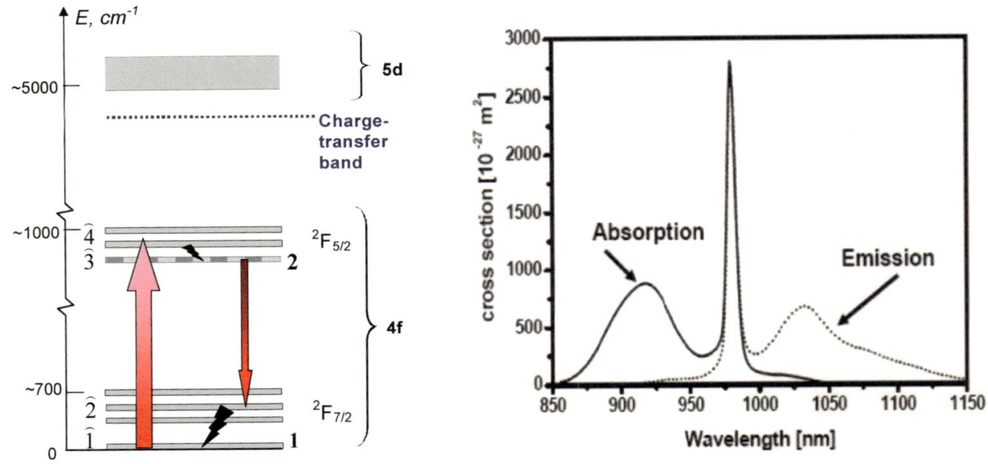

(a) (b)

Fig. 1. The atomic manifold level system (a) (Antipov et al., 2003) and emission and absorption cross section of ytterbium in silica host (b) (Paschotta et al., 1997). **1, 2** and $\hat{1}, \hat{2}, \hat{3}, \hat{4}$ indicate the levels for two- and four- level laser models, respectively.

Remaining in the frame of the two level laser approximation, we have to use only $p_1(\nu)$ and $p_2(\nu)$, i.e. the Yb-ion polarizabilities in the ground ($^2F_{7/2}$) and excited ($^2F_{5/2}$) states, respectively. According to equation (1), the dominating contributions to these polarizabilities are expected from the transitions, whose resonant frequencies are closer to the testing frequency $\nu_{1l}, \nu_{2l} \approx \nu_T$ and/or from the non-resonance transitions with the strongest oscillator forces. In Yb-doped materials, the well-allowed UV transitions to the 5d-electron shell and the charge-transfer transition are characterized by the oscillator forces that are several orders of magnitude higher than the forces of optical transitions inside the 4f-electron shell. In IR spectrum band, the polarizability difference $\Delta p(\nu) = p_2 - p_1$ is expressed from equation (1) as a sum of contributions from the near-resonance transitions (between the ground and excited states) and non-resonance UV transitions. Far from optical resonances, i.e. within Yb-doped fiber IR transparency band ($\lambda_T > 1.2\ \mu m$), the PD could be represented by two major terms with different dependencies on the testing wavelength λ_T:

$$\Delta p(\lambda_T) \approx \frac{e^2}{4\pi^2 c^2 m}\left\{ A\left[\frac{1}{1-\lambda_R^2/\lambda_T^2} + o\left(\frac{\lambda_R^2}{\lambda_T^2}\right)\right] + B\left[1 + o\left(\frac{\lambda_U^2}{\lambda_T^2}\right)\right]\right\} \tag{2}$$

where $A \equiv \sum_{l\in R} f_{2l}\lambda_{2l}^2 - f_{1l}\lambda_{1l}^2$; $B \equiv \left(f_{2U}\lambda_{2U}^2 - f_{1U}\lambda_{1U}^2\right)$; $\lambda_R \sim 1\mu m$ is a typical wavelength of the resonance transitions, f_{1U}, λ_{1U}, f_{2U}, λ_{2U} are oscillation forces and wavelengths of well-allowed charge-transfer transitions to the 5d-electron shell from the ground and excited state, respectively. The free space wavelengths λ_{1U}, λ_{2U} are located in the UV spectrum band around $\lambda_U \sim 0.1\mu m$ and obviously $\lambda_{1U} \neq \lambda_{2U}$. So, the origin of non-resonance contribution to the PD (the second term, $\sim B$) is the difference in the probabilities of the transitions at the testing wavelength λ_T from the ground and excited states to the 5d-electron shell or the charge-transfer transition.

The average refractive index of the medium is expressed through the polarizability of individual ions by the Lorentz-Lorenz formula (Born & Wolf, 2003). In the unpumped fiber media, where all ytterbium ions are in the lowest $^2F_{7/2}$ state, they all own the same polarizability p_1; the only possible transitions accounted by equation (1) are those that start from this ground state (q = 1). In the pumped state, a part of the ytterbium ions are excited to the metastable $^2F_{5/2}$ state, with the steady state population of the excited state δN_2 that depends on the pump power launched into the fiber. In this case, transitions from both the ground (q = 1) and excited (q = 2) states are accounted for ions possessing polarizabilities p_1 and p_2, respectively. The index change induced by pumping can then be found as a difference of the index contributions accounted for the pumped and unpumped media:

$$\delta n^e = \frac{2\pi F_L^2}{n_0} \Delta p \delta N_2 , \tag{3}$$

where n_0 is the refractive index of host glass; $F_L = (n_0^2 + 2)/3$ is the Lorentz factor.
The phase shift corresponding to the electronic RIC detected at the test wavelength λ_T in the fiber of length L is evaluated from the equation (3) by its integration over the fiber volume with $\rho_T(r)$ as a weight function:

$$\delta\varphi = \frac{4\pi^2}{\lambda_T} \int_0^L \int_0^\infty \delta n^e(z,r) \rho_T(r) r dr dz \approx \frac{\bar{\eta}\rho_T(0)}{\lambda_T} \left[4\pi^2 \frac{F_L^2}{n_0} \Delta p \right] \delta \mathbf{N}_2^\Sigma \tag{4}$$

where $\delta \mathbf{N}_2^\Sigma = 2\pi \int_0^L \int_0^\infty \delta N_2(z,r) r dr dz$ is the pump-induced change in the number of the excited Yb^{3+} ions in the whole fiber volume, $\rho_T(r)$ is the normalized power radial distribution of the probe light; r is the polar coordinate describing the fiber cross-section, z is the linear coordinate along the fiber.
The parameter $\bar{\eta}\rho_T(0)$ approximates the efficiency of the probe mode interaction with the population changes $\delta N_2(r)$ induced in the doped fiber area. Here we intentionally separated the factor $\rho_T(0)/\lambda_T$ that includes the major part of the dependency on λ_T and the correction factor $\bar{\eta}$ taking into account the distributed character of the population changes $\delta N_2(r)$ within the doped core area.
The factors $\rho_T(0)/\lambda_T$ and $\bar{\eta}$ could be evaluated from the step-index fiber approach (Snyder & Love, 1983). Assuming the testing wavelength λ_T to be within the IR fiber transparency band, one could express the power mode distribution $\rho_T(r)$ as:

$$\rho_T(r) = \begin{cases} \dfrac{1}{\pi a^2} \left(\dfrac{v}{V} \dfrac{J_0(ur/a)}{J_1(u)} \right)^2 & for \;\; r < a \\ \dfrac{1}{\pi a^2} \left(\dfrac{u}{V} \dfrac{K_0(vr/a)}{K_1(v)} \right)^2 & for \;\; r > a \end{cases} \tag{5}$$

where J_n, K_n are Bessel functions, $a = w\left[1.3 + 0.864\,(\lambda_S/\lambda_c)^{3/2} + 0.0298(\lambda_S/\lambda_c)^6\right]^{-1}$ is the effective core radius (Jeunhomme, 1983), $V = 2.05\lambda_c/\lambda_T$ is the fiber dimensionless modal parameter, w is the modal field diameter at $\lambda_S \approx 1.06\,\mu m$, and λ_c is the fiber cut-off wavelength, $u(\lambda_T)$ and $v(\lambda_T)=\sqrt{V^2-u(\lambda_T)^2}$ are the modal phase parameters that are a solution of the characteristic equation $v\mathrm{J}_0(u)\mathrm{K}_1(v)=u\mathrm{K}_0(v)\mathrm{J}_1(u)$.

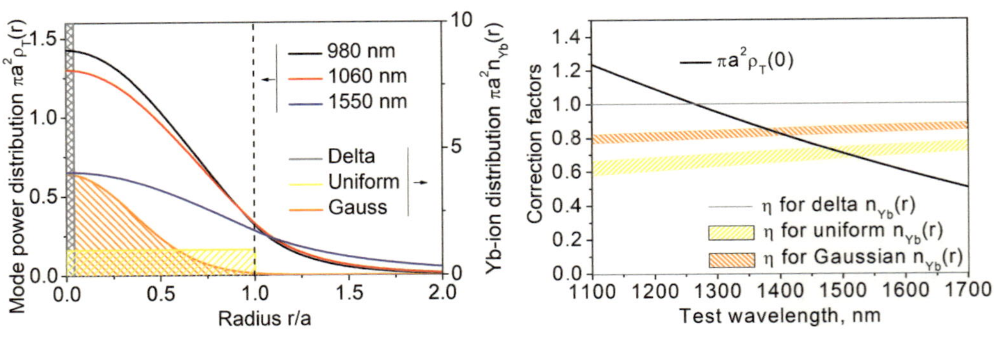

Fig. 2. The normalized mode power distributions at different wavelengths in comparison with examples of Yb-ion distributions: delta-like $n_{Yb}(r)=\delta(r)/\pi a^2$, uniform $n_{Yb}(r)=1/\pi a^2$ and Gaussian $n_{Yb}(r)==4\exp\left[-(2r/a)^2\right]/\pi a^2$ (a). The coefficient $\rho_T(0)$ (equations (5)) and correction factors $\bar{\eta}$ (equation (6)) calculated for different Yb-ion distributions.

Figure 2 explains the dependencies of the parameter $\bar{\eta}\rho_T(0)$ on the testing wavelength λ_T and on the Yb-ion distribution within the fiber core. One can see from figure 2 (a) that the longer testing wavelength λ_T corresponds to the wider distribution $\rho_T(r)$ providing the smaller overlap between the signal power and the doped fiber core area. As a result the parameter $\rho_T(0)$ (figure 2 (b)) decreases with the testing wavelength. As far as all changes $\delta N_2(r)$ are assumed to occur close to the fiber axis, the factor $\bar{\eta}\to 1$. However, the wider Yb-ion distribution $N(r)$ decreases the overlap between the signal power and the doped fiber core area (figure 2(a)) resulting in $\bar{\eta}<1$.
The factor $\bar{\eta}$ could be estimated from equation (4) to lie within two bounds:

$$\int_0^\infty n_{Yb}(r)\rho_T(r)rdr < \rho_T(0)\bar{\eta} < \frac{\int_0^\infty n_{Yb}(r)\rho_P(r)\rho_T(r)rdr}{\int_0^\infty n_{Yb}(r)\rho_P(r)rdr} \tag{6}$$

where $\rho_P(r)$ is the normalized mode radial power distribution at the pump wavelength λ_P and $n_{Yb}(r)=N(r)\Big/2\pi\int_0^\infty N(r)rdr$ is the normalized distribution of Yb-ions over the fiber core.

The typical values of the correction factor calculated for uniform and Gaussian distributions of the Yb-ion dopands are $\bar{\eta}\approx 0.7$ and $\bar{\eta}\approx 0.85$, respectively. The factors $\bar{\eta}$ determined from inequalities (6) perform a typical error of ~10% (figure 2(b)).

Equation (4) predicts that the phase shift is proportional to the pump-induced change in the whole number of the excited ions in the fiber. Therefore, the dynamics of both of them is governed by similar rate equations:

$$\frac{d\delta \mathbf{N}_2^{\Sigma}}{dt} = \frac{P_{in} - P_{out}}{h\nu_P} - \frac{P_{ASE}}{h\nu_S} - \frac{\delta \mathbf{N}_2^{\Sigma}}{\tau_{sp}} \tag{7}$$

$$\frac{d\delta\varphi}{dt} = K\left[P_{in} - P_{out} - \frac{\lambda_S}{\lambda_P}P_{ASE}\right] - \frac{\delta\varphi}{\tau_{sp}} \tag{8}$$

where $K = 4\pi^2 F_L^{\ 2}\Delta p\overline{\eta}\rho_T(0)\lambda_P / n_0 hc\lambda_T$, h is the Plank constant, ν_P , ν_S are the average frequencies of the pump and the amplified spontaneous emission (ASE) (or lasing), P_{in} , P_{out} are the input and output (residual) powers at ν_P , P_{ASE} is the emitted power at ν_S and an ideal quantum efficiency for the Yb-doped fiber is assumed.
The validity of the equation (8) has been confirmed in the experiment to be discussed in part 3 of the Chapter.

2.3 Fundamentals of thermally induced RICs in Yb-doped fibers

The described electronic RIC is not the only mechanism responsible for the induced phase shift in Yb-doped fibers and governed by the population inversion. The thermalization of the pump power leading to temperature rise may result in thermally-induced index changes in the fiber media. Moreover, thermal expansions of the fiber core along its length and its radius may cause phase modulation. However, one could show that the transverse expansion of the single-mode fiber core gives a small contribution to the thermally-induced phase shift in comparison with the volume RIC and core longitudinal expansion. The refractive index change due to temperature rise δT is expressed as

$$\delta n^T = \left(\frac{\partial n}{\partial T}\right)\delta T + \delta n_{ph} = \left[\left(\frac{\partial n}{\partial T}\right) + 2n_0^3\alpha^T C'\right]\delta T \tag{9}$$

where $\partial n/\partial T$ is the thermal RIC coefficient, δn_{ph} accounts for the photoelastic effect, C' denotes a photoelastic constant averaged over polarizations, and α^T is the thermal longitudinal expansion coefficient.
In the simplest approximation of four-level fiber amplifier (see figure 1(a)), the thermal load is achieved through radiation-less transitions only. The corresponding heat equation expresses the evolution of the temperature $T(r,z,t)$ within the fiber:

$$\frac{\partial T}{\partial t} - \frac{\kappa}{\rho C_p}\Delta T = \frac{1}{\rho C_p}\left(h\nu_{\hat{4}\hat{3}} N_{\hat{4}} w_{\hat{4}\hat{3}} + h\nu_{\hat{2}\hat{1}} N_{\hat{2}} w_{\hat{2}\hat{1}}\right) \tag{10}$$

where κ and ρ are the thermal conductivity and density of the host glass; C_p is the thermal capacity, $N_{\hat{4}}$ is the population of the upper pump level " $\hat{4}$ " (the Stark-depleted sub-level of the ($^2F_{5/2}$) state) of the Yb^{3+}-ions; $v_{\hat{4}\hat{3}}$, $w_{\hat{4}\hat{3}}$ are the frequency and rate of the transition from the pump level " $\hat{4}$ " to the upper laser level " $\hat{3}$ "; $N_{\hat{2}}$ is the population of the lower laser level " $\hat{2}$ " (the depleted sub-level of the ($^2F_{7/2}$) ground state); $v_{\hat{2}\hat{1}}$ and $w_{\hat{2}\hat{1}}$ are the frequency and rate of the transition from the low laser level " $\hat{2}$ " to the ground state " $\hat{1}$ ".

Since the radiation-less relaxation time is negligible in comparison with the upper level " $\hat{3}$ " lifetime τ_{sp}, one can estimate thermal load from the following rate equations:

$$N_{\hat{2}} w_{\hat{2}\hat{1}} \approx \gamma_{ASE} N_{\hat{3}}^2 + \frac{N_{\hat{3}}}{\tau_{sp}}$$
$$N_{\hat{4}} w_{\hat{4}\hat{3}} \approx \frac{\sigma_{\hat{1}\hat{4}} N_{\hat{1}}}{h\nu_{\hat{1}\hat{4}}} P_P(z,t)\rho_P(r) \tag{11}$$

where $\sigma_{\hat{1}\hat{4}}$, $\nu_{\hat{1}\hat{4}}$ are the pump absorption cross-section and frequency; γ_{ASE} is the constant characterizing ASE in the fiber.

The heat dissipation occurs from the fiber core to the lateral surface of the clad with a sink into the air. The temperature distribution in the fiber is expressed by the Newton formula:

$$\kappa\nabla_{\xi} T + H(T - T_0) = 0 \tag{12}$$

where T_0 is the temperature on the external surface of the fiber; ξ is clad surface normal vector; ∇_{ξ} is normal derivative and H is the heat-transfer coefficient accounted independently for core, cladding and counting parts.

To evaluate the thermally induced RIC in the fiber one has solve numerically the non-stationary heat equations (9 - 11) in combination with the boundary condition (12) on the cooled surfaces. The thermal fiber stresses and deformations could be taken into account through solving of the thermo-elastic steady-state problem with the thermal field distribution satisfying the heat equations (9 - 12).

2.4 Comparison of the electronic- and thermal- contributions to the RICs in Yb-doped fibers (a typical example).

The fundamental relations reported in the previous sections allow us to compare contributions of the electronic and thermal mechanisms to the RIC effects induced in Yb-doped fibers under diode pumping. To be more concrete in our estimations, we have to engage here typical fiber parameters and pumping conditions related to the low-power all-fiber RIC experiments discussed further in the Chapter.

Let us consider properties of a standard commercially available Yb-doped optical fiber characterized by the core diameter, $2a \approx 3\ \mu m$; the cladding diameter, $2R \approx 125\ \mu m$; the peak absorption coefficient at $\sim 980\ nm$, $\alpha \approx 230\ m^{-1}$ ($\sim 1000\ dB/m$). Through the absorption and emission cross-sections of Yb-ions at 980 nm, $\sigma_{12}^{(p)} \approx \sigma_{21}^{(p)} \approx 2700\,10^{-27}\ m^2$, the concentration of Yb-ions in such fiber core is estimated to be $N_0 \approx 8.5 10^{19} cm^{-3}$. For further estimations we will use the following important characteristics of pure glass silica: the refractive index, $n_0 \approx 1.5$; the Lorentz factor, $F_L \approx 1.41$, the PD value (as it is measured in the section 3.5) $\Delta p \approx 1.2\times 10^{-26} cm^3$; the coefficient $\partial n/\partial T \approx 1.2\,10^{-5}\ K^{-1}$; the glass density, $\rho = 2203\ kg\ m^{-3}$; the specific heat capacity, $C_P = 840\ J\ K^{-1}kg^{-1}$ and thermal conductivity $\kappa = 1.38\ W\ m^{-1}K^{-1}$.

The maximum electronic changes induced in the fiber core under diode pumping at 980 nm could be roughly estimated from equation (3):

$$\delta n^e = \frac{2\pi F_L^{\ 2}}{n_0}\Delta p N_0 \xi \approx 4.5\times 10^{-6} \tag{13}$$

where $\xi \approx 0.5$ is a maximal percentage of the excited ions available when $\sigma_{12}^{(p)} \approx \sigma_{21}^{(p)}$.
On other hand, the thermally induced RICs are expressed by equation (9) as:

$$\delta n^T = \left(\partial n/\partial T\right)\delta T \approx 1.2\,10^{-5}\ \delta T \tag{14}$$

where δT is a change of the core temperature due to pumping.
From estimations (13) and (14), one can conclude that the electronic and thermal contributions become equal, when the temperature of the fiber core increases by $\delta T \approx 0.35\,K$.

For estimation of the real core temperature variation achieved during the pump pulse excitation, we can assume a perfect thermal conductivity between the fiber core and fiber cladding. The duration of the excitation pulse is believed to be $\tau_P \prec \tau_{cl}$, where $\tau_{cl} \approx \rho C_p R^2/\kappa \sim 5\,ms$ is a typical time of temperature diffusion through the cladding. With these approximations the temperature on the clad surface during the pulse duration does not change significantly.
The growth of temperature in the fiber core center within the pulse duration can be expressed by a solution of the equation (10) in an approximation of the hard-tube fiber:

$$\delta T(t,r) = \frac{2}{R^2}\sum_{n=1}^{\infty}\int_0^{\infty} r'dr'\int_0^t Q(t',r')\exp\left(-C_n^2\frac{t-t'}{\tau_{cl}}\right)\frac{J_0\left(C_n\frac{r'}{R}\right)J_0\left(C_n\frac{r}{R}\right)}{J_1^2(C_n)}dt' \tag{15}$$

where $C_n > 0$ are the roots of $J_0(C_n)=0$, $J_n(x)$ denotes Bessel functions, $Q(t,r)$ is the right side of equation (10), i.e. the power income to the fiber core consumed for heating.
In each point of the fiber, the heating income $Q(t,r)$ depends on the pumping wavelength, the pump power and the degree of population inversion at the given moment of time. The pumping at 980 nm populates the lowest Shtark sub-level "$\hat{3}$" of the $^2F_{5/2}$ state directly. Therefore, the principal fiber heating occurs due to radiation-less relaxation of Yb-ions from the sub-levels "$\hat{2}$" to the ground state "$\hat{1}$" of the laser level $^2F_{7/2}$.
The power income from (10) can be expressed as

$$Q \approx \frac{h\nu_{\hat{2}\hat{1}}N_{\hat{2}}w_{\hat{2}\hat{1}}}{\rho C_p} \approx \frac{h(\nu_P - \nu_S)N_{\hat{3}}}{\rho C_p \tau_{sp}} \tag{16}$$

with the population $N_3(t,r,z)$ of the sublevel "$\hat{3}$" to be:

$$N_3(t,r,z) \approx \alpha\int_0^t \frac{P_P(z,t')\rho_P(r)}{h\nu_P}\exp\left(-\frac{t-t'}{\tau_{sp}} - \int_{t'}^t \frac{P_P(z,t'')}{\tau_{sp}P_P^{sat}}dt''\right)dt' \tag{17}$$

where $P_P^{sat} \approx \pi a^2 h\nu_p/\tau_{sp}\left(\sigma_{21}^{(p)}+\sigma_{12}^{(p)}\right) \approx 0.85\,mW$ is the fiber saturation power at 980 nm.

Under the conditions that $P_P(z,t) \equiv P_0$, the equations (16) – (17) give us an upper estimation of the heating induced in the fiber by the rectangular pulses with amplitude P_0 used in the experiment. Taking into account that $P_0 >> P_P^{sat}$, the quantity Q averaged over the fiber core cross-section and the pulse duration is estimated to be:

$$\bar{Q} \approx \alpha \frac{P_p^{sat}\left(\nu_p - \nu_s\right)}{\pi a^2 \rho C_p \nu_p} \tag{18}$$

The upper estimation of the temperature deviation δT in the fiber core centre achieved for the pump pulse excitation τ_P is derived from (15) by setting $Q(t',r') \equiv \bar{Q}$ and then formally assuming $t = \tau_P \to \infty$:

$$\delta T < \Delta T_{st} \approx \frac{a^2}{R^2} \tau_{cl} Q \left(2\ln\left(\frac{R}{a}\right) + 1 \right) \approx \frac{\alpha P_{sat}\left(\nu_p - \nu_s\right)}{2\pi\kappa\nu_p} \ln\left(\frac{R}{a}\right) \approx 0.01K \tag{19}$$

One can see from (19) that δT is at least of one order of magnitude lower than the core temperature increase expected for the case of equal contributions of thermal and electronic mechanisms. Following the same procedure we could estimate from equations (3) and (15) – (19) the ratio between the electronic and thermal RIC contributions during the excitation:

$$\frac{\Delta n^e}{\Delta n^T} > \frac{2\pi F_L^{\ 2}}{n_0} \frac{1}{a^2} \frac{\Delta p \tau_{sp} \kappa}{\left(2\ln\left(\frac{R}{a}\right) + 1 \right) (\partial n / \partial T) h(\nu_P - \nu_S)} \approx 35 \tag{20}$$

Just after the end of the pulse excitation the heating of the fiber core continues, since the sublevel " $\hat{3}$ " is still populated. In accordance with equation (3), relaxation of the population from the sublevel " $\hat{3}$ " occurs exponentially with a decay time $\tau_{sp} \approx 0.85\, ms$, so the quantity Q has to relax in the same manner. The exponential relaxation of the heating source Q causes similar relaxation of the fiber core temperature, because thermalization of the heating in the fiber core area occurs for the time $\tau_{cor} \approx \rho C_p a^2 / \kappa \sim 3\ \mu s$ much smaller then the relaxation time ($\tau_{cor} << \tau_{sp}$). As a result, the exponential RIC relaxation after pulse excitation is contributed by two synchronized components of the electronic and thermal RICs, Δn_{rel}^e and Δn_{rel}^T , respectively. The ratio between them is estimated from expressions (15)-(17) as:

$$\frac{\Delta n_{rel}^e}{\Delta n_{rel}^T} \approx \frac{2\pi\left(F_L^2 / n_0\right) \Delta p \tau_{sp}}{\left(\partial n / \partial T\right)\left(h(\nu_p - \nu_s) / \rho C_p\right) \tau_{cor}} \approx 3.5 \times 10^2 \tag{21}$$

In conclusion, the estimations (20 - 21) ensure us that the RIC effects measured in our experiment is almost of pure electronic nature. These experiments are under consideration in the next part of the Chapter.

2.5 Conclusion to the part 2

Summarizing this part, we have presented theoretical analysis of the electronic and thermal RIC effects taking place in Yb-doped optical fibers. We have clarified their joined population inversion nature and explained the principal difference between the two mechanisms. We derived key equations of the dynamical effects that will be fruitfully employed in the next sections for detailed analysis of numerous experimental observations. Importantly, the thermal contribution to the RICs observed in Yb-doped fibers is shown to be negligible under condition of the all-fiber low-power experiments considered below.

3. Pump-induced refractive index changes in Yb-doped optical fibers

In this part, we experimentally characterize the RICs induced by optical pulses at 980 nm in single-mode ytterbium-doped optical fibers and report details of the effect observation (Fotiadi et al., 2007, 2008b, 2008c). The RIC dynamics is shown to agree with the two level laser model discussed early in section 2.2. The only material model parameter, the absolute PD value, is measured in the spectral range of 1460-1620 nm for different fiber samples and is found to be independent on the fiber geometry and on the ion concentration. The PD dispersion profile highlights a predominant far-resonance UV rather than near-resonant IR transitions contribution to the RIC.

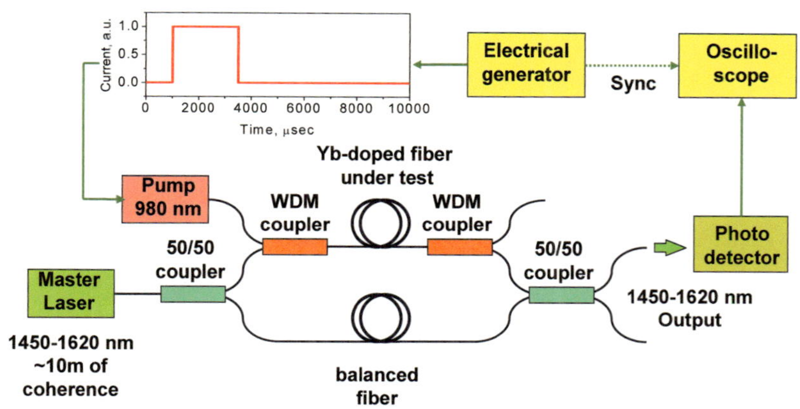

Fig. 3. Experimental setup for testing of the Yb-doped fiber samples.

3.1 Introduction to the part 3

The electronic RICs in pumped rare-earth-doped optical fibers is associated with changes of population of the ion states with different polarizabilities (Digonnet et al., 1997). This effect, which is important both for laser crystals and glasses, has been intensively investigated during the last decade. The nature of the polarizability difference (PD) in rare-earth ions is also widely discussed. Some authors believe that the main contribution to the PD expressed by equation (2) comes from the first term responsible for near-resonance IR transitions (Desurvire, 1994; Bochove, 2004; Barmenkov et al., 2004; Garsia et al, 2005). An alternative model suggests the predominant contribution from the second term responsible for strong UV transitions located far from the resonance (Digonnet et al., 1997), similar to those observed in laser crystals (Antipov et al., 2003, 2006; Margerie et al., 2006; Messias et al., 2007). Correct understanding of these electronic phenomena in Yb-doped fibers (Arkwright et al., 1998) is very important for numerous fiber applications. The pump-induced RICs could significantly affect the fiber laser behavior. The enhanced nonlinear phase shift could be employed for coherent beam combining discussed in the Chapter, optical switching (Wu et al., 1995), all-fiber adaptive interferometry (Stepanov et al., 2007).

In the following paragraphs, we discuss our original RIC observations in commercial single-mode Yb-doped optical fibers pumped at ~980 nm by a laser diode. The experiment is performed in a Mach-Zehnder interferometer configuration operating in IR spectrum band far from the absorption and emission Yb^{3+} ion resonances. The main objectives are to explore the RIC dynamics and to measure the values of the PD in standard Yb-doped fibers in spectral range 1460 - 1620 nm.

3.2 Experimental setup

The experimental setup is shown in figure 3. The Yb-doped fibers under investigation are introduced in one arm of the all-fiber spliced Mach-Zehnder interferometer. The tested fibers are pumped to the core from a standard laser diode operating at $\lambda_p \approx 980$ nm with the power up to ~145 mW. The CW-radiation of diode laser "Tunics" with a coherence length ~10 m is used as the test wave. The passed test signal is detected at the interferometer output by a fast photodiode. The test wavelength λ_T is tunable within the range from 1460 to 1620 nm.

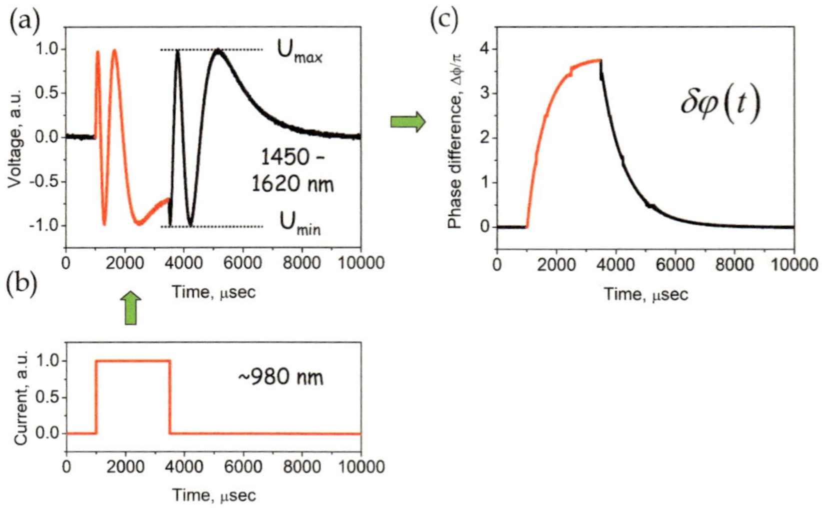

Fig. 4. A sample of the induced phase shift $\delta\varphi(t)$ recovery: the laser driver current pulse profile (a), the recorded oscilloscope trace (b), and reconstructed phase trace (c).

The RIC signal is measured as the photodiode response to a single rectangular pump pulse of tunable 10 μs- 10 ms duration (figure 4) applied to the tested fiber.
The induced phase shift $\delta\varphi(t)$ is recovered from the oscilloscope trace $U(t)$ as:

$$\begin{aligned}\delta\varphi(t) &= \varphi(t) - \varphi(0)\\ \varphi(t) &= (-1)^k \arcsin\left(\frac{2U(t) - U_{\max} - U_{\min}}{U_{\max} - U_{\min}}\right) + \pi k\end{aligned} \tag{22}$$

where $k = 0, 1, 2...$ provides continuity of $\delta\varphi(t)$.
Four fiber samples of a different lengths, index profile geometries and Yb^{3+}-ion concentrations have been examined. All fibers are single-mode at the pump and test wavelengths. Fiber A (fabricated in Fiber Optics Research Center, Russia) and Fibers B, C, D

(Yb-198, -118, -103, of CorActive, Canada) have rectangular and circular cladding profiles, respectively, with the cladding side/diameter $\sim 125\ \mu m$. All fibers are aluminum silicate with a Gaussian distribution of Yb^{3+}-ion concentration in the core. The fiber parameters are shown in Table 1.

Parameter	Fiber A	Fiber B	Fiber C	Fiber D
Peak absorption at 980 nm (α_P), dB/m	~ 900	~ 1073	~ 245	~ 35
Mode-field diameter at λ_S (w), μm	~ 4.5	~ 3.6	~ 4.5	~ 3.6
Cut-off wavelength (λ_c), nm	~ 810	~ 870	~ 680	~ 816
Equivalent core radius (a), μm	~ 1.6	~ 1.4	~ 1.3	~ 1.3
Coefficient K at 1550 nm, rad ms^{-1}mW^{-1}	$\sim 0.043\pi$	$\sim 0.056\pi$	$\sim 0.067\pi$	$\sim 0.067\pi$

Table 1. Specifications of Yb-doped single-mode fibers tested during the experiment.

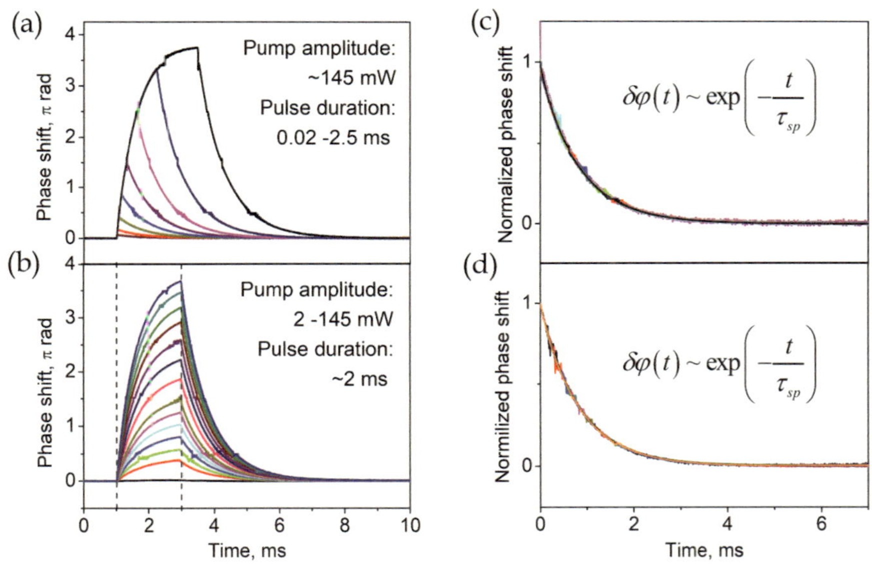

Fig. 5. Phase shifts induced in Fiber A by pulses of different pulse duration (a) and pulse amplitude (b). Only relaxing parts of the same curves normalized to the maximal values are shown in the delayed scale (c, d). The test wavelength is ~1550 nm, the fiber length is ~2m.

3.3 Relaxation of the phase shift after pump pulse excitation

The recorded phase shifts up to $\sim 4\pi$ shown in figure 5 highlight strong RIC effect induced in Fiber A by pump pulses of different amplitudes and durations. In every cases, the effect exhibits smooth saturation at some steady-state level. The saturation depends on the pulse amplitude and on the pulse duration (figure 5(a, b), respectively), or more precisely, on the pulse energy. Decaying parts of the phase traces describe relaxation of the refractive index after the end of the pulse excitation. They are perfectly fitted by an exponential decay function $\varphi(t) \sim \exp(-t/\tau_{sp})$ with the relaxation time constant equal to the Yb-ion excited state life-time $\tau_{sp} \approx 850\,\mu s$, almost the same for all fiber samples. All normalized curves shown in figure 5 (c, d) in the delayed time scale are similar, although they relate to different pumping conditions. Such relaxation behavior corresponds to the electronic mechanism of RIC predicted by equation (8) explained in section 2.2. No other features that might be attributed to the thermally induced RICs have been observed in the experiment.

3.4 Dynamics of pump induced phase shifts in highly doped fibers

Under conditions of rectangular pump pulse excitation, total pump absorption and low amplified spontaneous emission (ASE), equation (8) is resolved analytically to the solution:

$$\delta\varphi(t) = K\tau_{sp}\left[1-\exp\left(-\frac{t}{\tau_{sp}}\right)\right]P_0 \tag{23}$$

where P_0 is the pump pulse amplitude.

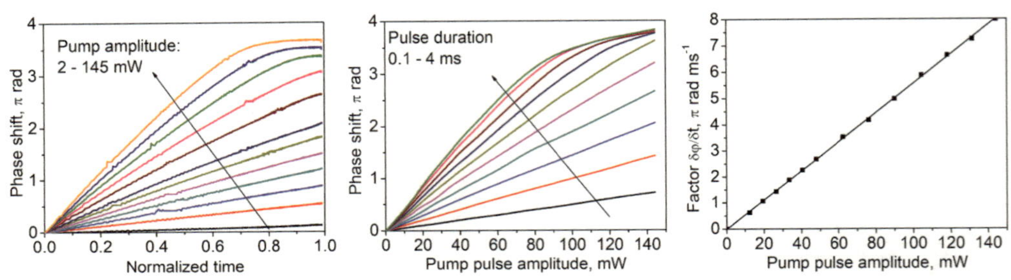

(a) (b) (c)

Fig. 6. Phase shifts induced in Fiber B by ~4 ms rectangular pulses as a dependence on the normalized time $\tau = 1-\exp(-t/\tau_{sp})$ (a) and pulse amplitude (b). The slope $\delta\varphi/\delta t|_{t\to 0}$ of the phase characteristics as a dependence on the pulse amplitude (c). The test wavelength is ~1550 nm, the fiber length is ~2m.

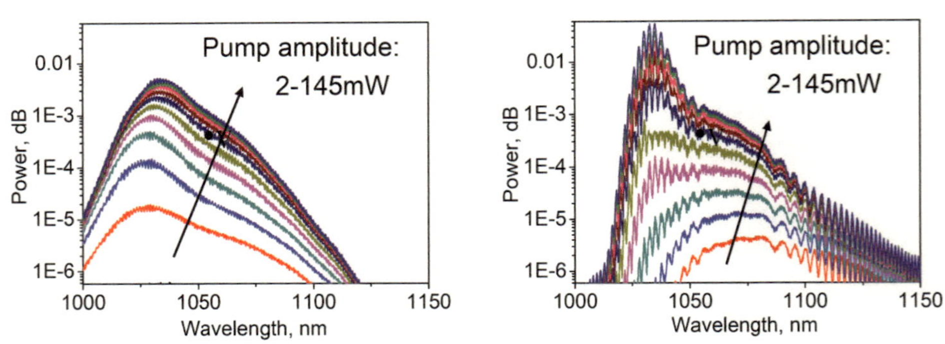

(a) (b)

Fig. 7. Optical spectra of the amplified spontaneous emission (ASE) recorded in Fiber B in backward (left) and forward (right) directions in respect to the pump propagation.

In highly doped fiber samples A, B and C, the pump power is completely absorbed by the fiber length. At low pump pulse energies, the experimental traces shown in figure 6 perfectly reproduce the predictions of equation (23). One can verify the exponential character of the phase growth to its steady-state level (figure 6(a)), and the linear dependence on the pump pulse amplitude (figure 6(b)). As the pass gain and ASE level remain low, the slopes $\delta\varphi/\delta t|_{t\to 0}$ can be measured for different pulse amplitudes (figure 6(c)). Their linear fit gives us the factor $K = 0.056\pi$ rad ms^{-1}mW^{-1} that is the only unknown material parameter in equation (8).

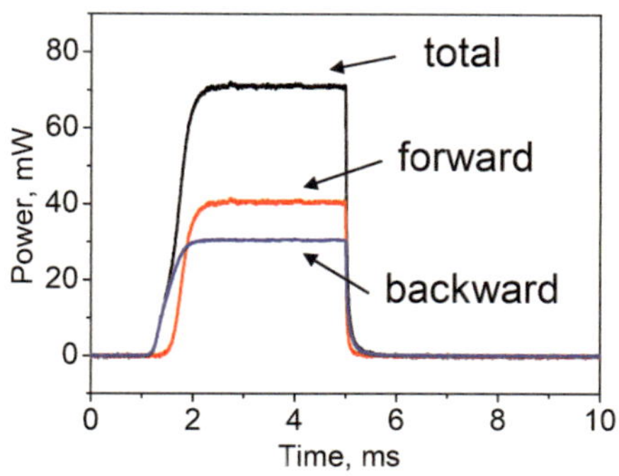

(a)

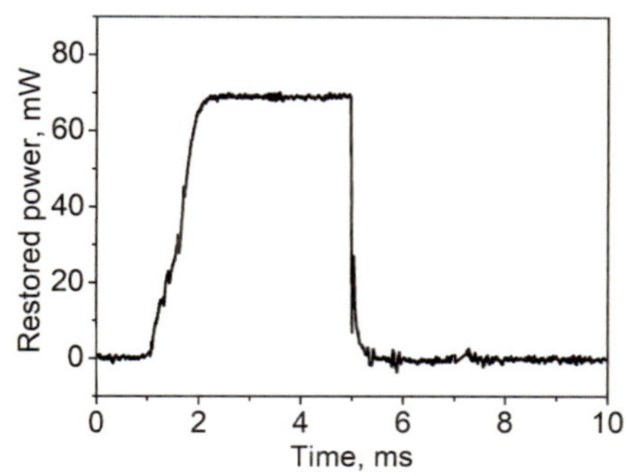

(b)

Fig. 8. The recorded (left) and reconstructed from equation (24) (right) ASE powers.

At higher excitation levels, an error between the recorded traces and the formula (23) appears and increases as the ASE power increases. The spectra of the optical emission at both fiber ends explain the ASE to be the reason of the saturation of the monotonic growth of the phase shift at high pump energies (figure 7).

In general case of all available pulse energies, the phase trace dynamics is driven by the differential equation (8) of the two-level RIC model (section 2.2). To validate the model at high pump power levels, the output powers from both fiber ends have been recorded during the pulse excitation simultaneously with recording of the test signal. Due to absorption saturation and re-absorption processes, the ASE pulse recorded in the forward direction is slightly delayed in respect to the backward ASE pulse thus causing a specific break in the leading edge of the total power (figure 8(a)).

We should compare the recorded trace of the total ASE power with the prediction of the equation (8) allowing reconstruction of the ASE power from the recorded phase traces though a simple mathematical procedure:

$$P_{ASE}(t) \approx P_{in} - K^{-1}\left[\frac{d\delta\varphi(t)}{dt} + \frac{\delta\varphi(t)}{\tau_{sp}}\right] \tag{24}$$

From figure 8, one can see that the reconstructed ASE profile is in a good quantitative agreement with the recorded total ASE power trace. Surprisingly, the model is able to reproduce the specific break of the original ASE pulse. This result proves our electronic RIC model to be eligible for whole range of the pulse energies used in the experiment.

3.5 Evaluation of the polarizability difference (PD)

Factor K in equations (23, 24) is the only model parameter relating the induced phase shift and the pump pulse parameters at a given test wavelength. Phase dynamics observed in two different sample lengths of Fiber C shows the factor K to be independent on the fiber length (figure 9(a)). Reduction in the fiber length causes reduction of the saturation energy, but does not affect the slope in phase changes at low pump pulse energies. At the same experimental conditions, the phase shift dependencies observed with different fiber samples reveal different slopes and other distinct features (figure 9(b)). The Fiber D with lowest Yb-ion concentration exhibits lower saturation power than others fibers, but the slope in this fiber is nearly the same as the slope relating to Fiber C and is larger than the slope relating to Fiber B and Fiber A. Approximation by equation (23) gives different factors K for different fibers presented in Table 1. However, the mutual ratios of the factors are in good agreement with our RIC model that predicts the factor K independent on Yb-ion concentration and inversely proportional to the square of the fiber core radius $\bar{a}$. The measured factors K allow us to estimate the polarizability difference at $1550\,\mathrm{nm}$ which proves to be the same for all tested fibers. Evaluation of $\bar{\eta}\rho_T(0)$ for a step-index fiber and Gaussian Yb-ion distribution with radius equal to $\bar{a}$ $(\bar{\eta} \approx 0.85)$ presented in figure 2 gives $\Delta p_{1550} \approx 1.2 \times 10^{-26}\,\mathrm{cm}^3$. We expect ~20% error in this value due to uncertainty of the doped area size. Notice that the expression for K given in (Fotiadi et al., 2008c) contains an unfortunate error causing overestimation of the PD value reported therein.

To measure the PD dispersion in the spectral range of 1450-1620 nm, Fiber B was additionally characterized at different testing wavelengths λ_T. The phase traces shown in figure 10(a) highlight significant differences in the phase slopes $\delta\varphi/\delta\tau$ referenced to different λ_T. Qualitatively similar results have been reported earlier for the two-core Yb-doped fibers (Arkwright et al., 1998). However, in our case the measured dispersion of the factor $K(\lambda_T)$ is found to be coincide with the dependence $\sim \rho_T(0)/\lambda_T$ as it is expressed from the step-index approach (figure 10(b)). As far as the polarizability difference (PD) $\Delta p(\lambda_T)$ is directly proportional to $\sim K^{-1}(\lambda_T)\rho_T(0)/\lambda_T$, the strong wavelength dependencies of $K(\lambda_T)$ and $\sim \rho_T(0)/\lambda_T$ are compensating each other, providing the PD to be nearly independent on the test wavelength in the measured spectrum range.

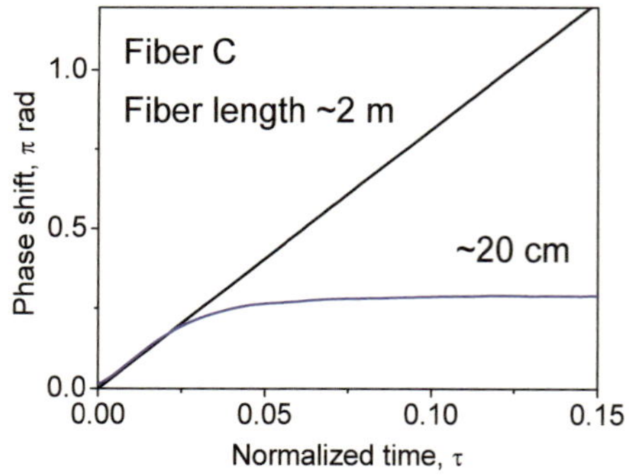

(a)

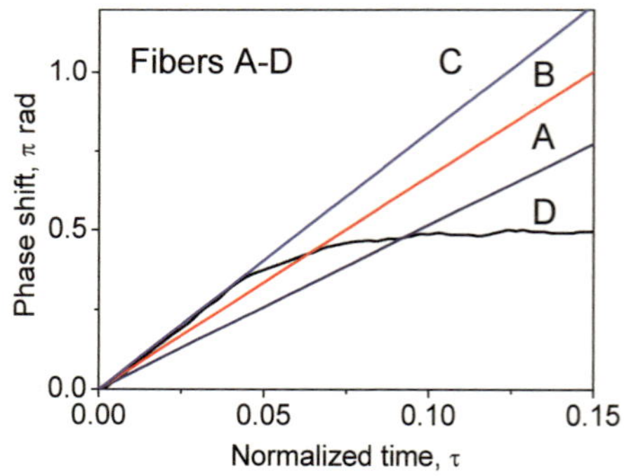

(b)

Fig. 9. Phase shift induced by 145-mW pulses in different fiber samples: in the same fiber of different lengths (a) and different fibers of ~2m length.

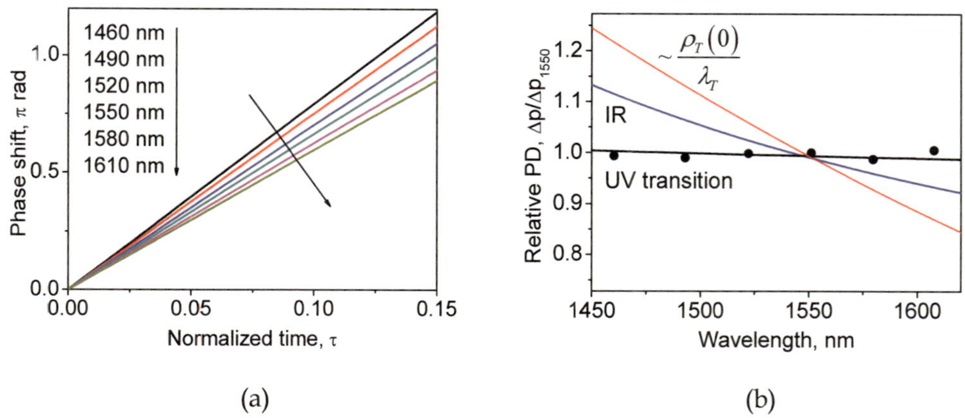

Fig. 10. Phase shift induced by 145-mW pulses in Fiber B at different testing wavelengths (a). The relative polarizability difference dispersion (points) in comparison with wavelength dependencies expressed by the first (resonance) and the second (non-resonance) terms of equation (3), and the dependence $\sim \rho_T(0)/\lambda_T$ (b).

The experimentally observed dispersion $\Delta p(\lambda_T)$ has to be compared with the Lorentz-line dispersion profiles predicted by equation (2) for PD contributions of near-resonant IR and far-resonance UV transitions of Yb^{3+} ions at $\sim 1\,\mu m$ and $\sim 0.4\,\mu m$, respectively. The reconstructed PD profile matches the UV-line wing and has the same value in all investigated spectral range.

3.6 Conclusion to the part 3

We have reported a strong RIC effect observed in different samples of commercial single-mode Yb-doped fibers. The effect induced in the fiber by optical pumping at its absorption resonance wavelength could be observed within the whole IR transparency band. The RIC exhibits a typical excited population dynamics in accordance with the electronic RIC mechanism and could be perfectly described by the two-level population model discussed in section 2.2. The PD value, the only material parameter of the model, is measured in the experiment and found to be independent on the Yb-fiber performance specifications. The PD dispersion curve highlights the dominant non-resonant contribution of the UV transitions.

4. All-fiber coherent combining of er-doped amplifiers through refractive index control in Yb-doped fibers

In this part, we discuss a simple all-fiber solution for coherent beam combining of rare-earth-doped fiber amplifiers. The RIC effect explained in the previous sections is used now for an active phase control in the fiber configuration. Algorithm based on the electronic RIC model supports straightforward implementation of the effect into a feedback loop. Combining of two 500-mW Er-doped amplifiers in a single-mode fiber is demonstrated with optical control by ~120-mW laser diode. The experimental optical loop employing RIC phase control is able to operate against the acoustic phase noise within the dynamical range of $\sim \pi$ and with a rate of $\sim 2.6\pi\ rad/ms$ suitable for combining of ~50 amplifiers similar to those used in our work.

4.1 Introduction

Being the most low-threshold nonlinear phenomena in optical fibers, Stimulated Brillouin scattering (SBS) reduces power levels available with CW narrow-band fiber sources, in particular leading to the self-pulsing and other laser instabilities (Fotiadi et al., 1998, 2002, 2004, 2006). A way to overcome this limitation is to use coherent combining of single-mode amplifiers each operating below the SBS threshold. The idea of this method is to split a single highly coherent beam into many beams which are then amplified by a parallel array of similar power amplifiers and finally recombined to a high power diffraction limited beam without significant broadening of the initial spectrum (Fan, 2005).

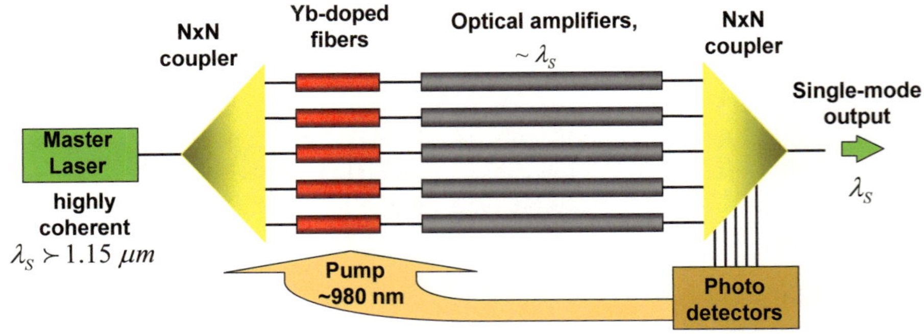

Fig. 11. Multiamplifier laser system with coherent beam combining based on RIC in YDF. The operating wavelength $\lambda_S \succ 1.15\ \mu m$ is selected within an Yb-fiber transparency spectrum band.

To achieve constructive interference after amplification, i.e. to collect the power from all channels in one single-mode fiber, the fiber amplifiers must be phase-matched together. Different approaches to the constructive phase-matching have been discussed: one can take advantage of the self-organization properties of multi-arm cavities (Bruesselbach et al., 2005b), of multi-core fibers delivering supermodes (Huo & Cheo, 2005), of digital holography (Bellanger et al., 2008), of SBS phase conjugation (Kuzin et al., 1994; Ostermeyer et al, 2008). A straightforward infallible beam combining can be provided by active phase control of each fiber amplifier by an attached phase modulator (Augst et al., 2004). However, piezo-electrical or electro-optical modulators are not perfect choice due to obvious disadvantages of parasitic resonances or integrating bulk and fiber components. In our all-fiber solution (figure 11) (Fotiadi et al, 2008a, 2009c), the rare-earth-doped fiber amplifiers operating at λ_S within an Yb-doped fiber transparency band are supplied by the sections of Yb-doped fibers (YDF) operating as optically controlled phase modulators. The principle of operation employs the RICs induced in YDF by optical pumping at 980 nm discussed in part 3. The method is applicable for Raman, Brillouin, neodymium-, erbium-, thulium-, or holmium- doped fiber amplifiers (Bruesselbach et al., 2005a; Goodno et al., 2009; Taylor et al., 2009; Fotiadi et al. 1989).

In the next sections, we verify the validity of our concept demonstrating the coherent combining of two 500-mW EDFAs. Two-level RIC model discussed in section 1.2 enables a simple algorithm with natural implementation of the RIC effect into an active phase control loop. The ability of the method to operate again the acoustic phase noise with the rate of $\sim 2.6\pi\ rad/ms$ is experimentally confirmed.

4.2 Experimental setup

The experimental setup (figure 12) is not so different from the setup used in the previous part, except the two amplifiers introduced into the interferometer. A master laser diode in combination with a 15-dBm preamplifier delivers single mode radiation at 1.55 µm with a coherence length of ~10 m. The first fiber coupler splits the laser emission into two arms, which are then amplified by two single mode EDFAs specified with 500-mW output. No spectral broadening of the amplified radiation has been observed at such power level. The EDFAs are supplied by thermo-electric controllers used for low-frequency phase noise elimination.

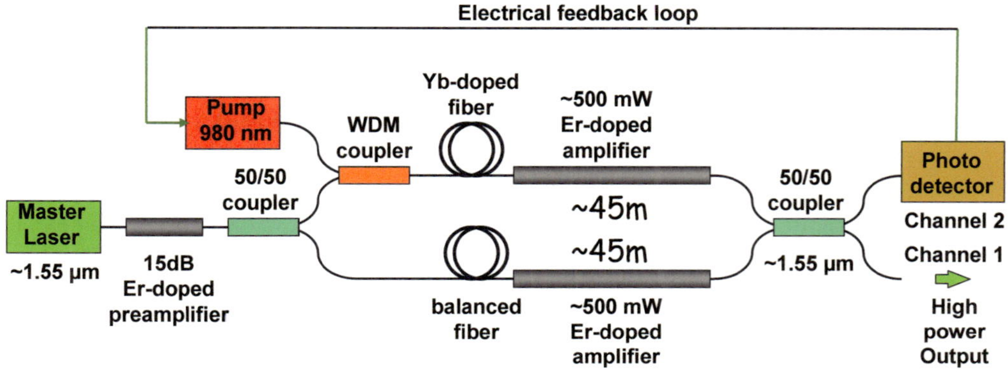

Fig. 12. Experimental setup used for demonstration of all-fiber coherent combining.

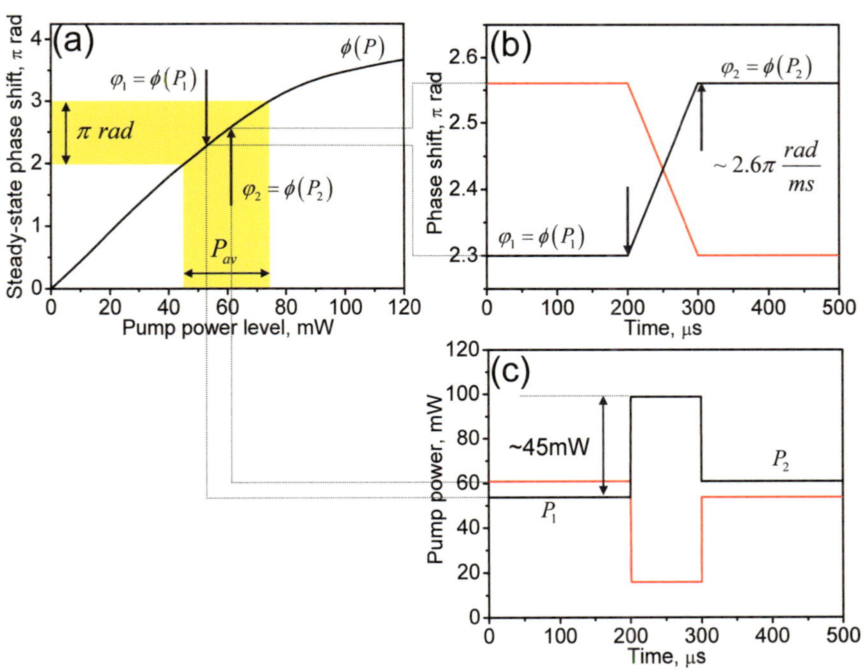

Fig. 13. Experimental steady-state (a) and dynamical (b) phase characteristics. Phase response (b) on a double jump of the diode power (c): positive (black) and inverted (red).

For fast phase adjustment, one of the arms is supplied by a 2-m length of YDF directly spliced with the amplifier. The YDF has an independent input for 120-mW laser diode operating at 980 nm. The use of the highly doped Fiber B (see, Table 1) ensures maximal RIC effect due to total absorption of the pump radiation inside the fiber. Since the RIC is directly proportional to the population density of the excited Yb-ions (see equation (3)), the phase shift induced in the fiber is determined by the laser diode power that could be used to maintain a constructive phase-matched coupling of two intense laser arms in a single-mode fiber (Channel 1). The power emitted through the Channel 2 is used for operation of the feedback loop discussed in next paragraphs. The relationship between the Channel 2 power and phase mismatching is regulated by equation (22).

4.3 Operation algorithm

The phase control operation algorithm is based on the steady-state and dynamical characteristics of the electronic RIC discussed in the previous part. The steady-state characteristic $\varphi = \phi(P)$ (figure 13(a)) is evaluated as the phase response to Heaviside step pulses of different amplitude P. One could tune the phase in the range up to $\sim 3.75\pi\ rad$ by a simple adjustment of the laser diode power. Commonly, the phase switching from $\varphi_1 = \phi(P_1)$ to $\varphi_2 = \phi(P_2)$ is provided by switching the diode power from P_1 to P_2, but this procedure requires several milliseconds to proceed.

Fast dynamical switching is possible within a part of the range covered by the steady-state curve $\varphi = \phi(P)$, for an example, within the range of $\sim \pi\ rad$ marked in figure 13(a). In the general case of total pump absorption and negligible spontaneous emission, we can extend equation (23) to all linear part of the steady-state curve, so the phase response $\Delta\varphi(t)$ on a single positive or negative jump of the diode power ΔP is expressed as:

$$\Delta\varphi(t) = K\tau_{sp}\left[1 - \exp\left(-\frac{t}{\tau_{sp}}\right)\right]\Delta P \tag{25}$$

where $K \approx 0.056\ \pi rad\ ms^{-1} mW^{-1}$ at $\lambda_T \approx 1.55\ \mu m$ (see Fiber B in Table 1).

In accordance with equation (25) a fast phase tuning (figure 13(b, c)) from $\varphi_1 = \phi(P_1)$ to $\varphi_2 = \phi(P_2)$ is achieved through two consecutive switchings of the diode power, first, from the level P_1 to a level $P_1 + \Delta P_0$ and then to the level P_2, where ΔP_0 is the power jump (positive or negative) available with the laser diode within the used tuning range. The switching time of such dynamical phase change is equal to the time τ between two opposite jumps of the diode power and is expressed for small phase steps as $\tau = (\varphi_2 - \varphi_1)/K\Delta P_0 << \tau_{sp}$. The higher ΔP_0, the faster phase tuning is. However, higher ΔP_0 is available within smaller tuning ranges. For tuning range of $\sim \pi\ rad$ ΔP_0 is limited by ~45mW that corresponds to an adjustment rate $\sim 2.6\pi\ rad/ms$.

The mission committed to the feedback loop circuit (figure 14) is to support the maximal power level emitted at 1.55μm through the Channel 1 (see figure 12). Therefore, the power emitted through the Channel 2 has to be kept as low as possible. This power recorded by the photodetector is used in the feedback circuit driven by 2.86-MHz acquisition card (National Instruments, NI PCI-6251). The period of data acquisition $\tau = 25\mu s$ is synchronized with by the card analogue output (alone or in combination with a standard pulse generator) that forces the laser driver to emit 50 μs -period meander signal with an amplitude

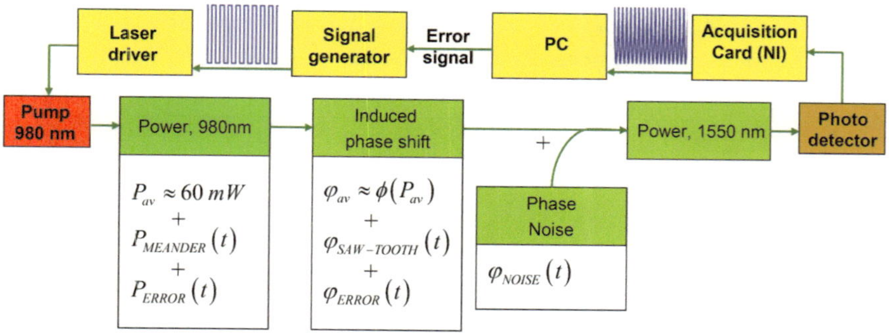

Fig. 14. The principal scheme of the electrical feedback loop operation.

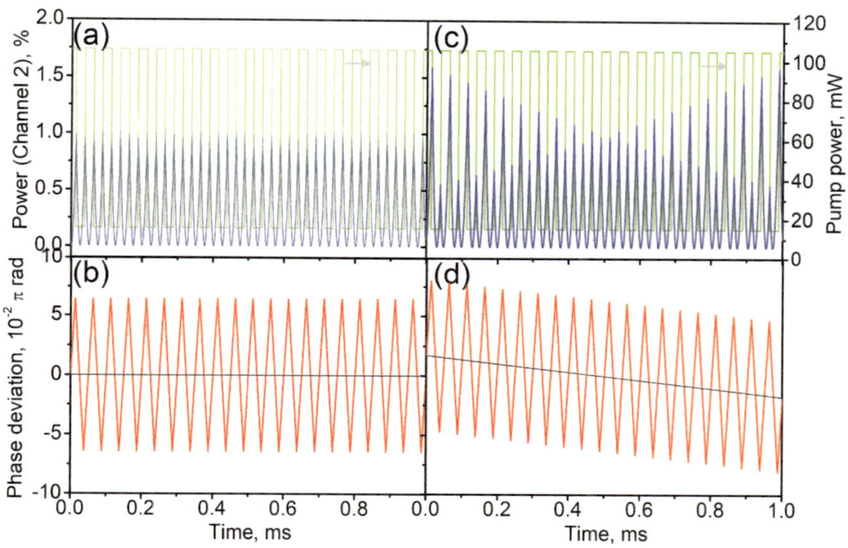

Fig. 15. Simulated feedback loop operation for cases of compensated (a, b) and uncompensated (c, d) phase noise: (a, c) laser diode (grey) and Channel 2 (black) powers; (b, d) the reconstructed phase deviation (grey) and reconstructed error signal (black); $P_{av} \approx 60mW$.

$\Delta P_0 = \pm 45\, mW$ and a controllable DC level P_{av} within the range of 45-75 mW (figure 15(a)). The modulated diode power induces a fast sawtooth modulation of the phase with a DC level of $\varphi_{av} = \phi(P_{av})$ and the excursion of $\Delta\varphi_0 = K\Delta P_0 \tau \approx 0.06\pi\ rad$ (figure 15(b)). Such phase modulation leads to ~100% amplitude modulation of the power in Channel 2 (figure 15(c, d)), while the modulation of the high power radiation in Channel 1 is negligible (~1%).

The signal acquired by the photodetector is the result of the superposition between the phase noise and the periodic phase modulation. In the case of right phase matching (when the phase noise is completely compensated), this signal is a perfect sawtooth signal, because the peaks associated with the positive and negative phase changes are of the same amplitude (figure 15(a)). In contrast, the presence of uncompensated noise causes the signal peaks to spread in two series. Importantly, the phase mismatching φ_{NOISE} is directly proportional to the difference between neighboring peaks (figure 15(c)). The error signal, produced by a PC from the acquired data, controls the phase φ_{av} through the control of the laser diode current: to compensate the phase mismatching, it produces a smooth correction $\delta P_{av} \to -P_{ERROR} = -\varphi_{NOISE}/K\tau$ to P_{av} (figure 14).

4.4 Experimental results

The phase control algorithm described in the previous section has been applied to the experimental setup (figure 12) demonstrating reliable operation against the phase noise.

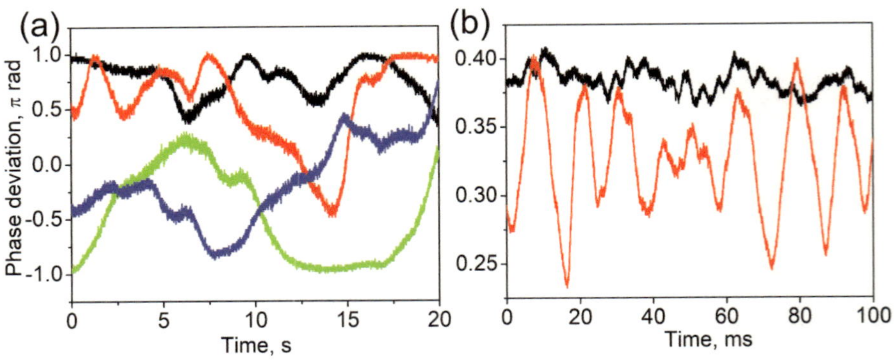

Fig. 16. Time series of the typical amplifier phase noise: (a) temperature noise, (b) acoustic noise: natural (black), caused by a flick given on the amplifier (grey).

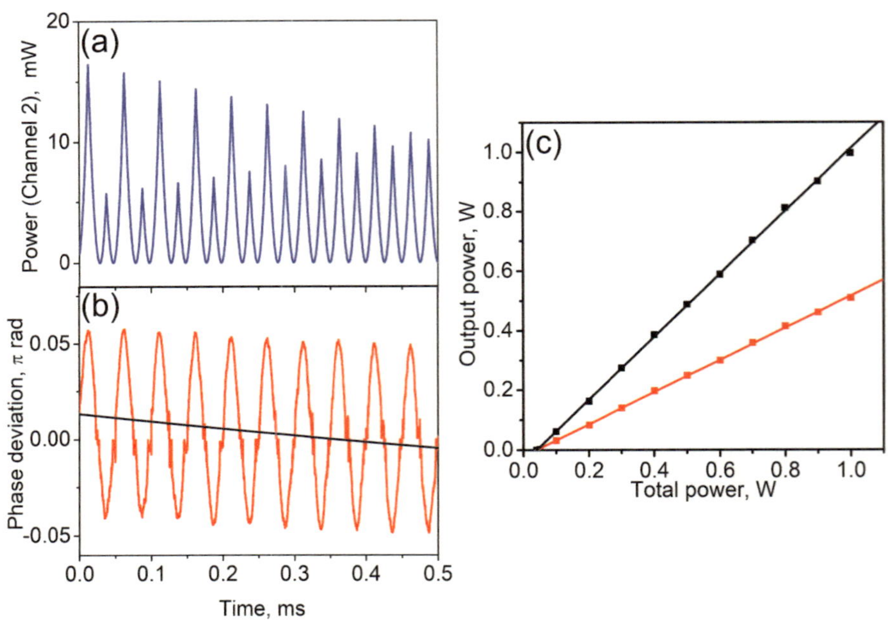

Fig. 17. Experimental operation of two combined amplifiers against the noise: (a) the photodiode signal; (b) the reconstructed phase (grey) and generated error signals (black); the laser system power characteristics without (grey) and with (black) active phase control.

Figure 16 presents the operation of the laser configuration without an active feedback. The phase traces have been recorded for a steady-state condition that is achieved after both amplifiers have reached thermal equilibrium (after 2-3 minutes of their operation with a power of 500 mW). Two classes of traces highlight phase fluctuations, associated with thermal variations (figure 16 (a)) and environmental acoustic vibrations (figure 16 (b)). These two kinds of noise contribute to different time domains and show different scales of the phase excursion that have to be compensated. Thermally induced phase noise is large fluctuations with amplitude up to several $\sim\pi$ *rad* attained for several seconds. For compensation of these fluctuations, no advanced fast technique is needed: the ordinary thermo-controllers connected with the feedback loop provide perfect suppression of low-frequency thermal noise. In contrast, acoustic phase noise associated mainly with mechanical resonances - noisy equipment, cooling fans, etc. - dominates with excursion rates of ~1ms and much smaller excursion amplitude, $\sim 0.01\pi$ *rad*. This should be considered as a minimum requirement for servo loop bandwidth since these measurements were taken in a quiet laboratory setting. Noisier environments would require a commensurate reduction of the feedback loop time, but we have checked that even a flick given on the amplifier causes a phase deviation with the amplitude no larger than $\sim 0.02\pi$ *rad* attending with a rate of $\sim 0.02\pi$ *rad/ms*. These observations give target parameters for the active phase control to be employed for compensation of the environmental acoustic noise in the system comprising two fiber amplifiers. Note, that a typical phase rate value to be compensated is more than 100 times lower than the capability of the proposed technique as just estimated. However, as the number of parallel amplifiers increases, the phase compensation rates have to be increased proportionally. The algorithm could be applied to the multi-amplifier system through time-division multiplexing.

The coherent combining of two 500-mW Er-doped amplifiers resulting in ~1 W power decoupled through the single-mode fiber output has been successfully demonstrated. Typical Channel 2 power and phase traces (figure 17(a, b)) exhibit the features similar to that shown in figure 15(c, d). The absolute power variations are about 10 - 20 mW, i.e. ~1 - 2 % of the total power emitted by two amplifiers. One can see how, during the given time series, the initial phase mismatch caused by a flick given on the amplifier is compensating due to operation of the feedback loop: the peaks initially separated in two series join together highlighting constructive interference achieved in Channel 1. The power characteristics of the combined system (figure 17(c)) give clear evidence that more than 95% of the radiation generated in two fiber amplifiers is efficiently decoupled through the single mode output.

4.5 Conclusion to the part 4

In conclusion, the reported result gives us the basis of the work towards development of the multichannel system shown in figure 11. The method is proved to operate against the noise with a rate of $\sim 2.6\pi$ *rad/ms* that potentially serves combining of ~50-100 amplifiers with noise properties like those shown in figure 16. It opens the potential to produce high-power narrow-band radiation through the complete near-infrared employing the YDF phase control in combination with variety of rare-earth-doped and Raman fiber amplifiers, in particular, based on large-mode-area fibers. Attractive features of these sources are their compactness, reliability, all-fiber integrated format.

5. Conclusions and recent progress

In this Chapter refractive index changes (RICs) induced in Yb-doped optical fibers by resonance pumping is considered as a side effect of the population inversion mechanism that generally manages operation of the fiber lasers and amplifiers. The fundamentals of the RIC phenomena and details of the effect dynamics are explained in terms of the two-level laser model. The effect is experimentally quantified for commercial aluminium silicate fibers demonstrating that the RIC dynamics follows the change of the population of the excited/unexcited ion states with a factor proportional to their polarizability difference (PD). The PD, the only material parameter used by the model, is measured for a number of fiber samples and found to be independent on the fiber geometry, the ion concentration and the tested wavelength in the range 1460 - 1620 nm. Continuation of this work is investigations of fibers with different matrixes of activators. In particular, our recent study of phosphate silicate Yb-doped fibers (Fotiadi et al., 2009a) highlighted their PD value to be not so different from the reported for aluminium silicate fibers, despite the Yb-ion life times in these fibers defer almost in two times. Deep understanding of these electronic phenomena in Yb-doped fibers is very important for numerous fiber applications.

In this Chapter, we propose a simple solution for coherent beam combining of rare-earth-doped fiber amplifiers. The RIC effect is employed for an active phase control in all fiber spliced configuration. A simple algorithm is developed on the based of the electronic RIC model allows straightforward implementation of the effect into a feedback loop. Combining of two 500-mW Er-doped amplifiers in a single-mode fiber is demonstrated with optical control by ~120-mW laser diode. The experimental optical loop is able to operate against the acoustic phase noise within the range of $\pi\ rad$ and with a rate of $\sim 2.6\pi\ rad/ms$ suitable for combining ~50 amplifiers similar to those used in our work. Recently we demonstrated that two wavelengths scheme improves the dynamical range and the response time of the phase control loop (Fotiadi et al., 2009b). Instead using the 980 nm laser diode alone, we proposed to use it in combination with a 1060nm laser. The use of both diodes gives a two times dynamical range or increases in several times the response of the phase control loop.

6. Acknowledgements

This research was supported by the Interuniversity Attraction Pole program VI/10 of the Belgian Science Policy and program "Scientific and Research-Educational Cadres for Innovation Russia" of Russian Federal Agency on Science and Innovation (contract 02.740.11.5093).

7. References

Agrawal, G. P. (1989). *Nonlinear Fiber Optics*, Academic, Boston, Mass..

Antipov, O. L.; Eremeykin, O. N.; Savikin, A. P.; Vorob'ev, V. A.; Bredikhin, D. V. & Kuznetsov, M. S. (2003). Electronic Changes of Refractive Index in Intensively Pumped Nd:YAG Laser Crystals, *IEEE J. Quantum Electron.*, vol. 39, pp. 910-918.

Antipov, O. L.; Bredikhin, D. V. ; Eremeykin, O.N.; Savikin, A. P. ; Ivakin, E.V. & Sukhadolau, A.V. (2006). Electronic mechanism of refractive index changes in intensively pumped Yb:YAG laser crystals, *Opt. Lett.*, vol. 31, pp.763-765.

Arkwright, J. W.; Elango, P.; Atkins, G. R.; Whitbread, T. & Digonnet, M. J. F. (1998). Experimental and Theoretical Analysis of the Resonant Nonlinearity in Ytterbium-Doped Fiber, *J. Lightwave Technol.*, vol. 16, pp. 798-806.

Arkwright, J. W.; Elango, P.; Whitbread, T. W. & Atkins, G. R. (1996). Nonlinear phase changes at 1310 nm and 1545 nm observed far from resonance in diode pumped ytterbium doped fiber. *IEEE Photon. Technol. Lett.*, vol. 8, pp. 408-410.

Augst, S. J.; Fan, T. Y. & Sanchez, A. (2004). Coherent beam combining and phase noise measurements of Yb fiber amplifiers, *Opt. Lett.*, vol. 29, pp. 474-476.

Barmenkov, Yu. O.; Kir'yanov, A. V. & Andres M. V. (2004). Resonant and thermal changes of refractive index in a heavily doped erbium fiber pumped at wavelength 980 nm, *Appl. Phys. Lett.*, vol. 85, pp.2466-2468.

Bellanger, C.; Brignon, A.; Colineau, J. & Huignard, J. P. (2008). Coherent fiber combining by digital holography, *Opt. Lett.*, vol.33, pp.293-295.

Bochove, E. (2004). Nonlinear refractive index of rare-earth-doped fiber laser, *Opt. Lett.*, vol.29, pp.2414-2416.

Born, M. & Wolf, E. (2003). *Principles of Optics.* University Press, Cambridge.

Bruesselbach, H.; Wang, Sh.; Minden, M.; Jones, D.C. & Mangir, M. (2005a). Power-scalable phase-compensating fiber-array transceiver for laser communications through the atmosphere, *J. Opt. Soc. Am. B*, vol. 22, pp. 347-353.

Bruesselbach, H.; Jones, D. C.; Mangir, M. S.; Minden, M. & Rogers, J. L. (2005b). Self-organized coherence in fiber laser arrays, *Opt. Lett.*, vol. 30, pp.1339-1341.

Desurvire, E. (1994). *Erbium-doped fiber amplifiers: Principles and Applications*, Willey, NY.

Digonnet, M. J. F.; Sadowski, R. W.; Shaw, H. J. & Pantell, R. H. (1997). Resonantly Enhanced Nonlinearity in Doped Fibers for Low-Power All-Optical Switching: A Review. *Opt. Fiber Technol.* Vol. 3, pp. 44-64.

Fan, T.Y. (2005). Laser Beam Combining for High-Power, High-Radiance Sources, *IEEE J. Sel. Top. Quantum Electron.*, vol. 11, pp. 567-577.

Feng, Y.; Taylor, L. & Calia, D.B. (2009). 25 W Raman-fiber-amplifier-based 589 nm laser for laser guide star, *Opt. Express*, vol. 17, pp. 19021-19026.

Fotiadi, A. A.; Kuzin, E. A.; Petrov, M. P. & Ganichev. A. A. (1989). Amplitude-frequency characteristic of an optical-fiber stimulated Brillouin amplifier with pronounced pump depletion. *Sov.Tech.Phys.Lett.*, vol.15, pp.434-436.

Fotiadi, A. A. & Kiyan, R. V. (1998). Cooperative stimulated Brillouin and Rayleigh backscattering process in optical fiber. Opt. Lett., vol. 23, pp.1805-1807.

Fotiadi, A. A.; Kiyan, R; Deparis, O.; Mégret, P. & Blondel, M. (2002). Statistical properties of stimulated Brillouin scattering in singlemode optical fibers above threshold, *Opt. Lett.*, vol.27, pp.83-85.

Fotiadi, A. A.; Mégret, P.; Blondel, M. (2004). Dynamics of self-Q-switched fiber laser with Rayleigh - stimulated Brillouin scattering ring mirror, *Opt. Lett.*, vol.29, pp. 1078-1080.

Fotiadi, A. A. & Mégret, P. (2006). "Self-Q-switched Er-Brillouin fiber source with extra-cavity generation of a Raman supercontinuum in a dispersion shifted fiber", *Opt.Lett.*, vol. 31, pp.1621-1623.

Fotiadi, A. A.; Antipov, O. L. & Mégret, P. (2007). Dynamics of pump/signal-induced index change in Yb-doped fiber amplifier, *Conference Digest: CLEO-Europe'2007,* IEEE, CJ3-4-THU.

Fotiadi, A. A.; Zakharov, N. G.; Antipov, O. L. & Mégret, P. (2008a). All-Fiber Coherent Combining of Er-Doped Fiber Amplifiers by Active Resonantly Induced Refractive Index Control in Yb-Doped Fiber, *in Conference on Lasers and Electro-Optics/Quantum Electronics and Laser Science Conference, OSA Technical Digest (CD)* (Optical Society of America, San Jose, CA, USA, 2008), CWB2.

Fotiadi, A. A.; Antipov, O. L. & Mégret, P. (2008b). Pump/Signal Induced Refractive Index Changes in Yb-Doped Fiber Amplifier: The Origin and Properties, *in Conference on Lasers and Electro-Optics/Quantum Electronics and Laser Science Conference and*

Photonic Applications Systems Technologies, OSA Technical Digest (CD) (Optical Society of America, San Jose, CA, USA, 2008), CFR3.

Fotiadi, A. A.; Antipov, O. L. & Mégret, P. (2008c). Dynamics of pump-induced refractive index changes in single-mode Yb-doped optical fibers, *Opt. Express*, vol.16, pp. 12658-12663.

Fotiadi, A. A.; Antipov, O. L.; Bufetov, I. A.; Dianov, E. M. & Mégret, P. (2009a). Comparative Study of Pump-Induced Refractive Index Changes in Aluminum and Phosphate Silicate Yb-Doped Fibers, *in Conference on Lasers and Electro-Optics/Quantum Electronics and Laser Science Conference, OSA Technical Digest (CD)* (Optical Society of America, 2009), JWA9.

Fotiadi, A. A.; Antipov, O. L. & Mégret, P. (2009b). All-fiber Coherent Combining of Er-doped Amplifiers via Refractive Index Control in Yb-doped Fibers by Two-wavelengths Optical Signal, *Conference Digest: CLEO-Europe'2009*, IEEE, CJ3-2-WED.

Fotiadi, A. A.; Zakharov, N. G.; Antipov, O. L. & Mégret, P. (2009c). All-fiber Coherent Combining of Er-doped Amplifiers through Refractive Index Control in Yb-doped Fibers, *Opt.Lett.*, vol. 34, in press.

Garsia, H.; Johnson, A. M.; Oguama, F. A. & Trivedi, S. (2005). Pump-induced nonlinear refractive index change in erbium and ytterbium-doped fibers: theory and experiment, *Opt. Lett.*, vol. 30, pp.1261-1263.

Goodno, G. D.; Book, L. D. & Rothenberg, J. E. (2009) Low-phase-noise, single-frequency, single-mode 608 W thulium fiber amplifier, *Opt. Lett.*, vol. 34, pp.1204-1206.

Huo, Y. & Cheo, P. K. (2005). Analysis of transverse mode competition and selection in multicore fiber lasers, *J. Opt. Soc. Am. B*, vol. 22, pp. 2345-2349.

Jeunhomme, L. (1983). *Single-Mode Fiber Optics*, Marcel Dekker, New. York.

Kuzin, E. A.; Petrov, M. P. & Fotiadi, A. A. (1994). Phase conjugation by SMBS in optical fibers. In: *Optical phase conjugation*, ed. by M.Gower, D.Proch. Springer-Verlag, pp.74-96.

Margerie, J.; Moncorgé, R. & Nagtegaele, P. (2006). Spectroscopic investigation of the refractive index variations in the Nd:YAG laser crystal, *Phys. Rev. B*, vol.74, pp.235108-10.

Messias, D.N.; Catunda, T.; Myers, J.D. & Myers, M.J. (2007). Nonlinear electronic line shape determination in Yb^{3+}-doped phosphate glass, *Opt. Lett.*, vol. 32, pp.665-667.

Ostermeyer, M.; Kong, et al. (2008). Trends in stimulated Brillouin scattering and optical phase conjugation, *Laser and particle beams*, vol. 26, pp. 297-362.

Paschotta R., Nilsson J.; Tropper, A. C.; & Hanna, D. C. (1997). Ytterbium-Doped Fiber Amplifiers, *IEEE J. Sel. Top. Quantum Electron.*, vol. 33, pp. 1049-1056.

Snyder, W. & Love, J. D. (1983). *Optical waveguide theory*, Chapman and Hall, London.

Stepanov, S. I.; Fotiadi, A. A.; Mégret, P. (2007). Effective recording of dynamic phase gratings in Yb-doped fibers with saturable absorption at 1064nm, *Opt. Express*, vol. 15, pp.8832-8837.

Taylor, L.; Feng, Y. & Calia, D. B. (2009). High power narrowband 589nm frequency doubled fibre laser source, *Opt. Express*, vol. 17, pp. 14687-14693.

Wu, B.; Chu, P. L. & Arkwright, J. W. (1995). Ytterbium-doped silica slab waveguide with large nonlinearity, *IEEE Photon. Technol. Lett.*, vol. 7, pp. 1450-1452.

12

Polarization Coupling of Light and Optoelectronics Devices Based on Periodically Poled Lithium Niobate

Xianfeng Chen*, Kun Liu, and Jianhong Shi
Shanghai Jiao Tong University,
China

1. Introduction

Ferroelectric crystals are important materials in many kinds of applications. Since the invention of so called quasi-phase matching technique, periodically domain-inverted ferroelectrics crystals with modulated second order nonlinear susceptibility has received much attention owing to its outstanding nonlinear optical properties(Magel et al.,1990; Xu et al.,1993; Zheng et al.,1998). In periodically domain-inverted ferroelectrics crystals, besides the nonlinear optical coefficients, other third-rank tensors, such as the electro-optic (EO) coefficient, are also modulated periodically because of the periodically reversed ferroelectric domains. Since the theoretical research on EO effect in PPLN was proposed by Y.Q Lu et al(Lu et al.,2000), this periodically modulation of the EO coefficient has given birth to some novel essential applications, especially when the voltage is applied along the transverse direction of the crystal. In this chapter, current research of transverse EO-based PPLN applications, such as tunable wavelength filter, polarization controller, electro-optical switch and laser Q-switch in periodically poled lithium niobate are summarized.

2. Solc-type wavelength filter

Tunable optical filters play an important role in optical signal processing and optical communication. In particular, a narrow flat-top bandpass filter, which not only enables a particular wavelength channel to pass through but also significant to keep signal stable, as well as a multiple-wavelength filter, which enables several wavelengths to pass through together at the same time, are widely needed in wavelength-division multiplexing systems. In this section, we present our research on tunable filter with single-wavelength, multiple-wavelength and flat-top passband based on PPLN crystal. Photovoltaic effect in PPLN is also investigated and based on which a tunable filter by use of UV light is demonstrated. Moreover, the latest researches on solc-type filter based on Ti:PPLN channel waveguide are also presented in the end of this section.

2.1 Principles

When an electric field is applied along the Y axis of a crystal with 3m symmetry, the new index ellipsoid deforms to make its principal axes rotated at an angle θ about the x axis with respect to the unperturbed principal axes. Since θ is proportional to the product of the electric field and the electro-optic coefficient, when an uniformed electric field applying along the Y axis of the crystal, the rotation angle of the original domain is opposite to that of the inverted domain because of the reversal of the spontaneous polarization(Lu et al.,2000). Thus, in a periodically domain inverted crystal with a uniformed electric field applied along the Y axis, a structure with alternating left and right rotation angle θ will be formed due to the periodic EO coefficient. This structure is similar to the well known folded Solc filter structure with alternating azimuth angles of the crystal axes. So we call it Solc layered structure.

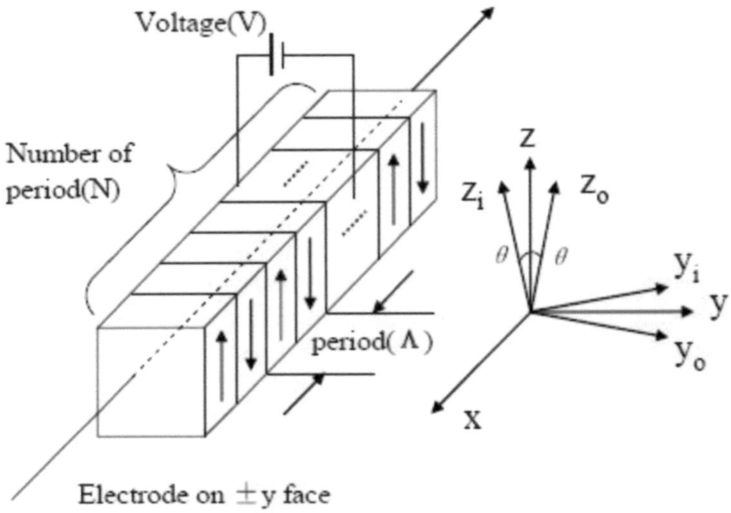

Fig. 1. Schematic diagram of a Solc layered structure based on electro-optic periodically domain inverted crystals. X, Y and Z are the principal axes of the unperturbed index ellipsoid. Λ is the period of this structure and N is the period number. The arrows inside the structure indicate the spontaneous polarization directions. Y_o, Z_o and Y_i, Z_i are the new perturbed principal axes of the original domains and the inverted domains with external electric field, respectively(Lu et al.,2000).

To have an insight into the filter process, polarization coupled-mode theory is established to track the polarization state of light propagation along PPLN. In PPLN the coupled-wave equations of the ordinary and extraordinary waves are:

$$\begin{cases} dA_1/dx = -i\kappa A_2 \exp(i\Delta\beta x) \\ dA_2/dx = -i\kappa^* A_1 \exp(-i\Delta\beta x) \end{cases} \quad (1)$$

with $\Delta\beta=(k_2-k_1)-G_m$, $G_m=2\pi m/\Lambda$ and $\kappa=-\frac{\omega}{2c}\frac{n_o^2 n_e^2 \gamma_{51} E_y}{\sqrt{n_o n_e}}\frac{i(1-\cos m\pi)}{m\pi}$ (m=1, 3, 5 ...), where A_1 and A_2 are the normalized amplitudes of the ordinary wave and the extraordinary wave, respectively, k_1 and k_2 are the corresponding wave vectors, G_m is the mth reciprocal vector corresponding to the periodicity of poling. Consider an input light with Y-axis polarization direction, the boundary condition of the equation is given by

$$\begin{cases} A_1(0)=1 \\ A_2(0)=0 \end{cases} \quad (2)$$

So the solution of coupled-mode equation (1) is:

$$\begin{cases} A_1(x)=\exp[i(\Delta\beta/2)x][\cos sx-i(\Delta\beta/2s)\sin sx] \\ A_2(x)=\exp[-i(\Delta\beta/2)x](-i\kappa^*)\sin sx/s \end{cases} \quad (3)$$

where s is given by $s^2=\kappa\kappa^*+(\Delta\beta/2)^2$.

When $\Delta\beta=(k_2-k_1)-G_m=0$, the reciprocal vector compensates for the wave-vector mismatch. We call this condition a QPM polarization conversion condition, which determines the fundamental wavelength of this kind filter. Under QPM condition, the transmission along the Z-axis is given by

$$T=\left|A_2(x)/A_1(0)\right|^2=\sin^2(Sx) \quad (4)$$

If $|Sx|=|\kappa|L=(2m+1)\pi/2$, (m=0, 1, 2...), then T=100% and the power of the input light is completely transferred from Y polarization to Z polarization. Thereby, with a polarizer along Z-axis at the output, the fundamental wavelength will pass it without any loss, while at the same time wavelengths which don't satisfy this condition will be forbidden by the analyzer. The working mechanism of solc-type filter can also be easily understood as follows: Because each domain serves as a half-wave plate with respect to the fundamental wavelength. After passing through the stack of half-wave plates, the optical plane of polarization of the fundamental wavelength rotates continually and emerges finally at an angle of 2Nθ, where N is the number of plates. Therefore, when $2N\theta=90^\circ$ at the filter output, light of the fundamental wavelength does not experience loss in passing through the crossed analyzer.

2.2 Device of tunable Solc-type wavelength filter

2.2.1 Tunable single-wavelength filter

The first observation of a Solc filter based on PPLN operating in the optical communication wavelength range was proposed by our group in 2003 (Chen et al., 2003). Similar to the traditional folded type Solc filter, a PPLN crystal is placed between two crossed polarizer. This sample is with dimensions of 28 mm×5 mm×0.5 mm consists of four gratings with periods from 20.2 to 20.8μm and a width of 1 mm. A measured transmission power versus wavelength for 20.6μm and 20.8μm period is shown in Fig.3, from which a typical transmission spectrum of traditional Solc-type filter can be seen. The amplitude modulation of the transmission power by applying an electric voltage along the Y axis of the PPLN is also observed and the result is shown in Fig.4. In above discussion, we have known that the fundamental wavelength is determined by equation $\Delta\beta=(k_2-k_1)-G_m=0$, which can be simplified to $\lambda_0=\Lambda(n_o-n_e)$. Because the refractive indices of the ordinary wave and the extraordinary wave are temperature dependent, yet the fundamental wavelength can be shifted to a series of different wavelengths by changing the temperature. The experiment

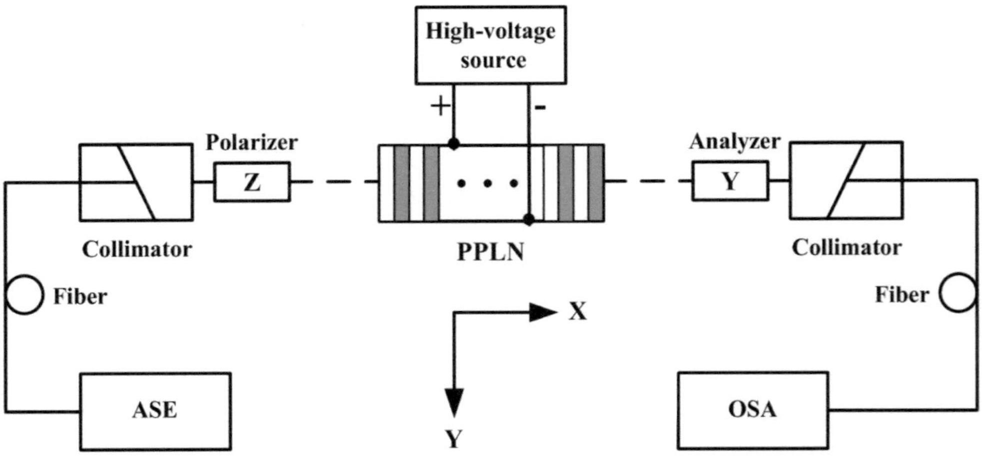

Fig. 2. Experimental setup for a PPLN single-wavelength filter. A PPLN crystal, which is Z cut, is placed between two crossed polarizer, the first of which is along Z direction and the second Y direction. The light propagates along the X direction and a uniform electric field is applied along the Y axis of the PPLN sample. ASE, amplified spontaneous emission; OSA, optical spectrum analyzer(Chen et al.,2003).

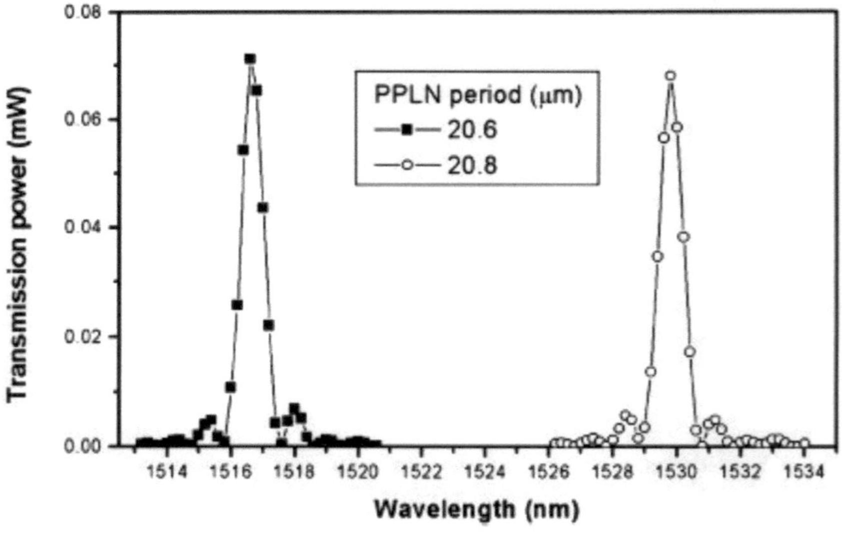

Fig. 3. Measured transmission power versus wavelength of the Solc filter in the 20.6- and 20.8-mm-period PPLN at a temperature of 24 ^{0}C. The FWHM of the transmission spectrum is~0.8 nm(Chen et al.,2003).

result on wavelength modulation is showed in Fig.5, where -0.415nm/^{0}C tuning rate is achieved. Wavelength tuning by temperature is interesting in dense wavelength division multiplexer optical fiber communication systems, where it can be employed as a wavelength-tunable filter for all-optical wavelength routing.

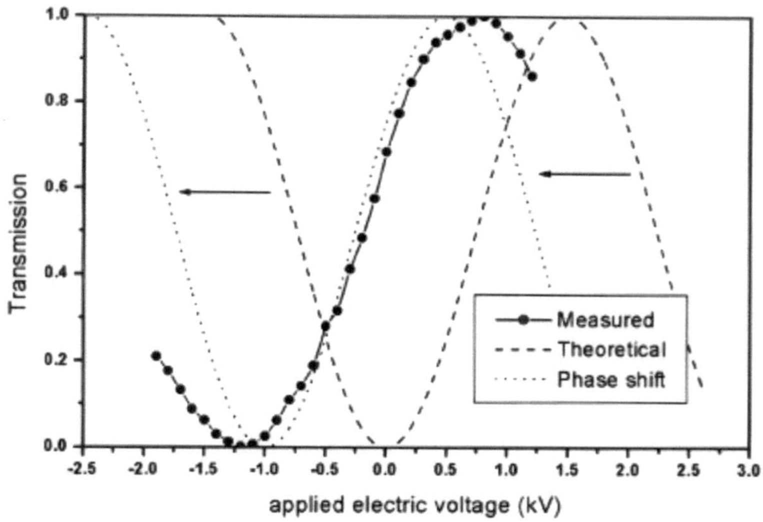

Fig. 4. Experimental measurement of the normalized transmission of the Solc filter (20.8 mm) as a function of the applied voltage at 24 ^{0}C for a given wavelength of 1529.80 nm. The solid curve is the experimental measurement, the dashed curve represents the theoretical values calculated from the Sellmeier equation(Jundt,1997), and the dotted curve is phase shifted from the dashed curve for comparison with the experiment(Chen et al.,2003).

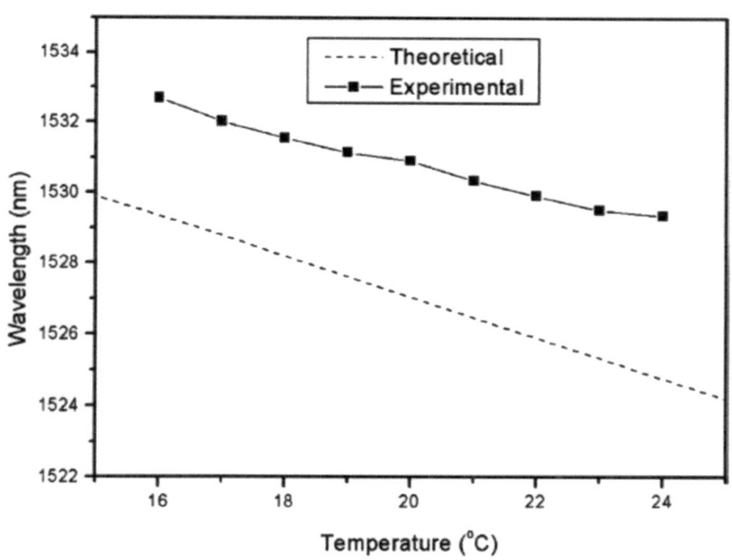

Fig. 5. Experimental measurement of the central wavelength of the Solc f ilter (20.8 mm) as a function of the temperature without the applied external electric voltage. The dashed line represents the theoretical values calculated from the Sellmeier equation(Jundt,1997), and the solid curve is the experimental measurement(Zhu et al.,2003).

2.2.2 Tunable multiple-wavelength filter

In previous section, we have discussed that the central wavelength of a PPLN Solc-type filter is determined by the period of the domain inversion and the ordinary and extraordinary refractive indices. Since the indices are temperature dependent, the central wavelength of such filter can be shifted by changing the working temperature of the whole PPLN. If we apply a temperature distribution along the PPLN through a local-temperature-control technique to control and reshape the second harmonic (SH) curve, then the PPLN will be divided into several temperature sections and each section will give a central wavelength determined by its local temperature. By combining the contributions from all the temperature sections in the PPLN, a multi-wavelength filter can be realized. By properly controlling the temperature of each section, we can construct a tunable multi-wavelength filter for any wavelength according to the practical needs of different applications(Wang et al.,2007).

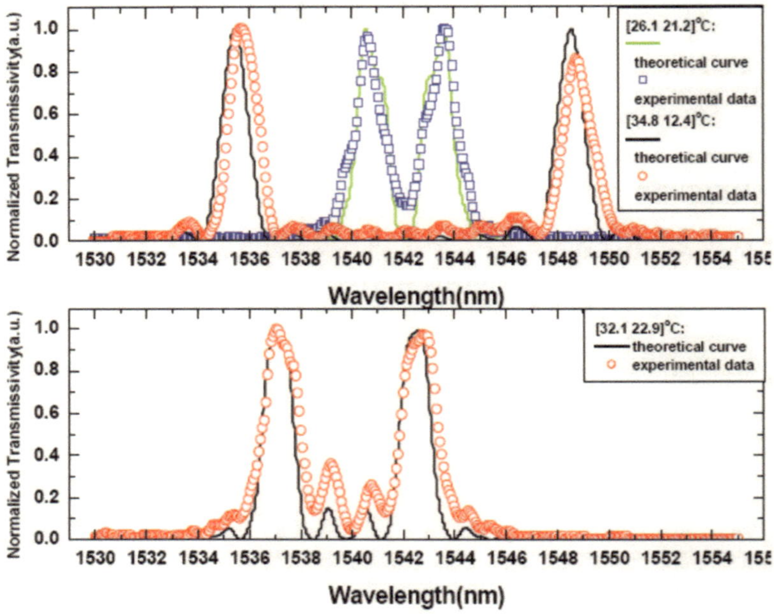

Fig. 6. Tunable double-wavelength filter realized in PPLN by applying a two-section pattern temperature distribution along the sample. In the first figure, under a [26.1 21.2] ^{0}C pattern temperature distribution, the double peaks were at 1540.616nm and 1543.63nm, and under a [34.8 12.4] ^{0}C pattern temperature distribution, the double peaks were tuned to 1538.562nm and 1545.6845nm, for the symmetrical case. In the second figure, under a [32.1 22.9] ^{0}C pattern temperature distribution, the double peaks were set to 1537.078nm and 1542.763nm, for the arbitrary case(Wang et al.,2007).

The multiple-wavelength filter can also been designed by optimizing the sequences of the opposite domains in aperiodic poled lithium niobate (APLN) using simulated annealing (SA) algorithm(Gu et al.,2004). Here, Jones matrix, a 2×2-matrix method, is employed to track the polarization state of light propagation along APLN (Yariv & Yeh,1984). The sequence of the positive and negative domains is optimized with the target that the transmission of the prescribed wavelengths are equal with maximum values at the same time. Here, we chose the objective function in the SA method as

$$F=\sum_{\alpha} T(\lambda_{\alpha})-\{\max[T(\lambda_{\alpha})]-\min[T(\lambda_{\alpha})]\} \quad (5)$$

where the symbol max[...] (min[...]) manifests to take their maximum (minimum) value along all the quantities including into [...]. λ_{α} is the transmission wavelength selected as the objective wavelength in the function. In our calculation, the length of each domain is set to be 5 μm. The length L of lithium niobate is 5 cm and the number of blocks is N=10000. The wavelength dependent refractive indices of extraordinary wave and ordinary wave are calculated at 20 ^{0}C from the Sellmeier equation(Jundt,1997). Wavelengths 1548.5, 1550.1, 1551.7, and 1553.3 nm are chosen as the filtering wavelengths of APLN filter. These wavelengths are from ITU standard wavelength table for DWDM optical communication, with frequency spacing of 200 GHz. Fig.7 shows the filtering results of two, three and four 200-GHz-spaced DWDM wavelengths by simulated annealing algorithm.

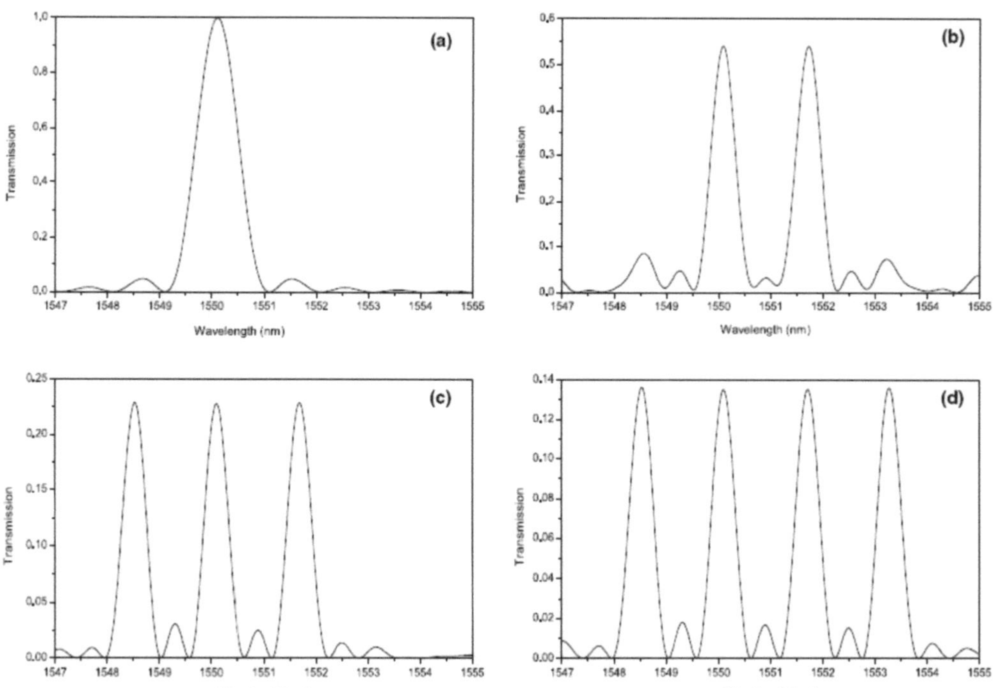

Fig. 7. Calculated transmission spectra of narrowband multiple wavelengths filters at 20 ^{0}C. The length L of lithium niobate is 5 cm. (a) a spectrum of PPLN Solc filter. (b)- (d) transmission spectrums with two, three and four peaks in APLN filters optimized by SA method. The number of domains is N =10000 and the applied electric field in Y -direction is 70 V/mm. The peak wavelengths are 1548.5, 1550.1, 1551.7 and 1553.3 nm, respectively (Gu et al.,2004).

The first experimental observation of narrowband multi-wavelength filters based on aperiodically poled lithium niobate crystals was reported by C. H. Lin et.al(C.H.Lin et.al.,2007). In the experiment, they simultaneously transmitted 8 ITU-standard wavelengths through a 5-cm long EO APLN filter with a transmittance of >90% and a bandwidth of ~0.45 nm for each wavelength channel. Moreover , the transmission spectrum of the EO APLN filter can be conveniently tuned by temperature at a rate of ~0.65 nm/^{0}C in the telecom C band.

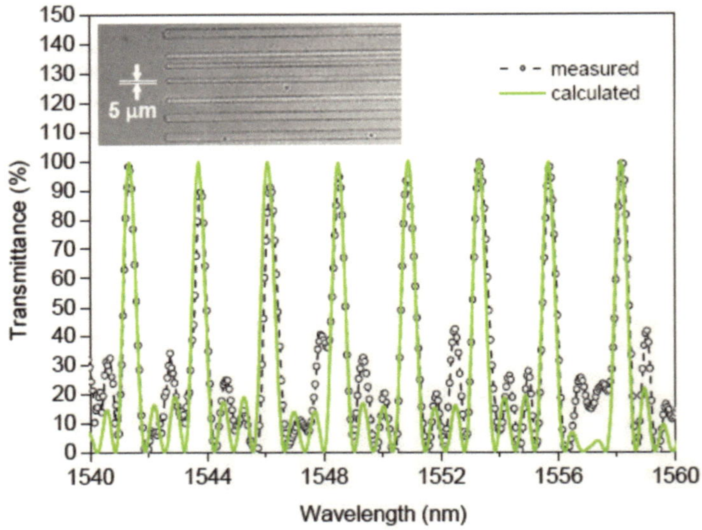

Fig. 8. Calculated (solid line) and measured (open circles) transmission spectra of the 5-cm long EO APLN filter. The two spectral curves agree reasonably well, except that the side lobes in the experimental curve are more apparent. The inset shows a microscopic image of an HF-etched z surface of the fabricated APLN crystal. The label in the inset indicates the width of the unit domain block used in this APLN crystal, i.e., $\Delta x = 5$ µm(C.H.Lin et al.,2007).

2.2.3 Flat-top wavelength filter

Generally, the passband of this kind filter is extremely narrow. Tiny shift of the fundamental wavelength because of unstable temperature or any other factors will contribute to great loss. Thereby, a flat-top filter which can avoid the interference of the outside and maintain signal quality is necessary.

To begin with this section, we would like to introduce the spectrum waveform similarity (SWS) rule discovered in PPLN firstly(Liu et al.,2009a). The SWS rule can be described as follows: if the product of the domain N and the rocking angle θ is fixed, the shape of the spectrum waveform remains similar, as N varies. To get a clear picture of this rule, Fig.9 will be helpful. In figure 9, "A" represents the product of the domain N and the rocking angle θ. With "A" fixed at pi/2, the shape of the spectrum waveform remains similar when N is changing, despite the FWHM correspondingly increases. Based on the spectrum waveform similarity (SWS) rule, we come to a conclusion that the shape of the spectrum waveform is determined by the product A. By observing the spectrums at different value of A, our study reveals a critical point (A=2.24) at which the spectrum surprisingly evolved into flat-top type (Fig.10). The 1 dB width of the flat-top waveform can be fitted as follows:

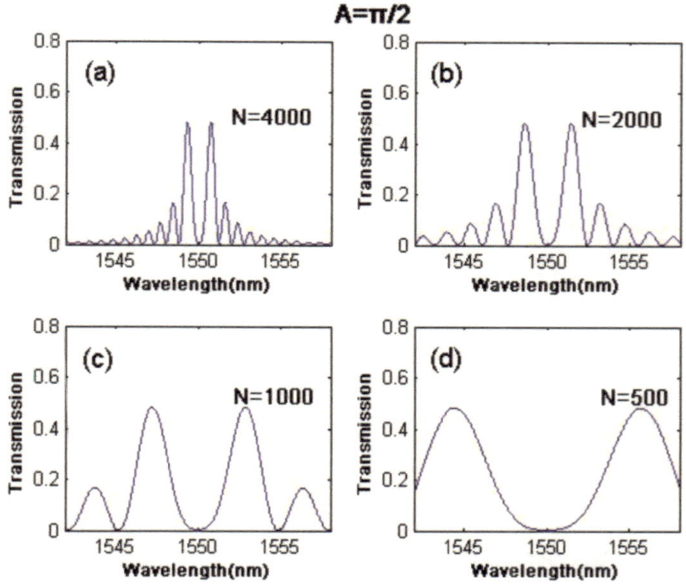

Fig. 9. The transmission spectrums by use of numeral calculation in PPLN with the product A fixed at $\pi/2$. (a) The transmission spectrum with N at 4000; (b) The transmission spectrum with N at 2000; (c) The transmission spectrum with N at 1000; (d) The transmission spectrum with N at 500(Liu et al.,2009a).

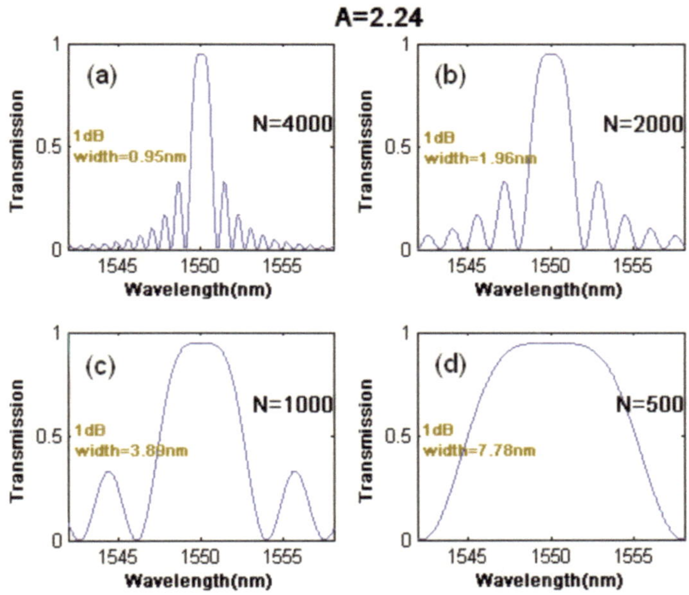

Fig. 10. The transmission spectrums by use of numeral calculation in PPLN with the product A fixed at 2.24. (a) The transmission spectrum with N at 4000; (b) The transmission spectrum with N at 2000; (c) The transmission spectrum with N at 1000; (d) The transmission spectrum with N at 500(Liu et al.,2009a).

$$\Delta\lambda_{1db} \approx 2.5\lambda_0/N \quad (6)$$

where λ_0 is the fundamental wavelength of the filter.

In order to confirm this theoretical discovery about flat-top spectrum in PPLN, further investigation has been performed experimentally(Liu et.al,2009b). The PPLN sample, employed in the experiment, with a dimension of $30(L)\times10(W)\times0.5(T)mm^3$ consists of 2857 domains with the period of 21 (μm). In above discussion, we have known the flat-top spectrum requires $N\theta=2.24$ which is greater than the single wavelength Solc-type filter ($2N\theta=\pi/2$). Thereby, as to a given PPLN (N fixed), higher electric field is requested than usual. During the experiment, the transmission of the fundamental wavelength attends to maximum at 3 (kV/cm) (A) and gradually declines when the electric field keeps on arising subsequently. Consistent with the theoretical result, the spectrum evolves into flat-top type with higher electric field at the critical point 4.2 (kV/cm) (B). The corresponding transmission spectrums at A and B are, respectively, shown in Fig.11. A flat-top waveform with 1 (nm) flat-top width is obtained in the experiment.

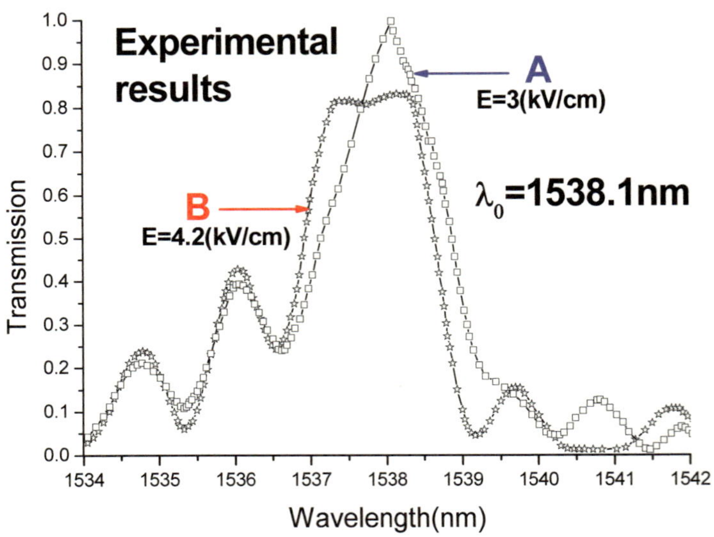

Fig. 11. The transmission spectrums at electric fields of A and B. The curve with square symbol represents the spectrum with 3 (kV/cm) electric field (A) and the curve with star symbol represents the spectrum with 4.2 (kV/cm) electric field (B) (Liu et al.,2009b).

Interestingly, as to a given PPLN, by extending the external electric field to a higher value, the theoretical results reveal that new critical points of flat-top will be found every 2.4 (kV/cm) electric field. Besides, the width of each flat-top transmission spectrum is proportionally increased with the increment of external electric field. Thereby, tunable passband can be achieved by switching the external electric field between different critical points which is shown in Fig.12.

In this section, both the theory and the experiment show that there is a critical voltage at which a flat-top transmission spectrum can be achieved. To have an insight into the flat-top process, more studies should proceed in future. We believe this phenomenon may be also found in other materials and structures similar to PPLN. As the flat-top spectrum is able to keep signal stable, this finding may improve the possibility of the practical application of Solc-type PPLN filter in optical networks and optical signal processing in future.

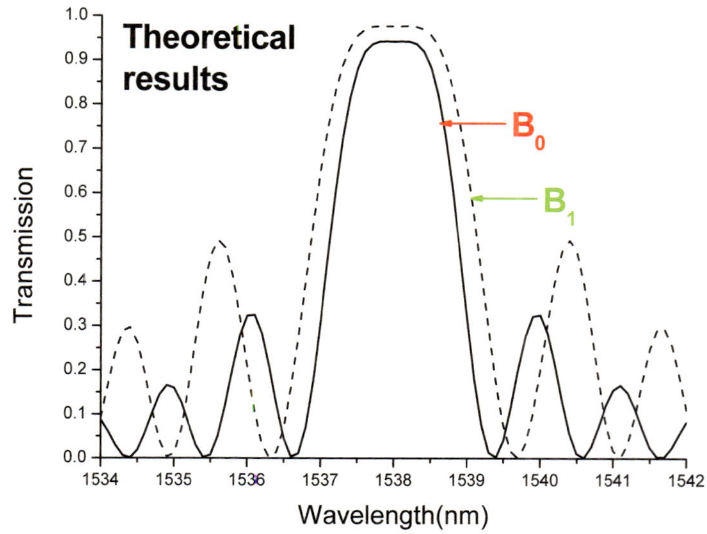

Fig. 12. The theoretical transmission spectrums at critical points B_0 and B_1. The solid curve represents the theoretical spectrum at B_0; The dash curve represents the theoretical spectrum at B_1(Liu et al.,2009b).

2.2.4 Photovoltaic effect in PPLN crystal

In the experiments of above sections, we also found that, without field applied, the device also has the function of Solc filter, which shows disagreement with the theory. Our study reveals that the photovoltaic effect (PVE) plays an essential role in the performance of non-field applied PPLN Solc filter(Chen et al.,2006). The PVE effect sometimes can not be neglected, especially in electric-optic devices. Our theoretical analysis shows that only the

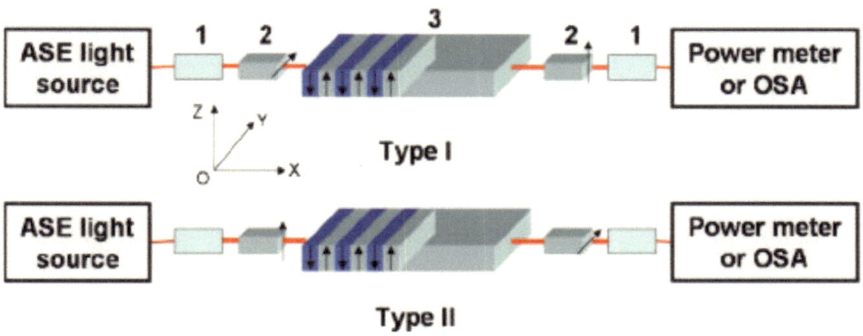

Fig. 13. Two types of device setups for PPLN Solc-type wavelength filter; the arrow on the polarizer indicates the polarization direction. The fact that type A takes on the filter behavior while type B does not reveals the test light has the PVE on the PPLN and it becomes a Solc filter when no external field exists. The numbered components are: 1, collimator; 2, crossed polarizer; 3, PPLN(Chen et al.,2006).

electric field E in the Y direction can enable the Solc filter. If the input polarizer is put in the Y direction, there will be a field formed in the Y direction that will make the device work as a Solc filter; Whereas, if the input polarizer is set in the Z direction, no field will cause the half-wave plate rocking and the PPLN will have no filter function. By observing the spectrums with two types of setups shown in Fig.13, the theoretical analysis is experimentally confirmed in Fig.14.

On the other hand, this PVE effect inspires us a new thought to build the light controlled optical devices or to measure the PVE coefficients (Shi et al., 2006). Fig.15 shows measured

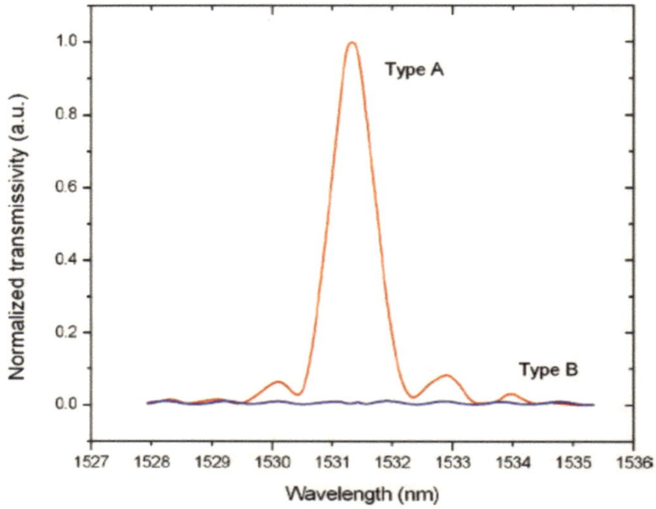

Fig. 14. The transmission curve of the type A setup of device is in accord with the Solc filter function while the type B setup almost has no pass peak(Chen et al.,2006).

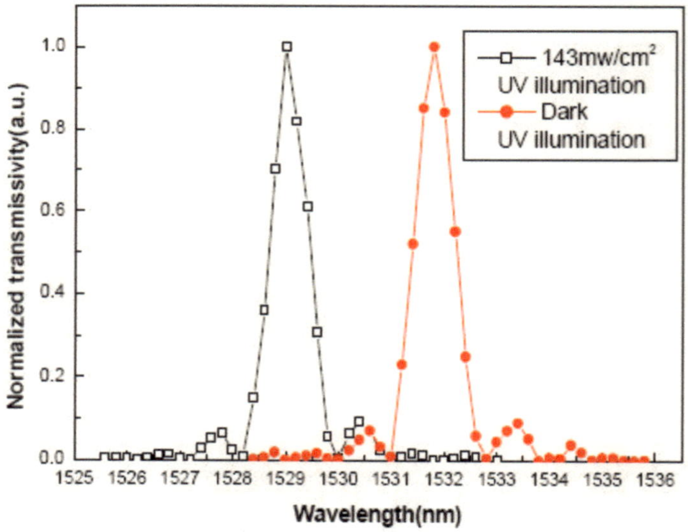

Fig. 15. The spectrum of the PPLN filter under 143mW/cm^2 UV illumination (empty square symbol) and without UV light illumination (red circle symbol) (Shi et al.,2006).

output power against wavelength for the PPLN of period 20.8μm with and without the illumination of the UV light, respectively. The UV light intensity is about 143mw/cm^2. The passband of the filter moves from 1531.9nm to 1529.1nm.4). It costs several seconds until spectrum of the filter becomes stable. However, the peak value and the shape of the pass band are unchanged. It means the Y field of the PVE is not changed. This suggests that no Y direction field is induced when a non-polarized light propagates in Z direction.

2.2.5 Solc-type filter in Ti:PPLN channel waveguide

In this section, we will present the latest research on solc-type filter based on Ti:PPLN channel waveguide which has a significant advantage of low driven voltage. The filtering mechanism of this kind filter is same to the filter based on PPLN crystal introduced above.
In this kind filter, the gap between the electrodes can be as short as (μm) level, therefore, only several voltages are enough for filtering or switching a light(Lee et al.,2007). The experimental setup to perform the wavelength filtering based on the Ti:PPLN channel waveguide is shown in Fig.16. An optical signal from a wavelength-swept fiber laser based on a semiconductor optical amplifier and a Fabry–Perot tunable filter was collimated and end-fire coupled into the Ti:PPLN waveguide, which is placed between two crossed polarizer by a 10x objective lens. The polarization direction of input beam was adjusted parallel to the Y axis of the Ti:PPLN device, and the output signal was observed by an optical spectrum analyzer. The filtering results of one-mode guiding condition and two-mode guiding condition are shown in Fig. 17. In one-mode condition, the measured FWHM of the filter was about 0.21 nm, and in two-mode condition, the origin of one peak is the TEM00-like mode, and the other is the TEM01-like mode. The wavelength difference of the two peaks is about 0.8 nm.

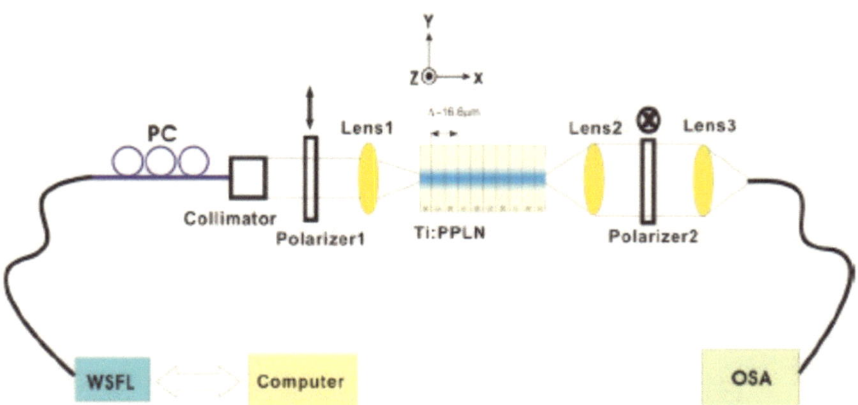

Fig. 16. (Color online) Experimental setup for the waveguide-type solc-type filter based on the Ti:PPLN channel waveguide. WSFL, wavelength-swept fiber laser; OSA, optical spectrum analyzer; PC, polarization controller(Lee et al.,2007).

Similar to the light controlled optical solc-type filter by use of PVE effect in PPLN crystal, Y.L. Lee et.al demonstrate a tunable all-optical solc-type wavelength filter based on Ti:PPLN channel waveguide(Lee et al.,2008). Fig.18 shows the centre wavelength of the filter as a function of the UV illumination intensity. As the intensity increases, the centre wavelength of the filter shifts to a shorter wavelength. The measured wavelength tuning rate of the filter

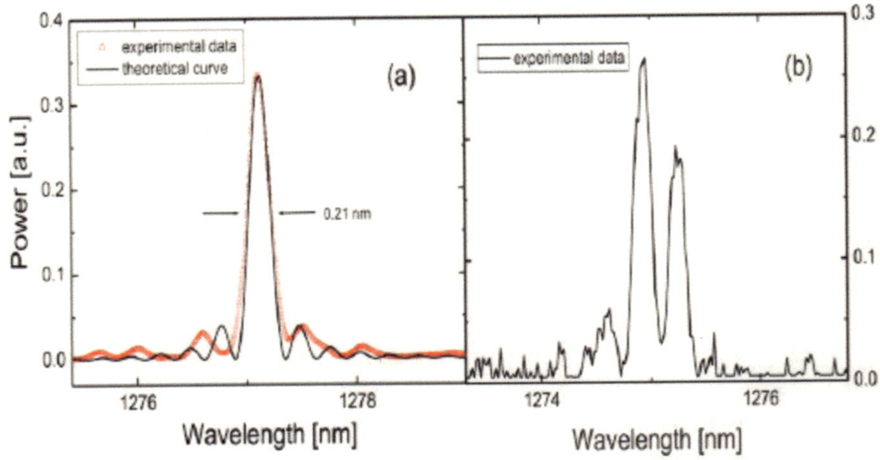

Fig. 17. (Color online) Optical spectra of the Šolc filter based on the Ti:PPLN waveguide depending on guiding modes at room temperature: (a) one-mode guiding condition, (b) two-mode guiding condition(Lee et al.,2007).

was about -26.42 nm/W cm^{-2} which shows more slant than that of a bulk PPLN Solc-type filter (-19.23 nm/W cm^{-2}) (Shi et al.,2006). The right axis of Fig.18 indicates the amount of change in the refractive index difference between n_o and n_e as a function of UV intensity. These results indicate that a Ti:PPLN solc-type filter gives a wider wavelength tuning range than that of a bulk PPLN solc-type filter.

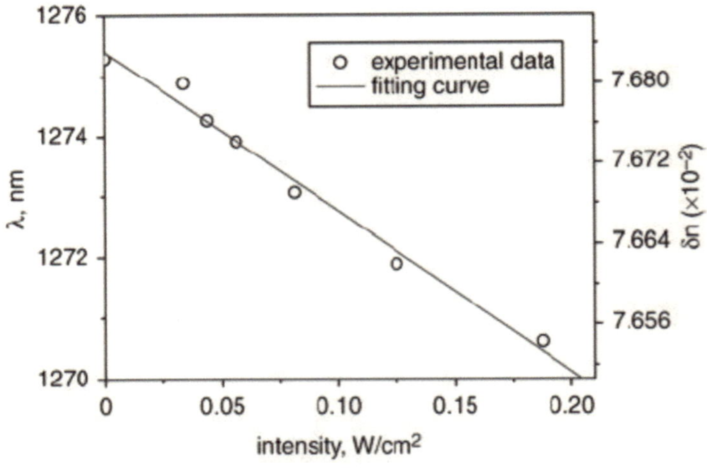

Fig. 18. Centre wavelength of Ti:PPLN solc-type filter aganst UV illumination intensity(Lee et al., 2008)

3. Polarization controller

In this section, we will introduce our study on precise linear polarization-state controller based on PPLN(Liu et al.,2009c), which was also inspired by Y.Q. Lu et.al.'s study(Lu et al.,2000), where they theoretically proposed the electro-optical effect in PPLN. The working mechanism is the same with the Solc-type filter. It is known that with respect to the fundamental wavelength of solc-type filter, each domain serves as a half-wave plate. Therefore, after passing through the stack of half-wave plates, the optical plane of polarization of the fundamental wavelength rotates continually and emerges finally at an angle of $2N\theta$. As θ is controlled with external electric field, the final rotation angle is as well dominated by the electric field. Consequently, an EO linear polarization-state controller for the fundamental wavelength can be realized. In the following discussion, we use another name of "operating wavelength" to represent the fundamental wavelength which is more appropriate than the latter.

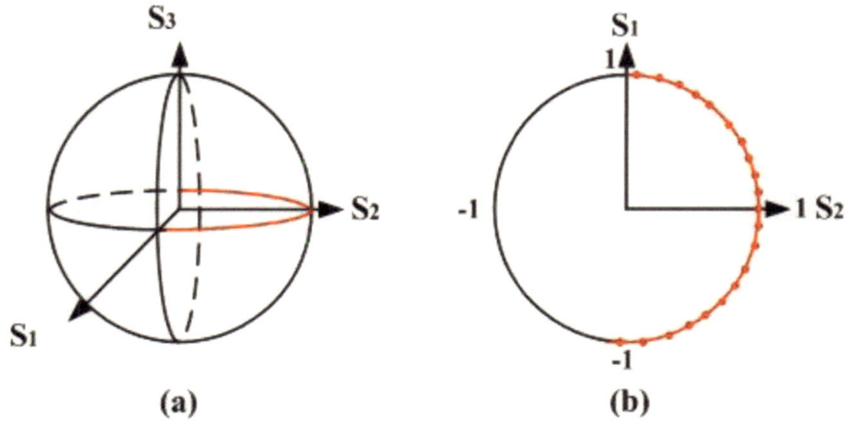

Fig. 19. Experimental measurement of the state of polarization of the output light (λ=1543.47nm) on a Poincare sphere when varying the applied electric field from 0 to 3 (kV/cm) (Liu et al.,2009c).

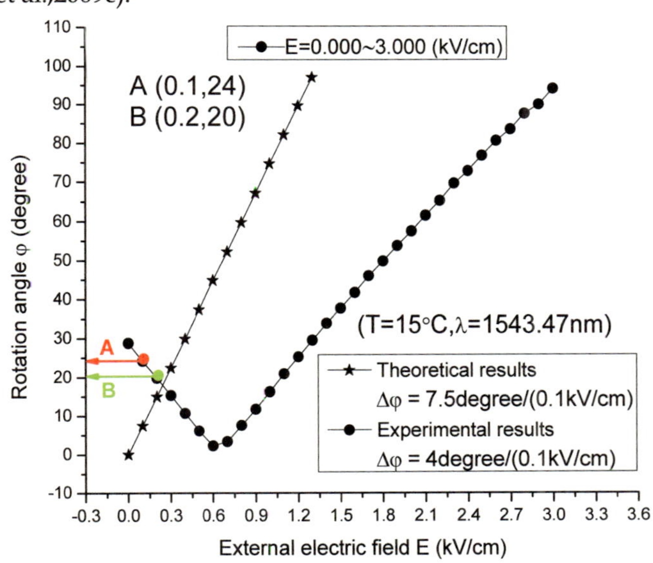

Fig. 20. Experimental measurement of the rotation angle of the output light (λ=1543.47nm) when varying the applied electric field from 0 to 3 (kV/cm) (Liu et al.,2009c).

As for the PPLN sample employed in the experiment, the operating wavelength is found out to be λ=1543.47nm. Then, experimental measurement of the state of polarization of the output light (λ=1543.47nm) is shown on a Poincare sphere in Fig.19 when varying the applied electric field. The experiment shows the state of polarization remains on the equatorial plane when changing the electric field which indicates the output light remains linear state of polarization. The corresponding rotation angle at each electric field is shown in Fig.20, where the electric field intensity is tuned from 0 to 3 (kV/cm), with the step of 0.1 (kV/cm) each time and the rotation angle linearly varies between 0-100 degrees.

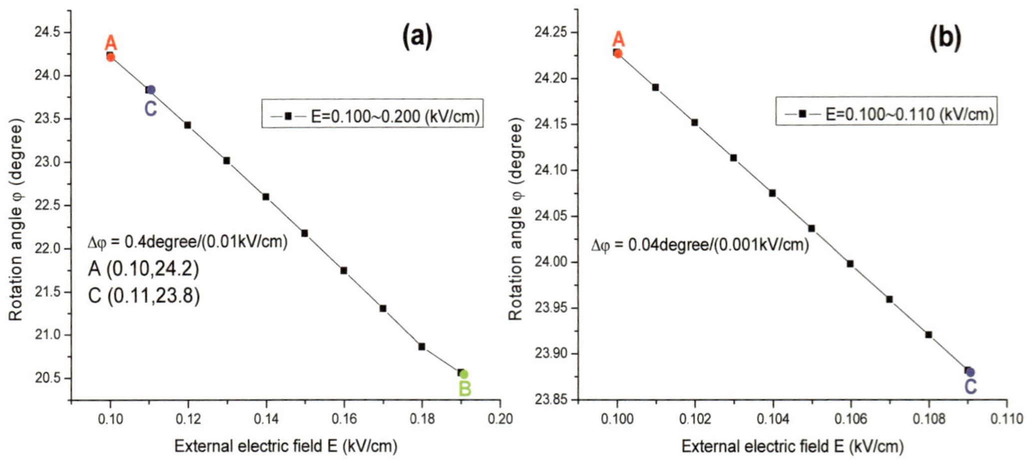

Fig. 21. Experimental measurement of the rotation angle of the output light (λ=1543.47nm) when varying the applied electric field with smaller step. (a) The external electric field varies from 0.100 to 0.200 (kV/cm); (b) The external electric field varies from 0.100 to 0.110 (kV/cm) (Liu et al.,2009c).

To obtain higher precision, we reduce the electric field step to 0.01 (kV/cm) and 0.001 (kV/cm), respectively. Higher precision of 0.4 degree and 0.04 degree is finally achieved, limited by the accuracy of the measurement system, which are shown in Fig.21. Moreover, the operating wavelength can be tuned to a series of different wavelengths by changing the temperature.

The device operating as a precise linear polarization-state controller may find many applications where high precision control of linear polarization-state light is requested ranging from polarization analysis to monitoring of network performance, material birefringence, and measurement of polarization mode dispersion, swept-wavelength measurement, medical imaging and fiber sensor systems.

4. Electro-optical switch

High speed optical switch is one of the key components in optical communication. In this section, with the same experimental setup of solc-type filter, we change to with focus on the application of a precise EO switch. The first observation of an EO switch in PPLN is shown in Fig.22, where high extinction ratio with 43 dB is gained in the experiment.

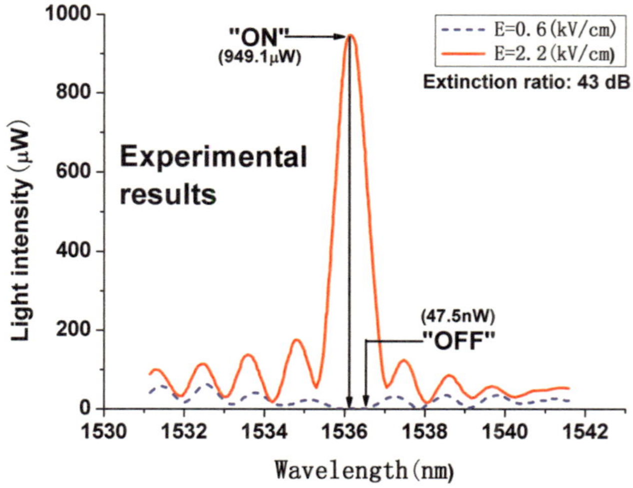

Fig. 22. The experimental transmission spectrums at electric fields of 2.2 (kV/cm) and 0.6 (kV/cm); The sold curve represents the switching state "ON", and the dash curve represents the state "OFF".

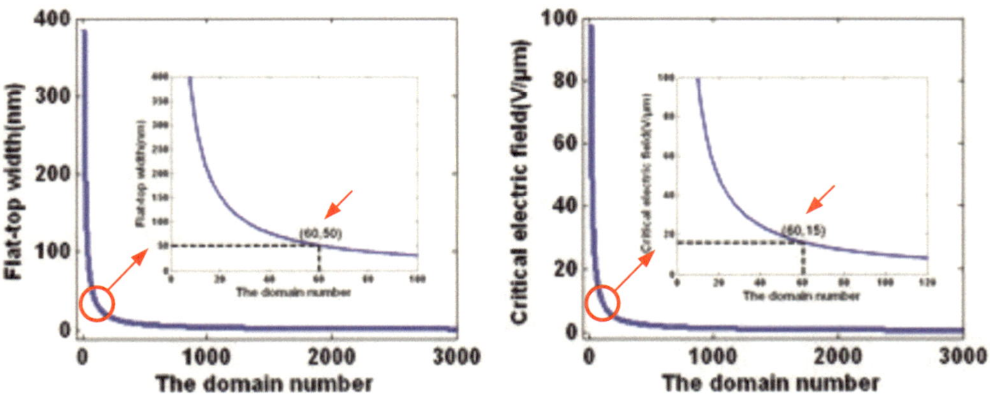

Fig. 23. (a) The flat-top width as a function of the domain number; (b) The critical electric field as a function of the domain number; The transmission spectrum at the domain number of 60 is showed in Fig.24

However, an obstacle to the application of such switch is the narrowband and wavelength limit. Inspired by the discovery of flat-top spectrum in PPLN in 2.2.3 section, we further find out a critical electric field at which the spectrum evolves into broadband. Fig.23 shows the flat-top width and the critical electric field as a function of the domain number. Considering a given PPLN with 60 domains, the transmission spectrum at the critical electric field is shown in Fig.24, which has a broadband with 50 nm flat-top width. The broadband here can efficiently extend the switch's operating wavelengths. However, although sharp increase of the flat-top width can be learned from Fig.23 (a), higher width than 50 nm is impossible in practice, because the corresponding critical electric field will synchronously rapidly rise in Fig.23 (b), which will damage the PPLN.

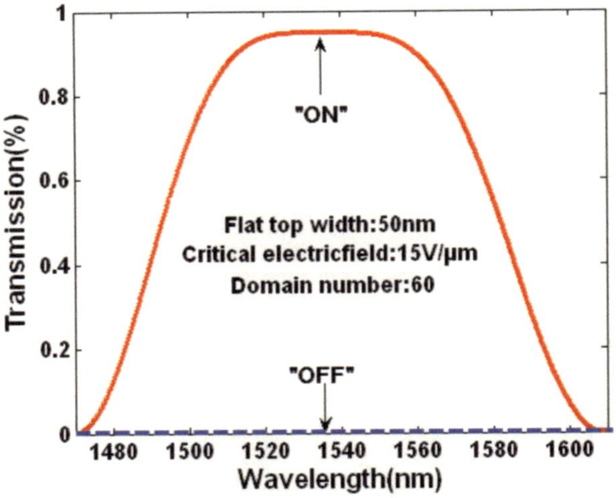

Fig. 24. The transmission spectrum at the critical electric field of 15 (V/μm) when the domain number is 60

5. Laser Q-switch

A class of Q-switched lasers employs a fast intra-cavity polarization rotator for switching the laser polarization between high- and low-loss laser modes. The two loss modes can be established by using a polarization-dependent loss element such as EO PPLN crystal. In this section, we present another significant application on laser Q switch laser based on EO PPLN crystal, which were proposed by Y. H. Chen et.al(Chen & Huang,2003) and Y. Y. Lin et.al(Lin et al.,2007).

In the year of 2003, Y. H. Chen et.al successfully employed a PPLN Pockels cell as a laser Q-switch with a switching voltage as low as ~100 V(Chen & Huang,2003). The PPLN Pockels cell consists of a quarter-wave plate and an EO PPLN crystal having the same crystal orientation as a typical PPLN wavelength converter.

Fig. 25 shows the continuous-wave (CW) performance of the Nd:YVO4 laser with and without a 100-V voltage applied to the EO PPLN crystal. Without the applied voltage, the Nd:YVO4 laser had a CW threshold of 400-mW pump power, attributable to the high output coupling loss (13%). With a 100-V voltage applied to the EO PPLN crystal, the laser's CW threshold was increased to ~1.4-W pump power as expected from the polarization rotation effect.

To show laser Q-switching operation, the EO PPLN is drove by using a 7-kHz, 100-V voltage pulser with a 300-ns pulse width. At 1.2-W pump power, 60-mW average power or 0.74-kW peak power with 8.6-mJ pulse energy in an 11.6-ns laser pulse width is measured in the experiment (Fig .26).

Despite the low switching voltage and excellent integrated QPM performance, the PPLN Pockels cell was sensitive to temperature and generated appreciable green-laser power at 532 nm in the Nd:YVO4 laser, both due to the QPM structure in the PPLN. It is desirable to have a "green-free" and temperature-insensitive laser Q-switch that still retains the advantages of low switching voltage and excellent integrated QPM performance. In the year

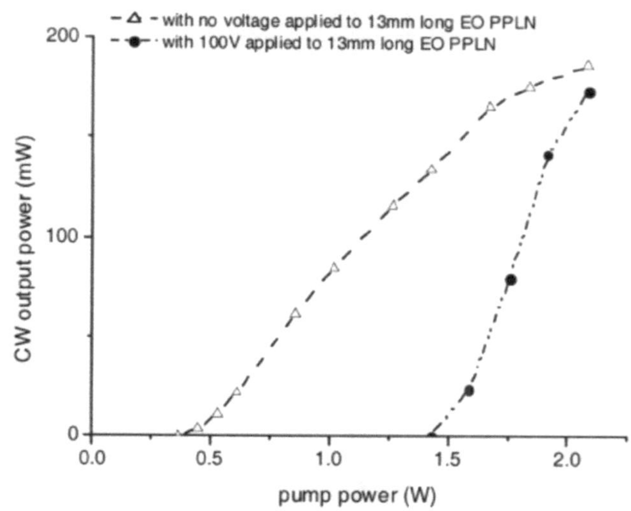

Fig. 25. CW performance of the Nd:YVO4 laser with (solid curve) and without (dashed curve) a 100-V voltage applied to the EO PPLN. With the 100-V voltage, the laser threshold was increased from 400 mW to 1.4 W because of the polarization-rotation effect from the EO PPLN crystal(Chen & Huang,2003).

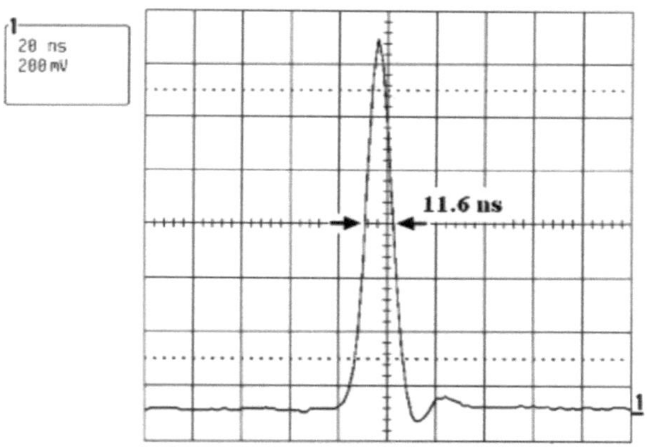

Fig. 26. Measured Q-switched pulse from the EO PPLN Q-switched Nd:YVO4 laser at 1.2-W pump power. The pulse width was 11.6 ns, and the peak power was ~0.74 kW(Chen & Huang,2003).

of 2007, Y. Y. Lin et.al reported such a laser Q-switch which had overcome the previous disadvantages using a novel EO PPLN Bragg modulator, producing 7.8 ns, 201 μJ pulses at a 10 kHz repetition rate when pumped by a 19.35 W diode laser at 808 nm(Y.Y.Lin et al.,2007). The schematic setups of two different kind laser Q-switches are shown in Fig.27.
Figure 28 shows the measured Q-switched pulse energy versus the diode pump power of the second kind laser Q-switch. At 19.35 W pump power, the Q-switched output pulse at 1064 nm has 201μJ energy and 7.8 ns width, corresponding to a peak power of 26 kW. The error bar in the plot shows that the pulse-to-pulse energy jitter was less than 5% over the range of our measurement. The inset shows the temporal profile of the Q-switched output

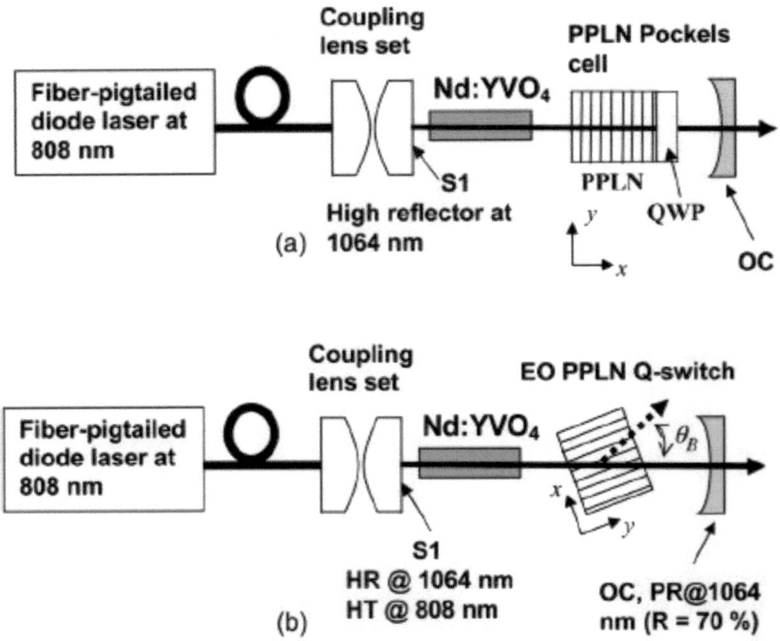

Fig. 27. Schematic of the actively Q-switched Nd:YVO4 laser using (a) a PPLN Pockels cell and (b) an EO PPLN Bragg modulator as a laser Q-switch. The laser cavity is formed by the surface S1 and the output coupler (OC). In (b), the Bragg angle θ_B is exaggerated for clarity. QWP, quarter-wave plate; HR, high reflection; HT, high transmission; PR, partial reflection(Y.Y.Lin et al,2007).

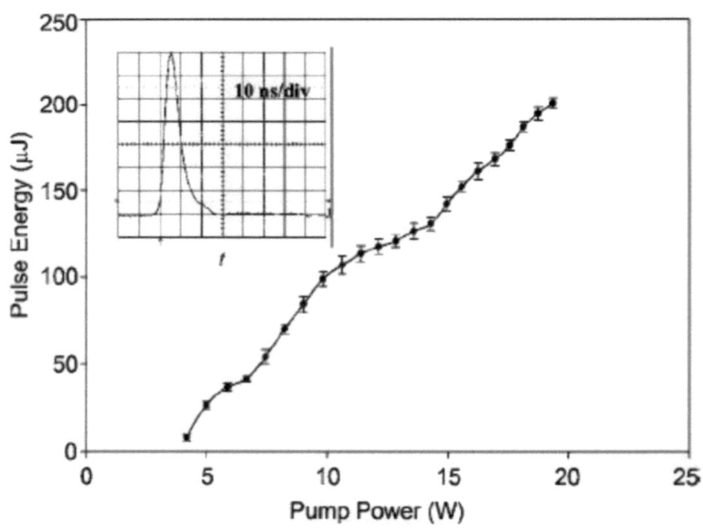

Fig. 28. Output pulse energy of the actively Q-switched Nd:YVO4 laser versus pump power. With 19.35 W pump power at 808 nm, the 1064 nm laser pulse has an energy of 201 ~J and a width of 7.8 ns, corresponding to a laser peak power of 26 kW. The inset shows the measured temporal profile of the Q-switched laser pulse(Y.Y.Lin et al.,2007).

pulse. In the experiment, no noticeable change was observed on the laser performance when the EO PPLN grating was heated from room temperature to 180 °C; In addition, no green-laser power produced from the non-phase-matched second-harmonic generation in the EO PPLN grating was observed in the experiment as well. The experiment successfully confirmed the advantages of this kind Q-switch.

6. Conclusion

In this chapter, we have reviewed essential applications by use of polarization coupling in periodically poled lithium niobate during the past five years. Because of the outstanding properties of PPLN, we believe that PPLN will remain to be an important contemporary realm of research.

7. References

Chen, X.F. ; Shi, J. H.; Chen, Y. P.; Zhu, Y. M.; Xia, Y. X. & Chen, Y. L. (2003). Electro-optic Solc-type wavelength filter in periodically poled lithium niobate, Opt. Lett. 28.

Chen, Y. H. & Huang, Y. C. (2003). Actively Q -switched Nd:YVO4 laser using an electro-optic periodically poled lithium niobate crystal as a laser Q -switch, Opt. Lett. 28, 1460-1462.

Chen, L.J.; Shi, J.H.; Chen, X.F. & xia, Y.X. (2006). Photovoltaic effect in a periodically poled lithium niobate Solc-type wavelength filter, Appl. Phys. Lett. 88, 121118.

Gu, X.; Chen, X.F.; Chen, Y.P.; Zeng, X.L.; Xia, Y.X. & Chen, Y.L. (2004). Narrowband multiple wavelengths filter in aperiodic optical superlattice, Opt. Commun.237, 53–58.

Jundt, D. H. (1997). Temperature-dependent Sellmeier equation for the index of refraction, no, in congruente lithium niobate," Opt. Lett. 22, 1553-1555.

Lu, Y. Q.; Wan, Z. L.; Wang, Q.; Xi, Y. X. & Ming, N. B. (2000). Electro-optic effect of periodically poled optical superlattice $LiNbO_3$ and its applications, Appl. Phys. Lett. 77, 3719.

Lin, C. H.; Chen, Y. H.; Lin, S. W.; Chang, C. L.; Huang, Y. C. & Chang, J. Y. (2007). Electro-optic narrowband multi-wavelength filter in aperiodically poled lithium niobate, Opt. Exp. 15, 9859-986.

Lee, Y. L.; Yu, N. E.; Kee, C. -.; Ko, D. -.; Lee, J.; Yu, B. -.; Shin, W.; Eom, T. J. & Noh, Y. -. (2007). Wavelength filtering characteristics of Solc filter based on Ti:PPLN channel waveguide, Opt. Lett. 32, 2813-2815.

Lee, Y. L.; Yu, N.E.; Kee, C.-S.; Ko, D.-K.; Noh, Y.-C.; Yu, B.-A.; Shin, W.; Eom, T.-J.; Oh, K. & Lee, J. (2008). All-optical wavelength tuning in Solc filter based on Ti:PPLN waveguide, Electron. Lett. 44, 30-32.

Lin, Y.Y.; Lin, S.T.; Chang, G.W.; Chiang, A.C.; Huang, Y.C. & Chen, Y.H. (2007). Electro-optic periodically poled lithium niobate Bragg modulator as a laser Q-switch, Opt. Lett. 32, 545-547.

Liu, K.; Shi, J.H.; Zhou, Z.E. & Chen, X.F. (2009a). Electro-optic Solc-type flat-top bandpass filter based on periodically poled lithium niobate, Opt. Commun.282, 1207–1211.

Liu, K.; Shi, J.H. & Chen, X.F. (2009b). Electro-optical flat-top bandpass Solc-type filter in periodically poled lithium niobate. Opt. Lett. 34, 7.

Liu, K.; Shi, J.H. & Chen, X.F. (2009c). Linear polarization-state generator with high precision in periodically poled lithium niobate, Appl. Phys. Lett., 94, 101106.

Magel, G. A.; Fejer, M. M. & Byer, R. L. (1990). Quasi-phase-matched second-harmonic generation of blue light in periodically poled $LiNbO_3$, Appl. Phys. Lett. 56, 108-110.

Shi, J. H.; Wang, J. H.; Chen, L. J.; Chen, X. F. & Xia, Y. X. (2006). Tunable Šolc-type filter in periodically poled LiNbO3 by UV-light illumination, Opt. Exp. 14, 6279.

Wang, J. H.; Shi, J. H.; Zhou, Z. E. & Chen, X. F. (2007). Tunable multi-wavelength filter in periodically poled LiNbO3 by a local-temperature-control technique, Opt. Exp. 15,1561.

Xu, C. Q.; Okayama, H. & Kawahara, M. (1993). 1. 5μm band efficient broadband wavelength conversion by difference frequency generation in a periodically domain-inverted $LiNbO_3$ channel waveguide, Appl. Phys. Lett. 63, 3559-3561.

Yariv, A. & Yeh, P. (1984). Optical Waves in Crystal: Propagation and Control of Laser Radiation (Wiley, NewYork).

Zheng, J. J.; Lu, Y. Q.; Luo, G.P. ; Ma, J.; Lu, Y. L.; Ming, N. B.; He, J. L. & Xu, Z. Y. (1998). Visible dual-wavelength light generation in optical superlattice through Er:$LiNbO_3$ unconversion and quasi-phase-matched frequency doubling, Appl. Phys. Lett.72, 1808–1810.

Zhu, Y. M.; Chen, X. F.; Shi, J. H.; Chen, Y. P.; Xia, Y. X. & Chen, Y. L. (2003). Wide-range tunable wavelength filter in periodically poled lithium niobate, Opt. Comm., 228,139-143.

13

All-Optical Wavelength-Selective Switch by Intensity Control in Cascaded Interferometers

Hiroki Kishikawa, Nobuo Goto and Kenta Kimiya
The University of Tokushima
Japan

1. Introduction

Wavelength-division-multiplexing (WDM) technology has been developing in broadband optical networks. The wavelength range used in fiber transmission systems has been increasing by employing various kinds of fiber amplifiers. In order to make it possible to handle signals of multiple wavelengths efficiently in network nodes, wavelength-selective processing in switching, routing and buffering systems will be required. In such processing systems, integrated-optic switches become one of fundamental key devices. In particular, high-speed wavelength-selective switches are required for WDM packet processing.
Wavelength-selective switches are classified from viewpoint of control method into electrically controlled devices and optically controlled ones. The former includes collinear acoustooptic (AO) devices (Goto & Miyazaki, 1990). Although multiple signals at different wavelengths can be independently switched with a single AO switch, it provides a response time only at an order of micro seconds. Optical circuits consisting of arrayed waveguide gratings (AWGs) and wavelength-insensitive switches are also regarded as a device in this class (Goh et al., 2008). The switching response depends on the control mechanizm of the wavelength-insensitive switches. The response time in thermo-optic switches is an order of milli seconds (Suzuki et al., 1998). On the other hand, optically controlled switches include devices using phase shift induced by optical nonlinear Kerr effect (Doran & Wood, 1988) and phase shift in semiconductor optical amplifiers (SOAs) (Nakamura et al., 2000) and in quantum dots (Kitagawa et al., 2009). These switches using phase shifting, in general, cannot provide wavelength-selectivity. Therefore, to realize wavelength-selective switching of the latter class, combination with multiplexers and demultiplexers such as AWGs is required.
As a switch of the latter class, the authors have proposed a new type of wavelength-selective switch where optical amplitudes in waveguide arms of interferometers are controlled by Raman amplifiers (Kishikawa & Goto, 2005; Kishikawa & Goto, 2006; Kishikawa & Goto, 2007a; Kishikawa & Goto, 2007b). In this article, the operation principle of wavelength-selective switching with the proposed devices is described. Computer simulation by finite-difference beam-propagation-method (FD-BPM) is performed to verify the switching operation.

2. Wavelength-selective switching by light control

Optical switching circuits discussed in this article consist of directional couplers, waveguide-type Raman amplifiers and attenuators as shown in Fig. 1. Designing of the switching circuits indicated by the dotted rectangular is theoretically discussed in the next section. In the switching circuits, the signal light intensity is controlled with waveguide-type Raman amplifiers. We consider that WDM optical signals at wavelengths λ_1, λ_2 and λ_3 are incident into the upper waveguide of the circuit, and the two signals at λ_1 and λ_3 are to be switched to OUT B, whereas the signal at λ_2 is to be forwarded to OUT A. To induce this routing operation, optical pumping signals at λ_1' and λ_3' are injected to Raman amplifiers, where λ_i', i=1,2,3, is shorter than λ_i by an amount of a Raman shift λ_R, that is, $\lambda_i' = \lambda_i - \lambda_R$. The Raman shift depends on the material of the amplifier, whose typical value is approximately 100 nm. By injecting these pumping signals, the Raman amplifiers amplify wavelength-selectively the signals at λ_1 and λ_3. In some structures of the circuits, control signal λ_2' may be injected to another Raman amplifier placed in the circuits. We will find the structure of switching circuits with the required conditions for the amount of amplification. Since the Raman amplifiers are controlled by pumping light, all-optical switching can be realized.

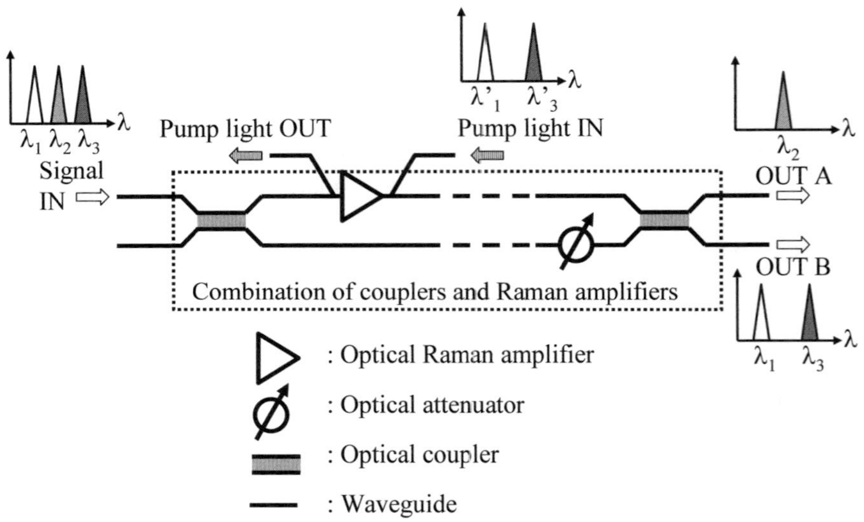

Fig. 1. Wavelength-selective switching circuit using a combination of optical couplers, optical Raman amplifiers and optical attenuators.

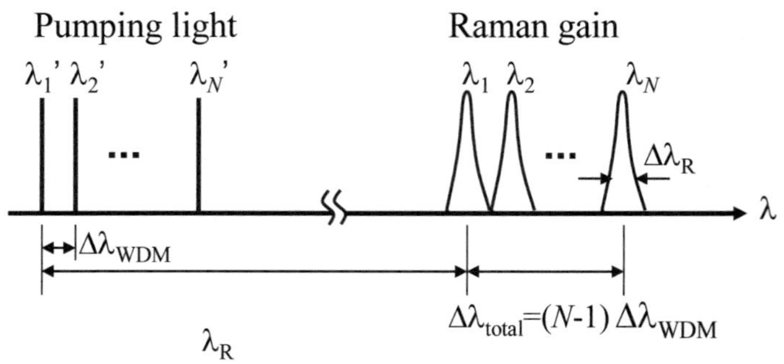

Fig. 2. Wavelength-selective amplification by stimulated Raman scattering with waveguide-type Raman amplifiers.

The wavelength-selective amplification can be performed in wide wavelength range by employing stimulated Raman scattering as shown in Fig. 2. The Raman gain bandwidth Δf_R in frequency for a specific pumping wavelength is tens GHz, for example, Δf_R =20 GHz with pulse-pumped maximum gain of 20 dB and continuous wave (cw)-pumped gain of 3.7 dB at λ =960 nm in GaP waveguides (Suto et al., 2002). In a Si waveguide, Δf_R =75 GHz with cw-pumped gain of 2.3 dB at wavelength 1574 nm (Rong et al., 2004), and pulse-pumped gain of 20 dB at wavelength 1673 nm (Raghunathan et al., 2005) were reported.
We denote the wavelength range available for this device as $\Delta\lambda_{total}$. Then, the WDM number handled with this switch is expressed as $N_{WDM} = \Delta\lambda_{total} / \Delta\lambda_R = c\,\Delta\lambda_{total} / \Delta f_R \lambda^2$, where c is the light velocity and $\Delta\lambda_R$ is the Raman gain bandwidth in wavelength. If $\Delta\lambda_{total}$ is as wide as 40 nm with the center wavelength of λ=1550 nm, N_{WDM} can be as large as 65 with Si Raman amplifiers.

2. Theoretical analysis of switching

2.1 Key elements

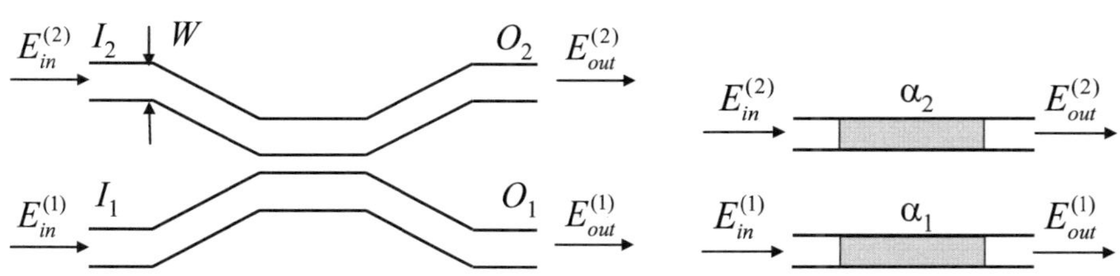

(a) directional coupler (b) parallel waveguides with amplifiers

Fig. 3. Key elements consisting the proposed wavelength-selective switches.

This section describes theoretical analysis with ideal device models. Key elements consisting the proposed switch are a 3-dB directional coupler and parallel independent waveguides containing amplifiers or attenuators as shown in Fig. 3. In the 3-dB directional coupler shown in (a), the electric field amplitude $E_{out}^{(i)}$, i=1, 2, of the optical output signals are related with $E_{in}^{(i)}$ of the optical input signals from the coupled mode theory, by eliminating the common phase term along the propagation, as

$$\begin{pmatrix} E_{out}^{(1)} \\ E_{out}^{(2)} \end{pmatrix} = \begin{pmatrix} \cos(\pi\kappa/4\kappa_0) & -j\sin(\pi\kappa/4\kappa_0) \\ -j\sin(\pi\kappa/4\kappa_0) & \cos(\pi\kappa/4\kappa_0) \end{pmatrix} \begin{pmatrix} E_{in}^{(1)} \\ E_{in}^{(2)} \end{pmatrix}, \tag{1}$$

where $j=\sqrt{-1}$, κ and κ_0 are the coupling coefficient and the coefficient at a designed wavelength for 3-dB coupling. The value κ depends on the field overlapping of optical individual waveguides. The wavelength dependence of κ therefore depends on the waveguide structure. This wavelength dependence of the 3-dB directional coupler causes the wavelength dependency in the proposed switch. At the designed wavelength for the 3-dB coupling, eq.(1) is simplified by using $\kappa = \kappa_0$, as

$$\begin{pmatrix} E_{out}^{(1)} \\ E_{out}^{(2)} \end{pmatrix} = \frac{1}{\sqrt{2}} \begin{pmatrix} 1 & -j \\ -j & 1 \end{pmatrix} \begin{pmatrix} E_{in}^{(1)} \\ E_{in}^{(2)} \end{pmatrix}. \tag{2}$$

Optical waves propagated along the independent parallel waveguides with amplification coefficient α_1 and α_2 are expressed as

$$\begin{pmatrix} E_{out}^{(1)} \\ E_{out}^{(2)} \end{pmatrix} = \begin{pmatrix} \alpha_1 & 0 \\ 0 & \alpha_2 \end{pmatrix} \begin{pmatrix} E_{in}^{(1)} \\ E_{in}^{(2)} \end{pmatrix}. \tag{3}$$

2.2 Switch A

We have proposed two switch architectures. Figure 4 shows the device structure of one of the switch architectures, named as switch A. The switch consists of two cascaded interferometers. In the first interferometer, each arm has a Raman amplifier with amplification coefficient α_A and α_B for the lower and upper arms, respectively. An attenuator with attenuation coefficient β is placed in the upper arm of the second interferometer. The Raman amplifiers are pumped by backward traveling light coupled with polarization-selective or wavelength-selective couplers.

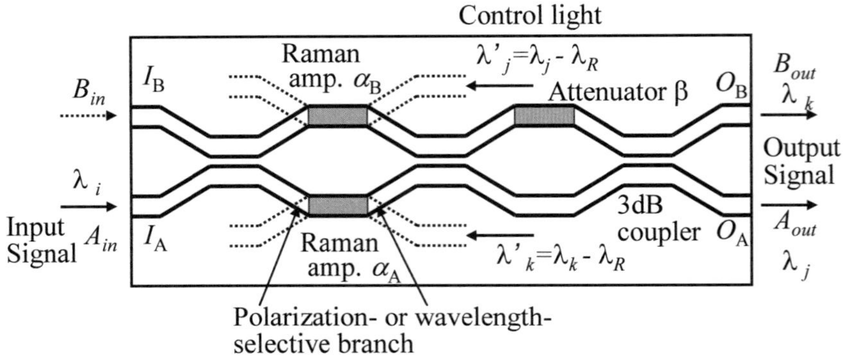

Fig. 4. Schematic diagram of the proposed wavelength-selective switch A consisting of two cascaded interferometers.

Optical WDM signals at wavelength λ_i, i=1, ..., N, are incident at input port I_A. When optical WDM signals at λ_k, $k \in \{1,...,N\}$, are to be switched to output port O_B, pumping WDM lights at λ_k' are coupled to the lower Raman amplifier to amplify the signals with gain coefficient α_A. The wavelength λ_k' satisfies the condition $\lambda_k' = \lambda_k - \lambda_R$. On the contrary, when the WDM signals at λ_j are to be switched to output port O_A, pumping WDM lights at λ_j' are coupled to the upper Raman amplifier to amplify the signals with gain coefficient α_B. Since the Raman amplifiers made of crystalline waveguides can wavelength-selectively amplify optical signals with narrow gain bandwidth $\Delta\lambda_R$, multiple WDM signals can be simultaneously and wavelength-selectively switched with this single switch.
Using eqs.(2) and (3), the output fields through the switch are related to the input fields as

$$\begin{pmatrix} A_{out} \\ B_{out} \end{pmatrix} = \left(\frac{1}{\sqrt{2}}\right)^3 \begin{pmatrix} 1 & -j \\ -j & 1 \end{pmatrix} \begin{pmatrix} 1 & 0 \\ 0 & \beta \end{pmatrix} \begin{pmatrix} 1 & -j \\ -j & 1 \end{pmatrix} \begin{pmatrix} \alpha_A & 0 \\ 0 & \alpha_B \end{pmatrix} \begin{pmatrix} 1 & -j \\ -j & 1 \end{pmatrix} \begin{pmatrix} A_{in} \\ B_{in} \end{pmatrix}. \tag{4}$$

Since the incident optical signal is coupled into the lower port I_A, we assume

$$\begin{pmatrix} A_{in} \\ B_{in} \end{pmatrix} = \begin{pmatrix} E_{in} \\ 0 \end{pmatrix}. \tag{5}$$

By substituting eq. (5) into eq. (4), the outputs are obtained as

$$\begin{pmatrix} A_{out} \\ B_{out} \end{pmatrix} = \frac{E_{in}}{2\sqrt{2}} \begin{pmatrix} \alpha_A - \alpha_B - \beta(\alpha_A + \alpha_B) \\ -j[\alpha_A - \alpha_B + \beta(\alpha_A + \alpha_B)] \end{pmatrix}. \tag{6}$$

It is expected from eq. (6) that B_{out}=0 for some values of α_A, α_B and β, on the other hand, A_{out}=0 for some other values of these parameters. Since the number of these unknown parameters is three for two equations, the solutions for these parameters cannot be determined. However, by considering physical limitations in these parameters, we obtain the solutions as

$$\begin{pmatrix} A_{out} \\ B_{out} \end{pmatrix} = \begin{cases} \begin{pmatrix} -E_{in} \\ 0 \end{pmatrix} & when \begin{cases} \alpha_A = 1 \\ \alpha_B = 1+\sqrt{2} \\ \beta = 1/(1+\sqrt{2}) \end{cases} \\ \begin{pmatrix} 0 \\ -E_{in} \end{pmatrix} & when \begin{cases} \alpha_A = 1+\sqrt{2} \\ \alpha_B = 1 \\ \beta = 1/(1+\sqrt{2}) \end{cases} \end{cases}. \tag{7}$$

It is found from eq.(7) that the switch operates as a 1x2 spatial switch whose output port is selected by proper amplification and attenuation effects. It is noted that the value β is constant in these two conditions. The incident light is routed to output port O_A by amplifying the upper amplifier with $\alpha_B = 1+\sqrt{2}$. The amplification is not required in the other amplifier α_A. On the contrary, the light is routed to output port O_B by amplifying the lower amplifier with $\alpha_A = 1+\sqrt{2}$. A drawback of this architecture is that either amplifier has to be activated for an input signal. That is, there is no default single output path without pumping.

2.3 Switch B

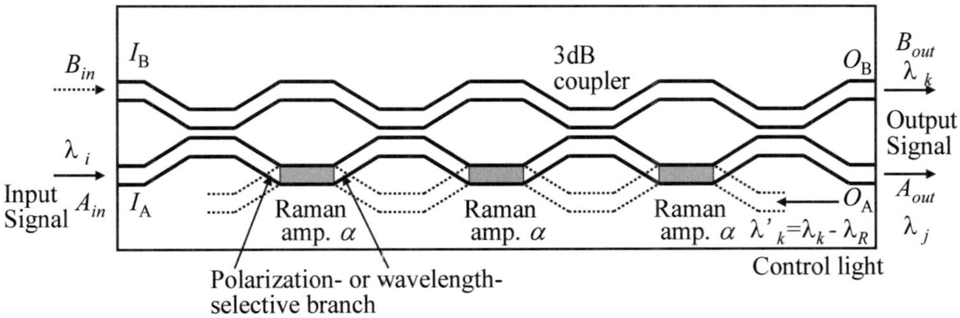

Fig. 5. Schematic diagram of the proposed wavelength-selective switch B consisting of three cascaded interferometers.

An alternative architecture of the switch consists of cascaded three interferometers as shown in Fig. 5. Three Raman amplifiers are placed in the lower arms of the three interferometers.

The feature of this architecture is that optical incident signals are routed to output port O_A without any pumping light for amplification, and are switched to output port O_B only when three amplifiers are activated by pumping. Since the three amplifiers are activated to have the same amplification coefficient α, the control for switching is simpler than that in switch A. The output fields through this switch are related to the input fields as

$$\begin{pmatrix} A_{out} \\ B_{out} \end{pmatrix} = \left(\frac{1}{\sqrt{2}}\right)^4 \begin{pmatrix} 1 & -j \\ -j & 1 \end{pmatrix} \begin{pmatrix} \alpha & 0 \\ 0 & 1 \end{pmatrix} \begin{pmatrix} 1 & -j \\ -j & 1 \end{pmatrix} \begin{pmatrix} \alpha & 0 \\ 0 & 1 \end{pmatrix} \begin{pmatrix} 1 & -j \\ -j & 1 \end{pmatrix} \cdot \begin{pmatrix} \alpha & 0 \\ 0 & 1 \end{pmatrix} \begin{pmatrix} 1 & -j \\ -j & 1 \end{pmatrix} \begin{pmatrix} A_{in} \\ B_{in} \end{pmatrix}. \tag{8}$$

Since the incident optical signal is coupled into the lower port I_A, we assume

$$\begin{pmatrix} A_{in} \\ B_{in} \end{pmatrix} = \begin{pmatrix} E_{in} \\ 0 \end{pmatrix}. \tag{9}$$

By substituting eq. (9) into eq. (8), the outputs are obtained as

$$\begin{pmatrix} A_{out} \\ B_{out} \end{pmatrix} = \frac{E_{in}}{4} \begin{pmatrix} \alpha^3 - 3\alpha^2 - \alpha - 1 \\ -j[\alpha^3 - \alpha^2 - \alpha + 1] \end{pmatrix}. \tag{10}$$

For no amplification at each Raman amplifier, that is α=1, output electric fields are calculated to be A_{out}= - E_{in} and B_{out}=0. It means that the incident optical wave is routed to output port O_A without amplification. In contrast, to switch the wave from output O_A to output O_B, the output electric fields have to be A_{out}=0 by proper amplification. The condition is given as a solution of an equation $\alpha^3-3\alpha^2-\alpha-1=0$. This solution for α can be satisfied with α=3.383. With this value of α, B_{out} is calculated to be $-j(\alpha+1)(\alpha-1)^2 E_{in} = -j6.222\, E_{in}$. It means that the incident optical wave is amplified and routed to output O_B. This switching operation is summarized as

$$\begin{pmatrix} A_{out} \\ B_{out} \end{pmatrix} = \begin{cases} \begin{pmatrix} -E_{in} \\ 0 \end{pmatrix} & when\ \ \alpha = 1 \\ \begin{pmatrix} 0 \\ -j6.222E_{in} \end{pmatrix} & when\ \ \alpha = 3.383 \end{cases}. \tag{11}$$

The difference of the output intensity by a factor of $(6.222)^2$ can be compensated by employing an attenuator at output port O_B or an amplifier at output port O_A.

3. Numerical simulation of switching

3.1 Switch A

The switching characteristics can be analyzed using the equations described in the previous section. Wavelength dependence is also found by using eq. (1) instead of eq.(2) in the similar manner as described above. In this section, on the contrary, we verify the switching operation by numerical simulation to clarify the wave propagation along the waveguides of the switch. We use FD-BPM simulation with seventh-order Pade approximation (Hadley, 1992).

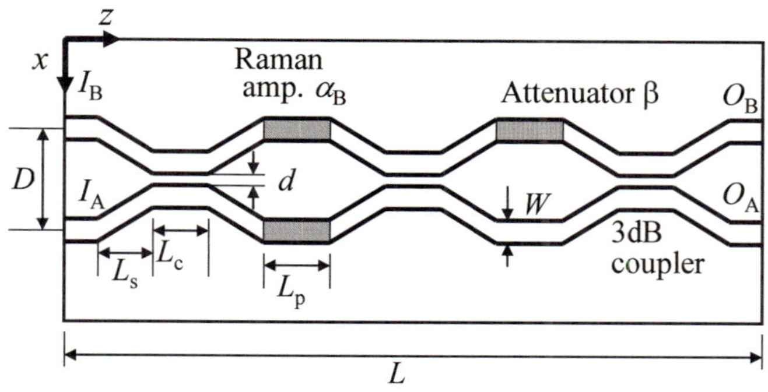

Fig. 6. Two-dimensional model of switch A for FD-BPM simulation.

We consider a two-dimensional slab waveguide model as shown in Fig. 6. The step length and the step width in the lateral direction are Δz=1.0 μm and Δx=0.05 μm, respectively. As the waveguide material, we consider silica waveguides. The core and cladding regions have refractive indices of n_c=1.461 and n_s=1.450, respectively. The width of the waveguide is W=3.0 μm. The interval and the length of the coupled region are set to be d=3.2 μm and L_c=100 μm to realize 3-dB coupling at wavelength 1550 nm. The length of the parallel waveguides corresponding to the Raman amplifiers and the attenuator is L_p=50 μm. The branching waveguides have the length of L_s=1025 μm. The distance between the two input ports is D=26.5 μm. The total length is L=6.65 mm.

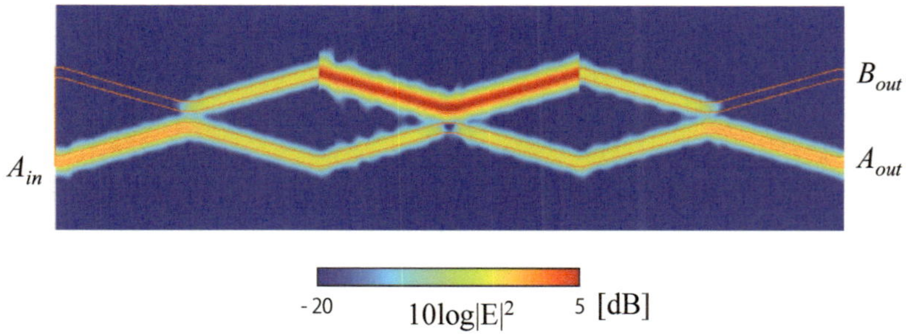

(a) $\alpha_A = 1,\ \alpha_B = 1+\sqrt{2}, \beta = 1/(1+\sqrt{2})$

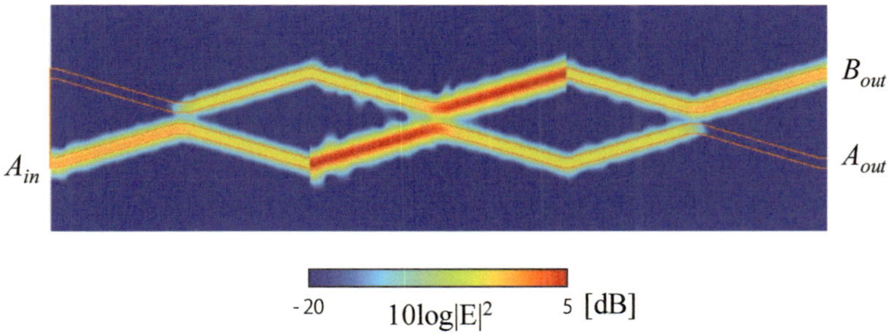

(b) $\alpha_A = 1+\sqrt{2},\ \alpha_B = 1, \beta = 1/(1+\sqrt{2})$

Fig. 7. Distribution of optical fields with two different switching conditions in swith A.

The switching operation at λ=1550 nm is confirmed as shown in Fig.7, where squared electric fields $|E|^2$ are plotted. In this simulation, Raman amplification is equivalently simulated just by increasing the optical electric field by multiplying the gain coefficient α_A or α_B at a place located at the end of the waveguide region for the amplifier. In a similar manner, the attenuator is modelled by multiplying the optical electric field by β at the beginning of the attenuator.
The wavelength dependence of the output intensity with the switch is plotted in Fig. 8. At the designed wavelength λ=1550 nm, the switching extinction ratio is larger than 35 dB. The wavelength range to achieve an extinction ratio larger than 20 dB is approximately 40 nm, though the 10-dB extinction ration is obtained over 100 nm. This wavelength dependence is caused by the wavelength dependence of the directional couplers.

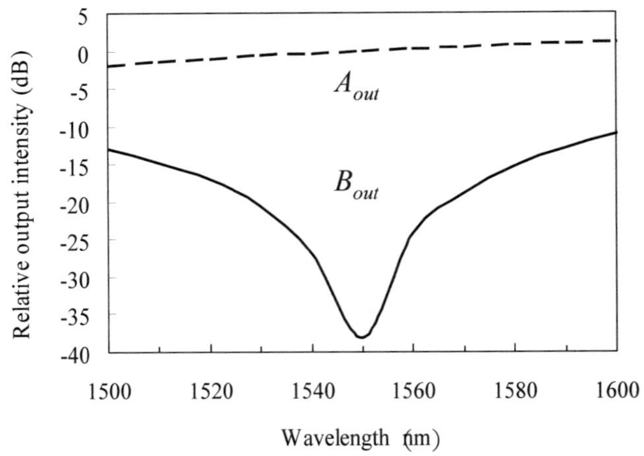

(a) $\alpha_A = 1,\ \alpha_B = 1+\sqrt{2}, \beta = 1/(1+\sqrt{2})$

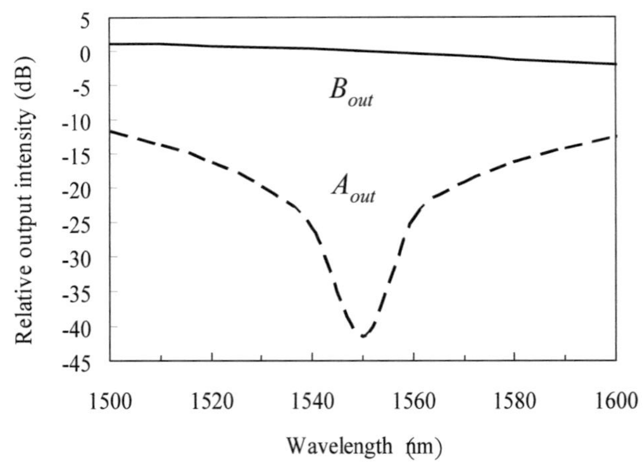

b) $\alpha_A = 1+\sqrt{2},\ \alpha_B = 1, \beta = 1/(1+\sqrt{2})$

Fig. 8. Wavelength dependence of switched outputs for the switch A designed at wavelength λ=1550nm.

3.2 Switch B

The switching operation with switch B is also verified by FD-BPM simulation. The model used in the simulation is shown in Fig. 9. The total length of the switch is L=8.85 mm.

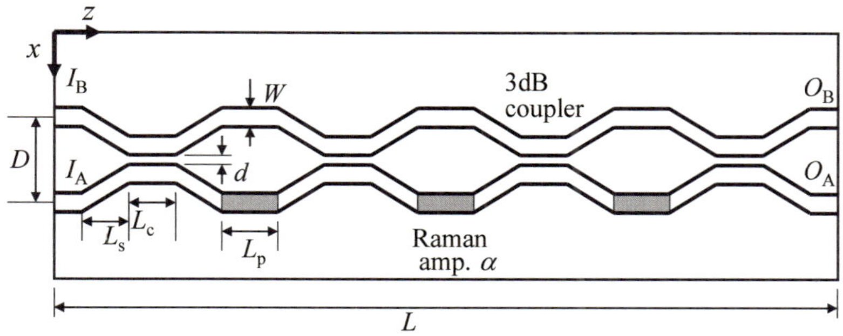

Fig. 9. Two-dimensional model of switch B for FD-BPM simulation.

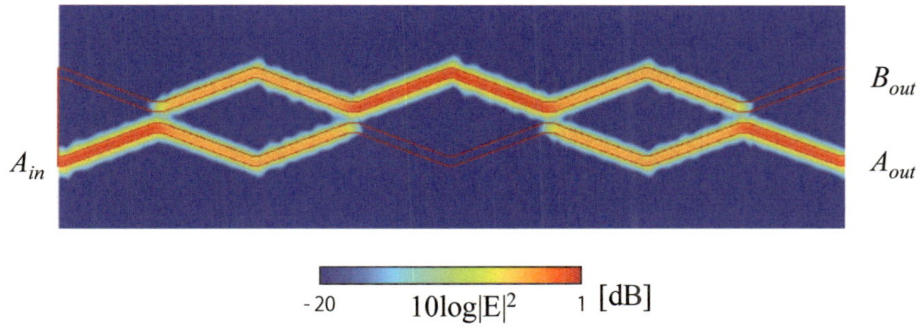

(a) $\alpha = 1$

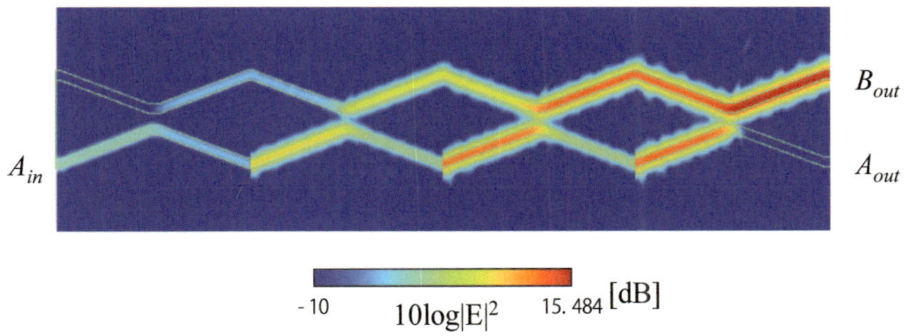

(b) $\alpha = 3.383$

Fig. 10. Distribution of optical fields with two different switching conditions in swith B.

The switching operation at λ=1550 nm is confirmed as shown in Fig. 10. Although the output intensities from output port O_A and O_B are different, switching is successfully simulated.

The wavelength dependence is shown in Fig. 11. At the designed wavelength λ=1550nm, the switching extinction ratio is larger than 25 dB. The wavelength range to achieve the extinction ratio larger than 20 dB is approximately 30 nm, though the 10-dB extinction ration is obtained over 80 nm.

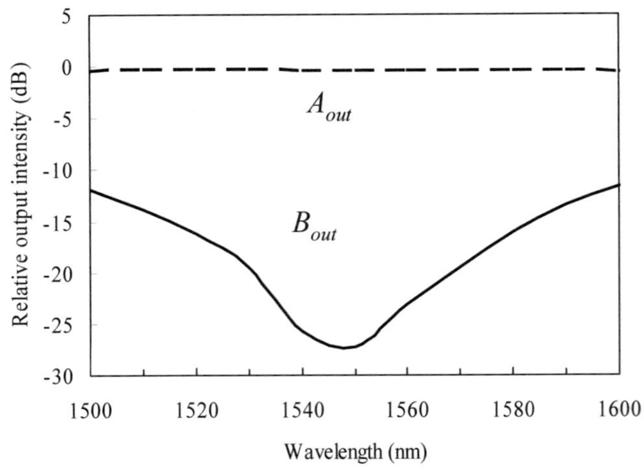

(a) $\alpha = 1$

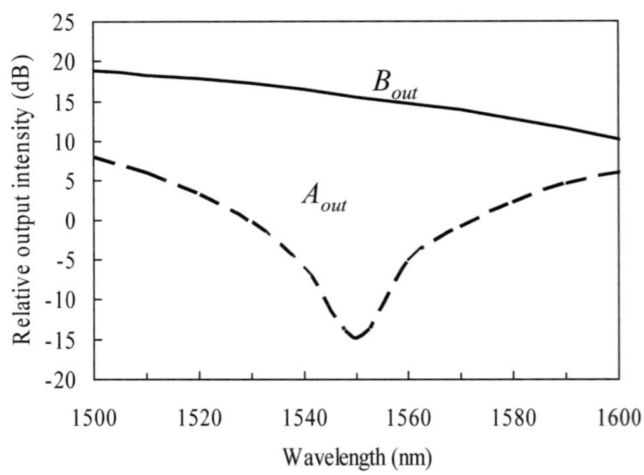

(b) $\alpha = 3.383$

Fig. 11. Wavelength dependence of switched outputs for the switch B designed at wavelength λ=1550nm.

4. Improvement of wavelength dependency

Waveguide-type Raman amplifiers do not depend on wavelength bands to be used because stimulated Raman scattering which is the base effect of Raman amplification can occur at any wavelength bands. Meanwhile, 3dB couplers have wavelength dependency in general, that is, the function of dividing an incident optical wave into two waves at the rate of 50:50 is available at some particular wavelength bands. The main cause of the wavelength dependency is the wavelength dependence of the coupling coefficient κ in eq.(1). For improving the characteristics of wavelength dependency of the switch and utilizing it at any wavelength bands, wavelength-independent (or wavelength-flattened) optical couplers should be employed. Fiber-type wavelength-independent couplers, that can be used for 50:50 of the dividing rate at wavelength bands such as 1550 nm± 40 nm and 1310 nm ± 40 nm, have already been on the market. However, waveguide-type wavelength-independent couplers have advantage from the viewpoint of integrating the switch elements.

An alternative for improving wavelength dependence is to replace the directional couplers by asymmetric X-junction couplers (Izutsu et al., 1982; Burns & Milton, 1980; Hiura et al., 2007). The asymmetric X-junction coupler has basically no dependence on wavelength and helps to improve the wavelength dependency of the proposed switch (Kishikawa et al., 2009a; Kishikawa et al., 2009b).

5. Another issue in implementation

Phase shift of the signal pulse experienced in the waveguide-type Raman amplifiers should be discussed because it can impact the operation of the switch. The phase shift is induced from refractive index change caused by self-phase modulation (SPM), cross-phase modulation (XPM), free carriers generated from two-photon absorption (TPA) (Roy et al., 2009), and temperature change. Although the structure of the switch becomes more complex, the effect of SPM and TPA-induced free carriers can be cancelled by installing the same nonlinear waveguides as those of the waveguide-type Raman amplifiers into counter arms of the Mach-Zehnder interferometers of the switch. The influence of XPM and temperature change involved with high power pump injection can also be suppressed by injecting pump waves, having the same power and different wavelengths that do not amplify the signal pulse, into the counterpart nonlinear waveguides.

6. Conclusion

We proposed a novel all-optical wavelength-selective switching having potential of a few tens of picosecond or faster operating speed. We discussed the theory and the simulation results of the switching operation and the characteristics. Moreover, the dynamic range over 25dB was also obtained from the simulation results of the switch. This characteristics can be wavelength-selective switching operation. More detailed analysis and simulation taking the nonlinearity of Raman amplifiers into account are required to confirm the operation with actual devices.

Although the principle and the fundamental verification were performed with the switches consisting of directional couplers, the idea can be similarly applied to switches consisting of other components such as asymmetric X-junction couplers to increase the wavelength range.

8. References

Doran, N. J. & Wood, D. (1988). Nonlinear-Optic Loop Mirror, Optics Lett., vol.13, no.1, pp.56-58, Jan. 1988.

Burns, W. K. & Milton, A. F. (1980). An Analytic Solution for Mode Coupling in Optical Waveguide Branches, IEEE J. Quantum Electron., vol.QE-16, no.4, pp.446-454, Apr. 1980.

Goh, T., Kitoh, T., Kohtoku, M., Ishii, M., Mizuno, T. & Kaneko, A. (2008). Port Scalable PLC-Based Wavelength Selective Switch with Low Extinction Loss for Multi-Degree ROADM/WXC, The Optical Fiber Communication Conference and the National Fiber Optic Engineers Conference (OFC/NFOEC2008), San Diego, OWC6, Mar. 2008.

Goto, N & Miyazaki, Y. (1990). Integrated Optical Multi-/Demultiplexer Using Acoustooptic Effect for Multiwavelength Optical Communications, IEEE J. on Selected Areas in Commun., vol.8, no.6, pp.1160-1168, Aug. 1990.

Hadley, G. R. (1992). Wide-Angle Beam Propagation Using Pade Approximant Operators, Opt. Lett., vol.17, no.20, pp.1426-1428, Oct. 1992.

Hiura, H., Narita, J. & Goto, N. (2007). Optical Label Recognition Using Tree-Structure Self-Routing Circuits Consisting of Asymmetric X-junction, IEICE Trans. Commun., vol.E90-C, no.12, pp.2270-2277, Dec. 2007.

Izutsu, M., Enokihara, A. & Sueta, T. (1982). Optical-Waveguide Hybrid Coupler, Opt. Lett., vol.7, no.11, pp.549-551, Nov. 1982.

Kishikawa, H. & Goto, N. (2005). Proposal of All-Optical Wavelength-Selective Switching Using Waveguide-Type Raman Amplifiers and 3dB Couplers, J. Lightwave Technol., vol.23, no.4, pp.1631-1636, Apr. 2005.

Kishikawa, H. & Goto, N. (2006). Switching Characteristics of All-Optical Wavelength-Selective Switch Using Waveguide-Type Raman Amplifiers and 3-dB Couplers, IEICE Trans. Electron., vol.E89-C, no.7, pp.1108-1111, July 2006.

Kishikawa, H. & Goto, N. (2007a). Optical Switch by Light Intensity Control in Cascaded Coupled Waveguides, IEICE Trans. Electron., vol.E90-C, no.2, pp.492-498, Feb. 2007.

Kishikawa, H. & Goto, N. (2007b). Designing of Optical Switch Controlled by Light Intensity in Cascaded Optical Couplers, Optical Engineering, vol.46, no.4, pp.044602-1-10, Apr. 2007.

Kishikawa, H., Kimiya, K., Goto, N. & Yanagiya, S. (2009a). All-Optical Wavelength-Selective Switch Controlled by Raman Amplification for Wide Wavelength Range, Optoelectronics and Communications Conf., OECC2009, Hong Kong, TuG3, July 2009.

Kishikawa, H., Kimiya, K., Goto, N. & Yanagiya, S. (2009b). All-Optical Wavelength-Selective Switch by Amplitude Control with a Single Control Light for Wide Wavelength Range", Int. Conf. on Photonics in Switching, PS2009, Pisa, PT-12, Sept. 2009.

Kitagawa, Y., Ozaki, N., Takata, Y., Ikeda, N., Watanabe, Y., Sugimoto, Y. & Asakawa, K. (2009). Sequential Operations of Quantum Dot/Photonic Crystal All-Optical Switch With High Repetitive Frequency Pumping, J. Lightwave Technol., vol.27, no.10, pp.1241-1247, May 2009.

Nakamura, S., Ueno, Y., Tajima, K., Sasaki, J., Sugimoto, T., Kato, T., Shimoda, T., Itoh, M., Hatakeyama, H., Tamanuki, T. & Sasaki, T. (2000). Demultiplexing of 168-Gb/s Data Pulses with a Hybrid-Integrated Symmetric Mach-Zehnder All-Optical Switch, IEEE Photon. Tech. Lett., vol.12, no.4, pp.425-427, Apr. 2000.

Raghunathan, V., Boyraz, O & Jalali, B. (2005). 20dB On-Off Raman Amplifiation in Silicon Waveguides, Conf. Lasers and Electro-Optics (CLEO2005), Baltimore, CMU1, May 2005.

Rong, H., Liu, A., Nicolaescu, R., Paniccia, M., Cohen, O. & Hak, D. (2004). Raman Gain and Nonlinear Optical Absorption Measurements in a Low-Loss Silicon Waveguide, Appl. Phys. Lett., vol.85, no.12, pp.2196-2198, Sept. 2004.

Roy, S., Bhadra, S. K. & Agrawal, G. P. (2009). Raman Amplification of Optical Pulses in Silicon Waveguides: Effects of Finite Gain Bandwidth, Pulse Width, and Chirp, J. Opt. Soc. Am. B, vol. 26, no. 1, Jan. 2009.

Suto, K., Saito, T., Kimura, T., Nishizawa, J. & Tanabe, T. (2002). Semiconductor Raman Amplifier for Terahertz Bandwidth Optical Communication, J. Lightwave Technol., vol.20, no.4, pp.705-711, Apr. 2002.

Suzuki, S., Himeno, A. & Ishii, M. (1998). Integrated Multichannel Optical Wavelength Selective Switches Incorporating an Arrayed-Waveguide Grating Multiplexer and Thermooptic Switches, J. Lightwave Technol., vol.16, no.4, pp.650-655, Apr. 1998.

14

Nonlinear Optics in Doped Silica Glass Integrated Waveguide Structures

David Duchesne[1], Marcello Ferrera[1], Luca Razzari[1],
Roberto Morandotti[1], Brent Little[2], Sai T. Chu[2] and David J. Moss[3]
[1]INRS-EMT,
[2]Infinera Corporation,
[3]IPOS/CUDOS, School of Physics, University of Sydney,
[1]Canada
[2]USA
[3]Australia

1. Introduction

Integrated photonic technologies are rapidly becoming an important and fundamental milestone for wideband optical telecommunications. Future optical networks have several critical requirements, including low energy consumption, high efficiency, greater bandwidth and flexibility, which must be addressed in a compact form factor (Eggleton et al., 2008; Alduino & Paniccia, 2007; Lifante, 2003). In particular, it has become well accepted that devices must possess a CMOS compatible fabrication procedure in order to exploit the large existing silicon technology in electronics (Izhaky et al., 2006; Tsybeskov et al., 2009). This would primarily serve to reduce costs by developing hybrid electro-optic technologies on-chip for ultrafast signal processing. There is still however, a growing demand to implement all-optical technologies on these chips for frequency conversion (Turner et al., 2008; Venugopal Rao et al., 2004), all-optical regeneration (Salem et al., 2008; Ta'eed et al., 2005), multiplexing and demultiplexing (Lee et al., 2008; Bergano, 2005; Ibrahim et al., 2002), as well as for routing and switching (Lee et al., 2008; Ibrahim et al., 2002). The motivation for optical technologies is primarily based on the ultrahigh bandwidth of the optical fiber and the extremely low attenuation coefficient. Coupled with minimal pulse distortion properties, such as dispersion and nonlinearities, optical fibers are the ideal transmission medium to carry information over long distances and to connect optical networks. Unfortunately, the adherence of the standard optical fiber to pulse distortions is also what renders it less than perfectly suited for most signal processing applications required in telecommunications. Bending losses become extremely high in fibers for chip-scale size devices, limiting its integrability in networks. Moreover, its weak nonlinearity limits the practical realization (i.e. low power values and short propagation lengths) of some fundamental operations requiring nonlinear optical phenomena, such as frequency conversion schemes and switching (Agrawal, 2006).
Several alternative material platforms have been developed for photonic integrated circuits (Eggleton et al., 2008; Alduino & Panicia, 2007; Koch & Koren, 1991; Little & Chu, 2000), including semiconductors such as AlGaAs and silicon-on-insulator (SOI) (Lifante, 2003;

Koch and Koren, 1991; Tsybeskov et al., 2009; Jalali & Fathpour, 2006), as well as nonlinear glasses such as chalcogenides, silicon oxynitride and bismuth oxides (Ta'eed et al., 2007; Eggleton et al., 2008; Lee et al., 2005). In addition, exotic and novel manufacturing processes have led to new and promising structures in these materials and in regular silica fibers. Photonic crystal fibers (Russell, 2003), 3D photonic bandgap structures (Yablonovitch et al., 1991), and nanowires (Foster et al., 2008) make use of the tight light confinement to enhance nonlinearities, greatly reduce bending radii, which allows for submillimeter photonic chips. Despite the abundance of alternative fabrication technologies and materials, there is no clear victor for future all-optical nonlinear devices. Indeed, many nonlinear platforms require power levels that largely exceed the requirements for feasible applications, whereas others have negative side effects such as saturation and multi-photon absorption. Moreover, there is still a fabrication challenge to reduce linear attenuation and to achieve CMOS compatibility for many of these tentative photonic platforms and devices. In response to these demands, a new high-index doped silica glass platform was developed in 2003 (Little, 2003), which combines the best of both the qualities of single mode fibers, namely low propagation losses and robust fabrication technology, and those of semiconductor materials, such as the small quasi-lossless bending radii and the high nonlinearity. This book chapter primarily describes this new material platform, through the characterization of its linear and nonlinear properties, and shows its application for all-optical frequency conversion for future photonic integrated circuits. In section 2 we present an overview of concurrent recent alternative material platforms and photonic structures, discussing advantages and limitations. We then review in section 3 the fundamental equations for nonlinear optical interactions, followed by an experimental characterization of the linear and nonlinear properties of a novel high-index glass. In section 4 we introduce resonant structures and make use of them to obtain a highly efficient all-optical frequency converter by means of pumping continuous wave light.

2. Material platforms and photonic structures for nonlinear effects

2.1 Semiconductors

Optical telecommunications is rendered possible by carrying information through waveguiding structures, where a higher index core material (n_c) is surrounded by a cladding region of lower index material (n_s). Nonlinear effects, where the polarization of media depends nonlinearly on the applied electric field, are generally observed in waveguides as the optical power is increased. Important information about the nonlinear properties of a waveguide can be obtained from the knowledge of the index contrast ($\Delta n = n_c - n_s$) and the index of the core material, n_c. The strength of nonlinear optical interactions is predominantly determined through the magnitude of the material nonlinear optical susceptibilities ($\chi^{(2)}$ and $\chi^{(3)}$ for second order and third order nonlinear processes where the permittivity depends on the square and the cube of the applied electromagnetic field, respectively), and scales with the inverse of the effective area of the supported waveguide mode. Through Miller's rule (Boyd, 2008) the nonlinear susceptibilities can be shown to depend almost uniquely on the refractive index of the material, whereas the index contrast can easily be used to estimate the area of the waveguide mode, where a large index contrast leads to a more confined (and thus a smaller area) mode. It thus comes to no surprise that the most commonly investigated materials for nonlinear effects are III-V semiconductors, such as silicon and AlGaAs, which possess a large index of refraction at the telecommunications wavelength (λ = 1.55 μm) and where waveguides with a large index contrast can be formed. For third order nonlinear phenomena such as the Kerr effect[1], the strength of the nonlinear interactions can be estimated through the nonlinear parameter $\gamma = n_2\omega/cA$ (Agrawal, 2006), where n_2 is the nonlinear index coefficient determined solely from material properties, ω is the angular frequency of the light, c is the speed of light and A the effective area of the mode, which will be more clearly defined later. The total cumulative nonlinear effects induced by a waveguide sample can be roughly estimated as being proportional to the peak power, length of the waveguide and the nonlinear parameter (Agrawal, 2006). In order to minimize the energetic requirements, it is thus necessary either to have long structures and/or large nonlinear parameters. Focusing on the moment on the nonlinear parameter, in typical semiconductors, the core index $n_c > 3$ (~3.5 for Si and ~3.3 GaAs) leads to values of n_2 ~10^{-18} - 10^{-17} m^2/W, to be compared with fused silica (n_c = 1.45) where n_2 ~2.6 x 10^{-20} m^2/W. Moreover, etching through the waveguide core allows for a large index contrast with air, permitting photonic wire geometries with effective areas below 1 um^2, see Fig. 1. This leads to extremely high values of γ ~ 200,000$W^{-1}km^{-1}$ (Salem et al., 2008; Foster et al., 2008) (to be compared with single mode fibers which have γ ~ 1$W^{-1}km^{-1}$ (Agrawal, 2006)). This large nonlinearity has been used to demonstrate several nonlinear applications for telecommunications, including all-optical regeneration at 10 Gb/s using four-wave mixing and self-phase modulation in SOI (Salem et al., 2008; Salem et al., 2007), frequency conversion (Turner et al., 2008; Venugopal Rao et al., 2004; Absil et al., 2000), and Raman amplifications (Rong et al., 2008; Espinola et al., 2004).

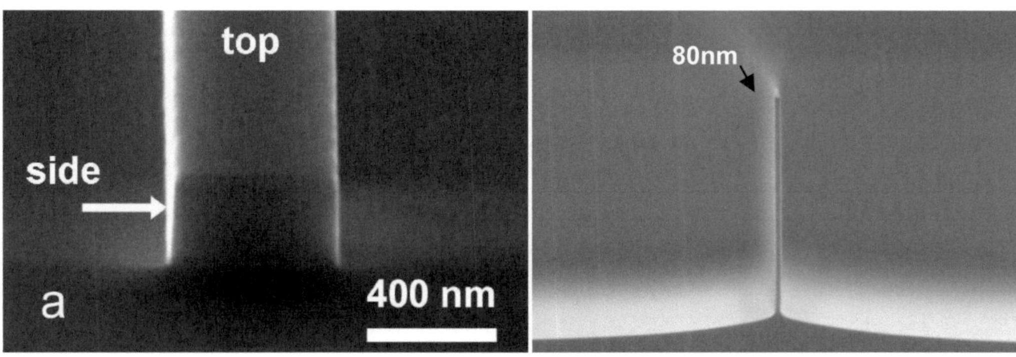

Fig. 1. (left) Silicon-on-insulator nano-waveguide (taken from (Foster et al., 2008)) and inverted nano-taper (80nm in width) of an AlGaAs waveguide (right). Both images show the very advanced fabrication processes of semiconductors.

There are however major limitations that still prevent their implementation in future optical networks. Semiconductor materials typically have a high material dispersion (a result of being near the bandgap of the structure), which prevents the fabrication of long structures. To overcome this problem, small nano-size wire structures, where the waveguide dispersion dominates, allows one to tailor the total induced dispersion. The very advanced fabrication technology for both Si and AlGaAs allows for this type of control, thus a precise waveguide

[1] We will neglect second order nonlinear phenomena, which are not possible in centrosymmetric media such as glasses. See (Boyd, 2008) and (Venugopal Rao et al., 2004; Wise et al., 2002) for recent advances in exploiting $\chi^{(2)}$ media for optical telecommunications.

geometry can be fabricated to have near zero dispersion in the spectral regions of interest. Unfortunately, the small size of the mode also implies a relatively large field along the waveguide etched sidewalls (see Fig. 1). This leads to unwanted scattering centers and surface state absorptions where initial losses have been higher than 10dB/cm for AlGaAs (Siviloglou et al., 2006; Borselli et al., 2006; Jouad & Aimez, 2006), and ~ 3 dB/cm for SOI (Turner et al.,2008).
Another limitation comes from multiphoton absorption (displayed pictorially in Fig. 2 for the simplest case, i.e. two-photon absorption) and involves the successive absorption of photons (via virtual states) that promotes an electron from the semiconductor valence band to the conduction band. This leads to a saturation of the transmitted power and, consequently, of the nonlinear effects. For SOI this has been especially true, where losses are not only due to two-photon absorption, but also to the free carriers induced by the process (Foster et al., 2008; Dulkeith et al., 2006). Moreover, the nonlinear figure of merit (= $n_2/a_2\lambda$, where a_2 is the two photon absorption coefficient), which determines the feasibility of nonlinear interactions and switching, is particular low in silicon (Tsang & Liu, 2008).
Lastly, although reducing the modal area enhances the nonlinear properties of the waveguide, it also impedes coupling from the single mode fiber into the device; for comparison the modal diameter of a fiber is ~10μm whereas for a nanowire structure it is typically 20 times smaller. This leads to high insertion losses through the device, necessitating either expensive amplifiers at the output, or of complicated tapers often requiring mature fabrication technologies and sometimes multi-step etching processes (Moerman et al., 1997) (SOI waveguides make use of state-of-the-art inverse tapers which limits the insertion losses to approximately 5dB (Almeida et al., 2003; Turner et al., 2008)).

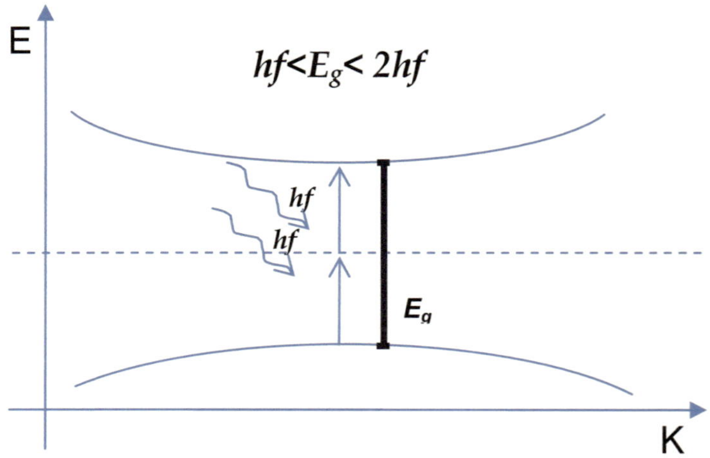

Fig. 2. Schematic of two-photon absorption in semiconductors. In the most general case of the multiphoton absorption process, electrons pass from the valence band to the conduction band via the successive absorption of multiple photons, mediated via virtual states, such that the total absorbed energy surpasses the bandgap energy.

2.2 High index glasses

In addition to semiconductors, a number of high index glass systems have been investigated as a platform for future photonic integrated networks, including chalcogenides (Eggleton et al., 2008; Ta'eed et al., 2007), silicon nitride (Gondarenko et al., 2009) and silicon oxynitride (Worhoff et al., 2002). Chalcogenides in particular have been shown to have extremely high nonlinear parameters approaching $\gamma \sim 100{,}000 W^{-1}km^{-1}$ in nanotapers (Yeom et al., 2008), which has been used to demonstrate demultiplexing at 160 Gb/s (Pelusi et al., 2007). However, all of these platforms suffer from shortcomings of one form or another. Fabrication processes for chalcogenide glasses are still under development (Li et al., 2005; Ruan et al., 2004) and while they generally possess a very high nonlinear figure of merit - significantly better than silicon, for example - it can be an issue for some glasses (Lamont et al., 2006). Photosensitivity and photo-darkening, while powerful tools for creating novel photonic structures, can sometimes place limits on the material stability (Shokooh-Saremi et al., 2006). Whereas other high-index glasses, such as silicon oxynitride, have negligible nonlinear absorption (virtually infinite figure of merit), high temperature annealing is required to reduce propagation losses, making the entire process non-CMOS compatible.

A high-index, doped silica glass material called Hydex® (Little, 2003), was developed by Little Optics in 2003 as a compromise between the attractive linear features of silica glass and the nonlinear properties of semiconductors. Films are first deposited using standard chemical vapour deposition. Subsequently, waveguides are formed using photolithography and reactive ion etching, producing waveguide sidewalls with exceptionally low roughness. The waveguides are then buried in standard fused silica glass, making the entire fabrication process CMOS compatible and requiring no further anneal. The typical waveguide cross section is 1.45 x 1.5 μm^2 as shown in Fig. 3. The linear index at λ = 1.55 µm is 1.7, and propagation losses have been shown to be as low as 0.06 dB/cm (Duchesne et al., 2009; Ferrera et al., 2008). In addition, fiber pigtails have been designed for coupling to and from Hydex waveguides, with coupling losses on the order of 1.5dB. The linear properties of this material platform has already been exploited to achieve filters with >80dB extinction ratios (Little et al., 2004), as well as the optical sensing of biomolecules (Yalcin et al., 2006).

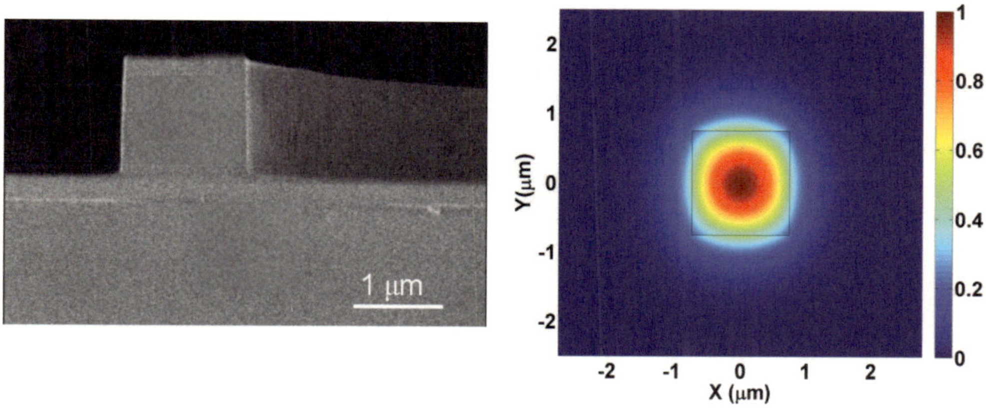

Fig. 3. Scanning electron microscopy picture of the high-index glass waveguide (prior to upper SiO_2 deposition), and electromagnetic field distribution of the fundamental mode.

As will be shown in the subsequent sections below, this material platform also has a moderate nonlinearity, and coupled with long or resonant structures, can be used to generate significant nonlinear effects with low power requirements. In the next section we produce the necessary equations governing light propagation in a nonlinear media, followed by a characterization method for the nonlinearity, and explain the possible applications achievable by exploiting resonant and long structures.

3. Light dynamics in nonlinear media

In order to completely characterize the nonlinear optical properties of materials, it is worthwhile to review some fundamental equations relating to pulse propagation in nonlinear media. In general, this is modelled directly from Maxwell's equations, and for piecewise homogenous media one can arrive at the optical nonlinear Schrodinger equation (Agrawal, 2006; Afshar & Monro, 2009):

$$\frac{\partial \psi}{\partial z} + \beta_1 \frac{\partial \psi}{\partial t} + i\frac{\beta_2}{2}\frac{\partial^2 \psi}{\partial t^2} + HOD + \frac{\alpha_1}{2}\psi = i\gamma|\psi|^2\psi - \frac{\alpha_2}{2A}|\psi|^2\psi - HOL \tag{1}$$

Where ψ is the slowly-varying envelope of the electric field, given by: $E = \psi'(z,t)F(x,y)\exp(i\beta_0 z - i\omega_0 t)$, where ψ' has been normalized such that $|\psi|^2$ represents the optical power. ω_0 is the central angular frequency of the pulse, β_0 the propagation constant, β_1 is the inverse of the group velocity, β_2 the group velocity dispersion, a_1 the linear loss coefficient, a_2 the two-photon absorption coefficient, γ (= $n_2\omega_0/cA$) the nonlinear parameter, t is time and z is the propagation direction. Here $F(x,y)$ is the modal electric field profile, which can be found by solving the dispersion relation:

$$\nabla^2 F + \frac{\omega^2 n^2}{c^2}F = \beta^2 F \tag{2}$$

The eigenvalue solution to the dispersion relation can be obtained by numerical methods such as vectorial finite element method (e.g. Comsol Multiphysics). From this the dispersion parameters can be calculated via a Taylor expansion:

$$\beta = \beta_0 + \beta_1(\omega - \omega_0) + \frac{\beta_2}{2}(\omega - \omega_0)^2 + \frac{\beta_3}{6}(\omega - \omega_0)^3 + \ldots \tag{3}$$

The effective area can also be evaluated:

$$A = \frac{\left[\iint_\infty |F|^2\, dxdy\right]^2}{\iint_\infty |F|^4\, dxdy} \tag{4}$$

In arriving to eq. (1), we neglected higher order nonlinear contributions, non-instantaneous responses (Raman) and non-phase matched terms; we also assumed an isotropic cubic medium, as is the case for glasses. These approximations are valid for moderate power values and pulse durations down to ~100fs for a pulse centered at 1.55 μm (Agrawal, 2006). The terms *HOL* and *HOD* refer to higher order losses and higher order dispersion terms, which may be important in certain circumstances (Foster et al., 2008; Siviloglou et al., 2006). Whereas eq. (1) also works as a first order model for semiconductors, a more general and exact formulation can be found in (Afshar & Monro, 2009). Given the material dispersion properties (found either experimentally or from a Sellmeier model (Sellmeier, 1871)), the only unknown parameters in Eq. (1) are the nonlinear parameter γ (or n_2 to be more precise), the linear propagation loss coefficient α_1 and the nonlinear loss term α_2.

The solution to the nonlinear Schrodinger equation has been studied in detail (Agrawal, 2006; Kivshar & Malomed, 1989). Here we present the solution to this equation for doped silica glass at low and high power regimes. This allows a complete characterization of the waveguide properties which will be extremely useful in studying nonlinear applications such as frequency conversion.

As will be shown below, one of the several advantages of high-index doped silica glass is in its mature fabrication technology which allows for long waveguides with minimal losses. As can be readily seen in Fig. 4, long spiral waveguides of more than 1m of length can be contained in a 2.5 x 2.5 mm^2 area.

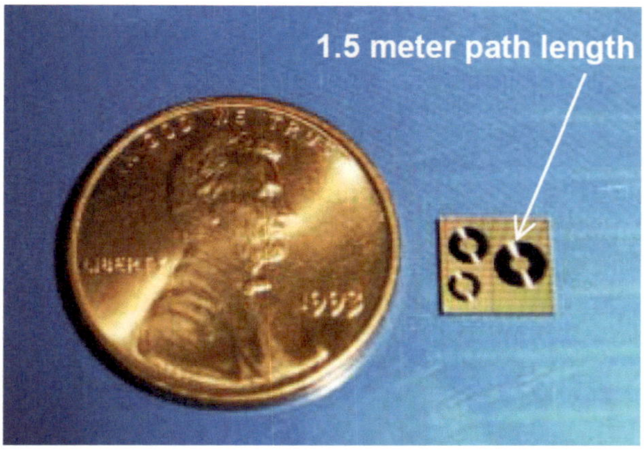

Fig. 4. A 1.5 meter long waveguide confined on a photonic chip smaller than the size of a penny.

3.1 Low power regime

At low power, dispersive terms dominate thus leading to temporal pulse broadening. In this limit, the nonlinear Schrodinger equation reduces to:

$$\frac{\partial \psi}{\partial z} + \beta_1 \frac{\partial \psi}{\partial t} + i\frac{\beta_2}{2}\frac{\partial^2 \psi}{\partial t^2} + HOD + \frac{\alpha_1}{2}\psi = 0 \tag{5}$$

This equation transforms to a simple linear ordinary differential equation in the Fourier domain, and assuming an input unchirped Gaussian pulse of width T_0 the solution (neglecting *HOD* terms) is given by (Agrawal, 2006):

$$\psi = \frac{T_0}{\sqrt{T_0^2 - i\beta_2 z}} \exp\left(-\frac{\alpha_1}{2}z\right) \exp\left(-\frac{(t-\beta_1 z)^2}{2\left(T_0^2 - i\beta_2 z\right)}\right) \tag{6}$$

The pulse is seen to acquire a chirp, leading to temporal broadening via dispersion. The analogue in the spectral domain is that the pulse acquires a quadratic phase, which can serve as a direct measurement of the dispersion induced from the waveguide. A well known experimental technique for reconstructing the phase and amplitude at the output of a device is the Fourier Transform Spectral Interferometry (FTSI) (Lepetit et al., 1995). Using this spectral interference technique, the dispersion of the 45cm doped silica glass spiral waveguide was determined to be very small (on the order of the single mode fiber dispersion, β_2~22ps^2/km), and not important for pulses as short as 100fs (Duchesne et al., 2009). This is extremely relevant, as 3 critical conditions must be met to allow propagation through long structures (note that waveguides are typically <1cm): 1) low linear propagation loses, so that a useful amount of power remains after propagation; 2) low dispersion value so that ps pulses or shorter are not broadened significantly; and 3) long waveguides must be contained in a small chip for integration, as was done in the spiral waveguide discussed. This latter requirement also imposes a minimal index contrast Δn on the waveguide, such that bending losses are also minimized. Moreover, as will be discussed further below, having a low dispersion value is critical for low power frequency conversion.

3.2 Nonlinear losses

In order to see directly the effects of the nonlinear absorption on the propagation of light pulses, it is useful to transform Eq. (1) to a peak intensity equation, $I = |\psi|^2 / A$, as follows:

$$\frac{dI}{dz} = \frac{\psi^*}{A}\frac{\partial \psi}{\partial z} + \frac{\psi}{A}\frac{\partial \psi^*}{\partial z} = -\alpha_1 I - \alpha_2 I^2 - \sum_n \alpha_n I^n , \tag{7}$$

where we have neglected dispersion contributions based on the previous considerations. We have also explicitly added the higher order multiphoton contributions (three-photon absorption and higher), although it is important to note that these higher order effects typically have a very small cross section that require large intensity values [see chapter 12 of (Boyd, 2008)]. Considering only two-photon absorption, the solution is found to be:

$$I = \frac{\alpha I_0 \exp(-\alpha z)}{\alpha + \alpha_2 I_0 \left(1 - \exp(-\alpha z)\right)} \tag{8}$$

From this one can immediately conclude that the maximal output intensity is limited by two-photon absorption to be $1/\alpha_2 z$; a similar saturation behaviour is obtained when considering higher order contributions. Multiphoton absorption is thus detrimental for high intensity applications and cannot be avoided by any kind of waveguide geometry (Boyd, 2008; Afshar & Monro, 2009).

Experimentally, the presence of multiphoton absorption can be understood from simple transmission measurements of high power/intensity pulses. Pulsed light from a 16.9MHz Pritel fiber laser, centered at 1.55µm, was used to characterize the transmission in the doped silica glass waveguides. An erbium doped fiber amplifier was used directly after the laser to achieve high power levels, and the estimated pulse duration was approximately 450fs. Fig. 5 presents a summary of the results, showing a purely linear transmission up to input peak powers of 500W corresponding to an intensity of 25GW/cm^2 (Duchesne et al., 2009). This result is extremely impressive, and is well above the threshold for silicon (Dulkeith et al., 2006; Liang & Tsang, 2004; Tsang & Liu, 2008), AlGaAs (Siviloglou et al., 2006), or even Chalcogenides (Nguyen et al., 2006). Multiphoton absorption leads to free carrier generation, which in turn can also dramatically increase the losses (Dulkeith et al., 2006;

Liang & Tsang, 2004; Tsang & Liu, 2008). For the case of two-photon absorption, the impact on nonlinear signal processing is reflected in the nonlinear figure of merit, $FOM = n_2 / \lambda\alpha_2$, which estimates the maximal Kerr nonlinear contribution with limitations arising from the saturation of the power from two-photon absorption. In high-index doped silica glass, this value is virtually infinite for any practical intensity values, but can be in fact quite low for certain chalcogenides (Nguyen et al., 2006) and even lower in silicon (~0.5) (Tsang & Liu, 2008).

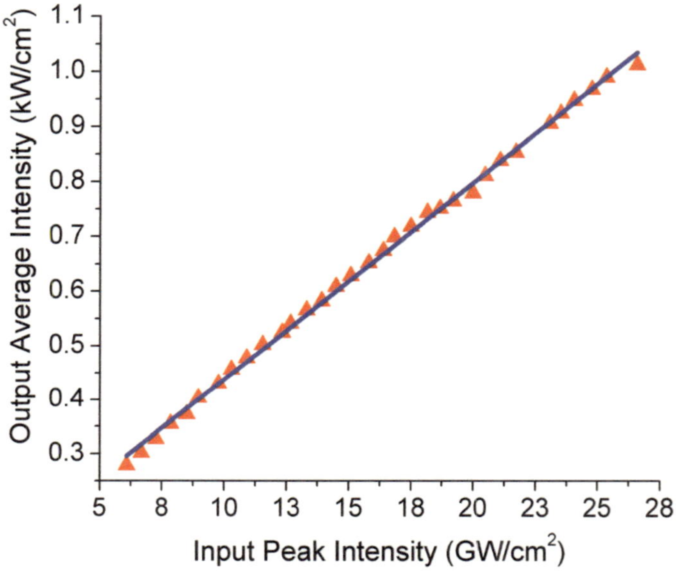

Fig. 5. Transmission at the output of a 45cm long high-index glass waveguide. The linear relation testifies that no multi-photon absorption was present up to peak intensities of more than 25GWcm2 (~500W).

By propagating through different length waveguides, we were able to determine, by means of a cut-back style like procedure, both the pigtail losses and propagation losses to be 1.5dB and 0.06dB/cm, respectively. Whereas this value is still far away from propagation losses in single mode fibers (0.2dB/km), it is orders of magnitude better than in typical integrated nanowire structures, where losses >1dB/cm are common (Siviloglou et al., 2006; Dulkeith et al., 2006; Turner et al., 2008). The low losses, long spiral waveguides confined in small chips, and low loss pigtailing to single mode fibers testifies to the extremely well established and mature fabrication process of this high-index glass platform.

3.3 Kerr nonlinearity

In the high power regime, the nonlinear contributions become important in Eq. (1), and in general the equation must be solved numerically. To gain some insight on the effect of the nonlinear contribution to Eq. (1), it is useful to look at the no-dispersion limit of Eq. (1), which can be readily solved to obtain:

$$\psi = \psi_0 \exp\left[i\gamma |\psi_0|^2 \alpha_1^{-1} \left(1 - \exp(-\alpha_1 z)\right)\right] \tag{9}$$

The nonlinear term introduces a nonlinear chirp in the temporal phase, which in the frequency domain corresponds to spectral broadening (i.e. the generation of new frequencies). This phenomenon, commonly referred to as self-phase modulation, can be used to measure the nonlinear parameter γ by means of recording the spectrum of a high power pulse at the output of a waveguide (Duchesne et al., 2009; Siviloglou et al., 2006; Dulkeith et al., 2006). The nonlinear interactions are found to scale with the product of the nonlinear parameter γ, the peak power of the pulse, and the effective length of the waveguide (reduced from the actual length due to the linear losses). For low-loss and low-dispersion guiding structures, it is thus useful to have long structures in order to increase the total accumulated nonlinearity, while maintaining low peak power levels. It will be shown in the next section how resonant structures can make use of this to achieve impressive nonlinear effects with 5mW CW power values. For other applications, dispersion effects may be desired, such as for soliton formation (Mollenauer et al., 1980).

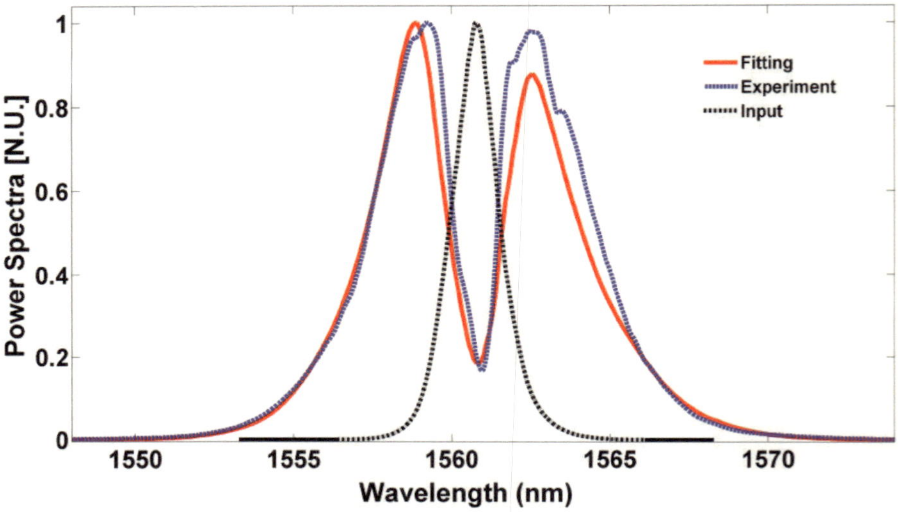

Fig. 6. Input (black) and output spectra (blue) from the 45cm waveguide. Spectral broadening is modelled via numerical solution of Eq. (1) (red curve).

Experimentally, the nonlinearity of the doped silica glass waveguide was characterized in (Duchesne et al., 2009) by injecting 1.7ps pulses (centered at 1.55μm) with power levels of approximately 10-60W. The output spectrum showed an increasing amount of spectral broadening, as can be seen in Fig. 6. The value of the nonlinearity was determined by numerically solving the nonlinear Schrodinger equation by means of a split-step algorithm (Agrawal, 2006), where the only unknown parameter was the nonlinear parameter. By fitting experiments with simulations, a value of γ = 220 $W^{-1}km^{-1}$ was determined, corresponding to a value of n_2 = 1.1 x 10^{-19} m^2/W (A = 2.0 μm^2). Similar experiments in single mode fibers (Agrawal, 2006; Boskovic et al., 1996), semiconductors (Siviloglou et al., 2006; Dulkeith et al., 2006), and chalcogenides (Nguyen et al., 2006) were also performed to characterize the Kerr nonlinearity. In comparison, the value of n_2 obtained in doped silica glass is approximately 5 times larger than that found in standard fused silica, consistent with Miller's rule (Boyd, 2008). On the other hand, the obtained γ value is more than 200 times larger, due to the much smaller effective mode area of the doped silica waveguide in contrast to the weakly guided single mode fiber. However, semiconductors and chalcogenides nanotapers definitely have the upper hand in terms of bulk nonlinear parameter values, where $\gamma \sim 200{,}000\ W^{-1}km^{-1}$ have been reported (Foster et al., 2008; Yeom et al., 2008), due to both the smaller effective mode areas and the larger n_2, as previously mentioned.

From Eq. (9), there are 2 ways to improve the nonlinear interactions (for a fixed input power): 1) increasing the nonlinear parameter, or 2) increasing the propagation length. To increase the former, one can reduce the modal size by having high-index contrast waveguides, and/or using a high index material with a high value of n_2. Thus, for nonlinear applications, the advantage for doped silica glass waveguides lies in exploiting its low loss and advanced fabrication processes that yield long winding structures, which is typically not possible in other material platforms due to nonlinear absorptions and/or immature fabrication technologies.

4. Resonant structures

Advances in fabrication processes and technologies have allowed for the fabrication of small resonant structures whereby specific frequencies of light are found to be "amplified" (or resonate) inside the resonator (Yariv & Yeh, 2006). Resonators have found a broad range of applications in optics, including high-order filters (Little et al., 2004), as oscillators in specific parametric lasers (Kippenberg et al., 2004; Giordmaine & Miller, 1965), thin film polarization optics, and for frequency conversion (Turner et al., 2008; Ferrera et al., 2008). For the case of nonlinear optics, disks (whispering gallery modes) and micro-ring resonators have been used in 2D for frequency conversion (Grudinin et al., 2009; Ibrahim et al., 2002), whereas microtoroids and microsphere have been explored in 3D (Agha et al., 2007; Kippenberg et al., 1991). The net advantage of these structures is that, for resonant frequencies, a low input optical power can lead to enormous nonlinear effects due to the field enhancement provided by the cavity. In this section we examine the specific case of waveguide micro-ring resonators for wavelength conversion via parametric four wave mixing. Micro-ring resonators are integrated structures which can readily be implemented in future photonic integrated circuits. First a brief review of the field enhancement provided by resonators shall be presented, followed by the four-wave mixing relations. Promising experimental results in high-index doped silica resonators will then be shown and compared with other platforms.

4.1 Micro-ring resonators

Consider the four port micro-ring resonator portrayed in Fig. 7, and assume continuous wave light is injected from the Input port. Light is coupled from the input (bus) waveguide into the ring structure via evanescent field coupling (Marcuse, 1991). As light circulates around the ring structure, there is net loss from propagation losses, loss from coupling from the ring to the bus waveguides (2 locations), and net gain when the input light is coupled from the bus at the input to the ring. Note that this is in direct analogy with a standard Fabry-Perot cavity, where the reflectivity of the mirrors/sidewalls has been replaced with coupling coefficients. Using reciprocity and energy conservation relations at the coupling junction, the total transmission from the Input port to the Drop port is found to be (Yariv & Yeh, 2006):

$$I_{Drop} = I_0 \frac{k^2 \exp(-\alpha L/2)}{1+(1-k)^2 \exp(-\alpha L) - 2(1-k)\exp(-\alpha L/2)\cos(\beta L)} \tag{10}$$

Where I_0 is the input intensity, L the ring circumference (=$2\pi R$), and k is the power coupling ratio from the bus waveguide to the ring structure. A typical transmission profile inside such a resonator is presented in Fig. 8, where we have also defined the free spectral range and the width of the resonance, Δf_{FW}. Resonance occurs at frequencies $f_{res} = mc/nL$, where m is an even integer, and n is the effective refractive index of the mode, whereas the free spectral range is given by $FSR = c/nL$. In general the ring resonances are not equally spaced with frequency, as dispersion causes a shift in the index of refraction. The coupling coefficient can be expressed in terms of experimentally measured quantities:

$$k = 1 - \left(\rho + 1 - \sqrt{\rho^2 + 2\rho}\right)\exp(\alpha L/2) \tag{11}$$

where $\rho \approx \frac{1}{2}\left(\pi \frac{\Delta f_{FW}}{FSR}\right)^2$. At resonance, the local intensity inside the resonator is enhanced due to constructive interference. This intensity enhancement factor can be expressed as:

$$IE = \frac{k}{\left[(1-k)\exp(-\alpha L/2) - 1\right]^2} \tag{12}$$

These equations have extremely important applications. From Eq. (10) the transmission through the resonator is found to be unique at specific frequencies, hence the device can be utilized as a filter. Even more importantly for nonlinear optics, for an input signal that matches a ring resonance, the intensity is found to be enhanced, which can be utilized to observe large nonlinear phenomena with low input power levels (Ferrera et al., 2008). In the approximation of low propagation losses, Eq. (12) results in $IE \approx \pi \cdot Q \cdot FSR / f_0$, which implies that the larger the ring Q-factor ($Q = f_0/\Delta f_{FW}$), the larger the intensity enhancement.

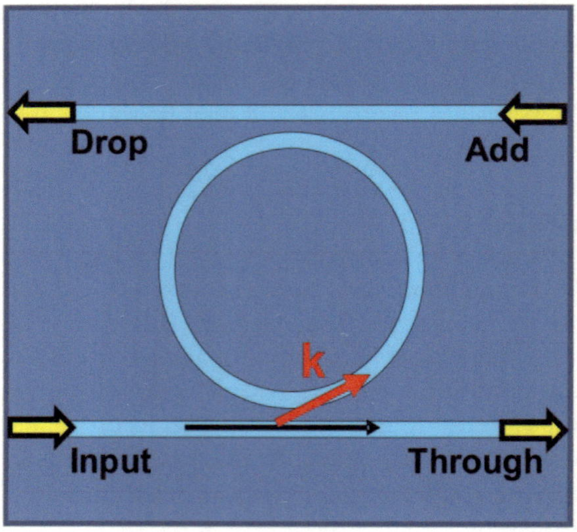

Fig. 7. Coupling coefficients and schematic of a typical 4-port ring resonator.

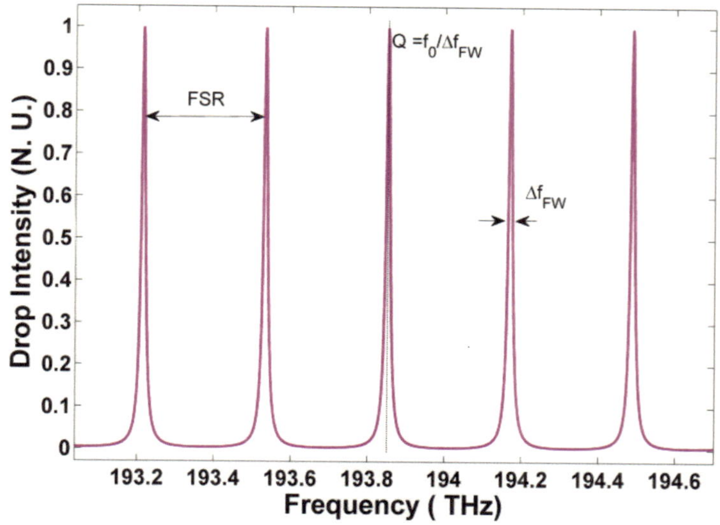

Fig. 8. Typical Fabry-Perot resonance transmission at the drop port of a resonator (input port excited). Here a FSR of 500GHz and a Q of 25 were used.

4.2 Four-wave mixing

Section 3 discussed third order nonlinear effects following the propagation of a single beam in a Kerr nonlinear medium. In this case the nonlinear interaction consisted of generating new frequencies through the spectral broadening of the input pulse. In general, we may consider multiple beams propagating through the medium, from which the nonlinear Schrodinger equation predicts nonlinear coupling amongst the components, a parametric process known as four wave mixing. This process can be used to convert energy from a strong pump to generate a new frequency component via the interaction with a weaker signal. As is displayed in the inset of Fig. 9, the quantum description of the process is in the simultaneous absorption of two photons to create 2 new frequencies of light. In the semi-degenerate case considered here, two photons from a strong pump beam (ψ_2) are absorbed by the medium, and when stimulated by a weaker signal beam (ψ_1) a new idler frequency (ψ_3) is generated from the parametric process. By varying the signal frequency, a tunable output source can be obtained (Agha et al., 2007; Grudinin et al., 2009). To describe the interaction mathematically, we consider 3 CW beams $E_i = \psi'(z)F(x,y)\exp(i\beta_i z - i\omega_i t)$, from which the following coupled set of equations governing the parametric growth can be derived (Agrawal, 2006):

$$\frac{\partial \psi_1}{\partial z} + \frac{\alpha}{2}\psi_1 = 2i\gamma|\psi_2|^2 \psi_1 + i\gamma\psi_2^2\psi_3^* \exp(i\Delta\beta z) \tag{13a}$$

$$\frac{\partial \psi_2}{\partial z} + \frac{\alpha}{2}\psi_2 = i\gamma|\psi_2|^2 \psi_2 + 2i\gamma\psi_1\psi_2^*\psi_3 \exp(-i\Delta\beta z) \tag{13b}$$

$$\frac{\partial \psi_3}{\partial z} + \frac{\alpha}{2}\psi_3 = 2i\gamma|\psi_2|^2 \psi_3 + i\gamma\psi_2^2\psi_1^* \exp(i\Delta\beta z) \tag{13c}$$

Where $\Delta\beta = 2\beta_2 - \beta_3 - \beta_1$ represents the phase mismatch of the process. In arriving to these equations we have assumed that the pump beam (ω_2) is much stronger than the signal (ω_1) and idler (ω_3), and that the waves are closely spaced in frequency so that the nonlinear parameter $\gamma = n_2\omega_0/cA$ is approximately constant for all three frequencies (the pump frequency should be used for ω_0). The phase mismatch term represents a necessary condition (i.e. $\Delta\beta = 0$) for an efficient conversion, and is the optical analogue of momentum conservation. On the other hand, energy conservation is also required and is expressed as: $2\omega_2 = \omega_1 + \omega_3$.

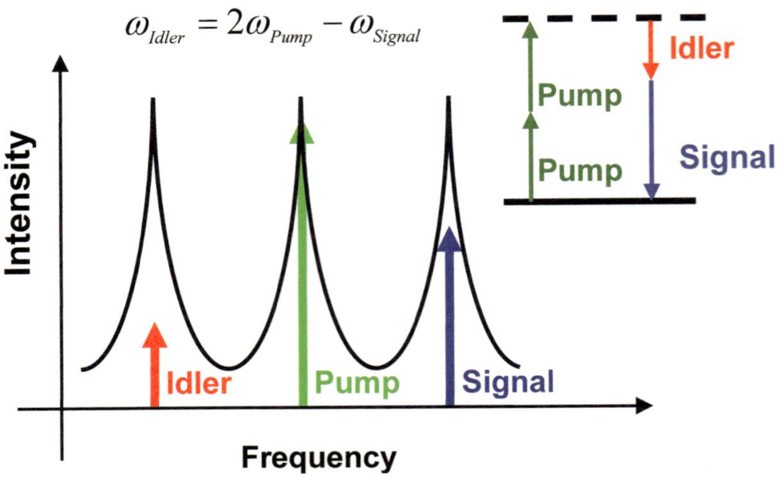

Fig. 9. Typical spectral intensity at the output of the resonator. (Inset) Energy diagram for a semi-degenerate four-wave mixing interaction.

The growth of the idler frequency can be obtained by assuming an undepleted pump regime, whereby the product $\psi_2 \exp(-az/2)$ is assumed to be approximately constant, and by solving the Eqs. (13) (Agrawal, 2003; Absil et al., 2000) we obtain:

$$P_3(z) = P_1\gamma^2 P_2^2 L_{eff}^2 \tag{14a}$$

$$L_{eff}^2 = z^2 \exp(-\alpha z)\left|\frac{1-\exp(-\alpha z + i\Delta\beta z)}{\alpha z - i\Delta\beta z}\right|^2 \tag{14b}$$

Where P_1, P_2 and P_3 here refer to the input powers of the signal, pump and idler beams respectively. The conversion is seen to be proportional to both the input signal power, and the square of the pump power. Again, we see that the process scales with the nonlinear parameter and is reduced if phase matching (i.e. $\Delta\beta = 0$) is not achieved. Various methods exist to achieve phase matching, including using birefringence and waveguide tailoring (Dimitripoulos et al., 2004; Foster et al., 2006; Lamont et al., 2008), but perhaps the simplest way is to work in a region of low dispersion. As is shown in (Agrawal, 2006), the phase mismatch term can be reduced to:

$$\Delta\beta \approx \beta_2(\omega_2 - \omega_1)^2 \tag{15}$$

and is thus directly proportional to the dispersion coefficient (note that at high power levels the phase mismatch becomes power dependant; see (Lin et al., 2008)).
For micro-ring resonator structures, Eq. (14) can be modified to account for the power enhancement provided by the resonance geometry. When the pump and signal beams are aligned to ring resonances, and for low dispersion conditions, phase matching will be obtained, and moreover, the generated idler should also match a ring resonance. In this case we may use Eq. (12) with Eq. (14) to give the expected conversion efficiency:

$$\eta \equiv \frac{P_3(L)}{P_1} = \gamma^2 P_2^2 L_{eff}^2 \cdot IE^4 , \tag{16}$$

where P_3 is the power of the idler at the drop port of the ring, whereas P_1 and P_2 are the input powers both at the Input port, both at the Add port, or one at the Add and the other at the Input (various configurations are possible). The added benefit of a ring resonator for four-wave mixing is clear: the generated idler power at the output of the ring is amplified by a factor of IE^4, which can be an extremely important contribution as will be shown below.
Four-wave mixing is an extremely important parametric process to be used in optical networks, and has found numerous applications. This includes the development of a multi-wavelength source for wavelength multiplexing systems (Grudinin et al., 2009), all-optical reshaping (Ciaramella & Trillo, 2000), amplification (Foster et al., 2006), correlated photon pair generation (Kolchin et al., 2006), and possible switching schemes have also been suggested (Lin et al., 2005). In particular, signal regeneration using four-wave mixing was shown in silicon at speeds of 10Gb/s (Rong et al., 2006). In an appropriate low loss material platform, ring resonators promise to bring efficient parametric processes at low powers.

4.3 Frequency conversion in doped silica glass resonators

The possibility of forming resonator structures primarily depends on the developed fabrication processes. In particular, low loss structures are a necessity, as photons will see propagation losses from circulating several times around the resonator. Furthermore, integrated ring resonators require small bending radii with minimal losses, which further require a relatively high-index contrast waveguide. The high-index doped silica glass discussed in this chapter meets these criteria, with propagation losses as low as 0.06 dB/cm, and negligible bending losses for radii down to 30 μm (Little, 2003; Ferrera et al., 2008).

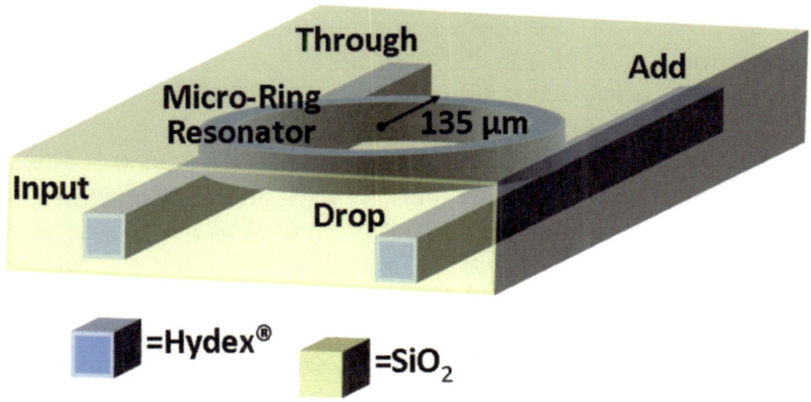

Fig. 10. Schematic of the vertically coupled high-index glass micro-ring resonator.

Two ring resonators will be discussed in this section, one with a radius of 47.5 μm, a Q factor ~65,000, and a bandwidth matching that for 2.5Gb/s signal processing applications, as well as a high Q ring of ~1,200,000, with a ring radius of 135μm for high conversion efficiencies typically required for applications such as narrow linewidth, multi-wavelength sources, or correlated photon pair generation (Kolchin et al., 2006; Kippenberg et al., 2004; Giordmaine & Miller, 1965). In both cases the bus waveguides and the ring waveguide have the same cross section and fabrication process as previously described in Section 2.2 and 3 (see Fig. 3). The 4-port ring resonator is depicted in Fig. 10, and light is injected into the ring via vertical evanescence field coupling. The experimental set-up used to characterize the rings is shown in Fig. 11, and consists of 2 CW lasers, 2 polarizers, a power meter and a spectrometer. A Peltier cell is also used with the high Q ring for temperature control.

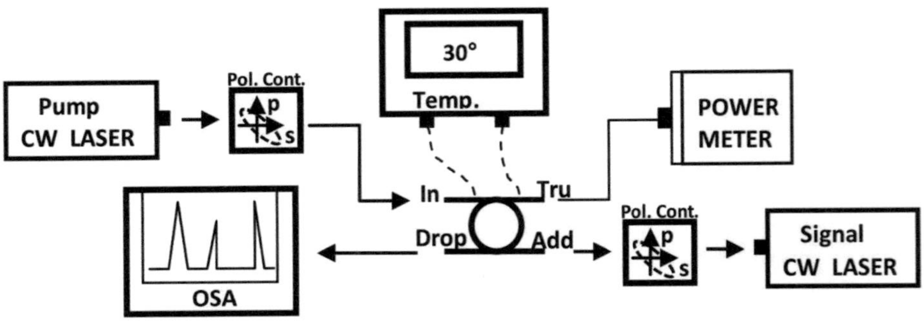

Fig. 11. Experimental set-up used to characterize the ring resonator and measure the converted idler from four wave mixing. 2 tunable fiber CW lasers are used, one at the input port and another at the drop port, whose polarizations and wavelengths are both set with inline fiber polarization controllers to match a ring resonance. The output spectrum and power are collected at the drop and through ports. A temperature controller is used to regulate the temperature of the device.

4.3.1 Dispersion

As detailed above, dispersion is a critical parameter in determining the efficiency of four-wave mixing. In ring resonators the dispersion can be directly extracted from the linear transmission through the ring. This was performed experimentally by using a wavelength tunable CW laser at the Input port and then recording the transmission at the drop port. The transmission spectral scan for the low Q ring can be seen in Fig. 12, from which a free spectral range of 575GHz and a Q factor of 65,000 were extracted (=200GHZ and 1,200,000 for the high Q).

As was derived in the beginning of Section 4, the propagation constants at resonance can be found to obey the relation: $\beta = m/R$, and thus are solely determined by the radius and an integer coefficient m. From vectorial finite element simulations the value of m for a specific resonance frequency can be determined, and hence the integer value of all the experimentally determined resonances is obtained (as they are sequential). This provides a relation between the propagation constant β and the angular frequency of the light ω. By fitting a polynomial relation to this relation, as described by Eq. (3), the dispersion of the ring resonator is obtained. Fig. 13 presents the group velocity dispersion in the high Q ring (due to the smaller spectral range, a higher degree of accuracy was obtained here in comparison with the low Q) obtained by fitting a fourth order polynomial relation on the experimental data (Ferrera et al., 2009). It is important to notice that the dispersion is extremely low for both the quasi-TE and quasi-TM modes of the structure, with zero dispersion crossings at λ = 1560nm (1595nm) for the TM (TE) mode. At 1550nm we obtain an anomalous GVD of β_2= -3.1 ± 0.9 ps^2/km for the TM mode and -10 ± 0.9 ps^2/km for the TE mode. These low dispersion results were expected from previous considerations (see Section 3.1), and are ideal for four wave mixing applications.

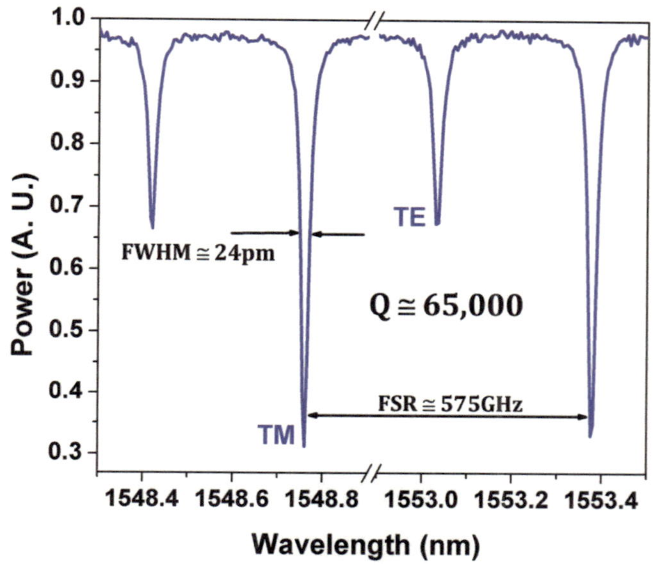

Fig. 12. Input-Drop response of the low Q (65,000) micro ring resonator.

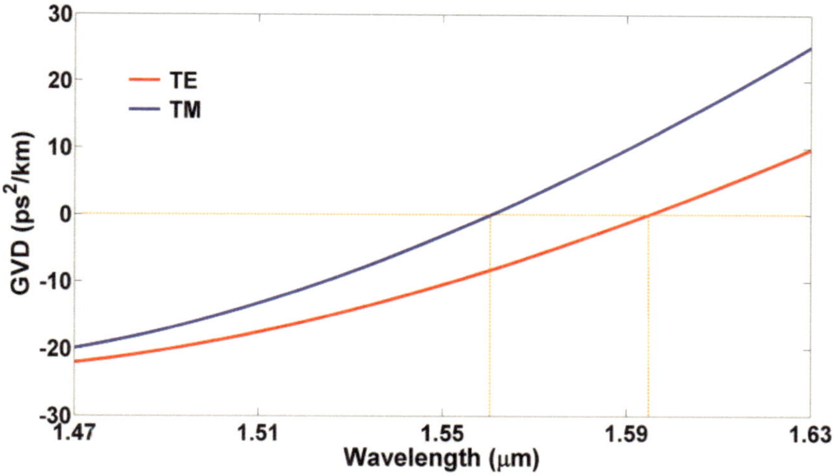

Fig. 13. Experimentally obtained dispersion (after fitting) of both the TE and TM fundamental modes. The zero-GVD points are found to be at 1594.7nm and 1560.5nm for TE and TM, respectively.

In addition, the dispersion data can be used to predict the bandwidth over which four-wave mixing can be observed. In resonators, the linear phase matching condition for the propagation constants is automatically satisfied as the resonator modes are related linearly by : $\beta = m/R$ (Ferrera et al., 2009; Lin et al., 2008). Rather, energy conservation becomes the new phase matching conditions, expressed as $\Delta\omega = 2\omega_2 - \omega_1 - \omega_{3r}$, where ω_1 and ω_2 are aligned to resonances by construction, but where the generated idler frequency $\omega_3 = 2\omega_2 - \omega_1$ is not necessarily aligned to its closest resonance ω_{3r} (Ferrera et al., 2009; Lin et al., 2008). We define the region where four-wave mixing is possible through the relation $|\Delta\omega| = |\omega_3 - \omega_{3r}| < \omega_{3r} / 2Q$, such that the generated idler is within the bandwidth of the resonance. This condition for the high Q ring is presented in Fig. 14, where the region in black represents absence of phase matching, whereas the colored region represents possible four-wave mixing (the blue region implies the lowest phase mismatch, and red the highest). The curvature in the plots is a result of high order dispersive terms. It can be seen that the four wave mixing can be accomplished in the vicinity of the zero dispersion points up to 10THz (80nm) away from the pump. This extraordinary result comes from the low dispersion of the resonator, which permits appreciable phase matching over a broad bandwidth at low power. However, it is important to note that four-wave mixing can always be accomplished in an anomalous dispersion regime given a sufficiently high power (Lin et al., 2008).

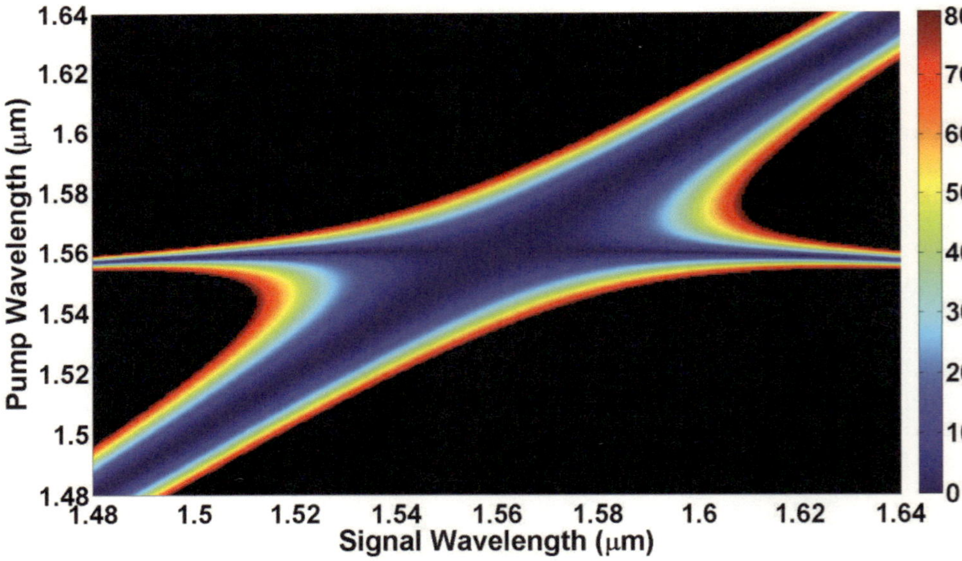

Fig. 14. Phase matching diagram associated to four-wave mixing in the high Q micro-ring resonator (interpolated). The regions in black are areas where four wave-mixing is not possible, whereas the coloured regions denote possible four-wave mixing with the colour indicating the degree of frequency mismatch (blue implies perfect phase matching; colour scale is $\Delta\omega$ in MHz).

4.3.2 Four-wave mixing in microring resonators

The 3 GHz bandwidth of the low Q resonator can be used for telecommunication applications such as signal regeneration and amplification (Salem et al., 2008; Ta'eed et al., 2005; Salem et al., 2007). We note that the device reported here was primarily designed for linear filter applications at 2.5 Gb/s, leaving room for further optimization for higher-bit-rate nonlinear applications. Four-wave mixing was obtained in this ring using only 5mW of input pump power at a resonance of 1553.38 nm, while the signal was tuned to an adjacent resonance at 1558.02nm with a power of 550 μW. Fig. 15 depicts the recorded output spectrum showing the generation of 2 idler frequencies: one of 930pW at 1548.74nm, and a second of 100pW at 1562.69nm. The latter idler is a result of formally exchanging the role of the pump and signal beams. The lower output idler power for the reverse process is a direct result of Eq. (16), where the conversion is shown to be proportional to the pump power squared. As is reported in (Ferrera et al., 2008), this is the first demonstration of four wave mixing in an integrated glass structure using CW light of such low power. This result is in part due to the relatively large γ factor of 220 $W^{-1}km^{-1}$ (as compared to single mode fibers) and, more importantly, due to the low losses, resulting in a large intensity enhancement factor of $IE^4 \sim 1.4 \times 10^7$, which is orders of magnitude better than in semiconductors, where losses are typically the limiting factor (Turner et al., 2008; Siviloglou et al., 2006). Recent results in SOI have also shown impressive, and slightly higher, conversion efficiencies using CW power levels. However, as can be seen in (Turner et al., 2006), saturation due to two-photon absorption generated free carriers limits the overall process, whereas in silica-doped glass it has been shown that no saturation effects occur for more than 25 GW/cm^2 of intensity, allowing for much higher conversion efficiencies with an increased pump power (note that the pump intensity in the ring is only ~0.015GW/cm^2 at resonance for 5mW of input power) (Duchesne et al., 2009).

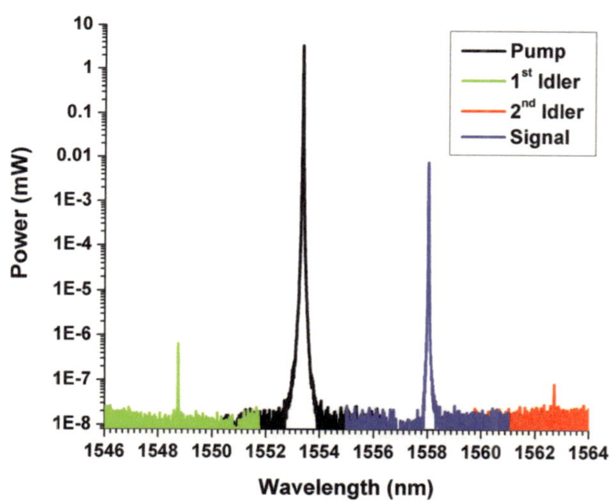

Fig. 15. Wavelength conversion in the low Q ring resonator.

The predicted frequency conversion, Eq. (16), was also verified experimentally. Firstly, for a fixed input pump power of 20mW the signal power was varied and the expected linear relationship between idler power and signal power was obtained, with a total conversion efficiency of 25×10^{-6}, as is shown in Fig. 16. Moreover, the reverse situation in which the pump power was varied for a fixed signal power also demonstrated the expected quadratic dependence, validating the approximations leading to Eq. (16). Lastly, by tuning the signal wavelength slightly off-resonance and measuring the conversion efficiency, it was experimentally shown, Fig. 17, that these results were in quasi-perfect phase matching, as predicted from Fig. 14.

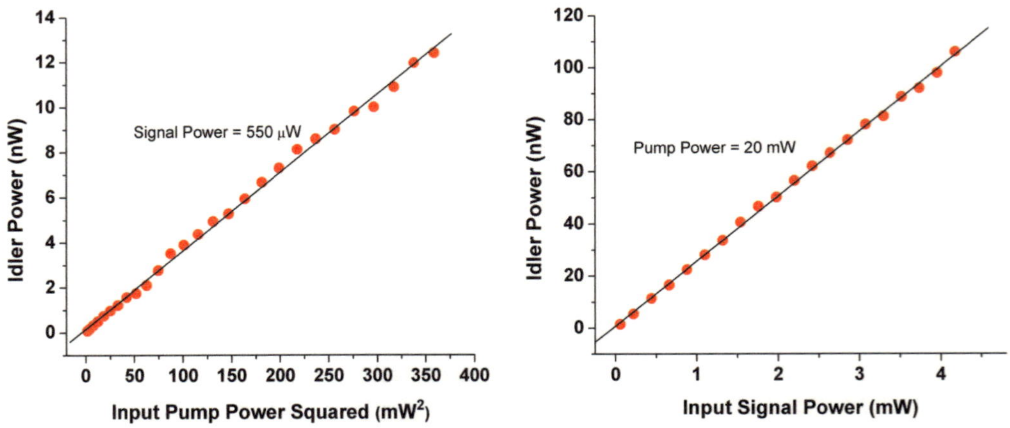

Fig. 16. Linear and quadratic relation of the idler power versus input signal power and input pump power, respectively.

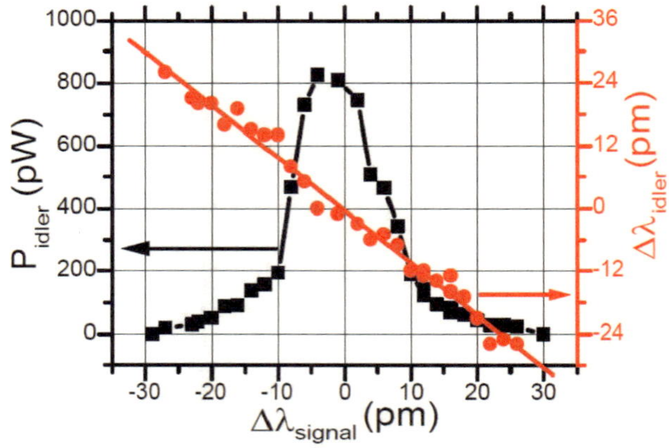

Fig. 17. Idler detuning curve, showing that dispersion is negligible in the system.

Experiments were also carried out in the high Q=1,200,000 ring resonator, which is for applications other than telecommunications, such as the realization of a narrow line source (Kawase et al., 2001). The advantage of this ring is the tremendous intensity enhancement factor *IE* =302.8, which amounts to a conversion efficiency enhancement as high as 8.4 x 10^9. Fig. 18 summarises the results of two different experiments where the pumps are placed to adjacent resonances, and when they are placed 6 free spectral ranges away from each other, respectively. In both cases the conversion efficiency was estimated to be -36 dB with only 8.8mW of input power. Moreover, a cascade of four-wave mixing processes can be seen whereby the pump and 1st idler mix to generate a 3rd idler (the numbers refer to Fig. 18).

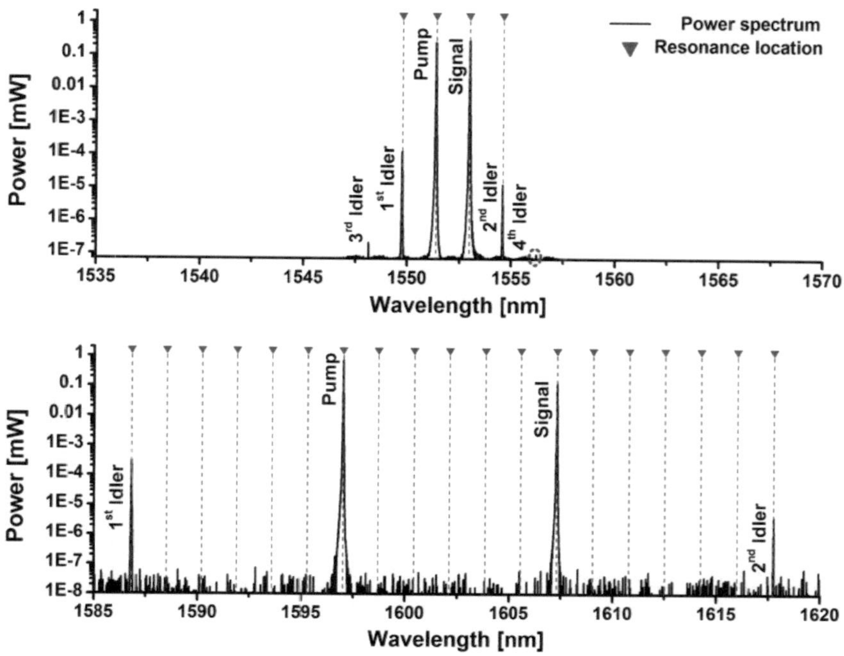

Fig. 18. Four wave mixing across several resonances in the high Q resonator. A conversion efficiency of -36db is obtained with only 8.8mW of external CW power.

5. Perspectives and conclusions

In this chapter we have presented a novel high-index doped-silica material platform for future integrated nonlinear optical applications. The platform acts as a compromise between the attractive linear properties of single mode fibers, namely low propagation losses and robust fabrication process, and those of semiconductors and other nonlinear glasses, and this by having a relatively large nonlinear parameter. Moreover, it outshines other high index glasses in its ability to have very low loss waveguides of 0.06 dB/cm without high temperature anneal, allowing for a complete CMOS compatible fabrication process. A small cross sectional area combined with a high index contrast also allows for tight bends down to 30 µm with negligible losses, permitting long spiral or resonant structures on chip. We have shown that although semiconductors possess a much larger nonlinearity γ, the low losses and robust fabrication allows for long and resonant structures with large and appreciable nonlinear effects that would otherwise not be possible in most semiconductors, or saturate with increasing input powers for others. In particular, we have presented and described measurement techniques to characterise the linear and third order nonlinearities, with specific applications to parametric four wave mixing.

Apart from the imminent applications in future photonic integrated circuits, these results may also pave way for a wide range of applications such as narrow linewidth, and/or multi-wavelength sources, on-chip generation of correlated photon pairs, as well as sources for ultra-low power optical hyperparametric oscillators.

6. References

Absil, P. P., Hryniewicz, J. V.; Little, B. E.; Cho, P. S.; Wilson, R. A.; Joneckis, L. G. & Ho, P.-T. (2000). Wavelength conversion in GaAs micro-ring resonators. *Opt. Lett.*, Vol. 25, No, 8, 554-556.

Afshar V., S. & Monro, T. M. (2009). A full vectorial model for pulse propagation in emerging waveguides with subwavelength structures part I: Kerr nonlinearity. *Opt. Express*, Vol. 17, No. 4, 2298-2318.

Agha, I. H.; Okawachi, Y.; Foster, M. A.; Sharping, J. E. & Gaeta, A. L. (2007). Four-wave-mixing parametric oscillations in dispersion-compensated high-Q silica microspheres. *Phys. Rev. A*, Vol. 76, No. 4, 043837.

Agrawal, G. P. (2006). *Nonlinear Fiber Optics*. Academic Press, ISBN 978-0123695161, San Diego.

Alduino, A. & Paniccia, M. (2007). Interconnects: Wiring electronics with light. *Nat. Photon.*, Vol. 1, 153-155.

Almeida, V. R.; Panepucci, R. R. & Lipson, M. (2003). Nanotaper for compact mode conversion. *Opt. Lett.*, Vol. 28, No. 15, 1302-1304.

Bergano, N. S. (2005). Wavelength division multiplexing in long-haul transoceanic transmission systems. *J. Lightw. Technol.*, Vol. 23, No. 12, 4125-4139.

Borselli, M.; Johnson, T. J. & Painter, O. (2006). Measuring the role of surface chemistry in silicon microphotonics. *Appl. Phys. Lett.*, Vol. 88, No. 13, 131114.

Boskovic, A.; Chernikov, S. V.; Taylor, J. R.; Gruner-Nielsen L. & Levring, O. A. (1996). Direct continuous-wave measurement of n2 in various types of telecommunication fiber at 1.55 μm. *Opt. Lett.*, Vol. 21, No. 24, 1966-1968.

Boyd, R. W. (2008). *Nonlinear Optics*. Academic Press, ISBN 978-0123694706, San Diego.

Ciaramella, E. & Trillo, S. (2000). All-optical reshaping via four-wave mixing in optical fibers. *IEEE Photon. Technol. Lett.*, Vol. 12, No.7 , 849-851.

Dimitropoulos, D.; Raghunathan, V.; Claps, R. & Jalali, B. (2004). Phase-matching and nonlinear optical processes in silicon waveguides. *Opt. Express*, Vol. 12, No. 1, 149–160.

Duchesne, D.; Ferrera, M.; Razzari, L.; Morandotti, R.; Little, B. E.; Chu, S. T. & Moss, D. J. (2009). Efficient self-phase modulation in low loss, high index doped silica glass integrated waveguides. *Opt. Express*, Vol. 17, No. 3, 1865-1870.

Dulkeith, E.; Vlasov, Y. A.; Chen, X.; Panoiu, N. C. & Osgood, R. M. Jr. (2006). Self-phase-modulation in submicron silicon-on-insulator photonic wires. *Opt. Express*, Vol. 14, No. 12, 5524-5534.

Eggleton, B. J.; Radic, S. & Moss, D. J. (2008). Nonlinear Optics in Communications: From Crippling Impairment to Ultrafast Tools, In: *Optical Fiber Telecommunications V: Components and Sub-systems*, Kaminow, I. P.; Li, T. & Willner, A. E., 5th Ed., Chap. 20, Academic Press, ISBN 978-0123741714, Oxford.

Espinola, R.; Dadap, J.; Osgood, R. Jr.; McNab, S. & Vlasov, Y. (2004). Raman amplification in ultrasmall silicon-on-insulator wire waveguides. *Opt. Express*, Vol. 12, No. 16, 3713-3718.

Ferrera, M.; Razzari, L.; Duchesne, D.; Morandotti, R.; Yang, Z.; Liscidini, M.; Sipe, J. E.; Chu, S.; Little, B. & Moss, D. J. (2008). Low power continuous-wave nonlinear optics in doped silica glass integrated waveguide structures. *Nat. Photonics*, Vol. 2, 737-740.

Ferrera, M.; Duchesne, D.; Razzari, L.; Peccianti, M.; Morandotti, R.; Cheben, P.; Janz, S.; Xu, D.-X.; Little, B. E.; Chu, S. & Moss, D. J. (2009). Low power four wave mixing in an integrated, micro-ring resonator with Q = 1.2 million. *Opt. Express*, Vol. 17, No. 16, 14098-14103.

Foster, M. A.; Turner, A. C.; Sharping, J. E.; Schmidt, B. S.; Lipson, M. & Gaeta, A. L. (2006). Broad-band optical parametric gain on a silicon photonic chip. *Nature*, Vol. 441, 960-963.

Foster, M. A.; Turner, A. C.; Lipson, M. & Gaeta, A. L. (2008). Nonlinear optics in photonic nanowires. *Opt. Express*, Vol. 16, No. 2, 1300-1320.

Giordmaine, J. A. & Miller, R. C. (1965). Tunable coherent parametric oscillation in LiNbO3 at optical frequencies. *Phys. Rev. Lett.*, Vol. 14, No. 24, 973 - 976.

Gondarenko, A.; Levy, J. S. & Lipson, M. (2009). High confinement micron-scale silicon nitride high Q ring resonator. *Opt. Express*, Vol. 17, No. 14, 11366-11370.

Grudinin, I. S.; Yu, N. & Maleki, L. (2009). Generation of optical frequency combs with a CaF2 resonator. *Opt. Lett.*, Vol. 34, No. 7, 878-880.

Ibrahim, T. A.; Van, V. & Ho, P.-T. (2002). All-optical time-division demultiplexing and spatial pulse routing with a GaAs/AlGaAs microring resonator. *Opt. Lett.*, Vol. 27, No. 10, 803-805.

Izhaky, N.; Morse, M. T.; Koehl, S.; Cohen, O.; Rubin, D.; Barkai, A.; Sarid, G.; Cohen, R. & Paniccia, M. J. (2006). Development of CMOS-compatible integrated silicon photonics devices. *IEEE J. Sel. Top. Quantum Electron.* Vol. 12, No. 6, 1688 – 1698.

Jalali, B.; Fathpour, S. (2006). Silicon photonics. *J. Lightw. Technol.*, Vol. 24, No. 12, 4600–4615.

Jouad, A. & Aimez, V. (2006). Passivation of air-exposed AlGaAs using low frequency plasma-enhanced chemical vapor deposition of silicon nitride. *Appl. Phys. Lett.*, Vol. 89, No. 9, 092125.

Kawase, K.; Shikata, J.; Imai, K. & Ito, H. (2001). Transform-limited, narrow-linewidth, terahertz-wave parametric generator. *Appl. Phys. Lett.*, Vol. 78, No. 19, 2819-2821.

Kippenberg, T. J.; Spillane, S. M. & Vahala, K. J. (2004). Kerr-nonlinearity optical parametric oscillation in an ultrahigh-Q toroid microcavity. *Phys. Rev. Lett.*, Vol. 93, No. 8, 083904.

Kivshar, Y. S. & Malomed, B. A. (1989). Dynamics of solitons in nearly integrable systems. *Rev. Mod. Phys.*, Vol. 61, 763 - 915.

Koch, T. L. & Koren, U. (1991). Semiconductor photonic integrated circuits. *IEEE J. Quantum. Electron.*, Vol. 27, No. , 641-653.

Kolchin, P.; Du, S.; Belthangady, C.; Yin, G. Y.; Harris, S. E. & Ginzton, E. L. (2006). Generation of narrow-bandwidth paired photons: use of a single driving laser. *Phys. Rev. Lett.*, Vol. 97, No. 11, 113602.

Lamont, M. R. E.; Rochette, M.; Moss, D. J. & Eggleton, B. J. (2006). Two-photon absorption effects on self-phase-modulation-based 2R optical regeneration. *IEEE Photon. Technol. Lett.*, Vol. 18, No. 10, 1185-1187.

Lamont, M. R. E.; Kuhlmey, B. T. & Martijn de Sterke, C. (2008). Multi-order dispersion engineering for optimal four-wave mixing. *Opt. Express*, Vol. 16, No. 10, 7551-7563.

Lee, J. H.; Kikuchi, K.; Nagashima, T.; Hasegawa, T.; Ohara, S. & Sugimoto, N. (2005). All-fiber 80-Gbit/s wavelength converter using 1-m-long bismuth oxide-based nonlinear optical fiber with a nonlinearity gamma of 1100 $W^{-1}km^{-1}$. *Opt. Express*, Vol. 13, No. 8, 3144-3149.

Lee, B. G.; Chen, X.; Biberman, A.; Liu, X.; Hsieh, I.-W.; Chou, C.-Y.; Dadap, J. I.; Xia, F.; Green, W. M. J.; Sekaric, L.; Vlasov, Y. A.; Osgood, R. M. & Bergman, K. (2008). Ultrahigh-bandwidth silicon photonic nanowire waveguides for on-chip networks. *IEEE Photon. Technol. Lett.*, Vol. 20, No. 6, 398-400.

Lepetit, L.; Chériaux, G. & Joffre, M. (1995). Linear techniques of phase measurement by femtosecond spectral interferometry for applications in spectroscopy. *J. Opt. Soc. B*, Vol. 12, No. 12, 2467-2474.

Li, W. T.; Ruan, Y. L.; Luther-Davies, B.; Rode, A. & Boswell, R. (2005). Dry-etch of As2S3 thin films for optical waveguide fabrication. *J. Vacuum Sci. Technol.*, Vol. 23, No. 6, 1626-1632.

Liang, T. K. & Tsang, H. K. (2004). Role of free carriers from two-photon absorption in Raman amplification in silicon-on-insulator waveguides. *Appl. Phys. Lett.*, Vol. 84, No. 15, 2745-2747.

Lifante, G. (2003). *Integrated Photonics*. John Wiley & Sons, ISBN 978-0470848685, England.

Little, B. E. & Chu, S. T. (2000). Towards very large-scale integrated photonics. *Opt. Photon. News*, Vol. 11, No. 11, 24-29.

Little, B. E. (2003). A VLSI photonics platform. *Optical Fiber Communication Conference*, 444-445, Atlanta, Georgia, March 2003, Optical Society of America.

Little, B. E.; Chu, S. T.; Absil, P. P.; Hryniewicz, J. V.; Johnson, F. G.; Seiferth, F.; Gill, D.; Van, V.; King, O. & Trakalo, M. (2004). Very high-order microring resonator filters for WDM applications," *IEEE Photon. Technol. Lett.*, Vol. 16, No. 10, 2263-2265.

Lin, Q.; Jiang, R.; Marki, C. F.; Mckinstrie, C. J.; Jopson R.; Ford J.; Agrawal G. P. & Radic S. (2005). 40-Gb/s optical switching and wavelength multicasting in a two-pump parametric device. *IEEE Photon. Technol. Lett.* Vol. 17, No. 11, 2376-2378.

Lin, Q.; Johnson, T. J.; Perahia, R.; Michael, C. P. & Painter, O. J. (2008). A proposal for highly tunable optical parametric oscillation in silicon micro-resonators. *Opt. Express*, Vol. 16, No. 14, 10596-10610.

Marcuse, D. (1991). *Theory of dielectric optical waveguides*. Academic Press, ISBN 978-0124709515, Boston.

Moerman, I.; Van Daele, P. P. & Demeester, P. M. (1997). A review on fabrication technologies for the monolithic integration of tapers with III–V semiconductor devices. *IEEE J. Sel. Top. Quantum. Electron.*, Vol. 3, No. 6, 1308-1320.

Mollenauer, L. F.; Stolen, R. H. & Gordon, J. P. (1980). Experimental observation of picosecond pulse narrowing and solitons in optical fibers. *Phys. Rev. Lett.*, Vol. 45, No. 13 , 1095 - 1098.

Nguyen, H. C.; Finsterbusch, K.; Moss, D. J. & Eggleton, B. J. (2006). Dispersion in nonlinear figure of merit of As2Se3 chalcogenide fibre. *Electr. Lett.*, Vol. 42, No. 10, 571-572.

Pelusi, M. D.; Ta'Eed, V. G.; Lamont, M. R. E.; Madden, S.; Choi, D.-Y.; Luther-Davies, B. & Eggleton, B. J. (2007). Ultra-high nonlinear As2S3 planar waveguide for 160 Gb/s optical time-division demultiplexing by four-wave mixing. *IEEE Photon. Technol. Lett.*, Vol. 19, No. 19, 1496–1498.

Rong, H.; Kuo, Y.-H.; Liu, A.; Paniccia, M. & Cohen, O. (2006). High efficiency wavelength conversion of 10 Gb/s data in silicon waveguides. *Opt. Express*, Vol. 14, No. 3, 1182-1188.

Rong, H.; Xu, S.; Cohen, O.; Raday, O.; Lee, M.; Sih, V. & Paniccia, M. (2008). A cascaded silicon Raman laser. *Nat. Photon.* Vol. 2, 170-174.

Ruan, Y. L.; Li, W. T.; Jarvis, R.; Madsen, N.; Rode, A. & Luther-Davies, B. (2004). Fabrication and characterization of low loss rib chalcogenide waveguides made by dry etching. *Opt. Express*, Vol. 12, No. 21, 5140-5145.

Russell, P. (2003). Photonic crystal fibers. *Science*, Vol. 299, No. 5605, 358-362.

Salem, R.; Foster, M. A.; Turner, A. C.; Geraghty, D. F.; Lipson, M. & Gaeta, A. L. (2007). All-optical regeneration on a silicon chip. *Opt. Express*, Vol. 15, No. 12, 7802–7809.

Salem, R.; Foster, M. A.; Turner, A. C.; Geraghty, D. F.; Lipson, M. & Gaeta, A. L. (2008). Signal regeneration using low-power four-wave mixing on silicon chip. *Nat. Photon.*, Vol. 2, 35–38.

Sellmeier, W. (1871). *Annalen der Physik und Chemie*, Vol. 143, 271.

Shokooh-Saremi, M.; Ta'eed, V. G.; Baker, N. J.; Littler, I. C. M.; Moss, D. J.; Eggleton, B. J.; Ruan, Y. & Luther-Davies, B. (2006). High-performance Bragg gratings in chalcogenide rib waveguides written with a modified Sagnac interferometer. *J. Opt. Soc. Am. B*, Vol. 23, No. 7, 1323-1331.

Siviloglou, G. A.; Suntsov, S.; El-Ganainy, R.; Iwanow, R.; Stegeman, G. I.; Christodoulides, D. N.; Morandotti, R.; Modotto, D.; Locatelli, A.; De Angelis, C.; Pozzi, F.; Stanley, C. R. & Sorel, M. (2006). Enhanced third-order nonlinear effects in optical AlGaAs nanowires. *Opt. Express*, Vol. 14, No. 20, 9377-9384.

Ta'eed, V. G.; Shokooh-Saremi, M.; Fu, L.; Moss, D. J.; Rochette, M.; Littler, I. C. M.; Eggleton, B. J.; Ruan, Y. & Luther-Davies, B. (2005). Integrated all-optical pulse regeneration in chalcogenide waveguides. *Opt. Lett.*, Vol. 30, No. 21, 2900–2902.

Ta'eed, V. G.; Baker, N. J.; Fu, L.; Finsterbusch, K.; Lamont, M. R. E.; Moss, D. J.; Nguyen, H. C.; Eggleton, B. J.; Choi, D.-Y.; Madden, S. & Luther-Davies, B. (2007). Ultrafast all-optical chalcogenide glass photonic circuits. *Opt. Express*, Vol. 15, No. 15, 9205–9221.

Tong, L. M.; Gattass, R. R.; Ashcom, J. B.; He, S. L.; Lou, J. Y.; Shen, M. Y.; Maxwell, I. & Mazur, E. (2003). Subwavelength-diameter silica wires for low-loss optical wave guiding. *Nature*, Vol. 426, 816–819.

Tsang, H. K. & Liu, Y. (2008). Nonlinear optical properties of silicon waveguides. *Semicond. Sci. Technol.*, Vol. 23, No. 6, 064007.

Tsybeskov, L.; Lockwood D. J. & Ichikawa M. (2009). Silicon photonics: CMOS going optical," *Proceedings of the IEEE*, Vol. 97, No. 9, 1161-1165.

Turner, A. C.; Foster, M. A.; Gaeta, A. L. & Lipson, M. (2008). Ultra-low power parametric frequency conversion in a silicon microring resonator. *Opt. Express*, Vol. 16, No. 7, 4881–4887.

Venugopal Rao, S.; Moutzouris K. & Ebrahimzadeh, M. (2004). Nonlinear frequency conversion in semiconductor optical waveguides using birefringent, modal and quasi-phase-matching techniques. *J. Opt. A: Pure Appl. Opt.*, Vol. 6, 569–584.

Wise, F.; Qian, L. & Liu, X. (2002). Applications of cascaded quadratic nonlinearities to femtosecond pulse generation. *J. Nonlin. Opt. Phys. Mat.*, Vol. 11, No. 3, 317-338.

Wörhoff, K.; Hilderink, L. T. H.; Driessen, A. & Lambeck, P. V. (2002). Silicon oxynitride. A versatile material for integrated optics applications. *J. Electrochem. Soc.*, Vol. 149, No. 8, F85-F91.

Yablonovitch, E.; Gmitter, T. J. & Leung, K. M. (1991). Photonic band structure: The facecentered-cubic case employing nonspherical atoms. *Phys. Rev. Lett.*, Vol. 67, No. 17, 2295–2298.

Yalcın, A.; Popat, K. C.; Aldridge, J. C.; Desai, T. A.; Hryniewicz, J.; Chbouki, N.; Little, B. E.; King, O.; Van, V.; Chu, S.; Gill, D.; Anthes-Washburn, M.; Unlu, M. S. & Goldberg, B. B. (2006). Optical Sensing of Biomolecules Using Microring Resonators. *IEEE J. Sel. Top. Quantum Electron.*, Vol. 12, No. 1, 148-155.

Yariv, A. & Yeh, P. (2006). *Photonics: Optical Electronics in Modern Communications*. Oxford University Press, ISBN 978-0195179460, USA.

Yeom, D.-I.; Mägi, E. C.; Lamont, M. R. E.; Roelens, M. A. F.; Fu, L. & Eggleton, B. J. (2008). Low-threshold supercontinuum generation in highly nonlinear chalcogenide nanowires. *Opt. Lett.*, Vol. 33, No. 7,660–662.

15

Advances in Femtosecond Micromachining and Inscription of Micro and Nano Photonic Devices

Graham N. Smith[1], Kyriacos Kalli[2] and Kate Sugden[1]
[1]*Photonics Research Group, Aston University*
[2]*Nanophotonics Research Laboratory, Cyprus University of Technology*
[1]*United Kingdom*
[2]*Cyprus*

1. Introduction

Femtosecond micromachining and inscription uses focussed femtosecond laser pulses to fundamentally change materials through the interaction of the pulse and material. The fabrication of intricate microstructures on the surface of opaque materials, or within the bulk volume of optically transparent glassy or polymeric materials, is widely recognized as an important development in the innovation of advanced components in fields such as medicine (stent production) and photonics (micro-devices, sensors). This chapter sets out to introduce the topic of femtosecond laser micromachining and inscription of different materials and will focus on the practical considerations of the subject.

A femtosecond, often shortened to fs, corresponds to 10^{-15} of a second and in femtosecond laser machining specialised lasers generate extremely high intensity pulses, where each pulse has a temporal duration that typically ranges between 50 to 500fs, which subsequently interact with materials. This interaction has a number of interesting properties that can be utilised in shaping and permanently changing the physical, chemical and optical properties of the material. Recent advances in the development of femtosecond laser sources have accelerated the growth of this field resulting in many exciting novel applications.

The extremely short temporal profiles of femtosecond laser pulses, when combined with tightly focusing optics, produce sub-micron scale material changes that are associated with the parts of the beam where the optical intensity is extremely high, typically reaching $10^{13}W/cm^2$. Under such conditions, intensity-dependent non-linear absorption processes, such as multi-photon absorption and avalanche ionisation take place, leading to permanent structural and chemical changes in a myriad of materials (N. H. Rizvi, nd). This is particularly important for transparent materials since they have electronic band gaps too large to bridge with a single photon process and therefore the linear absorption of laser light does not occur. As a result of this there is minimal interaction between the laser and material; however, with a tightly focused femtosecond laser beam the interaction can be triggered and the non-linear effects dominate. Any structuring therefore only occurs around the laser focus and has strong spatial confinement. The induced structural changes can take many forms; index change and void creation are two common examples and can be used to form true 3-dimensional structures. This is only possible because of the non-linear behaviour of the pulse-material interaction at the focus.

In this chapter we discuss the following subject areas: the key concepts of femtosecond pulse and material interaction; the materials that are used as machinable substrates; we offer a review of the types of femtosecond lasers that are commercially available; we introduce some techniques applied in the field to date and a review of some of the current applications for this type of technology.

2. Fundamental considerations

Having first looked at the fundamentals of how femtosecond lasers interact with materials, we will look at the properties of a femtosecond laser beam and discuss a number of parameters that strongly affect the micromachining process.

2.1 Pulse energy deposition mechanisms

The nature of the interaction between a material and a femtosecond pulse is largely dependent on the energy bandgap of the material and the energy of the incident photons. This determines whether single or multiphoton (figure 1) absorption will dominate. If the photon energy is greater than the bandgap then single photon absorption dominates. In this instance a photon is absorbed and an electron in the valence band is promoted to the conduction band leaving a hole. It is worth noting that in some indirect bandgap semiconductors single photon absorption can also occur but this process requires assistance from phonon interactions. When the bandgap is greater than the photon energy then multiphoton absorption becomes important. Multiphoton absorption relies on a number of photons arriving and being absorbed by the electron within a short space of time and is the key absorption mechanism in transparent insulating materials such as fused silica (amorphous SiO_2), where the band gap is large. Multiphoton absorption can also be important for laser absorption in materials where single photon absorption is inhibited by band filling, for example in indirect gap semiconductor materials such as silicon (Sundaram & Mazur, 2002; J. P. Callan et al., 2001).

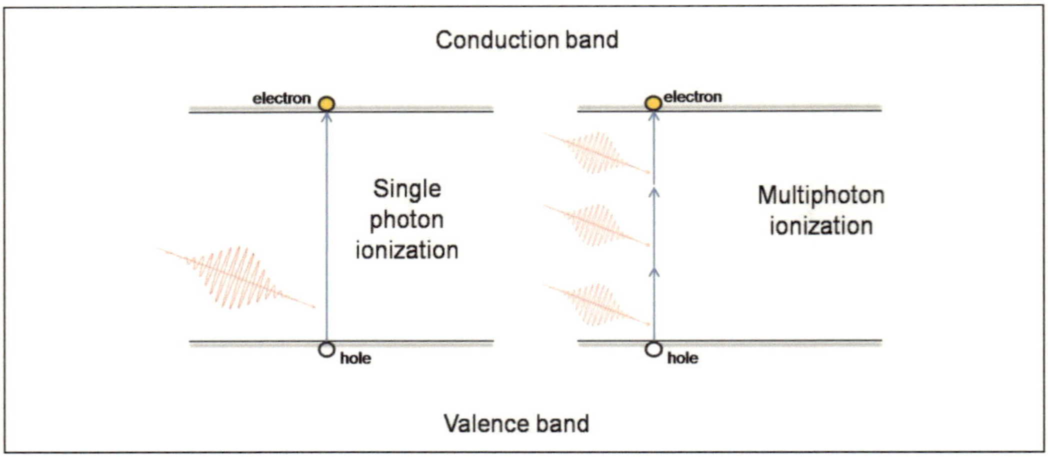

Fig. 1. A schematic of single and multiphoton absorption of incident photons (shown in red) causing electrons to be promoted to the conduction band from the valence band.

When a femtosecond laser beam is focussed onto a transparent material very little absorption occurs until the light intensity is high enough for multiphoton absorption to occur. The multiphoton ionisation that triggers the multiphoton absorption results from several photons being resonantly absorbed causing an electron to be promoted from the valence band to the conduction band.

In solids, scattering effects cause a loss of phase coherence between the excitation and incident photons. This means that the coherence dependent effects (i.e. ionisation) are observed in timescales of between 10-100 fs. It is only after this time that any created carriers become free. These carriers are then able to absorb more energy through free carrier absorption. This process does not increase the free carrier population only the energy although in extreme cases impact ionization can be triggered creating additional excited carriers.

Once excited the electrons and holes, through carrier-carrier and carrier-phonon scattering, spread and re-organise themselves in the conduction and valence bands. The two processes involved in the redistribution of carrier energy vary significantly in their mechanisms. Carrier-carrier exchange is a two body electrostatic interaction between two carriers. In this process there is no loss or gain in the total energy in a given system. This type of scattering takes 100s of femtosecond to reach an even distribution (the Fermi-Dirac state) through dephasing. Carrier-phonon scattering involves the free carriers losing and gaining energy and momentum through the emission and absorption of phonons. This transfer can be both inter and intra valley as well as within the bands themselves. The interaction does not change the number of carriers, however, their energy decreases due to spontaneous phonon emission with the energy being transferred to the substrate lattice. The two processes also have different timescales, although they can scatter concurrently. The carrier-phonon scattering lasts several picoseconds as the individual phonons carry limited energy and thus to dissipate the energy to thermal equilibrium takes longer.

Once the electrons, holes and lattice are in thermal equilibrium there will still be an excess of free carriers, as compared to when the entire material is in thermal equilibrium. These excess carriers are removed through either recombination (Auger and radiative) of electron-hole pairs or by carrier diffusion from the region of exposure. These processes take 10-100s of nanoseconds to occur. Once both thermal equilibrium is restored and the excess free carriers have been removed the material is often changed structurally and or thermally. If the material threshold has been exceeded, in other words if taken to temperatures above melting/boiling points, then melting and vaporization can occur. This occurs in the ns to μs timescale after initial exposure. The material becomes extremely hot but is still solid until a form of nucleation occurs and it can locally turn into a liquid or gaseous state. At this point there is an expansion into the surrounding media. The remaining energy, from the trailing edge of incident pulse(s), is converted to kinetic energy creating ions and allowing atoms and molecules to gain sufficient energy to break bonds. If this occurs on the surface and the particles leave the surface of the substrate this is called ablation. If this occurs within a transparent material it is often referred to as void creation where there is an absence of material often with compacted areas surrounding it. The extent to which the expansion occurs is dependent on the thermal diffusion rate of the material as it determines the rate of cooling of the exposed area. If melting or other processes have occurred resolidification and condensation are the mechanisms for returning the material to thermal equilibrium. This, however, may not leave the material in its' original state and this is where femtosecond micromachining comes in with the aim of manipulating the modification to create useful devices in a highly controlled manner.

To look at multiphoton absorption in a little more depth let us consider the inscription of dielectric crystals. As in the case of most glasses, the material bandgap is significantly greater than the incident photon energy. This ratio between the two is known as the multiphoton absorption order, K and for glass exposed with a femtosecond laser in the IR spectrum is typically 4-6. It is possible to reduce this value to 2 by using a source operating in the UV range thus obtaining two-photon absorption. This happens because of the relative increase in the individual photon energy that occurs at the lower wavelength. The dependence on having a sufficient flux of incident photons to initiate multiphoton absorption and ionisation implies a non-linear process. This is due to its' energy loss rate being proportional to the Kth power of the peak pulse intensity. This leads to the threshold pulse intensity being given by:

$$I_{MPA} = \left(\frac{K\hbar\omega\rho_{cr}}{t_p\beta_K}\right)^{1/K} \quad (1)$$

where ρ_{cr} is the plasma breakdown intensity, t_p is the pulse width, β_K is the multiphoton absorption co-efficient and ω is the laser frequency. It should be noted that as K increases the influence of the other parameters decreases. It is also only the section of the plasma field induced by multiphoton absorption that creates a modified region. This region is typically half the size of that which is heated. This can be exploited to give sub diffraction limited nanoscale modulations in index and ablation (S. K. Turitsyn et al., 2007).

2.2 Typical pulse properties

Each femtosecond laser has its' own characteristic pulse properties; however, there are some fundamental properties that they share. Most are Gaussian in both the temporal and spatial domain although it should be noted that $sech^2$ is most common for mode-locked lasers. The pulses typically have a low M^2 factor (ISO Standard 11146, 2005); more details on this are covered in section 4. This allows the pulses to be shaped easily through a number of techniques (N. Sanner et al., 2005). However, complex pulse shaping is not typically applied in most systems. Most pulse shaping is done with the use of spatial light modulators. These are typically 2D liquid crystal displays that are adjusted in the Fourier domain, although they are progressing to the use of micro-mirrors, which modulate the beam. In the case of the displays often the pulses are monitored and the liquid crystal programmed to act as a filter for certain areas until the desired pulse shape is achieved. These systems are usually attached to the lower repetition rate amplified laser systems although technological progress means that refresh rates towards the MHz range are close to being achieved (E. Frumker & Y. Silberberg, 2007).

2.3 Control of ablation area

When a pulse passes through an objective lens it becomes a stretched Gaussian profile in the spatial domain. Through the correct choice of lens it is possible to control the energy profile of the pulse and its spatial footprint. The spot diameter is given by:

$$\text{Spot diameter} = 1.22 * \frac{\lambda}{\text{NA}} \quad (2)$$

where λ is the wavelength and NA is the numerical aperture of the focusing lens objective.

At the focus, if the material's threshold energy is exceeded, the pulse will interact with the material and be absorbed. The use of lenses to focus and control the laser induced plasma is a very effective method of spatially shaping the resultant profile. Also by changing the pulse energy it is possible to control the spatial dimensions of the pulse-material interaction, figure 2. This allows the pulses to be spatially confined to a sub micron focal spot under the right conditions enabling a range of complex structures to be micromachined. (R. W. Applegate et al., 2006; P. Bado et al., 2006; S. Kawata et al., 2001; B. Yu, 2008).

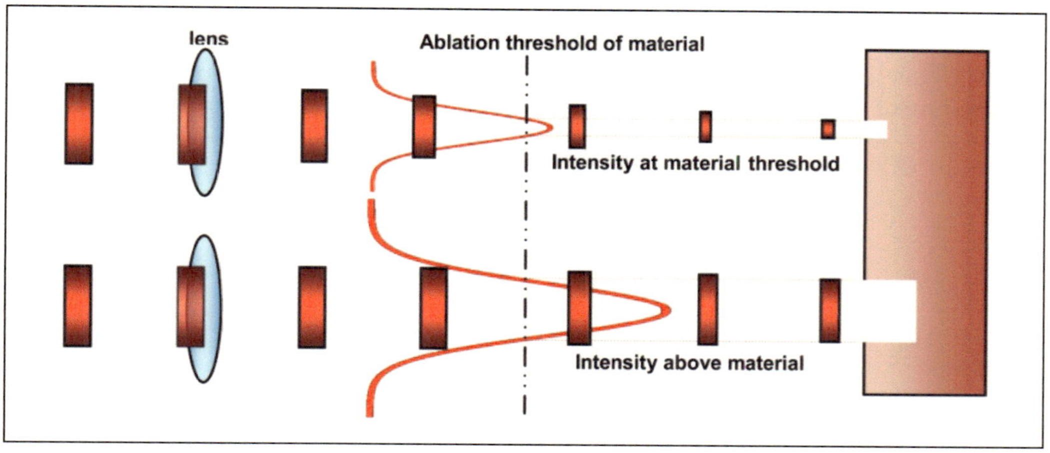

Fig. 2. A schematic showing that control of the laser pulse energy, relative to the material ablation threshold, can spatially control the structural change, in this case ablation.

Focussing can occur through the use of lenses and /or by using a low numerical aperture (NA) lens and allowing the material to self focus. The most common approach is to use an objective lens to create a focal spot because it is a more controlled process and works for lower power laser pulses. The threshold for the Kerr self focusing effect, where the refractive index seen by the beam is intensity dependent and is given by $n = n_0 + n_2 I$, is sufficiently high that for most materials this technique requires a significant amount of energy incident to create the fluence levels required (M. Sheik-Bahae et al., 1990; G. P. Agrawal, 2006). Above the threshold the Kerr effect occurs and the pulse experiences self focusing until it collapses. This effect at most energy levels is balanced by multiphoton absorption, plasma absorption and the plasma self-defocusing, due to the lower refractive index it creates. It is often considered preferable to avoid the Kerr effect when micromachining as it causes the laser focus to be deviated from its point of intended path. Self focusing occurs when the power, P, becomes greater than the critical power, P_{Cr} which can be calculated from the equation:

$$P_{Cr} = \frac{\lambda_0^2}{2\pi n_1 n_2} \tag{3}$$

where λ_0 is the laser wavelength, n_1 is the linear refractive index and n_2 is the nonlinear refractive index.

The time between pulses is a function of the repetition rate of a laser. The time between subsequent pulses interacting with the media makes a significant difference to the heating and subsequent cooling of the focal volume. The two forms of sources, oscillators and amplifiers, have very different repetition rates and so have quite different heat accumulation and diffusion characteristics.

The average power of a laser is determined by the pump source, this then produces either large amplified pulses separated by ms or smaller oscillator pulses with pulse separations of tens of ns, see figure 3. This range in repetition rates has a significant effect on the interaction with the material. The heat from a typical focal spot in a femtosecond micromachining system has a diffusion time of a μs in the case of fused silica. This means that in high repetition rate systems, when the repetition rate is such that the pulse separation is of the order of ns, the material in the focal region gains heat and is thermally loaded by successive pulses. If this thermal loading causes the focal volume to be above the temperature of the material melting point then permanent structural changes take place.

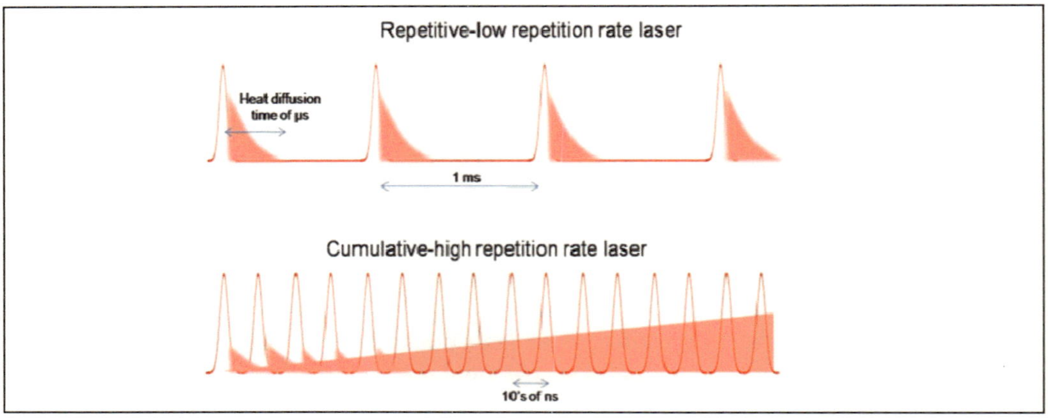

Fig. 3. A schematic showing the difference in low and high repetition rate lasers effect on the focal spots thermal accumulation and diffusion.

In contrast, amplified sources have much lower repetition rates and as such the heat has time to diffuse from the focal volume before the next pulse arrives. This tends to make the regions of change more confined to the focal volume than with oscillator laser sources. This dependence can be manipulated through the use of different lenses or combination of lenses but often the resultant structure has a shape associated with the focal spot. This fine dependence on the lens focussing is less stringent with high repetition rate sources as the thermal effects tend to impose a more spherical shape.

3. Materials

One of the least understood, yet most crucial aspects of femtosecond micromachining, is the precise nature of the interaction of the laser pulses with the substrate materials. The material properties affect the plasma as it interacts with it. Materials can be divided into two groups for classification, transparent and non-transparent. Figure 4 shows a range of both types of materials and their ablation/damage thresholds. The values for the material thresholds are taken from published work and are for a number of different wavelengths and pulse durations (P. Mannion et al., 2002; K. Nagashima, 2004; T. V. Kononenko et al., 2008; S. Baudach, 2000; G. Olivié et al., 2008; J. Kim & S. Na, 2007; A. Baum et al., 2007; J. Kruger et al., 2007; J. S. Yahng et al., 2009; N. Sanner et al., 2009). The values can be used to illustrate the general trends but should not be taken as finite points.

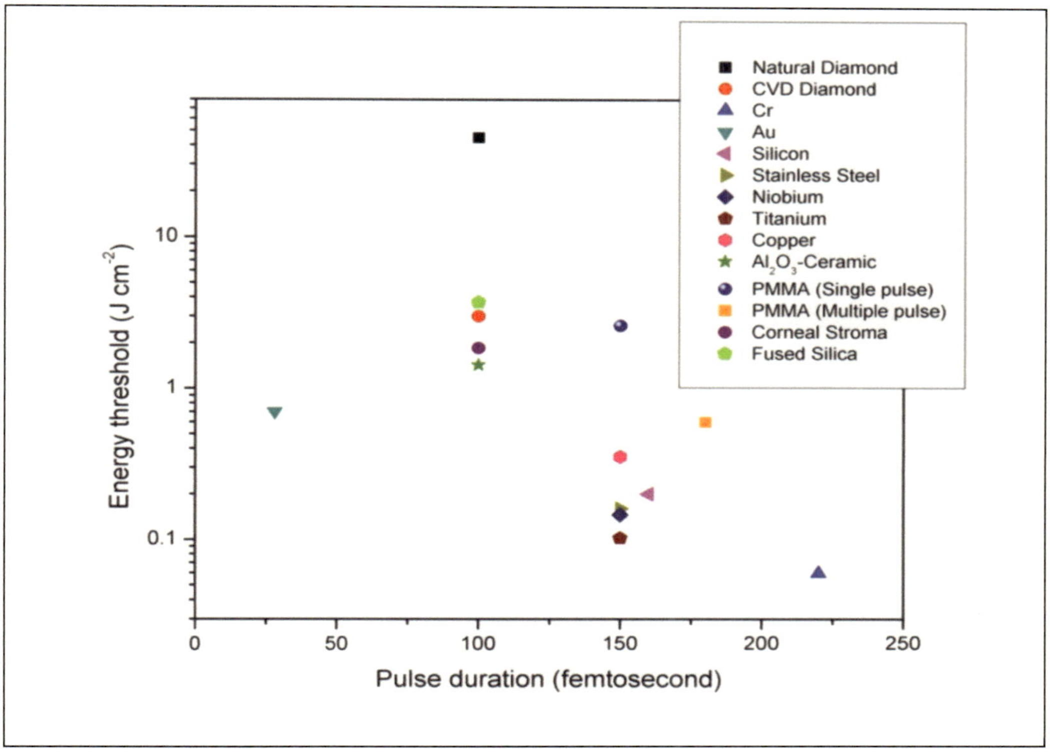

Fig. 4. A plot of the energy thresholds for modification vs pulse duration, for a range of materials, with the values taken from a literature survey.

Most material thresholds have a small range of values attributed to them arising from the laser parameters and techniques being applied being different between groups conducting the research. A good example of this is given in the paper by Sanner et al (N. Sanner et al., 2009), where the range of values quoted for the ablation threshold for fused silica is given as 2-12 J cm^{-2}. This range is by no means an anomaly or the widest range published, for example the value for natural diamond is given as 10-80 J cm^{-2} (T. V. Kononenko et al., 2008) where the value fluctuates widely across a specimen surface. The reason for the range in values can be explained by considering that there are a number of ways of measuring the thresholds and even defining them (N. Sanner et al., 2009), there are also differences in the exact composition of the materials used and their source, as well as differences in the experimental set-up. The variation in material and the dependence on small variations of impurities and dopants can lead to significantly different thresholds. The values also do not reflect the range of repetition rates that are available and the subsequent change in energy threshold (S. M. Eaton et al., 2005; C. B. Schaffer, 2003). There is also a dependence on the wavelength of the laser as this will determine the physical mechanisms involved from multiphoton absorption, impact ionization, free carrier absorption and many others.

It should be noted that the chart also reflects the work carried out using the lasers that have been practically available in laboratories around the world. The vast majority of these systems have a repetition rate of about 1kHz and are Ti:Sapphire amplifier lasers. This will potentially bias the results of the plot as the lasers that may be best suited to a given material may not be reflected in the work carried out to date. A good example of this may be

Poly(methyl methacrylate) (PMMA) where at the correct wavelength a particularly short pulse may suit the material in which the chain length of the molecules is long and the thermal sensitivity is high. The short pulse duration would potentially enable the confinement of the resultant shockwaves to be significantly greater.

3.1 Non-transparent materials

As one would expect, the use of femtosecond lasers with non-transparent materials is typically limited to working on the surface in either ablation or modification (E. D. Diebold, 2009). There are a number of non-transparent materials that have been studied as shown in figure 4. Each material has a number of different properties that affect the energy threshold. For non-transparent materials, the band gap is the most important contributor when looking to determine the energy threshold. This will, as detailed in section 2.1, determine the process by which the photons interact with the material. The interaction of femtosecond pulses with non-transparent materials causes excitation of the majority of electrons, formally in bound states, creating a plasma-like state which immediately creates disorder in typical lattice structures. This means that the process is typically considered to be athermal as it occurs before the lattice has come to an equilibrium state through the carriers. There is some debate as to exactly how athermal the process is but there is little disagreement that the reduction of the heat affected zone (often abbreviated to HAZ) is significant as compared to the longer pulse durations that had previously been studied (S. Valette et al., 2005). The potential for the use of non-transparent materials, either on their own, in multi-layered combinations or, in combination with dielectric materials, for a range of applications holds a great deal of future development capacity and will be a subject of continued research for a number of years to come.

3.2 Transparent materials

Transparent materials also rely on non-linear absorption and subsequent ionization to create either ablation at the surface or index modification within the bulk. The exact mechanisms are not fully determined. However, work on modelling the processes involved is beginning to compare well with practical results and illuminate the mechanisms (J. S. Petrovic, 2007). The mechanism for the energy transport from pulse to material is commonly thought to be through nonlinear absorption. A femtosecond pulse of sufficient energy when focused inside a material will optically breakdown. At this point a proportion of the energy is transferred to the lattice via excitation of the electrons in a multiphoton absorption process. These electrons reach thermal equilibrium within several femtoseconds and seed the subsequent avalanche ionisation which absorbs the energy in the tail of the pulse. It is assumed that the energy that does not go into the ionisation is stored as kinetic energy in the electrons. The recombination time for electrons in fused silica is of the same order as most of the pulse durations as shown in figure 4 at 170fs (Q. Sun et al., 2005). It is commonly assumed that the mechanical and thermal changes on the time scale of a pulse are negligible. In the next picosecond a proportion of the energy in the electrons is transferred to the lattice. During this time there is plastic deformation occurring which propagates at the speed of sound covering about a micron in 0.1ns. This pressure or shockwave separates from the hot focal volume in a matter of nanoseconds. This is then followed by thermal diffusion acting in the microsecond timescale where the thermal energy diffuses out of the focal volume. Beyond critical energies these processes give rise to permanent changes in structure that creates one of two commonly accepted forms in the shape of voids (A. Martinez et al., 2006) or smooth index changes (N. T. Nguyen et al., 2006). The difference between the two depends on whether the pulse is above or below the damage threshold of the material concerned.

Void creation is achieved when the focusing is very tight and the pulse energies used are high. This has the effect of creating huge pressures at the point of focus and subsequently on the surrounding material. The surrounding material experiences a pressure wave moving away from the peak intensity forming a rarefied region of material followed by a region of densification. When the pressures created, through the focusing and pulse energy, are great enough a void may be created (E. N. Glezer & E. Mazur, 1997).

Smooth index changes are achieved with pulses below the structural damage threshold. They too create rarefied areas; however, as the pressure exerted by the focal volume is much smaller the resultant index change gradient is smoother leading to structures that can be used for waveguiding. The effect of the pulse repetition rate on the material-pulse interaction is shown to have a significant effect on the profile of the index change at higher repetition rates (typically in the MHz region). This is shown to be due to the thermal diffusion of the material being longer than the pulse to pulse separation (C. B. Schaffer et al., 2003). This means that the material is thermally loaded when the next pulse is incident at the focal region. This effect has not been observed at lower repetition rates in the kHz range.

It should be noted that there are competing mechanisms for explaining the effects of pulse material interaction such as tunnelling ionization; however, work by Keldysh has shown that they are just two boundaries of the same process under different conceptual approaches (L. V. Keldysh, 1965).

3.3 Ablation in transparent materials

When the conditions for void creation are met and the energy incident in the focal region is sufficient to create a plasma with sufficient energy for Coulomb repulsion to create voids, ablation may occur. The only difference is the depth at which the changes occur. If near enough the surface the plasma will inevitably weaken the surface wall allowing the pressure on the side walls to be reduced in due course. This causes the removal of material from a substrate and is characterised as ablation.

3.4 Heat affected zone

The nature of the timescales involved in both transparent and non-transparent material interactions with femtosecond pulses, and the subsequent structural changes, offer up insight into the advantages of micromachining with ultrashort pulsed lasers. The electron-phonon scattering time, of approximately 1 picosecond, is greater than the pulse duration and so the pulse has ended before the excitation of the ions has begun. This means that theoretically the heat diffusion outside of the focal area is minimized (S. Valette et al., 2005). This extra control on the locale of micromachined change adds a level of precision not afforded to lasers of sub-picosecond pulse duration. This reduction in the heat affected zone is often important in many applications. The reduction can lead to fewer, if any, micro-stress fractures being created when machining materials which, for example, are used for heavy strain loading thus reducing the chances of fracture. Also, to induce changes in a material using a femtosecond laser it is not necessary to have dopants to seed the absorption process. The effect of the nonlinear ionization can be used, with great care and refined technique, to produce a confined and reproducible material change that can be put to practical use.

4. Types of femtosecond laser

The results of a market survey on femtosecond laser sources are presented in table 1 (Amplitude Systemes; Coherent; High Q laser; Kapteyn-Murnane Laboratories; Raydiance). There are a number of ways of evaluating femtosecond lasers, each parameter set leading to a different profile of applications that they suit. The parameters chosen for this summary are wavelength, energy per pulse, pulse width, repetition rate, average power and the key amplification medium of the source. There are, of course, other parameters that are also fundamental but for reasons of space have not been included for example, the M^2 value, peak power, beam diameter and polarization. These parameters should not be overlooked, however, when considering which laser source is applicable to a given task.
The M^2 factor, also known as the beam quality or beam propagation factor, is commonly used as a measure of the beam quality (ISO Standard 11146, 2005). For diffraction limited Gaussian beam the divergence is thus given by:

$$\Theta = M^2 \left(\frac{\lambda}{\pi \omega_0} \right) \tag{4}$$

where Θ is the half angle beam divergence, λ is the wavelength and ω_0 is the beam radius at the beam waist. A diffraction limited beam has an M^2 of 1 and is considered Gaussian. The level of divergence that a beam has will limit the degree to which it can be focused and since the M^2 factor is a measure of this it is a good indication of how a laser beam will focus given a certain numerical aperture objective lens. When combined with the optical power of a laser it determines the laser fluence achievable. Typically femtosecond lasers will have an M^2 value of between 1 and 1.5, in most applications a value closer to 1 is usually preferable.
The peak power is a parameter that is typically considered when deciding what material you wish to ablate or into which you wish to inscribe a structure. The key being that the peak power must exceed the material threshold energy so that work may be carried out. This can be done by comparing the many sources of material thresholds (see figure 4 for published results) against the specifications of the systems supplied. The peak pulse power, ppp, is given by:

$$ppp = F.\frac{Average\ power}{Repetition\ rate.\,\tau_p} \tag{5}$$

where τ_p is the pulse duration and F is a factor determined by the shape of the pulse; for a $sech^2$ pulse the factor is 0.88, for a Gaussian it is 0.94.
The beam diameter needs to be taken into consideration when designing the optics that a femtosecond laser system will require or, of course, vice versa. It is important to control the beam divergence and aperture size of the lens as through miscalculation of these much of the beam quality, wavefront, energy profile and peak energy may be lost or reduced.
Most of the sources in table 1 are linearly polarised. Although the polarisation is not generally an issue, care should be taken when designing the system to make sure that the polarisation for any beam is known at any given point. As an example, when using an autocorrelator to monitor the laser, an orthogonal polarisation state may be required to that which your laser outputs. There is also work on the polarisation state relative to the direction of beam-substrate translation; this is covered in more detail in section 5.3.
There are currently two key types of femtosecond laser commonly commercially available. They may be grouped into amplifier and oscillator systems. The details as to how they work

will not be covered here but may be found in references (T. Südmeyer, 2008; S. Backus, 1998), however, differences in terms of the output pulses will be discussed.
The wavelength ranges of the types of laser are determined by a number of factors. There are two main bands covered by the fibre lasers at 1030-1045nm and around 1550-1560nm. The two bands correspond to the dopant used in the lasers cavities. The conventional C-band erbium window is at 1530-1565nm and ytterbium sources operate at around 1030-1050nm (S. B. Poole, 1985). There is a third much smaller group of fibre lasers operating at around 800nm. Bulk amplifiers and oscillators, are also governed by the amplification material chosen. They typically use Ti:Sapphire and ytterbium and as such commonly operate at wavelengths around 800nm and 1030-1050nm.
The energy per pulse is a parameter to be considered in a similar way to the peak power. The pulse energy required will depend on both the material and the chosen application. Machining of a crystal for instance will typically require a much greater energy per pulse, for example energies up to and above 80 Jcm^{-2} (T. V. Kononenko et al., 2008) for natural diamond, while for index change in PMMA energies above 0.6 Jcm^{-2} (A. Baum et al., 2007) cause permanent change. The energy per pulse of the types of laser are detailed in table 1. The oscillators typically have energies in the range of 1-100s nJ per pulse, whereas the fibre lasers offer energies in the μJ range and amplifier pulse energies typically fall in the mJ range. The choice of pulse energy for a given application is critical as most materials have a small window of energies between the desired effect, say index change, and damage. The other consideration is that to control the energy, and other parameters, incident on a sample is significantly easier when not having to operate at the extreme limits of attenuators or with insufficient laser energy after the losses experienced through the system.

Type	Wavelength (nm)	Energy per pulse	Pulse width (fs)	Repetition rate	Average power (W)	Key material	Company
Amplifier	800	0.3-6mJ	25-130	1 or 5 kHz		Ti:Sapphire	Coherent
Amplifier	1030	10μJ-1mJ	400-500	0-100kHz		Diode pumped	Amplitude
Amplifier-multi pass	Tuneable	0.25-20mJ	<25-100	1-20 kHz	3.0 to 50	Ti:Sapphire	Kmlabs
Amplifier-Regen	1035	0.4-1mJ	350-500	1-100kHz	1.0 to 2.0	Ytterbium	High Q
Amplifier-Regen	1055	1mJ-5μJ	650	10-100kHz	0.100 to 0.200	Nd:Glass	High Q
Amplifier-Regen		3mJ	60-160	100/250	0.3 to 0.75	Ti:Sapphire	Coherent
Fibre	780		100	50MHz	0.02	Erbium doped	IMRA
Fibre-Amplifier	810		<150	75MHz	0.1	Erbium doped	IMRA
Fibre	1030	0.5-10 μJ	100-700	30-2MHz	20		Amplitude
Fibre	1043	2μJ	500	200KHz	0.4	Ytterbium doped	IMRA
Fibre	1043	10	700	200KHz	3W	Ytterbium doped	IMRA
Fibre	1045	10μJ	700	100kHz		Ytterbium doped	IMRA
Fibre	1552	1-10 μJ	<800	1Hz-300kHz	3		Raydiance
Fibre	1552	5-50 μJ	<800	1Hz-100kHz	5		Raydiance
Fibre	1560		100	50MHz	0.06	Erbium doped	IMRA
Oscillator	800	3-10nJ	<12	80-100MHz	0.250 to 0.900	Ti:Sapphire	Kmlabs
Oscillator	800		<20	80MHz	>0.300	Ti:Sapphire	Coherent
Oscillator	1030	20-500nJ	200-500	10-50MHz	1 to 5	diode pumped	Amplitude
Oscillator	1060	1.4nJ	<150	72MHz	>0.100	Nd:Glass – diode pumped	High Q
Oscillator	1030-1053	2-300nJ	~350	10-80MHz (fixed)	0.150 to 3	Ytterbium	High Q
Oscillator	790-870		100-300	73MHz	>0.200	Ti:Sapphire	High Q
Oscillator-Cavity dumped	800	>60nJ	100	0.5-1MHz	0.06	Ti:Sapphire	High Q
Oscillator-Cavity dumped	1040	>500nJ	<400	0.5-1MHz	0.5	Ytterbium	High Q
Oscillator-Cavity dumped	800	30nJ	<15	1kHz-4MHz			Kmlabs

Table 1. Table showing the market survey of femtosecond sources and basic properties

Femtosecond pulses are considered ultrashort and as table 1 shows they range greatly in practical terms. There are effectively two or three classifications of pulse duration. There are the extremely short pulsed lasers, with pulses typically in the 10s of femtosecond duration which are most commonly, although not exclusively, oscillator lasers. The next region is about 100-350 fs that are often amplifier lasers. The final group is from 350-800 fs and is largely occupied by fibre and amplifier lasers. The pulse duration makes a significant difference to the pulse-material interaction and the pulse energy required.

Repetition rates of commercially available systems range greatly from single kHz through to 100MHz. The range leads to a significant difference in the applications of each. There is some evidence to suggest that better quality waveguides, for instance, are written with lasers operating in the MHz regime rather than kHz (S. M. Eaton et al., 2005). On the contrary often for micro-machining ablation lower repetition rates in the 1-300 kHz range tend to be chosen because they have higher pulse energies which are above the ablation threshold. For these lower repetition rate systems there is also less thermal loading due to the pulse train spacing. Repetition rates and the resultant thermal loading, or absence, offers clear advantages of one repetition rate over another for a specific task.

In conclusion the parameters of a chosen laser will strongly influence the effectiveness of work in particular area. The parameter windows are relatively small for high quality results in any given application.

5. Techniques employed

There are several different techniques employed when making micro-machined devices through inscription and ablation. Some of them are techniques applied to both regimes and others are applied more specifically to one or the other. Typically using a laser to perform micro-machining involves complex physical processes and is dependent on fine parameters of the material and laser. Theoretical models exist and are touched upon in other sections of this chapter, however, they are often considered to be only a guideline and require refinement for optimal processing when using a practical system. In this section some of the basic methods and techniques applied to micro-machining are explained.

5.1 The basic system

Systems tend to either operate by having the sample fixed and the laser beam moving or by fixing the sample to a moving stage, or set thereof, and having a fixed objective lens, figure 5. There is also the option to use galvanometric systems where the beam is manipulated using mirror(s) and obviously a combination of all three. Each of the layouts has its own pros and cons depending on the main purpose of use, for instance when the desired sample is small and is required to be machined quickly then galvanometric systems can be most advantageous, however, when operating over a larger area these systems suffer from spherical plane effects and correcting for these often leads to a loss of sharpness in the focusing. This is especially important for femtosecond work where the depth of focus, due to the nonlinear nature, is so small.

Often the most practical systems use a partially fixed objective, where the objective is also on a stage but often remains stationary when working at a given depth in the sample, and use mechanical or air bearing stages to move the sample. These are often programmed by computer linked drive control units. The majority of stages operate some version of CNC (Computer Numerical Control) system (Smid, 2005) each of which have their own protocols, however, the techniques used are applicable to most if not all systems of this sort.

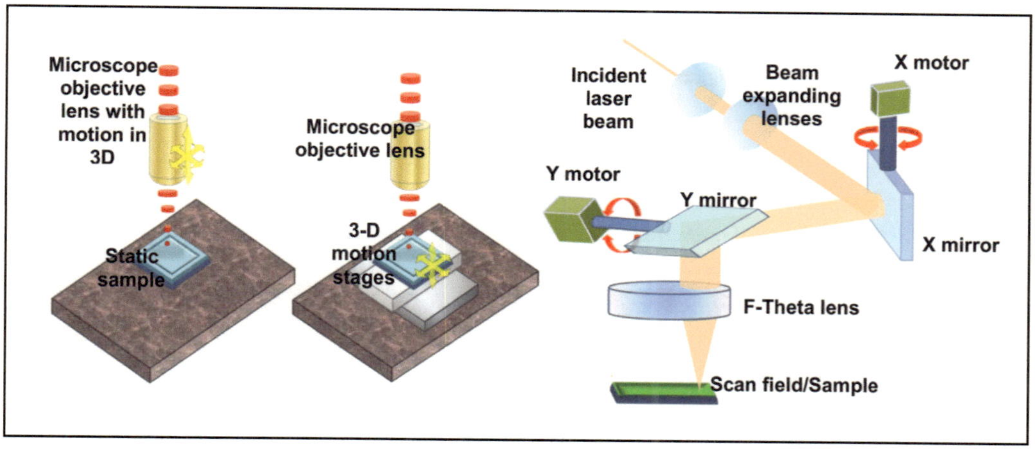

Fig. 5. A schematic of three types of focusing arrangement from left to right a static sample with moving objective, a moving stage with static objective lens and a galvanometric set up with motion controlled by mirror angle.

5.2 Common terminology & basic techniques

There are a number of terms applied to certain types of machining that describe the fundamental technique applied to working on a work piece and these are defined in table 2. The first of which we will consider is percussion drilling. This is a process of firing a number of pulses on a given area, each pulse removing a very small volume of material, thus leading to the creation of a hole. Typical laser repetition rates over 1kHz allow removal rates to be viable for use. This technique is used for the creation of small holes through or in materials. In general the material removal rates are relatively constant for small depths (to ~100 µm) after which the removal rate operates as the square root of the depth. Thus the time taken to double the depth is typically in excess of four times that of the initial hole. This occurs because as the beam penetrates to the bottom of a hole energy is lost to the material and the nature of Gaussian beam paths, after focusing, means that the energy available at the bottom decreases as the wing that is clipped is inversely proportional to the aspect ratio increase. Typically helical trepanning produces some of the smoothest side walls and most uniform holes but takes longer and tends to be best applied to smaller artefacts.

5.3 Other considerations

There are a number of other parameters and components to be aware of that can be critical to the finish and quality of a desired object. It is important to consider the desired aspect ratio or etch depth, the NA and working distance of the lens, the position of the focus in the sample, the beam polarisation, the speed of the moving parts and an inspection mechanism.

The aspect ratio is defined by the ratio of depth to width of an artefact, for example, a microfluidic channel or hole through a ceramic. The etch depth is the effective write depth of an inscribed feature such as a waveguide or diffractive element inside the bulk of the material. To optimise both of these parameters the choice of lens, power and beam shaping are fundamental. If aiming to write a deep slot into a substrate one would typically choose a lens with a low NA and long working distance so that it could operate at a distance and over a range of positions without being coated with the debris created by the plasma and

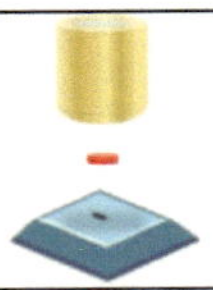	**Single shot drilling -** The process of using a single laser pulse to drill, this is rarely used.
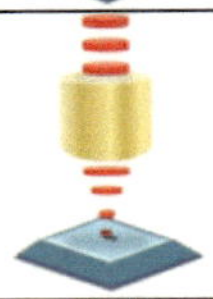	**Percussion drilling -** The use of a number of laser pulses at a repetition rate spacing above that of the length of the pulse used to remove material. Can lead to surface spatter which can lead to micro-cracking, deformation of hole shape and achieving high aspect ratios is often difficult.
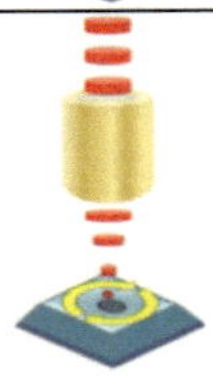	**Trepanning -** This is essentially percussion drilling with circular motion, often a pilot hole is drilled and then a spiral motion followed by circular finishing. The technique suffers from the same drawbacks as percussion drilling. The hole size formed by this motion is, to within the radius of the plasma, the diameter of motion. The holes produced by trepanning are generally more circular and accurate than a percussion drilled hole, however, they are larger in size.
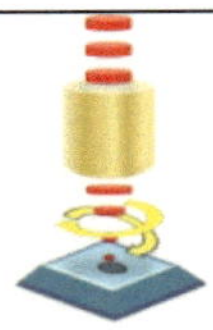	**Helical drilling -** The process of quantizing the ablation steps reaching breakthrough only after a number of passes described by a spiral motion. This often has a more circular geometry than trepanning and also minimises the load placed on the opposite face to that of the focus. It tends to also give less recast, however, takes significantly longer to process.
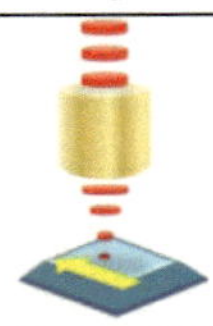	**Cutting -** Cutting through a sample using a series of pulses through motion of the beam or sample, often multiple passes are required. **Etching / Milling -** Removing a defined depth of material through control of pulse energy and/or number of pulses per location.
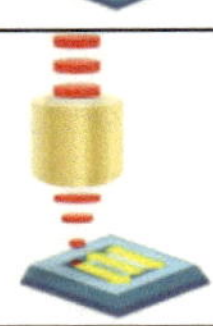	**Rastering -** The motion of moving back and forth over an area with lines separated by a given pitch. By varying the pitch this can lead to the removal of material from an area or in trenches. Typically these form square wave patterns although other forms are also used.

Table 2. A Table of the common techniques and a brief description of their mechanism.

avoiding contact with the material. Ideally most of the work should be done with a static z component and the right choice of lens, however, there are times when stepping the lens towards the sample is necessary to achieve a specific depth or profile. The position of the focus required to ablate a slot, when scanning, is typically not at the midpoint of the desired slot depth. Through experience it comes out at typically 1/3 of the depth but the exact position will change depending on the sample and other parameters. There are also issues to do with shielding by the walls when looking to achieve high aspect ratio side walls. This is because the pulses wings are clipped reducing the power of the pulse.

The speed of any scanned motion, as with repetition rate, will affect the rate of removal of material. This is because the fluence will be varied by the change in the speed of motion as the number of pulses per unit volume will be less. A variation in the repetition rate would have the converse effect. That is to say that if the pulse rate increases by a factor of 2 that the removal rate would increase linearly, assuming constant pulse energy. Whereas a doubling of the speed would half the removal rate or create a series of dots rather than a line.

There are two types of polarisation that can be used, linear and circular polarisation. The polarisation is believed to affect the write quality of inscribed lines such as waveguides. The current thinking is that a polarisation orthogonal to the direction of write for straight waveguides, or circular for curved ones, is preferable and results in smoother tracks (M. Ams et al., 2006). Polarisation parallel to the direction of write is not favourable since it produces less smooth tracks. There are other techniques employed such as combining cylindrical lenses with the regular microscope objectives to refine the width of written lines.

The ability to fully inspect and align a sample pre- and post-inscription or ablation is of fundamental advantage to any system. The use of confocal systems and inspection methods to inspect during writing has also developed considerably in recent years (J. Li, et al., 2008). A standalone camera can also be used to monitor the sample. The exact design and components used will not be uniform across all systems but the importance and advantage gained by their inclusion are extremely significant to the complexity of the fabricated devices.

Some examples of femtosecond micromachining are shown in figure 6. The images illustrate some of the common effects observed, both good and bad, from femtosecond micromachining. By reducing the separation between the slots it is possible to reduce the wall thickness and create extremely high aspect ratio structures. Figure 6 also shows entry and exit holes. The entry holes in this example are slightly rounded which can be corrected for by adjusting the focus position. The third image shows how both good and bad set up parameters affect the resultant finish quality.

Fig. 6. Slots machined in stainless steel shim 0.178 mm thick; (LHS) entry side showing gradual reduction in slot separation, (left middle) exit of the same slots, (right middle) slot showing what happens when the parameters and finishing are correct and wrong, (RHS) showing the high aspect ratio structures remaining after ablation.

5.4 Computer Aided Design (CAD) & rapid prototyping

There are a number of applications of femtosecond micro-machining where the complexity and rapid prototyping required are less suited to programming the motion line by line. This is clearly shown in the complexity of the microfluidic device illustrated by figure 7 below.

To code this line by line would be extremely time consuming and to change something like the machining pitch could take considerable effort going through the code line by line. In these situations the use of CAD software packages can be a significant advantage in being able to vary the parameters (such as pitch, write speed and scaling) quickly and design complex structures that would otherwise take significantly longer. Although it is not impossible to code some of the more complex structures the plausibility and economy of doing so when the software packages are available becomes more weighted in favour of the automated approach (G. Smith, 2008).

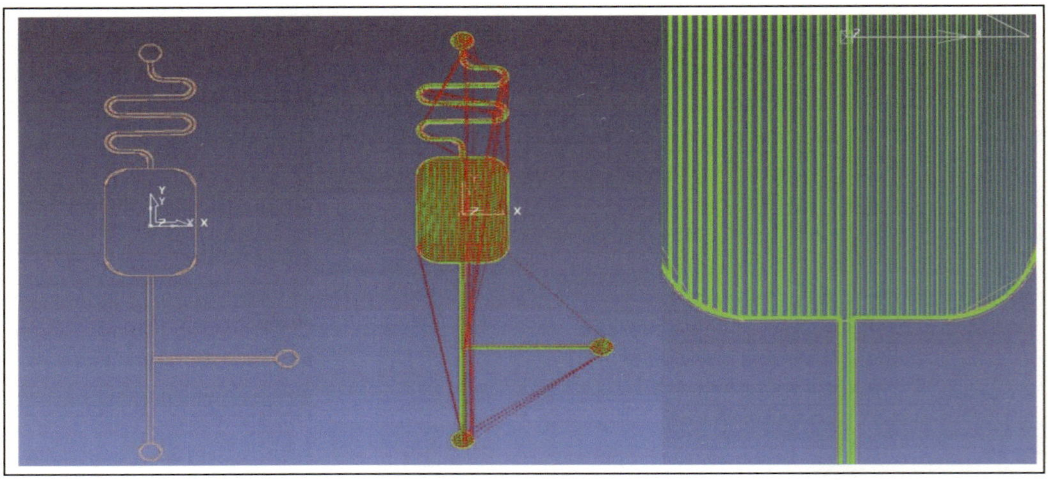

Fig. 7. Computer Aided design images, from left to right 1) An plan view of a computer designed microfluidic device, 2) the machine path lines shown for workpiece with green representing the path of the laser ablation and red being the skimming non ablation transit, 3) a close up of the tool path for ablation showing rastering and a finishing edge pass.

5.5 Post processing

There are a few post processing techniques that are important in relation to femtosecond micromachining. The most common technique is to wet etch using either hydrofluoric acid (commonly abbreviated to HF and is hydrogen fluoride in water solution) or ammonium bifluoride (ABF is chemically NH_4HF_2 and a diluted version of HF in a salt form, although used in water solution). The whole process involves inscribing the material (below the ablation threshold) using the laser focal spot then placing the substrate in the acid. The acid preferentially etches the inscribed areas at a rate of 50:1 in fused silica (K. Sugioka et al., 2007) and as such removes the inscribed area selectively. This technique offers the ability to make smoother structures in transparent materials with smaller features and higher aspect ratios. It is also possible to fabricate subsurface channels that would otherwise take a sequence of layer deposition stages or lithographic techniques. There is a downside, in that the use of these chemicals adds additional processes and time over direct ablation and involves the handling of hazardous chemicals. Figure 8 shows work done in optical fibre. The fibre has been exposed by femtosecond laser inscription below the damage threshold then wet etched using HF producing very narrow, high aspect ratio channels through the fibre core.

The use of heat treatment, cycled and constant, may be important for femtosecond micromachined structures. In theory, the thermally induced stresses created by the shockwaves propagating in the material around the plasma can be thermally annealed out through heating the substrates post inscription. Heat treatment thermally relaxes the material such that the stress is released and the permanent change of the inscription is all that is left. This effect is still the subject of study and its ability to offer further understanding of the plasma-material interaction will most likely be of fundamental impact (S. Juodkazis et al., 2004).

Fig. 8. Micro-channels fabricated in standard fibre using fs inscription and chemical etching (Y. Lai et al., 2006).

6. Applications

The numerous properties of femtosecond pulse interactions with a range of materials have led to a diverse range of novel applications. For example, the ability to micromachine in 3 dimensions in transparent media due to the nonlinear interaction has opened up possibilities that were previously not available without the addition of dopants and short wavelength laser exposure. There are also a number of applications that would simply not be possible without the use of femtosecond lasers for micromachining. Having said this there should be a note of caution as while there are numerous advantages to the technology it should not be considered as the only solution to all applications. Instead the advantages should be utilised for specific purposes.

6.1 Periodic structures

Because of the short pulse duration and the high refractive index changes that can be induced femtosecond lasers can be used to produce period structures in transparent materials. More specifically, they have been used to fabricate fibre Bragg gratings (Y. Kondo et al., 1999). These structures are written into or near the core of an optical fibre and reflect light at a wavelength determined by the periodicity of the structure.

Two approaches to the fabrication of these structures have been optimised over the last few years in the femtosecond domain: the point-by-point method (E. Wikszak et al., 2004; A. Martinez et al., 2004, K. Kalli et al., 2009) and the phase mask method (K. A. Zagorul'ko et al., 2003). Both methods had previously been used for the UV fabrication (with either CW or conventional pulsed lasers) of fibre Bragg gratings however the femtosecond regime provides some key differences due mainly to the localisation of the fringes which allows, for example, multiple gratings to be positioned in unique positions around a single core, as shown in figure 9. This can be highly advantageous from a device design point of view as, for example, it enables the production of a single fibre Bragg grating device that can be used as a directional bend sensor. Gratings can also be inscribed through the hole structure of microsctructure optical fibres using femtosecond lasers. Kalli et al have shown that with a suitable fibre design it is possible to use femtosecond pulses to penetrate the holes of the microstructure fibre without significant breakup of the femtosecond laser pulse during inscription.

In planar samples femtosecond lasers have been used to inscribe diffraction gratings which can in turn be used to fabricate fibre Bragg gratings (G. N. Smith et al., 2009). A photograph of one of these is shown in figure 9 showing first, second and third order phase masks. The work to date demonstrates the proof of concept and flexibility for the use of femtosecond lasers to make complex and reproducible phase masks. This approach has the potential to rival e-beam fabrication of phase masks and has the advantage of being a single step fabrication process that uses no chemicals.

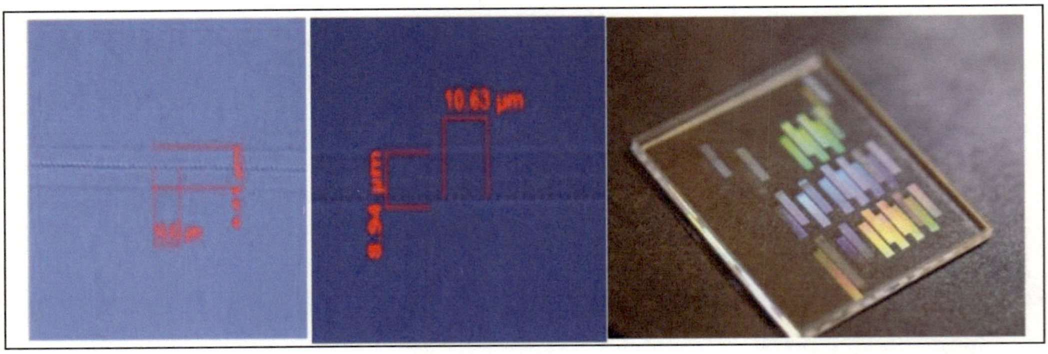

Fig. 9. Femtosecond inscribed fibre Bragg gratings in (LHS) the centre of the fibre core and (middle) on the edge of the fibre core, (RHS) photograph of a femtosecond phase mask inscribed with fs laser underneath the surface of the UV grade fused silica (G. N. Smith et al., 2009).

6.2 Micromachining of planar glass

Microfluidic device, incorporating high aspect ratio micron scale channels, can be directly machined. These devices are developed as lab-on-chip devices for purposes such as measuring a specific particle to particulate sorting and counting (D. N. Schafer, 2009). The advantage is that they only require tiny amounts of a fluid to function thus reducing costs of development of chemicals, allowing more information to come from smaller samples at increased speed of prototyping and development. Some of typical structures that are employed are shown in figure 10. They show bends, micropump holes, joints and high aspect ratio structures in both planar and fibre samples all of which can be easily adapted and machined using femtosecond micromachining giving advantages for rapid prototyping (G. Smith et al., 2008).

There are a number of methods for making these devices. The most common is to inscribe a structure in the material and then expose it to hydrofluoric acid. Another is to ablate structures or create voids in the presence of what are known as wetting fluids (Y. Iga et al., 2004). This works in the same way as you would use fluid with a standard milling process to remove the debris from a machined area. A third method is dry ablation, however, the results often lead to sidewalls that suffer from turbulent flow (rather than the ideal lamina flow) due to the surface roughness.

6.3 Waveguiding

There has been a great deal of interest in the use of femtosecond lasers to make waveguides. They have been used to make a number of things from straight connectors and curved waveguides to more complex structures like splitters, beam shapers, amplifiers and

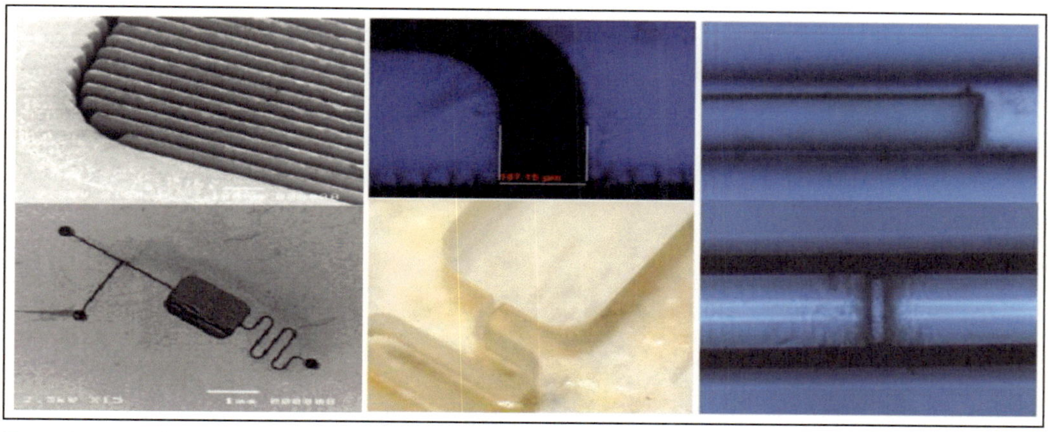

Fig. 10. Microfludic devices - (top LHS) SEM image of micro-groves to enhance fluid mixing (bottom LHS) SEM image of test structure, (top middle) microscope image showing smooth channel bend from microfluidics device, (bottom middle) photograph of larger scale structure showing high aspect ratio of fluid guides, (top RHS) slot ablated along the fibre axis in optical fibre using fs laser to within 5μm of the fibre core, (bottom RHS) slot ablated perpendicular to the fibre axis.

interferometers (A. A. Said, 2004; A. Szameit et al., 2006; A. M. Kowalevicz et al., 2005; K. Minoshima et al., 2001). There have been other avenues where the properties have been utilised such as the image reconstruction using a waveguide array (A. Szameit et al., 2009). This and other applications rely on the 3D write capability of femtosecond lasers allowing the creation of complex structures that are otherwise typically built layer by layer. The only pre-requisite is to create permanent index change localised to the area of write, typically the desired effect is a positive index change although other structures are also possible, thus forming a guide for the light to travel along. There are normally areas around the waveguides where the pulses have interacted with the media through the wings of a pulse or through heat shockwaves etc. These are best reduced through optimisation of the material and laser parameters used. Waveguides have typically been written in planar glass or crystalline samples, however, using femtosecond laser it is possible to inscribe waveguide structures in optical fibre. Figure 11 shows an example of this written at Aston University using a femtosecond laser in standard single mode optical fibre. The guide ends close to the edge of the fibre core and couples light from the evanescent field out of the fibre. This shows the potential to include complex waveguide based structures in fibres which could have a range of telecommunications and sensing applications.

6.4 Other applications

Femtosecond lasers have been used for numerous other applications, some of which are briefly described here to provide an illustration of the scope and potential of femtosecond lasers.

Optical data storage uses micron sized defects, typically index variations, in substrates used for the storage of data in a highly dense arrangement. This has now been accomplished in 3 dimensions and in a rewriteable format (K. Miura et al., 2002). The ability to write the points in 3 dimensions is something that can only be achieved through the use of the nonlinear femtosecond processing. The other key advantage of using a femtosecond laser process is

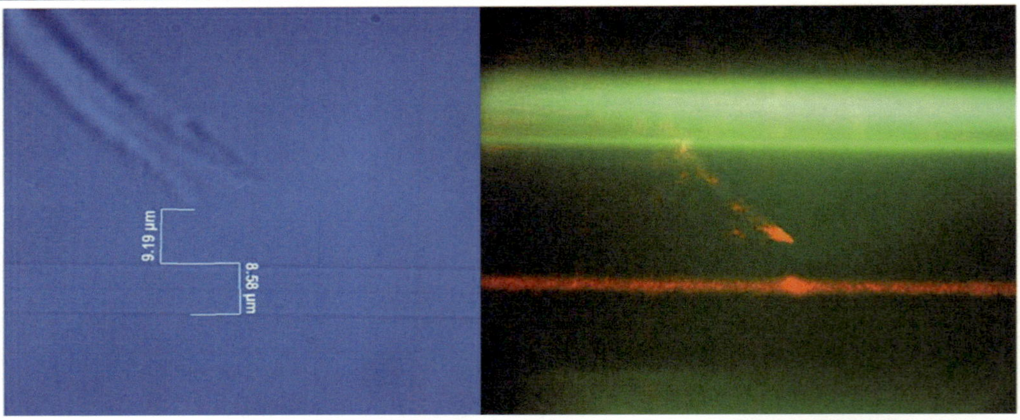

Fig. 11. (LHS) microscope image showing the separation of the fibre core and start of written guiding structure, (RHS) dark field microscope image of light being outcoupled from fibre core by a waveguide/lightpipe structure written with a femtosecond laser.

that the high point density that can be achieved through the spatial confinement of the femtosecond pulse-material interaction. This more localised index change means that the point defects can be more closely arranged giving higher density data storage without undesirable, peripheral modified regions that would otherwise make data recovery impossible.

Photopolmerisation exploiting the two photon polymerisation affect using femtosecond lasers has been used to create submicron features in polymer after chemical rinsing (R. Guo et al., 2006). This has produced a range of intricate structures that have a wide range of applications from terahertz lenses and photonic crystal structures and to non-photonics applications for example to produce medical stents. The femtosecond lasers unique ability to polymerise liquids and resins allows more intricate and complex artefacts than any other source to be produced. This ability to form fine structures is due to the small volume of interaction and spatial confinement femtosecond lasers.

There are a number of medical applications for femtosecond micromachining ranging from eye surgery and dentistry to cell transfection (C. T. A. Brown et al, 2008; C. McDougall et al., 2009; S. H. Chung & E. Mazur, 2009). Femtosecond lasers are used to perform submicrometer resolution surgery with minimal invasion and also use multiphoton absorption processes initiated by the femtosecond pulses to permeate cell membranes in a transient manner. One example of this is the injection of gold nanoparticles into mammalian cells using optical transfection combined with optical tweezing techniques. Interest in the use of femtosecond lasers in dental surgery has been generated because of a few key qualities. The short pulse durations provide crack free machining (an important attribute for successful cement fillings), the removal rates are comparable to those of mechanical drills and much improved over picoseconds lasers and reduced pain levels incurred are all fundamental reasons for ongoing development in this field.

Laser assisted in-situ keratomileusis (LASIK) is the surgical manipulation of the cornea (keratomileusis is Greek for cornea flap cutting). Using lasers to do this has become a common surgical procedure with high success rate and is now the most commonly performed refractive surgical procedure. The unique nature of the femtosecond pulse-matter interaction and the lack of damage caused to surrounding areas, due to better spatial confinement and lower thermal loading, has led to femtosecond lasers being developed to replace the other lasers and perform as minimally invasive, accurate and precise scalpels on a daily basis (J. F. Bille, 2008).

7. Conclusions

The use of focussed femtosecond laser pulses to fundamentally change materials through the interaction of the pulse and material offers new opportunities in device design. This is especially true for fabrication of intricate microstructures within the bulk volume of optically transparent glassy or polymeric materials. But it also can give significant advantages for the micromachining of surface structures in opaque materials in terms of feature size and aspect ratio.

Although femtosecond laser micromachining and inscription has been studied for several decades recent significant improvements in the range of lasers available have accelerated the technology into a range of diverse fields. The lasers available today offer vastly improved peak powers and reliability making commercial exploitation more viable. The advantages of using the nonlinear interaction of light with solid materials are being explored in a number of exciting ways, both in science and engineering, with new avenues opening up as new materials, sources and techniques are developed.

The capacity for making use of the short pulse durations, nonlinear absorption and other characteristics discussed above to create complex three dimensional structures both on the surface and within materials has attracted much recent research effort. However, there is much more potential through the combination of techniques and the development of further knowledge, simulation and modelling that will likely lead to future applications and fields that are only in their infancy at present.

The unique capabilities of femtosecond micromachining make it preferential in a great number of applications. The capacity to locally modify and create permanent change in a range of both transparent and non-transparent materials is of fundamental importance not only to photonics but to a growing number of manufacturing processes. The industrialisation of micromachining processes will be of great significance in the future success of solar cell and flexible organic light emitting diodes (OLEDs) in the manufacture of large sheets that need highly localised and complex machining patterns cut at speed. The most prominent current technology that will be able to facilitate this is the use of femtosecond lasers.

The reliability of the current generation of femtosecond sources compared to earlier models means that these lasers are rapidly being accepted as an option for commercial fabrication. With the continued development in the supporting technologies associated with femtosecond lasers such as the improvement in pump sources, development and commercialisation of more efficient glass compounds, the pulse-material interaction being more fully understood and the delivery systems and techniques being refined, there is a promising future for femtosecond micromachining to expand into more fields and become a common part of manufacturing and photonics industries.

8. References

Agrawal, G. P. (2006). Nonlinear Fiber Optics, 4th edn., Academic Press, New York

Ams, M.; Marshall, G.D. & Withford, M.J. (2006). Study of the influence of femtosecond laser polarisation on direct writing of waveguides, *Opt. Express* 14, 2006, pp.13158-13163

Applegate Jr, R.W.; Squier, J.; Vestad, T.; Oakey, J.; Marr, D.W.M.; Bado, P.; Dugan, M.A. & Said, A. A. (2006). Microfluidic sorting system based on optical waveguide integration and diode laser bar trapping, *Lab on a Chip,* 6, 2006, pp. 422–426

Backus, S.; Durfee III, C.G.; Murnane, M.M. & Kapteyn, H.C. High power ultrafast lasers, *Rev. Sci. Instrum.* 69, 1998, 1207

Bado, P.; Said, A.A.; Dugan, M. & Bellouard, Y. (2006). Laser-based fabrication of a displacement sensor with an integrated high-accuracy position sensor, *Proc. SPIE 6108*, 2006

Baudach, S.; Bonse, J.; Krüger J. & Kautek, W. (2000). Ultrashort pulse laser ablation of polycarbonate and polymethylmethacrylate, *Appl. Surf. Sci.,* 555, 2000, pp.154-155,

Baum, A.; Scully, P. J.; Perrie, W.; Sharp, M.; Watkins, K. G.; Jones, D.; Issac, R. & Jaroszynski, D. A. (2007). NUV and NIR Femtosecond Laser Modification of PMMA, *Laser Precision Microfabrication Conference LPM 2007,* April 2007, Vienna, Austria

Bille, J.F. (2008). Femtosecond Lasers in Ophthalmology: Surgery and Imaging in *Ultrashort Laser Pulses in Biology and Medicine*, Springer Berlin Heidelberg, 2008.

Brown, C.T.A.; Stevenson, D.J.; Tsampoula, X.; McDougall, C.; Lagatsky, A.A.; Sibbett, W.; Gunn-Moore, F.J. & Dholakia, K. (2008) Enhanced operation of femtosecond lasers and applications in cell transfection, *J. Biophoton.* 1, No. 3, 2008, pp.183–199

Callan J.P. (2001). Ultrafast dynamics in highly excited GaAs, Chapter 4 of *Ultrafast physical processes in semiconductors*, vol.67, 2001, pp 167-203

Chung, S.H. & Mazur, E. (2009). Surgical applications of femtosecond lasers, *J. Biophoton.* 1–16, 2009

Diebold, E.D.; Mack, N.H.; Doorn, S.K. & Mazur, E. (2009). Femtosecond Laser-Nanostructured Substrates for Surface-Enhanced Raman Scattering, *Langmuir,* 25 (3), 2009, pp 1790–1794

Eaton, S.M.; Zhang, H.; Herman, P.R.; Yoshino, F.; Shah, L.; Bovatsek, J. & Arai, A.Y. (2005). Heat accumulation effects in femtosecond laser-written waveguides with variable repetition rate, Optics Express, 13, 2005, pp. 4708-4716

Frumker E. & Silberberg, Y. (2007). Femtosecond pulse shaping using a two-dimensional liquid-crystal spatial light modulator, *Opt. Lett.* 32, 2007, pp. 1384-1386

Glezer, E. N. & Mazur, E. (1997). Ultrafast-laser driven micro-explosions in transparent materials. *Appl.Phys. Lett.* 71, 1997, pp. 882–884

Guo, R.; Xiao, S.; Zhai, X.; Li, J.; Xia, A. & Huang, W. (2006). Micro lens fabrication by means of femtosecond two photon photopolymerization, *Opt. Express* 14, 2006, pp. 810-816

Iga, Y.; Ishizuka, T.; Watanabe, W.; Itoh, K.; Li, Y. & Nishii, J. (2004). Characterization of micro-channels fabricated by in-water ablation of femtosecond laser pulses, *Japanese Journal of Applied Physics Part 1,* 43, 2004, pp. 4207-4211

ISO Standard 11146, Lasers and laser-related equipment – *Test methods for laser beam widths, divergence angles and beam propagation ratios,* http://www.iso.org/iso/home, (2005)

Juodkazis, S.; Yamasaki, K.; Mizeikis, V.; Matsuo S. & Misawa, H. (2004). Formation of embedded patterns in glasses using femtosecond irradiation, Applied Physics A: Materials Science & Processing, Volume 79, Numbers 4-6 / September, 2004

Kalli, K.; Geernaert, T.; Koutsides, C.; Komodromos, M.; Nasilowski, T.; Urbanczyk, W.; Wojcik, J.; Berghmans, F. & Thienpont, H. (2009). Point-by-point Bragg grating inscription in single-mode microstructure fibre using NIR femtosecond laser, *Twentieth International Conference on Optical Fibre Sensors 2009*, 5-9 October 2009, Edinburgh, UK

Kawata, S.; Sun, H-B.; Tanaka, T. & Takada, K. (2001). Finer features for functional microdevices, *Nature* 412, 2001, pp. 697-698

Keldysh, L.V. (1965). *Soviet Physics JETP* 20, 1965, 1307

Kim, J. & Na, S. (2007). Metal thin film ablation with femtosecond pulsed laser" *Optics & Laser Technology*, Volume 39, Issue 7, October 2007,Pages 1443-1448

Kondo, Y.; Nouchi, K.; Mitsuyu, T.; Watanabe, M.; Kazansky, P.G. & Hirao, K. (1999). Fabrication of long-period fiber gratings by focused irradiation of infrared femtosecond laser pulses, *Optics Letters*, vol. 24, May 15 1999, pp. 646-648

Kononenko, T.V.; Meier, M.; Komlenok, M.S.; Pimenov, S.M.; Romano, V.; Pashinin, V.P. & Konov, V.I. (2008). Microstructuring of diamond bulk by IR femtosecond laser pulses, *Appl. Phys. A,* 90, 2008, pp. 645–651

Kowalevicz, A. M.; Sharma, V.; Ippen, E. P.; Fujimoto, J. G. & Minoshima, K. (2005). Three-dimensional photonic devices fabricated in glass by use of a femtosecond laser oscillator, *Opt. Lett.*, 30, (2005), 1060-1062

Kruger, J.; Dufft, D.; Koter, R. & Hertwig, A. (2007). Femtosecond laser-induced damage of gold films, *Applied surface science*, 2007, pp:7815-7819

Lai, Y.; Zhou, K.; Zhang, L. & Bennion I. (2006). Microchannels in conventional single-mode fibers, *Optics Letters*, Vol. 31, Issue 17, 2006, pp. 2559-2561

Li, J.; Zhang, H.; Eaton, S.M.& Herman, P.R. (2008). 5-D Spectroscopic Microscopy for Intelligent Femtosecond Laser Writing of Optical Waveguides, in *CLEO/QELS Conference and PhAST, OSA Technical Digest (CD),* Optical Society of America, 2008, paper JWE3

Mannion, P.; Magee, J.; Coyne, E.; O'Connor, G.M. (2002). Ablation thresholds and feature quality in ultrafast laser micro-machining of common metals in air, to appear in the *proceedings of SPIE OPTO Ireland*, September 2002, Galway, Ireland

Martinez, A.; Dubov, M.; Khrushchev, I. & Bennion, I. (2004). Direct writing of fibre Bragg gratings by femtosecond laser, *Electronics Letters*, vol. 40, Sep 16th 2004, pp. 1170-1172

Martinez, A.; Dubov, M.; Khrushchev, I. & Bennion, I. (2006). Structure of Fibre Gratings Directly Written by Infrared Femtosecond Laser", *2006 Conference on Lasers and Electro-Optics (CLEO)*, 2006, JTuD13

McDougall, C.; Stevenson, D.J.; Brown, C.T.A.; Gunn-Moore, F. & Dholakia, K. (2009) Targeted optical injection of gold nanoparticles into single mammalian cells, *J. Biophoton.* 1–8, 2009

Minoshima, K.; Kowalevicz, A. M.; Hartl, I.; Ippen, E. P. & Fujimoto, J. G. (2001). Photonic device fabrication in glass by use of nonlinear materials processing with a femtosecond laser oscillator. *Opt. Lett.* 26, 2001, pp. 1516–1518

Miura, K.; Qiu, J.R.; Fujiwara, S.; Sakaguchi, S. & Hirao, K. (2002). Three-dimensional optical memory with rewriteable and ultrahigh density using the valence-state change of samarium ions. *Appl. Phys. Lett.* 80, 2002, pp. 2263–2265

Nagashima, K.; Hashida, M.; Katto, M.; Tsukamoto, M.; Fujita, M. & Izawa, Y. (2004). Femtosecond laser ablation of Al2O3 ceramics, *The Transactions of the Institute of Electrical Engineers of Japan*, 124, 2004, pp. 388-392. (in Japanese)

Nguyen, N.T.; Saliminia, A.; Chin, S.L. & Vallee, R. (2006). Control of femtosecond laser written waveguides in silica glass, *Appl. Phys. B*, 85, 2006, 145

Olivié, G.; Giguère, D.; Vidal, F.; Ozaki, T.; Kieffer, J.C.; Nada, O. & Brunette, I. (2008). Wavelength dependence of femtosecond laser ablation threshold of corneal stroma," *Opt. Express* 16, 2008, pp. 4121-4129

Petrovic, J.S.; Mezentsev, V.; Schmitz, H. & Bennion, I. (2007). Model of the femtosecond laser inscription by a single pulse, Optical and Quantum Electronics, Volume 39, Numbers 10-11 / August, 2007

Poole, S.B.; Payne, D.N. & Fermann, M.E. (1985). Fabrication of low loss optical fibers containing rare earth ions, *Electron. Lett.* 21, 1985, 737

Rizvi, N.H. (no date). Femtosecond Laser Micromachining:Current Status and Applications, http://www.exitech.org/pdfFiles/Femtosecond%20Laser%20Micromachining%20Current%20Status%20and%20Applicat.pdf, no date

Said, A.A.; Dugan, M.; Bado, P.; Bellouard, Y.; Scott A. & Mabesa Jr., J.R. (2004). Manufacturing by laser direct-write of three dimensional devices containing optical and microfluidic networks, In: *Photon Processing in Microelectronics and Photonics III*, Proc SPIE 5339, 194-204

Sanner, N.; Utéza, O.; Bussiere, B.; Coustillier, G.; Leray, A.; Itina T. & Sentis, M. (2009). Measurement of femtosecond laser-induced damage and ablation thresholds in dielectrics, Applied Physics A: Materials Science & Processing, Volume 94, Number 4, March 2009, pp.889-897

Sanner, N.; Huot, N.; Audouard, E.; Larat, C.; Huignard, J-P. & Loiseaux, B. (2005) Programmable focal spot shaping of amplified femtosecond laser pulses, *Opt. Lett.* 30, 2005, pp. 1479-1481

Schafer, D.N.; Gibson, E.A.; Salim, E.A.; Palmer, A.E.; Jimenez, R. & Squier, J. (2009). Microfluidic cell counter with embedded optical fibers fabricated by femtosecond laser ablation and anodic bonding, *Opt. Express* 17, 2009, pp. 6068-6073

Schaffer, C.B.; Garcia, J.F.; & Mazur, E. (2003). Bulk heating of transparent materials using a high-repetition rate femtosecond laser, *Appl. Phys. A* 76, 2003, pp. 351-354

Sheik-Bahae, M. *et al.* (1990). Dispersion and band-gap scaling of the electronic Kerr effect in solids associated with two-photon absorption, *Phys. Rev. Lett.* 65 (1), 1990, 96

Smid, P. (2006). *CNC Programming Techniques: An Insider's Guide to Effective Methods and Applications*, First edition, Industrial Press Inc, USA

Smith, G.N.; Kalli, K.; Bennion, I. & Sugden, K. (2009). Demonstration of inscription and ablation of phase masks for the production of 1st, 2nd, and 3rd order FBG gratings using a femtosecond laser, *Advanced Fabrication Technologies for Micro/Nano Optics and Photonics II. Proceedings of the SPIE*, Volume 7205, 2009, pp. 720511-720511-11

Smith, G.; Karnakis, D.; Knowles, M.; Ferguson, A.; Bennion, I. & Sugden, K. (2008). Fabrication of High Aspect Ratio Microfluidic Devices Using Direct FS Ablation, in *CLEO/Q ELS Conference and PhAST, OSA Technical Digest (CD)*, Optical Society of America, 2008

Südmeyer, T.; Marchese, S.V.; Hashimoto, S.; Baer, C.R.E.; Gingras, G.; Witzel, B. & Keller, U. (2008). Femtosecond laser oscillators for high-field science, Nature Photonics 2, 2008, pp. 599 - 604

Sugioka, K.; Hanada, Y. & Midorikawa, K. (2007). 3D integration of microcomponents in a single glass chip by femtosecond laser direct writing for biochemical analysis, *Applied Surface Science* 253, 2007, pp. 6595-6598

Sun, Q.; Jiang, H.; Liu, Y.; Wu, Z.; Yang, H. & Gong, Q. (2005). Measurement of the collision time of dense electronic plasma induced by a femtosecond laser in fused silica. *Opt. Lett.* 30, 2005, 320

Sundaram, S.K. & Mazur, E. (2002). *Nature Mater.* 12, 2002, 217

Szameit, A.; Dreisow, F.; Heinrich, M.; Pertsch, T.; Nolte, S.; Tünnermann, A.; Suran, E.; Louradour, F.; Barthélémy A.; & Longhi, S.; (2008). Image reconstruction in segmented femtosecond laser-written waveguide arrays, *Appl. Phys. Lett.* 93, (2008), 181109

Szameit, A.; Burghoff, J.; Pertsch, T.; Nolte, S.; Tünnermann, A.; & Lederer, F. (2006). Two-dimensional soliton in cubic fs laser written waveguide arrays in fused silica, *Opt. Express* 14, 2006, pp.6055-6062

Turitsyn, S.K.; Mezentsev, V.K., Dubov, M.; Rubenchik, A.M.; Fedoruk, M.P. & Podivilov, E.V. (2007). Sub-critical regime of femtosecond inscription, *Optics Express*, vol. 15, Oct 29 2007, pp. 14750-14764

Valette, S.; Le Harzic, R.; Huot, N.; Audouard, E. & Fortunier, R. (2005). 2D calculations of the thermal effects due to femtosecond laser-metal interaction, *Applied Surface Science*,Volume 247, Issues 1-4, 2005, Pages 238-242

Wikszak, E.; Burghoff, J.; Will, M.; Nolte, S.; Tuennermann, A. & Gabler, T. (2004). Recording of fiber Bragg gratings with femtosecond pulses using a "point by point" technique in *CLEO. vol. 96 San-Francisco*: Optical Society of America, 2004, Paper CThM7

Yahng, J.S.; Nam, J.R. & Jeoung, S.C. (2009).The influence of substrate temperature on femtosecond laser micro-processing of silicon, stainless steel and glass, *Optics and Lasers in Engineering*, v. 47, iss. 7-8, 2009, pp. 815-820,

Yu, B.; Lu, P.; Dai, N.; Li, Y.; Wang, X.; Wang, Y. & Zheng, Q. (2008). Femtosecond laser-induced sub-wavelength modification in lithium niobate single crystal, *J. Opt. A: Pure Appl. Opt.* 10, 2008

Zagorul'ko, K.A.; Kryukov, P.G.; Dianov, E.M.; Dragomir, A. & Nikogosyan, D.N. (2003) Fibre-Bragg-grating writing in single-mode optical fibres by UV femtosecond pulses, *Quantum Electronics*, vol. 33, Aug 2003, pp. 728-730

8.1 Laser suppliers

For details of the laser specifications go to the links below;

Amplitude systemes: http://www.amplitude-systemes.com/

Coherent: http://www.coherent.com/Lasers/
High Q: http://www.highqlaser.at/en/products/
IMRA: http://www.imra.com/
Kapteyn-Murnane Laboratories: http://www.kmlabs.com/
Raydiance-Inc: http://www.raydiance-inc.com/our-products

16

Magneto-optical Devices for Optical Integrated Circuits

Vadym Zayets and Koji Ando
Nanoelectronics Research Institute, National Institute of Advanced Industrial Science and Technology (AIST)
Japan

1. Introduction

Magneto-optical materials have two unique properties, which make them important for a variety of optical applications. The first property is non-reciprocity. The time inverse symmetry is broken in magneto-optical materials. Therefore, properties of magneto-optical materials are different for two opposite directions of light propagation and optical non-reciprocal devices like the optical isolator and the optical circulator can be fabricated only by utilizing magneto-optical materials. The second important property of the magneto-optical materials is a memory function. If the material is ferromagnetic, the data can be memorized by means of two opposite directions of the residual magnetization. Both the reading and writing of such memory can be done by magneto-optical effect.
The optical isolator is an important component of optical networks. It is transparent in one direction and blocks light in opposite direction. Due to the imperfect matching between optical components in the network, the unwanted back reflection always exists and it severely disturbs the network performance. To avoid this, the optical components have to be protected by an optical isolator. Also, the isolator is important to cut the back-travelling amplified spontaneous emission in the case of serially-connected amplifiers.
Today there is a big demand to integrate all optical components into an opto-electronics chip. In fact, the isolator is one of few components, which have not yet been integrated into commercial chips. It is because of difficulties to integrate magneto-optical materials into a semiconductor-made chip. To solve this, we proposed to use (Cd,Mn)Te as a magneto-optical material for such isolator. The (Cd,Mn)Te exhibits a huge Faraday effect and can be grown on a semiconductor substrate. For (Cd,Mn)Te waveguide grown on GaAs substrate we achieved a high Faraday rotation of 2000 deg/cm, a high isolation ratio of 27 dB, a low optical loss of 0.5 dB/cm, and a high magneto-optical figure-of-merit of 2000 deg/dB/kG in a wide 25-nm wavelength range (Debnath et al., 2007). These values meet or exceed similar values of commercial discrete isolators.
We predicted theoretically (Zaets & Ando, 1999) and proved experimentally (Zayets & Ando, 2005) the effect of non-reciprocal loss in hybrid semiconductor/ferromagnetic metal waveguides. This effect can be utilized for new designs of waveguide optical isolator. Because the structure of this isolator is similar to that of laser diode, such a design is beneficial for the integration. The bistable laser diode with non-reciprocal amplifier was proposed to be used for high-speed optical logic (Zayets & Ando, 2001).

We proposed the non-volatile high-speed optical memory, which utilizes the magnetization reversal of nanomagnet by spin-polarized photo-excited electrons. It was demonstrated experimentally that one selected pulse from a train of two optical data pulses with interval of 450 fs can solely excite the spin-polarized electrons without a disturbance from the unselected optical data pulse (Zayets & Ando, 2009). This proves feasibility for proposed memory to record data train with rate of 2.2 TBit/sec.

2. $Cd_{1-x}Mn_xTe$ waveguide optical isolator

The conventional bulk-type optical isolator consists of a 45-degree Faraday rotator placed between two polarizers [Fig.1]. The angle between axes of entrance polarizer and exit polarizer is 45 degrees. In forward direction the polarization of light is 45 degree rotated by the Faraday rotator to be along the axis of the exit polarizer. Therefore, the light can pass through the isolator in forward direction. In backward direction, the direction of polarization rotation is opposite to that in forward direction due the non-reciprocal nature of the magneto-optical effect. At the entrance polarizer, the polarization is 90 degrees to the polarizer axis and the light is fully blocked.

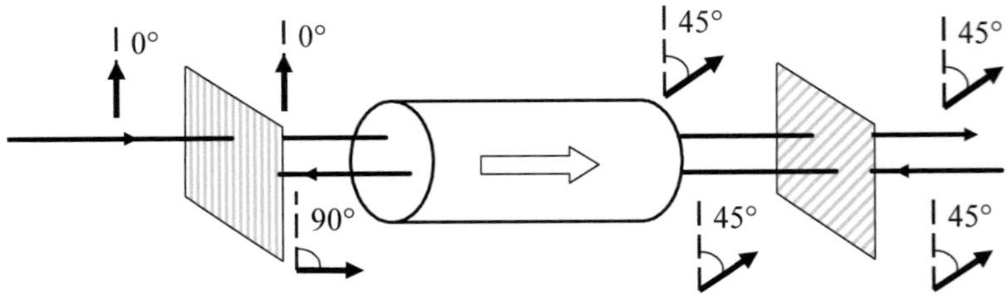

Fig. 1. Design of free-space optical isolator. The Faraday rotator is placed between entrance polarizer (left side) and exit polarizer (right side). Upper diagrams show polarization in forward direction. Lower diagrams show polarization in backward direction.

In present optical networks, ferrimagnetic garnet oxide crystals such as $Y_3Fe_5O_{12}$ (YIG) and $(GdBi)_3Fe_5O_b$ are used as magneto-optical materials for discrete optical isolators. Because most of the active optical elements (such as the laser diode, optical amplifier, modulator, and optical gate) are produced on GaAs or InP substrates, it is desirable to integrate monolithically all optical components on these types of substrate, but integration of the isolator is a difficult task. Waveguide optical isolator based on the garnet film has been reported (Ando et al., 1988). But the garnet-made isolators have not been monolithically integrated with semiconductor optoelectronic devices, because these oxide crystals can not be grown on semiconductor substrates.

Paramagnetic semiconductor $Cd_{1-x}Mn_xTe$ is promising as a magneto-optical material for integrated optical isolators and circulators. $Cd_{1-x}Mn_xTe$ shares the zinc-blende crystal structure with the typical semiconductor optoelectronic materials such as GaAs and InP; thus its film can be grown directly on GaAs and InP substrates. $Cd_{1-x}Mn_xTe$ also exhibits a huge Faraday effect (its Verdet constant is typically 50-200 deg/cm/kG) (Furdyna 1988) near its absorption edge because of the anomalously strong exchange interaction between the sp-band electrons and the localized d-electrons of Mn^{2+}. Furthermore, the tunability of its absorption edge from 1.56 to 2.1 eV with Mn concentration makes the $Cd_{1-x}Mn_xTe$ magneto-optical waveguide compatible with (Al,Ga,In)P:GaAs optoelectronic devices operating in the wavelength range of 600-800 nm. For longer-wavelength (λ=800-1600 nm) optoelectronic devices, $Cd_{1-x-y}Mn_xHg_yTe$ can be used. Bulk optical isolators using these materials are now commercially available (Onodera et al. 1994).

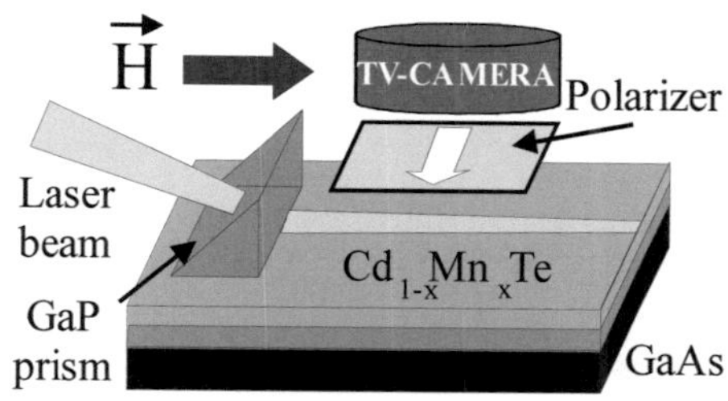

Fig. 2. Experimental set-up to evaluate magneto-optical TE-TM waveguide mode conversion

For the $Cd_{1-x}Mn_xTe$ to be used as a material for the waveguide isolator, several conditions should be satisfied. For a practical $Cd_{1-x}Mn_xTe$ waveguide isolator, the isolation ratio should exceed 20 dB, insertion loss should be below 1 dB and operation wavelength range should be wider than 20 nm. This performance can only be achieved with a magneto-optical waveguide having a mode conversion ratio above 95 % and a figure-of-merit above 100 deg/dB. Below we will show that using advanced waveguide structure and optimized fabrication technique, this conditions can be achieved in $Cd_{1-x}Mn_xTe$ waveguide grown on GaAs substrate.

The $Cd_{1-x}Mn_xTe$ has about 12% lattice mismatch with GaAs. The growth conditions of $Cd_{1-x}Mn_xTe$ on GaAs substrate should be well optimized. Otherwise, the high density of dislocation in $Cd_{1-x}Mn_xTe$ film causes high optical loss in $Cd_{1-x}Mn_xTe$ waveguide (Zaets et al.,1997) and low value of Faraday rotation. The $Cd_{1-x}Mn_xTe$ waveguide was grown by molecular beam epitaxy (MBE) on GaAs (001) substrate. We optimized the growth conditions and fabricated the $Cd_{1-x}Mn_xTe$ waveguide in the following way. In the beginning, GaAs substrate was thermally cleaned at 400^0 C under atomic hydrogen flux to remove oxides from GaAs surface. Before initiating the growth, the GaAs substrate was kept for 30 minutes under Zn flux to prevent the formation of the undesired Ga_2Te_3 compound. At first, a thin 10 nm ZnTe film was grown on the GaAs substrate to initialize the (001) growth. Following a 1-μm thick CdTe buffer layer, $Cd_{1-x}Mn_xTe$ waveguide was grown. It consists of a 3-μm-thick $Cd_{0.73}Mn_{0.27}Te$ waveguide cladding and a 1-μm-thick $Cd_{0.77}Mn_{0.23}Te$ waveguide core. The waveguide core was sandwiched between two 500-nm-thick $Cd_{1-x}Mn_xTe$ (x=0.27-0.23) graded-refractive-index clad layers, for which the Mn concentration was changed linearly with thickness. We used the $Cd_{0.73}Mn_{0.27}Te$ layer as a cladding layer, since GaAs is an optical absorber with a higher refractive index than that of $Cd_{1-x}Mn_xTe$, a single $Cd_{1-x}Mn_xTe$ layer on GaAs does not work as a waveguide. One needs transparent cladding layers with smaller refractive index. $Cd_{0.73}Mn_{0.27}Te$ satisfies these conditions because $Cd_{1-x}Mn_xTe$ with higher Mn concentration has a smaller refractive index and wider optical band gap. The graded-refractive-index clad layers are essential for $Cd_{1-x}Mn_xTe$ waveguide to achieve high magneto-optical TE-TM waveguide mode conversion and high optical isolation.

Figure 2 illustrates the experimental setup for evaluating optical propagation loss and TE-TM waveguide mode conversion (Zaets & Ando, 2000). A GaP prism was used to couple the laser light from tunable Ti:sapphire laser (λ=680 -800 nm) into a $Cd_{1-x}Mn_xTe$ waveguide. A cooled CCD TV-camera collected light scattered normally from the film surface. A linear polarizer was placed in front of the TV camera with its polarization axis perpendicular to the light propagation direction. With this configuration, only the TE mode component of waveguiding light can be detected by the high-sensitivity TV camera. In the absence of a magnetic field, a scattered light streak was seen when the TE mode was excited (Fig. 3(a)), but it was not seen when TM mode was excited (Fig. 3 (b)). Also, weak dot-like scattering on defects was seen in both cases.

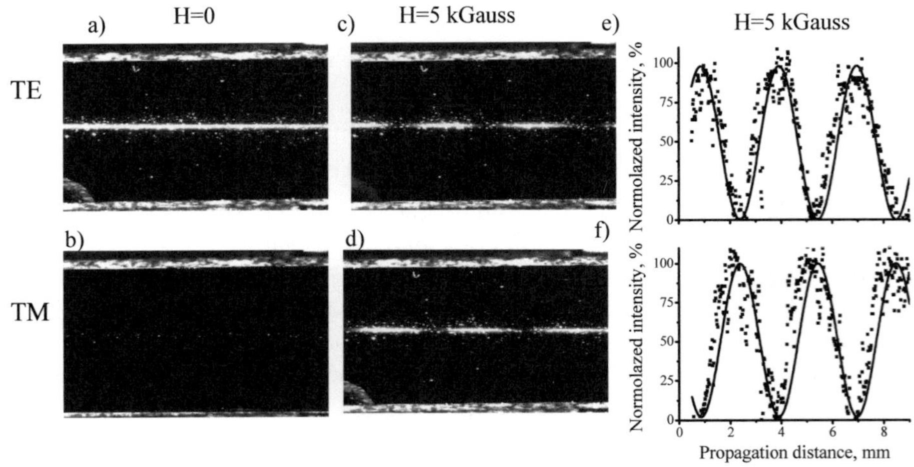

Fig. 3. TM-TE mode conversion ratio in $Cd_{1-x}Mn_xTe$ waveguide at λ=730 nm. (Zayets & Ando, 2004)

For the evaluation of the magneto-optical TE-TM waveguide mode conversion, a magnetic field was applied in parallel to the light propagation direction. A light streak with a periodically modulated intensity was observed for both TE mode excitation (Fig. 3 (c)) and TM mode excitation (Fig. 3 (d)). Figures 3 (e)-(f) show the measured intensity of the modulated streak along the propagation length. The intensity was normalized to input intensity. The oscillations maxima in the case of TE excitation (Fig. 3 (e)) correspond to the oscillations minima in the case of TM excitation (Fig. 3 (f)) and vice versa. Under an applied magnetic field the polarization of the waveguide mode rotates because of Faraday effect. If the TE-TM mode phase mismatch is not zero, the eigenmodes of the waveguide are elliptically polarized and the rotation between TE and TM polarizations is not complete. As seen from Figs. 3 (c)- 3 (f), the $Cd_{1-x}Mn_xTe$ waveguide with the graded index cladding layer shows almost complete mode conversion.

The $Cd_{1-x}Mn_xTe$ waveguide with graded buffer layers has low optical loss, high TE-TM mode conversion efficiency (more than 98 %) and high isolation ratio (more than 20 dB). However, high isolation ratio was obtained in narrow about 3 nm wavelength range. For practical application of the isolator the operation wavelength range should be at least 20 nm. For the operation of the optical isolator, the rotational angle of Faraday rotator should be 45^0 (Fig.1) for any operational wavelength. $Cd_{1-x}Mn_xTe$ is a diluted magnetic semiconductor. It has a high value of Faraday rotation, but it is high only near its bandgap and near the bandgap the dispersion of Faraday rotation is significant as well. Of course, $Cd_{1-x}Mn_xTe$ is a paramagnetic material and at each wavelength the Faraday rotation can be tuned to 45^0 by the changing magnetic field. However, such tuning is not practical for real applications because a practical isolator needs a permanent magnet with a fixed magnetic field. Below we will show that it is possible to achieve practically dispersion-free Faraday rotation in wide wavelength range by combining in a waveguide $Cd_{1-x}Mn_xTe$ bulk material and a $Cd_{1-x}Mn_xTe$ quantum well (QW).

The Faraday effect in a $Cd_{1-x}Mn_xTe$ QW is greater than that of bulk $Cd_{1-x}Mn_xTe$ and it is not as dependent on wavelength. However, due to the two-dimensional nature of the QW, its optical properties become significantly different for light polarized in the plane of the QW and perpendicular to the QW. Therefore, for a waveguide composed of only a single QW, there is a big difference between propagation constants of TE and TM modes. Due to TE-TM mode phase mismatch, the linearly polarized light can be easily converted to elliptically polarized light, which reduces the performance of the isolator. Therefore, a waveguide composed of only a single QW cannot be used for the isolator application.

In order to make a high performance isolator, we need a large, wavelength independent Faraday effect and small phase mismatch between TE and TM modes. For that purpose, we proposed using an optical waveguide that combines $Cd_{1-x}Mn_xTe$ bulk material and a single QW (Debnath et.al. 2004)

Figure 4 shows the (Cd,Mn)Te/(Cd,Zn)Te QW waveguide structure. There are two buffer layers of ZnTe (10 nm) and CdTe (1 μm) and a $Cd_{0.71}Mn_{0.29}Te$ (3 μm) waveguide clad layer. The waveguide core layer was sandwiched between two $Cd_{1-x}Mn_xTe$ (0.5 μm) graded layers in order to reduce TE-TM mode phase mismatch. The waveguide core consists of a $Cd_{0.76}Mn_{0.2}Te/Cd_{0.75}Zn_{0.25}Te$ single QW and a 1-μm-thick $Cd_{0.75}Mn_{0.25}Te$ layer, where thickness of the $Cd_{0.76}Mn_{0.24}Te$ well varies between 20–100 Å and the thickness of $Cd_{0.75}Zn_{0.25}Te$ barrier is 100 Å.

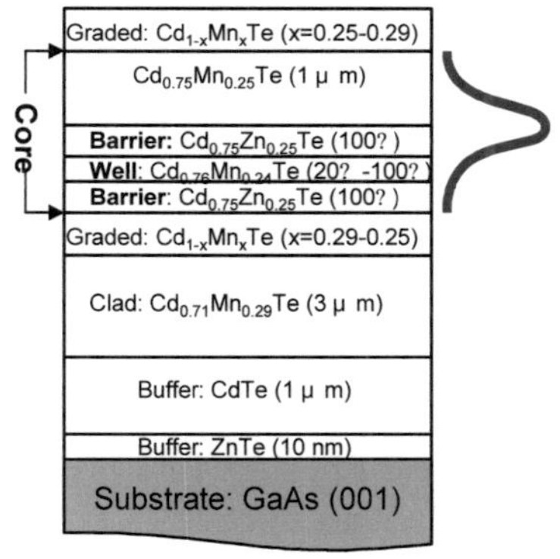

Fig. 4. Structure of a (Cd,Mn)Te waveguide with (Cd,Mn)Te/(Cd,Zn)Te QW. The waveguiding light intensity distribution is shown in the right side . (Debnath et al, 2007)

Figure 5 shows a spatially modulated light streak of the waveguide mode at two different wavelengths (760 and 785 nm) for the waveguides with QW and without QW. The high contrast between the minima and maxima of the light intensity oscillations shows that complete mode conversion is attained for both waveguides. The distance between peaks corresponds to 180 degrees of the rotation. For the waveguide without QW [Figs. 5 (c) and 5 (d)], there is a big difference of the rotational period for these two wavelengths. However, for the waveguide with QW [Figs. 5(a) and 5 (b)], there was no such difference. This means that, for the waveguide with QW, the Faraday rotation at these two wavelengths is the same. Also, for the waveguide with QW, the oscillation period is much shorter than that of the waveguide without QW. This corresponds to the larger Faraday rotation in the waveguide with QW.

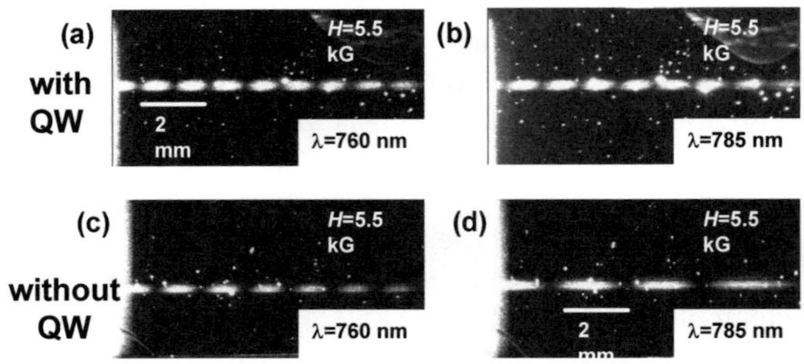

Fig. 5. Spatially modulated light streak from waveguide TE mode for CdMnTe waveguide with QW (a), (b) and waveguide without QW (c), (d) at λ = 760 nm (a), (c) and λ = 785 nm (b), (d) under magnetic field of 5.5 kG . (Debnath et al, 2007)

Figure 6 compares the Faraday effect in $Cd_{1-x}Mn_xTe$ waveguide with QW and without QW at H=5.5 kG. In the case of the waveguide with QW, the Faraday rotation is very high (~1800 deg/cm) and it is almost constant in a wide wavelength range. Figure 7 shows the wavelength range within which more than 95% conversion efficiency was obtained for the waveguide with single QW as a function of well width. For well widths of 20–40 Å, the operational wavelength range is as wide as 25-nm. However, for thicker well widths of 70–100 Å, the operational wavelength range sharply decreases. Analysis shows that the expansion of the wavelength range for thinner QW waveguides was due to the reduction of the mode phase mismatch, to as low as 50 deg/cm, whereas this value rose to more than 500 deg/cm for thicker QW waveguides. Thinner QW waveguides have high Faraday rotation (≈ 2000 deg/cm) and small phase mismatch (≈ 50 deg/cm) This is the reason why thinner QW waveguides provided a wider operational wavelength range of complete mode conversion. From this result we conclude that, for the practical optical isolator application, only waveguides with a single QW thinner than 40 Å can be used.

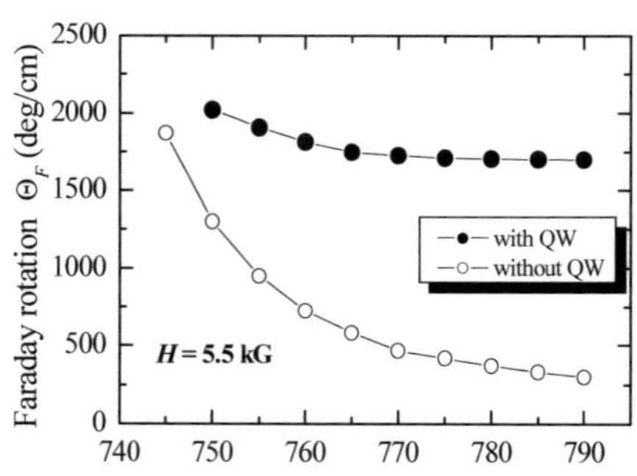

Fig. 6. Faraday rotation in $Cd_{1-x}Mn_xTe$ waveguide with QW and without QW . (Debnath et al, 2007)

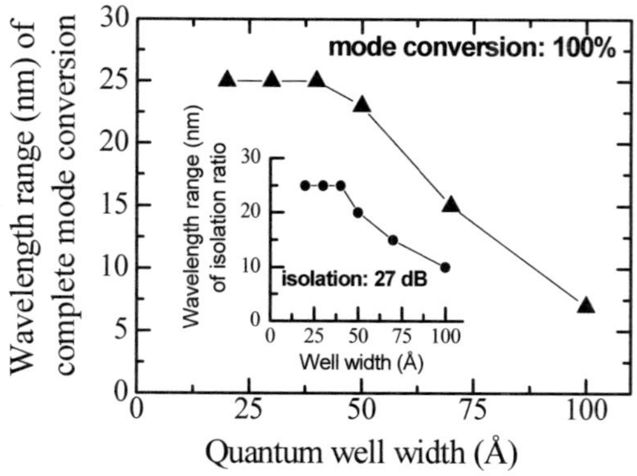

Fig. 7. Wavelength range, within which the complete mode conversion is obtained, as a function of QW width. Inset shows the isolation ratio. (Debnath et al, 2004)

For an integrated optical isolator, the CdMnTe magneto-optical waveguide has to be integrated with a reciprocal polarization rotator and a polarizing beam splitter. Both these components can be fabricated utilizing passive optical waveguides. The material of the waveguides is not essential for the operation of these components. Therefore, it is better to use the same passive waveguides as utilized for optical interconnection in a photonic circuit, where the isolator should be integrated. Figure 8 shows an example of a waveguide-type reciprocal polarization rotator. It is a passive optical waveguide in which the top is cut at an angle of 45 degrees. TM and TE modes are not eigenmodes in this waveguide. Therefore, there is a conversion between TM and TE mode along mode propagation. The length of this waveguide can be adjusted to achieve the desirable angle of polarization rotation. The waveguide type reciprocal polarization rotators were demonstrated utilizing Si waveguide (Brooks et al., 2006], AlGaAs waveguides (Huang et al., 2000) and GaInAsP/InP waveguides (Kim et al., 2009). Figure 11 shows an example of a waveguide-type polarizing beam splitter. It is a 2x2 waveguide splitter. In any waveguide splitter, the coupling efficiency between an input ports and an output ports depends on the value of mode propagation constant. Generally, in an optical waveguide the propagation constants of TM and TE modes are different. Therefore, it is possible to adjust the splitter so that the TM mode couples from port 1 into port 4 and the TE mode couples from port 1 into port 3. The waveguide-type polarizing beam splitters were demonstrated utilizing Si waveguide (Fukuda et al., 2006) and InGaAsP–InP waveguides (Augustin et al., 2007).

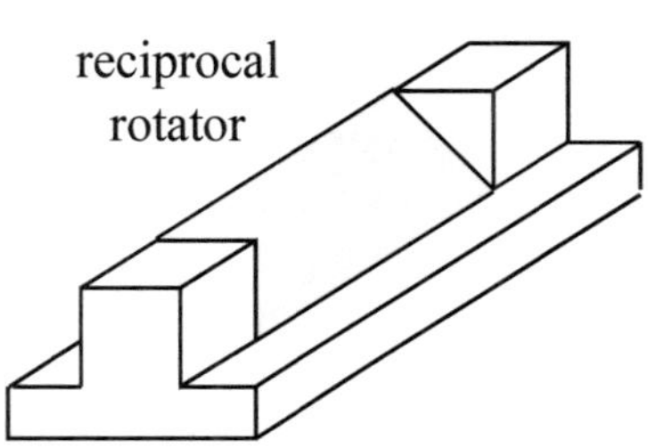

Fig. 8. Waveguide-type reciprocal polarization rotator.

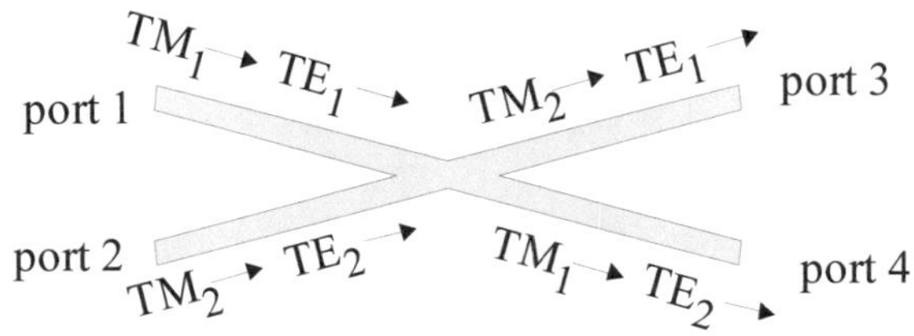

Fig. 9. Waveguide-type polarization beam splitter.

Figure 10 shows the design of a waveguide-type polarization-independent optical isolator. It consists of two polarizing beam splitters connected by two arms. Each arm consists of an 45-degree reciprocal rotator and a 45-degree $Cd_{1-x}Mn_xTe$ -made Faraday rotator. There is an optical absorber at port 2 to absorb backward travelling light. In the forward direction, the direction of polarization rotation in the Faraday rotator is the same as that in the reciprocal rotator and the total rotation angle by the reciprocal rotator and the Faraday rotator is 90 degrees. The light of both polarizations propagates through the isolator from input 1 port to output port 3. In the backward direction, the direction of polarization rotation in the Faraday rotator is opposite to that in the reciprocal rotator due to the non-reciprocal nature of the Faraday rotator. In this case, the total rotation angle is zero. The light propagates from output port 3 to the port 2, where there is an absorber. Therefore, the input port 1 is isolated. The optical paths for each polarization are shown in Fig. 10. A waveguide-type polarization-independent optical circulator can be fabricated utilizing the same design. In this case the correspondence between the input and output ports is: port 1-> port 3, port 2 ->port 4, port 3 -> port 2, port 4 -> port 1.

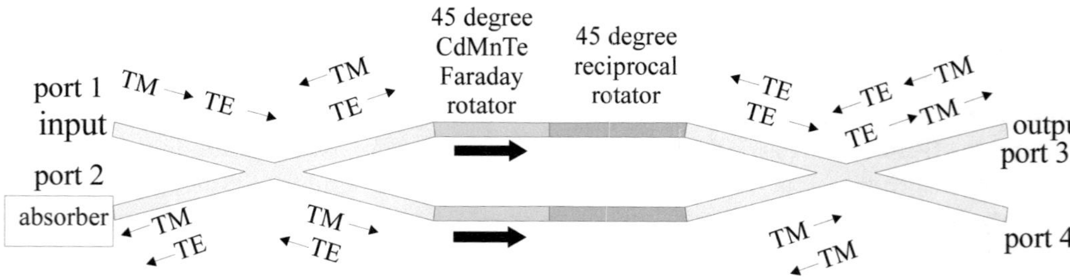

Fig. 10. Polarization-independent waveguide-type optical isolator / circulator. Polarization transformations for both propagation directions are shown.

In conclusion, the high performance of $Cd_{1-x}Mn_xTe$ waveguide isolator grown on GaAs substrate was demonstrated. Complete TE-TM mode conversion, a high Faraday rotation of 2000 deg/cm, a high isolation ratio of 27 dB, a low optical loss of 0.5 dB/cm, and a high magneto-optical figure-of-merit of 2000 deg/dB/kG were achieved in a wide 25-nm wavelength range. These values are comparable or better to that of commercial discrete isolators. The propagation of waveguide mode in $Cd_{1-x}Mn_xTe$ waveguide is very similar to the light propagation in magneto-optical bulk media. Therefore, non-reciprocal elements such as an optical isolator, circulator and polarization independent isolator can be fabricated by $Cd_{1-x}Mn_xTe$ waveguides using a similar scheme as is used for free space components. Therefore, using $Cd_{1-x}Mn_xTe$ all these components can be integrated with semiconductors optoelectronical components.

3. Ferromagnetic metal/semiconductor hybrid isolator

Both types of isolators, either made of $Cd_{1-x}Mn_xTe$ or made of garnets, required high-crystal quality materials in order to have a low optical loss and a high value of Faraday rotation. In the case of $Cd_{1-x}Mn_xTe$, the magneto-optical film can be directly grown on a semiconductor substrate. The defect density in $Cd_{1-x}Mn_xTe$ film should be kept low until the end of the microfabrication process. For the fabrication of integrated optical circuits it is more convenient to use common fabrication technique like sputtering, e-beam evaporation and lift-off. Ferromagnetic metals, like Fe, Co or Ni are very attractive for this purpose. They have high magneto-optical constants and the microfabrication of these metals is simple and well established for optoelectronic integrated circuits. For example, Cr and Co is often used as a metal for Ohmic contact to p-GaAs and p-InGaAs.

We predicted theoretically (Zaets & Ando, 1999) and proved experimentally (Zaets & Ando, 2005) the effect of non-reciprocal loss in hybrid semiconductor/ferromagnetic metal waveguides. This effect can be use for a new design of waveguide optical isolator. For this design, the magnetization of the ferromagnetic metal was perpendicular to the light propagation direction and lay in the film plane (Voigt configuration). In this case a large difference exists in values of loss/gain for TM modes propagating in opposite directions. Thus, an amplifier covered by a ferromagnetic metal can itself function as an optical isolator. This ferromagnetic-metal/ semiconductor hybrid isolator can be beneficial for monolithic integration of the optical isolator with semiconductor optoelectronic devices, because its structure is very similar to the structure of a laser diode and its fabrication process is almost the same as that of a laser diode. Therefore, the isolator can be integrated utilizing the present technology for a semiconductor laser diode and a semiconductor amplifier.

The effect of the non-reciprocal loss is unique for the hybrid waveguides. In the case of the light propagation in a bulk MO material the non-reciprocal effect (variation of optical properties for opposite directions of light propagation) occurs only when the magnetization of the material is parallel to the light propagation (Faraday effect and magnetic circular dichroism). There is no non-reciprocal effect, if the light propagates perpendicularly to the magnetization. On the contrary, in a MO waveguide, even if the magnetization is perpendicular to the light propagation direction and lies in the film plane, the TM mode has a non-reciprocal change of the propagation constant.

Figure 11 compares the magneto-optical effect in a bulk material and in an optical waveguide covered by ferromagnetic metal. In the bulk material, when the light propagates along magnetic field, there are two magneto-optical effects: magnetic circular dichroism (MCD) and Faraday effect. In magnetic materials the spin-up and spin-down bands are split along the magnetic field. Due to the conservation law of the time inverse symmetry, the light of the left circular polarization interacts only with the spin-up band and the light of the right circular polarization interacts only with the spin-down band. (In general, the light of elliptical polarization interacts solely with one spin band). Therefore, there is a difference of refractive index (Faraday effect) and of absorption (MCD effect) for the left and right circular polarizations, when the light propagates along the magnetic field. When the magnetic field is perpendicular to the light propagation, there is no linear magneto-optical effect. The case of a waveguide covered by magnetic metal is different. Inside of the metal the optical field is evanescent. As it will be shown below, in the case of evanescent field, the polarization rotates in xz-plane perpendicularly to the waveguide mode propagation direction (Fig. 11) even without magnetic field or anisotropy. Therefore, if the magnetic field is applied perpendicularly to the waveguide mode propagation direction, since the polarization of the optical field is elliptical along this direction; there are MCD and Faraday effects. Both these effects contribute to the non-reciprocal loss.

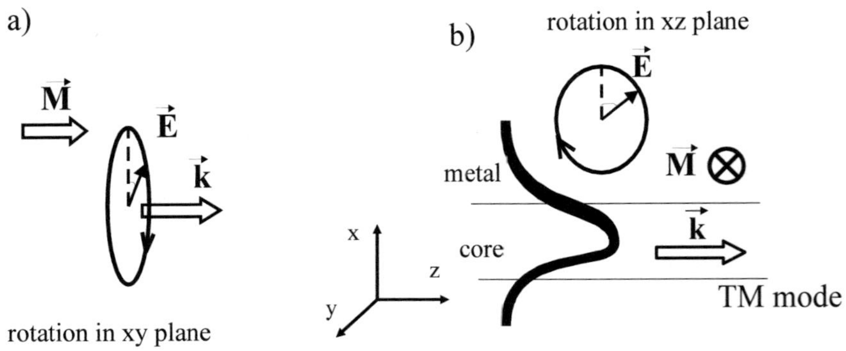

Fig. 11. Magneto-optical effect a) in bulk b) in waveguide covered by ferromagnetic metal.

The field of an electromagnetic wave is transverse. That means that the electrical field of the wave is always perpendicular to the wave vector. In the case of the light propagation in the bulk along z axis the electrical field can be described as

$$\vec{E} \sim e^{i(k_z z-\omega t)} + c.c. \quad \text{and} \quad E_z = 0 \tag{1}$$

For circular polarized wave, the polarization, which rotates in xy-plane, can be described as

$$\frac{E_x}{E_y} = \pm i \tag{2}$$

where + and – correspond to the right and left circular polarized waves.
Next, let us consider the waveguide covered by a magnetic metal. The electrical field in the metal can be described as

$$\vec{E} \sim e^{-k_x x+i(k_z z-\omega t)} + c.c. = e^{i(ik_x x+k_z z-\omega t)} + c.c \tag{3}$$

Since the electrical field of the electromagnetic wave is transverse $\vec{k} \perp \vec{E}$, for the TM mode ($E_y = 0$), we have

$$ik_x E_x + k_z E_z = 0 \quad \text{or}$$

$$\frac{E_x}{E_z} = i\frac{k_z}{k_x} \tag{4}$$

Comparing Eq. 4 and Eq. 2, it can be clearly seen that optical field in the metal is elliptically polarized and the polarization rotates in the xz-plane. Also, from Eq. (4) it can be seen that the ellipticity changes polarity for opposite directions of mode propagation (k_z -> $-k_z$). Therefore, if the magnetization of the metal is along the y-direction, the absorption will be different for two opposite mode propagation directions due to MCD effect. Also, the Faraday effect contributes to the non-reciprocal loss in waveguide. Due to the Faraday effect the effective refractive index of the metal is different for the left and right circular polarizations. The absorption by the metal is directly proportional to the amount of light inside the metal and this in turn depends on the difference of refractive indexes between the metal and the waveguide core. If there is a change of refractive index due to the Faraday effect, it causes the change of optical absorption of waveguide mode by the metal.

The effect of non-reciprocal loss has another simple explanation. The propagation of waveguide mode can be considered as a plane wave, which propagates in the waveguide core and experiences multi reflections from boundaries of waveguide. If the magnetization of the metal is perpendicular to the plane of reflection, the plane wave experiences the transverse Kerr effect. The transverse Kerr effect states that if the reflection plane of the light is perpendicular to the magnetization of the metal, the absorption by the metal is different for two opposite directions of the magnetization. Therefore, the plane wave experiences different absorption for opposite propagation directions.

The optical isolation of the amplifier covered by ferromagnetic metal can be calculated by solving Maxwell's equations for multilayer structure. As an example of the waveguide isolator operating at a wavelength of 790 nm, we consider a $GaAs_{0.9}P_{0.1}/Al_{0.3}Ga_{0.7}As$ quantum-well (QW) optical amplifier covered by a Co layer (Fig. 12). To reduce the absorption by the Co layer, a buffer layer of p-$Al_{0.7}Ga_{0.3}As$ is inserted between the absorbing Co layer and $GaAs_{0.9}P_{0.1}/Al_{0.3}Ga_{0.7}As$ QW amplifying core layer. The optical field of a waveguide mode exponentially penetrates through the buffer layer into the Co layer.

magneto-optic Co layer
p-$Al_{0.7}Ga_{0.3}As$, 250 nm
$GaAs_{0.9}P_{0.1}/Al_{0.3}Ga_{0.7}As$ QW amplifying core layer
n-$Al_{0.7}Ga_{0.3}As$, 1 μm
n-GaAs substrate

Fig. 12. GaAsP/AlGaAs optical amplifier covered by Co

Figure 13 shows the dependence of the optical loss/gain for the forward and backward propagating modes as a function of internal gain of the $GaAs_{0.9}P_{0.1}$ active layer. Depending

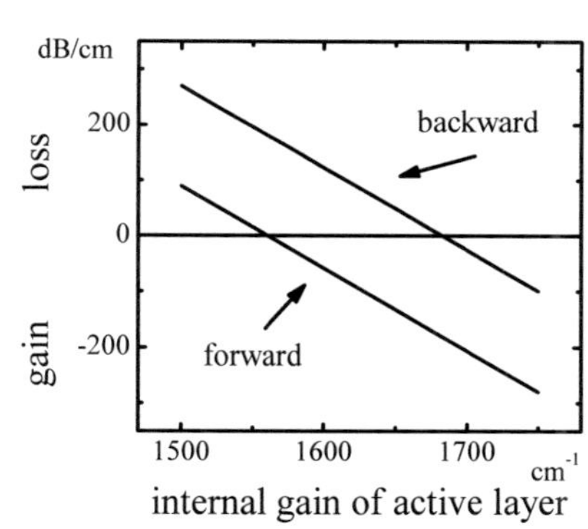

Fig. 13. Waveguide loss/gain as a function of the internal gain of $GaAs_{0.9}P_{0.1}$ active layer (Zaets&Ando, 1999).

on the value of the internal gain, the waveguide can operate as a non-reciprocal amplifier (internal gain >1680 cm-1) or as a non-reciprocal absorber (internal gain <1560 cm-1) or as an isolator (1560 cm-1> internal gain >1680 cm-1). The isolation ratio is almost constant against internal gain and it is about 180 dB/cm.

Because the waveguide mode interacts with the MO Co layer by its exponential tale through the buffer layer, the non-reciprocal loss/gain depends on the thickness of the buffer layer. Figure 14 shows a dependence of the isolation ratio and the internal gain of $GaAs_{0.9}P_{0.1}$ active layer as function of the thickness of the p-$Al_{0.7}Ga_{0.3}$ As buffer layer when the value of loss for the mode propagating in forward direction is kept to be zero. The thinner the buffer layer is, the larger isolation ratio can be obtained, although the higher amplification is necessary to compensate the loss.

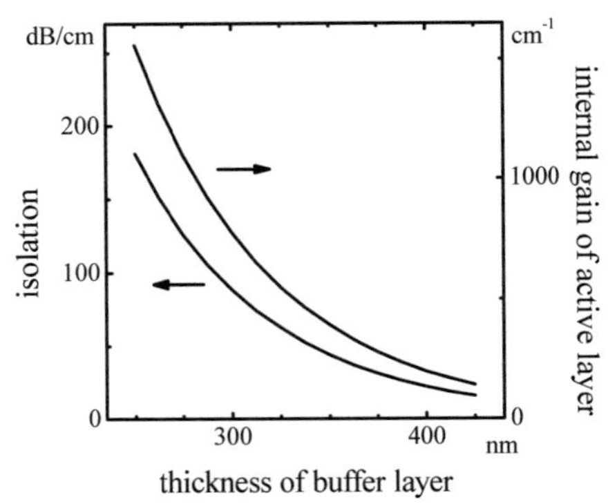

Fig. 14. Isolation ratio and the required internal gain of $GaAs_{0.9}P_{0.1}$ active layer vs. thickness of *p*-$Al_{0.7}Ga_{0.3}As$ buffer layer. The loss/gain of the mode propagating in forward direction is fixed to be zero. (Zaets&Ando, 1999).

Figure 15 compares isolation ratio of structure of Fig.12 with Co, Fe and Ni cover layer. Wavelength dependence of optical constants of Co, Fe and Ni are taken into account, while

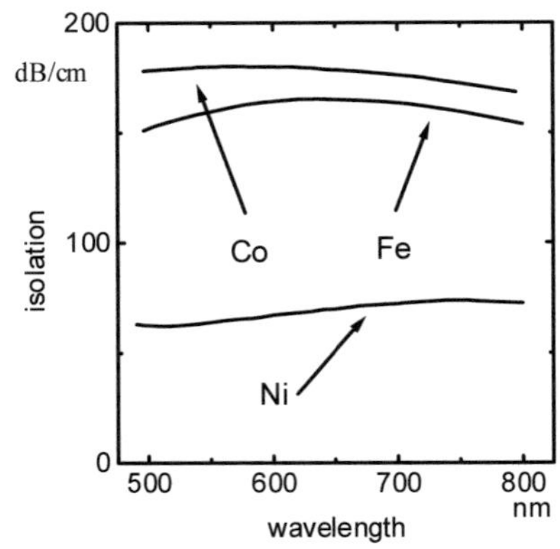

Fig. 15. Isolation ration vs. wavelength. Wavelength dependence of optical constants of Co, Fe and Ni are taken into account, while optical constants of amplifier is assumed to be independent on wavelength.

optical constants of amplifier are assumed to be independent of wavelength. Among these materials the cobalt is the most suitable for the proposed isolator. The isolation ratio only slightly changes vs. wavelength. Therefore, the wavelength dependence of the isolator is determined only by the bandwidth of the amplifier.

The proposed design of the isolator does not require the periodical reverse of magnetization or the phase matching between TE and TM modes or the interferometer structure. It provides the high isolation ratio in a wide range of the internal gain. The isolation ratio is proportional to the isolator length. These all merits make this design suitable for the monolithical integration of optical isolator with optoelectronic devices.

It should be noted that MO materials (ferromagnetic metals, magnetic semiconductors and magnetic garnets) show the highest values of the MO effect in a wavelength region where they show large absorption. Thus, the combination of the absorbing MO materials with the optical amplifiers is a viable way to achieve smaller sizes of non-reciprocal devices, which is suitable for integrated circuits.

For the experimental verification of the effect of non-reciprocal loss, we fabricated passive optical waveguides covered by a ferromagnetic metal. The directional dependence of absorption by the metal is a reason for the isolation in this structure. The optical gain in the isolator is used only to compensate the loss. To avoid side effects due to the optical amplification, we studied a passive waveguide covered by a ferromagnetic metal.

Two identical waveguides were fabricated, where only the material of buffer layer between Co layer and waveguide core was different. Figure 16(a) shows the structure of a $Ga_{1-x}Al_xAs$ waveguide covered by Co with SiO_2 buffer layer and Fig. 16 (b) shows the waveguide with AlGaAs buffer layer. The $Ga_{1-x}Al_xAs$ waveguide was grown with molecular-beam-epitaxy (MBE) on a GaAs (001) substrate. Following a 2500-nm-thick $Ga_{0.55}Al_{0.45}As$ clad layer and a 900-nm-thick $Ga_{0.7}Al_{0.3}As$ core layer, a buffer layer of 12-nm-thick SiO_2 or 120-nm-thick $Ga_{0.55}Al_{0.45}As$ was grown. The 10-μm-wide 600-nm-deep rib waveguide was wet etched. A 100 nm of Co layer and a 100 nm of Au layer were deposited on the buffer layer. A protection layer of 100-nm-thick SiO_2 with 8-μm-wide window was used to avoid light absorption at the sidewalls of the waveguide.

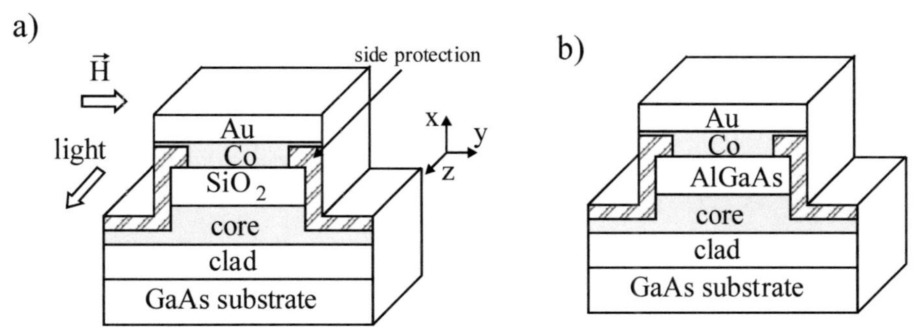

Fig. 16. AlGaAs optical waveguide covered with Co. Either a) SiO_2 or b) $Al_{0.45}Ga_{0.55}As$ was used as the buffer layer. Waveguide light propagates in the core layer and slightly penetrates into the Co layer.

For the evaluation of non-reciprocal loss, laser light (λ=770 nm) was coupled into the waveguide with a polarization-maintaining fiber. The output light was detected by a CCD camera. A polarizer was placed in front of the CCD camera. The magnetic field was applied perpendicularly to the light propagation direction and in the film plane with an electromagnet.

Figure 17 shows the transmission coefficient of TM mode as a function of applied magnetic field for the waveguide with SiO_2 buffer and the waveguide with $Ga_{0.55}Al_{0.45}As$ buffer. A clear hysteresis loop of the transmission coefficient was observed with a coercive force of 35 Oe. The transmission coefficient of TE mode showed no dependence on the magnetic field. The magnetization curve measured by a superconducting quantum interface device (SQUID). For both samples magnetization curve was identical and the coercive force was 35 Oe. The observation of the hysteresis loop of the transmission coefficient of TM mode with the same coercive force is that of Co proves the TM mode transmission depends on magnetization of the Co. Considering time-inversion symmetry, the difference of transmission in the same direction of light propagation for two opposite directions of magnetic field is equal to the difference in transmission for opposite directions of light propagation in one direction of magnetic field. Therefore, the amplitude of the hysteresis loop of the transmission corresponds to the isolation provided by the waveguide. As can be seen from Fig. 17, the isolation direction for a waveguide with a SiO_2 buffer is different from that for a waveguide with a $Ga_{0.55}Al_{0.45}As$ buffer. Therefore, the isolation direction depends not only on the magnetization direction of the ferromagnetic metal, but on the waveguide structure as well.

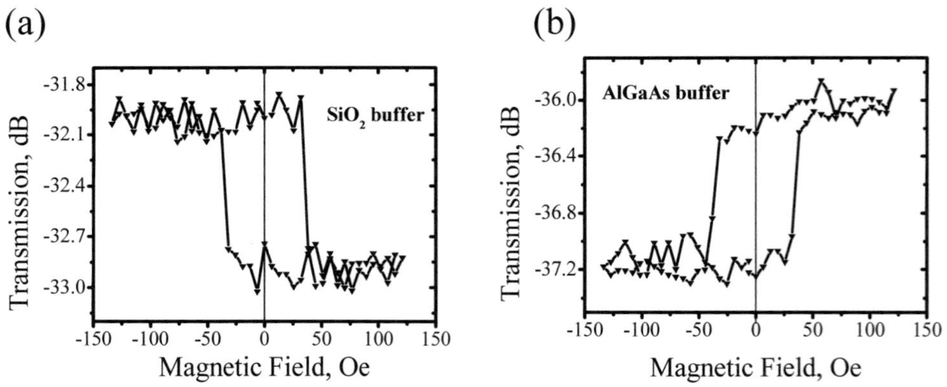

Fig. 17. Optical transmission in 1.1-mm-long $Ga_{1-x}Al_xAs$ optical waveguides covered by Co at λ=770 nm as a function of applied magnetic field a) with SiO_2 as the buffer layer and b) with $Ga_{0.55}Al_{0.45}As$ as the buffer layer. (Zayets &Ando (2005)).

These results can be explained by considering two contributions to the non-reciprocal loss by MCD and Faraday effects. We defined the figure-of-merit (FoM) for this isolator as a ratio of the non-reciprocal absorption to the total absorption by the metal. The mode propagation constants, non-reciprocal loss, and FoM were rigorously calculated from a direct solution of Maxwell's equations for the planar waveguide. In addition, both MCD and Faraday contributions to FoM were roughly estimated by estimating the light energy dissipation resulting from each contribution. For the waveguide with SiO_2 buffer, FoM was calculated to be 7.95%, where MCD and magneto-reflectivity contributions were estimated as -8.01 % and 15.86%, respectively. For the waveguide with $Ga_{0.55}Al_{0.45}As$ buffer, FoM was calculated to be -7.19%, where MCD and Faraday contributions were estimated as -8.01 % and 1.11 %, respectively. The sign of the contributions is different. The magnitude of the MCD contribution is almost same for both waveguides. On the contrary, the magnitude of Faraday contribution is significantly different. That is a reason that opposite isolation directions were observed in the waveguides with SiO_2 and $Ga_{0.55}Al_{0.45}As$ buffers. For both waveguides the sums of the MCD and Farady contributions are approximately equal to the FoM calculated from the rigorous solution of Maxwell equations.

Next, we verify the isolation in the optical amplifier covered by the ferromagnetic layer. Figure 18 (a) showed the fabricated isolator. The optical amplifier consists of 8 $In_{0.46}Ga_{0.54}As/In_{0.26}Ga_{0.74}P$ tensile strained quantum wells grown on n-InP substrate. The p-InP was used as buffer layer and InGaAs as a contact layer. The amplifier has maximum amplification at λ=1566 nm. Figure 18 (b,c) shows the intensity of amplified spontaneous emission (ASE) as a function of applied magnetic field measured at both edges of the isolator, when the weak injection current of 10 mA was used. At each edge, a hysteresis loop was observed for ASE , however the polarity of the loop was different. It is because there is difference in gain for the light propagating in forward and backward directions in the amplifier.

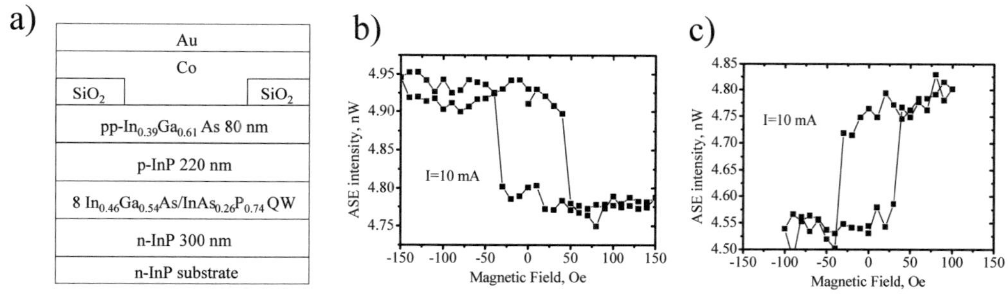

Fig. 18. a) InGaAs/InAsP optical amplifier with Co contact.Amplified spontaneous emission measured b)from left and c) right edge under small 10 mA injection current as a function of applied magnetic field.

Optical isolation of 147 dB/cm for TE mode was experimentally observed InGaAsP/InP optical amplifier at λ=1550 nm, in which one sidewall of waveguide was covered by TiO_2/Fe (Simizu & Nakano, 2004). Optical isolation of 147 dB/cm for TM mode at λ=1550 nm was reported for InGaAsP/InP covered by MnAs (Amemiya et al.,2007) and the optical isolation of 99 dB/cm for TM mode at λ=1300 nm was reported for AlGaInAs/InP optical amplifier covered by CoFe (Van Parys et al., 2006). In all these cases, to compensate the loss induced by the metal, the high injection current of more than 100 mA (current density more than 3 KA/cm^2) is required. The required current is too high for the isolator to be used in practical devices. In order to be suitable for integration, the injection current for the isolator should be reduced. It is only possible if the figure-of-merit for non-reciprocal loss could be increased. As we showed above, there are two contributions into the non-reciprocal loss: Faraday effect and MCD. In the case of Fe or Co, these contributions have almost the same amplitude and opposite sign. The value of Faraday effect is roughly proportional to the real part of off-diagonal element of permittivity tensor of the metal and the value of MCD is roughly proportional to imaginary part of off-diagonal element. If some ferromagnetic metal could be found, for which mutual signs between imaginary and real part of off-diagonal tensor element would be different to that of Fe and Co, it would make both contributions of the same sign and would significantly enlarge the figure-of-merit for the non-reciprocal loss.

4. Spin-photon memory

Photonics devices benefit from unique non-reciprocal properties of magneto-optical materials. If the material is ferromagnetic, the data can be stored in this material by means of two directions of residual magnetization. An ability to memorize data is another unique property of magneto-optical materials and it can be used for new designs of high-speed optical memory. Zayets & Ando (2009) proposed a new type of high-speed optical memory. Non-volatile data storage and high-speed operation are major advantages of this memory. Figure 19 shows the proposed design. The memory consists of micro-sized memory cells integrated on a semiconductor wafer. A bit of data is stored by each cell. Each cell consists of semiconductor-made photo detector and nanomagnet made of a ferromagnetic metal. The nanomagnet has two stable magnetization directions. The data is stored as a magnetized direction in the nanomagnet. For the data recording, the magnetization direction must be reversed by optical pulse. The circularly-polarized optical pulse is absorbed in the semiconductor detector creating spin-polarized electrons. Under applied voltage these spin polarized electrons are injected from the detector into the nanomagnet. The spin transfer torque is a consequence of the transfer of spin angular momentum from a spin-polarized current to the magnetic moment of a nanomagent. If the torque is sufficient, the magnetization turns and the data is memorized. Due to the optical selection rule, the spin-polarized electrons can be created only by the circular polarized optical pulse The linear polarized light excites equal amount of electrons of both up and down spins, therefore there is no net spin polarization, the current injected into nanomagnet is not spin polarized and there is no spin torque.

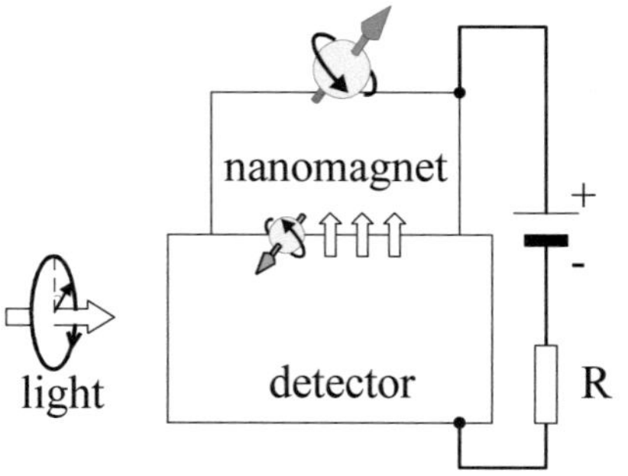

Fig. 19. Design of spin-photon memory. Zayets&Ando (2009)

Figure 20 shows integration of two memory cells and explains principle of high speed recording. There are two waveguide inputs. One input is for data pulses and one input is for the clock pulse. The clock pulse is used to select for recording a single pulse from sequence of the data pulses. Polarization of data pulses and clock pulse are linear and mutually orthogonal. Optical paths were split that each memory cell is illuminated by the data pulses and the clock pulse. The lengths of waveguides are adjusted so that the phase difference between clock and data pulses is $\lambda/4$ at each memory cell. At the first memory cell the clock pulse came at the same time with first data pulse. Therefore, these two pulses combined into one circularly polarized pulse. Since only first pulse is circularly polarized, only this pulse excites spin polarized electrons, changes magnetization and is memorized. All other data pulses are linearly polarized, they do not excite spin polarized electrons and they have no effect on the magnetization. For the second memory cell, the clock pulse is slightly delayed relatively to the data pulses and it comes together with second data pulse. Only the second pulse is circularly polarized and can be memorized by the second memory cell. Therefore, each data pulse can be memorized by individual memory cell. The closer the pulses can be placed relatively to each other, the more data can be transformed through one line and the faster recording speed of memory can be achieved. The minimum interval between pulses, at which a pulse can be recorded without any influence of nearest pulse, determines the recording speed of the memory.

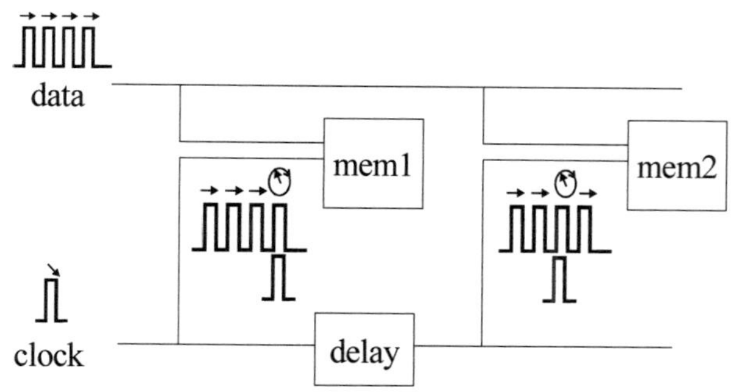

Fig. 20. The scheme for the integration of memory cells. Pulse diagrams explains the method for the high speed recording. Zayets & Ando (2009)

It was shown (Heberle et al., 1996) that spin-polarized excitons can by excited by two linearly cross-polarized pulses even the pulses arrived at different time. If the exciton dephasing time is longer than the interval between pulses, the excitons excited by two pulses can coherently interfere and create spin polarization.

For the successful demultiplexing by method proposed on Fig. 20, the interval between data pulses should be at least longer than electron dephasing time. Also, the spin polarization created by circularly polarized pulse should not be destroyed by following linear polarized pulses. Next, we verified the proposed demultiplexing method at the speed of 2.2 TBit/sec. For this purpose, we studied dynamics of excitation of spin polarized electrons in Si-doped GaAs ($n=7x10^{16}$ cm-3) at 80° K.

Figure 21 shows the experimental setup. A mode-locked Ti:sapphire laser (λ=820 nm) provides 160 fs linearly-polarized pump and probe pulses. Polarization of the pump was rotated by a $\lambda/2$ waveplate and split by polarization beam splitter (PBS) into clock pulse and data pulse of linear and mutually orthogonal polarizations. The data pulse was split into two pulses. The second data pulse was 165 λ (~450 fs) delayed. Clock and data pulses were combined together by another PBS and focused on the sample. The linearly polarized probe beam was 100 ps delayed relatively to the pump and focused on the same spot on the sample. The spin polarization of electrons excited by pump beam was estimated from Kerr rotation angle of the probe beam. The rotation angle of reflected probe beam was measuring by polarizer (P) and photo detector (det). Intensity of each data pulse was 1-10 μJ/cm2. Intensity of the probe pulse was 10 times smaller. The data pulses and clock pulse were phase locked. When the clock pulse and one of data pulse nearly coincided, the positive and negative angles of Kerr rotation were observed for delays of $(n+1/4)\lambda$ and $(n-1/4)\lambda$, respectively. When the pulses were away from each other, the Kerr rotation was not observed.

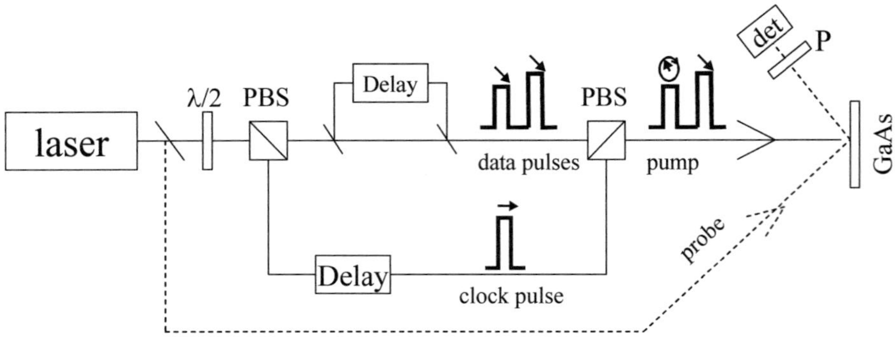

Fig. 21. Experimental setup for study the recording speed for spin-photon memory

Figure 22 shows spin polarization of the excited electrons as a function of delay between clock pulse and first data pulse. For each point the delay length was equal to (n+1/4)λ, where n is an integer. The maximum of spin polarization was observed when the clock pulse coincides with first data pulse. The spin polarization decreases when the clock pulse is delayed out of the first data pulse, and again increases as the clock pulse coincides with the second data pulse. Fig. 46 clearly shows that the spin polarization excited by the second data pulse can be separately distinguishable from the spin polarization excited by the first data pulse. This means that from two closely placed optical data pulses, only one pulse can trigger the recording without influence of another pulse. The interval between the data pulses is 450 fs. It corresponds to the recording speed of 2.2 TBit/sec. Notice that the detection of spin polarization was done 100 ps after data pulse arrived. This means that the lifetime of the spin polarized electrons is sufficient long to inject the spin polarized electrons into the nanomagnet for the magnetization reversal.

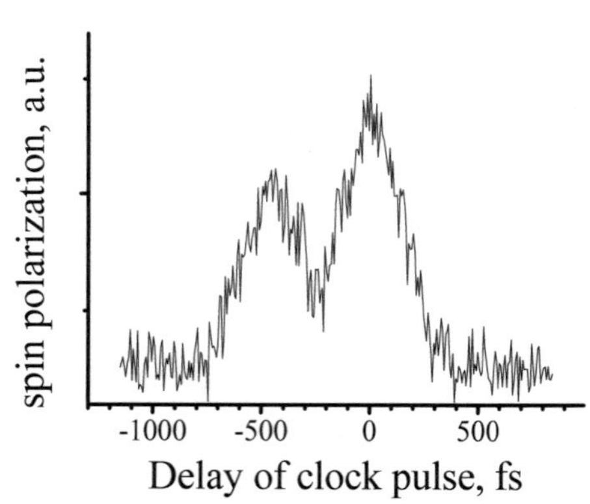

Fig. 22. Spin polarization of electrons excited by combined beam of clock pulse and two data pulse as a function of delay of clock pulse. For each point, the delay length was (n+1/4)λ, where n is integer. Reading of spin polarization was done 100 ps after arrival of pump pulses. Zayets & Ando (2009)

For the memory operating at speed 2.2 TBit/sec, the delay of clock pulse between cells (Fig. 21) should be 165·λ (~450 fs), the initial magnetization of all nanomagnets should be spin down and the delay of data pulses relatively to the clock pulse should be $(1/4+m\cdot 165)\cdot \lambda$, where m is a number of the data pulse. As result of Fig. 22, in this case in each cell only one data pulse excites spin polarization and memorized there. Other data pulses have no influence on spin polarization of that cell. In our experiment, the shortest interval between the data pulses, when the spin polarization excited by each pulse can be individually distinguished, is about 450 fs. For shorter interval the spin polarization is overlapped and the spin polarization created by preceding data pulse is reduced by next data pulse, which is causing overwriting of data of preceding pulse by next pulse in the cell. Therefore, the memory can not operate at speed faster than 2.2 TBit/sec.

By these experiments we demonstrated that from sequence of short-interval pulses it is possible to select only a single pulse for excitation of spin polarized electrons. That proves high recording speed of this memory. For the memory to be fully functional, the magnetization reversal by spin polarized photo-excited electrons should be realized. It requires injection of sufficient amount of spin-polarized current from detector into the nanomagnet (Slonczewski, 1996), Also, the time for the injection of the photo current from the detector into nanomagnet need to be adjusted so that it should be enough long to accomplish the magnetization reversal of the nanomagnet, but still it should be shorter than electron spin lifetime in the detector. The time, which takes for the magnetization of nanomagnet to turn between two stable direction, is about 500-1000 ps (Krivorotov, 2005). There are several semiconductors, in which the electron spin lifetime is longer or comparable with that time. For example, the spin life time is 100 ns in GaAs at T=4 K (Kikkawa & D. D. Awschalom, 1999) and 100 ps at room temperature (RT) (Hohage et al., 2006) 10 ns in GaAs/AlGaAs quantum well (QW) at RT (Adachi et al., 2001) and several ns in ZnSe QW at RT (Kikkawa et al.,1997). For the design of Fig. 19, the photo current injected in nanomagnet decays with time constant, where C is capacity of detector and R is resistance of close loop. The should be comparable with magnetization reversal time of the nanomagnet and smaller than the spin life time in the detector.

To estimate the energy of optical pulse required for magnetization reversal, we assumed that τ_{RC} is 500 ps, the efficiency of photon to spin conversion is 40 % and the critical current for the magnetization reversal is 5 mA (Kubota et al., 2005) To generate such current, the required energy of optical pulse should be about 3 pJ. It is in the range of the pulse energies which typically used in case of all-optical switching. Therefore, the memory may be suitable for the use in present optical communication systems.

Figure 23 shows the multiplexing scheme for this memory. For each memory cell, the optical waveguide passes under the nonomagnet. Due to the effect of non-reciprocal loss, the absorption of light in waveguide will be different for two opposite magnetization directions of the nanomagnet. The mode-locked laser, which could be integrated on the chip, provides the short pulses. The waveguide path is split, so that a pulse is split and it illuminates every memory cell. At output, all waveguide paths are combined. Each path has different length, so that each pulse reaches the output with a different delay and the amplitude of the pulse corresponded to the magnetization of each cell. Therefore, at the output the serial train of pulses will form with intensities proportional to the data stored in each memory cell.

This memory is compact, integratable, compatible with present semiconductor technology and it has fast operation speed. If realized, it will advance data processing and computing technology towards faster operation speed.

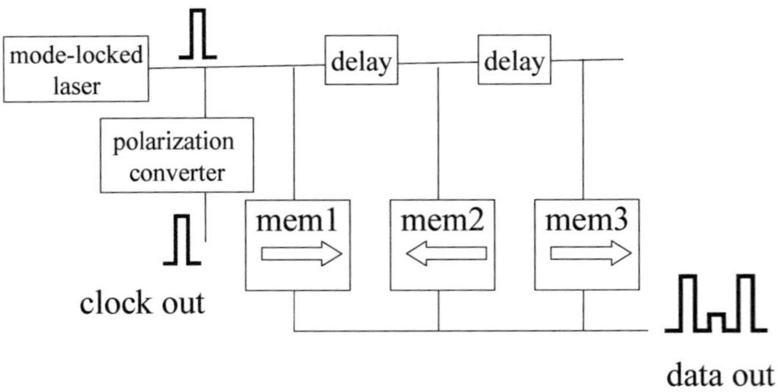

Fig. 23. Multiplexing scheme for spin-photon memory.

5. Conclusion

We have demonstrated that magneto-optical materials are important for optoelectronics, because of their unique non-reciprocal and memory properties. At present, there is an industrial demand for a waveguide-type optical isolator and a high-speed non-volatile optical memory. The use of magneto-optical materials is unavoidable for the fabrication of these devices.

6. Acknowledgement

This work was supported by NEDO. We thank R. Akimoto, Y. Suzuki, S. Yuasa, H. Saito, J.-C. Le Breton, H. Kubota, A. Fukushima, and K. Yakushiji, and V. Ganzha for helpful discussions, C. Debnath for the development of (Cd,Mn)Te waveguide isolator and K. Moriyama for technical assistance.

7. References

T. Adachi, Y. Ohno, F. Matsukura, &H. Ohno (2001), " Spin relaxation in n-modulation doped GaAs/AlGaAs (1 1 0) quantum wells " *Physica E* vol. 10, 36-39, 2001.

T. Amemiya; H. Shimizu; M. Yokoyama; P. N. Hai, M. Tanaka & Y. Nakano (2007), " 1.54-μm TM-mode waveguide optical isolator based onthe nonreciprocal-loss phenomenon: device design to reduce insertion loss " *Applied Optics* vol. 46, No. 23, 5784 , Aug. 2007.

K. Ando; T. Okoshi, and N. Koshizuka, (1988) "Waveguide magneto-optic isolator fabricated by laser annealing," *Applied Physics Letters,* Vol. 53, 4-6 ,1988.

L.M. Augustin; R. Hanfoug, R.; J.J.G.J van der Tol,.; W.J.M de Laat & M.K Smit (2007) "A Compact Integrated Polarization Splitter/Converter in InGaAsP–InP", *IEEE Photonics Technology Letters* , Vol. 19, No. 17, 1286 – 1288, Sept. 2007 .

C. Brooks,P.E. Jessop, H. Deng , D.O. Yevick, G. Tarr (2006), "Passive silicon-on-insulator polarization-rotating waveguides ", *Optical Engineering* Vol. 45, No. 4, 044603, 2006.

M. C. Debnath; V. Zayets & K. Ando, (2007). "(Cd,Mn)Te/(Cd,Zn)Te quantum-well waveguide optical isolator with wide wavelength operational bandwidth" *Applied Physics Letters*, vol. 91, No.4, 043502 , July 2007.

H. Fukuda; K.Yamada,; T. Tsuchizawa; T. Watanabe;,H. Shinojima & S. Itabashi (2006) "Ultrasmall polarization splitter based on silicon wire waveguides" *Optics Express* Vol. 14, No. 25, 12401, Dec. 2006.

J. K. Furdyna (1988), "Diluted Magnetic Semiconductors," *Journal of Appled Physics* vol. 64, R29- R64, 1988.

A. P. Heberle; J. J. Baumberg; E. Binder; T. Kuhn; K. Kohler & K. Ploog (1996), "Coherent Control of Exciton Density and Spin" *IEEE Journal of Selected Topics in Quantum Electronics* vol. 2, No.3, 769-775 , Sep.1996.

P. E. Hohage, G. Bacher, D. Reuter & A. D. Wieck (2006), " Coherent spin oscillations in bulk GaAs at room temperature" *Applied Physics Letters,* vol. 89, No. 23, 231101, Dec. 2006.

J.Z. Huang, R. Scarmozzino, G. Nagy, M.J. Steel, R.M. Osgood (2000), "Realization of a compact and single-mode optical passive polarization converter ", *IEEE Photonics Technology Letters,* Vol. 12, No. 3, 317-319 , Mar 2000

J. M. Kikkawa & D. D. Awschalom (1999), " Resonant Spin Amplification in n-Type GaAs" *Physics Review Letters* vol. 80, No.19, 4313-4316, May 1999.

J. M. Kikkawa, I. P. Smorchkova, N. Samarth, & D. D. Awschalom (1997), " Room-Temperature Spin Memory in Two-Dimensional Electron Gases " *Science* 277, 1284-1287, Aug. 1997.

S.H. Kim, R. Takei, Y. Shoji & T. Mizumoto (2009) "Single-trench waveguide TE-TM mode converter ", OPTICS EXPRESS vol. 17, No. 14, 11267-11273. Jul.2009.

I. N. Krivorotov; N. C. Emley; J. C. Sankey; S. I. Kiselev D. C. Ralph & R. A. Buhrman (2005)," Time-Domain Measurements of Nanomagnet Dynamics Driven by Spin-Transfer Torques" *Science* vol. 307, 228-231, Jan. 2005.

H. Kubota; A. Fukushima; Y. Ootani; S. Yuasa; K. Ando; H. Maehara; K. Tsunekawa; D. D. Djayaprawira; N. Watanabe & Y. Suzuki (2005), " Evaluation of Spin-Transfer Switching in CoFeB/MgO/CoFeB Magnetic Tunnel Junctions " *Japanese Journal of Applied Physics* Vol. 44, No. 40, 2005, pp. L 1237–L 1240, Sep. 2005.

K. Onodera; T. Masumoto & M. Kimura (1994)" 980 nm compact optical isolators using Cd/sub 1-x-y/Mn/sub x/Hg/sub y/Te single crystals for high power pumping laser diodes," *Electronics Letters* Vol. 30, 1954- 1955, 1994.

J. C. Slonczewski (1996), "Current-driven excitation of magnetic multilayers" *Journal of Magnetism and Magnetic Materials* vol. 159, L1-L7, 1996.

H. Shimizu &Y. Nakano (2004), "First demonstration of TE mode nonreciprocal propagation in an InGaAsP_InP active waveguide for an integratable optical isolator," *Japanese Journal of Applied Physics* Part 2, vol. 43, L1561-L1563, Nov. 2004.

W. Van Parys,B. Moeyersoon, D. Van Thourhout, R. Baets, M. Vanwolleghem,B.Dagens, J. Decobert, O. Le Gouezigou, D. Make, R.Vanheertum & L. Lagae (2006), " Transverse magnetic mode nonreciprocal propagation in an amplifying AlGaInAs/InP optical waveguide isolator" *Applied Physics Letters,* vol.88, No.7, 071115, Feb. 2006.

H. Yokoi; T. Mizumoto; T. Takano & N. Shinjo (1999), "Demonstration of an optical isolator by use of a nonreciprocal phase shift," *Applied Optics* vol.38, 7409-7413, 1999.

W. Zaets; K. Watanabe & K. Ando (1997) "CdMnTe magneto-optical waveguide integrated on GaAs substrate," *Applied Physics Letters,* vol. 70, No. 19, 2508-2510, Mar. 1997.

W. Zaets & K. Ando (1999), "Optical waveguide isolator based on non-reciprocal loss/gain of amplifier covered by ferromagnetic layer", *IEEE Photonics Technology Letters* vol. 11, No.8, 1012-1014, Aug. 1999.

W. Zaets & K. Ando (2000), "Magneto-optical mode conversion in Cd1-xMnxTe waveguide on GaAs substrate," *Applied Physics Letters,* vol. 77, No. 11, 1593- 1595, Sep. 2000.

W. Zaets & K. Ando (2001), "Magnetically Programmable Bistable Laser Diode With Ferromagnetic Layer" *IEEE Photonics Technology Letters.* vol. 113, No.3, 185-187, Mar. 2001.

V. Zayets; M. C. Debnath & K. Ando (2004), "Complete magneto-optical waveguide mode conversion in Cd 1-xMnxTe waveguide on GaAs substrate," *Applied Physics Letters,* vol. 84, No. 4, 565- 567, Jan. 2004.

V.Zayets & K. Ando (2005), "Isolation effect in ferromagnetic-metal/semiconductor hybrid optical waveguide" *Applied Physics Letters* , vol. 86, No. 26, 261105, June 2005.

V. Zayets & K.Ando (2009), "High-speed switching of spin polarization for proposed spin-photon memory" *Applied Physics Letters* , vol. 94, vol.12, 121104 , Mar. 2009. Reprinted with permission from *American Institute of Physics* Copyright 2004, 2005, 2007, 2009 American Institute of Physics.

17

Tunable Hollow Optical Waveguide and Its Applications

Mukesh Kumar, Toru Miura, Yasuki Sakurai and Fumio Koyama
Microsystem Research Center, Tokyo Institute of Technology
Japan

1. Introduction

The prime focus of this chapter is a hollow optical waveguide for integrated optics. Starting with a review of various novel hollow waveguide structures, physics and technology of tunable hollow waveguide have been presented with theory and experiments. With the flexible structure of hollow waveguide, a number of tunable optical functions can be realized; some of them are presented in the chapter. After giving a brief review on various hollow waveguides, we describe the basic structure and propagation characteristics of the tunable hollow waveguide. The possibility of two-dimensional (2D) confinement with a 3D hollow waveguide has been discussed. A widely tunable Bragg reflector based on hollow waveguide is then presented and the possibility of on-chip polarization control using 3D hollow waveguide has been discussed which is followed by the description of a novel tapered 3D hollow waveguide Bragg reflector for an adjustable compensation of polarization mode dispersion (PMD). The chapter closes with a discussion on future prospects of using tunable hollow waveguide in reconfigurable integrated photonic circuits and in other applications. The integration Technology for widely tunable hollow waveguide devices has been introduced and a novel hollow waveguide with vertical and lateral periodicity has then been shown with a high index contrast grating which can add extra design flexibility to control the polarizations and can be more fabrication tolerant.

2. Novel hollow waveguides

Guiding light in lower refractive index media has been interesting for many applications. The novel concepts for guiding light in a waveguide of a core of refractive index lower than that of cladding were proposed in form of Bragg fibers [1], Omni waveguides [2], and antiresonant reflecting optical waveguide (ARROW) [3]. Usually, optical guiding is accomplished by confining light to a region with high refractive index surrounded by cladding material with lower refractive index. Hollow waveguide with an air core offer the potential to minimize the dependence of light transmission on air core transparency, thus various hollow waveguide structures such as photonic band-gap fibers have been studied extensively as shown in Fig. 1.

The index of air is lower than those of typical solid cladding materials, thus it is impossible to confine light due to total internal reflections in hollow waveguide structures. Until now, various ways [4-10] to achieve low index guiding have been proposed and demonstrated, which include Bragg waveguides (Photonic band-gap waveguide), metallic waveguides, ARROW, and slot waveguide. Recently, silica hollow core photonic band-gap fiber with a minimum attenuation of 1.72 dB/km has been reported [9]. These hollow core waveguides have a lot of advantages including low nonlinearity, low optical damage threshold and strong optical confinement. Thus, there are various applications such as high power lasers/optical guiding, precise micromachining and gas/liquid sensing as shown in Fig. 2 [11-13].

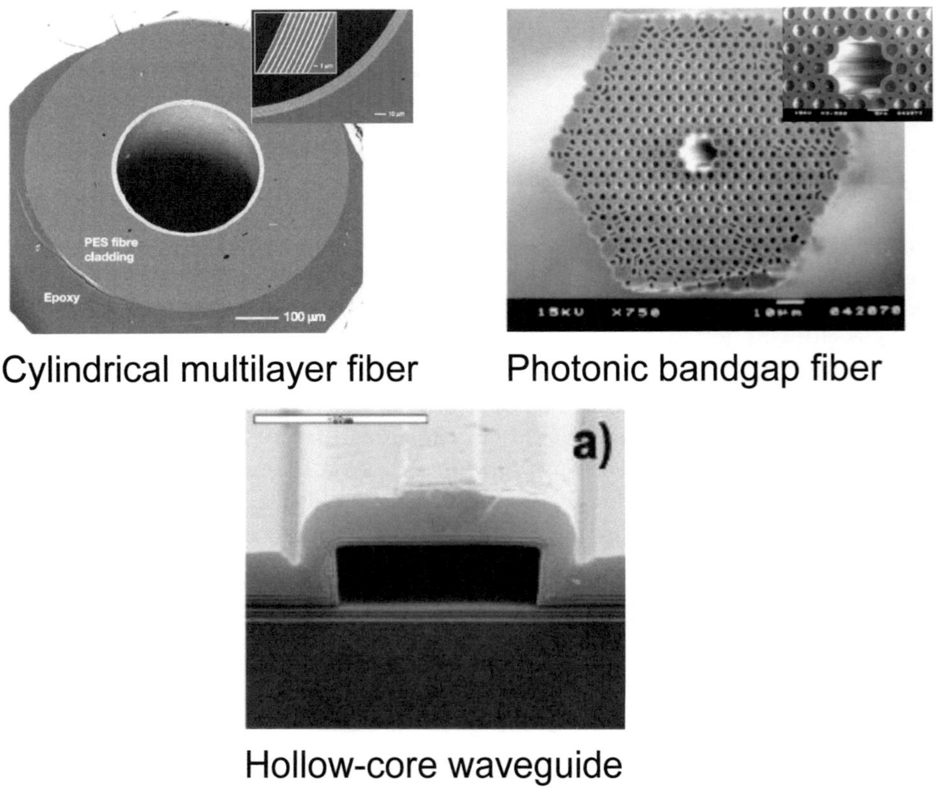

Fig. 1. Various air-core guiding structures [1-3].

Another advantage of hollow core waveguides is temperature insensitivity [7, 14-16]. For many applications in optical fiber communications and sensing, the sensitivity of the phase delay of optical guided mode to temperature is critical and it would be advantageous to decrease this sensitivity. In hollow waveguides, the guided mode is almost entirely confined to the air core region, whereas in a conventional waveguide it is entirely contained in solid materials. Since the temperature dependence of refractive index of air is smaller than that of solid material, the phase delay in hollow waveguides will have much smaller temperature dependence than that in conventional waveguides. In addition, a combination of vertical and lateral periodicities as top and bottom mirrors in hollow waveguide can reduce its polarization dependence [17] and thereby can make it more useful. Also, ultra low-loss hollow waveguide can be realized by using high index contrast grating [18] as top and bottom mirrors with periodicity being in the direction of propagation [19].

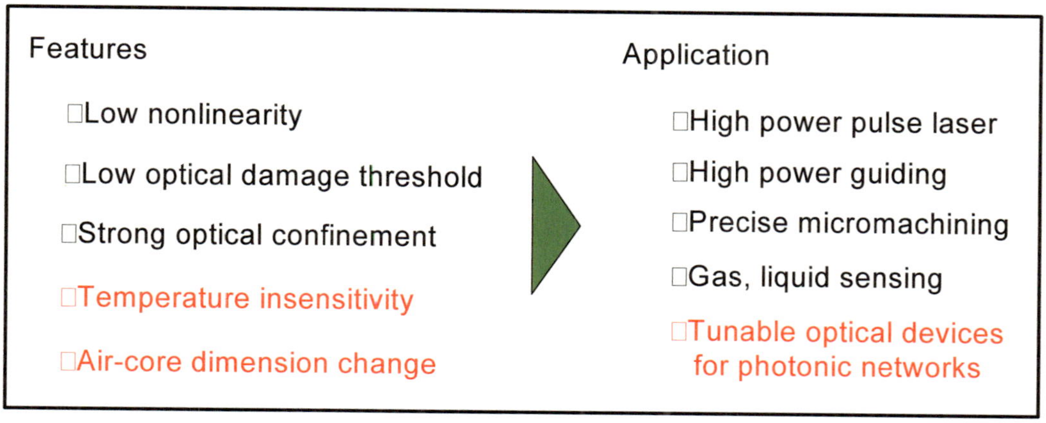

Fig. 2. Application of air-guiding waveguides [11-13].

3. Tunable hollow waveguide

3.1 Tunable and high confinement hollow structure

It has been difficult to utilize the conventional tuning schemes such as thermo-optic or electro-optic effects in hollow waveguides with an air core. Thus, applications to tunable optical devices were hardly discussed at all. In 2001, a novel tunable hollow waveguide structure with variable air core for tunable optical device applications was proposed by T. Miura and F. Koyama [7]. The combination of both hollow waveguides and MEMS actuators enable us to add the tuning functions to hollow waveguides. In tunable hollow waveguides, the light is guided in an air gap between the two multilayer mirrors. Multilayer mirrors with MEMS actuator driven by electrostatic force or thermal bimorph stress bring dynamic mechanical displacement of an air core. The propagation characteristics of hollow waveguide are strongly dependent on air-core dimension, thus one can achieve wide tuning operation of optical properties by air-core dimension control. One-dimensional (1D) and two-dimensional (2D) confinements have been proposed and demonstrated with tunable hollow waveguide structures based on this concept as shown in Fig. 3 [20-23]. In 1D confinement hollow waveguides, vertical confinement of light is achieved by Bragg reflections due to multilayer cladding. To obtain horizontal optical (2D) confinement in 3D hollow waveguides, we use total internal reflections due to effective refractive index

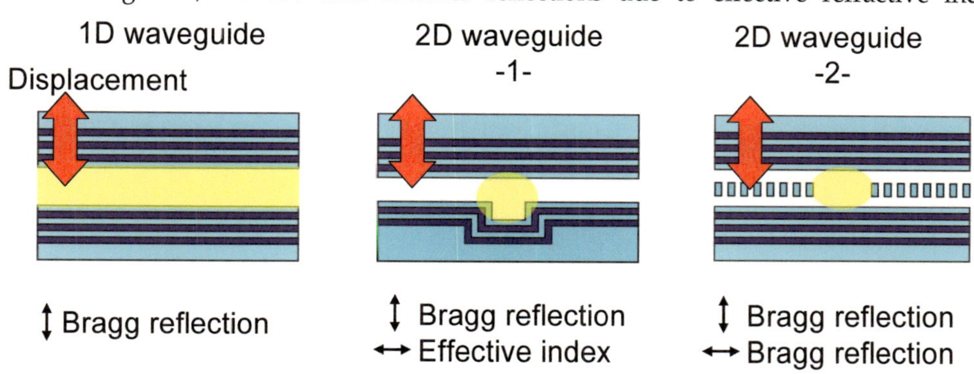

Fig. 3. Tunable Hollow optical waveguide structures [38]-[40].

differences generated by the air-gap steps or Bragg reflections due to periodic cladding structures just like vertical confinement. Tunable hollow waveguide structures are suitable for the temperature insensitive and widely tunable optical devices.

3.2 Propagation characteristics of tunable hollow waveguide

For the simplicity of the consideration, we assume the hollow waveguide composed of metal cladding. Metal-coated hollow waveguide is widely used in microwave or mili-wave frequency region. If the propagation loss of metal-coated 3D hollow waveguide is in an acceptable level, it can be used in our hollow waveguide devices. We investigate the propagation characteristics of metal coated 3D hollow waveguide in optical frequency region [24].

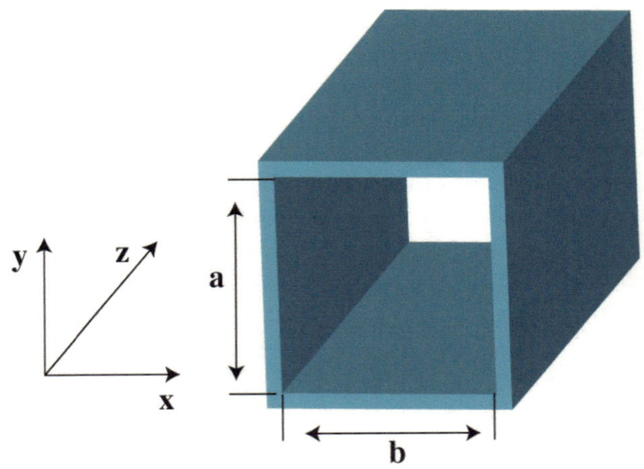

Fig. 4. Assumed 3D hollow waveguide model for analysis.

Assumed 3D hollow waveguide model for analysis is shown in Fig. 4. We have to derive the power flowing in the 3D hollow waveguide for determination of an attenuation factor. The flowing power Φ, which can be obtained by integrating the Poynting vector over transverse waveguide area of xy-plane, is expressed by

$$\Phi = \frac{1}{2} \cdot \mathrm{Re}\left[\int_0^a \int_0^b \left[E \times H^* \right]_z dxdy \right] \tag{1}$$

where $[E \times H^*]_z$ is the z-component of $[E \times H^*]$. We focus on TE_{10} mode that is expected as the mode of the lowest propagation loss. The power flow of TE_{10} mode can be expressed by

$$\Phi_{TE_{10}} = \frac{1}{4}\eta\left(\frac{2a}{\lambda}\right)\sqrt{1-\frac{\lambda}{2a}} \cdot a \cdot b \cdot H_0^{\,2} \tag{2}$$

where wave impedance: $\eta = (\mu/\varepsilon)^{1/2}$ and $E_x = 0$ is used.

The main factor of the propagation loss of a metallic hollow waveguide is ohmic loss in a metallic wall. If the electric conductivity is finite value, the electromagnetic wave penetrates into the inside of the metallic wall. The depth of this penetration, called *skin depth,* is expressed by

$$\delta = \frac{1}{\sqrt{\pi \cdot f \cdot \mu \cdot \sigma}} = \sqrt{\frac{2}{\omega \cdot \mu \cdot \delta}} \text{ [m]} \tag{3}$$

where f and δ is frequency and electric conductivity of metal, respectively.
Then faradic current is caused by the penetrate electromagnetic wave as shown in Fig. 5. Assuming magnetic field to be H_t, the surface current density can be expressed by

$$i_s = H_t \text{ [A/m]}. \tag{4}$$

It is thought that the surface current is flowing uniformly in the penetrate depth of δ. Therefore the ohmic loss for the unit area is expressed by

$$p = \frac{1}{2} i_s^{\,2} R_s = \frac{1}{2} i_s^{\,2} \sqrt{\frac{\pi \mu f}{\sigma}} \text{ [}\Omega\text{]} \tag{5}$$

where R_s is surface resistance,

$$R_s = \frac{1}{\delta \cdot \sigma} = \sqrt{\frac{\pi \mu f}{\sigma}} \text{ [W/m2]}. \tag{6}$$

The total ohmic loss *P* can be obtained by sum of the ohmic loss *p* along all surrounding metallic wall for the unit length. Therefore the attenuation factor for the power flow can be expressed by

$$\alpha_p = \frac{P}{\Phi} \text{ [Neper/m]}. \tag{7}$$

The attenuation factor for the electric or magnetic field flow is also expressed by

$$\alpha_c = \frac{\alpha_p}{2} = \frac{P}{2\Phi}. \tag{8}$$

We derive the ohmic loss in surrounding metallic wall for TE10 mode.
The all of the surrounding ohmic loss for TE_{10} mode is given by the sum of p_1, p_2 and p_3. p_1 is the ohmic loss at the surface of y = 0, b resulting from the faradic current i_x, which is caused by H_z at the surface of metallic wall. p_2 is the ohmic loss at the surface of x = 0, a resulting from the faradic current i_y, which is caused by H_z at the surface of metallic wall. p_3 is the ohmic loss at the surface of y = 0, b originated from the faradic current i_z, which is caused by H_x at the surface of metallic wall. Each component of magnetic field can be obtained by solving Maxwell's equations as shown in [24].

$$\begin{aligned} H_x &= j\frac{a}{\pi} \cdot \beta_{01} \cdot H_0 sin\left(\frac{\pi x}{a}\right) \cdot \exp\left(-j\beta_{01} z\right) \\ H_y &= 0 \\ H_z &= H_0 \cdot \cos\left(\frac{\pi x}{a}\right) \cdot \exp\left(-j\beta_{01} z\right) \end{aligned} \tag{9}$$

where β_{01} is the propagation constant of TE_{01} mode.

$$\beta_{10} = k\sqrt{1-\left(\frac{\lambda}{2a}\right)^2} \tag{10}$$

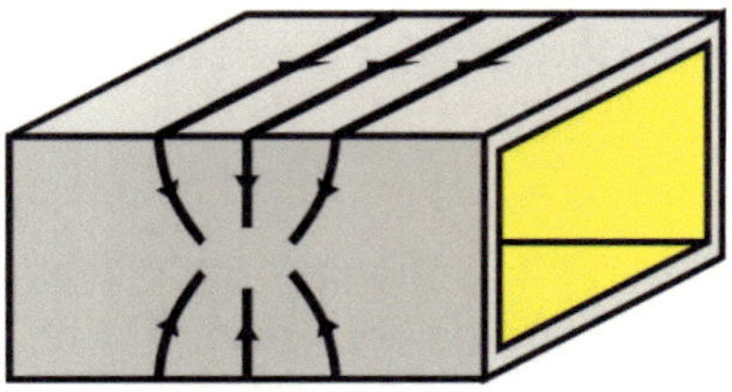

Fig. 5. Surface current caused by the penetration of electromagnetic waves.

The ohmic loss p_1, p_2 and p_3 can be expressed by

$$\begin{aligned} p_1 &= \int_0^a \left[\frac{1}{2}\left|i_x\right|_{y=0}\right|^2 R_s + \frac{1}{2}\left|i_x\right|_{y=b}\right|^2 R_s\right] dx = \frac{1}{2} R_s H_0 a \\ p_2 &= \int_0^b \left[\frac{1}{2}\left|i_y\right|_{x=0}\right|^2 R_s + \frac{1}{2}\left|i_y\right|_{x=a}\right|^2 R_s\right] dy = H_0^{\,2} R_s b \\ p_3 &= \int_0^a \left[\frac{1}{2}\left|i_z\right|_{y=0}\right|^2 R_s + \frac{1}{2}\left|i_z\right|_{y=b}\right|^2 R_s\right] dx = \frac{a^3}{2\pi^2}\beta_{01}^{\,2} H_0^{\,2} R_s \end{aligned} \tag{11}$$

Therefore the total ohmic loss is expressed by

$$P_{TE_{10}} = p_1 + p_2 + p_3 = \frac{1}{2} R_s H_0 \left[a + 2b + \frac{a^3}{\pi^2}\beta_{01}^{\,2}\right]. \tag{12}$$

The attenuation factor for power flow can be described as

$$\alpha_{P_{TE10}} = \frac{P_{TE_{10}}}{\Phi_{TE_{10}}} = \frac{2R_s\left\{a + 2b + \frac{a^3}{\pi^2}\beta_{10}^{\,2}\right\}}{\eta ab\left(\frac{2a}{\lambda}\right)^2 \sqrt{1-\left(\frac{\lambda}{2a}\right)^2}} \tag{13}$$

We can estimate the propagation loss of metallic hollow waveguide for the optical frequency region using eq. (13). The propagation loss for 20 x 20 μm hollow waveguide is as large as about 95 dB/cm at a wavelength of 1550 nm with gold walls. Even if the waveguide width is set to 100 μm, the propagation loss of about 19 dB/cm is expected. This propagation loss is too large to use in the optical devices.

The propagation constant of the light guided in the waveguide, shown in Fig. 6 can be expressed both for TE and TM mode as

$$\beta_{mn} = \sqrt{k^2 - \left[\left(\frac{m\pi}{D}\right)^2 + \left(\frac{n\pi}{W}\right)^2\right]} \tag{14}$$

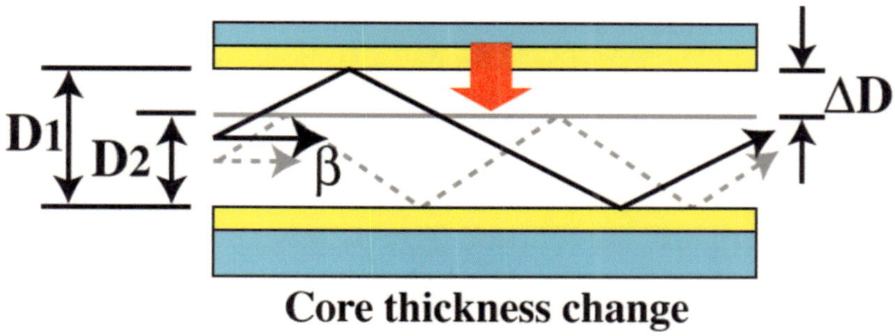

Fig. 6. Conceptual diagram of propagation constant tuning.

where m and n are mode number, k is wave vector in vacuum. This equation can be derived from Maxwell's equation [24]. We are now focusing on the case of the slab hollow waveguide; the width of the waveguide D becomes infinite. Therefore the propagation constant is independent of the mode number of n, since the light cannot form the mode for lateral direction. The propagation constant can be expressed as

$$\beta_m = k\sqrt{1-\left(\frac{m\lambda}{2D}\right)^2} \tag{15}$$

We rewrite this equation for a fundamental mode as

$$\beta_1 = k\sqrt{1-\left(\frac{\lambda}{2D}\right)^2} \tag{16}$$

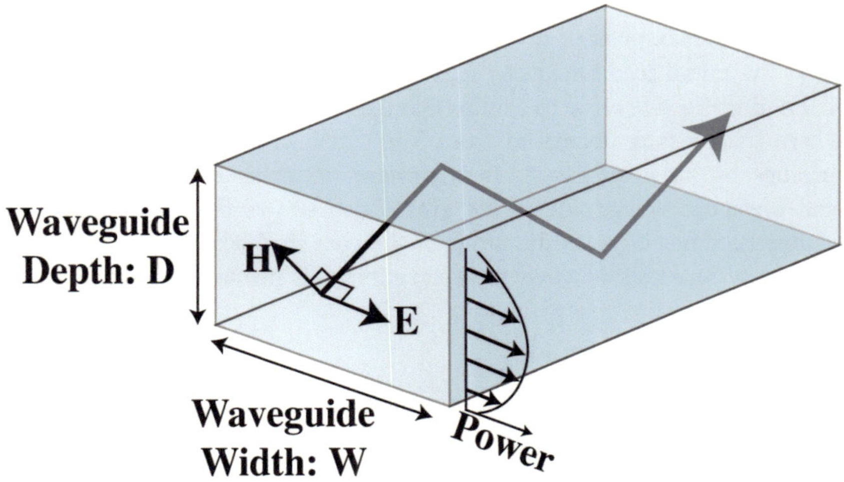

Fig. 7. Assumed model for analysis of propagation characteristics of hollow waveguide.

This equation shows the dependence of propagation constant of the guided fundamental mode on the air core thickness of the waveguide. From this equation, we can say that the propagation constant will be larger for larger values of waveguide thicknesses; the propagation constant is very sensitive to any change in the core thickness for the narrower cores. The propagation constant in the hollow waveguide is smaller than that in free space, in other words, the effective refractive index becomes smaller than unity. It is to be noted that the phase velocity of the guided mode seems faster than that of the free space. However the propagated mode doesn't propagate faster than the light propagated in free space. The group velocity of the guided mode remains slower than that in the free space.

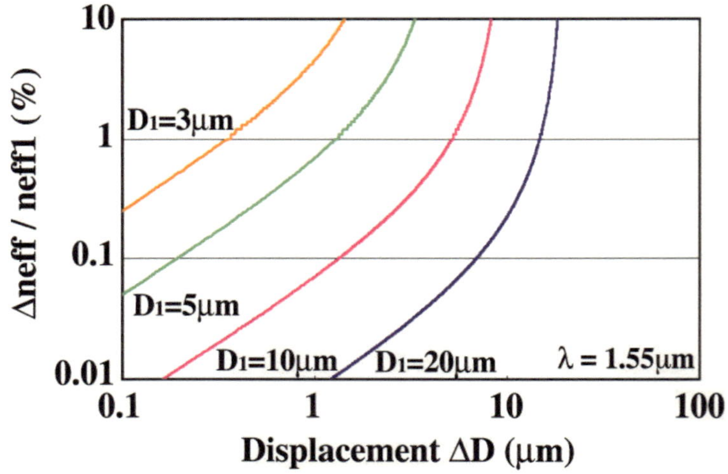

Fig. 8. Calculated effective refractive index change with the displacement of core thickness.

The calculated the effective refractive index change, namely change in propagation constant, is shown in Fig. 8. The effective refractive index change is plotted as a function of the displacement of the core thickness, when the core thickness changes from "initial thickness" to "initial thickness - ΔD" for the different initial thickness of 5, 10, 15 and 20 μm. The effective refractive index change diverges, when the displacement of the core thickness approaches to the initial core thickness. This can be understood from eq. (16). From Fig. 8, we can see that the effective refractive index change of few % can be expected. For example, when the core thickness is decreased from 5 μm to 3 μm, the effective refractive index change becomes as large as 2.2 %. It is difficult to achieve such a large number in conventional dielectric waveguides. This graph also shows that the change in effective refractive index becomes large as the range of the core thickness deviation becomes small. Therefore a narrow core hollow waveguide can effectively provide larger tuning.

4. Tunable 3D hollow waveguide

Hollow waveguides offer various interesting features including high power transmission, temperature insensitivity, low nonlinearity and strong optical confinement in comparison with conventional waveguides consisting of solid materials. In addition, waveguide type devices provide us the advantages in terms of miniaturization and integration of optical components and modules. These characteristics are very suitable for optical communication and sensing applications. Especially, the requirements of temperature insensitivity and wide tunability in optical components will become important along with the progress in future photonic networks. Our group has demonstrated a number of tunable optical devices based on slab hollow waveguides [25-27]. As shown in Fig. 3, the lack of lateral confinement in slab hollow waveguide is not suitable for practical applications in integrated photonic circuits, especially for two port devices where spot size mismatch may be an issue. A 3D hollow waveguide can solve this problem and can be flexibly designed for number of functionalities [28].

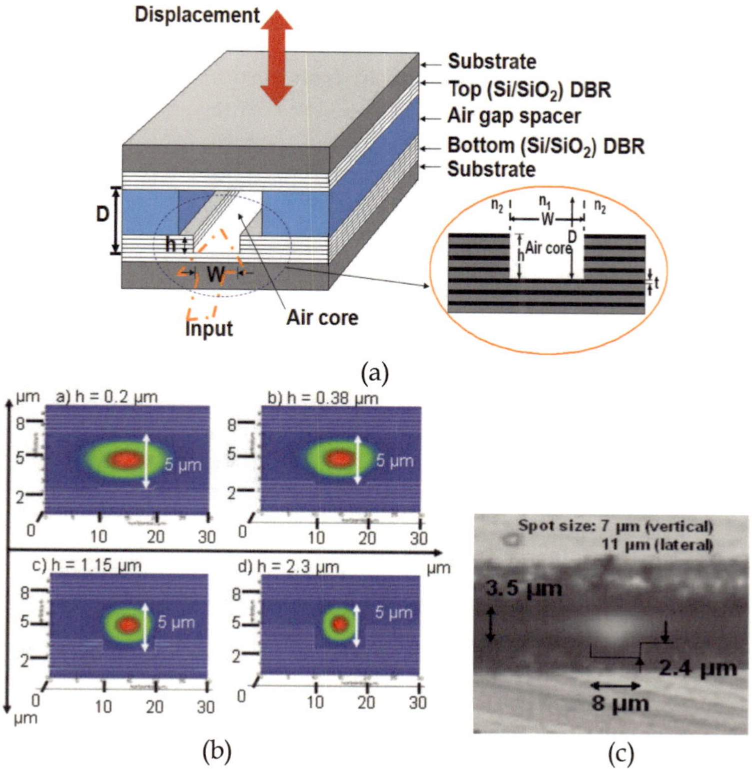

Fig. 9. (a) Schematic of tunable 3D hollow waveguide. All the layers in top and bottom DBR mirrors are quarter wavelength thick except the first silicon layer in bottom DBR just under the air core which we call phase matching layer with thickness t, shown in inset. (b) Computed effect of step-height on mode-field distributions at an air core thickness D = 5 μm and step width W = 10 μm, and (c) measured near field pattern for TE-mode.

The schematic of tunable 3D hollow waveguide is shown Fig. 9(a). The light is guided in the air with the help of two multilayer mirrors called distributed Bragg reflectors (DBRs). The vertical optical confinement is achieved by making the layers of DBR quarter wavelength thick for oblique incidence of light. For the lateral optical confinement, a groove is formed in the bottom DBR. The air core thickness will be larger in the region of this groove than in the lateral region. In view of eq. (16), larger air core thickness will have larger effective index. If the effective index of air core is n_1 and that of lateral air cladding is n_2 where $n_1 > n_2$ because of the difference in air gaps in lateral direction. The guided wave exhibits a difference in effective indices in the lateral direction, resulting in a lateral optical confinement by *quasi* total internal reflection. Thus, light can be laterally confined in the core region with a groove in one of the DBRs. In addition, the presence of groove, we call it a *step*, formed in one of the DBRs provides extra degree of freedom to control the propagation characteristics of hollow waveguide which can be tailored for a number of applications.

The lateral optical confinement and hence the lateral spot size can be varied with the step height *h*. The variation in TE-mode field distribution has been studied in Fig. 9(b) at an air core thickness of 5 μm with step width of 10 μm. Fig. 9(c) shows measured near field pattern for TE mode of the waveguide where the top DBR mirror of the 3D hollow waveguide with 2.4 μm deep and 8 μm wide groove is fabricated by electron-beam evaporation with 11 pairs of Si/SiO_2 on silicon substrate followed by inductive coupled plasma (ICP) -reactive ion etching (RIE). A strong optical confinement, in both vertical and lateral directions, can be achieved with the 3D hollow waveguide with a step formed in one of the two DBRs. The lateral optical confinement in hollow waveguides carries multifold benefit of efficient external coupling and wider tuning [29]. In following sections we will discuss the tailoring of hollow waveguide characteristics to realize important optical functionalities.

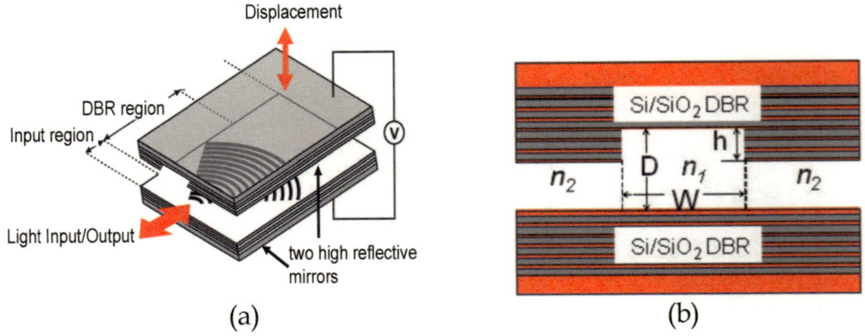

Fig. 10. (a) Schematic of tunable Bragg reflector based on 3D hollow waveguide and (b) cross section of 3D hollow waveguide.

5. Tunable Bragg reflector based on 3D hollow waveguide

5.1 Bragg grating loaded hollow waveguide

The wide tuning in propagation constant, discussed in section 3, of hollow waveguide with variable can be utilized to make various tunable optical devices. One example of this can be a Bragg grating loaded hollow waveguide which shows a tuning in Bragg wavelength by varying the air core [25] which can be an important photonic building block for the realization of in-plane photonic integrated circuits. Fig. 10 shows a tunable Bragg reflector based on a 3D hollow waveguide. The top DBR is having a step formed in it for the lateral optical confinement and the bottom DBR is loaded with a circular grating.

The top and bottom DBRs provide vertical confinement in the air gap and the step in top DBR provides lateral confinement. The circular Bragg grating on the bottom DBR gives high reflection for Bragg wavelength which is decided, for first-order grating in slab hollow waveguide [25], by the following equation

$$\lambda_{Bragg} = 2 \cdot \left(\beta / k_0 \right) \cdot \Lambda \tag{17}$$

where Λ is grating pitch. Putting eq. (16) into (17) gives

$$\lambda_{Bragg} = 2 / \sqrt{(1/D^2) + (1/\Lambda^2)} \tag{18}$$

This eq. shows that Bragg wavelength can be tuned with a variable air core thickness.

5.2 Fabrication and characterization

The two DBRs used in hollow waveguide are fabricated by electron beam (EB) evaporation. A 2.3 μm deep and 10 μm wide step is formed in top DBR by dry etching followed by selective wet etching. To confirm the optical confinement in hollow waveguide, a measurement set up discussed in [25] is used. The measured near field pattern for TE mode was shown in Fig. 9(c). To form the Bragg reflector, a circular Bragg grating is fabricated using EB lithography followed by dry etching. The grating depth, pitch and length are 500 nm, 820 nm and 1 mm, respectively. The total device length is 3 mm.

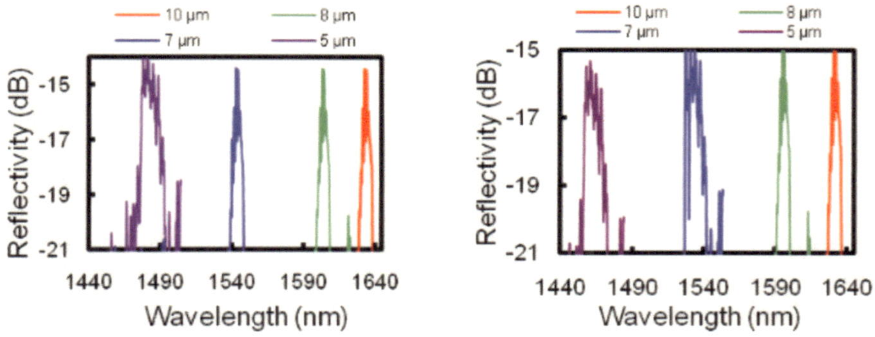

Fig. 11. Measured reflected spectra at various air core thicknesses of tunable Bragg reflector based on 3D hollow waveguide for (a) TE mode and (b) TM mode.

The same measurement set up of [25] is used to demonstrate the tuning in 3D hollow waveguide. As is clear from eq. (18), a variation in air core thickness D leads to a change in Bragg wavelength. The air core has been varied using an external piezoelectric actuator. The measured reflected spectra of TE and TM modes, at various air core thicknesses, are respectively shown in Fig. 11(a) and (b). The reflected spectra are measured at air core thicknesses of 10, 8, 7 and 5 μm. A clear blue-shift in the reflected spectra of the Bragg reflector, for both TE and TM modes, has been observed by decreasing the air core thickness from 10 μm to 5 μm. It is evident that a change in air core thickness can change the Bragg wavelength and hence a tunable Bragg reflector can be realized with hollow waveguide.

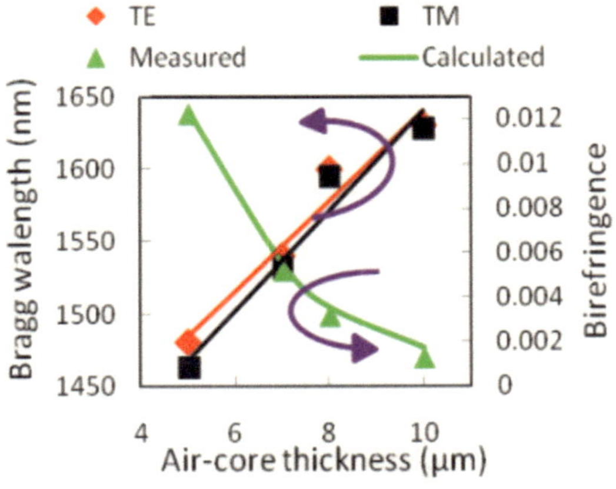

Fig. 12. Right y-axis: measured tuning in Bragg wavelengths with a variable air core for TE and TM modes. Left y-axis: measured tuning in birefringence by varying the air core thickness.

5.3 Tuning in Bragg wavelength and in birefringence

The measured Bragg wavelength as a function of air core thickness is shown in Fig. 12. By varying the air core thickness from 10 μm to 5 μm, a Bragg wavelength tuning of 152 nm for TE mode and that of 164 nm in TM mode are measured. Thus a grating loaded hollow waveguide can be used to provide giant tuning in Bragg wavelength which can find applications in realizing various tunable optical functions such as tunable filter, tunable lasers and so on. Further large tuning range can be achieved with narrower air core. The displacement of air core from smaller values of air core thicknesses results in larger tuning because of the tighter confinement. The reported tuning range is much larger than that in the tunable photonic devices based on the conventional dielectric waveguides using electro-optic and thermo-optic effects. Also, the tuning range is larger than that of slab hollow waveguide for the same displacement of air core because of the lateral optical confinement due to the presence of the step in top DBR [29]. Fig. 12 also shows how the modal birefringence can be tuned with variable air core thickness. The birefringence is defined by B = $(\lambda_{TE} - \lambda_{TM}) / \lambda_{TE}$, where λ_{TE} and λ_{TM} are the Bragg wavelengths of TE and TM modes, respectively. By varying the air core thickness from 5 μm to 10 μm, the birefringence decreases from 0.012 to 0.0012, providing a 900 % tuning and thereby making the 3D hollow waveguide a potential candidate to realize on-chip polarization control. The tunable birefringence can be useful for making tunable polarization manipulating devices.

6. Adjustable first-order polarization mode dispersion compensation

6.1 Giant birefringence from hollow waveguide

In long haul optical transmission systems, the degradation of transmitted optical signal from material and waveguide dispersion is a crucial problem. The dispersion, including polarization mode dispersion (PMD), is also a potential data rate limiting factor [30]. PMD occurs because of many intrinsic and extrinsic parameters including fiber core asymmetries, pressure, tension, vibration and temperature fluctuations, which lead to different transmission speeds of orthogonal polarizations causing random pulse distortion and pulse broadening in data stream; PMD becomes severe at increased data rates over 40 Gbit/s and higher [31]. To avoid the degradation of signal in fiber optic links, PMD compensation is necessary. A large birefringence and a tunable differential group delay (DGD) are needed to realize an adjustable PMD compensator, which can be achieved with a chirped Bragg grating filter to cancel the PMD in optical fibers [32]. A long fiber Bragg grating with some complicated tuning scheme is usually required to achieve large birefringence and a wide tuning in DGD [33]. An on-chip control of DGD with the tuning scheme other than the thermal tuning can be useful to realize a compact, temperature insensitive and low power consumption PMD compensator. Based on a tapered slab HWG, a planar Bragg reflector can provide chirping and an adaptable dispersion compensator can be realized [27]. In this section, we show that a giant birefringence can be realized with a 3D HWG by optimizing the step-height. A tapered-3D-HWG-Bragg-reflector has been proposed in which DGD can be tuned by varying the taper angle.

The schematic side-view of tunable Bragg reflector based on a 3D HWG is shown in Fig. 13(a), in which a step, of height h and width W, has been formed in a top DBR while the bottom DBR is loaded with a SiO_2 grating. The cross-section of the 3D hollow waveguide was shown in Fig. 10(b). The quarter wavelength thick top and bottom multilayer (DBR) mirrors provide strong optical confinement in vertical direction. Fig. 13(b) shows the effect of step height on the birefringence which is defined by B = (β_{TE} - β_{TM}) / β_{TE}, where β is propagation constant. The birefringence increases with increasing step-height because of the increase in lateral effective index contrast which was shown in Fig. 9(b). The calculated birefringence at a step-height of 2.3-μm is 0.012 with 5-μm thick air core. This giant birefringence, which is two orders of magnitude larger than that with other conventional fiber based technologies, can provide us compact devices with efficient functionality for PMD compensation etc.

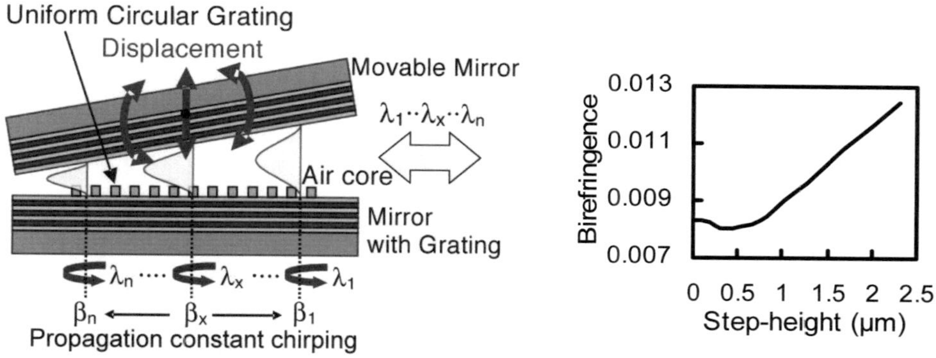

Fig. 13. (a) Schematic of tapered 3D hollow waveguide Bragg reflector (b) calculated birefringence (of straight hollow waveguide) versus step height *h* at an air core thickness *D* = 5 μm and step width *W* = 10 μm.

For the fabrication of tunable Bragg reflector based on 3D hollow waveguide, the top and bottom Si/SiO_2 multilayer mirrors have been prepared by electron beam (EB) evaporation. The 2.3-μm deep and 10-μm wide step is formed, in the top DBR mirror, by dry etching followed by selective wet etching. The grating has been fabricated, on the bottom DBR by electron-beam lithography followed by dry etching. The grating depth, pitch and length are 500 nm, 860 nm and 1 mm, respectively; and the total device length is 3 mm. The larger step-height, discussed above, carries multifold benefits of increased birefringence and reduced spot size because of increase in lateral index contrast. Also, The grating coupling coefficient is largest around 2.3-μm, which can broaden the reflection-band of the Bragg reflector [28], an important requirement to define a constant wavelength of operation of the device for two orthogonal polarizations.

6.2 Demonstration of tunable DGD

The dispersion and hence the group delay in hollow waveguides can be enhanced by introducing a taper. The large birefringence causes a delay between TE and TM modes of the 3D HWG; and the variable taper-angle can enhance the group delays and delay difference between these two orthogonal polarizations. The measured reflected spectra of TE and TM modes at various taper angles are shown respectively in Fig. 14(a) and (b). The broadband reflection of both the polarizations is obtained which is because of the combined effect of optimized 3D hollow waveguide and tapering. An overlap between the reflection-bands of TE and TM modes has also been observed. The taper angle has been varied from -0.038° to 0.057°. The taper-angle 0° refers to the straight 3D hollow waveguide (without taper). Increasing the taper-angle has the effect on reflection bandwidths of TE and TM modes; the bandwidth increases by increasing the taper-angle, because of the increased spatial chirping in Bragg wavelength. The net insertion loss remains less than 2.5 dB for 3-mm long device.

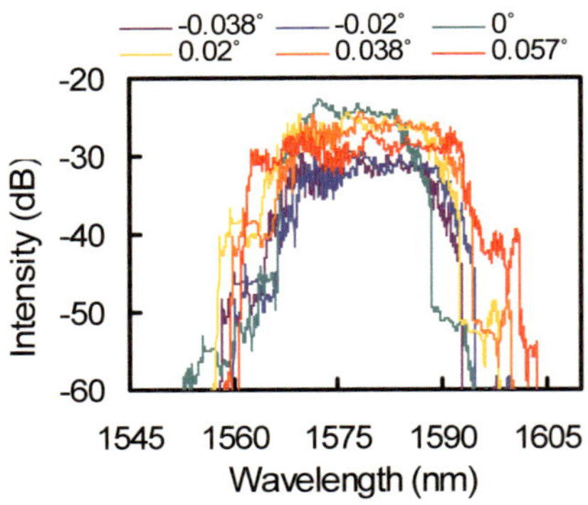

(a)

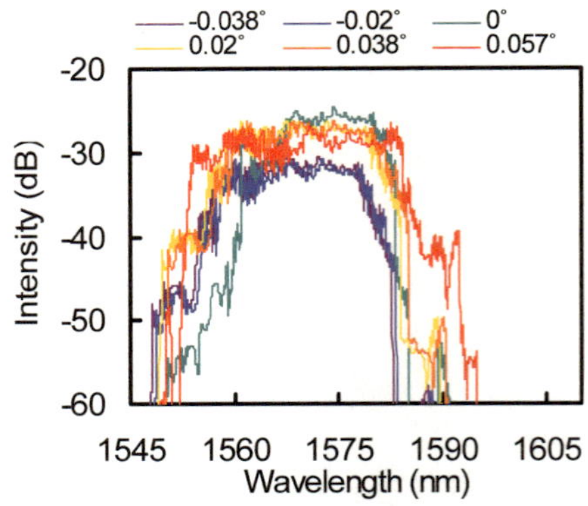

(b)

Fig. 14. Measured reflection spectra of tapered 3D hollow waveguide Bragg reflector for (a) TE mode and (b) TM mode.

In hollow waveguides, because of air core guiding, the angle of incidence of fundamental mode is smaller than Brewster's angle which makes the orthogonal polarizations out of phase causing large polarization dependence. Also, the presence of a step in one of the multilayer mirrors affects the orthogonal polarizations and polarization dependence can be further increased by varying the step-height. As shown in Fig. 13(b), by increasing the step-height, birefringence can be increased; an ultrahigh birefringence of 0.012, much larger than that of the slab HWG (step-height h = 0), can be achieved in a straight 3D HWG with a 2.3-μm high and 10-μm wide step at a 5-μm thick air core. The measured birefringence of the 3D hollow waveguide Bragg reflector as a function of taper-angle is shown in Fig. 15(a); where birefringence is defined by $B = (\lambda_{TE} - \lambda_{TM}) / \lambda_{TE}$, where λ is the center Bragg wavelength. A small variation in birefringence has been observed by varying the taper angle. The birefringence is less dependent on the taper angle; the birefringence remains around 0.01, which is in agreement with the Fig.13(a), and is almost two-order of magnitude larger than the birefringence of fiber Bragg gratings.

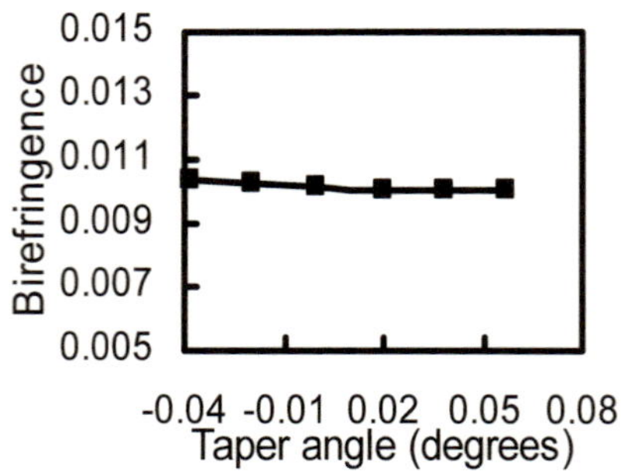

(a)

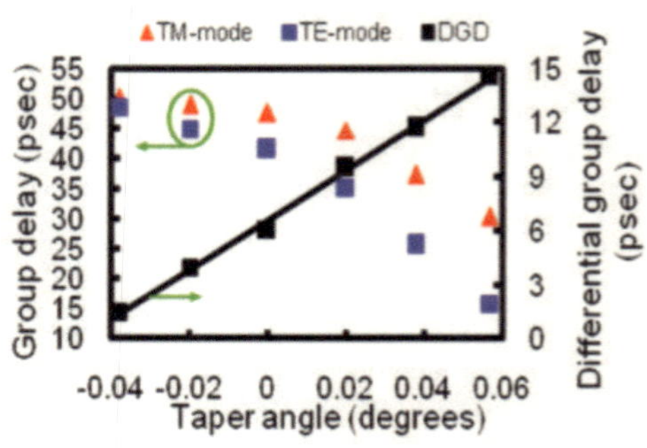

(b)

Fig. 15. (a) Measured birefringence versus taper-angle and (b) measured group delays of TE and TM modes, and DGD as a function of taper-angle.

The measured group delays of TE and TM modes of the Bragg reflector based on tapered 3D hollow waveguide, at various taper-angles, are shown in Fig. 15(b). By increasing the taper-angle, group delay, for both the polarizations, increases because of the enhancement of chirping in propagation constant of the guided mode. The increasing taper-angle has larger effect on TE polarization than that on TM polarization. Thus at larger taper-angles, the difference between group delays of TE and TM modes becomes larger which causes a tuning in DGD, where DGD is defined as the difference of group delays between TE and TM modes. Fig. 15(b) also shows tuning in DGD as a function of taper-angle; increasing the taper angle from -0.038° to 0.057°, DGD increases from 1.5-psec to 14.6-psec, providing us a 13.1-psec tuning in DGD, with a 3-mm long compact device. The large birefringence and tunable DGD of the proposed compact 3D HWG Bragg reflector makes it a good candidate for dynamic adjustable compensation of PMD in high bit rate and high speed transmission in optical fiber links. The giant birefringence can be used in tunable polarization manipulating devices based on a tunable 3D hollow waveguide.

7. Future prospects

7.1 Integration technology for tunable hollow waveguide devices

Hollow waveguides offer various interesting features including temperature insensitivity and ultra wide tuning. The design of hollow waveguide is flexible which carries the possibility of implementation of other functionalities in hollow waveguides. For efficient coupling of a hollow waveguide with a single mode fiber, a spot size converter may be used [26]. Fig. 16 shows how the tunable hollow waveguide can be implanted to realize reconfigurable photonic integrated circuit using high index core loaded waveguides for WDM applications. A spot size converter can be used for coupling to ultra-widely tunable functional devices consisting of hollow waveguide with sub-micron air core. And then, high index core loaded non-taper waveguides can be used as connecting waveguides between tunable devices or input/output waveguides to hollow waveguide devices. This kind of circuit is shown in Fig. 16.

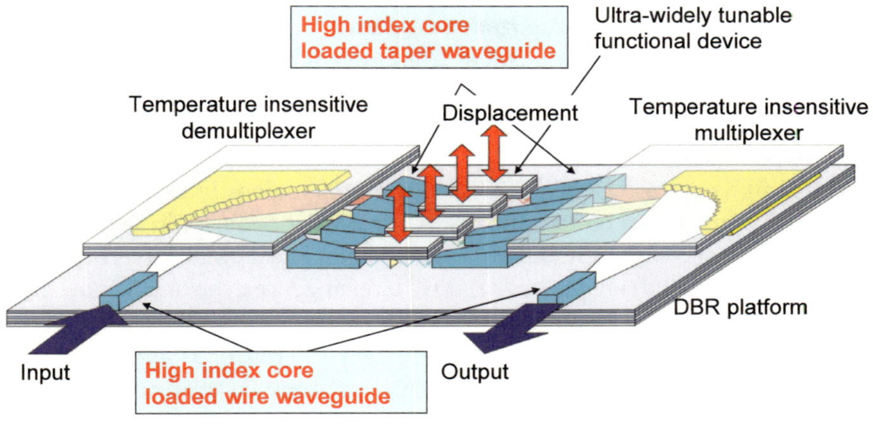

Fig. 16. Image of the scheme of reconfigurable photonic integrated circuit with tunable hollow waveguide using a high index core loaded waveguides for WDM applications.

The limiting factor of the tuning range has resulted from the alignment accuracy of air gap between two multilayer mirrors used as hollow waveguide cladding. This difficulty could be overcome by replacing a PZT actuator with a monolithic MEMS actuator. For this purpose a tunable hollow waveguide Bragg reflector with monolithically integrated MEMS electro-thermal actuator is shown in Fig. 17 [34]. This concept for hollow waveguide integration would enable to realize the tunable optical devices such as tunable lasers and tunable filters etc. with ultra-wide tuning and temperature insensitivity at the same time.

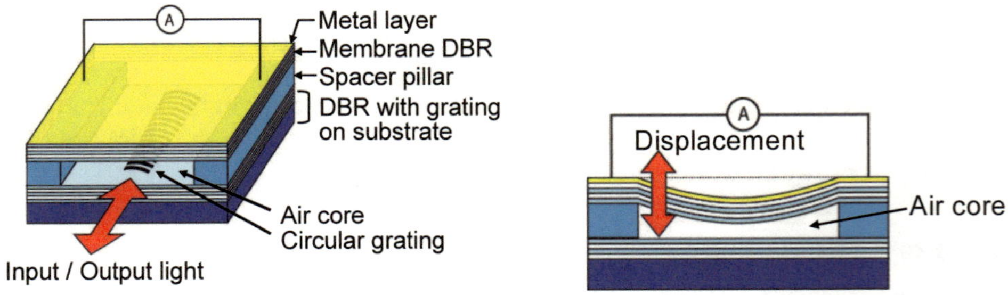

(a) (b)

Fig. 17. Schematic structure of tunable hollow waveguide Bragg reflector with integrated electro-thermal actuator. (a) Bird's-eye and (b) cross-sectional views.

7.2 Hollow waveguide with DBR and high index contrast grating: Extra degree of freedom for on-chip polarization control

As discussed in previous sections, the strong vertical optical confinement in hollow waveguides (HWGs) can be achieved using top and bottom DBRs as cladding, having quarter wavelength thick layers designed for oblique incidence of light, where light is guided in the air gap in between the two DBRs. The reflectivity of DBRs in the HWG is highly polarization dependent because of the periodicity only in one direction. The polarization dependence of a HWG becomes very large at a narrow air core where a HWG offers giant tuning of over 10% in propagation constant.

The birefringence of the HWG largely comes from the guiding DBR having periodicity only in the one direction. If a lateral periodicity (perpendicular to both, the direction of propagation and the direction of DBR periodicity) is introduced in the HWG, the polarization dependence of the HWG can be reduced. High reflectivity DBRs can be replaced with a high contrast grating (HCG) mirror which gives broadband high reflection [28, 29]. The HCG mirror can be introduced, as a lateral periodicity, in hollow waveguides to reduce its birefringence. The combined (and opposite) effect of DBR (vertical periodicity) and HCG (lateral periodicity) enable us to reduce birefringence even at a narrow air core.

In this section, a hollow waveguide with vertical and lateral periodicity is proposed which consists of a DBR mirror as vertical periodicity and an HCG mirror as lateral periodicity. We show that the combination of DBR and HCG mirrors can provide strong optical confinement and can reduce the birefringence of the hollow waveguide with narrow air-cores.

The schematic of the proposed hollow waveguide is shown in Fig. 18(a). It consists of a DBR as a bottom mirror and a high index contrast grating (HCG) as a top mirror of a hollow waveguide. Light can be confined vertically in an air-core utilizing the high reflectivities of

DBR and HCG. The structure consists of a 5 pairs Si/SiO_2 bottom DBR mirror where each layer in DBR is quarter wavelength thick, which is designed for oblique incidence. Because of the high refractive index contrast between Si and SiO_2, only 5-pairs Si/SiO_2 are enough to achieve a high reflectivity. The top HCG mirror consists of a high index contrast silicon/air grating. In an HCG mirror, many in-plane reflections add up and give rise to vertical high reflection [18]. The design parameters for the hollow waveguides are: air core thickness D, grating thickness t_g, grating pitch Λ, grating teeth width w and grating duty cycle C. The calculated mode field distributions of TE mode in the hollow waveguide is shown in Fig. 18(b). A strong vertical confinement is observed in the hollow waveguide. The propagation loss at Λ = 1.2 µm, C = 0.41 and t_g =0.45 µm is 0.55dB/cm foe TE mode and that of TM mode is 2.7 dB/cm, at an air core thickness of 5µm which is acceptable for millimeter long waveguide devices. The set of the optimized parameters with D = 5 µm, t_g =0.45 µm, Λ = 1.2 µm, w = 0.5 µm and C = 0.41 is called "set A" for convenience.

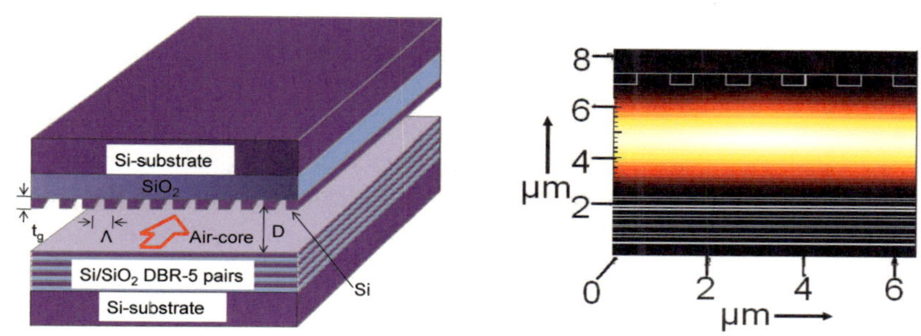

Fig. 18. (a) Schematic of hollow waveguide with DBR and HCG and (b) computed TE-mode distribution.

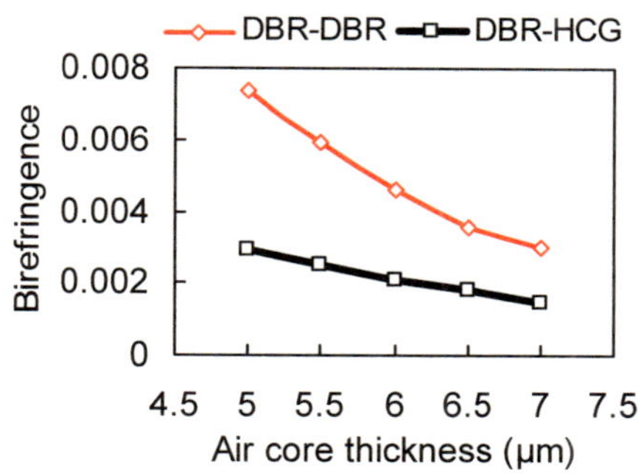

Fig. 19. birefringence as a function of D at the grating parameters in "set A", also shown birefringence of hollow waveguide with DBR on the both sides.

Hollow waveguides show large polarization dependence because of the shallow incidence angle of a fundamental mode. To realize ultra wide tuning in hollow waveguides, we need to use a hollow waveguide with a narrower air core; but the birefringence at narrower cores remains large. The polarization characteristics of HCG hollow waveguides are shown in terms of birefringence in Fig. 19 at the grating parameters of "set A". The birefringence is defined as $B = (\beta_{TE} - \beta_{TM}) / \beta_{TE}$, where β is propagation constant. In Fig. 19, the effect of an air core on the birefringence is shown where the birefringence of conventional hollow waveguides (having DBRs on top and bottom) is also shown for comparison. The result shows that the birefringence of the HCG hollow waveguide is smaller than that of the conventional hollow waveguide at narrower air cores. By increasing the air core thickness, the birefringence of the hollow waveguides decreases for both the cases. At an air core thickness of 5 μm, the birefringence of the HCG hollow waveguide is 2.8×10^{-3}; which is almost 2.6-times smaller than the birefringence of a conventional hollow waveguide. To achieve ultra wide tuning, the HCG hollow waveguide with narrower air-cores can be realized with smaller birefringence. The reason for this small birefringence is because of the opposite nature of conventional DBR mirror and HCG mirror; their combined (and opposite) interaction with the orthogonal polarizations results in the reduced birefringence. In case of the DBR-DBR HWG, the large birefringence comes from the field penetration of a TM mode into the DBR. The propagation constant of a TM mode is larger than that of a TE mode. On the other hand, a HCG gives us stronger optical confinement for a TM mode in an air core than DBRs. Thus, the propagation constant of a TM mode in the HCG-DBR HWG is smaller than that in the DBR-DBR HWG while the propagation constant of a TE mode remains almost same for both the cases. The difference in the field penetration results in a reduced birefringence.

The birefringence and the polarization dependence loss can be reduced with the further optimizations of the grating and the waveguide parameters [17]. The introduction of HCG into hollow waveguides can extend the fabrication tolerance of the structure because of the large fabrication tolerance of HCG and it can provide the lateral optical confinement as well [17]. Also, because of the ease of fabrication of HCG, it is possible to design it for one particular polarization, making polarization control very flexible. With the flexibility of HCG-design, it is possible to build hollow waveguides with widely tunable polarization nature for variety of applications. With the collaboration of HCG and the inherent properties of hollow waveguides such as temperature insensitivity and wide tuning, it will be possible to design novel tunable optical functions with large design tolerances.

8. References

[1] Pochi Yeh and Amnon Yariv, "Bragg reflection waveguides", *Opt. Commun.* 19, 427 (1976).

[2] M. Ibanescu, Y. Fink, S. Fan, E.L. Thomas and J.D. Joannopoulos, "The coaxial omniguide - a novel all- dielectric waveguide", *Science* 89 (2000) 415.

[3] M.A. Duguay, Y. Kokubun, T. Koch, and L. Pfeiffer, "Antiresonant reflecting optical waveguides in SiO_2-Si multilayer structures," *Appl. Phys. Lett.* 49, 13-15 (1986).

[4] R. F. Cregan, B. J. Mangan, J. C. Knight, T. A. Birks, P. St. J. Russell, P. J. Roberts, and D. C. Allan: *Science* 285 (1999) 1537.

[5] H. Schmidt, D. Yin, J. P. Barber, and A. R. Hawkins: *IEEE J. Select. Topics Quantum Electron.* 11 (2005) 519

[6] M. Miyagi, A. Hongo, and S. Kawakami: *IEEE J. Quantum Electron.* QE-19 (1983) 136.

[7] T. Miura, F. Koyama, Y. Aoki, A. Matsutani, and K. Iga: Jpn. J. Appl. Phys. 40 (2001) L688.

[8] T. Miura, F. Koyama, and A. Matsutani: IEEE Photonics Technol. Lett. 15 (2003) 1240.

[9] J. Roberts, D. P. Williams, F. Couny, M. Lawman, M. Mason, S. Coupland, R. Flea, and H. Sabert: Proc. Optical Fiber Communications Conf., California, 2004, PDP24.

[10] V.R. Almeida, Q. Xu, C.A. Barrios, and M. Lipson, "Guiding and confining Light in void nanostructure," *Optics Letters* 29 pp. 1209-1211, 2004.
[11] D. G. Ouzounov, F. R. Ahmad, D. Muller, N. Venkataraman, M. T. Gallagher, M. G. Thomas, J. Silcox, K. W. Koch, And A. L. Gaeta: *Science* 301 (2003) 1702.
[12] J. D. Shephard, J. D. C. Jones, D. P. Hand, G. Bouwmans, J. C. Knight, P. St. J. Russell, and B. J. Mangan: *Opt. Express* 12 (2004) 717.
[13] J. C. Petersen, T. Sorensen, T. P. Hansen, and H. R. Simonsen: *Opt. Express* 12 (2004) 4080.
[14] V. Dangui, H. K. Kim, M. J. F. Digonnet, and G. S. Kino: Proc. Optical Fiber Communications Conf., Anaheim, 2005, OTuI4.
[15] Y. Sakurai, Y. Yokota, A. Matsutani, and F. Koyama: Proc. Optical Fiber Communications Conf., Anaheim, 2005, OME30.
[16] Y. Sakurai, Y. Yokota, A. Matsutani, and F. Koyama: Appl. Phys. Lett. (2005) 071111.
[17] M. Kumar, C. Chase, V. Karagodsky, T. Sakaguchi, F. Koyama, and C. J. Chang-Hasnain, "Low birefringence and 2-D optical confinement of Hollow waveguide with distributed Bragg reflector and high index contrast grating", *IEEE Photonics Journal 1*, No. 2, 135 (2009).
[18] M. C. Y. Huang, Y. Zhou, and C. J. Chang-Hasnain, "A surface-emitting laser incorporating a high-index-contrast subwavelength grating", *Nature Photonics* 1 (2007) 119.
[19] Y. Zhou, V. Karagodsky, Bala Pesala, F. G. Sedgwick, and C. J. Chang-Hasnain, "A novel ultra-low loss hollow-core waveguide using subwavelength high-contrast gratings," Optics Express 17 (2009) 1508.
[20] T. Miura, F. Koyama, and A. Matsutani: *Jpn. J. Appl. Phys.* 41 (2002) 4785.
[21] T. Miura, F. Koyama, and A. Matsutani: *Jpn. J. Appl. Phys.* 41 (2002) 5607.
[22] T. Miura and F. Koyama: Proc. Optoelectronics and Communications Conference/ International Conference on Optical Internet, Yokohama, 2004, 892.
[23] M. Kumar and F. Koyama, "Investigation and theoretical analysis of low-loss 3D hollow waveguides for widely tunable photonic devices", *Jpn. J. Appl. Phys.* 47 (2008) pp 3471-3475.
[24] T. Sekiguchi: *Electro-Magnetic Wave* (Asakura-shoten, Tokyo, 1976).
[25] Y. Sakurai and F. Koyama: "Tunable hollow waveguide Bragg reflectors with variable air core," *Optics Express* 12 (2004) 2851.
[26] Y. Sakurai and F. Koyama: "Modeling of a low-loss spot size converter for hollow waveguide with sub-wavelength air core", *IEICE Elect. Express* 1 (2004) 115-119.
[27] Y. Sakurai, A. Matsutani, and F. Koyama, *Applied Physics Letters*. 88 (2006) 121103.
[28] M. Kumar, T. Sakaguchi, and F. Koyama, "Giant birefringence and tunable differential group delay in Bragg reflector based on tapered three-dimensional hollow waveguide", *Applied Physics Letters* 94, Issue 6, 061112 (2009).
[29] M. Kumar, T. Sakaguchi, and F. Koyama, "Wide tunability and ultralarge birefringence with 3D hollow waveguide Bragg reflector", *Optics Letters* 34 (2009) 1252.
[30] T. Takahashi, T. Imai, and M. Aiki, Electron. Lett. 30, 348 (1994).
[31] Xia Zhang, Yuehui Xia, Yongqing, and Xiaomin Ren, Optics Commun. 214, 123 (2002).

[32] S. Lee, R. Khosravani, J. Peng, V. Grubsky, D. S. Starodubov, A. E. Willner, and J. Feinberg, IEEE Photon. Technol. Lett. 11, 1277 (1999).
[33] X. Yi, C. Lu, X. Pang, W. -D. Zhong, F. Wei, and Y. Fang, Opt. Express 11, 2634 (2003).
[34] F. Koyama, T. Miura and Y. Sakurai, IEICE Trans. Electron. J88-C (2005) 388.

18

Regenerated Fibre Bragg Gratings

John Canning[1], Somnath Bandyopadhyay[2], Palas Biswas[2], Mattias Aslund[1], Michael Stevenson[1] and Kevin Cook[1]

[1]*Interdisciplinary Photonics Laboratories (iPL), Madsen Building F09, School of Chemistry, University of Sydney, NSW, 2006*

[2]*Fibre Optics Laboratory, Central Glass and Ceramic Research Institute (CGCRI), Council of Scientific & Industrial Research, Kolkata - 700032,*

[1]*Australia*

[2]*India*

1. Introduction

Silica remains the key optoelectronic and photonic medium, the essence of nearly all modern optical transport systems. Engineering of silica in its various forms ranges from 1 to 3-dimensional waveguide and periodic structures, including recent interest in 3-D photonic crystals. Most of the processing methods involve complex vapour deposition and various co-dopants, which have an advantage of overcoming the lack of finesse involved with general formation of glass structure through high temperature processing and quenching. Nevertheless, to obtain micron or sub-micron precision over the processing of glass for device purposes, invariably post processing methods are commonly used, ranging from etching of systems with dopants, often through patterned masks, to laser processing using UV to mid IR lasers. Concrete examples of micron scale laser processing of glass include direct written waveguides, Bragg gratings in waveguides and optical fibres and photonic crystals. The drawback with these post-processing techniques is that they often produce glass that is structurally less stable than the starting phase. For many applications the thermal stability of laser induced glass changes determines the limits in which they can operate - an excellent example which will form the basis for this chapter, is the optical fibre Bragg grating.

Fibre Bragg gratings are used in many industrial and technological applications. Within standard telecommunications applications, for example, type I fibre Bragg gratings that can operate to 80°C for 25 years are required - such gratings can in principle operate for lengthy periods up to 300°C. Gratings that can operate at temperatures well above standard telecommunication requirements are critical to the success of many real time sensing applications. In the oil and gas industries, an alternative application, although standard oil bores are typically quoted as having an environment no more than ~(180-250)°C [Schroeder et al. 1999; Kersey 2000], variations can occur and the increasing depth of the next generation bores suggest sensors that can operate to 400°C or more are desirable for long term or permanent operation. In industries involving high temperature furnaces, such as aluminium smelting or coal based power stations, it would be of interest to be able to monitor temperatures in excess of 1000°C. Similar temperature requirements span many other types of industries: structural health monitoring of buildings need to be able to operate in temperatures above 400°C in the event of a serious fire [Cardoza et al. 2007], engine turbines in various vehicle formats, particularly aircraft, can reach temperatures well above this, whilst the integration into next generation composite structures [Epaarachchi et al. 2009] in many of these applications often need to be carried out in temperatures above (300-400)°C. Another particularly important industry is the fibre laser sector [Canning 2006 and refs therein]. Presently, fibre Bragg gratings are spliced onto the ends of the fibre using specially designed matched photosensitive cores. However, some fibre mismatch, along with power tolerance issues within doped glass, remains. One approach to dealing with these and to reduce overall costs, is to write gratings directly into the active medium. Unfortunately, we have recently demonstrated that, in Yb^{3+} -doped fibre lasers at least, the UV induced grating index of a type I grating within the doped fibre can be readily annealed at moderate powers (~13W) [Åslund et al. 2005]. Even using femtosecond produced type II gratings, annealing occurs when the internal fields exceed kW, as within a Q-switched fibre laser [Åslund et al. 2008,2009b]. Until these issues are resolved, gratings will continue to be spliced onto the ends of the gain medium.

Within all these applications, the same stringent fabrication capabilities imposed on telecommunications are also increasingly desirable as sensor system and components become more complex than simple low reflection filters. Numerous distributed components raise the challenging specter of cross-talk between devices. Complex filter properties such as apodisation, chirping, phase shifts and more are increasingly in demand. Therefore, ideally there is a need to produce high temperature gratings that retain the best features of the current workhorse, the type I grating. There is a new type of grating attracting worldwide attention that promises to deliver this: the regenerated fibre Bragg grating. It is essentially formed from the initial type I "seed" grating, precipitating through thermal processing with a structure that is set by the laser written seed. For many other applications requiring similar levels of holographic precision by laser processing, the need for tailored and controlled *ultra stabilised* index change remains equally vital. To demonstrate new developments towards these goals we, however, concentrate in this chapter on the 1-D periodic structure of the optical fibre Bragg grating, noting that the general processes are much more widely applicable.

1.1 Historical background

Previous studies have already established that the operable temperature of FBGs can be increased by several means, including tailoring the glass composition [Shen et al. 2007; Butov et al. 2006], pre-processing with seed irradiation [Åslund & Canning 2000; Canning et al. 2001], the formation of type-I*n* (or type II*A*) [Xie et al. 1993; Dong et al. 1996; Groothoff & Canning 2004] gratings and type-II [Archambault et al. 1993; Hill et al. 1995] gratings, including those inscribed using femtosecond IR lasers [Grobnic et al. 2006]. For a general review on photosensitivity and grating types, see [Canning 2008a]. Another variant with superior high temperature stability is the so-called "chemical composition grating (CCG)" [Fokine 2002] where a periodic index modulation can be regenerated after erasure of the UV induced type-I grating written in H-loaded germanosilicate fibre, which happens to contain fluorine, if annealed ~1000°C. The prediction was a local reduction, or increase, of fluorine in the UV-exposed zones at that high temperature through diffusion of hydrogen fluoride. A subsequent study on annealing of type-I gratings at high temperature, however, has shown that the presence of fluorine is not necessary for this regeneration of index modulation [Trpkovski et al. 2005]. So-called chemical composition gratings (CCG) are found in Er-doped fiber with other dopants such as Ge, Al and Sn – as the number of possible diffusing materials increased, a number of researchers settled on oxygen diffusion (through OH) leading to stoichiometric changes. Very recently, the general phenomenon of regeneration has been found in simple H-loaded germanosilicate fibre [Zhang & Kharizi et 2007]. We soon recognized a more general implication of this result - rather than rely on a diffusive interpretation and subsequent polarisability change as the basis for writing gratings, an alternative approach to engineering the index change based on glass structural transformation arising from relaxation of high internal pressures and high temperature processing was proposed. Whilst much work remains to verify details of this approach, it implicitly did not rely on the fibre glass dopants at all (other than maximizing the seed grating strength) and directly led to the development of regenerated gratings in standard photosensitive fibres with transmission rejection >10%/cm and which can tolerate temperatures as high as 1295°C [Bandyopadhyay et al. 2008; Canning et al. 2008b]. The use of hydrogen was important, but not necessarily essential, to obtain index modulation of useful magnitude for very high temperature operation, since it permitted enhanced localisation of the pressure differences between processed and unprocessed regions. The model proposed in Canning et al. 2008b is independent of this and recent work demonstrates that regeneration can be achieved without hydrogen, although for much lower temperature operation [Linder et al. 2009]. In this chapter, we briefly review the hypothesis and demonstrate thermal processing of UV-induced templates that retain the nano-scale precision of the template whilst stabilising the glass change to unprecedented levels. Specifically, we concentrate on studying the thermal properties of regenerated gratings for ultra high temperature operation and show complex behaviour until properly stabilised. There is a linear growth in grating strength with length. Further, we regenerated two types of complex structures (superposed twin grating and a Moiré grating) to demonstrate that all the properties introduced by the seed grating are retained with nm resolution, suggesting that this method could form an advanced processing method for creating holographic structures that go beyond 1-D filters and which have ultra-high thermal stability.

2. Thermal stabilisation of seed index change

Prior to regeneration, the relaxation processes describing normal index change involved with type I grating writing is also characterised by a complex distribution of many relaxations with different timescales and different thermal stabilities. This is characteristic of glass preparation within an amorphous regime. In this section we show quite clearly that there is a distinct regeneration process threshold which suggests, upon regeneration there is an overwhelming preference for fewer relaxation processes. Even below this threshold, thermal stabilisation leads to a single relaxation. We postulate that this reduction is uncharacteristic of an amorphous system and that the glass may very well be transforming into a crystalline polymorph.

2.1 Type I gratings

Standard type I fibre Bragg gratings have become ubiquitous in telecommunication and related devices. These are thermally stabilised by annealing to operate over the telecommunications standards between -25 °C and +80 °C for up to 25 years, as predicted by numerous accelerated aging (annealing) models. Generally speaking experimental results have been consistent with these predictions. For new applications, however, the thermal operation of the fibre Bragg gratings needs to be stabilised to much higher temperatures but for many applications these are *temperatures which are still below that characterised by regeneration*. For example, in the oil, gas and mining industries, including seismic exploration and acoustic detection underground and underwater, the grating is expected to operate to temperatures as high as 400 °C. The options available, within certain limitations, are numerous. They include type 1*n* (type IIa) and type I*p* (type Ia) gratings that can operate between 500-800 °C depending on preparation [Canning 2008a and refs therein]; type II damage gratings [Canning 2008] produced by irradiation above the damage threshold of the glass using UV lasers or multiphoton excitation with longer wavelengths and more recent regenerated gratings that are based on glass structural change achieved using type I seed gratings as a template [Bandyopadhyay et al. 2008; Canning et al. 2008b]. The problem with these various methods is that the gratings produced often compromise the ideal properties offered by type I gratings. For example, type I*n* (and type I*p*) require overly long fluencies that often result in not only weaker gratings but a degradation in profile. Thermally more stable type II gratings are associated with large diffractive losses that make them poor choices for distributed systems and given the role of severe structural change, often on the nanoscale, the long term performance remains poorly understood. Regenerated gratings, on the other hand, operate at the highest temperatures ever achieved (up to 1295 °C) and have losses comparable to type I gratings. Unlike any other high temperature gratings, they can also preserve the complex functionality (shown later) of an advanced type I grating making them appear to be the ideal choice. Unfortunately, the local refractive index change is an order of magnitude less than the original type I seed grating making strong gratings with complex profiles, such as apodisation, difficult to achieve. As gratings now become more sophisticated within the new sensing environment, standard weak gratings used for simple distributed reflection systems are increasingly insufficient. For those applications operating well below the extremes of regeneration, is it possible to retain the advantages of type I gratings? Is there a genuinely distinct threshold between the two? To address these sorts of questions, thermal stabilisation of ordinary type I gratings is revisited with some new ideas based on the description of relaxation and as well on the role of hydrogen in glass. In doing so, we predict and demonstrate, through isochronal annealing, thermal stabilisation of type I gratings to temperatures as high as 600°C simply by annealing at increasing temperatures. In fact, evidence that the thermal stabilisation can be tuned to suit the ideal grating strengths required is presented. A specific aim is for good performance at 400°C, as demanded by many of the industries mentioned earlier. Strikingly, in this work, type I gratings written with pulsed 193nm are characterised with a regeneration threshold temperature of 800°C, about 100°C lower than that obtained using CW 244nm gratings.

2.2 Method of thermal stabilisation

Four Bragg gratings (I-IV, L = 10mm) were inscribed into a H_2 doped conventional photosensitive germanium and boron co-doped (~33 mol % Ge; ~10mol % B) step index fibre using the 193nm output from an ArF laser. The energy used was well below the damage threshold ensuring gradual positive index growth and normal type I grating behaviour. No rollover in growth was observed ruling out any type 1*n* contributions. The typical grating strengths were ~65 dB, (inset in Fig. 1, shows a typical grating spectra prior to thermal treatment). For a uniform grating, numerical simulation indicates an index modulation of $\Delta n_{mod} \sim 1 \times 10^{-3}$ so these are strong gratings.

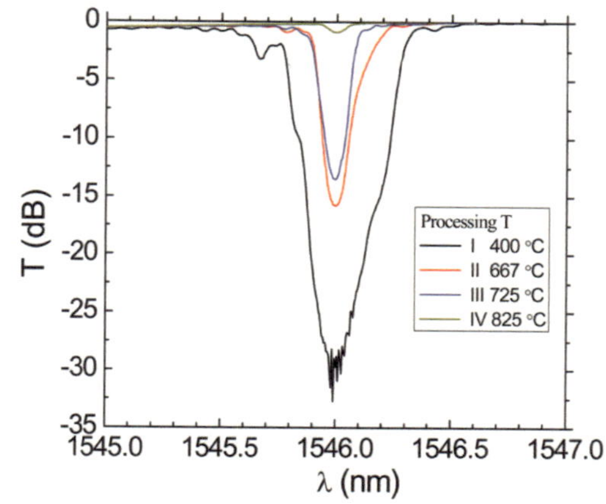

Fig. 1. Transmission spectra of fibre Bragg gratings I-IV with different thermal processing history prior to isochronal annealing.

These gratings were then individually pre-annealed as follows:

I. Ramped in 5 mins to 400 °C, held for 10 min, then removed from heater

II. Ramped in 15 mins to 667 °C, held for 10 mins, then removed from heater

III. Ramped in 15 mins to 667 °C, held for 10 mins, ramped in 10 mins to 745 °C, held for 10 mins, then removed from heater

IV. Ramped in 15 mins to 667 °C, held for 10 mins, ramped in 10 mins to 745 °C, held for 10 mins, ramped in 10 mins to 825 °C, held for 10 mins, then removed from heater

The pre-annealing sequence recipe is important to achieving this stabilisation and there is room for further optimisation. The final grating spectra for each grating are shown in Fig. 2. The large grating strength reduction observed above 800 °C suggests a threshold effect has been reached. As we shall see, this closely corresponds to the temperature at which structural change begins to take place and regeneration occurs.

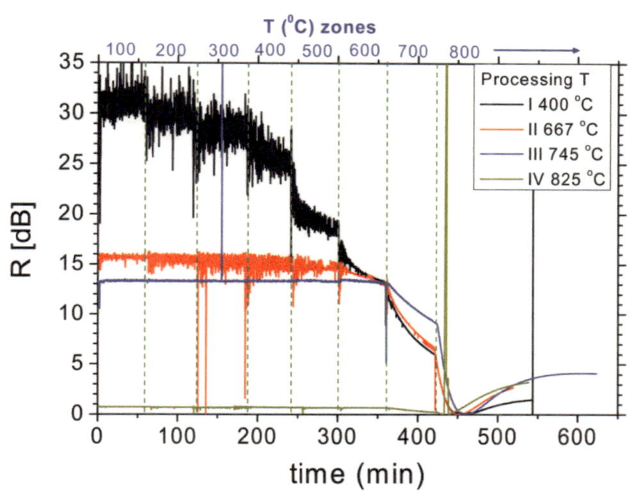

Fig. 2. Isochronal annealing of the four samples as described in the text.

2.3 Isochronal annealing

The four gratings were then sequentially isochronally annealed in a high temperature oven in steps of 100 °C for 1hr. A Gaussian fit to the transmission spectra was used to track the grating strength as a function of time (shown in Fig 2). The origin of the observed noise has not been identified but seems to be highest for gratings stronger than 90%. The following observations for each sample are made:

I. No stabilisation as a function of temperature is observed with rapid step decay between temperatures. Partial stabilisation at low temperatures is observed. This behaviour is not dissimilar to that observed for the hypersensitisation of gratings at lower temperatures - 400 °C is higher than the temperature required to form Ge-OH (>300 °C) but lower than that required to form Si-OH (>500 °C) [Sørenson et al. 2005]. Whilst there is some growth and decay observed at lower temperatures, clear exponential decay is observed at 400 °C and beyond.

II. Stabilisation up to and including 500 °C is observed. This is consistent with Si-OH formation and hypersensitisation to these temperatures as previously reported [Sørenson et al. 2005]. At 600 °C and beyond exponential decay is observed.

III. Stabilisation is observed up to and including 600 °C. At 700 °C and beyond exponential decay is observed. Pre-annealing to the higher temperature suggests significant improvement in stabilisation versus only a 2-3dB reduction in grating strength.

IV. The behaviour here is only slightly improved over that of III, suggesting an optimal was being approached with III. It comes at the expense of a very large grating strength reduction (>10dB) suggesting a threshold limit has been exceeded.

From the above results it is clear there is a large scope to adjust the pre-annealing conditions to optimise the thermal stability with the grating strength as required for a specific application. This means there is significant scope for fine tuning grating writing to enable strong gratings of any type to be produced at temperatures suitable for 400 °C operation, those typically used in the oil, gas (including gas sequestration monitoring) and seismic industries. This tuneability is very reminiscent of the thermal tuning we have demonstrated previously for the generation of OH-style gratings using thermal hypersensitisation [Sørenson et al. 2005]. However, it is important to note that if the amorphous nature of the glassy network characterises these processes, similar processes should be possible in glass without hydrogen and, in many cases, irrespective of the processing conditions (including femtosecond induced changes).

An important observation is the reduced threshold for regeneration. In these experiments there is evidence of a threshold-like effect above which regeneration occurs. This seems to be ~800 °C for pulsed 193nm which is at least 100 °C less than that reported earlier for CW 244nm. Further, there is a time dependence for the onset of this regeneration at 800 °C which we believe is probably fibre strain dependent. It also suggests there remains scope for reducing this threshold still further – this is consistent with the changes in local pressure affected by changes in applied strain along the fibre (through Poisson's ratio).

Another striking feature of the thermal stabilisation process is the subsequent single exponential decay readily fitted to the isochronal annealing of the stabilised gratings. Again, the similarities with previous thermal hypersensitisation work are noted [Sørenson et al. 2005; Canning & Hu 2001]. This suggests a single relaxation process is invoked, which is not usual for an amorphous network which is typically is characterised by a distribution of spectral and temporal relaxations that can all be described by single exponentials, though not all necessarily unique (the general idea of a distribution of Debye relaxations extends to all forms of physical relaxation of viscoelastic solids - see Gross 1968). Generally, the relaxation process is summarised as:

$$N(t) = N_0 \exp[-(t/\tau)^{\beta}] \qquad (2.3.2)$$

Where N is the parameter being measured, N_0 is the initial value, t is the time, and $1/\tau$ is the rate of relaxation; β determines the distribution of relaxation times. In strong glass formers such as silica, β is found to be temperature independent and ~0.3 or so. When $\beta = 1$ there is no distribution but a single Debye relaxation. Figure 3 shows the annealing decay profile at 700°C of the grating stabilised at 745°C. A single exponential fits this data. What is unusual is that disordered materials have $\beta < 1$; the implication is that the glass has become far more ordered, perhaps crystalline. If we assume phase separated glass, with the change is confined only to that of silica, an examination of the phase diagram for silica would suggest α-quartz, or an analogue, as the likely candidate for a crystalline polymorph. This is an extremely interesting prospect because α-quartz is a known piezoelectric material, raising the possibility of all-fibre piezo-based optical devices where the material is introduced periodically or with any profile.

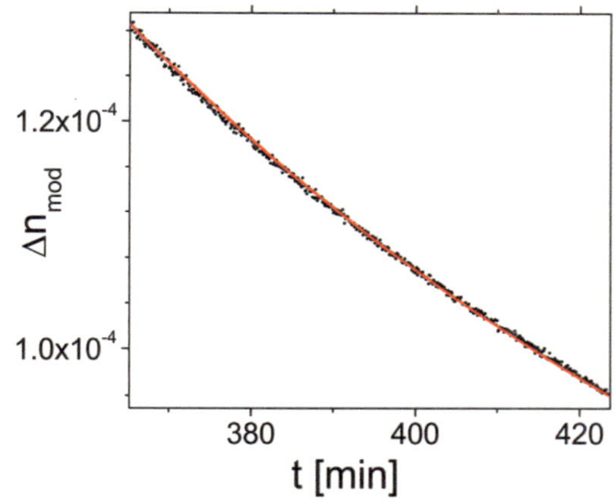

Fig. 3. Decay at 700°C of grating stabilised at 745°C. Single exponential fit also shown.

An approximation of the expected lifetime of a material is often extrapolated from an Arrhenius fit of the rate constant $k = 1/\tau$ where $k = A\exp(-E_a/RT)$. Using similar parameters to those applied in previous work (Inglis 1997; Baker et al. 1997), an estimate of the 3dB lifetime at each temperature for a grating stabilised at 745°C can be obtained, as shown in Figure 4. Although the lifetime falls of rapidly with temperature, these are extremely remarkable results - a grating operating continuously at 400°C can be expected to last for ~10 years, whilst at 350°C >500 years and at 300°C >300,000 years. Further, there remains scope to continue optimising this process.

In conclusion, for many applications, conventional fibre Bragg gratings can be thermally stabilised to operate at temperatures in excess of that required for telecommunications, all the way to 600°C. Beyond this point, regeneration occurs and in this work where pulsed 193nm was used to write the gratings, appears to be characterised by a distinct threshold at 800°C. In contrast, when using CW 244nm light, the threshold seems to be >900°C,

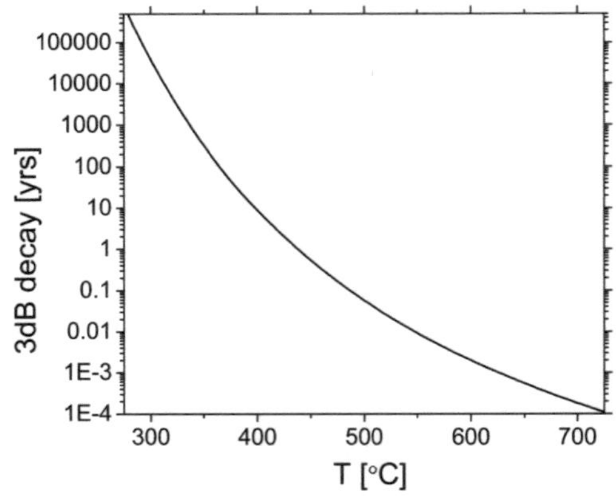

Fig. 4. Lifetime prediction based on 50% (3dB) decay of grating stabilised at 745°C.

suggesting an intensity dependence, and related to this a penetration depth, if the same excitation pathway is assumed. The tuneability offered by thermolytically adjusting, through annealing, the various relaxation processes means its possible to fabricate application specific gratings that require moderate thermal operation to this temperature whilst retaining strong, and sometimes complex (e.g. apodised) reflection spectra, generally not possible by other means. This overcomes one of the current problems associated with weak gratings strengths through regeneration. The versatility of this approach is consistent with the description earlier of an intrinsic distribution of stability regimes and relaxation kinetics that exist within a complex polyamorphous system such as silica. The complex distribution of results accessible by adjusting various parameters from temperature to laser intensity along with temporal processing sheds new insight into annealing generally and more specifically raises interesting questions about the meaning of accelerated ageing tests. This work sets the scene for a more careful evaluation of the processes after which stretched exponential decays reflecting dispersive relaxations, appear less meaningful - above threshold or the onset of regeneration, potentially the onset of crystallisation or a polymorphic (polyamorphic) transition.

3. Regeneration

Glass is generally considered to be metastable material with a large distribution of relaxation processes spread in time. As described above, the majority of these relaxations can each be described with a distribution of exponential decays, or Debye relaxations, in dielectric media – therefore, it is not uncommon that the spread of these processes within a random network is described by a stretched exponential distribution governed mostly by one parameter, β [Angell 1995a; Inglis 1997]. It also implies that the thermal processing history and the quenching rate can play a critical role in determining the final index change of the glass. Stretched exponential fits have been successfully used, for example, to determine the aging process of UV-induced periodic changes in glass that make up a fibre Bragg grating [Baker et al. 1997; Inglis 1997]. However, the success of this approach belies the fact that a comprehensive understanding of glass and glassy materials, as a result of the huge distribution of processes in materials that are not governed by an overwhelming and simple periodicity, remains one of the greatest challenges in condensed matter physics. From a practical perspective, for example, although polymorphism of the glassy state

[Mishima et al. 1984, 1985] (or polyamorphism [Wolf et al. 1992, Angell 1995b]) is generally accepted, its role in determining glass structure is rarely invoked when analyzing both thermal and glass transformation for actual device engineering. In photonics, for example, when laser processing is involved, there remains an almost exclusive preference to interpret refractive index changes to defect, or atomic, diffusion as well as direct polarisability changes from localised defect creation or annihilation. On the other hand, as we suggested above the possibility of creating functional polymorphs, such as a-quartz, exists. This introduces the general proposition of how to induce such polymorphic transitions within glass and within an optical waveguide, in this case fibre.

The original premise for our approach to developing high temperature gratings was based on recognizing that the conditions for potentially achieving a pressure-induced transformation of glass, not dissimilar to that reported for ice by Mishima et al. [1994,1995] is seemingly present within an ordinary optical fibre. Optical fibre drawing generally leads to a core glass (often with some germanate to raise the index above the core) which is under tension upon quenching since its thermal expansion coefficient is higher than that of the surrounding glass, which tends to cool first. However, if drawn under appropriate conditions which exploit Poisson's ratio and the fact that vitreous silica tends to have a molar volume larger in the solid state than the liquid state, it is possible to achieve compressive stress. In most typical drawing conditions it is tensile with effective internal pressures that can range from a few tens of MpA to those that exceed 100MpA. These effective internal pressures should allow for an ability to select from within a range of silicate based polyamorphs of varying density and refractive index - (we are not aware of a detailed investigation on this topic as yet). Nonetheless, the fact is that the pre-existing pressures within an optical fibre at the core/cladding interface can be very large and should play a role in a number of phenomena, including photosensitivity and glass. It is almost certain that the rollover from type I gratings to type In (or type IIa) is explained by this as is the observation of a tunable rollover threshold based on longitudinally applied strain (for a review see Canning 2008a).

The only missing ingredient from a post-induced regenerative phase transformation is temperature. If the phase diagram of silica is considered, for a pressure of ~100Mpa and a temperature of 900°C, tridymite (thought to be metastable), and β -quartz are stable. At temperatures above 1470°C, cristabolite is stable - below this it is unstable. Therefore, a simple interpretation of what might occur with thermal processing is that metastable tridymite is formed (perhaps cristabolite at higher temperatures) which the converts to α–quartz upon cooling back to room temperature, since this is the only crystal state stable at ~100Mpa and 25°C. At lower temperatures, as suggested earlier, α–quartz or an analogue involving dopants is feasible and can account for the stabilisation of type I gratings to temperatures in excess of 700°C. For temperatures above this, the conditions must exist for substantially more stable polymorph, or alternatively, polyamorph. In the regimes we are operating at, it is likely that introduced impurities such as hydrogen, will play an additional role given the local strain produced when OH forms, for example. OH formation can lead to stress relaxation in the core and therefore a small structural density change.

The question is how to measure such a transformation - the simplest approach we have is to use a periodically stressed structure which can be characterised optically. A fibre Bragg grating, as we have described, is such a structure. The induced periodic stress of a fibre Bragg grating inscribed with great precision using UV lasers has a modulation close to that required for our thought experiment to be implemented - a periodic modulation ~50Mpa has been measured in a grating without hydrogen with a spatial resolution of 0.3mm [Raine et al. 1999]. Average UV-induced index changes in excess of 100Mpa have also been reported near the core cladding interface [Raine et al. 1999; Belhadj et al. 2008], suggesting that it is indeed possible to achieve a periodic structure that will seed a periodic regeneration of a different density glass or crystal.

If the possibility of continuous polymorph transformation above the regeneration threshold is accepted, then it is most unlikely the annealing of laser induced changes is going to lead to complete recovery of the glass system - that is, we expect some memory of the UV written "seed" grating to be retained (contrary to a single Debye relaxation below threshold), in particular at the core/cladding boundary. In order to maximise the possibility of achieving high temperature regeneration, hydrogen loading is used for three reasons: *(a)* optimise the induced stress differences between processed and unprocessed regions in the Bragg structure since UV-induced OH causes periodic relaxation of the stresses; and *(b)* minimise the refractive index contribution of the UV written grating to the polarisability changes associated with OH formation; and *(c)* reduce the chance of eventually relaxing to crystallised α quartz which has the potential of changing to other structures each time the grating is heated to higher temperatures (this can explain why the results of Linder et al. 2009, and indeed most type I*n* (type IIa), are unstable above 600°C).

Therefore, the experiment is to write a standard grating in photosensitive fibre loaded with hydrogen and post-anneal this gradually to 900°C and beyond. If this model is correct, then it should be possible to thermally process a fibre and transform the grating from an ordinary type I grating, for example, to an ultra-stable regenerated grating far exceeding the properties of the thermally stabilised gratings described in the previous section. To maximise the chance of the process working, the original experiments focused on strong fibre Bragg gratings written with hydrogen. The results reported in [Canning et al. 2008b] verify that a new grating structure which was stable up to 1295°C before the fibre broke could be obtained. Despite the deterioration and breaking of the fibre in an open environment at such temperatures, the remaining regenerated grating remained intact when broken pieces were analysed, consistent with a stronger glass than the original glass. The nature of this glass is not yet resolved but preliminary scanning electron microscopy (SEM) analysis suggests a reduced Si concentration within the fibre. Figure 5 shows the results - no change in Ge concentration is revealed, pointing towards a transformation of silica and not germanosilicate. The results are also hinting that that change is indeed located more at the interface - however, more comprehensive work is necessary to understand these results further. Indirect support for the results appears when a detailed study of the role of germanium is investigated - despite varying the concentration of GeO_2 from ~3mol% (standard telecommunications fibre) to >20mol%, for equal seed grating strengths, the regenerated grating of similar strength is obtained. This suggests strongly that the regeneration is largely independent of the germanium concentration - it is not obvious how this fits into a general distribution of relaxations in an amorphous network but germanium dioxide, whilst having many structural analogies to silicon dioxide, has significant thermal history differences because of weaker bonding and a much lower melting point (~1100°C).

The likelihood of a more complex polyamorphic transition spectrum is increased when conditions for the threshold transformation are shown to be sensitive to a number of parameters, including writing intensity: using CW 244nm written gratings in standard

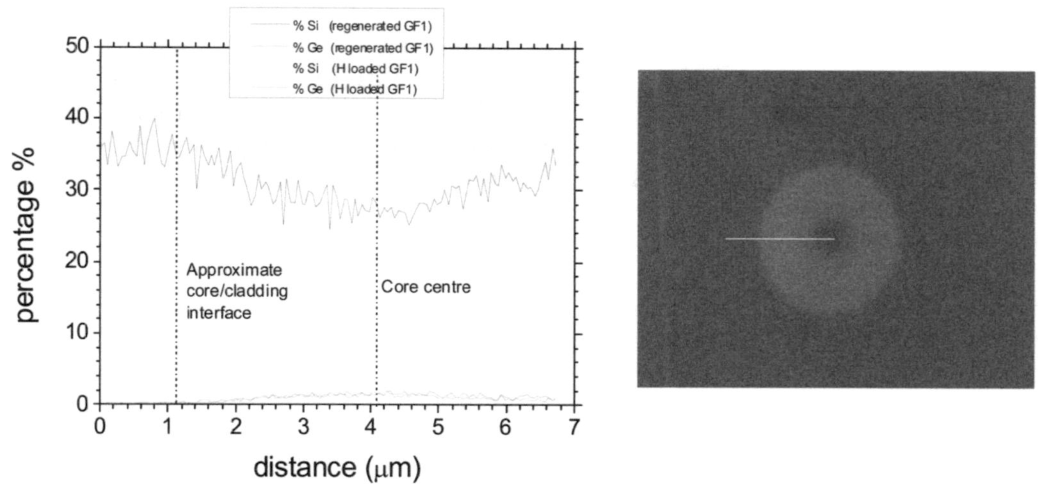

Fig. 5. X-ray backscatter analysis of germanosilicate core fibre with hydrogen loading and no grating, and another identical fibre containing a regenerated grating. No change in Ge concentration is observed whilst there is ~10% reduction in Si, consistent with a density drop in the silica (Preliminary data collected by Matthew Kolibac, 2009, Electron Microscope Unit, University of Sydney).

photosensitive fibre requires ~900°C to be reached, whilst for gratings written with 193nm pulsed light it appears that temperature can be reduced to 800°C [Åslund et al. 2009a; Canning et al. 2009]. More interestingly, perhaps, below the regeneration "threshold", the annealing phenomena we have observed is much closer to the idealised spectrum of polymorphic transformation: we have proposed and demonstrated "tuning" of the thermal stability of fibre Bragg gratings so that they can be tailored to operate at arbitrary temperatures, well in excess of type I gratings but below that of regenerated gratings [Åslund et al. 2009a; Canning et al. 2009].

Here, we focus on regenerated gratings within hydrogen loaded standard photosensitive optical fibres since this gives the most pronounced stabilisation to date. Since our work, others [Linder et al. 2009] have shown that regeneration can be obtained at lower temperatures in gratings written without hydrogen - the low thermal stability of these gratings is consistent with type 1*n* gratings. If crystallisation is occurring it is likely to be via the mechanism we described above to α–quartz, which is stable at ~100Mpa close to the observed 600°C. Whatever the state of glass, it is clearly distinct from that obtained at higher temperatures with hydrogen - if the general model described earlier holds, it is possible that at higher temperatures, a secondary regeneration process is observed in fibres without hydrogen (since it is not essential to the basic tenet). Interestingly, it appears that the procedures involved with thermal stabilisation may be important to achieving the subsequent regeneration - straight heating up to the regeneration temperature produced very poor results and in many instance none at all. This may indicate that the initial structural change which occurs is important to the subsequent regeneration process.

In the following sections, we describe how regenerated gratings are fabricated and characterise their annealing behaviour. An important practical consideration is the quality and resolution achievable with these gratings. We explore this by examining the regeneration of much more complex structures than simple uniform gratings.

3.1 Fabrication of seed gratings

In order to precipitate the structural change associated with regenerated gratings, a seed grating is necessary. In practice, we have confirmed that the stronger the seed grating, the stronger the final regenerated grating. For this case, a conventional Bragg grating inscribed into hydrogen loaded (24hrs, 200atm, 70°C) relatively highly germanium doped step index fibre with no boron (r_{core} ~ 2μm, [GeO_2 ~ 10.5mol%], $\Delta n_{co/cl} = 0.012$), using the 244nm output from a frequency doubled Ar^+ laser (P ~ 50W/cm^2, $f_{cumulative}$ ~ (6-12)kJ/cm^2, the same as that reported in Bandyopadhyay et al. 2008 and Canning et al. 2008b). Figure 6 shows the transmission and reflection of a very strong type I Bragg grating, readily exceeding the noise floor in transmission of our tunable laser and power meter setup (res: 1pm). Ignoring the slight quadratic chirp in the transmission profile, the simulation spectra for a uniform grating, based on transfer matrix solution of the coupled mode equations, was fitted to the bandwidth to estimate the index modulation achieved: Δn_{mod} ~ 1.6 x 10^{-3}, consistent with a grating >120dB in strength.

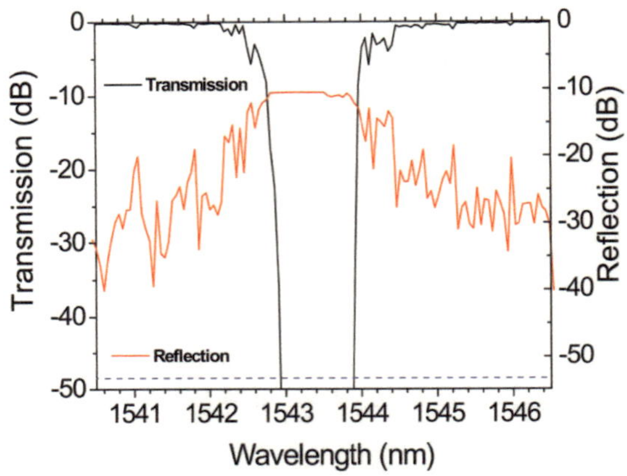

Fig. 6. Transmission and reflection spectra of the "conventional" seed grating. Noticeably, the large side lobes of this structure obscure the stitching errors expected from the phase mask used. The dashed line represents the noise floor.

3.2 Fabrication of regenerated gratings

Using a processing procedure identical to that optimised in Bandyopadhyay et al. 2008 and Canning et al. 2008b, ultra strong seed gratings were thermally processed sequentially with a standard recipe. At 950°C the onset of regeneration is observed, and over time as the seed grating disappears completely, the regenerated grating appears. The final transmission and reflection spectra of the regenerated grating, obtained from the seed grating shown in figure 6, are shown in figure 7. It is more than 50dB in strength and below the noise floor. Numerical simulation indicates an index modulation of Δn_{mod} ~ 1.55 x 10^{-4}, which is substantial. However, the average index may likely be greater than this since the Bragg wavelength, λ_B, tends to be shifted to longer wavelengths to that of the seed grating at room temperature, indicating that the fringe contrast is not optimal, potential scope for improvement.

In order to study the growth and annealing properties, a second regenerated grating was made from a weaker seed grating, written with a cumulative fluence ~30% less than that of the first grating, so that the full transmission spectrum can be observed within the noise

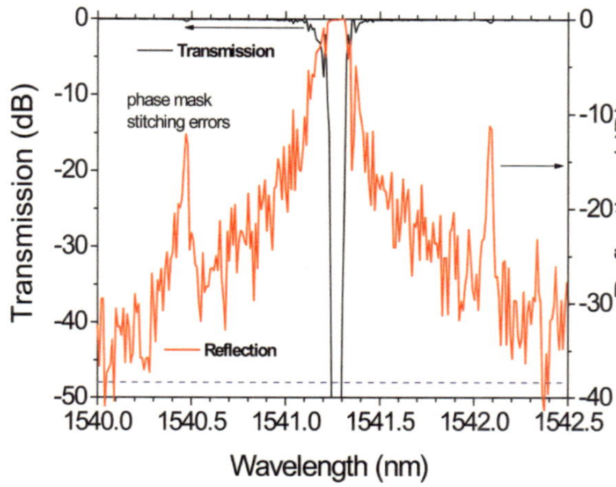

Fig. 7. Transmission and reflection spectra of the regenerated grating. Noticeably, the stitching errors of the phase mask are clearly visible indicating all the relative phase information has been retained. The dashed line represents the noise floor.

floor of the tunable laser and power meter setup. Figure 8 shows a close-up of the grating formation over time at ~950°C - in this example, the structure is less uniform with a quadratic chirp present. This chirp is an exaggerated copy of the seed grating quadratic chirp so the complex profile of the seed grating was preserved. When tension is removed during regeneration, λ_B is the same between seed and regenerated grating, also consistent with our previously reported observations [Bandyopadhyay et al. 2008; Canning et al. 2008b].

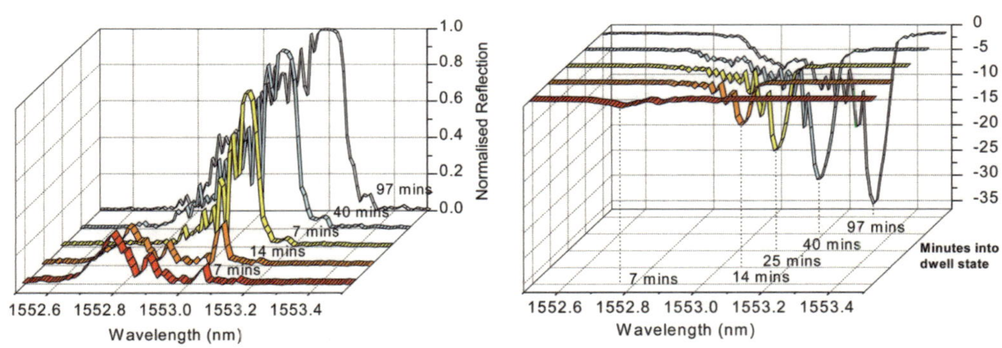

Fig. 8. Reflection (normalised) and transmission spectra during formation and growth of regenerated grating. Significant, amplified quadratic chirp is observed.

4. Annealing

The regenerated grating, shown in figure 7, was then cycled back to room temperature, back up to 1100°C and back to room temperature. For the short exposure times of ~10mins at each temperature every 100°C, no changes are observed in the grating spectra within the noise floor. The second regenerated grating was used to determine the longer term performance at 1000°C and 1100°C. The regenerated grating, ~42dB in strength, was cooled back to ~200°C before being taken up to 1000°C to show that no decay occurs at lower temperatures. The grating was then ramped to 1100°C over ~50 minutes. It was then allowed to sit at 1100°C for an additional 4 hours and 10 minutes. During the ramping period the grating decays with a single exponential to ~19dB. This stabilised to a final rejection of ~18dB within ten minutes at 1100°C. The grating strength after this point remained constant over the remaining exposure period. Figure 9 (a) shows a full summary of seed grating decay, regeneration formation and then the annealing results whilst figure 9 (b) shows a contour image diagram of the decay process over time showing the Bragg wavelength shift with temperature, the exponential decay in contour form and the subsequent stable and steady performance after an hour.

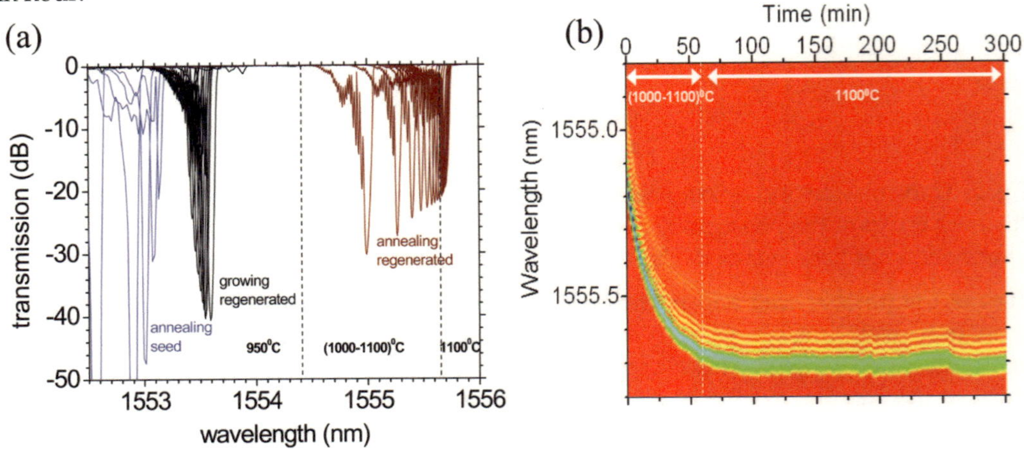

Fig. 9. (a) Summary of seed grating decay, regenerated grating formation and annealing; (b) detailed examination of the annealing of the grating between 1000 and 1100 °C.

When comparing with our previous work, we have shown that by simply extending the length of the seed grating we can regenerate longer gratings that increased in strength approximately linearly after stabilisation. In the previous work, the regenerated grating was reported to be ~2.3dB over 0.5cm, or a coefficient of 4.6dB/cm within the same fibre used here and for the same seed grating writing conditions. The reported value for 5cm in this work after stabilisation is ~18dB, less than the 23dB expected - this is because the seed grating fluence of this grating was actually less than that used previously so as to be able to observe the peak regeneration, ~42dB, prior to stabilisation. (The first grating broke during experiments so additional stabilisation could not be pursued). Within experimental error, there is no evidence to suggest a non-linear regenerated grating strength with length in this fibre.

The observation that the regenerated grating is stabilised after an initial decay process suggests two contributions to the grating strength, the second contribution clearly enabling extremely strong thermal resistance of the grating (up to 1295°C as demonstrated in Canning et al. 2008b). The single exponential process supports the notion that the first decay process is indeed a singular one. It too is quite stable after formation at least at temperatures below the regeneration temperature of ~950°C. Given the localisation of the changes to the periodic scale of the seed grating, the likely contributions are related to stresses not only at the core-cladding interface but indeed between processed regions. What is important to note is that the decrease in grating strength does not appear to be accompanied by any average index decrease (i.e. λ_B continues to be red-shifted and no blue shift is observed though its difficult to separate this because of the slow thermal equilibration of the oven used). Based on this observation, the appearance of two contributions may be an artifact of gradual fringe relaxation until the one contribution is stabilised - this suggests there is a longitudinal component (through stress between processed and unprocessed regions).

4.1 Dependence on germanium dioxide concentration [GeO_2]

With regards to the mechanism of formation, we have ruled out the role of fluorine in the cladding in this process by observing nearly identical regeneration (actually slightly reduced) and annealing within fibres that have no fluorine at all. The SEM observation in figure 5 reveals an apparent reduction in [Si] after regeneration, consistent with a decrease in silica density (and therefore oxygen as well). In contrast, no change in Ge concentration is observed. This suggests that the concentration of GeO_2 is not a direct factor in regeneration. To explore this, gratings were written in fibres with varying concentration, from ~3 mol% used in standard telecommunications grade optical fibre to ~30 mol% in highly photosensitive fibres. In all these cases, for gratings with the same seed index modulation (strength), the regenerated gratings are identical in strength. What is important is the initial seed grating strength and not [GeO_2]. This supports the null result from the SEM data.
The regenerated grating strength obtained is shown to be determined by a number of factors including seed grating strength and as well the fibre V parameter determined from both numerical aperture (NA) and core radius. More complex results appear to be obtained when additional dopants are employed. Early results indicate that when a fibre has 3 times the concentration of GeO_2 than the fibre used here, but is loaded with B_2O_3 to reduce the numerical aperture and allow a larger core radius, we find the regenerated grating strength is less than half. B_2O_3 is also a glass softener and can significantly change the internal frozen-in stresses of the fibre during drawing. This supports the idea that stress is important in the regeneration process (and indeed below it). More detailed scientific studies are currently underway to explore the changes involves.

5. Regeneration of complex gratings

In order to determine whether this process can be applied beyond simple Bragg grating writing as a realistic approach to the production of complex gratings and patterns and structures that can operate at high temperature whilst retaining the complexity of a nano-scaled device, we explored the impact of the regeneration process on two complex grating structures: (1) structure consisting of two superposed gratings with $\lambda_1 \sim 1548.73$ nm and $\lambda_2 \sim 1553.56$ nm, i.e. with $\Delta\lambda \sim 4.8$nm; and (2) a dual channel grating produced by writing a Moiré grating. In a Moiré grating, the refractive index variation along the length of the grating is also different where a uniform period, Λ_B, is modulated by a low spatial frequency sinusoidal envelope of period, Λ_e, that produce two sidebands (essentially a phase shifted structure built up from a periodic distribution of identical phase shifts). Given the sensitivity of the Moire grating to any perturbation in phase, the preservation of the transmission notch and overall profile will be indicative of nanoscale resolution in the regenerated structure.
For the superposed gratings (L ~ 5 mm) were inscribed into a H_2 loaded (24hrs, P = 100atm, T = 100°C) GeO_2 doped core silica fibre ([GeO_2] ~ 10%, fabricated at CGCRI) using a pulsed KrF exciplex laser (248 nm, pulse duration = 20 ns, $f_{\text{pulse}} \sim 70$ mJ/cm^2, repetition rate = 200

Hz). The Moiré grating was written into a fibre which was similar to that used for the superposed grating but also had boron to increase the seed photosensitivity. Regeneration is carried out with an identical recipe to that described earlier but inside a short fibre micro-heater. The hot zone of this heater is supposedly uniform over 5mm only (the exact variation along this length is not known but we suspect a Gaussian profile), and this dictates the grating length.

5.1 Superposed gratings

Sample #1 was prepared by superposing two seed gratings with Bragg wavelengths $\lambda_1 \sim$ 1548.73 nm and $\lambda_2 \sim$ 1553.56 nm, i.e. with $\Delta\lambda \sim$ 4.8nm. Each of the seed gratings was of moderate strength with transmission loss at $\lambda \sim$ -20 dB (grating with λ_1 being slightly stronger than that at λ_2). The superposition of two gratings leads to a compound form of the local index modulation described as [Bao et al. 2001]:

$$\Delta n(z)=2\Delta n_0 Cos\left(\frac{\pi(2\Lambda_{B1}+\Delta\Lambda)}{\Lambda_{B1}(\Lambda_{B1}+\Delta\Lambda)}z+\frac{\Delta\Phi}{2}\right)Cos\left(\frac{\pi\Delta\Lambda z}{\Lambda^2}-\frac{\Delta\Phi}{2}\right) \qquad (1)$$

Λ_{B1} and Λ_{B2} are the periods of the gratings with $\Lambda_{B2} = \Lambda_{B1}+\Delta\Lambda$ and $\Delta\Phi$ is the initial phase difference of the gratings. It is clear from this expression any non-uniformity introduced by the thermal annealing process will result in a spread of $\Delta\Phi$ and broadening of the peaks. The structure was then thermally processed as described earlier until regeneration was complete. The results are summarised in figure 10. Within experimental uncertainty, the Bragg wavelength separation remains the same (~4.8nm) although, as expected the annealing has led to a decrease in average index so that the Bragg wavelengths are blue-shifted. This reduction leads to a change in the phase distribution and the regenerated gratings have a more asymmetric profile, shown in the inset of figure 10 (c). This is consistent with a very weak Gaussian, or quadratic, chirp on the grating. The origin for this chirp almost certainly arises from the hot zone temperature distribution of the micro-heater rather than any intrinsic grating property.

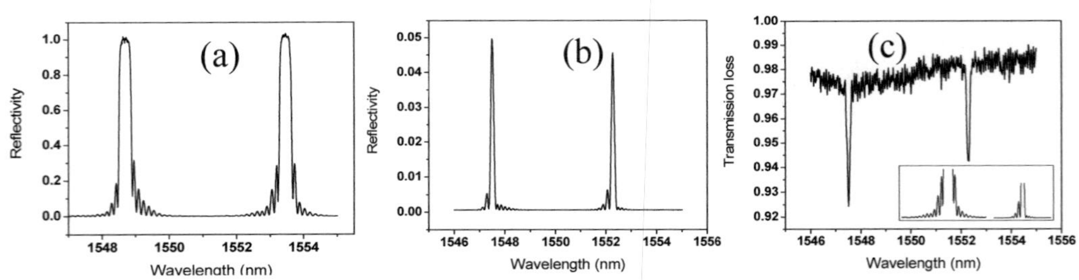

Fig. 10. Spectrum of dual over-written gratings. (a) Normalised reflection spectrum of the seed; (b) and (c) reflection and transmission spectrum of the regenerated grating respectively represented in absolute scale. Inset: close-up of side lobe structure of right hand peaks of seed and regenerated grating for comparison.

5.2 Moiré gratings

In a Moiré grating, a uniform period, Λ_B, is modulated by a low spatial frequency sinusoidal envelope of period, Λ_e, (figure 11) that produces two sidebands. The structure is equivalent to two gratings with stopgaps that overlap sufficiently to produce a resonant phase shift–

like structure in the stop gap of the superstructure. A similar profile is obtained by placing phase shifts with a low frequency period along a uniform grating.

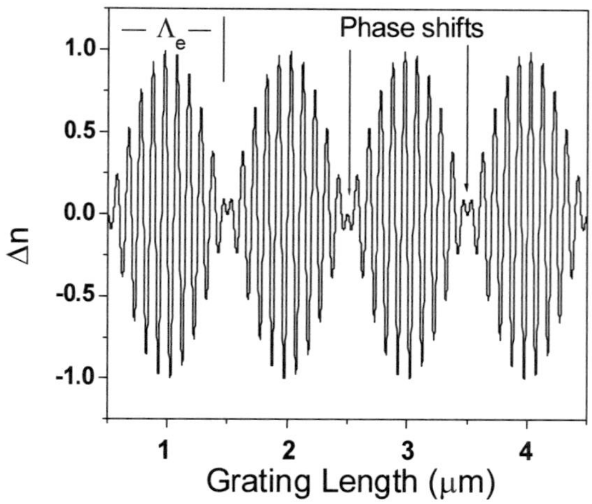

Fig. 11. Index modulation introduced into the seed Moiré grating.

The position dependent index amplitude modulation profile can be described as [Ibsen et al. 1998]:

$$\Delta n(z) = 2n\Delta n_0 F(z) Cos\left(\frac{2\pi Nz}{\Lambda_B}\right) Cos\left(\frac{2\pi Mz}{\Lambda_e}\right) \tag{2}$$

where N and M are integer and $2n\Delta n_0$ is the UV induced index change, $F(z)$ is the apodisation profile. On simplifying, eq.(2) directly leads to the resultant spatial frequencies at the sum and difference frequencies where two Bragg reflections will occur and may be represented as:

$$\Delta n(z) = n\Delta n_0 F(z)\left\{\cos\left(\frac{2\pi N}{\Lambda_B}\left[1+\frac{M\Lambda_B}{N\Lambda_e}\right]z\right)+\cos\left(\frac{2\pi N}{\Lambda_B}\left[1-\frac{M\Lambda_B}{N\Lambda_e}\right]z\right)\right\} \tag{3}$$

The new reflection has two effective bands separated in wavelength, $\Delta\lambda$, as:

$$\Delta\lambda = \frac{\lambda_B^{\,2}}{2n_{eff}\Lambda_e} \tag{4}$$

Based on the principle mentioned above a dual seed grating with a 100 GHz separation i.e. $\Delta\lambda \sim 0.8$ nm was written. Selected $\Lambda_B = 533.17$ nm produces a grating with $\lambda_B \sim 1554$ nm. Modulating Λ_B with $\Lambda_e = 1028$ μm we could generate two channels Bragg wavelengths, $\lambda_1 = 1553.51$nm and $\lambda_2 = 1554.34$ nm respectively. The effective refractive index of the fibre is $n_{eff} = 1.4573$. A precisely controlled scanning beam writing setup was used to produce π-phase shifts at specific locations of the grating to generate the required low frequency sinusoidal modulation of the index profile. A summary of the induced profile is shown in figure 6. The seed grating reflection profile is shown in figure 12 (a) and the regenerated grating reflection and transmission profiles are shown in figures 12 (b) & (c). Unlike the superposed gratings, where the sidebands of the grating are a result of the interference between end reflections of the grating and therefore susceptible to temperature gradients in the micro heater hot zone, the interference in the phase shift region is a result of the distributed interference between the grating and super period of the phase shifts. This means the structure is less sensitive to overall gradients on a macro scale. Importantly, the interference in the phase shift region is preserved after regeneration indicating that despite the very large macro heating process involved in creating the regenerated grating, the structure retains full memory of the seed grating, indicating that there is no evidence of a diffusive process that would alter the phase relationship anywhere over the grating length. Full preservation on a nanoscale is maintained through regeneration - this is a remarkable result.

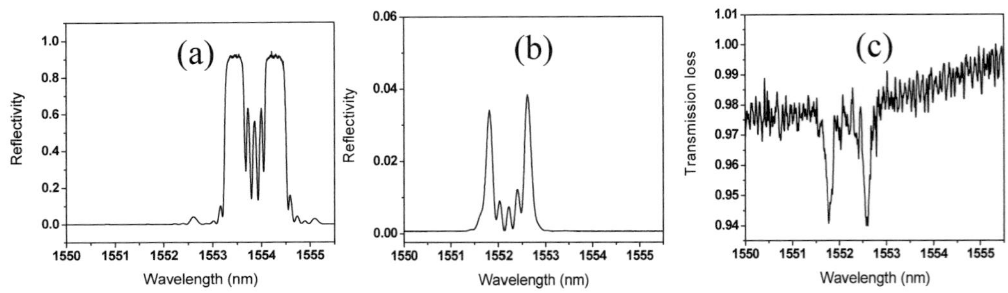

Fig. 12. Spectrum of Moiré grating. (a) Normalised reflection spectrum of the seed, (b) and (c) reflection and transmission spectrum of the regenerated grating respectively represented in absolute scale.

6. Conclusion

Strong regenerated gratings (~18dB, L = 5cm) that can withstand temperatures in excess of 1200°C have been produced. These gratings have a number of potential applications from monitoring furnace temperatures in various fields, to high intensity optical field-resistant gratings for high peak power fibre lasers. Below the regeneration threshold, stabilisation of type I gratings offers a realistic prospect of balancing grating strength with practical temperature operation up to 700°C. Remarkably, single exponential relaxation, consistent with annealing of a regular rather than an amorphous structure, is observed during isochronal annealing of thermally stabilised type I gratings.

Retaining the complex functionality available to type I gratings has also been demonstrated. In particular, complex regenerated gratings (L = 0.5cm) were produced. Two dual channel filter designs - a superposed grating and a Moiré grating - were fabricated with more than 4% transmission. The regenerated superposed structure showed signs of a small chirp possibly arising from the slightly Gaussian profile of the micro heater hot zone employed. This suggests that regenerated gratings can be thermally post-tailored during regeneration from the seed grating on a macro scale. In contrast, despite the significantly reduced strength and the reduced average index change (measured as a shift to shorter wavelengths) the regenerated Moiré grating exactly preserved the interference profile within the central transmission notch of the grating spectrum, and therefore the embedded phase information of the seed grating.

Preliminary material analysis using SEM backscattered x-rays and comparing results between fibres of different compositions indicates an independence of GeO_2 concentration. The SEM results suggest there is a reduction of Si and no reduction of Ge, consistent with a stress (pressure) driven silica transformation. This drop in Si is across the core though the changes, experimental variation not withstanding, appear higher at the core/cladding interface - periodic longitudinal stress profiles between processed regions within the seed grating maybe equally important to any interface effects. The regenerated grating strength appears to be mainly dependent on the initial seed grating modulation, or grating strength. This is also consistent with the extraordinary localisation of the thermally induced change with the original seed grating.

Given that much of the general process involves glass re-quenching under a different thermal history, both the thermal stabilisation and the regenerative processes described here are unlikely to be confined to silica fibres loaded with or without hydrogen (as recent results indicate). Rather, both processes have huge scope to be applied to many numerous materials systems - for example, stresses (and therefore equivalent pressures) at interfaces can be controlled by many means including the use of different thin film layers [Canning 2001]. It opens the way of using nanoscale precision laser processing to introduce nanoscale patterns and structures in materials which can then be thermally processed, with unique recipes given each environment, for additional stability. We believe regeneration has the potential to greatly expand advanced holographic processing of systems and templates by extending the lifetime and operational thresholds of the materials to a level not previously thought possible. This will be particularly important within applications involving very high intensity optical fields such as pulsed lasers.

Finally, the use of an optical fibre grating as the test bed for exploring thermal annealing of glasses generally has been proposed, offering a novel way to study the complex relaxations possible in glasses, including mixed systems.

7. Acknowledgements

We would like to acknowledge various colleagues who have helped prepare this work, including Jacob Fenton (a summer student at iPL) and Matthew Kolibac who have worked on various aspects of regenerated gratings. Funding from the Australian Research Council (ARC) and an International Science Linkage Grant from the Department of Industry, Innovation, Science and Research (DIISR), Australia and the Council of Scientific and Industrial Research (CSIR), India, under the 11th five year plan is acknowledged.

8. References

Åslund, M. & Canning, J. (2000). "Annealing properties of gratings written into UV-presensitized hydrogen-out diffused optical fiber", *Opt. Lett.* 25, 692-694

Åslund, M.; Jackson, S.D.; Canning, J.; Groothoff, N.; Ashton, B. & Lyytikainen K. (2005). "High power Yb^{3+} doped air-clad fibre laser using a Bragg grating written into the active medium", *Australian Conference on Optics and Lasers and Spectroscopy (ACOLS 2005)*, Roturua, New Zealand, paper WeA2

Åslund, M. L.; Jovanovic, N.; Jackson, S. D.; Canning, J.; Marshall, G. D.; Fuerbach, A. & Withford M. J. (2008). "Photo-annealing of femtosecond laser written Bragg gratings", *Australian Conference on Optical Fibre Technology & Opto-Electronics and Communications Conference, (ACOFT/OECC 08)*, Darling Harbour, Sydney

Åslund, M. L.; Canning, J.; Stevenson, M. & Cook, K. (2009a). "Thermal stabilisation of type I grating", *IEEE Photonics Society Annual Meeting*, Turkey

Åslund, M. L.; Jovanovic, N.; Canning, J.; Jackson, S. D.; Marshall, G. D.; Fuerbach, A. & Withford M. J. (2009b). "Photo-annealing of femtosecond laser written Bragg gratings", Accepted to *Photonic Technology Letters*

Angell, C. A. (1995a). "Formation of Glasses from Liquids and Biopolymers", *Science* 267, 1924

Angell, C. A. (1995b). "The old problems of glass and the glass transition, and the many new twists", *Proc. Natl. Acad. Sci. USA* 92, 6675-6682

Archambault, J.L.; Reekie L. & Russell, P. St. (1993). "100% reflectivity Bragg reflectors produced in optical fibres by single excimer laser pulses", *Electron. Lett.* 29, 453-455

Baker, S. R.; Rourke, H. N.; Baker, V. & Goodchild, D. (1997). "Thermal decay of fiber Bragg gratings written in boron and germanium codoped silica fiber", J. Lightwave Tech. 15 (8), 1470-1477

Bao, J.; Zhang, X.; Chen, K. & Zhou, W. (2001). "Spectra of dual overwritten Bragg grating", *Optics Commun.* 188, 31-39

Bandyopadhyay, S.; Canning, J.; Stevenson M. & Cook, K. (2008). "Ultrahigh-temperature regenerated gratings in boron-codoped germanosilicate optical fiber using 193 nm", *Opt. Lett.* 33 (16), 1917-1919

Belhadj, N.; Park, Y.; LaRochelle, S.; Dossou, K. & Anzana J. (2008). "UV-induced modifications of stress distribution in optical fibres and its contribution to Bragg grating birefringence", *Opt. Express* 16 (12) 8727-8741

Butov, O.V.; Dianov E. M. & Golant, K.M. (2006). "Nitrogen-doped silica core fibres for Bragg grating sensors operating at elevated temperatures," *Meas. Sci. Technol.* 17, 975-979

Canning, J.; Sommer, K. & Englund, M. (2001). "Fibre gratings for high temperature sensor applications", *Meas. Sci. Tech.* 12, 824-828

Canning, J. (2001)."Birefringence control in planar waveguides using doped top layers", *Opt. Comm.* 191, (3-6), 225-228

Canning, J. & Hu, P-F. (2001). "Low temperature hypersensitisation of phosphosilicate waveguides in hydrogen", *Opt. Lett.*, 26 (16), 1230-1232

Canning, J. (2006). Fibre lasers and related technologies, *Opt. & Las. In Eng.*, 44, 647-676

Canning, J. (2008a). Fibre Gratings and Devices for Sensors and Lasers, *Lasers and Phot. Rev.* 2 (4), 275-289

Canning, J.; Stevenson, M.; Bandyopadhyay, S. & Cook, K. (2008b). "Extreme Silica Optical Fibre Gratings", *Sensors* 8, 6448-6452

Canning, J.; Åslund, M.; Stevenson, M. & Cook, K. (2009). Australian Conference on Optical Fibre Technology (ACOFT 2009), Adelaide, Australia

Cardozo da Silva, J.C.; Martelli, C.; Kalinowski, H.J.; Penner, E.; Canning J. & Groothoff, N. (2007). "Dynamic analysis and temperature measurements of concrete cantilever beam using fibre Bragg gratings", *Opt. & Las. in Eng.* 45 (1), 88-92

Dong, L.; Liu W.F. & Reekie L. (1996). "Negative index gratings formed by 193nm laser", *Opt. Lett.* 21 (24), 2032-2034

Epaarachchi, J.; Canning, J. & Stevenson M. (2009). "An investigation of response of embedded near infrared fibre Bragg grating (FBG) sensors (830nm) in glass fibre composites under fatigue loading", Accepted to *J. of Composite Materials* Online at: http://jcm.sagepub.com/cgi/content/abstract/0021998309346382v1

Fokine, M. (2002). "Formation of thermally stable chemical composition gratings in optical fibers", *J. Opt. Soc. Am. B* 19, 1759-1765

Grobnic, D.; Smelser, C.W.; Mihailov, S.J. & Walker, R.B. (2006). "Long term thermal stability tests at 1000ºC of silica fibre Bragg gratings made with ultrafast laser radiation," *Meas. Sci. Technol.* 17, 1009-1013

Groothoff, N. & Canning, J. (2004). "Enhanced type IIA gratings for high-temperature operation", *Opt. Lett.* 29, 2360-2362

Gross, B. (1968). *Mathematical Structure and Theories of Viscoelasticity*, (Paris: Herman)

Hill, P.; Atkins, G.R.; Canning, J.; Cox, G.; Sceats, M.G. (1995). "Writing and visualisation of low threshold type II Bragg gratings in stressed optical fibres", *Appl. Opt.* 33 (33), 7689-7694

Ibsen, M.; Durkin, M.K. & Laming, R.I. (1998). "Chirped Moiré fiber gratings operating on two wavelength channels for use as dual-channel dispersion compensators," *IEEE Photon. Tech.. Lett.* 10, 84–86

Inglis, H. G. (1997). ''Photo-induced effects in opticals fibres,'' Ph.D. dissertation (School of Chemistry, University of Sydney, Sydney, NSW, Australia, 1997).

Kersey, D. (2000). "Optical fiber sensors for permanent downwell monitoring in the oil and gas industry", *IEICE Trans.* E83-C (3), 400-404

Linder, E.; Chojetski, C.; Brueckner, S.; Becker, M.; Rothhardt, M. & Bartelt, H. (2009) "Thermal regeneration of fibre Bragg gratings in photosensitive fibres", *Opt. Express* 17, 12523-12531

Mishima, O.; Calvert, L. D. & Whalley, E. (1984). " 'Melting ice' I at 77 K and 10 kbar: a new method of making amorphous solids," *Nature (London)* 310, 393-395

Mishima, O.; Calvert, L. D. & Whalley, E. (1985). "An apparently first-order transition between two amorphous phases of ice induced by pressure," *Nature (London)* 314, 76-78

Raine, K.W.; Feced, R.; Kanellopoulos, S.E. & Handerek, V.A. (1999). "Measurement of Axial Stress at High Spatial Resolution in Ultraviolet-Exposed Fibers," *Appl. Opt.* 38, 1086-1095

Schroeder, R.J.; Yamate, T. & Udd, E. (1999). "High pressure and temperature sensing for the oil industry using fibre Bragg ratings written into side hole single mode fibre", *Proc. SPIE* 3746, 42-45

Shen, Y.; He, J.; Qiu, Y.; Zhao, W.; Chen, S.; Sun, T. & Grattan, K.T. (2007). "Thermal decay characteristics of strong fiber Bragg gratings showing high temperature sustainability", *J. Opt. Soc. Am. B* 24, 430-438

Sørenson, H. R.; Canning, J.; Kristensen, M. (2005). "Thermal hypersensitisation and grating evolution in Ge-doped optical fibre", *Opt. Express,* 13 (7), 2276-2281

Trpkovski, S.; Kitcher, D.J.; Baxter, G.W.; Collins, S.F. & Wade, S.A. (2005). "High temperature-resistant chemical composition gratings in Er^{3+}-doped optical fiber," *Opt. Lett.* 30, 607-609

Wolf, G. H.; Wang, S.; Herbst, C. A.; Durben, D. J.; Oliver W. J.; Kang, Z. C. & Halvorsen, C. (1992) in *High Pressure Research: Application to Earth and Planetary Sciences*, eds.

Manghnani, Y. S. & Manghnani, M. H. (Terra Scientific/Am. Geophys. Union. Washington, USA), pp.503-517

Xie, W.X.; Niay, P.; Bernage, P.; Douay, M.; Bayon, J.F.; Georges, T.; Monerie, M. & Poumellec, B. (1993). "Experimental evidence of two types of photorefractive effects occurring during photo inscriptions of Bragg gratings within germanosilicate fibers", *Opt. Commun.* 104, 185-195

Zhang, B. & Kahriziet, M. (2007). "High temperature resistance fiber Bragg grating temperature sensor fabrication", *IEEE Sensor J.* 7, 586-590

19

Optical Deposition of Carbon Nanotubes for Fiber-based Device Fabrication

Ken Kashiwagi and Shinji Yamashita
Department of Electronic Engineering, The University of Tokyo
Japan

1. Introduction

Since carbon nanotubes (CNTs) have been applied to a passive mode-locker, or a saturable absorber (Set et al., 2004a; Set et al., 2004b), applications of CNTs in photonics field have been intensively investigated. Their quasi-one-dimensional structures produce their distinctive characteristics, their strong third-order nonlinearity and ultrafast recovery time, shorter than 1ps (Chen et al., 2002; Ichida et al., 2002). These characteristics are attractive for future ultrafast photonic networks and can be used in ultrafast photonic devices, such as ultrafast all-optical switches and all-optical logic gates.

Since CNTs are nano-sized material, one of the largest problems to realize CNT-based devices is handling issue. Optical deposition method has an advantage in efficiency over the other handling methods, such as spraying (Set et al., 2004a; Set et al., 2004b), direct synthesis (Yamashita et al., 2004), and polymer embedding methods (Sakakibara et al., 2005). These methods mostly require complicated processes, large-scale setups, and dissipate significant numbers of CNTs. A simpler and more cost effective handling technique of CNTs is required for mass productive CNT-based optical devices. We proposed optical CNT deposition technique to solve the problem. The light injection into CNT-dispersed solution from an optical fiber end deposited CNTs onto core region of the optical fiber end. We realized a passively mode-locked fiber laser using the CNT-deposited fiber as a saturable absorber, or a passive mode-locker.

However, the technique requires very precise control of the light injection power to deposit uniform and less scattering CNT layer, because highly uniform CNT solution, which has very small CNT entanglements, is required. Smaller CNT entanglements require higher injection power. High power injection makes the CNT layer around the core, not on the core. The upper limit of optical intensity depends on the flow speed caused by the injected light. Additional technique is, accordingly, needed to optimize injection power for each solution. We introduced in-situ optical reflectometry to monitor the deposition process, and area-selectively deposited CNTs onto core regions of optical fiber ends. The area-selectivity was confirmed by field emission scanning electron microscope (FE-SEM).

The technique has been applied only to deposition of CNTs onto fiber ends, though it has potentially versatile applications. We propose and demonstrate optically induced deposition of CNTs around microfibers by injecting light through the fibers. We also demonstrate a passively mode-locked fiber laser using a CNT-deposited microfiber fabricated by the technique.

In this chapter, we report optical deposition of CNTs onto optical fibers to realize simple and low cost fabrication of CNT-based fiber-structure devices. This chapter is organized as follows. In section 2, we explain CNT-based optical devices, including fundamental properties of CNTs, CNT fabirication methods, and structures of CNT-based optical devices. We propose the technique to deposit CNTs using light injection in section 3. In section 4, we show the technique with optical reflectometry to enhance the performance of the technique in terms of CNT-use efficiency and repeatability. We report application of the technique to CNT deposition around microfibers in section 5. Finally, we summarize this chapter in section 6.

2. Optical device based on carbon nanotubes

2.1 Carbon nanotubes fabrication methods

Graphite and diamond have been well known allotropes of carbon. In 1985, the third allotrope of carbon, fullerene, was discovered by Kroto et al. (Kroto et al., 1985), and carbon based nanomaterial research field emerged. CNTs are the forth new group of carbon materials which have nearly identical one-dimensional cylindrical structures, and their structures are assumed to be rolled sp^2-bonded graphene sheets. Graphene sheet is a sheet which carbon atoms are hexagonally bonding with each other. Since the discovery of CNTs by Iijima in 1991 (Iijima, 1991), theoretical studies of CNTs have been intensively examined. The studies brought out the distinctive physical properties of CNTs, such as their electronic density of states (eDOS), and metallic and semiconducting distinction that depends only on their structure, chirality (Saito et al., 1992). There are two types of CNTs in terms of their structures, single walled nanotubes (SWNTs) have only single cylinder, and multi walled nanotubes (MWNTs) have two or more cylinders. The CNTs which have optically interesting properties are the SWNTs.

Experimental studies on CNTs became possible after the establishment of production methods in the late 1990s, such as laser-furnace (Guo et al., 1995), arc-discharge (Journet et al., 1997), and catalytic CVD methods with supported catalysts (Dai et al., 1996) and floated catalysts (Satishkumar et al., 1998). These methods are used in the mass synthesis of CNTs in which the CNTs are obtained as soot adhered on the wall of the production chambers. Since the as-synthesized soot contains amorphous carbons, multi-shelled graphites, fullerenes, and/or catalyst metal particles as impurities, a purification process is generally required prior to use (Bandow et al., 1998; Bai et al., 2004; Chiang et al., 2001; Colomer et al., 1999; Morishita & Takarada, 1999; Park et al., 2006). The high pressure carbon mono-oxide (HiPco) method (Nikolaev et al., 1999) has been one of the commercialized fabrication methods for the mass production of CNTs.

The development of production methods of CNTs enabled experimental studies for specific applications. One of the applications is saturable absorbers for ultrashort pulse generation.

2.2 Optical characteristics of carbon nanotubes

In the previous subsection, we reviewed fabrication methods of CNTs. In this subsection, we briefly explain characteristics of CNTs, especially optical characteristics.

CNTs typically have 0.6 ~ 2 nm diameter and 1 μm length so that they have nearly identical 1D structures. CNT structures and their physical properties are determined by their diameter and chiral angle (rolling direction of graphene sheet). Fig. 1 shows an unrolled hexagonal lattice of a CNT, a graphene sheet. The chiral vector ($\boldsymbol{C}_h$) connects the two points which become the identical point when we roll up the sheet to make it a cylinder, or a CNT. The vector is called the "chiral vector".

The chiral vector (C_h) is expressed using unit vectors $\boldsymbol{a}_1$ and $\boldsymbol{a}_2$ and two integers n and m ($0 \leq m \leq n$) as

$$\mathbf{C}_h = n\mathbf{a}_1 + m\mathbf{a}_2 \equiv (n,m) \tag{1}$$

The type of CNT is completely determined by the two integers (n,m) through the definition given in Eq. (1). For example, (6,2) chiral vector is shown in Fig. 1. Originating from their 1D structure, wave vectors of CNTs are discretely existing. This discretion form a sharp divergence in their electron density of state (eDOS), called "van Hove singularities," that is typical characteristic of 1D materials.

Their eDOS structures are determined by their chirality. We can classify CNTs into two groups, metallic and semiconducting. In general, $\mathrm{mod}(n-m, 3) = 0$ CNTs are metallic and $\mathrm{mod}(n-m, 3) \neq 0$ CNTs are semiconducting. One of the most important characteristics of CNTs is that the difference between metallic and semiconducting properties is determined by their chirality.

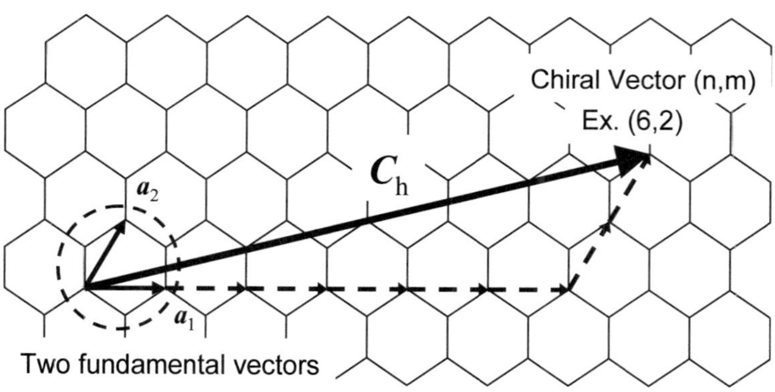

Fig. 1. Schematics of 2D graphene sheet

Semiconducting CNTs absorb the light whose energies are same as their bandgaps. The bandgap energies are controllable by choosing appropriate CNTs diameter, since the bandgap energies are almost inversely proportional to their diameters. The CNTs work as saturable absorbers in the absorption band. Saturable absorbers are materials or devices which change their absorbance depends on power of incident light. They absorb the light which has low intensity, whereas the absorbance decreases due to the saturation of absorption in the case of high intensity light. This phenomenon can be understood as eDOS occupation in conduction bands of saturable absorber material (Fig. 2). If the low intensity light incidents, electrons in valence band are excited up to conduction band, and photons are highly absorbed. On the other hand, if the optical intensity is high, some photons are not absorbed because the eDOS in conduction band is occupied with other electrons which excited by the light. Thus, the optical intensity dependent transparency, the saturable absorption reveals.

This kind of intensity dependent attenuation enables the high-intensity components of an optical pulse to pass through saturable absorbers, while the lower intensity components of the pulse, such as the pulse wings, pedestals, and background CW radiation, not to pass

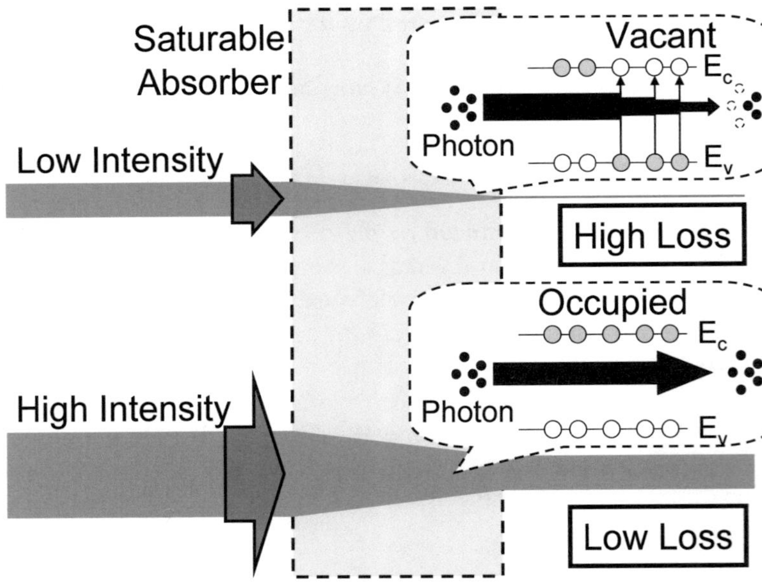

Fig. 2. Schematics of saturable absorption (E_v and E_c denotes energy levels of valence band and conduction band, respectively)

(Fig. 3). When a saturable absorber is inserted in a laser cavity, amplified spontaneous emission (ASE) noise of a gain medium is shaped to be a pulse train. In every round trip, light pass the saturable absorber as high intensity noise with low loss and low intensity noise with high loss, resulting in high intensity contrast. Finally, light start to oscillate in pulsed state.

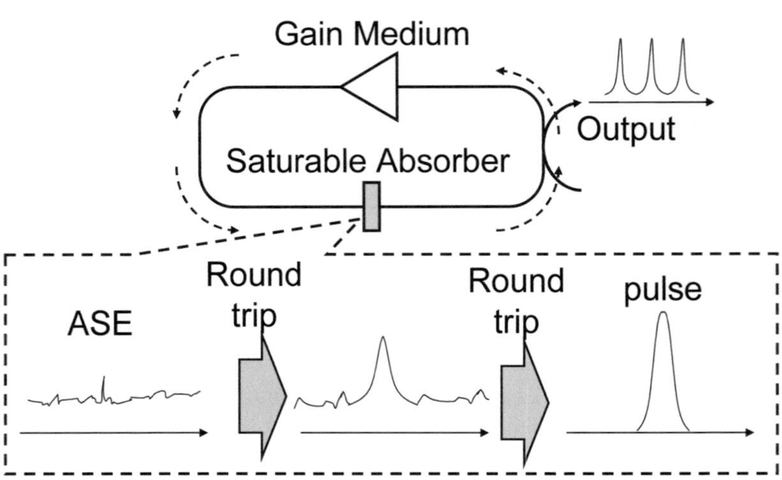

Fig. 3. Schematics of noise suppression by saturable absorber

For ultrashort pulse generation, a saturable absorber with a fast recovery time is required for stabilizing laser mode-locking, while a slower recovery time could facilitate laser self-starting. Recovery time of CNTs have been measured to be shorter than 1 ps, and CNTs are suitable material for ultrashort pulse generation. However, the recovery time of CNTs only consisted by semiconducting CNTs is not so fast (around 30 ps) (Rubtsov et al., 2004). The ultrafast response time of CNTs are based on bundles and/or entanglements of semiconducting and metallic CNTs because the electrons which are excited by photons in semiconducting CNTs couple to metallic CNTs, resulting in ultrafast recovery time of semiconducting CNTs. As-synthesized CNT samples consist of CNTs which have several different chiralities, including both metallic and semiconducting CNTs. Therefore, the CNT samples inherently have ultrafast recovery time shorter than 1 ps.

It is difficult to selectively fabricate CNTs which have certain chirality. However, the mixture of several types of CNTs have two advantages. One of the advantages is ultrashort recovery time which we mentioned above. Another advantage is its wide absorption band. Different chiralities of CNTs have different absorption band, and, consequently, their mixture effectively has very wide absorption band. The wide saturable absorption band is required for passively mode-locked laser whose output pulse width is ultrashort, such as femtosecond regime.

2.3 CNT based optical device structures and fabrication methods

After CNT optical devices were first demonstrated, three types of CNTs based device structures have been proposed. They are depicted in Fig. 4, transmission type, reflection type and fiber end type. These device structures were confirmed to have good performance to provide CNT-based optical devices. However, the fabrication process had some problems in terms of efficiency.

The three types of devices was first fabricated and demonstrated by spraying method (Set et al., 2004a; Set et al., 2004b). The spraying method is common method to evaluate and characterize CNT samples. In the method, we first prepare a CNT-dispersed solution. CNTs tend to be entangled with each other, and few kinds of solvents can be used to disperse them in high uniformity and fewer entanglements. DMF is one of the most commonly used solvents where CNTs can be dispersed. After the preparation, we spray substrates and fiber ends with the solution, and evaporate the residual DMF by heat. The largest advantage of this method is simplicity: we can fabricate optical devices using simple setup. However, the efficiency of CNT-use is poor. The sprayed solution spread around the target position where we want to deposit CNTs.

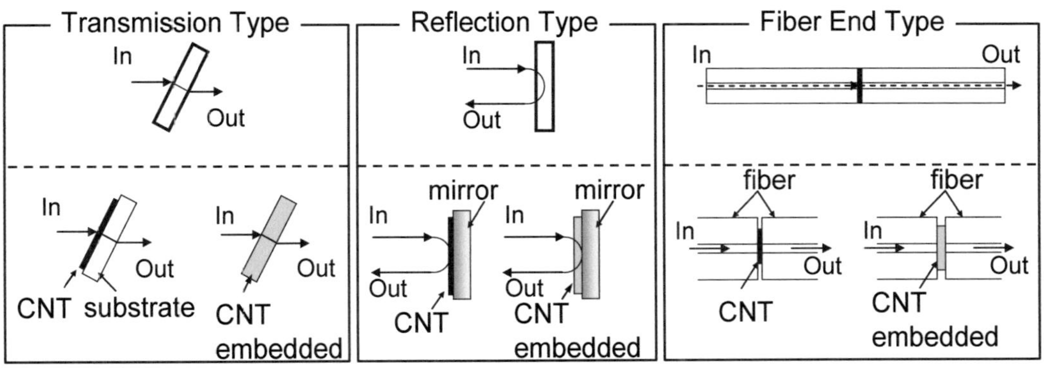

Fig. 4. Schematics of three types of carbon nanotubes optical devices

After the proposal, direct synthesis method (Yamashita et al., 2004) and polymer embedding method (Sakakibara et al., 2005) have been reported. Though embedding CNTs into polymer materials have been already demonstrated (Sakakibara et al., 2003), the report was the first application to the passive mode-locker in the fiber-end-type configuration. The polymer embedding method can remove the impurities before embedding CNTs into polymer materials. The method requires setups for polymer material processing, and the large number of CNTs is not settled through optical pass of devices. In the direct synthesis method, CNTs were directly synthesized onto optical fiber ends, and the fiber produced passively mode-locking of a fiber laser. However, there is no method to remove impurities in as-synthesized CNTs, and consequently ultra-high performance CNT fabrication setup is required.

From the next section, we show our proposal of optical deposition of CNTs. Our proposal offers very simple CNT-based optical device fabrication method, and efficiency of CNT-use drastically increases.

3. Optical deposition of carbon nanotubes

3.1 Optical deposition of carbon nanotubes onto fiber ends

There is a problem in handling CNTs because CNTs tend to entangle with each other and they are difficult to be dispersed in common solvents. In optical device applications, CNTs are conventionally used by spraying, directly synthesizing onto a device, or embedding into a polymer material. These processes are mostly complicated, large-scale setups are required and CNTs are not efficiently used in these methods. To minimize the dissipation number of CNTs, we have proposed and demonstrated optical deposition of CNTs. The method enables area-selective deposition of CNTs only onto a core region of an optical fiber end. This technique will drastically improve the efficiency of CNT usage and the fabrication costs of CNT-based photonic devices.

Our experimental setup is very simple as shown in Fig. 5. It is composed of just two equipments, a laser diode and an erbium doped fiber amplifier (EDFA). We first prepared a purified CNT-dispersed DMF solution. Light produced from a laser diode which had the wavelength of 1560 nm and the optical power of -10 dBm was amplified upto about 20 dBm by a high-power EDFA. The light was incident into the solution through a cleaved fiber end. We observed CNT deposition conditions on the fiber end facets using a microscope for several different optical powers. Microscopic Raman spectroscopy was used to confirm the existence of CNTs at selected area. To find the dependence of the numerical aperture (NA) of the optical fiber, we used two types of optical fibers having different NAs, single-mode fibers (SMFs) and dispersion-shifted fibers (DSFs).

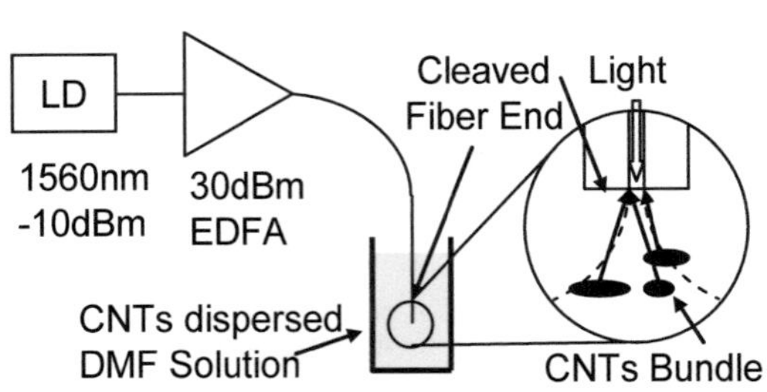

Fig. 5. Experimental setup for optically depositing carbon nanotubes onto fiber end

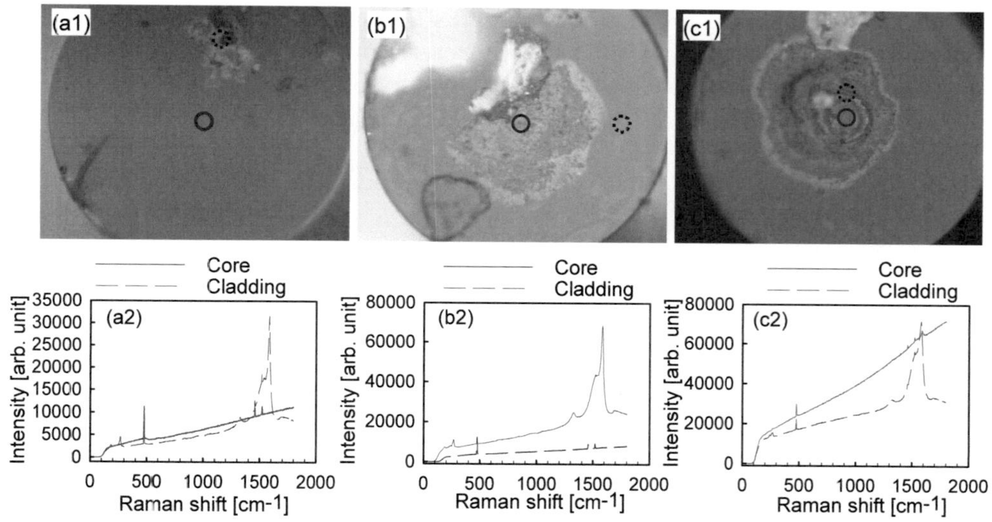

Fig. 6. Microscope images and Raman spectra of CNT-deposited SMF ends. Injected light power: (a) 20.0, (b) 21.5, and (c) 22.0 dBm.

In Fig. 6, microscope images and Raman spectra of CNTs deposited fiber ends of standard SMFs are shown. Fig. 6 (a), (b), (c) correspond to incident optical power of 20 dBm, 21.5 dBm and 22.0 dBm, respectively. Solid circles in Fig. 6 (a1), (b1), (c1) indicate core areas of SMFs, and corresponding Raman spectra are shown with solid curves in Fig. 6 (a2), (b2), (c2). Broken circles are the areas where CNTs are deposited outside of the core, and corresponding Raman spectra are shown with broken curves.

There are three major peaks in the Raman spectra of CNTs. The most typical peak of CNT appears at around 250 cm^{-1}, and it due to the radial breathing mode (RBM), which is a vibration mode of CNTs in the radial direction. We confirmed the presence of CNTs from the peak. The optical power of 20.0 dBm was not enough to trap the CNTs, whereas Fig. 6 (a) shows the existence of CNTs that may be accordingly attached to the cladding region of the fiber end. In Fig. 6 (b), CNTs were area-selectively deposited onto the fiber end only on the core region, and the Raman spectrum shows the existence of CNTs. By increasing the optical power up to 22.0 dBm, CNTs were not deposited onto the core region but deposited around the core.

In the case of DSF, an optical power of 19.0 dBm was enough to deposit CNTs. Fig. 7 shows a microscope image of the fiber end facet and its Raman spectrum at the core center. By increasing the optical power up to 21.5 dBm, CNTs were deposited around the core. Stronger confinement of light in DSF reduced the required optical power for CNT deposition by 2.5 dB, and widened the margin to ±2 dB.

The principle of this technique is not yet confirmed, but we presume that one possible mechanism is the optical tweezer effect, which is caused by the optical intensity diversion of the light in a solution. Another possible mechanism would be the flow of solution due to the injected light. The light might thermally induce convection and swirl nearby the core, and entangled CNTs were attached.

It is possible to deposit CNTs onto a fiber ferrule end as shown in Fig. 8. We deposited CNTs onto an end facet of a standard SMF with a ferrule using an optical intensity of 21.5

dBm, which is the same experimental condition as that of the fiber shown in Fig. 6 (b). Raman spectra in Fig. 8 show that CNTs were area selectively deposited only onto the core region. This technique will reduce the alignment cost after CNT deposition.

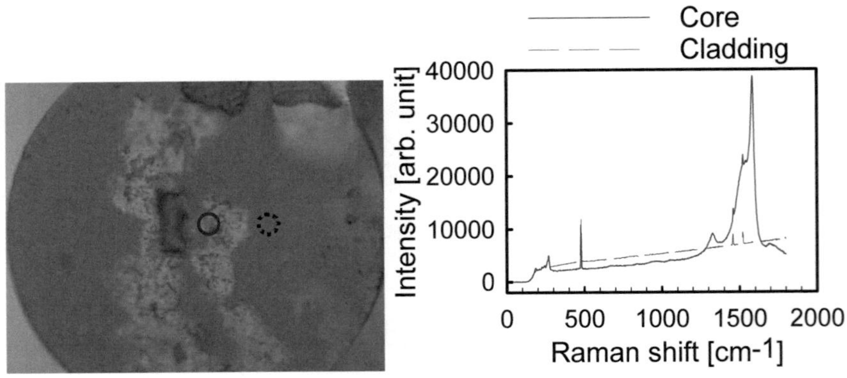

Fig. 7. Microscope image and Raman spectrum of CNTs deposited onto DSF end using 19.0 dBm injected light.

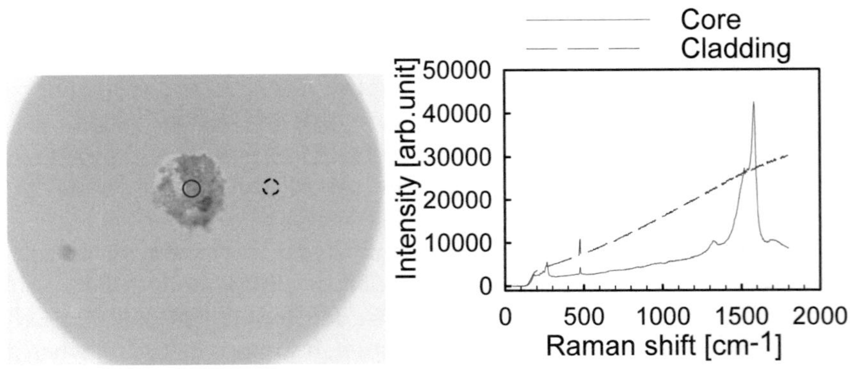

Fig. 8. Microscope image and Raman spectrum of CNT-deposited SMF end with ferrule using 21.5 dBm injection light.

3.2 Passive mode locking of fiber laser using optically deposited CNTs

In previous subsection, we have explained the method to deposit CNTs only onto the core region of optical fiber ends. We have deposited CNTs onto optical fiber end with ferrule intended to use as a saturable absorber. In this subsection, to ensure that the performance of the technique is sufficient for fabricating optical devices, we report a passively mode-locked fiber laser that employs the fiber, which had optically deposited CNTs on its fiber end.

The experimental setup is shown in Fig. 9. An EDFA was used as the laser gain medium and an isolator was inserted to prevent back reflection in the cavity to ensure one-directional lasing. We controlled the polarization state using a polarization controller (PC). The total dispersion in the laser cavity was adjusted to be nearly zero by inserting a 20-m-long SMF. The output light came out from a 3 dB coupler. A CNT-deposited fiber on its end with ferrule was inserted as an alignment-free passive mode-locker. The insertion loss of the fiber was about 4.2 dB. We measured an optical spectrum using an optical spectrum analyzer and an autocorrelation trace using a second-harmonic generation (SHG) autocorrelator.

By injecting 200 mA to the pump laser of the EDFA and controlling the polarization state of the light inside the laser cavity, we achieved passively mode-locked oscillation. An optical spectrum of the laser output measured with a 0.1 nm resolution is shown in Fig. 10 (a). The 3 dB spectral width was 3.2 nm. The SHG autocorrelation trace with a 50 fs resolution is shown in Fig. 10 (b) and had a full-width at half-maximum (FWHM) of 630 fs. Assuming a transform-limited sech^2 pulse waveform, the pulse width is calculated to be as short as 400 fs. The pulse width was almost independent of EDFA gain and was highly dependent on the dispersion of the laser cavity.

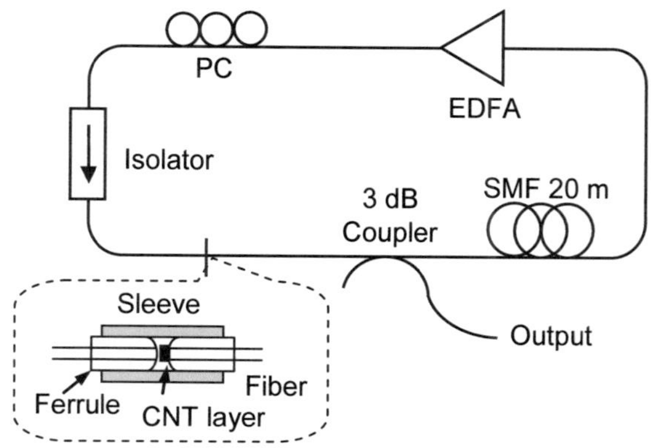

Fig. 9. Experimental setup for passively mode-locked fiber laser using a CNT-deposited fiber as a saturable absorber.

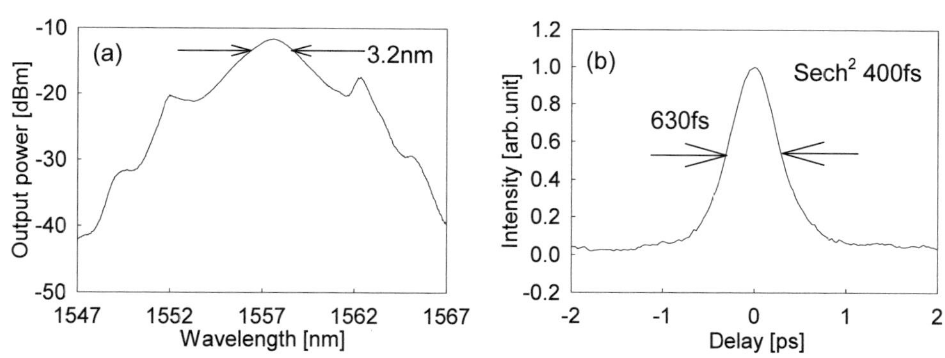

Fig. 10. Optical output characteristics of passively mode-locked fiber laser in Fig. 9 (a) Optical spectrum (resolution 0.1 nm). (b) SHG autocorrelation trace (resolution 50 fs).

4. In-situ monitoring of optical deposition of carbon nanotubes onto fiber end

The method described in the previous section requires only a lightsource to deposit CNTs onto a core region of an optical fiber end. Light injection from the fiber end into CNT-dispersed solution preferentially deposits CNTs onto core regions of optical fiber ends, resulting in efficient CNT-use. Optical pass alignment, therefore, is not necessary for this technique.

However, the technique requires very precise control of the light injection power to deposit uniform and less scattering CNT layer because highly uniform CNT solution, which has very small CNT entanglements, is required. Smaller CNT entanglements require higher injection power. With the increase of the light intensity, flow induced by the light injection becomes too high speed to trap the CNTs onto the core. High power injection deposits CNTs around the core, not on the core, as we have already depicted in Fig. 6. The upper limit of optical intensity depends on the flow speed caused by the injected light. Additional technique is, accordingly, needed to optimize injection power for each solution. In this section, we employed optical reflectometry to simplify the optimization process and deposit CNTs onto very small areas.

The experimental setup for the in-situ optical reflectometry is shown in Fig. 11. Light at a wavelength of 1560 nm from a laser diode was used for both the optical deposition and the optical reflectometry. The light was amplified by a high-power EDFA, and subsequently was split into two by a 10:90 coupler. The 10 % of the light was monitored for reference by a power meter after 20 dB attenuation. The light of 90 % was injected from a cleaved fiber end into a DMF solution, where purified CNTs were uniformly dispersed. The power of the reflected light from the fiber end was measured by another power meter through a circulator. The reference and the reflected light powers were measured at every 500 msec. The refractive indices of DMF and silica-glass are 1.42 and 1.44, respectively. Since the refractive index difference between DMF and silica-glass was small, the reflection was suppressed before CNT deposition. On the contrary, semiconducting CNTs had the refractive indecies of around 3.0, though the refractive indecies of CNTs depend on their chiralities(Margulis & Gaiduk, 2001).

The reflectivity drastically increased after the first deposition of an entanglement. The deposition was achieved by the optimization of the injection power with monitoring the reflection. Even if we repeat the experiment by changing the injection power with the highly uniform CNT solution, we could not deposit the CNTs only onto the core without the reflectometry. It was because there was very small margin of the injection power when we used the highly uniform CNT solution. Moreover, the solution condition, especially the sizes of the CNT bundles, changed in time and this prevented us from the preferential deposition without the reflectoemtry. The optical reflectometry offered the detection capability of the starting time of CNT deposition to the system and, consequently, controllability of the number of CNTs by adjusting the light injection duration after the deposition started. Subsequent to the process, we took microscope images and field emission scanning electron microscope (FE-SEM) images of the fiber ends.

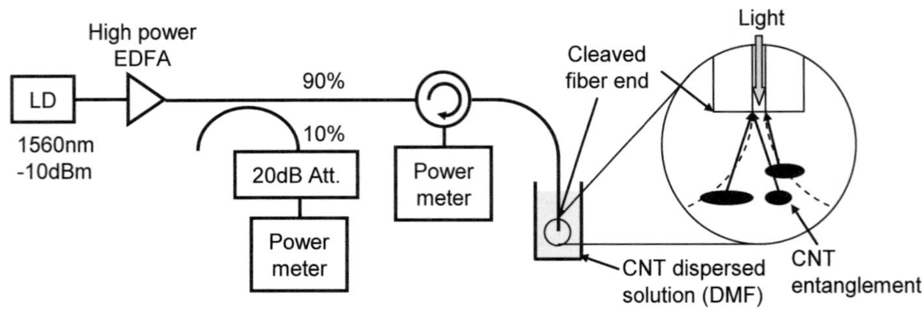

Fig. 11. Experimental setup for optically manipulated CNT deposition with optical reflectometry monitoring.

Fig. 12 shows in-situ optical reflectometry data series which describe the deposition processes of thin and thick layers. Once the EDFA was turned on, its output was kept constant around at 20 dBm. It took a short period of time (10 - 100 s) to start the deposition. The start times were not the same in the two cases because the solution was not completely uniform. The reflectivity was as small as -40 dB because of the small refractive index difference between DMF and the fiber. Fig. 12 clearly show the drastic increase of the reflectivities about 20 dB due to the first entanglement deposition. The thicknesses were controlled by changing the light injection period after the increase of reflectivity. To deposit a thin layer, the EDFA was turned off at 8 seconds after the start of deposition as shown in Fig. 12 (a). In the case of the thick layer deposition, the EDFA kept on for about 4 minutes after the deposition starting. During the deposition, the fluctuation of reflectivity gradually decreased (Fig. 12 (b)). It indicates that the CNT layer gradually became uniform.
The introduction of optical reflectometry produced important functionalities to the system: deposition starting time detection and in-situ layer uniformity evaluation. The thickness of CNT layer could be roughly adjusted by controlling duration of injection even without the reflectometry. Since entangled CNTs which flowed to a fiber end were trapped and deposited, CNT deposition did not start just after the light injection. This was the reason why it was difficult to control the number of CNTs. First deposition of CNT entanglement drastically increased reflectivity at the fiber end due to the high index contrast between CNTs and silica-glass. It, then, became a seed of a CNT layer and the deposition continued because of strong Van der Waals force among CNTs. The duration of the deposition, therefore, controlled the layer thickness by introducing optical reflectometry. As the layer became more uniform, the reflection became less affected by the solution flow. The fluctuation of the reflectivity gradually decreased.
Optical microscope images of the thin and thick layers are shown in Fig. 13. The figures describe the thickness difference between the two layers. To directly observe the existence of CNTs, the layer thicknesses and the sizes of the deposition regions, we took the FE-SEM images of the fiber ends shown in Fig. 14 and Fig. 15. The thin CNT layer was clearly observed only at the core region in Fig. 14, whereas the deposited layer was very thick in Fig. 15. The thick layer was preferentially deposited only onto ~15 μm diameter circular region. This result was achieved using the highly uniform solution solution and the precise optimization of injection power which was enabled by the in-situ optical reflectometry.

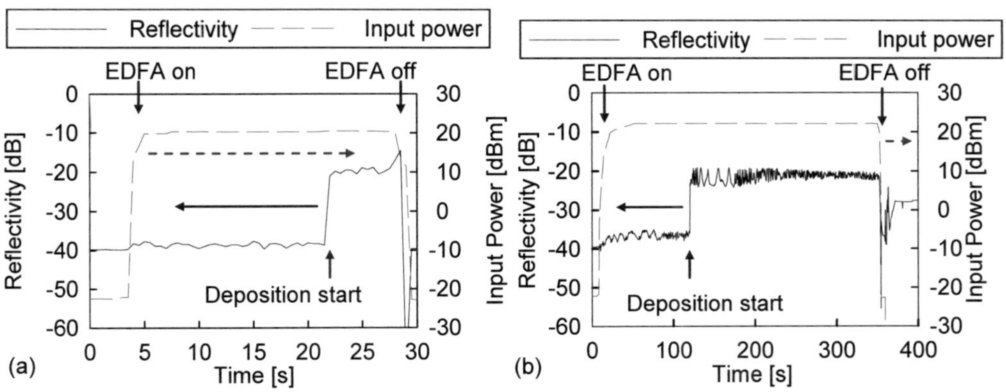

Fig. 12. Data series of optical reflectometry of CNT deposition (a)thin layer deposition process (b)thick layer deposition process.

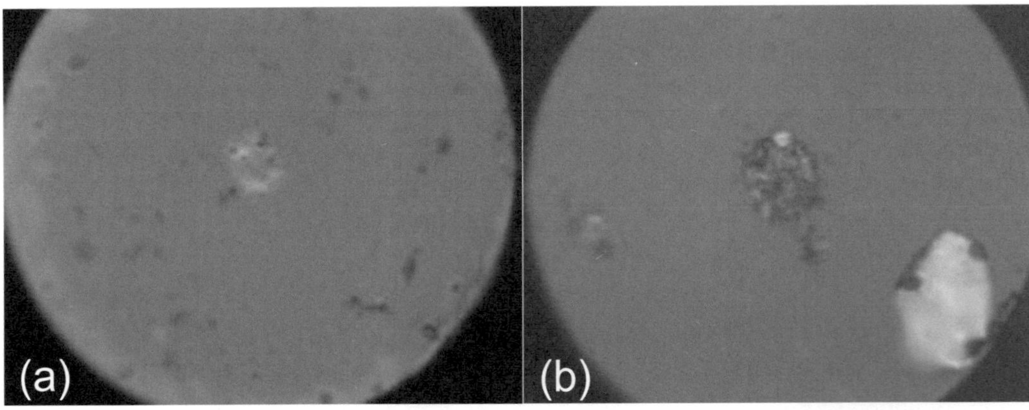

Fig. 13. Optical microscope images of fiber ends with CNT layers on the core region (a) thin CNT layer (b) thick CNT layer.

Fig. 14. FE-SEM images of the fiber end with the thin layer (a) whole fiber end
(b) magnified around the core region

Fig. 15. FE-SEM images of the fiber end with the thick layer (a) whole fiber end
(b) magnified around the core region.

5. Optical deposition of CNTs around microfiber

In section 3 and 4, we reported the optical deposition of CNTs onto the optical fiber ends. In this section, we show one aspect of versatility of the deposition method, CNT deposition around a microfiber. Evanescent coupling between CNTs and propagation mode of a microfiber is one way to realize polarization insensitve CNT device as already reported (Kieu & Mansuripur, 2007; Song et al., 2007).

First, we fabricated a microfiber using an experimental setup shown in Fig. 16. A bare standard single mode fiber (SMF) was set on two fiber holders, which were fixed on two translation stages. The fiber ends were connected to an erbium doped fiber amplifier (EDFA) and a power meter, respectively. We fabricated a microfiber by stretching the fiber with heat produced by a flame. During the fabrication process, we monitored the insertion loss of the fiber using the power meter. The fiber was tapered down so that its taper waist diameter became ~6 μm.

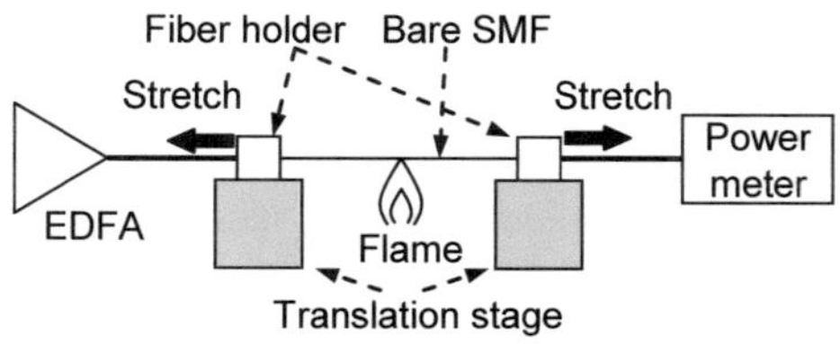

Fig. 16. Schematic of fabrication of microfiber

Next, we deposited CNTs around the microfiber using a setup shown in Fig. 17. The fabricated microfiber was immersed into a CNT-dispersed DMF droplet on a slide glass. Light from a laser diode at a wavelength of 1560 nm and at -10 dBm optical power was amplified up to 13 dBm by an EDFA and consequently injected into the microfiber. The output power was monitored by a power meter to detect start of CNT deposition, and consequently to control the deposition time.

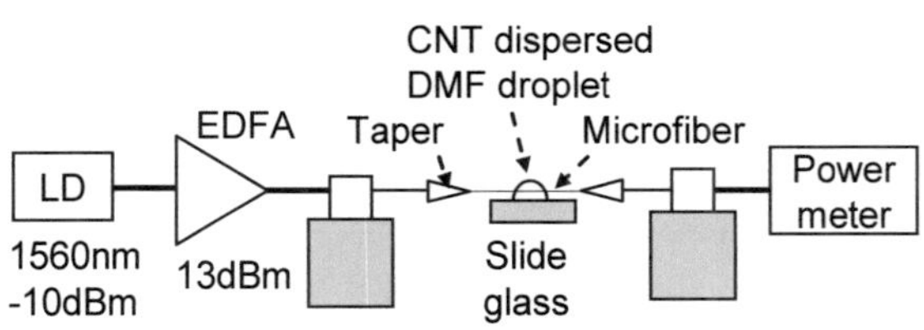

Fig. 17. Schematic of optical deposition of CNT around microfiber

Fig. 18 (a) shows a microscope image of a CNT-deposited microfiber, whose waist diameter was ~6 μm. We found that the CNTs start to be deposited around the microfiber at the incident power of 13 dBm. To ensure the existence of CNTs, we performed microscopic Raman spectroscopy. The dotted circle in Fig. 18 (a) was the area where microscopic Raman spectroscopy was performed. The microscopic Raman spectrum in Fig. 18 (b) confirms that the CNTs were certainly deposited around the microfiber. The optical deposition technique was confirmed to be applicable not only to deposition onto fiber ends but also to deposition around microfibers. We could detect the start of the CNT deposition by the drop of the output power due to scattering and absorption induced by deposited CNTs. We stopped the light injection about 5 sec after the deposition started. The excess loss induced by the tapering was 0.2 dB. The CNT deposition increased insertion loss by 5.8 dB and the total loss was 6 dB.

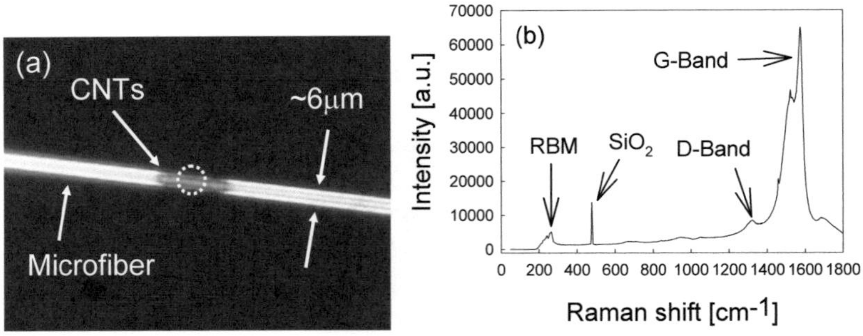

Fig. 18. (a) Microscope image of CNT-deposited microfiber (waist ~6 μm), dotted circle indicates area where microscopic Raman spectrum was measured, (b) Microscopic Raman spectrum of CNT-deposited microfiber.

We presume the similar phenomenon occurred both in the case of optical CNT deposition onto fiber ends and around microfibers. In the case of the CNT deposition onto fiber ends, swirl and convection caused by the light injection to a CNT-dispersed solution might deposit CNTs. The deposition mechanism might be the combination of the convection and the optical tweezer effect. On the contrary, in the present experiment, the light injection from air into the droplet thermally caused swirl and convection at the boundary. The swirl and convection induced by the light injection might deposit the CNTs around the microfibers. This phenomenon occurs only at the light input side boundary, and enables us to area-selectively deposit them onto desired position. Another possible mechanism is the optical tweezer effect, which can traps micro- and nano-sized objects by the optical intensity diversion in the solution. Evanescent field had the optical intensity diversion, and mode field mismatch between the air section and the DMF section caused the optical intensity diversion by scattering. The diversion might trap the CNTs and deposited onto the microfiber.

In order to confirm the saturable absorption property of the fabricated device, we applied it to mode-lock a fiber laser. We inserted the fiber, mentioned above, in a fiber laser cavity to achieve a passive mode-locking. The experimental setup is shown in Fig. 19. An EDFA produced gain of the laser. The polarization state inside the laser cavity was controlled by a polarization controller (PC). An isolator eliminated back-reflection inside the laser cavity, mainly occurred at the tapering region and the CNT-deposited part. The laser output came out from a 10% port of a 10:90 coupler. We measured optical spectrum and second harmonic generation (SHG) autocorrelation trace of the laser output using an optical spectrum analyzer and a SHG autocorrelator. We measured pulse repetition rate using a photodiode and an oscilloscope.

Fig. 20 shows an optical spectrum, an autocorrelation trace and a pulse train of the fiber laser output. The laser had the center wavelength at 1565 nm and the 3-dB bandwidth of 3.7 nm. The full width half maximum (FWHM) of the autocorrelation trace was 1.61 ps, which correspond to inferred pulse duration of 1.14 ps (assuming Gaussian pulse profile). The time bandwidth product (TBP) of the pulse was 0.528. Comparing to an unchirped transform-limited value of Gaussian pulse (0.441), the result indicates the pulse had chirp. The chirp was due to the residual dispersion in the laser cavity. The repetition rate of the fiber laser was 1.54 MHz according to the time interval of the output pulse train.

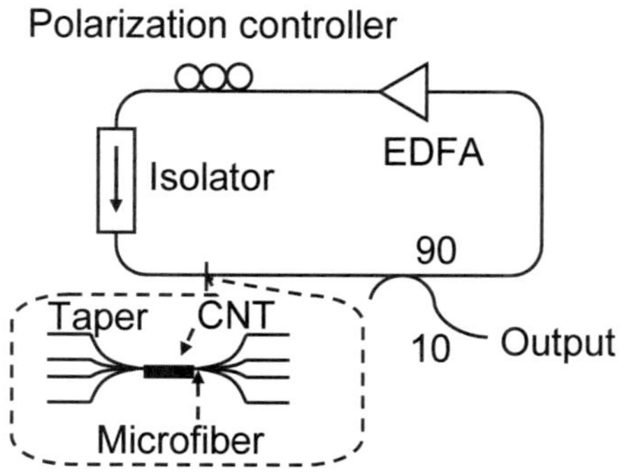

Fig. 19. Schematic of a passively mode-locked fiber laser using a CNT-deposited microfiber.

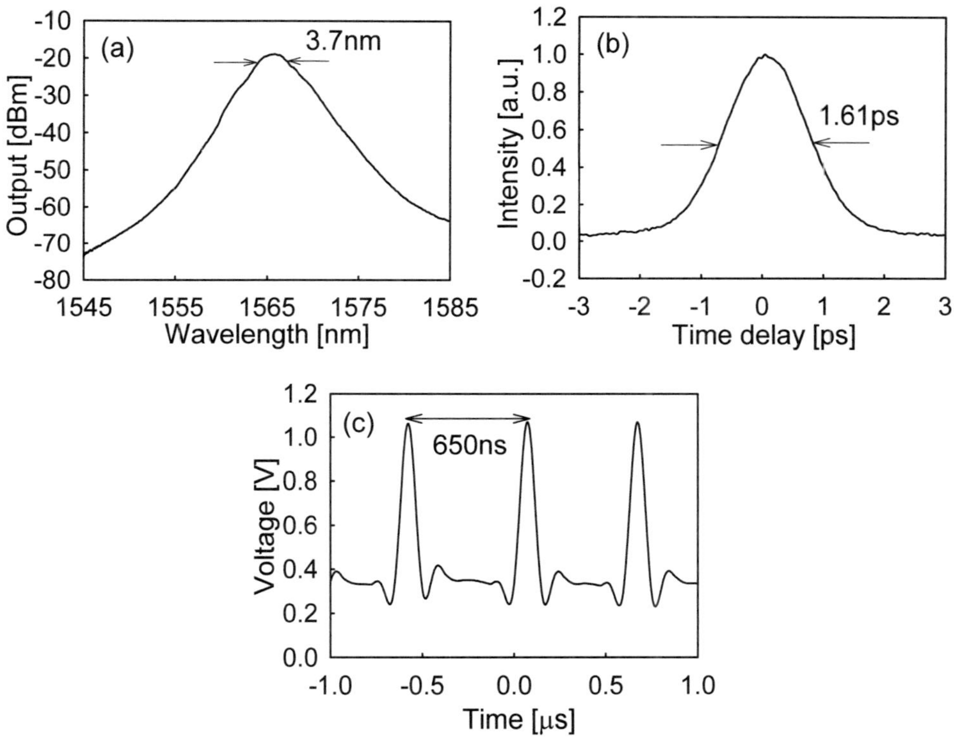

Fig. 20. Passively mode-locked fiber laser output (a) Optical spectrum of the fiber laser (resolution: 0.1 nm) (b) Autocorrelation trace of laser output (resolution: 50 fs, corresponding pulse width: 1.14 ps (assuming Gaussian pulse) (c) Output pulse train of the fiber laser (repetition rate: 1.54 MHz)

6. Conclusion

In this chapter, we have proposed and demonstrated the optical deposition method of CNTs. After brief introduction of this chapter (section 1), we explained the general characteristics, optical properties, and optical devices based on CNTs in section 2. From section 3, we reported optical deposition of CNTs. In section 3, we deposited CNTs by injected light from the optical fiber end into a CNT-dispersed solution. In subsection 3.1, we area-selectively deposited CNTs only onto the core region of optical fiber ends, such as a bare SMF and a bare DSF, by controlling the injection power. The area-selectivity was confirmed by microscopic Raman spectroscopy. A SMF with ferrule could be used as a deposition object, and it produced passive mode-locking of a fiber laser after the deposition (subsection 3.2).

The result confirmed the technique to have good performance of CNT-based device fabrication. However, the result showed that we need strict power control of the injection light within 2 dB for SMF. To solve this problem, we employed optical reflectometry into the method for in-situ monitoring of the deposition in section 4. The reflectometry offered easy optimization method for the injection power control. We deposited CNTs onto very small area (~15 μm diameter circular region) using highly uniform CNT-dispersed solution.

In section 5, we showed that the optical deposition method was also applicable to microfibers. We achieved CNT deposition around a microfiber, which were immersed in a CNT-dispersed DMF droplet, by injecting light through the microfiber. We confirmed the fiber could be used as a passive mode-locker in a fiber ring laser.

We believe our proposed method, optical deposition of CNTs, will contribute to development of ultrafast photonic applications based on CNTs in device fabrication aspect.

7. References

Bai et al., 2004. Bai, X., Li, D., Du, D., Zhang, H., Chen, L. & Liang, J. (2004), 'Laser irradiation for purification of aligned carbon nanotube films', *Carbon* 42(10), 2125 - 2127.

Bandow et al., 1998. Bandow, S., Asaka, S., Zhao, X. & Ando, Y. (1998), 'Purification and magnetic properties of carbon nanotubes', *Applied Physics A: Materials Science & Processing* 67(1), 23 – 27.

Chen et al., 2002. Chen, Y.-C., Raravikar, N., Schadler, L., Ajayan, P., Zhao, Y.-P., Lu, T.-M., Wang, G.-C. & Zhang, X.-C. (2002), 'Ultrafast optical switching properties of single-wall carbon nanotube polymer composites at 1.55 μm', *Applied Physics Letters* 81(6), 975–977.

Chiang et al., 2001. Chiang, I., Brinson, B., Huang, A., Willis, P., Bronikowski, M., Margrave, J., Smalley, R. & Hauge, R. (2001), 'Purification and characterization of single-wall carbon nanotubes (swnts) obtained from the gas-phase decomposition of co (hipco process)', *Journal of Physical Chemistry B* 105(35), 8297 - 8301.

Colomer et al., 1999. Colomer, J.-F., Piedigrosso, P., Fonseca, A. & Nagy, J. (1999), 'Different purification methods of carbon nanotubes produced by catalytic synthesis', *Synthetic Metals* 103(1-3 pt 3), 2482 - 2483.

Dai et al., 1996. Dai, H., Rinzler, A., Nikolaev, P., Thess, A., Colbert, D. & Smalley, R. (1996), 'Single-wall nanotubes produced by metal-catalyzed disproportionation of carbon monoxide', *Chemical Physics Letters* 260(3-4), 471 - 475.

Guo et al., 1995. Guo, T., Nikolaev, P., Thess, A., Colbert, D. & Smalley, R. (1995), 'Catalytic growth of single-walled nanotubes by laser vaporization', *Chemical Physics Letters* 243(1-2), 49 – 54.

Ichida et al., 2002. Ichida, M., Hamanaka, Y., Kataura, H., Achiba, Y. & Nakamura, A. (2002), 'Ultrafast relaxation dynamics of photoexcited states in semiconducting single-walled carbon nanotubes', *Physica B: Condensed Matter* 323(1-4), 237 - 238.

Iijima, 1991. Iijima, S. (1991), 'Helical microtubules of graphitic carbon', *Nature* 354(6348), 56–58.

Journet et al., 1997. Journet, C., Maser, W. K., Bernier, P., Loiseau, A., de la Chapelle, M. L., Lefrant, S., Deniard, P., Lee, R. & Fischer, J. E. (1997), 'Large-scale production of single-walled carbon nanotubes by the electric-arc technique', *Nature* 388(6644), 756–758.

Kieu & Mansuripur, 2007. Kieu, K. & Mansuripur, M. (2007), 'Femtosecond laser pulse generation with a fiber taper embedded in carbon nanotube/polymer composite', *Opt. Lett.* 32(15), 2242–2244.

Kroto et al., 1985. Kroto, H., Heath, J., O'Brien, S., Curl, R. & Smalley, R. (1985), 'C_{60}: Buckminsterfullerene', *Nature* 318(6042), 162–163.

Margulis & Gaiduk, 2001. Margulis, V. & Gaiduk, E. (2001), 'Nature of near-infrared absorption in single-wall carbon nanotubes', *Physics Letters A* 281(1), 52 - 58.

Morishita & Takarada, 1999. Morishita, K. & Takarada, T. (1999), 'Scanning electron microscope observation of the purification behaviour of carbon nanotubes', *Journal of Materials Science* 34(6), 1169 - 1174.

Nikolaev et al., 1999. Nikolaev, P., Bronikowski, M., Bradley, K., Rohmund, F., Colbert, D., Smith, K. & Smalley, R. (1999), 'Gas-phase catalytic growth of single-walled carbon nanotubes from carbon monoxide', *Chemical Physics Letters* 313(1-2), 91 - 97.

Park et al., 2006. Park, T.-J., Banerjee, S., Hemraj-Benny, T. & Wong, S. S. (2006), 'Purification strategies and purity visualization techniques for single-walled carbon nanotubes', *Journal of Materials Chemistry* 16(2), 141 - 154.

Rubtsov et al., 2004. Rubtsov, I., Russo, R., Albers, T., Deria, P., Luzzi, D. & Therien, M. (2004), 'Visible and near-infrared excited-state dynamics of single-walled carbon nanotubes', *Applied Physics A: Materials Science and Processing* 79(7), 1747 - 1751.

Saito et al., 1992. Saito, R., Fujita, M., Dresselhaus, G. & Dresselhaus, M. (1992), 'Electronic structure of chiral graphene tubules', *Applied Physics Letters* 60(18), 2204 - 2206.

Sakakibara et al., 2005. Sakakibara, Y., Rozhin, A. G., Kataura, H., Achiba, Y. & Tokumoto, M. (2005), 'Carbon nanotube-poly(vinylalcohol) nanocomposite film devices: Applications for femtosecond fiber laser mode lockers and optical amplifier noise suppressors', *Japanese Journal of Applied Physics, Part 1: Regular Papers and Short Notes and Review Papers* 44(4 A), 1621 - 1625.

Sakakibara et al., 2003. Sakakibara, Y., Tatsuura, S., Kataura, H., Tokumoto, M. & Achiba, Y. (2003), 'Near-infrared saturable absorption of single-wall carbon nanotubes prepared by laser ablation method', *Japanese Journal of Applied Physics, Part 2: Letters* 42(5 A), 494 - 496.

Satishkumar et al., 1998. Satishkumar, B., Govindaraj, A., Sen, R. & Rao, C. (1998), 'Single-walled nanotubes by the pyrolysis of acetylene-organometallic mixtures', *Chemical Physics Letters* 293(1-2), 47 - 52.

Set et al., 2004*a*. Set, S. Y., Yaguchi, H., Tanaka, Y. & Jablonski, M. (2004*a*), 'Laser mode locking using a saturable absorber incorporating carbon nanotubes', *Journal of Lightwave Technology* 22(1), 51 - 56.

Set et al., 2004*b*. Set, S. Y., Yaguchi, H., Tanaka, Y. & Jablonski, M. (2004*b*), 'Ultrafast fiber pulsed lasers incorporating carbon nanotubes', *IEEE Journal on Selected Topics in Quantum Electronics* 10(1), 137 - 146.

Song et al., 2007. Song, Y.-W., Morimune, K., Set, S. Y. & Yamashita, S. (2007), 'Polarization insensitive all-fiber mode-lockers functioned by carbon nanotubes deposited onto tapered fibers', *Applied Physics Letters* 90(2), 021101.

Yamashita et al., 2004. Yamashita, S., Inoue, Y., Maruyama, S., Murakami, Y., Yaguchi, H., Jablonski, M. & Set, S. (2004), 'Saturable absorbers incorporating carbon nanotubes directly synthesized onto substrates and fibers and their application to mode-locked fiber lasers', *Optics Letters* 29(14), 1581 - 1583.

20

High Power Tunable Tm^{3+}-fiber Lasers and Its Application in Pumping Cr^{2+}:ZnSe Lasers

Yulong Tang and Jianqiu Xu
Shanghai Institute of Optics and Fine Mechanics
Chinese Academy of Science, Shanghai 201800, China

1. Introduction

1.1 Research background

Interest in the Tm^{3+}-doped fiber laser originates from its emission band in the range of 1400-2700nm lying between the bands of Er^{3+} and Nd^{3+} ions. Since the advent of double-cladding configuration of fiber and the recent technological development of high-power laser diodes, output power (performance) of Tm^{3+}-doped fiber lasers has scaled exponentially. Up to date, the maximum output achieved in Tm^{3+}-doped fiber lasers has been comparable with that from Yb^{3+}-doped fiber lasers. Laser beam in the 2~3μm wavelength range has wide applications. First, it is a good candidate in laser microsurgery due to high absorption of water in this spectral region thus can provide high-quality laser tissue cutting and welding. In addition, this wavelength-range laser has potential applications in environment monitoring, LIDAR, optical-parametric-oscillation (OPO) pump sources, and so on [1-4].

For obtaining laser emission in the mid-infrared wavelength region, the Tm^{3+}-doped fiber is an excellent candidate due to several unique advantages it possesses. First, the Tm^{3+}-doped fiber has a strong absorption spectrum that has good overlap with the emission band of commercially available AlGaAs laser diodes, which have been significantly developed and are being developed with an unprecedented speed. Second, the specific energy-level structure of Tm^{3+} ions provides the Tm^{3+}-doped fiber laser a special advantageous energy-transfer process-the $^3H_4+^3H_6\rightarrow^3F_4+^3F_4$ cross relaxation process. In this process, two excited-state ions can be obtained with depletion of just one absorbed pump photon. With an appropriately high doping level, the cross relaxation process can offer a quantum efficiency close to two, which greatly improves the efficiency of the Tm^{3+}-doped fiber laser. Thirdly, the Tm^{3+}-doped fiber has a very broad emission band, spanning over more than 400 nm. This feature offers the Tm^{3+}-doped fiber laser an especially high-degree wavelength tunability, which is very useful in applications such as spectroscopy, atmospheric sensing and so on.

1.2 Host material of Tm^{3+}-doped fiber

For laser ions, the combination of the energy gaps between the excited level and the one just below it and the maximum phonon energy (MPE) plays an important role in the nonradiative relaxation rate, which in turn has a significant influence on the laser efficiency originated from the excited states. For Tm^{3+} ions, different host materials show a great difference in the MPE value. Two most common host materials used for Tm^{3+} fibers are silica and fluoride glass. Their MPE differs about several times, being 1100 cm^{-1} (silicate) and 550 cm^{-1} (fluorides), respectively [5]. Large MPE of the silicate glass fiber limits its infrared transparency range less than 2.2 μm and improves its multiphonon relaxation rates. Therefore, fluoride fibers are preferred as the host material for Tm^{3+} ions to achieve comparatively longer-wavelength emission.

In Tm^{3+}-doped fiber lasers, the $^3F_4 \rightarrow ^3H_6$ transition is usually exploited to achieve the ~2 μm high-power laser output. This transition can produce a very wide emission band, providing a broad tuning range for lasers and a wide optical bandwidth for amplifiers. However, the relaxation of the $^3F_4$ level is predominantly nonradiative. The measured lifetime of the $^3F_4$ level for Tm^{3+} doped silica fiber is just 0.2 ms [6] showing a high nonradiative rate thus low quantum efficiency. Therefore, Tm^{3+}-doped fiber lasers usually have high laser thresholds. On the contrary, Tm^{3+}-doped fluoride fiber lasers have comparatively lower thresholds due to a low MPE. The high nonradiative rate, however, does not impair laser slope efficiency, because stimulated emission will dominate nonradiative relaxation once the laser has been risen above threshold. Due to high damage threshold and the very effective modified chemical vapor deposition (MCVD) technique for fiber fabrication, Tm^{3+} silica fibers are usually chosen to construct high-power 2-μm fiber lasers.

1.3 Fiber design for cladding pumping

Tm^{3+}-doped fibers can be either core pumped or cladding pumped. In the past, the fiber laser was usually core pumped. The fiber core areas are generally <100 μm^2, which limit the power scalability because this method depend on expensive high-beam-quality pump sources. Since the invention of double cladding fiber configuration with a larger cladding area >10000 μm^2, together with a high numerical aperture (NA) of 0.3-0.55, output power of Tm^{3+}-doped fiber lasers can be greatly improved by use of high-power diode-arrays as pumping sources.

In the design of fibers for cladding pumping, the core of the fiber is usually made small (such as less than 5 microns) to guide a single-transverse mode (LP_{01}). The cladding generally has a much larger cross section (several-hundred-micron diameter) for high-power launching, and the shape of the cladding can be flexible with novel consideration. The shape of the inner cladding of the fiber has a great impact on the absorption efficiency of launched pump light. In the past, the inner cladding is used to be circularly symmetric, which can be drawn with ease and is compatible with the pig-tail fiber of the pump LD. However, the circular symmetry will make large portion of the pump light to be skew light, greatly reducing the absorption efficiency of doping ions. In order to improve the utility efficiency of pump sources, and take the pump light shape into account (compatible with the inner cladding shape), various double cladding fiber structures are invented, as shown in Fig. 1. By using these double cladding fibers, the pump efficiency is significantly enhanced.

1.4 Tm^{3+}-doped fiber laser cavity structure

The fiber laser resonator can be very simple when it is in free-running regime. The commonly adopted three kinds of fiber laser resonators, defined as Fabry-Perot resonator, are shown in Fig. 2. Fig 2 (a) is the simplest cavity, the pump light is launched into the fiber through a dichroic mirror, which is high reflective for laser light. Laser oscillation forms between this dichroic mirror and the distal-end fiber facet (~4% Fresnel reflection). In this cavity configuration, addition of output coupler or pump-light reflector will improve the

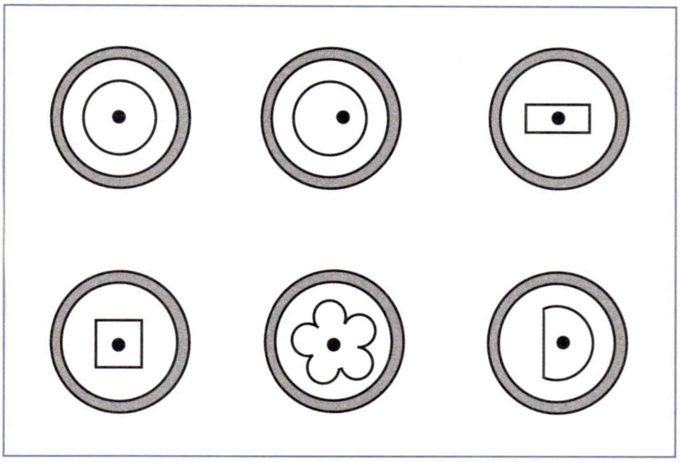

Fig. 1. Various shape of the double cladding fiber cross section

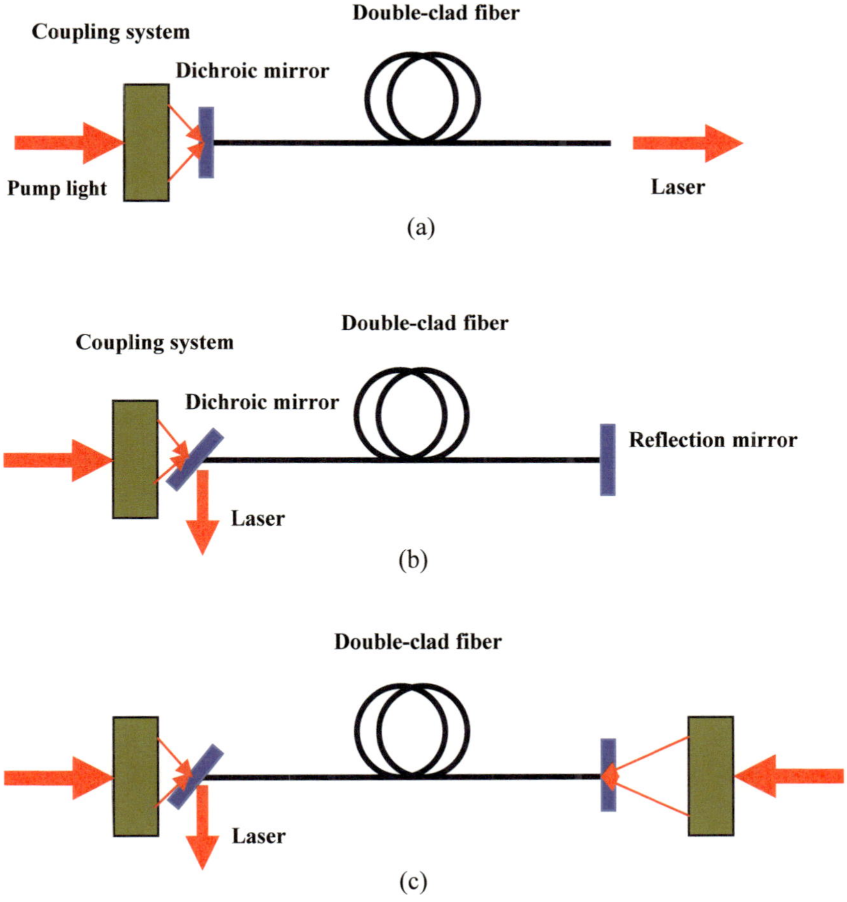

Fig. 2. Tm^{3+}-doped fiber laser resonators (a) Single-end forward pumping; (b) Single-end backward pumping; (c) Double-end pumping.

optical efficiency of the fiber laser. In Fig 2 (b), the pump light is launched into the fiber from the output end, and the dichroic mirror is set at 45 degree with respect to the fiber axis for extracting laser output. At the distal fiber end, a signal high reflection mirror is added to form the laser resonator. The cavity of Fig 2 (c) is used to further improve the pump power that can be launched into the fiber for the aim of power scaling.

In order to realize wavelength tuning, and at the same time obtain narrow-width laser spectrum, a bulk gratings can be put in the cavity as the output coupler. However, wavelength-tuning with bulk grating is inconvenient, and brings laser instability. The commonly used wavelength-tuning fiber laser resonator is constructed by using fiber Bragg gratings as the feedback device and output coupler, as shown in Fig. 3. This kind of fiber laser resonator can provide not only wavelength tuning, but also narrow spectral width and high stability.

Fig. 3. Complete-fiber resonator with fiber Bragg grating.

1.5 Spectral characteristics of Tm^{3+}-doped silica fiber

Our aim is to focus on the development of high-power ~2μm fiber laser, so the description about the spectral characteristics is only confined to the Tm^{3+}-doped silica fiber. The absorption spectrum of Tm^{3+}-doped silica fiber is shown in Fig. 4 [7]. This spectrum has strong absorption near 790 nm, which has good overlap with the emission band of present fully developed AlGaAs diode lasers. This feature of the Tm^{3+}-doped fiber laser makes the pump process comparatively easier and less expensive, offering an exciting potential for power scaling in the 2-μm wavelength range.

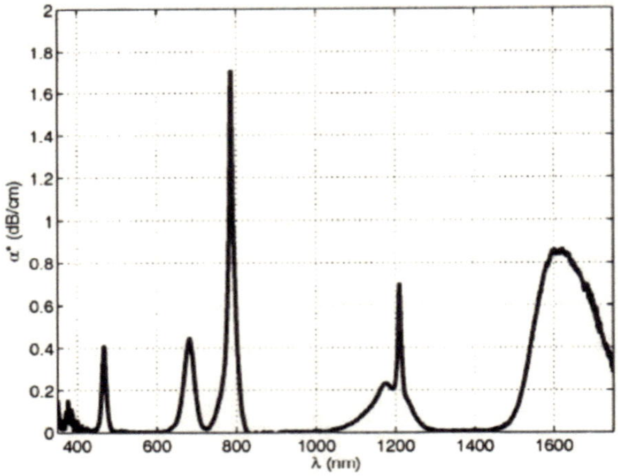

Fig. 4. Absorption spectrum of Tm^{3+}-doped silica fiber

The simplified energy-level diagram of Tm^{3+} ions is shown in Fig. 5. The pump light at ~790 nm excites Tm^{3+} ions from $^3H_6$ to $^3H_4$, which then nonradiatively decay to the upper laser level of $^3F_4$ with a fluorescence lifetime of 0.55 ms [8]. The transition from $^3F_4 \rightarrow ^3H_6$ will radiates photons at wavelength of ~2μm. Due to large Stark splitting of the lower laser level (the ground state level), the Tm^{3+}-doped fiber laser is a quasi-four-level system.

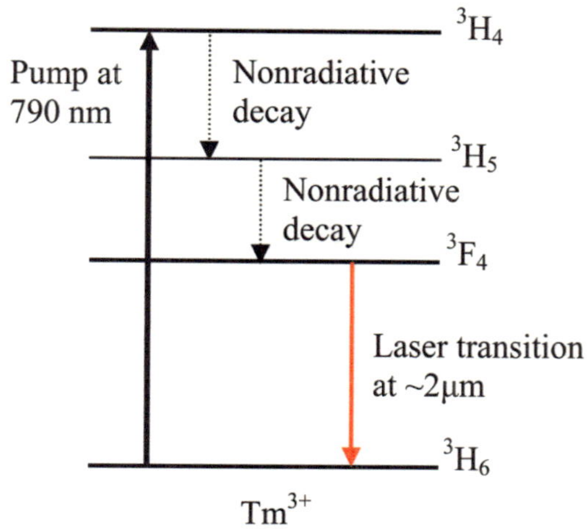

Fig. 5. The simplified energy-level diagram of Tm^{3+} ions

The special energy-level structure of Tm^{3+} ions provides the Tm^{3+}-doped silica fiber laser the advantageous **cross relaxation** process ($^3H_4+^3H_6 \rightarrow ^3F_4+^3F_4$). With this process, high quantum efficiencies can be achieved in Tm^{3+}-doped fiber lasers by appropriately improving the ion doping concentration. The cross relaxation process, as shown in Fig 6, can produce >100% quantum efficiencies for Tm^{3+}-doped fiber lasers [9-10]. In experiment, slope efficiency larger than 60% has been achieved (quantum efficiency>150%) in Tm^{3+}-doped fiber lasers [11-12].

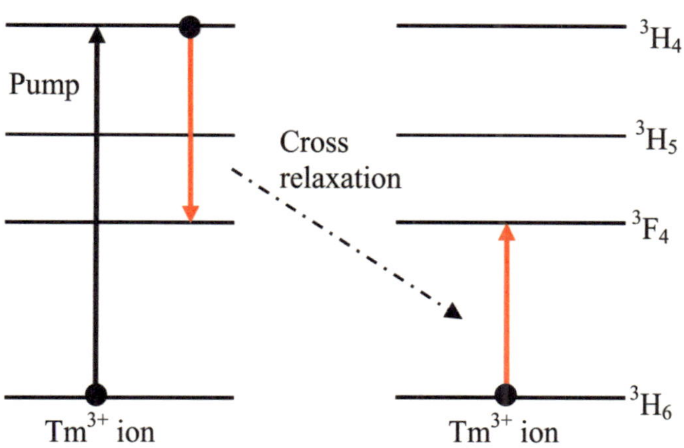

Fig. 6. Cross relaxation between Tm^{3+} ions.

In Tm^{3+}-doped silica fiber lasers, high-degree Stark splitting of the $^3H_6$ level by local electric field produces a broad emission band (>400 nm), as shown in Fig. 7 [13]. This special broad emission band makes Tm^{3+}-doped fibers very suitable for wavelength tuning.

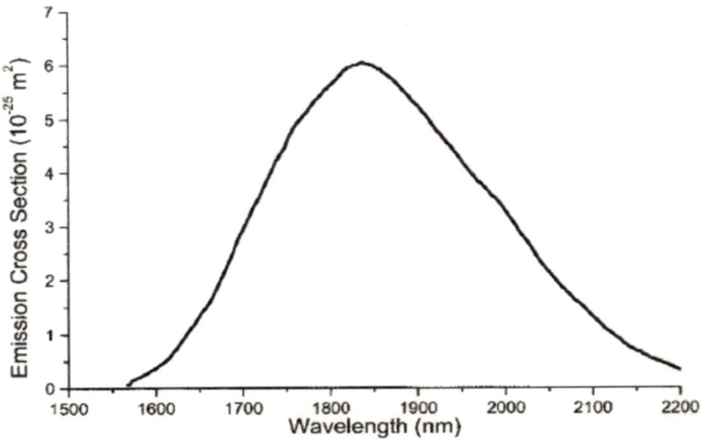

Fig. 7. The emission spectrum of the $^3F_4 \rightarrow {}^3H_6$ transition in Tm^{3+}-doped silica fiber

1.6 Pump sources

With the fully development and decreasing cost of AlGaAs diode lasers, diode lasers are the primary and ultra most preferred pump source in pumping ~2μm Tm^{3+}-doped fiber lasers at present. The 790-nm diode lasers pump Tm^{3+} ions to the $^3H_4$ level, which can lead to the advantageous cross relaxation process, greatly enhancing the quantum efficiency of Tm^{3+}-doped fiber lasers. From the absorption band of Tm^{3+} ions, light sources at other wavelength can also be used as the pump. As shown in Fig. 8, apart from 790nm, the pump sources include the 1.064μm and 1.319μm output from Nd:YAG lasers, 1.09 μm from Yb-doped fiber lasers, and 1.57μm Er^{3+}-doped fiber lasers.

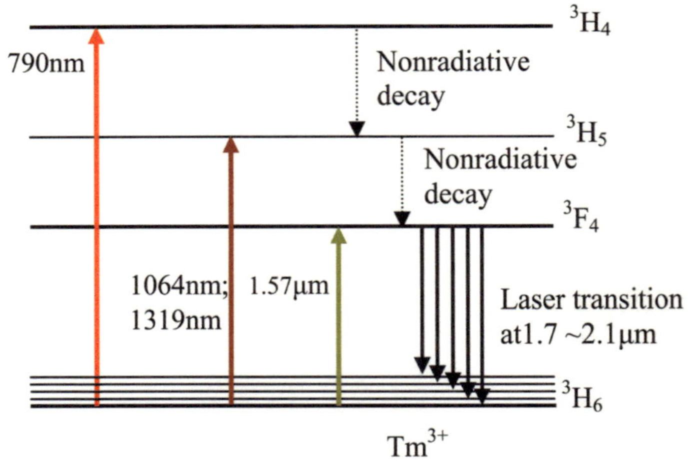

Fig. 8. The schematic of Tm^{3+} silica fiber with different pump wavelength.

The advantage of the 1.064 μm and 1.09 μm pumping is that Nd:YAG lasers and Yb-doped fiber lasers have been fully developed, which can be obtained with ease and low cost. This wavelength is used for the $^3H_5$-band pumping. Due to strong ESA, this band pumping leads to slope efficiencies less than 30% [14-15].

When the 1.319μm Nd:YAG laser is used as the pump source for Tm^{3+}-doped fiber lasers, the Tm^{3+} ions are pumped to the $^3H_5$ level. The small absorption cross section of the pump light limits the slope efficiency of the 1.319μm-pumped Tm^{3+}-doped fiber laser. Besides, at this wavelength, excited-state absorption of the pump is strong. Therefore, the slope efficiency of Tm^{3+}-doped fiber lasers with this kind of pump source is less than 25% [16].

When the Tm^{3+}-doped fiber laser is pumped at 1.57μm wavelength, the system is defined as a resonant pump structure due to the pump level ($^3F_4$) is also the upper laser level. Such a pump scheme has a small Stokes shift, thus can provide high operation efficiencies for Tm^{3+}-doped fiber lasers. With this pump technique, M. Meleshkevich reported a slope efficiency of 60% with respect to absorbed pump power [17]. However, this pump scheme cannot make use of the advantageous "cross relaxation" as the 790-nm pumping. Therefore, the highest efficiency is also less than that has been achieved in 790-nm-pumped Tm^{3+}-doped fiber lasers.

1.7 Recent power scaling of CW Tm^{3+}-doped fiber laser

The first ~2μm laser output from the $^3F_4 \rightarrow {}^3H_6$ transition in Tm^{3+}-doped silica fiber laser was achieved by Hanna et al in 1988 with a 797-nm dye laser as the pump source [18]. With the advent of AlGaAs diode lasers, diode-pumped Tm^{3+}-doped silica [19] and fluoride [20] fiber lasers were both realized in 1990.

The first high-power double-clad Tm^{3+}-doped silica fiber laser was reported by Jackson in 1998. The maximum output power is 5.4 W, and the slope efficiency is 31% [21]. In 2000, the output power was improved to 14 W with a slope efficiency of 46% [22]. In 2002, the 793-nm laser diode pumped Tm^{3+}-doped fiber laser has achieved a power output of 85 W with the slope efficiency of 56%. At the same power level, 75 W 2μm laser output was realized in an Yb-Tm codoped fiber laser, where ytterbium ions are used as sensitizing ions to facilitate pumping with 976 nm diodes. The slope efficiency however is just 32% due to the lack of "cross relaxation" process compared with those directly 790-nm pumped Tm^{3+}-doped fiber lasers [24]. In 2007, G. Frith reported a Tm^{3+}-doped fiber laser with output power improved to 263 W and the slope efficiency of 59% [25]. In 2009, the 2 μm output power from the Tm^{3+}-doped fiber laser was significantly enhanced to 885 W with a slope efficiency of 49.2% [26]. In addition, single frequency output has also been over 600 watts [27]. Further improving the 2-μm laser output over 1-kilowatt level, comparable with Yb-doped fiber lasers, is just a matter of time.

2. Short-length Tm^{3+}-doped silica fiber lasers

2.1 Introduction

Compared with long fiber lasers, short fiber lasers have many unique advantages. First, short fiber lasers can provide narrow-linewidth laser emission due to large axial mode spacing. From the equation [28]

$$\Delta \nu = \frac{c}{2nL}, \tag{1}$$

the shorter the fiber length, the larger the axial mode spacing. In a same broad gain spectrum, short fiber length leads to fewer axial modes obtaining enough gain to overcome cavity loss achieving oscillation. Therefore, the laser spectrum will be much narrower with shorter fiber length. Second, in pulsed laser operation, short fiber lasers are preferable to produce short pulse-width emission with high peak power. The laser pulse width can be expressed by [29]

$$\tau_p = \frac{r\eta(r)}{r-1-\ln r} \times \frac{2L}{c\delta_0}. \tag{2}$$

where, δ_0 is the single-pass cavity loss, $\eta(r)$ is the energy extraction efficiency, r is the pump strength, L is the cavity optical length. We can see that the laser pulse width is proportional to the fiber length. Therefore, short fiber laser is preferred to achieved short pulse duration, thus achieve high peak power. In addition, short fiber lasers are free from the undesired nonlinear effects such as stimulated Raman, Brillouin scattering and so on.

Short-length Tm^{3+}-doped fiber lasers can be used to achieve single frequency output [7, 30-31], as well as moderate-level laser output [32] in the mid-infrared region. Such kind of compact mid-infrared laser device has great potentials in communication and integrated optical systems. Power scaling of short Tm^{3+}-doped fiber lasers will improve the utility of 2μm fiber lasers and speed the development of compact fiber laser systems.

In this section, short thulium-doped fiber lasers with high-power output are introduced. Besides, the performance dependence of such kind of laser on temperature and cavity parameters are discussed.

2.2 Experiment and CW output

The double-cladding thulium-doped fiber has a 30 μm diameter, 0.22 N.A. core doped with ~2wt.% Tm^{3+}. The pure-silica cladding, coated with a low-index polymer, has a ~410 μm diameter and a N.A. of 0.46. The cladding absorption coefficient at the pump wavelength (790 nm) is ~7dB/m. These fiber specifications of large fiber cross section, large ratio of core diameter to cladding diameter, high N. A. and high doping concentration make it possible to achieve high output power from short length of fibers. The pump source for the experiment is a single high-power LD bar, operating TM mode centered at 790 nm, shifting to ~793 nm at comparatively higher operating temperature. With this pump source, the maximum power launched into the fiber is near 12 W.

The laser cavity configuration is shown in Fig. 9 [33]. The laser pumping beam is reshaped first by a micro-prism stack, and then focused into a circular spot of ~0.5×0.5 mm diameter with a cylindrical lens and an aspheric lens. The focused pump beam is launched into the thulium-doped fiber through a dielectric mirror. The pump end of the fiber is butted directly to the dielectric mirror with high reflectivity (>99%) at 1850~2100 nm and high transmission (>97%) at 760~900 nm. The distal end of the fiber is directly butted to the output coupler with ~90% reflectivity at 1800~2000 nm to build a Fabry-Perot laser resonator. During the course of the experiments, the fiber is clipped between two copper sheets, which are closely fixed onto a water-cooling heat sink. The combination of a dielectric mirror (R>99%@1850~2100nm, T>99%@7 90 nm) and a Ge filter is used to extract the laser light and block the unabsorbed pump light.

Figure 10 shows the output power of a 7-cm length fiber laser as a function of launched pump power without water-cooling (40°C) [33]. When the cavity is constructed with an output coupler of 10% transmission, the laser has a threshold pump power of 135 mW and a maximum output power of 1.09 W. When the pump power is over 10 W, rollover of the output power occurred, which probably stemmed from large amount of heat generated in the fiber due to large quantum defect. The slope efficiency of the fiber laser is about 9.6% with respect to launched pump power, which is much higher than that of silica based DFB fiber lasers [7, 32], but lower than that from a thulium-doped germinate glass fiber laser [31]. The comparatively low slope efficiency can be accounted by poor pump absorption due to relatively low doping concentration and short fiber length. Another alternative explanation can be related to the background loss of the silica-based fiber at wavelengths longer than 1.8 μm [34]. When the output coupler mirror is removed and only the 4% Fresnel reflection of the fiber end is used to complete the laser cavity, the maximum output power and slope efficiency decrease to 0.96 W and 8.7%, respectively. However, below 2.3 W of pump power, the fiber-end output coupler leads to slightly higher output power. Different output coupling seems to have little influence on the threshold pump power, which remains around 135 mW.

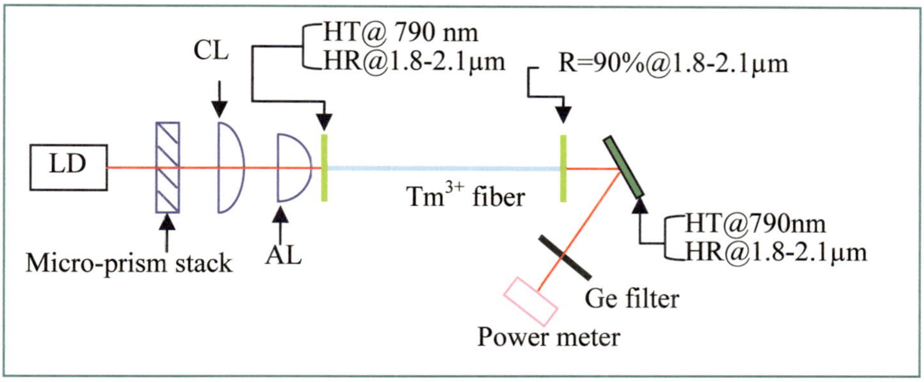

Fig. 9. Schematic diagram of the experimental setup. HT: high transmission; HR: high reflection; CL: cylindrical lens; AL: aspheric lens.

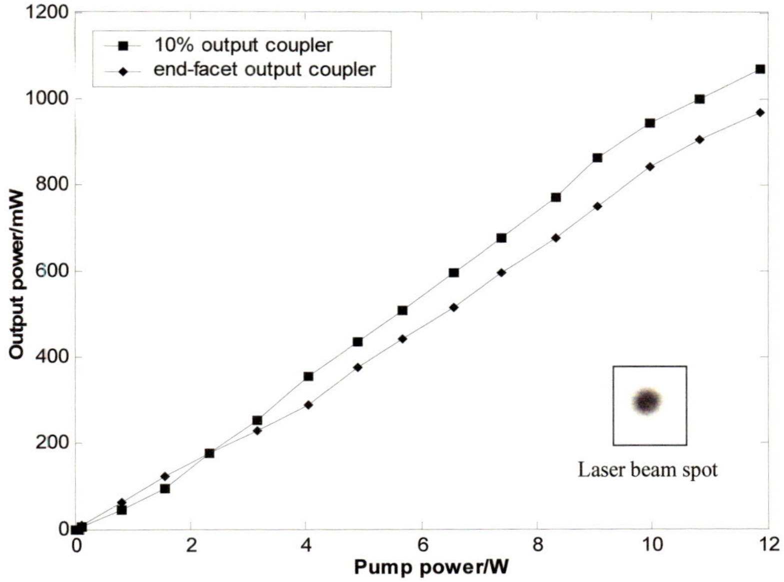

Fig. 10. Output characteristics of a 7 cm-length fiber laser with 10% output coupler and fiber-end-face coupler (R~4%). Inset is the laser beam spot.

In order to find out the influence of the spacing between the rear fiber end and the output coupler, a 6-cm-long fiber is used. The air spacing is denoted by L_S. This fiber laser is still operated with just air convectional cooling (40°C). The output characteristics are shown in Fig. 11 [33]. When L_s =0, a maximum output power of 496 mW and a slope efficiency of 4.6% can be obtained. When L_s =0.5 and 1 mm, the laser operation efficiency decreases greatly: the maximum output power reduces to 328 mW, and the slope efficiency decreases to 2.95%. This can be accounted by large reduction of the cavity Q factor by strong fiber-end diffraction loss. As far as L_s =0.5 mm and L_s =1 mm are compared, no significant different performance can be observed.

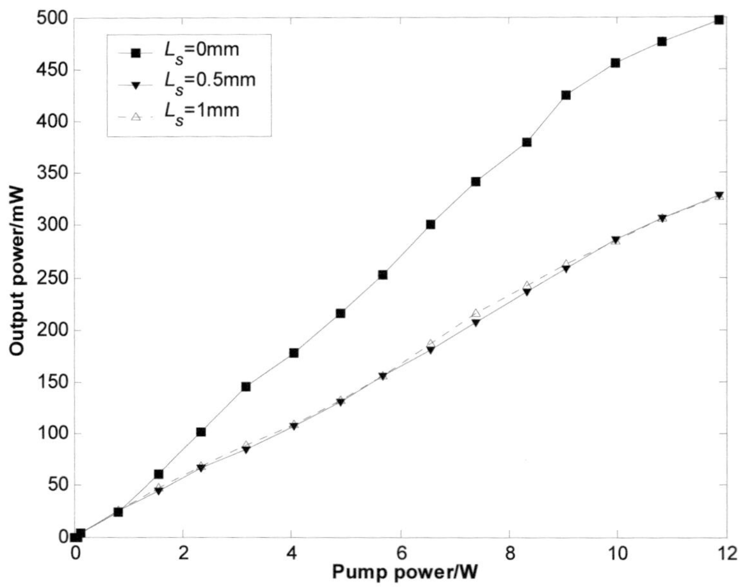

Fig. 11. Output characteristics of a 6 cm-length fiber laser with different air spacing between distal fiber end and output mirror.

2.3 Influence of temperature

In Tm^{3+}-doped fiber lasers, the ground-state level is Stark splitted into many sub levels. The Tm^{3+}-doped fiber laser is a quasi-four-level system, so the operation temperature has an important impact on the laser efficiency. For simplicity, we assume the ground state level of the Tm^{3+} ion is spited into just two sub levels N_1 and N_2. N_1 is the level where pump absorption starts from, and N_2 is the lower laser level. Between these two Stark levels, the population follows the Boltzman distribution [35]

$$\frac{N_2}{N_1} = e^{-\frac{E_2 - E_1}{kT}}, \tag{3}$$

where, E_1 and E_2 are energies of these two levels, k is the Boltzman constant and T is temperature. From the above expression, different temperatures will lead to different population distributions between N_1 and N_2, which in turn has an influence on the population inversion for the laser transition. Therefore, the laser operation temperature will have an important impact on laser efficiency.

A 5.5-cm long Tm^{3+}-doped fiber laser is constructed to observe the impact of the fiber operation temperature on the laser efficiency. The operation temperature of the fiber is controlled by changing the temperature of the circulating water through the copper heat-sink. Without water-cooling, the operation temperature of the fiber is around 40°C. With water-cooling, the operation temperature can be adjusted from ~0°C to 40°C. The output characteristic of the 5.5-cm-lenght fiber laser with different operation temperatures is plotted in Fig. 12 [33]. Both the maximum output power and slope efficiency exhibit a notable increase as the operation temperature of the fiber laser decreases from 40°C to 10°C. The maximum output grows from 345 mW to 531 mW, and the slope efficiency rises from 4% to 5.8%. The threshold pump power drops to 85 mW at operation temperature of 10°C. At high pump levels, a rollover of the output can be observed for all operation temperatures, which is accounted by saturation of pump absorption and distortion of the fiber end by high light intensity. At the maximum output level, the laser output fluctuation is less than 5% (as shown in the inset of Fig. 12), showing high stability of short-length Tm^{3+}-doped fiber lasers.

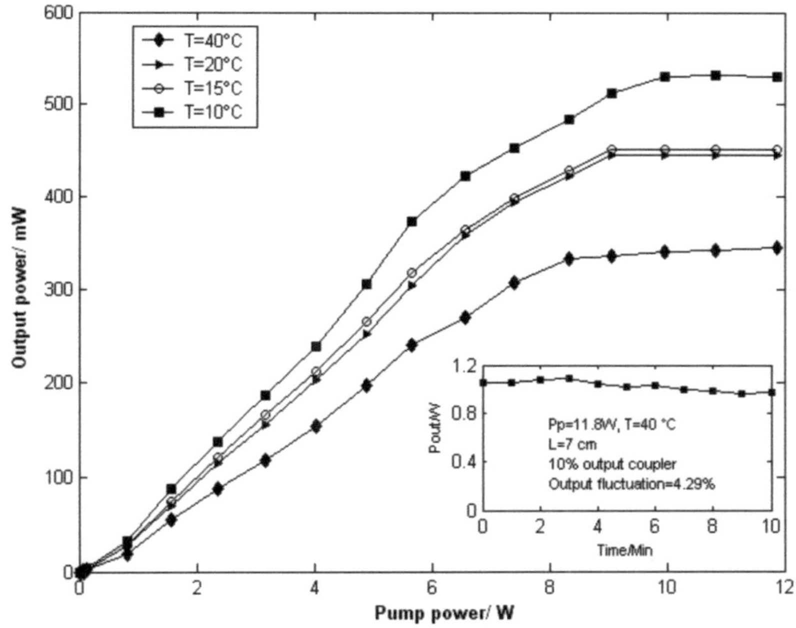

Fig. 12. Output characteristics of a 5.5 cm-length fiber laser with different operation temperatures; inset is the output power versus time (minute) at constant pump (11.8W) and operation temperature (40°C).

2.4 The laser spectra

Fig. 13 is the laser spectrum of the 5cm-length Tm^{3+}-doped fiber laser measured at pump power of 10 W (output of ~500mW). It is clear that the short fiber laser operates in few longitudinal modes. This is because that the fiber length is small, leading to a large longitudinal mode spacing, only few modes can obtain enough gain to initiate laser oscillation. The main spectral peak situates at 1970 nm, with a FWHM of ~3 nm. This spectral band width is much narrower than that obtained in the longer fiber laser [23], whose laser band width is tens of nm. Therefore, it is much easier to achieve single-frequency laser output with short-length Tm^{3+}-doped fibers. The existence of another small spectral peak indicates the mode competition in the short Tm^{3+}-doped fiber laser.

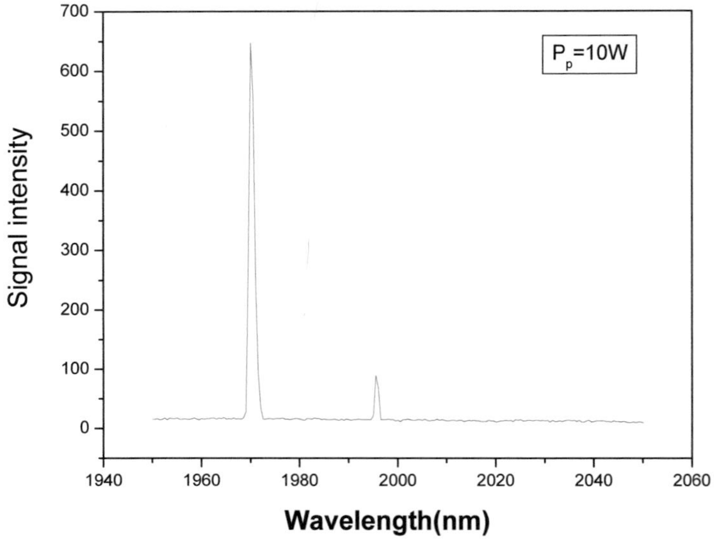

Fig. 13. Laser spectrum of the 5cm-length Tm^{3+}-doped fiber laser

The laser spectra of this short fiber laser under different pump power levels are shown in Fig. 14. When the pump power is 0.8 W, the laser only shows one spectral peak. The single-spectral-peak operation preserves with improvement of the pump to 2 W. When the pump is increased to 5W, laser operation changes from single spectral peak to two spectral peaks. At high pump level, mode hopping occurs, e.g. the main spectral peak changes to 1997 nm at the pump power of 8 W. The mode hopping phenomenon is probably due to changes of the cavity configuration parameters induced by the variation of the circumstance (such as fiber temperature or deformation of the cavity mirror).

3. High power ~2 μm Tm^{3+}-doped silica fiber laser and its wavelength tunability

The Tm^{3+}-doped fiber laser pumped by diode lasers has a great potential to scaling its output comparable to or even over the Yb^{3+}-doped fiber laser. This is because that Tm^{3+}-doped fiber can be pumped at around 790 nm, where efficient diode lasers are readily available. Another reason is that the Tm^{3+}-doped fiber laser can benefit from the cross-relaxation process, obtaining two excited Tm^{3+} ions for one pump photon. Therefore, the Tm^{3+}-doped fiber laser can provides, in theory, a maximum efficiency of 82% (a maximum

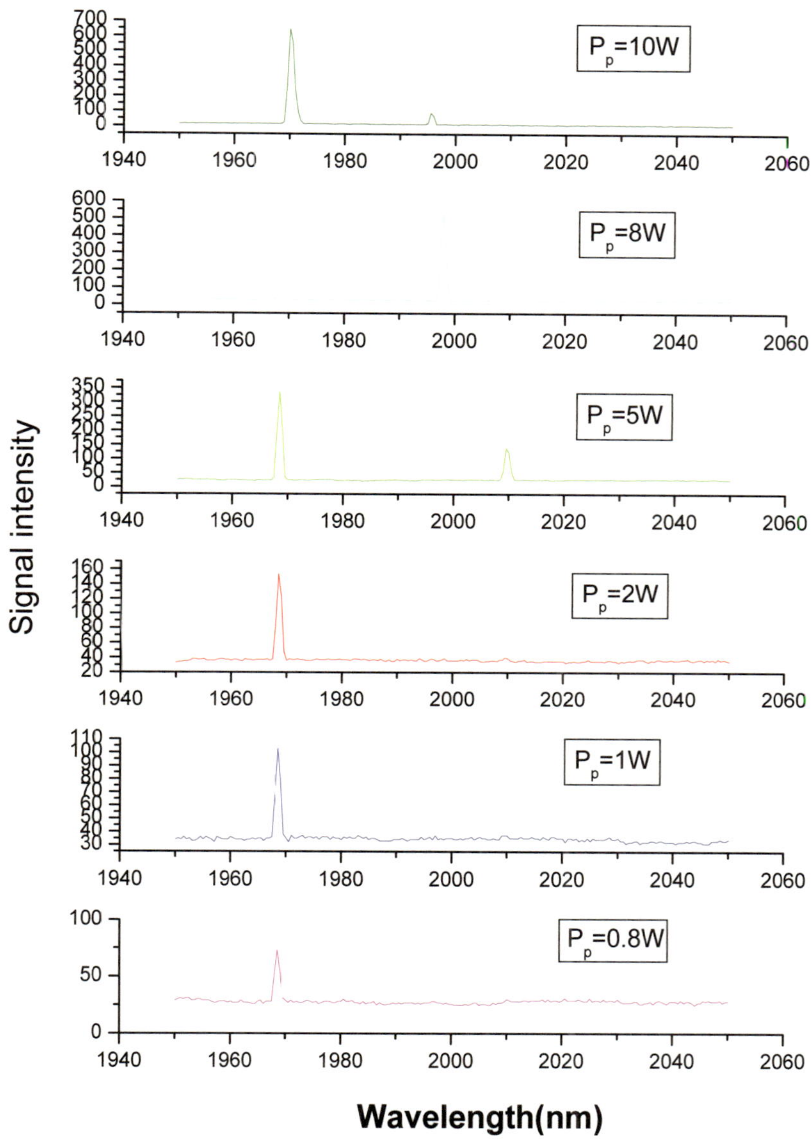

Fig. 14. Laser spectrum of the 5cm-length Tm^{3+}-doped fiber laser at different pump levels

quantum efficiency of 200%). Since 1998, the output power of double-clad Tm^{3+}-doped fiber laser has scaled steadily, especially when Al was codoped into the fiber to improve the Tm^{3+} doping level [36-37]. At present, the maximum CW output from the 2μm Tm^{3+}-doped fiber laser has arrived at 885W [26]. With further improvement of the pump diodes and pumping techniques, the 2μm output from the Tm^{3+}-doped fiber laser will be over 1 kilowatts and rise to tens of kilowatts.

3.1 Power scalability of Tm^{3+}-doped fiber laser

In this section, the power scalability and the limitation of Tm^{3+}-doped fiber lasers is analyzed and discussed. Detailed analysis shows that Tm^{3+}-doped fiber lasers have higher power scaling capabilities than that of Yb fiber lasers, i.e. higher power can be achievable from Tm^{3+}-doped fiber lasers than Yb fiber lasers. Thereafter, a 100-W CW Tm^{3+}-doped silica fiber laser with 60% slope efficiency is demonstrated experimentally and a design of twisted slab-like Tm^{3+}-doped fiber with the modal area of 5000μm^2 is proposed for high-power Tm^{3+}-doped fiber lasers.

3.1.1 Introduction

Compared with solid-state lasers, fiber laser has many advantages, such as good heat management (large surface to volume ratio), high beam quality, compactness and high stability [38-39]. Yb-doped fiber lasers have always been the research interest due to its merits such as high output and high slope efficiency. Since the invention of double-clad fiber, the output power of Yb fiber laser follows the Moore's law of fiber laser-Payne's law, i.e. the laser output doubles every two years. The 2μm laser wavelength from the Tm^{3+}-doped fiber laser is "eye-safe", which has wide applications in medical treatment, optoelectric Lidar and pump sources for mid-infrared lasers. Although Tm^{3+}-doped fiber lasers appear later than the Yb fiber laser [40], and the output power is lower than that of the latter, the output of the Tm^{3+}-doped fiber laser also follows Payne's law.

3.1.2 Limitation and advantages for power scalability of Tm^{3+}-doped fiber lasers

1. Energy levels of the Tm^{3+}-doped fiber laser

In Tm^{3+}-doped fiber lasers, the 2μm laser light comes from the transition $^3F_4 \rightarrow {}^3H_6$. the energy level diagram of Tm^{3+} ions is shown in Fig. 15 [41]. Due to strong Stark splitting of the $^3F_4$ level, the Tm^{3+}-doped laser is a quasi-four-level laser system. Besides, ions have a much smaller absorption cross section of 1.14×10^{-21}cm^2 [40] than that of Yb^{3+} ions (7.7×10^{-21}cm^2). Therefore, the Tm^{3+}-doped fiber laser requires more bright pump diode lasers to over its threshold. For the $^3F_4$ upper laser level, the emission cross section is $\sigma_{emi}=0.6\times10^{-21}$cm^2 [40]. The saturation intensity can be calculated from the expression [42]

$$I_{sat} = \frac{h\nu}{\sigma_{emi}\tau_f}, \tag{1}$$

where τ_f is the lifetime of the upper laser level.

From the calculation, we obtain the saturation intensity I_{sat} = 0.5kW/cm^2.However,the present output laser intensity of Tm^{3+}-doped fiber lasers has arrived at 2GW/cm^2, far over the saturation intensity. Therefore, high extraction efficiencies near quantum limitation can be achievable. In practice, the accessible laser efficiency is influenced by the physical-chemical features (spectrum, clustering, solubility and re-absorption, et al) of Tm^{3+} ions doped in silica fibers. The doping level of Tm^{3+} in silica fibers is usually lies in the 5000ppm level, and the fiber length is less than 13 meters. These conditions play a basic limitation in the output and energy-storing capability for the Tm^{3+}-doped silica fiber laser.

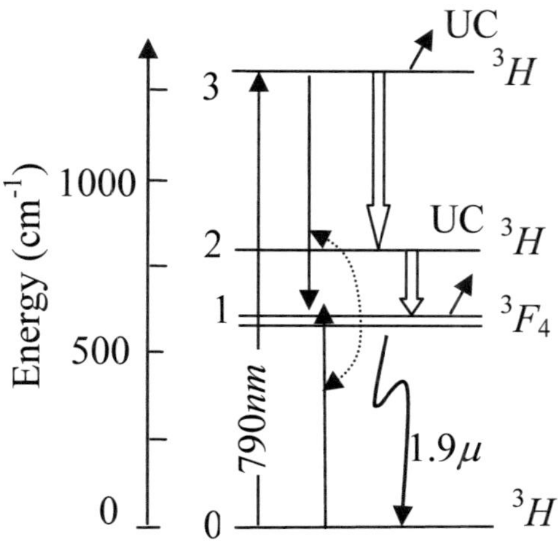

Fig. 15. Diagram of the energy levels of Tm^{3+} ions

2. Optical damage

Optical damage to fiber is always a factor should be considered when high power output is expected from fiber lasers, because the optical damage of fiber set a upper limit for achieved power. Optical damages to fiber mainly include the material damage by instant light intensity and the photo-darkening effect by long-time illumination. In order to obtain single-transverse-mode operation, the normalized frequency of fiber should be

$$V = \frac{\pi \cdot d \cdot (N.A.)}{\lambda} < 2.4 \tag{2}$$

Where, d is the fiber diameter, and N.A. is the fiber numerical aperture. Limited by the fiber drawing technique and bend loss, the N.A. of fiber is generally less than 0.06. Therefore, the diameter of single-mode fiber cannot be too large. To obtain the same V, 2μm-fiber can have a diameter twice that of 1μm-fiber, thus a cross section of 4 times larger. Therefore, the Tm^{3+}-doped fiber is more preferred to achieve single-mode high power laser output.

Damage to bulk fiber

The damage to silica fiber by high-power laser is a complicate process, including many factors such as local heating induced glass fracture, glass evaporation, and non-linear effects, et al. By using 1-μm laser in measurement, damage threshold of silica fiber was obtained from 16GW/cm^2 to 400GW/ cm^2, showing great discrepancy. When Yb^{3+} ions are doped, the damage threshold will decrease to be 0.5~0.8 times that of the pure silica fiber [43]. As far as the fiber facet is considered, the damage threshold will drop to 1/4 times that of pure silica fiber due to the impact of surface plasma, being about 40GW/cm^2 [44]. For the passive components in fiber lasers, such as fiber gratings, coated mirrors and fiber couplers, the damage threshold will be further decreased. At present, the maximum light intensity achieved in 1μm fiber laser is 2GW/cm^2, much lower than the fiber damage threshold.

As shown in Fig. 16, the silica fiber has much stronger absorption at 2μm wavelength than at 1μm wavelength. Therefore, the silica fiber damage threshold will be lower at 2μm wavelength (~40GW/cm^2) than at 1μm wavelength. On the other hand, the surface plasma effect will be smaller by 2μm laser, which will improve the damage threshold to some extent. In experiment, the 2μm laser light intensity achieved in the Tm^{3+} fiber has arrived at $2GW/cm^2$, being the same order of magnitude with that of 1μm laser.

As far as short-pulse (ps to ns regime) lasers are concerned, when neglecting self-focusing and other factors, the damage threshold of silica fiber follows the integrating law, which can also be applied to silica fiber laser [43].

$$E_{thres} \approx \sqrt{2/\pi P \Delta t}\ , \tag{3}$$

where, P is peak power and Δt is the pulse width. The damage threshold peak power P= $475GW/cm^2$ is nearly unchanged [43], indicating that heating effect is not the mail factor leading to fiber damage. For fs-level laser pulses, self-focusing will be the primary source of fiber laser, and eq. (3) cannot be used.

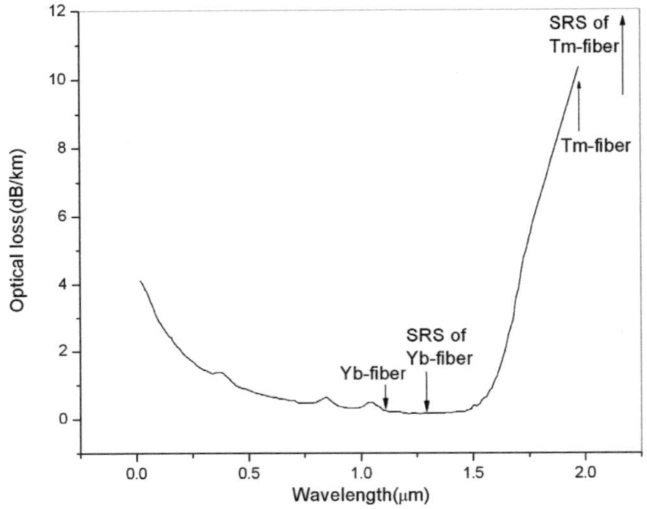

Fig. 16. Optical loss of Yb and Tm lasers, and their respective SRS emissions in silica fiber

Photo-darkening

Photo-darkening is the phenomenon that, after long time of laser operation, high-energy photons will induce color-center and defects in silica fiber, thus lead to growth of laser absorption and decrease of laser efficiency. Photo-darkening mainly stems from the radiation of UV and visible light. After 4 hours' radiation of 5 mW laser light at 488 nm, photo-darkening can be clearly observed. The photo-darkening-induced absorption primary occurs in the UV and visible spectral range. However, due to the fact that silica fiber itself has very weak absorption at the 1~2μm wavelength range, even slight photo-darkening will reduce the efficiency of fiber lasers [45].

Photo-darkening is extensively observed in Tm^{3+}-doped fiber lasers. Therefore, photo-darkening was considered a hamper limiting long-time operation of Tm^{3+}-doped fiber lasers pumped by 790 nm diode lasers. However, present experiment has shown that Tm^{3+}-doped fiber laser can operate stably over 2000 hours [46], close to that of Yb fiber lasers. Employing pre-radiation with UV light will decrease photo-darkening effect, thus enhance the lifetime of Tm^{3+}-doped fiber lasers.

3. Thermal issues

Larger ratio of surface to volume of fiber provides fiber lasers a good heat management system [47]. The heat generated in fiber core through multiphonon relaxation can be effectively dissipated by radiation and convection from the outer surface of the fiber. Most important is that, due to the guiding effect of fiber, even large temperature gradient exists will not deteriorate the light beam quality severely. However, with enhancement of output power, thermal issues of the Tm^{3+}-doped fiber laser become serious and should be considered carefully.

Heat effects

Just as the Yb fiber laser, the key factor deciding how high power the Tm^{3+}-doped fiber laser can endure is the melting temperature of the fiber core. By adopting positive cooling, the heat resistance ability of the Tm^{3+}-doped fiber laser is about 100 W/m.

The 790nm-pump Tm^{3+}-doped fiber laser has the "one-for-two" energy-transfer process-cross relaxation [23], there exist three transitions leading to occurrence upper laser level ions from absorbing a pump photon. They are (1) $n_3 \rightarrow 2 \cdot n_1$ (cross relaxation); (2) $n_3 \rightarrow n_2 \rightarrow n_1$; and (3) $n_3 \rightarrow n_1$. Taking into account the up-conversion and re-absorption, the quantum efficiency of achieving upper laser level ions is

$$\eta = [2\eta_{3101} + \eta'_{3101} + \eta_{3201}]\eta_{las} - \eta_{up} - \eta_{reabs} \tag{3}$$

Neglecting some weak effects, the expression can be simplified to

$$\eta = 2\eta_{31} + (1 - \eta_{31})(\beta_{31} + \beta_{32}) \tag{4}$$

From the parameters of Tm^{3+}-doped silica fiber, we can obtain $\eta = 0.74$, which is comparable with that of Nd:YAG laser. The quantum efficiency of Tm^{3+}-doped fiber laser is also related to the doping concentration, operation temperature and other cavity parameters. Based on the heat management capability, it is concluded that the Tm^{3+}-doped fiber laser can provide a output potential of about 300W/m.

Efficiency decrease by operation temperature

The ground state energy level of Tm^{3+} ions is consisted of a series of sublevels, forming a continuous energy band with a broad bandwidth of 5770 cm^{-1} [48]. This feature provides the Tm^{3+}-doped fiber laser a quasi-four level laser system, possessing high temperature stability. At kilowatt level, the reduction of population inversion of Tm^{3+}-doped fiber laser due to temperature growth is less than 1%.

Based on a 5.5cm-length Tm^{3+}-doped fiber laser, the influence of temperature on the laser's efficiency has been observed, as shown in Fig. 12. When the temperature is increased from 10°C to 40°C, the output and slope efficiency decrease about 30% [33]. The primary reason leading to the reduction of output power is the 0.3nm/°C wavelength shift of the pump diode laser, which making the pump light cannot overlap well with the Tm^{3+} absorption band, decreasing the pump absorption efficiency. If only the absorbed pump power is taken into account, the power reduction is just 5%. Another alternative explanation is that the fluorescence spectrum of Tm^{3+}-doped fiber is broadened with temperature, decreasing the gain per unit-length of fiber. As far as long fiber lasers are concerned, the influence of these two factors is weak. Our 100-W Tm^{3+}-doped fiber laser experiment shows that, when the fiber length is 10 meters, the reduction of output power and slope efficiency induced by the temperature growth from 10°C to 40°C is less than 5%.

4. Non-linear optical effects

The silica glass itself is non-symmetric, so no intrinsic second-order non-linear effect exists [49]. However, the existence of third-order non-linear effect such as stimulated Raman scattering (SRS) and stimulated Brillouin scattering (SBS) has significant impact on high-power CW lasers, especially single-frequency output is expected. Self-phase-modulation (SPM) and cross-phase-modulation (XPM) has negligible effects on CW-mode lasers, but has obvious impact on ultra-short pulsed lasers. Self-focusing can greatly decrease the damage threshold of fiber. The parameter of fiber material that is closely related with third-order non-linear effects is the second-order refractive index n_2, which differs greatly when different doping and fiber drawing techniques are used. For silica fiber operating at 2μm, the second-order refractive index n_2 is about 2.5×10^{-20} m^2/W [49].

SRS and SBS

Both SRS and SBS are elastic scattering processes. SRS originates from the elastic scattering of optical phonons, and has a frequency shift of ~ 13THz, and is hardly influenced by the band width of input laser; SBS stems from the elastic scattering of acoustic phonons, and has a frequency shift of ~ 17GHz, and can be significantly influenced by the band width of input laser. When the band width of the input laser is larger than 0.5GHz, the SBS threshold will be significantly improved, and the SRS will be the dominant limitation for power scaling of fiber lasers.
In fiber, the power threshold of SRS and SBS can be expressed as [49]

$$P_0 \approx C_1 A_{eff} / g_R L_{eff} \tag{5}$$

Where, C_1 is the threshold parameter (C_{SRS}=16 and C_{SBS}=21), A_{eff} and L_{eff} is effective mode area and interaction length, respectively. At the wavelength of 2μm, the effective SRS coefficient $g_R=5\times10^{-14}$m/W [40], is just a half of the one at the wavelength of 1μm; while the SBS coefficient $g_R=5\times10^{-11}$m/W [40], is the same as the one at the wavelength of 1μm. In fiber, the Stocks loss is also important to SRS. From Fig. 16, we can see that the Stocks loss will higher than the pump when laser wavelength is longer than 1.55μm. The optical loss at 2μm (~10dB/km) is 10 times that at 1μm, so the SRS threshold in 2μm fiber lasers will be 20 times that in 1μm fiber lasers. In Tm^{3+}-doped silica fiber lasers, the SRS threshold is 30~100MW/μm^2.

Self-focusing

Self-focusing is the light beam focusing phenomenon induced by the intensity-related change of refractive index of the material. Due to long interaction length in fiber lasers, even a weak self-focusing effect will induce laser beam focusing, thus damage the fiber. The threshold for self-focusing in fiber is expressed by [50]

$$P_{self} = \frac{0.15\lambda^2}{n_0 n_2}. \tag{6}$$

The threshold P_{self} is proportional to the square of laser wavelength. At 2μm, Both n_0 and n_2 are slightly smaller than that at 1μm. Therefore, the self-focusing threshold at 2μm is more than 4 times that at 1μm. For 2μm fiber lasers, the self-focusing threshold is about 16MW, with no relation to fiber diameter.

SPM and XPM

In fiber lasers, SPM and XPM change the phase of ultra-short laser pulses, playing a severe role in the reshaping of pulses. The phase change can be calculated by

$$B = \frac{2\pi}{\lambda} \int n_{2I}(z) I(z) dz \,, \tag{7}$$

where, n_{2I} is non-linear refractive index, I is light intensity and λ is laser wavelength.
Generally, in order to keep a small phase distortion in the propagation of ultra-short pulses, the B integral should less than π. With the same n_2, 2μm laser pulses have a smaller B integral than that of 1μm. Solving the propagation equations of laser pulse in fiber, the frequency change induced by SPM and XPM can be expressed by a non-linear character parameter

$$\gamma = \frac{2\pi n_2}{\lambda A_{eff}} \,. \tag{8}$$

The 2μm fiber laser has a comparatively smaller n_2 and a comparatively larger A_{eff} than that of 1μm lasers. Therefore, the 2μm fiber laser has a non-linear character parameter about 1/10 times that of 1μm fiber laser, providing a much higher peak-power enduring capability of ultra-short pulses.

3.1.3 Power scaling techniques for the Tm^{3+}-doped fiber laser

Experimental results

For achieving high power laser output, the double-clad Tm^{3+}-doped silica fiber has a large doped core diameter of 27.5 μm with the N.A. of 0.20 and. High Tm^{3+} ions doping concentration of 2.5 wt.% is chosen to facilitate the cross relaxation process. A small portion of Al^{3+} ions were also doped into the fiber to suppress the energy transfer upconversion (ETU) processes. The pure silica inner cladding, coated with a low-index polymer, has a 400-μm diameter and the N.A. of 0.46. The hexagonal cross section of the inner clad helps to improve pump absorption.
Fig. 17 is the experimental setup for realizing the high-power Tm^{3+}-doped fiber laser. Two high-power fiber-coupled LD arrays operating at ~793 nm was used as the pump source. The pump beam was launched into the fiber by aspheric-lens coupling system through dichroic mirrors. The right-hand side dichroic mirror is used to guide the 2 μm laser light out. The launched efficiency was as high as ~90%, and the largest pump power of 180 W can be launched into the fiber. The pump end of the fiber was butted directly to the dichroic mirror with high reflectivity (>99.7%) at 2.0 μm and high transmission (>97%) at 790 nm. Both fiber ends were cleaved perpendicularly to the axis and polished carefully. The ends of the fiber were clamped tightly in water-cooled copper heat-sinks, and the remaining fiber was immersed into water to achieve maximum efficiency.

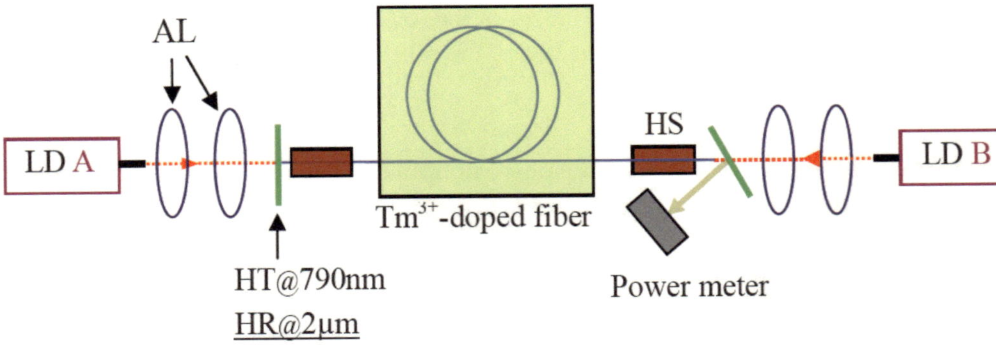

Fig. 17. Experimental setup of the Tm^{3+}-doped fiber laser. LD: laser diode; AL: aspheric lens; HT: high transmission; HR: high reflection; HS: heat sink.

Fig.18 shows the 100-W level Tm^{3+}-doped silica fiber laser achieved in our laboratory. When pumped by 790 nm diode laser, the laser output is 102 W with a slope efficiency of 58.6%. In this Tm^{3+}-doped fiber laser, the power density is just 30kW/cm^2, far lower than the previously mentioned power density limit. Therefore, power scaling this Tm^{3+}-doped fiber laser to over 1 kW power level is only limited by obtaining high brightness pump diode lasers [51].

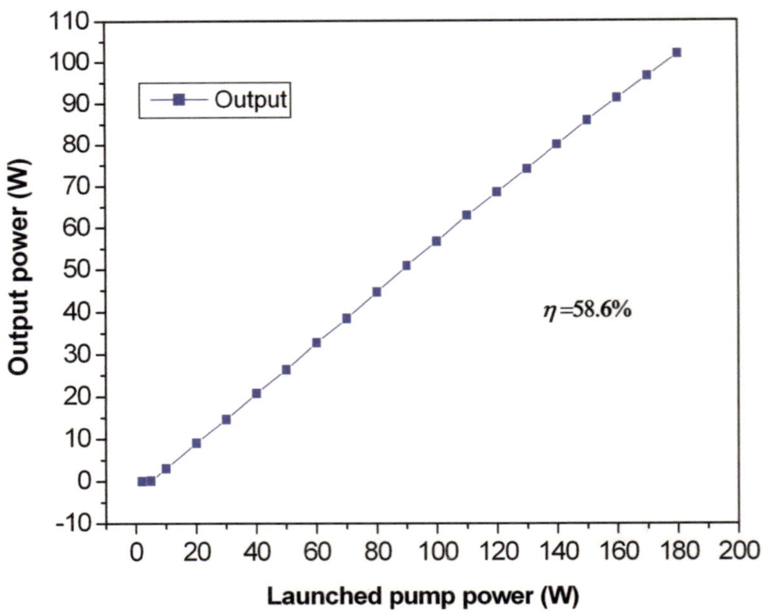

Fig. 18. Output characteristics of the Tm^{3+}-doped fiber laser.

Fig. 19 is the laser output spectrum. The laser spectrum is centered at 2070 nm, and has a full band width of about 25 nm. Due to a broad gain bandwidth, the fiber laser operates in the multi-mode regime. Therefore, there are many spectral peaks in the measured spectrum.

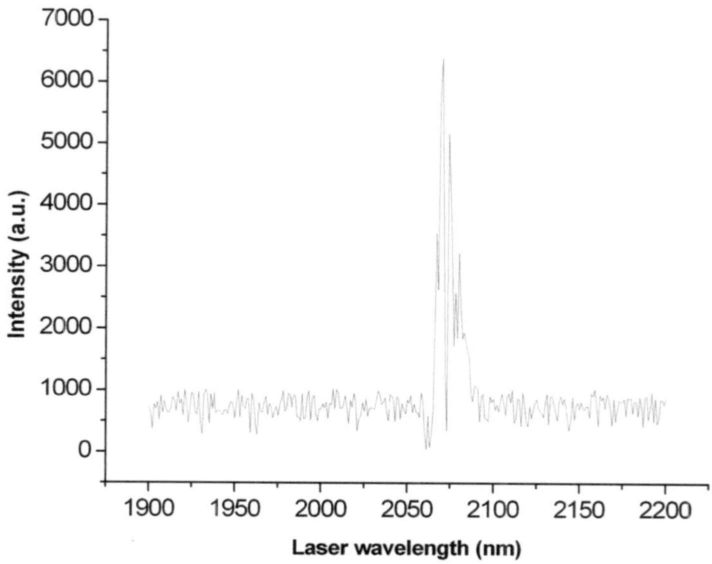

Fig. 19. Laser spectrum of the Tm^{3+}-doped fiber laser.

A novel power-scaling method

There are two technical ways to improve the output power of the 2μm Tm^{3+}-doped fiber laser over 1kW: one is by using large mode area (LMA) fibers or multi-mode fibers through special fiber-structure design; another way is adopting different laser cavity configurations such as multi-stage cavity (different cladding-size fibers are used in different stages), master oscillator power amplifier (MOPA) structure or laser beam combination.

From the analysis in the previous section, we found that the limitation of power scaling of fiber lasers is the finite fiber diameter. In order to obtain LMA fiber together with high beam quality, photonic crystal fiber (PCF) can be used. Another alternative is designing the fiber with a W-shape refractive index. By using this method, we design a slab-core fiber, achieving a core mode area larger than 5000μm^2 [52]. In such kind of fiber, laser in the width direction is multi-mode, so the key thing is to control the beam quality in this direction. Using the V-groove method can improve the beam quality in the width direction, but will decrease the mechanical feature of the fiber. Therefore, we proposed the rotating slab-core fiber configuration, as shown in Fig. 20. The slab-core rotates at a angular rate of θ, the trace of the slab-core edge can be described as

$$\begin{aligned} x &= d\cos(\theta z) \\ y &= d\sin(\theta z) \end{aligned} \tag{9}$$

Where, d is the slab-core width and z is the coordinate in the length direction.

With different θ, the rotating strength of the slab core will be changed. The winding of the fiber core induce much higher loss for higher-order modes (suppressing the higher-order modes), thus laser oscillation occurs for lower-order modes. Fig. 20 shows the loss of different orders of modes as a function of the rotating rate of the slab core. It can be seen that higher-order modes can be effectively suppressed when the rotating rate of the fiber core is less than $2\pi/5\text{cm}^{-1}$.

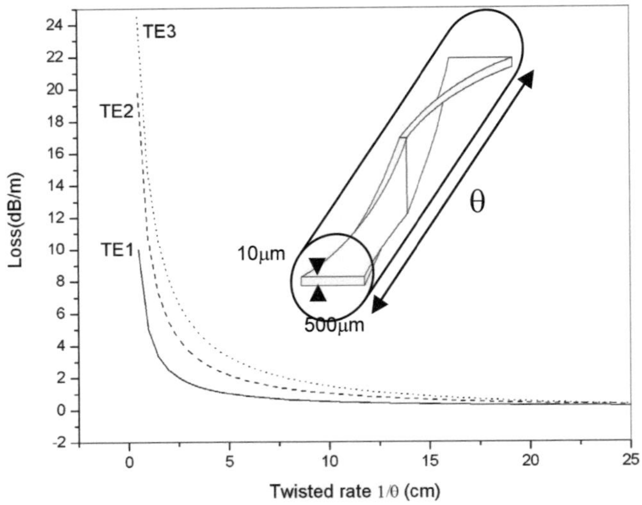

Fig. 20. Mode loss of the rotating rectangular-core fiber versus the angular rotating rate.

Conclusion and prospect

The Tm^{3+}-doped silica fiber laser as a kind of high efficient 2μm laser source, its further power scalability is not limited by the intrinsic physical mechanism of fiber. At present, the limitation for power scaling lies in obtaining high bright laser diodes and ultra-low-loss Tm^{3+}-doped silica fibers. Compared with the 1μm Yb fiber lasers, Tm^{3+}-doped fiber laser has a higher power scalability. Besides, the Tm^{3+}-doped fiber laser has a longer lifetime of the upper laser level and a much broader emission band, providing itself a much better advantage in achieving pulsed laser output. The Tm^{3+}-doped fiber laser has great potentials in high-power "eye-safe" laser output and high-peak-power pulsed laser.

3.2 Wavelength tuning of high-power Tm^{3+}-doped fiber laser

3.2.1 Introduction

The large degree of Stark splitting of the $^3H_6$ ground state , as shown in Fig. 8, provides the $^3H_4 \rightarrow {}^3H_6$ transition in the Tm^{3+}-doped silica fiber with a very broad emission spanning >400 nm. This feature provides the Tm^{3+}-doped fiber laser the great potential of being wavelength tuning in a large spectral range. The emission spectrum of the $^3F_4 \rightarrow {}^3H_6$ transition in Tm^{3+}-doped silica fiber is shown in Fig. 7 [18].

Apart from being doped into silica fiber, the Tm^{3+} ions can also achieve very broad emission band when doped into other host materials. For instance, an emission band of ~300 nm can be obtained when Tm^{3+} is doped into ZBLAN fiber, as shown in Fig. 21 [53]. Therefore, Tm^{3+}-doped fiber lasers have a particular advantage in wavelength tuning regime.

According to the character of fiber laser, many methods can be used to achieve the wavelength tuning function. As far as the Tm^{3+}-doped fiber laser is concerned, the wavelength tuning techniques include temperature-tuning, length-tuning, birefringence-tuning and grating-tuning et al. Among these tuning techniques, the most widely used and most skillful is the grating wavelength tuning technique. Along with the mature of the fiber Bragg grating, the grating tuning technique is becoming more and more important.

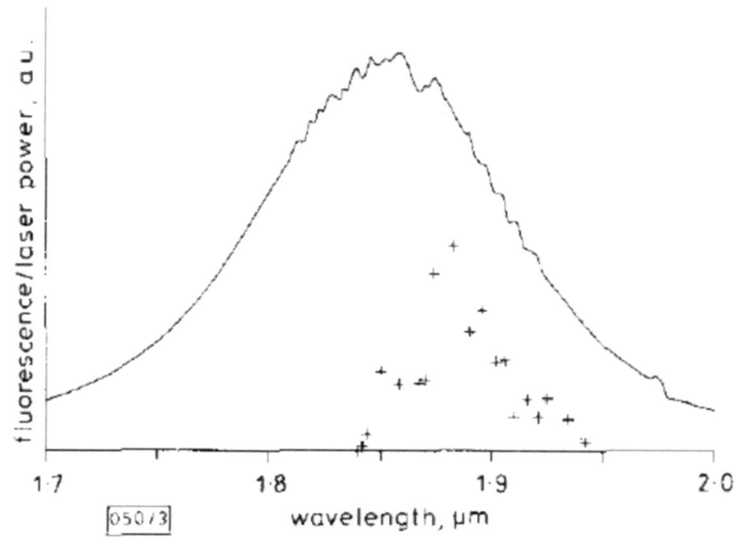

Fig. 21. The fluorescence spectrum of $^3F_4 \rightarrow {}^3H_6$ transition in Tm^{3+}-doped ZBLAN fiber

Several kinds of wavelength tuning techniques in Tm^{3+}-doped fiber lasers:

1. Fiber-length tuning.

Due to the quasi-four-level system feature, the Tm^{3+}-doped fiber laser can be wavelength tuned by changing the fiber length. When the fiber length is elongated, re-absorption of the signal light will increase, leading to red-shift of laser wavelength. This tuning method is simple and convenient for manipulating, but the tuning range is limited. The broadest tuning wavelength spanning is less than 100 nm [54-55]. The most dominated shortcoming of this wavelength tuning technique is that laser wavelength cannot be tuned continuously. In the tuning process, the replacement of fiber requires re-adjusting the laser cavity, complicating the tuning work. This tuning method has little potential in practical applications.

2. Birefringence-tuning.

This wavelength tuning method is based on changing the birefringence characteristic of the signal light in the cavity. By using a birefringence filter, the Tm^{3+}-doped fiber laser has been tuned over a 200 nm spectral range [56]. Although this method can provide a wide tuning range, the tuning laser configuration is rather complicated, and very inconvenient for tuning. Besides, this technique is confined by the free-spectral range of the birefringence filter. Therefore, this method is far from practical application.

3. Temperature-tuning.

Due to the circumstance-field impact, the ground-state level of Tm^{3+} ions is Stark splitted into many sub levels. As one of the Stark sub levels, the lower laser level has a population distribution significantly influenced by the circumstance temperature (according to the Boltzman distribution). This leads to the wavelength shift with temperature. Electrical oven has been used to heat the Tm^{3+}-doped fiber laser for wavelength tuning [57], and a tuning range of 18 nm was achieved when the fiber temperature was changed during a 109°C range. With a Peltier plate, a wavelength tuning range of 40 nm was realized with the tuning rate of ~2nm/°C in a 6-meter-length Tm^{3+}-doped fiber [58]. This tuning technique is simple and convenient, but the tuning range is also narrow. The melting point of the fiber polymer cladding set a upper limit for the temperature, and low temperature operation cannot be practically used, which limits the wide application of this tuning method.

4. Grating-tuning.

At present, the grating tuning method is the most fully developed and widely used. This is primarily due to the fast development of the grating fabrication technique. By using the grating-tuning technique, Tm^{3+}-doped fiber laser has achieved tuning range over 200 nm [59-61]. Compared with the above mentioned three methods, the grating tuning technique can provide a broader tuning range with a much narrower linewidth. This method is, up to date, the most mature wavelength tuning technique.

3.2.2. High-power Tm3+-doped fiber laser tuned by a variable reflective mirror

Due to the quasi-four-level system feature, the Tm^{3+}-doped fiber laser can be wavelength tuned by changing the transmittance of the output coupler. With a variable reflective mirror (VRM) as the output coupler, high-power Tm^{3+}-doped fiber laser can be wavelength tuned over a range of >200 nm [47]. The combination of high power and wavelength tuning of the Tm^{3+}-doped fiber laser provides an excellent kind of laser source in the ~2 μm spectral range.

In the experiment, the double-clad Tm^{3+}-doped silica fiber has a doped core with the N.A. of 0.20 and diameter of 27.5 μm. High Tm^{3+} ions doping concentration of 2.5 wt.% is essential to facilitate the CR energy transfer process. A small portion of Al^{3+} ions were also doped into the fiber to suppress the energy transfer upconversion (ETU) processes, which may cause the quenching of the $^3F_4$ multiplet lifetime. The pure silica inner cladding, coated with a low-index polymer, has a 400-μm diameter and the N.A. of 0.46. The hexagonal cross section of the inner clad helps to improve pump absorption. The absorption coefficient at the pump wavelength (790 nm) is ~2.8 dB/m.

Fig. 22 shows the experimental setup [47]. High-power LD arrays operating at 790 nm and TM mode was used as the pump source. The outputs from two LD arrays were polarizedly combined to form a single pump beam. This pump beam was reshaped by a micro-prism stack at first, and then focused into a circular spot using a cylindrical lens and an aspheric lens. Through a dichroic mirror, the pump light was launched into the fiber. The launched efficiency was measured through a 4-cm-long Tm^{3+}-doped fiber. The largest pump power of 51 W can be launched into the fiber. The pump end of the fiber was butted directly to the dichroic mirror with high reflectivity (>99.7%) at 2.0 μm and high transmission (>97%) at 790 nm. Both fiber ends were cleaved perpendicularly to the axis and polished carefully. The output coupler was formed by a VRM or the bare fiber-end facet. The transmission of the VRM can be changed continuously from 5% to 80% (the reflection R is changed from ~94.8% to 18.4%) at 2 μm by simply horizontally displacing the VRM with a one-dimensional stage. The ends of the fiber were clamped tightly in water-cooled copper heat-sinks, and the remaining fiber was immersed into water to achieve maximum efficiency. During the experiment, both cavity mirrors were carefully adjusted with five-dimensional holders.

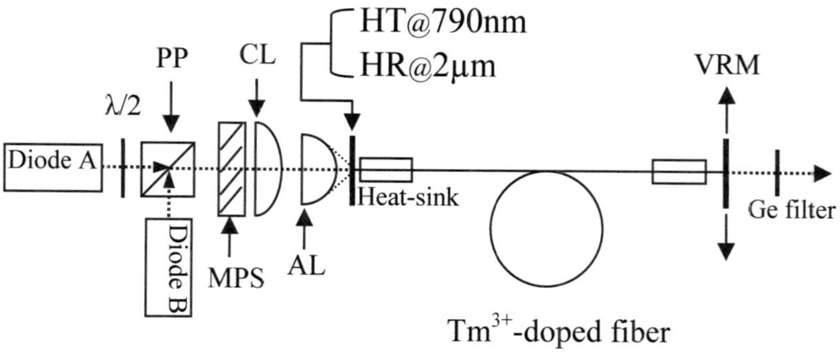

Fig. 22. Schematic of the experimental setup. PP: polarizing plate; MPS: micro-prism stack; CL: cylindrical lens; AL: aspheric lens; HT: high transmission; HR: high reflection; VRM: variable reflective mirror.

The lasing characteristics obtained with relative higher output couplings in a 4-m long fiber laser are shown in Fig. 23 [47]. When the VRM was moved away from the fiber end and the bare fiber-end facet was used as the output coupler (T≈96%), the laser reached threshold at a launched pump power of 5.9 W, and produced a maximum output power of 32 W at 1949 nm for 51-W launched pump power, corresponding to a slope efficiency of 69% and a quantum efficiency of 170%. The high efficiency was attributed to high Tm^{3+}-doping concentration, suppression of ETU with Al^{3+} ions [38], and efficient fiber-cooling. With T=80% output coupling, a slightly lower output power of 29.8 W was generated at 1970 nm, and the slope efficiency with respect to launched pump power was ~65%. When the output coupling decreased to 60%, the output power dropped to 27.4 W at 1994 nm with a slope efficiency of ~58%. In all these cases, the output power increased linearly with the launched pump power, suggesting that the laser can be power scaled further by increasing the pump power. The power stability of the laser output, monitored by an InAs PIN photodiode and a 100 MHz digital oscilloscope, was less than 1% (RMS) at ~30 W power levels.
After carefully optimization the position of the coupler, the fiber laser was wavelength tuned by simply horizontally moving the VRM coupler. The peak wavelength of the laser spectrum is taken as the laser wavelength. Fig. 24 shows the dependence of the laser wavelength on the output coupling [47]. When the output coupling decreased from ~96% to 5% in the 4-m long fiber laser, the laser wavelength was tuned from 1949 to 2055 nm with a tuning range of 106 nm. The nearly linear dependence provides a basic knowledge to choose the wavelength from Tm^{3+}-doped silica fiber lasers. The phenomenon can be explained by the enhanced re-absorption of laser in the high-Q cavity. Since the photon lifetime in the cavity is increased with higher reflective mirrors, the photon travels more round-trips, and undergoes more re-absorption before escapes from the cavity.
Employing different fiber lengths from 0.5 m to 10 m, as shown in Fig. 24, the laser can be tuned from 1866 to 2107 nm. The total tuning range is over 240 nm at above-ten-watt levels. A typical laser spectrum obtained with the 4-m fiber at coupling of T=15% and 16-W output power is shown as inset in Fig. 24. The laser spectra under different couplings and fiber lengths hold nearly identical features. The spectrum has a bandwidth (FWHM) of ~15 nm and several lasing peaks. The multi-peak spectrum indicates the laser operated in multiple longitudinal modes.

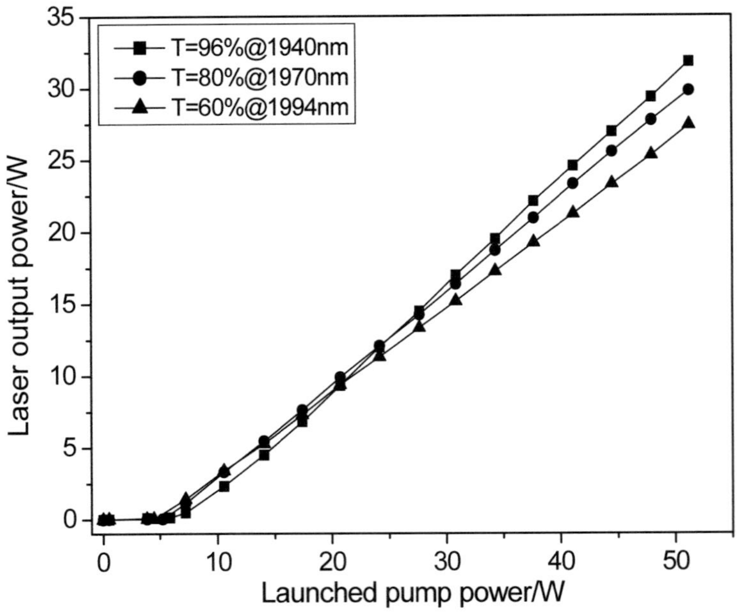

Fig. 23. Laser output power versus launched pump power with three high output couplings.

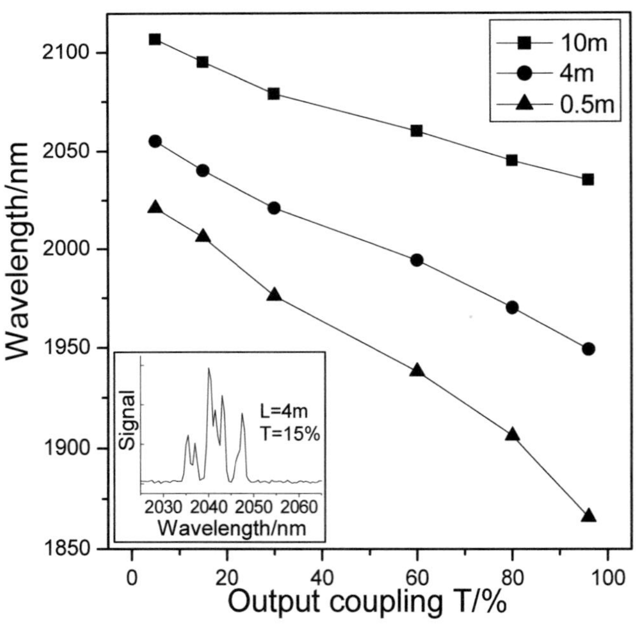

Fig. 24. Laser peak wavelength as a function of output coupling; inset is the laser spectrum obtained with the 4-m fiber at coupling of *T*=15%.

The maximum output power and launched threshold pump power as functions of the output coupling are shown in Fig. 25 [47]. When the output coupling decreases from ~96%

to 5%, the threshold pump power reduces almost linearly from 5.9 to 1.0 W, and the maximum output power drops from 32 W to 9.0 W. The sharp decreasing of the output power with <15% output coupling is mainly due to low output transmission and increased re-absorption of laser light. Between the output coupling of 20% and 96%, the laser output power exceeds 20 W over a tuning range of 90 nm from 1949 to 2040 nm (see Fig. 24). This presents the potential of Tm^{3+}-doped silica fiber lasers to generate multi-ten-watt output over a hundred-nanometer tuning range.

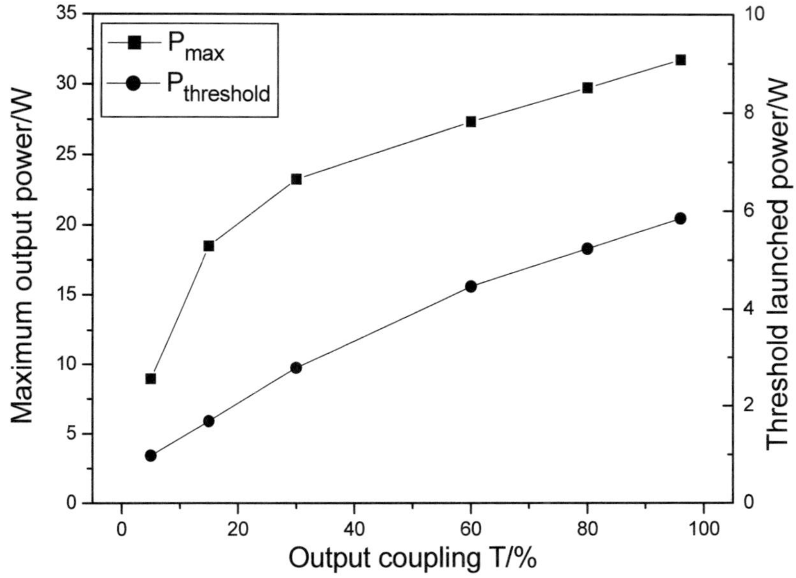

Fig. 25. Maximum laser output power and threshold launched power as functions of the output coupling.

3.2.3 Conclusion

At present, high-power widely tunable Tm^{3+}-doped silica fibers must make use of high-power diode lasers as the pump source. Due to the comparatively low damage threshold of grating and difficulty in fabricating 2μm grating, wavelength tuning high-power Tm^{3+}-doped fiber laser with fiber Bragg grating is still unpractical. Using the variable reflective output coupler to tune high-power 2μm fiber lasers is a feasible alternative. The combination of high power, high efficiency, and wide tunability of Tm^{3+}-doped fiber lasers will provide a great opportunity for applications of eye-safe lasers.

4. Self-pulsing and passively Q-switched Tm^{3+}-doped fiber laser

Due to its special energy-level structure and the wave-guiding effect of fiber, Tm^{3+}-doped fiber lasers can produce fluent dynamical behaviors, including self-pulsing, self-mode-locking and et al [62-63]. On the other hand, the particular broad emission band of Tm^{3+} ions provides the potential to achieve ultra-short pulses from the Tm^{3+}-doped fiber laser.

4.1 Self-induced pulsing in Tm^{3+}-doped fiber lasers with different output couplings

1. Introduction

It's well known that self-pulsing can be achieved in any lasers with an adequate saturable absorber [64]. Erbium-doped fiber lasers have demonstrated a large variety of dynamical behaviors, including self-pulsing operations [65], static and dynamic polarization effects [66], antiphase and chaotic dynamics [67]. The dynamic behaviors have been attributed to the presence of ion-pairs or clusters acting as a saturable absorber [68-69], bidirectional propagation in "high-loss cavity" and Brillouin scattering effects in the fiber [70]. Ion pair concentration can play an important role in self-pulsing dynamic behaviors [71].

It has been shown that the Tm^{3+}-doped fiber laser can operate successively in continuous-wave (CW) mode, self-pulsing mode and quasi-CW mode with increase of pump power [62]. Self-mode-locking phenomenon has also been observed in the Tm^{3+}-doped fiber laser, which was supposed to stem from saturable absorption or strong interactions between the large number of longitudinal modes oscillating in the cavity [63].

2. Experimental observation

In order to understand the mechanism and features of self-pulsing in Tm^{3+}-doped fiber lasers, different output couplers are used to construct the fiber laser cavity. Self-pulsing behavior was observed under various pumping rates.

The experimental arrangement for observing self-pulsing operation is shown in Fig. 26 [72]. The 2 μm Tm^{3+}-doped fiber laser is pumped by a single CW-diode laser, operating TM mode centered at 790 nm, shifting to~793 nm at comparatively higher operating temperature. With this pump source, the maximum power launched into the fiber was near 12 W.

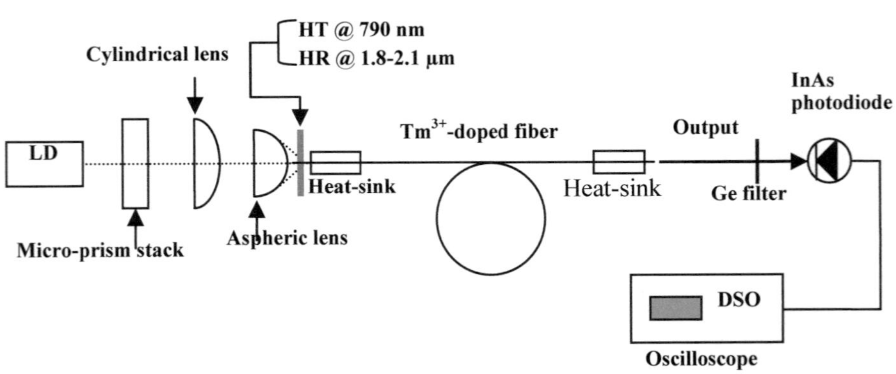

Fig. 26. Experimental arrangement of LD-pumped Tm^{3+}-doped fiber laser

The double-clad MM-TDF with ~10 m length (Nufern Co.) had a 30 μm diameter, 0.22 N.A. core doped with Tm^{3+} of ~2 wt.% concentration (the V value is about 9.42 when laser wavelength is of ~2 μm). The pure-silica cladding, coated with a low-index polymer, had a 410 μm diameter and a NA of 0.46. The fiber has an octagon-shape clad, which helps to improve the pump absorption. The fiber ends were perpendicularly cleaved and carefully polished carefully to ensure flatness, so that the loss was minimized.

The laser pumping beam was reshaped first by a micro-prism stack, and then focused into a circular spot of ~0.5×0.5 mm diameter with a cylindrical lens and an aspheric lens. The focused pump beam was launched into the thulium-doped fiber through a dielectric mirror. The pump end of the fiber is butted directly to the dielectric mirror with high reflectivity (>99%) at 1850~2100 nm and high transmission (>97%) at 760~900 nm. The Fabry-Perot laser cavity was formed between the dielectric mirror and the output-end fiber facet (with Fresnel reflection of ~3.55% providing feedback for laser oscillation). Both ends of the fiber were held in metallic heat-sinks, and the remaining fiber was wrapped on a water-cooling metallic drum to prevent possible thermal damage to the fiber.

The threshold pump power of the long fiber laser with the output coupler of the fiber-end facet is about 5.8 W. Various self-pulsing regimes obtained with increasing pump level are shown in Fig. 27 [72]. When the pump power is near the threshold (P=6 W), the laser delivers a regular train of pulses, as shown in Fig. 28(a). The pulse duration is 7.2 μs, and the frequency is 42 kHz. When the pump power is increased to P=7 W, the pulse width narrows to 6.5 μs and the pulse frequency grows to 63 kHz, as seen in Fig. 28(b). At high pump levels, a second set of pulses began to appear as shown (the arrow point to) in Fig 28(b) and (c). This is due to that the high peak power confined in the fiber core may favor the excitation of a Brillouin backscattered wave, especially in the "high-loss cavity" configuration (high output coupling) [70].

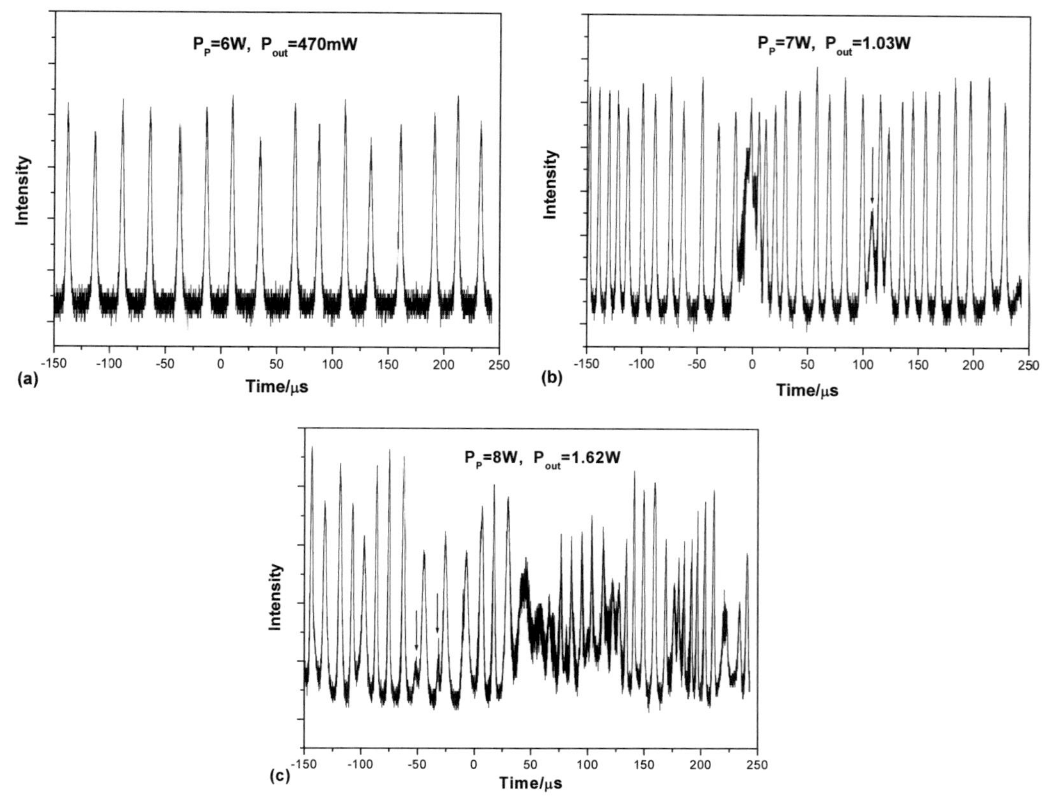

Fig. 27. Output intensity time trace of 10 m fiber laser with end-facet output coupler for (a) Pp=6 W, (b) Pp =7 W, (c) Pp =8 W.

When the pumping level is high enough, the laser output becomes quasi-CW, as shown in Fig. 28 [72]. This result is in agreement with that obtained in previously studies [62, 69]. In the case of 10 W of pump power, the pulse repetition rate increases to 132 kHz, but the pulse width randomizes. At this time, the laser operates in a similar self mode-locking state [63, 71].

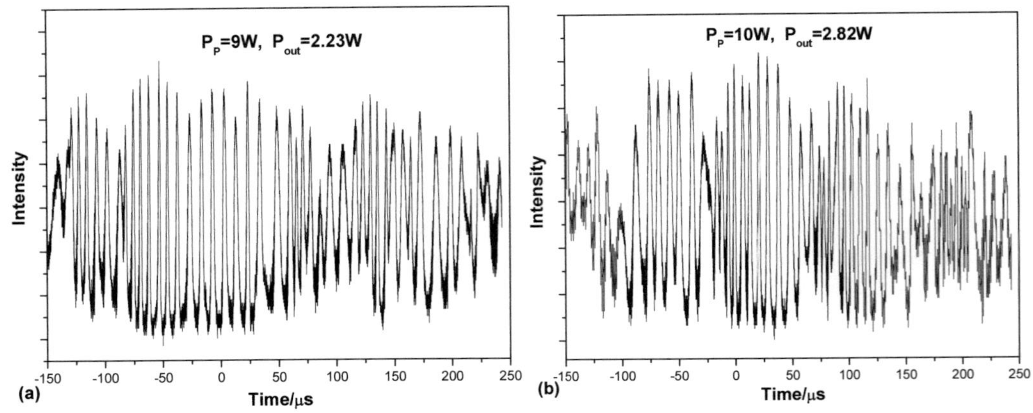

Fig. 28. Quasi-CW operation for pumping power (a) Pp=9 W, (b) Pp =10 W.

With the fiber-end coupler, the pulse width and frequency as functions of pump power are indicated in Fig. 29 [72]. The pulse width decreases, but the pulse repetition rate increases, near linearly with enhanced pump power. At high pump levels, e.g. over 9 W, the pulse width begins saturating. Therefore, it seems hard to derive short pulse duration through self-pulsing in Tm^{3+}-doped fiber lasers.

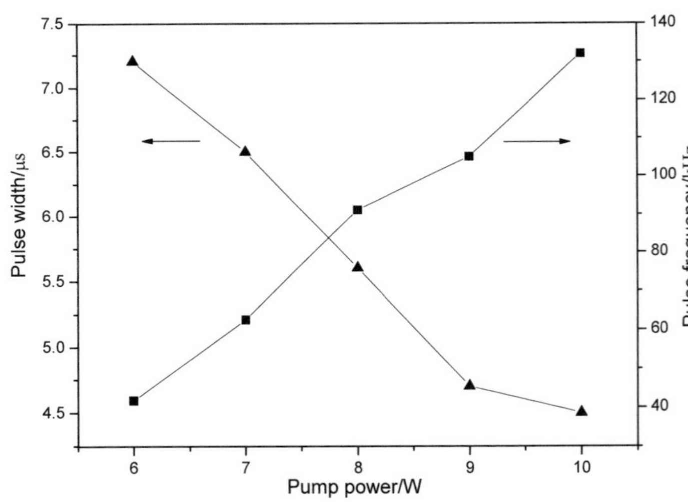

Fig. 29. Pulse width and pulse frequency versus pump power.

When a dielectric mirror with T=10% at 2 μm is used as the output coupler, the dynamics behavior is somewhat different from that obtained with the fiber-end coupler, as indicated in Fig. 30 [72]. For this cavity configuration, the threshold pump power is about 3 W. Near the threshold, a regular train of pulses is observed, as shown in Fig. 30(a). The pulse duration is around 18 μs, and the pulse frequency is about 21 kHz. Increasing the pump power to 4 W, the pulse duration decreases to 16 μs and the frequency increases to 37 kHz, respectively. However, when the pump power is further increased to 5 W and 6 W, only the pulse frequency shows a definite changing trend, becoming higher and higher. The pulse width indicates an indefinite advancing trend: some become broader and some become narrower. The irregularity of the pulse increases significantly with pump power enhanced.

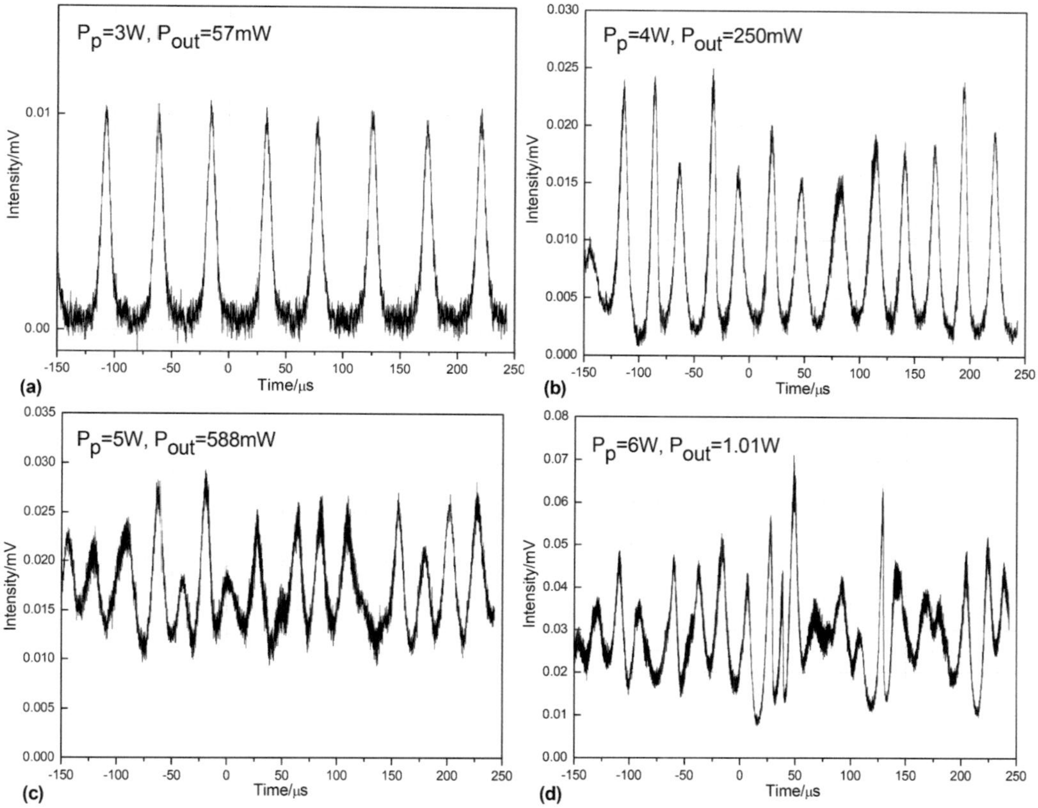

Fig. 30. Output intensity time trace of 10 m fiber laser with 10% output coupler for (a) P_p =3 W, (b) P_p =4 W, (c) P_p =5 W, (d) P_p =6 W.

When a dielectric mirror with T=5% at 2 μm is used as the output coupler, the dynamics behavior is completely different from the previous results, as indicated in Fig. 31 [72]. For this cavity configuration, the threshold pump power is also about 3 W. However, even near the threshold, the pulse train is very irregular, as shown in Fig. 31(a). The pulse duration is around 23 μs, and the pulse frequency is about 28 kHz. Increasing the pump power to 4 W, the laser output becomes near-CW. With 5 W of pump power, the output is completely CW. This clearly demonstrates that the self-pulsing behavior of heavily doped fiber lasers can be suppressed by using low-transmission output couplers.

The dependence of the pulse width and frequency on the output coupler transmission (T) is shown in Fig. 32 [72]. The pulse width and pulse frequency were obtained near respective pump threshold. It is clear that the pulse width decreases near linearly with T. This is because that the pulse width scales similar to the photon cavity lifetime [73]. A laser cavity with a lower T has a longer photon cavity lifetime due to less output loss, thus has broader pulse duration. The pulse frequency first decreases and then increases with increasing T. Considering that the threshold pump power is different for different cavities, we normalized pulse frequency to pump power. As shown in Fig. 32(b), the normalized pulse frequency increases with decreasing T. When T<10%, the pulse frequency grows sharply, transforming to CW operation.

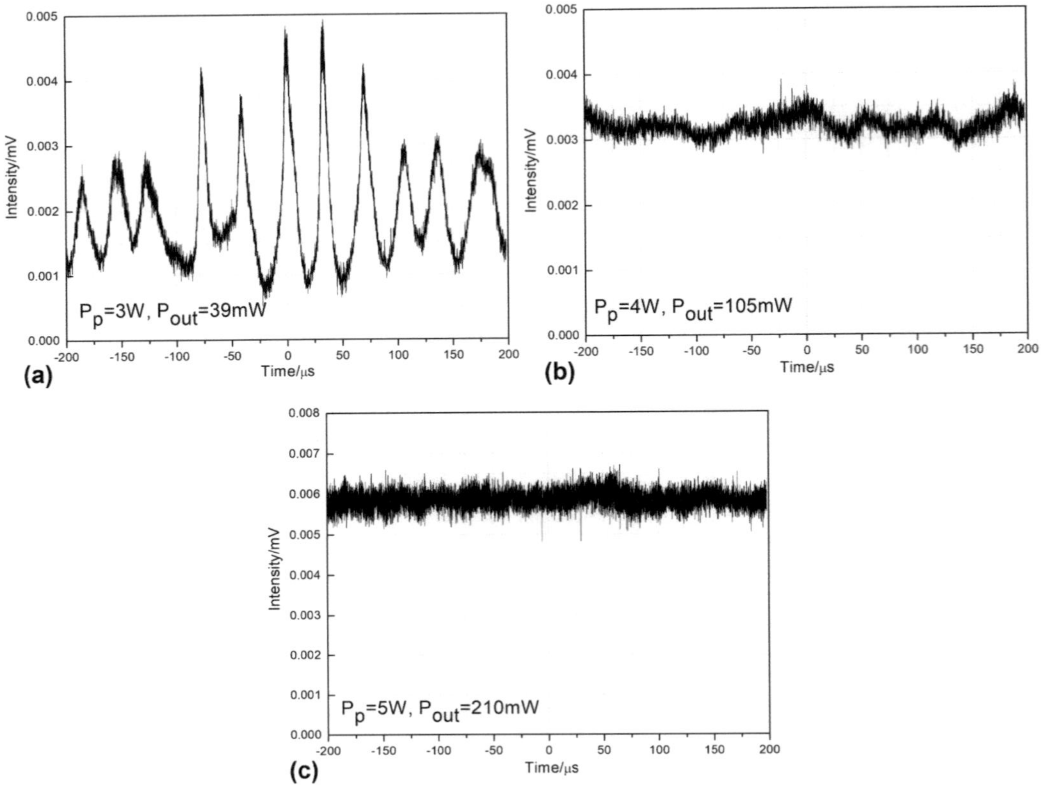

Fig. 31. Output intensity time trace of 10 m fiber laser with 5% output coupler for (a) Pp =3 W, (b) Pp =4 W, (c) Pp =5 W.

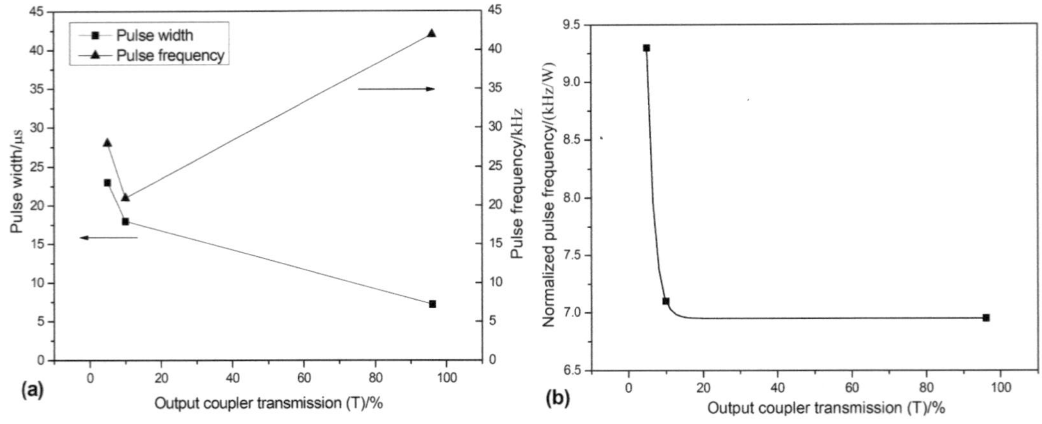

Fig. 32. Output coupler transmission dependence of (a) pulse width and frequency, and (b) normalized pulse frequency near pump threshold.

3. Several possible loss mechanisms for self-pulsing formation

In heavily Tm^{3+}-doped fibers, the distance between Tm^{3+} ions decreases, leading to the formation of ion pairs or clusters, this in turn strengthens the interaction between ions. Such interactions occur among the ions doped in fibers, leading to several energy-transfer processes, one of which is dubbed as the up-conversion process [74].

For Tm^{3+}-doped fiber lasers, as shown in Fig. 35, the pump light at 790 nm excites the ions from $^3H_6$ state to $^3H_4$ state, which quickly relaxes to the upper laser level $^3F_4$. In Tm^{3+}-doped fibers, the up-conversion processes include $^3F_4$, $^3F_4 \rightarrow ^3H_4$, $^3H_6$ and $^3F_4$, $^3H_5 \rightarrow ^3H_6$, $^3F_3$, as shown in Fig 32 (1) and (2). This effect results in one ground ion and one up-converted ion, which quickly relaxes to the $^3F_4$ level. Consequently, this energy transferring process losses one potential stimulated photon. High Tm^{3+} ion doping concentration leads to high ion-pair and ion-cluster concentration, thus induces large quenching effect.

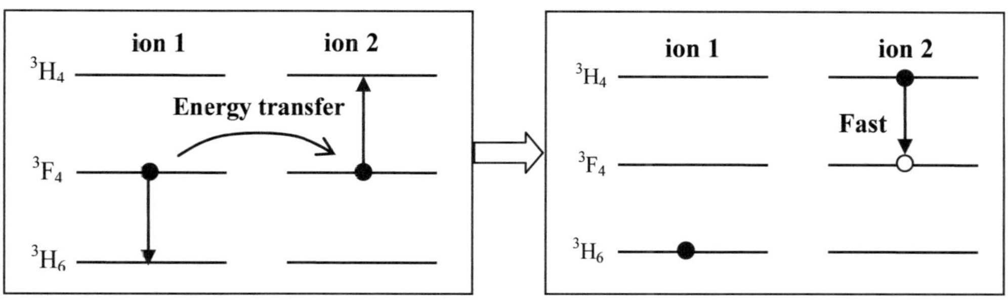

Fig. 33 (1). Up-conversion energy transfer process between the same energy levels in Tm^{3+}-doped fiber lasers

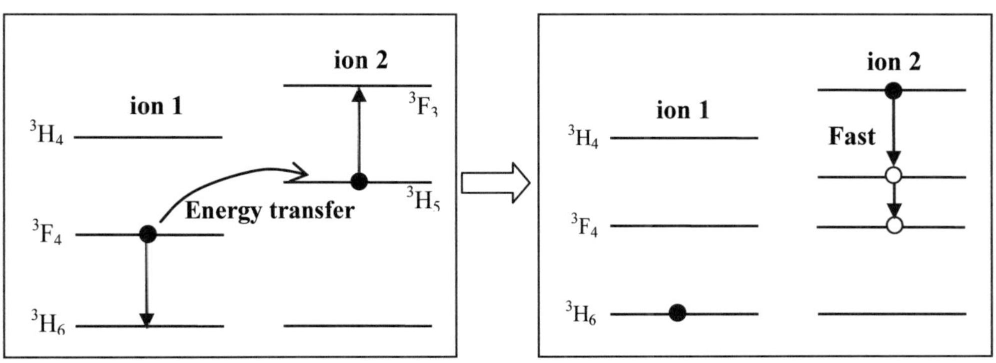

Fig. 33 (2). Up-conversion energy transfer process between different energy levels in Tm^{3+}-doped fiber lasers

Another photon loss mechanism may be due to laser self-absorption (ground-state absorption) through the $^3F_4$, $^3H_6 \rightarrow ^3H_6$, $^3F_4$ energy transfer process, as shown in Fig. 34. When one excited ion and a ground-state ion stay near enough, the excited ion will transfer its energy to the latter and relaxes to ground state, while the latter ion will absorbs the energy and transits to higher levels. When such process occurs repeatedly between a large number of ions, the energy migration process happens, acting as a loss mechanism. These above mentioned energy-transfer processes all have the possibility to act as saturable absorbers.

In the Tm^{3+}-doped fiber laser, self-pulsing is a commonly observed phenomenon, which is considered as an output instability. The true mechanism leading to the formation of this interesting phenomenon is still unclear. In the following section, the origin of self-pulsing in the Tm^{3+}-doped fiber laser will be discussed.

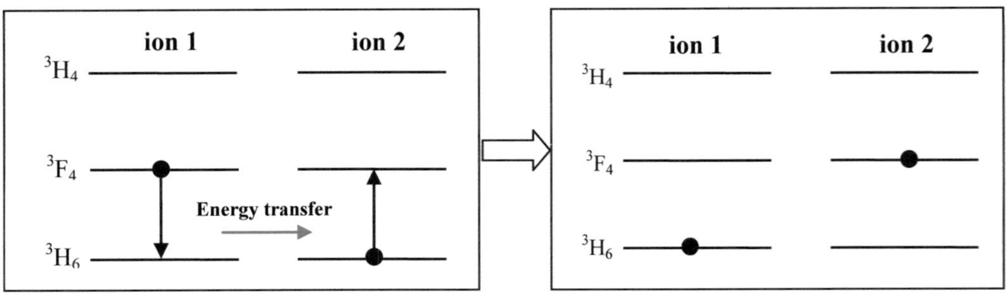

Fig. 34. schematic diagram of self-absorption process in heavily Tm^{3+}-doped fiber lasers

4.2 Theoretical modeling and simulation of Self-pulsing in Tm^{3+}-doped fiber laser

4.2.1 Effects of Excited-state Absorption on Self-pulsing in Tm^{3+}-doped Fiber Lasers

Introduction

Followed various experimental observations, many mechanisms have been proposed to explain the origin of self-pulsing in Tm^{3+}-doped fiber lasers. Some of them are controversial, and consistent agreement has not been satisfied. The in-depth understanding for self-pulsing formation in Tm^{3+}-doped fiber lasers is required.

In this section, mechanisms of self-pulsing in Tm^{3+}-doped fiber lasers are theoretically investigated by taking into account several important energy-transfer processes. A simplified model is constructed to explain the self-pulsing characteristics in Tm^{3+}-doped fiber lasers.

Numerical model

The four lowest energy manifolds of trivalent thulium ions are sketched in Fig. 35. The pump transition, laser transition, and different energy transfer mechanisms including cross relaxation, energy transfer up-conversion and spontaneous decay are indicated. The energy manifolds were numbered 1-4 and these denominations will be used throughout this paper. The rate equations for the local population densities of these levels are as follows [75-77]:

$$\begin{aligned}\frac{dN_4}{dt} &= R(z,t)(N_1 - N_4) - k_{4212}N_1N_4 \\ &+k_{2124}N_2^2 - \frac{N_4}{\tau_4} + \sigma_{sa}c\phi(z,t)(N_3 - N_4)\end{aligned}, \tag{1}$$

$$\begin{aligned}\frac{dN_3}{dt} &= k_{2123}N_2^2 + \beta_{43}\frac{N_4}{\tau_4} - \frac{N_3}{\tau_3} \\ &-\sigma_{sa}c\phi(z,t)(N_3 - N_4)\end{aligned}, \tag{2}$$

$$\frac{dN_2}{dt} = 2k_{4212}N_1N_4 - 2(k_{2124}+k_{2123})N_2^2 + \beta_{42}\frac{N_4}{\tau_4} + \beta_{32}\frac{N_3}{\tau_3} - \frac{N_2}{\tau_2} - c\phi(z,t)\sigma_e(N_2 - \frac{g_2}{g_1}N_1) + c\phi(z,t)\sigma_{ga}(N_1 - N_2) \quad , \tag{3}$$

$$\frac{d\phi}{dt} = c\phi(z,t)\sigma_e(N_2 - \frac{g_2}{g_1}N_1) + m\frac{N_2}{\tau_2} - \sigma_{ga}c\phi(z,t)(N_1 - N_2) - r_c\phi(z,t) \quad , \tag{4}$$

$$N_1 = N_{tot} - N_2 - N_3 - N_4 \, , \tag{5}$$

$$R(z,t) = R(0,t)\cdot e^{-\alpha_p \cdot z} \quad , \tag{6}$$

where N_i are the populations of four energy manifolds $^3H_6$, $^3F_4$, $^3F_5$, $^3H_4$, and N_{tot} is the total density of Tm^{3+} ions. R is the pump rate, and ϕ is the average photon density of the laser field. σ_e is the stimulated emission cross section of signal light, σ_{ga} and σ_{sa} are the absorption cross sections of ground state and excited state, respectively. Where g_1 and g_2 are the degeneracies of the upper and lower laser levels, τ_i is the level lifetimes of four manifolds, and r_c is the signal photon decay rate. β_{ij} are branch ratios from the i to j level, m is the ratio of laser modes to total spontaneous emission modes. The coefficients k_{ijkl} describe the energy transfer processes: k_{4212} and k_{3212} are the cross relaxation constants, and k_{2124} and k_{2123} are the up-conversion constants. The coefficient α_p is the pump absorption of the fiber, which is calculated by $\alpha_p = \sigma_{ap} \cdot N_{tot}$, where σ_{ap} is the pump absorption cross section. In the simulation, the phonon-assisted ESA process of $^3F_4$, $^3H_5 \rightarrow {^3H_6}$, $^3H_4$ and ground-state absorption (GSA) through the $^3F_4$, $^3H_5 \rightarrow {^3H_6}$, $^3F_3$ energy transfer process are considered.
The corresponding parameters for Tm^{3+} ions doped in silica host are listed in Table 1 [12, 77-79].

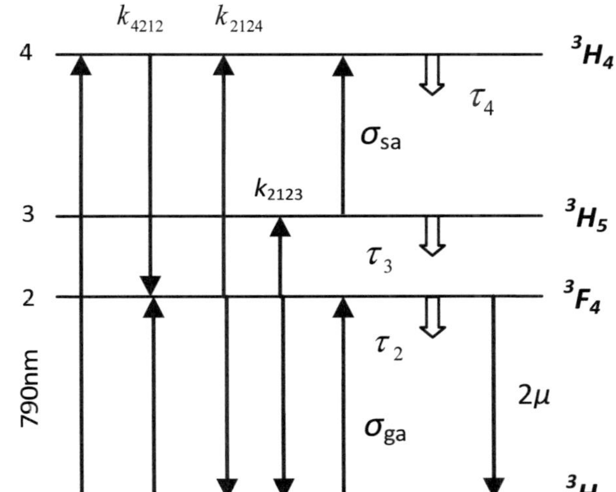

Fig. 35. Schematic of the four lowest energy manifolds in Tm^{3+} ions.

Parameter	numerical value
k_{4212}	1.8×10^{-16} cm^{3}s^{-1}
k_{2123}	1.5×10^{-18} cm^{3}s^{-1}
k_{2124}	1.5×10^{-17} cm^{3}s^{-1}
τ_i	$\tau_4 = 14.2\mu s$ $\tau_3 = 0.007\mu s$ $\tau_2 = 340\mu s$
β_{ij}	$\beta_{43} = 0.57$ $\beta_{42} = 0.051$ $\beta_{32} \approx 1$
σ_e	2.5×10^{-21} cm^2
σ_{sa}	Variable (4×10^{-21} cm^2)
σ_{ga}	variable
m	8×10^{-7}
r_c	9.7×10^{6} s^{-1}
σ_{ap}	1×10^{-20} cm^2
N_{tot}	1.37×10^{20}cm^{-3}

Table 1. The parameters in the rate equations

Theoretical calculation

As can be seen from table 1, the lifetime of level N_3 (0.007 μs) is much shorter than that of level N_2 (340 μs), we can simplify the energy manifolds to three levels. In the above rate equations, we assume the relaxation from N_3 to N_2 is very fast so that N_3~0. Let $N_{23} = N_2+N_3$ and add Eq. (2) and (3), we get

$$\begin{aligned}\frac{dN_{23}}{dt} &= 2k_{4212}N_1N_4 - (2k_{2124} + k_{2123})N_{23}^2 + (\beta_{43} + \beta_{42})\frac{N_4}{\tau_4} - \frac{N_{23}}{\tau_2} \\ &-c\phi[\sigma_e(N_{23} - \frac{g_2}{g_1}N_1) + \sigma_{sa}(N_{23} - N_4) - \sigma_{ga}(N_1 - N_{23})]\end{aligned} \quad . \tag{7}$$

By replaced Eq. (2) and (3) with Eq. (7), the rate equations Eq. (1–6) are simplified to a three-level system. All important energy transfer processes, ESA, and GSA are kept in the simplified rate equations. The simplified model is sufficiently to investigate the dynamic characteristics involved these processes.

Suppose the laser operating in the steady-state (or CW, continuous-wave) regime, the rate of change of the photon density and population must be equal to zero,

$$\frac{d\phi}{dt} = 0 \ , \tag{8}$$

$$\frac{dN_i}{dt} = 0 . \tag{9}$$

Neglecting GSA, the rate equations (1), (4) and (7) lead to

$$\begin{aligned} &R(N_1 - N_4) - k_{4212}N_1N_4 + k_{2124}N_{23}^2 - \frac{N_4}{\tau_4} , \\ &+\phi c\sigma_{sa}(N_{23} - N_4) = 0 \end{aligned} \tag{10}$$

$$c\phi\sigma_e(N_{23} - \frac{g_2}{g_1}N_1) + m\frac{N_{23}}{\tau_2} - r_c\phi = 0 , \tag{11}$$

$$\begin{aligned} &2k_{4212}N_1N_4 - (2k_{2124} + k_{2123})N_{23}^2 + (\beta_{42} + \beta_{43})\frac{N_4}{\tau_4} - \frac{N_{23}}{\tau_2} \\ &-\phi[c\sigma_e(N_{23} - \frac{g_2}{g_1}N_1) + c\sigma_{sa}(N_{23} - N_4)] = 0 \end{aligned} . \tag{12}$$

From Eq. (10-12) and Eq. (5), we can solve ϕ, N_1 and N_{23} with the prerequisite that $\phi > 0$, $N_{tot} > N_1$ and $N_{23} > 0$.

Solving the equations, we find that there is a certain range of pump rate R (defined as ΔR), where the steady-state solution for the rate equations can not be found, as shown in Fig. 36 [80]. In this range, the laser will not be operated in the continuous-wave state. With increase or decrease of pump power out of the range ΔR, the operation of Tm^{3+}-doped fiber lasers undergoes phase transition (changes to CW operation). Such a case is in good agreement with the experimental observation in the self-pulsing operation in Tm^{3+}-doped fiber lasers.

The non-CW range ΔR is calculated as varying the ESA cross section and the cross relaxation parameter k_{4212}. The variation of ΔR as a function of the ESA cross section is shown in Fig. 37 [80]. It is clear that the ESA cross section has an important impact on the self-pulsing operation of Tm^{3+}-doped fiber lasers. The non-CW range ΔR increases with the larger ESA cross section, especially, increases exponentially when the ESA cross section is larger than 3×10^{-21} cm^2. When the ESA cross section is less than 1×10^{-21} cm^2, the range ΔR shrinks sharply, and goes to zero with a small value of ESA cross section. The CW operation of Tm^{3+}-doped fiber lasers can sustain for any pump rate when the ESA cross section is sufficiently small. On the other hand, with a larger ESA cross section, the CW operation will always be broken in certain pump range.

The influence of the cross relaxation on the self-pulsing of Tm^{3+}-doped fiber lasers is evaluated. The non-CW range ΔR is calculated as a function of cross relaxation strength k_{4212} as shown in Fig. 38 [80]. Large values of k_{4212} will obviously enlarge the range ΔR. However, even when the cross relaxation k_{4212} is decreased to zero, the breaking of CW operation still preserves, implying that the cross relaxation energy-transfer process is not the key process in the formation of self-pulsing in Tm^{3+}-doped fiber lasers.

In order to investigate exactly the revolution of the photon density in Tm^{3+}-doped fiber lasers, numerical simulation based on complete rate equations Eq. (1-6) is carried out in the following section.

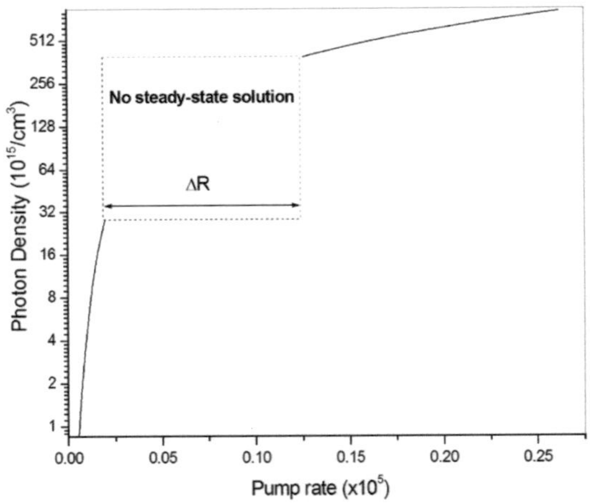

Fig. 36. Photon density as a function of pump rate R.

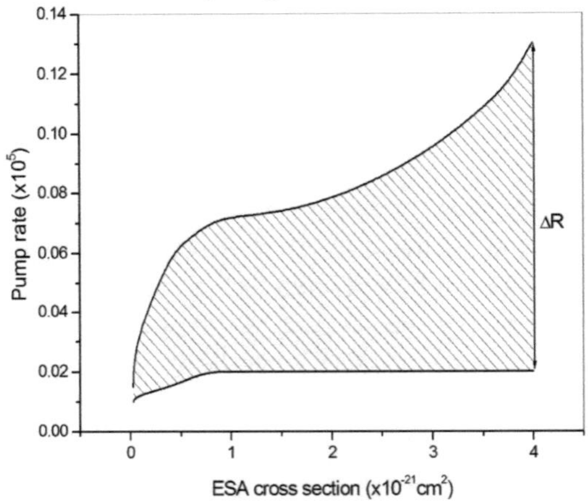

Fig. 37. The non-CW pump range ΔR as a function of the ESA cross section.

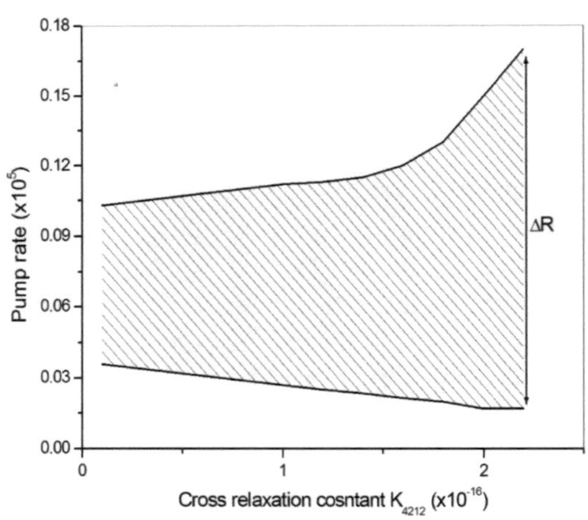

Fig. 38. The non-CW pump range ΔR as a function of the cross relaxation strength.

Simulation results

In order to compare with our experiments, the fiber laser is made up of a 10-m long Tm^{3+}-doped silica fiber with the doping concentration of ~2 wt.%. One fiber end is attached with a dichroic mirror, which is high reflective (R=100%) at laser wavelength and anti-reflective at pump wavelength. Another fiber-end facet is used as the output coupler with signal light transmission of *T*~96%. The pump light is coupled into fiber through the dichroic mirror and the pump rate is set to be 8×10^3 $cm^{-3}s^{-1}$.

In the simulation, the fiber is divided into 100 gain segments. The coupled rate equations are solved in every segment sequentially. The output of previous segment is used as the input of the next segment. The photon intensity in the last segment transmitted through the fiber end is assumed to be the laser output intensity. The returned light is used as the input for the next calculation cycle.

Four energy-transfer processes: cross relaxation, energy transfer up-conversion, GSA and ESA are calculated separately to analyze their influence on the formation of self-pulsing.

A. Cross relaxation

In this sub-section, only the cross relaxation process is taken into account and the processes of energy-transfer up-conversion, GSA and ESA are all neglected. The impact of cross relaxation is evaluated by varying the value of the parameter k_{4212}. The simulation results are shown in Fig. 39 [80]. Stable CW laser operation preserves over a very large region of k_{4212} from 1.8×10^{-20} to 1.8×10^{-12} cm^3s^{-1}. Further decreasing or increasing the cross relaxation strength does not change the nature of the stable CW laser operation. Clearly, the cross relaxation process is not the determinate process leading to self-pulsing formation. With the increase of k_{4212}, the decay of the laser relaxation oscillation will be lengthened, and the laser intensity be increased. A strong cross relaxation parameter may be helpful for improving the slope efficiency of heavily-doped Tm^{3+}-doped fiber lasers.

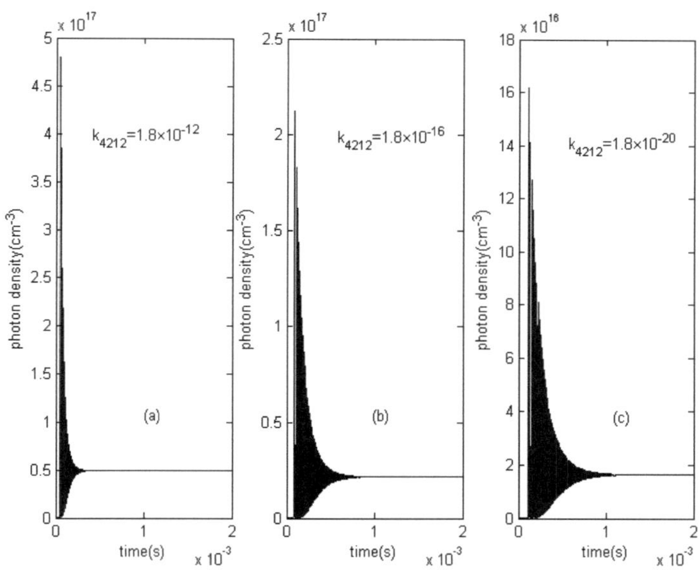

Fig. 39. Laser photon density dynamics characteristics with different cross-relaxation strength k_{4212}.

B. Energy-transfer up-conversion process – k_{2123} and k_{2124}

In this sub-section, the energy-transfer up-conversion process $^3F_4$, $^3F_4 \rightarrow {}^3H_6$, $^3H_5$ (k_{2123}) and $^3F_4$, $^3F_4 \rightarrow {}^3H_6$, $^3H_4$ (k_{2123}) are taken into account. The simulation results are shown in Fig. 40 and 41 [80].

The behaviors of the parameters k_{2123} and k_{2124} are very similar. The up-conversion processes $^3F_4$, $^3F_4 \rightarrow {}^3H_6$, $^3H_5$ or $^3F_4$, $^3F_4 \rightarrow {}^3H_6$, $^3H_4$ consume the population inversion. When the energy-transfer up-conversion is too strong, *i.e.*, $k_{2123} > 1.5\times10^{-17}$ or $k_{2124} > 1.5\times10^{-16}$ cm^3s^{-1}, the laser

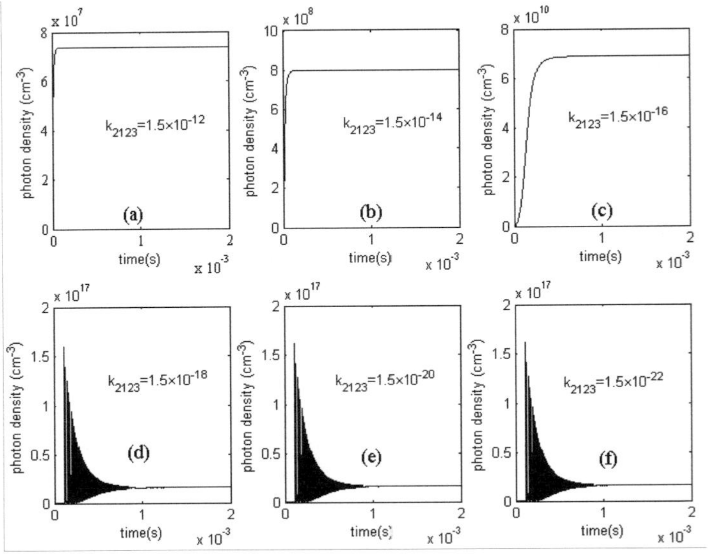

Fig. 40. Laser photon density dynamics characteristics with different energy-transfer up-conversion strength k_{2123}.

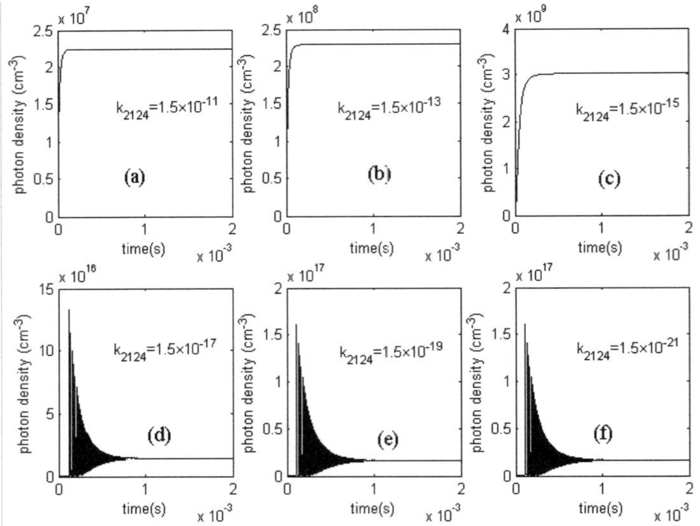

Fig. 41. Laser photon density dynamics characteristics with different energy-transfer up-conversion strength k_{2124}.

relaxation oscillation is suppressed when the pump rate is 8×10^3 $cm^{-3}s^{-1}$, as shown in Fig. 40(a)-(c) and 41(a)-(c). The photon density is clamped in a very low level. The laser threshold is run up by the stronger up-conversion.

When the parameters $k_{2123}<1.5\times10^{-18}$ or $k_{2124}<1.5\times10^{-17}$ cm^3s^{-1}, the laser relaxation oscillation occurs again. The smaller the parameters are, the longer the relaxation oscillation suspends. No matter which values of the parameters (from 1.5×10^{-6} cm^3s^{-1} to zero) are chosen, no self-pulsing phenomenon is observed. The up-conversion process does not directly connect to the self-pulsing operation in Tm^{3+}-doped fiber lasers.

In the practical Tm^{3+}-doped system, the values of k_{2123} and k_{2124} are around 10^{-17} - 10^{-18} cm^3s^{-1}. The main influence of up-conversion is increasing the laser threshold.

C. Ground-state absorption (GSA)

The GSA is also called as the re-absorption in the Tm^{3+}-doped fiber lasers because the laser will be re-absorbed by the ions in the ground state when it propagates along the fiber. The GSA looks like the saturable absorption at the first sight, and had been thought as a possible mechanism for the self-pulsing formation. However, because the photon absorbed by the GSA will be re-emitted back, the laser can not be switched off by the GSA. In such a situation, it is impossible to form the self-pulsing by the GSA.

The GSA process $^3H_6\rightarrow{}^3F_4$ can be thought as a reverse process of the laser transition $^3H_6\rightarrow{}^3F_4$. The photon resonates between the levels $^3H_6$ and $^3F_4$ back and forth, which effectively extends the lifetime of N_2 ($^3F_4$). Consequently, the laser threshold is lowered with a relative large GSA cross section.

In Fig. 42 [80], the revolution of photon density is plotted for various GSA cross section σ_{ga}. Obviously, the laser can generate only when the GSA cross section σ_{ga} is less than the emission cross section σ_e. As the GSA cross section σ_{ga} is taken the value from 1×10^{-21} to 1×10^{-23} cm^2, stable CW operation always occurs after the relaxation oscillation. The final photon density decreases with the smaller GSA cross section σ_{ga}.

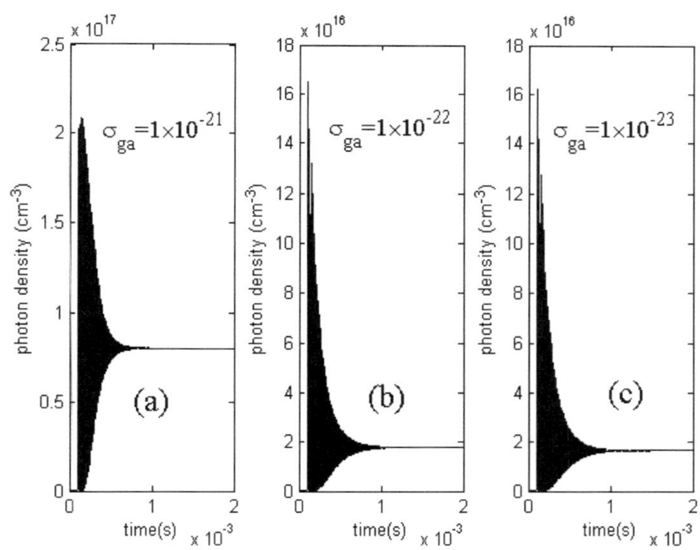

Fig. 42. Laser photon density dynamics characteristics with different ground-state absorption strength σ_{ga} (cm^2).

D. Excited-state absorption (ESA)

As the theoretical analysis in the previous section, the ESA is the key process in the self-pulsing operation of Tm^{3+}-doped fiber lasers. In this sub-section, the cross relaxation k_{4212}, up-conversion k_{2123} and k_{2124}, and GSA cross section σ_{ga} are set to be zero, and only the ESA process $^3H_5 \rightarrow {}^3H_4$ is taken into account. The evolution of photon densities for various ESA cross sections σ_{sa} are shown in Fig. 43 [80]. When the ESA cross section σ_{sa} is chosen in the range from 4×10^{-21} to 4×10^{-19} cm^2, it is clear to observe stable, regular self-pulsed trains. This verifies the theoretical predication that the ESA process is the key reason leading to the self-pulsing dynamics in the Tm^{3+}-doped fiber lasers. The pulse width is about several microseconds and the pulse frequency is tens of kilohertz, showing excellent agreement with the previous experimental results.

When the ESA cross section σ_{sa} is much lower, the ESA is too weak to hinder accumulation of the population in the level $^3H_5$ (N_3), and CW operation occurs after relative long relaxation oscillation as shown in Fig. 43 (c) and (d). With the increase of the ESA cross section σ_{sa}, the decay time of the relaxation oscillation becomes longer and longer, and finally, the relaxation oscillation evolves to a stable self-pulsed train. On the other hand, when the ESA cross section is very large, a great number of population in the level $^3H_5$ (N_3) is depleted by the ESA. Consequently, the population inversion in the level $^3F_4$ (N_2) is not enough to sustain the laser oscillation.

As shown in Fig. 43 (a) and (b), the self-pulse repetition rate and pulse width are reduced as increase of the ESA cross section. Although the self-pulsing is induced by the ESA, the pulse properties are influenced by the cross relaxation, up-conversion, and GSA.

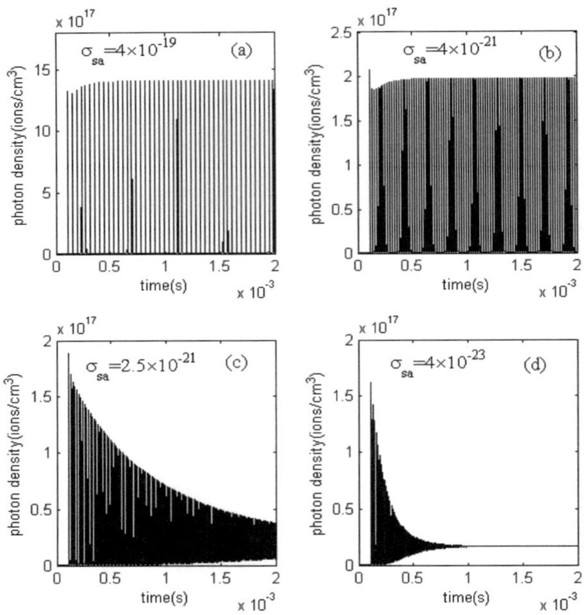

Fig. 43. Laser photon density dynamics characteristics with different ESA strength σ_{sa} (cm^2).

Conclusion

Based on theoretical analysis and numerical simulation, the ESA (excited-state absorption) process is clarified as the key reason leading to the formation of self-pulsing in Tm^{3+}-doped fiber lasers. Increasing the ESA cross section can reduce both the self-pulse frequency and pulse width. With increase of the pump rate, the operation of Tm^{3+}-doped fiber lasers will undergo phase transition (CW to self-pulsing, and then back to CW). The laser threshold and efficiency are influenced by the cross relaxation, up-conversion and the GSA.

4.2.2. Theoretical study on self-pulsing in Tm^{3+}-doped fiber lasers

1. Introduction

In the previous section, explicit theoretical analysis has proved that the ESA process is responsible for self-pulsing formation in fiber lasers [80]. In this section, the self-pulsing characteristics in Tm^{3+}-doped fiber lasers are theoretically investigated by changing several key parameters-the output coupling, pump rate and active ion doping concentration. Besides, how to optimize these corresponding parameters for obtaining expected laser pulse frequency and pulse width, and potential applications of self-pulsing are discussed.

2. Simulation results

The simulation process and the adopted rate equations and corresponding parameters are described in the previous section.

2.1. Influence of output coupling T on the self-pulsing characteristics

In this section, the ESA cross section is kept at 4×10^{-21} cm^2, and the impact of the output coupling strength on the self-pulsing characteristics is simulated. It is found that too low pump rate will not lead to self-pulsing with any output coupling T. The minimum pump rate required to initiate self-pulsing operation is defined as self-pulsing pump threshold R_{threP}. The self-pulsing threshold R_{threP} as a function of output coupling is shown in Fig. 44 (a) [81]. The self-pulsing threshold increases first moderately and then quickly with the output coupling. This is due to that high output coupling causes high cavity loss. After

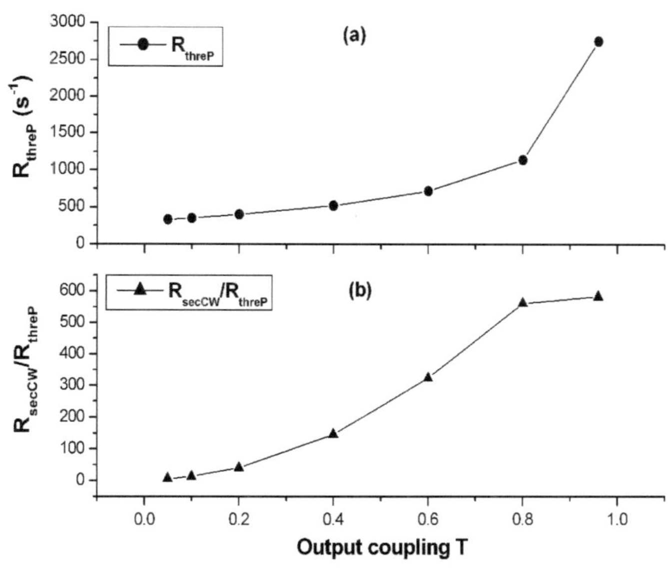

Fig. 44. Self-pulsing threshold (a) and the ratio of second CW threshold to self-pulsing threshold (b) as a function of output coupling.

being initiated, the self-pulsing state will preserve stably for a certain pump range. Further improving the pump rate, self-pulsing operation will be transferred to continuous-wave (CW) state. Because the laser operates in CW mode both before and after the occurrence of self-pulsing, the pump rate that renders the laser to CW mode after the self-pulsing regime is defined as the second CW threshold R_{secCW}. The ratio of the second CW threshold (R_{secCW}) to the self-pulsing threshold (R_{threP}) will indicate the capability of a fixed output coupling to support self-pulsing operation in the fiber laser. This threshold ratio R_{secCW}/R_{threP} as a function of output coupling is shown in Fig. 44 (b) [81]. It is clear that the ratio (R_{secCW}/R_{threP}) increases near linearly with output coupling. Therefore, in order to achieve self-pulsing operation over a large power range, high output-coupling cavity configurations should be adopted. Stable operation of self-pulsing over a large power range will offer a new alternative to obtain pulsed laser output. Therefore, self-pulsing modulation has the potential to become a novel Q-switching technique.
Fig. 45 shows the numerically calculated self-pulsing train of the Tm^{3+}-doped fiber laser with output coupling of T=96% at the threshold pump rate [81]. This regular self-pulsing train has a pulse width and repetition rate of 7.68 μs and 16.89 kHz, respectively. Self-pulsing begins after the pump power being switched on for 0.4 ms.

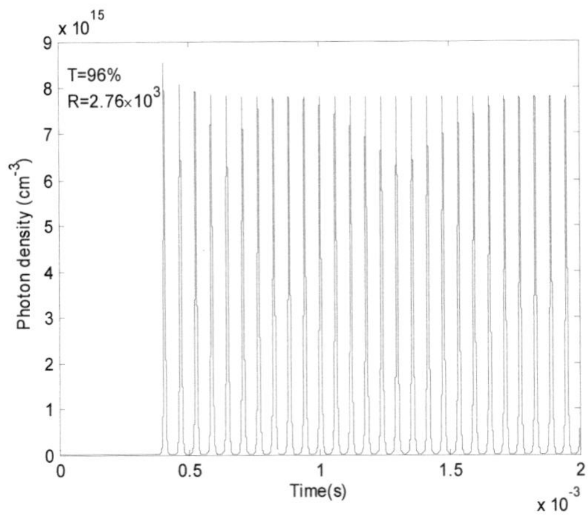

Fig. 45. Self-pulsing train at corresponding threshold pump rate with output coupling T=96%.

At respective self-pulsing threshold pump rates, the laser pulse width and pulse frequency as a function of the output coupling are shown in Fig. 46 [81]. Increasing the output coupling, the laser pulse frequency and pulse width respectively increases and decreases near linearly. As shown in Fig. 46, higher output couplings need higher pump rates to initiate self-pulsing. Higher pump rate induces higher population inversion and higher gain, leading to higher self-pulsing repetition rate. Narrower pulse width at higher coupling is due to the reduction of cavity lifetime resulted from higher coupling loss. This offers a clue that, in order to achieve shorter pulse width and higher repetition rate simultaneously in self-pulsing fiber lasers, higher output couplings should be adopted.
In order to further the understanding about the influence of output coupling on the self-pulsing features, we carry out simulation with different output coupling strengths at a identical pump rate of $R=4\times10^3$. With this pump rate, self-pulsing operation can be achieved

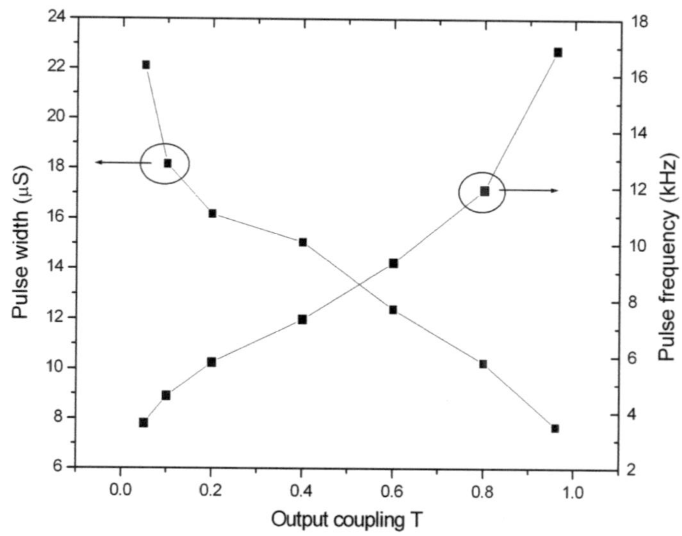

Fig. 46. Pulse width and frequency for different couplings at respective threshold pump rate.

between T=96% and T=10%. When T<10%, self-pulsing cannot be sustained. Detailed values about the pulse width and pulse repetition rate are shown in Fig. 47 [81]. The output coupling of T=10% provides the maximum pulse frequency of ~75 kHz. As the output coupling increases from T=10% to T=96%, the pulse frequency drops down sharply first and then more slowly, reaching a minimum value of 30 kHz. Higher pulse frequency at lower output coupling is due to that lower output coupling provides less cavity loss thus a faster

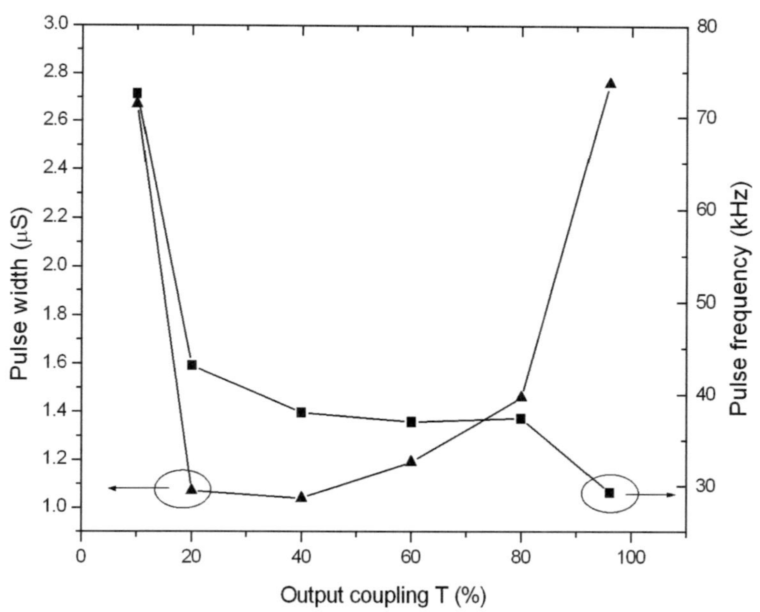

Fig. 47. Pulse width and frequency for different output couplings at pump rate $R=4\times10^3$.

gain recovery. Therefore, in making use of self-pulsing to obtain high-frequency pulses, lower output couplings are preferred. As the output coupling increases from T=10% to T=96%, the pulse width decreases sharply first, then arrives at a minimum value, thereafter increases steadily and then significantly. Increasing the output coupling reduces the population inversion, but also the photon lifetime in the cavity, which play opposite roles on the pulse width. Therefore, an optimum output coupling (~40%) exists that produces the shortest pulse duration of ~1 μs.

2.2. Influence of pump rate R on the self-pulsing characteristics

In this section, fiber-end facet is considered as the output coupler (T=96%), the ESA cross section is kept at 4×10^{-21} cm^2, and the pump rate is changed to investigate its impact on the self-pulsing characteristics. Increasing the pump rate, the laser operation undergoes several stages. First, the CW operation occurs at a comparatively low pump rate. Thereafter, self-pulsing begins when the pump rate arrives at the self-pulsing threshold $R_{threP}=2.76\times10^3$. The self-pulsing operation maintains for some pump power range, and then transfers to CW mode as the pump rate further reaches the second CW threshold R_{secCW}. In the self-pulsing regime, the pulse width and pulse frequency as a function of pump rate are depicted in Fig. 48 [81]. As the pump rate increases, the pulse frequency increases first slowly, and then rapidly. On the contrary, the pulse width decreases sharply first, and then saturates to about 200 ns, finally shows a little rise again. At the self-pulsing threshold, the pulse width and pulse frequency are 7.68 μs and 16.89 kHz, respectively. As the pump rate increases to 5.72×10^3, the pulse width and pulse frequency decreases and increases to 1.83 μs and 38.3 kHz, respectively. The maximum pulse frequency approaches 900 kHz, and the minimum pulse width is about 200 ns. An interesting phenomenon is that the pulse width keeps nearly constant at the minimum value over a very large pump range (as the double arrow line denotes). From this phenomenon, several conclusions can be arrived at. First, the self-pulsing must originate from some intra-ionic processes. This excludes such mechanisms as

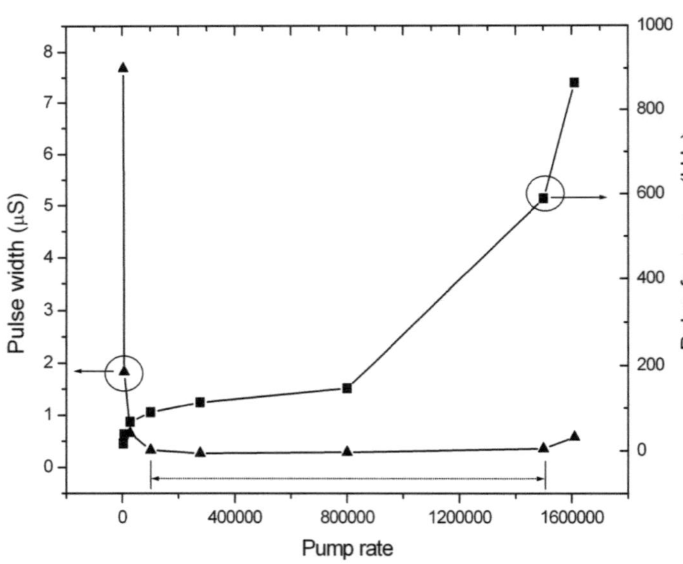

Fig. 48. The pulse width and pulse frequency versus the pump rate for T=96% output coupling.

Brillouin scattering and interaction between longitudinal modes from accounting for self-pulsing formation. Secondly, stable self-pulsing operation can be realized with high output power, showing a great power scalability of this pulsing technique. Thirdly, there is a limitation in the achievable minimum pulse width. Further narrowing the pulse width may require combining with other modulation techniques, such as nonlinear polarization control et al.

2.3. Influence of active-ion doping concentration N_{tot} on the self-pulsing characteristics

In Tm^{3+}-doped fiber lasers, appropriately high Tm^{3+} doping concentration can strengthen the cross relaxation process, which significantly enhances the quantum efficiency of the laser. However, too high Tm^{3+} doping concentration will form Tm^{3+} ion clusters thus present concentration quenching. Besides, high doping level can induce strong energy up-conversion processes, leading to reduction of population inversion. So in fact, there is an appropriate Tm^{3+} doping concentration, which provides the fiber laser the maximum laser efficiency.

In this section, simulation is carried out with constant output coupling T=96% and constant pump rate $R=4\times10^3$. Self-pulsing operation is observed by changing the particle density N_{tot} (Tm^{3+} ion density). As the particle density increases from 1.37×10^{20} to 1.37×10^{21} cm^{-3}, the self-pulsing threshold R_{threP} decreases more than ten times. The pulse width and repetition rate as a function of N_{tot} are shown in Fig. 49 [81]. As the particle density increases from 1.37×10^{20} to 1.5×10^{21} cm^{-3}, the pulse frequency grows from 30 kHz to ~145 kHz near linearly. Higher doping concentration leads to higher cross relaxation, energy up-conversion and signal light re-absorption. The combination of these processes can speed the recovery of population inversion after a pulse output, thus improve the pulse repetition rate. Increasing the particle density, the laser pulse width decreases sharply first, and then slowly. Therefore, comparatively higher doping concentration is preferred to simultaneously

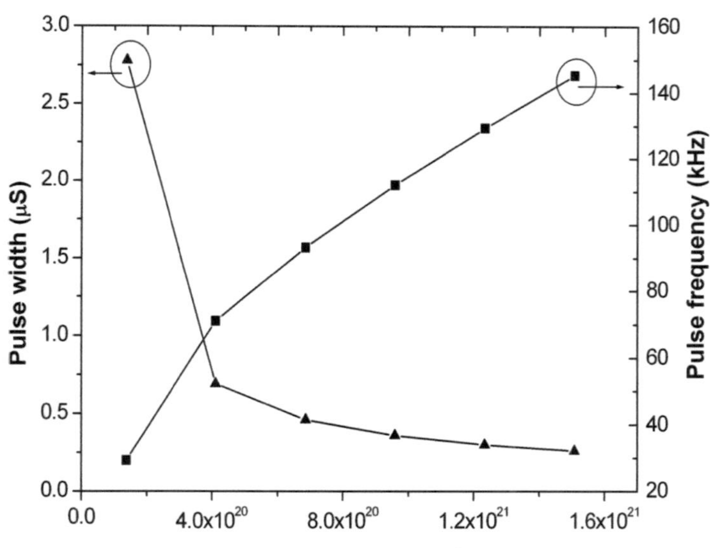

Fig. 49. Pulse width and frequency as a function of doping concentration at pump rate $R=4\times10^3$.

achieve high pulse frequency and narrow pulse width in self-pulsing fiber lasers. However, especially high active-ion doping in fibers is impracticable. In addition, too high doping concentration will cause ion clustering, thus decrease laser efficiency. Therefore, there is a tradeoff between obtaining narrow pulse width and achieving high laser efficiency. Further narrowing the pulse duration requires combining self-pulsing with other modulation techniques.

3. Conclusion

Through numerical simulation, self-pulsing behavior of Tm^{3+}-doped fiber lasers is theoretically analyzed, and the influence of pump rate, output coupling and ion doping concentration is explicitly studied. In order to achieve self-pulsing over a large pump range, high output couplings are preferred. At a fixed pump level, higher output coupling decreases pulse frequency, but narrowest pulse width can only be achieved with an optimum output coupling. For a given output coupling, higher pump strength increases the pulse frequency, but decreases and saturates the pulse width. Reduction of total cavity loss helps narrowing the pulse width. Increasing doping concentration significantly decreases the self-pulsing threshold. In self-pulsing Tm^{3+}-doped fiber lasers, high pulse frequency and narrow pulse width can be simultaneously achieved with appropriately higher active-ion doping concentration.

Understanding the self-pulsing characteristics and methods as how to make use of it will help improving the performance and utility of 2-μm Tm^{3+}-doped fiber lasers. The excellent features of self-pulsing provide it the potential to become a new competitive Q-switching technique. Self-pulsing has great prospects in applications such as constructing self-switching pulsed lasers, improving pulsed lasers' output power, and simplifying the cavity configuration of ultra-short-pulse-width laser systems.

4.3 Passively Q-switching Tm^{3+}-doped fiber laser with Cr^{2+}:ZnSe

1. Introduction

Thulium-doped fiber lasers have an especially broad emission band, providing great potential for achieving ultra-short mid-infrared pulses. For obtaining pulsed laser output, various modulation techniques, can be used, e.g., gain-switching, Q-switching, cavity dumping and mode-locking. Gain-switched Tm^{3+}-doped silica fiber lasers have generated high pulse energy over 10 mJ in several hundred nanosecond pulse duration [82-83], while Q-switched Tm^{3+}-doped silica double-clad fiber lasers have produced several-kW peak power with the pulse durations around 100 ns [84-85]. Ultra-short 2-μm pulses have been achieved from mode-locked thulium-doped fiber lasers with the pulse durations as less as several hundred femto-seconds [86].

Compared with the positive Q-switching techniques, Passive Q-switching is an attractive approach to construct simple, robust, and cost-efficient pulsed lasers.

Many saturable materials have been utilized as the modulation elements in the 2-μm spectral range, including Ho^{3+}:CaF_2, Co^{2+}:ZnSe, and Cr^{2+}:ZnSe crystals [87]. Tm^{3+}-doped silica fiber lasers near 2 μm passively Q-switched by a Cr^{2+}:ZnSe crystal have been demonstrated with the pulse duration of 330 ns and peak power of 15 W [87]. Passive Q-switching Tm^{3+}-doped silica fiber lasers by a Ho^{3+}-doped silica fiber has also been investigated with the pulse width of ~1 μs [88].

In this section, short-length passively Q-switched Tm^{3+}-doped silica fibers (5-50 cm) are described with polycrystalline Cr^{2+}:ZnSe microchips as the saturable materials.

2. Experimental setup

The polycrystalline Cr^{2+}:ZnSe, prepared by CVD technique and doped by thermal diffusion method, has a relaxation time of 5 μs [89], which is significantly shorter than the lifetime of the upper laser level of Tm^{3+} ions (3F4→3H6) of ~335 μs in silica [12]. The Cr^{2+} doping concentration of the polycrystalline is ~7×10^{18} cm^{-3}. Double-clad large-mode-area fibers are used to increase the extractable energy from short-length fibers. The core, doped with approximate 2.5 wt.% Tm^{3+} ions, is 27.5 μm in the diameter and 0.20 of numerical aperture (NA). The pure silica inner cladding, coated with a low-index polymer, has a 400-μm diameter and NA of 0.46.

Fig. 50 shows the experimental setup [90]. A high-power laser-diode array operating at 790 nm is used as the pump source. The pump beam is reshaped to a square beam pattern by a micro-prism stack, and then focused into the fiber using a cylindrical lens and an aspheric lens. The pump beam is launched into the fiber through a dichroic mirror with high reflectivity (>99.7%) at 2.0 μm and high transmission (>97%) at 790 nm. The output end of the fiber is butted directly to the polycrystalline Cr^{2+}:ZnSe. Whole fiber is clamped in an air-cooled copper heat-sink. At the same time, the polycrystalline Cr^{2+}:ZnSe microchip is fixed separately on a water-cooled heat-sink to prevent it from thermal-induced fracture.

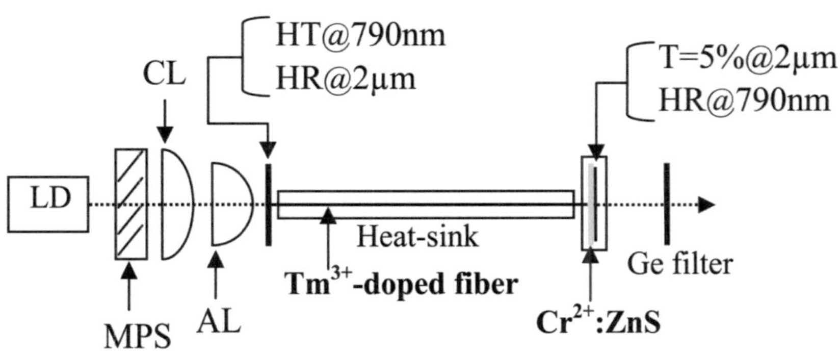

Fig. 50. Schematic of the experimental setup; LD: laser diode, MPS: micro prism stack, CL: cylindrical lens, AL: aspheric lens, HT: high transmission, HR: high reflection.

Laser output power is measured with a pyroelectric power meter after the leaky pump light being blocked by a Ge filter (T=0.1% at 790 nm). Laser spectra are examined by a mid-infrared spectrograph with a resolution of 0.2 nm and a TEC-cooled InAs detector (J12 series). An InAs PIN photodiode and a 500 MHz digital oscilloscope are used to measure the laser temporal characteristics.

3. Results and discussion

The bleaching experiment is carried out with an uncoated polycrystalline Cr^{2+}:ZnSe microchip with thickness of 1 mm. A 2-μm Tm^{3+}-doped silica fiber laser with output power up to 3 W is used as the test source. The 2-μm laser was focused onto the crystal with an objective lens of 11-mm focal length and NA of 0.25. The beam diameter at the focus is measured to be 30 μm by using the knife edge method. Figure 51 shows the power transmission for various laser intensities onto the crystal [90]. The transmission nearly remains a constant ~0.6 as the incident intensity is above 164 kW/cm². The saturation intensity I_s ~164 kW/cm² is higher than that from Cr^{2+}:ZnSe crystal [87]. The higher saturation intensity is probably attributed to uncoated surfaces and additional scattering loss in the polycrystal. From the saturation intensity I_s and fiber diameter, the laser power to saturate the polycrystal is calculated to be ~1.3 W.

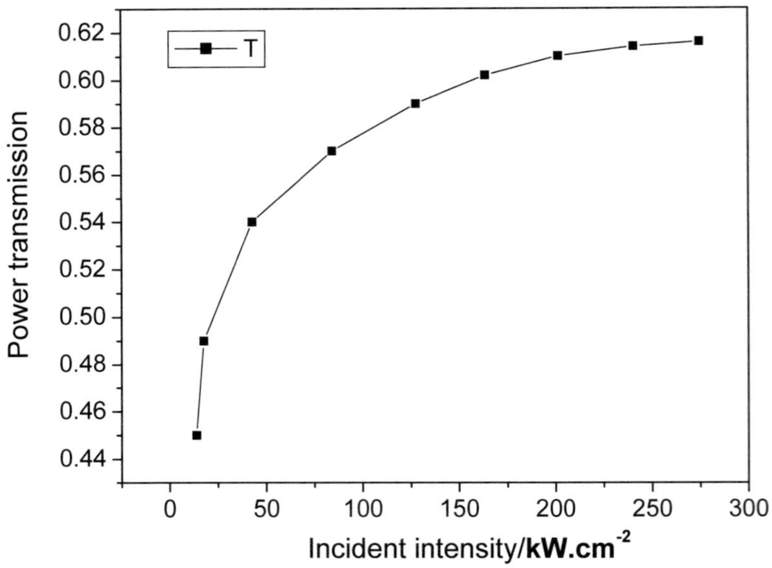

Fig. 51. Power transmission characteristics by 1 mm thick Cr^{2+}:ZnSe polycrystalline showing saturation intensity I_s=164 kW/cm².

One surface of the polycrystalline microchip is precisely polished and the other one is coated with partial reflectivity (R=95%) at ~2 μm wavelength and high reflectivity (R>99.8%) at 790 nm. To Q-switch the fiber laser, different thicknesses of polycrystalline Cr^{2+}:ZnSe microchips are chosen for various fiber lengths. The selected thickness of polycrystalline Cr^{2+}:ZnSe microchip for various fiber lengths is shown in Table 2 [90].

Fiber length (cm)	5	10	20	50
Thickness of Cr^{2+}:ZnSe (mm)	0.3	0.4	0.5	1

Table 2. Parameters of the fiber length and thickness of Cr^{2+}:ZnSe

The characteristics of a 5-cm long Q-switched fiber laser are shown in Fig. 52 and 53 [90]. The fiber laser has a threshold pump power of ~3 W. Maximum average output power is around 50 mW, corresponding to a slope efficiency of ~0.5% with respect to ~10 W launched pump power. The output power increases nearly linearly with increasing launched pump power. When the pump power higher than 10.5 W, the dielectric film on the Cr^{2+}:ZnSe is damaged. The low output power and slope efficiency arise primarily from insufficient pump absorption in short fibers and reflection loss at the uncoated surface of the polycrystalline Cr^{2+}:ZnSe microchips.

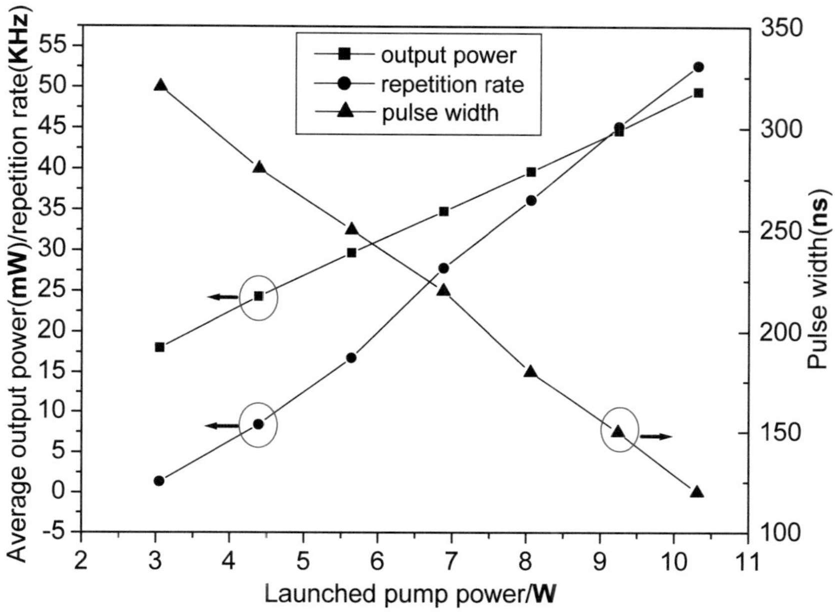

Fig. 52. Average output power, pulse repetition rate and pulse width as a function of launched pump power for a 5 cm-length fiber.

When the polycrystalline Cr^{2+}:ZnSe microchips are replaced by a partially reflecting mirror, the short fiber lasers are operated in the CW regime. The performance of CW operation of short Tm^{3+}-doped fiber lasers has been described in the previous publication [33]. The maximum output power of a 5-cm long fiber laser is ~290 mW, corresponding to a slope efficiency of 2.9%. The Q-switched fiber lasers can extract about 17% energy of the CW operation.

Pulse width of 120 ns is obtained from the 5-cm long fiber laser, which presents the shortest pulse duration achieved from passive Q-switched Tm^{3+}-doped fiber lasers. It is expected that sub-hundred-nanosecond pulses can be accessible by shortening the fiber length to ~1 cm. The pulse repetition rate increases linearly and reaches the maximum value of 53 kHz with the pump power of 10.3 W.

Peak power shown in Fig. 53 is obtained by dividing the pulse energy by the pulse width [84]. As the pump power is increased, both the pulse energy and peak power decrease. The decrease of pulse energy and peak power can be accounted by the faster increase of repetition rate than that of the output power (see Fig. 52). The maximum pulse energy and peak power are ~14 μJ and 45 W, respectively, which decreases to ~1 μJ and ~8 W at higher power level.

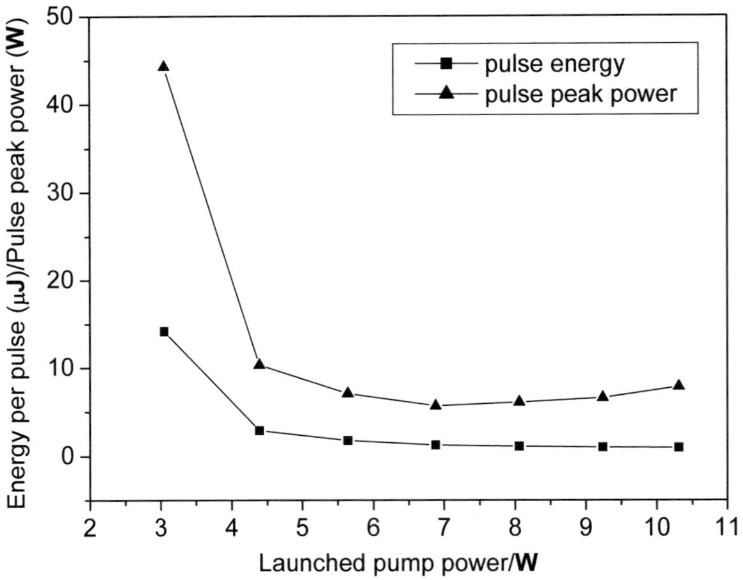

Fig. 53. Energy per pulse and pulse peak power as a function of launched pump power for the 5 cm-length fiber.

With 8-W launched pump power, a typical pulse train and individual pulse obtained from the 50-cm long fiber laser are shown in Fig. 54. As shown in Fig. 54(a), the intensity stability of the Q-switched pulse train (84 kHz) is over 90%, and the pulse spacing stability is near 90%. The single laser pulse in Fig. 56(b) shows a smooth and typical pulse shape with the pulse duration of ~450 ns. Mode-locking phenomenon, investigated by T. A. King et al [87] in passive Q-switching long Tm^{3+}-doped fibers, has not been observed in our short Tm^{3+}-doped fibers.

Average output power, repetition rate and pulse width, obtained at pump level of 10.3 W, as a function of fiber length are shown in Fig. 55 [90]. As a result of longer cavity round-trip time, the pulse duration becomes broader by using the longer fiber. When the fiber is longer than 20 cm, the pulse duration is broadened to around 300 ns. The pulse repetition rate increases linearly with fiber length at first, and then shows a sharp augmentation when the fiber longer than 50 cm. The highest average output power and repetition rate generated from a 50 cm-long fiber laser, are ~200 mW and 530 kHz, respectively.

Pulse energy and peak power obtained for various fiber lengths at pump power of 3 and 10.3 W, are shown in Fig. 56 [90]. As the fiber length is changed from 5 to 50 cm, the pulse energy displays a moderate decrease with increasing fiber length. On the other hand, the peak power reduces by one order of magnitude. The decrease of peak power comes mainly from the broadening of the pulse width as shown in Fig. 55. When the fiber length changed from 5 to 20 cm, the pulse width increases much faster, leading to a remarkable decrease of the peak power.

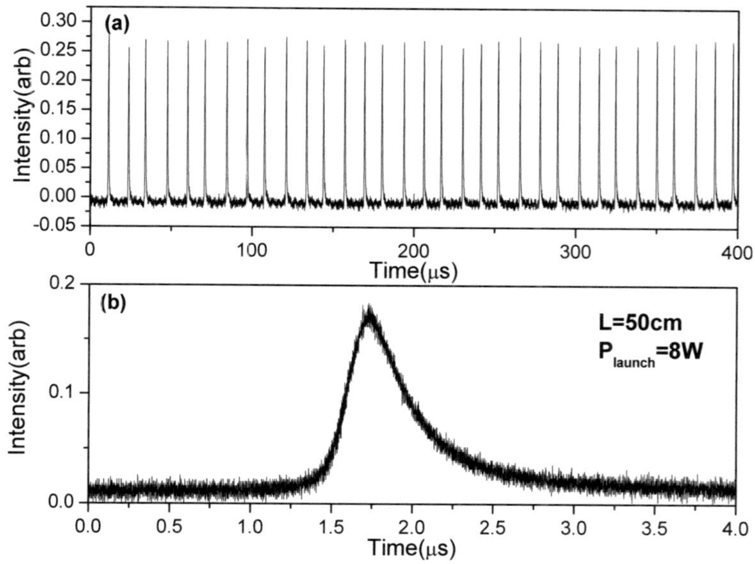

Fig. 54. The pulse train and a single pulse for a 50 cm-length fiber at 8 watts of launched pump power.

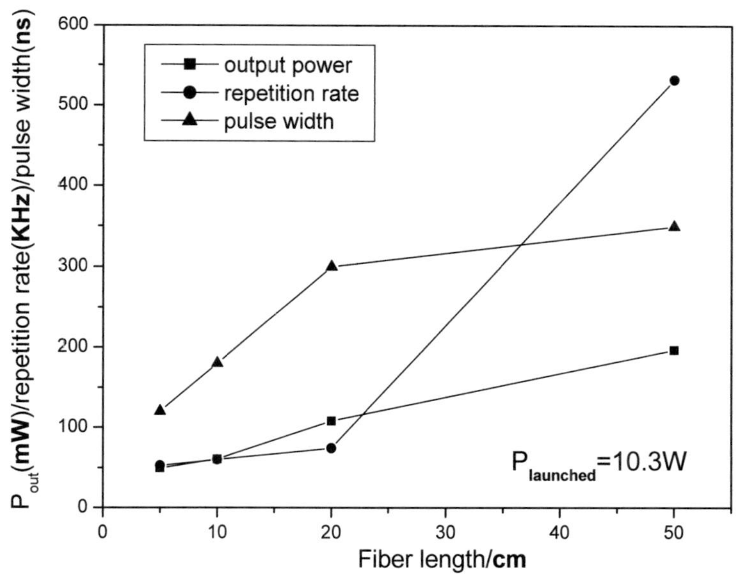

Fig. 55. Maximum average output power, repetition rate, and minimum pulse width as a function of fiber length at 10.3 W of launched pump power.

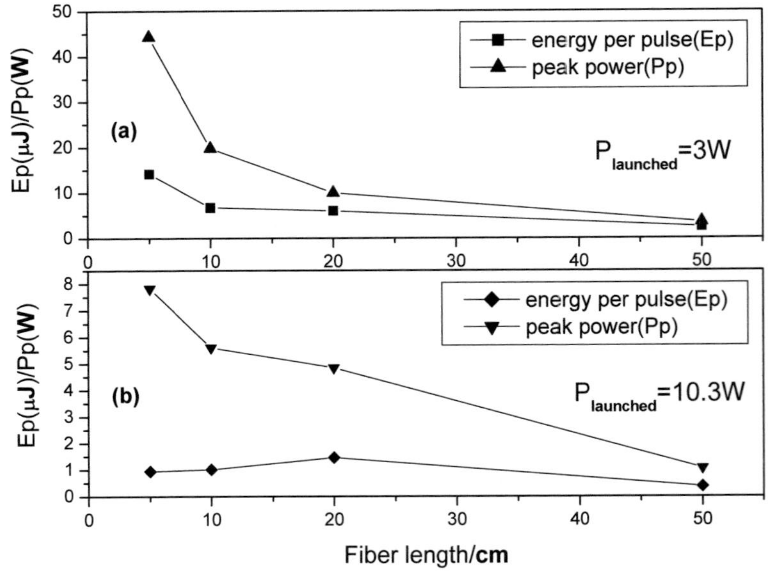

Fig. 56. Pulse energy and pulse peak power as a function of fiber length at launched pump power of 3 W (a) and 10.3 W (b).

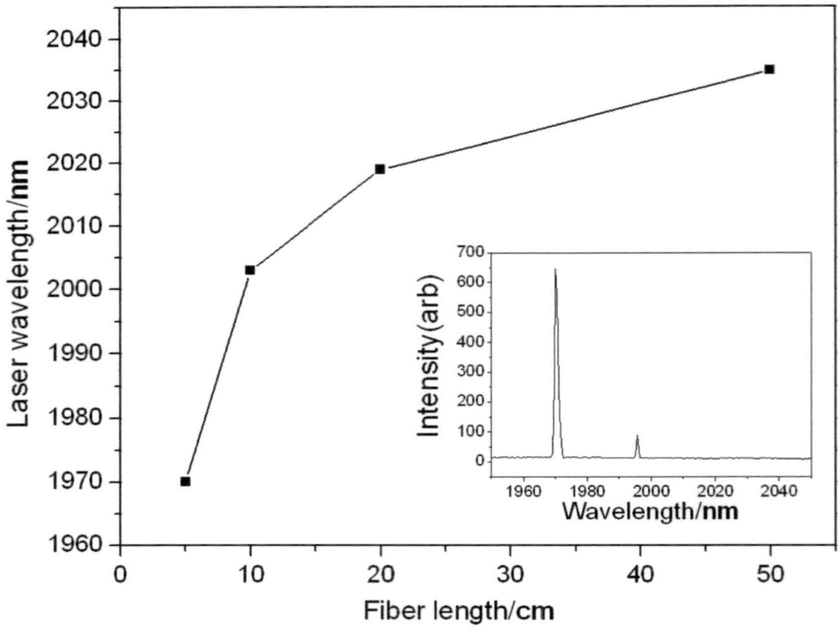

Fig. 57. Laser wavelength as a function of fiber length; inset is the laser spectrum obtained with the 5 cm-length fiber at 8 W of pump power.

Laser wavelength of the fiber laser as a function of fiber length is shown in Fig. 57 [90]. When the fiber length alters from 5 to 50 cm, the laser wavelength shows a parabolic increase from 1970 to 2035 nm. The elongation of wavelength is a result of the re-absorption of laser light in the Tm^{3+}-doped fibers. From the inset of Fig. 57, the fiber laser was operated in multiple modes with the bandwidth (FWHM) of around 2 nm for the main peak.

4. Conclusion

Polycrystalline Cr^{2+}:ZnSe microchips can be used to passively Q-switch the thulium-doped silica fibers effectively. With this technique, pulse duration as short as 120 ns and repetition rate as high as 530 kHz can be achieved from several-centimeter-length Tm3+ fibers. The moderate pulse energy and peak power (14 μJ and 45 W) can be further improved by increasing the thulium doping concentration or adopting high-gain, high absorption glass fibers. The combination of polycrystalline microchips, short-length fibers and LDA pump source provides a simple and reliable laser device. Besides, the easy-growth feature of polycrystalline Cr^{2+}:ZnSe adds the laser system outstanding advantages of convenience and economy. Due to its wide absorption spectrum from ~1.5 to 2.1 μm, polycrystalline Cr^{2+}:ZnSe can also be used to passively Q-switch Er^{3+} lasers and Tm^{3+}:Ho^{3+}-doped lasers. Such short pulse width, 2-μm lasers have wide applications in trace gas detecting and communications.

5. Tm^{3+}-doped fiber laser pumped Cr^{2+}:ZnSe solid-state laser

Introduction

The application requirement for room-temperature tunable mid-IR sources like, e.g. spectroscopic, medical, and remote sensing has provoked increased interest for long-wavelength, i.e. above 2 μm, tunable directly diode-pumped solid-state lasers.

Divalent transition-metal doped II-VI chalcogenides laser materials have such interesting features as: room-temperature operation between 2 and 5μm, broad tunability, negligibly low excited state absorption, high emission cross section, possibility of direct diode pumping and CW operation. Infrared transition metal ion-doped solid-state lasers, have become the critical components in optical frequency standards, space-based remote sensing systems, and have potential application in femtochemistry and attosecond science[91-94].

For achieving laser output over 2 μm, Cr^{2+}:ZnSe crystal is a good candidate. The low maximum phonon frequency in ZnSe (~250 cm-1) leads to a low nonradiative decay rate and high fluorescence quantum yield (close to unity) [95]. This provides Cr:ZnSe with the highest gain among vibronic lasers and enables efficient room-temperature operation. The remarkable characteristics of the ultrabroadband (~1000 nm) Cr^{2+}:ZnSe crystal, such as the high emission cross section of the order of 10^{-18} cm^2 [96], the negligibly low excited state absorption (ESA) [97], the fairly good chemical and mechanical stability and the thermal conductivity nearly as high as in sapphire, gives this material enormous potential as a laser medium for diode-pumped tunable MIR lasers.

Remarkable progress has been achieved in Cr^{2+}-doped zinc chalcogenides, and in particular of Cr:ZnSe as broadly tunable continuous wave (CW) laser operating around 2.5μm. Diode-pumped room-temperature CW Cr^{2+}:ZnSe laser was first demonstrated by Sorokin et al. in 2002 [93]. Tuning below 2μm was first demonstrated by Umit and Alphan in 2006 [98].

Record of the highest average pulse power of 18.5W was achieved by CTI in 2004 [99]. Sorokina et al have achieved a wavelength tuning range over 1100 nm and ps-level laser pulses in Cr^{2+}:ZnSe lasers [100-101]. Among all solid-state laser materials, the gain spectrum of Cr^{2+}:ZnSe is the broadest. This feature of the Cr^{2+}:ZnSe laser make it possible to produce 1~2 optical-cycle ultra-short laser pulses, which will find great potential applications in the femtosecond laser regime.

5.1 Watt-level CW ceramic Cr^{2+}:ZnSe laser pumped by Tm^{3+} fiber laser

In this section, watt-level CW laser, based on 1.7-mm ceramic Cr^{2+}:ZnSe disk in the mid-infrared range around 2.4 μm is introduced.

Material Characterization

The Cr^{2+}:ZnSe disk is doped by means of PVT. The pure ZnSe poly-crystal and Cr powder was mixed in a vacuum tube under temperature of 950°C. After diffusion for 200 hours, Cr^{2+} ion-doping concentration reached $10^{19}/cm^3$. The disk has a diameter of 15mm and a thickness of 1.7mm. The diagram of energy levels of Cr^{2+} in Cr:ZnSe crystal is shown in Fig. 58 [102]. The energy levels associated with laser operation are the $^5T_2$ ground state and the ^5E exited state. Transitions to/from the other nearby excited states, the $^3T_1$ and $^3T_2$ states, are spin forbidden and are weak compared to the transition from $^5T_2$ to ^5E [103]. There is a possibility of excited state absorption (ESA) due to the ^5E-$^3T_1$ transition, but it was shown to have a small cross section in Cr^{2+}:ZnSe. The wavelength of the strongest line comes from the ^5E-$^3T_1$ transition. The II-VI materials themselves are fully transparent over the wavelength range of interest. Altogether this provides Cr^{2+}:ZnSe the highest gain among vibronic lasers and enables efficient broadband room-temperature operation[101,104].

The absorption spectra of Cr:ZnSe, shown in Fig. 59 [105], is centered at 1776nm and has a bandwidth of 400 nm, which makes a wide pumping range.

A double-clad Tm^{3+}-doped silica fiber laser for excitation, an Omni-λ500 spectrometer and PD (InAs Photodiode) detector for wavelength measurements, the super fluorescence

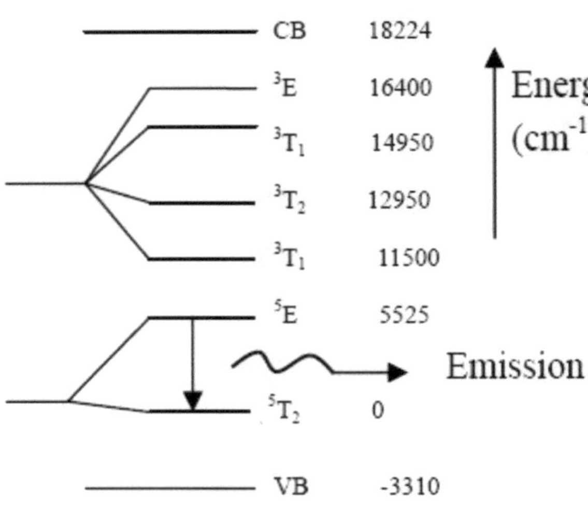

Fig. 58. Cr^{2+} Energy levels in Cr^{2+}:ZnSe as described by Grebe et al.

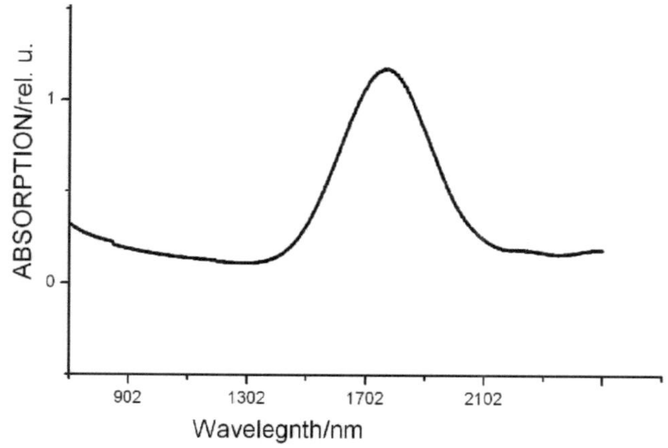

Fig. 59. Cr:ZnSe disk absorption spectra

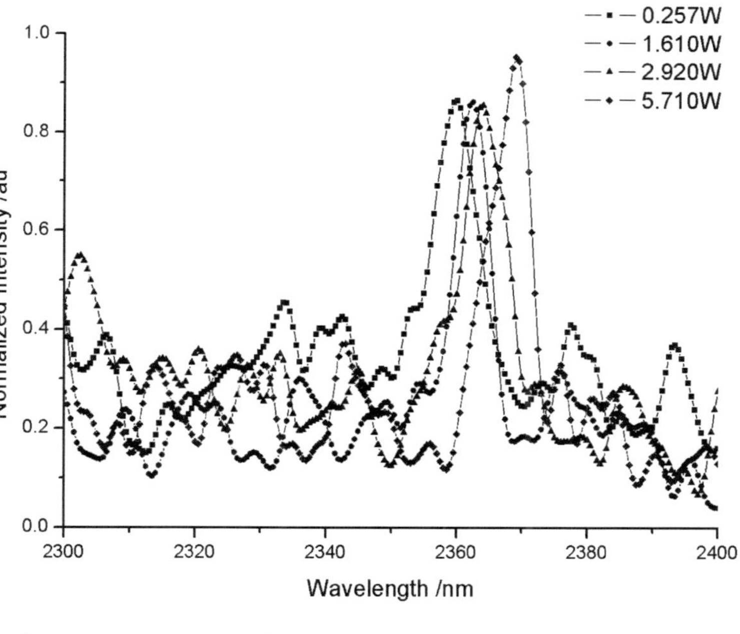

Fig. 60. Super fluorescence spectra of Cr^{2+}:ZnSe pumped by different power

spectra of Cr:ZnSe pumped by different power is shown in Fig. 60 [105]. The peak emission increases from 2.360μm to 2.369μm with increase of pump power and the bandwidth of super fluorescence was about 20 nm which decreased by several nanometers when pump intensity increased. Rotating the angle of incidence (0° for pump beam launched on the crystal surface vertically), the intensity and peak emission wavelength of super fluorescence varied significantly, shown in Fig. 61.

Experimental setups

Cr^{2+}:ZnSe disk has been coated with tens of microns thick PR (Partial-Reflective) film on one surface, with measured transmission of 5% at 2.4μm, and AR (Anti-Reflective) film at both 1.9μm and 2.4μm on the other. The PR film has a transmittance of lower than 3% at 1.9μm.

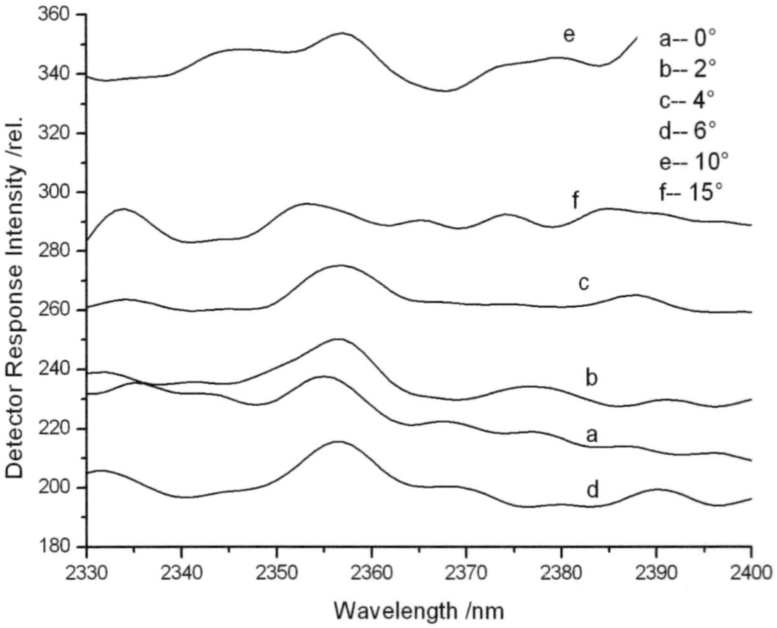

Fig. 61. Super fluorescence spectra of Cr^{2+}:ZnSe pumped in different angles

The laser was pumped by a double-clad Tm^{3+}-doped silica fiber laser. The pump beam was focused on the disk into a 0.5-mm-diameter spot.

The laser setup (Fig. 62) consists of the coated Cr^{2+}:ZnSe disk, whose surface coated with a PR film, was used as the output window, a concave HR mirror with a radius of curvature of 500mm as the rear mirror. The Tm^{3+}-doped silica fiber laser beam was directly delivered onto the disk with the angle no more than 10° to the optical axis. The fiber of the Tm^{3+}-doped fiber laser had a core diameter of 30μm and pure silica inner-cladding of size 250μm with a low refractive index (n=1.375) polymer outercladding of size 400μm, and was pumped at one end by beam-shaped diode-bars at 790nm delivering a maximum combined pump power of ~80W. The Tm^{3+}-doped fiber laser produced a maximum output of ~8 W and the peak wavelength was 2050nm. The ZnSe disk was mounted on a copper holder for cooling. The laser operated at room temperature in continuous wave regime, producing maximum ~110 mW of output power at 5% output coupling. Another laser arrangement was set by inserting two focusing lens of 13mm and 22.6mm focal length and a reflective mirror to improve the beam overlap on the ZnSe disk surface, shown in Fig. 63 [105]. In the second arrangement of laser setup, the laser performance was effectively improved.

Experimental results

The resonator is shown in Fig. 62. The plat-concave cavity length is ~100mm, with no extra etalons. The cw laser output characters are given in Fig. 64. With 5% output coupler, the threshold power was ~600mW of the incident power. In this arrangement of laser, the pump fiber was directly headed to the surface of Cr:ZnSe disc with no focusing. So, the pump laser

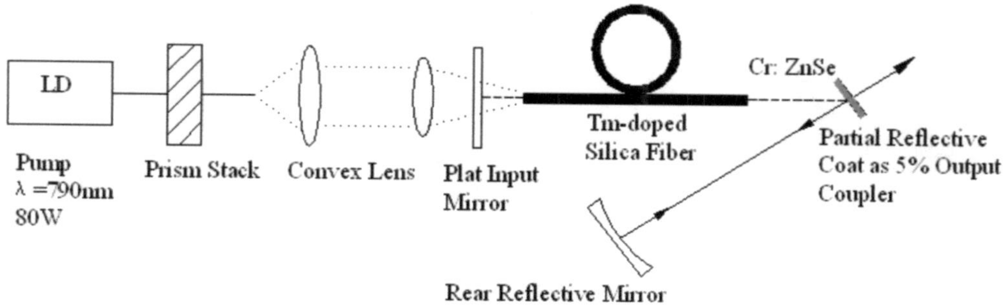

Fig. 62. Schematic diagram of the ceramic Cr:ZnSe laser setup

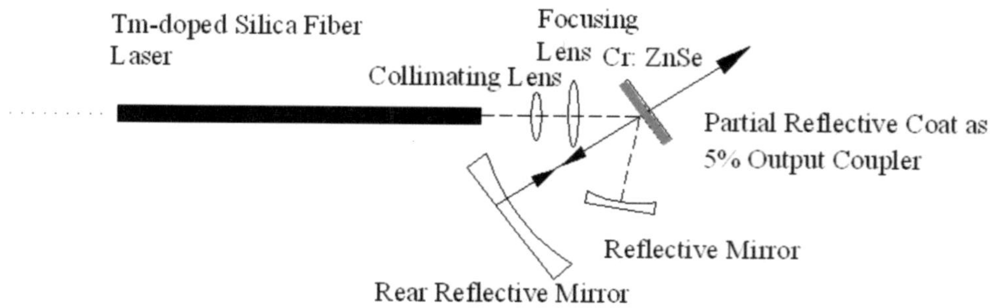

Fig. 63. The second arrangement of Cr:ZnSe laser setup

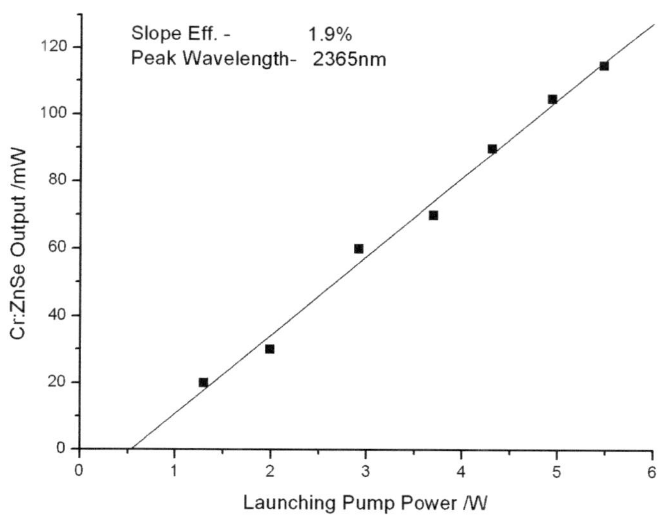

Fig. 64. Output power versus launching power of the Cr^{2+}:ZnSe laser in the first arrangement

spot was as big as 3 mm in diameter which is as big as the laser output window. Highest output laser power was ~110mW with 4.4W of the launching power, corresponding to a conversion efficiency of 1.9%, which is quite low and is mostly due to the loss of the pump power. To raise the slope efficiency, the pump laser beam should be focused to a much smaller spot at the entrance surface of active medium--Cr:ZnSe. And an extra rear reflective mirror should be placed to reduce the loss of the pump power.

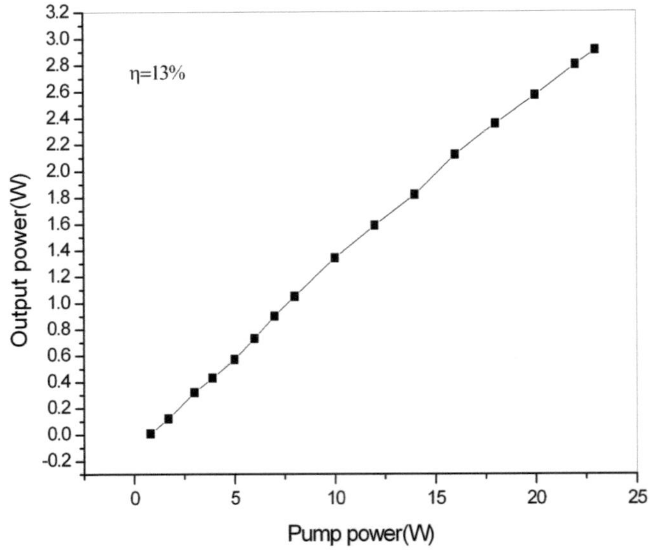

Fig. 65. Output power versus launching power of the Cr^{2+}:ZnSe laser in the second arrangement

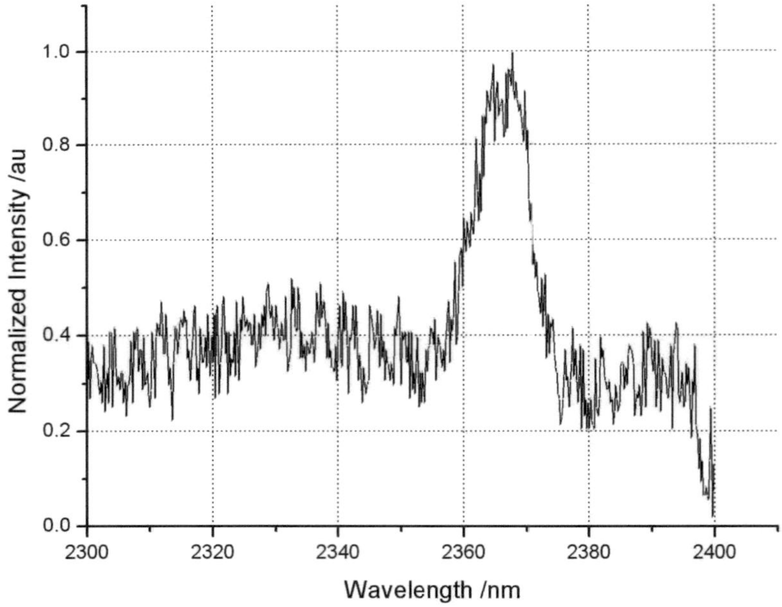

Fig. 66. Emission spectra of the Cr^{2+}:ZnSe laser

In the second arrangement of the laser, the collimated pump beam was focused on the disc into a 0.5-mm-diameter spot by an uncoated 22.6mm lens, and the cavity length was decreased to 26mm. The laser output characteristic and spectrum are shown in Fig. 65 and Fig. 66. As shown in Fig. 66, maximum output power of 2.91 W was achieved with a slope efficiency of 13% with respect to the launching pump power. The efficiency is far lower than the limit of ~56% of the modeling result. It is mostly caused by the coatings at the surfaces of ZnSe disk, which were optimized at 2.4 μm and have very narrow bandwidth. At the peak wavelength of our laser around 2.367 μm, the reflectivity and transmittance were lower than the optimal values. Another result of this narrow-band coating is the relative narrow laser linewidth. As shown in Fig. 66 [105], the measured laser linewidth was only 10 nm, which was much narrower than that reported by other authors [91, 94 and 99]. Tightly focusing pump beam into the ZnSe disk can improve the laser output power. However, it poses a hazard of damage on the surface coatings. Further work of optimizing beam delivery, focusing optics and recoating ZnSe surface relative to the peak wavelength of laser will surely decrease the threshold and increase the output power.

5.2 Wavelength tunability of Cr^{2+}:ZnSe laser

The Cr^{2+}:ZnSe laser has good wavelength tenability. In our experiment, the band width of the laser light from the Cr^{2+}:ZnSe laser is about 10 nm. Therefore, we studied the tunability character of the Cr^{2+}:ZnSe laser by adopting the angle-tuning technique.

In the experiment, the incident angle of the pump beam on the ZnSe crystal will change with rotating this crystal. The anti-reflection film coated on the crystal acts as a partial-reflection mirror due to its non-hundred-percent transmittance. Therefore, the two surfaces of the ZnSe crystal form a quasi Fabry-Poret etalon, which has a selection effect on laser modes and exert the wavelength-tuning function. In the Fabry-Poret cavity, light experiences multi-pass reflection between the two parallel surfaces M_1 and M_2, leading to interference. Light beams leaving the M2 with the same phase will interfere coherently. The wavelength walking through the etalon after the multi-pass reflection will possess a phase delay as

$$\delta = \frac{2\pi}{\lambda} 2nd \cos\Theta \,, \tag{13}$$

where, nd is the optical length of the crystal, $\Theta = \Theta' / n$ is the refractive angle, λ is the light wavelength. The transmittance of the crystal-etalon is

$$T = \left[1 + \frac{4r}{(1-r)^2} \sin^2\left(\frac{\delta}{2}\right) \right]^{-1} , \tag{14}$$

where, r is the reflectance of the crystal surface. When the optical length difference of the transmitted light beams is multi times of λ, the transmittance will be maximum [106]:

$$T_{\max} = 1, \quad 2nd \cos\Theta = m\lambda, \quad m = 1,2,3\cdots .$$

As shown in Fig. 67 [107], when the ZnSe crystal is rotated, the incident angle of the pump beam will change (the incident angle is referred to the angle between the normal of the crystal surface and the axis of the cavity). Consequently, the resonant wavelength will change correspondingly, leading to output laser wavelength tuned over 100 nm range (2272→2383nm). Besides, under different rotating angles of the crystal, the overlap between the pump beam and the laser beam will be different, leading to different laser threshold for different wavelength output. When the incident angle is 0°, the pump beam is normal to the crystal surface, the pump and laser beam modes have a good overlap. At this case, the output power is the highest, and the laser spectrum has only one peak band. When the incident angle is increased to 2° (by tilting the crystal), laser wavelength red-shifts a little. As the incident angle is enhanced to 4°, the modes overlap become worse, so the sidebands appear in the spectrum. Further increasing the incident angle (10°→15°), the selectivity of the cavity decreases seriously, leading to occurrence of 2, 3, and 5 spectral peaks. Besides, the band width of the laser spectrum is broadened. Detailed experimental results are shown in table 3 [107].

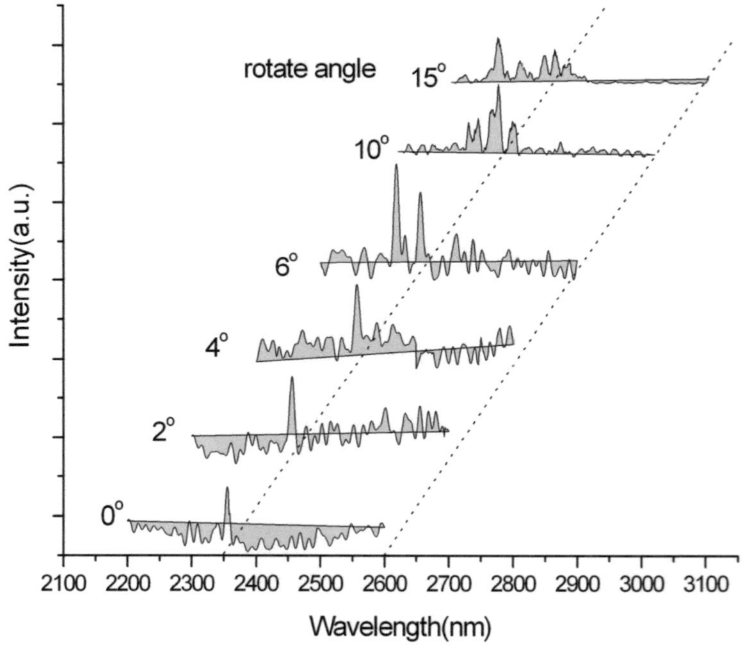

Fig. 67. laser wavelength change with different input angles of the pump light.

Rotating angle	Spectral peak position	FWHM
0°	2355nm	~10nm
2°	2356nm	~13nm
4°	2357nm(small sidebands)	~11nm
6°	2318nm(two peaks)	~10nm
10°	2319nm (three peaks)	~18nm
15°	2272,2306,2345,2360,2383nm (five peaks)	~15nm

Table 3. Cr^{2+}:ZnSe angle-tuning spectra

6. Conclusion and prospect of 2 μm Tm^{3+}-doped fiber laser

Based on the high-degree development of high-brightness laser diodes and optimizing of Tm^{3+} fiber fabrication technique, and further understanding about the spectral properties of Tm^{3+} ions, Output power and performance of the Tm^{3+}-doped fiber laser can be improved to a higher level. Due to its so many specific advantages, the Tm^{3+}-doped fiber laser has great potential in the development toward high-power output, wide wavelength tunability, narrow pulse duration, and high peak power. With further enhancement of the performance and quality of the Tm^{3+}-doped fiber laser, this kind of ~2-μm laser device will has wide applications in medicine, machining, environment detecting, LIDAR, optical-parametric-oscillation (OPO) pump sources, and so on.

There are several directions for the development of Tm^{3+}-doped fiber laser in the future.

- High power regime;
- Pulsed mid-infrared laser output, high-peak power, including femto-second laser pulse operation;
- Single frequency (narrow linewidth) and single mode operation;
- Wide tunable mid-infrared laser output (including multi-color wavelength laser output simultaneously).

7. References

[1] P. Myslinski, X. Pan, C. Barnard, J. chrostowski, B. T. Sullivan, and J. F. Bayon, "Q-switched thulium-doped fiber laser," Opt. Eng. 32 (9), 2025-2030 (1993).

[2] L. Esterowitz, "Diode-pumped holmium, thulium, and erbium lasers between 2 and 3 μm operating CW at room-temperature," *Opt. Eng.*, 29 (6): 676-680 (1990).

[3] R. C. Stoneman and L. Esterowitz, "Efficient, broadly tunable, laser-pumped Tm-YAG and Tm-YSGG CW lasers," *Opt. Lett.*, 15 (9): 486-488 (1990).

[4] S. W. Henderson, P. J. M. Suni, C. P. Hale, S. M. Hannon, J. R. Magee, D. L. Bruns, and E. H. Yuen, "Coherent laser-radar at 2 μm using solid-state lasers," *IEEE Trans. Geosci. Remote Sens.*, 31 (1): 4-15 (1993).

[5] I.T. Sorokina, K.L. Vodopyanov (Eds.): Solid-State Mid-Infrared Laser Sources, Topics Appl. Phys. 89, 219–255 (2003).

[6] Hanna, D. C., R. M. Percival, R. G. Smart, A. C. Tropper, "Efficient and tunable operation of a Tm-doped fibre laser," Opt. Commun. 75: 283-286 (1990).

[7] S.Agger, J.H.Povlsen and P.Varming, "Single-frequency thulium-doped distributed-feedback fiber laser," Optics Letters, 29(13): 1503-1505 (2004).

[8] H.W. Gandy, R.J. Ginther and J.F. weller, Stimulated emission of Tm^{3+} radiation in silicate glass, J. Appl. Phys. 38: 3030-3031 (1967).]

[9] R.G. Smart, J.N. Carter, A.C. Tropper and D.C. Hanna, "continuous-wave oscillation of Tm-doped fluorozirconate fibre laser at around 1.47m, 1.9 and 2.3 when pumped at 790nm", Opt. Commun. 82: 563-570 (1991).

[10] Jihong Geng, Jianfeng Wu, and Shibin Jiang, "Efficient operation of diode-pumped single-frequency thulium-doped fiber lasers near 2 μm," Opt. Lett., 32 (4): 355-357 (2007).

[11] N. Y. Voo, J. K. Sahu, and M. Ibsen, "345-mW 1836-nm Single-Frequency DFB Fiber Laser MOPA," IEEE Photon. Technol. Lett. 17, 2550 (2005).

[12] Jianqiu Xu, Mahendra Prabhu, Jianren Lu, Ken-ichi Ueda, and Da Xing, "Efficient double-clad thulium-doped fiber laser with a ring cavity," Applied Optics, 2001, 40(12): 1983-1988.

[13] Stuart D. Jackson and Terence A. King, "Theoretical Modeling of Tm-Doped Silica Fiber Lasers, J. Lightwave Technology", 17(5): 948-956 (1999).

[14] Yuen H. Tsang, Daniel J. Coleman, Terence A. King, "High power 1.9 µm Tm3+-silica fibre laser pumped at 1.09 µm by a Yb3+-silica fibre laser," Optics Communications 231 (2004).

[15] D. C. Hanna, M. J. McCarthy, I. R. Perry, P. J. Suni, "Efficient high-power continuous-wave operation of monomode Tm-doped fibre laser at 2 µm pumped by Nd YAG laser at 1.064 µm," Electron. Lett., (1989) 25 (20): 1365-1366.

[16] P.S. Golding, S.D. Jackson, P.-K. Tsai, B.C. Dickinson, T.A. King, "Efficient high power operation of a Tm-doped silica fiber laser pumped at 1.319 mm," Optics Communications 175: 179–183 (2000).

[17] M. Meleshkevich, N. Platonov, D. Gapontsev, A Drozhzhin, V. Sergeev, and V. Gapontsev, "415W Single-mode CW Thulium fiber laser in allfiber format," in *Proc. Eur. Conf. Lasers Electro-Opt., 2007 Int. quantum Electron. Conf. (CLEOE-IQEC 2007)*, Munich, Germany, Jun. 17–22, p. 1.

[18] D.C. Hanna, I.M. Jauncey, R.M. Percival, I. R. Perry, R.G. Smart, P. J. Suni, J. E. Townsend, A. C. Tropper: Continuous-wave oscillation of a monomode thulium-doped fibre laser, Electron. Lett. 24, 1222 (1988).

[19] W. L. Barnes, J.E. Townsend: Highly tunable and efficient diode pumped operation of Tm^{3+} doped fibre lasers, Electron. Lett. 26, 746 (1990).

[20] J. N. Carter, R.G. Smart, D.C. Hanna, A.C. Tropper: CW diode-pumped operation of 1.97 µm thulium-doped fluorozirconate fibre laser, Electron. Lett. 26, 599 (1990).

[21] Jackson, S.D., and King, T.A.: 'High-power diode-cladding-pumped Tm-doped silica fiber laser', *Opt. Lett.*, 1998, 23, pp. 1462–1464.]

[22] R.A. Hayward, W.A. Clarkson, P.W. Turner, J. Nilsson, A.B. Grudinin and D.C. Hanna, Efficient cladding-pumped Tm-doped silica fibre laser with high power singlemode output at 2µm, Electron. Lett., 36 (8): 711-712 (2000)

[23] G. Frith, D.G. Lancaster and S.D. Jackson, "85W Tm^{3+}-doped silica fibre laser," Electron. Lett., 41, 687-688 (2005).

[24] Y. Jeong, P. Dupriez, J. K. Sahu, J. Nilsson, D. Shen, and W. A. Clarkson, "Thulium-ytterbium co-doped fiber laser with 75 W of output power at 2 µm," SPIE (2004) 5620: 28-35]

[25] Evgueni Slobodtchikov, Peter F. Moulton, Gavin Frith, "Efficient, High-Power, Tm-doped Silica Fiber Laser," 2007ASSP-MF2.

[26] P. F. Moulton, G. A. Rines, E. V. Slobodtchikov, K. F. Wall, G. Frith, B. Samson, and A. L. G. Carter, "Tm-Doped Fiber Lasers: Fundamentals and Power Scaling," IEEE J. Sel. Topics in Quantum Electronics, 15 (1): 85-92 (2009).

[27] G. D. Goodno, L. D. Book, and J. E. Rothenberg, "low-phase-noise single-frequency single mode 608 W thulium fiber amplifier," Opt. Lett. 34 (8): 1204-1206 (2009)

[28] Orazio Svelto, "principles of Lasers", P142, 4th edit. 1998, Plenum Press, New York.
[29] A. E. Siegman, LASERS, 1986, P492, P1017, Miller/Scheier Associates, Palo Alto, CA.
[30] Jianfeng Wu, Shibin Jiang, Tao Luo, Jihong Geng, N. Peyghambarian, and Norman P. Barnes. "Efficient Thulium-Doped 2-μm Germanate Fiber Laser," *IEEE Photon. Technol. Lett.* 18 (2): 334-336 (2006).
[31] Jihong Geng, Jianfeng Wu, and Shibin Jiang, "Efficient operation of diode-pumped single-frequency thulium-doped fiber lasers near 2 μm," *Opt. Lett.*, 32 (4): 355-357 (2007).
[32] N. Y. Voo, J. K. Sahu, and M. Ibsen, "345-mW 1836-nm Single-Frequency DFB Fiber Laser MOPA," *IEEE Photon. Technol. Lett.* 17, 2550 (2005).
[33] Yulong Tang, Yong Yang, Xiaojin Cheng and Jianqiu Xu. Short Tm3+-doped fiber lasers with watt-level output near 2 μm [J]. *Chinese Opt. Lett.*, 2008, 6 (1): 44-46.
[34] S. D. Jackson and S. Mossman, "Efficiency dependence on the Tm3+ and Al3+ concentrations for Tm3+-doped silica double-clad fiber lasers," *Appl. Opt.*, 42, no. 15, pp. 2702-2707, 2003.
[35] Orazio Svelto, "principles of Lasers", P5, 4th edit. 1998, Plenum Press, New York.
[36] S.D. Jackson and S. Mossman, "Efficiency dependence on the Tm3+ Al3+ concentrations for Tm3+ -doped silica double-clad fiber lasers," *Appl. Opt.*, vol. 42, no. 15, pp. 2702–2707, 2003.
[37] S. D. Jackson, "Cross relaxation and energy transfer upconversion processes relevant to the functioning of 2 μm, Tm3+ -doped silica fiber lasers," *Opt. Commun.*, vol. 230, pp. 197–203, 2004.
[38] Ping Yan, Shupeng Yin, and Mali Gong. 175-W continuous-wave master oscillator power amplifier structure ytterbium-doped all-f iber laser [J]. *Chinese Opt. Lett.*, 2008, 6 (8): 580-582.
[39] Dong Xue, Qihong Lou, Jun Zhou, Lingfeng Kong, Jinyan Li, and Shiyu Li. A 110-W fiber laser with homemade double-clad fiber [J]. *Chinese Opt. Lett.*, 2005, 3 (6): 345-347.
[40] S. D. Jackson and T. A. King. High-power diode-cladding-pumped Tm-doped silica fiber laser. *Opt. Lett.*, 1998, 28(18): 1462-1464.
[41] S. D. Jackson and T. A. King. Dynamics of the output of heavily Tm-doped double-clad silica fiber lasers [J]. *J. Opt. Soc. Am. B*, 1999, 16(12): 2178-2188.
[42] Walter Koechner, "solid-state laser engineering," 5th edit. p95, Springer series, 1999.
[43] A. V. Smith, B. T. Do, G. R. Hadley and R. L. Farrow. Optical Damage Limits to Pulse Energy From Fibers [J]. *IEEE J. Selected Topics in Quantum Electron.*, 2009, 15(1): 153-158.
[44] W. H. Lowdermilk and D. Milam. Laser-Induced Surface and Coating Damage [J]. *IEEE J. Quantum Electron.*, 1981, QE-17(9): 1888-1903.
[45] S. Yoo, C. Basu, A. J. Boyland, C. Sones, J. Nilsson, J. K. Sahu and D. Payne. Photo darkening in Yb-doped aluminosilicate fibers induced by 488 nm irradiation [J]. *Opt. Lett.*, 2007, 32(12): 1626-1628.
[46] http://www.slideshare.net/nufchas/power-scaling-790nmpumped-tmdoped-devices-from-191-to-213-m .

[47] Yulong Tang, Yong Yang, and Jianqiu Xu. High Power Tm3+-Doped Fiber Lasers Tuned by a Variable Reflective Output Coupler [J]. *Research Letters in Optics*, 2008, 2008: 1-3.

[48] X. Zou, and H. Toratani, "Spectroscopic properties and energy transfers in Tm3+ singly- and Tm3+/Ho3+ doubly-doped glasses," Journal of non-crystalline solids 195, 113-124 (1996).

[49] G. P. Agrawal (1995). Nonlinear Fiber Optics, Academic, San Diego, CA.

[50] W. Torruellas, Y. Chen, B. McIntosh, J. Farroni, K. Tankala, S. Webster, D. Hagan, M. J. Soileau, M. Messerly and J. Dawson. High peak power Ytterbium doped fiber amplifiers [J]. Proc. of SPIE 2006, 6102:61020N-1-61020N-7.

[51] Jianqiu Xu, Yulong Tang, Yong Yang, and Yin Hang. High power tunable Tm^{3+}-fiber lasers and its application in pumping Cr^{2+}:ZnSe lasers [C]. CLEO/QELS 2008, Paper JTuA3.

[52] Xin Ye, Tao Fang, Zhimin Wang, Shixun Dai, Jianqiu Xu. Nd:glass belt lasers with improved beam quality [J]. *Chinese Opt. Lett.*, 2005, 5 (9): 527-529.

[53] Allain, J.Y., M.Monerie, H.Poignant. "Tunable CW lasing around 0.82 1.48 1.88 and 2.35 μm in thulium-doped fluorozirconate fibre", Electronics Letters, 25 (24): 1660-1662 (1989).

[54] S. D. Jackson and T. A. King, "High-power diode-cladding-pumped Tm-doped silica fiber laser", Optics Letters, 23 (18): 1462-1464 (1998).

[55] T. Yamamoto, Y. Miyajima and T. Komukai, "1.9 μm Tm^{3+}-doped silica fibre laser pumped at 1.57 μm" , Electron. Lett. 30 (3), 220-221 (1994).

[56] D. C. Hanna, R. M. Percival, R. G. Smart, and A. C. Tropper, "Efficient and tunable operation of a Tm^{3+}-doped fiber laser" , Opt. Commun. 75 (3-4), 283-286 (1990).

[57] C. Ghisler, W. Luthy and H. P. Weber, "Tuning of a Tm^{3+} Ho^{3+} Silica Fiber Laser at 2μm", IEEE J. Quantum Electron., 31 (11): 1877-1879 (1995).

[58] R. L. Shubochkin, V. A. Kozlov, A. L. G. Carter, and T. F. Morse, "Tunable Thulium-Doped All-Fiber Laser" , IEEE Photon. Tech. Lett. 10 (7), 944-945 (1998).]

[59] D. Y. Shen, J. K. Sahu, and W. A. Clarkson, "High-power widely tunable Tm^{3+}: fiber lasers pumped by an Er, Yb co-doped fibre laser at 1.6 μm," Opt. Express 14 (13), 6084-6090 (2006).

[60] W. L. Barnes, J. E. Townsend, "Highly tunable and efficient diode pumped operation of Tm^{3+} doped fibre lasers", Electron. Lett. 26 (11), 746-747 (1990).

[61] W. A. Clarkson, N. P. Barnes, P. W. Turner, J. Nilsson, and D. C. Hanna, "High-power cladding-pumped Tm^{3+}-doped silica fiber laser with wavelength tuning from 1860 to 2090 nm," Opt. Lett. 27 (22), 1989-1991 (2002).

[62] Ashraf F. El-Sherif, Terence A. King, "Dynamics and self-pulsing effects in Tm^{3+}-doped silica fibre lasers," Opt. Commun. 208, 381-389(2002).

[63] F.Z. Qamar and T.A. King, "Self-induced pulsations, Q-witching and mode-locking in Tm-silica fibre lasers", J. Mod. Opt. 52 (7), 1031-1043 (2005).

[64] T. Erneux, "Q-switching bifurcation in a laser with a saturable absorber," J. Opt. Soc. Am. B 5 (5), 1063-1069 (1988).

[65] P. Le Boudec, M. Le Flohic, P. L. Frangois, F. Sanchez, and G. Stephan, "Self-pulsing in Er^{3+}-doped fibre laser," Opt. Quantum Electron. 25, 359-367 (1993).

[66] R. Leners, P. L. Francois, and G. Stephan, "Simultaneous effects of gain and loss anisotropies on the thresholds of a bipolarization fiber laser," Opt. Lett. 19 (4), 275-277 (1994).
[67] P. Le Boudec, C. Jaouen, P. L. Fransois, and J.-F. Bayon, F. Sanchez, P. Besnard, and G. Stephan, "Antiphase dynamics and chaos in self-pulsing erbium-doped fiber lasers," Opt. Lett. 18 (22), 1890-1892 (1993).
[68] P. L. Boudec, P. L. Francois, E. Delevaque, J. F. Bayon, F. Sanchez, G. M. Stephan, "Influence of ion pairs on the dynamical behaviour of Er^{3+}-doped fibre lasers," Opt. Quantum Electron. 25, 501-507 (1993).
[69] S. Colin, E. Contesse, P. Le Boudec, G. Stephan, F. Sanchez, "Evidence of a saturable-absorption effect in heavily erbium-doped fibers," Opt. Lett. 21 (24), 1987-1989 (1996).
[70] A. Hideur, T. Chartier, C. Ozkul, F. Sanchez, "Dynamics and stabilization of a high power side-pumped Yb-doped double-clad fiber laser," Opt. Commun. 186, 311-317 (2000).
[71] F.Z. Qamar and T.A. King, "Self- mode-locking effects in heavily doped single-clad Tm^{3+}-doped silica fibre lasers", J. Mod. Opt. 52 (8), 1053-1063 (2005).
[72] Yulong Tang, Jianqiu Xu, "Self-induced pulsing in Tm3+-doped fiber lasers with different output couplings," Photonics and OptoElectronics Meetings (POEM2008), Nov. 2008, Wuhan, China.
[73] Michel J. F. Digonnet, "Rare-Earth-Doped Fiber Lasers and Amplifiers," second ed. New York, Basel, 382 (2001).
[74] F. Sanchez, P.L. Boudec, P.L. Francois and G. Stephan, "Effects of ion pairs on the dynamics of erbium-doped fiber lasers," Phys. Rev. A 48 (3), 2220-2229 (1993).
[75] W. Koechner, "Solid-State Laser Engineering", Fifth Edition, springer-Verlag Berlin Heidelberg New York, pp.17–27, 1999.

[76] Gunnar Rustad and Knut Stenersen, "Modeling of Laser-Pumped Tm and Ho Lasers Accounting for Upconversion and Ground-State Depletion", *IEEE Journal of Quantum Electronics*, vol. 32, no. 9, pp. 1645–1656, September 1996.
[77] Igor Razdobreev and Alexander Shestakov, "Self-pulsing of a monolithic Tm-doped YAlO3 microlaser", *Physical Review* A, vol. 73, no. 5, pp. 053815 (1-5), 2006.
[78] S.D. Jackson and T.A. King, "Theoretical modeling of Tm-doped silica fiber lasers," J. Lightwave Tech. vol. 17, no. 5, 948-956, 1999.
[79] B. M. Walsh, N. P. Barnes, "Comparison of Tm:ZBLAN and Tm: silica fiber lasers; Spectroscopy and tunable pulsed laser operation around 1.9 μm," Appl. Phys. B, vol. 78, 325-333 (2004).
[80] Yulong Tang, Jianqiu Xu, "Effects of Excited-state Absorption on Self-pulsing in Tm^{3+}-doped Fiber Lasers,"IEEE J. of Quant. Electronics, (2009) being submitted.
[81] Yulong Tang, Jianqiu Xu, "Theoretical study on self-pulsing in Tm^{3+}-doped fiber lasers," Optics communications, (2009) being submitted.
[82] B. C. Dickinson, S. D. Jackson, and T. A. King, Opt. Commun. 182 (2000) 199.
[83] Y. J. Zhang, B. Q. Yao, Y. L. Ju, and Y. Zh. Wang, Opt. Express 13 (2005) 1085.
[84] A. F. El-Sherif, T. A. King, Opt. Commun. 218 (2003) 337.

[85] A. F. El-Sherif, T. A. King, Opt. Lett. 28 (2003) 22.
[86] R. C. Sharp, D. E. Spock, N. Pan, and J. Elliot, Opt. Lett. 21 (1996) 881.
[87] F. Z. Qamar, T. A. King, Opt. Commun. 248 (2005) 501.
[88] S. D. Jackson, Appl. Opt. 46 (2007) 3311.
[89] V.I. Levchenko, V.N. Yakimovich, L.I. Postnova, V.I. Konstantinov, V.P. Mikhailov, N.V. Kuleshov, J. Crystal Growthk, 198/199 (1999) 980.
[90] Yulong Tang, Yong Yang, Jianqiu Xu and Yin Hang, "Passive Q-switching of short-length Tm^{3+}-doped silica fiber lasers by polycrystalline Cr2+:ZnSe microchips," Optics Communications 281 (2008) 5588–5591.
[91] R. H. Page, K. I. Schaffers, L. D. DeLoach, G. D. Wilke, F. D. Patel, J. B. TassanoJr., S. A. Payne, W. F. Krupke, K.-T. Chen, and A. Burger, IEEE JOURNAL OF QUANTUM ELECTRONICS, 33, 11 (1997).
[92] G. J. Wagner, T. J. Carrig, R. H. Jarman, R. H. Page, K. I. Schaffers, J.-O. Ndap, X. Ma, and A. Burger, Trends in Optics and Photonics, ADVANCED SOLID-STATE LASERS 26, 8 (1999).
[93] E. Sorokin and I. T. Sorokina, APPLIED PHYSICS LETTERS 80, 3 (2002).
[94] E. Sorokin, S. Naumov, and I. T. Sorokina, IEEE JOURNAL OF SELECTED TOPICS IN QUANTUM ELECTRONICS 11, 23 (2005).
[95] I.T. Sorokina, K.L. Vodopyanov (Eds.): Solid-State Mid-Infrared Laser Sources, Topics Appl. Phys. 89, p271 (2003).
[96] L.D. DeLoach, R.H. Page, G.D. Wilke, S.A. Payne, W.P. Krupke: Transition metal-doped zinc chalcogenides: spectroscopy and laser demonstration of a new class of gain media, IEEE J. Quantum Electron. 32, 885–895 (1996)
[97] A.V. Podlipensky, V. G. Shcherbitsky, N.V. Kuleshov, V. I. Levchenko, V. N. Yakimovich, M. Mond, E. Heumann, G. Huber, H. Kretschmann, S. Kuck: Efficient laser operation and continuous-wave diode pumping of Cr^{2+}:ZnSe single crystals, Appl. Phys. B. 72, 253–255 (2001).
[98] U. Demirbas and A. Sennaroglu, OPTICS LETTERS 31, 3 (2006).
[99]. T. J. Carrig, G. J. Wagner, W. J. Alford, and A. Zakel, Proceedings of SPIE 5460, 9 (2004).
[100] I. T. Sorokina, E. Sorokin, A. D. Lieto, M. Tonelli, R. H. Page, and K. I. Schaffers, JOURNAL B OF OPTICAL SOCIETY OF AMERICA 18, 5 (2001).
[101] I. T. Sorokina, OPTICAL MATERIALS 26, 18 (2004).
[102] G. Grebe, G. Roussos, and H.-J. Schultz, JOURANAL OF PHYSICS C: SOLID STATE PHYSICS, 6 (1976).
[103] G. Goetz, H. Zimmerman, and H.-J. Schultz, Zeitschrif fur Physik B, 18 (1993).
[104] J. B. McKay, Thesis, THE AIR FORCE AIR UNIVERSITY, 2003.
[105] YANG Yong, TANG Yu-Long, XU Jian-Qiu and HANG Yin, "Tm-Doped Fibre Laser Pumped Cr2+:ZnSe Poly-Crystal Laser," CHIN.PHYS.LETT. Vol. 25, No. 1 (2008) 116.
[106] W. Koechner, Solid-State Laser Engineering, Vol. 1, 6 ed. (Springer, 2006).
[107] Yong Yang, Yulong Tang, Jianqiu Xu and Yin Hang, "Study on Laser Output and Tuning Ability of Cr^{2+}:ZnSe," Chinese Lasers, 35 (10): 1495~1499 (2008) (Chinese).

21

2 µm Laser Sources and Their Possible Applications

Karsten Scholle, Samir Lamrini, Philipp Koopmann and Peter Fuhrberg
LISA laser products OHG
Germany

1. Introduction

The wavelength range around 2 µm which is covered by the laser systems described in this chapter is part of the so called "eye safe" wavelength region which begins at about 1.4 µm. Laser systems that operate in this region offer exceptional advantages for free space applications compared to conventional systems that operate at shorter wavelengths. This gives them a great market potential for the use in LIDAR and gas sensing systems and for direct optical communication applications. The favourable absorption in water makes such lasers also very useful for medical applications. As it can be seen in figure 1, there is a strong absorption peak near 2 µm which reduces the penetration depth of this wavelength in tissue to a few hundred µm.

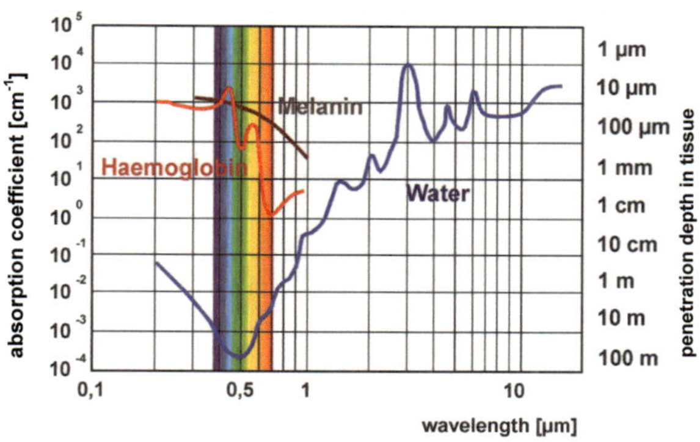

Fig. 1. Absorption and penetration depth in water and other biological tissue constituents for different wavelengths

Due to the strong absorption in water, the main constituent of biological tissue, substantial heating of small areas is achieved. This allows for very precise cutting of biological tissue. Additionally the bleeding during laser cutting is suppressed by coagulation, this makes 2 µm lasers ideal for many surgical procedures.

Furthermore 2 µm lasers are well suited to measure the health of planet earth. They can be used directly for measuring the wind velocity and for the detection of both water vapour and carbon dioxide concentration. Wind sensing is very important for weather forecasting, storm tracking, and airline safety. Water vapour and carbon dioxide detection is useful for weather and climate prediction and for the analysis of the green house effect.

2. Solid state laser systems around 2 μm

In the wavelength range around 2 μm the most interesting transitions for high power continuous wave (cw) and pulsed laser operation exist in the trivalent rare earth ions Tm^{3+} and Ho^{3+}. Using these ions laser emission was achieved in many different host crystals and glass fibres. For cw operation the thulium lasers are most interesting; however for pulsed and q-switched operation holmium lasers are more attractive due to the higher gain of the holmium doped crystals. The first experiments with Tm^{3+} and Ho^{3+} doped crystals were already carried out in the 1960s (Johnson, 1963). For both ions the relevant laser transition for the 2 μm emission ends in the upper Stark levels of the ground state. Therefore both lasers can be described as quasi three level lasers with a thermally populated ground state (Svelto, 1998; Koechner, 2006). Thulium lasers have the great advantage that the Tm^{3+} ions can be directly excited with commercially available laser diodes around 800 nm. To achieve efficient laser operation at 2.1 μm holmium can only be excited directly around 1.9 μm or by exploiting an energy transfer process from thulium or ytterbium.

2.1 Thulium lasers systems

With thulium doped crystals laser emission on many different transitions was reached so far. The laser emission around 2.0 μm is resulting from a transition that starts in the $^3F_4$ manifold and ends in a thermally populated Stark level of the $^3H_6$ ground state. The first Tm:YAG laser at 2 μm using this transition was realised in 1965 (Johnson et al., 1965). It was a flash lamp pumped laser which operated at 77 K. It took some years until the first pulsed laser operation at room temperature was realised in 1975 using Cr,Tm:YAG (Caird et al., 1975). Shortly after the development of the first laser diodes in the wavelength range around 800 nm continuous wave diode pumped laser operation at room temperature was shown (Huber et al., 1988; Becker et al., 1989). Until now thulium laser emission around 2 μm was demonstrated in many different host materials and there are some thulium based laser systems commercially available (LISA laser products OHG; IPG Photonics Corp.).

The energy level scheme of Tm^{3+} with the relevant energy transfer processes for this laser transition is shown in figure 2. The scheme of Tm:YAG is shown, as YAG is the most commonly used host material for thulium lasers. The figure also shows the Stark splitting of the ground state, which is important for the thermal population of the lower laser level of the 2 μm laser transition. In the figure one can see that the thulium ions can be excited around 800 nm from the ground state to the $^3H_4$ energy level. The upper laser level $^3F_4$ is then populated by a cross relaxation process (CR) that occurs between two thulium ions. In this non-radiative process for one ion an electron relaxes from the $^3H_4$ level to the $^3F_4$ level and for a second ion an electron is excited from the ground state to the $^3F_4$ level (French et al., 1992; Becker et al., 1989). This excitation process yields two excited ions for each absorbed pump photon. Therefore the quantum efficiency is nearly two when the cross relaxation process is highly efficient. Thus, instead of a maximum efficiency of 41 %, one can obtain an efficiency of 82 %, in theory. The efficiency of the cross relaxation process depends on the doping concentration of the thulium ions since the involved dipole-dipole interaction depends on the ion spacing. It is also possible to pump the $^3F_4$ energy level directly between 1700 nm and 1800 nm, but there are no well developed pump sources commercially available. A comparison between this direct excitation and the excitation exploiting the cross relaxation process was made by Peterson et al. (Peterson et al., 1995).

The efficiency of the laser process can be lowered by some energy transfer processes and by excited state absorption (ESA). Both possible upconversion processes that start from the upper laser level are phonon assisted. Barely any losses result from the upconversion process UC 1 which starts from the upper laser level, because this is the reverse process of the cross relaxation. More losses result from the upconversion process UC 2, because in this case the excitation of one ion is lost and another ion is excited into the $^3H_5$ level. The $^3H_5$ energy level has a very short lifetime and it is mostly depopulated by a non radiative process ($^3H_5 \rightarrow {}^3F_4$) which generates heat inside the crystal. Also excited state absorption which can start from the upper laser level $^3F_4$ or the upper level of the cross relaxation process ($^3H_4$) causes losses for the laser. The influence of these processes is usually low, due to the required phonon assistance. Only at high pump powers a slight blue fluorescence can be observed that starts from the $^1G_4$ manifold which is situated at approximately 21000 cm^{-1}.

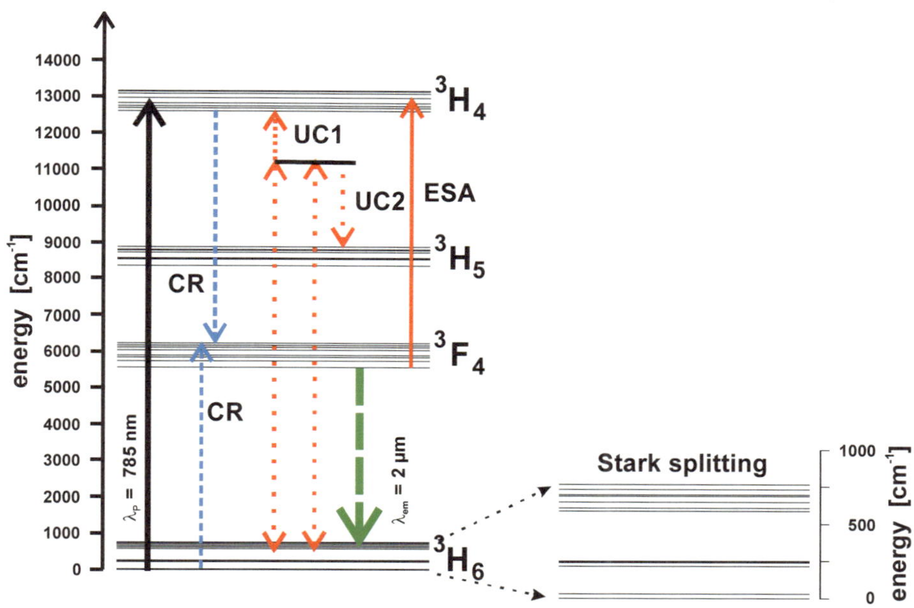

Fig. 2. Tm:YAG energy scheme with the relevant transitions for the 2 μm laser emission and the Stark splitting of the ground state

Thulium lasers have been realised in a wide variety of host crystals and fibre materials, in table 1 important parameters for the 2 μm laser transition are listed for a selection of crystals used for high power lasers. Further information about different thulium doped crystals can be found in the literature (Kaminskii, 1996; Sorokina & Vodopyanov, 2003). The absorption cross section σ_{abs} for the strongest absorption peak of the transition from the ground state to the $^3H_4$ manifold is given. The typical emission wavelength for the free running laser λ_{em} and the emission cross section for this transition σ_{em} are shown. For a comparison of the suitability of the different host materials for the laser operation the thermal conductivity λ_{th} and the lifetime of the upper laser level τ are listed.

The thermal conductivity of the host material is very important for the laser operation. The generated heat in the laser crystal has to be dissipated and removed efficiently to achieve high output powers. As it can be seen in table 1 the thermal conductivity of YLF is very low and it is rather high for Sc_2O_3, for YAG it is in between. The thermal conductivities are measured with un-doped crystals, but normally the thermal conductivity is reduced significantly with higher thulium doping concentrations (Gaumé et al., 2003). The heat removal from the laser material can be increased by using special geometries like thin disks or slabs instead of the standard rod geometry.

The lifetime of the upper laser level τ also depends on the thulium doping concentration of the crystal. In table 1 the lifetimes are given for very low doping concentrations, with higher thulium doping concentrations the lifetimes are often strongly reduced (Scholle et al., 2004). The main reason for this effect is the increased energy migration between the Tm^{3+} ions which supports the energy transfer to crystal impurities. The longest lifetimes of 15.6 ms of the upper laser level were measured for Tm:YLF crystals, which are about 1.5 times longer than in Tm:YAG and up to four times longer then for Tm:Lu_2O_3 and Tm:Sc_2O_3 crystals. Longer lifetimes allow larger energy storage in the upper laser level which is essentially important for q-switching operation.

laser host material	σ_{abs} (10^{-21} cm^2)	λ_{em} (nm)	σ_{em} (10^{-21} cm^2)	λ_{th} (W m^{-1} K^{-1})	τ (ms)	reference
YAG	7.5	2013	1.8	13	10	Heine, 1995
YLF	σ pol 3.6 π pol 8.0	1910 1880	2.35 3.7	6	15.6	Payne et al., 1992 Walsh et al., 1998
Lu_2O_3	3.8	2070 1945	2.3 8.5	13	3.8	Koopmann et al., 2009a
Sc_2O_3	5.0	1994	8.4	17	4.0	Fornasiero et al., 1999
Y_2O_3	5.0	2050 1932	2.1 8.1	14		Ermeneux et al., 1999
LuAG	5.7	2023	1.66	13	10.9	Scholle et al., 2004
$YAlO_3$		1936	5.0	11	4.8	Payne et al., 1992
silica fibre	4.5	1860	3.9		6.6	Agger & Povlsen, 2006
germanate f.	6	1840	4.1		5.3	Turri et al.,2008

Table 1. Properties of widely used thulium doped laser crystals for high power applications. Absorption cross section σ_{abs}; free running laser emission wavelength λ_{em}; emission cross section σ_{em}; thermal conductivity λ_{th}; lifetime of the upper laser level τ.

As mentioned, thulium 2 µm lasers can be pumped around 800 nm, exploiting the cross relaxation process to populate the upper laser level. Tm:YAG has one of the highest absorption cross sections in this wavelength region, but the main absorption peak is located at 785 nm. Figure 3 shows the absorption spectra of Tm:YAG, Tm:Lu_2O_3 and Tm:YLF. Tm:YLF has a natural birefringence, therefore the spectra for π and σ polarisation are shown.

A challange for most of the thulium doped crystals is that the available diodes around 800 nm where mainly developed for Nd:YAG pumping at 808 nm. So the available diodes in the range from 785 - 795 nm are more expensive and possess lower brightness and output powers compared to those operating close to 808 nm. Therefore most of the Tm doped crystals can not be pumped at the strongest absorption peak, only thulium doped sesquioxides like Lu_2O_3 and some vanadates have strong absorption lines near 808 nm. Due to the weak absorption larger crystals or multi pump pass set-ups have to be used to achieve sufficient pump light absorption.

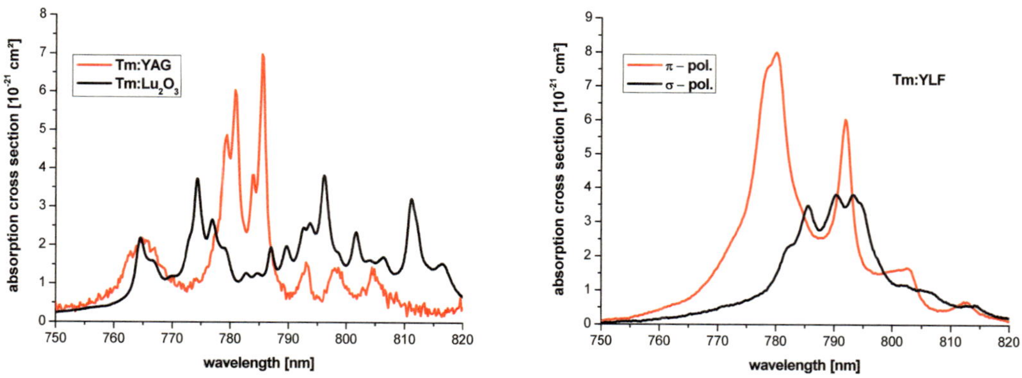

Fig. 3. Absorption cross sections of the $^3H_6$ → $^3H_4$ transition for Tm:YAG, Tm:Lu_2O_3, (Koopmann et al. 2009b) and Tm:YLF.

As can be seen in table 1, the different host materials provide the possibility to access many wavelengths in the range between 1840 nm and 2100 nm with thulium lasers. In the table the emission cross sections are shown for the typical free running laser transition, but thulium has a very broad and strongly structured emission spectrum in most crystals. As an example the emission spectra for Tm:YAG, Tm:Lu_2O_3 and Tm:YLF are shown in figure 4. One can see that the emission cross sections of Tm:YAG are much lower than for Tm:YLF and most other crystals. Low emission cross sections lead to low gain, therefore in YAG a co-doping of thulium and holmium was often used in the past since holmium has six times larger emission cross sections in YAG.

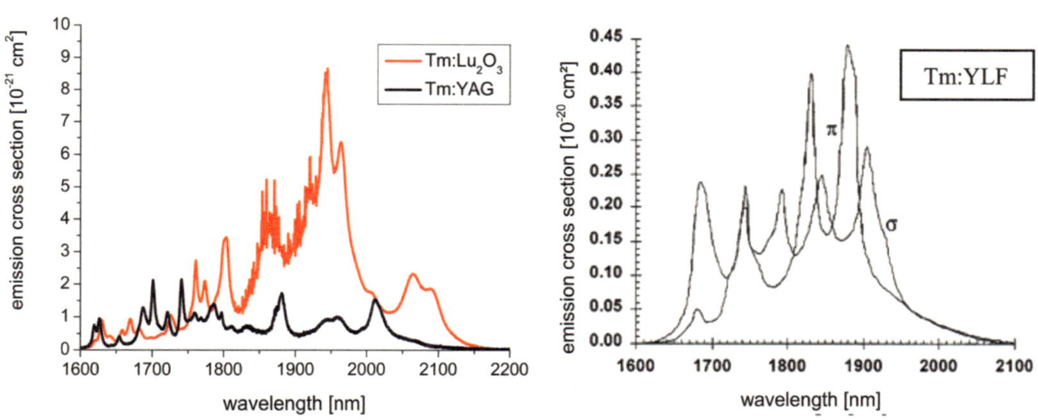

Fig. 4. Left side: Emission cross sections of Tm:YAG and Tm:Lu_2O_3 for the transitions from the $^3F_4$ manifold to the ground state (Koopmann et al., 2009b). Right side: Emission cross sections for π and σ polarisation of Tm:YLF (Budni et al., 2000).

The broad emission spectra of thulium doped crystals enable very large wavelength tuning ranges for thulium laser systems. This is very useful for a couple of laser applications. Additionally a broad gain spectrum allows the generation of extremely short laser pulses in mode-locked laser operation. Wavelength tuning is achieved by integration of wavelength selective elements into the laser resonator. Mostly prisms, diffraction gratings or birefringent filters under Brewster angle are used for wavelength tuning (Svelto, 1998). Tuning ranges of over 200 nm were achieved in different thulium doped crystals so far, for instance the tuning curves for Tm:LuAG and Tm:Lu_2O_3 are shown in figure 5. Tm:LuAG can only be sufficiently operated from 2010 nm to 2040 nm. With Tm:Lu_2O_3 efficient laser operation is possible from 1900 nm to 2110 nm. Especially the tuning range up to 2.1 µm with high output powers makes Tm:Lu_2O_3 attractive, since this laser can be an alternative to Ho:YAG lasers that emit around this wavelength.

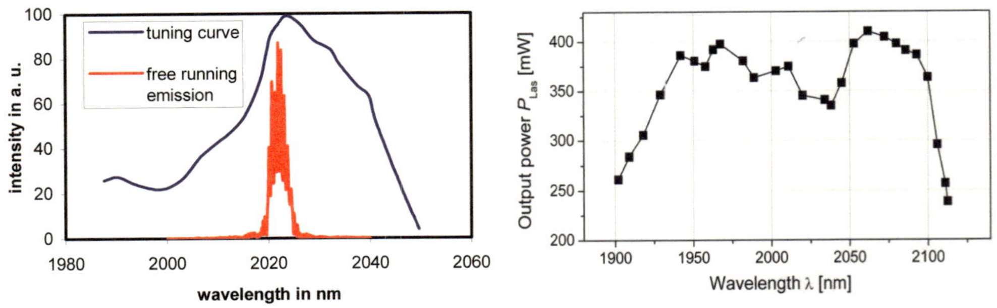

Fig. 5. Tuning curves of Tm:LuAG (Scholle et al., 2004) and Tm:Lu_2O_3 (Koopmann et al., 2009a)

Some of the important parameters of thulium doped crystals for laser operation around 2 µm have been discussed, but there are still some more aspects, which are important for the realisation of a high power thulium laser. A very important point is the crystal quality. To achieve high output powers, high quality crystals with very few impurities and defects are necessary. Additionally the achievable crystal size is important. In big crystals the generated heat can be distributed over a larger area. For rod or slab lasers much larger crystals than for thin disc lasers are required. The best qualities and largest crystal sizes are achieved today with Tm:YAG and Tm:YLF, but these crystals are not the best choice regarding thermal conductivity or emission cross section. Therefore in the future other host crystals like Lu_2O_3 or Sc_2O_3 can become important when larger high quality crystals become available.

Although thulium lasers have been realised in many different host crystals, high power lasers with output powers of some tens of watts or even more have been demonstrated only with YAG and YLF as laser host materials so far. In table 2 a short overview of some recently published thulium crystal and fibre laser systems is shown. The first high power thulium lasers were realised with solid state systems, nowadays the fibre laser system deliver the highest output powers.

The first cw diode pumped thulium 2 µm laser with output powers > 100 W was demonstrated in 1997 (Honea et al., 1997). To achieve these high output powers, Tm:YAG rods with undoped YAG end caps were used. The end caps were diffusion bonded to the doped rod to optimise the cooling of the rods and to reduce the thermal stress on the end surfaces. The laser rod was end-pumped by a diode bar operating at 805 nm using a fused silica lens duct to couple the pump light into the rod of 3 mm in diameter. With such a set-up and a 2 % thulium doped rod in a short plane-concave laser cavity up to 115 W of output power at room temperature were achieved. The laser showed a high slope efficiency of about 52 %, however the beam quality was poor at high powers ($M^2 = 23$). Nowadays Tm:YAG lasers emitting at 2.0 µm are commercially available. For instance LISA laser products OHG offers the RevoLix 120 Watt laser system, which is a medically approved Tm:YAG laser system. This system is used for non-invasive surgery, where the laser light needs to be delivered by a fibre. The system can deliver up to 120 W output power through different application fibres with very small core diameters.

The highest output power achieved so far with Tm:YLF rods is 55 W (Schellhorn, 2008). This high power was achieved using two 3.5 % doped Tm:YLF rods in one folded laser cavity that were pumped with four laser diodes. Each rod was pumped from both ends by a diode emitting at 792 nm. With this set-up a slope efficiency of 49 % and a beam quality of $M^2 < 3$ was observed. With a single rod the maximum output power was limited to 30 W. Due to the low thermal conductivity and the low fracture limit of Tm:YLF the rod geometry is not very well suited for high power Tm:YLF lasers. With a slab geometry further power scaling is possible, which is due to the better thermal management (So et al. 2006). The highest output power of a Tm:YLF slab reported so far is 148 W (Schellhorn et al., 2009). This was achieved with a 2 % doped Tm:YLF slab double-end-pumped by two laser diode stacks emitting at 790 nm. With an optical to optical conversion efficiency of 26.7 % 148 W of cw output power at 1912 nm were achieved at room temperature.

laser host material	λ_p (nm)	λ_{em} (nm)	cw output power (W)	slope eff. (%)	reference
YAG	805	2013	115	52	Honea et al., 1997
YAG	800	2013	120		LISA laser products OHG *
YLF	792	1910	55	49	Schellhorn, 2008
YLF	790	1912	148	32.6	Schellhorn et al., 2009
Lu_2O_3	796	2070	1.5	61	Koopmann et al., 2009a
germanate f.	800	1900	64	68	Wu et al., 2007
silica fibre	793	2050	110	55	Frith et al., 2007
silica fibre	1567	1940	415	60	Meleshkevich et al., 2007
silica fibre	790	2040	885	49.2	Moulton et al., 2009

Table 2. Brief overview of recently published continuous wave thulium solid state and fibre laser results (λ_p = pump wavelength; λ_{em} = emission wavelength). * RevoLix 120 Watt commercial system from LISA laser products OHG

Nowadays nearly the same maximum output powers can be reached with Tm:YAG and Tm:YLF lasers. The slope efficiencies of the Tm:YAG systems are higher than for the Tm:YLF ones, but the beam quality of the Tm:YLF systems is better due to the weaker thermal lenses which occur in YLF. Further power scaling of the output power from both systems should be possible especially with the slab geometry. So far the maximum reported output powers were limited by the available pump powers, not by fracture of the laser crystals. Great potential for power scaling is also exhibited by Tm:Lu_2O_3. The properties of the crystal and the recently reported results with record high slope efficiencies indicate the large potential of this crystal.

In recent years a lot of research has been performed on the improvements of fibre lasers and great advances were made in the power scaling. Since the late 1980s for many years single-mode diode pumped fibre lasers that emitted a few tens of milliwatts were used because of their large gain and the feasibility of single-mode continuous wave lasing. The most well-known application of these fibre lasers is in the telecom market around 1550 nm where erbium-doped fibre lasers and amplifiers are used. The modern high-power fibre lasers are built mostly with double-clad fibres that have a small inner core that is doped with the laser active ions and is surrounded by a much larger cladding. These fibres can be pumped by high-power multimode diodes or even diode bars or stacks. The pump light is guided in the cladding by total internal reflection between the cladding and the coating and it is only absorbed when it passes through the doped core of the fibre. This fibre design concept allows the efficient conversion of multimode laser diode radiation into fibre laser radiation with very high brightness.

Today, the highest output powers of fibre lasers (some kW) have been demonstrated with ytterbium doped silica fibres that operate in the wavelength region centred around 1080 nm. Lasers in this wavelength range are not "eye safe" which is a problem for a lot of laser applications. But nowadays also thulium doped fibres almost reach the kW output power level. A selection of the latest publications reporting on high power thulium fibre lasers around 2.0 μm is shown in table 2. The fibre geometry has the great advantage that the heat that is generated during the laser process is dissipated over a large area if a long absorption distance is used. The absorption length of a fibre system can be adjusted not only by the doping concentration and the pump wavelength, also the core to cladding ratio can be used. In 2007 Frith et al. reported on a highly efficient thulium fibre laser with up to 110 W of cw output power. They used newly designed large mode area fibres which yield a low numerical aperture (NA) of the doped core (NA = 0.06). This enables the usage of large core diameters by still retaining single transversal mode laser operation. Therefore Frith et al. could build up a laser using a fibre with a core diameter of 20 μm, a cladding diameter of 400 μm, and a fibre Bragg grating as a highly reflective mirror. With this concept and pumping of one fibre end through the fibre Bragg grating 110 W of narrow line width (full width at half maximum (FWHM): 3 nm) output power were achieved with a slope efficiency of 55 %. In the same year Wu et al. reported on the high power operation of a thulium doped germanate fibre (Wu et al., 2007). They achieved 64 W of output power in a one-end pumped configuration with an only 20 cm long piece of fibre. With respect to the launched 800 nm pump power an extremely high slope efficiency of 68 % was measured. In a dual-end pumped configuration the maximum output power could be increased to 104 W, but the slope efficiency was reduced to 52.5 %. Also in 2007 a 415 W thulium fibre laser that was inband pumped at 1567 nm was presented (Meleshkevich et al., 2007). A double clad single mode thulium fibre was used, which was end pumped by an assembly of 18 cw erbium fibre lasers. By the usage of fibre Bragg gratings an all-fibre set-up was realised which yielded an output beam with $M^2 < 1.1$ and a slope efficiency of 60 %. Using a thulium doped fibre with a core diameter of 35 μm (numerical aperture 0.2) and a cladding diameter of 625 μm Moulton et al. achieved a cw output power of up to 885 W at room temperature (Moulton et al., 2009). This is the highest output power achieved with a single Tm-doped fibre so far. The laser showed multi mode emission with a slope efficiency of about 49.2 %. To achieve this high output power a 7 m long piece of fibre was used that was pumped from both ends with fibre coupled laser diode sources emitting at 793 nm. So actually the highest thulium fibre laser output powers were achieved with diode pumping around 800 nm, but also the approach with resonant pumping around 1570 nm has the potential to reach such high powers. The overall efficiency of the 800 nm pumped systems is better than for the resonant pumping due to the limited efficiency of the Yb,Er:fibre lasers used. An all fibre system should be possible with both concepts, although so far this was only presented for the resonant pumping at 1570 nm. Actually there are two companies that are offering thulium fibre lasers commercially. One offers systems that are pumped around 800 nm (Nufern) and the other is using the resonant pumping concept (IPG Photonics Corp.).

Actually the highest thulium laser output powers are achieved with fibre lasers, but also crystal lasers can reach more than 100 W of output power. The fibre lasers also yield a better beam quality than the crystal lasers, which is an enormous advantage for some applications. Both systems show approximately the same slope efficiencies, nevertheless for high output powers the optical to optical efficiency of the fibre lasers is higher. For applications with output powers in the range of 100 W to 200 W crystal lasers are still a good alternative to fibre lasers, especially since these systems are well developed and commercially available.

2.2 Holmium laser systems

Until recently there were no laser diodes available in the wavelength ranges which allow direct pumping of Ho^{3+} ions. Therefore the first holmium lasers were realised in co-doped systems. Usually thulium co-doping is used, because one can exploit the cross relaxation process of the thulium ions for the excitation of the upper laser level in holmium. There are two energy transfer processes which lead to the population of the upper laser level ($^5I_7$) of the holmium ions. The first transfer is the fast spatial energy migration among the Tm^{3+} ions and the second one is the energy transfer from the thulium $^3F_4$ to the holmium $^5I_7$ level. The net energy transfer can be determined as

$$f_{Ho} = \frac{N_{Ho} \cdot \sum_i g_i \cdot e^{-\frac{E_i}{k \cdot T}}}{N_{Ho} \cdot \sum_i g_i \cdot e^{-\frac{E_i}{k \cdot T}} + N_{Tm} \cdot \sum_j g_j \cdot e^{-\frac{E_j}{k \cdot T}}}$$

where f_{Ho} is the net transfer efficiency, N_{Tm} is the concentration of Tm^{3+} ions, the sums over i and j are sums over the Ho^{3+} $^5I_7$ and Tm^{3+} $^3H_4$ crystal field splittings, respectively, E_i and E_j are the energy levels of these splittings, g_i and g_j are their degeneracies, k is the Boltzmann constant, and T is the temperature (Fan et al., 1988). In YAG as host crystal, about 50 % of the excitation energy is transferred to the holmium ions at room temperature when using a thulium co-doping. The rest of the energy is stored in the excited thulium ions (Storm, 1988). YAG is the most common host crystal for Ho^{3+} because of its high values of specific heat, heat conductivity and optical quality (Rothacher et al., 1998). The energy scheme of Ho:YAG with the relevant transitions for the 2.1 μm laser emission and the Stark splitting of the ground state are shown in figure 6. The 2.1 μm emission emerges from a transition which starts in the upper laser level $^5I_7$ and terminates in the thermally populated sublevels of the $^5I_8$ ground state manifold. Both involved energy levels show a strong Stark splitting, the higher level consists of 14 sublevels from 5228 cm^{-1} to 5455 cm^{-1} and the ground state splits up into 11 sublevels from 0 cm^{-1} to 535 cm^{-1} (Kaminskii, 1996).

The thermal population of the lower laser level at room temperature for the free running Ho:YAG laser is about 2 %, which is nearly the same as for the Tm:YAG laser. But the upper laser level of the holmium laser is also thermally populated and this population is much lower than for thulium lasers. At room temperature in YAG only 10 % of the holmium ions, which are excited to the $^5I_7$ manifold, populate the Stark level which is the upper laser level. For thulium this number is about 46 %. Therefore the temperature dependence of holmium lasers is stronger than the one of thulium lasers. The upconversion process, in which one holmium ion gets excited into the $^5I_5$ or the $^5I_6$ manifold, is a non resonant process (see figure 6). It is a phonon assisted process, for which two closely spaced holmium ions that are excited into the $^5I_7$ manifold are necessary. Therefore this process becomes important when the population density of the $^5I_7$ energy level is high. Thus the upconversion process is most important for the q-switched operation, when the energy storage in the upper laser level is high.

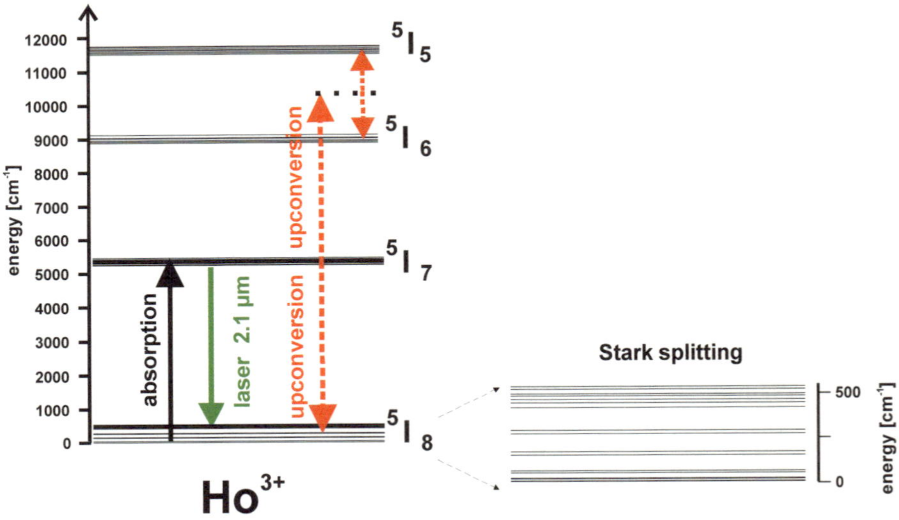

Fig. 6. Ho:YAG energy scheme with the relevant transitions for 2.1 μm laser operation

The first holmium laser on the 2.1 μm transition used a thulium co-doping and YAG as a host material. It was a flash lamp pumped pulsed laser that was built up in 1965 (Johnson et al., 1965; Johnson et al., 1966). This laser needed to be cooled with liquid nitrogen. Room temperature, lamp pumped, pulsed operation of Ho:YAG and Ho:YLF lasers was demonstrated in 1970 and 1971, respectively (Remski et al., 1970; Chicklis et al., 1971). The investigated crystals were sensitised with Er^{3+} and Tm^{3+}, leading to an optimised absorption of the pump light. The first room temperature cw laser operation of a holmium laser with thulium co-doping was demonstrated in 1985, using a krypton laser as pump source (Duczynski et al., 1986). Shortly after the development of diode lasers around 800 nm many innovative approaches of pumping thulium co-doped holmium lasers were successfully realised. The first cw holmium lasers utilising diode pumping were demonstrated in 1986 for YAG as host material and for YLF in 1987 (Fan et al., 1987; Kintz et al., 1987). Intra-cavity pumping of Ho:YAG was demonstrated for the first time in 1992 (Stoneman et al., 1992). The Ho:YAG crystal was embedded in a Tm:YAG laser cavity and acted as output coupler.

The following section focuses on the latest results of different methods for in-band pumping (direct pumping into the upper laser level $^5I_7$) of Ho:YAG crystals, which is the most promising approach to reach the highest output powers. In-band pumping of most holmium crystals is possible in the wavelength range around 1.9 μm. The latest results of continuous wave and q-switched holmium laser operation will be shown and reviewed. The co-doping of thulium and holmium in crystals and fibres has significant drawbacks. The probability of the upconversion process that populates the $^5I_5$ and the $^5I_6$ level is increased by the co-doping and the thermal load in the crystal is higher, even when the cross relaxation process of the thulium ions is exploited very well. Due to the fact that the emission wavelength of 2.1 μm addresses a wide variety of applications that require short laser pulses with high pulse energies or high continuous wave powers, many high power holmium laser systems were realised in recent years. Most of these laser systems use thulium crystal or fibre lasers for pumping the holmium ions. An overview of some of the latest published results is shown in table 3.

laser material	pump source	λ_p (μm)	cw power (W)	pulse energy (mJ)	λ_{em} (nm)	slope. eff. (%)	reference
Ho:YAG	Tm:YLF	1.95	1.6		2090	21	Schellhorn et al., 2003
Ho:YAG	Tm:YLF	1.9		50	2090		Budni et al., 2003
Ho:YAG	Tm fibre	1.905	6.4		2097	80	Shen et al., 2004
Ho:YLF	Tm fibre	1.94	43	40	2050	42	Dergachev et al., 2005
Ho:YAG	Tm fibre	1.908	10	15	2100	52	Moskalev et al., 2006
Ho:YAG	Tm:YLF	1.908	9.4		2090	40	Schellhorn, 2006
Ho:YAG	Tm:YLF	1.91	14		2100	16	So et al., 2006b
Ho:YAG	diode	1.91	40	3.5	2120	57	Scholle & Fuhrberg 2008
Ho:YLF	Tm fibre	1.94	12.4	10.9	2065	47	Bollig et al., 2009
Ho:YAG	Tm fibre	1.908	18.7		2090	80	Mu et al., 2009

Table 3. Overview of some recently published cw and pulsed holmium laser results in the literature (λ_p = pump wavelength; λ_{em} = emission wavelength)

Thulium fibre lasers are an excellent pump source for holmium lasers. They offer a nearly diffraction limited beam quality and a very narrow emission bandwidth. The wavelength of the fibre lasers can be tuned to the maximum absorption of the holmium ions by using fibre Bragg gratings. Due to these benefits the highest slope efficiencies for in-band pumped holmium lasers were achieved with thulium fibre lasers as pump sources. Up to 80 % of slope efficiency for cw laser operation were demonstrated by Mu et al. and Shen et al. with Ho:YAG lasers. In 2004 Shen et. al. reported on a room temperature Ho:YAG laser pumped by a cladding-pumped Tm:silica-fibre laser (Shen et al., 2004). The emission wavelength of the Tm:silica-fibre laser was tuned with an external grating to the absorption maximum of Ho:YAG. With the available 9.6 W of pump power, 6.4 W of unpolarized output power of the Ho:YAG system were reached in a short plane-concave resonator. The optical to optical conversion efficiency of the system was 67 %. Mu et. al. used an adhesive free bonded YAG/Ho:YAG/YAG laser composite crystal in a water cooled heat sink (Mu et al., 2009). The front facet of the composite crystal acted as a plane highly reflective mirror for the holmium laser and the back side had a high reflectivity coating for the pump wavelength to achieve a multi pass of the pump light. Thus the resonator was built with the front facet of the crystal and a concave output coupler. The pump source, an unpolarized thulium doped fibre laser (FWHM 0.7 nm), was tuned to the strongest absorption line of Ho:YAG at 1907.65 nm. The maximum output power of this system was 18.7 W at 24.3 W of pump power, which results in an optical to optical conversion efficiency of 77.6 %, which is the highest efficiency reported so far.

In 2006, Moskalev et. al. demonstrated a q-switched Ho:YAG laser pumped by a commercially available thulium fibre laser (Moskalev et al., 2006). They used a 50 mm long Ho:YAG rod with a doping concentration of 0.5 % that was conductively cooled. With a plane concave folded cavity a maximum output power of 10 W in cw operation were demonstrated with a corresponding slope efficiency of 52 %. With an acousto-optic modulator inside the resonator, q-switched laser operation with a maximum output power of 15 mJ at 100 Hz repetition rate was shown.

The highest cw output powers and pulse energies of a holmium laser achieved with thulium fibre laser pumping were reached with YLF as host material for the holmium ions. Dergachev et al. reached 43 W of cw output power and 40 mJ of q-switched pulse energy with a Ho:YLF laser that was pumped at 1940 nm with a commercially available 100 W thulium fibre laser (Dergachev et al., 2005). These high powers were achieved using two holmium crystals in one cavity. The crystals were pumped by one fibre laser using a polarisation beam splitter to spread the pump power. The 40 mJ of pulse energy were reached for repetition rates below 400 Hz, at 1 kHz 28 mJ of pulse energy were reached. Using a single Ho:YLF crystal Bollig et al. reached a slope efficiency of 47 % and 10.9 mJ of pulse energy at 1 kHz repetition rate (Bollig et al., 2009). With an additional Ho:YLF amplifier that was pumped by the pump light transmitted from the first holmium crystal, up to 23.7 mJ of pulse energy at 1 kHz repetition rate were reached. This amplifier system had a slope efficiency of 47 % and yielded a beam quality of $M^2 < 1.1$.

Another attractive pump source for holmium lasers are Tm:YLF lasers. Tm:YLF lasers are highly efficient and offer high pump powers with good beam quality. These laser systems can also be tuned to the maximum absorption peaks of holmium in the wavelength range between 1.9 µm and 2.0 µm. In 2003, Budni et al. demonstrated a high pulse energy q-switched Ho:YAG laser with 50 mJ of output energy and an M^2 of about 1.2, that was pumped by a Tm:YLF laser (Budni et al., 2003). They used a folded plane concave cavity, where the pump light was coupled into the holmium crystal trough a thin-film polariser that also acted as a folding mirror. Using output couplings of up to 70 % damage free q-switched operation with pulse lengths of 14 ns and peak powers of up to 3.6 MW were achieved. A Ho:YAG thin-disk laser with an output power of up to 9.4 W was realised by M. Schellhorn using two polarisation coupled Tm:YLF lasers as pump sources (Schellhorn, 2006). With a 0.4 mm thick, 2 % doped Ho:YAG crystal an optical to optical efficiency of 36 % was reached using 24 passes of the pump light through the Ho:YAG crystal.

Using Tm:YLF lasers as pump sources also intra-cavity pump schemes are possible, where the holmium crystal is placed inside the thulium laser resonator. In this case the holmium laser acts as an output coupler for the thulium laser. A schematic set-up for an intra-cavity side pumped holmium laser is shown in figure 7.

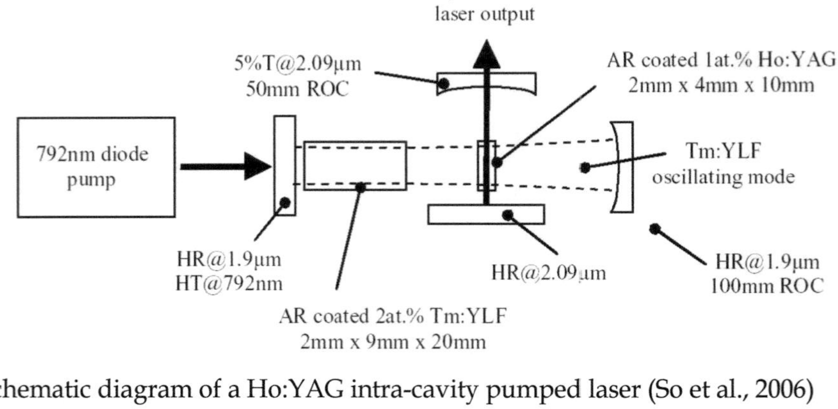

Fig. 7. Schematic diagram of a Ho:YAG intra-cavity pumped laser (So et al., 2006)

Using intra-cavity pumping a low holmium doping concentration can be used, since the pump light intensity is very high inside the laser resonator. In 2003 a compact Ho:YAG laser intra-cavity pumped by a diode-pumped Tm:YLF laser was realised (Schellhorn et al., 2003). At room temperature a maximum average holmium laser output of 1.6 W with a slope efficiency of 21 % with respect to the incident diode pump power was achieved. In 2006 So et al. demonstrated an intra-cavity side-pumped Ho:YAG laser system (So et al., 2006b). A high-power Tm:YLF slab laser (9 x 1.5 x 20 mm^3) with an optimised thulium doping concentration of 2 at. % pumped by a laser diode stack at 792 nm served as the pump source. The maximum output power of the slab laser itself at a wavelength of 1.91 μm was 68 W. The corresponding slope efficiency was 44 %. The intra-cavity side-pumped Ho:YAG slab (4 x 10 x 2 mm^3) had a doping concentration of 1 at. %. With the Ho:YAG slab 14 W of output power at 2.1 μm with a slope efficiency of 16 % were achieved with this set-up, which is shown in figure 7.

In all the laser systems described above the holmium lasers were pumped by thulium lasers, whose emission wavelengths were tuned to the most efficient absorption peaks of the holmium ions. These thulium laser systems are pumped with commercially available laser diodes around 800 nm. Exploiting the cross relaxation process of the thulium ions these systems reach optical to optical conversion efficiencies between 40 % and 60 %. This pumping concept yields a quite complex overall set-up and it limits the overall efficiency of the laser systems. Taking into account an electrical to optical efficiency of 50 % for the laser diodes around 800 nm the maximum overall system efficiency with these pump systems for the realised holmium lasers is about 15 %. Direct in-band pumping with laser diodes around 1.9 μm is therefore an attractive alternative to develop simple and compact holmium laser systems with high overall efficiencies.

The first directly diode in-band pumped holmium laser was realised in 1995 (Nabors et al., 1995). With six 1.9 μm laser diodes using angle multiplexing and polarisation beam combining nearly 0.7 W of output power were reached. With respect to the absorbed pump power a slope efficiency of 35 % was achieved. This demonstrated that efficient in-band pumping of Ho:YAG lasers by laser diodes at 1.9 μm is generally possible, although the emission bandwidth and the beam quality of laser diodes is inferior to thulium laser systems. Nabors et al. used a mix of GaInAsSb and InGaAsP diode lasers, each with about 0.7 W of output power, which were available in 1995. In recent years diode lasers based on GaSb material systems (AlGaIn)(AsSb) were significantly improved. They cover the wavelength range from 1.85 μm to 2.35 μm. In section 3 these diodes will be described in detail. The improvements of these high power diodes make them very interesting for the excitation of holmium lasers. The newly developed GaSb-based diode bars and stacks provide enough pump power to realise high power holmium laser systems.

Using a GaSb-based laser diode stack which consists of ten bars with an output power up to 158 W Scholle et al. presented the first high power Ho:YAG laser that was in-band pumped by such diodes in 2008 (Scholle et al., 2008). A continuous wave output power of 40 W with a slope efficiency of 57 % was demonstrated. Also q-switched operation with an acousto optic modulator (AOM) inside the laser cavity was investigated. A maximum q-switched output power of 3.5 mJ - limited by damage of the optical components - at a repetition rate of 1 kHz with 33 % output coupling was reached. The pulse durations were around 150 ns.

Using a diode stack as pump source yields some challenges for the laser set-up. In figure 8 one can see that the absorption peaks for the excitation from the ground state to the upper laser level in Ho:YAG (left side) and Ho:YLF (right side) are narrow. The maximum absorption in Ho:YAG can be found around 1910 nm ($\sigma_{abs} = 9 \times 10^{-21}$ cm²) and the corresponding FWHM is only 7 nm. For Ho:YLF the maximum absorption depends on the polarisation, for π polarisation the maximum is at 1940 nm ($\sigma_{abs} = 10 \times 10^{-21}$ cm²) and for σ polarisation it is at 1945 nm and significantly lower ($\sigma_{abs} = 6 \times 10^{-21}$ cm²).

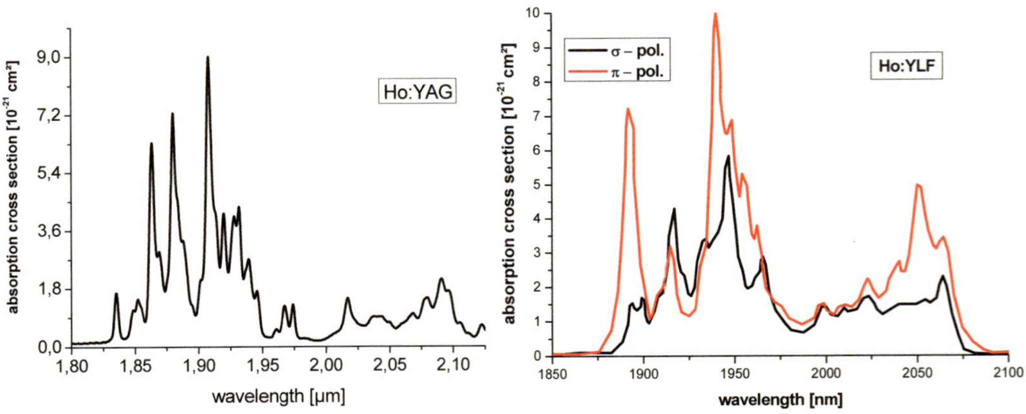

Fig. 8. Absorption cross sections of the $^5I_8 \rightarrow {}^5I_7$ transition for Ho:YAG and Ho:YLF (Walsh et al., 1998)

As shown in figure 14 the FWHM of the emission spectrum of a multi-bar stack is about 25 nm for high output powers, so the effective absorption coefficient in Ho:YAG for the diode stack is approximately 0.35×10^{-20} cm², when the centre wavelength of the emission is at 1910 nm. Taking this into account the absorption length of a 0.5 % doped Ho:YAG crystal is about 40 mm and for a 1 % doped crystal it is 20 mm, respectively. Another problem for pumping with diode stacks is the absorption minimum around 1895 nm, since the emission of the stack shifts over this absorption minimum when the power of the stack is continuously increased (see figure 13). When using a double pass pump scheme a significant part of the pump light can be coupled back into the diode stack, when the diode emits at this absorption minimum. This can lead to the destruction of the diode. The poor beam quality of the diode stack makes it necessary to use special optics for the collimation of the light and also for focussing into the laser rod. In our experiments the pump light of the used GaSb stack was only fast axis collimated. The slow axis divergence was compensated by a cylindrical lens inside the anti-reflective coated multi lenses focussing optic. With this optic pumping of laser rods with a diameter of 3 mm is possible. Inside the Ho:YAG rod the pump light is guide by total reflection on the polished surface of the rod.

Figure 9 shows the set-up used for the holmium laser experiments. The Ho:YAG rod and the multi-bar laser diode stack were water cooled to 15 °C with a common cooling circuit. All Ho:YAG rods used were anti reflective (AR) coated for the pump and the laser wavelength on both sides. The plane-plane resonator used for the q-switching experiments had a length of about 150 mm, for the cw experiments a shorter resonator with 80 mm in length was used. The first resonator mirror has high reflective (HR) coating for 2.1 µm and an AR coating for the pump light at 1.9 µm. After the Ho:YAG rod a pump light reflector was integrated to achieve a double pass of the pump light. For q-switching an acousto optic modulator cut under Brewster angel was integrated in the resonator. Output couplers with transmissions from 3.8 % to 30 % were tested so far for cw and q-switched laser operation. Until now laser rods with doping concentrations of 0.5 % and 1 % were investigated. Figure 10 shows the laser output curves of a 62 mm long 0.5 % doped Ho:YAG rod for different output couplings on the left side. The highest output powers of up to 40 W were reached using 7 % output coupling. All curves show a drop of the holmium output power at an incident pump power of about 80 W because of the mentioned minimum in the absorption of Ho:YAG around 1895 nm (see fig. 8). The laser threshold is around 40 W and the maximum output power was achieved with 135 W of incident pump power on the crystal. This yields an optical to optical efficiency of about 33 %. The slope efficiency is about 40 % with respect to the incident pump power on the Ho:YAG crystal and 57 % with respect to the absorbed pump power. The 1 % doped crystal showed a slightly inferior performance. On the right side of figure 10 the best input output curve of a 1 % doped crystal is shown in comparison to the results of the 0.5 % doped crystal. The threshold of the 1 % doped crystal is higher (60 W) and the slope efficiency with respect to the absorbed pump power is nearly the same as for the lower doped crystal.

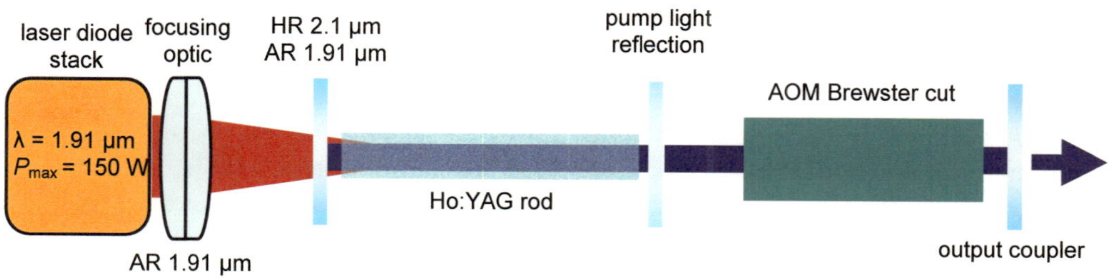

Fig. 9. Experimental set-up for the holmium laser experiments with q-switching

In recent experiments with the q-switched laser operation pulse energies up to 5 mJ were achieved with a pulse length of 180 ns. With an output coupling of 30 % the q-switched laser operated just above the laser threshold. The 5 mJ were observed for repetition rates of up to 250 Hz; at 1 kHz repetition rate only 3 mJ of pulse energy were observed.

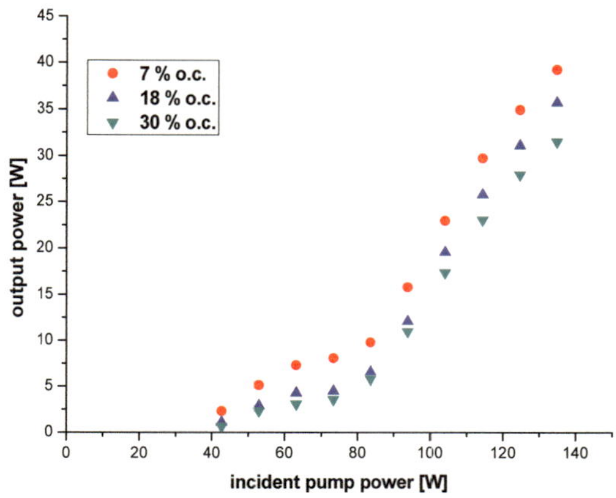

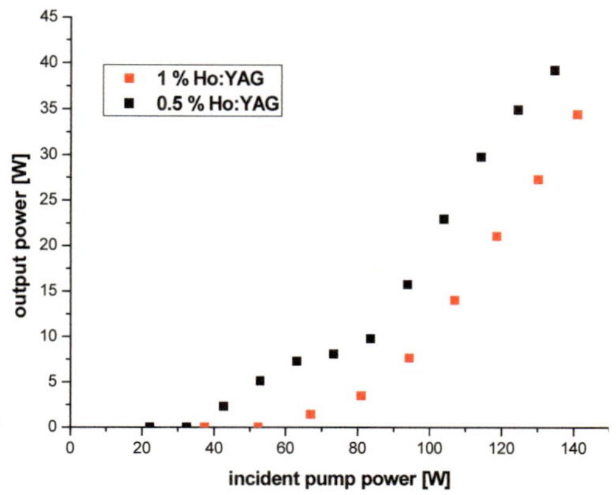

Fig. 10. Left side: Characteristic curves of a 0.5 % doped Ho:YAG rod with a length of 62mm for different output couplings (o.c.). Rigth side: Comparison of the output power for the 0.5 % and the 1 % doped crystal using 7 % o.c..

The presented results show the great potential of the in-band pumping of holmium crystals. Maximum cw output powers of up to 43 W and maximum pulse energies of up to 50 mJ were shown. The achieved slope efficiency 80 % with the fibre laser pumping shows that there are no significant upconversion losses for in-band pumping. The achieved results with the diode pumping demonstrated the feasibility of this sources for the pumping of Ho:YAG, but pumping Ho:YLF should also be possible with these diodes. In general the performances achieved with Ho:YLF and Ho:YAG are comparable. Further power scaling of the output powers should be possible with both systems when pump sources with higher powers become more easily available. A second option to achieve higher output powers is the realisation of multi stage amplifier systems.

3. GaSb-based diode laser systems around 2 µm

3.1 Laser diodes and diode stacks

In the last years great progress was made in the development of laser diodes emitting in the wavelength range around 2 µm. The efficiency and the output power could be significantly increased. Additionally the beam quality was improved (lower fast axis beam divergence) and the lifetime at room temperature operation was extended. Therefore these diodes can now address an increasing number of applications and they will become more and more commercially available. Depending on the emitted wavelengths and output powers one can use these compact and low-cost coherent sources for instance in gas sensing, free-space telecommunication, and medical applications or as pump sources for solid state laser systems. In principle GaSb-based laser diodes can be realised with emission wavelengths between 1.85 µm and 2.35 µm for room temperature operation (Rattunde et al., 2000).

In 1963 I. Melngailis demonstrated the first mid-infrared semiconductor laser based on the InAs material system (Melngailis, 1963). The laser operated at low temperatures and emitted light at 3.1 µm. The first room temperature mid-infrared laser was developed by Caneau et al. in 1985 (Caneau et al., 1985). This (GaInAsSb)(AlGaAsSb) double-heterostructure laser operated at 2.2 µm. Ten years later Baranov et al. reported on high temperature operation of the same semiconductor structures emitting at 2.1 µm (Baranov et al., 1994). All these systems delivered very low output powers, a breakthrough with respect to the upscaling of the laser output power for 2 µm diodes was achieved in 1992 (Choi et al., 1992). For the first time quantum-well-structured GaSb-based laser diodes were presented. This new design approach led to laser output powers up to 190 mW and it showed improved power exploitation in comparison with the mentioned double-heterostructures. From this invention it took about 10 years to achieve further power scaling of laser diodes that operate in the 2 µm wavelength range. The next invention to achieve higher output powers was a p-side down mounted broad-area GaSb-based laser diode (Rattunde et. al., 2002). The direct gap band-edge profile of a laser structure with an emission wavelength around 2 µm that was realised by Rattunde et al. is shown in figure 11. In such a diode, three compressively strained 10 nm thick $Ga_{0.78}In_{0.22}Sb$ quantum wells which are separated by 20 nm wide lattice matched $Al_{0.29}Ga_{0.71}As_{0.02}Sb_{0.98}$ barriers compose the active region. The active region itself is located between two 400 nm thick $Al_{0.29}Ga_{0.71}As_{0.02}Sb_{0.98}$ confinement layers which were sandwiched between 2.0 mm wide lattice matched $Al_{0.84}Ga_{0.16}As_{0.06}Sb_{0.94}$ n- and p-doped cladding layers.

Due to the broad area the differential and thermal resistance of this laser was reduced. This provided a maximum output power of up to 1.7 W at room temperature. The great potential of this semiconductor material system was recognised instantly and Shterengas et al. soon fabricated the first linear array with this type of diodes (Shterengas et. al., 2004). Linear laser arrays on a 1-cm-wide bar with 19 100 μm wide emitters and 1-mm-long cavities were built and characterised. With this first high-power diode laser linear array 10 W of output power at 2.3 μm were achieved for continuous wave laser operation at room temperature. The spectral width of the linear array output was about 20 nm and the achieved peak wall-plug efficiency was already near 9 %. But these first high power diodes and bars had one big disadvantage, which was the large fast axis beam divergence. This large divergence of 67° FWHM results from the broad-area design of the diode laser and it considerably limits the applications for this type of diodes. With this large divergence fibre coupling is extremely difficult and also the practical use as a pumping source for solid state lasers requires a FWHM ≤ 45° in order to collimate the emitted beam of the laser diodes with state-of-the-art collection optics.

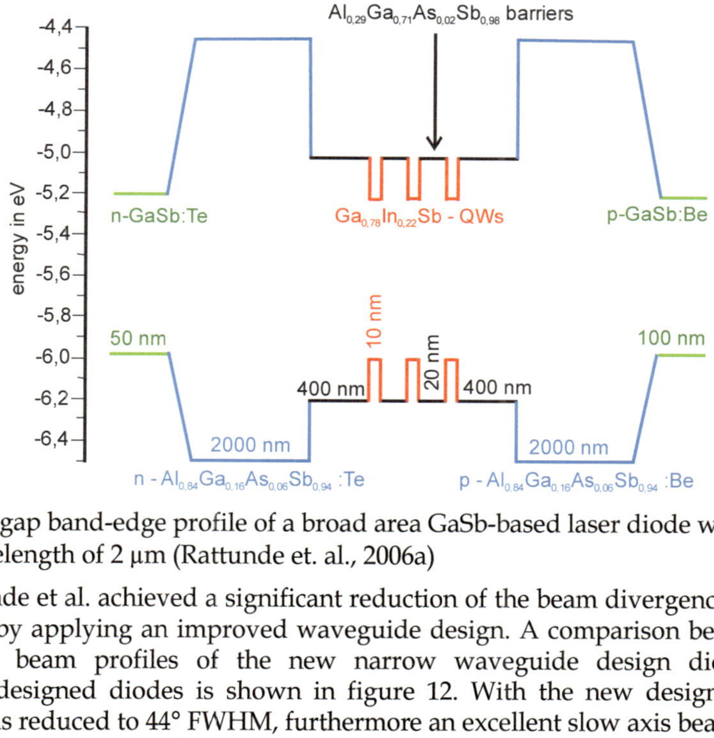

Fig. 11. Direct gap band-edge profile of a broad area GaSb-based laser diode with an emission wavelength of 2 μm (Rattunde et. al., 2006a)

In 2006 Rattunde et al. achieved a significant reduction of the beam divergence of the GaSb-based diodes by applying an improved waveguide design. A comparison between the fast axis far field beam profiles of the new narrow waveguide design diodes and the conventional designed diodes is shown in figure 12. With the new design the fast axis divergence was reduced to 44° FWHM, furthermore an excellent slow axis beam quality was achieved (Rattunde et al., 2006b). The newly designed laser diodes exhibited a maximum cw output power of 1.96 W at a wavelength of 2.0 μm.

In 2006 Kelemen et al. developed and characterised high-power 1.91 μm (AlGaIn)(GaSb) quantum-well diode laser single emitters and linear arrays with the new narrow waveguide design (Kelemen et al., 2006). A single emitter showed a maximum output power of nearly 2 W with a slope efficiency of 0.32 W/A at room temperature. The 1-cm-long tested linear array consisted of 19 emitters. The maximum cw output power of this bar was limited to 16.9 W by the thermal rollover. The slope efficiency of the bar was about 0.31 W/A and the maximum wall plug efficiency was 26 %. The wavelength shift of the bar was measured to be 1.4 nm/K and 1.2 nm/W, so the thermal shift and the shift on power dissipation are slightly higher than for a single emitter (1.2 nm/K and 8.6 nm/W).

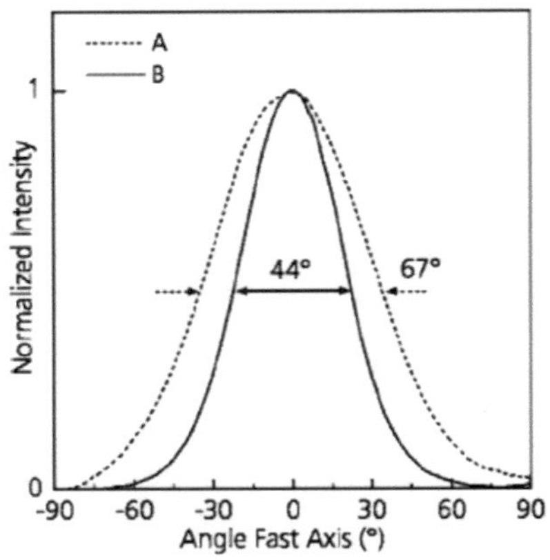

Fig. 12. Measured fast axis far-field beam profiles of a broad area diode (doted line) and a diode with reduced beam divergence (continuous line) (Rattunde et al., 2006b)

In 2007 LISA laser products OHG bought a high power multi-bar stack that was build with ten linear arrays of the type that was characterised and described by Kelemen et al. before. This first GaSb-based diode stack is directly water cooled and integrated into a sealed housing. The single emitters used have a stripe-width of 150 µm and a pitch of 500 µm. The rear facets are coated with a highly reflective double-stack of Si and SiO_2 films (> 95 % reflectivity) and the front facets are coated by a single layer of SiN (3 % reflectivity). The emitters were collimated only in the fast axis with micro optics which are integrated in the sealed housing. The used waveguide concept which leads to a low fast axis beam divergence of 44° FWHM allows a collimation with common pumping diode optics. In order to focus the emitted beam of the laser diode stack into a laser crystal the remaining slow axis divergence has to be compensated with a cylindrical lens.

Important aspects for the holmium pumping are the current-power characteristic, output spectra and the spectral shift of the laser diode stack. Figure 13 shows the current-power characteristic at a cooling temperature of 15 °C for continuous wave operation on the left side and the shift of the central wavelength with respect to the output power of the diode stack on the right side.

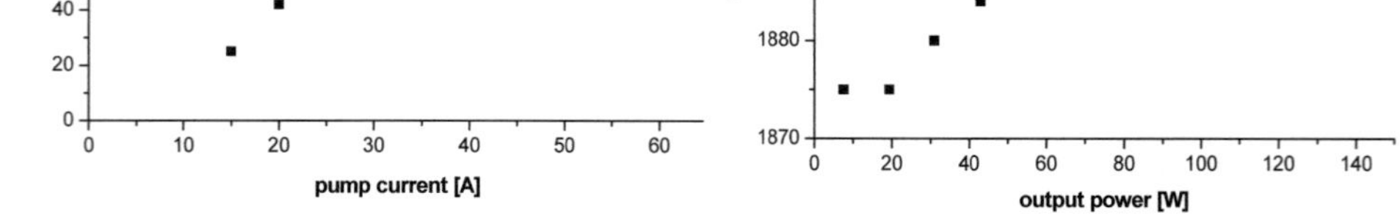

Fig. 13. Left diagram: Characteristic curve of the GaSb multi-bar stack at 15 °C. Right diagram: Shift of the central wavelength with respect to the output power.

The threshold current of the laser diode stack is 8 A and the maximum continuous wave output power is 158 W leading to a slope efficiency of 2.95 W/A. The right side of figure 13 shows that the central wavelengths of the laser diode stack shifts over 40 nm from laser threshold to the maximum output power (0.26 nm/W). This large wavelength shift is a significant disadvantage for the pumping of holmium lasers, since the absorption of holmium ions in most host crystals changes dramatically over this wavelength range. One can see in figure 8 that the main absorption peak of Ho:YAG at 1910 nm has a line width of only 7 nm and also Ho:YLF has a line width of only about 15 nm at the absorption maximum around 1940 nm.
An additional problem of the laser diode stack is the very broad emission spectrum. In figure 14 the emission spectra of the laser diode stack at different pump powers (left side) and an output spectrum of a single emitter (right side) are shown. Each spectrum is normalised. On the left side one can see that the emission of the diode stack shifts to longer wavelengths with higher pumping currents and in addition, the structure of the spectrum becomes more inhomogeneous. This effect causes the broadening of the emission spectrum with higher pumping currents from 8 nm to 20 nm.

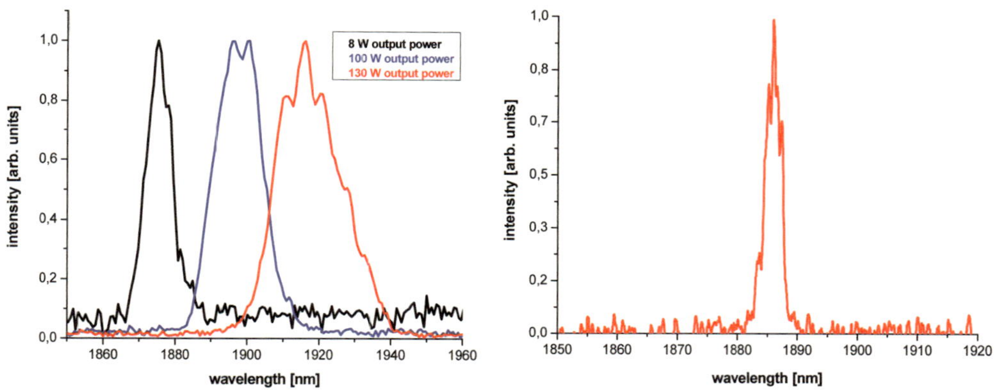

Fig. 14. Left side: Normalised output spectra of the diode stack for different output powers. Right side: Output spectrum of a single emitter at low output power.

Due to the big improvements of the GaSb-based diodes in the recent years these laser sources have now become interesting for many purposes. Current research and development therefore focuses on the optimisation of these lasers towards specific applications. Single emitters can be used as cheap sources for sensing or spectroscopy. Using external resonators single-mode cw operation can be realised with the diodes, which makes it possible to use them as seed sources for high power lasers (Jacobs et al., 2004). As shown in section 2, laser bars or stacks can be used for solid state laser pumping. But these lasers become also interesting for other industrial or medical applications, especially when the efficiency and the beam quality can be further increased. Nowadays GaSb-based diodes are becoming commercially available, for instance by m2k Laser.

3.2 Optically pumped GaSb-based semiconductor disk lasers

The poor spatial quality of the output of conventional GaSb-based 2 μm laser diodes limits their applications. Using a different design for the semiconductor and optical pumping, a very high brightness can be achieved. Optically pumped semiconductor disc lasers (OPSDL) also known as vertical-external-cavity surface emitting lasers (VECSEL) are optically pumped similar to solid state thin disc lasers (Kuznetsov et al., 1997). The OPSDL chip, formed by a sequence of epitaxially grown semiconductor layers, simultaneously acts as the cavity mirror and the gain region. A pump laser is focused onto the surface of the chip with a spot-size typically several tens or hundreds of microns in diameter. To complete the laser resonator an arrangement of external mirrors and other components is used. A typical OPSDL set-up with a two mirror cavity is shown on the left side of figure 15. To achieve optimal cooling of the OPSDL chip a heat-spreader is often mounted on the surface (Schulz et al., 2008). A typical OPSDL structure with heat-spreader and the heat flow is shown on the right side of figure 15. The OPDSDL design offers several advantages: The emission wavelength can be varied by the design of the OPSDL structure in the same way as for conventional semiconductor lasers. Mode matching of pump and laser beam by variation of the external resonator design often provides diffraction limited output beams. The external resonator also allows flexible control of the laser properties by the insertion of intra-cavity elements like frequency stabilization, passive mode-locking, or frequency doubling.

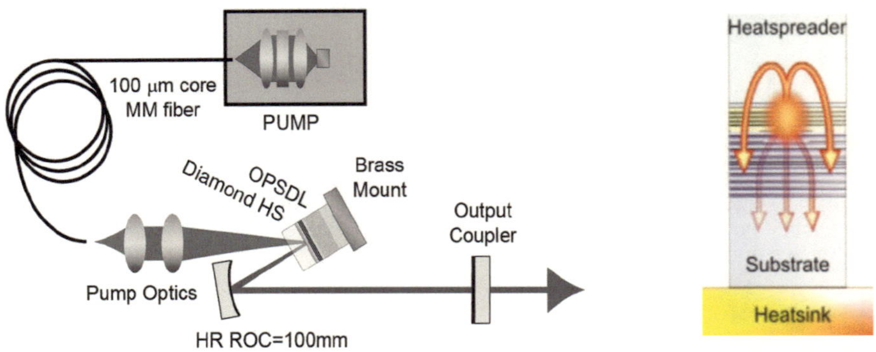

Fig. 15. Schematic set-up of an OPSDL with a folded external cavity (left side) and the heat flow in an OPSDL with a heat-spreader (right side) (Schulz et al. 2008).

GaSb-based OPSDLs can be realised within the wavelength range from 2.0 μm to 2.5 μm. In the 2 μm wavelength range either SiC or diamond are used as heat-spreader. Due to the lower thermal conductivity of SiC the output powers of lasers with SiC heat-spreaders are lower than for diamond ones. On the left side of figure 16 typical input-output curves of an GaSb-based OPSDL with a SiC heat-spreader emitting at 2020 nm are shown for different cooling temperatures. As a pump source a commercial fibre coupled laser diode emitting at 976 nm and a resonator similar to the one shown in figure 15 with about 4 % output coupling was used. The maximum output power of this laser was 2.5 W at 9 °C and 2.0 W at 18 °C. The strong temperature dependence is a result of the high heat load in the chip caused by the large quantum defect between the pump and the laser wavelength. The highest output powers of up to 5 W were achieved with GaSb-based OPSDLs operating at 2.0 μm using a diamond heat-spreader (Hopkins et al., (2008). At longer wavelengths the maximum output powers are lower so far.

By integrating a quartz birefringent filter into the collimated arm of the folded resonator, wavelength tuning and single mode laser operation can be achieved. With an 8 mm thick quartz birefringent filter a maximum single mode output of 0.8 W at 2005 nm with an OPSDL chip with a SiC heatspreader were achieved. On the right side of figure 16 the maximum single mode output powers achieved with this chip for different wavelengths are shown. The large wavelength tuning was achieved using the birefringent filter and the variation of the pump power for wavelength tuning. The wavelengths tuning was not continuously due to the etalon effect of the uncoated SiC heatspreader used in this experiment. The highest tuning range achieved with a birefringent filter and single mode laser operation of a GaSb-based OPSDL was 70 nm around 2.3 μm (Hopkins et al., 2007). An OPSDL with a diamond heat spreader was used which reached a maximum single mode output power of 860 mW when cooled to -15 °C. Stable single-mode laser operation of an OPSDL could also be demonstrated by using a volume Bragg grating (VBG) as output coupler (Scholle et al., 2009). Up to 0.8 W were achieved at 2013 nm by using the VBG with about 98 % reflectivity as output coupler.

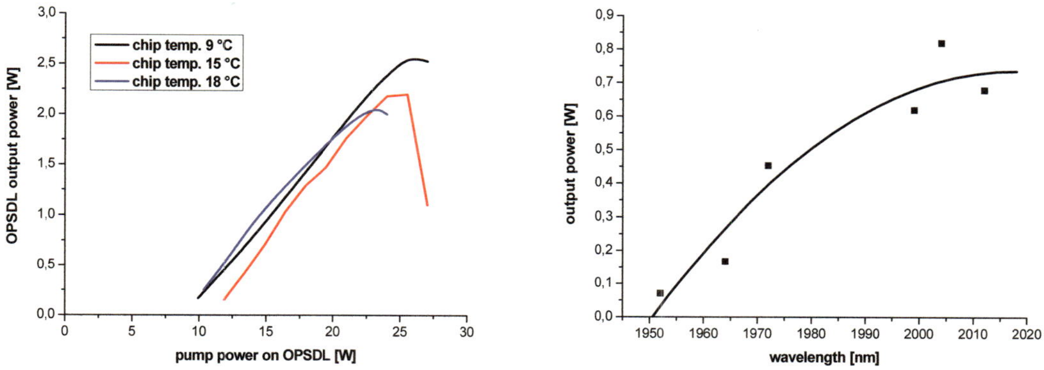

Fig. 16. Input-output curve of a GaSb-based OPSDL with a SiC heatspreader (left side) and maximum single mode output of this OPSDL achieved with a birefringent filter for different wavelengths (right side).

OPSDL combine some of the biggest advantages of semiconductor and solid state lasers. They can be designed for different wavelengths like semiconductor lasers and they can provide high beam qualities like solid state lasers. This makes OPSDLs attractive for a wide spectrum of applications. Until now the maximum output powers of GaSb-based OPSDLs are quite low, but they can be increased in the future. The possibility of wavelength selection and single mode operation makes OPSDLs an ideal seed source for secondary amplifier or laser stages. The major disadvantage for OPSDLs is the lack of energy storage due to the short carrier lifetimes.

4. Most attractive laser applications in the 2 μm wavelength range

4.1 Laser sensing and spectroscopy

The 2 μm wavelength range is called "eye safe", since laser radiation of this wavelength is absorbed in the vitreous body of the eye and does not reach the retina. Therefore the threshold for untreatable eye damage is much higher than for shorter wavelengths. Laser systems that operate in the "eye safe" wavelength range have great market potential especially in free space applications where eye safety is very important. So these laser systems are ideal for the usage in LIDAR (LIght Detection And Ranging) systems. LIDAR systems operate very similar to radar systems except that aerosol particles suspended in the air provide the return signal.

In the wavelength range at 2 µm there are absorption lines of a number of atmospheric gasses (e.g. H_2O, CO_2, N_2O) which can be detected and analysed in this spectral region. There are different detection techniques for chemical sensing based on short, medium and long detection paths. Short range detection means in the immediate location of the laser system and the detector, often with both in the same compact housing. Medium range refers to a short open path between the source and the detector (some metres). This includes systems with an integrated measurement system that use a multi-pass Heriotstyle cell or chamber containing the gas of interest. Long range here refers to any detection made over an appreciable distance (100 m to several km). These systems use the backscattered light for the measurement. To gain position and direction information, pulsed sources of appreciable energy are required for long range detection. In the 2 µm wavelength range there are many atmospheric transmission windows where very long detection ranges can be realised.

One of the key technologies opened up by tunable laser sources in the 2 µm region is the sensitive detection of some lighter atmospheric gasses and molecules. For this type of detection there are a number of lower power cw techniques appropriate. Most of these rely on differential optical absorption spectroscopy DOAS (Platt & Stutz, 2008). The systems detect a specific gas by exploiting the specific absorption of the gas at some wavelength. The backscattered light at this wavelength is compared with the level of returned light at another to confirm the presence of the gas of interest while eliminating other potential contaminants. In most cases, it is reasonable to compare the absorption of the species of interest with a known absorption line of another species like water vapour for example. There are many methods of detection with varying complexity and suitability for different applications; many of these employ a tunable source to allow the probe wavelength to scan through the wavelengths of interest. Remote chemical sensing with compact and robust laser sources in the 2 µm wavelength range has good potential in the chemical and petroleum industries in terms of safety, quality control, and regulatory enforcement as well as in medical and environmental applications. Recently the sensing of a number of chemical markers in breath analysis (such as ammonia - NH_3) has been shown to provide early diagnosis of chronic medical conditions.

The wind velocity can be measured using a Doppler-LIDAR system which operates with heterodyne detection. Aerosol particles travel essentially as fast as the surrounding air. When they scatter the incident laser radiation, they impart a slight Doppler shift to the returning radiation. Using heterodyne detection, the Doppler shift can be measured and thus the wind velocity at the place of the scattering can be determined. To achieve high detection ranges and high resolutions most of the Doppler-LIDAR systems are realised in a master oscillator – power amplifier (MOPA) configuration. A low-power narrow line width master laser with a fixed or tunable wavelength is used to seed a high power laser that adopts the characteristics of the master laser. A high short-term stability of the laser frequency is necessary to reach a high resolution for the wind velocity measurement. At a wavelength of 2.0 µm a resolution of 1 m/s of the wind velocity can be reached only when the frequency stability of the laser is better than 1 MHz.

Ground-based 2 µm Doppler-LIDAR systems can be used at airports to detect wake vortex formation during take-off and landing of aircraft to improve the safety. In the European project I-WAKE (Instrumentation system for on-board wake vortex and other hazards detection, warning and avoidance) a thulium solid state laser based 2 µm MOPA system was developed for an eye safe air-borne Doppler-LIDAR. The goal of the system was to detect wake-vortex formation during approach and landing and to detect wind shear and clear-air turbulences during the flight to increase the capacities of the airports and the flight safety (I-WAKE, 2001).

4.2 Medical applications

There are several aspects which make the 2 μm wavelength range a promising candidate for highly precise surgical applications for both soft and hard tissue. The most important property is the high absorption in water combined with minimal penetration depth within human tissue. The superficial mid-infrared tissue ablation effects lead to submicron ablation rates which result in minimal damage around the exposed area. The second important aspect is the coagulation effect caused by the 2 μm laser radiation, which suppresses the bleeding during operations. In the most cases, solid state lasers are used for well established medical procedures such as precise tissue ablation, ophthalmic surgery or dentistry. Figure 18 shows the main medical applications and their anatomic regions within the whole human body. For the applications printed in red pulsed laser systems are used, for the others cw laser radiation is more practical.

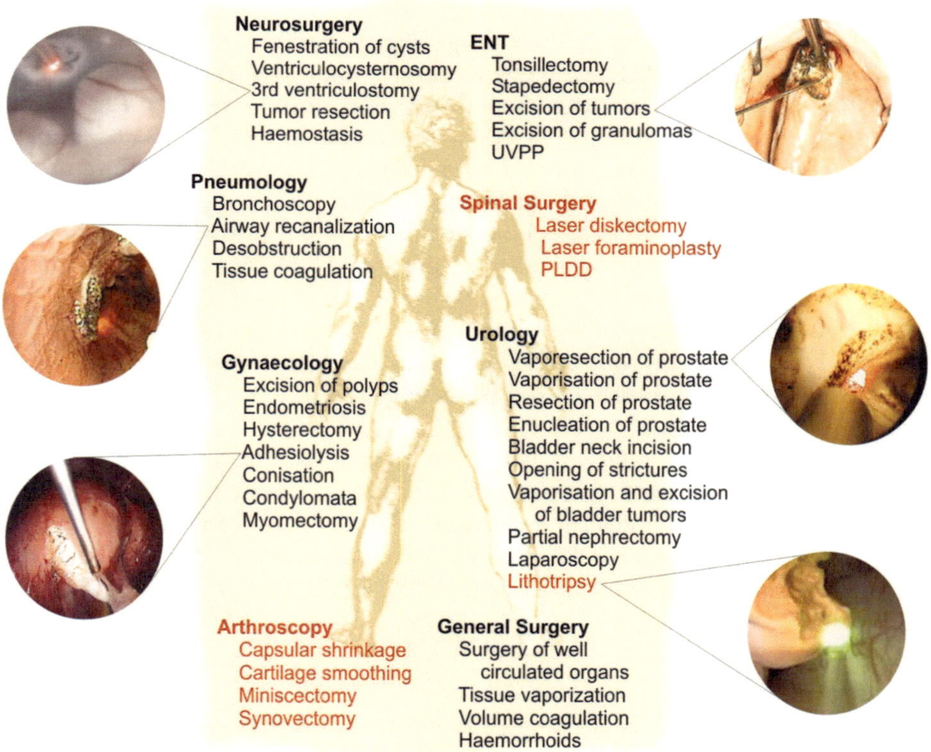

Fig. 18. Overview with key-hole pictures of different possible surgery operations which can be carried out with laser systems operating in the 2 μm wavelength range. For the applications marked in red pulsed laser systems are used.

Depending on the diagnostic findings and the sort of therapy, most of the medical procedures are carried out either with a Ho:YAG laser or Tm:YAG laser systems because of the penetration depth and the already mentioned absorption of this specific laser radiation in water. The Ho:YAG laser operates at 2.1 µm and the Tm:YAG laser operates at 2 µm. Although their wavelengths are close-by their penetration depths distinguish clearly. In soft tissue the Ho:YAG laser has a penetration depth of 300 µm and the Tm:YAG laser penetrates only 100 µm. There are four fundamental photoablation mechanisms which cause these differences depending on the wavelength: photochemical processes, photothermal processes, photomechanical processes and photoelectrical processes. The first type describes laser stimulation effects on biochemical and molecular interactions. The second mechanism is the actual photoablation. The tissue is removed by vaporization and superheating of tissue fluids. The photomechanical process breaks apart the tissue structures and leads to tissue removal by laser-induced shock-wave generation. The fourth mechanism, the photoelectrical process, achieves tissue removal by electrically charged ions and particles.

The first laser lithotripsy was reported in 1981 (Orii et al., 1981). A cw Nd:YAG laser emitting at 1.064 µm was used to successfully clear pigmented common bile duct (CBD) stones from humans. A big drawback of this wavelength is the strong generation of wall heat which leads to damage of the bile ducts. Therefore many investigations were carried out characterising other laser wavelengths. The pulsed dye laser operating at 504 nm gained in importance and led to many successful interventions, but nevertheless did not become widely accepted because of its limited range of applications and high costs. Today, interventions made by Ho:YAG laser systems show complete stone fragmentation in over 90 % of all lithotripsies and the stones of all compositions can be reduced better than with other laser wavelengths (Teichman et al. 2001). In the case of the Ho:YAG laser the power required for lithotripsy is the same for all stones. In contrast, the required laser power of the dye laser had to be changed substantially among different stone compositions. The reason is that the fragmentation by Ho:YAG lithotripsy is caused primarily by photothermal mechanism which transfers the laser energy directly to the stone. Therefore potentially hazardous cavitation bubbles or shock waves are not caused by the Ho:YAG laser. Teichman et al. also investigated the potential of the lithotripsy of gallstones by a Ho:YAG laser system. The laser energy was guided endoscopically by a fibre to the targeted stones. They conclude that Ho:YAG laser lithotripsy is effective and safe when used to fragment gallstones.

In a study Bach discusses about the feasibility and efficacy of a Tm:YAG laser system (RevoLix, LISA laser products OHG, Germany) for prostatectomy (Bach et al. 2009). Three surgeons used the 70-W 2-µm cw Tm:YAG laser system which was coupled with a 550 µm core fibre for the intervention of 208 patients. The study showed that the Tm:YAG laser assisted prostatectomy is feasible and effective, even in patients with potentially impaired detrusor function.

4.3 Material processing

The 2 µm wavelength range is also very attractive for material processing especially for plastics. The most interesting aspect is the processing of plastic materials that are transparent in the visible wavelength range. Figure 19 shows a schematic diagram of the welding process for two transparent plastics and two pictures of plastic welds prepared with a thulium 2 µm laser. Most of the relevant plastic materials show sufficient absorption around 2.0 µm to allow direct processing with lasers operating at this wavelength. Using the absorbed energy of a 2.0 µm laser cutting, welding, and marking are easily possible. Using "standard" 1 μm lasers this is only possible with additives inside the plastic that increase the absorption of the laser light, because the plastics are still highly transparent in this wavelength range. These additives make the fabrication process more complicated and the addition is sometimes prohibited for example when the plastics are used in medical applications. Laser systems for such applications have to deliver high cw output powers with a good beam quality and they have to be highly reliable for industrial operation. Possible applications in this field also include the processing and fabrication of transparent bio-fluidic chips (precise welding or generating of micro-channels) for biological or medical mass-screening experiments.

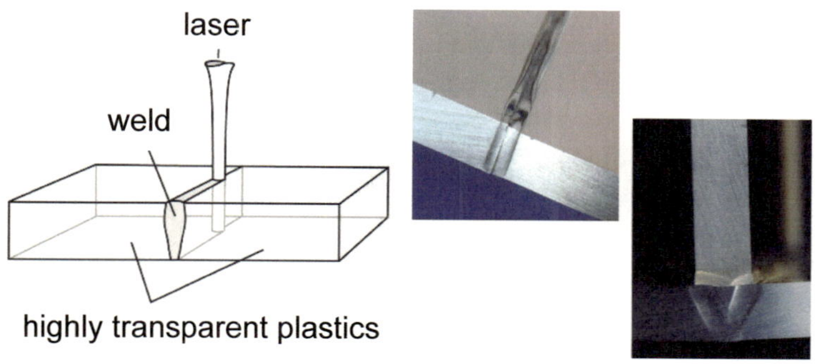

Fig. 19. Left side: Schematic diagram for welding two highly transparent plastics with a butt joint. Right side: Welding examples prepared with a 2 μm Tm:YAG laser by Prolas GmbH, Aachen (butt joint 4 mm PMMA and T-bonding 3 mm PMMA)

4.4 Further possibilities

Besides the described applications 2 μm laser systems are also interesting for free space optical (FSO) communication, military and security applications and for pumping of laser sources for the mid infra-red region. FSO communication solutions are used for low-cost metro networks. Suitably robust and cheap laser systems working at emission wavelengths in the first water absorption window at 2 μm are attractive for these systems. Due to the eye safe wavelength higher powers can be used as at 1 μm and long ranges can be achieved for operation at an atmospheric transmission window. Combined with the availability of small area fast detectors, this offers real prospects for such systems.

There are a number of security and military applications for 2 μm laser systems with different requirements on the brightness, compactness or cost-effectiveness of the systems. While many heavier elements and more complex explosive compounds (TNT, TATP) and chemo-biological agents (Sarin, Soman) may be directly detected at longer wavelength ranges (7 to 9 and 9 to 12 μm respectively) it is worth noting that the presence of explosives may be betrayed by increased levels of lighter constituent compounds or 'markers' either through out gassing or decomposition or residue from manufacture (such as ammonia or acetone), which can be detected at a wavelength of 2 μm (Day et al., 2007). For instance, current airport swab spot checks look for residue markers - this could be done remotely and passively with a photonic chemical sensor providing mass throughput screening. Detecting chemical and biological warfare agents and related solvents is obviously of increasing importance for ensuring public and armed-forces safety.

Another potential application in the security context is the deployment of secure short-range communication networks on a battlefield. This can be well achieved at eye-safe wavelengths within an atmospheric transmission window. Here it is advantageous to provide low-power, tunable, and pulsed operation from the source to ensure covert, spread-spectrum communications with little backscattered or dispersed light out of the line-of-sight path, therefore ensuring secrecy and security.

2 µm laser systems are also ideal pump sources for laser systems that emit at longer wavelengths in the mid-infrared. Especially the pumping of Cr^{2+} lasers and optical parametric oscillators (OPOs) is of interest. Cr^{2+} lasers can be pumped and operated continuously and they are broadly tunable. They can cover the whole wavelength range from 2000 nm to 3100 nm (Sorokina & Vodopyanov, 2003). Efficient pumping of Cr2+ lasers has already been demonstrated with diode lasers and thulium lasers. OPOs allow broad continuous wavelength tuning, high peak powers, and high conversion efficiencies. They use a nonlinear optical frequency down-conversion mechanism in which one pump photon decays into two photons with lower energy (longer wavelength). The sum of the energies of the two new photons is equal to the energy of the pump photon. The two new photons do not have the same energy; typically the photon with the higher energies is called "signal" and the other one "idler". For high power applications in the mid-infrared, usually OPOs based on ZGP are used. ZGP OPOs can ideally be pumped with holmium lasers around 2.1 µm (Budni et al. 1998; Vodopyanov, 2003).

5. Conclusion

In this chapter the recent progress in the development of crystal lasers, fibre lasers and semiconductor lasers operating in the wavelength range close to 2 µm are discussed. For the crystal and fibre lasers the focus is put on the improvements, developments, and recently achieved high power laser results of thulium and holmium doped systems. Thulium laser systems emitting around 2.0 µm can be pumped very efficient nearby 800 nm, when the cross relaxation process, which populates the upper laser level, is exploited very well. Output powers close to 1 kW and slope efficiencies of up to 68 % have recently been reported. Concerning the holmium laser systems the most recently achieved results for inband pumping at 1.9 µm with different sources are reviewed and discussed. Using inband pumping q-switched pulses of up to 55 mJ of pulse energy and of 43 W of cw output power have been demonstrated.

Additionally the latest improvements of GaSb-based laser diodes and stacks are outlined. These diodes can cover the whole wavelength range from 1.85 µm to 2.35 µm and they get more and more commercially available. The potential of efficient holmium pumping with a diode stack is shown.

In the last section the potential applications for the different 2 µm laser sources are presented. Applications in spectroscopy, sensing, surgery, and material processing are discussed in detail.

6. References

Agger, S. D. & Povlsen, J. H. (2006). Emission and absorption cross section of thulium doped silica fibers, *Opt. Express,* Vol. 14, No. 1, pp. 50-57

Bach, T.; Herrmann, T. R. W.; Haecker, A.; Michel, M. S. & Gross, A. (2009). Thulium:yttrium-aluminium-garnet laser prostatectomy in men with refractory urinary retention, *BJU International,* Vol. 104, No. 3

Baranov, A. N.; Fouillant, C.; Grunberg, P.; Lazzari, J. L.; Gaillard, S. & Joullié, A. (1994). High temperature operation of GaInAsSb/AlGaAsSb double-heterostructure lasers emitting near 2.1 μm, *Appl. Phys. Lett.*, Vol. 65, 5

Becker, T.; Clausen, R. & Huber, G. (1989). Spectroscopic and Laser Properties of Tm-doped YAG at 2 μm. *OSA Proceedings on Tunable Solid-State Lasers,* 5, 150

Bollig, C.; Strauss, H. J.; Esser, M. J. D.; Koen, W.; Schellhorn, M.; Preussler, D.; Nyangaza, K.; Jacobs, C.; Bernhardi, E. H. & Botha, L. R. (2009). Compact Fibre-Laser-Pumped Ho:YLF Oscillator-Amplifier System, *CLEO EUROPE 2009*

Budni, P. A.; Pomeranz, L. A.; Lemons, M. L.; Schunemann, P. G.; Pollak, T. M. & Chicklis, E. P. (1998). 10 W mid-IR holmium pumped $ZnGeP_2$ OPO, *Advanced Solid State Lasers, Tech. Digest,* Opt. Soc. Am., Washington, DC, pp. 90-92

Budni, P. A.; Lemons, M. L.; Mosto, J. R. & Chicklis, E. P. (2000). High-Power/High-Brightness Diode-Pumped 1.9-μm Thulium and Resonantly Pumped 2.1-μm Holmium Lasers, *IEEE J. of Selected Top. In Quant. Electr.,* Vol. 6, No. 4, pp. 629634

Budni, P. A.; Ibach, C. R.; Setzler, S. D.; Gustafson, E. J.; Castro, R. T. & Chicklis, E. P. (2003). 50-mJ, Q-switched, 2.09-μm holmium laser resonantly pumped by a diode-pumped 1.9-μm thulium laser, *Optics Letters,* Vol. 28, No. 12, pp. 1016-1018

Caird, J. A.; DeShazer, L. G. & Nella, J. (1975). Characteristics of room-temperature 2.3-μm laser emission from Tm^{3+} in YAG and $YALO_3$. *IEEE J. Quantum Electron.* , 11

Caneau, C.; Srivastava, A. K.; Dentai, A. G.; Zyskind, J. L. & Pollack, M. A. (1985). Room-temperature GaInAsSb/AlGaAsSb DH injection lasers at 2.2 μm, *Electron. Lett.*, Vol. 21, p 815

Chicklis, E. P.; Naiman, C. S., Folweile, R. C.; Gabbe, D. R.; Jenssen, H. P. & Linz, A. (1971). High-Efficiency Room-Temperature 2.06-MUM Laser Using Sentitized HO3+-YLF, *Appl. Phys. Lett,* 19, 4

Choi, H. K. & Eglash, S. J. (1992). High-power multiple-quantum-well GaInAsSb/AlGaAsSb diode lasers emitting at 2.1 μm with low threshold current density, *Appl. Phys. Lett.,* Vol. 61, 10

Day, T.; Pushkarsky, M.; Weida, M.; Arnone, D. & Pritchett, R. (2007). Miniaturized, Tunable, External Cavity Quantum Cascade Lasers, *8th International Conference on Mid-Infrared Optoelectronics: Materials and Devices (MIOMD-VIII),* Bad Ischl, Austria

Dergachev, A.; Moulton, P. F. & Drake, T. E. (2005). High-power, high-energy Ho:YLF laser pumped with Tm:fiber laser, *Advaced Solid State Photonics 2005* (The Optical Society of America, Washington, DC, 2005), post-deadline paper SS1-6

Duczynski, E. W.; Huber, G.; Ostroumov, V. G. & Shcherbakov, I. A. (1986). Cw double cross pumping of the $^5I_7$-$^5I_8$ laser transition in Ho^{3+}-doped garnets, *Appl. Phys. Lett.,* 48, 23, pp. 1562-1563

Ermeneux, F. S.; Sun, Y.; Cone, R. L.; Equall, R. W.; Hutchenson, R. L. & Moncorge, R. (1999). Efficient cw 2 μm Tm^{3+}:Y_2O_3 Laser, *Adv. Solid-State Lasers,* Vol. 26, pp. 497-502

Fan, T. Y.; Huber, G.; Byer, R. L. & Mitzscherlich, P. (1987). Continuous-wave operation at 2.1 μm of a diode-laser-pumped, Tm-sensitized Ho:$Y_3Al_5O_{12}$ laser at 300 K, *Optics Lett.,* Vol. 12, No. 9, pp 678-679

Fan, T. Y.; Huber, G.; Byer, R. L. & Mitzscherlich, P. (1988). Spectroscopy and Diode Laser-Pumped Operation of Tm, Ho:YAG. *IEEE J. Quantum Electro.*, 24, 6

Fornasiero, L.; Berner, N.; Dicks, B.-M.; Mix, E.; Peters, V.; Petermann, K. & Huber, G. (1999). Broadly tunable Laseremission from $Tm:Y_2O_3$ and $Tm:Sc_2O_3$ at 2µm, *Conference on Advanced Solid-State Lasers, OSA Technical Digest,* paper WD5

French, V. A.; Petrin, R. R.; Powell, R. C. & Kotka, M. (1992). Energy-transfer processes in Y3Al5O12:Tm,Ho, *Phys. Rev. B,* 46, p. 8018

Frith, G. P., Samson, B., Carter, A. Farroni, J. & Tankala, K. (2007). Highpower (110), high efficiency (55%) FBG fibre laser operating at 2 µm, Proc. SPIE 6453, 64532B

Gaumé, R.; Viana, B; Vivien, D.; Roger, J.-P. & Furnier, D. (2003). A simple model for the prediction of thermal conductivity in pure and doped insulating crystals, *Appl. Phys. Lett.*, Vol. 83, No.7, p.1355

Heine, F. (1995). Diodengepumpte Festkörperlaser für Kommunikationstechnologie und Fernerkundung, *Dissertation,* Institut für Laser-Physik, Universität Hamburg

Honea, E. C.; Beach, R. J.; Sutton, S. B.; Speth, J. A.; Mitchell, S. C.; Skidmore, J. A.; Emanuel, M. A. & Payne, S. A. (1997). 115-W Tm:YAG Diode-Pumped Solid-State Laser. *IEEE J. of Quant. Electr.*, Vol. 33, No. 9, pp. 1592-1600

Hopkins, J.-M.; Hempler, N.; Rösener, N.; Schulz, N.; Rattunde, M.; Manz, C.; Köhler, K.; Wagner, J. & Burns, D. (2008). High-Power, (AlGaIn)(AsSb) semiconductor disk laser at 2.0 µm, *Opt. Lett.*, Vol. 38, No. 2, pp. 201-203

Huber, G.; Duczynski, E. W. & Petermann, K. (1988). Laser pumping of Ho-, Tm-, Er-doped garnet lasers at room temperature, *IEEE J. Quantum Electro.*, 24, pp. 924-933

I-WAKE Consortium (2001). Instrumentation systems for on-board WAKE-vortex and other hazards detection warning and avoidance. *DOW, Growth Project GRD1-2001-4076*

Jacobs, U. H., Scholle, K., Heumann, E., Huber, G. Rattunde, M. & Wagner, J. (2004). Room-temperature external cavity GaSb-based diode laser around 2.13 mm, *Appl. Phys. Lett.*, Vol. 85, No.4, pp. 5825-5826

Johnson, L. F. (1963). Optical maser characteristics of rare-earth ions in crystals, *J. Appl. Phys.*, 34, pp. 897-909

Johnson, L. F.; Geusic, J. E.; & Uitert, L. G. V. (1965). Coherent oscillations from Tm^{3+}, Ho^{3+}, Yb^{3+} and Er^{3+} ions in yttrium aluminum garnet, *Appl. Phys. Letters,* 7 (5), 127

Johnson, L. F.; Geusic, J. E. & Uitert, L. G. V. (1966). Efficient, high-power coherent emission from Ho3+ ions in yttrium aluminum garnet, assisted by energy transfer. *Appl. Phys. Letters* 8, 200

Kaminskii, A. A. (1996). Crystalline Lasers: Physical Processes and Operating Schemes, *CRC Press*, 978-0849337208, Boca Raton

Kelemen, M. T.; Weber, J.; Ratunde, M.; Kaufel, G.; Schmitz, J.; Moritz, R.; Mikulla, M. & Wagner, J. (2006). High-Power 1.9-µm Diode Laser Arrays With Reduced Far-Field Angle, *IEEE Photonics Technology Letters,* 18, 4

Kintz, G. J.; Esterowitz, L. & Allen, R. (1987). Cw diode-pumped Tm3+:YAG 2.1µm room-temperature laser *Electr. Lett.*, Vol. 23, No. 12, p. 616

Koechner, W. (2006). Solid-State Laser Engineering, Springer Series in Optical Sciences, Vol.1, p. 748, *Springer-Verlag GmbH,* 978-0-387-29094-2, Berlin, Heidelberg

Koopmann, P.; Peters, R.; Petermann, K. & Huber, G. (2009a). Highly Efficient, Broadly Tunable $Tm:Lu_2O_3$ Laser at 2 µm, *CLEO EUROPE 2009,* CA10.3

Koopmann, P.; Peters, R.; Petermann, K. & Huber, G. (2009a). Hocheffizienter Laserbetrieb von Tm^{3+}:Lu_2O_3 bei einer Wellenlänge von 2 μm, *DPG Frühjahrstagung 2009 in Hamburg*, Q8.4

Kuznetsov, M.; Hakimi, F.; Spraque, R. & Mooradian, A. (1997). High-Power (>0.5 W CW) Diode-Pumped Vertical-External-Cavity Surface-Emitting Semiconductor Lasers with Circular TEM_{00} Beams, *IEEE Photonics Technol. Lett.*, Vol. 9, pp. 1063

Meleshkevich, M.; Platonov, N.; Gapontsev, D.; Drozhzhin, A.; Sergeev, V. & Gapontsev, V. (2007). 415W Single-Mode CW Thulium Fiber Laser in all-fiber format, *CLEOE-IQEC 2007*, Munich

Melngailis, I. (1963). Maser action in InAs diodes, *Appl. Phys. Lett.*, Vol.2, 9

Moskalev, I. S.; Fedorov, V. V.; Mirov, S. B.; Babushkin, A.; Gapontsev, V. P.; Gapontsev, D. V. & Platonov, N. (2006). Efficient Ho:YAG Laser Resonantly Pumped by Tm-Fiber Laser, *Advanced Solid-State Photonics,*(Optical Society of America, 2006), TuB10

Moulton, P. F.; Rines, G. A.; Slobodtchikov, E. V. Wall, K. F.; Frith, G.; Samson, B: & Carter, A. L. G. (2009). Tm-Doped Fiber Lasers: Fundamentals and Power Scaling, *IEEE J. of Selected Top. in Quant. Electr.*, Vol. 15, No. 1, pp 85-92

Mu, X.; Meissner, H. E. & Lee, H. C. (2009). Thulium Fiber Laser 4-Pass End-Pumped High Efficiency 2.09-μm Ho:YAG Laser, *OSA/CLEO/IQEC 2009*

Nabors, C. D.; Ochoa, J.; Fan, T. Y.; Sanchez, A.; Choi, H. K. & Turner, G. W. (1995). Ho:YAG Laser Pumped By 1.9 μm Diode Lasers, *IEEE J. Quantum Elect*, Vol. 31, 9

Orii, K.; Nakahara, A.; Ozaki, A.; Sakita, T. & Iwasaki, Y. (1981). Choledocholithomy by YAG laser with a choledochofiberscope: case reports of two patients, *Surgery*, 90

Payne, S. A.; Chase, L. L.; Smith, L. K., Kway, W. L. & Krupke, W. F. (1992). Infrared Cross-Section Measurements for Crystals Doped with Er3+, Tm3+, and Ho3+, *IEEE J. of Quant. Electr.*, Vol. 28, No. 11, pp. 2619-2630

Peterson, P.; Gavrielides, A. & Sharma, M. P. (1995). Diode-pumped Tm:YAG solid-state lasers with indirect and direct manifold pumping, *Appl. Phys. B*, 61, 195

Platt, U. & Stutz, J (2008). Differential Optical Absorption Spectroscopy, *Springer-Verlag GmbH*, 978-3-540-21193-8, Berlin, Heidelberg

Rattunde, M.; Mermelstein, C.; Simanowski, S.; Schmitz, J.; Kiefer, R.; Herres, N.; Fuchs, F.; Walther, M. & Wagner, J. (2000). Temperature dependence of thershold current for 1.8 to 2.3 mm (AlGaln)(AsSb)-based QW diode lasers, *IEEE 27th International Symposium on Compound Semiconductors (ISCS'00)*, Monterey, 437

Rattunde, M.; Mermelstein, C.; Schmitz, J.; Kiefer, R.; Pletschen, W.; Walther, M. & Wagner, J. (2002). Comprehensive modeling of the electro-optical-thermal behaviour of (AlGaln)(AsSb)-based 2.0 μm diode lasers, *Appl. Phys. Lett.*, Vol. 80, 22

Rattunde, M.; Schmitz, J.; Mermelstein, C.; Kiefer, R. & Wagner, J. (2006a). III-Sb-based Type-I QW Diode Lasers, In: *Mid-infrared Semiconductor Optoelectronics* , Krier, A., pp. 131-157, Springer Series in Optical Sciences, ISBN 1-84628-208-X, London

Rattunde, M.; Schmitz, J.; Kaufel,G.; Kelemen, M. T.; Weber, J. & Wagner, J. (2006b). GaSb-based 2.X mm quantum-well diode lasers with low beam divergence and high output power, Appl. Phys. Lett., Vol. 88

Remski, R. L. & Smith, D. J. (1970). Temperature dependence of pulsed Laser Threshold in YAG-ER3+,TM3+,HO3+, *IEEE J. Quantum Electro.*, 6, 11

Rothacher, Th.; Lüthy, W. & Weber, H. P. (1998). Spectral properties of a Tm:Ho:YAG laser in active mirror configuration, *Appl. Phys. B*, 66, pp. 543–546

Schellhorn, M.; Hirt, A. & Kieleck, C. (2003). Ho:YAG laser intracavity pumped by a diode-pumped Tm:YLF, *Opt. Lett*, Vol. 28, 20

Schellhorn, M. (2006). Performance of a Ho:YAG thin-disc laser pumped by a diode-pumped 1.9 µm thulium laser, *Appl. Phys. B*, 85, 4

Schellhorn, M. (2008). High power diode-pumped Tm:YLF laser, *Appl. Phys. B*, Vol. 91, p. 71

Schellhorn, M.; Ngcobo, S. & Bollig, C. (2009). High-power diode - pumped Tm:YLF slab laser, *Appl. Phys. B*, Vol. 94, pp. 195-198

Scholle, K.; Heumann, E. & Huber, G. (2004). Single mode Tm and Tm,Ho:LuAG lasers for LIDAR applications, *Laser Phys. Lett.*, Vol. 1, No. 6, pp. 1-5

Scholle, K. & Fuhrberg, P. (2008). In-band pumping of high-power Ho:YAG lasers by laser diodes at 1.9 µm, *OSA/CLEO/IQEC 2008*

Scholle, K.; Lamrini, S.; Fuhrberg, P.; Rattunde, M. & Wagner, J. (2009). Wavelength stabilization and mode selection of a GaSb-based semiconductor disk laser at 2000 nm by using a Volume Bragg Grating, *CLEO EUROPE 2009*, CB7.4M

Schulz, N.; Hopkins, J.-M.; Rattunde, M.; Burns, D. & Wagner, J. (2008). High-Brightness long-wavelength semiconductor disk lasers, *Laser & Photonics Rev.*, Vol. 2, No. 3

Shen, D. Y.; Abdolvand, A.; Cooper, L. J. & Clarkson, W. A. (2004). Efficient Ho:YAG laser pumped by a cladding-pumped tunable Tm:silica-fibre laser, *Appl. Phys. B*, 79

So, S.; Mackenzie, J. I.; Shepherd, D. P. & Clarkson, W. A.; Betterton, J. G. & Gorton, E. K. (2006a). A power-scaling strategy for longitudinally diode-pumped Tm:YLF lasers, *Appl. Phys. B*, Vol. 84, pp. 389-393

So, S.; Mackenzie, J. I.; Shepherd, D. P. & Clarkson, W. A. (2006b). Intra-cavity side-pumped Ho:YAG laser, *Optics Express*, Vol. 14, No. 22

Sorokina, I. T. & Vodopyanov, K. L. (2003). Solid-State Mid-Infrared Laser Sources, *Springer-Verlag GmbH*, 3-540-00621-4, Berlin-Heidelberg

Stoneman, R. C. & Esterowitz, L. (1992). Intracavity-pumped 2.09-µm Ho:YAG laser, *Opt. Lett.*, Vol. 17, 10

Strom, M. E. (1988). Laser characteristics of a q-switched Ho:Tm:Cr:YAG, *Appl. Opt.*, 27, (20)

Svelto, O. (1998). Principles of Lasers Fourth Edition, *Springer-Verlag GmbH*, 978-0306457487, Berlin, Heidelberg

Teichman, J. M. H.; Schwesinger, W. H., Lackner, J. & Cossman, R. M. (2001). Holmium: YAG laser lithotripsy for gallstones, *Surg. Endosc.*, Vol. 15, pp. 1034-1037

Turri, G.; Sudesh, V.; Richardson, M.; Bass, M.; Toncelli, A. & Tonelli, M. (2008). Temperature-dependent spectroscopic properties of Tm3+ in germinate, silica, and phosphate glasses: A comparative study, *J. of Appl. Phys.*, Vol. 103

Vodopyanov, K. (2003). Pulsed MID-IR Optical Parametric Oscillators, Solid-State Mid-Infrared Laser Sources, Sorokina, I. T. & Vodopyanov, K. L. *Springer-Verlag GmbH*, 3-540-00621-4, Berlin-Heidelberg

Walsh, B. M.; Barnes, N. P. & Di Bartolo, B. (1998). Branching ratios, cross sections, and radiative lifetimes of rare earth ions in solids: Application to Tm^{3+} and Ho^{3+} ions in $LiYF_4$, *J. of Appl. Phys.*, Vol. 83, No. 5, pp. 2772-2787

Wu, J.; Yao, Z.; Zong, J. & Jiang, S. (2007). Highly efficient high-power thulium-doped germanate glass fiber laser, *Optics Letters*, Vol. 32, No. 6, pp. 638-640

22

Designer Laser Resonators based on Amplifying Photonic Crystals

Alexander Benz[1], Christoph Deutsch[1], Gernot Fasching[1], Karl Unterrainer[1], Aaron M. Maxwell[2], Pavel Klang[2], Werner Schrenk[2] and Gottfried Strasser[2]
[1]*Photonics Institute and Centre for Micro- and Nanostructures, Vienna University of Technology*
[2]*Institute of Solid-State Electronics and Centre for Micro- and Nanostructures, Vienna University of Technology*
Austria

1. Introduction

The design and realisation of lasers involves two main steps: the development of the gain medium and the development of the optical resonator. Naturally, both have a major impact on the device performance. The light that passes through the active medium and lies within a certain frequency range is amplified. The resonator confines the light and defines the way it travels through the gain medium.

The focus of the present work is the design and realisation of micro resonators based on photonic crystals (PhCs). The key feature of PhCs is the possibility to design the dispersion relation for electromagnetic waves by patterning on a wavelength scale Joannopoulus et al. (2008). The optical properties of the resonator can be manipulated without almost any restrictions. Essential device parameters such as emission frequency, lasing far-field or cavity quality-factor (Q-factor) are balanced simultaneously. The PhCs are incorporated directly into the active region of a terahertz (THz) quantum-cascade laser (QCL) Köhler et al. (2002). The use of THz-QCLs allows for an easy experimental realisation of the desired resonator geometry. The large emission wavelength on the order of 100 μm allows for a simple processing using optical lithography and sets high tolerances for the fabrication.

We present two different schemes for PhC laser resonators in theory and experiment. The first one uses a bulk active region which is surrounded by a PhC-mirror. The light is confined by the mirror and amplified in the central gain region. The spatial separation of the two main laser components allows for a simpler fabrication. However, it comes at the expense of larger devices. The second resonator type uses the PhC directly as the gain medium. It is fabricated from the active region of a THz-QCL, making the bulk gain region redundant. The device performance is greatly enhanced in terms of emission frequency control.

2. Active region of the THz quantum-cascade lasers

The basic building block of every QCL is a superlattice, materials with different bandgaps are grown on top of each other. If the thickness of the individual layers is in the range of a few nm's, a further quantisation of the electron states can be achieved. The theoretical fundamentals are from the 1970's mainly by R. Tsu and L. Esaki Tsu & Esaki (1973); Tsu et al. (1974). The idea to use intersubband transitions as an active laser medium has been proposed in 1971 by R. F. Kazarinov and R. A. Suris Kazarinov & Suris (1971). The essential point is the possibility to design the electron wave functions by carefully selecting the thickness of the individual layers. We are able to control all the parameters in the active region, such as the dipole matrix element, the upper state life time or the transition energy independently of the used material system. All these parameters become independent of the used material system. To increase the gain further, identical cascades are grown on top of each other. Electrons from the ground state in one cascade are injected into the upper laser state of the next cascade, allowing for electron recycling. All these advantages and the opportunities in design have made QCLs the preferred monolithic source in the MIR and THz spectral region, covering wavelengths from 3 to 300 μm.

The physical nature of the QCLs has another direct consequence, which is essential for this work. QCLs are unipolar devices, only one carrier type is involved in the current transport and the light generation. This makes QCLs insensitive to surface effects allowing for the realization of almost any resonator geometry. In contrast, the light generation in classical bandgap lasers is based on an electron-hole recombination across the bandgap. Therefore, an increase in device surface leads to a strong increase in surface leakage currents which in the end limits the device performance.

The active region we are using is based on a gallium-arsenide (GaAs) / aluminium-gallium-arsenide (AlGaAs) heterostructure. The barriers have an Al-content of 15 % leading to a typical height of 110 meV. A calculated bandstructure at lasing field is shown in Fig. 1(a), one cascade consists of only four wells. The optical transition in this design is vertical, leading to stronger coupling of the upper and lower laser state and reduced sensitivity to interface roughness. The 2.7 nm thin barrier, in between the double well, has the strongest influence on the transition energy. An increased thickness leads to a red-shift in the emission frequency as the anti-crossing between the upper and lower laser states is reduced.

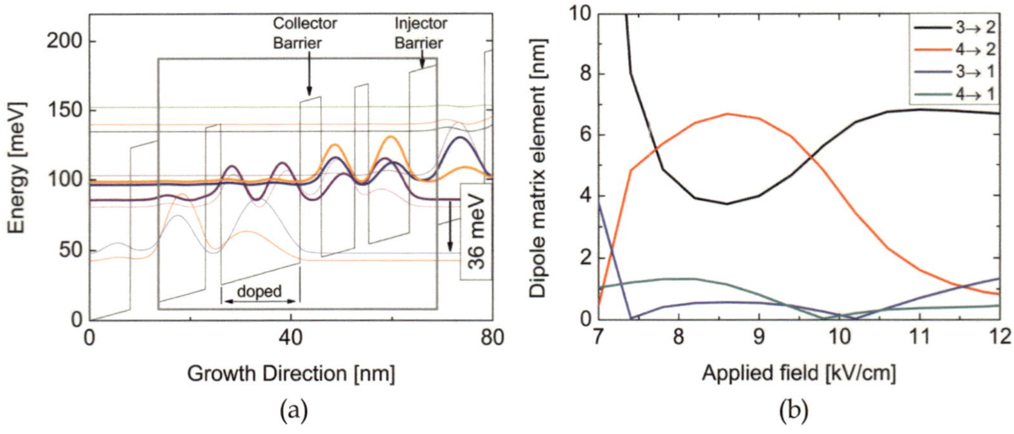

Fig. 1. (a) Calculated bandstructure at the lasing field of 9.8 kV/cm. The lasing states are marked with bold lines (purple=2, blue=3, orange=4). One cascade is marked with the grey box. (b) Field dependence of the dipole matrix elements. Only the transitions 3→2 and 4→2 have a relevant coupling strength, the other transitions are optically not active.

The lower laser state is depopulated via a resonant LO phonon emission allowing us to achieve scattering times of around 0.5 ps Williams et al. (2002). Thereby, a stable population inversion can be achieved. The large separation of 36 meV between the lower laser state and the ground state reduces the thermal back filling strongly. Nevertheless, the operating temperature of THz-QCLs is still limited to cryogenic levels. The highest operating temperature reported currently is 186 K Kumar et al. (2009).

One should keep in mind that the optical parameters of the active region depend on the applied field. The coupling strength between the upper and the lower laser state varies with the field. For values below 9.5 kV/cm the transition $4 \rightarrow 2$ is the dominant one, for higher fields $3 \rightarrow 2$, as illustrated in Fig. 1(b). The emission energy shows a blue-shift with increasing field for all the transitions due to the Stark effect. A larger applied field leads to a stronger separation of the electron states.

3. Plasmonic waveguides

The waveguide is responsible for the confinement of the optical mode, it guarantees the overlap with the gain region. For traditional bandgap lasers, waveguides based on total internal reflection are used. The gain region has a higher refractive index than the surrounding medium. However, this concept only works for waveguide thicknesses larger than the optical wavelength involved. For THz-QCLs this is almost impossible to achieve. The wavelengths are in the range of 60 to 300 μm in general. With the available growth techniques such as molecular beam epitaxy or metal-organic chemical vapour deposition, it is not practical to grow a heterostructure with the desired thickness and still to maintain the precision required. Typically, we are limited to 10 or 15 μm thick heterostructures. Such waveguides cannot rely on total internal reflection. The solution is the use of plasmonic waveguides.

To be more precise, the waveguides for THz-QCLs rely on surface plasmons. As any other surface wave, surface plasmons are bound to an interface and decay exponentially into both media. For propagation, it is necessary that the refractive index has different sign on both sides of the interface. A conventional semiconductor which is undoped or low doped has a positive refractive index. Metals or highly doped semiconductors, which act quasi-metallic, have a negative refractive index in the THz spectral region. Therefore, if the active region is sandwiched between two, thin metallic or quasi-metallic layers, an efficient waveguide is formed. This concept is well known from the microwave technology, where micro strip waveguides are used successfully Käs & Pauli (1991).

In this work we use the so-called double-metal (DM) waveguide Kohen et al. (2005). The entire structure consists of two gold layers and the active region in between. For an electrical contact between the metal and the semiconductor we use 100 nm thin highly doped n^+-contact layers. As the THz-mode hardly penetrates the metal, a confinement of almost 100 % is realised. Simulations for a 15 μm thin and 100 μm wide DM waveguide are shown in Fig. 2. As the waveguide is much thinner the wavelength, only the first order mode in vertical direction can propagate. This mode shows an almost constant mode profile along the vertical axes. It has to be stressed that also the lateral confinement is excellent. The mode hardly leaks out of the waveguide into the surrounding air. Another advantage is the high reflectivity of 90 % for untreated facets due to the impedance mismatch between the waveguide and the free space Maineult et al. (2008).

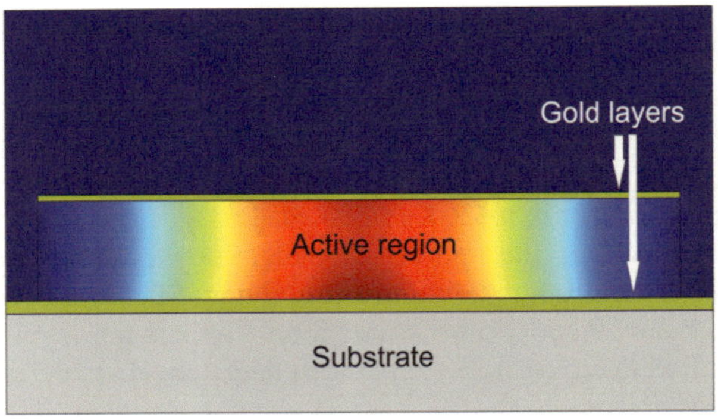

Fig. 2. Two-dimensional simulation of a DM waveguide using *Comsol Multiphysics* (www.comsol.com). The mode profile is almost constant in the vertical direction. Despite the finite width of the ridge, there is no significant leakage in the lateral direction.

4. Photonic crystal theory

Section 2 described the design of the active region itself. The heterostructure designed can be interpreted as artificial crystal for electrons manipulating the electron wave functions. PhCs on the other hand can be seen as artificial crystals for light. The full dispersion relation for electromagnetic waves becomes designable. The combination of QCLs and PhCs allows us to design all aspects of the laser in terms of gain medium and optical resonator.

4.1 Mathematical description of photonic crystals

We will describe the optical properties for a classical system, the quantisation of the electromagnetic field can be neglected in this case. We use the Maxwell's Equations Jackson (1999) with some significant simplifications.

- There are no free charges or currents present in the system.
- We restrict ourselves to purely dielectric media: dispersion is neglected, non-linearity is neglected, the refractive index is independent of time and the relative permeability is set to 1. Therefore, we are able to write the relative permittivity $\epsilon(\vec{r})$ solely as a function of position.
- The involved fields vary harmonically in time, which allows us to replace any derivative in respect to time $\partial/\partial t$ by $-i\omega$.

All this simplifications allows us to decouple the Maxwell's Equation resulting in two independent equations for the electric field $\vec{E}(\vec{r})$ and for the magnetic field $\vec{H}(\vec{r})$

$$\nabla \times \left(\frac{1}{\epsilon(\vec{r})} \nabla \times \vec{H}(\vec{r})\right) = \left(\frac{\omega}{c}\right)^2 \vec{H}(\vec{r}), \tag{1}$$

$$\nabla \times \nabla \times \vec{E}(\vec{r}) = \left(\frac{\omega}{c}\right)^2 \epsilon_r(\vec{r}) \vec{E}(\vec{r}). \tag{2}$$

For mathematical convenience normally the equation for the magnetic field is solved and then used to calculate the electric field. We want to obtain the solution for the PhC, which is a periodic system. According to Bloch's Theorem, the solution has to have the same periodicity as the crystal, giving us the basic solution to our problem:

$$\vec{H}_k = exp\left(i\vec{k}\cdot\vec{r}\right)\vec{u}_k\left(\vec{r}\right), \tag{3}$$

where $\vec{k}$ is the Bloch wave vector and $\vec{u}_k(\vec{r})$ the so-called Bloch function. We know that almost any periodic function can be written as a sum of harmonic functions. This defines the basic form for our unknown Bloch function:

$$\vec{u}_k\left(\vec{r}\right) = \sum_G \vec{c}_G\left(\vec{k}\right) exp\left(i\vec{G}\cdot\vec{r}\right), \tag{4}$$

where $\vec{G}$ is the sum over all reciprocal lattice vectors and $\vec{c}_G$ are the Fourier coefficients. After lengthy calculations, we have simplified a set of partial differential calculations into a set of linear equations. Now the Fourier coefficients become accessible:

$$\sum_G -\epsilon_{G'-G}^{-1}\cdot\left(\vec{k}+\vec{G}'\right)\times\left(\vec{k}+\vec{G}\right)\times\vec{c}_G = \frac{\omega^2}{c^2}\vec{c}_{G'}. \tag{5}$$

The Fourier transform of the inverse dielectric function ϵ_G^{-1} can be calculated very efficiently with the Fast-Fourier Transformation giving us an easy-to-use tool for the design of PhCs.

4.2 Limitations of the mathematical description

As already mentioned, due to all these simplifications the results are valid only for 'perfect' systems. The PhC is infinitely large and consists of perfectly linear and loss-less materials. However, these limitations are not very critical for our devices. Due to the excellent coupling between the mode and PhC, even a small resonator behaves similar to the ideal PhC. The frequency dependent gain can be implemented by perturbation theory using a complex permittivity in the form $\epsilon = \epsilon' + i\delta$. For realistic THz-QCL it is not even necessary as the error due to the gain is negligible Nojima (1998b).

4.3 Design of photonic crystals

The PhC used for the resonator experiments has always the same basic structure, a schematic is shown in Fig. 3(a). It consists of isolated, free-standing pillars which are surrounded by air. Such structures typically show full bandgaps for TM-polarised light Johnson et al. (1999), which is the polarisation of light emitted by a QCL. The PhC is embedded in a DM waveguide for the real devices. For the simulations we use 2D simulations of infinitely high rods which significantly reduces the computational effort. As the waveguide allows only one vertical mode with an almost constant mode profile to propagate, there is virtually no dependence in the vertical direction. Experiments by Schartner et al. (2006) show that 2D-PhCs in plasmonic waveguides can be simulated using only 2D simulations and achieving very accurate results.

We use a ratio r/a of 0.3 for the calculations, where r is the radius of the pillars and a the period of the PhC. The refractive index for the pillars is set to 3.65, an experimental value for GaAs at 3 THz Yasuda & Hosako (2008). For now, the system is set ideal, i.e. the system is loss-less and infinitely large. The calculated bandstructure in Fig. 3(b) shows the expected full bandgaps for TM-modes. The first one spans from 0.21 to 0.3 [fa/c], the second one from 0.39 to 0.51 [fa/c], where f is the frequency and c the speed of light. Frequencies which lie inside the bandgap cannot propagate through the PhC. The only possible solution for the

Maxwell's Equations are waves with a complex $\vec{k}$-vector, in other words exponentially decaying waves.

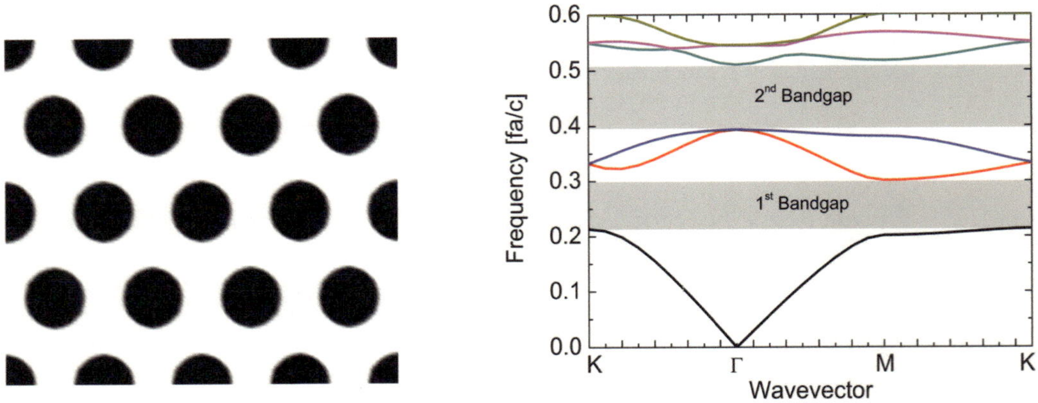

(a) (b)

Fig. 3. (a) Schematic of the PhC used. The isolated high-index pillars are arranged in a hexagonal lattice and surrounded by air. (b) The band structure for the ideal PhC. The first eight bands are presented, the full bandgaps for TM-polarised light are clearly visible.

5. Photonic crystal mirror devices

The first set of devices consists of two parts, a bulk gain region is surrounded by a PhC mirror. A schematic of the resonator is illustrated in Fig. 4. The PhC works as a frequency selective mirror allowing for tuning the emission frequency from the gain maximum into the bandgap of the PhC. The parameters of the PhC are identical to the previous section. The period is varied from 22 to 35 μm while r/a is 0.3. The entire device is embedded into a DM waveguide ensuring a strong interaction between the mode inside the gain region and inside the PhC. The thickness of the waveguide is equal for the different periods, it is defined by the thickness of the active region of 15 μm.

5.1 Simulation results for photonic crystal mirrors

For the simulation of the real device, the plane-wave expansion method (PWE) method is not well suited. The bulk gain region can be seen as a defect which breaks the periodicity of the PhC. So-called 'super-cells' allow for an incorporation of the defect while maintaining periodic boundaries by an artificial increase of the computational cell. However, the artificial increase leads to band folding, the number of possible bands is increased making the identification of individual bands difficult Feng & Arakawa (1997); Kuzmiak & Maradudin (1998); Zhi et al. (2003).

To avoid these problems FDTD-calculations using the open-source package *MIT Electromagnetic Equation Propagation* (MEEP) are performed for our resonators Farjadpour et al. (2006). We restrict ourselves again to 2D-simulations of a hexagonal core which is surrounded by two rows of pillars. The core is large enough to support modes on its own. Therefore, we have also simulated reference cavities consisting of only the core, the pillars have been removed. Now the modes visible can be assigned either to the hexagonal cavity

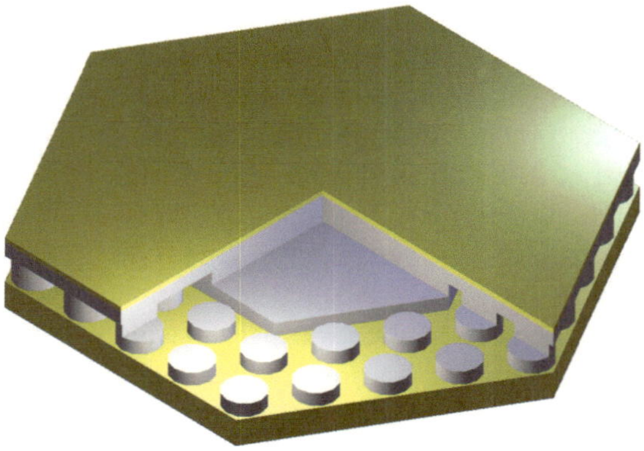

Fig. 4. Schematic of the PhC-mirror device Benz et al. (2007). The bulk gain region is surrounded by a PhC which acts as a frequency selective mirror. The device is embedded in a DM-waveguide to ensure a strong modal coupling between the gain region and the PhC.

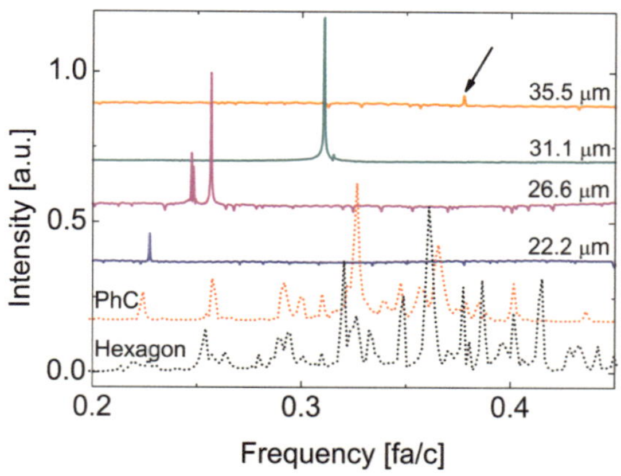

Fig. 5. Simulated and measured spectra for the PhC-mirror devices Benz, Deutsch, Fasching, Unterrainer, Andrews, Klang, Schrenk & Strasser (2009). The dotted lines represent simulations for the bare cavity and the full device respectively. The solid lines show measured spectra for different periods of the PhC.

or to the PhC. The calculated spectra in Fig. 5 show clearly that additional modes appear due to the PhC. For higher frequencies, the number of possible modes increases strongly, as the wavelength becomes smaller while the resonator size remains unchanged. Due to the strong optical coupling with the PhC, it is enough to use only two rows of pillars to achieve the frequency control.

5.2 Experimental results for photonic crystal mirrors

We have used the same active region to process devices with different periods, the corresponding spectra are shown in Fig. 5. The predicted modes are clearly visible in the measurements, corresponding to frequencies of 0.224, 0.256, 0.309 and 0.378 [fa/c]. One should keep in mind that the exact position of the frequencies also depends on the size of the core as the geometric path is changed for the modes. Nevertheless, the mode position overlaps nicely with the calculated bandgaps. This behaviour is expected, as the PhC-mirrors shows the lowest losses for frequencies within the bandgap. The devices do not show single-mode emission, in general, as broad stop bands are used. Due to the inhomogeneously broadened gain with a typical full-width at half-maximum (FWHM) gain bandwidth of 130 GHz, multi-mode emission becomes easily possible Kröll et al. (2007).

The device with a period of 31.05 μm has to be treated separately from the other devices. The laser emits between the first two bandgaps at 0.31 [fa/c]. This corresponds to a mode at the K-point. Such modes at high symmetry point appear at flat-band regions in the PhC-bandstructure due to very low group velocity. The long interaction time with the PhC results in strong feedback. These regions are especially interesting for resonators without any bulk gain region. As they do not require any states inside the bandgap and still allow for an excellent frequency selection.

6. Active photonic crystal laser

The last resonator type that we are going to investigate is based purely on isolated pillars. The specifications for the PhC are identical to the previous sections, the same PhC geometry, ratio r/a, refractive index and waveguide. It is important to stress that now the bulk gain region has been removed, as shown in Fig. 6. The sub-wavelength, isolated pillars have to provide the optical feedback and the required gain. The pillars are fabricated directly from the active region of a THz-QCL and again embedded in a DM waveguide. Thereby, we are able to achieve a very simple and efficient pumping scheme. Only the pillars underneath the top contact are pumped and the electric field distribution is perfectly homogeneous.

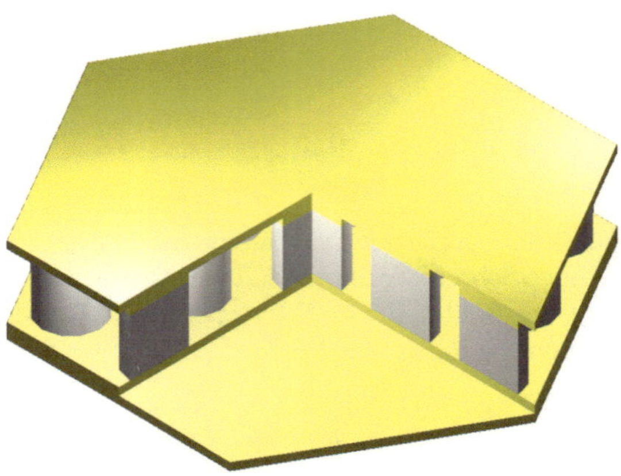

Fig. 6. Schematic of the active PhC laser Benz, Deutsch, Fasching, Andrews, Unterrainer, Klang, Schrenk & Strasser (2009). The pillars are fabricated directly from the active region of a THz-QCL, there is no need for an additional bulk gain region. This type of cavity relies on PhC band edges rather than on defect states.

6.1 Simulation results for active photonic crystals

As there is no defect incorporated, the real device is very close to the ideal PhC. The only limitation is the finite size which will be treated later. Starting from the PWE calculations we see immediately that lasing inside the bandgap is not possible as there are no allowed states. Lasing will occur at the flat band region at the K-point. However, to establish lasing at all it

is important to know the mode distribution inside the resonator. It is easy to calculate the electric field for the ideal PhC. The corresponding energy distribution shows that 95 % of the modal energy overlaps with the active pillars. This is necessary as only the pillars are pumped, the surrounding air is, in the best case, transparent. The high modal overlap in combination with the strongly reduced group velocity is predicted to give strong gain enhancement in this type of structure Nojima (2001; 1998a); Sakoda (1999).
According to simulations, finite size arrays of isolated pillars support lasing modes on their own without the need for any further mirrors Nojima (1999). For our FDTD-calculations we arranged 37 pillars in a hexagonal shape. Even this small number of pillars already supports a few lasing modes, the calculated spectrum is shown in Fig. 7(a). The mode at 0.19 [fa/c] corresponds to the M-point and has a Q-factor of 60, the one at 0.21 [fa/c] to the K-point and has a Q-factor of 1000. For frequencies above 0.3 [fa/c] additional modes corresponding to higher lying bands show up. In between, the PhC bandgap is nicely visible. These calculations predict that these devices should operate in stable single-mode emission at the K-point. One should keep in mind that the simulations overestimate the Q-factor of the real device. The main limitation is the loss in the gold and the thin n^+-layers. These effects are naturally ignored in the 2D-simulations. The reduction of the Q-factor due to interface roughness, which normally limits the Q-factor for PhC in the visible Asano et al. (2006); Srinivasan & Painter (2002; 2003), is not a problem as the processing imperfections are much smaller than the wavelength.

6.2 Experimental results for active photonic crystals

As expected, the devices show a stable single-mode emission defined by the PhC. Fig. 7(a) shows the emission of devices with a 26.6 and 31.1 μm period respectively. Both devices are lasing around 0.23 [fa/c] which corresponds to the K-point or the mode with the significantly higher Qfactor. We attribute the discrepancy between experiment and simulation to imperfections in the processing, uncertainties in the refractive index of the active region and the effects of the DM waveguide.
Apart from the possibility to predict the emission frequency, one should keep in mind that this concept shows a huge tuning range for lasing. The devices with a 26.62 μm period are lasing at 2.56 THz, which corresponds almost to the gain maximum of 2.6 to 2.7 THz Benz, Deutsch, Fasching, Andrews, Unterrainer, Klang, Schrenk & Strasser (2009). Using a 31.05 μm period shifts the emission frequency to 2.25 THz. Spectra for both devices at different applied fields are shown in Fig. 7(b). This corresponds to a possible tuning range of 400 GHz which is significantly larger the typical FWHM gain bandwidth of THz-QCLs of only 130 GHz Kröll et al. (2007). This is a strong evidence for the theoretically predicted gain enhancement in this type of structures. Additionally, it also shows the large potential for this kind of resonator. It combines an excellent frequency control with a huge tuning range.

7. Conclusion

The design and the realisation of laser resonators is a challenging task. In general, it is necessary to balance aspects such as quality factor, laser far-field, threshold and output power. Incorporating PhCs allows for a full control of the dispersion relation of the resonator. This new flexibility can be used to fulfil all the requirements at the same time.

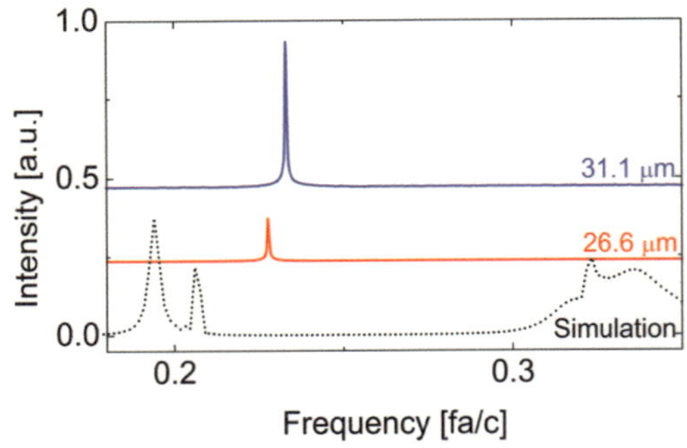

(a)

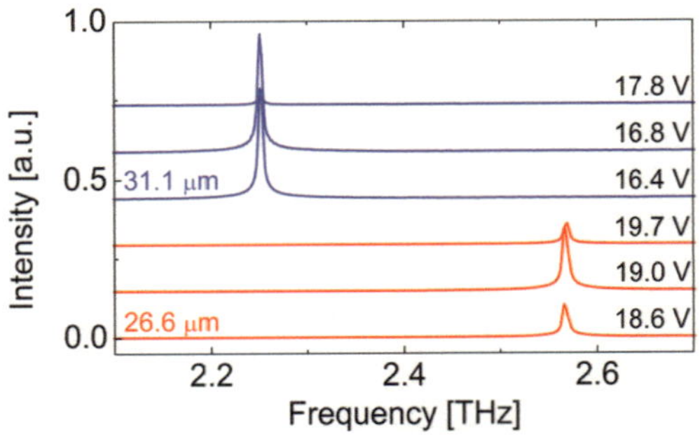

(b)

Fig. 7. (a) Simulated and recorded spectra for the active PhC laser Benz, Deutsch, Fasching, Andrews, Unterrainer, Klang, Schrenk & Strasser (2009). The realised devices are both lasing at the K-point, close to the predicted emission frequency. (b) Measured spectra for different periods of the PhC as a function of the applied bias. Both devices show a stable single-mode emission which is independent of the voltage.

Here, we have presented the design of PhCs for THz-QCLs, which are an ideal system to study the behaviour experimentally. The large wavelength allows for simple processing and large tolerances. The unipolar nature of QCLs makes the devices insensitive to surface leakage. Therefore, it is possible to directly integrate the PhC into the active laser medium.

The devices presented can be split into two different concepts. The first type of devices are based on PhC mirrors. A bulk gain region is surrounded by a PhC mirror. The emission of the device is tuned into the bandgap of the PhC, corresponding to the lowest mirror losses. These devices show a multi-mode emission in general due to the broad stop bands used. Nevertheless, this is an easy concept to be realised. The second type of devices consists of the active pillars only, there is no bulk gain region. The sub-wavelength pillars are used to create the required gain and the optical feedback. The strong modal confinement in lateral and vertical directions allows us to built resonators with dimension comparable to the emission wavelength. The excellent optical properties of the resonator also allow us to achieve a tuning range of 400 GHz, which is significantly wider than the typical FWHM gain bandwidth of THz-QCLs of only 130 GHz. In addition, the number of modes can be reduced significantly as narrow band edges are used which are surrounded by wide forbidden bands.

The simulations are in excellent agreement with the experimental results. We are able to precisely predict the emission frequencies for both types of cavities. Using the PWE method, we can determine the band structure of the ideal crystal including the position of the bandgaps. The group velocity and the modal overlap for any mode can be calculated. The FDTD-simulations allow us to simulate the entire resonator. There are no restrictions on the symmetry of the resonator or the linearity of the used materials. One simulation run generates the entire frequency information. The Q-factors can be determined easily. Especially for lower frequencies and smaller devices, the predictions are excellent, as the number of possible modes remains limited.

This work was partly supported by the Austrian Scientific Fund FWF (SFB ADLIS, SFB IRON, DK CoQuS), the Austrian nano initiative project (PLATON), the EC (TERANOVA) and the Austrian Society for Microelectronics (GMe).

8. References

Asano, T., Song, B.-S. & Noda, S. (2006). Analysis of experimental q factors (~1 million) of photonic crystal nanocavities, *Opt. Express* 14: 1996.

Benz, A., Deutsch, C., Fasching, G., Andrews, A. M., Unterrainer, K., Klang, P., Schrenk, W. & Strasser, G. (2009). Active photonic crystal terahertz laser, *Opt. Express* 17: 941.

Benz, A., Deutsch, C., Fasching, G., Unterrainer, K., Andrews, A. M., Klang, P., Schrenk, W. & Strasser, G. (2009). Photonic crystal mode terahertz lasers, *J. Appl. Phys.* 105: 122404.

Benz, A., Fasching, G., Deutsch, C., Andrews, A. M., Unterrainer, K., Klang, P., Schrenk, W. & Strasser, G. (2007). Terahertz photonic crystal resonators in double-metal waveguides, *Opt. Express* 15: 12418.

Farjadpour, A., Roundy, D., Rodriguez, A., Ibanescu, M., Bermel, P., Joannopoulos, J. D., Johnson, S. G. & Burr, G. (2006). Improving accuracy by subpixel smoothing in fdtd, *Opt. Lett.* 31: 2972.

Feng, X.-P. & Arakawa, Y. (1997). Defect modes in two-dimensional triangular photonic crystals, *Jpn. J. Appl. Phys.* 36: 120.

Jackson, J. D. (1999). *Classical Electrodynamics*,Wiley, New York.

Joannopoulus, J. D., Johnson, S. G., Winn, J. N. & Meade, R. D. (2008). *Photonic crystals - Molding the flow of light*, Princeton University Press.

Johnson, S. G., Fan, S., Villeneuve, P. R., Joannopoulos, J. D. & Kolodziejski, L. A. (1999). Guided modes in photonic crystal slabs, *Phys. Rev. B* 60: 5751.

Käs, G. & Pauli, P. (1991). *Mikrowellentechink - Grundlage, Anwendungen, Messtechnik*, Franzis Verlag Gmbh, München.

Kazarinov, R. F. & Suris, R. A. (1971). Possibility of amplification of electromagnetic waves in a semiconductor with a superlattice, *Fizika i Tekhnika Poluprovodnikov* 5: 797.

Kohen, S., Williams, B. S. & Hu, Q. (2005). Electromagnetic modeling of terahertz quantum cascade laser waveguides and resonators, *J. Appl. Phys.* 97: 053106.

Köhler, R., Tredicucci, A., Beere, H., Lienfield, E., Davis, A., Ritchie, D., Iotti, R. & Rossi, F. (2002). Terahertz semiconductor heterostructure laser, *Nature (London)* 417: 156.

Kröll, J., Darmo, J., Dhillon, S. S., Marcadet, X., Calligaro, M., Sirtori, C. & Unterrainer, K. (2007). Phase-resolved measurements of stimulated emission in a laser, *Nature (London)* 449: 698.

Kumar, S., Hu, Q. & Reno, J. L. (2009). 186 K operation of terahertz quantum-cascade lasers based on a diagonal design, *Appl. Phys. Lett.* 94: 131105.

Kuzmiak, V. & Maradudin, A. A. (1998). Localized defect modes in a two-dimensional triangular photonic crystal, *Phys. Rev. B* 57: 15242.

Maineult, W., Gellie, P., Andronico, A., Filloux, P., Leo, G., Sirtori, C., Barbieri, S., Peytavit, E., Akalin, T., Lampin, J.-F., Beere, H. E., & Ritchie, D. A. (2008). Metal-metal terahertz quantum cascade laser with micro-transverse-electromagnetic-horn antenna, *Appl. Phys. Lett.* 93: 183508.

Nojima, S. (1998a). Enhancement of optical gain in two-dimensional photonic crystals with active lattice points, *Jpn. J. Appl. Phys.* 37: 565.

Nojima, S. (1998b). Polarization anisotropy of optical gain in two-dimensional photonic crystals with active lattice point, *Jpn. J. Appl. Phys.* 37: 6418.

Nojima, S. (1999). Single-mode laser oscillation in semiconductor gain photonic crystals, *Jpn. J. Appl. Phys..* 38: 867.

Nojima, S. (2001). Optical-gain enhancement in two-dimensional active photonic crystals, *J. Appl. Phys.* 90: 545.

Sakoda, K. (1999). Enhanced light amplification due to group-velocity anomaly peculiar to two- and three-dimensional photonic crystals, *Opt. Express* 4: 167.

Schartner, S., Golka, S., Pflügl, C., Schrenk,W., Andrews, A. M., Roch, T. & Strasser, G. (2006). Band strucuture mapping of photonic crystal intersubband detectors, *Appl. Phys. Lett.* 89: 151107.

Srinivasan, K. & Painter, O. (2002). Momentum space design of high-q photonic crystal optical cavities, *Opt. Express* 10: 670.

Srinivasan, K. & Painter, O. (2003). Fourier space design of high-q cavities in standard and compressed hexagonal lattice photonic crystals, *Opt. Express* 11: 579.

Tsu, R. & Esaki, L. (1973). Tunneling in finite superlattice, *Appl. Phys. Lett.* 22: 562.

Tsu, R., Koma, A. & Esaki, L. (1974). Optical properties of semiconductor superlattice, *J. Appl. Phys.* 46: 842.

Williams, B., Callebaut, H., Kumar, S., Hu, Q. & Reno, J. (2002). 3.4 thz quantum cascade laser based on longitudinal-optical-phonon scattering for depopulation, *Appl. Phys. Lett.* 82: 1015.

Yasuda, H. & Hosako, I. (2008). Measurement of terahertz refractive index of metal with terahertz time-domain spectroscopy, *Jpn. J. Appl. Phys.* 47: 1632.

Zhi, W., Guobin, R., Shuqin, L. & Shuishend, J. (2003). Supercell lattice method for photonic crystal fibers, *Opt. Express* 11: 980.

23

High-power and High Efficiency Yb:YAG Ceramic Laser at Room Temperature

Shinki Nakamura
Ibaraki University
Japan

1. Introduction

Ceramic laser media fabricated by vacuum sintering (Ikesue et al., 1995), (Ikesue et al., 1996), and nanocrystalline (Lu et al., 2000) technology are very attractive materials because they have several remarkable advantages compared with single crystal laser materials. Ceramic samples with a large size can be easily fabricated, whereas this is extremely difficult for single crystals; multiplayer and multifunctional ceramic laser materials are possible because of the polycrystallinity of ceramics (Taira, 2007). Potentially, because of their short fabrication period and because they can be mass-produced, the cost of ceramic laser materials could be much lower than that of single crystals. Furthermore, no complex facilities and critical techniques are required for the growth of ceramics. Since 1995, Ikesue and coworkers have been developing several types of ceramic laser material (Ikesue et al., 1995), (Ikesue et al., 1996), and they found in 2000 that the output power from a 3.4 at.% Nd:YAG ceramic microchip laser is twice that from a Nd:YAG crystal microchip laser of the same size (Shoji et al., 2000). At a low doping concentration, it was found that the efficiency of a diode-end-pumped Nd:YAG ceramic laser is even higher than that of a Nd:YAG single crystal laser. Since 1998, Yanagitani and coworkers have been developing several types of ceramic lasers, and Lu et al. reported the Nd:YAG ceramic laser as one of them in 2001 (Lu et al., 2001). The mechanical properties of YAG ceramics were reported by Kaminskii et al. in ref. (Kaminskii et al., 2003). YAG ceramics had a 10% higher hardness than a YAG single crystal, and the fracture toughness of the YAG ceramics was more than threefold that of the YAG single crystal. Therefore, the ceramics had a higher resistance to thermal shock than the single crystal. Ytterbium (Yb^{3+})-doped materials are very attractive for diode-pumped solid-state lasers (DPSSLs) (Krupke, 2000). The Yb^{3+}-doped materials have high quantum efficiency and exhibit no concentration quenching simply because the Yb^{3+} ion has only two manifolds, namely, the ground state $^2F_{7/2}$ and the upper level $^2F_{5/2}$. Thus far, many articles about Yb:YAG crystal lasers have been published (Lacovara et al., 1991), (Hönninger et al., 1995), (Sumida et al., 1998). Yb:YAG has broad absorption and emission bands. The broad absorption band in the near-IR region is suitable for laser-diode (LD) pumping, and the broad emission band enables the generation of ultrashort pulses (Hönninger et al., 1995). However, an Yb:YAG laser is known as a quasi-three-level laser or a quasi-four-level laser, and a finite population exists at the Stark level of the lower manifold $^2F_{7/2}$, where laser transition terminates, which requires high-intensity pumping, a high-brightness pump source, and an efficient heat removal technique (Matsubara et al., 2007) , (Sato & Taira, 2004), (Shoji et al., 2004) to prevent reabsorption from the lower level of the laser. Takaichi et al. reported the absorption and emission spectra of a Yb:YAG ceramic (C_{Yb} = 1 at.%) and demonstrated laser oscillation, which was the first diode-end-pumped Yb:YAG ceramic laser (not Nd:YAG) with a 345 mW cw output power and a slope efficiency of 26% (Takaichi et al., 2003). Recently, Tsunekane and Taira have demonstrated a high-power diode-edge-pumped single-crystal Yb:YAG / ceramic undoped YAG composite microchip laser (Tsunekane&Taira, 2005), (Tsunekane&Taira, 2006). Early in 2007, a diode-edge-pumped, composite all-ceramic Yb:YAG (C_{Yb} = 10 at.%) microchip laser was demonstrated by Tsunekane and Taira, and a 414 W cw output power was obtained with a slope efficiency of 47% (Tsunekane&Taira, 2007). Very recently, Dong et al. have demonstrated a 2.7 W heavily doped (20 at.%) Yb:YAG ceramic laser with a slope efficiency of 52% (Dong et al., 2007) ; however, its two-pass-pumping miniature laser configuration was more complex than a simple conventional end-pumping configuration and its output power was not markedly high. Nakamura et al. demonstrated a 5.5 W cw Yb:YAG (9.8 at.%) ceramic laser with a slope efficiency of 52% using a simple end-pumping scheme (Nakamura et al., 2008a) with a 400 μm fiber-coupled LD. Dong et al. demonstrated a highly efficient (a slope efficiency of 79%) Yb:YAG ceramic laser (Dong et al., 2006) with a 100 μm fiber-coupled LD using an end-pumping scheme, but its output power was 1.7 W.

In section 2, we report a high-power (6.8 W) and high-efficiency tunable Yb:YAG ceramic laser demonstrated using an end-pumping scheme with a slope efficiency of 72% (Nakamura et al., 2008b) at room temperature (20°C).

Additionally, the tunability of the Yb:YAG ceramic laser for spectroscopy and femtosecond laser application. The broadest tunability is demonstrated in section 3. Finally, A diode-pumped passively mode-locked Yb:YAG ceramic laser is demonstrated in section 4. 417 and 286 fs pulses with average powers of 250 and 25 mW were obtained at 1030 nm using 1 and 0.1% output couplers, respectively. 233 fs pulses with an average power of 20 mW were also obtained at a center wavelength of 1048.3 nm using a 0.1% output coupler. To the best of our knowledge, this is the first demonstration of a diode-pumped mode-locked Yb:YAG ceramic laser.

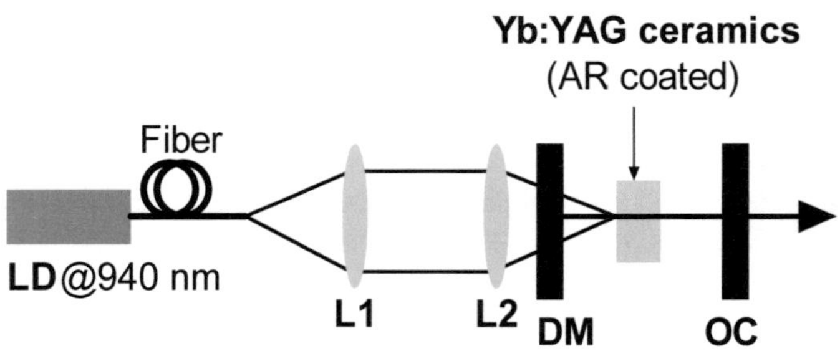

Fig. 1. Experimental setup for the high-power Yb:YAG ceramic laser (Nakamura et al., 2008a)

2. High-power and high efficiency Yb:YAG

2.1 Experimental setup of linear cavity

The experimental setup for the Yb:YAG ceramic laser is shown in Fig. 1. A 940 nm fiber-coupled LD (JENOPTIK Laserdiode, JOLD-30-FC-12) was used as a pumping source, the core diameter of the fiber was 200 μm, and the numerical aperture (NA) of the fiber was 0.22. The pumping beam was focused onto the ceramic with a ratio of 1:1 using the lenses L1 (f = 25 mm) and L2 (f = 25 mm). The diameter of the focused spot on the ceramic was ~200 μm. To obtain high efficiency and high power, a laser cavity consisting of a flat dichroic mirror (DM) and a flat output coupler (OC) as a linear resonator was used. The DM was antireflection (AR)-coated at 940 nm and had a high reflectivity at 1030 nm. The OC was partially-reflection-coated with a transmittance of T = 1, 5, and 10% at 1030 nm. An AR-coated ceramic Yb:YAG (C_{Yb} = 9.8 at.%, Konoshima Chemical) with dimensions of 5x10x1 mm^3 was used. A 1-mm-thick Yb:YAG ceramic plate was wrapped with indium foil and mounted in a water-cooled copper block that acted as a heat sink. Water was maintained at a room temperature of 20°C during laser oscillation. The cavity length was 20 mm, which was optimized, as shown in the later part of this section.

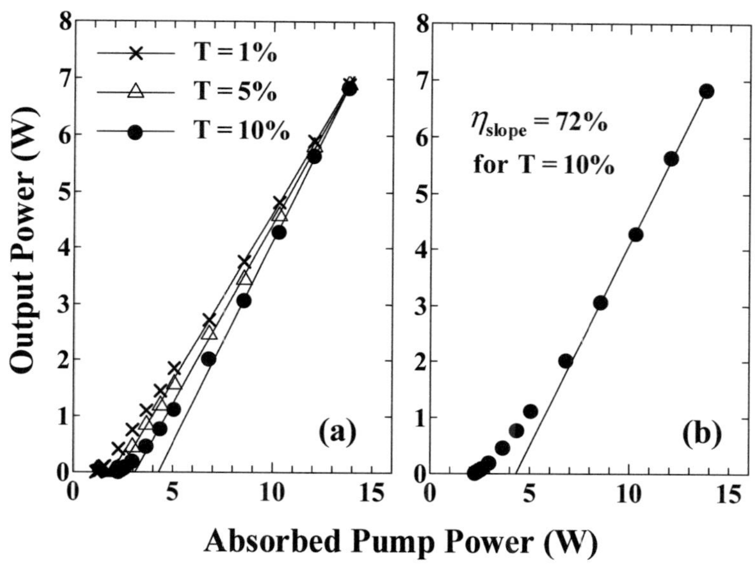

Fig. 2. Input-output power dependence of efficient Yb:YAG (C_{Yb} = 9.8 at%) ceramic laser: (a) with output couplers of T = 1, 5, and 10%, and (b) only T = 10% (Nakamura et al., 2008b).

2.2 Experimental results of linear cavity

Figure 2(a) shows the output power as function of the absorbed pump power in the cases for the three transmittances of the output couplers T = 1, 5, and 10%, (Nakamura et al., 2008b) and Fig. 2(b) shows the output power as function of the absorbed pump power only for the case of T = 10%(Nakamura et al., 2008b). The absorbed pump powers at the lasing threshold were 1.2, 2.0, and 2.3 W, and the maximum output powers of 6.9, 6.9, and 6.8 W for T = 1, 5, and 10%, respectively, were obtained at the absorbed pump power of 13.8 W. The round trip loss L in the resonator was estimated to be 0.09 by the lasing thresholds and the reflectivity of the output couplers (Koechner, 1992), which resulted in a small signal gain g_0 of 2.0 cm^{-1}, and a single pass gain G of 1.2 with the 1 mm thick gain medium. Each linear line was fit in Fig. 2(a) for T = 1, 5, and 10%. The slope efficiencies η_{slope} were 60, 64, and 72% for T = 1, 5, and 10%, respectively. Since we considered that T = 10% is best for obtaining the highest slope efficiency of 72%, we filled the data for the T = 10% case to Fig. 2(b). The maximum output power of 6.8 W for T = 10% was obtained at the absorbed pump power of 13.8 W, indicating that the efficiency of converting pumping optical power to output optical power, $\eta_{opt-opt}$, was 49%. The line of the best fit is shown in Fig. 2(b). The slope efficiency η_{slope} was 72% for T = 10%. The maximum output power of 6.8 W was determined to be fourfold higher than 1.7 W and the slope efficiency was determined to be 7% lower than 79% using the 100 μm fiber-coupled LD reported by Dong et al. (Dong et al., 2006). Our 6.8 W laser with the slope efficiency of 72% is expected to have a higher slope efficiency than the present result if the pumping source is replaced with a 100 μm fiber-coupled 25 W LD, for example, LIMO25-F100-DL940 (Lissotschenko Mikrooptick) while maintaining the high output power, because the pumping intensity would increase to a value of fourfold higher than that of a 200 μm fiber-coupled LD. In comparing our laser with the edge-pumped composite Yb:YAG ceramic laser (Tsunekane&Taira, 2007) developed by Tsunekane and Taira, we limit our discussion to the cw case; the laser power of 414 W obtained by Tsunekane and Taira is much higher than our result, but their slope and optical-optical conversion efficiency were 47 and 44%, which were 25 and 5% lower than our slope and optical-optical conversion efficiency of 72 and 49%, respectively. The transverse intensity profile of the Yb:YAG ceramic laser beam is shown in Fig. 3. The intensity distribution indicates that the beam is a Gaussian beam (a TEM_{00} mode beam). The beam image in Fig. 3 was as stable as the pumping LD and we found no amplitude instability.

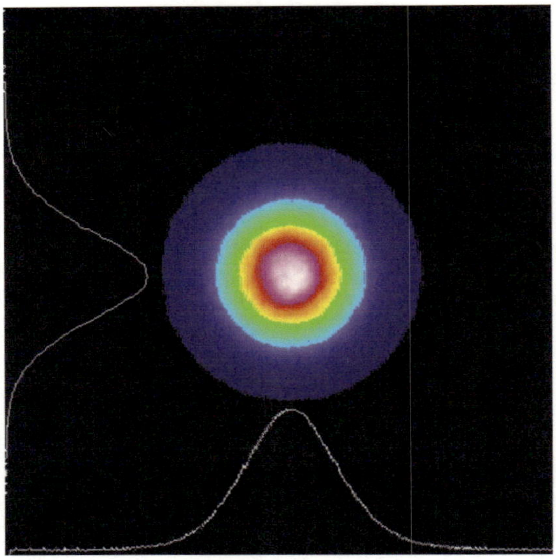

Fig. 3. Transverse intensity profile of the Yb:YAG ceramic laser beam (Nakamura et al., 2008b). The intensity distribution indicates that the beam is a Gaussian beam (a TEM_{00} mode beam).

These results of high output power, high efficiency, and good beam quality were obtained after the optimization of the cavity length. The cavity length was varied to obtain an optimum value for the highest efficiency and highest output power, and the focal length of the thermal lens for designing a tunable laser cavity configuration. Figure 4 shows the maximum output power as a function of the cavity length. Figure 4 shows that the optimum cavity length is less than 20 mm. This value is the appropriate cavity length for our laser, because there is no space to reduce the cavity length less than 20 mm. When we used the 400 μm fiber-coupled LD (Nakamura et al., 2008a), the optimum cavity length with the highest output power and highest slope efficiency was 25 mm, and reducing the length less than 25 mm yielded a worse result.

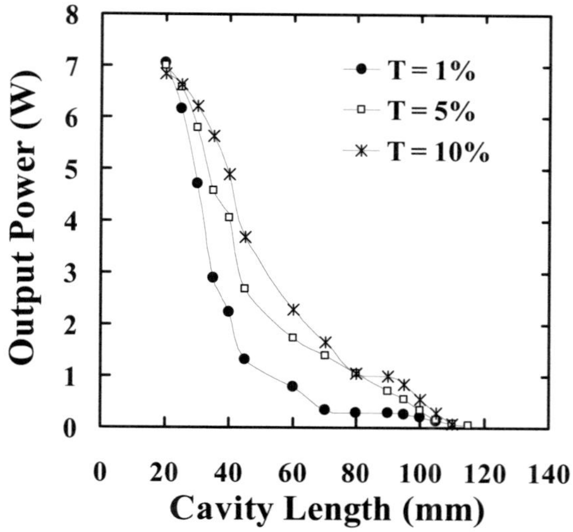

Fig. 4. Maximum output power as function of cavity length of efficient Yb:YAG ceramic laser (Nakamura et al., 2008b).

The focal length of the thermal lens in the ceramic Yb:YAG plate was considered for designing a tunable laser cavity configuration. Figure 4 also shows that the focal length of the thermal lens is 109 mm [120 mm (the cavity length) minus 11 mm (the distance of the ceramic Yb:YAG and the DM)], because the cavity becomes unstable, terminating the laser oscillation when the Fabry-Perot cavity length exceeds the thermal lens focal length. By considering this thermal lens, a tunable laser with a v-shape cavity including a concave mirror M (radius of curvature, ROC = 250 mm) and an SF10 dispersive prism was obtained, as shown in Fig. 5 in section 3.

2.3 Summary of linear cavity

A diode-end-pumped high-efficiency high-power Yb:YAG ceramic laser was demonstrated at a room temperature of 20°C with an Yb concentration of 9.8 at.%, a gain medium thickness of 1 mm, a pumping power of 13.8 W, an output coupler of T = 10%, and a cavity length of 20 mm. A 6.8 W cw output power was obtained with a slope efficiency of 72%. This is the relatively high efficiency of ceramic Yb:YAG lasers at room temperature. The beam quality was shown as a transverse intensity distribution indicating a Gaussian beam (a TEM_{00} mode beam).

3. Tunability of Yb:YAG ceramic laser

3.1 Tunable Yb:YAG laser

In previous reports on Yb:YAG ceramic lasers, no descriptions of the tunability of the lasers are given. However, there are some reports on the tunability of Yb:YAG crystal lasers.

(Brauch et al., 1995), (Saikawa et al., 2000), (Saikawa et al., 2002) In 2000, the widest tunability range from 1024.1 to 1108.6 nm was demonstrated for a 160 mW Yb:YAG crystal laser using a 0.1% output coupler and a birefringent filter by Saikawa et al. (Saikawa et al., 2000) Subsequently, Saikawa et al. reported a 180 mW Yb:YAG crystal laser with a tunability in the spectral range of 59 nm from 1021.9 to 1081.2 nm in 2002. (Saikawa et al., 2002) Very recently, we have reported the first study of a tunable Yb:YAG ceramic laser (Nakamura et al, 2008b) that exhibits continuous tunability with a spectral range of 63.5 nm from 1020.1 to 1083.6 nm for T = 1% at a maximum output power of 1.6 W. In this section, we investigated the broader tunability range of a 1.4 W Yb:YAG ceramic laser using a 1% output coupler and a prism; this range was twofold broader than that in our previous work. (Nakamura et al, 2008b)

Comparing the emission spectra of a Yb:YAG crystal (Dong et al., 2003) doped with 10 at.% Yb^{3+} and the ceramic Yb:YAG (Dong et al., 2006) with 9.8 at.% Yb^{3+} reported by Dong et al., we found that the ceramic Yb:YAG had a slightly higher emission intensity and the possibility of producing shorter-wavelength oscillation than the crystal Yb:YAG. To consider the difference in the emission spectra in the short-wavelength region, we employed a dichroic mirror (DM; Layertec No. 102927) with a high-reflectivity (HR) wavelength range from 1010 to 1200 nm and an antireflective range from 800 to 970 nm as a replacement for the previously used DM (Layertec No. 103542) (Nakamura et al, 2008b) with an HR range from 1020 to 1200 nm and an antireflective range from 808 to 980 nm. To the best of our knowledge, we obtained the broadest reported tunable range of 118.31 nm from 992.52 to 1110.83 nm for a Yb:YAG laser owing to this modification.

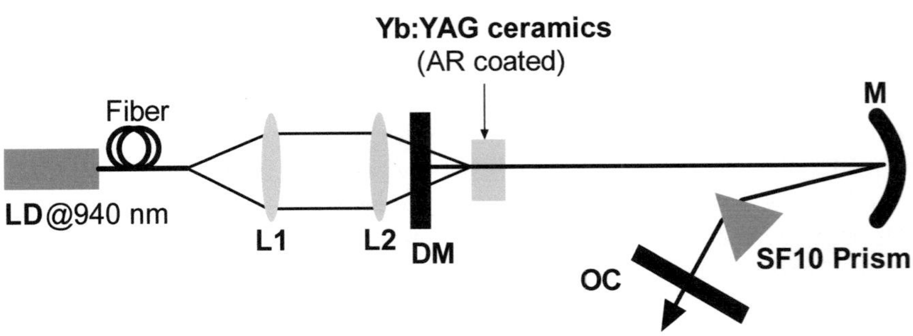

Fig. 5. Experimental setup for tunable Yb:YAG ceramic laser (Nakamura et al., 2008b). LD: fiber-coupled diode laser; L1, L2: focusing lenses; DM: flat dichroic mirror; M; a concave mirror (ROC = 250 mm); OC: output coupler.

3.2 Experimental setup of tunable Yb:YAG ceramic laser

The experimental setup for the Yb:YAG ceramic laser is shown in Fig. 5. A 940 nm fiber-coupled LD (JENOPTIK Laserdiode JOLD-30-FC-12) was used as a pumping source; the fiber has a core diameter of 200 μm and a numerical aperture (NA) of 0.22. The pumping beam was focused onto the ceramic at a ratio of 1:1 using the lenses L1 (f = 25 mm) and L2 (f = 25 mm). The diameter of the spot focused on the ceramic was ~200 μm. A laser cavity consisting of a flat DM and a flat output coupler (OC) with a mirror M and an SF10 prism was used. The DM was antireflection (AR)-coated at 940 nm and had an HR at 1030 nm. In detail, we employed a DM (Layertec No. 102927) with an HR wavelength range from 1010 to 1200 nm and an AR range from 800 to 970 nm, which was a replacement for the previously used DM (Layertec, No. 103542) (Nakamura et al, 2008b) with an HR range from 1020 to 1200 nm and an AR range from 808 to 980 nm. The OC was partially reflection-coated with transmittances of T = 0.1, 1, 5, and 10% at 1030 nm. An AR-coated ceramic Yb:YAG (C_{Yb} = 9.8 at.%, Konoshima Chemical) with dimensions of 5x10x1 mm³ was used. The dimensions of the ceramics were selected to enable Brewster angled alignment for a mode-locked laser performance in future to reduce cost. A 1-mm-thick Yb:YAG ceramic plate was wrapped with indium foil and mounted in a water-cooled copper block that acted as a heat sink. The water was maintained at room temperature (20 °C) during laser oscillation. By considering the thermal lens effect (Nakamura et al, 2008b), a tunable laser with a V-shape cavity including a concave mirror M (radius of curvature, ROC = 250 mm) and an SF10 dispersive prism was obtained, as shown in Fig. 5. The SF10 dispersive prism was inserted into a V-shape resonator as the tuning element between the folded mirror M and the output coupler OC at the Brewster angle. The cavity length was 315 mm.

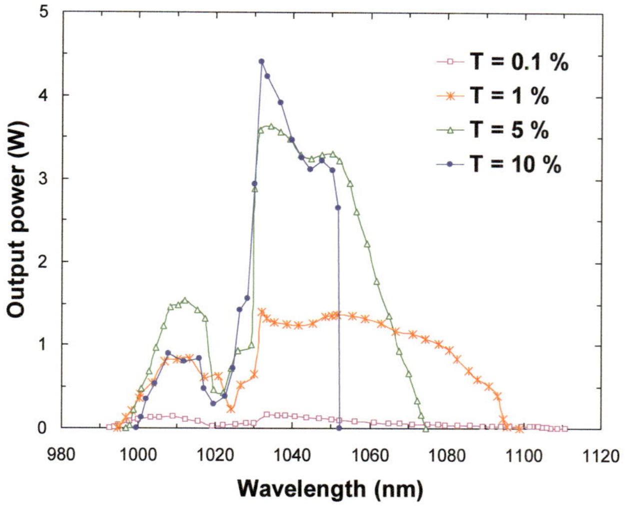

Fig. 6. Output power of ceramic Yb:YAG tunable laser as a function of oscillation wavelength for various output couplers (Nakamura et al., 2009).

3.3 Experimental results of tunable Yb:YAG ceramic laser

Figure 6 shows the dependence of output power on laser oscillation wavelength for output couplers of T = 0.1, 1, 5, and 10% when the absorbed pump power was 13.8 W. The averaged spectral linewidth of the output was 3.1 nm in these measurements. We achieved smoothly continuous tunability. A tunable range of 52.64 nm from 999.48 to 1052.12 nm was obtained with a maximum output power of 4.41 W at 1031.78 nm for T = 10%. A tunable range of 77.83 nm from 996.75 to 1074.58 nm and a maximum power of 3.63 W at 1034.14 nm were obtained for T = 5%. A maximum power of 1.40 W was obtained at 1033.42 nm and a tunable range of 104.52 nm from 994.35 to 1098.87 nm was obtained for T = 1%, which is broader than the tunability of 84.5 nm reported by Saikawa et al. (Saikawa et al., 2000). Finally, as shown in Fig. 6, T = 0.1% results in the largest tunable range of 118.31 nm from 992.52 to 1110.83 nm, which is broader than the tunability for T =1%, with a maximum output power of 163 mW at 1033.42 nm. To the best of our knowledge, this is the broadest tunability (Nakamura et al, 2009) reported for an Yb:YAG ceramic laser. The shortest wavelength of the tuning range in Fig. 6 is not limited to 1020 nm (Nakamura et al, 2008b) because of the broad dichroic coating range of the pumping mirror (Layertec No. 102927) with an HR wavelength range from 1010 to 1200 nm and an AR range from 800 to 970 nm. This tuning range of 118.31 nm for the Yb:YAG ceramic laser at 20 °C or 293 K is 3.29-fold broader than the 36.0 nm tuning range, from 1018 to 1054 nm, which was produced from a Yb:YAG crystal laser with a three-plate birefringent filter at 218 and 245 K. (Brauch et al., 1995). Furthermore, our tuning range of 118.31 nm from 992.52 to 1110.83 nm for the high-power ceramic Yb:YAG laser at 20 °C is twofold broader than that for the crystal Yb:YAG laser reported by Saikawa et al. (Saikawa et al., 2002), which has a tuning range of 59 nm from 1022 to 1081 nm at 18 °C. The widely tunable Yb:YAG crystal laser with birefringent filters reported by Saikawa et al. (Saikawa et al., 2000) had a tuning range of 84.5 nm from 1024.1 to 1108.6 nm; however, the highest output power was 180 mW, which is much lower than the maximum output power of 1.4 W of our ceramic laser with an SF10 prism for T = 1%, and our tuning range of 104.52 nm for T = 1% is 1.24-fold broader than their 84.5 nm tunable range.

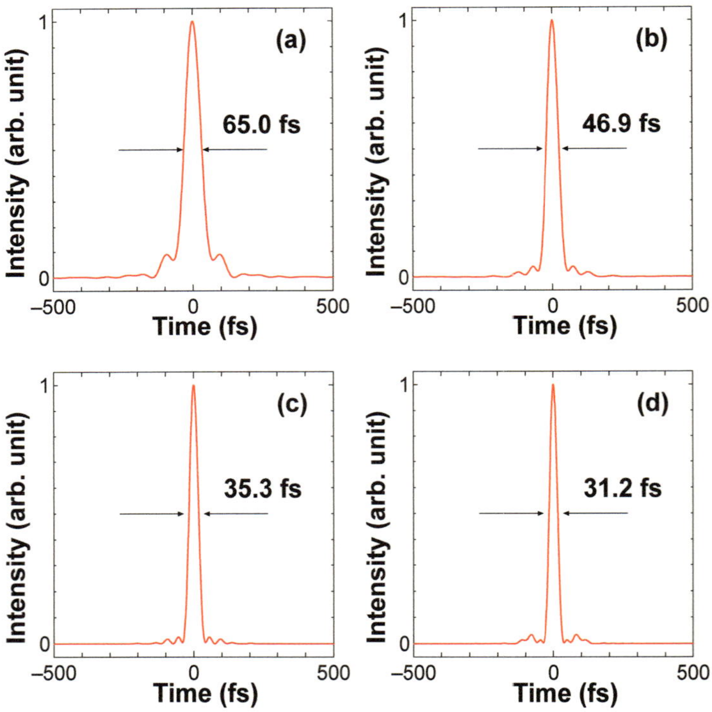

Fig. 7. Expected transform limited pulses numerically obtained by inverse Fourier transformation of Fig. 6 for T = (a) 10, (b) 5, (c) 1, and (d) 0.1% (Nakamura et al., 2009).

Figure 7 shows the expected transform-limited pulses obtained by an inverse Fourier transformation of Fig. 6 for T = (a) 10, (b) 5, (c) 1, and (d) 0.1%. The full width at half maximum (FWHM) of the numerically obtained pulse duration was t_p = 65.0, 46.9, 35.3, and 31.2 fs for Figs. 7(a)-7(d), respectively. These numerical results suggest that the ceramic Yb:YAG laser medium has the potential for a laser performance of less than 65 fs.

3.4 Summary of tunable Yb:YAG ceramic laser

A diode-end-pumped broadly tunable Yb:YAG ceramic laser was demonstrated at room temperature (20 °C) with Yb concentration of 9.8 at.%, a gain medium thickness of 1 mm, a pumping power of 13.8 W. Continuous tunability with a spectral range of 104.52 nm from 994.35 to 1098.87 nm for T = 1% at a maximum output power of 1.4 W and with a spectral range of 118.31 nm from 992.52 to 1110.83 nm for T = 0.1% at a maximum output power of 0.16 W was also obtained at room temperature. To the best of our knowledge, this is the broadest tunability reported for a Yb:YAG laser. This tunability is expected to be very attractive for femtosecond laser applications. Numerical results suggest that the ceramic Yb:YAG laser medium has the potential for a laser performance of less than 65 fs. The cost of Yb:YAG ceramic laser materials is much lower than that of single crystals because of their high-speed, large production, and mass production potential, is making them tremendously attractive for industrial applications.

4. Mode-locked Yb:YAG ceramic laser

4.1 Mode-locked Yb-doped lasers

Femtosecond mode-locked lasers are applied in various fields of physics, engineering, chemistry, biology and medicine, with applications including ultrafast spectroscopy, metrology, superfine material processing and microscopy. Specific and interesting properties of the femtosecond laser pulses have been used in these applications. For example, femtosecond pulses have a very precise time resolution, and their strong electric field induces important and unusual nonlinear effects. For those applications, high-power, high-efficiency and compact femtosecond lasers are required. Ceramic materials are attractive for satisfying these requirements. YAG ceramics have 10% higher hardness than a YAG single crystal, and the fracture toughness of YAG ceramics is more than threefold that of the YAG single crystal. Therefore, the ceramics have a higher resistance to thermal shock than the single crystal. Ytterbium (Yb^{3+}) also has interesting properties satisfying the above requirements. Its broad absorption and emission spectra allow the realization of a directly laser-diode (LD)-pumped femtosecond laser. Moreover, its small quantum defect, absence of excited-state absorption, upconversion and cross-relaxation reduce the thermal load and enable highly efficient operation. The emission and absorption spectra and thermal conductivity strongly depend on the host material.

Recently, various ceramic materials have been progressively investigated for use in ultrashort-pulse lasers (Shirakawa et al., 2003), (Tokurakawa et al., 2006), (Shirakawa et al., 2003), (Tokurakawa et al., 2007), (Tokurakawa et al., 2008). A diode-pumped femtosecond $Yb:Y_2O_3$ ceramic laser was demonstrated, and 615 fs pulses at a center wavelength of 1076.5 nm were obtained with a 420 mW average power (Shirakawa et al., 2003). A diode-pumped passively mode-locked $Yb:Lu_2O_3$ ceramic laser was demonstrated, for which 357 fs pulses at a center wavelength of 1033.5 nm with a 352 mW average power were obtained (Tokurakawa et al., 2006). A passively mode-locked femtosecond Yb^{3+}-doped $Y_3(Sc_{0.5}Al_{0.5})_2O_{12}$ (Yb:YSAG) ceramic laser pumped by a Ti:sapphire laser was also demonstrated, and 280 fs pulses at a center wavelength of 1035.8 nm with a 62 mW average power were obtained (Saikawa et al., 2007), but the laser was not diode-pumped. A diode-pumped Kerr lens mode-locked laser of $Yb^{3+}:Sc_2O_3$ ceramics was demonstrated with 92 fs pulses at a center wavelength of 1042 nm and an 850 mW average power, and 90 fs pulses at a center wavelength of 1092 nm and a 160 mW average power (Tokurakawa et al., 2007). A diode-pumped Kerr lens mode-locked laser with Yb^{3+}:Lu_2O_3 and undoped Y_2O_3 combined ceramics was demonstrated, and 65 fs pulses at a center wavelength of 1032 nm with a 320 mW average power were obtained (Tokurakawa et al., 2008). In previous reports on Yb:YAG ceramic lasers, there is no description of the mode locking of lasers. However, in some reports the femtosecond mode locking of Yb:YAG crystal lasers is discussed (Hönninger et al., 1999), (Aus der Au et al., 2000), (Neuhaus et al., 2008a), (Neuhaus et al., 2008b), (Uemura&Torizuka, 2008). The average power and pulse energy have been scaled up to 76 W and 25.9 μJ using a mode-locked thin-disk Yb:YAG laser (Neuhaus, 2008b). A Kerr lens mode-locked Yb:YAG crystal laser has been demonstrated, for which the pulse duration was as short as 100 fs at a center wavelength of 1051 nm (Uemura&Torizuka, 2008). The shortest pulse obtained using a diode-pumped ceramic laser without the Kerr lens effect was 357 fs (Tokurakawa et al., 2006).

In this section, we report a diode-pumped passively mode-locked Yb:YAG ceramic laser generating 286 fs pulses with an average power of 25 mW at a wavelength of 1033.5 nm using a 0.1% output coupler. 233 fs pulses with an average power of 20 mW were also obtained at a center wavelength of 1048.3 nm using a 0.1% output coupler. To the best of our knowledge, this is the first demonstration of a diode-pumped mode-locked Yb:YAG ceramic laser and the shortest pulse using diode-pumped ceramic lasers without Kerr lens mode locking. In our previous studies as shown in section 3, we measured the broadest tunability of Yb:YAG ceramics at room temperature (RT) (Nakamura et al., 2008b), (Nakamura et al., 2009). This broad tunability indicates the possibility of sub-100-fs pulse generation using Yb:YAG ceramics.

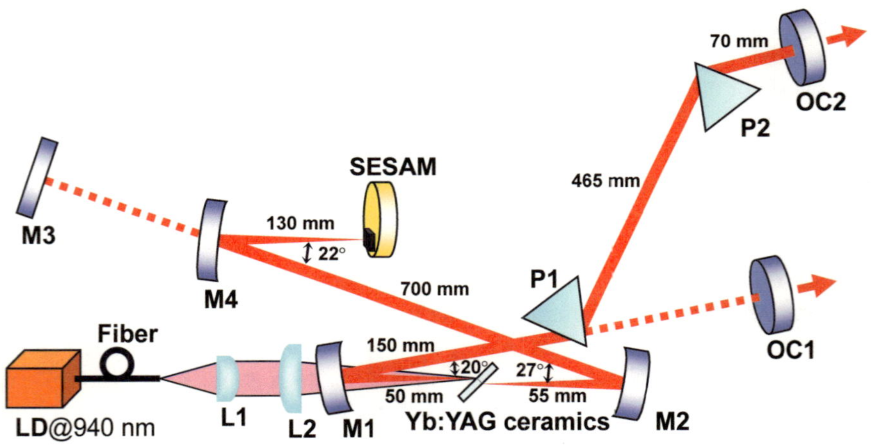

Fig. 8. Experimental setup of the mode-locked Yb:YAG ceramic laser (Yoshioka et al., 2009). LD: fiber-coupled diode laser. L1, L2: focusing lenses. M1-M4: high-reflectivity mirrors. OC1, OC2: output couplers. P1-P2: SF10 Brewster prisms. SESAM: semiconductor saturable absorber mirror. Dashed lines indicate the beam lines in cw operation. Solid lines indicate the beam lines in mode-locked operation.

4.2 Experimental setup of mode-locked Yb:YAG ceramic laser

The experimental setup for the mode-locked Yb:YAG ceramic laser is shown in Fig. 8. An x-fold cavity configuration was used. A 940 nm fiber-coupled LD was used as a pumping source. The core diameter of the fiber was 200 μm. The numerical aperture (NA) of the fiber was 0.22. The maximum pump power was 26.6 W. The pumping beam was imaged by relay to the ceramics using lens L1 (f = 50 mm) and lens L2 (f = 70 mm). The 1-mm-thick Yb:YAG (C_{Yb} = 9.8 at.%) ceramic plate was arranged at the Brewster's angle. The Yb:YAG plate was wrapped with indium foil and mounted in a water-cooled copper heat sink block. The copper block was cooled by flowing water at 20°C. The ceramic was placed between two high-reflectivity mirrors (M1, M2) that were anti-reflection (AR)-coated at 940 nm and had high reflectivity at 1030 nm with a 100 mm radius of curvature (ROC). The reference laser operated in cw mode with a 10% output coupler (OC1) and a high-reflectance end mirror (M3) along the dashed lines in Fig. 8. For passive mode locking, a 1 or 0.1% output coupler (OC2) and a semiconductor saturable absorber mirror (SESAM, BATOP) with 2% saturable absorption at 1030 nm, 70 μJ/cm² saturation fluence and 500 fs relaxation time constant were used in the respective arms. The total cavity length was 1620 mm. The laser beam was focused onto the SESAM by a concave mirror (M4, ROC = 250 mm). The distance among the mirrors and the folded angle of the mirrors are shown in Fig. 8. The astigmatism compensation was not considered. The spot sizes of laser mode in the laser crystal and on SESAM were estimated ~ 61 × 53 μm and ~ 450 × 330 μm, respectively. An SF10 Brewster prism pair (P1, P2) with 465 mm separation was inserted in the other arm to compensate for the dispersion. The total negative GDD of this cavity was about -2670 fs² per a round trip.

4.3 Experimental results of mode-locked Yb:YAG ceramic laser

We demonstrated mode locking for three cases. The first case was mode locking at 1030 nm using a 1% output coupler. The second case was mode locking at 1030 nm using a 0.1% output coupler. The third case was mode locking at 1050 nm region with cw oscillation at 1030 nm region using a 0.1% output coupler.

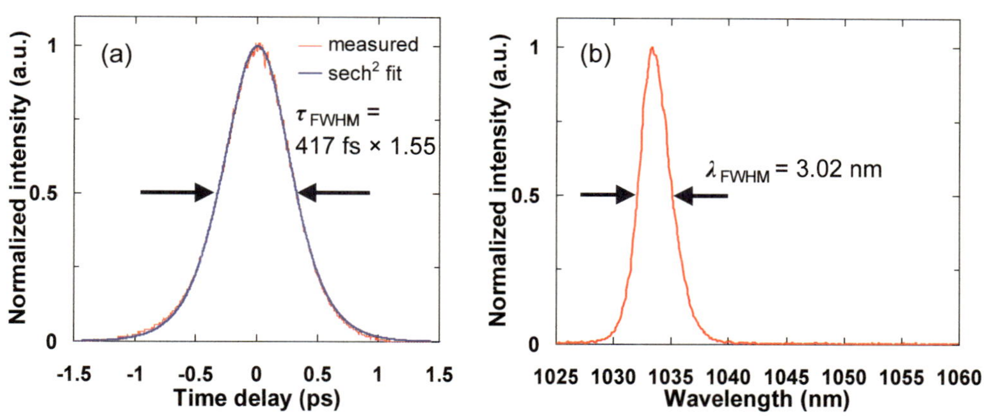

Fig. 9. (a) Measured autocorrelation trace and the sech² fitting, and (b) spectrum of the mode-locked pulses using a 1% output coupler (Yoshioka et al., 2009).

4.3.1 Mode locking at 1030 nm region using a 1% output coupler

Figure 9 shows the intensity autocorrelation trace and the spectrum of mode-locked pulses using a 1% output coupler. The average output power was 250 mW at a pump power of 26.6 W. The sech²-fitted pulse width was 417 fs and the spectral width was 3.02 nm at the center wavelength of 1033.3 nm, which results in a time-bandwidth product of 0.353, slightly above the Fourier limit for a sech² pulse (0.315). The repetition rate was 91 MHz. The pulse energy and peak power were 2.75 nJ and 6.60 kW, respectively. When the 1% output coupler was used, mode locking at 1050 nm region was not observed but cw oscillation at 1050 nm region or simultaneous cw dual-wavelength oscillation at 1030 and 1050 nm was observed.

4.3.2 Mode locking at 1030 nm region using a 0.1% output coupler

Figure 10 shows the intensity autocorrelation trace and the spectrum of mode-locked pulses using a 0.1% output coupler. The average output power was 25 mW at a pump power of 26.6 W. The sech²-fitted pulse width was 286 fs and the spectral width was 4.51 nm, and centered at 1033.5 nm. This results in a time-bandwidth product of 0.363, which is 15% above the Fourier limit for a sech² pulse (0.315). This indicates that the pulse has potential to be further shortened by extracavity compression. The repetition rate was 91 MHz. The pulse energy and peak power were 0.275 nJ and 0.960 kW, respectively. The pulse width of 286 fs obtained using a 0.1% output coupler (Fig. 10) was shorter than that of 417 fs obtained using a 1% output coupler (Fig. 9) because the intracavity power is increased using a 0.1% output coupler, which makes it possible for the laser to oscillate with a broad spectral range. However, the average output power decreased to 25 mW owing to the low transmission of the 0.1% output coupler.

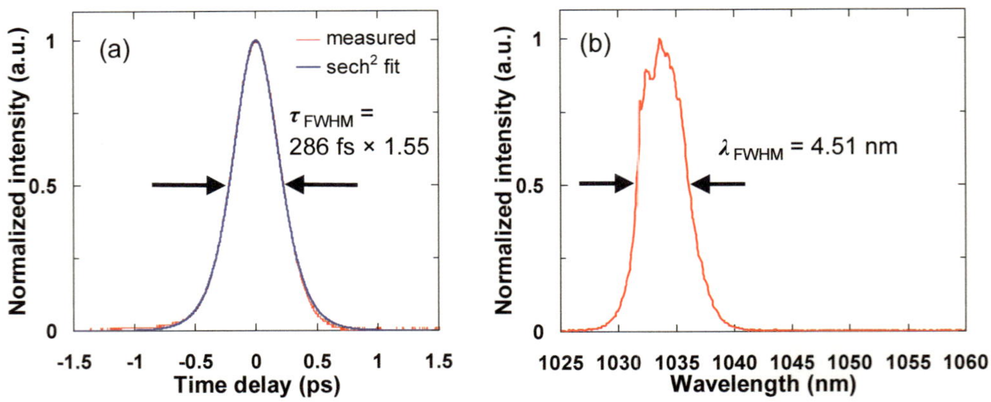

Fig. 10. (a) Measured autocorrelation trace and the sech² fitting, and (b) spectrum of the mode-locked pulses at 1030 nm region using a 0.1% output coupler (Yoshioka et al., 2009).

4.3.3 Mode locking at 1050 nm region using a 0.1% output coupler

When the SESAM angle was changed slightly from the conditions described in Sec. 4.3.2, mode locking at 1033.5 nm suddenly stopped and simultaneous cw dual-wavelength oscillation at 1030 and 1050 nm occurred. When the intensity of the cw spectral component at 1050 nm was increased and the laser mode radius focused on the SESAM was optimized, mode-locked pulses were generated again. Figure 11 shows (a) the intensity autocorrelation trace, (b) the spectrum of mode-locked pulses. The output spectrum in Fig. 11(b) was measured by injection into the fiber-coupled spectrometer directly from OC2. The average output power was 20 mW at a pump power of 26.6 W. The sech²-fitted pulse width was 233 fs and the spectral width was 5.20 nm and centered at 1048.3 nm, which results in a time-bandwidth product of 0.330, slightly above the Fourier limit for a sech² pulse (0.315). The repetition rate was 91 MHz. The pulse energy and peak power were 0.220 nJ and 0.946 kW, respectively. Figure 12 shows a pulse train of cw mode-locking in millisecond time scale. This indicates that the mode-locked pulses had a long-term stability. The pulse width of 233 fs at 1048.3 nm in Fig. 11 obtained using a 0.1% output coupler was the shortest among these results. This indicates that the laser has the potential to generate shorter pulses at 1050 nm region rather than 1030 nm region. However, the cw spectral component at 1032.4 nm was not quenched in our cavity during mode locking at 1048.3 nm. This occurred owing to the strong emission at 1030 nm.

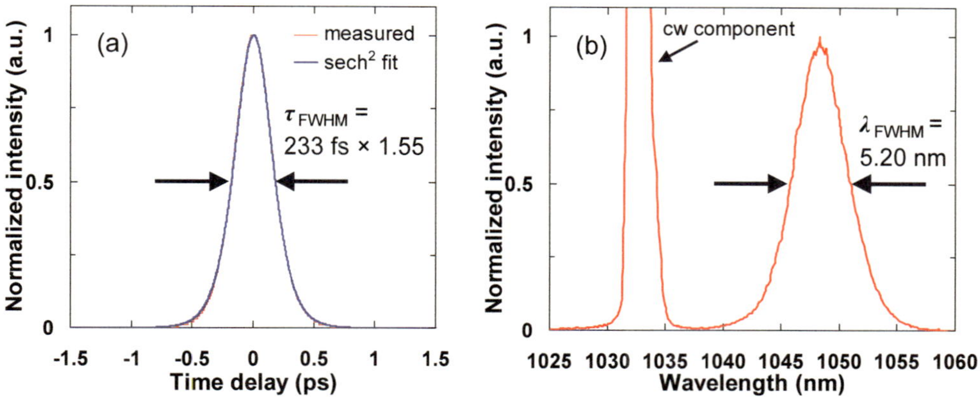

Fig. 11. (a) Measured autocorrelation trace and the sech² fitting, (b) spectrum of the mode-locked pulses at 1050 nm region using a 0.1% output coupler (Yoshioka et al., 2009).

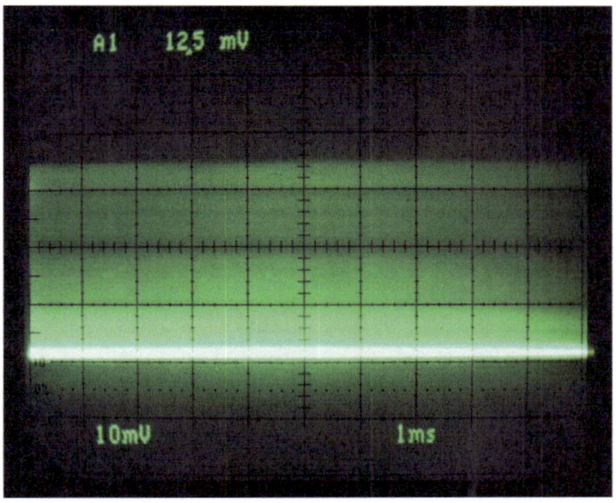

Fig. 12. Pulse train of cw mode-locking in millisecond time scale (Yoshioka et al., 2009).

4.4 Summary of mode-locked Yb:YAG ceramic laser

A diode-pumped femtosecond ytterbium laser with a host material of YAG ceramic was demonstrated. We successfully achieved passive mode locking at wavelengths of 1033.5 and 1048.3 nm. At 1033.5 nm, passive mode locking by a semiconductor saturable absorber mirror generated 286 fs pulses with an average power of 25 mW using a 0.1% output coupler. This is the shortest pulse width in the 1030 nm region. At 1048.3 nm, the shortest pulse of 233 fs with an average output power of 20 mW was generated using a 0.1% output coupler. To the best of our knowledge, this is the first mode-locked Yb:YAG ceramic laser, and the shortest pulse for diode-pumped ceramic lasers without Kerr lens mode locking was obtained.

5. Summary

A high-power efficient ceramic Yb:YAG laser was demonstrated at a room temperature of 20°C with an Yb concentration of 9.8 at.%, a gain medium of 1 mm, a pumping power of 13.8 W, an output coupler of T = 10%, and a cavity length of 20 mm. A 6.8 W cw output power was obtained with a slope efficiency of 72%.
A diode-end-pumped Yb:YAG ceramic laser with a broad tunability was demonstrated. The continuous tunability in the spectral range of 104.52 nm from 994.35 to 1098.87 nm for T = 1% at a maximum output power of 1.4 W and in the spectral range of 118.31 nm from 992.52 to 1110.83 nm for T = 0.1% at a maximum output power of 0.16 W was also obtained at room temperature. To the best of our knowledge, this is the broadest tunability of Yb:YAG lasers.
A diode-pumped passively mode-locked Yb:YAG ceramic laser was demonstrated. 417 and 286 fs pulses with average powers of 250 and 25 mW were obtained at 1030 nm using 1 and 0.1% output couplers, respectively. 233 fs pulses with an average power of 20 mW were also obtained at a center wavelength of 1048.3 nm using a 0.1% output coupler. To the best of our knowledge, this is the first demonstration of a diode-pumped mode-locked Yb:YAG ceramic laser.

6. Referrence

Aus der Au, J.; Spühler, G. J.; Südmeyer, T.; Paschotta, R.; Hövel, R.; Moser, M.; Erhard, S.; Karszewski, M.; Giesen, A. & Keller, U. (2000). 16.2-W average power from a diode-pumped femtosecond Yb:YAG thin disk laser. *Opt. Lett.*, Vol. 25, No. 11, (June, 2000) 859-861, ISSN0146-9592.

Brauch, U.; Giesen, A.; Karszewski, M.; Stewen, Chr. & Voss, A. (1995). Multiwatt diode-pumped Yb:YAG thin disk laser continuously tunable between 1018 and 1053 nm. *Opt. Lett.*, Vol. 20, No. 7 (Apr., 1995) 713-715, ISSN0146-9592.

Dong, J.; Bass, M.; Mao, Y.; Deng, P. & Gan, F. (2003). Dependence of the Yb^{3+} emission cross section and lifetime on temperature and concentration in yttrium aluminum garnet. *J. Opt. Soc. Am. B*, Vol. 20, No. 9 (Sep., 2003) 1975–1979, ISSN 0740-3224.

Dong, J.; Shirakawa, A.; Ueda, K.; Yagi, H.; Yanagitani, T.; Kaminskii, A. A. (2006). Efficient Yb^{3+}:$Y_3Al_5O_{12}$ ceramic microchip lasers. *Appl. Phys. Lett.*, Vol. 89, No. 9, (2006) id. 091114 (3 pages), ISSN 0003-6951.

Dong, J.; Shirakawa, A; Ueda, K; Yagi, H.; Yanagitani, T. & Kaminskii, A. A. (2007). Laser-diode pumped heavy-doped Yb:YAG ceramic lasers. *Opt. Lett.*, Vol. 32, No. 13, (Jul., 2007) 1890-1892, ISSN0146-9592.

Hönninger, C.; Zhang, G.; Keller, U. & A. Giesen, (1995). Femtosecond Yb:YAG laser using semiconductor saturable absorbers. *Opt. Lett.* Vol. 20, No. 23, (Dec., 1995) 2402–2404, ISSN 0146-9592.

Hönninger, C.; Paschotta, R.; Graf, M.; Morier-Genoud, F.; Zhang, G.; Moser, M.; Biswal, S.; Nees, J.; Braun, A.; Mourou, G.A.; Johannsen, I.; Giesen, A.; Seeber, W. & Keller, U. (1999). Ultrafast ytterbium-doped bulk lasers and laser amplifiers. *Appl. Phys. B*, Vol. 69 (1999) 3-17, ISSN 0946-2171.

Ikesue, A.; Kinoshita, T.; Kamata, K. & K. Yoshida (1995). Fabrication and Optical Properties of High-Performance Polycrystalline Nd:YAG Ceramics for Solid-State Lasers. *J. Am. Ceram. Soc.* Vol. 78 (1995) 1033-1040, ISSN 0002-7820.

Ikesue, A.; Kinoshita, T. & Yoshida, K. (1996). Effects of Neodymium Concentration on Optical Characteristics of Polycrystalline Nd:YAG Laser Materials. *J. Am. Ceram. Soc.* Vol. 79 (1996) 1921-1926, ISSN0002-7820.

Kaminskii, A. A.; Akchurin, M. Sh.; Alshits, V. I.; Ueda, K.; Takaichi, K.; Lu, J.; Uematsu, T.; Musha, M.; Shirakawa, A.; Gabler, V.; Eichler, H. J.; Yagi, H.; Yanagitani, T.; Bagayev, S. N.; Fernandez, J. & Balda, R. (2003). New results in studying physical properties of nanocrystalline laser ceramics. Crystallogr. Rep. Vol. 48 (2003) 515, ISSN 1063-7745.

Koechner, W. (1992). In: *Solid-State Laser Engineering*, 3rd ed., Koechner, W. (Ed.), p. 101, Springer, ISBN 3540650644, Berlin.

Krupke, W. F. ((2000). Ytterbium solid-state lasers. The first decade. *IEEE J. Sel. Top. Quantum Electron.* Vol. 6 (2000) 1287, ISSN 1077-260X.

Lacovara, P; Choi, H. K.; Wang, C. A.; Aggarwal, R. L. & Fan, T. Y. (1991). Room-temperature diode-pumped Yb:YAG laser. *Opt. Lett.*, Vol. 16, No. 14 (Jul., 1991) 1089-1091, ISSN0146-9592.

Lu, J.; Prabhu, M.; Song, J.; Li, C.; Xu, J.; Ueda, K.; Kaminskii, A. A.; Yagi, H. & Yanagitani, T. (2000). Optical properties and highly efficient laser oscillation of Nd:YAG ceramics. Appl. Phys. B Vol. 71 (2000) 469, ISSN 0946-2171.

Lu, J.; Murai, T.; Takaichi, K.; Uematsu, T.; Misawa, K.; Prabhu, M.; Xu, J.; Ueda, K.; Yanagitani, T.; Yagi, H.; Kaminskii, A. A. & Kudryashov, A. (2001). 72 W Nd:$Y_3Al_5O_{12}$ ceramic laser. Appl. Phys. Lett. Vol. 78 (2001) 3586-3588, ISSN 0003-6951.

Matsubara, S.; Ueda, T. ; Kawato, S. & Kobayashi, T. (2007). Highly Efficient Continuous-Wave Laser Oscillation in Microchip Yb:YAG Laser at Room Temperature. *Jpn. J. Appl. Phys.* Vol. 46 (2007) L132-L134, ISSN0021-4922.

Nakamura, S.; Matsubara, Y.; Ogawa, T. & Wada, S. (2008a). High-power High-Efficiency Yb^{3+}-doped $Y_3Al_5O_{12}$ Ceramic Laser at Room Temperature. Jpn. J. Appl. Phys. Vol. 47, No. 4, (Apr., 2008) 2149-2151, ISSN0021-4922.

Nakamura, S.; Yoshioka, H.; Matsubara, Y.; Ogawa, T. & Wada S. (2008b). Efficient tunable Yb:YAG ceramic laser. *Opt. Commun.* Vol. 281 (Sep., 2008) 4411-4414, ISSN 0030-4018.

Nakamura, S.; Yoshioka, H.; Ogawa, T. & Wada S. (2009). Broadly Tunable Yb^{3+}-doped $Y_3Al_5O_{12}$ Ceramic Laser at Room Temperature. *Jpn. J. Appl. Phys.*, Vol. 48, No. 6 (June, 2009) id. 060205 (3page), ISSN 0021-4922.

Neuhaus, J.; Kleinbauer, J.; Killi, A.; Weiler, S.; Sutter, D. & Dekorsy, T. (2008a). Passively mode-locked Yb:YAG thin-disk laser with pulse energies exceeding 13 μJ by use of an active multipass geometry. *Opt. Lett.* Vol. 33, No. 7 (Apr., 2008) 726-728, ISSN 0146-9592.

Neuhaus, J.; Bauer, D. ; Zhang, J. ;Killi, A.; Kleinbauer, J. ; Kumkar, M.; Weiler, S. ; Guina, M.; Sutter, D. H. & Dekorsy, T. (2008b). Subpicosecond thin-disk laser oscillator with pulse energies of up to 25.9 microjoules by use of an active multipass geometry. *Opt. Express,* Vol. 16, No. 25 (Dec., 2008) 20530-20539, ISSN1094-4087.

Saikawa, J.; Kurimura, S.; Pavel, N.; Shoji, I. & Taira, T. (2000). (2000). Performance of widely tunable Yb:YAG microchip lasers. In: *Advanced Solid-State Lasers (OSA Trends in Optics and Photonics, Vol. 34),* Ingeyan, H. ; Keller, U. ; Marshall, C. (Eds.), p. 106, OSA, ISBN 978-1557527929, Washington D. C.

Saikawa, J.; Kurimura, S.; Shoji, I. & Taira, T. (2002). *Tunable frequency-doubled Yb:YAG microchip lasers. Opt. Mater.* Vol. 19 (2002) 169, ISSN0925-3467.

Saikawa, J.; Sato, Y.; Taira, T. & Ikesue, A. (2007). Femtosecond Yb^{3+}-doped $Y_3(Sc_{0.5}A_{l0.5})_2O_{12}$ ceramic laser. *Opt. Mat.* Vol. 29 (2007) 1283-1288, ISSN0925-3467.

Sato, Y. & Taira, T. (2004). Saturation factors of pump absorption in solid-state lasers. *IEEE J. Quantum Electron.* Vol. 40 (2004) 270, ISSN 0018-9197.

Shirakawa, A.; Takaichi, K.; Yagi, H.; Bisson, J-F.; Lu, J.; Musha, M.; Ueda, K.; Yanagitani, T.; Petrov, T. S. & Kaminskii, A. A. (2003). Diode-pumped mode-locked Yb^{3+}:Y_2O_3 ceramic laser. *Opt. Express,* Vol. 11, No. 22 (Nov., 2003) 2911-2916, ISSN1094-4087.

Shoji, I.; Kimura, S.; Sato, Y.; Taira, T.; Ikesue, A. & Yoshida, K. (2000). Optical properties and laser characteristics of highly Nd^{3+}-doped $Y_3Al_5O_{12}$ ceramics. *Appl. Phys. Lett.,* Vol. 77 (2000) 939-941, ISSN 0003-6951.

Shoji, T.; Tokiwa, S.; Kawanaka, J.; Fujita, M. & Izawa, Y. (2004). Quantum-Defect-Limited Operation of Diode-Pumped Yb:YAG Laser at Low Temperature. *Jpn. J. Appl. Phys.,* Vol. 43 (Mar., 2004) L496-L498, ISSN 0021-4922.

Sumida, D. S.; Bruesselbach, H.; Byren, R.W.; Mangir, M. & Reeder, R. (1998). High-power Yb:YAG rod oscillators and amplifiers. *Proc. SPIE* Vol. 3265 (1998) 100, ISSN0277-786X.

Taira, T. (2007). RE3+-Ion-Doped YAG Ceramic Lasers. *IEEE J. Sel. Top. Quantum Electron.,* Vol. 13 (2007) 798-809, ISSN 1077-260X.

Takaichi, K.; Yagi, H.; Lu, J. ; Shirakawa, A.; Ueda, K. & Yanagitani, T. (2003). Yb^{3+}-doped $Y_3Al_5O_{12}$ ceramics – a new solid-state laser material. *Phys. Stat. Sol. A,* Vol. 200 (2003) R5-R8, ISSN 0031-8965.

Tokurakawa, M.; Takaichi, K.; Shirakawa, A.; Ueda, K.; Yagi, H.; Hosokawa, S.; Yanagitani T. & Kaminskii, A. A. (2006). Diode-pumped mode-locked Yb^{3+}:Lu_2O_3 ceramic laser. *Opt. Express,* Vol. 14, No. 26 (Dec., 2006) 12832-12838, ISSN1094-4087.

Tokurakawa, M.; Shirakawa, A.; Ueda, K.; Yagi, H.; Yanagitani, T. & Kaminskii, A. A. (2007). Diode-pumped sub-100 fs Kerr-lens mode-locked Yb^{3+}:Sc_2O_3 ceramic laser. *Opt. Lett.,* Vol. 32, No. 23 (Dec., 2007) 3382-3384, ISSN 0146-9592.

Tokurakawa, M.; Shirakawa, A.; Ueda, K.; Yagi, H.; Hosokawa, S.; Yanagitani, T. & Kaminskii, A. A. (2008). Diode-pumped 65 fs Kerr-lens mode-locked Yb^{3+}:Lu_2O_3 and nondoped Y_2O_3 combined ceramic laser. *Opt. Lett.,* Vol. 33, No. 12 (June, 2008) 1380-1382, ISSN 0146-9592.

Tsunekane, M. & Taira, T. (2005). High-Power Operation of Diode Edge-Pumped, Glue-Bonded, Composite Yb:$Y_3Al_5O_{12}$ Microchip Laser with Ceramic, Undoped YAG Pump Light-Guide. *Jpn. J. Appl. Phys.,* Vol. 44 (Sep., 2005) L1164-L1167, ISSN0021-4922.

Tsunekane, M. & Taira, T. (2006). 300 W continuous-wave operation of a diode edge-pumped, hybrid composite Yb:YAG microchip laser. *Opt. Lett.,* Vol. 31, No. 13 (Jul., 2006) 2003-2005, ISSN 0146-9592.

Tsunekane, M. & Taira, T. (2007). High-power operation of diode edge-pumped, composite all-ceramic Yb:$Y_3Al_5O_{12}$ microchip laser. *Appl. Phys. Lett.,* Vol. 90 (2007) 121101, ISSN 0003-6951.

Uemura, S. & Torizuka, K. (2008). Kerr-Lens Mode-Locked Diode-Pumped Yb:YAG Laser with the Transverse Mode Passively Stabilized. *Jpn. J. Appl. Express,* Vol. 1, No. 1 (Jan. 2008) id. 012007 (3 pages), ISSN0021-4922.

Yoshioka, H.; Nakamura, S.; Ogawa, T. & Wada, S. (2009). Diode-pumped mode-locked Yb:YAG ceramic laser. *Opt. Express,* Vol. 17, No. 11 (Jun., 2009) 8919-8925, ISSN1094-4087.

Polarization Properties of Laser-Diode-Pumped Microchip Nd:YAG Ceramic Lasers

Kenju Otsuka
Tokai University
Japan

1. Introduction

Laser-diode(LD)-pumped solid-state lasers have attracted much attention as compact, highly efficient light sources for various applications. A wide variety of materials have been studied to develop more efficient and high power microchip lasers with LD pumping. In end-pumping schemes, materials with a short absorption length for the LD pump beam are most promising for highly efficient operations because of the excellent match between the mode and pump beam profiles. In the past decade, new single-crystalline laser materials having increased absorption coefficients for LDs, such as Nd:$GdVO_4$ and Yb:YAG, have been reported. On the other hand, the recent advancement of sintering methods led to the development of polycrystalline ceramic laser materials that have transparency comparable to single crystals (Ikesue et al., 1995 a, b). Various ceramic laser materials, e.g., host crystals and active ions, have been sintered toward LD-pumped solid-state lasers. Intensive studies are going on toward high power lasers utilizing such advantages of ceramic lasers as the growth of samples with large arbitrary shapes and high doping of active ions.

From laser physics point of view, on the other hand, ceramic lasers, which consist of randomly distributed single crystals surrounded by grain boundaries, are interesting active materials for studying lasing properties in random media. This Chapter is devoted to the formation of transverse modes and polarization characteristics in Nd:YAG ceramic lasers, focusing on the effect of average grain sizes on lasing properties.

2. Formation of local transverse modes and associated dynamic effects

Systematic studies of Nd:YAG ceramic lasers for different doping levels showed that if we dope Nd^{3+}-ions highly, it is difficult to keep the large grain size and the number of grain boundaries will increase within the same path length accordingly (Ikesue et al., 1995 a, b; Shoji et al., 2000; Lu et al., 2000). In this section, our experimental results concerning lasing pattern formations due to grain boundaries and the resultant dynamic instabilities observed in Nd:YAG ceramic lasers (Kawai et al, 2004; Narita et al., 2005; Otsuka et al., 2006) are reviewed.

2.1 Segregation into local transverse modes and dynamic instability

Thin-slice Nd:YAG ceramic lasers consist of randomly distributed single-crystalline grains, whose directions of crystal axes surrounded by grain boundaries are independent, placed between closely-spaced reflective end surfaces. Since the thermal birefringence (i.e., depolarization) strongly depends on the direction of crystal axes of the grain and thus strong phase distortions occur over transverse directions with increasing the pump-power. Due to resultant random phase disturbances distributed inside the plane-parallel cavity, Hermite-Gauss modes cannot be formed and the lasing pattern is expected to be split into multiple localized transverse modes, namely "local modes" hereafter, having slightly different lasing frequencies resulting from slightly different optical cavity lengths, i.e., standing-wave conditions.

The experiment was carried out using mirror-coated thin-slice Nd:YAG ceramic lasers with various Nd concentrations sintered by Polytechno Corp. to clarify the formation of local modes, in which stable optical resonators are formed by the thermal lens effect due to LD pumping. The experimental setup is shown in Fig. 1(a). All the samples were 5 mm square and 1 mm in thickness, and end surfaces were coated by dielectric mirrors M_1 (99.9% reflection at l064 nm and high transmission at 808 nm) and M_2 (98% reflection at 1064 nm). A nearly collimated elliptical-shape pump beam from the LD with 808-nm wavelength was passed through a pair of anamorphic prisms to transform the elliptical shape into circular one, and then it was focused onto the surface of Nd:YAG ceramic lasers by microscope objective lenses with different magnifications, i.e., M = 10x (numerical aperture, N.A = 0.25) and 20x (N.A. = 0.4). Example scanning electron microscope (SEM) images of the etched surfaces, indicating single-crystalline grain structures, are shown in Fig. 1(b), in which a pump-beam size is depicted. Grain sizes are decreased with increasing Nd concentrations.

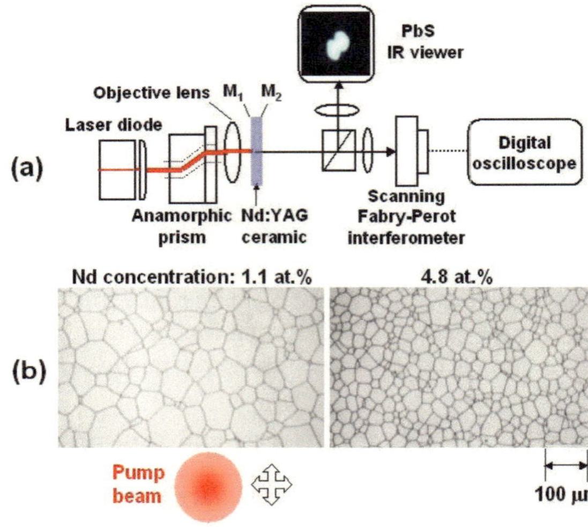

Fig. 1. (a) Experimental setup. (b) SEM image of the etched surfaces of Nd:YAG ceramic. The pump-beam spot size with M = 10x and 20x focusing lens was about 80 and 40 μm.

In order to identify expected splits into spatially-distributed local modes, we measured near-field patterns, corresponding optical spectra and output waveforms. Lasing patterns and the corresponding detailed oscillation spectra were measured using the same PbS infrared viewer (Hamamatsu Photonics, C-1000) and scanning Fabry-Perot interferometer (Burleigh, SA Plus, free-spectral range: 2 GHz; resolution: 6.6 MHz). Here, a single longitudinal-mode operation in the entire pump-power region was confirmed by a multi-wavelength-meter. Example results are shown in Figs. 2. A knife edge placed close to the laser was used to block a part of the lasing beam. Near-field patterns for the whole and partial beams and their optical spectra are shown in Figs. 2 (a)-(b). In this case, two local modes separated by 250 MHz are oscillating at their own frequencies. A high-speed intensity modulation at 250 MHz was observed as shown in a waveform of Fig. 2(c). The corresponding power spectrum peaks at 250 MHz and its harmonic of 500 MHz. Therefore, the periodic waveform is not a simple beat note, but it arises from the intensity modulation at the beat frequency.

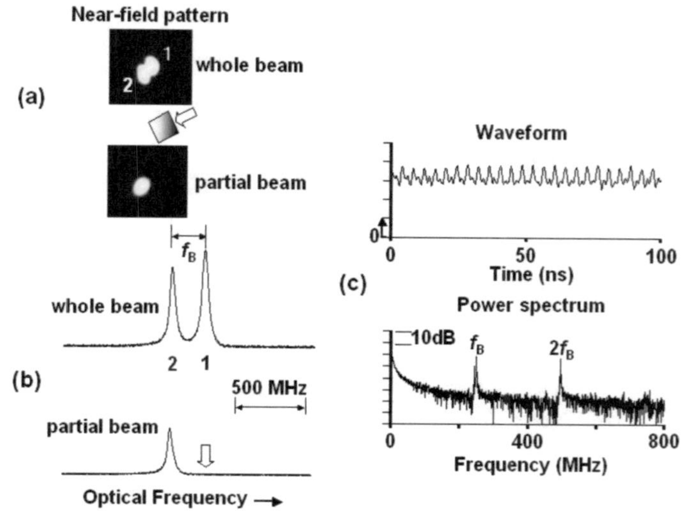

Fig. 2. (a) Near field patterns, (b) optical spectra, (c) intensity waveform and the corresponding power spectrum observed in the 1.1 at.%-doped Nd:YAG ceramic laser.

Various modulation patterns featuring periodic and breathing-type waveforms at beating frequencies of adjacent local modes were observed by slight changes in pump positions. Furthermore, the modulation amplitude was found to be greatly enhanced by a fine tuning of the pump position with a larger magnification objective lens. A variety of high-speed modulation waveforms at frequencies up to 1 GHz, which is much higher than the relaxation oscillation frequency of few MHz, was observed in the present experiment for slightly different pump positions, e.g., several tens of microns, at a fixed pump power. Example high-frequency modulation waveforms and corresponding power spectra observed for different pump-beam (i.e., lasing-beam) diameters observed in a 3.6 at.% doped sample are shown in Fig. 3. Complicated waveforms modulated by many high frequency beats among local modes was observed for a larger pump-beam diameter [(a)], while larger amplitude breathing-type waveforms involving small number of local modes took place for a tightly focused pump-beam [(b)] in which the threshold pump power was 92 mW and the slope efficiency was 20 %. With increasing lasing beam diameter larger by the use of smaller magnification lenses, the number of grain boundaries increases across the lasing beam increases. As a result, the number of interacting local modes, i.e., beat frequencies, increases accordingly, yielding complicated modulation waveforms.

Quasi-periodic and chaotic modulation waveforms around at the relaxation oscillation frequency were also observed as shown in Figs. 4(a)-(b) at the same pump power as Fig. 3, where the pump beam was tightly focused using the 20x (N.A. = 0.4) magnification lens.

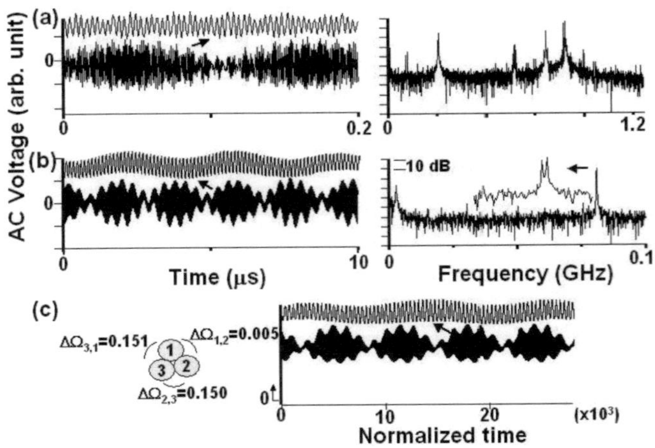

Fig. 3. Intensity waveforms and the corresponding power spectra. (a) High-speed multiple-frequency modulation, (b) Breathing-mode modulation. (c) Numerically reproduced breating-mode modulation.

The simultaneous measurement of optical spectra revealed that these instabilities were confirmed to take place when a beat-frequency among coupled local modes approached the relaxation oscillation frequency. Provided averaged grain size of 20 μm shown in Fig. 1(a), which is smaller than the measured focusing beam diameter of 40 μm, an extremely sensitive change in observed waveforms against a pump position change, i.e., grain structures, can be well understood since a slight change of pump position could results in a change in the spatial arrangement of local modes within the lasing beam, e.g., nearest neighbors, having different lasing frequencies at which beating modulations take place.

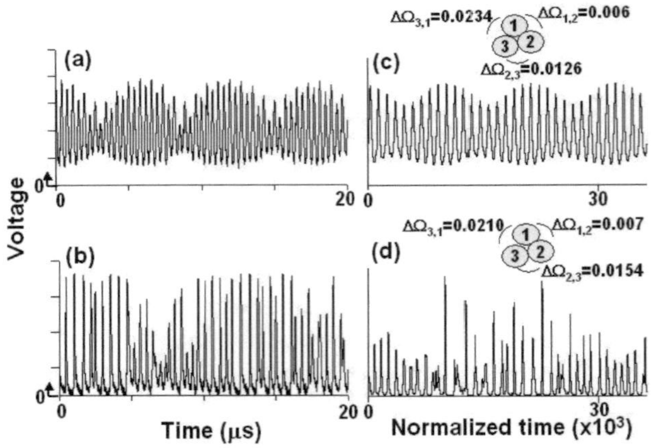

Fig. 4. Observed intensity waveforms (left) and numerical reproductions (right). (a), (c): Quasi-periodic relaxation oscillations, (b), (d): Chaotic relaxation oscillations.

2.2 Theoretical model

Here, the theoretical reproduction of observed dynamic instabilities is shown based on the model of coupled local-mode lasers (Otsuka et al. 2006).

The local modes are coupled through spatial field overlapping across the transverse direction. Assuming the nearest neighbor coupling among spatially-distributed local modes, the model equation of such coupled lasers are given as follows (Otsuka et al., 2000):

$$dN_i/dt = \{w - 1 - N_i - [1 + 2N_i]E_i^2\}/K, \tag{1}$$

$$dE_i/dt = N_iE_i + \eta_{i,j}\Sigma E_j \sin \Phi_{i,j}, \tag{2}$$

$$d\Phi_{i,j}/dt = \Delta\Omega_{i,j} - \eta_{i,j}\, \Sigma[(E_j/E_i) + (E_i/E_j)] \sin \Phi_{i,j}, \tag{3}$$

Here, $E_i = (g\tau)^{1/2}E(t)$ is the normalized field amplitude of i-th local mode, $N_i = gN_{th}\tau_p[N_i\ (t)\ /N_{th}-1]$ is the normalized excess population inversion of i-th local mode where N_{th} is the threshold population inversion. g is the differential gain coefficient, where gain is defined as $G = G_{th} + g(N_i\ (t) - N_{th})$. $w = P\ /P_{th}$ is the relative pump power normalized by the threshold, $K = \kappa\tau$ is the fluorescence lifetime normalized by the damping rate of the optical cavity, $\kappa = (2\tau_p)^{-1}$ (τ_p : photon lifetime), $\Phi_{i,j}$ is the phase difference between i-th local mode field and adjacent j-th local mode field, $\Delta\Omega_{i,j} = (\omega_j - \omega_i)/\kappa$ is the normalized frequency difference between i-th local mode field and adjacent j-th local mode field. $\eta_{i,j}$ is the coupling coefficient between adjacent local mode fields, and t is the time normalized by the damping rate of the optical cavity. In a weak coupling regime, $\Delta\Omega_{i,j} >> \eta_{i,j}$, phase-locking between local modes fails and beating modulations at $\Delta\Omega_{i,j}$ take place. Large K values in thin-slice solid-state lasers with short photon lifetimes guarantee efficient high-speed modulation far above relaxation oscillation frequencies (Otsuka et al., 2002). Numerical high-speed breathing-type modulation, quasi-periodic and chaotic relaxation-oscillation waveforms, ΣE_i^2, which correspond to Figs. 3(b), 4(a) and 4(b), are shown in Figs. 3(c), 4(c) and 4(d), respectively. Here, three coupled localized modes with $\Delta\Omega_{i,j}$ shown in figures are assumed.

2.3 Periodic spiking operations

Periodic spiking operations also appear with a precise tuning of the pump position. Such Q-switching type periodic spiking oscillations, which appeared even in the single-mode region near the lasing threshold, are shown to be brought about by a different dynamic origin from modal-beating mediate intensity modulations shown in Figs. 3-4. An example bifurcation diagram and typical waveforms obtained at a fixed pump position is shown in Fig. 5. The periodic spiking oscillation in the single-mode region **PS** [(b)] tended to coexist with high-speed beating modulations in two-mode region, featuring random switching between two dynamic states **RS** [(c)], and finally high-speed modulation dominated spiking oscillations as the pump power approached the threshold pump power in region **HS** [(d)]. Note that the spiking oscillation frequency decreased with in decreasing the pump power.

Self-induced periodic "spiking" modulations observed for a tight pump-beam focusing condition have been explained theoretically assuming saturable absorber type of inclusions in grain boundaries. The standard single-mode laser rate equation including saturable absorbers are given by: (Erneux, 1990)

$$dn_l/dt = (w_l - n_l - n_l s_l)/K, \tag{4}$$

$$dn_a/dt = (w_a - n_a - \alpha n_a s)/K, \tag{5}$$

$$ds/dt = 2(n_l - n_a - 1)s, \tag{6}$$

where n_l, n_a are normalized population densities for the laser and saturable absorber, respectively, s is the photon density, w_l, w_a are normalized pump rates and α is the saturation parameter. Observed periodic spiking oscillations, which appeared near the lasing threshold $w = P/P_{th} = w_l/(w_a + 1) < 2$, have been well explained by the linear stability analysis and numerical simulation of single-mode nonlinear equations described above. An example numerical result is shown in Fig. 6. It was found theoretically that the periodic spiking oscillation starts to appear just above the threshold pump power, i.e., $w = 1$, and the pulsation frequency increases as the pump power increases. In the real experiment, however, such a pulsation sometimes coexists with beating-type high-speed modulations shown in Fig. 5(d). Observed spiking oscillations and theoretical reproductions strongly suggest the possible inclusion of saturable absorber-type defects into grain boundaries in highly-doped Nd:YAG ceramics.

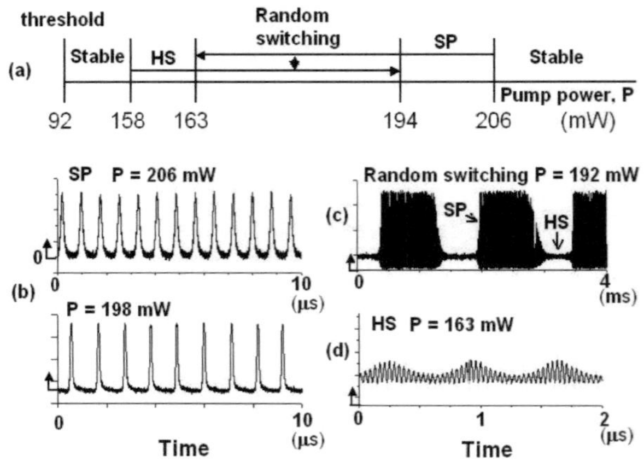

Fig. 5. (a) Bifurcation diagram as a function of the pump power observed at the fixed pump position. **HS**: high-speed modulation, **SP**: periodic spiking oscillation. (b) Observed Q-switching-type of spiking oscillation. (c) Observed random switching among SP and HS. (d) Observed high-speed modulation.

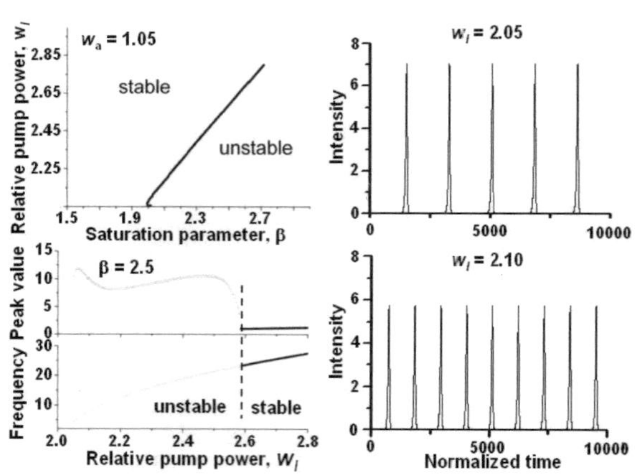

Fig. 6. Linear stability analysis and pulsation waveforms.

Random switching behaviors among different dynamic states similar to Fig. 5(c) have been reported and interpreted in terms of chaotic itinerancy (Otsuka, 1999; Ko et al., 2001).

2.4 Summary

In this section, self-induced dynamic instabilities, featuring modal-beating mediate high speed modulations, large-amplitude quasi-periodic and chaotic relaxation oscillations, have been demonstrated experimentally using thin-slice Nd:YAG ceramic lasers with coated end mirrors. The theoretical model of coupled local-mode lasers including the effect of the lateral field overlapping, i.e., interference, among adjacent local modes has been presented for understanding dynamic instabilities. With a generalized model of coupled local-modes (i.e., filament) lasers proposed here, a variety of modal-beating mediate dynamic instabilities have been well reproduced. Self-induced Q-switching type of periodic spiking modulations observed for a tight pump-beam focusing condition have been also observed and explained theoretically assuming saturable-absorber type of inclusions in grain boundaries.

3. Effect of average grain sizes on modal and polarization properties

The inherent segregation of transverse patterns into multiple local modes possessing different lasing profiles occurs in thin-slice Nd:YAG ceramic lasers, whose grain sizes are several tens of micrometers as described in the previous section. On the other hand, single-frequency linearly polarized emissions that were free from dynamic instabilities were achieved in microcrystalline Nd:YAG ceramic samples, whose average grain size was a few micrometers. In this section, we describe the effect of average grain size on transverse lasing patterns, detailed oscillation spectra, and polarization properties of LD-pumped Nd:YAG ceramic lasers (Ohtomo, 2007) to address the problems associated with making highly coherent miniature ceramic lasers possessing high beam quality and lasing frequency purity for applications besides high-power operation, e.g., laser metrology (Sudo et al. 2006).

3.1 Input-output characteristics

The experimental setup is shown in Fig. 7(a), in which an external cavity is used instead of mirror-coated thin-slice Fabry-Perot laser in the previous section to verify the effect of grain sizes on lasing pattern formations in the optical cavity, which should support a well-defined TEM_{00} mode. The experiment was carried out using two types of 1-mm-thick Nd:YAG ceramic sample having different average grain sizes, namely **A** and **B,** sintered by Asaka Riken Co., Ltd. and Konoshima Chemical Co., Ltd., respectively, where Nd concentrations are 1.2 at.% for **A** and 1.0 at.% for **B**. A linearly polarized collimated elliptical LD beam of 808 nm wavelength was transformed into a circular one and focused onto the ceramic samples using a microscope objective lens with a numerical aperture of 0.25. The pump-beam spot size averaged over the 1-mm-thick sample length was determined to be 80 μm, where the absorption coefficient for the LD light was 3.55 and 2.96 cm^{-1} for A and B samples, respectively. The sample was placed within a 5-mm-long semi-confocal cavity consisting of a flat mirror M_1 (99.8% reflective at 1064 nm and > 95% transmissive at 808 nm) and a concave mirror M_2 (99% reflective at 1064 nm) of 1-cm radius of curvature. We have used several ceramic samples belonging to the same lot of A or B. Typical example SEM images of etched surfaces and histograms of grain sizes are shown in Fig. 8.

Distinct differences in threshold pump powers and slope efficiencies were not found for both samples. The input-output characteristics of samples belonging to A were found to depend on samples and the pumped (i.e., lasing) beam position across the cross-sectional area of each sample, reflecting the grain structure within the lasing beam. While in samples belonging to B, such dependences were small. Typical examples are shown in Fig. 7(b).

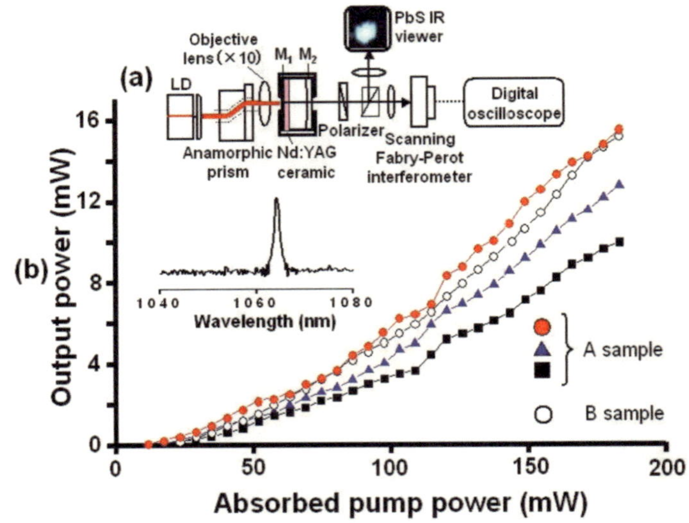

Fig. 7. (a) Experimental setup. (b) Input-output characteristics of large- and micro-grained ceramic lasers. A: large grain, B: micro-grain.

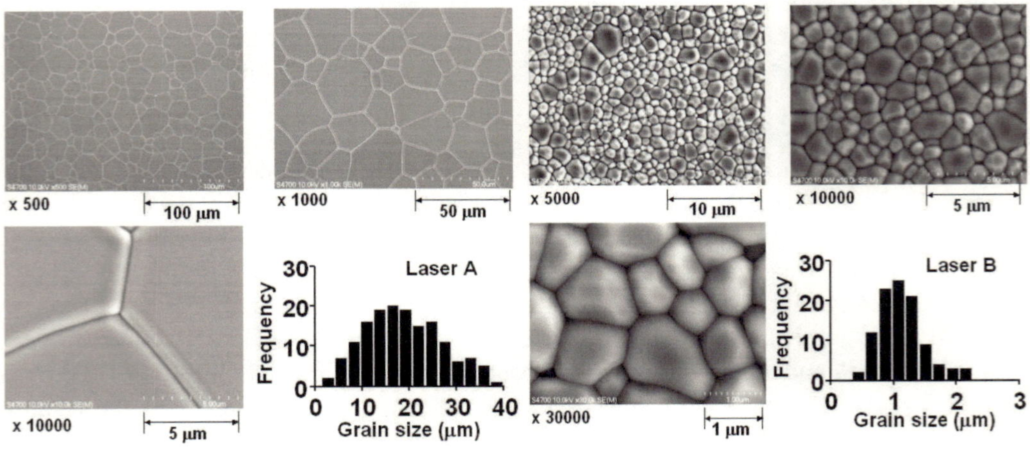

Fig. 8. SEM images of etched surfaces for several magnifications and measured histograms of grain sizes. Left: large grains, Right: micro-grains.

Larger undulations in output curves in samples A are considered to appear through structural changes of lasing patterns due to pump-dependent local-mode formations as will be discussed later. In all lasers, single-transition oscillations on the 1064-nm transition line were confirmed for the entire pump-power region by measurements using a multi-wavelength meter (resolution: 0.1 nm), as depicted in the inset of Fig. 7. This may result from the mode selection due to the etalon effect of the 1-mm-thick sample whose free-spectral range of $\Delta f = c/2nL = 56$ GHz (c: velocity of light, n = 1.82: refractive index, L = 1 mm: sample length) is comparable to the gain bandwidth (full width at half maximum) of Nd:YAG ceramic.

3.2 Lasing patterns and polarization states

Lasing patterns and the corresponding detailed oscillation spectra were measured using the same PbS infrared viewer and scanning Fabry-Perot interferometer as those in Fig. 1 after the output beam had passed through a polarizer. Example results for the two lasers are shown in Fig. 9, where grain structures (i.e., SEM images of thermally etched surfaces), near-field patterns and lasing optical spectra are also indicated.

The formation of local modes occurred in laser A just above the threshold pump power and the number of local modes increased with increasing pump power (i.e., transverse gain region). In the case of Fig. 9(a), a transverse mode belonging to the particular longitudinal mode was segregated into 4 local modes, possessing slightly different frequencies and polarization states. Note that the lasing beam consists of linearly polarized separated local modes with different polarization directions indicated by arrows. Laser A consists of single-crystal grains of several tens of micrometers in size whose crystal axes are randomly distributed, while the spot size of the cavity eigenmode averaged over the sample length was calculated to be 80 μm, in which the average grain size was determined to be 19.22 μm for the sample shown in Fig. 9(a). Therefore, several single-crystalline grains, whose sizes are orders of magnitude larger than the lasing wavelength, are included within a cavity eigenmode cross section on average, and the oscillation frequency and eigenpolarization state of the lasing beam cannot be uniquely determined because of position-dependent thermal birefringence across the beam. It thus appears that the lasing transverse mode should be split into local modes having different frequencies and polarization states, exhibiting dynamic instabilities as described in section **2**.

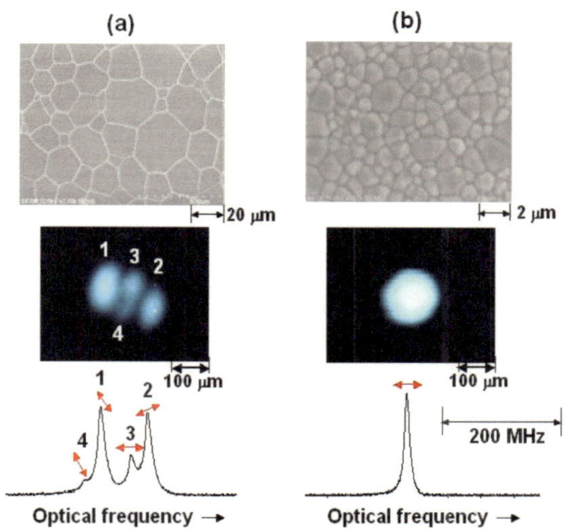

Fig. 9. SEM images of etched Nd:YAG ceramic surfaces, near-field patterns and optical spectra. (a) Large grain sample **A**, (b) micro-grained sample **B**. Polarization directions are indicated by red arrows.

In the case of laser B, on the other hand, linearly polarized oscillations were obtained in a single TEM_{00} transverse mode (i.e., Hermite-Gauss HG_{00} mode), as shown in Fig. 9(b). In this case, single-crystalline grain sizes are on the order of the lasing wavelength [i.e., average grain size of 1.16 μm for the sample shown in Fig. 9(b)], and local modes cannot be formed within the eigenmode (i.e., TEM_{00} mode) spot size. Consequently, segregations into local modes can be substantially suppressed, leading to stable single-mode operations without dynamic instabilities. An example far-field intensity profile of laser B is shown in Fig. 10, together with a Gaussian fitting curve, where a laser-beam profiler (Coherent, Beam Master-3Si) was used. The mode purity, which is the degree to which the intensity pattern reproduces the theoretical TEM_{00} mode (Arlt, 1998) was evaluated to be greater than 99% for the entire pump-power region.

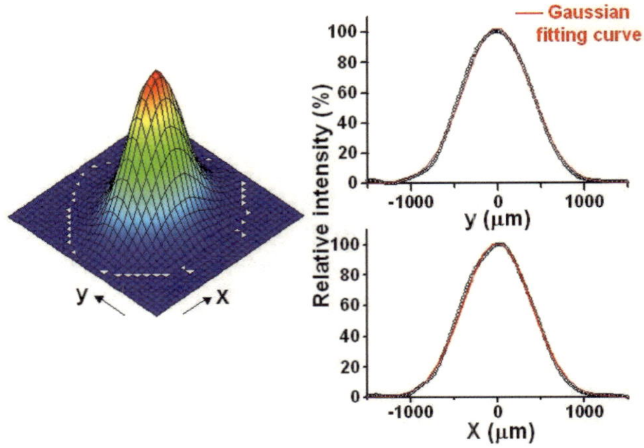

Fig. 10. Typical far-field lasing pattern of the micro-grained Nd:YAG ceramic laser.

Example pump-dependent optical spectra for other ceramic samples are shown in Fig. 11. Linearly polarized single-frequency oscillations, whose polarization direction coincided with that of the LD pump light, were obtained in laser B. In the high pump-power regime, the orthogonal linearly-polarized weak mode appeared as indicated by the double downarrow in this sample, featuring weak dual-polarization oscillations.

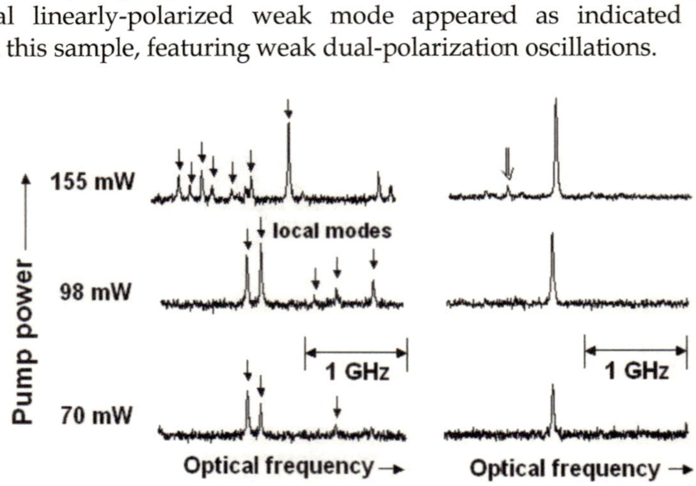

Fig. 11. Pump-dependent optical spectra. Left: large grains, Right: micro-grains

We tested 40 samples including Yb:YAG with different average grain sizes purchased from different suppliers and observed the formation of local modes in all the samples except for micro-grained samples with average grain sizes smaller than approximately 5 μm. The results are summarized in Table I. Stable linearly-polarized TEM_{00} mode operations without dynamic instabilities were observed in a mirror-coated 5 at.%-doped 2-mm-thick Yb:YAG micro-grained ceramic laser with average grain sizes of 3.2 μm (Ohtomo, 2007) and its application to the self-mixing laser metrology with enhanced optical sensitivity has been demonstrated (Ohtomo, 2009).

Doping (at. %)	1.1 (Nd)	2.4 (Nd)	4.8 (Nd)	1.2 (Nd)	4.8 (Nd)	5.0 (Yb)	1.0 (Nd)
Grain size (μm)	51.85	37.35	19.03	19.22	5.61	3.20	1.16
Lasing mode	LMs	LMs	LMs	LMs	LMs	TEM_{00}	TEM_{00}

Table I. Average grain size vs modal structure observed in various ceramic YAG samples purchased from different suppliers. LMs: local modes. The average grain size was determined by averaging the length along the longest direction of each grain.

3.3 Discussions

Let us discuss the polarization properties observed in LD-pumped miniature ceramic lasers, in comparison with those in single-crystalline lasers. In LD-pumped thin-slice single-crystalline Nd:YAG lasers, we observed linear-polarization oscillations in the low pump-power region, and dual-polarization oscillations (DPO) appeared in the wide pump-power region. A typical input-output characteristic observed in a 3-mm-diameter, 5-mm-thick mirror-coated Nd:YAG single-crystalline laser with LD pumping is shown in Fig. 12. Here, mirror coatings are the same as those shown in Fig. 7(a). The threshold pump power is larger than those of ceramic lasers shown in Fig. 7(b) because of the increased lasing beam spot size of 200 μm (i.e., mode volume). In a wide pump-power region, DPO is found to occur. The nonlinear complementary output characteristic was theoretically reproduced by the cross saturation of population inversion among orthogonally polarized modes (Otsuka et al., 2007). It was found that such polarization characteristics critically depend on the slight asymmetry in pumping geometry, i.e., a slight change in pump position with respect to the central axis of the Nd:YAG crystal. Such pump-dependent polarization has also been reported by Cabrera et al., 2005.

The intriguing question is why do linear or dual-polarization emissions exceed the unpolarized emissions expected in single-crystalline Nd:YAG with the isotropic fluorescence properties. A plausible physical origin might be attributed to the thermal birefringence inherent in Nd:YAG crystals, in which depolarization effect is minimum along the orthogonal directions across the center of the pump (i.e., lasing) beam stemming from the so-called *four-leaf rose* depolarization pattern (Koechner, 1970; Otsuka, 1971) as depicted in the left of Fig. 12. In short, in LD-pumped miniature laser configurations with small pump-beam (i.e., lasing) spot sizes like the present lasers, the central region of the lasing beam does not suffer from the depolarization effect in the case of orthogonal linearly polarized emissions. Therefore, polarization eigendirections are imposed by such a thermally induced crystal birefringence, and DPOs are considered to exceed unpolarized emissions, as expected in the case of small gain anisotropy.

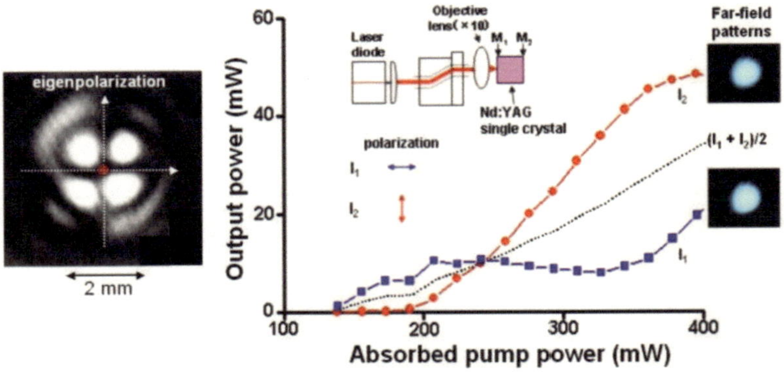

Fig. 12. Dual-polarization oscillation in the Nd:YAG single crystalline laser.

In micro-grained ceramic lasers, on the other hand, the averaging of thermal birefringence effect in Nd:YAG ceramics due to random distributions of crystal axes of grains reported by Shoji et al., 2002 is considered to become significant because of the extremely small grain sizes comparable to the lasing wavelength. Furthermore, in general, the thermally induced internal strain can be relaxed at grain boundaries in ceramics consisting of grains whose crystal axes are randomly distributed. With decreasing grain size, the suppression of thermal birefringence is expected to be more pronounced owing to increased grain boundaries. Indeed, depolarization patterns were observed to be substantially smoothed out in micro-grained ceramics. Therefore, micro-grained ceramic lasers are considered to act like highly isotropic lasers, being free from the thermal birefringence effect. Note that linearly polarized emissions predominate and DPO appears with increasing pump power in low-power LD end-pumped Nd-doped fiber lasers in the absence of thermal birefringence (Bielawski, 1992). Such a lasing polarization property has been well explained theoretically in terms of the polarization-selective gain, namely *pumping anisotropy*, in which the polarization-selective gain occurs because both the interactions between the polarized pump and the ions, and that between the ion and the fields along each direction of polarization depend on the local field experienced by the ion (Bielawski, 1992; Kravtsov et al., 2004). The observed polarization properties, which follow the pump polarization, in micro-grained ceramic lasers could be interpreted in terms of the pumping anisotropy in isotropic lasing media. This topic will be discussed later again in **4.2.**

4. Controlling modal and polarization properties

This section is devoted to the control of transverse lasing patterns and their polarization properties of Nd:YAG ceramic lasers by physical manipulations toward stable single-frequency linearly polarized emissions for various applications besides high-power applications.

4.1 Linearly-polarized Ince-Gauss mode operations with off-axis LD pumping

Here, let me show the unique method to prevent local-mode oscillations and achieve stable operations without dynamic instabilities, which has been realized by adjusting azimuthal cavity symmetry, in which forced Ince-Gauss (IG) mode operations take place in linearly-polarized single-frequency oscillations (Otsuka, 2007).

The experimental setup is the same as Fig. 7(a), where the two mirrors and the 1-mm-thick Nd:YAG ceramic sample of 1.2 at.% were assembled into one body. In the present experiment, we used larger grain samples categorized into **A** in the previous section sintered by Asaka Riken Co., Ltd. which exhibited segregations into local transverse modes and associate dynamic instabilities in the usual LD pumping scheme, i.e., on-axis pumping.

Far-field patterns and the corresponding detailed oscillation spectra were measured with the PbS infrared viewer and the scanning Fabry-Perot interferometer (free-spectral range: 2 GHz, resolution: 6.6 MHz), after the output beam had passed through a polarizer similar to Fig. 7(a). Examples of the results for different rotation angles of the polarizer obtained in the low pump-power region are shown in Fig. 13. It is obvious that a transverse mode belonging to the particular longitudinal mode was segregated into closely-spaced three local modes, **a**, **b**, and **c**, possessing slightly different frequencies and polarization states.

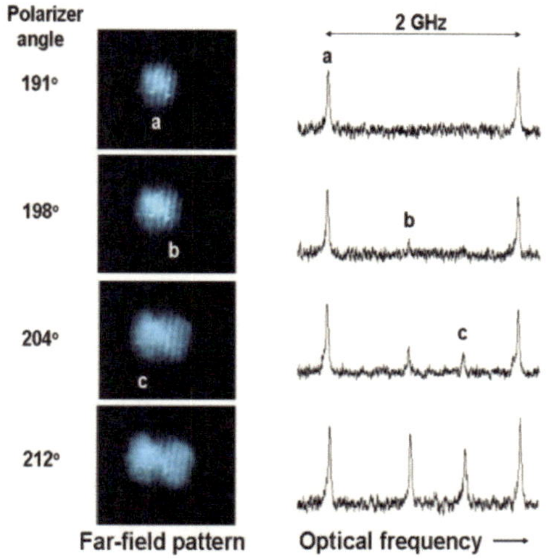

Fig. 13. Polarization-dependent far-field pattern change and the corresponding optical spectra.

As the pump power was increased, the number of local modes tended to increase due to the increase in active area in the transverse direction. The total and polarization-dependent far-field patterns are shown in Fig. 14, where polarization directions of individual local modes are indicated by arrows. Note that the lasing beam consists of linearly polarized separated 'clusters' of local modes with different polarization directions in this case. Dynamic instabilities resulting from beating-type modulations through field coupling among local modes described in section **2** were also observed in this particular sample.

Now let me show a simple method to achieve stable operations of Nd:YAG ceramic lasers to prevent segregations into local modes possessing different frequencies and polarization states by selective excitations of higher-order Ince-Gauss modes, whose mode volume can be larger enough than the grain size, with a controlled azimuthal cavity symmetry. Besides well-known Hermite-Gauss modes (HGMs) and Laguerre-Gauss modes (LGMs), Ince-Gauss modes (IGMs) is the third complete family of transverse eigenmodes. Ince-Gauss modes (IGMs) has been predicted as orthogonal solutions of the paraxial wave equation in elliptic

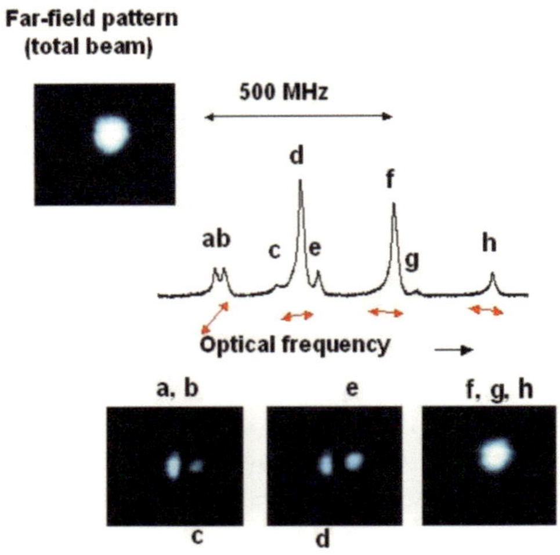

Fig. 14. Far-field patterns of total and partial beams, and their polarization states observed in the high pump-power regime.

coordinates and the IGMs with mode numbers p and m, and ellipticity ε are given by (M. Bandres & Gutierrez-Vega, 2004).

$$IG^{e}_{p,m}(\mathbf{r}, \varepsilon) = C[w_0/w(z)]C_p^m(i\xi,\varepsilon)C_p^m(\xi, \varepsilon)\exp[-r^2/w^2(z)]$$

$$\times \exp i\,[kz + \{kr^2/2R(z)\} - (p + 1)\psi_z(z)], \tag{7}$$

$$IG^{o}_{p,m}(\mathbf{r}, \varepsilon) = S[w_0/w(z)]S_p^m(i\xi,\varepsilon)S_p^m(\xi, \varepsilon)\exp[-r^2/w^2(z)]$$

$$\times \exp i\,[kz + \{kr^2/2R(z)\} - (p + 1)\,\psi_z(z)], \tag{8}$$

where the elliptic coordinate is defined in a transverse z plane as $x = f(z)\cos\xi\cos\eta$, $y = f(z)\sin\xi\sin\eta$ and $\xi \in [0, \infty]$, $\eta \in [0, 2\pi]$. $f(z)$ is the semifocal separation of IGMs defined as the Gaussian beam width, i.e., $f(z)= f_0 w(z)/w_0$, where f_0 and w_0 are the semifocal separation and beam width at the z=0 plane, respectively. $w(z)= w_0\,(1+ z^2/z_R^2)^{1/2}$ describes the beam width, $z_R = kw_0^2/2$ is the Rayleigh length, and k is the wave number of the beam. The terms C and S are normalization constants, and subscripts e and o refer to even and odd IGMs, respectively. C_p^m (. , ε) and S_p^m (. , ε) are the even and odd Ince polynomials of order p, degree m, and ellipticity parameter ε, respectively. In Eqs. (7) and (8), r is the radial distance from the central axis of the cavity, $R(z)=z+z_R^2/z$ is the radius of curvature of the phase front, and $\Psi_z(z) = \arctan(z/z_R)$. The parameters of ellipticity ε, waist w_0 and the semifocal seperation f_0 are not independent, but related by $\varepsilon = f_0^2/w_0^2$.

By adjusting azimuthal symmetry, a variety of higher-order IG mode oscillations, $IG_{p,m}$, were selectively excited, where the central axis of the resonator was tilted against the pump-beam axis as shown in Fig. 15(a) in the range of $0 < \theta < 30$ [mrad]. Such a tilt of the laser resonator assembled in one body against the pump-beam axis is considered to introduce an equivalent effect of the off-axis pumping (Schwarz et al., 2004) whose lateral shift ranges 0 - 150 μm from the center of the cavity. Example far-field lasing patterns observed for different azimuthal symmetries at a constant pump power of 293 mW are shown in Fig. 15(b), in

which each output power was decreased to 5 - 7 mW from the output power of 10 mW in local-mode operations with on-axis pumping. The formation of IG modes by adjusting the LD pump-beam focus to the brightest 'target' spot of IGM lasing patterns, i.e., off-axis pumping, has been demonstrated by the numerical simulation (Chu & Otsuka, 2007).
As for forced single IG mode oscillations, linearly-polarized single-frequency operations being free from dynamic instabilities have been achieved successfully, although the threshold pump power was increased to 80 - 100 mW depending on the tilt, i.e., astigmatic pumping, while higher-order HG modes were formed by a larger astigmatic pumping with further increased pump power. Linearly-polarized single-frequency operations were confirmed by optical spectra as shown in Fig. 15.

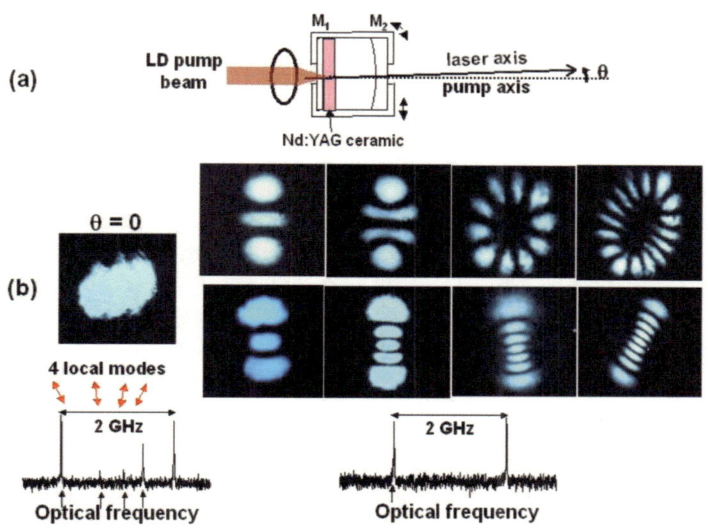

Fig. 15. (a) Azimuth laser-diode pumping scheme. (b) Far-field lasing patterns and the corresponding optical spectra. Left: On-axis pumping (θ = 0); Right: off-axis pumping.

We have tested various 1-mm-thick Nd:YAG ceramic samples with 1.2 at.%, 2.4 at.% and 3.6 at.% Nd concentrations in the same cavity configuration and forced single-frequency linearly-polarized IG mode operations were attained by adjusting azimuthal symmetry of the cavity against the LD-beam pump direction. Moreover, it is interesting to note that each single-frequency IG mode formed in the low pump-power region was maintained with increasing the pump power.
Single-mode IG patterns were formed stably in microchip Nd:YAG ceramic lasers with LD pump levels below 1 W presumably due to the less thermal lens effect. Analytical IG mode patterns with mode numbers [*p*,*m*] and ellipticities ε of the modes corresponding to observed patterns in Fig. 15 are shown in Fig. 16.
The mechanism here is that split of the fundamental TEM_{00} mode into local modes due to grain boundaries under the mode-matched pumping condition is suppressed by the selective excitation of higher-order IG modes whose mode volumes are much larger than the averaged grain size under astigmatic pump. In other words, distributed thermal birefringence effects are effectively smoothed out in IG mode operations, resulting in single-frequency linearly-polarized emissions similar to micro-grained Nd:YAG ceramic lasers with on-axis LD pumping demonstrated in section **3**.

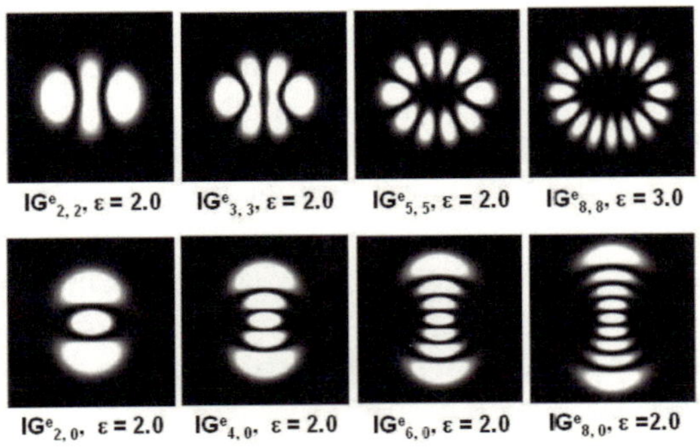

Fig. 16. Numerically reproduced IG mode patterns corresponding to observed IG mode patterns shown in Fig. 15(b).

4.2 Observation of Mathieu-Gauss mode operations in micro-grained ceramic lasers

In this section, the lasing pattern formation in micro-grained Nd:YAG ceramic lasers with azimuth LD pumping is discussed. Spontaneous Mathieu-Gauss mode oscillations are shown to appear instead of IG mode operations, which take place in larger grained Nd:YAG microchip ceramic lasers with azimuth laser diode pumping (Tokunaga et al., 2009). Here, non-diffractive beams (Durin, 1987), including Bessel-Gauss (BG) (Arlt & Dholakia, 2000) and Mathieu-Gauss (MG) modes (Gutierrez-Vega & Bandres, 2007), are promising for sophisticated micromachining and for trapping microparticles and manipulating them over longer distances than any other optical tweezers. The effect of fluorescence anisotropy or thermal birefringence on these pattern formations is also discussed for several laser materials.

A. Mathieu-Gauss and Bessel-Gauss mode operation

The laser cavity configuration is the same as Fig. 7(a) and the 1-mm-thick micro-grained Nd:YAG ceramic of 1 at.% Nd concentration was attached to M_1 and the cavity was assembled into one body as depicted in Fig. 17(a). The averaged grain size was measured to

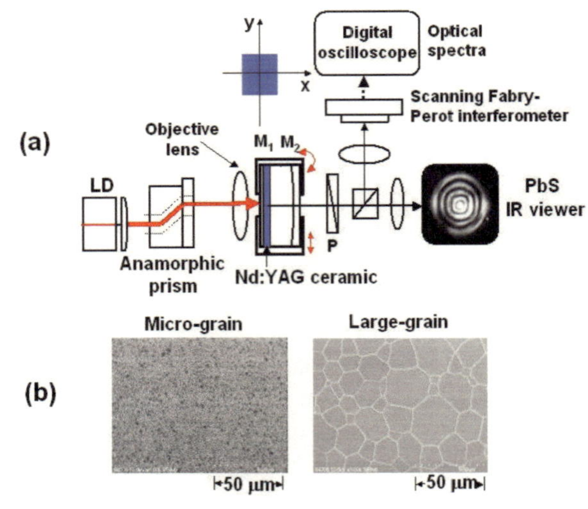

Fig. 17. (a) Experimental setup of micro-grained Nd:YAG ceramic lasers with azimuth LD pumping. (b) SEM images of etched surfaces.

be 1.1 μm. An SEM surface image of the micro-grained sample used in this experiment is shown in Fig. 17(b), together with that of large-grain sample. The collimated linearly-polarized LD beam was passed through an anamorphic prism pair and it was focused onto the sample by a microscope objective lens of NA = 0.25, where the focused beam diameter was about 80 μm. The laser exhibited a single-frequency TEM_{00}-mode oscillation, which is linearly polarized along the LD pump-beam polarization direction due to the reduced thermal birefringence for mode-matched on-axis pumping condition as mentioned in section **3**. By shifting or tilting the laser cavity slightly as depicted by arrows in Fig. 17(a), a variety of MG mode operations were observed, instead of Ince-Gauss (IG) modes, depending on the degree of effective off-axis pumping. Typical far-field patterns, including BG modes, are shown in Fig. 18. For the higher-order BG modes (BG_1, BG_2), an optical vortex having a topological charge of 1 and 2 was formed in the center.

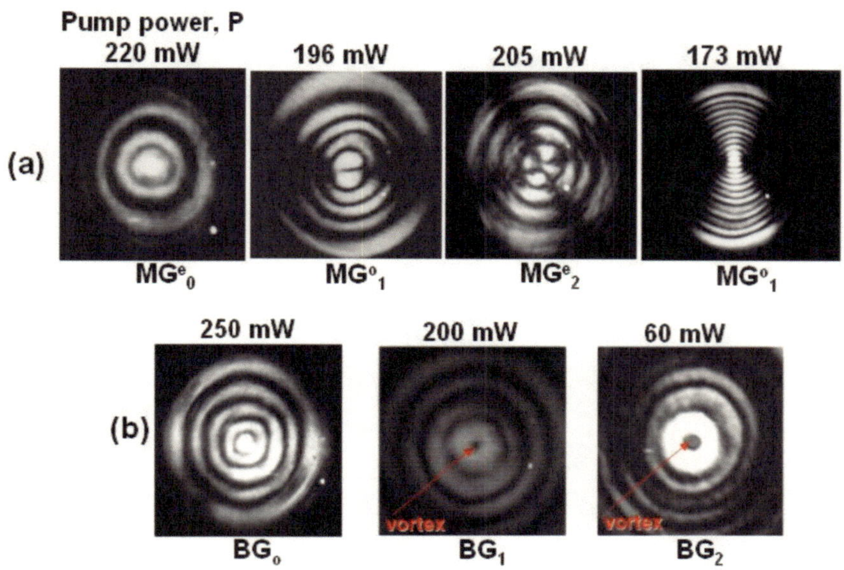

Fig. 18. Observed far-field lasing patterns. (a) Mathieu-Gauss laser beam. (b) Bessel-Gauss laser beam.

Numerically reproduced intensity patterns corresponding to Fig. 18 and the phase portraits are shown in Fig. 19. Here, the complex amplitude of the *m*-th order even and odd MG beams propagating along the positive *z* of an elliptic coordinate system $\boldsymbol{r}$ = (ξ, η, *z*) is given by (Gutierrez-Vega & Bandres, 2007):

$$MG^e{}_m(r) = \exp(-\frac{k_t{}^2 z}{2k\mu}) GB(r) Je_m(\xi, q) ce_m(\eta, q) \,, \tag{9}$$

$$MG^o{}_m(r) = \exp(-\frac{k_t{}^2 z}{2k\mu}) GB(r) Jo_m(\xi, q) se_m(\eta, q) \,. \tag{10}$$

Here, $Je_m(\cdot)$ and $Jo_m(\cdot)$ are the *m*-th order even and odd radial Mathieu functions, $ce_m(\cdot)$ and $se_m(\cdot)$ are the *m*-th order even and odd angular Mathieu functions, $GB(\boldsymbol{r}) = \mu^{-1}\exp(-r^2/\mu w_0{}^2)$ is

the fundamental Gaussian beam, $\mu(z) = 1 + iz/(kw_0^2)$, w_0 is the Gaussian width at the waist plane $z = 0$, and $k = 2\pi/\lambda$ is the longitudinal wave number. $q = k_t^2 f_0^2/4$ is the ellipticity parameter, which carries information about the transverse wave number k_t and the semiconfocal separation at the waist plane f_0.
Similarly, m-th order BG beams are given by

$$BG_m(r,\phi) = \exp\left(-\frac{k_t^2 z}{2k\mu}\right) GB(r) J_m\left(\frac{k_t r}{\mu}\right) \exp(-im\phi) . \tag{11}$$

Here, (r,ϕ) are the polar coordinates and $J_m(\cdot)$ is the m-th order Bessel function of the first kind.

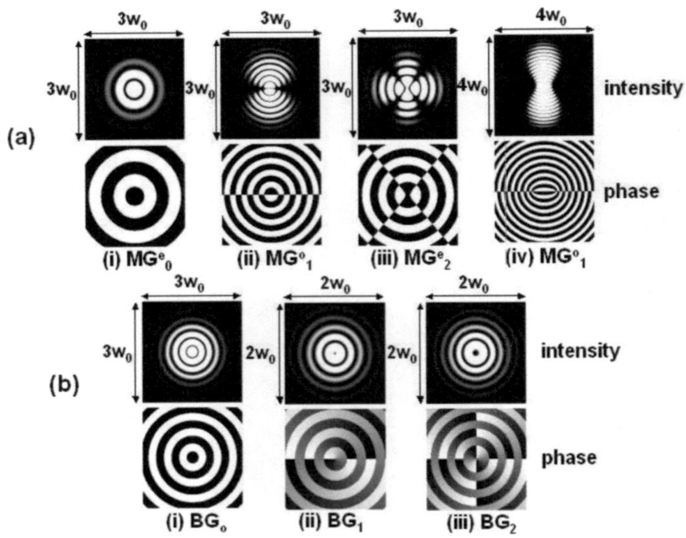

Fig. 19. Numerically reproduced intensity patterns corresponding to Fig. 18 and their phase portraits. (a) Mathieu-Gauss laser beam. (b) Bessel-Gauss laser beam. λ (wavelength) = 1064 nm, w_0 = 3 mm. Adopted parameter values (k_t, q) are (a)-(i): (2800/m, 0.2); (a)-(ii): (6000/m, 0.2); (a)-(iii): (4300/m, 0.5); (a)-(iv): (7500/m, 25); (b)-(i): (4500/m, 0); (b)-(ii): (5500/m, 0); (b)-(iii): (6500/m, 0).

Elliptical-polarization BG modes or dual-polarization MG modes appeared for small effective off-axis pumping. An example of polarization-dependent oscillation spectra is shown in Fig. 20(a). With larger off-axis pumping, linearly polarized single or double longitudinal MG mode operations were observed, where the longitudinal mode spacing coincided with 12.88 GHz, which corresponds to the inverse of two round-trip times as expected for BG and MG mode oscillations. An example oscillation spectrum consisting of two longitudinal modes is shown in Fig. 20(b).

B. Effect of fluorescence anisotropy on lasing pattern formation

We replaced the micro-grained Nd:YAG ceramic by $LiNdP_4O_{12}$ (LNP) and a-cut $Nd:GdVO_4$ crystals, which exhibit linearly polarized emission resulting from strong fluorescence anisotropy independently of the pump-beam polarization state. Under the same azimuth LD-pumping conditions as for micro-grained ceramic lasers, neither BG nor MG mode oscillations appeared. Instead, single-frequency linearly polarized IG mode operations on

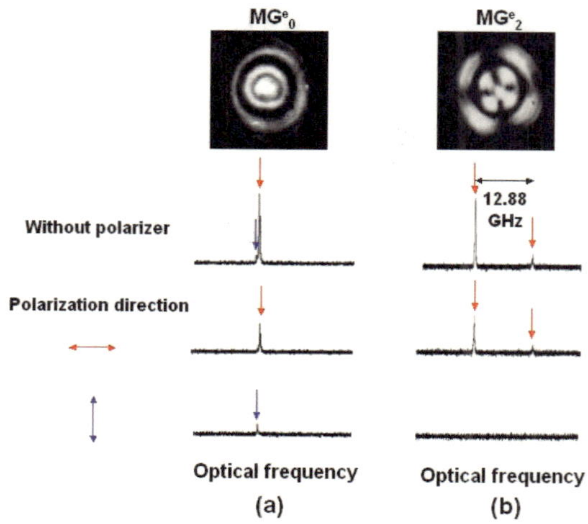

Fig. 20. Far-field lasing patterns and their polarization-dependent optical spectra.
(a) Dual-polarization Mathieu-Gauss beam with small off-axis pumping.
(b) Linear-polarization multi-longitudinal mode Mathieu-Gauss beam with large off-axis pumping.

elliptical coordinates were observed depending on the pump-beam position (Ohtomo et al., 2007), similar to large grain Nd:YAG ceramic lasers with spatially dependent thermal birefringence discussed in the previous subsection **4.1**. Examples are shown in Fig. 21. As for large-grain Nd:YAG ceramic lasers, neither BG nor MG mode oscillations appeared with azimuth LD pumping.

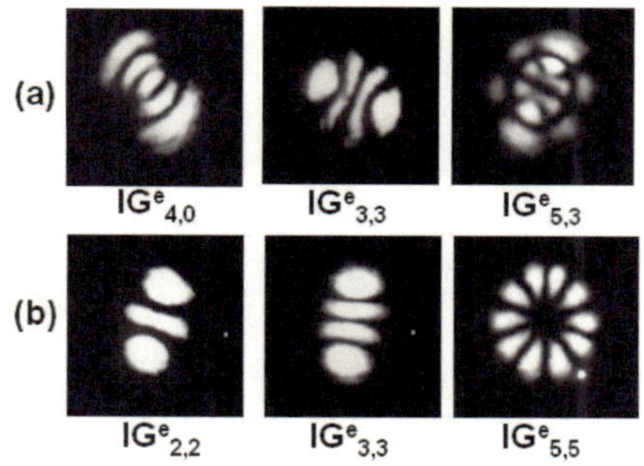

Fig. 21. Ince-Gauss mode operations with azimuth LD pumping. (a) $Nd:GdVO_4$ single crystal. (b) Large-grain Nd:YAG ceramic with average grain size of 19.2 μm.

C. Discussion

Laser oscillations in BG and MG modes are usually obtained in cavities with an axicon-type lens or mirror (Gutierrez-Vega, 2003; Alvarez-Elizondo, 2008) such that interference between conical lasing fields occurs within the laser cavity. In the present experiment, BG and MG mode oscillations were produced just by azimuth LD pumping. Let us offer a plausible explanation for MG mode oscillations in terms of effective off-axis pumping depicted in Fig. 22(a).

In the framework of vector lasers (Kravtsov, 2004), the angular amplification inhomogeneity has been shown to depend on the orientation of the polarization plane of laser radiation from that of pump radiation, in the form of $D(\theta, \Psi) = 2A_0\cos^2(\theta - \Psi)$ as depicted in Fig. 22(b),

and the polarization state is almost completely determined by the polarization of the pump radiation for an isotropic cavity with micro-grained Nd:YAG ceramic as described in section **3** (Ohtomo, 2007; Otsuka 2008). For azimuth LD pumping, the laser emission tends to occur such that its polarization direction follows the LD polarization direction within the pumped area.
Let us assume a small reflection loss difference at uncoated surfaces of the thermal lens between polarizations along radial and azimuth directions as depicted in Fig. 22(c). With the two effects combined, the laser polarization state may depend on the pump-beam position and size, i.e., gain area, if the LD polarization direction is fixed. For larger off-axis pumping, MG modes with a linear eigen-polarization are expected as a result of the stronger polarization discrimination effect and beam bending through the thermal lens as shown in Fig. 22(a). For small off-axis pumping, BG modes with orthogonal eigen-polarizations appear presumably because radial polarization components with a smaller reflection loss increase within the gain area.

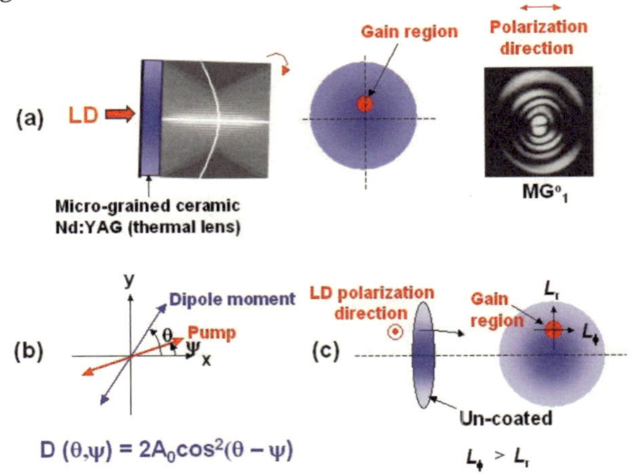

Fig. 22. (a) Conceptual illustration of the optical resonator containing a micro-grained Nd:YAG thermal lens with azimuth LD pumping. (b) Angle-dependent dipole moment induced by a linearly-polarized LD pump light. (c) Polarization-dependent reflection loss at un-coated surfaces.

In anisotropic lasers or large-grain Nd:YAG ceramic lasers, the laser polarization state is determined by fluorescence anisotropies or local thermal birefringence independently of the pump polarization, and neither BG nor MG mode oscillations take place.

5. Concluding remarks

In this Chapter, reviews were given on modal and polarization properties of microchip Nd:YAG ceramic lasers with laser-diode end pumping, featuring such effects as average grain sizes and azimuth pumping.
Segregations into multiple local-modes and the associated variety of dynamic instabilities occur in LD-pumped Nd:YAG samples with average grain size over several tens of microns resulting from the field interference effect among local-modes. The following results have been obtained for realizing stable single-frequency, linearly-polarized oscillations in Nd:YAG microchip ceramic lasers:

1. Micro-grained ceramics, whose average grain sizes are below 5 μm, can guarantee stable linearly-polarized TEM_{00} mode operations.

2. Large-grain ceramics, whose average grain sizes are larger than several tens of microns, can exhibit stable linearly-polarized oscillations in forced Ince-Gauss modes with azimuth/off-axis pumping.
3. Micro-grain ceramics can produce spontaneous Mathieu-Gauss and Bessel-Gauss lasing modes with azimuth/off-axis pumping.

6. References

Alvarez-Elizondo, M. B., Rodrlguez-Masegosa, R. & Gutierrez-Vega, J. C. (2008). Generation of Mathieu-Gauss modes with an axicon-based laser resonator. *Opt. Express* 16, 23 (2008) 18770-18775, eISSN 1094-4087.

Arlt, J., Dholakia, K., Allen, L. & Padgett, M. J. (1998). The production of multiringed Laguerre-Gaussian modes by computer-generated holograms. *J. Mod. Opt.* 45, 6 (1998) 1231-1237, ISSN 0950-0340.

Bandres, M. A. & Gutierrez-Vega, J. C. (2004). Ince–Gaussian modes of the paraxial wave equation and stable resonators. *J. Opt. Soc. Am. A* 21, 5 (2004) 873-880, ISSN 1084-7529.

Bielawski, S., Derozier, D. & Glorieux, P. (1992). Antiphase dynamics and polarization effects in the Nd-doped fiber laser. *Phys. Rev. A* 46, 5 (1992) 2811-2822, ISSN 1050-2947.

Cabrera, E., Calderon, O. G. & Guerra, J. M. (2005). Experimental evidence of antiphase population dynamics in lasers. *Phys. Rev. A* 72 (2005) 043824, ISSN 1050-2947.

Chu, S.-C. & Otsuka, K. (2007). Numerical study for selective excitation of Ince-Gaussian modes in end-pumped solid-state lasers. *Optics Express* 15 (2007) 16506-16519, eISSN 1094-4087.

Durin, J. (1987). Exact solutions for nondiffracting beams. I. The scalar theory. J. Opt. Soc. Am. *A*, 4, 4 (1987) 651-654, ISSN 1084-7529.

Erneux, T (1990). *Laser Bifurcations*, Northwestern University Press, Evanston, IL.

Gutierrez-Vega, J. C., Rodrlguez-Masegosa, R. & Chaves-Cerda, S. (2003). Bessel-Gauss resonator with spherical output mirror: geometrical- and wave-optics analysis. *J. Opt. Soc. Am. A* 20, 11 (2003) 2113-2122, ISSN 1084-7529.

Ikesue, A., Furusato, I. & Kamata, K. (1995a). Fabrication of polycrystalline, transparent YAG ceramics by a solid-state reaction method. *J. Am. Ceram. Soc.* 78, 1 (1995) 225-228, ISSN 0002-7820.

Ikesue, A., Kinoshita, T., Kamata, K. & Yoshida, K. (1995b). Fabrication and optical properties of high-performance polycrystalline Nd:YAG ceramics for solid-state lasers. *J. Am. Ceram. Soc.* 78, 4 (1995) 1033-1040, ISSN 0002-7820.

Kawai, R., Miyasaka, Y., Otsuka, K., Ohtomo, T., Narita, T., Ko, J.-Y., Shoji, I. & Taira, T. (2004). Oscillation spectra and dynamic effects in a highly-doped microchip Nd:YAG ceramic laser. *Opt. Express* 12, 10 (2004) 2293-2302, eISSN 1094-4087.

Kimura, T. & Otsuka, K. (1971). Thermal effects of a continuously pumped Nd^{3+}:YAG laser. *IEEE J. Quantum Electron.* QE-7, 8 (1971) 403-407, ISSN 00189197.

Ko, J.-Y., Otsuka, K. & Kubota, T. (2001). Quantum-noise-induced order in lasers placed in chaotic oscillation by frequency-shifted feedback. *Phys. Rev. Lett.* 86, 18 (2001) 4025-4028, ISSN 0031-9007.

Koechner, W. & Rice, D. K. (1970). Effect of birefringence on the performance of linearly polarized YAG:Nd lasers. *IEEE J. Quantum Electron.* QE-6,9 (1970) 557-566, ISSN 00189197.

Kravtsov, N. V., Lariontsev, E. G. & Naumkin, N. I. (2004). Dependence of polarisation of radiation of a linear Nd:YAG laser on the pump radiation polarization. *Quantum Electron.* 34, 9 (2004) 839-842, ISSN 1063-7818.

Lu, J., Prabhu, M., Xu, J., Ueda, K., Yagi, H., Yanagitani, T. & Kaminskii, A. (2000). Highly efficient 2% Nd:yttrium aluminum garnet ceramic laser. *Appl. Phys. Lett.* 77, 23 (2000) 3707-3709, ISSN 0003-6951.

Narita, T., Miyasaka, Y. & Otsuka, K. (2005). Self-Induced instabilities in Nd:$Y_3Al_5O_{12}$ ceramic lasers. *Jpn. J. Appl. Phys.* 37 (2005) L1168-L1170, ISSN 0021-4922.

Ohtomo, T., Kamikariya, K. & Otsuka, K. (2007). Effect of grain size on modal structure and polarization properties of laser-diode-pumped miniature ceramic lasers. *Jpn. J. Appl. Phys.* 46 (2007) L1043-L1045, ISSN 0021-4922.

Ohtomo, T., Kamikariya, K., Otsuka, K. & Chu, S.-C. (2007). Single-frequency Ince-Gaussian mode operations of laser-diode-pumped microchip solid-state lasers. *Opt. Express* 15, 17 (2007) 10705-10717, eISSN 1094-4087.

Ohtomo, T. & Otsuka, K. (2009). Yb:$Y_3Al_5O_{12}$ laser for self-mixing laser metrology with enhanced optical sensitivity. *Jpn. J. Appl. Phys.* 48 (2009) 070212, ISSN 0021-4922.

Otsuka, K. (1999). *Nonlinear Dynamics in Optical Complex Systems.* Kluwer Academic Publishers. Dordrecht/London/Boston (1999), Chapter 2, ISBN 07923-6132-6.

Otsuka, K., Kawai, R., Hwong, S.-L., Ko, J.-Y. & Chern, J.-L. (2000). Synchronization of mutually coupled self-mixing modulated lasers. *Phys. Rev. Lett.* 84, 14 (2000) 3049-3052, ISSN 0031-9007.

Otsuka, K., Ko, J.-Y., Lim, T.-S., and Makino, H. (2002). Modal interference and dynamical instability in a solid-state slice laser with asymmetric end-pumping. *Phys. Rev. Lett.* 87 (2002) 083903, ISSN 0003-6951.

Otsuka, K., Narita, T., Miyasaka, Y., Ching, C.-C., Ko, J.-Y. & Chu, S.-C. (2006). Nonlinear dynamics in thin-slice Nd:YAG ceramic lasers: Coupled local-mode model. *Appl. Phys. Lett.* 89, 8 (2006) 081117, ISSN 0003-6951.

Otsuka, K., Nemoto, K., Kamikariya, K., Miyasaka, Y., Ko, J.-Y. & Lin, C.-C. (2007). Chaos synchronization among orthogonally polarized emissions in a dual-polarization laser. *Phys. Rev. E* 76, 2 (2007) 026204, ISSN 1063-651X.

Otsuka, K., Nemoto, K., Kamikariya, K., Miyasaka, Y. & Chu, S.-C. (2007). Linearly polarized single-frequency oscillations of laser-diode-pumped microchip ceramic Nd:YAG lasers with forced Ince–Gaussian mode operations. *Jpn. J. Appl. Phys.* 46 (2007) 5865-5867, ISSN 0021-4922.

Otsuka, K. & Ohtomo, T. (2008). Polarization properties of laser-diode-pumped micro-grained Nd:YAG ceramic lasers. *Laser Phys. Lett.* 5, 9 (2008) 659-663, ISSN 1612-2011.

Schwarz, U. T., Bandres, M. A. & Gutierrez-Vega, J. C. (2004). Observation of Ince–Gaussian modes in stable resonators. *Opt. Lett.* 29, 16 (2004) 1870-1872, ISSN 0146-9592.

Shoji, I., Kurimura, S., Sato, Y., Taira, T., Ikesue, A. & Yoshida, K. (2000). Optical properties and laser characteristics of highly Nd3 + -doped Y3Al5O12 ceramics. *Appl. Phys. Lett.* 77, 7 (2000) 939-941, ISSN 0003-6951.

Shoji, I., Sato, Y., Kurimura, S., Lupei, V., Taira, T., Ikesue, A. & Yoshida, K. (2002). Thermal-birefringence-induced depolarization in Nd:YAG ceramics. *Opt. Lett.* 27, 4 (2002) 234-236, ISSN 0146-9592.

Sudo, S., Miyasaka, Y., Kamikariya, K., Nemoto, K. & Otsuka, K. (2006). Microanalysis of Brownian particles and real-time nanometer vibrometry with a laser-diode-pumped self-mixing thin-slice solid-state laser. *Jpn. J. Appl. Phys.* 45 (2006) L926-L928, ISSN 0021-4922.

Tokunaga, K., Chu, S.-C., Hsiao, H.-Y., Ohtomo, T. & Otsuka, K. (2009). Spontaneous Mathieu-Gauss mode oscillation in micro-grained Nd:YAG ceramic lasers with azimuth laser-diode pumping. *Laser Phys. Lett.* 6, 9 (2009) 635-638, ISSN 1612-2011.

25

Surface-Emitting Circular Bragg Lasers – A Promising Next-Generation On-Chip Light Source for Optical Communications

Xiankai Sun and Amnon Yariv
Department of Applied Physics, California Institute of Technology, Pasadena, California 91125, USA

1. Introduction

Surface-emitting lasers have been attracting people's interest over the past two decades because of their salient features such as low-threshold current, single-mode operation, and wafer-scale integration (Iga, 2000). Their low-divergence surface-normal emission also facilitates output coupling and packaging. Although Vertical Cavity Surface Emitting Lasers (VCSELs) have already been commercially available, their single-modedness and good emission pattern are guaranteed only for devices with a small mode area (diameter of ~μm). Attempts of further increase in the emission aperture have failed mostly because of the contradictory requirements of large-area emitting aperture and single modedness, which casts a shadow over the usefulness of VCSELs in high-power applications.

A highly desirable semiconductor laser will consist of a large aperture (say, diameter larger than 20 μm) emitting vertically (i.e., perpendicularly to the plane of the laser). It should possess the high efficiency typical of current-pumped, edge-emitting semiconductor lasers and, crucially, be single-moded. Taking a clue from the traditional edge-emitting distributed feedback (DFB) semiconductor laser, we proposed employing transverse circular Bragg confinement mechanism to achieve the goals and those lasers are accordingly referred to as "circular Bragg lasers."

There have been intensive research activities in planar circular grating lasers since early 1990s. Erdogan and Hall were the first to analyze their modal behavior with a coupled-mode theory (Erdogan & Hall, 1990, 1992). Wu et al. were the first to experimentally realize such lasers in semiconductors (Wu et al., 1991; Wu et al., 1992). With a more rigorous theoretical framework, Shams-Zadeh-Amiri et al. analyzed their above-threshold properties and radiation fields (Shams-Zadeh-Amiri et al., 2000, 2003). More recently, organic polymers are also used as the gain medium for these lasers due to their low fabrication cost (Jebali et al., 2004; Turnbull et al., 2005; Chen et al., 2007).

The circular gratings in the above-referenced work are designed radially periodic. In 2003 we proposed using Hankel-phased, i.e., radially chirped, gratings to achieve optimal interaction with the optical fields (Scheuer & Yariv, 2003), since the eigenmodes of the wave equation in cylindrical coordinates are Hankel functions. With their grating designed to follow the phases of Hankel functions, these circular Bragg lasers usually take three configurations as shown in Fig. 1: (a) circular DFB laser, in which the grating extends from the center to the exterior boundary x_b; (b) disk Bragg laser, in which a center disk is surrounded by a radial Bragg grating extending from x_0 to x_b; (c) ring Bragg laser, in which an annular defect is surrounded by both inner and outer gratings extending respectively from the center to x_L and from x_R to x_b. Including a second-order Fourier component, the gratings are able to provide in-plane feedback as well as couple laser emission out of the resonator plane in vertical direction.

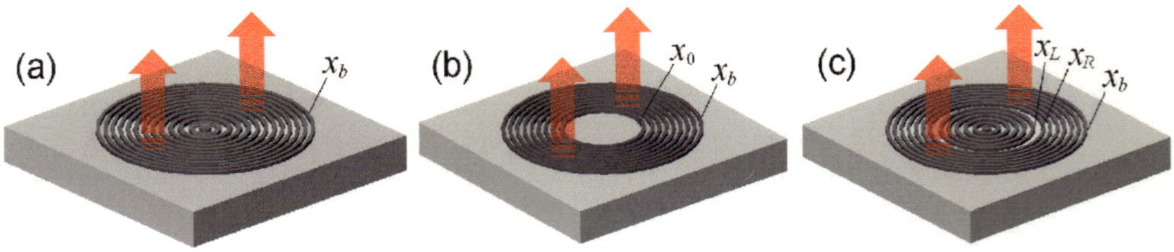

Fig. 1. Surface-emitting circular Bragg lasers: (a) circular DFB laser; (b) disk Bragg laser; (c) ring Bragg laser. Laser emission is coupled out of the resonator plane in vertical direction via the Bragg gratings

This chapter will present a comprehensive and systematic study on the surface-emitting Hankel-phased circular Bragg lasers. It is structured in the following manner: Sec. 2 focuses on every aspect in solving the modes of the lasers - analytical method, numerical method, and mode-solving accuracy check. Sec. 3 gives near-threshold modal properties of the lasers; comparison of different types of lasers demonstrates the advantages of disk and ring Bragg lasers in high-efficiency surface laser emission. Sec. 4 discusses above-threshold modal behavior, nonuniform pumping effect, and optimal design for different types of lasers. Sec. 5 concludes this chapter and suggests directions for future research.

2. Mode solving techniques

Taking into account the resonant vertical laser radiation, Appendix A presents a derivation of a comprehensive coupled-mode theory for the Hankel-phased circular grating structures in active media. The effect of vertical radiation is incorporated into the coupled in-plane wave equations by a numerical Green's function method. The in-plane (vertically confined) electric field is expressed as

$$E(x) = A(x)H_m^{(1)}(x) + B(x)H_m^{(2)}(x), \tag{1}$$

where $H_m^{(1)}(x)$ and $H_m^{(2)}(x)$ are the mth-order Hankel functions which represent respectively the in-plane outward and inward propagating cylindrical waves. A set of evolution equations for the amplitudes $A(x)$ and $B(x)$ is obtained:

$$\frac{\mathrm{d}A(x)}{\mathrm{d}x} = u(x)\cdot A(x) - v(x)\cdot B(x)\cdot e^{2i\delta\cdot x}, \tag{2}$$

$$\frac{\mathrm{d}B(x)}{\mathrm{d}x} = -u(x)\cdot B(x) + v(x)\cdot A(x)\cdot e^{-2i\delta\cdot x}, \tag{3}$$

where

$x = \beta\rho$: normalized radial coordinate with β being the in-plane propagation constant;
$\delta = (\beta_{design} - \beta)/\beta$: frequency detuning factor, representing a relative frequency shift of a resonant mode from the designed value;

$$u(x) = \begin{cases} g_A(x) - h_1, & \text{if } x \text{ is within a grating region} \\ g_A(x), & \text{if } x \text{ is within a no-grating region;} \end{cases}$$

$$v(x) = \begin{cases} h_1 + ih_2, & \text{if } x \text{ is within a grating region} \\ 0, & \text{if } x \text{ is within a no-grating region;} \end{cases}$$

$h_1 = h_{1r} + ih_{1i}$: grating's radiation coupling coefficient, representing the effect of vertical laser radiation on the in-plane modes;
h_2: grating's feedback coupling coefficient, which can always be chosen real;
$g_A(x) = g(x) - \alpha$: space-dependent net gain coefficient, the minimum value of which required to achieve laser emission will be solved analytically or numerically;
α: nonsaturable internal loss, including absorption and nonradiative scattering losses;
$g(x) = g_0(x)/[1 + I(x)/I_{sat}]$: intensity-dependent saturated gain profile;
$g_0(x)$: unsaturated gain profile; and
$I(x)/I_{sat}$: field intensity distribution in units of saturation intensity.

It should be noted that, although Eqs. (2) and (3) appear to be a set of coupled equations for in-plane waves only, they implicitly include the effect of vertical radiation due to h_1. As it will become clearer in Sec. 2.3, the vertical radiation can simply be treated as a loss term during the process of solving the in-plane laser modes.

2.1 Analytical mode solving method

When solving the modes at threshold with uniform gain (or pump) distribution across the device, the net gain coefficient g_A is x independent. The generic solutions of Eqs. (2) and (3) in no-grating regions are trivial:

$$A(x) = A_0 e^{g_A x}, \tag{4}$$

$$B(x) = B_0 e^{-g_A x}. \tag{5}$$

In grating regions, by introducing $\tilde{A}(x) = A(x)e^{-i\delta x}$ and $\tilde{B}(x) = B(x)e^{i\delta x}$, Eqs. (2) and (3) become:

$$\frac{d\tilde{A}(x)}{dx} = (u - i\delta)\tilde{A}(x) - v\tilde{B}(x), \tag{6}$$

$$\frac{d\tilde{B}(x)}{dx} = -(u - i\delta)\tilde{B}(x) + v\tilde{A}(x), \tag{7}$$

whose generic solutions lead to

$$A(x) = A(0)e^{i\delta x}\frac{\sinh[S(x-L)] + \mathbb{C}\cosh[S(x-L)]}{-\sinh[SL] + \mathbb{C}\cosh[SL]}, \tag{8}$$

$$B(x)=\frac{A(0)e^{-i\delta x}}{v}\cdot\frac{[(u-i\delta)-\mathbb{C}S]\sinh[S(x-L)]+[\mathbb{C}(u-i\delta)-S]\cosh[S(x-L)]}{-\sinh[SL]+\mathbb{C}\cosh[SL]},\tag{9}$$

where $S\equiv\sqrt{(u-i\delta)^2-v^2}$, $\mathbb{C}$ is a constant to be determined by specific boundary conditions, and L is a normalized length parameter (see Fig. 2). The determination of the constant $\mathbb{C}$ in Eqs. (8) and (9) requires the specific boundary conditions be applied to the grating under investigation.

We focus on two typical boundary conditions to obtain $\mathbb{C}$ and the corresponding field reflectivity in each case.

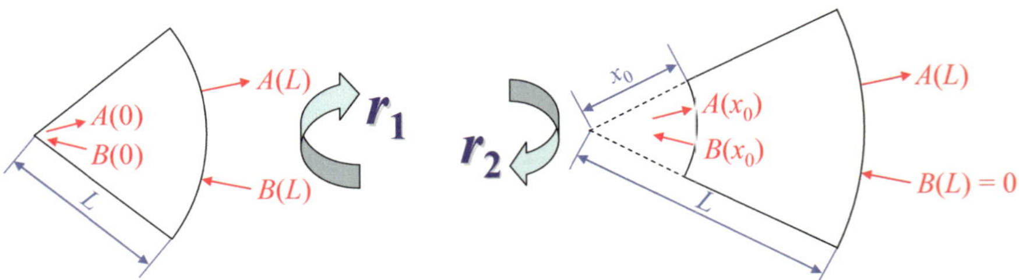

Fig. 2. Two types of boundary conditions for calculating reflectivities. (a) $A(0) = B(0)$, $r_1(L) = A(L)/B(L)$; (b) $B(L) = 0$, $r_2(x_0, L) = B(x_0)/A(x_0)$

Case I: As shown in Fig. 2(a), the grating extends from the center $x = 0$ to $x = L$. An inward propagating wave with amplitude $B(L)$ impinges from outside on the grating. The reflectivity is defined as $r_1(L) = A(L)/B(L)$. The finiteness of $E(x)$ at the center $x = 0$ requires $A(0) = B(0)$, leading to

$$\mathbb{C}=\frac{(u-v-i\delta)\sinh[SL]+S\cosh[SL]}{S\sinh[SL]+(u-v-i\delta)\cosh[SL]}$$

and to the reflectivity

$$r_1(L)=\frac{A(L)}{B(L)}=e^{2i\delta L}\frac{(u-v-i\delta)\sinh[SL]+S\cosh[SL]}{-(u-v-i\delta)\sinh[SL]+S\cosh[SL]}.\tag{10}$$

Case II: As shown in Fig. 2(b), the grating extends from $x = x_0$ to $x = L$. An outward propagating wave with amplitude $A(x_0)$ impinges from inside on the grating. The reflectivity is defined as $r_2(x_0, L) = B(x_0)/A(x_0)$. No inward propagating wave comes from outside of the grating, i.e., $B(L) = 0$. This condition leads to $\mathbb{C}=S/(u-i\delta)$ and to the reflectivity

$$r_2(x_0,L)=\frac{B(x_0)}{A(x_0)}=e^{-2i\delta x_0}\frac{v\sinh[S(L-x_0)]}{(u-i\delta)\sinh[S(L-x_0)]-S\cosh[S(L-x_0)]}.\tag{11}$$

It should be noted that, as seen from their definitions, the above reflectivities Eqs. (10) and (11) include the propagation phase.

With the obtained reflectivities for the two types of boundary conditions, it is easy to derive the laser threshold condition for each circular Bragg laser configuration.

1. Circular DFB laser:

The limiting cases $r_1(x_b) \to \infty$ or $r_2(0, x_b) = 1$ lead to the same result

$$\tanh[Sx_b] = \frac{S}{u - v - i\delta}. \tag{12}$$

2. Disk Bragg laser:

Considering the radially propagating waves in the disk and taking the unity reflectivity at the center, the threshold condition is $1 \cdot e^{2g_A x_0} \cdot r_2(x_0, x_b) = 1$, which reads

$$\frac{e^{2(g_A - i\delta)x_0} \cdot v \cdot \sinh[S(x_b - x_0)]}{(u - i\delta)\sinh[S(x_b - x_0)] - S\cosh[S(x_b - x_0)]} = 1. \tag{13}$$

3. Ring Bragg laser:

Considering the radially propagating waves in the annular defect, the threshold condition is $r_1(x_L) \cdot e^{2g_A(x_R - x_L)} \cdot r_2(x_R, x_b) = 1$, which reads

$$\frac{(u - v - i\delta)\sinh[Sx_L] + S\cosh[Sx_L]}{-(u - v - i\delta)\sinh[Sx_L] + S\cosh[Sx_L]} \cdot \frac{e^{2(g_A - i\delta)(x_R - x_L)} \cdot v \cdot \sinh[S(x_b - x_R)]}{(u - i\delta)\sinh[S(x_b - x_R)] - S\cosh[S(x_b - x_R)]} = 1. \tag{14}$$

The above threshold conditions Eqs. (12), (13), and (14) govern the modes of the lasers of each type and will be used to obtain their threshold gains (g_A) and corresponding detuning factors (δ). With these values, substituting Eqs. (4), (5), (8), and (9) into Eq. (1) and then matching them at the interfaces yield the corresponding in-plane modal field patterns. Despite their much simpler and more direct forms, these threshold conditions automatically satisfy the requirements that $E(x)$ and $E'(x)$ be continuous at every interface between the grating and no-grating regions (Sun & Yariv, 2009c).

2.2 Numerical mode solving method

When solving the modes at threshold with uniform gain (or pump) distribution across the device, g_A is independent of x so that Eqs. (2) and (3) can have analytical solutions Eqs. (4) and (5), or (8) and (9). In the case of using a nonuniform pump profile and/or taking into account the gain saturation effect in above-threshold operation, g_A becomes dependent on x and Eqs. (2) and (3) have to be solved numerically. The modes are then obtained by identifying those satisfying the boundary conditions.

As explained in Sec. 2.1, the same boundary conditions (BCs) apply to all the three types of circular Bragg lasers: (i) $A(0) = B(0)$; (ii) $B(x_b) = 0$; (iii) $A(x)$ and $B(x)$ continuous for $0 < x < x_b$. In Eqs. (2) and (3), $g_0(x)$ for a certain gain distribution profile can be parameterized with a proportionality constant, say, its maximal value g_0.

The mode solving procedure is as follows: Having BC(i), we start with an amplitude set $[A\ B] = A(0)[1\ 1]$ at the center, then numerically integrate Eqs. (2) and (3) along x to the exterior boundary x_b, during which both A and B values are kept continuous at every interface between grating and no-grating regions to satisfy BC(iii). After the integration, we have $B(x_b)$ whose absolute value marks a contour map in the 2-D plane of g_0 and δ. Now each minimum point in this contour map satisfies BC(ii) and thus represents a mode with corresponding g_0 and δ. Retrieving $A(x)$ and $B(x)$ for this mode and substituting them into Eq. (1) give the modal field pattern.

We can also calculate the modal pump level using the obtained g_0. Assuming a linear pump–gain relationship above transparency, the unsaturated gain $g_0(x)$ follows the profile of pump intensity $I_{pump}(x)$, and we may define the pump level $P_{pump} \equiv \int I_{pump}(x) \cdot 2\pi\rho \cdot d\rho =$

$P_0 \int g_0(x) \cdot x \cdot dx$, where P_0 having a power unit is a proportionality constant determined by specific experimental setup. For simple $g_0(x)$ profiles, P_{pump} can have analytical expressions as will be shown in Sec. 4.2, otherwise, numerical integration always remains a resort.

2.3 Mode-solving accuracy check

In this subsection we derive an energy relation on which the examination of mode-solving accuracy is based. This energy relation is a direct result of the coupled-mode equations (2) and (3) combined with the boundary conditions and thus is exact.

Similar to the procedure in (Haus, 1975), multiplying Eq. (2) by A^* and Eq. (3) by B^*, then adding each equation to its complex conjugate, one obtains

$$\frac{d|A|^2}{dx} = 2(g_A - h_{1r})|A|^2 - v \cdot A^* B \cdot e^{2i\delta x} - v^* \cdot AB^* \cdot e^{-2i\delta x}, \tag{15}$$

$$\frac{d|B|^2}{dx} = -2(g_A - h_{1r})|B|^2 + v \cdot AB^* \cdot e^{-2i\delta x} + v^* \cdot A^* B \cdot e^{2i\delta x}. \tag{16}$$

Subtracting Eq. (16) from Eq. (15) yields

$$\frac{d}{dx}\left(|A|^2 - |B|^2\right) = 2g_A\left(|A|^2 + |B|^2\right) - 2h_{1r}\left|Ae^{-i\delta x} + Be^{i\delta x}\right|^2. \tag{17}$$

Integrating Eq. (17) from $x = 0$ to $x = x_b$ and applying the boundary conditions $A(0) = B(0)$ and $B(x_b) = 0$ lead to

$$\underbrace{|A(x_b)|^2}_{\text{peripheral leakage}} + \underbrace{2h_{1r}\int_{\text{grating}}\left|Ae^{-i\delta x} + Be^{i\delta x}\right|^2 dx}_{\text{vertical laser emission}} = \underbrace{2\int_0^{x_b} g_A\left(|A|^2 + |B|^2\right)dx}_{\text{power generated in the gain medium}}, \tag{18}$$

which is interpreted as the energy conservation theorem for the surface-emitting circular Bragg lasers. This equation states that, in steady state, the net power generated in the gain medium is equal to the sum of peripheral leakage power and vertical emission power. Due to its exactness, we may use this relation to monitor the accuracy of mode solving by substituting into Eq. (18) the obtained modal $g_0(x)$, δ, $A(x)$, and $B(x)$ and comparing the left-hand and right-hand sides of the equation.

As an aside, it should be noted that all the power terms in Eq. (18) are in units of a saturation power defined by

$$P_{sat} \equiv E_{sat}^2\, 4D/\beta, \tag{19}$$

where E_{sat} is the saturation field which relates to the saturation intensity by $I_{sat} = cn\varepsilon_0|E_{sat}|^2/2$ (c, the speed of light; n, transverse effective index; ε_0, the vacuum permittivity), and D is the thickness (vertical dimension) of the laser resonator.

3. Near-threshold modal properties

3.1 Threshold, frequency detuning, and in-plane modal pattern

For numerical demonstration, we assume all the lasers possess a vertical layer structure as described in (Scheuer et al., 2005a) which was designed for 1.55 μm laser emission. The grating design procedure is detailed in Appendix B. The effective index n_{eff} is calculated to be 2.83 and the in-plane propagation constant $\beta = k_0 n_{\text{eff}} = 11.47\ \mu\text{m}^{-1}$. The circular grating is designed to follow the phase of Hankel functions with $m = 0$ to favor circularly symmetric modes. A quarter duty cycle is chosen to have both large feedback for in-plane waves while keeping a considerable amount of vertical emission. The coupling coefficients were found to be $h_1 = 0.0072 + 0.0108i$ and $h_2 = 0.0601$.

Since we would like to compare the modal properties of different types of lasers with a same footprint, a typical device size of $x_b = 200$ (corresponding to $\rho_b \approx 17.4\ \mu\text{m}$) is assumed for all. For the disk Bragg laser, the inner disk radius x_0 is assumed to be $x_b/2 = 100$. For the ring Bragg laser, the annular defect is assumed to be located at the middle $x_b/2 = 100$ and the defect width is set to be a wavelength of the cylindrical waves therein, yielding $x_L + x_R = x_b = 200$ and $x_R - x_L = 2\pi$. The calculated modal field patterns, along with the corresponding threshold gain values (g_A) and frequency detuning factors (δ), of the circular DFB, disk, and ring Bragg lasers are listed in Table 1.

Mode number		1	2	3	4	5
Circular DFB laser	Modal field	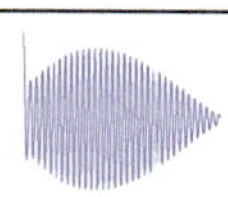	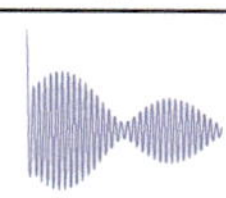	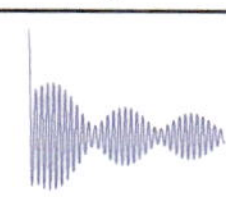	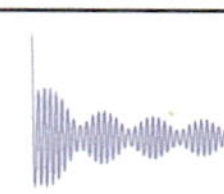	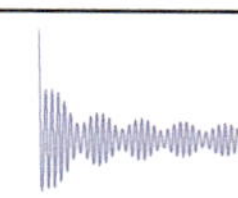
	g_A (10^{-3})	0.283	1.03	2.04	3.11	4.12
	δ (10^{-3})	61.8	66.6	74.1	83.6	94.6
Disk Bragg laser	Modal field	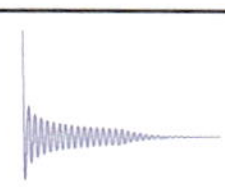	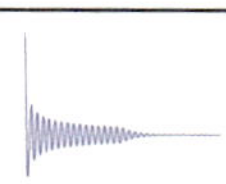	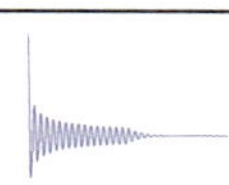	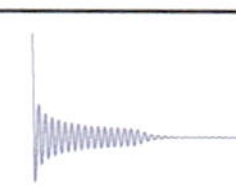	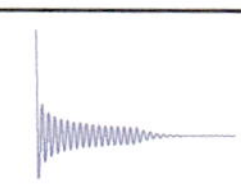
	g_A (10^{-3})	0.127	0.288	0.454	0.690	1.21
	δ (10^{-3})	49.8	21.2	-8.09	-37.4	-66.5
Ring Bragg laser	Modal field	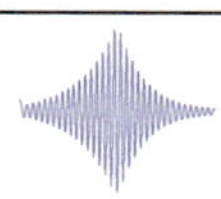	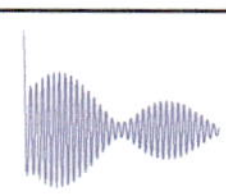	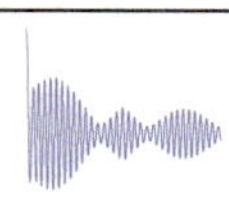	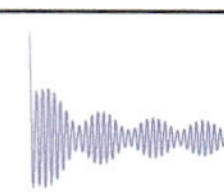	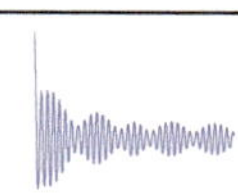
	g_A (10^{-3})	0.457	1.06	1.92	3.14	4.09
	δ (10^{-3})	55.9	66.9	71.0	84.4	91.6

Table 1. Modal field patterns, along with their threshold gains (g_A) and frequency detuning factors (δ), of the circular DFB, disk, and ring Bragg lasers. All the three types of lasers have an exterior boundary radius of $x_b = 200$. After (Sun & Yariv, 2008)

A comparison of these modal properties concludes the following features of the three laser structures:

1. All the displayed modes of the circular DFB laser are in-band modes on one side of the band gap (all $\delta > 0$). This is due to the radiation coupling induced mode selection mechanism (Sun & Yariv, 2007). Increased gain results in the excitation of higher-order modes.
2. All the displayed modes of the disk Bragg laser are confined to the center disk with negligible peripheral power leakage and thus possess very low thresholds and very small modal volumes as will be shown in Sec. 3.3.

3. All the displayed modes of the ring Bragg laser, with the exception of the fundamental defect mode, resemble their counterparts of the circular DFB laser. The defect mode has a larger threshold gain than the fundamental mode of the circular DFB laser, but the former possesses a much higher emission efficiency as will be shown in Sec. 3.3.

3.2 Radiation field and far-field pattern

As mentioned earlier, by implementing a second-order circular grating design, the gratings can not only provide feedback for the in-plane fields but also couple the laser emission vertically out of the resonator plane. As derived in Appendix A, Eq. (A12) relates the in-plane fields with the vertical radiation field in the grating regions. The radiation pattern at the emission surface is known as the near-field. For the grating design with m = 0, the near-field is expressed as

$$\Delta E = \left(s_1 A e^{-i\delta x} + s_{-1} B e^{i\delta x}\right)\left|H_0^{(1)}\right|, \tag{20}$$

where s_1 and s_{-1} at the emission surface can be obtained numerically according to Eq. (A13) for a given grating structure. Following the design procedure in Appendix B, both s_1 and s_{-1} at the emission surface were calculated to be 0.1725 − 0.0969i. Using the Huygens–Fresnel principle, the diffracted far-field radiation pattern of light from a circular aperture can be calculated under the parallel ray approximation ($|\mathbf{r}| \gg |\mathbf{r}'|$) (Hecht, 1998):

$$\begin{aligned} U(\mathbf{r}) &\propto \iint_{\text{aperture}} \Delta E(\rho,\varphi) \frac{\exp\left(ik|\mathbf{r}-\mathbf{r}'|\right)}{4\pi|\mathbf{r}-\mathbf{r}'|} \mathrm{d}\mathbf{r}' \approx \frac{e^{ikr}}{4\pi r} \iint_{\text{aperture}} \Delta E(\rho,\varphi) \exp\left(-ik(\mathbf{r}'\cdot\hat{\mathbf{r}})\right) \mathrm{d}\mathbf{r}' \\ &= \frac{e^{ikr}}{4\pi r} \int_{\varphi=0}^{2\pi} \int_{\rho=0}^{\rho_b} \Delta E(\rho) \exp\left[-ik\rho\sin\theta\cos(\varphi-\phi)\right] \rho \mathrm{d}\rho \mathrm{d}\varphi = \frac{e^{ikr}}{2r} \int_0^{\rho_b} \Delta E(\rho) J_0(k\rho\sin\theta) \rho \mathrm{d}\rho, \end{aligned} \tag{21}$$

where

$$\mathbf{r}' = \rho\cos\varphi\hat{\mathbf{x}} + \rho\sin\varphi\hat{\mathbf{y}}$$

is the source point and

$$\mathbf{r} = r\sin\theta\cos\phi\hat{\mathbf{x}} + r\sin\theta\sin\phi\hat{\mathbf{y}} + r\cos\theta\hat{\mathbf{z}}$$

is the field point. The far-field intensity pattern is then given by

$$I(\mathbf{r}) = U^*(\mathbf{r})U(\mathbf{r}) = |U(\mathbf{r})|^2 \tag{22}$$

and plotted in Fig. 3 for the fundamental mode of circular DFB, disk, and ring Bragg lasers.

In the far-field patterns, the different lobes correspond to different diffraction orders of the light emitted from the circular aperture. In the circular DFB and ring Bragg lasers, most of the energy is located in the first-order Fourier component thus their first-order diffraction peaks dominate. In the disk Bragg laser it is obvious that the zeroth-order peak dominates. These calculation results are similar to some of the experimental data for circular DFB and DBR lasers (Fallahi et al., 1994; Jordan et al., 1997).

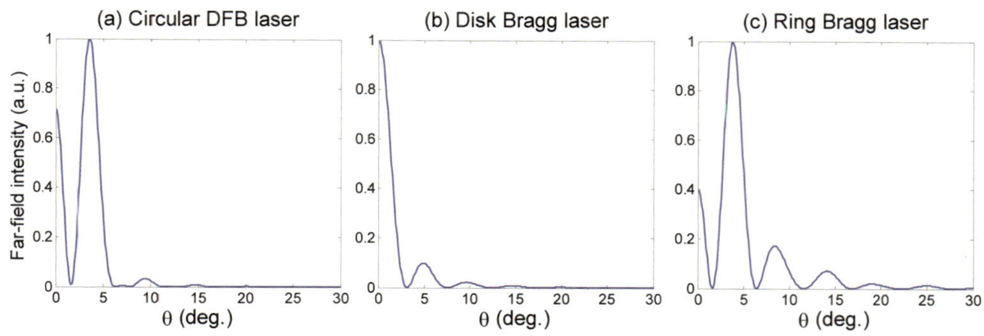

Fig. 3. Far-field intensity patterns of the fundamental mode of (a) circular DFB, (b) disk, and (c) ring Bragg lasers. After (Sun & Yariv, 2009a)

3.3 Single-mode range, quality factor, modal area, and internal emission efficiency

In the previous subsections we have compared the modal properties for devices with a fixed exterior boundary radius x_b = 200. In what follows we will vary the device size and investigate the size dependence of modal gains to determine the single-mode range for each laser type. Within each own single-mode range limit, the fundamental mode of these lasers will be used to calculate and compare the quality factor, modal area, and internal emission efficiency. Similar to the prior calculations with a fixed x_b, we still keep $x_0 = x_b/2$ for the disk Bragg laser and $x_L + x_R = x_b$, $x_R - x_L = 2\pi$ for the ring Bragg laser even as x_b varies.

Single-Mode Range

In the circular Bragg lasers, since a longer radial Bragg grating can provide stronger feedback for in-plane waves, larger devices usually require a lower threshold gain. The downside is that a larger size also results in smaller modal discrimination, which is unfavorable for single-mode operation in these lasers. As a result, there exists a range of the exterior boundary radius x_b values for each laser type within which range the single-mode operation can be achieved. This range is referred to as the "single-mode range." Figure 4 plots the evolution of threshold gains for the 5 lowest-order modes as x_b varies from 50 to 350. The single-mode ranges for the circular DFB, disk, and ring Bragg lasers are 50–250, 60–140, and 50–250, respectively, which are marked as the pink regions. Since single-mode

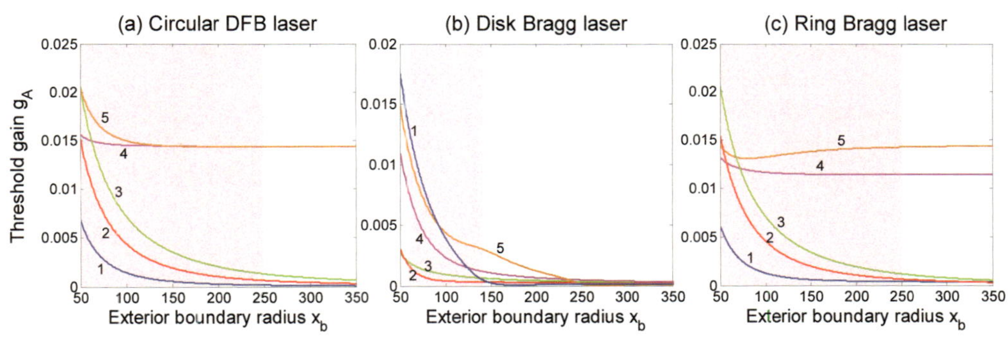

Fig. 4. Evolution of threshold gains of the 5 lowest-order modes of (a) circular DFB, (b) disk, and (c) ring Bragg lasers. The modes are labeled in accord with those shown in Table 1. The single-mode range for each laser type is marked in pink. After (Sun & Yariv, 2008)

operation is usually preferred in laser designs, in the rest of this subsection we will limit x_b to remain within each single-mode range and focus on the fundamental mode only.

Quality Factor

As a measure of the speed with which a resonator dissipates its energy, the quality factor Q for optical resonators is usually defined as $\omega\mathcal{E}/P$ where ω denotes the radian resonance frequency, $\mathcal{E}$ the total energy stored in the resonator, and P the power loss. In our surface-emitting circular Bragg lasers, the power loss P has two contributions: coherent vertical laser emission coupled out of the resonator due to the first-order Bragg diffraction, and peripheral power leakage due to the finite radial length of the Bragg reflector.
Jebali et al. recently developed an analytical formalism to calculate the Q factor for first-order circular grating resonators using a 2-D model in which the in-plane peripheral leakage was considered as the only source of power loss (Jebali et al., 2007). To include the vertical emission as another source of the power loss, a rigorous analytical derivation of the Q factor requires a 3-D model be established. This is much more complicated than the 2-D case. However, since we are interested in comparing different laser types, a relative Q value will be good enough. Considering that the energy stored in a volume is proportional to $\int|\mathbf{E}|^2\,\mathrm{d}V$ and that the outflow power through a surface is proportional to $\int|\mathbf{E}|^2\,\mathrm{d}S$, we define an unnormalized quality factor

$$Q' = \frac{\int_0^D \mathrm{d}z \int_0^{2\pi} \mathrm{d}\varphi \int_0^{\rho_b} |E(\rho,z)|^2\, \rho \mathrm{d}\rho}{\iint_{\text{grating}} |\Delta E(\rho,z=0)|^2\, \rho \mathrm{d}\rho \mathrm{d}\varphi + \int_0^D \mathrm{d}z \int_0^{2\pi} |E(\rho=\rho_b,z)|^2\, \rho_b \mathrm{d}\varphi} \quad (23)$$

$$= \frac{\int_0^D Z^2(z)\mathrm{d}z \cdot \int_0^{x_b} |E(x)|^2\, x\mathrm{d}x}{\int_{\text{grating}} |\Delta E(x,z=0)|^2\, x\mathrm{d}x + \int_0^D Z^2(z)\mathrm{d}z \cdot |E(x=x_b)|^2\, \beta x_b},$$

where $Z(z)$ denotes the vertical mode profile for a given layer structure [see Eq. (A3)] and D the thickness of the laser resonator. For a circularly symmetric mode, the angular integration factors are canceled out. The expressions for the in-plane field E and radiation field ΔE are given by Eqs. (1) and (20), respectively.
The unnormalized quality factor Q' Eq. (23) is obviously proportional to an exact Q and the former is more intuitive and convenient for calculational purposes. The Q' of the fundamental mode for the three laser types is calculated and displayed in Fig. 5. As expected, increase in the device size (x_b) results in an enhanced Q' value for all three types of lasers. Additionally, the disk Bragg laser exhibits a much higher Q' than the other two laser structures of identical dimensions. As an example, for x_b = 100, the Q' value of the disk Bragg laser is approximately 3 times greater than that of the circular DFB or ring Bragg lasers. This is consistent with their threshold behaviors shown in Table 1.

Modal Area

Based on the definition of modal volume (Coccioli et al., 1998), an effective modal area is similarly defined:

$$A_{\text{mode}}^{\text{eff}} = \frac{\iint |\mathbf{E}|^2\, x\mathrm{d}x\mathrm{d}\varphi}{\max\{|\mathbf{E}|^2\}}. \quad (24)$$

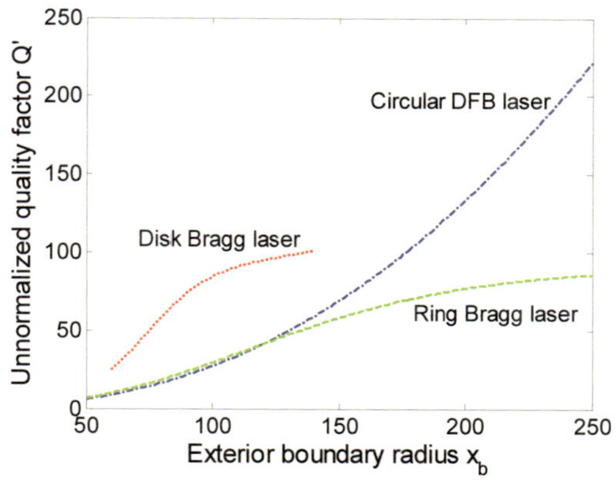

Fig. 5. Unnormalized quality factor of circular DFB, disk, and ring Bragg lasers. After (Sun & Yariv, 2008)

The modal area is a measure of how the modal field is distributed within the resonator. A highly localized mode having a small modal area can have strong interaction with the emitter. Figure 6 plots $A_{\text{mode}}^{\text{eff}}$ of the fundamental mode, within each single-mode range, for the three laser types. The top surface area of the laser resonator (πx_b^2) is also plotted to serve as a reference. The modal area of the disk Bragg laser is found to be at least one order of magnitude lower than those of the circular DFB and ring Bragg lasers. This is not surprising and can be inferred from their unique modal profiles listed in Table 1.

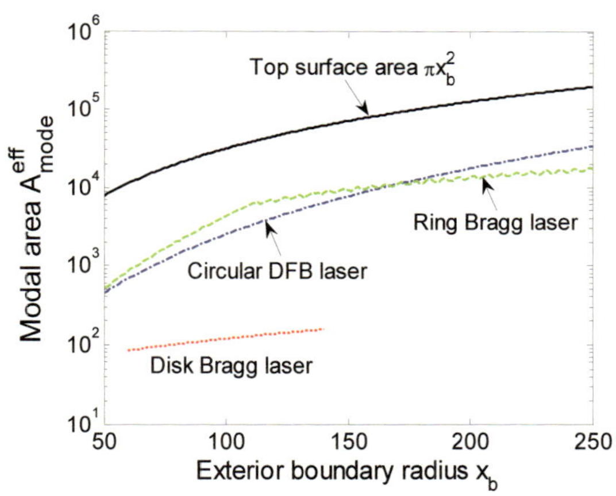

Fig. 6. Modal area of circular DFB, disk, and ring Bragg lasers. The top surface area of the laser resonator (πx_b^2) is also plotted as a reference. After (Sun & Yariv, 2008)

Internal Emission Efficiency

As mentioned earlier, the generated net power in the circular Bragg lasers is dissipated by two kinds of loss: vertical laser emission and peripheral power leakage. The internal emission efficiency η_{in} is thus naturally defined as the fraction of the total power loss which is represented by the useful vertical laser emission. Figure 7 depicts the η_{in} of the fundamental mode, within each single-mode range, for the three laser types. As expected, all the lasers possess a larger η_{in} with a larger device size. Comparing devices of identical dimensions, only the disk and ring Bragg lasers achieve high emission efficiencies. This is a result of their fundamental modes being located in a band gap while the circular DFB laser's fundamental mode is at a band edge, i.e., in a band. Band-gap modes experience much stronger reflection from the Bragg gratings, yielding less peripheral power leakage than in-band modes.

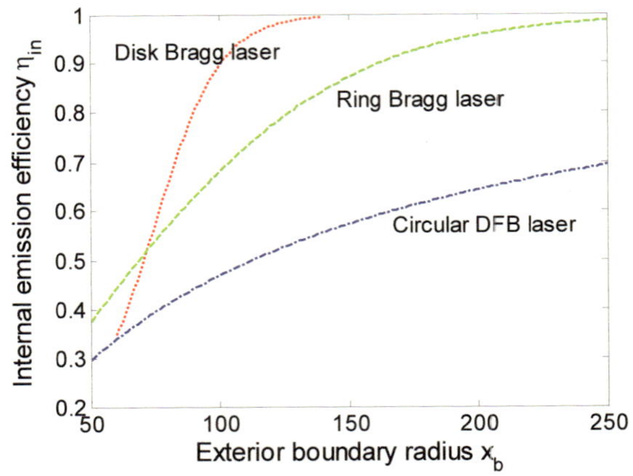

Fig. 7. Internal emission efficiency of circular DFB, disk, and ring Bragg lasers. After (Sun & Yariv, 2008)

Summary of Comparison

In this subsection, by varying the device size we have obtained the single-mode range and compared the quality factor, modal area, and internal emission efficiency of the three types of lasers. It is demonstrated that, under similar conditions, disk Bragg laser has the highest quality factor, the smallest modal area, and the highest internal emission efficiency, indicating its suitability in high-efficiency, low-threshold, ultracompact laser design, while ring Bragg laser has a large single-mode range, large modal area, and high internal emission efficiency, indicating its wide application as a high-efficiency, large-area laser.

4. Above-threshold modal analysis

In Sec. 3 we have solved the modes and compared the near-threshold modal properties of the three types of surface-emitting circular Bragg lasers. This section focuses on an above-threshold modal analysis which includes gain saturation effect. The coupled-mode equations (2) and (3) will be solved numerically with boundary conditions. The relation of surface emission power versus pump power will be simulated. The laser threshold and external emission efficiency will be compared for these lasers under different pump profiles. Lastly, with the device size varying in a large range, the evolution curve of pump level for several lowest-order modes will be generated and the optimal design guidelines for these lasers will be suggested.

4.1 Surface emission power versus pump power relation

The numerical mode solving recipe is described in detail in Sec. 2.2. Simply put, Eqs. (2) and (3) are integrated along x from $x = 0$ to $x = x_b$ with the initial boundary condition $[A\ B] =$

$A(0)[1\ 1]$. By identifying those satisfying the final boundary condition $B(x_b) = 0$ one finds the modes with corresponding g_0 and δ. The modal pump level is then given by $P_{\text{pump}} = \int g_0(x) \cdot x \cdot \text{d}x$ in units of a proportionality constant P_0. Explained in Sec. 2.3, the surface emission power P_{em} from the laser is just the second term on the left-hand side of Eq. (18). By varying the value of $A(0)$ at the beginning of the integration process, we are able to get the (P_{pump}, P_{em}) pairs which basically form the typical input–output relation for a laser mode.

As an example, we consider the circular DFB laser with $x_b = 200$ and the other structural parameters the same as those used in Sec. 3. The additional parameter used in the numerical integration, the nonsaturable internal loss α, is assumed to be 0.2×10^{-3} (already normalized by β) for typical III–V quantum well lasers. With the simulated (P_{pump}, P_{em}) pairs, the typical laser input–output relation is obtained for the fundamental mode and plotted in Fig. 8. The laser threshold P_{th} is defined as the pump level at the onset of surface laser emission. The external emission efficiency (or, energy conversion efficiency) η_{ex} is defined as the slope $\text{d}P_{\text{em}}/\text{d}P_{\text{pump}}$ of the linear fit of the simulated data points up to $P_{\text{em}} = 10P_{\text{sat}}$. As can be seen, the output power varies linearly with the pump power above threshold, which is in agreement with the theoretical and experimental results for typical laser systems [see, e.g., Sec. 9.3 of (Yariv, 1989)].

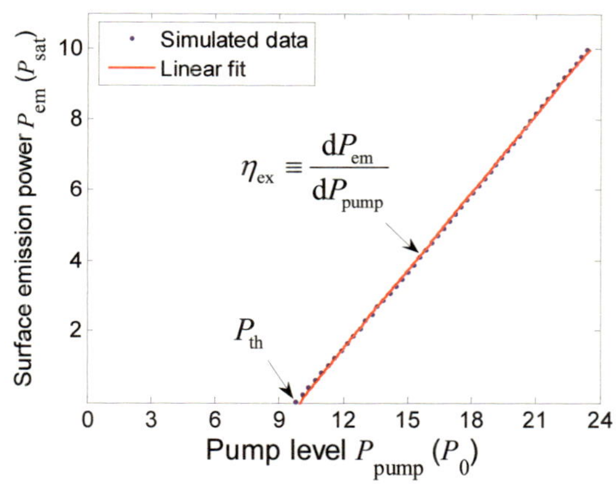

Fig. 8. Surface emission power P_{em} (in units of P_{sat}) versus pump power level P_{pump} for the fundamental mode of circular DFB laser ($x_b = 200$) under uniform pumping. The laser threshold P_{th} is defined as the pump level at the onset of surface laser emission. The external emission efficiency η_{ex} is defined as the slope of the linear fit of the simulated data points up to $P_{\text{em}} = 10P_{\text{sat}}$.

4.2 Nonuniform pumping effects

So far our studies on the circular Bragg lasers have assumed a uniform pumping profile and thus a uniform gain distribution across the devices. In practical situations, the pumping profile is usually nonuniform, distributed either in a Gaussian shape in optical pumping (Olson et al., 1998; Scheuer et al., 2005a) or in an annular shape in electrical pumping (Wu et al., 1994). The effects of nonuniform pumping have been investigated theoretically (Kasunic et al., 1995; Greene & Hall, 2001) and experimentally (Turnbull et al., 2005) for circular DFB lasers. In this subsection we will study and compare the nonuniform effects on the three types of surface-emitting circular Bragg lasers.

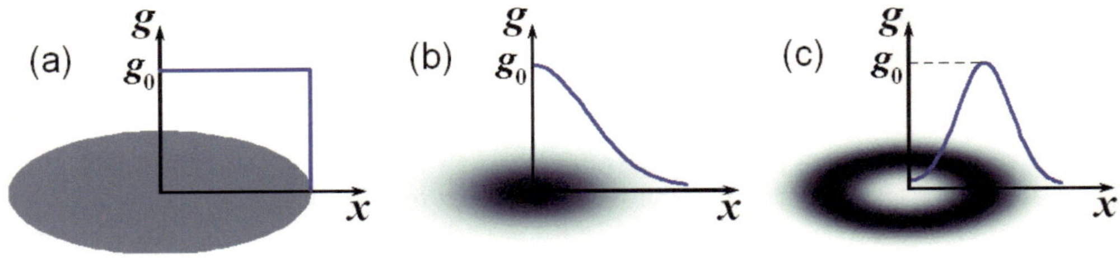

Fig. 9. Illustration of different pump profiles: (a) uniform; (b) Gaussian; (c) annular

Let us focus on three typical pumping profiles - uniform, Gaussian, and annular - as shown in Fig. 9. The pump level P_{pump} can be expressed analytically in terms of the pump profile parameters:

a. Uniform:

$$g_0(x) = g_0, \ 0 \le x \le x_b, \ P_{\text{pump}} = \int_0^{x_b} g_0 \cdot x \cdot \mathrm{d}x = \tfrac{1}{2} g_0 x_b^2, \tag{25}$$

b. Gaussian:

$$g_0(x) = g_0 \exp\left(-\tfrac{x^2}{w_p^2}\right), \ x \ge 0, \ P_{\text{pump}} = \int_0^{\infty} g_0 \exp\left(-\tfrac{x^2}{w_p^2}\right) \cdot x \cdot \mathrm{d}x = \tfrac{1}{2} g_0 w_p^2, \tag{26}$$

c. Annular:

$$g_0(x) = g_0 \left[\exp\left(-\tfrac{(x-x_p)^2}{w_p^2}\right) + \exp\left(-\tfrac{(x+x_p)^2}{w_p^2}\right) \right], \ x \ge 0,$$

$$P_{\text{pump}} = \int_0^{\infty} g_0 \left[\exp\left(-\tfrac{(x-x_p)^2}{w_p^2}\right) + \exp\left(-\tfrac{(x+x_p)^2}{w_p^2}\right) \right] \cdot x \cdot \mathrm{d}x = g_0 \left[w_p^2 \exp\left(-\tfrac{x_p^2}{w_p^2}\right) + \sqrt{\pi} w_p x_p \operatorname{erf}\left(\tfrac{x_p}{w_p}\right) \right], \tag{27}$$

where the error function $\operatorname{erf}(x) \equiv \frac{2}{\sqrt{\pi}} \int_0^x \exp(-t^2)\, \mathrm{d}t$.

To compare the nonuniform pumping effects, the typical exterior boundary radius $x_b = 200$ is again assumed for all the circular DFB, disk, and ring Bragg lasers. In addition, for the disk Bragg laser the inner disk radius is set to be $x_0 = x_b/2$, and for the ring Bragg laser the two interfaces separating the grating and no-grating regions are located at $x_L = x_b/2 - \pi$ and $x_R = x_b/2 + \pi$. Following the calculation procedure in Sec. 4.1, the threshold pump level P_{th} and the external emission efficiency η_{ex} of the fundamental mode of the three types of lasers were calculated with the uniform, Gaussian, and annular pump profiles, respectively, and the results are listed in Table 2. Without loss of generality, the Gaussian profile was assumed to follow Eq. (26) with $w_p = x_b/2 = 100$, and the annular profile was assumed to follow Eq. (27) with $x_p = x_b/2 = 100$ and $w_p = x_b/4 = 50$. The numbers shown in Table 2 indicate an inverse relation between P_{th} and η_{ex}. The lowest P_{th} and the highest η_{ex} are achieved with the Gaussian pump for the circular DFB and disk Bragg lasers and with the annular pump for the ring Bragg laser.

These observations can actually be understood with fundamental laser physics: In any laser system the overlap factor between the gain spatial distribution and that of the modal intensity is crucial and proportionate. In semiconductor lasers once the pump power is strong enough to induce the population inversion the medium starts to amplify light. The lasing threshold is determined by equating the modal loss with the modal gain, which is the

Pump profile	Circular DFB laser		Disk Bragg laser		Ring Bragg laser	
	P_{th}	η_{ex}	P_{th}	η_{ex}	P_{th}	η_{ex}
Uniform	9.760	0.7369	6.565	0.4374	13.162	0.9278
Gaussian	5.967	0.9961	2.373	0.8741	8.570	1.379
Annular	6.382	0.9742	5.855	0.7358	7.010	1.500

Table 2. Threshold pump level P_{th} (in units of P_0) and external emission efficiency η_{ex} (in units of P_{sat}/P_0) of circular DFB, disk, and ring Bragg lasers under different pump profiles. After (Sun & Yariv, 2009b)

exponential gain constant experienced by the laser mode. This modal gain is proportional to the overlap integral between the spatial distribution of the gain and that of the modal intensity. Therefore if one assumes that, to the first order, the gain is proportional to the excess pump power over the transparency, then the threshold pump level P_{th} is inversely proportional to the above overlap integral [see, e.g., Sec. 11.3 of (Yariv, 1989)]. On the other hand, since the rate of simulated emission per electron and thus the gain are proportional to the modal intensity as seen by the electron [see, e.g., Sec. 8.3 of (Yariv, 1989)], this leads to a direct proportion between the external emission efficiency η_{ex} and the overlap integral. The bottom line is that a larger overlap between the pump profile and the modal intensity distribution results in more efficient energy conversion in the gain medium which consequently leads to a lower P_{th} and a higher η_{ex}.

4.3 Considerations in optimal design

To obtain the optimal design for these circular Bragg lasers, we will again vary their device size in a large range and inspect their size-dependent behavior. Like what we have done in Sec. 3.3, we will vary the exterior boundary radius x_b for all the lasers while keeping $x_0 = x_b/2$ for the disk Bragg laser and $x_L = x_b/2 - \pi$, $x_R = x_b/2 + \pi$ for the ring Bragg laser.

Figure 10 shows the dependence of the pump level P_{pump} and the frequency detuning factor δ on the device size x_b for the 3 lowest-order modes, under uniform pump profile, of the three types of lasers. In each subfigure, the modes are numbered in accord with those shown in Table 1. For both P_{pump} and δ, dashed lines mark their values obtained at threshold and solid lines at $P_{em} = 10P_{sat}$.

Seen from the upper left and right subplots of Fig. 10, the circular DFB and ring Bragg lasers still possess large discrimination between the modes even when operated in above-threshold regime (e.g., at $P_{em} = 10P_{sat}$), which ensures them a large single-mode range of at least 50–250. Additionally, we have identified low-pump ranges for their Mode 1 at $P_{em} = 10P_{sat}$, which are 100–160 for the circular DFB laser and 80–130 for the ring Bragg laser. The low-pump range is another important factor in designing such lasers for high-efficiency, high-power applications. The existence of this low-pump range is a result of competition between the pumped area and the required gain level: although larger devices require a larger area to be pumped, their longer radial Bragg gratings reduce the needed gain because of stronger reflection of the optical fields from the gratings.

Seen from the upper middle subplot of Fig. 10, the P_{pump} for the disk Bragg laser exhibits interesting behaviors: (i) at $x_b = 200$, the order of Modes 1 and 2 exchanges from at threshold to above threshold due to the gain saturation effects; (ii) the single-mode range (for Mode 2) shifts from 60–140 at threshold to 90–175 at high surface emission level $P_{em} = 10P_{sat}$. Therefore the single-mode range for designing the disk Bragg laser should be the overlap of these two ranges, i.e., 90–140.

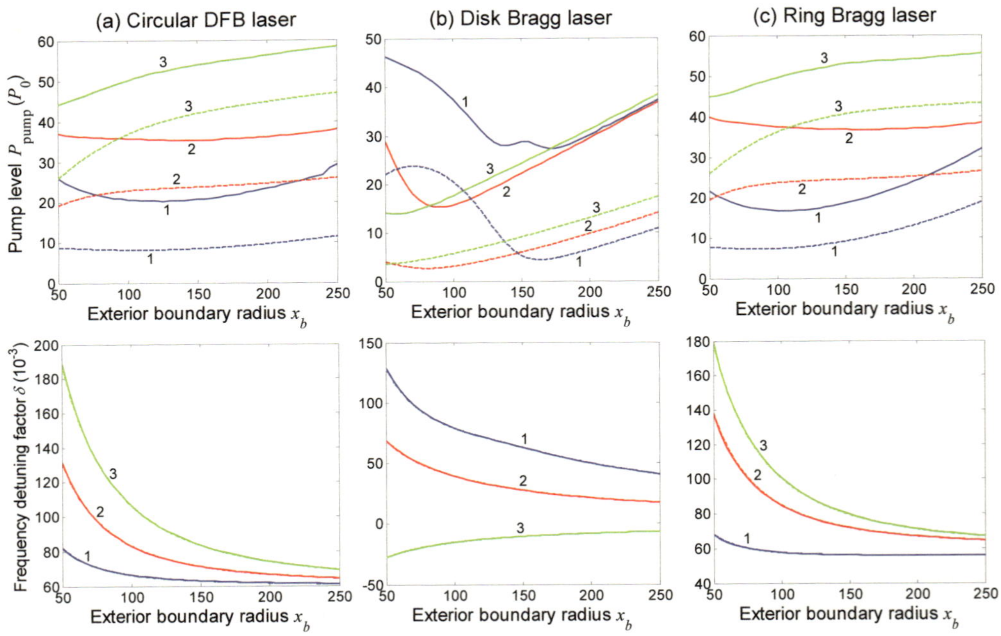

Fig. 10. Device-size-dependent pump level P_{pump} and frequency detuning factor δ of the 3 lowest-order modes, under uniform pump profile, of (a) circular DFB, (b) disk, and (c) ring Bragg lasers. x_b is the exterior boundary radius for all types of lasers. The inner disk radius x_0 of the disk Bragg laser is set to be $x_b/2$. The inner and outer edges of the annular defect of the ring Bragg laser are set to be $x_L = x_b/2 - \pi$ and $x_R = x_b/2 + \pi$, respectively. The modes are labeled in accord with those shown in Table 1. Dashed lines mark the values obtained at threshold and solid lines at $P_{em} = 10P_{sat}$. After (Sun & Yariv, 2009b)

Seen from the lower subplots of Fig. 10, all the laser modes have overlapped dashed and solid lines, which means their frequency detuning factors δ are unaffected by the surface emission level. This is because of δ being an intrinsic property of a laser mode.

5. Conclusion and outlook

In this chapter we have described and analyzed a type of on-chip microlasers whose surface emission is very useful for many applications. The main advantage of these lasers would be the relative high (say, more than tens of mW), single-mode optical power emitted broadside and coupled directly into a fiber or telescopic optics. Other areas of applications that can benefit from such lasers include ultrasensitive biochemical sensing (Scheuer et al., 2005b), all-optical gyroscopes (Scheuer, 2007), and coherent beam combination (Brauch et al., 2000) for high-power, high-radiance sources in communications and display technology. Furthermore, a thorough investigation of such lasers may also lead to a better understanding in designing and fabricating a nanosized analogue, if a surface-plasmon approach is employed.

Throughout this work we have been trying to make a small contribution to understanding of the circular Bragg lasers for their applications as high-efficiency, high-power, surface-emitting lasers. We have covered the basic concepts, calculation methods, near- and above-threshold modal properties, and design strategies for such lasers.

We have studied three typical configurations of such circular Bragg lasers - namely, circular DFB laser, disk Bragg laser, and ring Bragg laser. Following the grating design principle for linear DFB lasers, the gratings of circular Bragg lasers have to be in sync with the phases of optical waves in a circular (or cylindrical) geometry. Since the eigensolutions of wave equation in a circular geometry are Hankel functions, this leads to a varying period of the gratings in radial direction, i.e., radially chirped gratings. To obtain efficient output coupling in vertical direction, a second-order scheme has been employed, and a quarter duty cycle has proved to be a good choice.

After a series of comparison of the modal properties, it becomes clear that disk and ring Bragg lasers have superiority over circular DFB lasers in high-efficiency surface emission. More specifically, disk Bragg lasers are most useful in low-threshold, ultracompact laser design while ring Bragg lasers are excellent candidates for high-power, large-area lasers.

Considering above-threshold operation with a nonuniform pump profile, it has been numerically demonstrated and theoretically explained that a larger overlap between the pump profile and modal intensity distribution leads to a lower threshold and a higher energy conversion efficiency. To achieve the same level of surface emission, disk Bragg laser still requires the lowest pump power, even though its single-mode range is modified because of the gain saturation induced mode transition. Circular DFB and ring Bragg lasers find their low-pump ranges at high surface emission level. These results provide us useful information for designing these lasers for single-mode, high-efficiency, high-power applications.

Looking ahead, there is still more work to be done on this special topic. For example, it would be interesting to further investigate how the grating design effects on the modal far-field pattern and what design results in a pattern having all, or almost all, of the energy located in the zeroth-order lobe with narrow divergence. This will be useful for applications which require highly-directional, narrow-divergence laser beams. On the other hand, since this chapter is mainly theoretical analysis oriented, experimental work, of course, has to develop to verify the theoretical predictions. In the field of optoelectronics, a single-mode, high-power laser having controllable beam shape and compatible with on-chip integration is still being highly sought. Due to the many salient features that have been described, it is our belief that the surface-emitting circular Bragg lasers will take the place of the prevailing VCSELs and make the ideal on-chip light source for next-generation optical communications and many other areas.

Appendix A: Derivation of comprehensive coupled-mode theory for circular grating structure in an active medium

In the case that the polarization effects due to the waveguide structure are not concerned, we can start from the scalar Helmholtz equation for the z component of electric field in cylindrical coordinates

$$\left[\frac{1}{\rho}\frac{\partial}{\partial\rho}\left(\rho\frac{\partial}{\partial\rho}\right)+\frac{1}{\rho^2}\frac{\partial^2}{\partial\varphi^2}+k_0^2n^2(\rho,z)+\frac{\partial^2}{\partial z^2}\right]E_z(\rho,\varphi,z)=0, \tag{A1}$$

where ρ, φ, and z are respectively radial, azimuthal, and vertical coordinates, $k_0 = \omega/c = 2\pi/\lambda_0$ is the wave number in vacuum.

For an azimuthally propagating eigenmode, E_z in a passive uniform medium in which the dielectric constant $n^2(\rho, z) = \varepsilon_r(z)$ can be expressed as

$$E_z(\rho,\varphi,z) = E_z^{(m)}(\rho,z)\exp(im\varphi) = \left[AH_m^{(1)}(\beta\rho) + BH_m^{(2)}(\beta\rho)\right]Z(z)\exp(im\varphi), \tag{A2}$$

with m the azimuthal mode number, $\beta = k_0 n_{\mathrm{eff}}$ the in-plane propagation constant, and $Z(z)$ the fundamental mode profile of the planar slab waveguide satisfying

$$\left(k_0^2\varepsilon_r(z) + \frac{\partial^2}{\partial z^2}\right)Z(z) = \beta^2 Z(z). \tag{A3}$$

In a radially perturbed gain medium, the dielectric constant can be expressed as $n^2(\rho, z) = \varepsilon_r(z) + i\varepsilon_i(z) + \Delta\varepsilon(\rho, z)$ where $\varepsilon_i(z)$ with $|\varepsilon_i(z)| \ll \varepsilon_r(z)$ represents the medium gain or loss and $\Delta\varepsilon(\rho, z)$ is the perturbation profile which in a cylindrical geometry can be expanded in Hankel-phased plane wave series:

$$\begin{aligned}\Delta\varepsilon(\rho,z) &= -\Delta\varepsilon_0 \sum_{l=\pm1,\pm2} a_l(z)\exp\left(-il\Phi[H_m^{(1)}(\beta_{\mathrm{design}}\rho)]\right) \\ &= -\Delta\varepsilon_0 \sum_{l=\pm1,\pm2} a_l(z)\exp\left(-il\Phi[H_m^{(1)}(x)]\right)\exp(-il\delta\cdot x) \\ &= -\Delta\varepsilon_0\left[a_2(z)\frac{H_m^{(2)}}{H_m^{(1)}}e^{-2i\delta\cdot x} + a_{-2}(z)\frac{H_m^{(1)}}{H_m^{(2)}}e^{2i\delta\cdot x} + a_1(z)\frac{H_m^{(2)}}{\left|H_m^{(1)}\right|}e^{-i\delta\cdot x} + a_{-1}(z)\frac{H_m^{(1)}}{\left|H_m^{(1)}\right|}e^{i\delta\cdot x}\right].\end{aligned} \tag{A4}$$

In the above expression, $a_l(z)$ is the lth-order expansion coefficient of $\Delta\varepsilon(\rho, z)$ at a given z. x is the normalized radial coordinate defined as $x = \beta\rho$. $\delta = (\beta_{\mathrm{design}}-\beta)/\beta$ ($|\delta| \ll 1$), the normalized frequency detuning factor, represents a relative frequency shift of a resonant mode from the designed value.
To account for the vertical radiation, an additional term $\Delta E(x, z)$ is introduced into the modal field so that

$$E_z^{(m)}(x,z) = \left[A(x)H_m^{(1)}(x) + B(x)H_m^{(2)}(x)\right]Z(z) + \Delta E(x,z). \tag{A5}$$

Assuming that the radiation field $\Delta E(x, z)$ has an $\exp(\pm ik_0 z)$ dependence on z in free space, i.e.,

$$\left[\frac{1}{\rho}\frac{\partial}{\partial\rho}\left(\rho\frac{\partial}{\partial\rho}\right) - \frac{m^2}{\rho^2}\right]\Delta E = 0, \tag{A6}$$

substituting Eqs. (A4), (A5), and (A6) into Eq. (A1), introducing the large-radius approximations (Scheuer & Yariv, 2003)

$$\left|\frac{H_m^{(1,2)}(x)}{x}\right| \ll \left|\frac{\mathrm{d}H_m^{(1,2)}(x)}{\mathrm{d}x}\right|, \quad \frac{\mathrm{d}^n H_m^{(1,2)}(x)}{\mathrm{d}x^n} \approx (\pm i)^n H_m^{(1,2)}(x), \tag{A7}$$

neglecting the second derivatives of $A(x)$ and $B(x)$, and applying the modal solution in the passive unperturbed case, one obtains

$$2iZ\left(\frac{\mathrm{d}A}{\mathrm{d}x}H_m^{(1)}-\frac{\mathrm{d}B}{\mathrm{d}x}H_m^{(2)}\right)+i\frac{k_0^2\varepsilon_i}{\beta^2}\left(AH_m^{(1)}Z+BH_m^{(2)}Z\right)+\frac{1}{\beta^2}\left(k_0^2\varepsilon_r+ik_0^2\varepsilon_i+\frac{\partial^2}{\partial z^2}\right)\Delta E$$
$$=\frac{k_0^2\Delta\varepsilon_0}{\beta^2}\left[a_2\frac{H_m^{(2)}}{H_m^{(1)}}e^{-2i\delta\cdot x}+a_{-2}\frac{H_m^{(1)}}{H_m^{(2)}}e^{2i\delta\cdot x}+a_1\frac{H_m^{(2)}}{\left|H_m^{(1)}\right|}e^{-i\delta\cdot x}+a_{-1}\frac{H_m^{(1)}}{\left|H_m^{(1)}\right|}e^{i\delta\cdot x}\right]$$
$$\times\left(AH_m^{(1)}Z+BH_m^{(2)}Z+\Delta E\right). \tag{A8}$$

The phase-matching condition requires that the source and wave have close phase dependence. Grouping the terms with the same kind of Hankel functions leads to the following set of coupled equations:

$$2i\frac{\mathrm{d}A}{\mathrm{d}x}H_m^{(1)}Z+i\frac{k_0^2\varepsilon_i}{\beta^2}AH_m^{(1)}Z=\frac{k_0^2\Delta\varepsilon_0}{\beta^2}\left(a_{-2}BH_m^{(1)}e^{2i\delta\cdot x}Z+a_{-1}\frac{\Delta E}{\left|H_m^{(1)}\right|}H_m^{(1)}e^{i\delta\cdot x}\right), \tag{A9}$$

$$-2i\frac{\mathrm{d}B}{\mathrm{d}x}H_m^{(2)}Z+i\frac{k_0^2\varepsilon_i}{\beta^2}BH_m^{(2)}Z=\frac{k_0^2\Delta\varepsilon_0}{\beta^2}\left(a_2AH_m^{(2)}e^{-2i\delta\cdot x}Z+a_1\frac{\Delta E}{\left|H_m^{(1)}\right|}H_m^{(2)}e^{-i\delta\cdot x}\right), \tag{A10}$$

$$\left(k_0^2\varepsilon_r+\frac{\partial^2}{\partial z^2}\right)\Delta E=k_0^2\Delta\varepsilon_0\left(a_1A\left|H_m^{(1)}\right|e^{-i\delta\cdot x}Z+a_{-1}B\left|H_m^{(1)}\right|e^{i\delta\cdot x}Z\right). \tag{A11}$$

From Eq. (A11), ΔE can be expressed as

$$\Delta E=\left(s_1Ae^{-i\delta\cdot x}+s_{-1}Be^{i\delta\cdot x}\right)\left|H_m^{(1)}\right|, \tag{A12}$$

where

$$s_l(z)=k_0^2\Delta\varepsilon_0\int_{-\infty}^{+\infty}a_l(z')Z(z')G(z,z')\mathrm{d}z', \tag{A13}$$

and $G(z, z')$ is the Green's function satisfying

$$\left(k_0^2\varepsilon_r(z)+\frac{\partial^2}{\partial z^2}\right)G(z,z')=\delta(z-z'). \tag{A14}$$

Substituting Eq. (A12) into Eqs. (A9) and (A10), multiplying both sides by $Z(z)$, and integrating over z yield

$$\frac{\mathrm{d}A}{\mathrm{d}x}=\left(g_A-h_{-1,1}\right)A-\left(h_{-1,-1}+ih_{-2}\right)Be^{2i\delta\cdot x}, \tag{A15}$$

$$\frac{\mathrm{d}B}{\mathrm{d}x}=-\left(g_A-h_{1,-1}\right)B+\left(h_{1,1}+ih_2\right)Ae^{-2i\delta\cdot x}, \tag{A16}$$

where the gain coefficient

$$g_A\equiv-\frac{k_0^2}{2P\beta^2}\int_{-\infty}^{+\infty}\varepsilon_i(z)Z^2(z)\mathrm{d}z, \tag{A17}$$

the radiation coupling coefficients

$$h_{\pm1,\pm1} \equiv \frac{ik_0^2\Delta\varepsilon_0}{2P\beta^2}\int_{-\infty}^{+\infty} a_{\pm1}(z)s_{\pm1}(z)Z(z)\mathrm{d}z, \tag{A18}$$

the feedback coupling coefficients

$$h_{\pm2} \equiv \frac{k_0^2\Delta\varepsilon_0}{2P\beta^2}\int_{-\infty}^{+\infty} a_{\pm2}(z)Z^2(z)\mathrm{d}z, \tag{A19}$$

and the normalization constant

$$P \equiv \int_{-\infty}^{+\infty} Z^2(z)\mathrm{d}z. \tag{A20}$$

In the case of index grating, we can choose the phase of the grating such that $a_{-1} = a_1$, $a_{-2} = a_2$, then we denote $h_1 = h_{\pm1,\pm1}$, $h_2 = h_{\pm2}$. By defining $u = g_A - h_1$ and $v = h_1 + ih_2$, Eqs. (A15) and (A16) become

$$\frac{\mathrm{d}A(x)}{\mathrm{d}x} = u(x)\cdot A(x) - v(x)\cdot B(x)\cdot e^{2i\delta\cdot x}, \tag{A21}$$

$$\frac{\mathrm{d}B(x)}{\mathrm{d}x} = -u(x)\cdot B(x) + v(x)\cdot A(x)\cdot e^{-2i\delta\cdot x}. \tag{A22}$$

Specifically, in the unperturbed (i.e., no-grating) region where $\Delta\varepsilon = 0$, h_1 and h_2 vanish, and Eqs. (A21) and (A22) reduce to

$$\frac{\mathrm{d}A(x)}{\mathrm{d}x} = g_A(x)\cdot A(x), \tag{A23}$$

$$\frac{\mathrm{d}B(x)}{\mathrm{d}x} = -g_A(x)\cdot B(x). \tag{A24}$$

Appendix B: Grating design procedure and calculations of the numerical Green's function $G(z, z')$, coefficients h_1, h_2, and s_1

Since we have previously fabricated such Hankel-phased circular Bragg lasers in InGaAsP active semiconductor material (Scheuer et al., 2005a), we will use the layer structure therein as an example for our numerical study. The target lasing wavelength λ_0 is 1.55 μm. For simplicity, we approximate the complicated layer structure by an effective index profile comprising five layers as illustrated in Table B1. The vertical mode profile $Z(z)$ and the effective index n_{eff} can be obtained numerically using a 1-D mode solver. The effective index n_{eff} was calculated to be 2.83 and the in-plane propagation constant $\beta = k_0 n_{\mathrm{eff}} = 11.47$ μm^{-1}. Numerical calculations of the mode profile and effective index of the approximated layer structure indicate negligible deviations from those of the exact one.

We focus our analysis on the case of a partially etched grating with an etch depth of 185 nm. The numerical Green's function $G(z, z')$ that satisfies Eq. (A14) with the given layer structure was calculated for $-\infty < z < \infty$ and z' between the top and bottom surfaces of the laser resonator. A surface plot of the real part of $G(z, z')$ is displayed in Fig. B1.

Layer description	Refractive index n	Thickness
Upper cladding	1.54	∞
Third layer	3.281	60.5 nm
Second layer (active region)	3.4057	129 nm
First layer	3.281	60.5 nm
Lower cladding	1.54	∞

Table B1. Approximated layer structure for numerical study

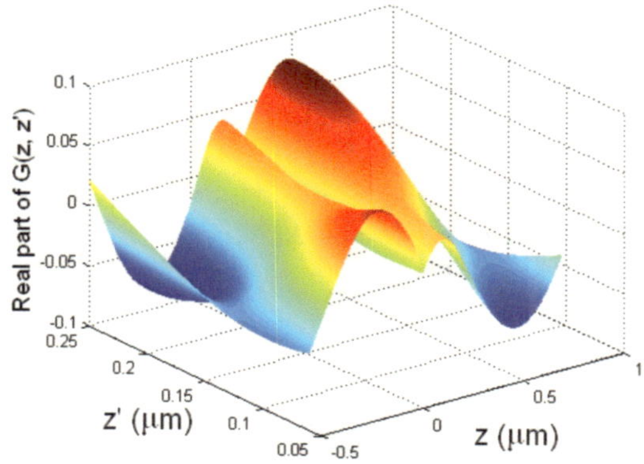

Fig. B1. Real part of the Green's function $G(z, z')$ with $-\infty < z < \infty$ and z' between the top and bottom surfaces of the laser resonator. After (Sun et al., 2008)

To favor a circularly symmetric laser mode, $m = 0$ is specially chosen in the grating phase design. In such case, a Hankel-phased grating modulation with rectangular profile

$$\Theta(\Phi[H_m^{(1)}(x)],\alpha)=\begin{cases}1, & \cos\left(\Phi[H_m^{(1)}(x)]\right)\geq\alpha\\ 0, & \cos\left(\Phi[H_m^{(1)}(x)]\right)<\alpha\end{cases} \tag{B1}$$

can be expanded in Fourier series as

$$\begin{aligned}&\frac{\arccos\alpha}{\pi}+\frac{2}{\pi}\sum_{l=1}^{\infty}\frac{\sin(l\arccos\alpha)}{l}\cos\left(l\Phi[H_m^{(1)}(x)]\right)\\&=\frac{\arccos\alpha}{\pi}+\frac{2}{\pi}\left[\sin(\arccos\alpha)\cos\left(\Phi[H_m^{(1)}(x)]\right)+\frac{\sin(2\arccos\alpha)}{2}\cos\left(2\Phi[H_m^{(1)}(x)]\right)+\cdots\right]\\&=d_c+\frac{1}{2\pi}\sin(2\pi d_c)\left[\exp(-i2\Phi[H_m^{(1)}(x)])+\exp(i2\Phi[H_m^{(1)}(x)])\right]\\&+\frac{1}{\pi}\sin(\pi d_c)\left[\exp(-i\Phi[H_m^{(1)}(x)])+\exp(i\Phi[H_m^{(1)}(x)])\right]+\cdots.\end{aligned} \tag{B2}$$

The expansion yields the coefficients

$$a_2=a_{-2}=\frac{\sin(2\pi d_c)}{2\pi}$$

and

$$a_1 = a_{-1} = \frac{\sin(\pi d_c)}{\pi}$$

where

$$d_c \equiv \frac{\arccos\alpha}{\pi} \quad (-1<\alpha<1, \;\; 0<d_c<1)$$

is the duty cycle of the Hankel-phased rectangular grating. It should be emphasized that the duty cycle has a significant role in determining the coupling coefficients h_1 and h_2 (Barlow et al., 2004). Figure B2 plots h_1 and h_2 as a function of the duty cycle d_c. A judicious choice would be d_c = 0.25 where h_2 is maximal and Re(h_1) is not small so that we can have a large ratio of vertical emission to power leakage, hence a high emission efficiency. Based on Eqs. (A18) and (A19), the grating's coupling coefficients were found to be h_1 = 0.0072 + 0.0108i and h_2 = 0.0601. Calculated from Eq. (A13), s_1 and s_{-1} at the emission surface (i.e., the top surface of the laser resonator) were found to be 0.1725 – 0.0969i.

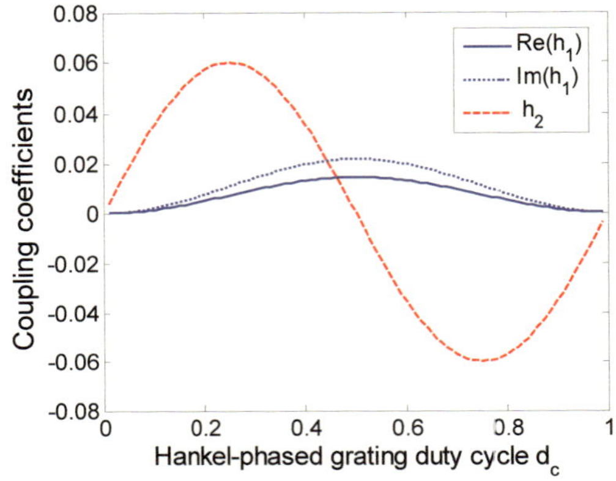

Fig. B2. Radiation coupling coefficient h_1 and feedback coupling coefficient h_2 as a function of the duty cycle d_c of the Hankel-phased rectangular grating. After (Sun et al., 2007)

6. References

Barlow, G. F.; Shore, A.; Turnbull, G. A. & Samuel, I. D. W. (2004). Design and analysis of a low-threshold polymer circular-grating distributed-feedback laser. *J. Opt. Soc. Am. B*, Vol. 21, No. 12, pp. 2142-2150

Brauch, U.; Loosen, P. & Opower, H. (2000). High-Power Diode Lasers for Direct Applications, In: *High-Power Diode Lasers: Fundamentals, Technology, Applications.* Diehl, R., (Ed.), pp. 303-368, Springer, Berlin/Heidelberg

Chen, Y.; Li, Z.; Zhang, Z.; Psaltis, D. & Scherer, A. (2007). Nanoimprinted circular grating distributed feedback dye laser. *Appl. Phys. Lett.*, Vol. 91, No. 5, pp. 051109

Coccioli, R.; Boroditsky, M.; Kim, K. W.; Rahmat-Samii, Y. & Yablonovitch, E. (1998). Smallest possible electromagnetic mode volume in a dielectric cavity. *IEE Proc.-Optoelectron.*, Vol. 145, No. 6, pp. 391-397

Erdogan, T. & Hall, D. G. (1990). Circularly symmetric distributed feedback semiconductor laser: an analysis. *J. Appl. Phys.*, Vol. 68, No. 4, pp. 1435-1444

Erdogan, T. & Hall, D. G. (1992). Circularly symmetric distributed feedback laser: coupled mode treatment of TE vector fields. *IEEE J. Quantum Electron.*, Vol. 28, No. 3, pp. 612-623

Fallahi, M.; Dion, M.; Chatenoud, F.; Templeton, I. M.; Barber, R. & Sedivy, J. (1994). Low threshold CW operation of circular-grating surface-emitting DBR lasers using MQW and a self-aligned process. *IEEE Photon. Technol. Lett.*, Vol. 6, No. 11, pp. 1280-1282

Greene, P. L. & Hall, D. G. (2001). Effects of radiation on circular-grating DFB lasers—part II: device and pump-beam parameters. *IEEE J. Quantum Electron.*, Vol. 37, No. 3, pp. 364-371

Haus, H. A. (1975). Gain saturation in distributed feedback lasers. *Appl. Opt.*, Vol. 14, No. 11, pp. 2650-2652

Hecht, E. (1998). *Optics*, 3rd edn., Addison-Wesley

Iga, K. (2000). Surface-emitting laser—its birth and generation of new optoelectronics field. *IEEE J. Sel. Top. Quantum Electron.*, Vol. 6, No. 6, pp. 1201-1215

Jebali, A.; Mahrt, R. F.; Moll, N.; Erni, D.; Bauer, C.; Bona, G.-L. & Bachtold, W. (2004). Lasing in organic circular grating structures. *J. Appl. Phys.*, Vol. 96, No. 6, pp. 3043-3049

Jebali, A.; Erni, D.; Gulde, S.; Mahrt, R. F. & Bachtold, W. (2007). Analytical calculation of the *Q* factor for circular-grating microcavities. *J. Opt. Soc. Am. B*, Vol. 24, No. 4, pp. 906-915

Jordan, R. H.; Hall, D. G.; King, O.; Wicks, G. & Rishton, S. (1997). Lasing behavior of circular grating surface-emitting semiconductor lasers. *J. Opt. Soc. Am. B*, Vol. 14, No. 2, pp. 449-453

Kasunic, K. J.; Wright, E. M. & Peyghambarian, N. (1995). Numerical modeling of inhomogeneously-pumped circular-grating DFB lasers. *Proc. SPIE*, Vol. 2398, No. 1, pp. 125-134

Olson, C.; Greene, P. L.; Wicks, G. W.; Hall, D. G. & Rishton, S. (1998). High-order azimuthal spatial modes of concentric-circle-grating surface-emitting semiconductor lasers. *Appl. Phys. Lett.*, Vol. 72, No. 11, pp. 1284-1286

Scheuer, J. & Yariv, A. (2003). Coupled-waves approach to the design and analysis of Bragg and photonic crystal annular resonators. *IEEE J. Quantum Electron.*, Vol. 39, No. 12, pp. 1555-1562

Scheuer, J.; Green, W. M. J.; DeRose, G. A. & Yariv, A. (2005a). InGaAsP annular Bragg lasers: theory, applications, and modal properties. *IEEE J. Sel. Top. Quantum Electron.*, Vol. 11, No. 2, pp. 476-484

Scheuer, J.; Green, W. M.; DeRose, G. & Yariv, A. (2005b). Ultra-sensitive biochemical sensor based on circular Bragg micro-cavities, *Proceedings of Conference on Lasers and Electro-Optics/Quantum Electronics and Laser Science and Photonic Applications Systems Technologies*, pp. CPDA7, Baltimore, MD, May 2005, Optical Society of America

Scheuer, J. (2007). Direct rotation-induced intensity modulation in circular Bragg micro-lasers. *Opt. Express*, Vol. 15, No. 23, pp. 15053-15059

Shams-Zadeh-Amiri, A. M.; Li, X. & Huang, W. P. (2000). Above-threshold analysis of second-order circular-grating DFB lasers. *IEEE J. Quantum Electron.*, Vol. 36, No. 3, pp. 259-267

Shams-Zadeh-Amiri, A. M.; Li, X. & Huang, W. P. (2003). Hankel transform-domain analysis of scattered fields in multilayer planar waveguides and lasers with circular gratings. *IEEE J. Quantum Electron.*, Vol. 39, No. 9, pp. 1086-1098

Sun, X. K. & Yariv, A. (2007). Modal properties and modal control in vertically emitting annular Bragg lasers. *Opt. Express*, Vol. 15, No. 25, pp. 17323-17333

Sun, X. K.; Scheuer, J. & Yariv, A. (2007). Optimal design and reduced threshold in vertically emitting circular Bragg disk resonator lasers. *IEEE J. Sel. Top. Quantum Electron.*, Vol. 13, No. 2, pp. 359-366

Sun, X. K.; Scheuer, J. & Yariv, A. (2008). Optimal design of vertically emitting circular Bragg disk resonator lasers, *Proceedings of SPIE*, pp. 689604, SPIE Photonics West 2008, San Jose, CA, Jan. 2008, SPIE

Sun, X. K. & Yariv, A. (2008). Surface-emitting circular DFB, disk-, and ring- Bragg resonator lasers with chirped gratings: a unified theory and comparative study. *Opt. Express*, Vol. 16, No. 12, pp. 9155-9164

Sun, X. K. & Yariv, A. (2009a). Surface-emitting circular DFB, disk-, and ring- Bragg resonator lasers with chirped gratings. II: nonuniform pumping and far-field patterns. *Opt. Express*, Vol. 17, No. 1, pp. 1-6

Sun, X. K. & Yariv, A. (2009b). Surface-emitting circular DFB, disk-, and ring- Bragg resonator lasers with chirped gratings. III: gain saturation effects and above-threshold analysis. *Opt. Express*, Vol. 17, No. 12, pp. 10119-10125

Sun, X. K. & Yariv, A. (2009c). A unified theory for surface emitting chirped circular grating lasers, *Proceedings of SPIE*, pp. 72180H, SPIE Photonics West 2009, San Jose, CA, Jan. 2009, SPIE

Turnbull, G. A.; Carleton, A.; Tahraouhi, A.; Krauss, T. F.; Samuel, I. D. W.; Barlow, G. F. & Shore, K. A. (2005). Effect of gain localization in circular-grating distributed feedback lasers. *Appl. Phys. Lett.*, Vol. 87, No. 20, pp. 201101

Wu, C.; Svilans, M.; Fallahi, M.; Makino, T.; Glinski, J.; Maritan, C. & Blaauw, C. (1991). Optical pumped surface-emitting DFB GaInAsP/InP lasers with circular grating. *Electron. Lett.*, Vol. 27, No. 20, pp. 1819-1821

Wu, C.; Svilans, M.; Fallahi, M.; Templeton, I.; Makino, T.; Glinski, J.; Maciejko, R.; Najafi, S. I.; Maritan, C.; Blaauw, C. & Knight, G. (1992). Room temperature operation of electrically pumped surface-emitting circular grating DBR laser. *Electron. Lett.*, Vol. 28, No. 11, pp. 1037-1039

Wu, C.; Makino, T.; Fallahi, M.; Craig, R. G. A.; Knight, G.; Templeton, I. & Blaauw, C. (1994). Novel circular grating surface-emitting lasers with emission from center. *Jpn. J. Appl. Phys.*, Vol. 33-Pt. 2, No. 3B, pp. L427-L429

Yariv, A. (1989). *Quantum Electronics*, 3rd edn., Wiley, New York

26

Novel Enabling Technologies for Convergence of Optical and Wireless Access Networks

Jianjun Yu[1], Gee-Kung Chang[2], Zhensheng Jia[2] and Lin Chen[3]
[1]NEC Laboratories America
[2]Georgia Institute of Technology,
[3]Hunan University,
[1,2]USA
[3]China

1. Introduction

New business models based on novel telecommunication technologies continue to dramatically change the way people live and work nowadays. People want to be always connected, with ultra-high speed data transfer, with high-quality video streaming and without face-to-face meeting. Thanks for the ~70 to 100% CAGR (cumulative annual growth rate) of mobile broadband growth, broadband mobile connectivity is now a necessity, which also leads to the increased capacity in the backbone and access networks. It is believed that wireless applications of HD-video are here and now and it is estimated that standard digital video will be unacceptable in future two years, HD-Video will be the new alternative to meet the increasing demand for broadband video services. Ultra uncompressed HD Video with (UHDV) 7680 x 4320 pixels (33 Mega pixels) plus 22.2 sound (24 channels in three layers) is expected to require larger than 40Gbps data speed. In addition to high-speed, symmetric and guaranteed bandwidth demands for future video services, next-generation access networks are driving the needs for the convergence of wired and wireless services to offer end users greater choice, convenience and variety in an efficient way. This scenario will require the delivery of voice, data and video services with mobility feature to serve the fixed and mobile users in a unified networking platform.

The most widely deployed access networks based on twisted-pair copper cable are approaching their upper limit of bandwidth-distance product (10 Mb/s*km). For distance under 1.5 km, for example, asymmetric digital subscriber loop (ADSL) can deliver about 8 Mb/s while the latest very-high-speed digital subscriber loop (VDSL) technology can deliver up to 26 Mb/s for distances under 1 km. Another dominant access medium is the hybrid of fiber and coax cables (HFC). The guaranteed bandwidth per subscriber is only 2.8-5.6 Mb/s for the downstream and 0.15-0.3 Mb/s for the upstream due to the bandwidth shared within a cell (500-1000 subscribers). It is quite obvious that the two technologies cannot meet the bandwidth demands for the future video services and will have limited lifetimes. With the trend to deploy optical fiber deeper and deeper and the development of highly recognized passive optical network (PON), it is expected that time-division multiplexing PON (TDM-PON) and wavelength-division multiplexing PON (WDM PON) will be the most promising candidates for next-generation access systems. A TDM PON, including asynchronous-transfer-mode (ATM) and broadband PON (A/BPON), Ethernet PON (EPON) and Gigabit PON (GPON), shares a single transmission channel to be a satisfactory solution for the near future bandwidth needs. A WDM PON provides point-to-point optical connectivity to multiple end users through a single feeder fiber and will be a future-proof access network. On the other hand, broadband wireless access (BWA) technologies have surged in popularity because they are more convenient, scalable and flexible for roaming connections. Most widely used technologies are local multipoint distribution service (LMDS) and multi-channel multipoint distribution service (MMDS). World Interoperability for Microwave Access (WiMAX) is another BWA technology being standardized by IEEE 802.16. These technologies can provide wireless connection but severely constrained by the required bandwidth especially for the video-centric services with high-definition TV (HDTV) quality.

To make full use of huge bandwidth offered by fiber and mobile feature presented via wireless scheme, the integration of wireless and optical networks is a potential solution for increasing capacity and mobility as well as decreasing costs in the access network. Thus, radio-over-fiber (ROF) based optical-wireless networks came into play and has emerged as an affordable alternative solution in environments such as conference center, airport, hotels, shopping malls - and ultimately to homes and small offices. The integration of optical and wireless means the convergence of two conventional technology field - radio frequency (RF) for wireless access and optical fiber for wired transmission. Long-range links are provided by optical fiber and the last tens of meters are accomplished by wireless. The requirement of more wireless bandwidth leads to spectral congestion at lower microwave frequencies in current wireless access networks. Millimeter-wave systems have the unique potential to resolve the scarcity of access bandwidth and the spectral congestion. In this situation, it is necessary to minimize the cost of the base station (BS) and shift the system complexity and expensive devices to the central office (CO) because the BS picocell has small coverage due to high atmospheric attenuation in the mm-wave band. Fig. 1 depicts the generic architecture of optical-wireless network and the enabling technologies we will discuss in the paper. At the central office (CO), the optical mm-wave signals are generated and mixed by using cost-efficient all-optical approaches. Optical networking technologies are leveraged to reach the longer transmission distance over single mode fiber (SMF) and integrate with WDM PON between the BS and CO. The BS design goal is to make the full-duplex operation and dual-service provision possible with a simple and low-cost way, which typically involves the choice of optical up-conversion technologies for overall architecture planning.

In this book chapter, we will demonstrate several key techniques for next-generation hybrid optical-wireless networks. Optical mm-wave signal generation and up-conversion with low-cost approaches are vital to future real deployment in access networks. Different schemes for optical mm-wave generation, up-conversion and transmission are compared. For all-optical up-conversion technique, the techniques based on external modulation (OCS and phase modulation) are the promising solution for downlink transmission in terms of required input power, transmission performance against dispersion in the fiber, support for WDM PON, mixing bandwidth, stability and system configuration. Using orthogonal

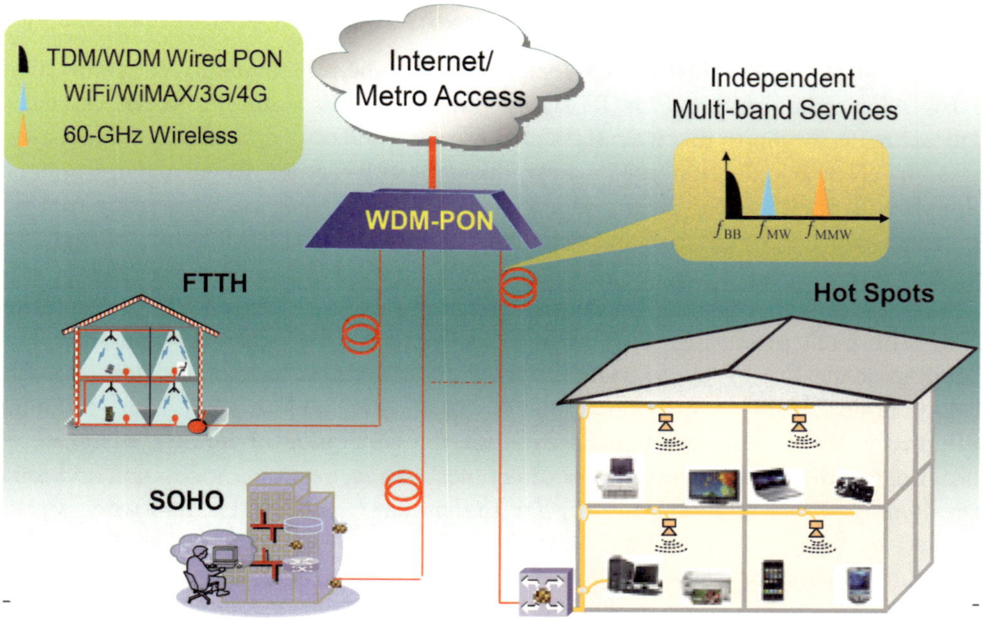

Fig. 1. A generic architecture of an optical-wireless network.

frequency division multiplexing (OFDM) modulation format, the spectral efficiency can be significantly increased and the fiber dispersion can be further mitigated. We will also demonstrate the possibility to seamlessly integrate RoF with WDM-PON networks for simultaneously serving multiple end users by leveraging the existing PON networks.

In the case of uplink, the network architecture consisting of a single light source at the CO and the reuse of the downstream wavelength at the base station BS is an attractive solution for consolidated bidirectional connection as it requires no additional light source and wavelength management at the BS. Several schemes for realizing full-duplex transmission are described in this chapter based on wavelength reuse to avoid the need of the light source at the BS. The delivery of 16Gb/s super broadband OFDM optical signals are also introduced in the transmission of 50m grade-index fiber.

To exploit the benefits of both wired and wireless technologies, carriers and service provides are actively seeking a convergent network architecture to deliver multiple services to serve both fixed and mobile users. This can be accomplished by using hybrid optical-wireless networks, which not only can transmit wireless signals at the BS over fiber, but also simultaneously provide the wired and wireless services to the end users. A super-broadband access testbed is presented to deliver both wired and wireless services simultaneously via an optical fiber or air, and an order of magnitude larger access bandwidth than the state-of-the-art Wi-Fi systems can be provided. A campus-wide field trial is demonstrated for RoF system to transmit uncompressed 270-Mbps standard definition (SD) and 1.485-Gbps high definition (HD) real-time video contents carried by 2.4-GHz radio and 60-GHz millimeter wave signals, respectively, between two on-campus research buildings distanced over 2.5-km standard single mode fiber (SMF-28) through the Georgia Institute of Technology's (GT) fiber network.

2. Optical mm-wave up-conversion for downstream

An intermediate frequency (IF) or baseband signal can be transmitted over optical fiber to the BS, where the baseband or IF signal are up-converted to mm-wave carriers for broadcasting in the air. In this way, the signal transmission over optical fiber is less severely affected by chromatic dispersion. However, this approach requires frequency up-conversion at the BS with a mm-wave mixer and a local oscillator (LO) signal. Generating mm-wave frequencies using electrical devices is challenging because of the bottleneck of high-speed electronic processing. Additionally, because of the requirement of numerous BSs in optical mm-wave access networks, this approach significantly increases the cost and complexity of the BS. The most promising solution is to use optical approaches for mm-wave signal generation at the CO. Over the past few years, many groups have conducted research to develop optical mm-wave generation, up-conversion and transmission techniques. Traditionally, three different methods exist for the generation of mm-wave signals over optical links with intensity modulation: direct modulation, external modulation and remote heterodyning. Direct modulation is by far the simplest, and 40GHz mm-wave signal is generated by using this modulation technique. However, due to the limited modulation bandwidth of the laser, there is no 60GHz mm-wave signal generation by using direction modulation. The configuration of external modulation is simple and can be used to generate optical mm-wave, but it has some disadvantages that limit its implementation at mm-wave because high cost for driving signals and low dispersion tolerance. The double-sideband (DSB) carrier modulation format severely suffers the chromatic dispersion when the signals

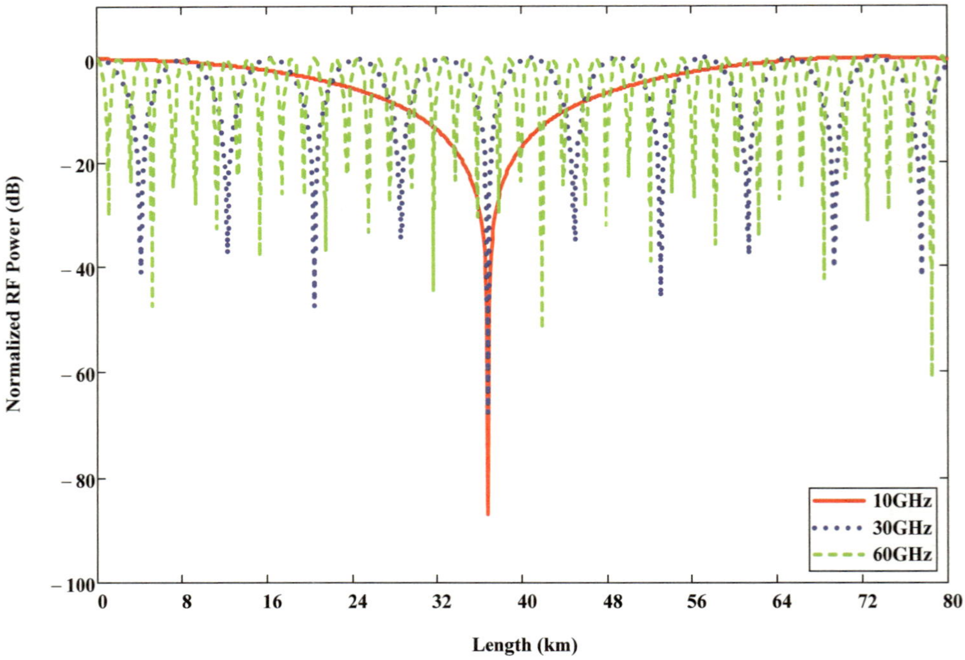

Fig. 2. Normalized received RF power for optical DSB modulation as a function of transmission distance in dispersive optical fiber (D: 17ps/km/nm, wavelength: 1.55um)

propagate along a dispersive fiber because two sidebands experience different amount of phase shift compared to original carrier. The received RF power of the mm-wave signal varies depending on the relative phase difference between two beating components, as shown in Fig.2. It can be seen that the RF power varies in a periodic manner along the fiber.
For optical heterodyning technique, two or more optical signals are simultaneously transmitted and are heterodyned in the receiver. However, it requires either a precisely biased electro-optic modulator or a complex laser to reduce the severe phase noise, which greatly adds to the cost and complexity of the system.
Recently, several new approaches for up-conversion of radio signals have been reported. These techniques, based on nonlinear effects in wave-guide device, exhibit low conversion efficiency and need very high input optical power. The scheme based on cross-gain modulation (XGM) in semiconductor optical amplifier (SOA) requires a large input power to saturate the gain of SOA. The scheme by using cross-phase modulation (XPM) in SOA Mach-Zehnder Interferometer (SOA-MZI) loosens the requirement for the input power, however, the complicated conversion structure and nonlinear crosstalk among multiple channels are major hurdle to greatly limit the signal quantity of wireless end users. Here some all-optical up-conversion schemes in direct modulation laser, highly-nonlinear dispersion-shifted fiber (HNL-DSF), electroabsorption modulator (EAM), external intensity, integrated modulator and phase modulator will be introduced.

2.1 Optical mm-wave generation by a DML

Fig. 3 shows the principle to generate optical mm-wave signals by using only one DML and wavelength reuse for uplink connection. In the central office, a wideband DML is used to generate optical mm-wave signal. The RF signals are generated by an electrical mixer which mixes the LO and downstream data. In the base station an optical interleaver with two

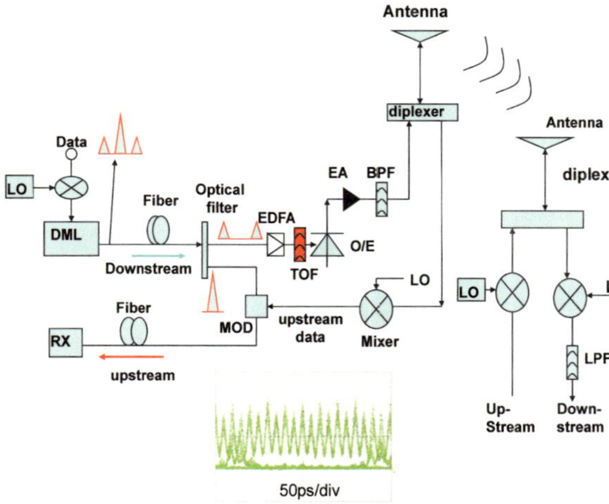

Fig. 3. Mm-wave signal generation by using one DML and ROF network configuration with both down and upstream connection and source-free in the base station. DML: direct-modulation laser, O/E: optical/electrical converter, RX: receiver, EA: electrical amplifier, LPF: low-pass filter, BPF: band-pass filter, MOD: modulator. The eye diagram (50ps/div) of optical mm-wave signal at 40GHz generated from the DML after one 25GHz optical inter-leaver is inserted.

output ports or cascaded an optical circulator and a fiber Bragg grating (FBG) is used to separate the optical carrier and the first-order sidebands. After the optical interleaver, the two peaks of the first-order sidebands will be beat to generate mm-wave signals with a double repetitive frequency of the RF signal when they are detected by a downlink receiver. The bandwidth of the commercial DML can be over 20GHz, so the repetitive frequency of the optical mm-wave can go up to 40GHz. The separated optical carrier can be used to carry the uplink optical signals. The base-band uplink data is used to drive an external modulator to generate optical uplink optical signals before it is transmitted over the fiber to the central office. In one experiment, one commercial DML with central wavelength of 1550nm biased at 57mA was used to generate lightwave. The output power biased at 57mA of the DML is 4dBm, and 3dB bandwidth of the DML is larger than 20GHz. The RF signals are generated by using an electrical mixer to combine the 20GHz RF clock (sinusoidal wave) and 2.5Gb/s electrical baseband signal (downstream data). Then the mixed RF signals are boosted to 2.7Vp-p to drive the DML. The repetition frequency of the optical mm-wave is 40 GHz as shown in the inserted figure in Fig. 2.

2.2 FWM in HNL-DSF

Four-wave mixing (FWM) is one of the important nonlinear effects to generate new waves or parametric amplification, especially for effectively ultra-fast response by using HNL-DSF. Relying on the third-order electric susceptibility and beating process with the frequency phase-matching condition when light of three or more different wavelengths is launched into HNL-DSF, it is possible to realize terahertz all-optical mixing or up-conversion as shown in Fig.4. Two pumping waves $\omega_{OCS} - \omega_{RF}$ and $\omega_{OCS} + \omega_{RF}$ are generated by using optical carrier suppression (OCS) technologies, where ω_{RF} is the RF sinusoidal clock frequency. The converted signal ω_{con} may then be determined by

$$\omega_{con} = (\omega_{OCS} - \omega_{RF}) + (\omega_{OCS} + \omega_{RF}) - \omega_s = 2\omega_{OCS} - \omega_s \tag{1}$$

where ω_s is the input signal light. The two pumping waves are set to coincide with the fiber zero-dispersion wavelength to generate beating grating in HNL-DSF efficiently, which will modulate the input signal ω_s to produce two sideband waves with the frequency shift

$$\omega_{con} \pm \left|(\omega_{OCS} - \omega_{RF}) - (\omega_{OCS} + \omega_{RF})\right| = \omega_{con} \pm 2\omega_{RF} \tag{2}$$

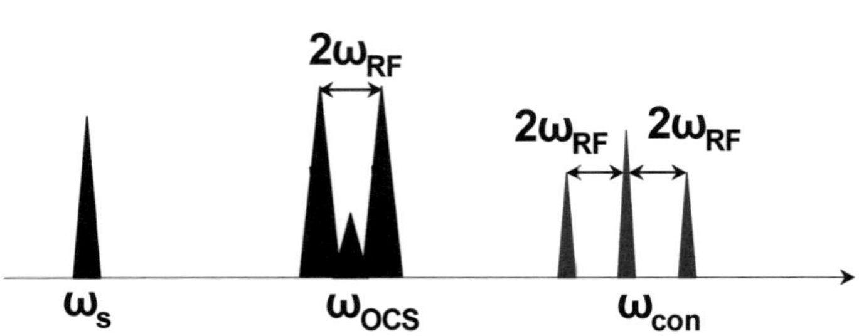

Fig. 4. Schematic of FWM based al-optical up-conversion.

The power is determined by the third-order nonlinearity susceptibility $\chi^{(3)}$ and the fiber parameters. Since FWM is independent of the signal bit rate and coding format, which can be used to realize simultaneous up-conversion of multiple WDM signals and easily integrated with passive optical access networks. In addition, the HNL-DSF has high Raman gain to assist the FWM process in terms of conversion efficiency and bandwidth. Table 1 shows transmission characteristics of the HNL-DSF used in these experiments. The fiber was made by OFS Denmark and has a nonlinear coefficient γ of 10 $W^{-1}km^{-1}$ with the length of 1 km.

Using this scheme, eight WDM signals with the channel spacing of 3.2 nm are generated from eight DFB lasers and modulated by a $LiNbO_3$ Mach-Zehnder (M-Z) modulator driven by 2.5-Gb/s baseband signals. The modulated WDM signals are then transmitted over 10-km SMF to de-correlate the signals before the up-conversion. OCS modulation scheme is used to generate a 40-GHz optical local-oscillator (LO) signal as the two FWM pumping signals. It is realized by driving a dual-arm M-Z modulator biased at V_π with two complementary 20-GHz radio frequency (RF) sinusoid waveforms. To increase the FWM conversion efficiency and broaden the conversion bandwidth, a backward Raman pump is utilized. The advantages for FWM based all-optical up-conversion is that FWM is transparent to the signal bit-rate and modulation formats, which is easy to realize up-conversion for different WDM signals. In addition, due to the ultra-fast nonlinear response of the fiber, it is possible to realize THz waveform all-optical mixing or up-conversion. Meanwhile, the HNL-DSF has higher Raman gain compared to standard DSF, which can be utilized to assist the FWM process.

Characteristics	Measured value
Attenuation coefficient	0.4 dB/km
Zero-dispersion wavelength	1561.0 nm
Dispersion slope (at zero-dispersion wavelength	0.03 $ps/nm^2/km$
Nonlinear coefficient	10 $W^{-1}km^{-1}$
Length	1 km
Raman Gain	4 - 8.5 dB

Table 1. Transmission characteristics of HNL-DSF.

To overcome polarization sensitivity when one single pump is employed, a novel scheme with dual-pumps is proposed and experimentally demonstrated. OCS can be realized when an intensity modulator (IM) is biased at null point, as shown in Fig. 1(b). If we assume that the modulator is driven by a RF sinusoidal wave signal with a repetitive frequency of f. After the IM, two subcarriers with wavelength spacing of $2f$ will be generated from a CW lightwave by carrier suppression scheme. The generated two lightwaves have the same polarization direction, optical power, and locked phase. The frequency spacing is exactly controlled by the RF frequency of the sinusoidal wave on the modulator. In our previous investigation, we have shown that the polarization-insensitive wavelength conversion can be realized when two pump lightwaves have the same polarization direction, and the converted signal locates the same side with the original signal. Another condition to meet the polarization-insensitive wavelength conversion is that the original signal should be far away from the pumps. This condition can be easily realized in the real system. The proposed scheme to realize all-optical up-conversion is shown in Fig. 5. To simplify the system, we only consider single channel input baseband signal. In our previous experiment, two pumps are generated from different laser source; therefore, their phase is not locked. Moreover, in order to maintain the same polarization direction, we need to use polarization controller (PC) or polarization maintaining fiber (PMF) to keep the two lightwaves have the same polarization direction. If the emitting wavelength of the laser is not stable, the wavelength of the converted signal will be drifted. Obviously, if we use OCS scheme to generate two pump lightwaves, these problems, like polarization direction and frequency spacing, can be effectively solved. As shown in Fig. 5(a), the single channel baseband signal with the dual-pumps generated by OCS are injected into the nonlinear medium such as nonlinear fiber or semiconductor optical amplifier. Due to FWM in the nonlinear medium, new optical components will be generated. To simplify this figure, we only consider the first-order FWM, and the high-order FWM is neglected. Also, we do not consider the FWM between the pumps. We can see that two converted signals are generated as shown in Fig. 5(b). The two converted signals have the same polarization direction, and their phase is locked. In this scheme, the optical signals similar to a double sideband (DSB) are generated which includes two converted signals and original signal after wavelength conversion. Then we use optical filter, an optical interleaver is optimal, to remove the original signal as shown in Fig. 5(c). In this way, the two converted new signals with channel spacing of $4f$ can be obtained after FWM process in the nonlinear medium. By using this scheme, 60GHz optical mm-wave signal, which carries 7.5Gb/s baseband signal, is generated, and one eye diagram is shown in Fig. 5(d).

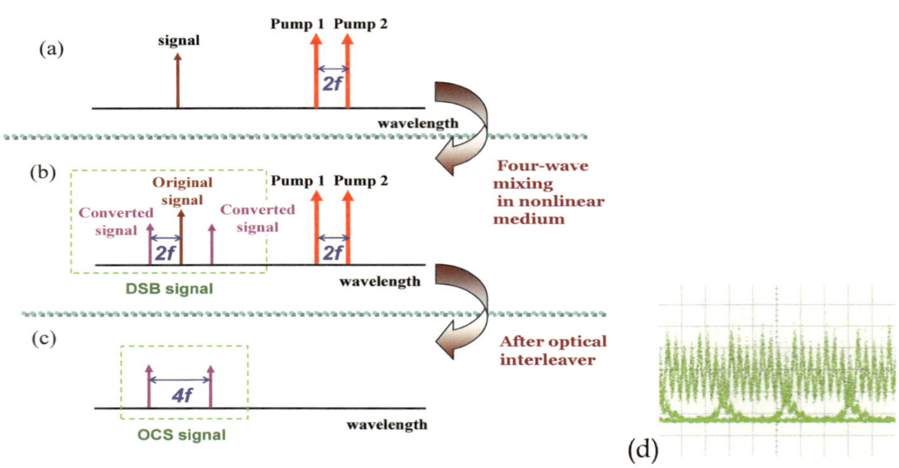

Fig. 5. All-optical up-conversion based on polarization-insensitive FWM in nonlinear medium. (d) Eye diagram of 10Gb/s signal carried by 60GHz mm-wave (50ps/div).

2.3 XPM in HNL-DSF

Cross-phase modulation (XPM) in the nonlinear optical loop mirror (NOLM) and straight pass in HNL-DSF is also possible to realize all-optical up-conversion for more wavelength channels without any interference and saturation effect limitation. A conceptual diagram by using OCS as the control signal is shown in Fig. 6(a) and (b). Regarding the NOLM architecture, the symmetry between the counter-propagation paths of the probe signal is broken due to XPM induced phase shift by the control signal. So the NOLM is changed into the mixer between the probe and control signal. For the straight-line structure, while propagating in the HNL-DSF, the intensity of the control signal modulates the electric field of the probe data signal via XPM effect. Therefore, the RF sinusoidal clock is imposed onto the probe signal as the sidebands. Since the NOLM suffers from the stability problems due to the sensitive polarization state in the loop, XPM-based up-conversion in straight-line HNL-DSF are simpler and more robust. By using this nonlinear effect with a straight-line architecture, 16 channels at 2.5Gb/s are upconverted to 40GHz carrier in recent experiment. It is noted that the optical carrier-to-sideband ratio (CSR) has influences on the system performance because the mm-wave signals are generated by the interplay between the optical powers in the carrier and sideband or the two sidebands themselves. CSR refers to carrier-to-sideband ratio, the ratio of the optical power in the optical carrier to that of the first-order sideband within a defined resolution bandwidth (here set as 0.01nm). While maintaining the CSR at 13 dB, the measured receiver. This scheme exhibits very good conversion performance at high data rate and can provide more wavelength channels by being extended into the whole optical fiber transmission band without any interference and saturation effect limitation.

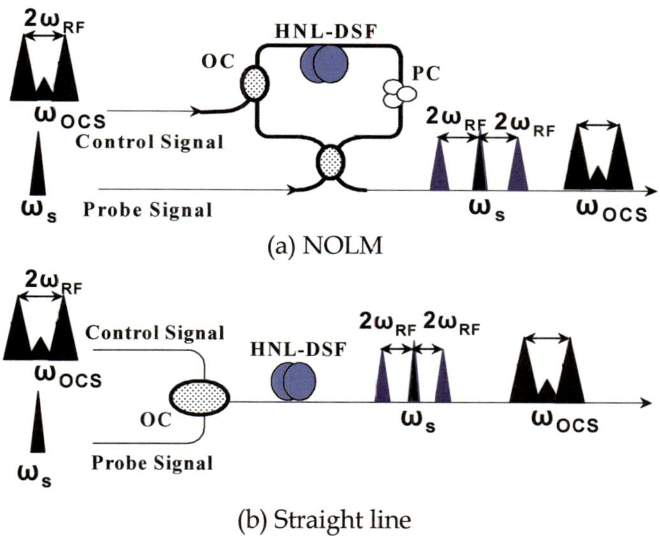

Fig. 6. Concept of using XPM-based all-optical up-conversion.

2.4 XAM in EAM

The principle of up-conversion based on EAM is similar to its wavelength conversion mechanism at high bit rate, the main difference from wavelength conversion is that the modulated data signal will be used to replace the CW source. The experimental setup is shown in Fig. 7.

An EAM (CyOptics: EAM 40) with 3-dB bandwidth of 32 GHz, fiber-to-fiber insertion loss of 8 dB and polarization sensitivity lower than 1 dB is used to realize the signal up-conversion. Optical LO signals are generated by using OCS modulation scheme with larger than 25 dB of the carrier suppression ratio. The electrical signal and waveform is also inserted as an inset in Fig. 7. This scheme has some advantages such as low power consumption, compact size, polarization insensitivity, easy integration with other devices and higher speed operation due to EAM inherent characteristics.

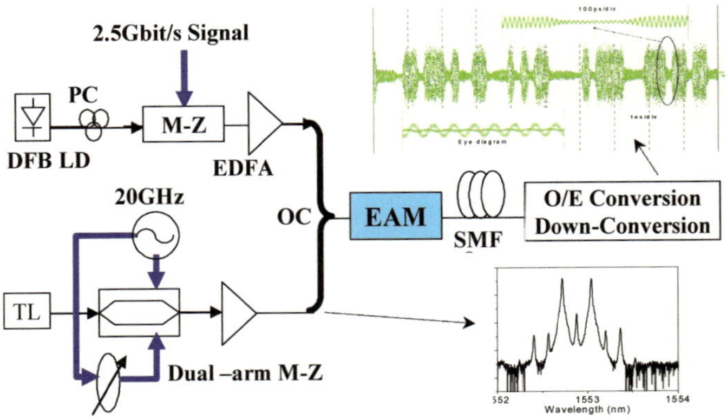

Fig. 7. Experimental setup of XAM-based all-optical mixer in EAM.

2.5 External intensity modulation

External modulation is another choice to up-convert mm-wave signals for ROF systems. Essentially, three different schemes exist for the generation and transmission of mm-wave signals over optical links with intensity modulation: double-sideband (DSB), single-sideband (SSB) and OCS modulation scheme. Fig. 8 shows optical mm-wave generation based on DSB, SSB and OCS modulation schemes respectively, and the corresponding optical spectra and eye diagrams after mixing with the 40-GHz RF signal are also inserted. Baseband data signal is generated by an M-Z modulator driven by 2.5-Gb/s PRBS electrical signal with a word length of 2^{31}-1. For DSB modulation scheme, the M-Z modulator 2 is biased at 0.5 V_π and the frequency of the driven RF signal is 40 GHz. The generated mm-

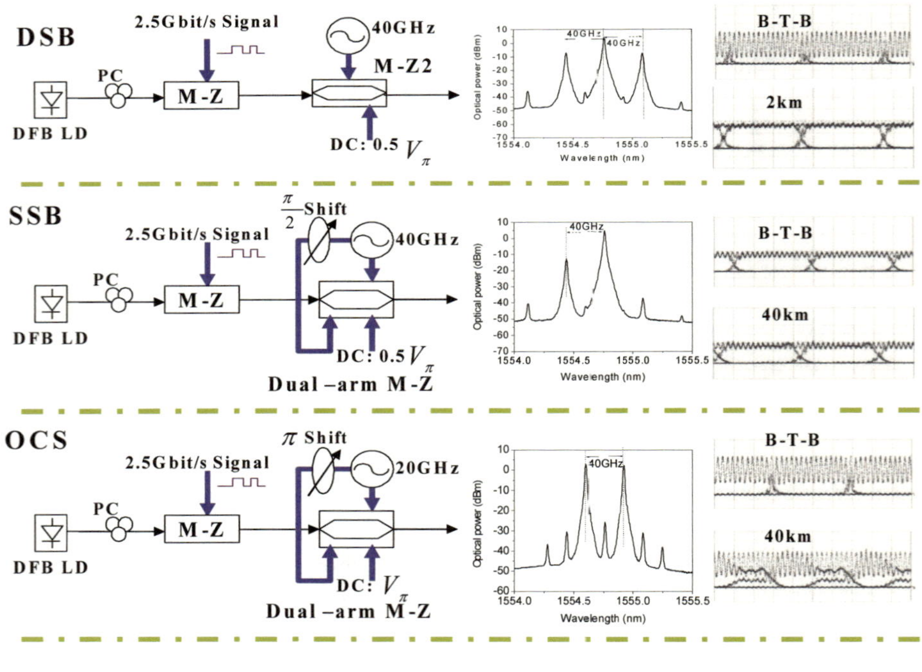

Fig. 8. Up-conversions using DSB, SSB and OCS techniques.

wave will occupy over 80-GHz bandwidth because it has two sidebands. Since the two sidebands have different velocities in SMF, the RF power at 40 GHz will disappear after transmitting over a certain length of SMF. As an example, the eye diagram after transmission over 2 km is inserted in Fig.8. It is seen that RF power at 40 GHz is almost faded, which leads to a large power penalty. The measured BER curves in Fig.9 show the power penalty is 17 dB at a BER of 10^{-10} after 2km transmission. These results indicate that DSB modulation based scheme is not suitable to a large area access network. A dual-arm M-Z modulator is employed to achieve SSB modulation. The two electrical RF signals to drive the dual-arm M-Z modulator has a phase shift of $\pi/2$, and the DC bias is at $0.5V_\pi$. The generated optical mm-wave will only occupy 40-GHz bandwidth, but the optical CSR is generally larger than 15 dB, which means it is full of DC components at the peak of center wavelength; hence it results in low receiver sensitivity. Fig. 9 shows the receiver sensitivity of back-to-back for SSB modulation is around 10 dB lower than that of DSB modulation. Although there is no power penalty after 20-km transmission, it is more than 5 dB after 40-km transmission due to fiber dispersion and large CSR. When the phase of the two electrical RF signals to drive the dual arm M-Z modulator is set to π difference and the DC bias is at the minimal intensity-output point or V_π, OCS modulation is realized. In this scheme, only 20-GHz RF signal is needed and the bandwidth for the M-Z modulator is also only 20 GHz, moreover the generated optical spectrum just occupies 40-GHz bandwidth. At BER of 10^{-10}, the receiver sensitivity of B-T-B mm-wave signal is -39.7 dBm, which is similar to that of the millimeter signal generation based on DSB modulation. There is no power penalty after transmission over 20 km, and the power penalty is less than 2 dB after 40-km transmission. The electrical eye diagrams after 10-km and 50-km transmission are shown as an inset (i) and (ii) in Fig. 9. These results show the mm-wave generated by OCS modulation can be used in large area access networks.

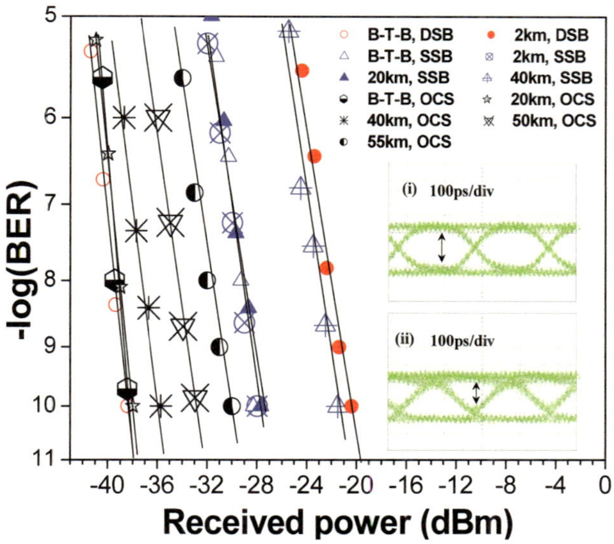

Fig. 9. BER curves with different mm-wave generation schemes.

Since the optical mm-wave has two peaks after OCS modulation, it will suffer from dispersion in fiber when transmission over SMF. The pulse width of the 2.5-Gb/s signal carried by the optical mm-wave is approximately 400 ps. The two peaks with a wavelength spacing of 0.32 nm will have a walk-off time of 400 ps caused by fiber dispersion after transmission over 74-km SMF with a group velocity dispersion (GVD) of 17 ps/nm/km, which means the eye will be fully closed after this distance. While considering the limited rise and fall times of the optical receiver and electrical amplifier, the maximum delivery distance will be shorter. Fig. 10 clearly shows the evolution of optical eye diagrams at different transmission distance. The un-flat amplitudes of the optical carriers at 40 GHz as shown in Fig. 10 (b) are caused by chromatic dispersion. Previous investigations show that fiber dispersion causes the amplitude fluctuation of the carrier but the RF power at 40 GHz does not disappear when the carrier is a pure dual-mode lightwave. Fig. 10 (d) shows that the eye is almost closed after the optical mm-wave is transmitted over 60 km.

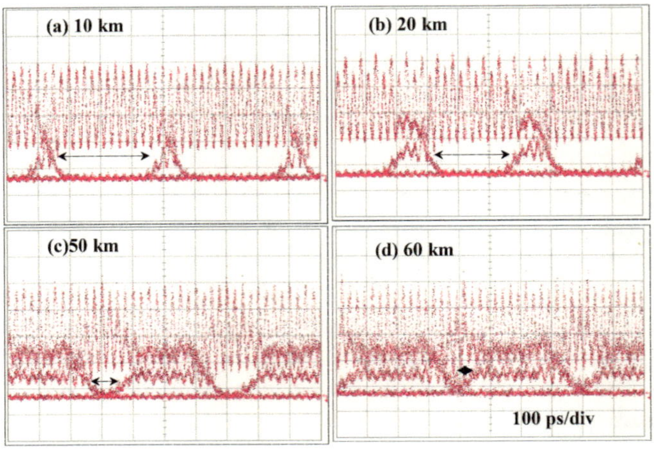

Fig. 10. Optical eye diagrams at different transmission distance.

By using the OCS modulation scheme, 32x2.5-Gb/s DWDM signals after transmission over 40 km are simultaneously up-converted to integrate with WDM PON networks. In this experiment, 32 DFB laser array is used to realize 32 wavelength signals from 1536.1 nm to 1560.9 nm with adjacent 100 GHz spacing. An AWG is used to combine the 32 CW lightwaves before modulation by an M-Z. The generated 32x2.5-Gb/s signals are transmitted over 40 km for simulation the metro optical network before they are up-converted by using a dual-arm M-Z based on OCS technique. The up-converted mm-waves are amplified by an EDFA to get a power of 20 dBm before transmission over variable length SMF. At the receiver, the desired channel is selected by using the identical O/E and down-conversion components as in forenamed setup. The measured power penalty is smaller than 2dB for all channels after transmission over 40km.

The up-conversion signals based on OCS modulation scheme have shown the best performance such as the highest receiver sensitivity, the highest spectral efficiency, and the smallest power penalty over long distance delivery compared to DSB and SSB modulation scheme.

There are some modified schemes to realize all-optical up-conversion based on external intensity modulator. One scheme is to use low RF frequency to drive an intensity modulator to generate optical mm-wave signal with frequency quadrupling technique. The principle is shown in Fig. 11. A $LiNbO_3$ intensity modulator (IM) is employed to generate optical mm-wave with low-frequency RF. Downstream data and RF signal at quarter of LO frequency are mixed by using subcarrier multiplexing (SCM) technique then to drive the IM. To realize an optical mm-wave carrier with four times of LO frequency, the modulator needs to be dc-biased at the peak output power when the LO signal is removed. If the repetitive frequency of the RF microwave source is f_0, the frequency spacing between the second-order modes is equal to $4f_0$ while the first-order modes are suppressed. As an example, the output optical spectrum shown in Fig. 11 is for the case of a 10 GHz modulation frequency. From the figure, it can be seen that the frequency spacing of the second-order modes is 40 GHz and the first-order sidebands are also suppressed. Taking the advantage of this property can dramatically lower the bandwidth requirements for the optical modulator and allows the use of a much lower frequency electrical drive signal. This can greatly reduce the cost of the system and makes it more practical to use.

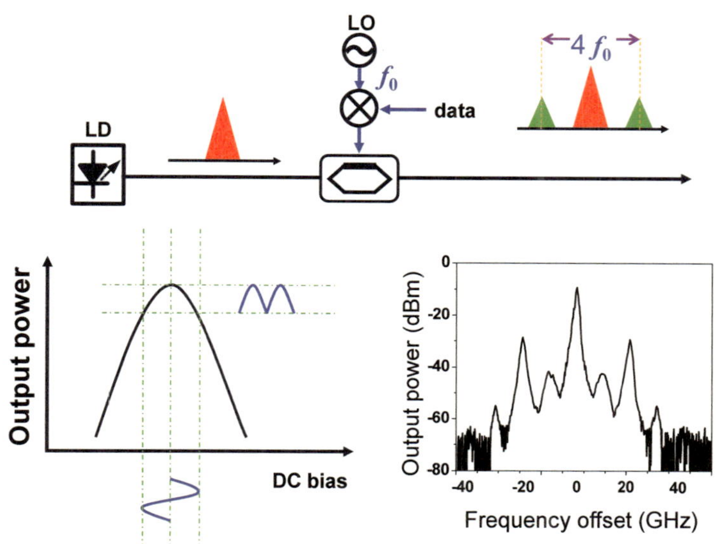

Fig. 11. Principle of frequency quadrupling scheme. Resolution of the optical spectrum is 0.01 nm.

Fig. 12 shows another scheme to generate optical mm-wave without optical filtering by using an integrated modulator. A conceptual diagram of optical carrier suppressed millimeter-wave signal generation using a frequency quadrupling technique without any optical filter. An external integrated MZM that consists of three sub-MZMs is key to generating optical millimeter-wave signals. One sub-MZM (MZ-a or MZ-b) is embedded in each arm of the main modulator (MZ-c). The optical field at the input of the integrated MZM is defined as (1) where is the amplitude of the optical field and is the angular frequency of the optical carrier. MZ-a and MZ-b are both biased at the full point. Electrical driving modulation signals sent into MZ-a and MZ-b are and , respectively. The odd sidebands are suppressed after passing through MZ-a and MZ-b. Since the MZ-c is biased at the null point, the optical carrier is suppressed when the lightwave passes through this integrated modulator, and only the second sidebands are left, hence a quadrupling mm-wave signal is generated.

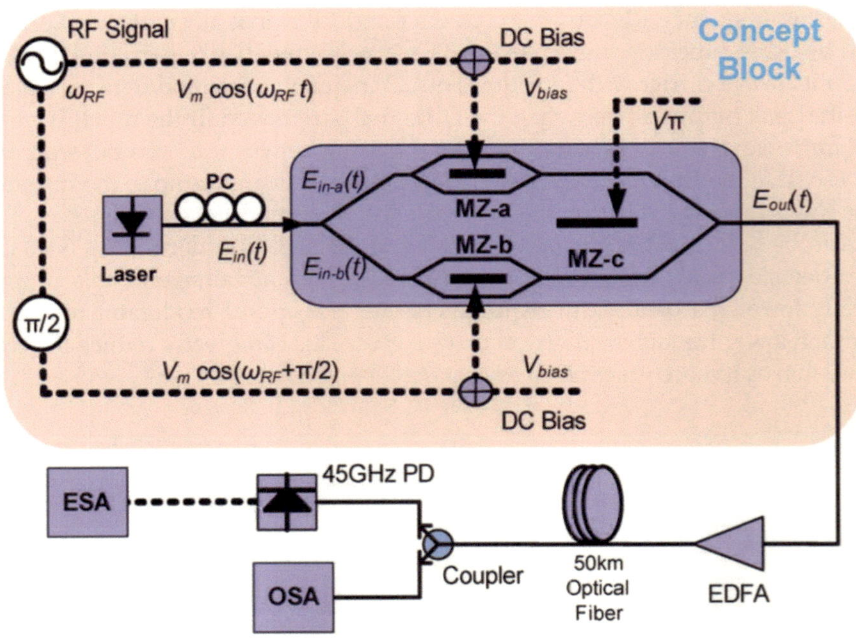

Fig. 12. Optical up-conversion using a frequency multiplication technique for WDM RoF systems. (MZ: Mach–Zehnder modulator; EDFA: erbium-doped fiber amplifier; OSA: optical spectrum analyzer; ESA: electrical spectrum analyzer).

2.6 External phase modulation

In addition to the intensity modulation, external phase modulation is also utilized to produce downstream optical mm-waves in optical-wireless networks. Figure 13 shows the principle of using phase modulator (PM) and subsequent interleaver for mm-wave generation. As a schematic illustration, the case of WDM signals with 100-GHz channel spacing as inset (i) is considered. When the WDM sources are modulated by a PM driven by a 20-GHz sinusoidal RF clock, the signal field of one channel at the output of PM can be written as a few sidebands:

$$E_{output1} = A_s \sum_{n=-\infty}^{+\infty} J_n(m_d)\cos[(\omega_s + n\omega_{RF})t + n\pi / 2] \tag{1}$$

where A_s is the amplitude of the original optical carrier, $J_n(m_d)$ is the nth Bessel function of the first kind, $m_d = \pi V_{RF} / V_\pi$ is the modulation depth of PM, V_{RF} is the driving voltage of the RF signals, $n\omega_{RF}$ is the generated sidebands. How many sidebands can be generated depends on the amplitude of the driven RF signal on the PM. Here we assume that only the first-order sideband is generated through optimizing the modulation depth m_d. The peak of the first sideband is 20 GHz away from the original optical carrier as shown as inset (ii). The interleaver, with one input and two out-ports of 25-GHz bandwidth, is used to suppress the optical carrier. When the central wavelengths of the WDM light sources can match up well to the interleaver, the optical carrier of each channel will be suppressed. The output signal of the interleaver is expressed as:

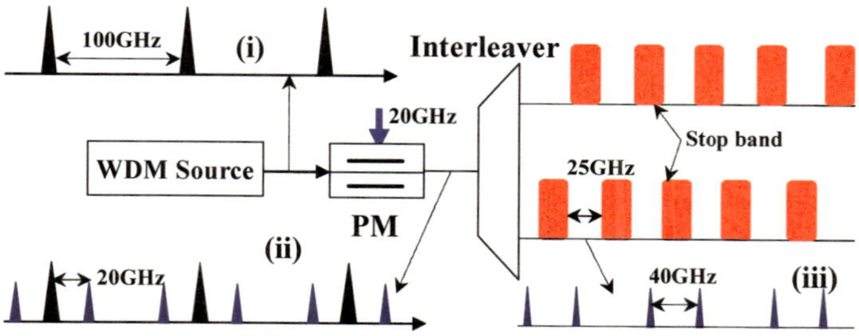

Fig. 13. Concept of using PM and interleaver for mm-wave generation.

$$E_{output2} = A_s \left\{ \sum_{n=-\infty}^{\infty} J_n(m_d)\cos\left[(\omega_c + n\omega_m)t + n\pi/2\right] - \alpha J_0(m_d)\cos\omega_c t \right\} \quad (2)$$

where α is the attenuation coefficient of the interleave filter at the peak of center frequency. The optical spectrum from output 1 of the interleaver is shown in inset (iii). In this way, optical mm-wave WDM channels are generated.

Eight-channel WDM optical mm-wave generation and transmission is demonstrated in one experiment, where eight channel signals at 2.5Gb/s are upconverted into 40GHz mm-wave by a phase modulator and optical filtering technique. Compared optical mm-wave generation by using an intensity modulator, this scheme also exhibits better performance on system stability due to the removal of any DC-bias controller.

3. OFDM-ROF system

In the previous section, all modulation formats for the baseband optical signal are optical On/Off keying signal. When this optical signal is carried by the optical mm-wave signal at high frequency, the transmission distance of optical mm-wave signals is quite limited by the fiber chromatic dispersion as shown in Fig. 6. On the other hand, orthogonal frequency division multiplexing (OFDM) modulation technology has been widely adopted in ADSL and RF-wireless systems such as IEEE 802.11a/g (Wi-Fi) and IEEE 802.16 (WiMAX). OFDM systems can provide excellent tolerance towards multipath delay spread and frequency-dependent channel distortion. Recently, optical transmission systems employing OFDM have gained considerable research interest because OFDM can overcome the effect of fiber chromatic dispersion and have the capability to use higher level modulation formats to increase spectral efficiency. So the combination of OFDM and ROF is naturally suitable for optical-wireless systems to extend the transmission distance over both fiber and air links.

The first experimental demonstration of a super-broadband OFDM-ROF system based on SCM and interleaver for all-optical mm-wave generation and up-conversion was presented in OFC 2006. In this experiment, the transmission of 4-QAM (QPSK) OFDM analog signals at 1 Gb/s on 40-GHz mm-wave carriers is achieved over 80-km standard single mode fiber (SSMF) without dispersion compensation with less than 0.5-dB power penalty at BER of 10^{-6}. Fig. 14 shows the schematic diagram of mm-wave OFDM-ROF system. At the central office (CO), the OFDM analog data and an RF clock at half of the local oscillator (LO) frequency are mixed by using SCM technology. The mixed signals are applied to drive a $LiNbO_3$ Mach-Zehnder modulator (LN-MZM) to create first-order sidebands. After transmission over SSMF, an interleaver is employed to separate the optical carrier from the first-order sidebands to generate optical mm-wave carrier at the BS. Then the boosted electrical mm-wave signal is down-converted through a mixer and retrieved by the OFDM receiver. The separated optical carrier is considered as the continuous wave (CW) and directly modulated by the uplink data and sent back to the CO.

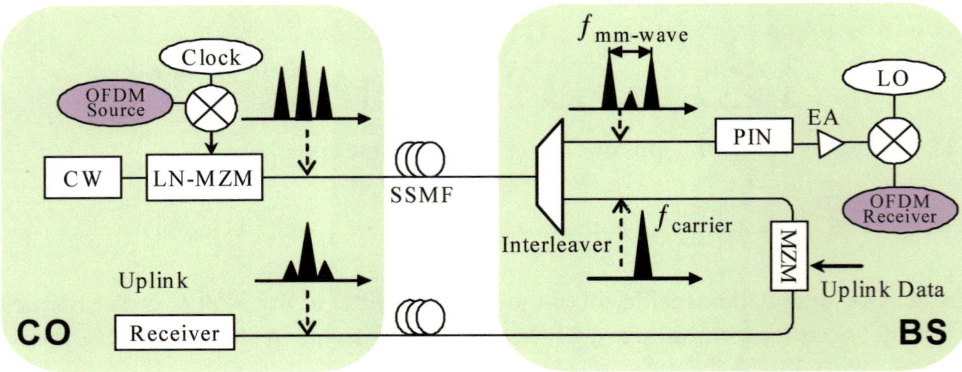

Fig. 14. Architecture of an mm-wave OFDM-ROF system based all-optical up-conversion.

Fig. 15 depicts the experimental setup. At the optical transmitter side, a CW lightwave from a tunable laser (TL) at 1559.7nm (2-MHz linewidth) is modulated by an MZM driven by the mixed OFDM analog signals. The 1-Gb/s OFDM baseband signals are calculated offline with Matlab program including mapping 2^{31}-1 PRBS into 256 4-QAM-encoded subcarriers, subsequently converting the OFDM symbols into time domain by using IFFT and then adding 6.4-ns cyclic prefix (CP). The digital waveforms are then downloaded to a Tektronix AWG 7102 arbitrary waveform generator (AWG) operating at 10 GS/s to produce 1-Gb/s analog OFDM signals, which is shown in OFDM source in Fig. 15. Among 256 OFDM subcarriers (FFT size), 200 channels are used for data transmission, 55 channels at high frequencies are set to zero for over-sampling, and one channel in the middle of the OFDM spectrum is set to zero for DC in baseband. The output waveforms are shown in Fig. 15 as inset (i). The 1-Gb/s OFDM signals are mixed with a 20-GHz sinusoidal wave to realize SCM for the mm-wave signals and then used to drive the MZM. The electrical spectrum of mixed signals and the optical spectrum of modulated optical signals are shown in Fig. 15 as inset (ii) and (a), respectively. The input power is 14 dBm before transmission over 80-km SSMF. At the optical receiver side, a 50/25-GHz optical interleaver with 35-dB channel isolation and two outputs is used to separate the optical carrier and the sub-carriers. The optical spectra of the separated optical carrier and mm-wave signals are shown in Fig. 15 as (b) and (c), respectively. The carrier is suppressed larger than 20 dB. The optical eye diagram of mm-wave signals is also shown in (b). Regarding the downlink connection, direct detection is made by a 50-GHz bandwidth PIN photodiode. The converted electrical mm-wave signal is then amplified by an electrical amplifier (EA) with 10-GHz bandwidth centered at 40 GHz. A 10-GHz clock is used in combination with a frequency multiplexer to produce 40-GHz electrical LO signal later mixed to down-convert the electrical signal to OFDM baseband form. The down-converted signals are sampled with a real-time digital oscilloscope (Tektronix TDS6154C). The received data are processed and recovered off-line with a Matlab program as an OFDM receiver. The sampling frequency is 1.25 GHz. The electrical spectrum of down-converted OFDM signals is shown in Fig. 15 as (iii). The spectrum fluctuations for different frequency components arise from the nonlinear response of TL, MZM and optical amplifier due to the large optical power. The measured power penalty for the downstream optical signal is smaller than 2dB.

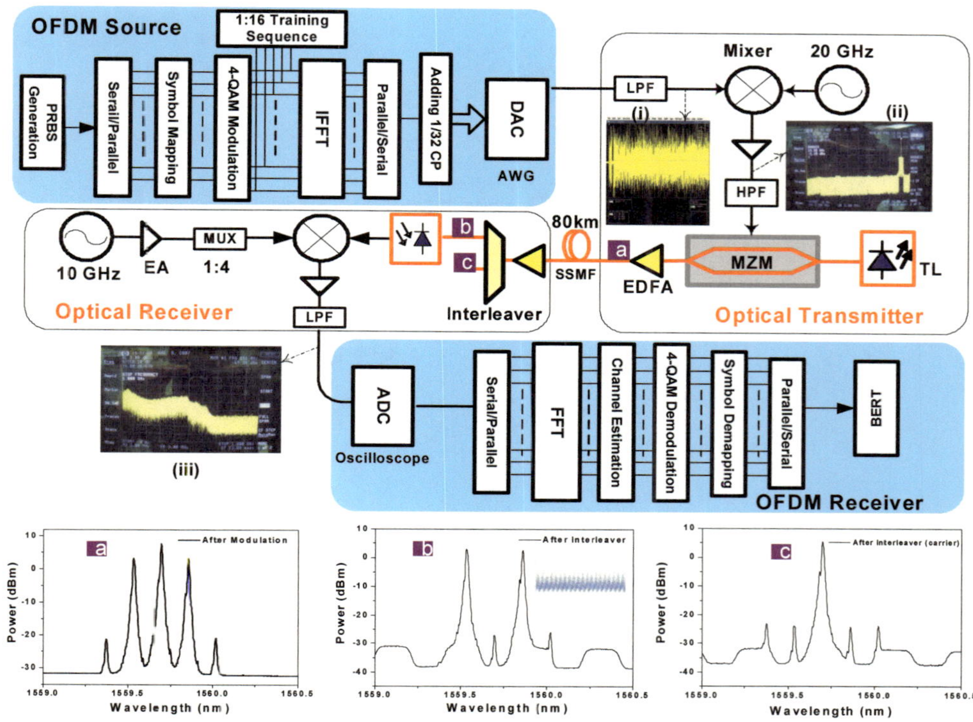

Fig. 15. Experimental setup for OFDM-ROF systems transmission over 80-km SSMF at 1Gb/s on 40-GHz mm-wave carrier. The electrical waveform and spectra are inserted. The optical spectra measured at point a, b, c are all inserted. The resolution for optical spectra is 0.01nm.

Usually, OFDM signal has a high peak to average power ratio (PAPR), and this high PAPR limits the transmission distance of optical OFDM. The PAPR of the optical OFDM can be reduced when the optical OFDM signal is generated by phase modulator because of its constant intensity of the optical OFDM signal. Recent experiment has confirmed this conclusion. In addition, advanced coding technique can be used to improve the receive sensitivity of the optical OFDM signal.

Figure 16 shows the principle of the OFDM-ROF architecture based on optical mm-wave signal generation by a phase modulator and turbo coding technique. The original data is encoded by turbo coding technique and then modulated in an OFDM block based on IFFT. The OFDM signal after mixing with the LO clock signal is used to drive a phase modulator to modulate lightwave from a CW lightwave source and generate optical mm-wave. Different kinds of signals have different PAPR. For example, OOK NRZ signal has a PAPR of 3dB, and OFDM intensity optical signal has more than 10dB. To reduce PAPR, one employs phase modulator for its constant amplitude. Comparing with the OFDM signal generated by an intensity modulator, optical OFDM signal generated by a phase modulator has two other benefits: 1) high OSNR due to relatively small insertion loss of a phase modulator, and 2) high stability without dc bias control system. Using advanced coding technique the receiver sensitivity can be improved over 2dB in the recent experiment.

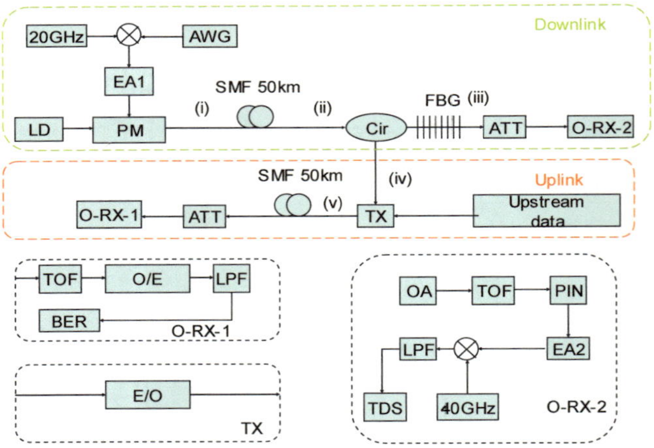

Fig. 16. Experimental setup of OFDM-RoF architecture based on phase modulator. AWG: arbitrary waveform generator, PM: phase modulator, EA: electrical amplifier, FBG: fiber Bragg grating, ATT: attenuator, O-RX: optical receiver, TX: transmitter, TOF: tunable optical fiber, LPF: low pass filter, BER: bit error tester, OA: optical amplifier, TDS: time domain scope (real time Osc.), O/E: optical/electrical converter, E/O: electrical/optical converter.

The experimental setup is depicted in Fig. 17. The center wavelength of the continuous lightwave (CW) generated by a distributed feedback laser-diode (DFB-LD) is 1543.72nm. The RF signal is generated by an electrical mixer to combine the 20GHz RF clock (sinusoidal wave) and 2.5-Gb/s OFDM signals. Then the mixed electrical signals are boosted to drive the phase modulator as shown in Fig. 16. The waveform and spectrum of OFDM source is depicted as inset (a) and (b) in Fig.16, respectively. In this experiment, the OFDM signal based on QPSK I/Q modulation scheme is generated by the Tektronix arbitrary waveform generator. Original data is encoded by recursive systematic convolution code (RSC1) firstly, and combine the interleaving original signal after RSC2. The length of interleaving is 1024. The generator vector of RSC is $g_0 = (1,0,11;1101)$. RSC encoding rate is 1/2. Then the combining data is punctured. Combining the original OFDM signal and puncturing data, encoding data is generated and modulated in the way of OFDM modulation. The Turbo coding rate is 2/3. In the OFDM frame there are 256 subcarriers, among them, 200 subcarriers are used for data and 56 subcarriers are set to zero as guard interval. The guard interval (cyclic prefix) in time domain is 1/32 which would be eight symbols every OFDM frame. The first and second order sideband is 20 and 50dB smaller than the carrier, respectively. The input power into transmission fiber is 13.2dBm. After 50-km SMF-28, the optical signals are divided into two parts including central optical carrier and two first order sidebands by a fiber Bragg grating (FBG) as depicted in Fig.16. The FBG is used to remove the central carrier and converted phase to intensity signals. An optical receiver (OR2) contains an EDFA as preamplifier and an optical band-pass filter (OBPF) with the 3dB

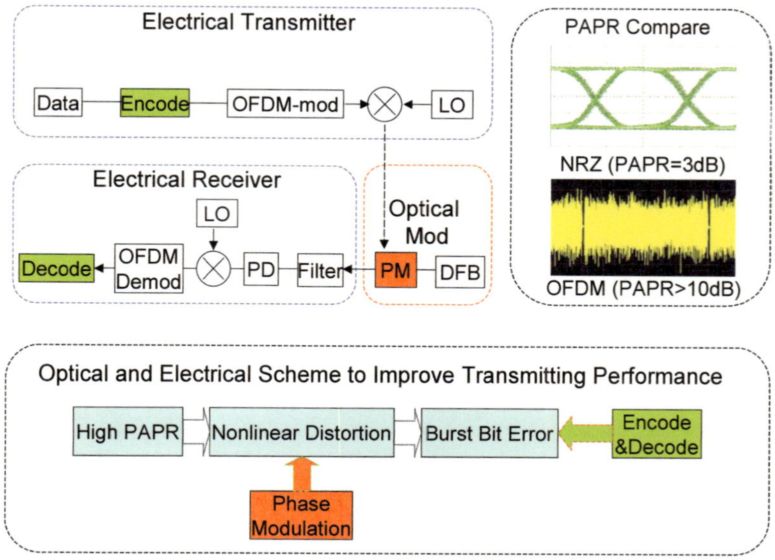

Fig. 17. Principle of the OFDM-ROF architecture based on a phase modulator with channel coding. Encode: Turbo encoding, LO: local oscillator, DFB: distributed feedback laser diode, PM: phase modulator, PD: photo-detector, Decode: Turbo decoding.

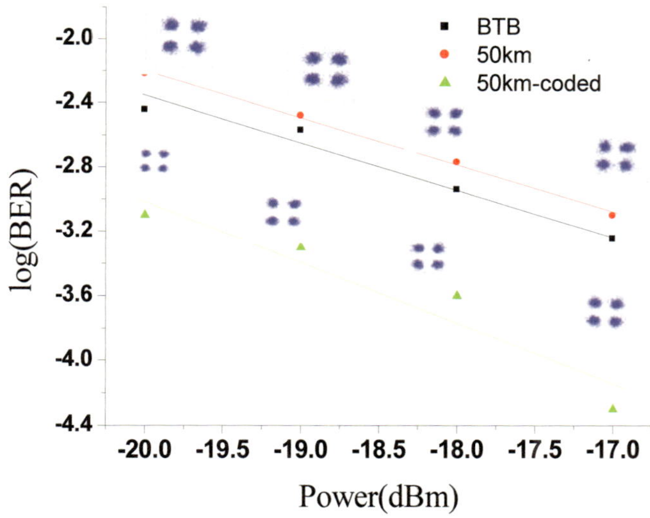

Fig. 18. BER curves for down-link OFDM optical signal.

bandwidth of 1nm as ASE noise block, respectively. Then the optical OFDM signals are converted to electrical signals by a PIN with the 3dB bandwidth of 50 GHz. After that, the converted electrical signals are down-converted to the baseband and sampled by a

Tektronix real-time oscilloscope and processed off-line. The center carrier is re-modulated by intensity modulator (IM) driven by 2.5Gb/s uplink PRBS data with a word length of 2^{31}-1. The received constellations of the OFDM signals before and after transmission over 50km downlink SMF-28 and the corresponding BER performance are shown in Fig. 18. A little expansion in the received constellation comes from the OSNR degradation. OFDM signals before and after transmission at a BER of $1x10^{-3}$ is -17.62 and -17.25 dBm. For upstream optical signal at 2.5Gb/s, there is no power penalty after transmission over 50km upstream fiber because the effect of the dispersion of 50km SMF-28 on the 2.5Gbit/s signals is very small.

4. Seamless integration of ROF with WDM-PON

Wavelength division multiplexed passive optical network (WDM-PON) has been regarded as a promising solution to meet access bandwidth requirements for delivering gigabits/sec data and video services to large number of users. The design of WDM-PON architecture is expected to be compatible with radio-over-fiber system without any change of optical line terminal (OLT) configuration to flexibly serve both fixed and mobile users. Here, we show two different architectures to realize this function of seamless integration of ROF with WDM-PON.

The first architecture is based on subcarrier multiplexing (SCM) technology to generate optical mm-wave signal and provide the lightwave source for up-stream reuse in WDM-PON network. The commercially available package of integrated SOA and EAM is used to be the uplink transmitter to increase the power margin through eliminating the need of RSOA and external modulators. Based on this scheme, the symmetric 2.5Gbit/s data signals per channel are transmitted over the same 40km single mode fiber (SMF) for both directions with less than 1dB power penalty.

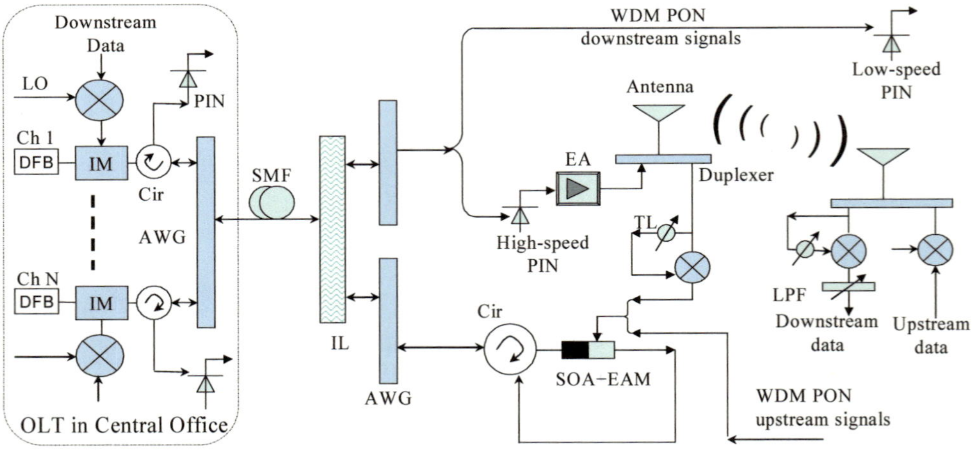

Fig. 19. WDM-PON access network compatible with ROF system. IL: interleaver, Cir: circulator, LPF: low-pass filter, TL: tunable line delay, IM: intensity modulator.

The network configuration is shown in Fig. 19. Each laser is modulated by means of SCM technique. A broadband downstream date after mixing with the LO clock signal are used to directly or externally modulate the lightwave. The WDM SCM signals are multiplexed by using an AWG before they are sent to the ONU after transmission over fiber. In the ONU or base-station, the WDM SCM signals are passed through an optical interleaver to realize the separation of the optical carrier and subcarrier signals where DI-MZ interferometer was employed. The optical interleaver has higher optical carrier suppression ratio and adjacent channel isolation compared to the DI-MZ interferometer. The separated optical carrier is then passed an AWG and delivered to the customer unit or base station before re-modulation by using an integrated SOA and EAM. Here SOA is used to provide the gain. Then the modulated upstream optical signals will be sent back to the OLT or central office after transmission over the same transmission fiber of the downstream signals before they detected by a 3R receiver.

For regular WDM-PON application, the separated subcarrier signals after AWG will be detected by a low-speed photo-diode before 3R receiver. For wireless service provided by ROF, the subcarrier signals are broadcasted by an antenna before detection by a high-speed receiver and amplification by a narrow-band electrical amplifier EA. At the customer unit, the wireless signal is received by an antenna and down-converted by a mixer to retrieve the baseband signals. For wireless upstream, the up-converted data will be sent back to the base station by wireless transmission. After amplification and down-conversion, it will modulate the EAM before it is sent back to the central office. In this scheme, we do not need to change the OLT configuration to make the WDM-PON compatible with ROF network.

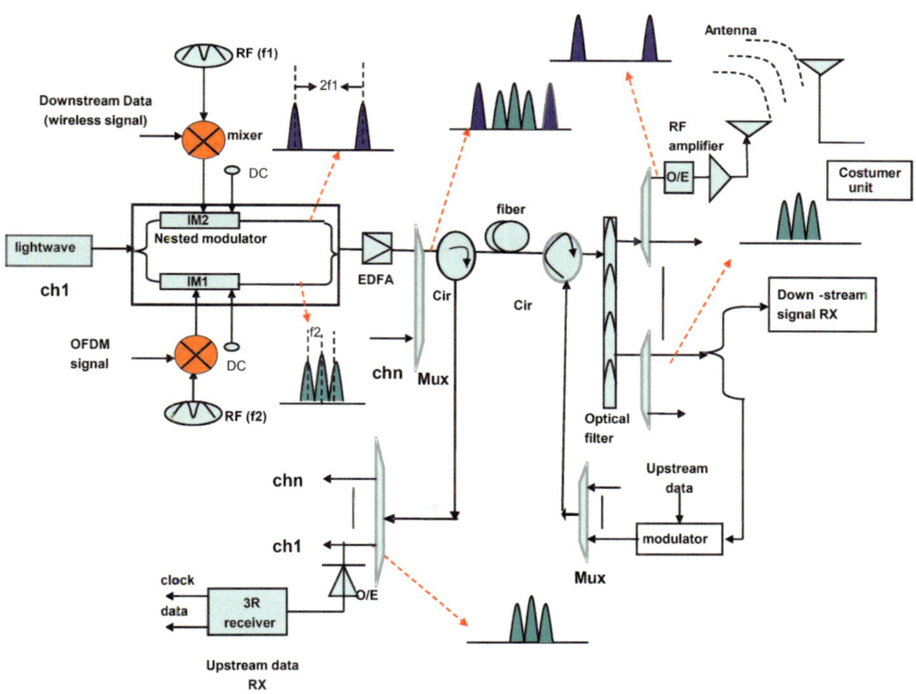

Fig. 20. Schematic diagram of the seamless integration of ROF with WDM-PON for providing triple service.

Other architecture is based on an external integrated modulator to generate optical mm-wave and provide lightwave for upstream reuse. Since the two arms in the integrated modulator do not affect each other, the generated mm-wave signal and the baseband optical signal have higher quality.

The proposed novel WDM-PON architecture is shown in Fig. 20. The CW lightwave is modulated by an external integrated modulator using x-cut $LiNbO_3$, consisting of two single-electrode Mach-Zehnder modulators (MZMs). The separated lightwave in the integrated modulator is modulated by intensity modulator (IM) IM-1 and IM-2 to generate SCM signal, respectively. The optical mm-wave is generated based on OCS modulation to carry the wireless signal by IM-1 while OFDM wired signal is based on DSB modulation by IM-2. The IM-1 and IM-2 has a 90 degree phase shift to avoid the interference between the up and bottom optical signal, hence there is no crosstalk between the mm-wave and baseband signal. The wireless signal and wired signal are combined by a waveguide coupler before all channels are multiplexed. After transmission over the optical fiber, a fiber filter is used to separate the OCS modulated downstream wireless signals and the double sideband (DSB) modulated downstream wired OFDM signals. The OCS modulated wireless downstream signals at different wavelengths are demultiplexed to the costumer unit by antenna before the O/E conversion. The DSB modulated downstream wired signals, after being demultiplexed, are sent to two paths. One part is converted to electrical OFDM wired signal by a low speed receiver. The other part is re-modulated by an IM driven by the upstream data. The WDM upstream data are demultiplexed before they are O/E converted.

5. Wavelength reuse for upstream

It has been expected that mm-wave bands would be utilized to meet the requirement for high bandwidth and overcome the frequency congestion in the optical-wireless networks. The negative side of mm-waves is the need for numerous BSs, which is a consequence of high RF propagation losses in the atmosphere. In this situation, it is necessary to minimize the cost of the BS and shift the system complexity and expensive devices to the CO. Hence, the overall architecture design and the scheme of RF signal generation, transmission for the uplink and downlink play the key roles on the successful deployment in the real networks. Fig. 21 shows two ROF systems with downstream and upstream connection. In Fig. 21(a), it shows a simple ROF system. The CW lightwave is modulated for up-conversion to generate optical mm-wave signal. The optical mm-wave signal is delivered to the base station by optical fiber. The optical fiber can be standard single mode fiber for long distance (>1km) transmission or multi-mode fiber for short distance transmission. The fiber loss can be compensated by optical amplifier. In the base station, the optical mm-wave signal is detected by a high-speed photodiode. Then the electrical mm-wave signal is boosted by a narrow-band electrical amplifier (EA) before it is broadcasted by antenna to the customer unit. The electrical upstream data obtained from the antenna is boosted by an narrow band EA. The boosted electrical upstream signal is used to drive an external modulator such as intensity modulator, electro-absorption modulator (EAM). The CW lightwave for upstream connection is generated from a lightwave source. The generated upstream mm-wave signal is delivered to the central office by upstream fiber. The upstream fiber and downstream can be the same fiber, however, it may have crosstalk between them. Obviously, in this architecture, an additional CW lightwave source is necessary at the base station. To simplify the base station, we can use the architecture shown in Fig. 21(b). Here the CW lightwave for upstream connection is extracted from the downstream optical signals. It means that a part of downstream signal is reused. This is quite similar to that technique used in WDM-PON architecture with source-free at remote node. For uplink connection, some methods have been recently proposed. However, most of them only demonstrated uplink connections over short transmission distances. Full-duplex operation using high RF carrier still raises difficulties that have to be addressed. The network architecture consisting in a single light source at the CO and the reuse of the downlink wavelength at the BS is an attractive solution for low-cost implementation as it requires no additional light source and wavelength management at the BS.

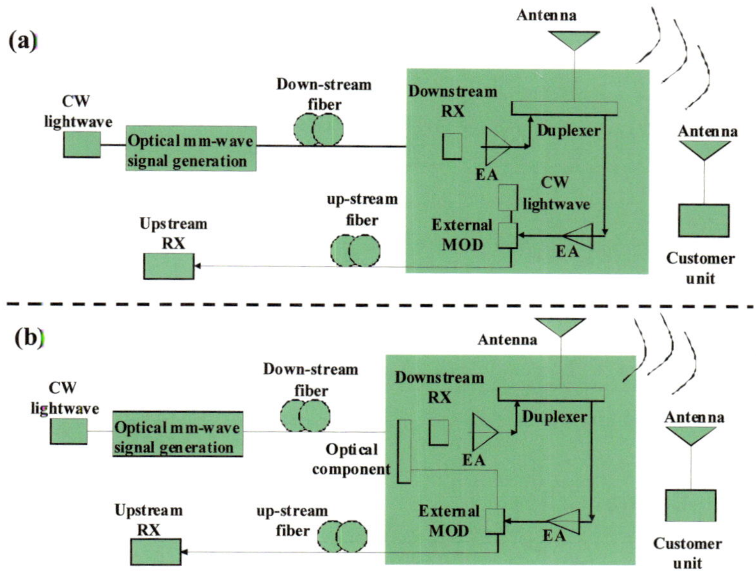

Fig. 21. ROF system with downstream and upstream connection.

5.1 IM with subsequent filter for downstream and OOK upstream generated by an external modulator

Figure 22 shows the principle to generate optical mm-wave by using DSB modulation along with optical filtering, and wavelength reuse for uplink connection. An external modulator such as $LiNbO_3$ modulator or electro-absorption modulator driven by RF signal can be used to generate DSB modulation signal. Usually the first and second sidebands will be generated after modulation. Here one can use an optical interleaver to separate the first order sideband from the optical carrier and the second order sideband signals. The two peak modes of the first order sideband will be beat to generate an optical mm-wave with a double repetitive frequency of the RF signal when they are detected by an optical receiver. The baseband down-link data is added on the optical mm-wave by using additional external intensity modulator. The separated optical carrier and the second order sideband signals are further separated by the second interleaver. After this, we can obtain a pure optical carrier. The pure optical carrier will be combined with the downlink optical mm-wave signal by using a 3dB optical coupler. In the base station, we use the third optical interleaver to separate the downlink optical mm-wave and the pure optical carrier. The separated downlink optical mm-wave signal is received by a high-speed detector, then, it is boosted by a narrow-band electrical amplifier before it is broadcasted by an antenna. The received uplink data from the antenna firstly are down-converted by using an electrical mixer. The down-conversion can be realized by using an electrical mixer without LO. The baseband uplink data is used to drive an external modulator to generate optical uplink data before it is transmitted over the fiber to reach the central office. Since the optical interleaver has periodic characteristic, this scheme for this ROF architecture including optical mm-wave generation and wavelength reuse for uplink connection can be used to generate the DWDM optical mm-wave and deliver them by sharing the same interleavers.

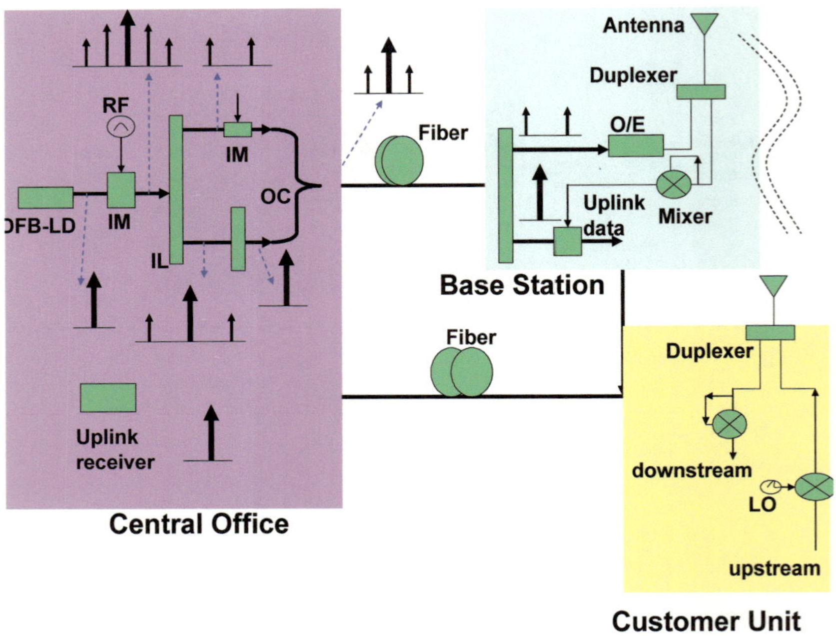

Fig. 22. Principle of DWDM optical mm-wave generation using phase modulator and optical interleaver. IL: Interleaver, IM: intensity modulator, OC: optical coupler, O/E: optical/electrical converter, LO: local oscillator.

In real network implementations, the diplexer connected with the antenna would act as a circulator to handle the up- and down-stream signals at the BS. The baseband upstream signals would be obtained after down-conversion of the end user's information coming from the diplexer in the BS. In this experiment, the same fiber length is used for both up-and down-streams. The uplink signal is detected by a low-frequency response receiver which also filters out the residual part of mm-wave signals.

5.2 OCS for downstream and reuse for upstream

The schematic diagram is shown in Fig. 23. OCS scheme is employed to generate optical mm-wave and up-convert baseband data signal simultaneously for the downstream. The original carrier is split prior to OCS operation and then coupled with optical mm-wave signal before they are transmitted to the BS. At the BS, a fiber Bragg grating (FBG) is used to reflect the carrier while passing the optical mm-wave signal to downlink receiver. The reflected carrier is acted as the CW and re-modulated with the symmetric upstream signal, then transmitted back to CO, where a low-cost receiver with low-frequency response detects the upstream signal.

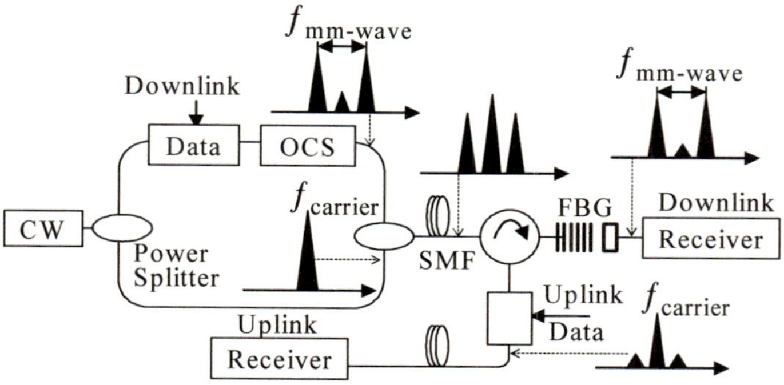

Fig. 23. A full-duplex ROF system based on OCS and reuse.

The experimental setup for the full-duplex ROF system is shown in Fig. 24. At the CS, a CW is generated by a TL at 1549.1 nm and split into two parts via a 50:50 optical PS. The first part is modulated via an M-Z modulator driven by 2.5-Gb/s signals. Optical mm-wave signals are generated by using OCS scheme. The optical spectrum and eye diagram after OCS are measured at point A, B and inset (i), (ii) in Fig. 24, respectively. Then the generated optical mm-wave is amplified by an EDFA to get a power of 6 dBm for transmission. The second part is sent directly for amplification by an EDFA to obtain a power of 9 dBm and then combined with the first part via OC before they are transmitted over SMF.

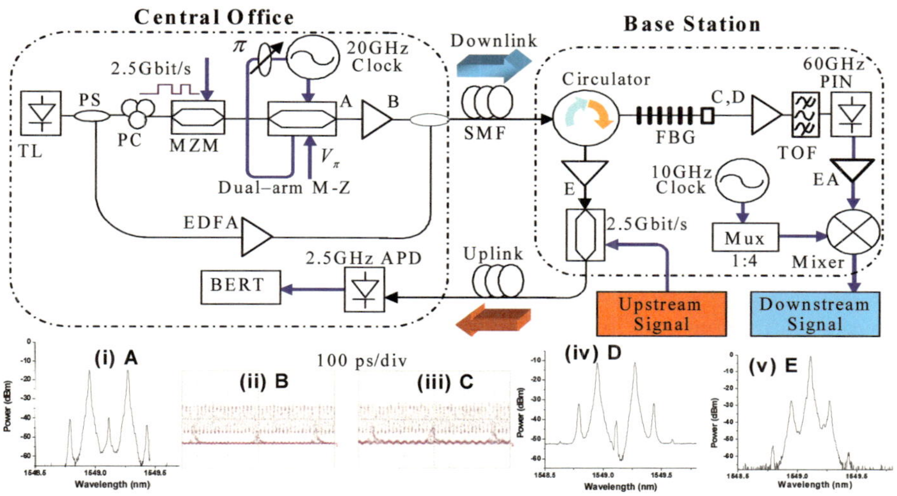

Fig. 24. Experimental setup for full-duplex ROF system based on optical carrier suppression and reuse.

At the BS, the FBG is used to take on two roles: one is to reflect the optical carrier to provide CW light source for uplink connection; the other is to pass the two sidebands generated by OCS simultaneously and as a consequence, it increased the carrier suppression ratio up to 30 dB due to sharp notch characteristics. This FBG filter has a 3-dB reflection bandwidth of 0.2 nm and reflection ratio larger than 50 dB at the reflection peak wavelength. The eye diagram, the spectrum of passing part and reflected part are measured at point C, D, E and inset (iii), (iv) and (v) in Fig.24. Then we use the identical O/E and down-conversion to retrieve the downstream signals. Fig. 25 shows that the evolution of dispersion impact on different distances for two directions signals. It is clearly seen that the eye still keeps open in despite of transmission over 40-km SMF, which is long enough for access network coverage. But the uplink might transmit longer distance because most of the component of high frequency is removed already by FBG filter before it is sent back to the CO. For the uplink, the reflected signal is amplified by an EDFA to compensate the insertion loss of the filter before it is modulated by symmetric 2.5-Gb/s electrical signal.

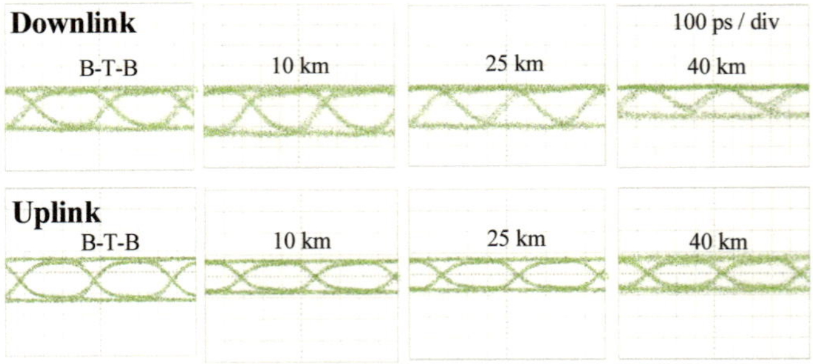

Fig. 25. Electrical eye diagrams after different length transmission.

5.3 PM with subsequent filter for downstream and directly modulated SOA for upstream

Fig. 26 shows the schematic diagram of full-duplex optical-wireless system architecture by using PM and SOA. An optical PM is driven by small RF signal (1/4 half-wave voltage of the PM) to create first-order sidebands while suppress the second-order components for increasing dispersion tolerance. An interleaver is employed to separate the optical carrier from the first-order sidebands to generate optical mm-wave carrier. After modulation by downlink data, the up-converted optical signal is combined with the original optical carrier and transmitted over the SMF. At the BS, an FBG is used to reflect the optical carrier while pass the optical mm-wave signal to O/E conversion. Then, the boosted electrical mm-wave signal is broadcasted by an antenna through a duplexer acting as a circulator to handle up- and down-stream signals at the BS. On the other hand, the upstream signals are down-converted through a mixer and tunable delay (TD) line without the need of LO signal.

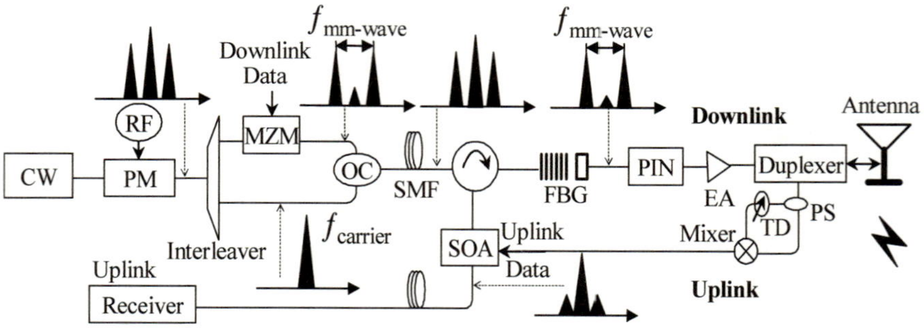

Fig. 26. Architecture of a full-duplex optical-wireless system by using PM and SOA.

The reflected optical carrier is considered as the CW source and directly modulated by baseband upstream signals in the SOA and sent back to the CS, where a low-cost and low-frequency response receiver is used for detection. In this scheme, the SOA performs the function of both amplification and modulation, which eliminates the requirement of optical amplifier and external modulator at the BS.

Fig. 27 depicts the experiment setup for the full-duplex optical-wireless system by using wavelength reuse and directly modulated SOA. At the CO, A CW is generated by a DFB-LD at 1534.4 nm and modulated by an optical PM driven by a 20-GHz RF sinusoidal wave with amplitude 1 V (half-wave voltage of the PM is 4 V). The optical spectrum (measured at point a) after modulation is shown in Fig. 27 as inset (a). A 50/25-GHz optical interleaver with 35-dB channel isolation is used to separate the remained optical carrier from the first-order sidebands. The generated optical mm-wave is then amplified and modulated by an IM driven by 2.5-Gb/s PRBS 2^{31}-1 electrical downstream signals. The separated optical carrier is directly sent to combine with up-converted baseband signals via OC before its transmission over 40-km SMF-28 with 8-dBm input power. The optical spectra of the separated optical carrier, optical mm-wave signal and combined signals are shown in Fig. 27 as inset (b), inset (c) and inset (d), respectively. At the BS, a FBG is used to reflect the optical carrier and transmit the optical mm-wave signals simultaneously. The FBG filter has a 3 dB reflection bandwidth of 0.2 nm and reflection ratio larger than 50 dB at the reflection peak wavelength. The spectra of reflected carrier and transmitted optical mm-wave signals are shown in Fig. 26 as inset (e) and inset (f). The downstream signals are down-converted through a mixer and TD line without requiring any LO signal.

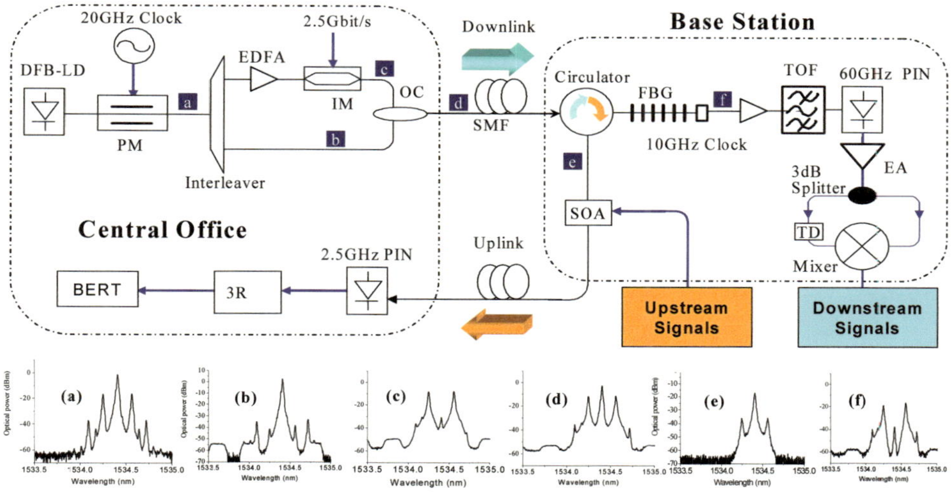

Fig. 27. Experimental setup for the full-duplex optical-wireless system by using wavelength reuse and directly modulated SOA. The optical spectra measured at point a, b, c, d, e, f are all inserted as insets.

For the uplink, the reflected optical carrier is directly modulated in an SOA driven by a 250-Mb/s (PRBS 2^7-1) electrical signal with amplitude 3.1-V and 165-mA bias. The gain is 10 dB in the 34-nm spectral width and the polarization sensitivity is smaller than 0.5 dB. In this experiment, the same fiber length is used for both up- and down-streams. The uplink signal is detected by a low-frequency response receiver which also filters out the residual part of the high-frequency mm-wave signal due to imperfect filtering by the FBG.

Fig. 28 shows the optical eye diagrams in B-T-B configuration and after the 40 km propagation in the SMF. The power fluctuations of the 40-GHz modulation arise from the chromatic dispersion in the fiber. The power penalty is measured. For the downlink, the power penalty at the given BER of 10^{-10} is 2.0 dB after 40-km transmission. For the uplink, the power penalty for the re-modulated optical carrier is around 0.5 dB over the same transmission distance. The eye diagrams of B-T-B configuration for both directions are also inserted in Fig. 28.

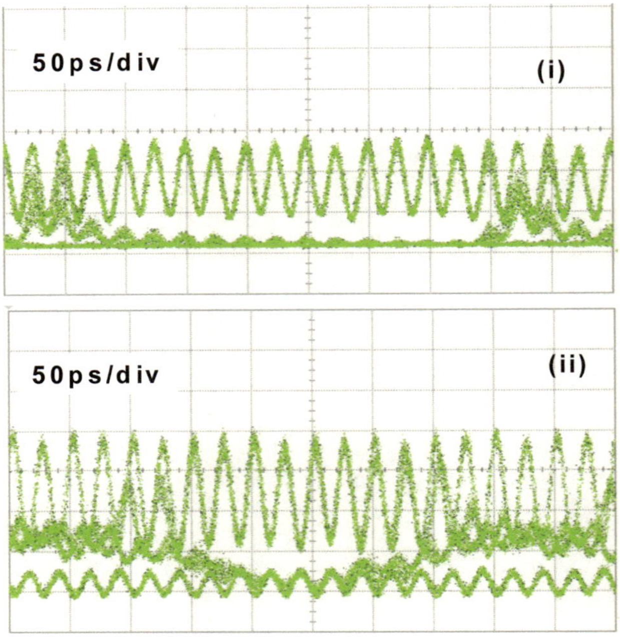

Fig. 28. Optical eye diagrams for downstream signals at (i) B-T-B and (ii) after 40km transmission.

5.4 Integrated modulator for optical mm-wave and baseband generation and wavelength reuse

The experimental setup for integrated 60-GHz bidirectional ROF and WDM-PON system using a single I/Q (nested) modulator is shown in Fig.28. At the central office, a CW lightwave is generated by a tunable laser (TL) at 1557.1nm. The I/Q modulator integrates two intensity Mach-Zehnder modulators and a phase shifter. The intensity modulator at the upper arm is driven by 10-Gb/s electrical signals with2Vpi (half-wave voltage of 5V) and PRBS length of 2^{31}-1. The modulator is biased at the null point to generate phase modulated optical signal (DPSK signal). The intensity modulator at the bottom arm is driven by a RF signal mixed with 30-GHz clock and 2.5-Gb/s baseband signal with the same PRBS length. This modulator is also biased at the null point. Through adjusting the DC bias on the phase shifter, we generate 90 degree phase difference between the upper and bottom arm optical signal. In this way, coherent crosstalk is reduced between the two arms. Optical signals are then boosted to 16 dBm before they are launched into the 20-km standard SMF-28. The optical spectrum before the EDFA is shown in Fig. 29 inset (a). After transmission, a 25-GHz optical interleaver is used to separate the optical wired signals (baseband signals) from the wireless signals (60-GHz optical mm-wave). The optical spectra of the two output ports of the IL are shown in Fig. 29 insets (b) and (d). The wired optical signal is divided into two parts. One is detected by a standard DPSK balanced receiver. Another part is reused for upstream connection. The eye diagram of this downstream DPSK signal at the BS is shown in Fig. 29 inset (i)., Another 50-GHz optical interleaver is employed to further suppress wired signal as shown in inset (c). The 60-GHz optical mm-wave signals are pre-amplified by a regular EDFA with a small-signal gain of 30 dB, and then filtered by a tunable optical filter (TOF) with the bandwidth of 0.8 nm to suppress ASE noise before O/E conversion via a PIN PD with a 3-dB bandwidth of 60 GHz. The converted electrical signal is then amplified by an electrical amplifier (EA) with a bandwidth of 10 GHz centered at 60 GHz. An electrical LO signal at 60 GHz is generated by using a frequency multiplier from 15- to 60- GHz. We use the electrical LO signal and an electrical mixer to down-convert the electrical mm-wave signal to the baseband. The optical eye diagrams of the 60-GHz optical mm-wave signals before and after down-conversion are shown in Fig. 29. The power penalty is negligible at a BER of $1x10^{-10}$. The wired BPSK optical signals are detected by a standard balanced receiver, which includes a one bit M-Z delay interferometer (MZ-DI, demodulator) and a balanced detector. There is no power penalty after transmission over

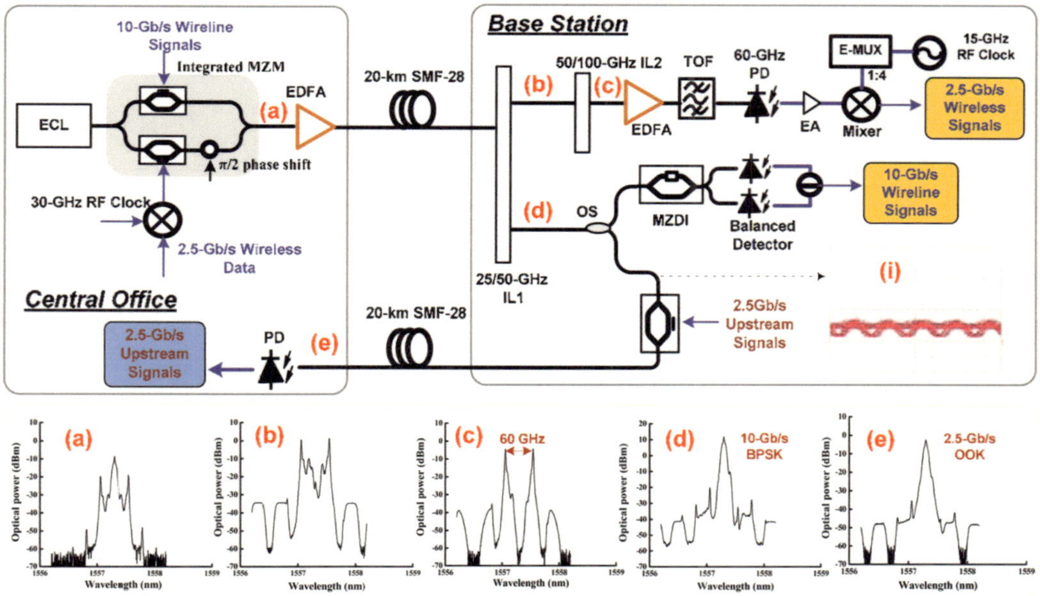

Fig. 29. Experimental setup. ECL: external cavity laser, MZM: Mach Zehnder modulator, IL: interleaver, TOF: tunable optical filter, EA: electrical amplifier, OS: optical splitter, MZDI: Mach Zehnder delay interferometer, resolution: 0.01nm.

20-km SSMF-28. The optical DPSK signal is re-modulated for upstream reuse by one intensity modulator driven by symmetric 2.5-Gb/s PRBS electrical signals with a word length of 2^{31}-1. In real networks, the diplexer connected with the antenna would be used to circulate the transmitting and receiving signal at the BS. The baseband upstream signals will be obtained after down-conversion of end user's information coming from the diplexer in the BS. In this experiment, after transmission over the same length SMF-28 as the downlink channel, the upstream optical signals are detected by a commercial APD receiver. The optical spectrum of the upstream optical signal after transmission over 20-km uplink fiber is shown in Fig. 29(e). The eye diagrams of optical upstream signals are shown in Fig. 30. Fig. 30 (e) is detected by a 60-GHz photodiode, in which the transit peak of the downstream DPSK signal exists. After detected by a 2-GHz photodiode, the transit peak of DPSK signal is removed as Fig. 30(f). Negligible power penalty is observed after transmission for the upstream signal.

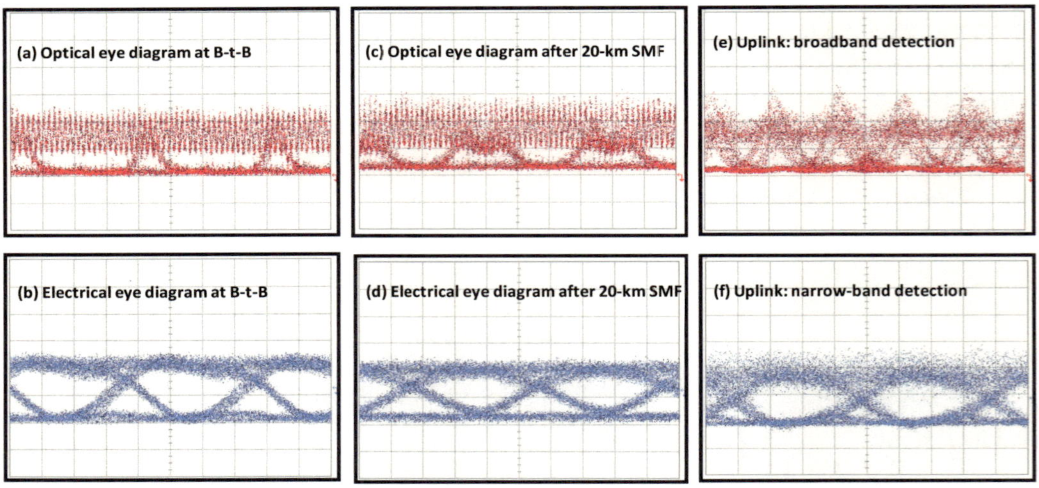

Fig. 30. Measured eye diagrams of downstream (a-d) and upstream (e-f) over 20-km SMF-28.

6. Optical OFDM signals over GI-POF

Perfluorinated graded-index polymer optical fibers (GI-POFs) can provide large bandwidth and low attenuation (60dB/km) at 850–1300nm, so it is a good replacement and a low cost alternative to traditional glass. With ease of use and affordability, GI-POFs make an excellent choice for the installation of high performance fiber networks. In addition, GI-POFs provide a higher transmission bandwidth than any other type of plastic optical fiber. Recently, a few 40Gb/s transmission experiments have been demonstrated. Until recently, all commercially available POFs have been fabricated from non-fluorinated polymers such as PMMA and, as a result, have had a refractive index that changes in steps. Although inexpensive, these fibers are characterized by large modal dispersion and typically operate at 530nm or 650nm, which is well outside of standard communication wavelengths (850nm or 1300nm), which is where high-speed transceivers are readily available. Due to the high attenuation in the near infrared, these fibers are restricted to low performance (<100Mb/s), short range (<50m) applications in the visible region. With the advent of an amorphous perfluorinated polymer, polyperfluoro-butenylvinylether (commercially known as

CYTOP®), the limitations presented by step-index POFs have been overcome. Perfluorinated fiber exhibits very low attenuation in the near infrared (~10dB/km) as shown in Fig. 5. Moreover, since the perfluorinated optical fiber can be constructed with a graded refractive index, it is capable of supporting bandwidths that are 100 times larger than those provided by conventional POFs. This is due to the interplay between high mode coupling, low material dispersion, and differential mode attenuation. Unlike conventional glass fibers, which suffer from high interconnection and receiver costs, perfluorinated GI-POFs are easy to install. To add a connector to a glass fiber, the fiber needs to be cleaved using an expensive, specialized tool. Then, epoxy is used to attach the fiber to the connector hardware. Finally, the assembled connector must be polished. In contrast, the GI-POF can be terminated using simple and inexpensive tools, connectors are crimped on, and polishing occurs in mere seconds, leading to a high quality optical link in a fraction of the time. Moreover, GI-POFs are compatible with standard multimode glass fiber transceivers. As an example, Table 2 lists the specification of the commercial GI-POF from Thorlabs.
Since 2006, a few world records by employing GI-POF have been achieved. The optical fiber communication conference (OFC) 2006, Georgia Tech's researchers reported that 30Gb/s on/off keying (OOK) signals are transmitted over 30m GI-POF. In ECOC 2007, Schollmann et al., reported the 40Gb/s OOK signals are delivered over 50m GI-POF with new designed multi-mode high-speed receiver. In OFC 2008, Yu in NEC Labs America reported 42.8Gb/s optical signal generated by chirped managed laser transmission over 100m GI-POF. In ECOC 2008, Yu in NEC Labs America demonstrated 16Gb/s OFDM signal transmission over 50m GI-POF. In OFC 2009, Yang reported 40Gb/s signal transmission over 100m GI-POF based on discrete multimode modulation. By using the new spectral efficiency modulation format, such as CML and OFDM, can furthermore increase the bandwidth of the GI-POF.

Fiber model	50SR	62SR	120SR
Attenuation at 850nm	<60dB/km		
Attenuation at 850nm	<60dB/km		
Bandwidth at 850nm	>300MHz·km		
Numerical aperture	0.190+0.015	0.190+0.015	0.190+0.015
Macrobend loss	<0.25dB	<0.35dB	<0.60dB
Zero dispersion wavelength	1200~1650nm		
Dispersion slope	<0.06ps/nm^2·km		
Core diameter	50+5um	62.5+/-5μm	120+/-10μm
Cladding diameter	490+/-5μm		
Temperature induced attenuation at 850nm (-20 to +70°C)	<5dB/km		

Table 2. Specification of Thorlab's GI-POF.

In this section we will experimentally demonstrate the transmission of upconverted 16Gbit/s OFDM signals on 24GHz microwave carrier over 50m GI-POF at 1310nm. The experimental setup of the proposed OFDM signals transmission over GI-POF is shown in Fig. 31. The lightwave from the DFB laser-diode (LD) at 1310nm with the output power around 10dBm is modulated by an intensity modulator (IM) driven by up-converted OFDM signals. The 16Gbit/s OFDM signals are generated by OFDM transmitter and then up-converted to 24GHz to realize RF-OFDM signals via an electrical mixer. The up-converted spectrum is inserted in Fig. 29. We can see that the bandwidth of the OFDM signal is 8GHz. The OFDM baseband signal is generated offline and uploaded into a Tektronix AWG7102. The waveforms produced by the arbitrary wave generator (AWG) are continuously output at a sample rate of 20GHz (8bits DAC, 4GHz bandwidth). The FFT size is 256, from which 200 channels are used for data transmission, 55 channels at high frequencies are set to zero for over-sampling, and one channel in the middle of the OFDM spectrum is set to zero for DC in baseband. 10 training sequences are applied for each 150 OFDM-symbol frame in order to enable phase noise compensation. At the output of the AWG, the low-pass filter (LPF) with 5GHz bandwidth is used to remove the high-spectral components. Subsequently, the RF-pilot tone is created by inserting a small DC offset before an analogue I/Q mixer is used to up-convert the OFDM signal from the baseband to an 8.5GHz intermediate frequency (IF). The electrical spectrum of the original signal is shown in Fig. 32(a) that was measured at the point (a) in Fig. 31. The IM is driven by the OFDM signals to create double sideband (DSB) optical signals. The bias and the power of the RF signals are carefully adjusted to obtain proper power ratio between the optical carrier and the first-order sideband signals. After IM, the signal was launched into 50m of commercially available GI-POF for transmission. The core of the GI-POF is 50 μm with 60dB/km attenuation at 1300 nm. The signal power launched and output of GI-POF was 5.5 and 2.5dBm. A PIN receiver is used in the receiver side with the bandwidth of 29GHz and a 50 m multimode-coupled input. Before low pass filter (LPF), a 24GHz electrical LO signal is mixed to down-convert the electrical signal to its baseband form. The down-converted signals are sampled

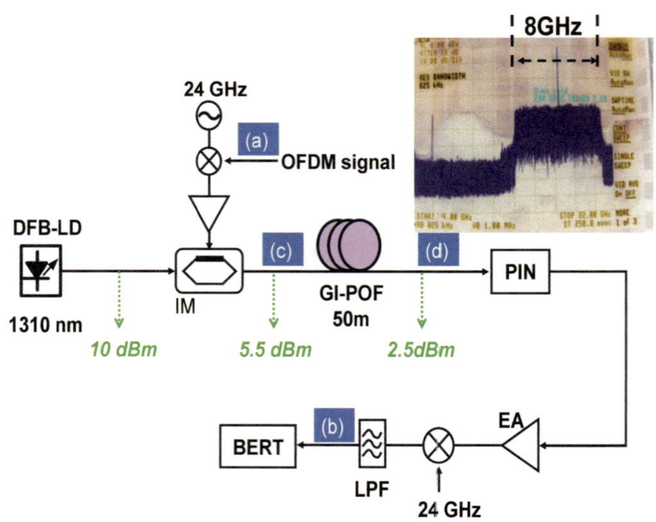

Fig. 31. Experimental configuration for 16Gb/s OFDM transmission over GI-POF. EA: electrical amplifier; IM: intensity modulator; GI-POF: graded-index plastic optical fiber; PIN: receiver; LPF: low pass filter. Inset: electrical spectrum of the OFDM signal after up-conversion.

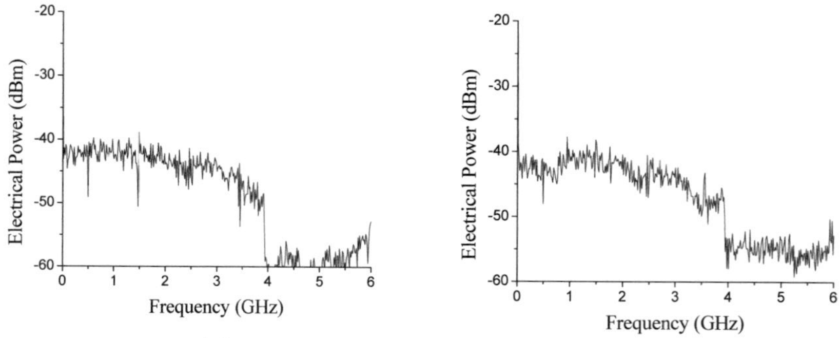

(a) (b)

Fig. 32. Received electrical spectra: (a) after arbitrary waveform generator, (b) after LPF; received optical spectra with 0.01nm resolution.

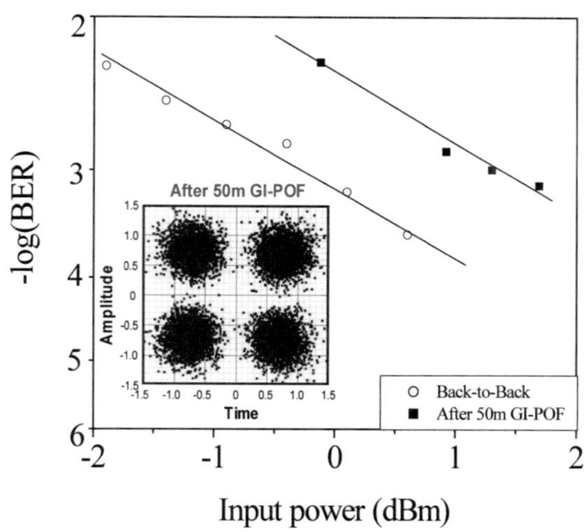

Fig. 33. Measured BER curves and the constellation figure of back-to-back and after 50m GI-POF.

with a real-time oscilloscope (Tektronix 6154C) and processed off-line. The electrical spectrum of down-converted signals is shown in Fig. 32(b). The measured BER of back-to-back and after transmission is shown in Fig. 33 and the constellation figure after 50m GI-POF is inserted. One million bits have been evaluated for all values of BER reported in this work. We can see that there exists signal degradation after 50m GI-POF. But the BER is still lower than $1x10^{-3}$, which is below the limitation of forward error correction (FEC) at $2x10^{-3}$. The main reason is the degradation of optical signal-to-noise- ratio (OSNR) from the fiber with an insertion loss of 3dB and modal dispersion.

7. Testbed implementation for delivery of wired and wireless services simultaneously

Currently, the wired and wireless services are separately provided by two independent physical networks. Wired networks based on fiber-to-the-home (FTTH) access technologies provide huge bandwidth to users but are not flexible enough to allow roaming connections. On the other hand, wireless networks offer mobility to users, but do not possess abundant bandwidth to meet the ultimate demand for video services with HDTV quality. Therefore, seamless integration of wired and wireless services for future-proof access networks will lead to convergence of ultimate high bandwidth for both fixed and mobile users in a single, low-cost transport platform. This can be accomplished by using our developed hybrid optical-wireless networks, which can not only transmit wireless signals at the BS over fiber, but also simultaneously provide the wired and wireless services to the end users due to the cascaded modulation scheme for downstream signals. Optical mm-waves can be generated by several all-optical up-conversion schemes that we have discussed in Section 2. No matter what kind of all-optical up-conversion scheme, one chooses, a part of base-band signal still exists in the whole electrical spectrum after all-optical up-conversion. To illustrate this point, we simulate the process of all-optical up-conversion schemes based on FWM in HNL-DSF and OCS modulation. As depicted in Fig. 34, it shows the electrical spectra after all-optical up-conversion. The LO frequency for all-optical up-conversion scheme based on FWM in fiber is 40 GHz and the baseband signal is 2.5-Gb/s NRZ signal. Fig.33 shows that there are

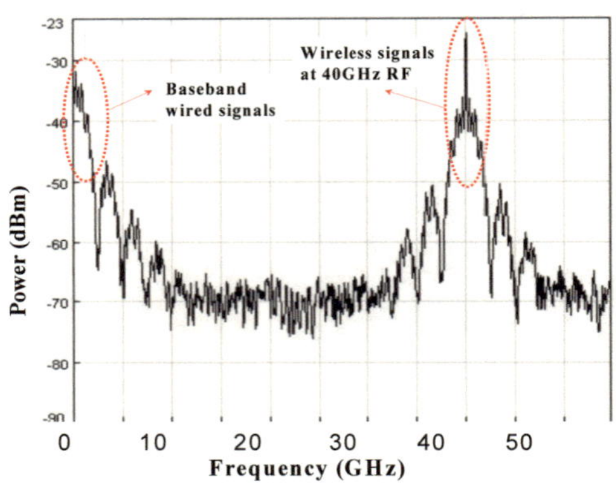

(a)

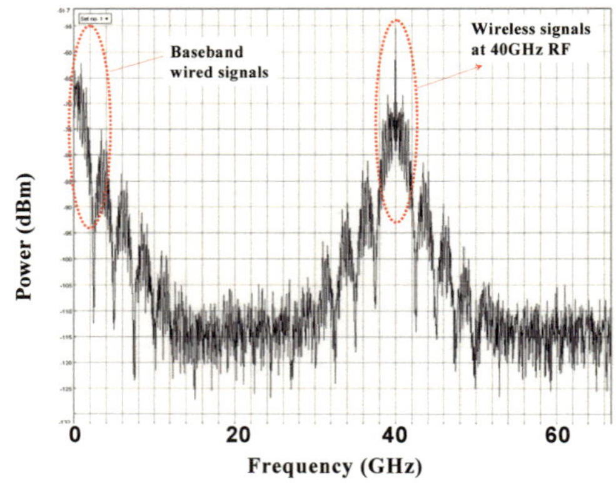

(b)

Fig. 34. Electrical spectra after all-optical up-conversion schemes. (a) FWM in HNL-DSF : LO signal is 40 GHz, (b) OCS modulation: RF driving frequency is 20 GHz.

two components of electrical signals after all-optical up-conversion, one part occupies the base-band, the other occupies high-frequency band near 40 GHz. Hence we propose a novel ROF network architecture to use the baseband signals for broadband optical access at 2.5 Gb/s at low cost and RF frequency for wireless connection.

Fig. 35 illustrates the architecture for concurrently providing super broadband wireless and wired services. The content providers or upstream networks send the data to the CO, where the multi-channel mm-wave carriers are generated through external modulator based on our developed all-optical up-converter. This up-conversion technique posses many advantages that allow data to be transmitted over wired and wireless medium in a single platform. First, the generation of the mm-wave carrier and the up-conversion of the original data channel are performed simultaneously in the optical domain. Second, as a result of this process, two identical data signals are generated concurrently: one at the electrical baseband and another at the RF-carrier frequency. The up-converted signals are multiplexed before they are transmitted over fiber to the BS where an AWG in WDM- PON is used to route the signals to customer premise. At the customer premise, the signal is divided by a PS into two. The first part is detected by a high-speed receiver and then electrically amplified using a narrow-band RF amplifier before broadcasted by an antenna as a wireless signal. The other part is sent directly to a wall-mounted optical port via fiber access, and a user can utilize a simple, low-cost receiver to detect the baseband data signal by filtering out the high frequency mm-wave signal. This newly hybrid system can allow wired and wireless transmission of the same content such as high-definition television, data and voice up to 100 times faster than current networks. The same services will be provided to customers who will either plug into the wired connection in the wall or access the same information through a wireless system. The customer premise can be conference centers, airports, hotels, shopping malls and ultimately homes and small offices.

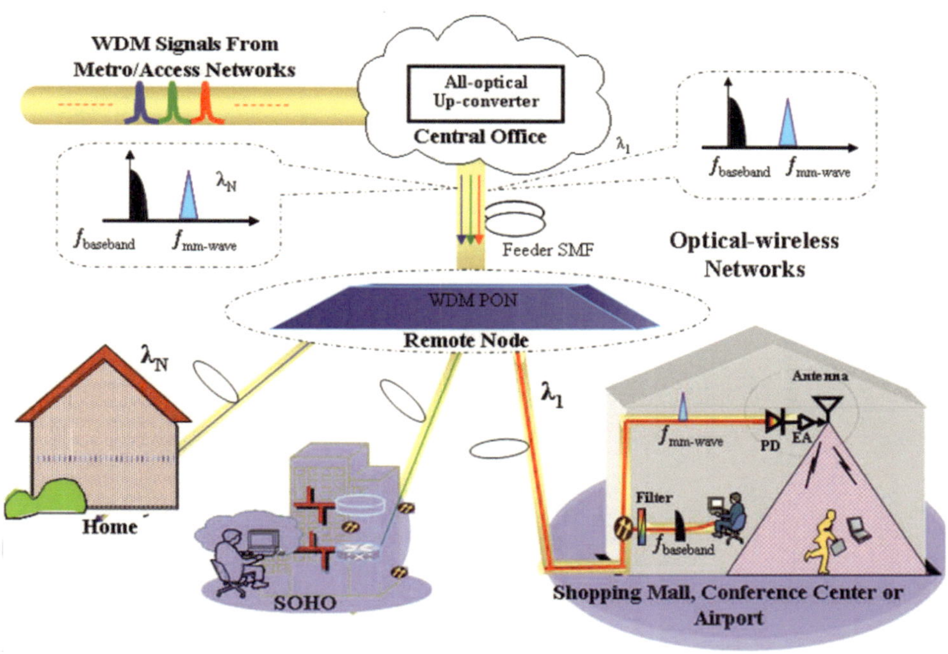

Fig. 35. Novel network architecture for providing dual-service in optical-wireless networks.

The experimental testbed setup for delivering wired and wireless services simultaneously is shown in Fig. 35. At the BS, the optical mm-wave is divided into two parts. The wired part is sent to a low-speed avalanche photodiode (APD) that has a 3-dB bandwidth of 2 GHz. Since the bandwidth of the APD receiver is limited at 2 GHz, the RF signal at high frequency is filtered out. In the case of the wireless part, the converted electrical signal is boosted by a RF electrical amplifier before it is broadcasted through a double ridge guide antenna with a gain of 19.2 dBi at frequency range of 18 to 40 GHz. After wireless transmission, the signal is received by another identical mm-wave antenna. The signal is down-converted through a mixer and TD line by using part of the incoming signal as the LO signal. The receiver sensitivities and eye diagrams at different air distances are shown in Fig. 37. The power

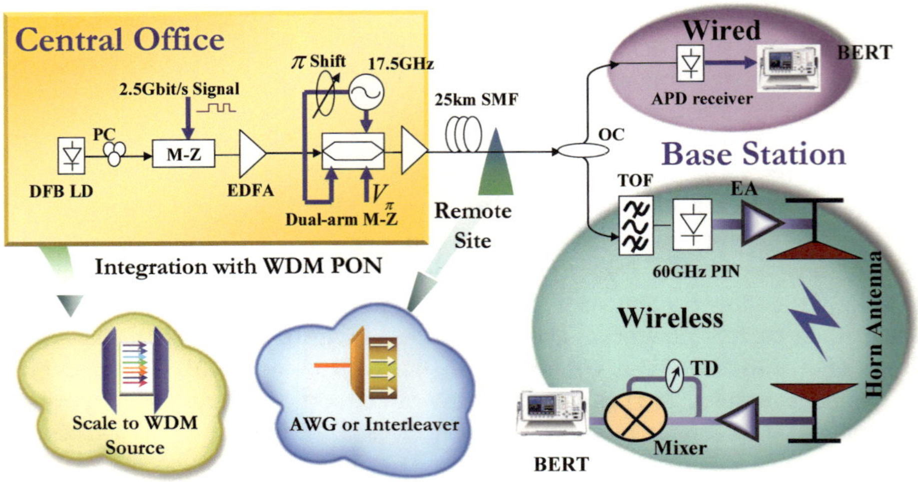

Fig. 36. Testbed setup for delivering wired and wireless services simultaneously.

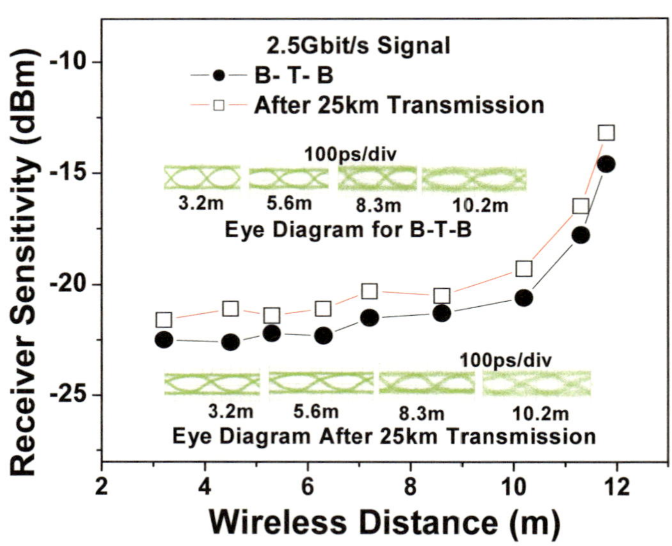

Fig. 37. Receiver Sensitivity for 2.5-Gb/s signal at different air distance.

penalty after transmission over 25-km SMF is less than 1.5 dB. The receiver sensitivity degrades quickly when the wireless signals are transmitted beyond 10-m indoor air space because the antenna as an isotropic point source in its radiation direction spreads its energy over a certain angle of surface area as the wave propagates in air space. The path loss is proportional to the reciprocal of distance square. Signal degradation via multiple reflections from the wall is a key factor that limits the maximum transmission distance for 2.5-Gb/s wireless signal in our testbed environment of office building hallway.

Fig. 38 shows the system implementation of the first field demonstration of delivering dual service uncompressed 270-Mbps standard definition (SD) and uncompressed 1.485-Gbps high definition (HD) video content using 2.4-GHz microwave and 60-GHz mm-wave radio signals, respectively, over Georgia Institute of Technology (GT) Campus fiber backbone network from *Centergy Research Lab* at 10th Street to *Aware Home Residential Lab* at 5th Street. The transmission distance is 2.5 km standard single-mode fiber (SMF-28). At the transmitter (*Centergy building*), all optical up-conversion method is used to perform simultaneous generation of 60-GHz mm-wave and up-conversion of 1.485-Gbps HD signals at wavelength 1554.0 nm. The all-optical mm-wave generation at 60-GHz is realized by using an optical phase modulator driven by 30-GHz sinusoidal electrical clock signal and an optical de-interleaver as described in the previous section. The HD signal at 1.485-Gbps is generated from the component output of commercially available Sony Blue-Ray Disc player and an analog-to-digital converter. For 2.4-GHz radio signal, we used electrical mixing and double-sideband optical modulation to up-convert 270-Mbps real-time SD video content generated from a commercially available Canon Camcorder before optically transmitted at wavelength of 1550 nm. At the receiver (Aware home), direct detection of optical 60-GHz mm-wave signal is performed by a 60-GHz bandwidth PIN photodiode to realize optical-to-electrical

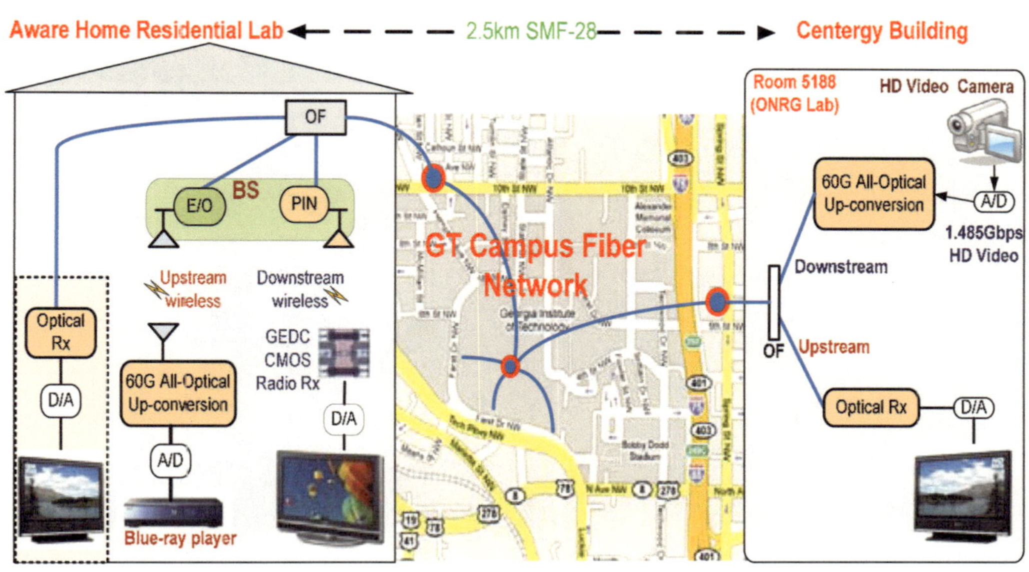

Fig. 38. Field trial demonstration setup of the SD/HD video content delivery using 2.4-GHz and 60-GHz mm-wave radio-over-fiber in the Georgia Tech (GT) campus fiber network.

conversion. The converted electrical mm-wave signal is then amplified by an electrical amplifier (EA) with 5 GHz bandwidth centered at 60 GHz and 3.55 before it is broadcasted through a double-ridge guide rectangular horn antenna with a gain of 25 dBi, frequency range of 50 to 75 GHz and 3 dB beam width of 7°. After the wireless transmission, the 60-GHz mm-wave signal is received by the end mobile terminal in order to perform the down-conversion and recover the 1.485-Gbps HD video signal. The down-conversion is performed by a 60-GHz balanced mixer using self-reflective mixing technique. Similarly, the 2.4-GHz radio signal is received by a 2.5-GHz PIN receiver at the BS and distributed over the wireless to the receiver antenna. The 270-Mbps SD signal is then recovered by down-conversion process. Fig. 39 shows the transmitter and receiver modules at two separate locations. We did not measure any BER in the field-trial, since we do not have any available electrical clock recovery module that is required to recover the clock at the distantly located receiver. However, the video quality displayed at the remote receiver TV screens located at

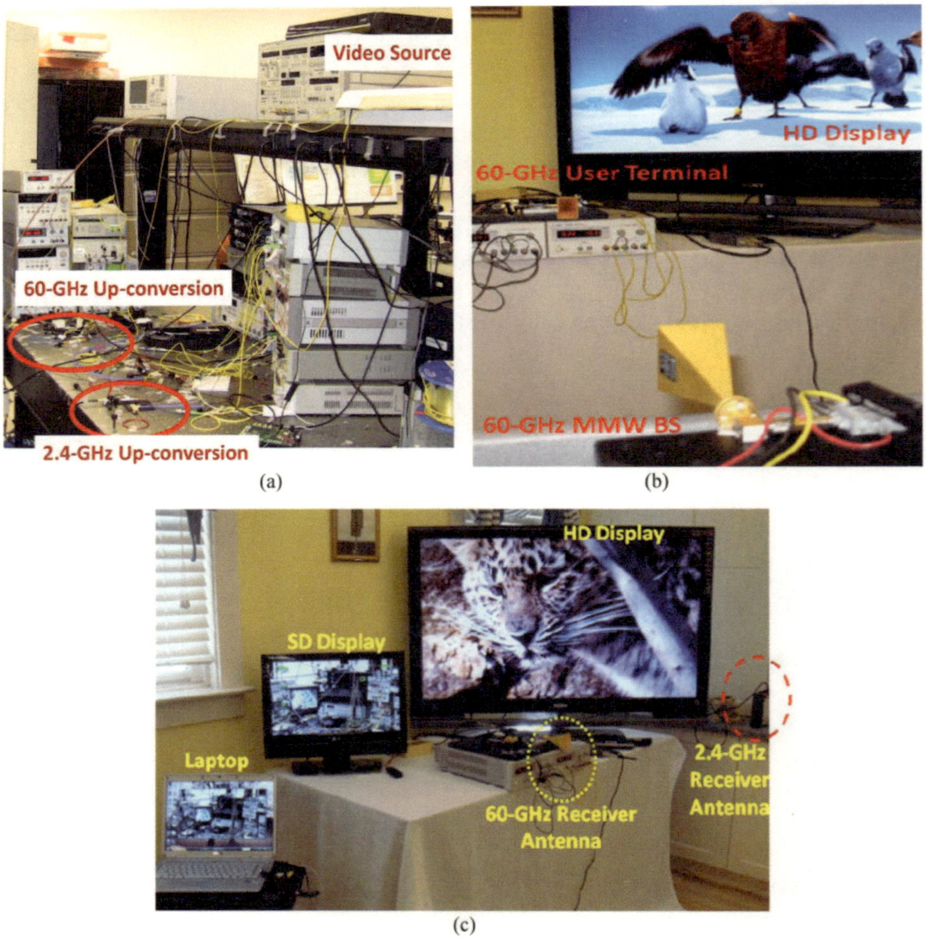

Fig. 39. Transmitter and receiver of the 3-screen, dual-service 2.4-GHz and 60-GHz RoF carrying 270-Mbps SD and 1.485-Gbps HD video content. (a) Transmitter at Centergy lab at 10th Street, (b) 60-GHz Wireless setup at residential lab at 5th Street, and (c) 3-screen receiver at residential lab.

the residential lab (2.5 km away from the transmitter at Centergy building) indicates the very good BER performance of the received signal.

ROF-based optical-wireless access networks have been considered the most promising solution to increase the capacity, coverage, bandwidth, and mobility as well as decreasing cost for next-generation access networks. This hybrid system explore the converged benefits of the optical and wireless technology to serve both fixed and mobile end users. However, there are many issues that still need to be addressed. The media access control (MAC) is a promising direction for future research. Although the analog nature of ROF links offers transparency for the transmission of the wireless protocol stack, the additional propagation delay introduced by the fier in the end-to-end logical link connection limits the maximum fiber lengths that can be accommodated to guarantee sufficiently low latency for adequate protocol performance. Another research area would be the FDM design for optical-wireless systems. OFDM is a good choice for wireless transmission, but the OFDM-ROF system has the stringent requirement of orthogonality between the sub-carriers. Since there are 7-GHz license-free spectrum, the traditional FDM can be used in the optical-wireless system to reduce the complexity of signal processing and increase the tolerance towards the frequency drift.

8. Acknowledgment

Author likes to thank Dr. T. Wang from NEC Laboratories America and Dr. M. Huang and Dr. A. Chowdhury from Georgia Institute of Technology for their great contribution for these studies.

9. References

M. Huchard, M. Weiss, A. Pizzinat, S. Meyer, P. Guignard, and B. Charbonnier (2008). Ultra-Broadband wireless home network based on 60-GHz WPAN cells interconnected via RoF, *J. of Lightwave Technol.*, 26(15), 2364-2372.

C. T. Lin, W. R. Peng, P. C. Peng, J. Chen, C. F. Peng, B. S. Chiou, and S. Chi (2006). Simultaneous Generation of Baseband and Radio Signals Using Only One Single-Electrode Mach–Zehnder Modulator With Enhanced Linearity, *IEEE Photon. Technol. Lett.*, 18, 2481–2483.

N. J. Gomes, A. Nkansah, and D. Wake (2008), Radio-Over-MMF techniques—part I: RF to microwave frequency systems, *J. of Lightwave Technol.*, 26(15), 2388-2395.

A. M. J. Koonen, and M. G. Larrodé (2008). Radio-Over-MMF techniques—part II: microwave to millimeter-wave systems, *J. of Lightwave Technol.*, 26(15), 2396-2408.

L. Chen, H. Wen and S. C. Wen (2006). A radio-over-fiber system with a novel scheme for millimeter-wave generation and wavelength reuse for up-link connection, *IEEE Photon. Technol. Lett.*, 18, 2056-2058.

A. Stöhr, A. Akrout, R. Buß, B. Charbonnier, F. v. Dijk, A. Enard, S. Fedderwitz, D. Jäger, M. Huchard, F. Lecoche, J. Marti, R. Sambaraju, A. Steffan, A. Umbach, and M. Weiß (2009). 60 GHz radio-over-fiber technologies for broadband wireless services, *J. of Optical Networking*, 8(5), 471-487.

J. Yu, et al. (2007). A cost-effective scheme to generate and de-multiplex multiple frequency millimeter-wave signals in a ROF network, *ECOC 2007*, 3.3.3.

J. Ma, J. Yu, C. Yu, X. Xin, Q. Zhang (2007). Transmission performance of the optical mm-wave generated by double sideband intensity-modulation, *Optics Communications*, 280(2), 317-326.

J. Ma, J. Yu, C. Yu, X. Xin (2007). Fiber dispersion influence on transmission of the optical millimeter-wave generated by LN-MZM intensity modulation, *IEEE/OSA Journal of Light. Technol.*, 25(11).

C. Bock, J. Prat, S. D. Walker (2005). Hybrid WDM/TDM PON using the AWG FSR and featuring centralized light generation and dynamic bandwidth allocation, *J. Lightw. Technol.*, 23(12), 3981-3988.

S.-J. Park, C.-H. Lee, K.-T. Jeong, H.-J. Park, J.-G. Ahn, K.-H. Song (2004). Fiber-to-the-home services based on wavelength-division-multiplexing passive optical network, *J. Lightw. Technol.*, 22(11), 2582-2591.

H. Bolcskei, A. J. Paulraj, K.V.S. Hari, R. U. Nabar, W. W. Lu (2001). Fixed broadband wireless access: state-of-the-art, challenges, and future directions, *IEEE Commun. Mag.*, 100-108.

B. Fong, N. Ansari, A. C. M. Fong, G. Y. Hong, P. B. Rapajic (2004). On the scalability of fixed broadband wireless access network deployment, *IEEE Commun. Mag.*, 42(9), S12-S18.

S. J. Vaughan-Nichols (2004). Achieving wireless broadband with WiMax, *Computer*, 37(6), 10-13.

A. Chowdhury, H. Chien, M. Huang, J. Yu and G. K. Chang (2008). Rayleigh Backscattering Noise-Eliminated 115-km Long-Reach Bidirectional Centralized WDM-PON With 10-Gb/s DPSK Downstream and Remodulated 2.5-Gb/s OCS-SCM Upstream Signal, *IEEE Photon. Technol. Lett.*, 20(24), 2081-2083.

J. Yu, Z. Jia, T. Wang, G. K. Chang (2007). Centralized Lightwave Radio-Over-Fiber System With Photonic Frequency Quadrupling for High-Frequency Millimeter-Wave Generation, *IEEE Photon. Technol. Lett.*, 19(19), 1499 - 1501.

J. Yu et al (2007). Demonstration of a Novel WDM Passive Optical Network Architecture to Provide Triple Play Services with Source-free Optical Network Units, *IEEE Photon. Technol. Lett.*, 19(8), 571-573.

Z. Jia, J. Yu, A. Chowdhury, G. Ellinas, G. K. Chang (2007). Simultaneous Generation of Independent Wired and Wireless Services Using a Single Modulator in Millimeter-Wave-Band Radio-Over-Fiber Systems, *IEEE Photon. Technol. Lett.*, 19(20), 1691 - 1693.

J. Yu, M. Huang, Z. Jia, A. Chowdhury, G.-K. Chang (2009). All-Optical Up-Conversion for 30x7.5Gb/s WDM Signals in a 60GHz ROF System, *OFC 2009*, OTuB4.

H.-C. Chien, A. Chowdhury, Y.-T. Hsueh, Z. Jia, S.-H. Fan, J. Yu, G.-K. Chang (2009). A Novel 60-GHz Millimeter-Wave over Fiber with Independent 10-Gbps Wired and Wireless Services on a Single Wavelength Using PolMUX and Wavelength-Reuse Techniques, *OFC 2009*, OTuB7.

Y.-T. Hsueh, Z. Jia, H.-C. Chien, A. Chowdhury, J. Yu, G.-K. Chang (2009). Generation and Transport of Independent 2.4 GHz (Wi-Fi), 5.8 GHz (WiMAX), and 60-GHz Optical Millimeter-Wave Signals on a Single Wavelength for Converged Wireless over Fiber Access Networks, *OFC 2009*, OTuJ1.

D. Qian, N. Cvijetic, Y.-K. Huang, J. Yu, T. Wang (2009). 22.4-Gb/s OFDM Transmission over 1000 km SSMF Using Polarization Multiplexing with Direct Detection, *OFC 2009*, OTuO7.

Z. Jia, H.-C. Chien, Y.-T. Hsueh, A. Chowdhury, J. Yu, G.-K. Chang (2009). Wirseless HD Services over Optical Access Systems: Transmission, Networking, and Demonstration, *OFC 2009*, JThA84.

J. Ma, J. Yu, C. Yu, Z. Jia, G.-K. Chang (2007). The influence of fiber dispersion on the code form of the optical mm-wave signals generated by single sideband intensity-modulation, *Optics Communications*, 271(2), 396-403.

D. Wake, M. Webster, G. Wimpenny, K. Beacham, L. Crawford (2004). Radio over fiber for mobile communications, in *Int. Topical Meeting Microwave Photonics*, 157–160.

K. Kitayama (2000). Architectural considerations of fiber radio millimeter wave wireless access systems," *Fiber Integr. Opt.* 19, 167-186.

L. Rosa, S. Selleri, G. Tartarini, P. Faccin, E.M. Fabbri (2006). Distortion Performance Prediction in Multi-Band Radio over Fiber Systems Exploiting Direct Laser Modulation, *36th European Microwave Conference*, 1292-1295.

K.E. Razavi, P.A. Davies (1997). Millimetre wave generation by filtering the FM-IM spectra of a directly modulated DFB laser, *IEEE MTT-S Int. Microw. Symposium Digest*, 3, 1707-1708.

L. A. Johansson, A. J. Seeds (2000). Millimetre-wave radio-over-fibre transmission using an optical injection phase-lock loop source, *Microwave Photonics*, 129-132.

B. Leesti, A. J. Zilkie, J. S. Aitchison, M. Mojahedi, R. H. Wang, T. J. Rotter, C. Yang, A. Stintz, and K. J. Malloy (2005). Broad-band wavelength up-conversion of picosecond pulses via four-wave mixing in a quantum-dash waveguide, *IEEE Photon. Technol. Lett.*, 17(5), 1046-1048.

J.-H. Seo, Y.-K. Seo, and W.-Y. Choi (2003). Nonlinear characteristics of an SOA photonic frequency up-converter, *Int. Topical Meeting on Microwave Photonics*, 109-112.

H.-J. Song, J.-S. Lee and J.-I. Song (2004). Signal up-conversion by using a cross-phase-modulation in all-Optical SOA-MZI wavelength converter", *IEEE Photon. Technol. Lett.*, 16(2), 593-595.

J. Ma, J. Yu, C. Yu, Z. Jia, X. Sang, Z. Zhou, T. Wang, G.K. Chang (2006). Wavelength conversion based on four-wave mixing in high-nonlinear dispersion shifted fiber using a dual-pump configuration, *J. Lightw. Technol.*, 24(7), 2851-2858.

J. Yu, J. Gu, X. Liu, Z. Jia and G. K. Chang (2005). Seamless integration of an 8x 2.5Gb/s WDM-PON and Radio-Over-Fiber using all-optical up-conversion based on Raman-assisted FWM, *IEEE Photon. Technol. Lett*, 17(9), 1986-1988.

W. Wang, H. N. Poulsen, L. Rau, D. J. Blumenthal (2003), Regenerative 80-Gb/s fiber XPM wavelength converter using a hybrid Raman/EDFA gain-enhanced configuration, *IEEE Photon. Technol. Lett*, 15(10), 1416-1418.

J. Yu, X. Zheng, C. Peucheret, A. T. Clausen, H. N. Poulsen, and P. Jeppesen (2000), All-optical wavelength conversion of short pulses and NRZ signals based on a nonlinear optical loop mirror, *J. Lightw. Technol.*, 18(7), 1007-1017.

W. Wang, H. N. Poulsen, L. Rau, H.-F. Chou, J. E. Bowers, D. J. Blumenthal (2005). Raman-enhanced regenerative ultrafast all-optical fiber XPM wavelength converter, *J. Lightw. Technol.*, 23(3), 1105-1115.

Z. Jia, J. Yu, G.-K. Chang (2005). All-optical 16x2.5 Gbit/s WDM signals simultaneous up-conversion based on XPM in an NOLM in ROF systems, *IEEE Photon. Technol. Lett*, 17(12), 2724-2726.

C. Lim, M. Attygalle, A. Nirmalathas, D. Novak, R. Waterhouse (2006). Analysis of optical carrier-to-sideband ratio for improving transmission performance in fiber-radio links, *IEEE Trans. on Microw. Theory and Tech.*, 54(5), 2181-2187.

P. S. Cho, D. Mahgerefteh, and J. Goldhar (1999). All-Optical 2R Regeneration and Wavelength Conversion at 20 Gb/s Using an Electroabsorption Modulator, *IEEE Photon. Technol. Lett.*, 11(12), 1662 -1664.

J. Yu, Z. Jia, T. Wang, and G. K. Chang (2007). A novel radio-over-fiber configuration using optical phase modulator to generate an optical mm-wave and centralized lightwave for uplink connection, *IEEE Photon. Technol. Lett.*, 19, 140–142.

W. Shieh, X. Yi, Y. Tang (2007). Experimental Demonstration of Transmission of Coherent Optical OFDM Systems, *OFC 2007*, OMP2.

A. M. Odlyzko (2003). Internet traffic growth: Sources and implications," in *Proc. SPIE: Optical Transmission systems and Equipment WDM Networking II*, 5427, 1-15.

J. Halpern and G. Garceau (2004). *Fiber: Revolutinizing the Bell's Telecome Networks.* New York: Bernstein/Telcordia Technologies Study.

B. Bing (2006). Broadband wireless access - the next wireless revolution, *Proceedings of the 4th Annual Communication Networks and Services Research Conference,* pp. 14.

D. Chrissan (2004). Uni-DSLT: One DSL for universal service, *Texas Instruments White Paper (Spay018).*

C. Lee, W. V. Sorin, B. Y. Kim (2006). Fiber to the home using a PON infrastructure, *J. Lightw. Technol.*, 24(12), 4568-4582.

T. Koonen (2006). Fiber to the home/fiber to the premises: what, where and when? *Proc. of the IEEE,* 94(5), 911-934.

J. Yu, D. Qian, M. Huang, Z. Jia, G. K. Chang, T. Wang (2008). 16Gbit/s radio OFDM signals over graded-index plastic optical fiber, *ECOC 2008*, P. 6. 16.

D. Qian, J. Yu, J. Hu, L. Zong, L. Xu, T. Wang (2008). 8x11.5Gbit/s OFDM transmisión over 1000km SSMF using convencional DFB lasers and direct-detection, *OFC 2008,* OMM3.

J. Yu, J. Hu, D. Qian, Z. Jia, G. K. Chang, T. Wang (2008). Transmission of micro-wave photonics generated 16Gbit/s super broadband OFDM signals in radio-over-fiber system, *OFC 2008*, OThP2.

M. Huang, J. Yu, H. Chien, A. Chowdhury, J. Chen, S. Chi, G. K. Chang (2008). A simple WDM-PON architecture to simultaneously provide triple-play services by suing one single modulator, *OFC 2008*, OTuI4.

Z. Jia, J. Yu, D. Qian, G. Ellinas, G. K. Chang (2008). Experimental demonstration for delivering 1Gbit/s OFDM signals over 80-km SSMF in 40GHz Radio-over-fiber access systems, *OFC 2008*, JWA108.

Z. Jia, J. Yu, L. Zong, G. K. Chang (2008). Transport of 8x2.5Gbit/s gíreles signals over optical millimeter wave through 12 straight-line WSSs and 160km fiber for advanced DWDM metro ntworks, *OFC 2008*, OMO3.

K. I. Kitayama and R. A. Grifin (1999). Optical downconversion from millimeter-wave to IF-Band over 50km long optical fiber link using an electroabsorption modulator, *IEEE Photon. Technol. Lett.*, 11(2), 287-289.

K. Nishimura, R. Inohara, M. Usami, S. Akiba (2005). All-optical wavelength conversion by electroabsorption modulator, *IEEE J. Sel. Topics in Quantum Electronics,* 11(1), 278 - 284.

J. Yu, Z. Jia, G.-K. Chang (2005). All-optical mixer based on cross-absorption modulation in electroabsorption modulator," *IEEE Photon. Technol. Lett.,* 17(11), 2421-2423.

D. Wake, C. R. Lima, P. A. Davies (1996). Transmission of 60GHz signals over 100km of optical fiber using a dual-mode semiconductor laser," *IEEE Photon. Technol. Lett.,* 8(4), 578-580.

U. Gliese, S. Norskov, T. N. Nielsen (1996). Chromatic dispersion in fiber-optic microwave and millimeter-wave links," *IEEE Trans. Microw. Theory and Tech.,* 44(10), 1716-1724.

G. Qi, J. Yao (2005). Optical generation and distribution of continuously tunable millimeter-wave signals using an optical phase modulator, *J. Lightw. Technol.,* 23(9), 2687-2695.

J. Yu, Z. Jia, L. Xu, L. Chen, T. Wang, G.-K. Chang (2006). DWDM optical millimeter-wave generation for radio-over-fiber using an optical phase modulator and an optical interleaver, *IEEE Photon. Technol. Lett.,* 18(13), 1418-1420.

M.G. Larrode, A. M. J .Koonen, J. J. V. Olmos, A. Ng'Oma (2006). Bidirectional radio-over-fiber link employing optical frequency multiplication," *IEEE Photon. Technol. Lett.,* 18(1), 265–267.

J.-H. Seo, C.-S. Choi, Y.-S. Kang, Y.-D. Chung, J. Kim, W.-Y. Choi (2006). SOA-EAM frequency up/down-converters for 60-GHz bi-directional radio-on-fiber systems, *IEEE Trans. Microw. Theory Tech.,* 54(2), 959-966.

P. J. Winzer, R.-J. Essiambre (2006). Advanced optical modulation format, *Proc. Of the IEEE,* 94(5), 952-985.

W. Huang, C.-K. Chan, L.-K. Chen, and F. Tong (2005). An optical network unit for WDM accesss networks with downstream DPSK and upstream remodulated OOK data using injection-locked FP laser, *IEEE Photon. Technol. Lett.,* vol. 15, no. 10, pp. 1476–1478, Oct. 2005.

Z. Jia, J. Yu, M. Haris, D. Boivin, G.-K. Chang (2006). A bi-directional radio-over-fiber system with all-optical up-converted DPSK for downstream and re-modulated OOK for upstream, *Proc. LEOS 2006,* TuV3.

J. J. V. Olmos, T. Kuri, and K.-I. Kitayama (2009). Half-duplex 12-channel dense WDM 2.6-GHz-band radio-over-fiber system employing a 1.5 GHz bandwidth reflective semiconductor optical amplifier, *J. of Optical Networking,* 7(12), 989-994.

I. G. Insua, D. Plettemeier, and C. G. Schäffer (2009). Broadband radio-over-fiber-based wireless access with 10 Gbits/s data rates, *J. of Optical Networking,* 8(1), 77-83.

Z. Jia, J. Yu, G.-K. Chang (2006). A full-duplex radio-over-fiber system based on optical carrier suppression and reuse, *IEEE Photon. Technol. Lett.,* 18(16), 1726–1728.

H. Shinohara (2005). Broadband access in Japan: rapidly growing FTTH market, *IEEE Commun. Mag.,* 43(9), 72-78.

G.-K. Chang, J. Yu, Z. Jia, J. Yu (2006). Novel optical-wireless access network architecture for simultaneously providing broadband wireless and wired services, *Proc. OFC 2006,* Anaheim, CA, OFM1.

Z. Xu, Y. Wen, M .Attygalle, X. Cheng, and W. Zhong (2006). Multiple channel carrier-reused WDM passive optical networks, *ECOC 2006,* Th 4.3.2.

M. Attygalle et al., (2006). Wavelength reuse scheme for source free optical network units in WDM passive optical networks, *ECOC 2006,* Th 3.5.3.

W.Hung, C. K. Chan, L. K. Chen, and F. Tong (2003). An optical network unit for WDM access networks with downstream DPSK and upstream remodulated OOK data using injection-lock FP laser, *IEEE Photon. Technol. Lett.*, 15(10), 1476-1478.

E. Wong, K.L. Lee, T.Anderson (2006). Directly Modulated Self-Seeding Reflective SOAs as Colorless Transmitters for WDM Passive Optical Networks, *OFC 2006*, PDP 49.

O. Akanbi et al (2005). A new bidirectional DWDM PON using optical carrier suppression and separation in the central office, *ECOC 2005*, We. 3. 3. 1.

M. Khanal, C. J. Chae, and R. S. Tucker (2005). Selective broadcasting of digital video signals over a WDM passive optical network", *IEEE Photon. Technol. Lett.*, 17(9), 1992-1994.

J. Yu, O. Akanbi, Y. Luo, L. Zong, Z. Jia, T. Wang and G. K. Chang (2007). A novel WDM-PON architecture with centralized lightwaves in OLT for providing triple play services, *OFC 2007*.

M. Attygalle, C. Lim, G. J. Pendock, A. Nirmalathas, and G. Edvell (2005). Transmission improvement in fiber wireless links using fiber Bragg gratings, *IEEE Photon. Technol. Lett.*, 17(1), 190-192.

A. Wiberg, P. Perez-Millan, M. V. Andres, P. A. Andrekson, P. O. Hedekvist (2005). Fiber-optic 40-GHz mm-wave link with 2.5Gb/s data transmission, *IEEE Photon. Technol. Lett.*, 17(9), 1938-1940.

G. H. Smith, D. Novak, Z. Ahmed (1997). Overcome chromatic-dispersion effects in fiber-wireless systems incorporating external modulators, *IEEE Trans. on Microw. Theory and Tech.*, 45(8), 1410-1415.

J. M. Fuster, J. Marti, J. L. Corral, V. Polo, F. Ramos (2000). Generalized study of dispersion-induced power penalty mitigation techniques in millimeter-wave fiber-optic links, *J. of Lightw. Technol.* 18(7), 933-940.

C. K. Sun, R. J. Orazi, S. A. Pappert (1996). Efficient microwave frequency conversion using photonic link signal mixing," *IEEE Photon. Technol. Lett.*, 8(1), 154-156.

J. Yu, Z. Jia, L. Yi, Y. Su, G.-K. Chang, T. Wang (2006). Optical millimeter-wave generation or up-conversion using external modulators, *IEEE Photon. Technol. Lett.*, 18(1), 2006.

J. Yu, Z. Jia, T. Wang, G. K. Chang and G. Ellinas (2007). Demonstration of a novel WDM-PON access network compatible with ROF system to provide 2.5Gb/s per channel symmetric data services, *OFC 2007*.

J. E. Mitchell (2004). Performance of OFDM at 5.8 GHz using radio over fiber link, Electronics Letters, 40(21), 1353- 1354.

W. Shieh, X. Yi, and Y. Tang (2007). Transmission experiment of multi-gigabit coherent optical OFDM systems over 1000km SSMF fiber, Electron. Lett., 43, 183-184.

H. Bao (2007). Transmission simulation of coherent optical OFDM signals in WDM systems, Optics Express, 15, 4410-4418.

Y. Tang, W. Shieh, X. Yi, and R. Evans (2007). Optimum design for RF-to-optical up-converter in coherent optical OFDM systems, *IEEE Photon. Technol. Lett.* 19, 483-485.

A. Kim, Y. H. Joo, and Y. Kim (2004). 60GHz wireless communication systems with radio-over-fiber links for indoor wireless LANs, *IEEE Transactions on Consumer Electronics*, 50, 517-520.

L. Chen, J. He, Y. Li. (2007). Simple ROF configuration to simultaneously realize optical millimeter-wave signal generation and source-free base station operation, *ECOC* 2007, 45-46.

T. Kawanishi, K. Higuma, T. Fujita, S. Mori, S. Oikawa, J. Ichikawa, T. Sakamoto, and M. Izutsu (2005). 40 Gbit/s versatile LiNbO lightwave modulator, *ECOC 2005*, Glasgow, Scotland, 2005, Th 2.2.5.

Z. Jia, J. Yu, Y.-T. Hsueh, H.-C. Chien, G.-K. Chang (2008). Demonstration of a symmetric bidirectional 60-GHz radio-over-fiber transport system at 2.5-Gb/s over a single 25-km SMF-28, *ECOC 2008*, Tu. 3. F. 5.

D. Qian, J. Yu, J. Hu, P. N. Ji, T. Wang (2008). 11.5-Gb/s OFDM transmission over 640km SSMF using directly modulated laser, *ECOC 2008*, Mo. 3. E. 4.

M.-F. Huang, J. Yu, Q. Dayou, G.-K. Chang (2008). Lightwave centralized WDM-OFDM-PON, *ECOC 2008*, Th. 1.F.5.

T. Sakamoto, T. Kawanishi, and M. Izutsu (2006), Continuous-phase frequency-shift keying with external modulation, *IEEE J. Sel. Topics Quantum Electron.*, 12(4), 589–595.

J. Yu et al. (2005). A novel optical frontend for ultra-high capacity of 32x2.5Gbit/s data delivery in radio-over-fiber systems, *ECOC 2005*, Th 4.5.4.

M. Attygalle et al. (2006). Wavelength reuse scheme for source free optical network units in WDM passive optical networks, *ECOC 2006*, Th 3.5.3.

M. R. Wiech et al. (2006). Bidirectional EDFA for future extra large passive optical networks, *ECOC 2006*, Mo. 4. 5. 7.

Z. Xu, X. Zhang and J. Yu (2007). Frequency up-conversion of multiple RF signals using optical carrier suppression for radio over fiber downlinks, *Optics Express*, 15(25), 16737-16747.

K. Kitayama, R. Griffin (1999). Optical downconversion from millimeter-wave to IF band over 50-km-long optical fiber link using an electroabsorption modulator, *IEEE Photon. Technol. Lett.*, 11, 287-289.

J. Ma, J. Yu, Z. Jia, C. Yu, Z. Zhou, and G.-K. Chang (2007). Optical mm-wave generation based on phase modulation along with optical filtering, *Microwave and Optical Technol. Lett.*, 49(8), 1787-1793.

S.L. Jansen, et al (2007). 20-Gb/s OFDM Transmission over 4,160-km SSMF Enabled by RF-Pilot Tone Phase Noise Compensation, in proc. *OFC 2007*, PDP 15.

S. Jansen, I. Morita, H. Tanaka (2007). 10-Gb/s OFDM with conventional DFB lasers, *ECOC 2007*, 5.2.2.

L. Xu, J. Hu, D. Qian and T. Wang (2008). Coherent optical OFDM systems using self optical carrier extraction, *OFC 2008*, OMu4.

G. K. Chang, Z. Jia, J. Yu (2006). Super-broadband optical wireless over optical fiber network architecture, *LEOS 2006*, TuV1.

J. Yu, Z. Jia, T. Wang, G.-K. Chang and G. Ellinas (2007). Demonstration of a novel WDM-PON access network compatible with ROF systems to provide 2.5Gb/s per channel symmetric data services, *OFC 2007*, OThM5.

J. Yu, Z. Jia and G.-K. Chang (2005). Seamless integration and large coverage delivery of 32x2.5Gbit/s DWDM signals in a radio-over-fiber network, *ECOC 2005*, Th4.5.4.

J. Ma, X. Xin, J. Yu, C. Yu, K. Wang, H. Huang, and L. Rao (2009). Optical millimeter wave generated by octupling the frequency of the local oscillator, *J. of Optical Networking*, 7(10), 837-845.

J. Chen, C.-T. Lin, P. T. Shih, W.-J. Jiang, S.-P. Dai, Y.-M. Lin, P.-C. Peng, and S. Chi (2009). Generation of optical millimeter-wave signals and vector formats using an integrated optical I/Q modulator, *J. of Optical Networking*, 8(2), 188-200.

C. Lim, A. Nirmalathas, M. Bakaul, K.-L. Lee, D. Novak, and R. Waterhouse (2009). Mitigation strategy for transmission impairments in millimeter-wave radio-over-fiber networks, *J. of Optical Networking*, 8(2), 201-214

A. Alphones (2009). Double-spread radio-over-fiber system for next-generation wireless technologies, *J. of Optical Networking*, 8(2), 225-234

Z. Jia, J. Yu, Y.-T. Hsueh, H.-C. Chien, A. Chowdhury, and G.-K. Chang (2009). Wireless high-definition services over optical fiber networks, *J. of Optical Networking*, 8(2), 235-243.

Á. A. de G. Vivero, O. Y. Alani, and J. M. Elmirghani (2009). Indoor airport radio-over-fiber network traffic model and performance analysis using load-balancing techniques, *J. of Optical Networking*, 8(3), 272-284.

S. L. Jansen, I. Morita, N. Takeda, H. Tanaka (2007). 20-Gb/s OFDM transmission over 4,160-km SSMF enabled by RF-pilot tone phase noise compensation, *OFC 2007*, PDP 15.

B. Schmidt, A. Lowery, J. Armstrong (2007). Experimental demonstration of 20 Gbit/s direct-detection optical OFDM and 12 Gbit/s with a colorless transmitter, *OFC 2007*, PDP 18.

H. Sasa, T. Niiho, K. Tanaka, K. Utsumi and S. Morikura (2003). Radio-over-fiber transmission performance of OFDM signal for dual-band wireless LAN systems, *MWP 2003*, 139-142.

A. Kim, Y. Joo and Y. Kim (2004). 60GHz wireless communication systems with radio-over-fiber links for indoor wireless LANs, *IEEE Trans. Consum. Electron.*, 50(2), 517-520.

J. Yu, M. Huang, Z. Jia, T. Wang, and G. K. Chang (2008). A novel scheme to generate single-sideband millimeter-wave signals by using low-frequency local oscillator signal, *IEEE Photon. Technol. Lett.*, 20(7), 478-480.

N. J. Gomes, M. Morant, A. Alphones, B. Cabon, J. E. Mitchell, C. Lethien, M. Csörnyei, A. Stöhr, and S. Iezekiel (2009). Radio-over-fiber transport for the support of wireless broadband services, *J. of Optical Networking*, 8(2),156-178.

Z. Jia, J. Yu, T. Hsuech, A. Chowdhury, H. Chien, J. A. Buck, G. K. Chang (2008). Multiband Signal Generation and Dispersion-Tolerant Transmission Based on Photonic Frequency Tripling Technology for 60-GHz Radio-Over-Fiber Systems, *IEEE Photon. Technol. Lett.*, 20(17), 1470 – 1472.

H. Yang, S. C. J. Lee, E. Tangdiongga, F. Breyer, S. Randel, and A. M. J. Koonen (2009). 40-Gb/s Transmission over 100m Graded-Index Plastic Optical Fiber based on Discrete Multitone Modulation, *in Proc. of OFC 2009*, PDPD8.

Photonic Crystal Multiplexer/Demultiplexer Device for Optical Communications

Sahbuddin Shaari[1] and Azliza J. M. Adnan[2]
[1]*Universiti Kebangsaan Malaysia, Bangi, Selangor,*
[2]*Telekom Reserach & Development Sdn. Bhd, Cyberjaya, Malaysia*

1. Introduction

Photonic crystals (PhCs) are periodic dielectric structures. They are called *crystals* because of their periodicity and *photonic* because they act on light. They can occur when the period (the separation of the periodic dielectric materials) is less than the wavelength of the light. If the choice of lattice geometries and compositional dielectric materials are suitable, it is possible all those reflection and refractions will cancel not only the light scattered sideways, but the light moving forward as well. Then, because the light has to go somewhere (energy is conversed), it has no choice but to go back, which in this case it is forbidden from entering the photonic crystals. This happens no matter what direction the light is coming from, in a certain range of wavelengths which called *photonic band gap*.
Fundamentally, PhCs are based on a concept of extended from conventional diffraction gratings and have unique analogy to solid state crystals. Thus the solid state physics theory is used for the analysis of PhCs. For example, one can calculate photonic band gaps (PBGs), impurity, defects and surface states, for PhCs structure. This allows the manipulation of light in dielectric mediums. For example, by carving a tunnel through the material, an optical "wire" can be created which no light can be deviate in the "wire". Also, by making a cavity in the center of the crystals, the beam of light could be caught and held which created an optical "cage". The abilities to trap and guide light have many potential applications in optical communications and computing, where one would like to make tiny optical "circuits" to help manage the ever-increasing traffic through the world's optical communications networks. Other devices, too, are made possible by this increased control over light: from more-efficient lasers and LED light sources, to opening new regimes for operating optical fibers, to cellular phones.

2. Application and fabrication of photonic crystals devices

2.1 The application of photonic crystals

Photonic crystals are attractive optical materials for controlling and manipulating the flow of light. One dimensional photonic crystals are already in widespread use in the form of thin-film optics with applications ranging from low and high reflection coatings on lenses and mirrors to colour changing paints and inks. Higher dimensional photonic crystals are of great interest for both fundamental and applied research, and the two dimensional ones are beginning to find commercial applications. K. Inoue have summarizes the relation of PhCs with other optics and various applications now being developed as shown in Fig.1. Considering that PhCs exhibit their functionalities by multidiffraction and multiscattering, a PhC can be said to be a kind of hologram in a broad sense. A novelty of PhCs, compared with conventional holograms, is the precise design method, namely the photonic band calculation, which is very effective for estimating their functionalities and performance.

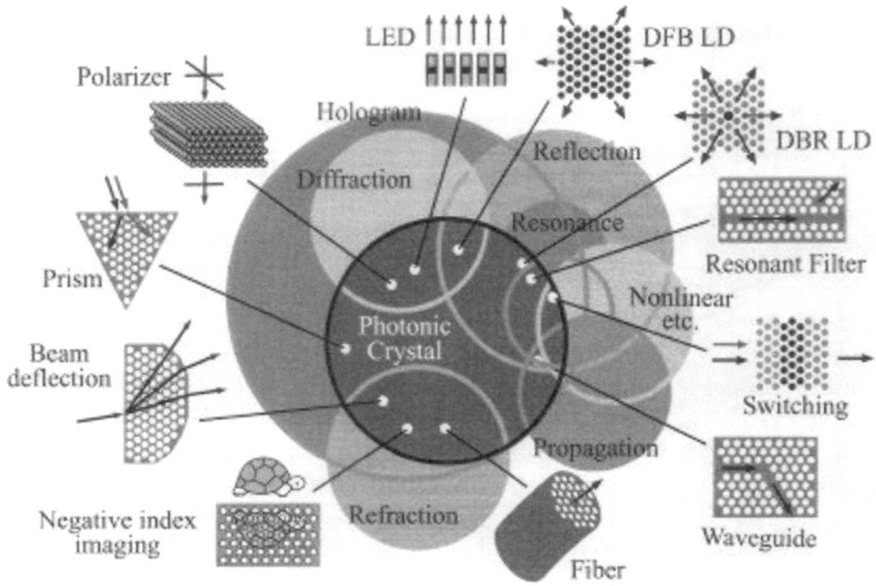

Fig. 1. Photonic crystal optics and various applications (Source: K. Inoue et al. 2004)

Many applications of two dimensional photonic crystals are directed towards integration of photonic devices. It is important to evolve into the photonic and electronics circuits, which will require high-density photonic and electronic circuits. The example of photonic crystals devices is LED, waveguide, resonator and photonic crystal fiber.

The introduction of defects into photonic lattices results in generation of a defect level and light is confined to be localized defect state. This high quality factor selection of one wavelength within the band gap can be used to make devices which combine advantages of thresholdless light emission and high reliability of LEDs with coherence and high efficiency of lasers. PhCs have also been used just as confining regions for lasers. Lasers have been fabricated with a 2D photonic lattice as mirrors instead of cleaved facets.

Artificial defects in PhCs and related formation on defect state within the band gap has been used in the realization of the devices such as filters and resonators. The confinement properties along with the defect properties can be used in various other applications such as sharp waveguide bends, fabry-perot cavity and reflection type lens.

Waveguiding of light within a photonic lattice with a defect region guiding light has been widely reported with near perfect transmission. The property is can be particularly useful for applications which need data transmission through bends whose feature size are in the nano meter rangers.

PhC also have been used as a clad of optical fibers for better confinement. One type of such a fiber uses the low effective refractive index of the PhC for optical confinement. Other such fibers use bragg reflection of PhCs or bragg reflection of a coaxial multilayer mirror. All such fiber predominantly use an air core and single mode PBG guidance has been demonstrated.

Anomalous dispersion and anistropic properties of PhCs have been used to fabricate optical components like prisms and polarizers. Wavelength dependent angular dispersion of PhCs is about hundred times more than in an ordinary prism. This phenomenon is termed as superprism phenomenon and is expected to be very useful in applications such as wavelength division multiplexing. A Si/SiO_2 multilayer with a zigzag cross section was fabricated and was found to work

The photonic band diagram indicates that there are three different frequency ranges for light, which can be utilized for real applications, as shown in Fig 2. The first one is the lowest frequency range below the first zone folding of the photonic band. The gradient of the lowest straight photonic band is determined by the effective refractive index n of the PhC and it is different for different polarizations. Classically, this characterization is called form birefringence. Since the PBG calculation precisely predicts the effective index for each polarization, the index is artificially controlled by the PhC structure. The second one is photonic band gap, which means the omni-directional stop band. It is one of the most unique properties of PhCs, and actually it was the main topic at the early stage of PhC research.

The photonic band gap can be used as a reflector for light that is to enter the PhC from arbitrary directions. It is applied to reflection-type devices, e.g., lasers and waveguides. The third one is the higher frequency range above the photonic band gap where complex photonic bands exist. The slope of a photonic band is proportional to the group velocity of the light. Therefore, the horizontal band at a band-edge means zero vg and the localization of light energy. In 2D and 3D PhCs, such a zero velocity group or a small velocity group appears not only at the band-edge but in many bands. They will be effective for the enhancement of various interactions of light with materials of the PhC. In addition, the 2D or 3D distribution of bands, the so called dispersion surface, provides unique light propagation in PhCs. Thus, this frequency range can be used in transmission-type functional devices.

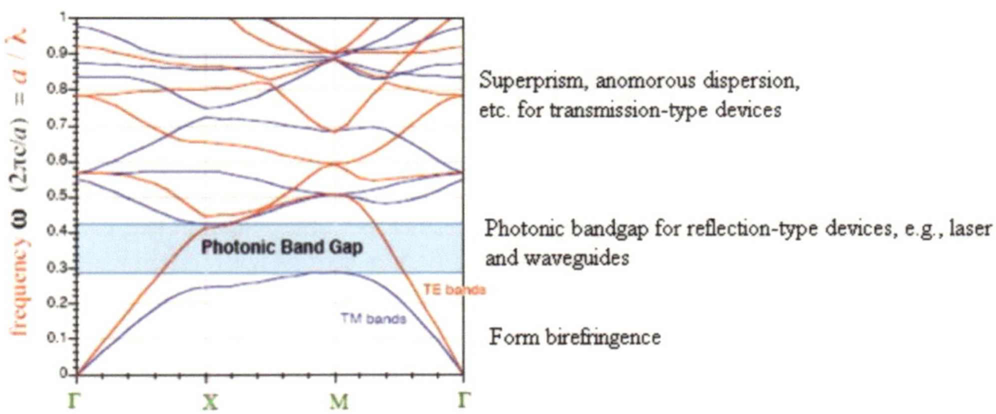

Fig. 2. Photonic band diagram for 2D PhC of holes in a triangular lattice and three frequency ranges for different application targets. (Source: K. Inoue et al. 2004)

2.2 Fabrication of photonic crystals waveguide

Fabrication of two dimensional photonic structures requires one to mark the planar pattern of unwanted areas on the surface of the semiconductor, and remove the material from these areas, thus creating holes and it is vice versa for creating pillars. Dry etching by reactive ions or wet electrochemical etching are most often employed for this purpose. Dry etching allows accurate control over the hole size and arrangement (with nanometer precision), but has limited maximum depth of etching. It is therefore used to fabricate low aspect ratio structures, such as thin waveguides. Electrochemical etching allows one to obtain very deep holes, and is suitable for fabricating high aspect ratio structures, but the size of etched holes can be controlled somewhat less predictably.

Marking of the unwanted areas is usually done by lithography. Since the most promising structures are those having photonic band gaps near-infrared and visible ranges, they must have features of comparable size, and conventional lithography cannot be applied. Thus, electron beam lithography (EBL) is used, which normally employs a scanning electron microscope with resolution around 1 nm, equipped with an electron beam drawing stage to transfer preset patterns on the surface of a semiconductor wafer, covered by photoresists, e.g. a layer of PMMA with 200-400 nm thickness. The pattern resulting after the photoresist development is used as a mask for subsequent etching.

The etching is done by reactive ions, accelerated towards the sample surface in a plasma discharge, and reacting with the material in the unmasked areas, thus destroying it. Typically, chlorine-based ($SiCl_4$ and Cl_2), or fluorine-based (CHF_3, CF_4, C_2F_6 and SF_6) reactive gases are used. For etching GaAs and AlGaAs structures, use of clorine-based gases has become standard, while fluorine-based etching chemistry is preferred for silicon. Reactive ion etching is highly anistropic, and therefore the initial mask pattern does not spread out laterally with increasing etching depth, and deep holes with almost parallel walls are obtained. The etching also destroys photoresists, albeit it is slower rate, which depends on the details of the etching process and photoresists parameters. Since etching must be stopped before mask becomes unrealiable, this factor limits the maximum achievable aspect ratio to about 10. To increase the aspect ratio, masks highly resistant to reactive etching (typically Au, Cr, SiO2 or Si_3N_4) are deposited and patterned on the surface using lithography, and etching is carried out with fluorine gases. As a result, photonic crystal structures with aspect ratios up to 20 were successfully fabricated.

3. Photonic band gap in 2-D photonic crystals

One of the most important features in photonic crystals is the photonic band gap (PBG), which is analogous to band gaps or energy gaps for electrons travelling in semiconductors. In the case of semiconductors, a band gap arises from the wave-like natures of electrons. Electrons as waves within a semiconductor experience periodic potential from each atom and are reflected by the atoms. Under certain conditions, electrons with certain wave vectors and energy constitute standing waves. The range of energy, named "band gap" in which electrons are not allowed to exist, appears. This phenomenon differentiates semiconductors from metal and insulators. In the similar manner, standing waves of electromagnetic waves can propagate through a periodic structure whose minimum features are less than the wavelength of light. In this case, the medium expels photons with certain wavelengths and wave vectors. Such a structure acts as an insulator of light, and this phenomenon is referred to as "photonic band gap".

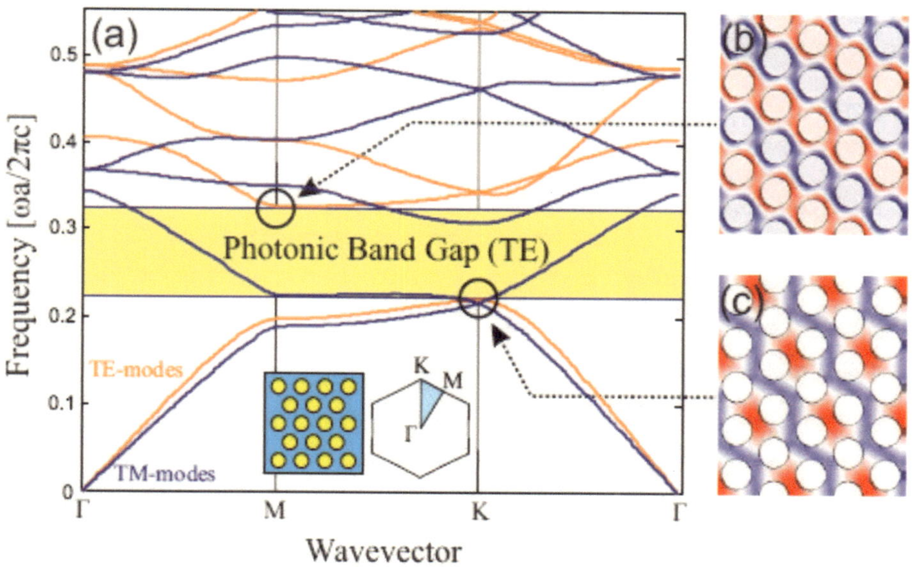

Fig. 3. (*a*) The photonic band structure for a triangular lattice of holes in a high index material. The frequencies for the two polarizations (TE modes in light gray, TM modes in dark gray) are plotted around the boundary of the irreducible Brillouin zone (shaded triangle in the inset). (*b*) The magnet field pattern of the TE mode corresponding to the second band at the first M point. (*c*) The magnetic field pattern of the TE mode corresponding to the first band at the K point. The grey levels indicated the amplitude of the magnetic field (dark-negative, light-positive). There is a band for the TE guided modes only.

A first characteristic optical property of PhC is the PBG. A two-dimensional triangular lattice with a hexagonal Brillouin zone exhibits a very high symmetry in the plane. Therefore, this structure is convenient for the formation of forbidden bands in all directions with the plane of periodicity. Fig. 3 shows the photonic band structure or dispersion diagram with the eigensolutions for a triangular lattice of holes in a high refractive index material. Both TE and the TM band structures are shown. The in-plane wavevector $\mathbf{k}_{//}$ goes along the edge of irreducible Brillouin zone, from Γ to M to K as shown in the inset in Fig. 3 (a). It is conventional to plot the frequency bands only extrema almost always occur along these boundaries. For TE modes (light gray lines in Fig.3 (a), there is no photonic band gap exists.

In order to understand in more detail the formation of photonic band gap for TE modes, the field patterns (magnetic field) at the lower and upper band edges corresponding to the high symmetry points K and M of the irreducible Brillouin zone was analyzed. At the lower band edge, the field associated with the lowest TE mode at K is strongly concerted in the high index material (Fig. 3 (a)) giving it a lower frequency. In contrast, the field pattern of the second mode at M, the upper band edge, has a nodal plane cutting through the high index material and therefore its energy is more concentrated in the air holes (Fig. 3(b)) giving it a higher frequency. For this reason, the bands above and below PBG are also referred to "air band" and "dielectric band", respectively. The PBG arises from this difference in field energy distribution. The higher the dielectric contrast in the periodic structure the larger is the PBG. Therefore, high index materials are essential for the realization of PhC structures.

Fig. 4 shows example of gap map plot for square, hexagon and circular scatterers pillars in honeycomb lattice. A gap map is a plot of the locations of the photonic band gaps of a crystal, as one or more of the parameters of the crystal are varied. The red and blue gaps show the TE band gap and TM band gap, respectively. Meanwhile, the green gaps show the absolute band gap which the TE and TM band gap overlap. As can be seen all three structures in honeycomb lattice have absolute band gap and the gaps all decrease in frequency as the filling fraction increases.

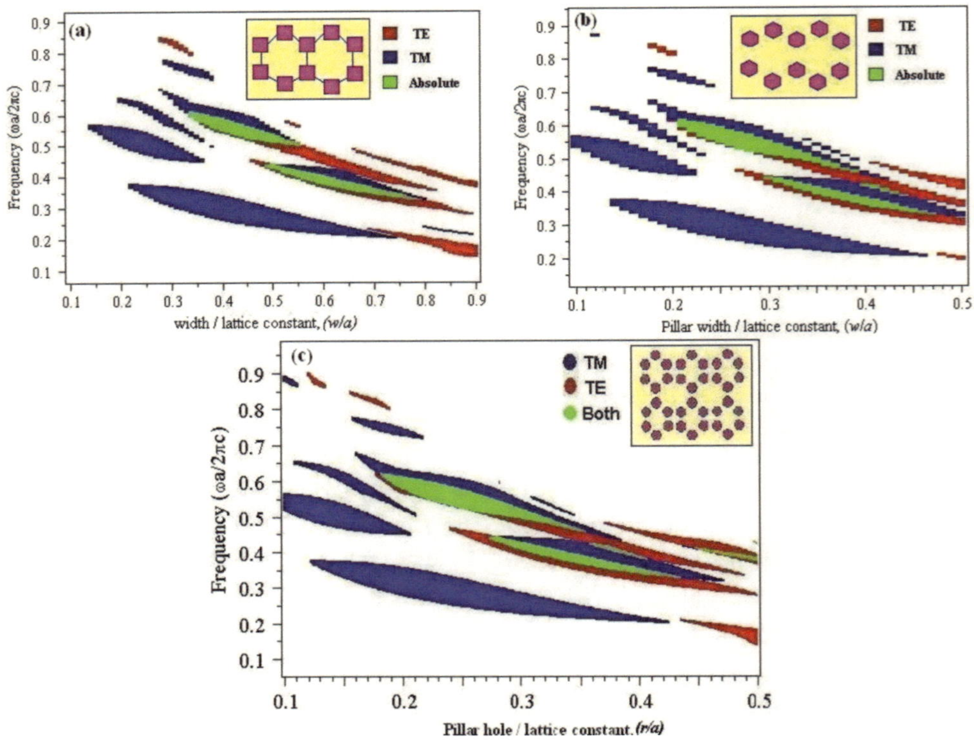

Fig. 4. Gap map for (a) Square scatterer pillars, (b) Hexagon scatterer pillar and (c) circular scatterer pillar in honey comb lattice

4. Defect engineering in photonic crystals waveguide

In the same way as for solid-state crystals, two main types of defects exist: cavities defects and extended defects. Cavities defects are associated to very local disruptions in the periodicity of the crystal, and their presence is revealed through the appearance of electromagnetic modes at discrete frequencies, which may be seen as analogous to isolated electronic states. Likewise, extended defects can be seen as analogous to dislocations of the crystal, and they may result in the appearance of transmission bands in spectral regions where a photonic band gap existed in the case of a perfectly periodic crystal.

Fig. 5 show the schematic illustration of possible defects in PhC. A single pillar from the crystal can be remove, or replace it with another whose size, shape, or dielectric constant is different than the original. Cavities and extended defects can be used to create a basic waveguiding component such as waveguide and bend waveguide.

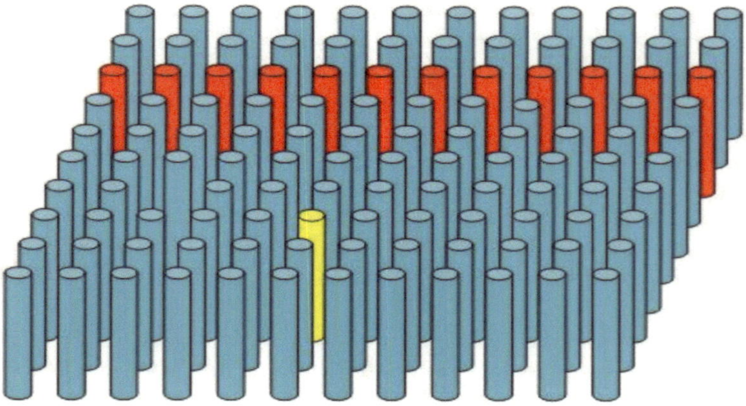

Fig. 5. Schematic illustrations of possible defects in PhC. Perturbing the line of pillar (red) might allow a localized state to exist. Perturbing one pillar in the bulk of the crystal (yellow) might allow a localized defect state to exist.

4.1 Cavities defects

Cavities defects can be created by locally modifying the refractive index, by changing the size of the patterns (substitution defect), by displacing one of the periodic patterns (interstitial defect) or by inserting a different pattern (dopant). Here again, the presence of a point defect may lead to discrete energy levels within the photonic gaps. Point defects in two-dimensional photonic crystals, correspond to localized electromagnetic modes in the case where the band gaps are omnidirectional, be it only for certain polarization. If this condition fulfilled, the electromagnetic field is actually found to be concentrated in the region of the defect and evanescent in the surrounding regions. By contrast, in the case where the band gap is not omnidirectional, a fraction of the electromagnetic energy will constantly leak away from the region of the defect towards directions along which the propagation is allowed. In this case, the presence of a point defect essentially leads to a peak in the density of electromagnetics states.

By removing a pillar from the lattice, we create a cavity that is effectively surrounded by reflecting walls. If the cavity has the proper size to support a mode in the band gap, then light cannot escape and we can pin the mode to the defect. In fact, a resonant cavity would be useful whenever one would like to control radiation within a narrow frequency range.

The important questions to address when designing a defect mode are how the defect shall be introduced into the structure, and which frequencies it will support as localized modes. First, one obvious way to introduce defect is allow one of the pillars of the rectangular lattice to grow or shrink in radius, calling the radius of the defect pillar is $r_{def,}$ the possibilities range from r_{def} = 0, corresponding to missing pillar in the structure, to around r_{def} = 0.5a µm, corresponding to an pillar that envelops one entire unit cell.

Next, we would like the defect to harbor modes of light that have frequencies within the band gap of the crystal. Fig. 6 shows the defect frequencies as the defect radius varies across the entire range in silicon rectangular pillars. The defect pillar is surrounded by perfect lattice at r = 0.18a. The photonic band gap is a white space between upper and lower green block which around frequency 0.30 to 0.445. From the plot, it shows that the bigger the defect pillar, the great quantity of defect modes occurred.

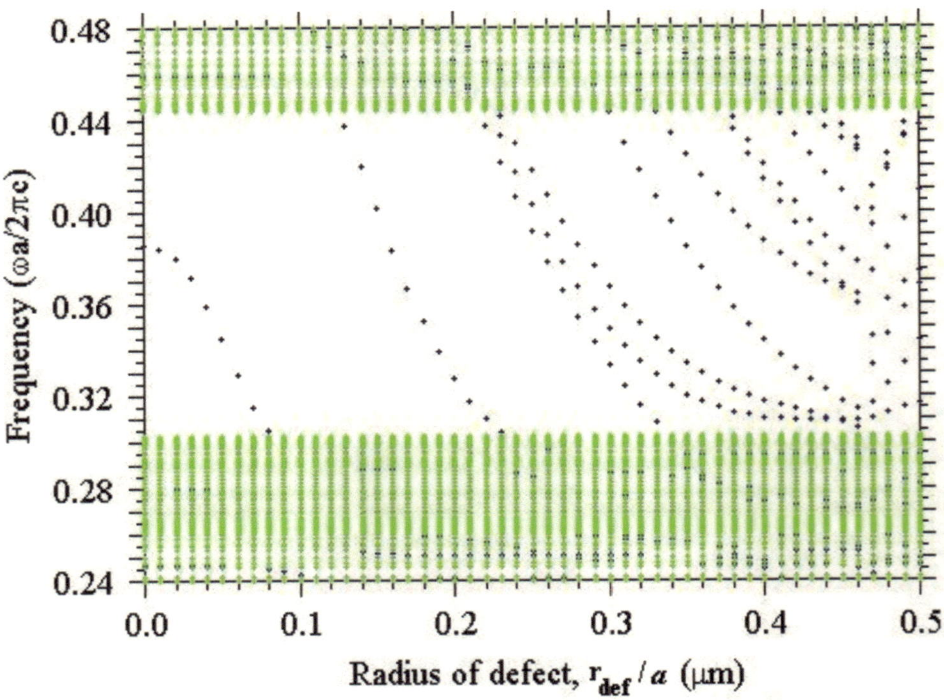

Fig. 6. A plot of the TM modes in rectangular silicon pillars in air with $r= 0.18a$. The photonic band gap is the white space between green block. The localized modes are shown as blue dotted lines. We create the defect cavity filling in a single silicon pillar.

Note that it is possible not only to create a defect mode with a frequency in the band gap, but also that the defect frequency sweeps continuously across the band gap as r_{def} is varied. In other words, we can "tune" the defect frequency or later called resonant frequency to any value within the band gap with a judicious choice of r_{def}. This complete tenability is an important feature of PhCs, it would be analogous to the ability to tune the properties of solids by somehow adjusting the radii of single dopant atoms.

Fig. 7 show the defect characteristics when a single missing pillar is involved at the centered of perfect circular pillar-type in rectangular lattice of PhC. Fig. 7 (a) represents the field distributions calculated for a defect created in a two-dimensional square lattice formed by dielectric pillars in air. The defect was created here by removing one pillar. The incident wave is assumed to be TM polarized. Removing one pillar introduce a peak into the crystal's density of states. In fig. 7 (b), the peak happens to be located in the photonic band gap which located in the yellow gap, then the defect-induced state must be evanescent-the defect mode cannot penetrate to the rest of the crystal, since it has frequency in the band gap. In this case a single missing pillar emits the resonant wavelength of 1.47 μm.

When several point defects of the same nature are present in a photonic crystal, and when the distance between these defects is large enough, their mutual influence can be neglected. In this case, everything happens as if an energy degeneracy of the system occurred several times. Indeed, while the electromagnetic modes associated to the different defects are localized in different region of the crystal, their field distributions are identical. By contrast, when the distance between the defects decreases, the coupling between these defects leads

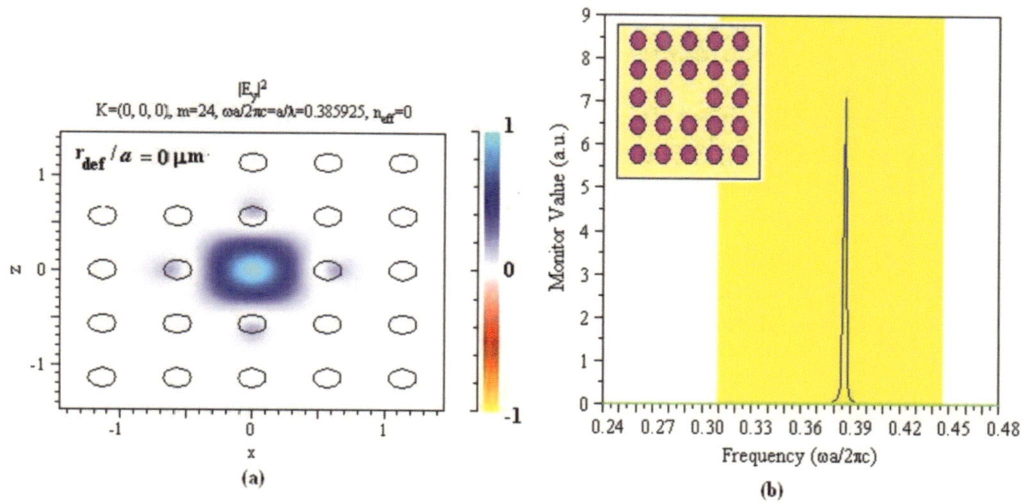

Fig. 7. Defect in a rectangular missing single pillar (*rdef* = 0). (a) The electric field patterns of the defect modes with defect frequency of 0.386. The panel at the most right side of (a) indicate the strength of the field. (b) The resonant frequency spectrum found from impulse simulation of the defect structure. The peak at 0.387 $\omega a/2\pi c$ represents a wavelength of 1.47µm.

to the formation of electromagnetic modes with different field distributions: in this case, the energy degeneracy is lifted. Fig.8 illustrates the effects induced by such a coupling through spectral measurement performed using FDTD. Two were introduced by removing two dielectric pillars. The transmission spectrum of the crystal was then measured for two different distances between the defects.

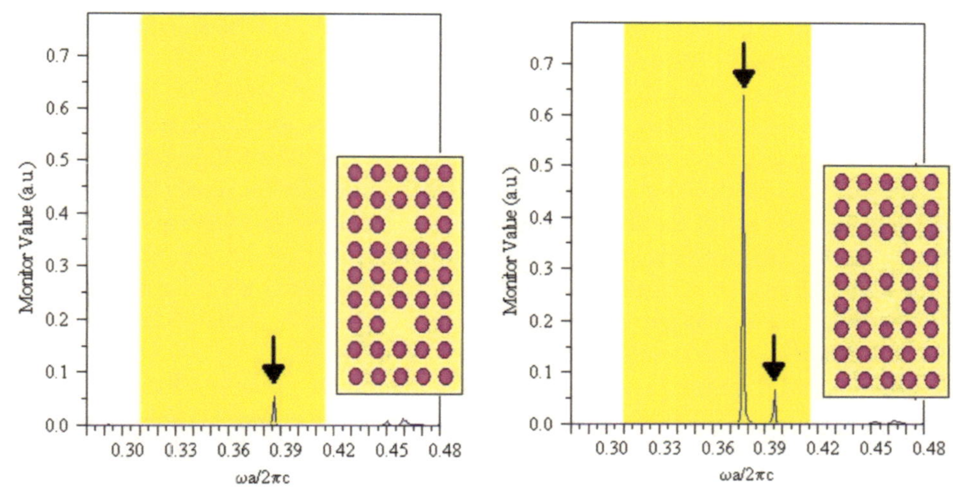

Fig. 8. Coupling between two point defects in a two-dimensional photonic crystal with a square symmetry. The crystal here is formed by a lattice of silicon pillars extending in the air. *Left*. Transmission spectrum measured in weak coupling regime. *Right*. Transmission spectrum measured in strong coupling regime.

When the two defects are distant from one another, as in the case in the left part of Fig. 8, a single transmission peak is observed at the high-frequency side of the TE band gap. This corresponds to an air defect, according to the terminology used in (Joannopolous, 1995). When the defects are at close distant from one another, the transmission maximum splits into two peaks, thereby revealing the existence of two different electromagnetic modes. Assuming the origin to be at the centre of the structure, the low-frequency mode presents a symmetric field distribution, while the high-frequency mode presents an anti-symmetric field distribution. This phenomenon is quite analogous to the situation occurring when a particle is released in a quantum well (Cohen-Tannoudji, 1973): the wave function of the fundamental state is symmetric whereas the wave function of the first excited state is anti-symmetric.

4.2 Extended defects

We can use cavities defects in photonic crystals to trap light, as we have seen in point defect. By using extended defects or line defects, we can also guide light from one location to another. The basic idea is to carve a waveguide out of an otherwise-perfect photonic crystal. Light that propagates in the waveguide with a frequency within the band gap of the crystal is confined to, and can be directed along the waveguide.

In Fig.9 (a) we show the band structure for the guide created by removing a row of pillars in the direction of the crystal, as shown in the inset. We find a single guided mode inside the band gap. The electric field of the mode has even symmetry with respect to the mirror plane along the guide axis. The mode itself bears a close resemblance to the fundamental mode of a conventional dielectric waveguide: it has sinusoidal profile inside the guide and decays exponentially outside.

In Fig.9 (b) the waveguide is made by removing three rows of pillars in the direction of the crystal (see the inset). There are now three guided modes inside the gap that can again be classified according to their symmetry with respect to the mirror plane along guide axis. The first and the second modes are even, whereas the third mode is odd.

It is generally true that the number of bands inside the band gap equals the number of rows of pillars removed when creating the guide. This can be understood from a simple counting of states in the crystal. If we decrease the dielectric constant of a single pillar in a prefect crystal, we pull up one defect state from the dielectric band. If we repeat this for a whole row of pillars, we pull up N localized states in an $N \times N$ crystal: one state at each k point for $\mathbf{k}$ along the guide. Analogously, when M rows of pillars are removed, we pull up M guided modes at each $\mathbf{k}$ from the dielectric band. Nevertheless, at some $\mathbf{k}$'s the modes may have frequencies outside the band gap and the entire band may not be contained in the gap, as is the case, for instance for the lowest guided mode band in Fig.9 (b).

Next we want to see what happen if we removed single row of pillar and then we put defect along the waveguide as shown in Fig.10 (a) which also known as couple cavity waveguide.

Fig. 10 (b) shows the transmission spectra of couple cavity waveguide over the wavelength as the radius of the cavity defect is varied. The radius is varied from $0.1a$ to $0.4a$. Obviously from the graph we can see that this structure can be made as filter. For example if we want to filter wavelength 1.31 μm and 1.55 μm separately, we can include radius defect of $0.3a$ along the waveguide to block wavelength 1.31 μm from entering the waveguide and wavelength 1.55μm is allow to enter the waveguide. The wavelength of 1.55 μm can be block from entering the waveguide by include the radius defect 0.2a. It can be noticed that when defect radius increases, the guided frequencies shift towards higher wavelengths. As the defect increases, the miniband is nearer to the dielectric band and also the modes can interact with in the bulk modes and, as a result of that, the transmission loss increases slightly.

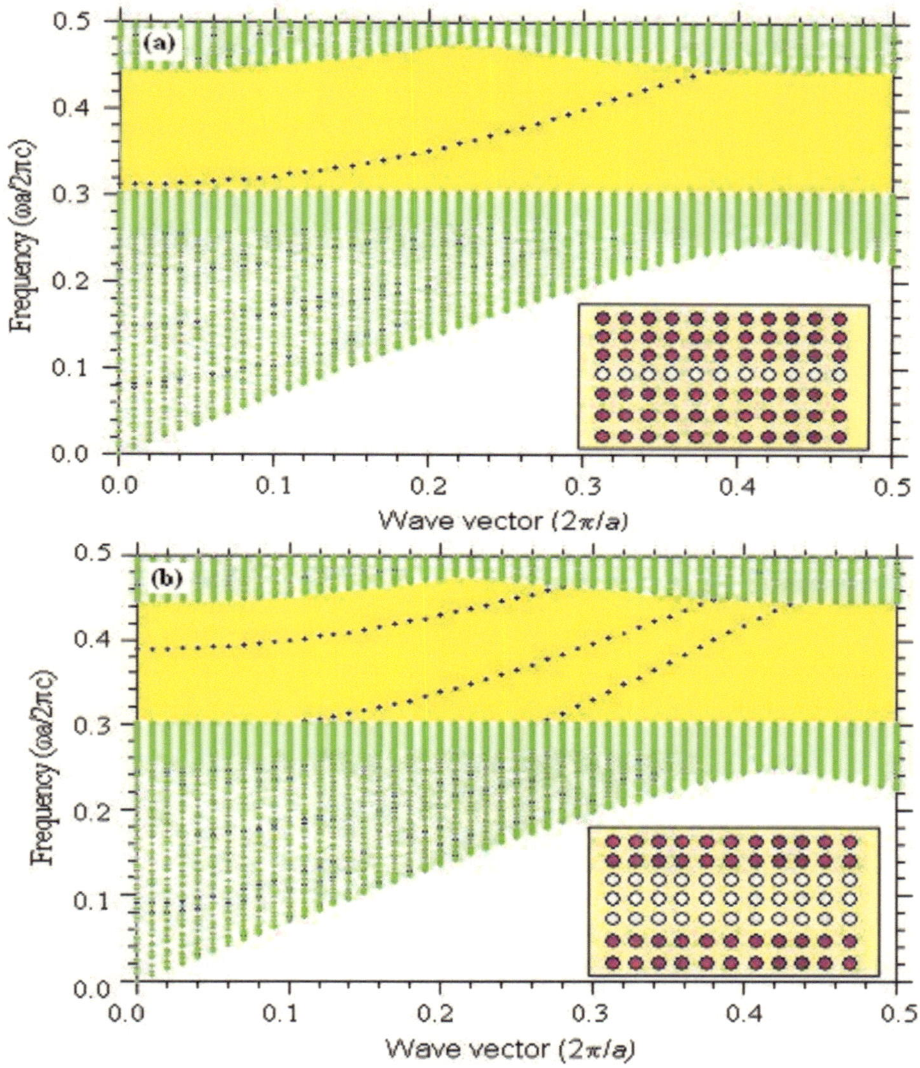

Fig 9. The projected band structure of TM modes for a waveguide in a square lattice of silicon pillars in air. The green region contains continuum of extended crystal states. The photonic band gap is colored yellow. The black dotted point is the band of guided modes that runs along the waveguide. (a) The waveguide is formed by removing one row of silicon pillar as shown in the inset. (b) . The waveguide is formed by removing three row of silicon pillar as shown in the inset.

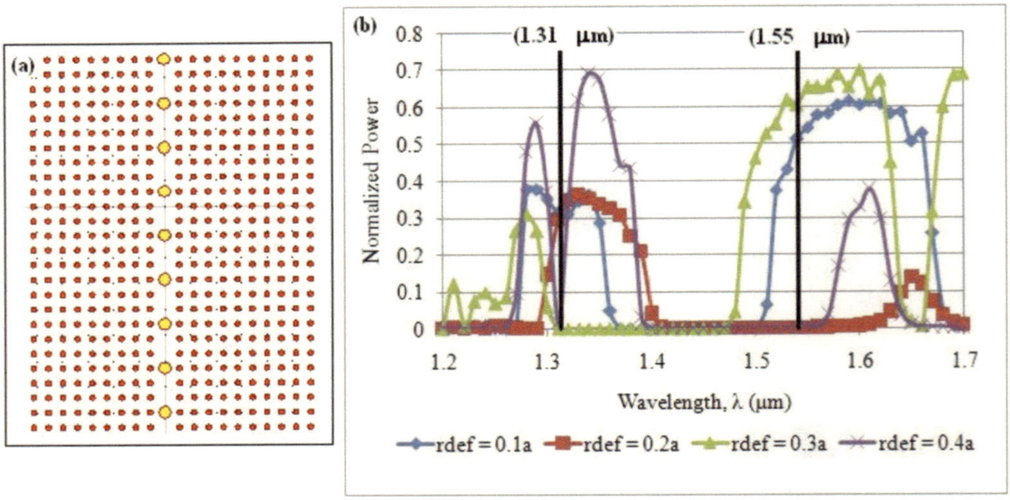

Fig. 10. (a) Couple cavity waveguide. (b) Transmission spectra of couple cavity waveguide as radius of the cavity are varied.

Fig.11 shows the projected band structured and dispersion curve for (a) coupling of two waveguide and (b) coupling of three waveguides. The PhC bands are shaded green and the bandgap is the gap between the two shaded green. In contrast to the waveguide modes in single missing row as shown in Fig.11 (a), there is at most one guided mode for all frequencies in the band gap. This is a property of common to most single line defect waveguides. In Fig.11 (a) there are two guided modes for coupling of two waveguide structure. At the small wavevector the guided modes is not couple together but as the wavevector increase, the two modes seem to couple together. The figure at the right inset of Fig.11 (a) show the enlarge point at the wavevector of 0.37 to 0.41. we can seen clearly that the two guided modes separate at wavevector 0.37, couple at point 0.385 and decouple back at point 0.39.

Fig.11 (b) shows the characteristics of guided modes for coupling of three waveguides. The right inside of Fig.11 (b) shows the two guide modes not couple to each other along the wavevector. A conclusion can be made, when two parallel identical PhC waveguides are brought close enough to have defect modes well coupled, the defect modes will split into two eigenmodes. The smaller the separation of waveguides, the larger the coupling and the more splitting in dispersion of the eigenmodes.

In this next section we want to show that photonic crystal can be use to guide light around the tight corners. In rectangular lattice, we can carve out a waveguide with a sharp 90 degree bend as shown in Fig. 12. Here we plot the displacement field of propagating TM mode as it travels around the corner. Even though the radius of the curvature of the bend is less than the wavelength of the light, very nearly all the light that goes in one end comes out the other.

Fig. 13 shows the loss over wavelength for 90°. The transmission loss for 90° bend along the telecommunication regime can be achieved less than 5 dB. The reflectivity at this sharp corner is around 0-15 dB. This figure proved that the PhC is a suitable material to guide light in very sharp corner with very small loss compare to conventional waveguide which loss usually 13-40 dB.

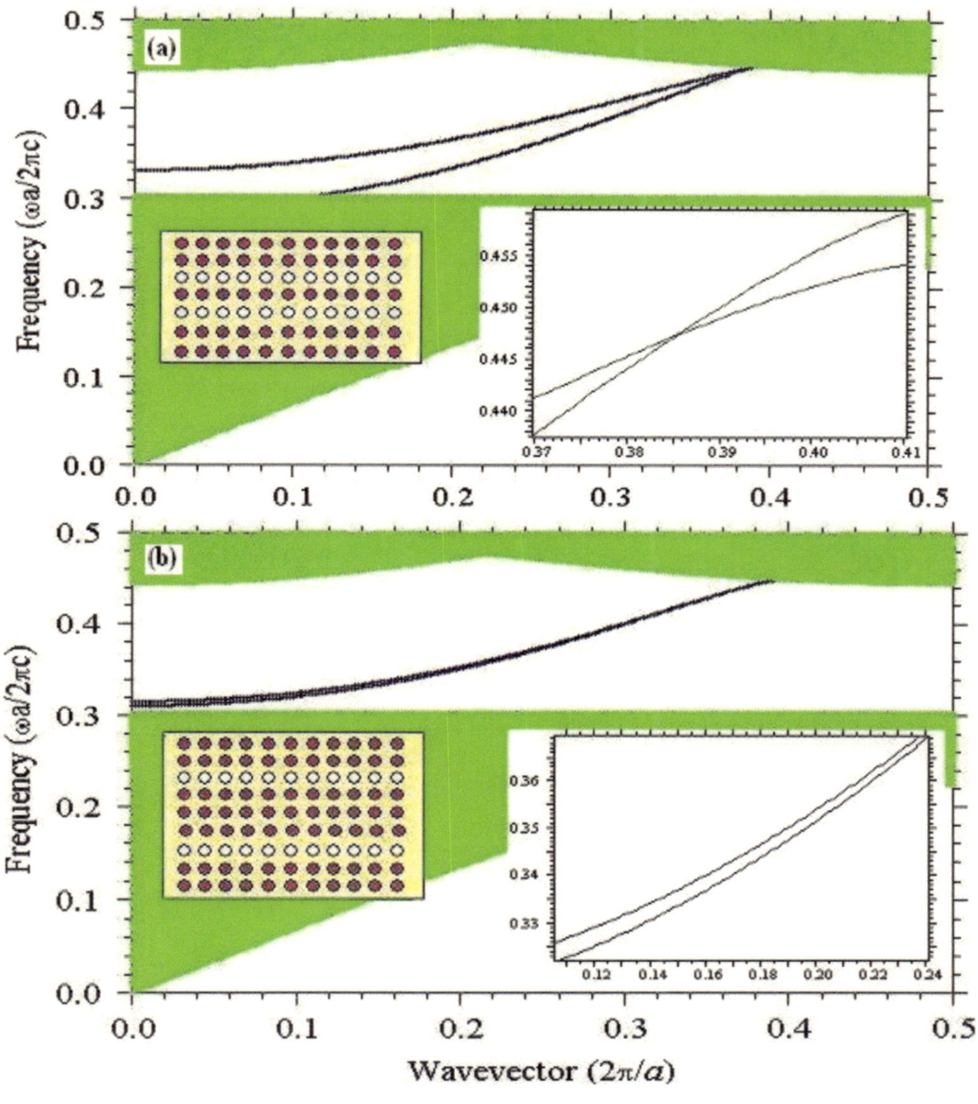

Fig. 11. Geometries and band structures for one (a) and three rows of dielectric pillars between two parallel waveguides. Enlarged parts of the band structures are shown in the insets of (a) and (b) to illustrate that the bands cross in (a) but do not cross in (b).

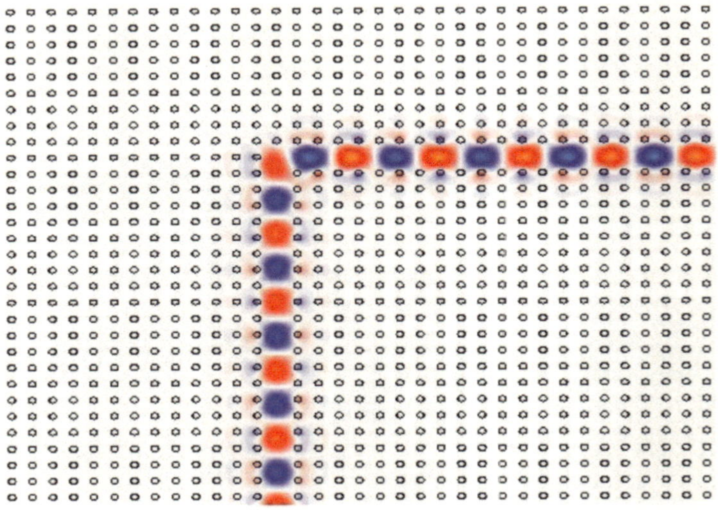

Fig. 12. The displacement field of a TM mode traveling around a sharp bend in a waveguide carved out of a rectangular lattice of dielectric pillars. Light is coming in from the bottom and exiting at the right.

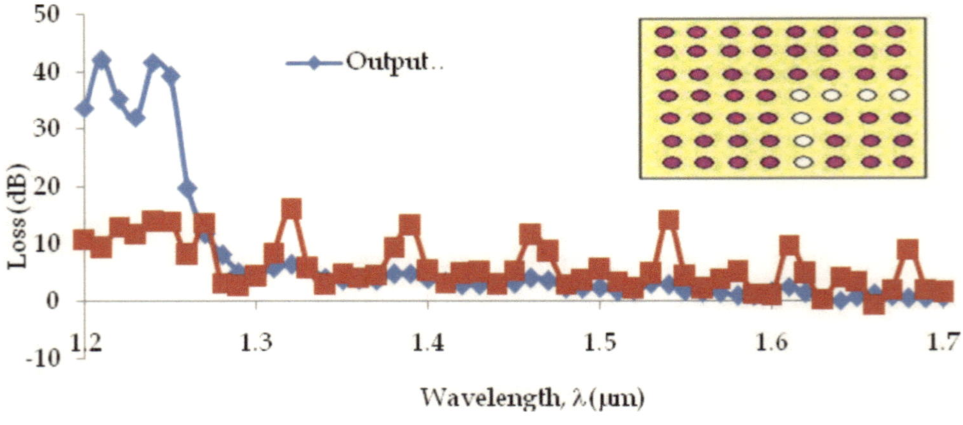

Fig. 13. Loss over wavelength for 90° bend.

5. Photonic crystals multiplexer/demultiplexer devices

A PhC with photonic band gap is a promising candidate as a platform on which to construct devices with dimensions of several wavelengths for future photonic integrated circuits. PhCs are particularly interesting, in all-optical systems to transmission and processing information due to the effect of localization of the light in the defect region of the periodic structure. Among the most important application areas of PhCs is low threshold single mode lasers (where PhCs are used as the optical confinement factor), wavelength filters, optical waveguide structures, WDM system devices, splitters and combiners. Wavelength filters of optical range based of two dimensional PBG structure can be created by the correct selection of geometrical and physical parameters.

In this subchapter, two wavelength demultiplexer/multiplexers (DEMUX/MUX) are designed. These DEMUX/MUXs consist of circular pillars in rectangular lattice surrounded by air with radius of circular scatterers is 0.18*a*. Device A which is 1310 nm / 1550 nm DEMUX/MUX device is based on splitting the wavelength channel using wide band filters. Finally, device B which split 1310 nm and 1550 nm wavelength channels was using MMI splitting mechanism.

5.1 Device A: 1310 nm /1550 nm Demultiplexer/Multiplexer

Fig. 14. shows the schematic diagram of 1310/1550 nm demultiplexer for rectangular lattice of silicon pillars. Two filters with radius defect inside it is placed at the T-junction. Filter 1 has radius defect of $r_1 = 0.11a$ which only filter wavelength 1310 nm and filter 2 has radius defect of $r_2 = 0.25a$ and absolutely only emit wavelength 1310 nm inside it.

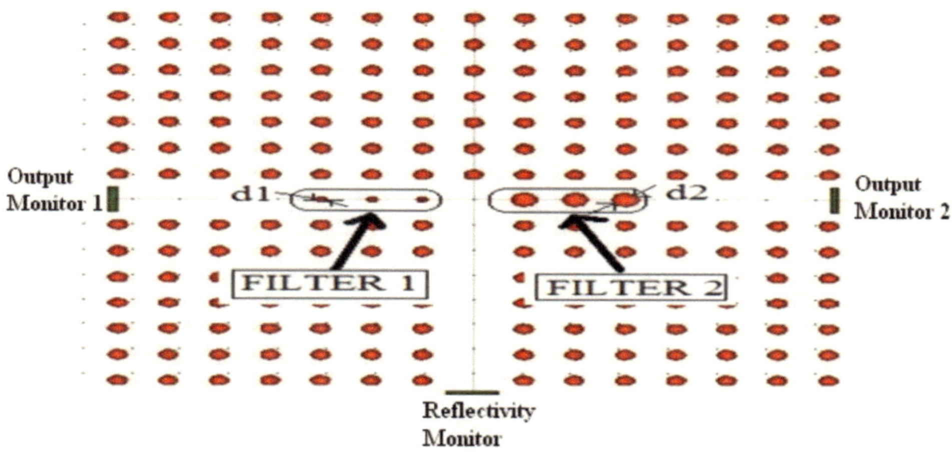

Fig. 14. Schematic diagram of the optical DEMUX/MUX based on a PhC waveguide with filter 1 and filter 2

The characteristics of defect pillars in filter 1 and filter 2 is investigated. The effect of number of pillar towards transmission power and reflectivity inside the filter was studied intensively. The output monitor of filter 1 is placed at left side of filter 1, meanwhile output monitor of filter 2 is placed at the right most of the waveguide and finally monitor for reflectivity is placed at the back of input signal as shown in fig. 14.

Fig. 15 shows the transmission characteristics at filter 1, filter 2 and reflectivity at varies number of defect pillars at input wavelength of 1310 nm. Basically when the wavelength of 1310 nm propagates inside the device A, it penetrate the filter 1. At filter 2, the maximum of 40% power is detected at monitor 2 and maximum power of 38% was reflect back at reflectivity monitor.

Fig. 15 (b) shows the normalized power inside the filter 2. Power occurred inside this filter is called crosstalk. For ideal case, power should be zero at this monitor 2 meanwhile power should transmit 100% at monitor 1. Unfortunately it did not happen in our device A because small amount of power dissipate into filter 2 and reflect back. The minimum number of defect pillars inside the filter caused the maximum crosstalk inside the device A.

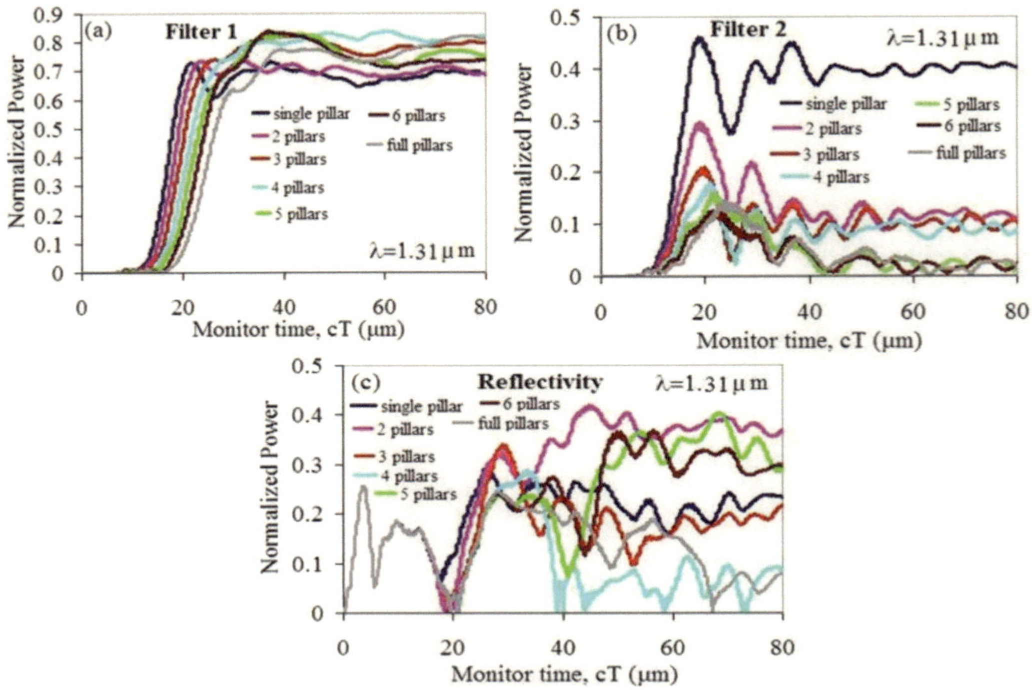

Fig. 15. Transmission characteristics inside filter 1, filter 2 and reflectivity when wavelength, λ of 1310 nm propagate inside device A.

A small amount of power was reflected back when all the pillars is occupied inside both filters as shown in fig.15 (c). Average of 24% power is reflected back when the wavelength of 1310 nm propagate inside device A.

Fig. 16 shows transmission characteristics at monitor 1, monitor 2 and reflectivity monitor when wavelength 1550 nm propagates inside device A. As can been seen, the filter at each arms work correctly because the power detected at monitor 1 is minimum meanwhile the power is maximum at monitor 2. This situation is vice versa when wavelength of 1310 nm propagates inside device A.

A maximum of 85% of power is transferred inside monitor 2 when wavelength 1550 nm propagates inside device A. The transmission graph in figure 16 (b) seem random and not depend on the number of defect pillars. This is because two pillars defect gave the most minimum power meanwhile the three single pillar give highest transmission.

The power that reflect back when wavelength 1550 nm propagates inside device is higher as shown in figure 16 (c). 78% of power is reflect back when the defect pillars inside the monitor is equal to two and five pillars.

From the investigation and analysis of the transmission characteristics when wavelength 1310 nm and 1550 propagates inside device A, an optimum design is proposed to split two wavelengths into two different output channels. The proposed design for splitting the wavelength 1310 nm and 1550 nm is using full defect pillars with diameter 124 nm in filter 1 and three pillars defect pillars with diameter of 285 nm in filter 2. With this design, it was found that it help to reduce reflectivity and boost up the power transmission in the device.

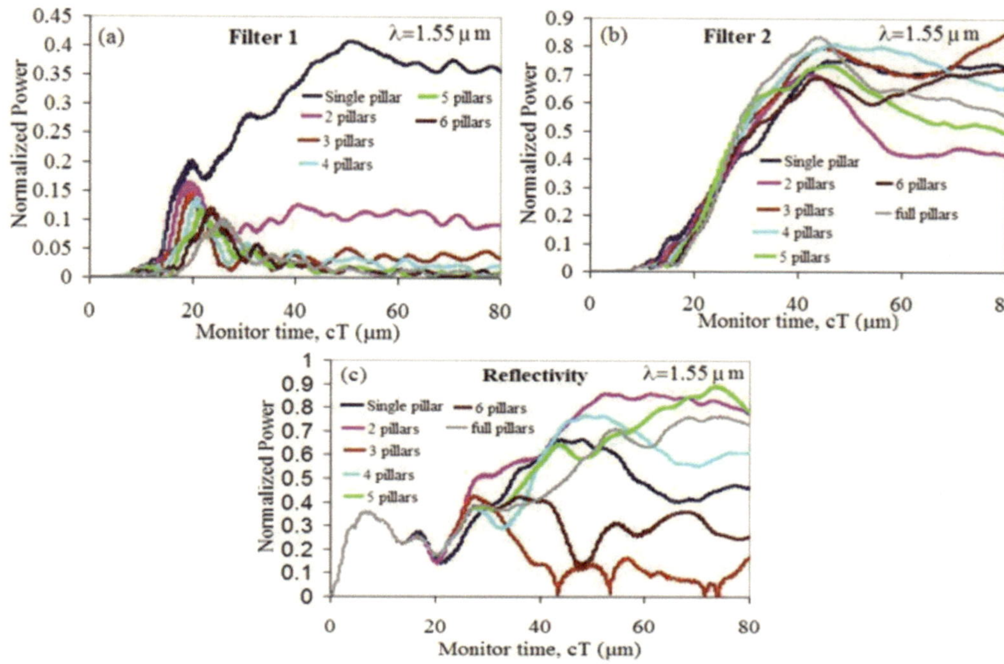

Fig. 16. Transmission characteristics inside filter 1, filter 2 and reflectivity when wavelength, λ of 1310 nm propagate inside device A.

The computed electromagnetic field distribution for device A is shown in Fig. 17 for two transmitted signals with wavelength 1310 nm (Fig. 17 (a)) and 1550 nm (Fig.17 (b)). Filter 1 has full defect pillars inside it and the diameter of defect pillars is smaller compare to diameter of surrounding pillars ($d_1 < d$) in the left arm. The filter1 only permit wavelength

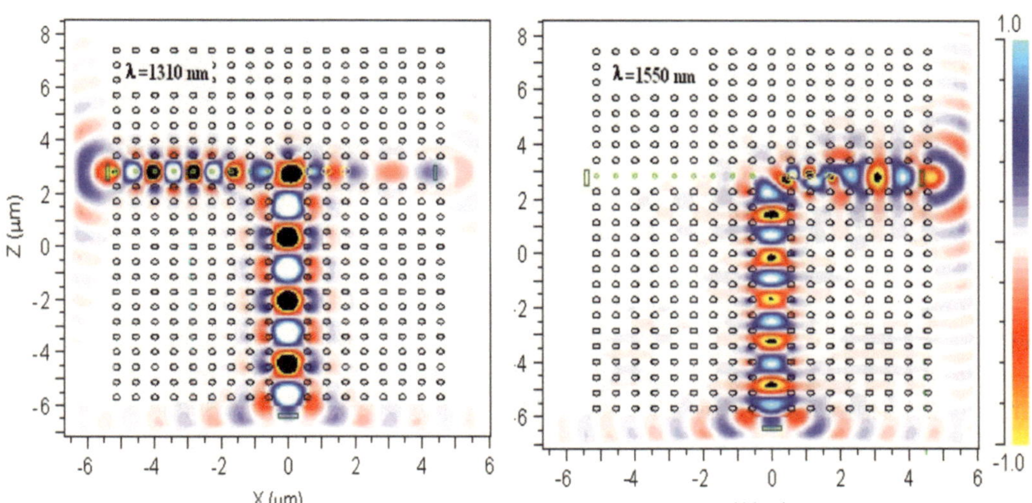

Fig. 17. Results of FDTD simulation of wavelength channel splitting for (a) source wavelength 1310 nm, and (b) source wavelength 1550 nm for device A

1310 nm to go through inside it and very small crosstalk was detected at monitor 2. Meanwhile filter 2 with diameter of defect pillars is bigger than the surrounding pillars (d_2>d) and three defect pillars inside it. Filter 2 in the right arm permit wavelength 1550 nm and reflects the wavelength of 1550 nm. Thus made the wavelength 1550 nm to turn to the left arm.

5.2 Device B: 1310 nm / 1550 nm demultiplexer/multiplexer based on Multimode Interference (MMI)

MMI based device is using the concept of interference phenomenon in it devices. Since the length of the device using an interference phenomenon is determined to be a common beat length for the multiple wavelengths, however, the device is quit long. To resolve this problem, in 2004, Kim et al. demonstrated that self imaging phenomenon also was valid in the PhC waveguide as well as in the dielectric waveguide (L.B. Soldano et al. 1995).

Self imaging is a property of multimode waveguides by which an input field profile is reproduced in a single or multiple images at periodic intervals along the propagation axis (L.B. Soldano et al. 1995). From the above definition, an input image can be reproduced in single or multiple images but, for the sake of simplicity, reproduction of only a single image is considered here. As shown in Fig.18, if an input field $\psi(0,y)$ is introduced into a multimode waveguide at x=0 with an asymmetric displacement d from the plane y=0, two kinds of images are reproduced at $x=L_m$ and $x=L_d$, depending on self-imaging conditions: one is a replica of the input field mirrored with respect to the plane y=0 at $x=L_m$ and the other is a direct replica of the input image at $x=L_d$.

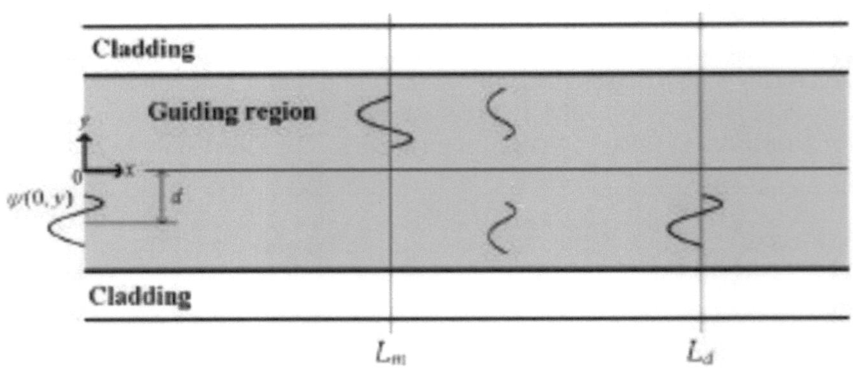

Fig. 18. Schematic illustration of a multi-mode waveguide. Input image is reproduced at $x=L_m$ and $x=L_d$.

For conventional waveguide, like the one presented above, self-imaging can be accepted without doubt or it may well be taken for granted. However, for a multi-mode PhC waveguide (PhCW), self imaging phenomena can still be observed. In order to prove that self imaging phenomena still valid in PhCW, a simulation which is similar to Fig.18 is simulate except that the conventional multi-mode waveguide is replaced with a PhC equivalent, as shown in Fig.19. In this structure, five consecutive rows are removed in otherwise perfect crystals to form a multi-mode PhCW.

As an access waveguide, a one-line defect PhCW is introduced into the multi-mode PhCW as shown in Fig.19. This access PhCW supports single-mode, and ensures that a well-confined input field is injected into the multi-mode PhCW for practical analysis.

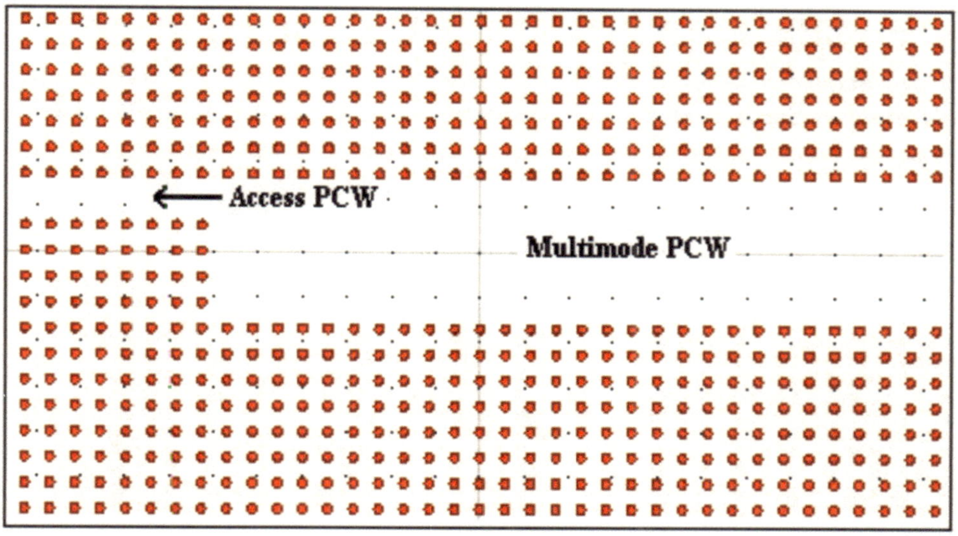

Fig. 19. Computational setup for observation of self imaging phenomena. The red dots represent dielectric rods (n=3.4) in air and their radius is 0.18a, where n is the refractive index of the rods and a is the lattice constant of the PhC.

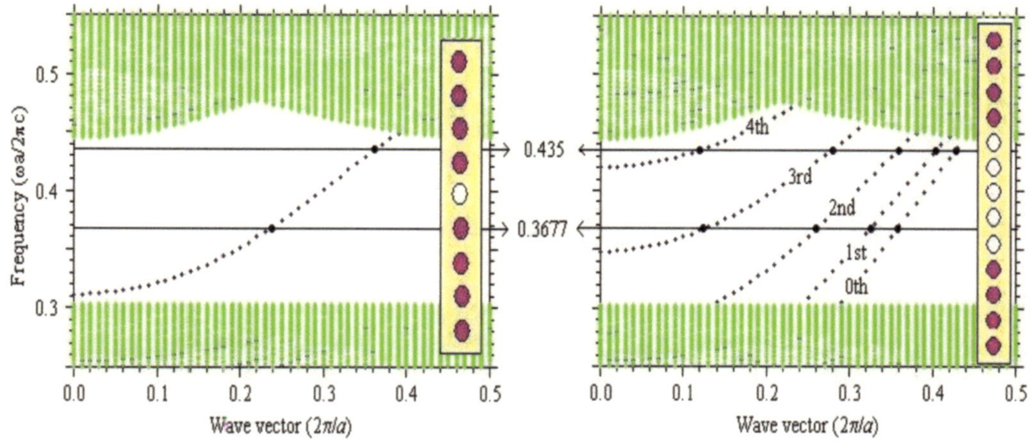

Fig. 20. (a) The dispersion curve for the access PhCW and the computational super-cell (inset). The access PhCW ensures single-mode operation from 0.312 (a/λ) to the top band gap. (b) The dispersion curve for the multi-mode PhCW and the computational super-cell (inset). The multi-mode PhCW supports 4 guided modes at 0.37(a/λ) and 5 guided modes at 0.43(a/λ)

Before launching an input field into the multi-mode PhCW, the property of guided modes in the PCWs should be understand because self-imaging is attributed to the multi-mode interference, which strongly depends on the number of modes, propagation constants, and modal field patterns. To confirm the number of guided modes supported by the access PhCW and by the multi-mode PhCW, the dispersion curves for two PhCWs are presented in Fig. 20. The dispersion curves are calculated by the plane wave expansion (PWE) method.

In this crystal, the band gap opens for the frequency range of 0.303-0.445(a/λ) for *E*-polarization (electric field parallel to the pillars), where λ is the wavelength in free space. While the frequency range of single-mode operation for the access PhCW extends from 0.312(a/λ) to the top of the band gap, as shown in Fig.20 (a), the multimode-mode PhCW supports from three to five guided modes for the same frequency range (Fig.20(b)). In the frequency range, the operating frequency of 0.3677(a/λ) is chosen where the multi-mode PhCW supports more than three guided modes (higher than 0.35(a/λ)), as presented in Fig.20 (b).

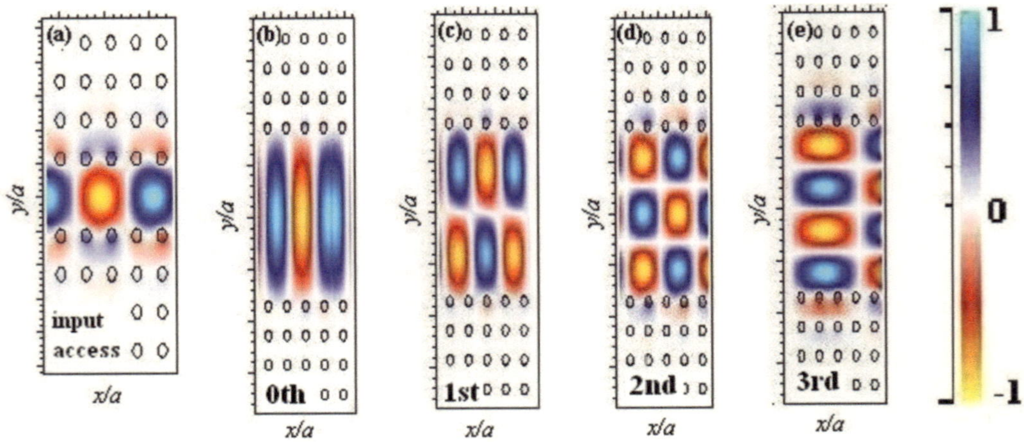

Fig. 21. Modal patterns of electric field in *z*-component for each mode at the operation frequency 0.37(a/λ) as presented in Fig.19. (a) Input image for access PhCW, (b) the 0th mode, (c) the 1st mode, (d) the 2nd mode, and (e) the 3rd mode at 0.37(a/λ).

To identify field patterns of guided modes in the access PhCW and the multi-mode PhCW, the modal field distribution are calculated at the opening frequency by the PWE method. Fig. 21 shows the *y*-component of the electric field at each calculation point, which are marked on dispersion curves in Fig. 20. The modal patterns have their own symmetry, even or odd, with respect to the propagation axis, hence the modal patterns can be selectively excited depending on the input position. However, in our case, since the access PhCW is introduced into the multi-mode PCW with an asymmetric position of $y=2/a$, as shown in Fig. 19, all the modes of the multi-mode PCW at the frequency of 0.37(a/λ) are excited by the input field (Fig. 20 (a)), which is confirmed by the overlap integral. Therefore, they all contribute to self-imaging.
The configuration of Fig. 19 is directly transferred to the FDTD computational domain for numerical experiment. The domain is surrounded by perfectly matched layers to absorb the outgoing waves. A continuous wave at the frequency of 0.37(a/λ) is launched into the access PhCW as shown in Fig. 22. The propagation shape inside the PhCW when 5 rows of pillar is removed is more likely to sinusoidal shape. It is likely that two kinds of images are reproduced by self imaging; one is expected to be mirrored replica at $x=10a$ (L_m) and the other to be a direct replica at $x=14a$ (L_d).

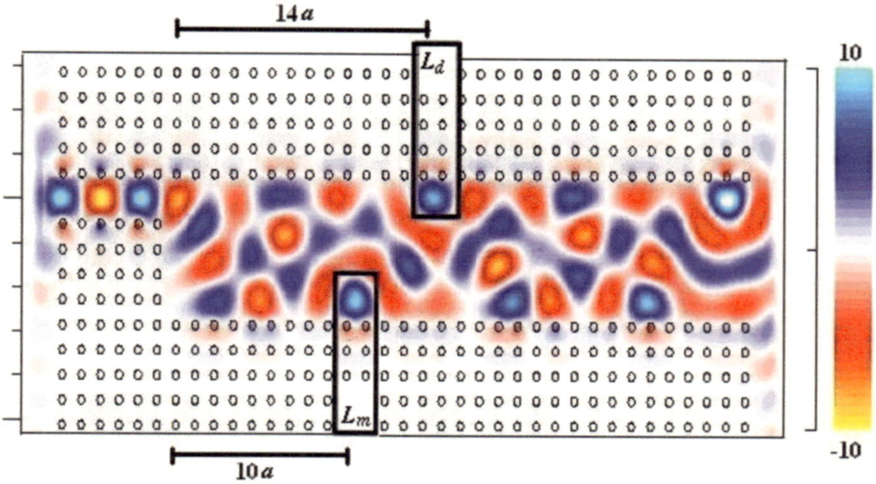

Fig. 22. Steady-state electric field distribution at 0.3677(a/λ).

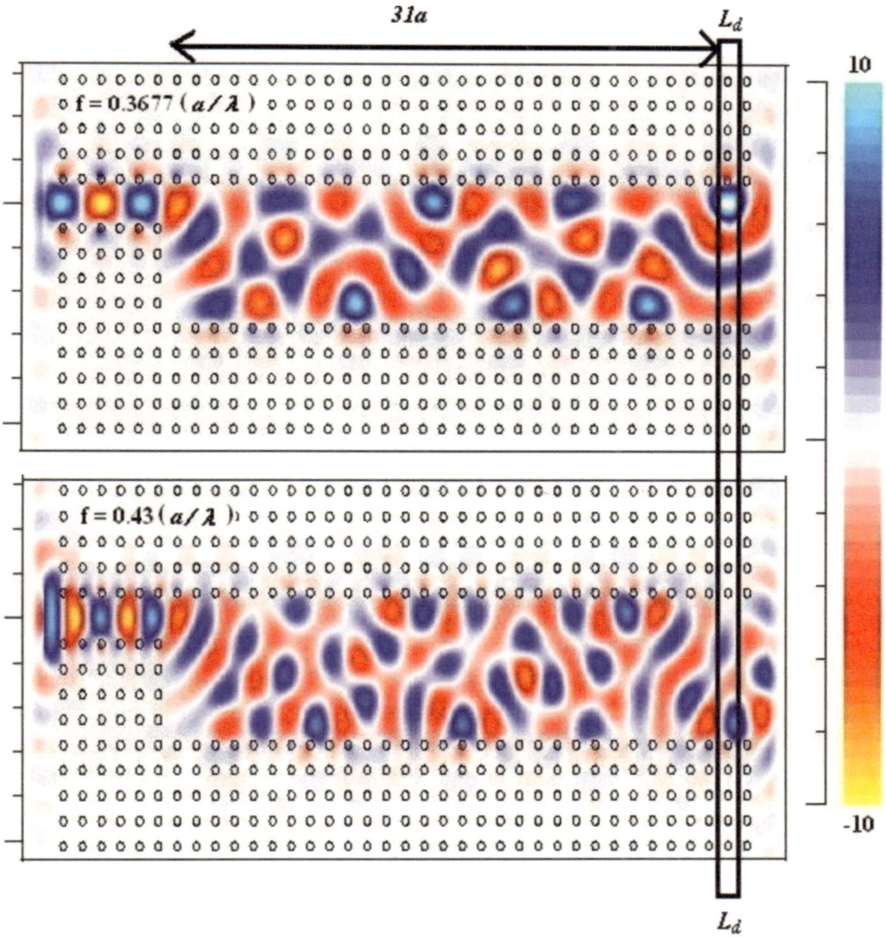

Fig. 23. Steady-state electric field distribution at 0.3677(a/λ) (*upper*) and at 0.43(a/λ) (*below*).

Fig. 23 shows steady-state electric field distribution at two different frequencies. As can be seen in both figures, the image inside the L_d box at 0.3677(a/λ) (upper) and at 0.43(a/λ) (below) is at different position but has same distant, L_d which is 31a. From this picture, a DEMUX/MUX based on MMI can be designed. If arms are placed at 31a, the two wavelengths can be guided inside two different output arms.

Fig. 24 shows a PhC DEMUX/MUX is designed by using self-imaging conditions. The objective is to separate two wavelengths (1310 nm and 1550 nm) so that a 1-to-2 structure is required for routing each wavelength to a corresponding output. As shown in Figure 23, two output PhCWs are added to the structure and the length, L_d of the multi-mode PhCW is set around 31a, since the direct replica at 0.3677(a/λ) and 0.43(a/λ) are imaged at that position.

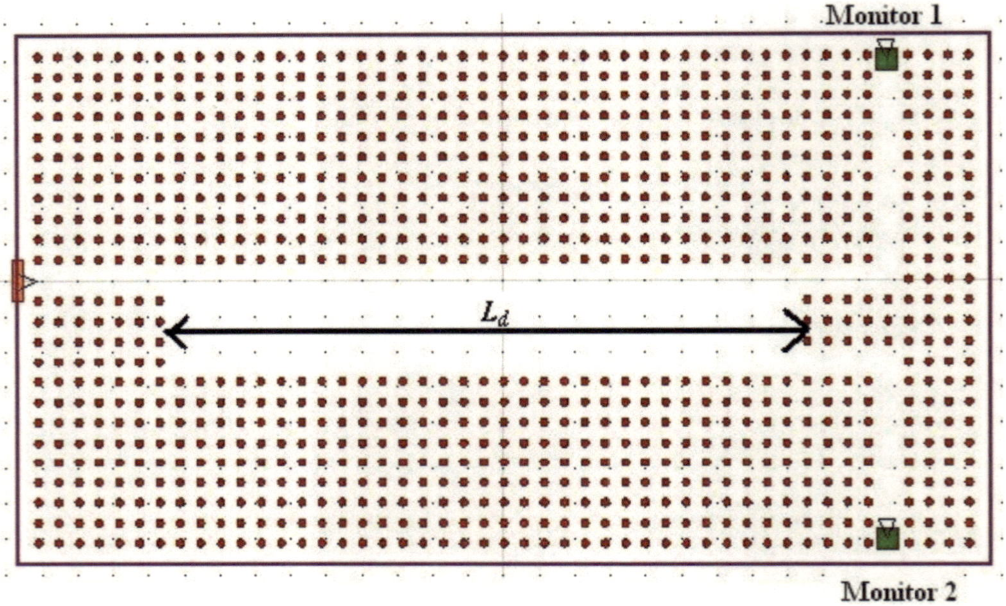

Fig. 24. The designed PhCW DEMUX/MUX MMI based.

This wavelength at 0.3677(a/λ) is intended to be routed to output port 1. Scanning computation of several frequency points on dispersion curves in Fig. 20 (b) makes it possible to determine another appropriate operating point of 0.43(a/λ) at which a mirrored replica is image at almost the same position as the direct replica is imaged (Fig. 23).

This design is directly applicable to a 1550/1310 nm DEMUX/MUX by setting the lattice constant, a, as 570nm. Unlike the previous situation at 0.37(a/λ), here there are five guided modes at 0.43(a/λ) to be considered-one more guided mode than at 0.37(a/λ), as shown in Fig. 20. However, the 4th mode is not excited, since it has odd symmetry with respect to the input field. Hence, four out of five modes are considered. Table 20 shows the final design parameters for device B with optimize value of L_m

Fig. 25 shows the steady-state electric field distributions as obtained by FDTD calculations after continuous waves at λ =1550 nm (Fig. 25 (left)) and λ=1310 nm (Fig.25 (right)) are launched into the access PhCW.

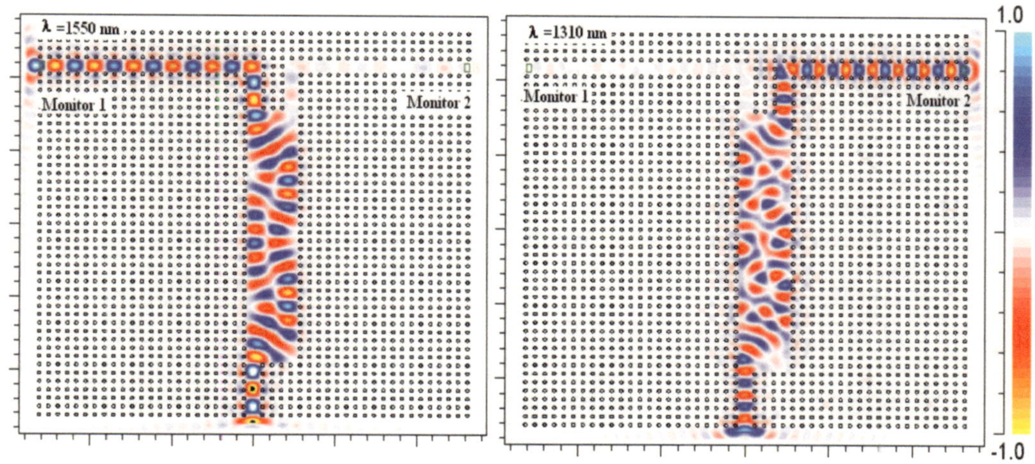

Fig. 25. Steady-state electric field distributions in the designed PhC wavelength DEMUX/MUX MMI based at 1550 nm (*left*) and at 1310 nm (*right*).

6. Conclusion

Several of frailty has been identified in designing the DEMUX/MUX in PhC. For example, the power transfer inside the device not transfers 100% at the output arms. Light propagates inside the devices might dissipate along the waveguide. To minimize this losses, the design need to be alter, such as incorporate the defect pillar inside the space between two pillars so that the light will reflect back when the light incident with the defect pillar.

Development of planar lightwave circuit (PLC) devices by combining the conventional waveguides and PhCW need to be study. One of the mechanisms how to minimize the coupling loss between conventional waveguide and PhCW is by introduces taper waveguide at the end of PhCW. By introducing taper waveguide, the bigger spot size from conventional waveguide will slowly shrink when enter the taper waveguide at the PhCW.

Most of the work in this thesis focuses on optical communication PLC technology. Application area of photonic crystal devices can be extends into other applications. PhC can also be implemented in other applications such as bio-sensing, imaging, illumination, etc. The investigation on how the existing devices can be used for such applications and new devices/materials will have to be developed to address these areas.

In the future, PhC components will by widely used in optical telecommunications. Since the invention of the concept of PhC in the end of of the 80's, their properties have been studied intensively. New applications have been proposed and realized. The only components that are so far in commercial use are the photonic crystal fibers. They have many extraordinary properties that cannot be achieved using conventional fibers and thus will have a profound effect on the fiber optics industry. In the near future, other PhC components, such as fiber lasers, will also be brought into market.

7. References

B. S. Song, S. Noda and T. Asano, 2003. *Science,* 300: 1537.

B. S. Song, S. Noda, T. Asano and Y. Akahane, 2005, *Nature Materials,* 2: 207.

Baba T., & Matsuzaki T. 1996. Fabrication and photoluminescence studies of GaInAs/InP 2-dimensional photonic crystals. *Japanese Journal of Applied Physics, Part 1 (Regular Papers & Short Notes)* 35(2B): 1348-1352.

Baba T., Inoshita K., Tanaka H.,Yonekura J., Ariga M., Matsutani A., Miyamoto T., Koyoma F. & Iga K. 1999. Strong enhancement of light extraction efficiency in GaInAsP 2D arranged microcolumns. *Journal of Lightwave Technology* 17(11):2113-2120.

Benisty H., Weisbuch C., Labilloy D., Rattier M., Smith C. J. M., Krauss T. F., de la Rue R. M., Hourde R., Oesterle U., Jouanin C.& Cassagne D. 1999. Optical confinement properties of two-dimensional photonic crystals. *Journal of Lightwave Technology* 17(11): 2063-2077.

Boroditsky M., Gontijo I., Jackson M., Vrijen R. Yablonovitch E., Krauss T., Chuan Cheng Cheng, Scherer A., Bhat R. & Krames M. 2000. Surface recombination measurements on III-V candidate materials for nanostructure light-emitting diodes. *Journal of Applied Physics.* 87(71): 3497-3504.

C. Manolatou, S. G. Johnson, S. Fan, P. R. Villeneuve, H. A. Haus & J. D. Joannopolous. 1999. High-density integrated optics. *IEEE J. Lightwave Technol.* 17: 1682-1692.

C. T. Lee & M. L. Wu. 2001. Apaxes-linked circle gratings for low-loss waveguide bends. *IEEE Photon. Technol. Lett* 13: 597-599.

Chul-Sik Kee, Jae-Eun Kim and Hae Yong Park. 1997. Absolute photonic band gap in a two-dimensional square lattice of square dielectric rods in air, *Rapid Communications Physical Review E*, Volume 56, number 6.

Chutinan A.& Noda S. 1999. Highly confined waveguides and waveguide bends in three-dimensional ph otonic crystal. *Applied Physics Letters.* 75(24): 3739-3741

Cregan R. F., Mangan B. J., Knight J. C., Birks T. A., Russell P. St. J., Roberts P. J. & Allan D. C. 1999. Size-mode photonic band gap guidance of light in air. *Science* 285(5433):1537-1539.

D. K. Armani, T. J. Kippenberg, S. M. Spillane and K. J. Vahala. 2003. Ultra-high-Q toroid microcavity on achip. *Nature* 421, 925-928.

D. L. Lee. 1986. *Electromagnetic principles of integrated optics* New York: John Wiley & Sons.

D. W. Vernooy, V. S. Ilchenko, H. Mabuchi, E. W. Streed, and H. J. Kimble. 1998. High-Q measurements of fused-silica microspheres in the near infrared. *Opt. Lett.*23, 247-249.

Djafar K., Mynbaev & Lowell L. Scheiner. 2001. Fiber-optic communications Technology New York: Prentice Hall

Eli Yablonovitch & T. J. Gmitter. 1991. Photonic band structure: the face centered cubic case employing nonspherical atoms. *Physical Review Letters* 67: 2295-2298.

Eli Yablonovitch. 1987. Inhibited spontaneous emission in solid State physics and electronics. *Physical Review Letters* 58: 2059-2062.

H. Nishihara, M Haruna & T. Suhara. 1989. *Optical integrated circuits.* New York: McGrawHill.

J. S. Foresi, P. R. Villeneuve, J. Ferrara, E. R. Theon, G. Steinmeyer, S. Fan, J.D. Joannopolous, L.C. Kimerling, H.I. Smith and E. P. Ippen, 1997, *Nature*, 390: 143.

K. Inoue & K. Ohtaka. 2004. *Photonic crystals: physics, fabrication and applications.* New York: Springer-Verlag Berlin Heidelberg.

K. M. Ho., C. T. Chan & C. M. Soukoulis. 1990. Existance of a photonic band gap in periodic dielectric structures. *Physical Review Letters* 65: 3152-3155.

K. M. Leung & Y. F. Lin. 1990. Full vector wave: calculation of photonic band structures in face centered cubic dielectric media.

K. Okomoto. 1999. *Integrated optical circuits and components: Design and applications*. New York: Marcel Dekker.

Knight T C., Birks T. A., Russell P. St. J. & de Sandro J. P. 1998. Properties of photonic crystal fiber and effective index material. *Journal of the Optical Society of America A (Optics, Image Science and Vision)* 15(3): 748-752.

Kosaka H., Kawashima T., Tomita A. Notomi M., Tamamura T., Sato T. & Kawakami S. 1998. Superprism phenomena in photonic crystals. *Physical Review B (Condensed Matter)* 58(1615): R10096-R10099.

Kristina Grifantini, 2008Nature's Photonic Crystal, Scientists find an elusive diamond structure in a Brazilian beetle. *Technology Review published by MIT*. www.technologyreview.com/Nanotech/20840/?nlid=1115&a=f

L.B. Soldano et al., 1995. Optical multi-mode interference devices based on self-imaging: principles and applications. *J. Lightwave Technol.* 13, 615-627.

M. Koshiba 1992. *Optical waveguide analysis*. New York:McGraw-Hill

M. Popovic, K. Wada, S. Akiyama, H. A. Haus & J. Miche. 2002. Air trenches for sharp silica waveguide bends. *J. Lightwave Technology* 20:1762-1772.

Mekis A., Chen J. C., Kurland I., Fan S., Villeneuve P. R. & Joannopolous J. D. 1996. High transmission through sharp bends in photonic crystal waveguides. *Physical Review Letters*. 77(18): 3787-3790.

O. Painter, R. K. Lee, A. Scherer, A. Yariv, J. D. O'Brien, P. D> Dapkus, and I. Kim, 1999, *Science*, 284: 1819.

O'Brien J., Painter O., Lee R., Cheng C. C., Yariv A., & Scherer A. 1996. Laser incorporating 2D photonic bandgap mirrors. *Electronics Letters*. 32(24): 2243-2244.

Ohtera Y., Sato T., Kawashima T. Tamamura T.& Kawakami S. 1999. Photonic crystal polarization splitters. *Electronics Letters* 35(15): 1271-1272.

Pete Vukusic & J. Roy Sambles. 2003. Photonic structures in bilogy. *Nature* 424(680): 852-855.

R. A. Jarvis, J. D. Love & F. Ladouceur. 1997. Bend-radius reduction in planar waveguide using UV post-tuning. *Electron Lett*.33:892-894.

R. Colombelli, K. Srinivasan, M. Troccoli, O. Painter, C. F. Gmacgl, D. M. Tennant, A. M. Sergent, D. L. Sivco, A. Y. Cho and F. capasso, 2003, *Science*, 302: 1374.

R. L. Espinola, R. U. Ahmad, F. Pizzuto, M. J. Steel & R. M. Osgood. 2001. A study of high-index contrast 90 degree waveguide bend structures. *Opt Express*. 8:517-528.

S. Lardenois, D. Pascal, L. Vivien, E. Cassan & S. Laval. 2003. Low-loss submicrometer silicon-on-insulator rib waveguides and corner mirrors. *Opt. Lett.* 28:1150-1152.

S. Noda, A. Chutinan and M. Imada, 2000. *Nature*, 407:608.

S. Ogawa, M. Imada, S. Yoshimoto, M. Okano and S. Noda, 2004. *Science*, 305: 227.

S. Y. Lin, V. M. Hietala, S. K. Lyo, and A. Zaslavsky, 1996, *Appl. Phys. Lett.*, 68,3233

Sajev John. 1987 Strong localization photon in certain disordered dielectric superlattices. *Physical Review Letters*. 58:2486-2489. *Physical Review Letters* 65: 2646-2649.

Shanshui Fan, Villeneuve P. R., Joannopolous J. D. & Schubert E. F. 1997. High extraction efficiency of spontaneous emission from slabs of photonic crystals. *Physical Review Letters*. 78(17): 3249-3247.

Y. Akahane, T. Asano, B. S. Song and S. Noda, 2003, *Nature*, 425: 944.

Yablonovitch E., Gmitter T. J., Meade R. D., Rappe A. M., Brommer K.D. & Joannopolous J.D. 1991. Donor and acceptor modes in photonic band structures. *Physical Review Letters* 38 (249): 3380-3383.

Yonekura J., Ikeda M.& Baba T. 1999. Analysis of finite 2D photonic crystals of columns and lightwave devices using scattering matrix method. *Journal of Lightwave Technology* 17(8): 1500-1508.

28

Improvement Scheme for Directly Modulated Fiber Optical CATV System Performances

Hai-Han Lu, Ching-Hung Chang and Peng-Chun Peng
National Taipei University of Technology
Institute of Electro-Optical Engineering
Taiwan

1. Introduction

In an optical CATV system, the signal was directly or externally modulated with lightwave before communicates. Directly modulating signal with LD is an economic method, but the transmission distance and performance are significantly limited by laser chirping issues. In another hand, the externally modulation schemes have been proofed to provide better outcomes by eliminating the laser chirping issue. Nevertheless, an expensive externally modulated transmitter is required causing an increased capital expenditures. In order to provide an economic structure with advanced transmission performance as in external modulation system, direct modulation method is often combined with other techniques or components to compose higher performance and lower cost CATV system. The split-band techniques for example have been proofed as a powerful assistant for direct modulation CATV systems. In this structure, the full channel loading is shared by additional LDs causing a wider optical linewidth to eliminate the SBS degradation. In addition, by increasing the wavelength numbers, major parts of CSO distortions from each transmission band are automatically removed from each channel. These outstanding techniques and impressive outcomes are consequently analyzed and discussed as parts of this chapter.
In parallel with the effective of the SBS degradation, the CNR is in direct proportion to the optical power, as well as the second-order harmonic distortion-to-carrier ratio (HD_2/C) and the third-order inter-modulation distortion-to-carrier ratio (IMD_3/C) values are in inverse proportion to the laser resonance frequency. Improving the utilized LD characteristic is therefore another useful methodology to upgrade the optical CATV system. External light-injection and optoelectronic feedback techniques in particularly have been experimentally proofed as efficient methods to enhance laser output power and laser resonance frequency (Lee et al., 2007; Lu et al., 2008). As a result, the CNR value will be proportionally increased with the enhanced LD intensity, as well as the HD_2/C and IMD_3/C will be proportionally reduced with the improved laser resonance frequency. The reduced HD_2/C and IMD_3/C values are subsequently leading to an improvement in CSO/CTB values.
Following with the modifications of modulation schemes and lightwave characteristics, compensating fiber dispersions is another research direction to promote fiber optical CATV systems. Numbers of compensation methodologies such as eliminating parts of optical dispersion by optical filter and cascading systems with negative dispersion fibers are

demonstrated as useful techniques to upgrade system performance. Since there is no modulating information in the optical carrier, putting an optical filter to eliminate the redundant spectra not only can increase spectra efficiency by reducing the spectral linewidth but also can ameliorate the CSO/CTB performance by reducing frequency dispersion. Similarly, cascading a section of negative dispersion fiber with CATV systems is also an efficient method to eliminate the fiber chromatic dispersion. By combining a negative dispersion fiber with single mode fiber (SMF), the total fiber dispersion is able to be removed to improve the system performance. Consequently, the relative fiber dispersion compensation techniques as well as the possible extensions and the future directions of the CATV systems are also discussed in this chapter.

2. Modulation promotion methods: split-band techniques

Long-distance transmission of fiber AM-VSB 80-channel CATV systems has been widely spread throughout the cable industry. Nevertheless, the maximum transmission distance of such systems is still limited by RF parameters and it is difficult to obtain good CNR, CSO, and CTB performances due to full channel loading (Lu & Lee, 1998). In order to provide an economic structure with advanced transmission performance as in external modulation system, direct modulation method is often combined with other techniques or components to compose higher performance and lower cost CATV system. The split-band techniques for example have been proofed as a powerful assistant for direct modulation CATV systems. In this structure, the full channel loading is shared by additional LDs causing a wider optical linewidth to eliminate the SBS degradation (Lee et al., 2007). In addition, by increasing the wavelength numbers, major parts of CSO distortions from each communication band are automatically removed from each communication channel.

Fig. 1 shows two AM-VSB 80-channel fiber optical CATV systems with two cascaded EDFA's (Lu & Lee, 1998). Fig. 1 (a) (referred to as system I) shows the conventional 80-channel fully loaded externally modulated system. Fig. 1 (b) (referred to as system II) shows a half-split-band directly modulated WDM system with 40 channels per transmitter.

In system I, channel 5~78 (77.25–547.25 MHz) are fed into an external modulator as the modulation signal through an RF predistortion circuit. The interaction of the 1550 nm DFB laser light and the RF-generated electric fields leads to a phase modulation via the electrooptical effect. The phase modulation was then converted to intensity modulation within the external modulator. During this converting process, the linewidth is broadening by the phase modulation causing a reduced SBS effect. While in system II, channel 5~40 (77.25–319.25 MHz; low bands) are fed into the first directly modulated transmitter, and channel 41~78 (325.25–547.25 MHz; high bands) are fed into the second one. Channel (55.25–67.25 MHz) are removed from system II to obtain best CSO value.

Fig. 2 (a)–(c) shows the measured CNR, CSO and CTB values under NTSC channel number for both system I and II respectively. Comparing with the system I, the CNR and CTB values in the system II are relatively degraded about 2 and 0.2 dB. Nevertheless, the CSO performance in the system II is much better than in system I especially at the high bands region.

This improvement, resulted from the use of half-split-band and WDM techniques, is caused by the reduction of fiber nonlinear and dispersion effects. For example, when the carrier frequency for CH_7 and CH_{13} is 175.25 and 211.25 MHz respectively, there will be a CSO distortion at:

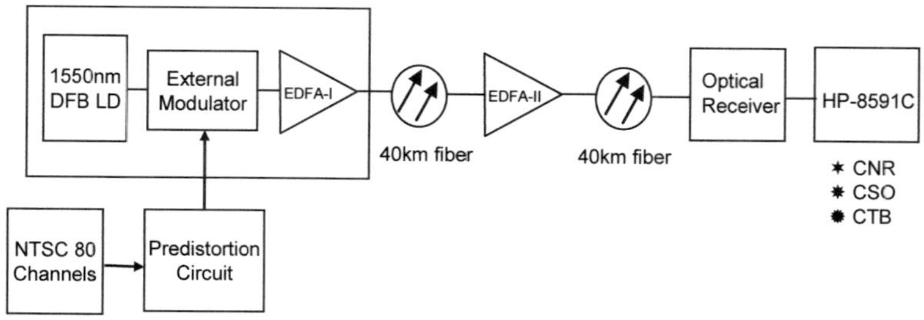

Fig.1(a)

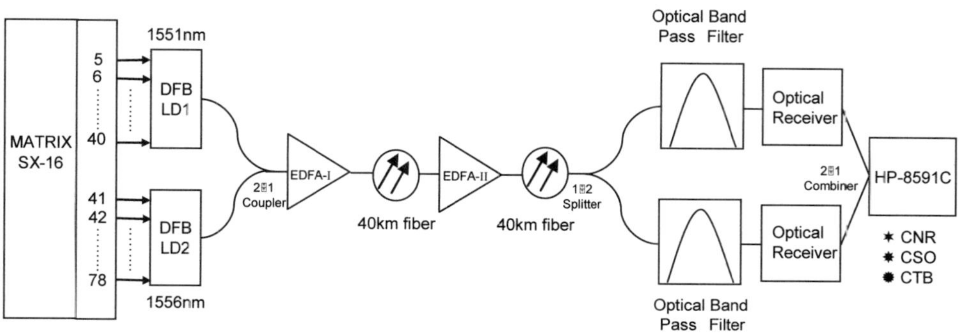

Fig.1(b)

Fig. 1. (a) Fully loading externally modulated transmitter system and (b) directly modulated transmitters using half-split-band techniques (Lu & Lee, 1998).

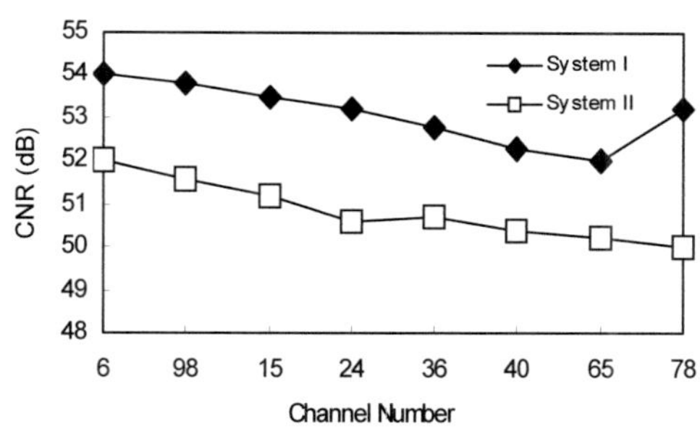

(a)

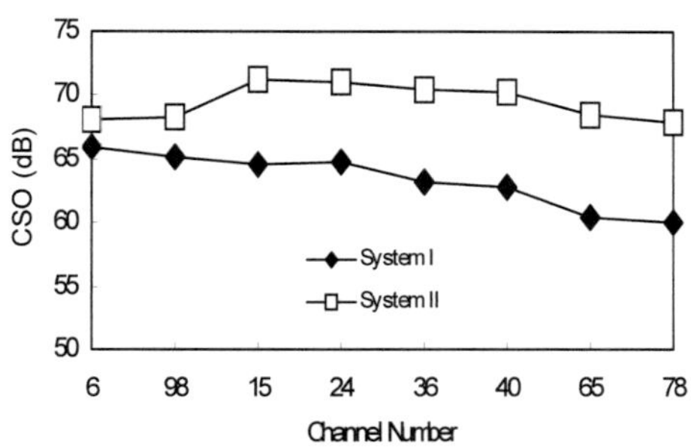

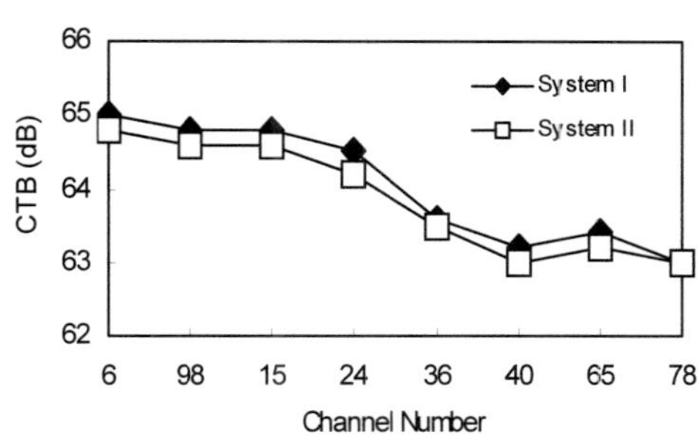

(b) (c)

Fig. 2. (a) Measured CNR, (b) CSO and (c) CTB values for systems I and II. (Lu & Lee, 1998).

$$175.25\text{MHz}\ (CH_7) + 211.25\text{MHz}\ (CH_{13}) = 386.25\ \text{MHz}. \quad (1)$$

Similarly, the CSO distortion caused by CH_{78} and CH_{50} is located at:

$$547.25\text{MHz}\ (CH_{78}) - 379.25\ \text{MHz}\ (CH_{50}) = 168\text{MHz}. \quad (2)$$

The above results present that the CSO distortions induced by low bands (CH_5 ~ CH_{40}) are located at high bands, and CSO distortions induced by high bands (CH_{41} ~ CH_{78}) are located at low bands. The half-split-band technique is therefore can remove major part of CSO distortions in system II to improve the transmission performance.

3. Lightwave enhancement schemes: light-injection techniques

In contrast to transform the modulation techniques, boosting up LD performance is another valuable technique to be recoded in here. External light-injection techniques and optoelectronic feedback techniques in particularly have been experimentally proofed as efficient methods to improve optical fiber CATV systems (Lee et al., 2007). By externally injecting a master light source into a slave laser, the output power and laser resonance frequency of the injection-locked slave laser can be greatly enhanced (Kazubowska et al., 2002; Lee et al., 2006). The CNR value is consequently proportionally increased with the enhanced LD intensity, as well as the HD_2/C and the IMD_3/C are proportion reduced with the laser resonance frequency. The reduced HD_2/C and IMD_3/C values in the light-injection optical CATV systems will lead to an improvement in CSO/CTB performance. As a result, the light-injection techniques can notably improve the lightwave features to transmit CATV programs.

3.1 Local light-injection techniques

Fig. 3 shows a directly modulated transport system employing external light injection technique to improve the overall system performance (Lu et al., 2003). The central wavelengths of the two DFB laser diodes are 1550.5(λ_1) and 1555.7(λ_2) nm respectively.

In the system, the frequency response of the DFB laser diode with and without external light injection is very different. In the free running case, the laser resonance frequency is ~5 GHz; with 3 dBm light injection, the laser resonance frequency is increased to ~15 GHz; and with 4.8 dBm light injection, the laser resonance frequency is further increased up to ~18.5 GHz. The 370% improvement in the laser resonance frequency has significantly demonstrated the advancement of the external light injection techniques to assist the fiber optical CATV systems.

Fig. 4 (a) shows the measured CNR values under NTSC channel number with and without external light injection respectively. It can be seen that the CNR values are boosted up with the increasing of the injected power level. For long-haul lightwave transmission systems, the CNR values are dominated mainly from signal-spontaneous beat noise (Way, 1998):

$$CNR_{sig-sp}^{-1} = \frac{8n_{sp}hv}{m^2 c_1^2 P_{in}}(1 - \frac{1}{G}) \quad (3)$$

where CNR_{sig-sp} is associated with the EDFA, n_{sp} is the population inversion factor, hv is the photon energy, m is the optical modulation index, c_1 is the input coupling loss to the EDFA,

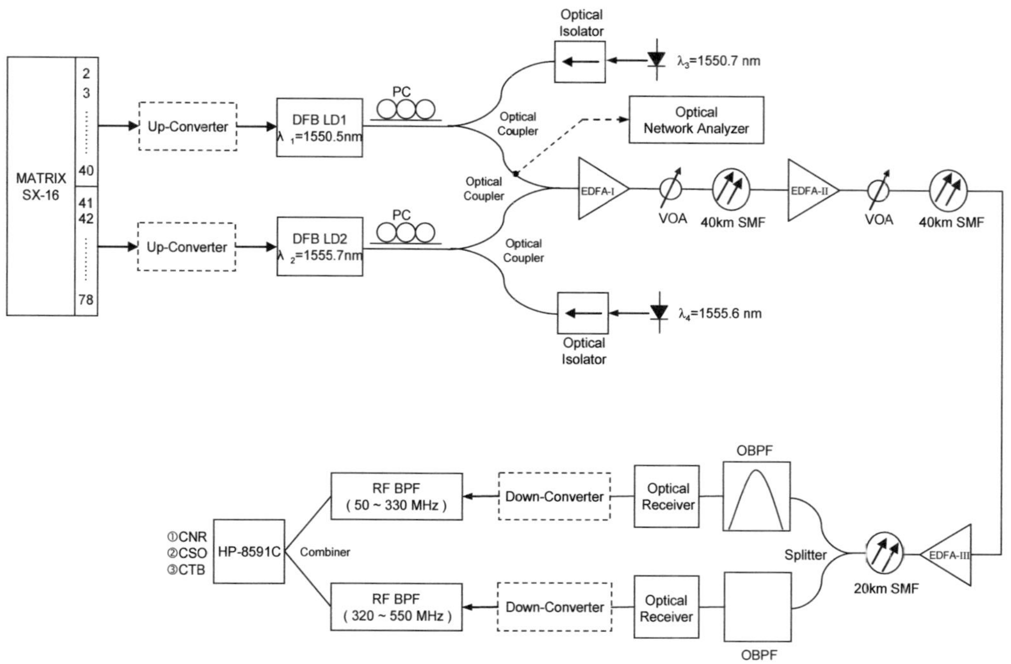

Fig. 3. Directly modulated transport system employing external light-injection technique (Lu et al., 2003)

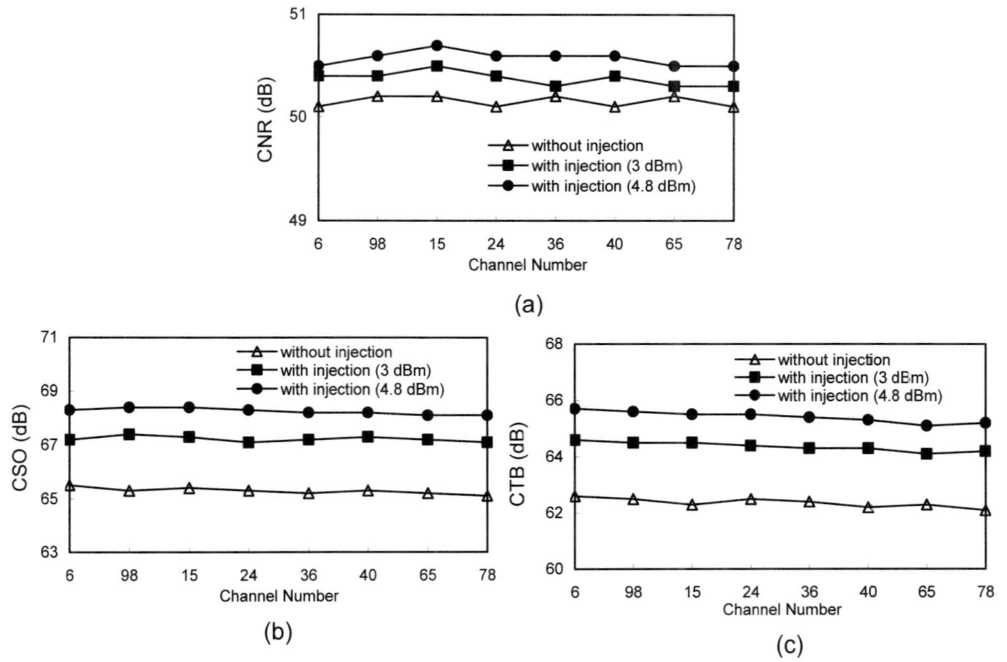

Fig. 4. (a) Measured CNR values with and without light injection. (b) The theoretical derived and experimental measured CSO and (c) CTB values.(Lu et al., 2003)

P_{in} is the EDFA optical input power, and G is the saturated gain of the EDFA. From the above equation, it is clear that the $CNR_{sig\text{-}sp}$ value depends critically on the optical input power P_{in}. When the injection power is increased, the power launched into the EDFA-I is enlarged. This is attributed by a fact that the external light injection will reduce the laser diode threshold current and then increase the optical output power of the laser diode (Kazubowska et al., 2002). As a result, the CNR performance is upgraded by the techniques.

In parallel with the CNR values, the CSO and CTB values can also be improved by using half-split-band and external light injection techniques. According to the analysis in (Lu & Lee, 1998), the CSO and CTB distortions can be expressed as:

$$CSO = HD_2 + 10\cdot\log N_{CSO} + 6\ (dB) \tag{4}$$

$$CTB = IMD_3 + 10\cdot\log N_{CTB} + 6\ (dB) \tag{5}$$

where HD_2 is the second order harmonic distortion, IMD_3 is the third order intermodulation distortion, N_{CSO} and N_{CTB} are the product counts of CSO and CTB respectively. Smaller N_{CSO} and N_{CTB} can be obtained from the smaller channel number. Therefore, the CSO and CTB values presenting in Fig. 4 (b) and (c) can satisfy the fiber optical CATV systems' requirements (>65/60 dB) (Lu & Lee, 1998).

In addition, it also can be observed that CSO and CTB improvements of ~2 and ~3 dB have been achieved with 3 and 4.8 dBm light injection respectively. This means that the external light injection technique not only can increase the laser resonance frequency, but also can reduce the HD_2/C and IMD_3/C. The HD_2/C and IMD_3/C can be expressed as (Helms, 1991):

$$IMD_3 / C = \left\{ \left[\left(\frac{f_1}{f_0}\right)^4 - \frac{f_1^2}{2 f_0^2} \right]^2 + \left(\frac{f_1}{f_0}\right)^2 \left[\frac{1}{4\pi f_0 \tau_n} - \left(\frac{f_1}{f_0}\right)^2 \left(2\pi f_0 \tau_p + \frac{3}{4\pi f_0 \tau_n} + \frac{3\varepsilon S_0}{2\pi f_0 \tau_p} \right) \right]^2 \right\} \tag{6}$$

$$HD_2/C = m^2 \cdot \left| FR(2 f_1) \right|^2 \cdot \left(\frac{f_1}{f_0}\right)^4 \tag{7}$$

where $FR(f)$ is the small-signal frequency response, f_1 is the modulation frequency, f_0 is the laser resonance frequency, S_0 is the photon density, τ_n is the recombination lifetime of carriers, τ_p is the photon lifetime, and ε is the gain compression parameter with respect to photon density. It is clear from Eq. 6 and 7 that both HD_2/C and IMD_3/C can become very small when $f_1 << f_0$. The use of external light injection technique lets the laser resonance frequency increased, and results in system with lower HD_2/C and IMD_3/C. Furthermore, it can be obviously seen from Eq. 4 and 5 that to reduce HD_2/IMD_3 will lead to CSO/CTB performance improvement. To show a more direct association among Eq. 6, 7 and the experimental results, the electrical spectra of the received signals with and without external injection are given in Fig. 5 (a) and (b). We can see that the flatness of the system with 4.8 dBm external injection is superior to that without external injection resulting in better CSO and CTB performances.

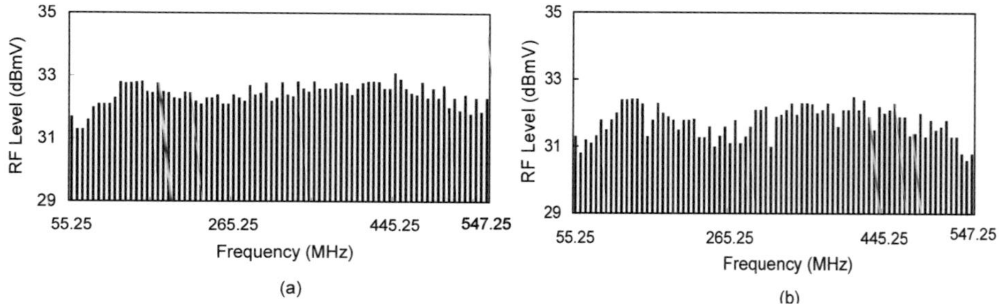

Fig. 5. (a) The electrical spectrum of the received signals with 4.8 dBm external injection and (b) without external injection into the laser (Lu et al., 2003).

3.2 Remote light-injection techniques

In this section, remote light injection technique is discussed. As presented in Fig. 6 (Lee et al., 2007), a total of 77 random phase continuous wave carriers from a multiple signal generator were used to simulate analog CATV channels ($CH_{2\text{-}78}$; 6 MHz/CH), and fed into two DFB LDs. The optical power was coupled into a 50-km SMF through an optical coupler. As to the remote light injection part, 1532.76 nm (λ'_1) and 1535.94 nm (λ'_2) lightwaves with 8 dBm power level are accurately chosen to inject through 3-port optical circulators (OCs). At the receiver end, the power levels of two DFB LDs in the free-running case are decreased obviously due to fiber transmission loss. However, these two power levels are able to be significantly increased by 8 dBm remote light injection and the optical spectra are slightly shifted toward longer wavelengths. This is because that the optimal injection locking condition is found when the detuning between λ_1 (λ_2) and the λ'_1 (λ'_2) is 0.12 nm, and the chaotic phenomenon is found when the detuning between λ_1 (λ_2) and the λ'_1 (λ'_2) is larger than 0.26 nm.

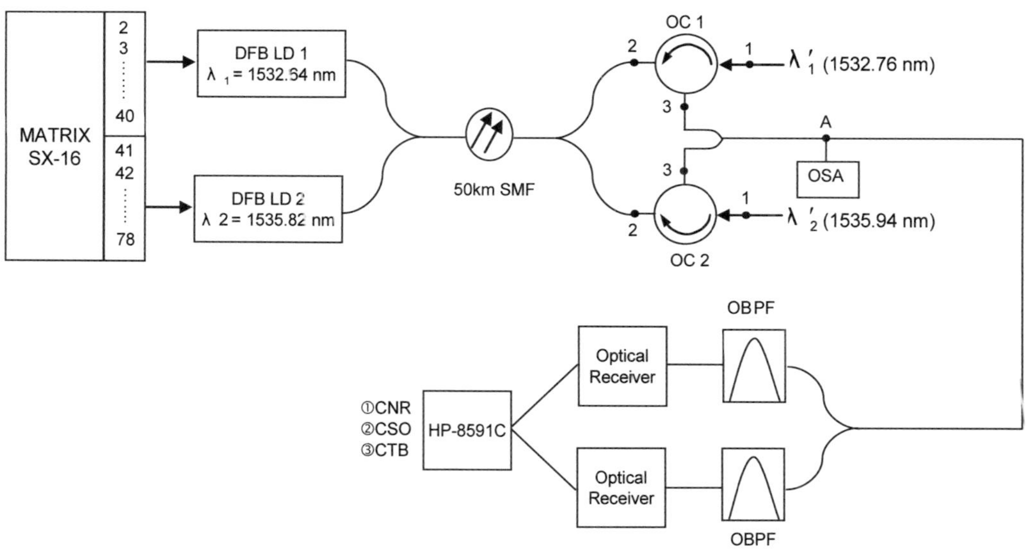

Fig. 6. Remote light-injection direct modulation fiber optical CATV transport systems (Lee et al., 2007).

The measured CNR, CSO and CTB values in the free running case and with 8 dBm remote light injection are presented in Fig. 7 (a), (b) and (c) respectively. It is obvious that the CNR value (≥ 50 dB) is increased largely as 8 dBm optical power is remotely injected. The CNR value depends critically on the received optical power level:

$$CNR = \left(CNR_{RIN}^{-1} + \left(CNR_{th}^{-1} + CNR_{shot}^{-1}\right)\right)^{-1} \tag{8}$$

where CNR_{RIN} results from the LD RIN; CNR_{th} (due to thermal noise) and CNR_{shot} (due to shot noise) are associated with the optical receiver. The summation of CNR_{th} and CNR_{shot} with 8 dBm remote light injection is higher than that in the free running case. This is due to a factor that the optical power is promoted by remote light injection causing a better CNR performance in the receiver end.

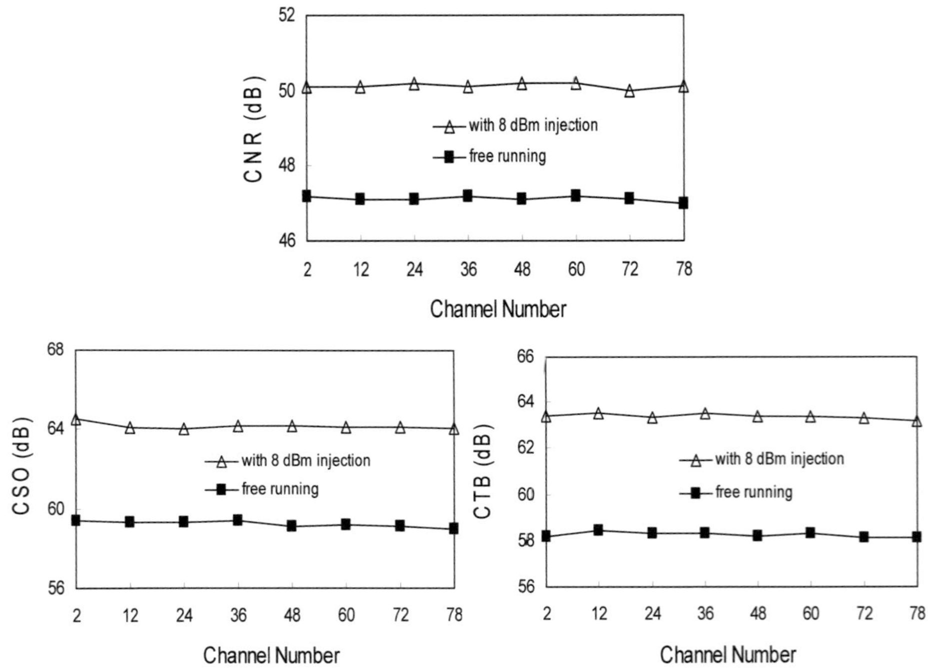

(b) (c)

Fig. 7. Measured (a) CNR (b) CSO and (c) CTB values under NTSC channel number (Lee et al., 2007)

As to the CNR performance, the CSO/CTB values (≥65/63 dB) of system with 8 dBm remote light injection are improved considerably. It can be observed from the results that large CSO and CTB improvements of about 6 and 5 dB have been achieved. The improvements are resulted from the use of the half-split-band and remote light injection techniques. CSO and CTB distortions are given by (Way, 1998):

$$CSO = 10\log\left[\frac{mD\lambda_c^{\,2}Lf}{4c}\sqrt{16(\Delta\tau)^2 + \frac{4\lambda_c^{\,4}L^2\pi^2 f^6}{c^2}}\right] + 10\log N_{CSO} + 6 \tag{9}$$

$$CTB = 10\log\left[\frac{9m^2D^2\lambda_c^{\,4}L^2f^2}{4c}\left(4(\Delta\tau)^2 + 4\pi^2 f\right)\right] + 10\log N_{CTB} + 6 \tag{10}$$

where m is the optical modulation index, D is the dispersion coefficient, λ_c is the optical carrier wavelength, L is the fiber length, f is the RF frequency, $\Delta\tau(=|D|\cdot L\cdot\Delta\lambda)$ is the fiber chromatic dispersion ($\Delta\lambda$ is the spectral width), and N_{CSO}/N_{CTB} are the product counts of CSO/CTB. By using half-split-band technique, smaller N_{CSO}/N_{CTB} can be obtained from smaller channel number; thereby, part of the CSO/CTB distortion will be removed dramatically in each split-band region. Moreover, the use of remote light injection technique decreases the frequency chirp of LD, letting system with lower fiber chromatic dispersion, and leading to an improvement of CSO/CTB performances.

3.3 Lower-frequency side-mode injection-locked techniques

In the early stage of developing light-injection techniques, researchers are firstly focused their eyes on main mode injection-locked. Nevertheless, this phenomenon was changed by the publications of low-frequency side mode injection-locked techniques (Lee et al., 2006; Seo et al., 2002), because the new method illustrates a much better improvement than main mode injection technique. Table 1 presents the SMSR values under lower-frequency side mode injection-locked of DFB LD at different wavelength detuning. It can be seen that the SMSR values of 40~48 dB are achieved when the locking range is -0.07 ~ +0.32 nm. As optimal injection locking happens, with a detuning of +0.12 nm, the maximum SMSR value of 48 dB is obtained. The injection-locked range for slave laser under light injection can be expressed as (Mondal et al., 2007):

$$-k\sqrt{\left(\frac{I_{in}}{I_m}\right)(1+\alpha)} \le \Delta\omega \le k\sqrt{\frac{I_{in}}{I_m}} \tag{11}$$

where k and α denote coupling coefficient between injected field and laser field; I_{in} and I_m are injected field and laser mode field intensity, and $\Delta\omega$ is the locking range. Within the locking range, the frequency of slave laser is locked nearly to that of the master laser. Furthermore, with light injection, because of the coherent summation of externally injecting and internally generated slave fields, the phase adds an additional dynamic variable. Consequently, a new resonant coupling between the field amplitude and phase appears and can dominate the laser resonance frequency.

Wavelength Detuning (nm) / SMSR (dB)							
-0.07 /40	-0.04/ 41.5	-0.00/ 43.1	0.03/ 44	0.07/ 45.6	0.07/ 45.6		
0.1/ 46.8	0.12/ 48	0.15/ 46.4	0.18/ 45	0.23/ 43.4	0.25/ 42.5	0.29/ 41.4	0.32/ 40

Table 1. The SMSR values under lower-frequency side mode injection-locked of DFB LD_1 at different wavelength detuning (Lu et al., 2008).

Fig. 8 (a), (b) and (c) show the measured CNR, CSO and CTB values under NTSC channel number for free-running, with 4.8 dBm main mode injection, and with 4.8 dBm lower-frequency side mode injection, respectively. The CNR value depends critically on the optical input power, so that the increased values between the scenarios of 4.8 dBm main and side modes injection are similar. Nevertheless, the performances of the measured CSO and CTB values are very different. According to the Eq. 6 and 7, the use of lower-frequency side mode injection locking technique can further increase the resonance frequency of the slave laser resulting in smaller values of HD_2/C and IMD_3/C. Consequently, better CSO and CTB performances are obtained in 4.8 dBm lower-frequency side mode injection scenario.

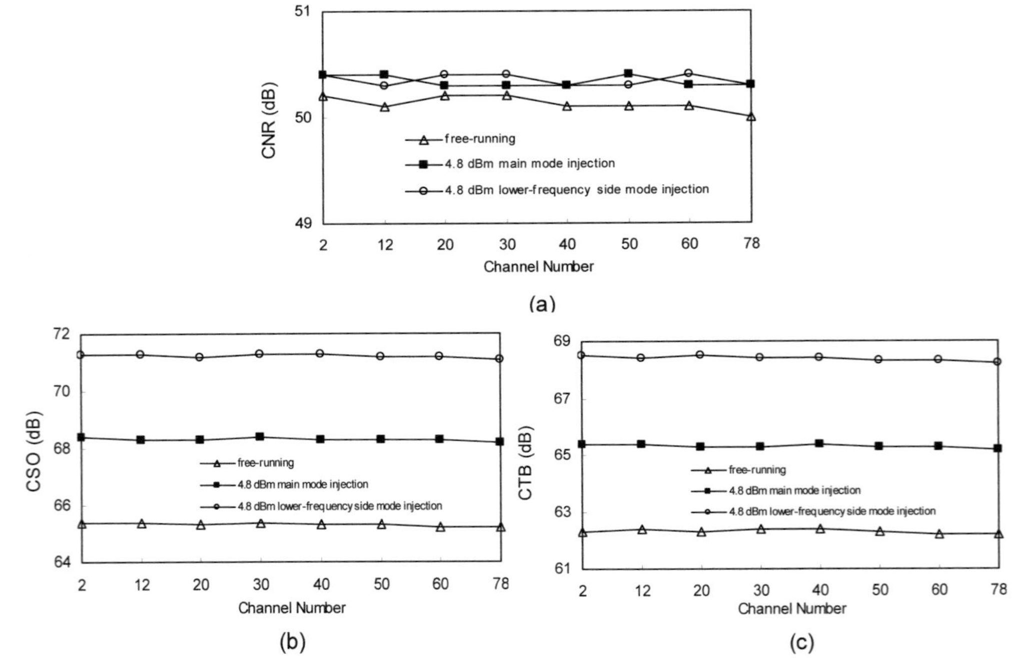

Fig. 8. Measured (a) CNR, (b) CSO and (c) CTB values under NTSC channel number

3.4 Hybrid local light-injection and optoelectronic feedback techniques

In parallel with the light-injection techniques, optoelectronic feedback technique has been used in high-speed digital optical communication systems to improve bit error rate (BER) performance (Attygalle & Wen, 2006). This technique, which can greatly enhance the laser resonance frequency (Li et al., 1995), is therefore can be integrated with light-injection methods to assist the transmission of fiber optical CATV systems as presented in Fig. 9 (Lu et al., 2006).

Experimentally, with the assisting of main mode injection and optoelectronic feedback, the laser resonance frequency is further improved to 25.2 GHz. The laser resonance frequency, f_0, is given by:

$$f_0^2 = \frac{g_0 S}{4\pi^2 \tau_\rho} \tag{10}$$

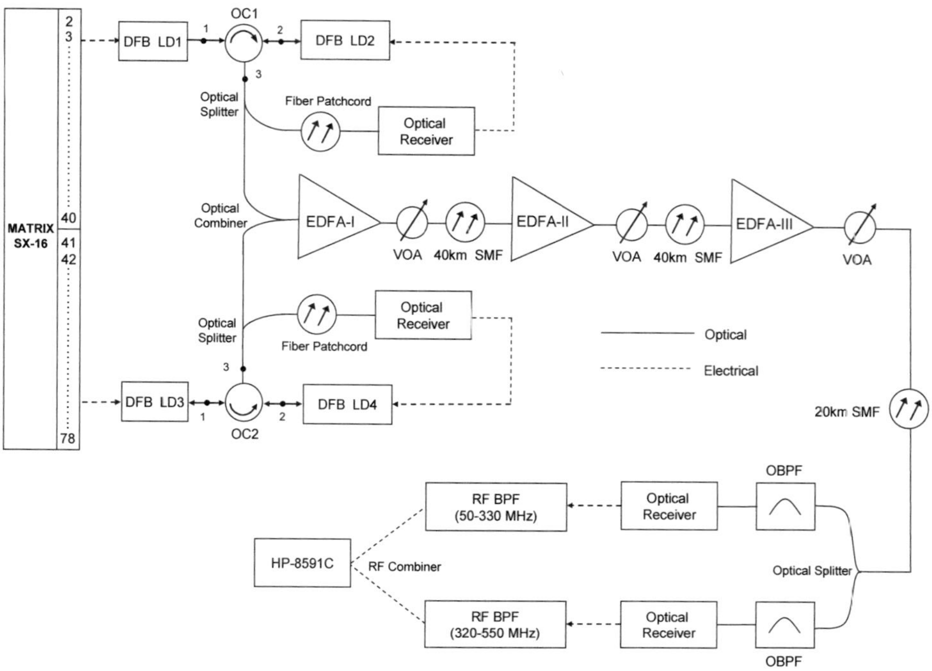

Fig. 9. Local light-injection and optoelectronic feedback techniques enhanced 100-km split-band directly modulated optical CATV system (Lu et al., 2006).

where g_0 is the gain coefficient, S is the photon density, and τ_p is the photon lifetime. Since f_0^2 is direct proportion to photon density, the increased photon density by the light-injection power will lead to a promotion of laser resonance frequency. Additionally, optoelectronic feedback techniques will further increase the stability of the laser, resulting in out-of-phase carrier re-injection. The re-injection can therefore assist the effect of the light-injection techniques, leading to an improvement of laser resonance frequency, as a result, presenting in an improvement of CSO/CTB values. The optoelectronic feedback techniques are hence an efficient method to enhance the performance of directly modulated optical CATV systems by promoting RF parameters.

4. Dispersions compensation schemes

Following with the modifications of direct modulation schemes and lightwave enhancement methods in fiber optical CATV systems, fiber chromatic dispersion is still another bottleneck needed to be solved out. Cascading CATV system with optical filter or a section of negative dispersion fiber (Lu et al., 2007) for example have been developed to overcome this issue. Since there is no useful modulating information in the redundant spectra, adding an optical filter to eliminate parts of these spectra not only can increase spectra efficiency but also can ameliorate the CSO/CTB performance. Similarly, cascading a section of negative dispersion fiber with a long-haul CATV system can also promote the system by eliminating the fiber chromatic dispersion. Consequently, the relative fiber dispersion compensation techniques are recorded and discussed in the following sections.

4.1 Downgrading dispersions by optical filter

Over a long-haul fiber transmission, fiber dispersion accumulates rapidly and leads to a worse system performance. To overcome this issue, utilizing optical filter to change the broad spectral linewidth into a narrow one has been demonstrated as a useful method in fiber optical transport systems (Lu et al., 2008). As presented in Fig. 10, the downstream optical signal was sent through a tunable optical band-pass filter (OBPF) and a Fabry-Perot (FP) etalon filter before received. The OBPF is applied to select the appropriate wavelength and the FP etalon filter is employed to narrow down the spectral linewidth as well as to compensate fiber dispersion.

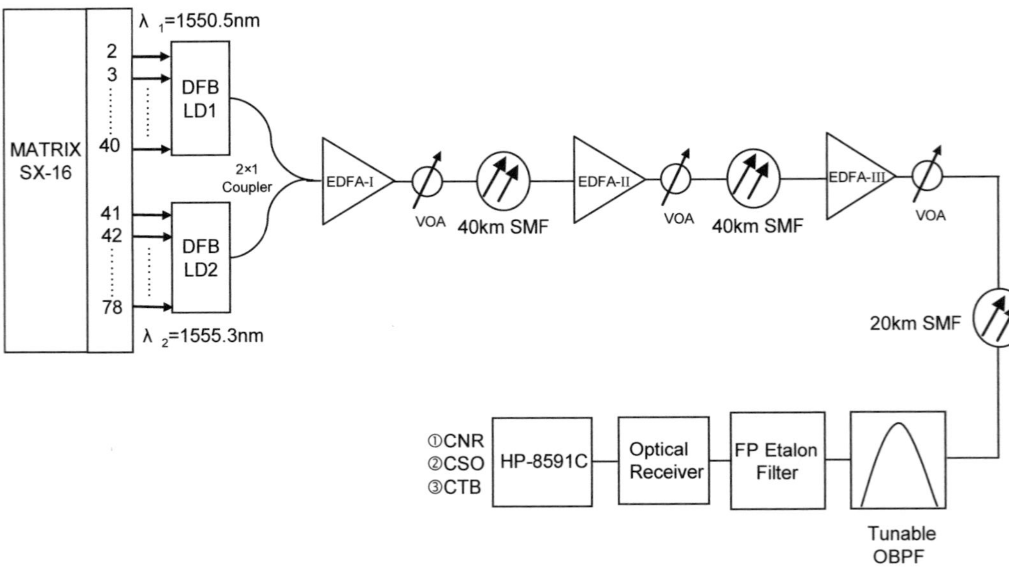

Fig. 10. Employing split-band technique and Fabry-Perot etalon filter to improve directly modulated fiber optical CATV system (Lu et al., 2008).

By eliminating the redundant spectra of optical signal, the spectral efficiency is enhanced and the dispersion is ameliorated resulting in better CSO/CTB performance. This optical filter is then worth deployed due to excellent optical characteristics such as sharp cutoff in the transmission spectrum. However, the wavelength misalignment between the selected optical wavelength and the optical filter will change optical power level launched into the fiber and degrade system performances. To avoid the wavelength misaligned by thermal effect, the filter needs to be utilized in a thermal package.

4.2 Dispersion compensated by special fiber

Different with cascading an optical filter to cut off optical spectral linewidth, adding a span of negative dispersion fibers, such as photonic crystal fiber (PCF), chirp fiber grating (CFG) and dispersion compensation fiber (DCF), into an optical CATV systems is experimentally demonstrated as another efficient method to compensate fiber dispersion (Ni et al., 2004). Fig. 11 for example demonstrates a 100-km optically amplified AM-VSB transmission system cascading with a span (3.6-km) of PCF disperion compensation fiber.

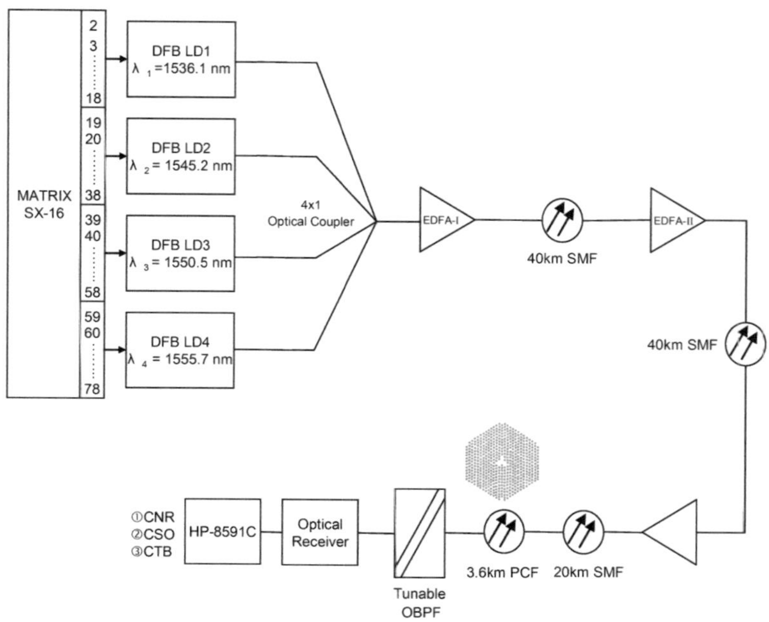

Fig. 11. A 100-km optically amplified AM-VSB transmission system with a length of PCF disperion compensation fiber (Lu et al., 2007).

In this system, the optical link with a transmission length of 100-km SMF has a total positive dispersion of 1700 ps/nm (17 ps/nm/km × 100 km). However, a length of 3.6-km PCF has a negative dispersion of -1710 ps/nm (-475 ps/nm/km × 3.6 km). By combining these two pieces together, the total dispersion is nearly eliminated (-10 ps/nm) leading to lower fiber-induced distortion and better CSO/CTB performance.

5. Extending applications of directly modulated fiber optical CATV systems

Following with an assistance of numerous techniques, the CATV service providers are able to offer high quality of CATV programs by cost effective optical fiber connection. Nevertheless, the potential of optical fiber is not fully utilized yet. Integrating other services, such as Internet access, WiMAX services and HDTV programs, with CATV transport systems would be quite useful to share the cost of deploying and maintaining optical fiber (Ying et al., 2007). Recently, passive optical networks (PONs) are promising way to obtain low cost and high capacity optical Internet access. DWDM in combination with PON has received considerable attentions due to its large capacity, network security, easy management, and upgrade- ability (Choi et al., 2005; Hann et al., 2004; Khanal et al., 2005). In parallel with the PON systems, radio-over-fiber (ROF) transport systems also present a potential to offer significant network flexibility, large transmission capacity and economic advantage to satisfy the increasing demand in wireless broadband services such as WiMAX (Masella & Zhang, 2006). Due to low attenuation and broad bandwidth characteristics of optical fiber, combining ROF and Internet access with CATV systems has subsequently attracted much attention to fully utilize the potential of optical fiber and to provide triple play services for clients. Fig. 12 for example presents a bidirectional HDTV/Gigabit Ethernet/CATV over DWDM-PON system. Services with 129 HDTV channels, 1.25 Gb/s

Gigabit Ethernet connection, and 77 CATV channels were successfully demonstrated over 40 km SMF links. Good performance of BER, CNR and CSO/CTB were achieved in this system.

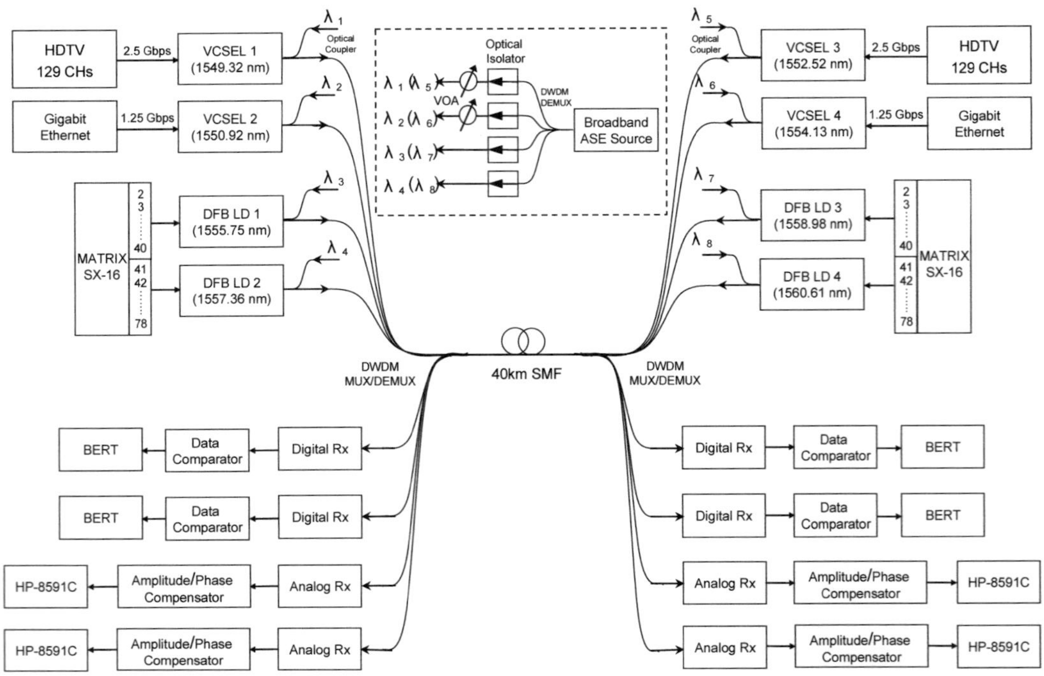

Fig. 12. HDTV/Gigabit Ethernet/CATV over bidirectional hybrid DWDM-PON (Lu et al., 2007).

6. Conclusion

Fiber optical CATV systems are recently enhanced by the introduction of 1550 nm technology. However the maximum transmission distance of the systems is still limited by RF parameters. Literarily, numbers of techniques such as split-band schemes, light-injection methods and dispersion compensation skills have been developed to extend the bottleneck in fiber optical CATV systems. Sharing full channel load from one LD to multiple LDs in split-band schemes has been demonstrated as an efficiency way to eliminate major part of CSO distortion from each optical band. Furthermore, improving laser resonance frequency and output power by light-injection techniques as well as compensating fiber dispersion by optical filters or by negative dispersion fiber are also presenting an advanced assistance in such systems. All of these techniques make a possibility to deploy a long-haul and cost-effective CATV system by direct modulation scheme. The main problem is that such systems do not fully utilize the potential of optical fiber. There is still plenty of capacity in fiber link waiting for people to dig out. As a result, combining fiber optical CATV systems with other applications, such as Internet access and WiMAX services, are discussed popularly in literature. The applications and characteristics of the mentioned techniques as well as the future directions of directly modulated fiber optical CATV system are consequently analyzed and illustrated in this chapter.

7. References

Attygalle, M., and Wen, Y. J. (2006): Injection-locked Fabry-Perot laser with electronic feedback for clock recovery from high-speed OTDM signals. *IEEE Photonics Technology Letters* 18, 478-480.

Choi, K. M., Baik, J. S., and Lee, C. H. (2005): Broad-band light source using mutually injected Fabry-Perot laser diodes for WDM-PON. *IEEE Photonics Technology Letters* 17, 2529-2531.

Hann, S., Kim, D. H., and S., P. C. (2004): Uni-lambda bidirectional 10/1.25 GbE access service based on WDM-PON. *Electronics Letters* 40, 194-195.

Helms, J. (1991): Intermodulation and harmonic distortions of laser diodes with optical feedback. *Journal of Lightwave Technology* 9, 1567-1575.

Kazubowska, A., Anandarajah, P., and Barry, L. P. (2002): Improved performance of a hybrid radio/fiber system using a directly modulated laser transmitter with external injection. *IEEE Photonics Technology Letters* 14, 233-235.

Khanal, M., Chae, C. J., and Tucker, R. S. (2005): Selective broadcasting of digital video signals over a WDM passive optical network. *IEEE Photonics Technology Letters* 17, 1992-1994.

Lee, C.-Y., Lu, H.-H., Yee, H., Lin, W.-I., Tzeng, S.-J., and Lai, P.-C. (2007): Directly modulated fiber optical CATV transport systems without optical amplification. *IEICE Electronics Express* 4, 282-287.

Lee, S. H., Kang, J. M., Choi, I. H., and Han, S. K. (2006): Linearization of DFB laser diode by external light-injected cross-gain modulation for radio-over-fiber link. *IEEE Photonics Technology Letters* 18, 1545-1547.

Li, G., Boncek, R. K., Wang, X., and Sackett, D. H. (1995): Transient and optoelectronic feedback-sustained pulsation of laser diodes at 1300 nm. *IEEE Photonics Technology Letters* 7, 854-856.

Lu, H.-H., Chuang, Y.-W., Chen, G.-L., Liao, C.-W., and Chi, Y.-C. (2006): Fiber-Optical Cable Television System Performance Improvement Employing Light Injection and Optoelectronic Feedback Techniques. *IEEE Photonics Technology Letters* 18, 1789-1791.

Lu, H.-H., Huang, H.-H., Su, H.-S., and Wang, M.-C. (2003): Fiber Optical CATV System-Performance Improvement by Using External Light-Injection Technique. *IEEE Photonics Technology Letters* 15, 1017-1019.

Lu, H.-H., and Lee, C.-T. (1998): Directly Modulated CATV Transmission Systems Using Half-Split-Band and Wavelength-Division-Multiplexing Techniques. *IEEE Photonics Technology Letters* 10, 1653-1655.

Lu, H.-H., Patra, A. S., Tzeng, S.-J., Peng, H.-C., and Lin, W.-I. (2008): Improvement of Fiber-Optical CATV Transport Systems Performance Based on Lower-Frequency Sidemode Injection-Locked Technique. *IEEE Photonics Technology Letters* 20, 351-353.

Lu, H.-H., Patra, A. S., Wu, H.-W., Tzeng, S.-J., Ho, W.-J., and Yee, H. (2008): Employing split-band technique and Fabry-Perot etalon filter to improve directly modulated fiber optical CATV system performances. *Optical Fiber Technology* 14, 227-231.

Lu, H.-H., Tsai, W.-S., Chine, T.-S., Cehn, S.-H., Chi, Y.-C., and Liao, C.-W. (2007): Bidirectional Hybrid DWDM-PON for HDTV/Gigabit Ethernet/CATV Applications. *ETRI Journal* 29, 162-168.

Lu, H.-H., Tzeng, S.-J., Chuang, Y.-W., Chen, G.-L., and Peng, H.-C. (2007): Fiber-optic CATV system performance improvement by using split-band technique and photonic crystal fiber. *Optics Communications* 271, 436-440.

Lu, H.-H., Tzeng, S.-J., Wang, M.-C., and Huang, H.-H. (2003): Fiber Optical CATV System Performance Improvement by Using Split-Band and Optical VSB Modulation Techniques. *IEICE Trans. Commun.* E86-B, 3296-3299.

Masella, B., and Zhang, X. (2006): A novel single wavelength balanced system for radio over fiber links. *IEEE Photonics Technology Letters* 18, 301-303.

Mondal, S. K., Roycroft, B., Lambkin, P., Peters, F., Corbett, B., Townsend, P., and Ellis, A. (2007): A multiwavelength low-power wavelength-locked slotted Fabry-Perot laser source for WDM applications. *IEEE Photonics Technology Letters* 19, 744-746.

Ni, Y., Zhang, L., An, L., Peng, J., and Fan, C. (2004): Dual-core photonic crystal fiber for dispersion compensation. *IEEE Photonics Technology Letters* 16, 1516-1518.

Seo, J. H., Seo, Y. K., and Choi, W. Y. (2002): Nonlinear distortion suppression in directly modulated distributed feedback lasers by sidemode optical injection locking. *Japanese journal of applied physics* 41, L136-L138.

Way, W. I. (1998): *Broadband hybrid fiber/coax access system technologies*. Academic Press. San Diego.

Way, W. I. (1998): *Broadband Hybrid Fiber/Coax Access System Technologies*. Academic Press. San Diego.

Ying, C.-L., Lu, H.-H., Tzeng, S.-J., Ma, H.-L., and Chuang, Y.-W. (2007): A hybrid WDM transport system based on mutually injection-locked Fabry Perot laser diodes. *Optics Communications* 276, 87-92.

Optical Beam Steering using a 2D MEMS Scanner

Yves Pétremand[1], Pierre-André Clerc[1], Marc Epitaux[2], Ralf Hauffe[3], Wilfried Noell[1] and N.F. de Rooij[1]
[1]*Ecole Polytechnique Fédérale de Lausanne (EPFL)*
[2]*formerly at Intel Corporation, Newark, CA*
[3]*Hymite GmbH, Berlin*
[1]*Switzerland*
[2]*USA*
[3]*Germany*

1. Introduction

Being currently performed on highly complex and expensive equipments, active optical alignment of single mode 10 Gb/s transmitters and receivers is proving to be the bottleneck process for high volume manufacturing (epitaux2005). In order to alleviate this production burden, new integrated technologies are required to align micro-components like micro-lenses and optical fibers with photonic devices. Although passive alignment using Silicon micro-machined V-grooves seems to be very attractive at first, it has only been successfully implemented for micro-lens which can tolerate up to several microns of misalignment (osenbach1998). Alternative solutions such as active MEMS alignment techniques have been employed mainly for switching applications. They are based on actuated micro-mirrors, causing incident collimated beams to be reflected onto another mirror or a focusing lens to illuminate a waveguide (syms2002, pezeshki2002). A feedback loop mechanism is usually employed to dynamically optimize the coupling. Mostly used to select optical channels, these micro-mirror active alignment methods require complicated collimator and pigtail configurations and lack the possibility of locking them at the optimal position.

To resolve the latter issue, we propose a MEMS XY scanner mounted with a micro-lens to actively perform optical alignment of fiber optics modules by steering a beam. Although similar techniques have been demonstrated (toshiyoshi2003, sunghoon2002), we describe a unique technology using a 2x2.7 mm silicon bulk micro-machining chip characterized by ±30 µm XY range of motion (petremand2004), electrostatic comb drives actuation and a silicon hybrid lens with alignment locking capability.

Presently, transceivers are mainly assembled into TO can systems (epitaux2005b, petremand2005). In the described work, the MEMS device is dedicated to be assembled into a vertical silicon optical bench. In this system, the beam propagates vertically through a stack of micro-machined silicon devices which comprise substrate, RF feed-through, hermetic sealing and optical alignment functions.

This chapter will first present the system concept and the MEMS device. Secondly, design and simulations performed with CoventorWare and CoventorWare-Architect will be shown. Thirdly, the fabrication process will be presented. Finally, the packaging technology will be explained and characterization results of the fabricated MEMS device as well as a fully assembled and functional system will be discussed.

2. Scanner design

Fig. 1 shows the concept of the described application of the MEMS device. By placing a lens on the 2D movable platform of the actuator, it is possible to steer the light beam coming out of the laser diode and couple it into a single mode fiber. Working in the infrared region of the spectrum (wavelength usually used in telecommunications), all the MEMS components can be micro-machined out of silicon.

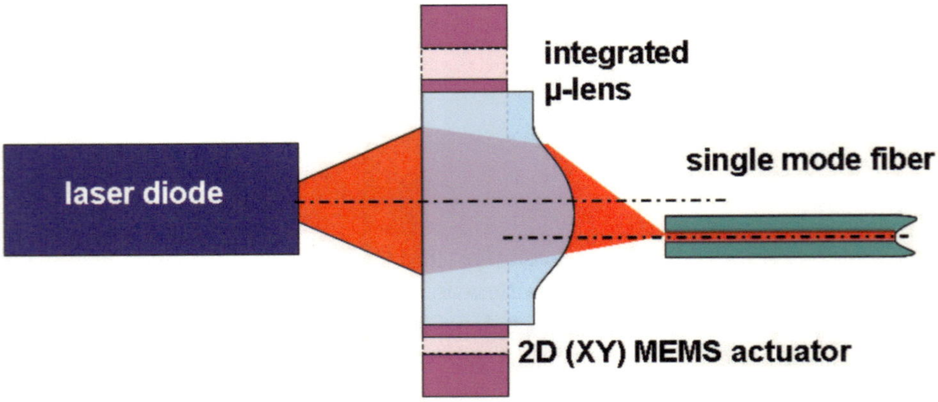

Fig. 1. Optical scheme of the light coupling concept.

Due to fabrication costs and available technology, it was chosen to assemble the silicon lens on top of the platform instead of integrating it directly on the MEMS. The advantages of such an assembly are the following:

- The cost of the device is reduced as one can manufacture more microlenses per surface unit by using a dedicated wafer.
- The cost of effective anti-reflecting coating on the lenses is reduced and facilitated.
- The technology is simplified as there is no need to protect the microlens during the HF release step of the process.

The principle of operation of the micromachined scanner is shown in Fig. 2. The 2D displacement of the hollowed platform of the device is provided by two pairs of electrostatic combdrive actuators (A, B, C and D, see Fig. 3) suspended by 4 springs to the chip base (petremand2004). The mobile platform is linked to the actuators by 4 compliant curved beams. These flexible arched beams are designed to convert two unidirectional (X) actuations into two dimensional (XY) displacements and as well as to amplify the movement in the direction orthogonal to the actuators by a factor of two. When the two actuators are moving accordingly by the same orientation and amount, the platform translates in the same orientation and by the same distance (displacement in X). Additionally when the two actuators are moving in opposite directions by the same amount, the platform translates in

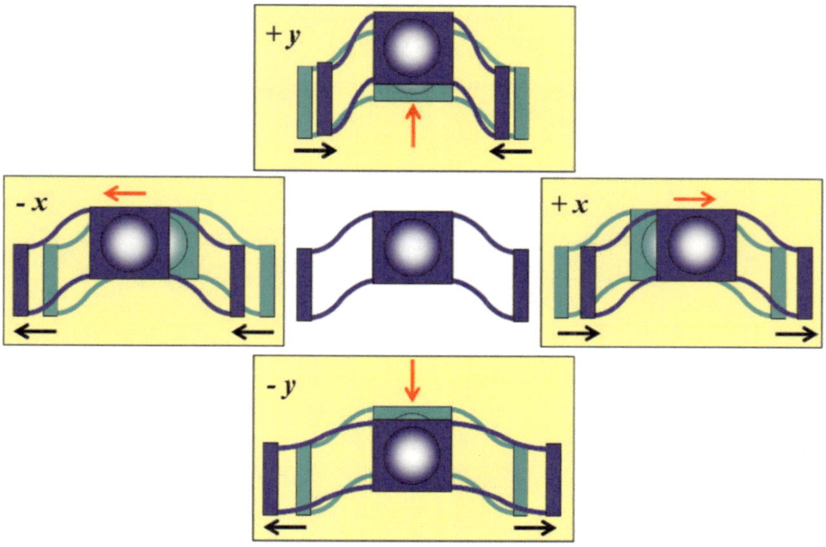

Fig. 2. Principle of operation of the MEMS scanner. By combining these basic displacements, any point of the scanning area can be reached.

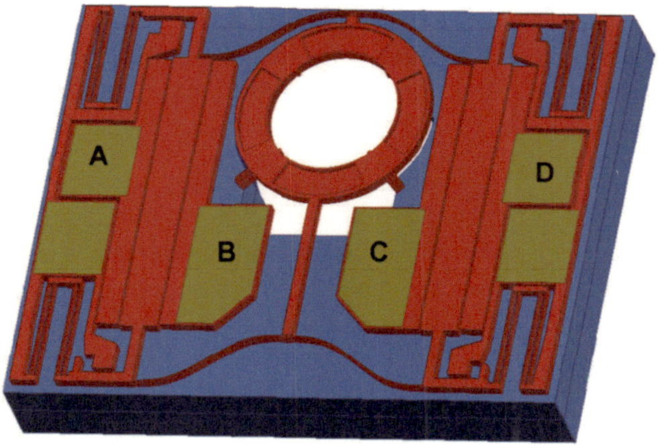

Fig. 3. CAD image of the device showing the two pairs of electrostatic combdrive actuators, the compliant suspension springs and the movable platform.

the orthogonal direction by either compressing or extending the compliant beams (displacement in Y). Fig. 2 shows the displacements along the two main axes (X) and (Y). With combining these basic displacements by applying different signal amplitudes to combdrive actuators, any off-axis points of the reachable area can be addressed.

Once the desired position reached, the movable platform can be locked in place by applying a voltage between the movable platform and the base of the chip (Fig. 4). Depending on the electrical current and the voltage applied, the locking is either temporary or permanent. In case of a temporary locking, the platform is held in place as long as the licking voltage is applied. For the permanent one, the platform stays at the desired position even after all the voltages are set back to zero.

The backside of the chip includes a through hole under the platform to accommodate the movable lens and transmit light across the MEMS chip.

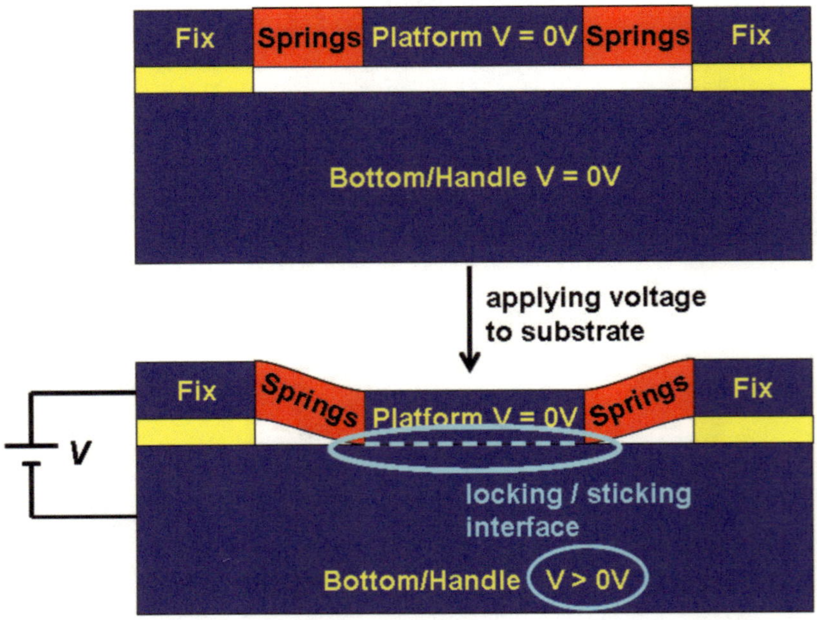

Fig. 4. Locking mechanism principle. Device layer is pulled down to the handle by applying a voltage between these two layers. Depending on the voltage/current applied, the resulting locking mechanism can be either temporary or permanent.

3. Simulations

Simulations were carried out to optimize the size and the shape of the compliant beams. By comparing straight and curved beams, for a given deformation, the maximal stress occurs in the same region of the beam but its value is different. In case of a straight beam, a maximal stress of 160 MPa was obtained while the maximal stress of the curved beam was 110 MPa. The curved beam shape allows a reduction of the maximal stress of more than 30%, which is very positive as the stress can be the cause of fatigue or breakage of the device.

In order to obtain a large enough scanning range in the Y direction even if the system is moved in the X direction, as the total displacement of the combdrives is limited, the shape of the curved springs was optimized to amplify the Y displacement with a factor of two. That way, if both actuators are displaced of a distance x in opposite directions, the corresponding perpendicular motion y of the platform is the following:

$$y \cong 2x \tag{1}$$

This result was obtained with a beam given by the following equation (karthe1991):

$$y(x) = \frac{\sqrt{2}}{4}x - \frac{\sqrt{2}L}{8\pi}\sin\left(\frac{2\pi}{L}x\right) \tag{2}$$

where L is the length of the beam.

Fig. 5 shows a schematic of the movement of the beam as force is applied on the movable end. Architect simulations results are shown in Fig. 6. For these simulations, a beam slightly simplified was used. In Architect, a model exists for the simulation of "freeform beams". These can be defined either by a polynomial equation or by points. The approximation of the shape described in equation 2 was performed with a set of 13 coordinates.

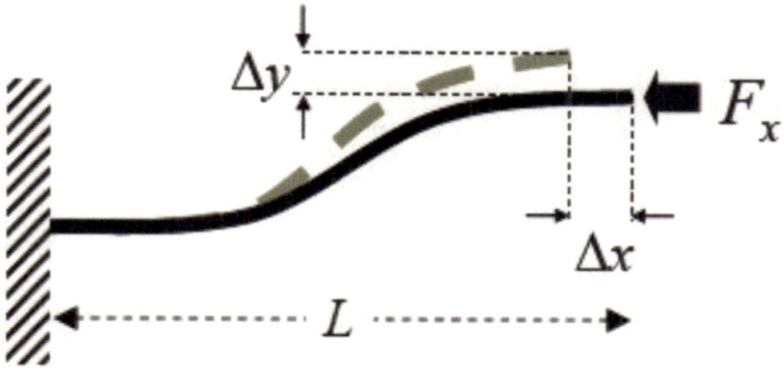

Fig. 5. Schematic view of the movement of the free end of the compliant curved beam as a force is applied to.

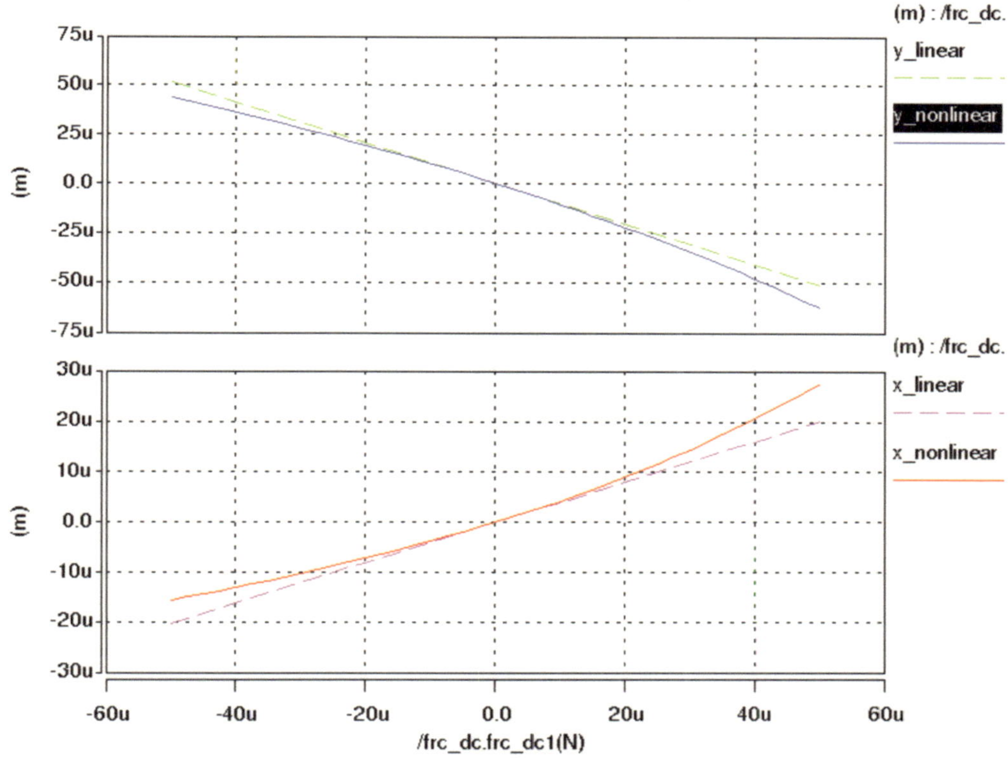

Fig. 6. Architect results of the simulation of the curved beam. In each graph, the linear and non-linear simulation modes are represented. The top image represents the displacement along the Y axis (Δy on Fig. 5) and the bottom one represents the displacement along the X axis (Δx on Fig. 5).

With such spring shape, one obtains the displacements of the platform shown on Fig. 7. This simulation visualizes the result of the shape of the beams but not their width. By changing the beam width, same behavior of the device was obtained but at different voltages.

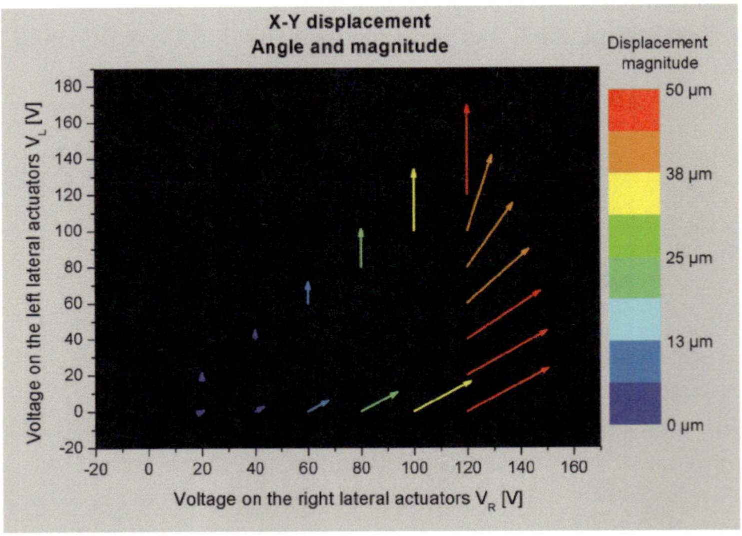

Fig. 7. Coventor simulation showing the displacement of the central platform in respect to voltages applied on two combdrives. This graph corresponds to one quarter of the total area reachable of the platform.

4. Fabrication process

For the fabrication four main process steps are performed: oxidation, metallization, DRIE with a delay mask process and etching of the sacrificial oxide layer. In between, to pattern the different layers, photolithographic steps have to be completed.

After metallization, PECVD oxide deposition is performed to avoid any contamination of gold during the rest of the process. As gold is known to be highly contaminant, it has to be buried in order not to expose the equipment used later in the process.

One challenge of this process is to perform a double bulk micromachining delay mask process. On the device layer side, shallow cavities are etched to allow the assembly of the lens on top of the MEMS. On the handle layer, deep openings on two levels had to be performed in order to create a cavity where the laser diode will be placed into.

The chips are finally released and singulated using HF vapor phase etching.

5. Packaging technology

The packaging technology approach is depicted in Fig. 8. The configuration is similar to a TO-can package with a coaxial optical connector. The TO header is replaced by a vertical hermetic Silicon Bench sub-assembly. In contrast to TO package assembly, the receptacle is simply passively bonded to this sub-assembly without requiring complex active alignment and welding equipment. The silicon sub-assembly is further illustrated in Fig. 8 b). It consists of a system composed of two micromachined silicon layers and a silicon microlens.

The first chip carries the active optical elements such as the laser diode, monitoring photodiode and other miscellaneous electrical components. As an edge emitting laser is used, a turning mirror is placed on the substrate to deflect the laser beam towards the other silicon sayers and the connector. The 2D MEMS scanner with the pre-mounted lens is assembled on top of the laser diode and the turning mirror. It is electrically connected to the first layer by wire bonding. This silicon sub-assembly is then hermetically sealed with a glass lid and flip-chip bonded on a flexprint containing the contacts to connect the system to the external elements.

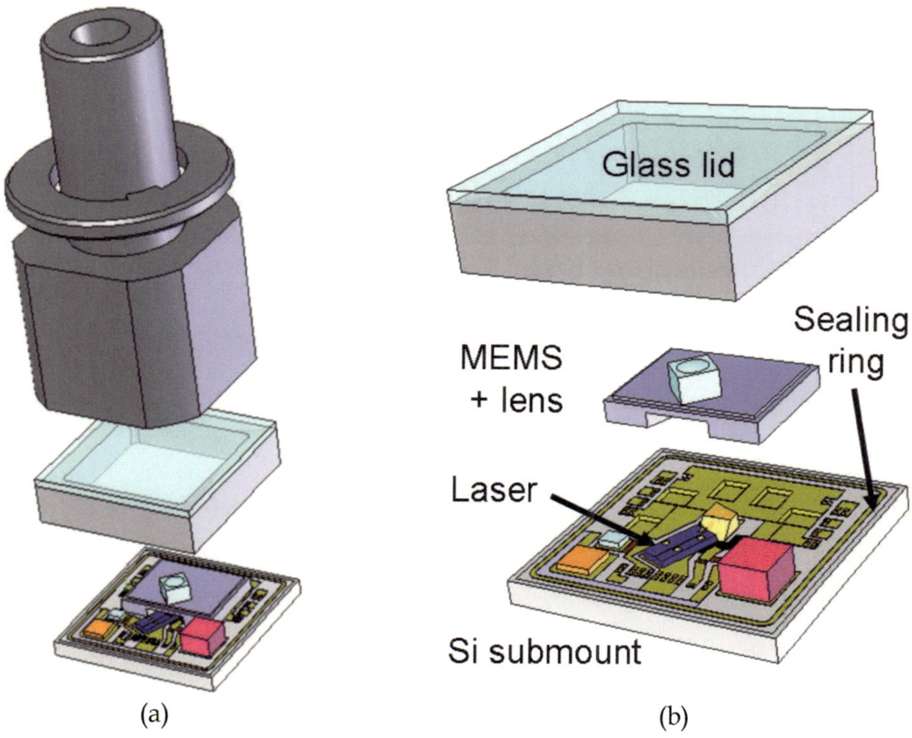

Fig. 8. Model of the system based on Vertical Optical Bench. a) General view of the packaging with the device module and the fiber ferrule. b) Detailed view of the device module with the glass lid, the silicon chip, the silicon submount containing the laser diode, the turning mirror and electronic components. The MEMS chip is partially hollowed out on the backside to allow space for the laser diode and turning mirror.

6. Experimental results

Fig. 9 shows a SEM image of the fabricated device with the silicon lens assembled on top of the movable platform. The bright parts of the image correspond to the metal pads defined to be connected by wire bonding.

Displacements of the platform were performed by applying voltages on the electrodes of the chips. Fig. 10 shows the maximal displacements achieved with a device with springs and compliant beams width designed at 3.5 µm. Depending on the width of the springs, the displacements obtained are the same but correspond to excitations with different amplitudes.

Fig. 9. MEMS device with a silicon lens assembled on top. This sub-assembly will be positioned on a submount over a laser diode (Fig. 8 b).

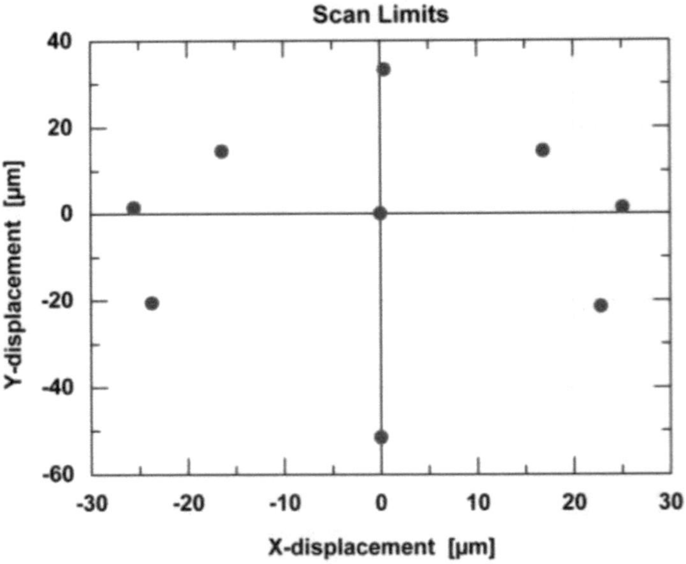

Fig. 10. Maximal displacements achieved with the scanning platform.

Resonance frequency was measured and compared to simulated values. Fig. 11 shows the resonance frequency calculated from the step response measurements performed with a Wyko DMEMS white light interferometer. It was performed by applying a Fourier transform (Fig. 11 b)) to the step response of the actuator (Fig. 11 a)). It gives a resonance frequency of 717.2 Hz. This measurement was carried out on a chip before assembly and packaging (i.e. without lens). This value can be compared to the simulated value shown in Fig. 12 where the resonance frequency was 698.83 Hz. These results are in very good agreement, the error between the measured and the simulated value being less than three percent.

Although the frequency values are very close, it is not the case of the damping coefficients. It can be seen on (Fig. 11 a)) that the measured damping value is very low. It is much lower than the simulated one. These differences are not understood yet and have to be further investigated.

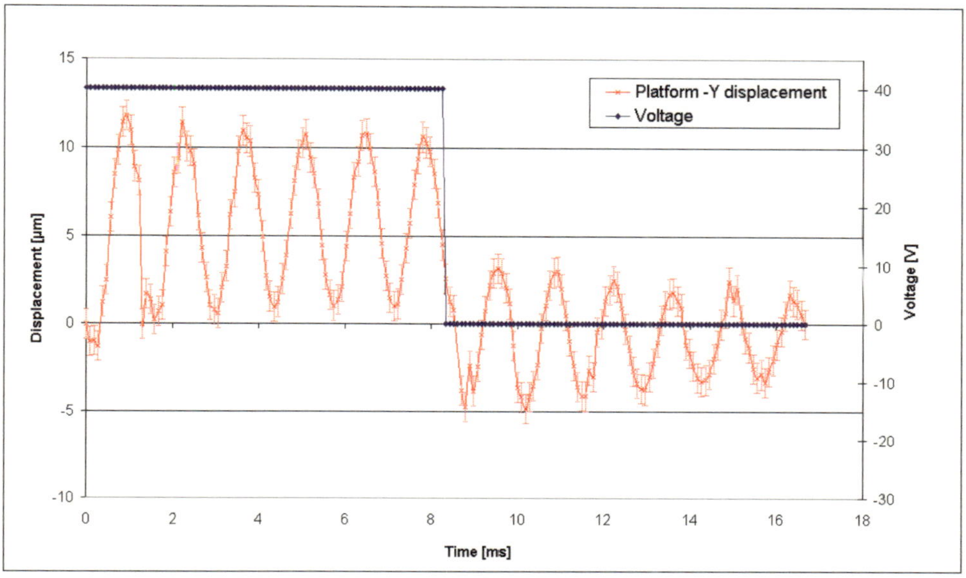

a)

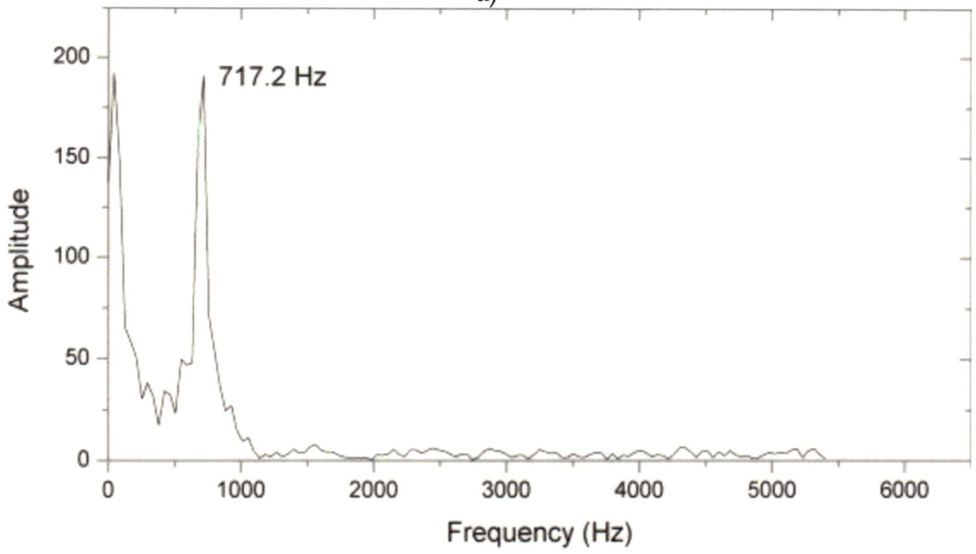

b)

Fig. 11. Determination of the resonance frequency by measuring the step response in "-Y" direction (Fig. 2) of a fabricated actuator. a) Measured displacement of the actuator and voltage applied on the combdrives. b) Fourier transform of the obtained step response.

The assembly of a complete system was then carried out (Fig. 13) and optical measurements of beam steering were performed. A complete assembly process flow had to be developed to assemble the different components of the system. Fig. 13 a) shows this procedure. First subassemblies are performed to glue the lens on the MEMS device and to bond the electronic components including the laser diode on the submount. Both are then assembled, wire bonded and protected by a cover glass. The measured and simulated displacements of the platform and the beam steering measurement angle were compared. By plotting the optical measurements and the simulations performed previously in the same graph, a good agreement of both (normalized) data sets can be observed (Fig. 14).

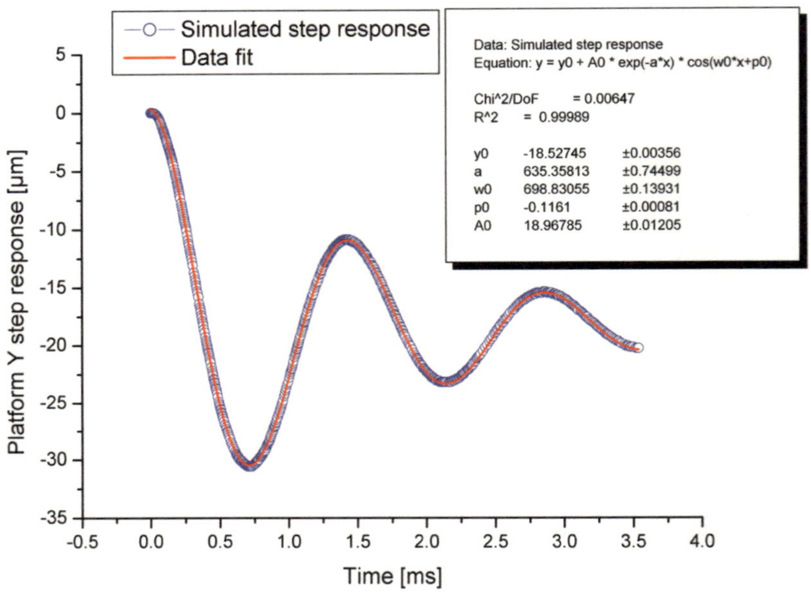

Fig. 12. Simulation of the step response of the actuator in "-Y" direction (Fig. 2). The fitted curve shows a frequency of 698.83 Hz. Excellent agreements between the simulated and the measured values (Fig. 11).

7. Conclusion

A bi-directional MEMS-based optical beam steerer was fabricated and assembled. The assembly of the different parts of the system was carried out. A displacement of several tens of micrometers has been demonstrated and optical beamsteering was shown with a system containing all the electronic and optical components. Resonance frequency and displacements measurements agree with the simulated ones. This displacement range is suitable for opto-electronic alignment applications.

8. Acknowledgments

The authors thank Süss Microoptics for providing the lenses, all the staff of the Comlab for their advice and help and Intel for funding the project.

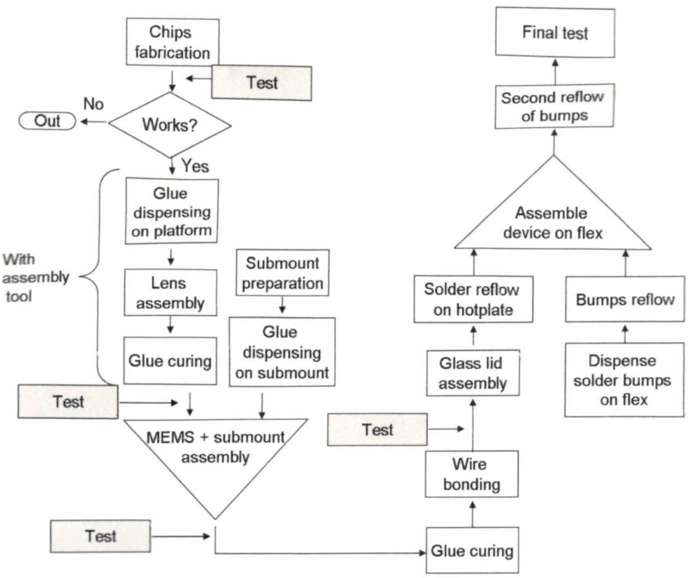

a)

b)

embly of the MEMS device. a) Flowchart of the assembly process. b) Fully ystem comprising flexprint, MEMS, microlens, electronic components, laser ass lid.